AF894138

STUDY GUIDE

for

COLLEGE ALGEBRA AND TRIGONOMETRY

Kolman/Shapiro

STUDY GUIDE

for

COLLEGE ALGEBRA AND TRIGONOMETRY

Kolman/Shapiro

by

James W. Snow

Whatcom Community College

Academic Press

A Subsidiary of Harcourt Brace Jovanovich, Publishers

New York London Toronto Sydney San Francisco

Academic Press, Inc.
111 Fifth Avenue, New York, New York 10003

United Kingdom Edition published by
Academic Press, Inc. (London) Ltd.
24/28 Oval Road, London NW1

ISBN: 0-12-417841-3

Manufactured in the United States of America.

PREFACE

In an attempt to assist you in your learning of mathematics, this study guide is written as a supplement to the basic text, not a replacement for it. You will find it easy to move back and forth between the text and the study guide since the language is very similar and the chapter and section numbers correspond.

In a very real sense this study guide is more of a solutions manual than it is a step-by-step guide through the material, since a large portion of the book is devoted to detailed solutions to the exercises found in the text. Did you ever just want to peek at an answer or a solution to see what it was, or to get a hint how to start? Probably so, but if you are like many students you thought that was cheating or not fair, or at least not a good thing to do. Right? Well, in my opinion, it is not only all right to do, but it is an excellent way to learn a lot of mathematics. After all, what do you do in math class when you don't understand the problem? You ask the professor to do it on the board, of course. And what does he then do? Most likely he will work it out to your satisfaction. So, indeed, I urge you to use the solutions as a learning device to help you see a way to approach a problem.

You will find the Study Guide and Solutions Manual consists of four major components:

1. The BASIC CONCEPTS you should learn from each unit;

2. WHAT YOU SHOULD KNOW HOW TO DO upon completion of each unit;

3. SOLUTIONS to selected problems from the text; and

4. A short CHAPTER QUIZ covering concepts and problem types. Answers to the quizzes are included.

There are a number of uses for a study guide and solutions manual such as this one; among these are:

1. Use as a learning tool as you proceed through the course;

2. Use as a learning tool to review just before a chapter test, a mid-term exam, and before the final examination;

3. Use as a reference book if you have difficulty understanding a concept or in solving a problem; and

4. Use as a quick overview before you begin the study of a section in the text.

Somewhere in this work you may find an error or two. They are probably mine, and I apologize for them in advance. I know all too well how very frustrating it is to be working along and then find out that 'the book is wrong.'

It is my sincere hope that by having this study guide and solutions manual available your learning of college algebra and trigonometry will be a bit easier and more enjoyable.

Acknowledgements: I wish to thank especially the editorial staff of Academic Press, Inc. who offered many constructive comments on the preparation of the manuscript, and also to Ms. Margaret Fogg who did a great job of typing the material under a rather tight time schedule.

Jim Snow

Bellingham, Washington

February, 1981

Table of Contents

Chapter | Page

1. The Foundations of Algebra 1
2. Equations and Inequalities 20
3. Functions 43
4. Exponential and Logarithmic Functions 64
5. Trigonometry: The Circular Functions 79
6. Angles and Triangles 98
7. Analytic Trigonometry 114
8. Analytic Geometry: The Conic Sections 140
9. Systems of Equations and Inequalities 156
10. Matrices and Determinants 180
11. Roots of Polynomials 201
12. Topics in Algebra 215

CHAPTER 1

The Foundations of Algebra

Section 1.1 The Real Number System

BASIC CONCEPTS you should learn from this unit.

1. Symbols and vocabulary of sets

2. The fundamental subsets of the real number system are Natural Numbers, Integers, Rational Numbers, Irrational Numbers.

3. The twelve axioms of a field

4. The four properties of equality

WHAT YOU SHOULD KNOW HOW TO DO upon completion of this unit.

1. Use set notation

2. Identify various numbers as being real, integers, rational, irrational

3. Identify the properties of real numbers

4. Identify and use the properties of equality.

SOLUTIONS to exercise set 1.1

2. Integers between -4 and 2 would be $\{-3, -2, -1, 0, 1\}$.

5. Only the positive integers of $S = \{1,2\}$.

8. Since x must be even, we have $\{2,8,10\}$.

11. F π is irrational, so $\pi/2$ is also irrational.

14. F $0.75 = \frac{75}{100}$, so is rational.

17. T $\sqrt{5}$ is an irrational and all irrationals and rationals are real.

20. T

23. Commutative property of addition

26. This is the closure property of addition of two real numbers.

29. Closure for multiplication

32. Zero is the additive identity element.

35. Multiplicative inverse property (property 10)

38. For example, let $a = 1$, $b = 3$; $\frac{1}{3} \neq \frac{3}{1}$.

41. Symmetric property: $a = b$, then $b = a$.

44. Substitution property

47. Given $a + c = b + c$

then $a + c + (-c) = b + c + (-c)$ — Theorem in text

and $a + 0 = b + 0$ — Additive inverse

thus $a = b$ — Additive identity

50. Using the hint, we have $b = \frac{1}{0}$ is the reciprocal of 0.

$1 = 0 \cdot \frac{1}{0}$ — Property 10 A number times its reciprocal = 1.

$1 = 0 \cdot b$ — Substitution

$1 = 0$ — See Theorem 3, (the right side will equal 0 because one factor equals 0.)

Section 1.2 Real Number Line

BASIC CONCEPTS you should learn from this unit:

1. Representation of real numbers geometrically on a number line
2. Properties of inequality, especially property (e): If $a < b$, then $ac > bc$ if $c < 0$. (Multiplying by a negative reverses the inequality.)
3. Concept of absolute value, and properties.

WHAT YOU SHOULD KNOW HOW TO DO upon completion of this unit:

1. Locate real numbers on a number line
2. Designate the relative size of two numbers using inequality symbols, as $-3 < 0.2$
3. Indicate sets of real numbers on a number line
4. Translate word statements into equalities and vice versa
5. Determine the absolute value of a given number
6. Find the distance between two points on a number line using the absolute value.

SOLUTIONS to Exercise Set 1.2:

2.

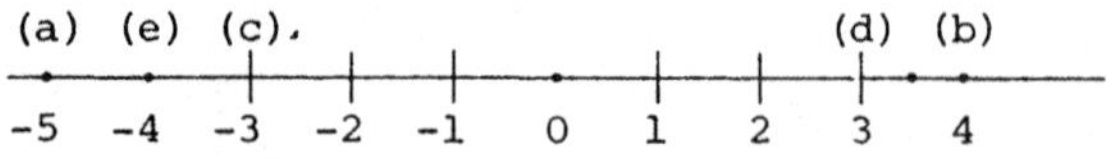

5. 4 would be to the left of 6.

8. -4 would be to the left of 0.

11.

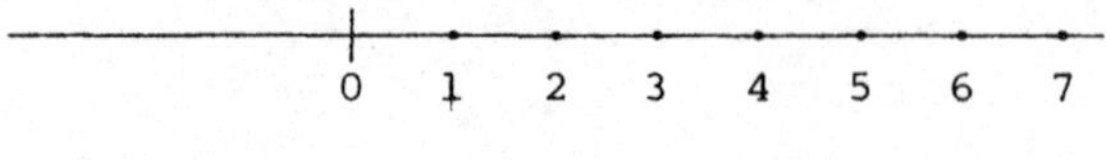

14. 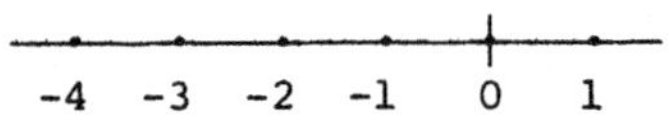

17. $a \geq 0$; that is, a is greater than or equal to zero.

20. $3 < a < 7$

23. $b \geq 5$.

26. Transitive property

29. Property (d); multiplying by a number greater than zero.

32. 2/3

35. -2

38. $|0| = 0$.

41. $\dfrac{|6|}{|-3|} = \dfrac{6}{3} = 2$.

44. $\dfrac{4 - 2 \cdot 4 \cdot 3}{1} = 4 - 24 = -20$.

47. $|-1 - (-3)| = |2| = 2$.

Section 1.3 Polynomials

BASIC CONCEPTS you should learn from this unit.

1. Definition of algebraic expression: variables and constants with the operations of +, -, x, ÷
2. Evaluation of an expression
3. Definition of an exponent
4. Definition of monomial and of polynomial as the sum of monomials
5. Degree of a monomial and polynomial
6. Properties of the real numbers applied to polynomials
7. Operations of addition and multiplication with polynomials.

WHAT YOU SHOULD KNOW HOW TO DO upon completion of this unit.

1. Define "Algebraic Expression."
2. Find the numeric value of an expression given the value(s) of the variable(s).
3. Recognize monomials and polynomials.
4. Add and multiply polynomials.

SOLUTIONS to Exercise Set 1.3:

2. $rst = 2 \cdot 3 \cdot 4 = 24.$

5. $\frac{r + s}{rt} = \frac{2 + 3}{2 \cdot 4} = \frac{5}{8}.$

8. $c = 37$, so we have $\frac{9}{5} \cdot 37 + 32 = \frac{333}{5} + 32 = \underline{98.6}.$

11. Substituting yields $(0.02)(2.5) + (0.314)(3.4)(2.81) + (2.25)(2.81)$

$= 0.05 + 2.999956 + 6.3225$

$= 9.372456.$

14. $|-3 + 2| + |-3 -2| = |-1| + |-5| = 6.$

17., $\frac{-|-2 -2|}{|-2 -1|} = \frac{-4}{3}$.

20. $x^3 \cdot x^5 = x^{3 + 5} = x^8.$

23. $\left(\frac{3}{2} x^3\right)\left(-2x\right) = -3x^4.$

26. (a) and (d) are not polynomials because of fractional exponents.

29. Leading coefficient is $\frac{3}{5}$. Degree is 4.

32. Degree is 4 (from the first term $4xy^3 = 4x^1y^3$).

35. Substituting, we obtain $3 \cdot 2^2 \cdot (-1)^2 + 2 \cdot 2(-1) - 2 + 2(-1) + 7$

$= 12 - 4 - 2 - 2 + 7 = 11.$

38. πr^2

41. The expression $55x + 45y + 60z$ represents the total cost of all stocks bought.

44. Remove parentheses: $4xy^2 + 2xy + 2x + 3 + 2xy^2 - xy + y - 2$; then combine terms to get $6xy^2 + xy + 2x + y + 1.$

47. Combine like terms: $- 2a^2bc + ab^2c - 2ab^3 + 3.$

50. $(2s - 3)(s^3 - s + 2) = 2s^4 - 2s^2 + 4s - 3s^3 + 3s - 6 = 2s^4 - 3s^3 - 2s^2 + 7s - 6.$

53. $(2y^2 + y)(- 2y^3 + y - 3) = - 4y^5 + 2y^3 - 6y^2 - 2y^4 + y^2 - 3y$

$= - 4y^5 - 2y^4 + 2y^3 - 5y^2 - 3y.$

56. $(2a^2 + ab + b^2)(3a - b^2 + 1) =$
$6a^3 - 2a^2b^2 + 2a^2 + 3a^2b - ab^3 + ab + 3ab^2 - b^4 + b^2$. There are no like terms to combine.

59. $2(3x - 2)(2x + 3) = (6x - 4)(2x + 3) = 12x^2 + 18x - 8x - 12 =$
$12x^2 + 10x - 12.$

62. A = total area - area of cutout; $A = xy - \left(\frac{x}{2}\right)^2.$

65. Mentally you would think $4x^2 + 6x + 2x + 3 = 4x^2 + 8x + 3.$

68. $(x + 4)(2x - 1) = 2x^2 - x + 8x - 4 = 2x^2 + 7x - 4.$

71. $(3x - 1)(3x - 1) = 9x^2 - 3x - 3x + 1 = 9x^2 - 6x + 1.$

74. $(3a + 2b)(3a + 2b) = 9a^2 + 6ab + 6ab + 4b^2 = 9a^2 + 12ab + 4b^2.$

Section 1.4 Factoring

BASIC CONCEPTS you should learn from this unit.

1. Factoring is a process of finding the factors of an expression.
2. Factoring a common factor
3. Factoring trinomials
4. Factoring the difference of two squares: $a^2 - b^2 = (a + b)(a - b).$

WHAT YOU SHOULD KNOW HOW TO DO upon completion of this unit.

1. Define factoring.
2. Factor polynomials of the types in this unit, that is, common factor, difference of two squares, trinomials.

SOLUTIONS to Exercise Set 1.4.

2. $\frac{1}{2}x + \frac{1}{4}y = \frac{1}{2}\left(x + \frac{1}{2}y\right)$ Common factor

5. $5bc + 25b = 5b(c + 5)$ Common factor

8. $3abc + 12bc = 3bc(a + 4).$

11. $x^2 + 4x + 3 = (x + \quad)(x + \quad)$
$= (x + 3)(x + 1).$

14. $y^2 + 7y - 8 = (y + \quad)(y - \quad)$
$= (y + 8)(y - 1).$

17. $y^2 - \frac{1}{9} = \left(y + \frac{1}{3}\right)\left(y - \frac{1}{3}\right)$ Difference of two squares.

20. $4b^2 - a^2 = (2b + a)(2b - a)$ Difference of two squares

23. $\frac{1}{16} - y^2 = (\frac{1}{4} + y)(\frac{1}{4} - y)$ Like Exercise 17

26. $a^2b^2 - \frac{1}{9} = (ab + \frac{1}{3})(ab - \frac{1}{3})$.

29. $x^2 + 11x + 24 = (x + \quad)(x + \quad)$
$= (x + 8)(x + 3)$.

32. $2x^2 + 7x + 6 = (2x \quad)(x \quad)$
$= (2x + 3)(x + 2)$.

35. $6x^2 + 13x + 6 = (3x + 2)(2x + 3)$

38. $9x^2 + 24x + 16 = (3x + 4)^2$ A square trinomial

41. $6a^2 - 5ab - 6b^2$ $(3a + 2b)(2a - 3b)$ (Hint: See Exercise #35).

44. $x^{12} - 1 = (x^6 - 1)(x^6 + 1)$ Difference of two squares. $(x^6 - 1)$ also is difference of two squares.
$= (x^3 + 1)(x^3 - 1)(x^6 + 1)$.

47. $8n^2 - 18n - 5 = (2n - 5)(4n + 1)$

50. $3y^2 + 6y - 45 = 3(y^2 + 2y - 15)$ Common factor first
$= 3(y + 5)(y - 3)$

53. $18x^2m + 33xm + 9m = 3m(6x^2 + 11x + 3)$
$= 3m(3x + 1)(2x + 3)$

56. $10r^2 - 5rs - 15s^2 = 5(2r^2 - rs - 3s^2)$
$= 5(2r - 3s)(r + s)$.

59. $b^4 + 2b^2 - 8 = (b^2 \quad)(b^2 \quad)$
$= (b^2 + 4)(b^2 - 2)$.

61. $6b^4 + 7b^2 - 3 = (3b^2 - 1)(2b^2 + 3)$.

64. $3(x + 2)^2(x - 1) - 4(x + 2)^2(2x + 7) = (x + 2)^2[3(x - 1) - 4(2x + 7)]$
$= (x + 2)^2(-5x - 31)$
$= -(x + 2)^2(5x + 31)$

BASIC CONCEPTS you should learn from this unit.

1. Definition of a rational expression; that is, the quotient of two polynomials

2. Multiplication and division are carried out much the same as is done with arithmetic fractions.

3. Addition and subtraction, too, are handled as in arithmetic, except that more care is needed in finding and using the lowest common denominator (LCD).

4. Remember that polynomials and thus rational expressions merely represent real numbers, so the same properties will apply.

WHAT YOU SHOULD KNOW HOW TO DO upon completion of this unit.

1. Multiply, divide, add, and subtract rational algebraic expressions.

2. Given two or more rational expressions, find the lowest common denominator.

3. Simplify complex fractions.

SOLUTIONS to Exercise Set 1.5

2. $\dfrac{y^2 - 25}{y + 5} = \dfrac{\cancel{(y + 5)}(y - 5)}{\cancel{(y + 5)}} = y - 5.$

5. $\dfrac{6x^2 - x - 1}{2x^2 + 3x - 2} = \dfrac{(3x + 1)\cancel{(2x - 1)}}{\cancel{(2x - 1)}(x + 2)} = \dfrac{3x + 1}{x + 2}.$

8. Invert divisor: $\dfrac{5\cancel{(x + 3)}}{8_2} \cdot \dfrac{4}{3\cancel{(x + 3)}} = 5/6.$

11. Invert divisor: $\dfrac{\cancel{(x + 2)}}{\cancel{3y}} \cdot \dfrac{\cancel{3y} \cdot 5y}{\cancel{(x + 2)}(x - 4)} = \dfrac{5y}{x - 4}.$

14. Factor all parts: $\dfrac{6x^2 + 7x - 2}{(4x + 1)\cancel{(x - 1)}} \cdot \dfrac{(5x + 2)\cancel{(x - 1)}}{(3x + 1)(x + 2)} = \dfrac{(6x^2 + 7 - 2)(5x + 2)}{(4x + 1)(3x + 1)(x + 2)}.$

17. Invert divisor: $\dfrac{\cancel{(x - 5)}(x + 3)}{1} \cdot \dfrac{(x^2 + 1)}{\cancel{(x - 5)}(x - 2)} = \dfrac{(x + 3)(x^2 + 1)}{(x - 2)}.$

20. Factor: $\dfrac{\cancel{(x + 3)}\cancel{(x - 3)}}{(3x - 1)\cancel{(2x + 1)}} \cdot \dfrac{\cancel{(2x + 1)}(x + 2)}{\cancel{(x + 3)}\cancel{(x + 1)}} \cdot \dfrac{(x - 2)\cancel{(x + 1)}}{x\cancel{(x - 3)}} = \dfrac{x^2 - 4}{3x^2 - x}.$

23. LCD = $\underline{2a}$

26. $(x^2 - 4) = (x + 2)(x - 2)$

$(x - 2) = (x - 2)$

Therefore, LCD $= (x + 2)(x - 2)$.

29. All denominators are needed: LCD $= x(x + 1)(x - 1)$.

32. $\dfrac{x}{x^2 - 4} + \dfrac{2}{4 - x^2} = \dfrac{x}{x^2 - 4} - \dfrac{2}{x^2 - 4} = \dfrac{(x - 2)}{(x + 2)(x - 2)} = \dfrac{1}{x + 2}$

35. $\dfrac{1}{a + 2} + \dfrac{3}{a - 2}$ The LCD $= (a + 2)(a - 2)$, so we have

$\dfrac{a - 2}{LCD} + \dfrac{3(a + 2)}{LCD} = \dfrac{a - 2 + 3a + 6}{LCD} = \dfrac{4a + 4}{(a + 2)(a - 2)}$

38. $\dfrac{4x - 1}{6x^3} + \dfrac{2}{3x^2} = \dfrac{4x - 1}{6x^3} + \dfrac{4x^4}{6x^3} = \dfrac{8x - 1}{6x^3}$ The LCD $= 6x^3$

41. $\dfrac{5x}{2(x + 3)(x - 3)} + \dfrac{4}{3(x^2 + 3)}$ The LCD $= 2 \cdot 3(x + 3)(x - 3)(x^2 + 3)$, so we have

$\dfrac{5x(3)(x^2 + 3) + 4(2)(x + 3)(x - 3)}{LCD} = \dfrac{15x^3 + 45x + 8x^2 - 72}{6(x + 3)(x - 3)(x^2 + 3)}$

44. $\dfrac{2x}{(2x + 1)} - \dfrac{x-1}{(2x + 1)(x - 2)}$ The LCD $= (2x + 1)(x - 2)$, so we have

$\dfrac{2x(x - 2) - (x - 1)}{LCD} = \dfrac{2x^2 - 5x + 1^{12}}{(2x + 1)(x - 2)}$.

47. $\dfrac{2x - 1}{(x + 3)(x + 2)} - \dfrac{x - 2}{(x + 3)(x + 1)}$ The LCD $= (x + 3)(x + 2)(x + 1)$, so we have

$\dfrac{(2x - 1)(x + 1) - (x - 2)(x + 2)}{LCD} = \dfrac{2x^2 + x - 1 - x^2 + 4}{LCD} = \dfrac{x^2 + x + 3}{(x + 3)(x + 2)(x + 1)}$

50. In this problem the LCD $= (x + 2)(x - 2)$, so we have

$\dfrac{2x(x - 2) + x(x + 2) - 1}{LCD} = \dfrac{3x^2 - 2x - 1}{x^2 - 4}$.

53. $\dfrac{(x + 1)}{(1 - \frac{1}{x})} \cdot \dfrac{x}{x} = \dfrac{x^2 + x}{x - 1}$.

56. $\dfrac{\dfrac{a}{a - b} - \dfrac{b}{a + b}}{a^2 - b^2} = \dfrac{a^2 + ab - ab + b^2}{a^2 - b^2} \cdot \dfrac{1}{a^2 - b^2} = \dfrac{a^2 + b^2}{(a^2 - b^2)^2}$.

59. $\dfrac{(\frac{a}{b} - \frac{b}{a})}{(\frac{1}{a} + \frac{1}{b})} \cdot \dfrac{ab}{ab} = \dfrac{a^2 - b^2}{b + a} = \dfrac{\cancel{(a + b)}(a - b)}{\cancel{(a + b)}} = a - b.$

62. $2 + \dfrac{3}{1 + \dfrac{2}{1 - x}} = 2 + \dfrac{3(1 - x)}{1 - x + 2} = 2 + \dfrac{3 - 3x}{3 - x}$

$= \dfrac{2(3 - x)}{3 - x} + \dfrac{3 - 3x}{3 - x} = \dfrac{9 - 5x}{3 - x}.$

65. $1 - \dfrac{1}{1 + \dfrac{1}{1 - \dfrac{1}{1 + x}}} = 1 - \dfrac{1}{1 + \dfrac{1 + x}{1 + x - 1}} = 1 - \dfrac{1}{1 + \dfrac{1 + x}{x}} = 1 - \dfrac{x}{x + 1 + x}$

$= 1 - \dfrac{x}{1 + 2x} = \dfrac{1 + 2x - x}{1 + 2x} = \dfrac{1 + x}{1 + 2x}$

Section 1.6 Integer Exponents

BASIC CONCEPTS you should learn from this unit.

1. The rules for positive integer exponents.
2. The meaning of the zero exponent; that is, $x^0 = 1$.
3. The meaning and use of negative integer exponents; for example, $x^{-2} = \dfrac{1}{x^2}$.

WHAT YOU SHOULD KNOW HOW TO DO upon completion of this unit.

1. Simplify expressions containing integer exponents.
2. Change an expression with negative exponents to one containing only positive exponents.

SOLUTIONS to Exercise Set 1.6

2. Multiply 2 times 5 to obtain the correct exponent

   Answer $= y^{10}$

5. $(2x)^4 = 2^4x^4 = 16x^4$.

8. x^{3m^2}

11. $-\dfrac{x^3}{y^3}$

14. $x^{12-8} = x^4$.

17. $x^{3m + m} = x^{4m}$.

20. $\left(\dfrac{3x^3}{y^2}\right)^5 = \dfrac{3^5x^{15}}{y^{10}}$

23. $\dfrac{(r^2)^4}{(r^4)^2} = \dfrac{r^8}{r^8} = 1.$

26. $\dfrac{(-2a^2b)^4}{(-3ab^2)^3} = \dfrac{16a^8b^4}{-27a^3b^6} = \dfrac{16a^5}{-27b^2}$

29. $(-2a^2b^3)^{2m} = (-2)^{2m}a^{4m}b^{6m}$

32. $(xy)^0 - 2^{-1} = 1 - \frac{1}{2} = \frac{1}{2}.$

35. $\dfrac{1}{3^{-4}} = 3^4 = 81.$

38. $-x^{-5} = -1/x^5.$

41. $5^{-3} \cdot 5^5 = 5^{-3+5} = 5^2 = 25.$

44. $(x^{-2})^4 = x^{-8} = 1/x^8.$

47. $\dfrac{2^2}{2^{-3}} = 2^{2-(-3)} = 2^5 = 32.$

50. $(x^4y^{-2})^{-1} = x^{-4}y^2 = \dfrac{y^2}{x^4}.$

53. $\left(-\frac{1}{2}x^3y^{-4}\right)^{-3} = \left(-\frac{1}{2}\right)^{-3}x^{-9}y^{12} = \dfrac{-8y^{12}}{x^9}.$

56. $\left(\dfrac{x^3}{x^{-2}}\right)^2 = \left(x^5\right)^2 = x^{10}.$

59. $(a - 2b^2)^{-1} = \dfrac{1}{a - 2b^2}$

62. $(a^{-1} + b^{-1})^{-1} = \dfrac{1}{a^{-1} + b^{-1}} = \dfrac{1}{\frac{1}{a} + \frac{1}{b}} = \dfrac{1}{\frac{b + a}{ab}} = \dfrac{ab}{a + b}.$

65. $\left(\dfrac{a}{b}\right)^{-m} = \dfrac{1}{\left(\frac{a}{b}\right)^m} = \dfrac{1}{\frac{a^m}{b^m}} = \dfrac{b^m}{a^m} = \left(\dfrac{b}{a}\right)^m$ Q.E.D.

68. $\left(\dfrac{7.65^{-1}}{7.65^2}\right)^2 = \left(\dfrac{1}{7.65^3}\right)^2 = \dfrac{1}{7.65^6} = 4.99 \times 10^{-6}.$

BASIC CONCEPTS you should learn from this unit.

1. Rational exponents and roots: $b^{1/n}$ is an $n^{\underline{th}}$ root of b; for example, $b^{1/3}$ is a cube root of b.

2. Meaning of $a^{m/n}$ as $(a^m)^{1/n} = (a^{1/n})^m$

3. Since $\sqrt{b} \cdot \sqrt{b} = b$ and since $b^{1/2} \cdot b^{1/2} = b$, $\sqrt{b} = b^{1/2}$. In general, $\sqrt[n]{b} = b^{1/n}$.

4. Meaning of a simplified radical.

5. Radicals can be written in exponent form, so that the rules of exponents can be used in simplifying.

6. Meaning of "Rationalize the denominator."

WHAT YOU SHOULD KNOW HOW TO DO upon completion of this unit.

1. Simplify expressions with rational exponents

2. Change an expression with rational exponents to one with radicals and vice versa

3. Simplify radical expressions including addition, multiplication, and division

4. Rationalize the denominator of a fraction containing a radical in the denominator.

SOLUTIONS to Exercise Set 1.7

2. $(-125)^{-1/3} = \left(-\frac{1}{125}\right)^{1/3} = -1/5.$

5. $\frac{2x^{1/3}}{x^{-3/4}} = 2x^{1/3 + 3/4} = 2x^{13/12}.$

8. $\frac{125^{4/3}}{125^{2/3}} = \frac{5^4}{5^2} = 5^2 = 25.$

11. $\left(\frac{x^{15}}{y^{10}}\right)^{3/5} = \frac{x^9}{y^6}.$

14. $x^{2/3} = \sqrt[3]{x^2}$

17. $(12x^3y^{-2})^{2/3} = \sqrt[3]{\frac{144x^6}{y^4}} = \frac{2x^2}{y^2}\sqrt[3]{18y^2}$

20. $\sqrt[5]{3^2} = 3^{2/5}$.

23. $\frac{1}{\sqrt[4]{\frac{4}{9}a^3}} = \frac{1}{(\frac{4}{9}a^3)^{1/4}} = \frac{1}{(\frac{2}{3})^{1/2}a^{3/4}}$.

26. $\sqrt{\frac{25}{4}} = \frac{5}{2}$.

29. $\sqrt{(-5)^2} = 5$.

32. $\sqrt{(-\frac{7}{2})^2} = \frac{7}{2}$.

35. Let $x = 2$, $y = 3$.

then $\sqrt{2^2 + 3^2} = \sqrt{13} \neq 2 + 3$.

$\sqrt{13} \neq 5$.

38. 10

41. $\sqrt[3]{y^7} = y^2\sqrt[3]{y}$.

44. $\sqrt{x^5y^4} = x^2y^2\sqrt{x}$.

47. $\sqrt[4]{16x^8y^5} = 2x^2y\sqrt[4]{y}$.

50. $\frac{4}{3\sqrt{11}} \cdot \frac{\sqrt{11}}{\sqrt{11}} = \frac{4\sqrt{11}}{33}$.

53. $\frac{4x^2}{\sqrt{2x}} \cdot \frac{\sqrt{2x}}{\sqrt{2x}} = \frac{4x^2\sqrt{2x}}{2x} = 2x\sqrt{2x}$.

56. $\sqrt[4]{48x^8y^6z^2} = 2x^2y\sqrt[4]{3y^2z^2}$ Note that $48 = 16 \cdot 3$; $\sqrt[4]{16} = 2$.

59. $3\sqrt{x} + 4\sqrt{x} = (3 + 4)\sqrt{x} = 7\sqrt{x}$.

62. $\sqrt{20} - 4\sqrt{45} + \sqrt{80}$

$2\sqrt{5} - 4 \cdot 3\sqrt{5} + 4\sqrt{5}$

$-6\sqrt{5}$

65. $2\sqrt{5} - (3\sqrt{5} + 4\sqrt{5}) = 2\sqrt{5} - 7\sqrt{5} = -5\sqrt{5}$.

68. $\sqrt{8}(\sqrt{2} - \sqrt{3}) = \sqrt{16} - \sqrt{24} = 4 - 2\sqrt{6}$.

71. $(\sqrt{2} - \sqrt{3})^2 = (\sqrt{2} - \sqrt{3})(\sqrt{2} - \sqrt{3}) = 2 - 2\sqrt{6} + 3 = 5 - 2\sqrt{6}$.

74. $\left(\sqrt[3]{2x} + 3\right)\left(\sqrt[3]{2x} - 3\right) = \left(\sqrt[3]{2x}\right)^2 - 9$

$= \sqrt[3]{4x^2} - 9.$

77. $\dfrac{-2}{\sqrt{3} - 4} \cdot \dfrac{\left(\sqrt{3} + 4\right)}{\left(\sqrt{3} + 4\right)} = \dfrac{-2\sqrt{3} - 8}{3 - 16} = \dfrac{2\sqrt{3} + 8}{13}$

80. $\dfrac{4}{2 - \sqrt{2y}} \cdot \dfrac{\left(2 + \sqrt{2y}\right)}{\left(2 + \sqrt{2y}\right)} = \dfrac{8 + 4\sqrt{2y}}{4 - 2y} = \dfrac{4 + 2\sqrt{2y}}{2 - y}$

83. $\dfrac{\left(\sqrt{2} + 1\right)\left(\sqrt{2} + 1\right)}{\left(\sqrt{2} - 1\right)\left(\sqrt{2} + 1\right)} = \dfrac{2 + 2\sqrt{2} + 1}{2 - 1} = 3 + 2\sqrt{2}.$

86. $\dfrac{2\sqrt{a}}{\sqrt{2x} + \sqrt{y}} \cdot \dfrac{\left(\sqrt{2x} - \sqrt{y}\right)}{\left(\sqrt{2x} - \sqrt{y}\right)} = \dfrac{2\sqrt{2ax} - 2\sqrt{ay}}{2x - y}$

89. $\sqrt{(-1)(-1)} = \sqrt{(-1)^2} = 1$ by agreement;

Section 1.8 Complex Numbers

BASIC CONCEPTS you should learn from this unit.

1. Definition of $i = \sqrt{-1}$; $i^2 = -1$.
2. Square roots of negative numbers.
3. Expansion of the number system to include numbers of the form $a + bi$. Note that if $b = 0$, we simply have the reals.
4. Adding and subtracting are done by adding (or subtracting) the real parts together and the imaginary parts together:

   $$(a + bi) + (x + yi) = a + x + (b + y)i$$

5. Multiplication is done as with real binomials except that when i^2 occurs we replace it with -1.

WHAT YOU SHOULD KNOW HOW TO DO upon completion of this unit.

1. Define complex numbers.
2. Simplify i^n expressions when n is any natural number.
3. Add, subtract, and multiply complex numbers.

SOLUTIONS to Exercise Set 1.8

2. $i^{27} = i^4 \cdot i^4 \cdot i^4 \cdot i^4 \cdot i^4 \cdot i^4 \cdot i^3 = i^3$ Recall that $i^4 = 1$.

$i^3 = i^2 \cdot i = -1$ Since $i^2 = -1$.

5. $-i^{33} = -i$ $i^{32} = \left(i^4\right)^8 = 1^8 = 1$

8. $-i^{39} = -i^3 = -i^2 \cdot i = i$

11. $-\frac{1}{2}$ as a complex number $= -\frac{1}{2} + 0i$

14. $-\sqrt{-5} = -\sqrt{-1} \cdot \sqrt{5} = -i\sqrt{5}$

17. $3 - \sqrt{-49} = 3 - i\sqrt{49} = 3 - 7i$

20. $-0.5 + \sqrt{-32} = -0.5 + \sqrt{-1} \cdot \sqrt{16} \cdot \sqrt{2} = -0.5 + 4i\sqrt{2}$

23. $(3x - 1) + (y + 5)i = 1 - 3i$ means that $3x - 1 = 1$ or $x = 2/3$, and $y + 5 = -3$ or $y = -8$.

26. $(y - 2) + (5x - 3)i = 5 = 5 + 0i$ means that $y - 2 = 5$ or $y = 7$ and $5x - 3 = 0$ or $x = 3/5$

29. Adding the real parts and then the imaginary parts gives $5 + i$.

32. $1\frac{1}{2} - 1\frac{2}{3}i$ or $\frac{3}{2} - \frac{5}{3}i$

35. $-2i(3 + 1) = (-2i)(3) + (-2i)(i)$

$= -6i - 2i^2 = -6i + 2 = 2 - 6i$

38. $(5 + i)(2 - 3i) = 10 - 15i + 2i - 3i^2$

$= 10 - 13i + 3 = 13 - 13i$

41. $(3 - 2i)(2 - i) = 6 - 3i - 4i + 2i^2$

$= 6 + 7i - 2 = 8 - 7i$

44. $x^2 - 2x + 5$ if $x = (2 - i)$, we have

$(2 - i)^2 - 2(2 - i) + 5 = 4 - 4i + i^2 - 4 + 2i + 5 = 5 - 2i + i^2 = 4 - 2i$

47. Prove that $(a + bi) + (c + di) = (c + di) + (a + bi)$.

The left side $= (a + c) + (b + d)i$ and since a,c,b,d are reals, this equals $(c + a) + (d + b)i$ by the commutative property of addition for Reals.

The right side also equals $(c + a) + (d + b)i$ again, by the definition of the addition of complex numbers. But this is exactly the same as the left and the statement is proved.

50. The sum of a number and its additive inverse is zero. So if $(-a - bi) + (a + bi)$ is zero, $-a - bi$ is the additive inverse of $a + bi$.

$$-a - bi + (a + bi) = (-a + a) + (-b + b)i$$
$$= 0 + 0i = 0$$

and the statement is proved.

SOLUTIONS to Review Exercises.

2. $\{-2\}$

5. F (Natural numbers are integers greater than zero.)

8. $(3a) + (-3a) = 0$ Additive inverse property

11. $9x \cdot 1 = 9x$ the multiplicative identity is 1.

14.

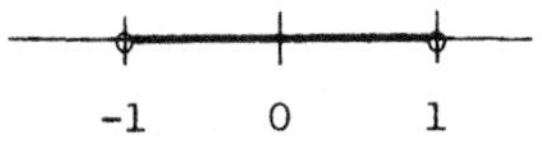

17. Substituting the given values of x and y produces
$(3.25)(12) + (0.15)(80) = 39 + 12 = \$51$

20. -7 is the leading coefficient; the degree is 5.

23. $3x(2x + 1)^2 = 3x(4x^2 + 4x + 1) = 12x^3 + 12x^2 + 3x$

26. $2a^2 + 3ab + ba + 9b = a(2a + 3b) + 3(2a + 3b)$
$= (a + 3)(2a + 3b)$

29. $6x^4 + 7x^2 - 3 = (3x^2 - 1)(2x^2 + 3)$

32. $\frac{\cancel{(a + b)}}{\cancel{(a + 2b)}} \cdot \frac{\cancel{(a + 2b)}(a - 2b)}{\cancel{(a + b)}(a - b)} = \frac{a - 2b}{a - b}$

35.
$$x = x$$
$$x^2 - x = x(x - 1)$$
$$(x - 1)^2 = (x - 1)(x - 1)$$
$$\text{LCD} = x(x - 1)(x - 1)$$

38. $2 + \frac{4}{a^2 - 4} = \frac{2a^2 - 8}{a^2 - 4} + \frac{4}{a^2 - 4} = \frac{2a^2 - 4}{a^2 - 4}$

41. $x^2 + \frac{\frac{1}{x} + 1}{x - \frac{1}{x}} \cdot \frac{(x)}{(x)} = x^2 + \frac{1 + x}{x^2 - 1} = x^2 + \frac{\cancel{(1 + x)}}{\cancel{(x + 1)}(x - 1)} = x^2 + \frac{1}{x - 1} = \frac{x^3 - x^2 + 1}{x - 1}$

44. $\left(\frac{x^3}{y^{-6}}\right)^{-4/3} = \frac{x^{-4}}{y^8} = \frac{1}{x^4 y^8}$

47. $\frac{2}{\sqrt{12}} = \frac{2}{\sqrt{4}\sqrt{3}} = \frac{2}{2\sqrt{3}} = \frac{1}{\sqrt{3}} = \frac{1}{3^{1/2}}$

50. $\frac{\sqrt{x}}{(\sqrt{x} + \sqrt{y})} \cdot \frac{(\sqrt{x} - \sqrt{y})}{(\sqrt{x} - \sqrt{y})} = \frac{x - \sqrt{xy}}{x - y}$

53. $(x - 2) + (2y - 1)i = -4 + 7i$ Then $x - 2 = -4$

$x = -2$

and $2y - 1 = 7$

$y = 4$

56. $(2 + i)^2 = 4 + 4i + i^2$

$= 3 - 4i$ since $i^2 = 1$

Chapter 1 - A SHORT QUIZ

1. A set is a ___?___ of objects.
2. Use set notation to describe the set of integers greater than -4 and less than or equal to 2.
3. Is the number $7i$ a rational number?
4. Which property of the real numbers is illustrated by $3xa + 7ya = (3x + 7)a$?
5. If $a < b$, which would be on the left if they were each plotted on a number line?
6. Draw a number line and graph the numbers represented by $-1.5 \le x \le 3$.
7. The absolute value of a number is sometimes defined as the ___?___ from zero.
8. State in words the meaning of $-80 < N \le 1$. (Assume that N is real.)
9. If $a < b$ and $d < 0$, then ad ___?___ bd.

10. Why is $2x^{2/3} + 4x - 1$ NOT a polynomial?

11. What is the degree of $7x^4 - 3x + 5x^5 - 2$?

12. Find the error.

$$(x + 5) - (2x - 2) = x + 5 - 2x - 2 = -x + 3$$

Correct the solution.

13. Evaluate $3x^2 - 2x + 4$ when $x = -2$.

14. Factoring is the process of ___?___ .

15. To factor this expression: $7a^2 - 63b^2$, what would you do first?

16. Factor: $3a^2 - 16a + 21$.

17. A polynomial is said to be ___?___ if it cannot be written as the product of two polynomials, each of positive degree.

18. Find the lowest common denominator for these three fractions.

$$\frac{x}{x^3 - x}, \quad \frac{5}{x}, \quad \frac{x + 3}{(x + 1)^2}$$

19. Divide as indicated.

$$\frac{3x}{x + 2} \div \frac{6x^2}{x^2 - x - 6}$$

20. Simplify this complex fraction.

$$\frac{x - \frac{1}{x}}{2 + \frac{1}{x}}$$

21. Write with a positive exponent.

$$\frac{1}{a^{-2}}$$

22. $x^0 =$ ___?___.

23. Write $\sqrt[3]{x^2}$ in exponent form.

24. $\left(b^{1/n}\right)^n = ?$

25. What is meant by "the principal square root"?

26. Simplify $\frac{1}{\sqrt{x}}$.

27. True or false: $\sqrt{9} + \sqrt{36} = \sqrt{45}$?

28. True or false: $\sqrt{9} \cdot \sqrt{36} = \sqrt{324}$?

29. Find the sum of $(3 - 2i) + (-6 - 7i)$.

30. Find the product of $i(3 - 2i)$.

Quiz Answers Chapter 1

1. collection

2. $\{-3,-2,-1,0,1,2\}$

3. No, it is complex.

4. The distributive property

5. a would be to the left of b.

6. 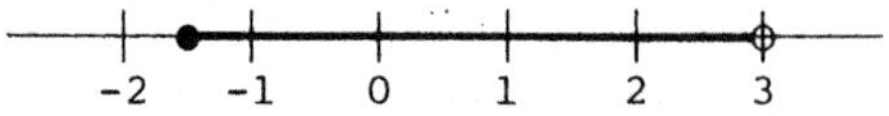

7. distance

8. N is greater than -80 and less than or equal to 1.

9. $>$

10. A rational exponent of $\frac{2}{3}$ is not allowed.

11. 5

12. The right side should be $x + 5 - 2x + 2 = -x + 7$.
    ↑ error (under the "+")

13. 20

14. Finding the numbers or expressions that, when multiplied, yield the given number or expression.

15. Factor out the common factor, 7.

16. $(3a - 7)(a - 3)$

17. prime or irreducible

18. $x(x - 1)(x + 1)^2$

19. $\frac{x - 3}{x}$

20. $\frac{x^2 - 1}{2x + 1}$

21. a^2

22. 1

23. $x^{2/3}$

24. b

25. The positive square root is the principal square root.

26. $\frac{\sqrt{x}}{x}$

27. False

28. True

29. -3 - 9i

30. 2 + 3i

CHAPTER 2

Equations and Inequalities

Section 2.1 Linear Equations in One Unknown

BASIC CONCEPTS you should learn from this unit.

1. A linear equation in one unknown is a statement of equality containing one variable that has the exponent one (1); for example, $3x \cdot 2 = 7$.

2. A solution to an equation is a numeric value of the variable that makes the statement true.

3. Linear equations can be solved by adding a number or expression or subtracting the same number or expression from both sides of the equation, and/or by multiplying or dividing both sides of the equation by a number or expression.

4. A linear equation is often called a first degree equation.

5. If it is necessary to multiply or divide a linear equation by an expression containing the unknown, the resulting equation may not be equivalent to the original one; hence, the solution must be checked in the original equation.

WHAT YOU SHOULD KNOW HOW TO DO upon completion of this unit.

1. Recognize linear equations.

2. Solve linear equations.

3. Verify supposed solutions.

4. Determine whether an equation is an identity or a conditional equation.

SOLUTIONS to Exercise set 2.1

2. $3 \cdot \frac{5}{2} - 4 = \frac{15}{2} - \frac{8}{2} = \frac{7}{2} \neq \frac{5}{2}$ so $\frac{5}{2}$ is not a solution.

5. $3x + 5 = -1,\ 3x = -6,\ x = -2$

8. $\frac{1}{2}s + 2 = 4,\ \frac{1}{2}s = 2,\ s = 4$

11. $0 = -\frac{1}{2}a - \frac{2}{3},\ \frac{1}{2}a = -\frac{2}{3},\ a = -\frac{4}{3}$

14. $2x - 1 = 3x + 2,\ 2x - 3x = 2 + 1,\ x = -3$

17. $2(3b + 1) = 3b - 4,\ 6b + 2 = 3b - 4,\ 3b = -6,\ b = -2$

20. $-3(x - 2) = 2(x + 4),\ -3x + 6 = 2x + 8,\ -5x = 2,\ x = -\frac{2}{5}$

23. $-4(2x + 1) - (x - 2) = -11,\ -8x - 4 - x + 2 = -11,\ -9x - 2 = -11,\ -9x = -9,\ x = 1$

26. $8 - 2kx = -3x,\ 8 = 2kx - 3x,\ 8 = x(2k - 3),\ x = \frac{8}{2k - 3}$

29. $\frac{x}{2} = \frac{5}{3}$ Multiply both sides by 6

$3x = 10$

$x = \frac{10}{3}$

32. $\frac{5}{a} - \frac{3}{2} = \frac{1}{4}$ (LCD = a), $4a\left(\frac{5}{a}\right) - 4a\left(\frac{3}{2}\right) = 4a\left(\frac{1}{4}\right)$, $20 - 6a = a$, $20 - 7a$, $a = \frac{20}{7}$

35. $\frac{1}{x - 2} + \frac{1}{2} = \frac{2}{x - 2}$ multiplying by the LCD, which is $2(x - 2)$, gives

$2(x - 2)\left[\frac{1}{x - 2} + \frac{1}{2} = \frac{2}{x - 2}\right] = 2 + x - 2 = 4$, $x = 4$,

check:

$\frac{1}{4 - 2} + \frac{1}{2} \stackrel{?}{=} \frac{2}{4 - 2}$, $\frac{1}{2} + \frac{1}{2} \stackrel{?}{=} \frac{2}{2}$, $1 = 1$

so that the solution is correct.

38. $\frac{3}{x - 1} + \frac{2}{x + 1} = \frac{5}{(x + 1)(x - 1)}$ LCD = $(x + 1)(x - 1)$

$(x + 1)(x - 1)\left[\frac{3}{x - 1} + \frac{2}{x + 1} = \frac{5}{(x + 1)(x - 1)}\right]$; $3x + 3 + 2x - 2 = 5$; $5x + 1 = 5$;

$5x = 4$; $x = \frac{4}{5}$

41. $\frac{4}{b} - \frac{1}{b + 3} = \frac{3b + 2}{(b + 3)(b - 1)}$ LCD = $b(b + 3)(b - 1)$

Multiplying both sides by the LCD gives

$4b^2 + 8b - 12 - b^2 + b = 3b^2 + 2b$

$7b = 12$

$b = \frac{12}{7}$

This, with a bit of work, does check!

44. $\frac{2x - 1}{x - 5} + 3 = \frac{3x - 2}{5 - x}$ <u>or</u> $\frac{2x - 1}{x - 5} + 3 = \frac{2 - 3x}{x - 5}$

Therefore, LCD = $x - 5$; multiplying, we have

$2x - 1 + 3x - 15 = 2 - 3x$; $8x - 16 = 2$; $8x = 18$; $x = \frac{9}{4}$. This also checks.

47. $2x + 1 = 3x - 1$; $2 = x$. Thus it is a conditional equation.

50. $5(x - 1) = 10$

$x - 1 = 2$ Divide by 5

$x = 3$ Add 1.

Yes, all are equivalent.

53. $3(x^2 + 2x + 1) = -6$

$x^2 + 2x + 1 = -2$ Divide by 3

$(x + 1)^2 = -2$ Simply factoring the left-hand side

Yes, all are equivalent.

Section 2.2 Applications

BASIC CONCEPTS you should learn from this unit.

1. "Applied" mathematics problems generally involve word sentences as well as numbers, so we have to "translate" such problems into mathematics. Helpful steps in this procedure are the following.
   - A. Read the problem carefully.
   - B. Assign a letter to represent the quantity to be found.
   - C. Write an equation that reflects the relationships stated in the problem.
   - D. Apply the rules for solving equations.

   (This sequence sounds easy, but many students make it hard by not doing the first two steps carefully.)

WHAT YOU SHOULD KNOW HOW TO DO upon completion of this unit.

1. Translate word statements into an algebraic equation.
2. Solve the equation using the methods of Section 2.1.
3. Substitute your answer in the original problem to check whether it makes sense.
4. Solve literal equations (often formulas) for a specified letter.

SOLUTIONS to Exercise Set 2.2

2. The number of sedans = n, so the number of station wagons would be $3n - 20$.

5. Let n = Fred's age now

   Then $n + 12$ = John's age now

   Also $n - 4$ = Fred's age 4 years ago

   $n + 12 - 4 = n + 8$ = John's age 4 years ago

   Equation: $n + 8 = 2(n - 4)$

8. n = larger number

   $n - 5$ = smaller number

   $n + (n - 5) = 11$

11. n = short piece

$n + 4$ = longer piece

The sum should be 12 meters.

$n + (n + 4) = 12$

14. Choose n = number of \$10 bills

Then $2n + 4$ = number of \$5 bills

and $n - 6$ = number of \$20 bills

The equation found by the <u>value</u> of the bills should equal \$460.

$$10n + 5(2n + 4) + 20(n - 6) = 460$$

17. Let n = number of \$6 tickets

$2n$ = number of \$5 tickets

$3n + 1$ = number of \$3 tickets

The equation found by the value of the tickets should equal \$503.

$$6n + 5(2n) + 3(3n + 1) = 503$$

20. Let n = amount invested in black and white film

$4000 - n$ = amount invested in color film

$0.12n$ = profit from black and white film

$0.21(4000 - n) = 0.12n - 150$

23. The distance traveled by each truck is, of course, equal when the faster truck overtakes the slower one. Using $d = rt$ as the formula for each truck and then setting these expressions equal, we have

t = time of the faster truck

$t - 2$ = time of the slower truck

$55t = 50(t - 2)$

distance = distance

Also, t will be the time required for the second truck to overtake the first.

27. Draw a diagram:

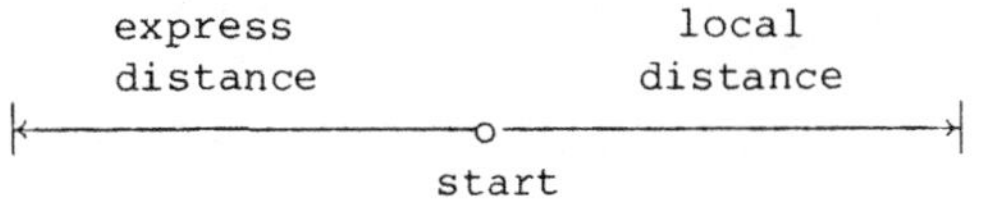

Use $d = rt$.

Let r = speed of the local

$2r$ = speed of the express

Then

$4r + 4(2r) = 480$

local distance + express distance = total distance

29. Let n = number of ounces of Ceylon tea

$8 - n$ = number of ounces of Formosa tea

Equation found by value of tea:

$$1.50n + 2.00(8 - n) = 1.85(8)$$

Ceylon value + Formosa value = mixture value

32. Let n = the number of kilograms of green beans. (This, again, is a "value" equation.)

$$1 \cdot n + 1.30(100) + 1.40(90) = 1.20(190 + n)$$

bean value + corn value + pea value = mix value

35. Let t = the number of hours with both computers.

$$\frac{t}{4} + \frac{t}{6} = 1$$

part by A + part by B = total job

38. Let t = the number of hours for hose B alone.

$$\frac{2}{3} + \frac{3}{t} = 1$$

part of fill done by A + part of fill done by B = total fill

41. Let r = speed of the plane in still air

$r + 20$ = speed with the wind

$r - 20$ = speed against the wind

Times are equal in this problem, so use $t = \frac{d}{h}$.

$$\frac{300}{r - 20} = \frac{400}{r + 20}$$

44. $V = \frac{1}{3}\pi r^2 h$; $3V = \pi r^2 h$; $\frac{3V}{\pi r^2} = h$

47. $A = \frac{1}{2}h(b + b')$; $2A = h(b + b')$; $\frac{2A}{h} = b + b'$; hence, $b = \frac{2A}{h} - b'$ or $b = \frac{2A - b'h}{h}$.

Section 2.3 Linear Inequalities

BASIC CONCEPTS you should learn from this unit.

1. A Linear Inequality is an inequality containing a variable to the first power.

2. The solution of a linear inequality is the set of all values of the variable that make the statement true.

3. Graphs of linear inequalities are generally intervals on a number line.

4. Procedures for solving an inequality are the same as those for solving equations, with this one exception: multiplying or dividing by a negative number reverses the sense of the inequality.

WHAT YOU SHOULD KNOW HOW TO DO upon completion of this unit.

1. Recognize a linear inequality.

2. GRAPH an inequality on a number line.

3. Solve an inequality algebraically, using addition, subtraction, multiplication. and division.

SOLUTIONS to Exercise Set 2.3

2. $-4 < x \leq 1$: $(-4, 1]$

5. $-12 \leq x \leq -3$: $[-12, -3]$

8. $x < 17$: $(-\infty, 17)$

11. $[5, 8]$: $5 \leq x \leq 8$

14. $[-3, 10]$: $-3 \leq x \leq 10$

17. $[0, \infty)$: $x \geq 0$

20. $x + 5 < 4$

$x + 5 - 5 < 4 - 5$

$x < -1$

23. $x - 3 \geq 2$

$x \geq 2 + 3$

$x \geq 5$

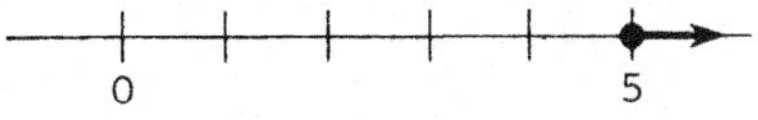

26. $-5 > b - 3$

$-5 + 3 > b$

$-2 > b$

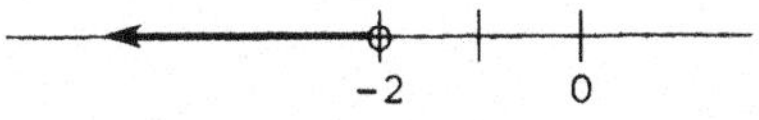

29. $2x \geq 0$

$x \geq 0$

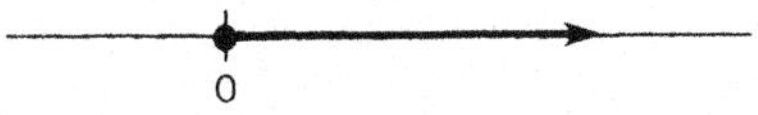

32. $3x - 2 > 4$

$3x > 6$

$x > 2$

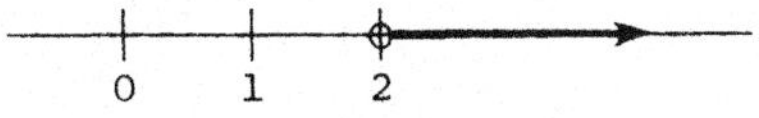

35. $\frac{4}{5 - 3x} < 0$

For this inequality to be true,
5-3x must be negative or

$5 - 3x < 0$

$5 < 3x$

$\frac{5}{3} < x$

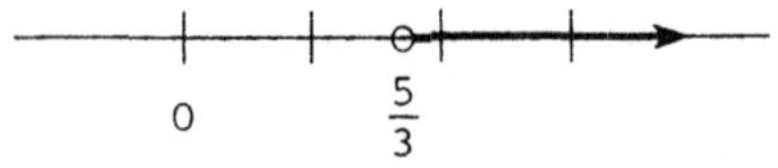

38. $\frac{1}{2}y - 2 \leq 2$

$\frac{1}{2}y \leq 4$

$y \leq 8$: $(-\infty, 8]$

41. $4(2x + 1) < 16$

$8x + 4 < 16$

$8x < 12$

$x < 3/2$: $\left(-\infty, \frac{3}{2}\right)$

44. $4(x - 3) \geq 3(x - 2)$

$4x - 12 \geq 3x - 6$

$x \geq 6$: $[6, \infty)$

47. $\frac{2}{3}(x + 1) + \frac{5}{6} \geq \frac{1}{2}(2x - 1) + 4$

$\frac{2}{3}x + \frac{2}{3} + \frac{5}{6} \geq x - \frac{1}{2} + 4$ Multiply both sides by the LCD, 6.

$4x + 4 + 5 \geq 6x - 3 + 24$

$4x + 9 \geq 6x + 21$

$-12 \geq 2x$

$x \leq -6$: $(-\infty, -6]$

50. $\frac{x}{5} - \frac{1 - x}{2} > \frac{x}{2} - 3$ Multiply both sides by 10.

$2x - 5 + 5x > 5x - 30$

$2x > -25$

$x > -25/2$: $\left(\frac{-25}{2}, \infty\right)$

53. $-2 < 4x \leq 5$

$-\frac{1}{2} < x \leq \frac{5}{4}$: $\left(-\frac{1}{2}, \frac{5}{4}\right]$

56. $5 \le 3x - 1 \le 11$

$6 \le 3x \le 12$ Add 2

$2 \le x \le 4$ Divide by 3

$[2, 4]$

59. $-8 < 2 - 5x \le 7$

$-10 \le -5x \le 5$ subtract 2

$2 \ge x \ge -1$ Divide by -5

or

$-1 \le x \le 2$: $[-1, 2]$

62. A is to be less than B:

$160 < 100 + 0.20m$

$60 < 0.20m$

$300 < m$ or more than 300 miles

65. Let n = number of copies sold

$12.50n > 19000 + 6n$

income from n copies > cost of n copies

$6.5n > 19000$

$n > 2923$ 2924 copies or more

Section 2.4 Absolute Value in Equations and Inequalities

BASIC CONCEPTS you should learn from this section.

1. Definition of Absolute Value: $|x| = \begin{cases} x & \text{if } x \ge 0 \\ -x & \text{if } x < 0 \end{cases}$

2. Solving equations and inequalities with Absolute Value expressions is done by using your understanding of absolute value <u>and</u> the rules learned from the preceding units for solving equations and inequalities. For example,

$|3x + 1| = 5$

$3x + 1 = 5$ $\quad x = \frac{4}{3}$

$-(3x + 1) = 5$ $\quad 3x + 1 = -5$ $\quad x = -2$

$|3x + 1| > 5$

$3x + 1 > 5$ $\quad x > \frac{4}{3}$

$-(3x + 1) > 5$ $\quad 3x + 1 > -5$ $\quad x > -2$

3. Graphs of equations or inequalities with Absolute Value generally consist of one or two intervals on a number line.

WHAT YOU SHOULD KNOW HOW TO DO upon completion of this unit.

1. Define Absolute Value.

2. Solve equations and inequalities containing Absolute Value expressions.

3. Graph equations and inequalities that include Absolute Value

SOLUTIONS to Exercise Set 2.4

2. $|r - 5| = \frac{1}{2}$; $r - 5 = \frac{1}{2}$ $\quad -(r - 5) = \frac{1}{2}$

$\underline{r = 5\frac{1}{2}}$, $\quad \underline{r = 4\frac{1}{2}}$

5. $|-3x + 1| = 5$

$-3x + 1 = 5$ or $+3x - 1 = 5$

$-3x = 4 \qquad 3x = 6$

$\underline{x = -\frac{4}{3}} \qquad \underline{x = 2}$

8. $\frac{1}{|x|} = 5$ Therefore, $1 = 5|x|$ or $|x| = \frac{1}{5}$.

So $\underline{x = \frac{1}{5}}$ or $\underline{x = -\frac{1}{5}}$

11. $|x + 1| > 3$

$x + 1 > 3$ or $-(x + 1) > 3$

$x > 2 \qquad x + 1 < -3$

$x < -4$

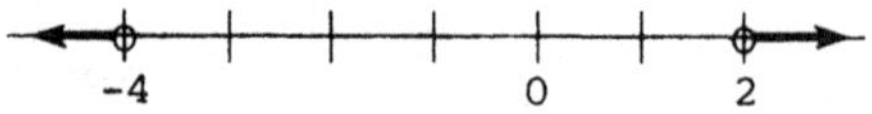

12. $|3x + 6| \leq 12$

$3x + 6 \leq 12 \qquad -3x - 6 \leq 12$

$3x \leq 6 \qquad -3x \leq 18$

$x \leq 2 \qquad x \geq -6$

14. Look out! Absolute Value cannot be negative! Therefore only values that satisfy $|3x + 2| \geq 0$ will do. What we're really saying here is that $|3x + 2| = -1$ is an invalid statement.

17. $|x - 3| \geq 4$

$x - 3 \geq 4 \qquad -(x - 3) \geq 4$

$x \geq 7 \qquad x - 3 \leq -4$

$x \leq -1$

$(-\infty, -1]$ or $[7, \infty)$

20. $\frac{|3x + 2|}{2} \le 4$

$|3x + 2| \le 8$

$3x + 2 \le 8$

$3x \le 6$

$x \le 2$

$-(3x + 2) \le 8$

$3x + 2 \ge -8$

$3x \ge -10$

$x \ge -\frac{10}{3}$

$[-\frac{10}{3}, 2]$

23. $\left|\frac{5 - x}{3}\right| > 4$:

$\frac{5 - x}{3} > 4$ or $\frac{x - 5}{3} > 4$

$5 - x > 12$

$x < -7$

$(-\infty, -7)$

$x - 5 > 12$

$x > 17$

$(17, \infty)$

26. Prove that $|x|^2 = x^2$.

For $x < 0$

$|x| = -x$

$(-x)^2 = x^2$

For $x = 0$

$|x| = x$

$x^2 = x^2$

For $x > 0$

$|x| = x$

$x^2 = x^2$

28. Let x = weekly income.

$|x - 300| \le 50$

(difference between weekly income and \$300) is less than or equal to (\$50)

Section 2.5 The Quadratic Equation

BASIC CONCEPTS you should learn from this unit.

1. Definition of a QUADRATIC EQUATION as one that can be written in the form $ax^2 + bx + c = 0$ where a, b, c are real numbers, but $a \neq 0$.

2. The methods of solving Quadratic Equations are

   A. Simply taking a square root;

   B. Factoring, then using the Zero Product Principle, that is, if $a \cdot b = 0$, either $a = 0$ or $b = 0$;

   C. Completing the square;

   D. Using the Quadratic Formula, $x = \frac{-b \pm \sqrt{b^2 - 4ac}}{2a}$

3. Quadratic equations have <u>two</u> roots which can be real or complex.

4. The <u>discriminant</u>, $(b^2 - 4ac)$, tells us the nature of the roots:

$b^2 - 4ac > 0$	$b^2 - 4ac = 0$	$b^2 - 4ac < 0$
two real roots	double root	two complex roots

5. Certain equations with radicals and exponents, although <u>not quadratic</u>, can be solved by the foregoing methods if a proper substitution of variable is made or by using the squaring process.

WHAT YOU SHOULD KNOW HOW TO DO upon completion of this unit.

1. Recognize an equation as quadratic.

2. Solve quadratic equations by various methods.

3. Solve certain types of equations that are quadratic in form.

SOLUTIONS to Exercise Set 2.5

2. $4x^2 - 64 = 0$, $4x^2 = 64$, $x = \pm 4$

5. $5y^2 - 25 = 0$, $5y^2 = 25$, $y = \pm\sqrt{5}$

8. $(4t + 1)^2 - 3 = 0$, $(4t + 1)^2 = 3$, $4t + 1 = \pm\sqrt{3}$, $t = \dfrac{-1\pm\sqrt{3}}{4}$

11. $x^2 - 3x + 2 = 0$, $(x - 2)(x - 1) = 0$; $x - 2 = 0$ and $x - 1 = 0$, so $x = 2, 1$.

14. $3r^2 - 4r + 1 = 0$, $(3r - 1)(r - 1) = 0$, $3r - 1 = 0$ and $r - 1 = 0$, so $r = \frac{1}{3}, 1$.

17. $y^2 - 4y = 0$, $y(y - 4) = 0$, $y = 0$, $y = 4$

20. $2s^2 - 5s - 3 = 0$, $(2s + 1)(s - 3) = 0$, $s = -1/2$, $s = 3$

23. $6x^2 - 5x + 1 = 0$, $(3x - 1)(2x - 1) = 0$; $x = 1/3$, $x = 1/2$

26. $t^2 - 2t = 15$

$$t^2 - 2t + 1 = 15 + 1$$
$$(t - 1)^2 = 16$$
$$t - 1 = \pm 4$$
$$t = 5$$
$$t = -3$$

29. $3x^2 + 8x = 3$

$$x^2 + \frac{8}{3}x + \frac{16}{9} = 1 + \frac{16}{9}$$
$$\left(x + \frac{4}{3}\right)^2 = \frac{25}{9}$$
$$x + \frac{4}{3} = \pm\frac{5}{3}$$
$$x = \frac{1}{3}, \ -3$$

32. $3x^2 - 4x = -3$

$$x^2 - \frac{4}{3}x + \frac{4}{9} = -1 + \frac{4}{9}$$

$$\left(x - \frac{2}{3}\right)^2 = -\frac{5}{9}$$

$$x - \frac{2}{3} = \frac{\pm i\sqrt{5}}{3} \;; \quad x = \frac{2 \pm i\sqrt{5}}{3}$$

35. $3x^2 + 2x = -1$

$$x^2 + \frac{2}{3}x + \frac{1}{9} = -\frac{2}{9}$$

$$\left(x + \frac{1}{3}\right)^2 = -\frac{2}{9}$$

$$x + \frac{1}{3} = \pm\frac{i\sqrt{2}}{3} \;; \quad x = \frac{-1 \pm i\sqrt{2}}{3}$$

38. $2x^2 + 3x + 3 = 0$

$$x = \frac{-3 \pm \sqrt{9 - 24}}{4}$$

$$= \frac{-3 \pm i\sqrt{15}}{4}$$

41. $5y^2 - 4y + 5 = 0$

$$y = \frac{4 \pm \sqrt{16 - 100}}{10} = \frac{4 \pm \sqrt{-84}}{10}$$

$$= \frac{2 \pm i\sqrt{21}}{5}$$

44. $2x^2 + 4x - 3 = 0$

$$x = \frac{-4 \pm \sqrt{16 + 24}}{4} = \frac{-4 \pm \sqrt{40}}{4}$$

$$= \frac{-2 \pm \sqrt{10}}{2}$$

47. $4u^2 + 3u = 0$

$$u = \frac{-3 \pm \sqrt{9 - 0}}{8}$$

$$= \frac{-3 \pm 3}{8} \qquad \underline{u = 0}, \quad \underline{u = -\frac{3}{4}}$$

50. $2t^2 + 2t + 3 = 0$

$$t = \frac{-2 \pm \sqrt{4 - 24}}{4} = \frac{-1 \pm i\sqrt{5}}{2}$$

53. $2x^2 + 5x + 4 = 0$

$$x = \frac{-5 \pm \sqrt{25 - 32}}{4} = \frac{-5 \pm i\sqrt{7}}{4}$$

56. $x^2 + 2 = 0 \quad x^2 = -2 \quad x = \pm\sqrt{-2} = \pm i\sqrt{2}$

59. $a^2 + b^2 = c^2 \quad b^2 = c^2 - b^2 \quad b = \pm\sqrt{c^2 - b^2}$

62. $A = \pi r^2 \quad r^2 = \frac{A}{\pi} \quad r = \pm\sqrt{\frac{A}{\pi}}$ or $\frac{\pm\sqrt{A\pi}}{\pi}$

65. $x^2 - 2x + 3 = 0$ Therefore, $b^2 - 4ac = 4 - 12 = -8$.
two complex roots

68. $2x^2 + x + 5 = 0$; $b^2 - 4ac = 1 - 10 = -9$
two complex roots

71. $3x^2 + 2x = 0$, $b^2 - 4ac = 4$
two real rational roots

74. $3x^2 + x - 5 = 0$, rearrange it, then $b^2 - 4ac = 1 + 60 = 61$
there are two real irrational roots

77. $3r^2 - 6r + 1 = 0$, $b^2 - 4ac = 6 - 12 = -6$
there are two complex roots

80. $4s^2 + 4s + 1 = 0$, $b^2 - 4ac = 16 - 16 = 0$
there is a double root (real).

83. $b^2 - 4ac$ is to be less than 0, so

$2x^2 + 3x + k = 0 \quad b^2 - 4ac = 3 - 8k$

and $3 - 8k < 0$, $\underline{k > 3/8}$

86. $x - \sqrt{13 - x} = 1$

$x - 1 = \sqrt{13 - x}$

$x^2 - 2x + 1 = 13 - x$

$x^2 - x - 12 = 0 \quad (x - 4)(x + 3) = 0 \quad x = 4$

Solution $x = -3$ will not check.

89. $\sqrt{3x + 4} - \sqrt{2x + 1} = 1$

$$\sqrt{3x + 4} = 1 + \sqrt{2x + 1}$$
$$3x + 4 = 1 + 2\sqrt{2x + 1} + 2x + 1$$
$$x + 2 = 2\sqrt{2x + 1}$$
$$x^2 + 4x + 4 = 4(2x + 1)$$
$$x^2 - 4x = 0$$
$$x = 0 , \quad x = 4$$

Both roots check.

92. $\sqrt{5x + 1} + \sqrt{4x - 3} = 7$

$$\sqrt{5x + 1} = 7 - \sqrt{4x - 3}$$
$$5x + 1 = 49 - 14\sqrt{4x - 3} + 4x - 3$$
$$x - 45 = -14\sqrt{4x - 3}$$
$$x^2 - 90x + 2025 = 196(4x - 3)$$
$$= 784x - 588$$
$$x^2 - 874x + 2613 = 0$$
$$(x - 3)(x - 871) = 0$$

Isn't that a dandy?

$x = 3$ checks.

$x = 871$ doesn't check.

95. $\frac{6}{x^2} + \frac{1}{x} - 2 = 0 ; \quad 6 + x - 2x^2 = 0 ; \quad (3 + 2x)(2 - x) = 0$

$x = -\frac{3}{2} \quad x = 2$

Substitute $u = \frac{1}{x}$.

$$6u^2 + u - 2 = 0$$
$$(3u + 2)(2u - 1) = 0 ,$$

$u = -\frac{2}{3}$ or $x = -\frac{3}{2}$

$u = \frac{1}{2}$ or $x = 2$

98. $3x^{4/3} - 4x^{2/3} - 4 = 0$

Let $u = x^{2/3}$; then

$$3u^2 - 4u - 4 = 0$$
$$(3n + 2)(u - 2) = 0$$

$u = -\frac{2}{3}$ $\qquad$ $u = +2$

$x^{2/3} = -\frac{2}{3}$ $\qquad$ $x^{2/3} = 2$

$\underline{x = \left(-\frac{8}{27}\right)^{1/2}}$ $\qquad$ $x = 8^{1/2} = \underline{\sqrt{8}}$

101. Prove that if r_1 and r_2 are roots to a quadratic equation, then

$$r_1 r_2 = \frac{c}{a} \quad \text{and} \quad r_1 + r_2 = -\frac{b}{a}$$

Proof:

If r_1 and r_2 are roots to a quadratic equation, then

$$r_1 = \frac{-b + \sqrt{b^2 - 4ac}}{2a} \quad \text{and} \quad r_2 = \frac{-b - \sqrt{b^2 - 4ac}}{2a}$$

$$r_1 r_2 = \left(\frac{-b + \sqrt{b^2 - 4ac}}{2a}\right)\left(\frac{-b - \sqrt{b^2 - 4ac}}{2a}\right)$$

$$= \frac{b^2 - b^2 + 4ac}{4a^2} = \frac{c}{a} \text{ as required}$$

also

$$r_1 + r_2 = \frac{-b + \sqrt{b^2 - 4ac}}{2a} + \frac{-b - \sqrt{b^2 - 4ac}}{2a}$$

$$= \frac{-2b}{+2a} = -\frac{b}{a} \text{ as required}$$

Section 2.6 Applications - Quadratic Equations

BASIC CONCEPTS you should learn from this unit.

1. Procedure for solving these word problems is similar to that for other word problems: Read carefully, choose a letter to represent the unknown, form an equation, and solve it.

2. Be alert to meaningless solutions in the context of the problem (negative age, for example).

3. Use of drawings is often helpful.

WHAT YOU SHOULD KNOW HOW TO DO upon completion of this unit.

1. Set up an equation of second degree to solve an applied problem.

2. Solve word problems similar to those illustrated and check your answers.

SOLUTIONS to Exercise Set 2.6

2. Let x = number of days for the designer

then $x + 16$ = number of days for the assistant

$$\frac{6}{x} + \frac{6}{x + 16} = 1 \qquad \text{Multiply both sides by } x(x + 16) \text{ gives}$$

$$6x + 96 + 6x = x^2 + 16x$$

$$x^2 + 4x - 96 = 0$$

$$(x + 12)(x - 8) = 0$$

$$x = -12, \qquad \text{Reject this root.}$$

$$x = 8 \text{ days}$$

$$x + 16 = 24 \text{ days}$$

5.

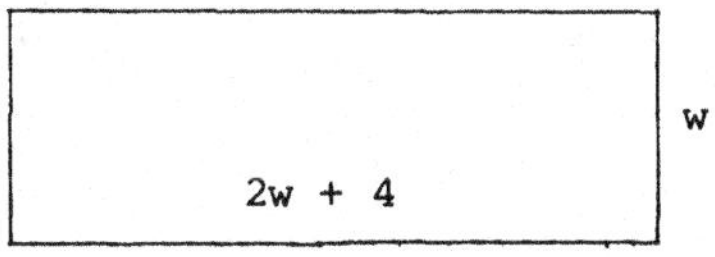

$$w = \text{width}$$
$$2w + 4 = \text{length}$$
$$\text{area} = w(2w + 4)$$
$$2w^2 + 4w = 48$$
$$w^2 + 2w - 24 = 0$$
$$(w + 6)(w - 4) = 0$$
($w = -6$; reject this root)

$w = 4'$
$2w + 4 = 12'$

8.

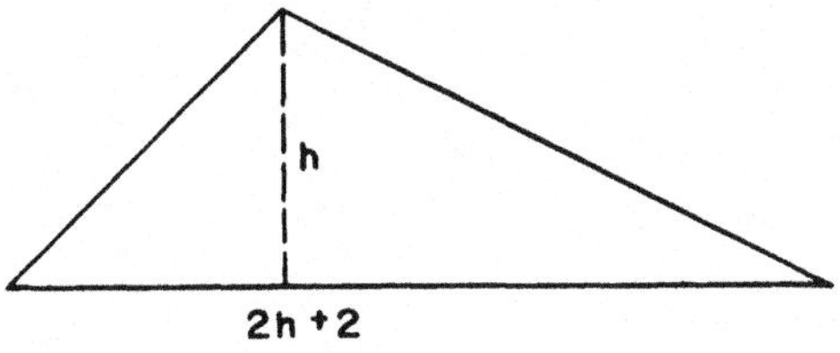

$$h = \text{height}$$
$$2h + 2 = \text{base}$$
$$A = \frac{1}{2}Bh$$
$$\frac{1}{2}(2h + 2) \cdot h = 12$$
$$(h + 1)h = 12$$
$$h^2 + h - 12 = 0$$
$$(h + 4)(h - 3) = 0$$

$h = 3'$
base $= 8'$

11. Let x = the number

$x + \frac{1}{x} = \frac{26}{5}$; $5x^2 + 5 = 26x$, $5x^2 - 26x + 5 = 0$, $(5x - 1)(x - 5) = 0$

so $\underline{x = \frac{1}{5}}$ or $\underline{x = 5}$

14. n = first odd number

$n + 2$ = next consecutive odd number

then $\frac{1}{n} + \frac{1}{n + 2} = \frac{8}{15}$ LCD $= 15n(n + 2)$

$$15n + 30 + 15n = 8n^2 + 16n$$
$$8n^2 - 14n - 30 = 0$$
$$4n^2 - 7n - 15 = 0$$
$$(4n + 5)(n - 3) = 0$$

hence $n = 3$ or $n = -\frac{5}{4}$ Reject this root

and $n + 1 = 5$

17. Let n = number of shares

cost of each share $= \frac{1200}{n}$

$$\frac{1200}{n - 30} = \frac{1200}{n} + 2 \qquad \text{LCD} = n(n - 30)$$
$$1200n = 1200n - 36000 + 2n^2 - 60n$$
$$2n^2 - 60n - 36000 = 0$$
$$n^2 - 30n - 18000 = 0$$
$$(n - 150)(n + 120) = 0$$

Therefore, $\underline{n = 150}$ shares

20. x = daily rate of pay

$\frac{480}{x}$ = number of days worked at \$x per day

$$\frac{480}{x} = \frac{480}{x - 8} + 2 \qquad \text{LCD} = x(x - 8)$$

$$480x - 3840 = 480x + 2x^2 - 16x$$

$$2x^2 - 16x + 3840 = 0 \; ; \qquad x^2 - 8x + 1920 = 0$$

$$(x - 48)(x + 40) = 0$$

hence, <u>x = \$48 per day</u>

Section 2.7 Second Degree Inequalities

BASIC CONCEPTS you should learn from this unit.

1. A <u>SECOND</u> DEGREE inequality has the form $ax^2 + bx + c \leq 0$ or $ax^2 + bx + c \geq 0$ or can be put in that form.
2. If factorable, second degree inequalities can be solved by analyzing the signs of the factors.
3. Inequalities of the form $\frac{ax + b}{cx + d} < 0$ can also be solved by analysis of signs.

WHAT YOU SHOULD KNOW HOW TO DO upon completion of this unit.

1. Recognize second degree inequalities.
2. Solve second degree inequalities by factoring and sign analysis.
3. Solve inequalities of the form $\frac{ax + b}{cx + d} < 0$ by sign analysis.

SOLUTIONS to Exercise Set 2.7

2. $x^2 + 3x - 4 \leq 0$

$(x + 4)(x - 1) \leq 0$

x + 4 ---0+++++++++++++++++++++++++++++++++++

x - 1 -----------------------0+++++++++++++++

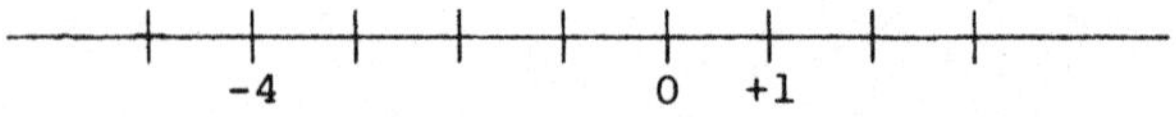

Signs are unlike when $-4 < x < 1$

final solution $-4 \leq x \leq 1$

5. $4x - 2x^2 < 0$

$2x(2 - x) < 0$

2x -----------0+++++++++++++++++++++++++++++

2 - x +++++++++++++++++++++0----------------

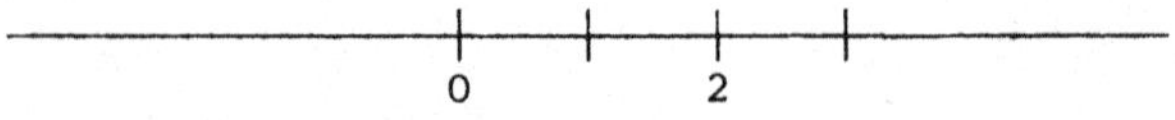

Signs are unlike when $x < 0$ <u>or</u> $x > 2$

8. $\frac{x - 6}{x + 4} \geq 0$

```
(x - 6) ------------------------------0++++++
(x + 4) ----0++++++++++++++++++++++++++++++++++
```

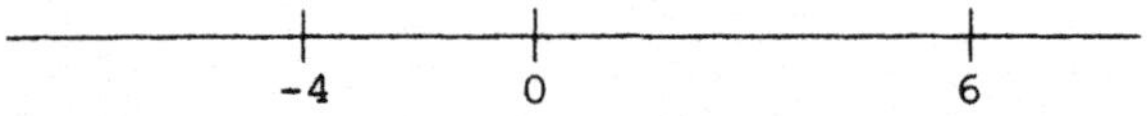

Solution when signs are the same $x < -4$ or $x \geq 6$

<u>or</u> $x - 6 = 0$

11. $\frac{3s + 2}{2s - 1} \geq 0$

```
3s + 2 ---------------0+++++++++++++++++++++
2s - 1 ----------------------0++++++++++++++
```

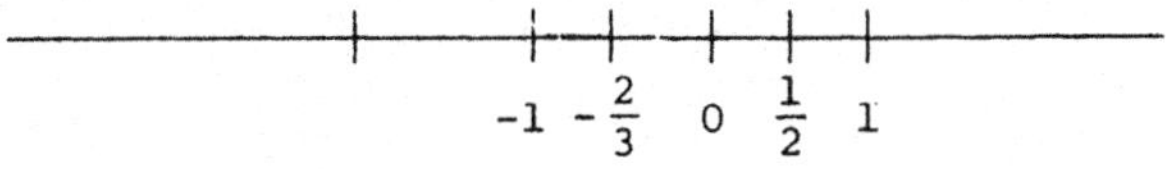

Solution: $s > \frac{1}{2}$ or $s \leq -\frac{2}{3}$

14. $(x - 4)(2x + 5)(2 - x) \leq 0$

```
  x - 4 ---------------------------------0++
 2x + 5 -------0+++++++++++++++++++++++++++++
  2 - x +++++++++++++++++++++++++0----------
```

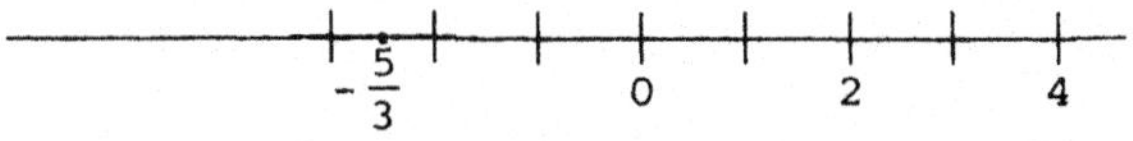

For $(x - 4)(2x + 5)(2 - x) \leq 0$ <u>one</u> or all three of the factors must be negative, <u>or</u> any one of the factors must be zero. Thus

$x \geq 4$ or $-5/2 \leq x \leq 2$

17. $2x^2 - 3x - 5 < 0$

$(2x - 5)(x + 1) < 0$

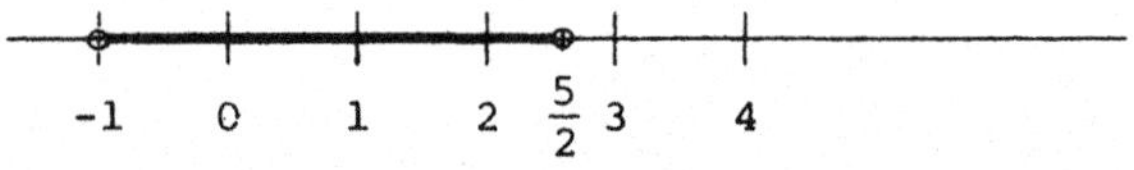

20. $\frac{3x + 2}{2x - 3} \geq 0$

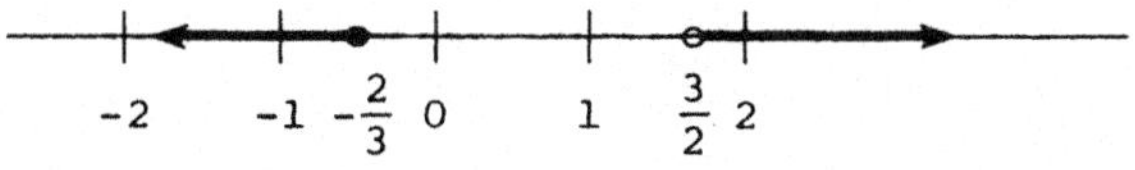

23. $6x^2 + 8x + 2 \geq 0$

$2(3x^2 + 4x + 1) \geq 0$

$2(3x + 1)(x + 1 \geq 0$

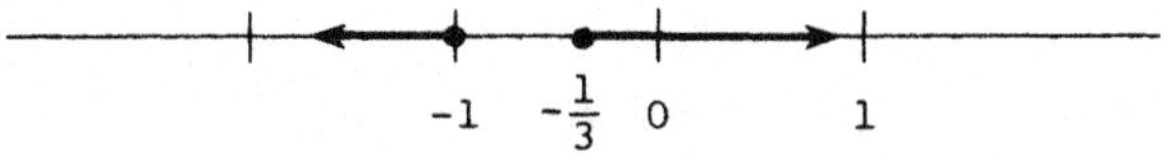

26. $(2x + 5)(3x - 2)(x + 1) < 0$

(2x + 5) ---0++++++++++++++++++++++++++++

(3x - 2) --------------------0+++++++++++

(x + 1) -----------0+++++++++++++++++++++

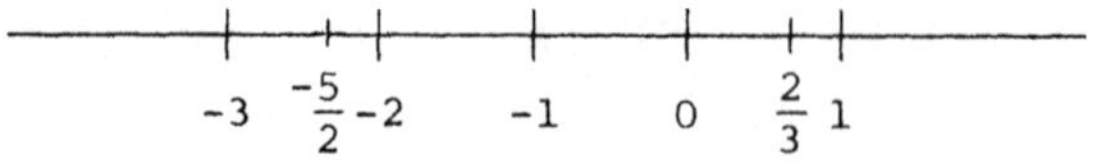

Solution

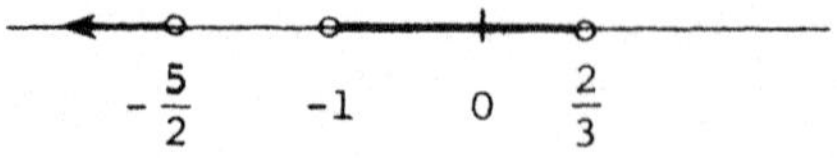

29. $\sqrt{(x - 2)(x + 1)}$ will be real when $(x - 2)(x + 1) \geq 0$;

that is, when $x \leq -1$ or $x \geq 2$.

32. $\sqrt{2x^2 + 3x + 1} = \sqrt{(2x + 1)(x + 1)}$ is real for

$x \leq -1 \quad x \geq -\frac{1}{2}$

SOLUTIONS to Review Exercises Chapter 2

2. $2(2x - 3) - 3(x + 1) = -9$

$4x - 6 - 3x - 3 = -9$

$x = 0$

5.

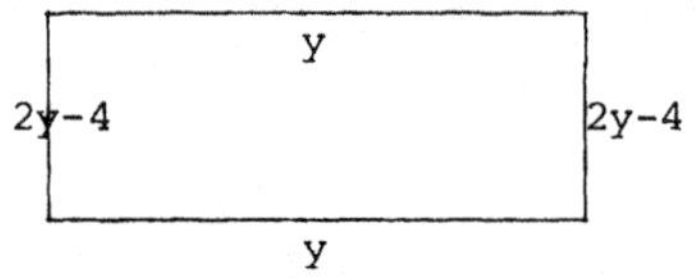

Let y = length

$y + y + 2y - 4 + 2y - 4 = 12$

$6y - 8 = 12$

$6y = 20 \qquad y = 10/3$

$2y - 4 = 8/3$

8. Let b = number of hours for B to do job alone

$\frac{2}{3} + \frac{2}{b} = 1$

$2b + 6 = 3b$

$6 = b$ Answer: 6 hours

11. $3 \leq 2x + 1$

$2 \leq 2x$

$x \geq 1$

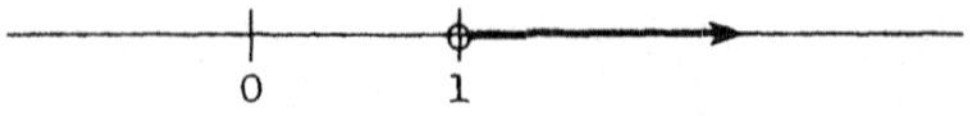

14. $\dfrac{-1}{2x - 5} < 0$ True only if $2x - 5 > 0$ or when $x > 5/2$

$\left(\dfrac{5}{2}, \infty\right)$

17. $|4x - 1| = 5$ $\qquad -(4x - 1) = 5$

$4x - 1 = 5$ $\qquad 4x - 1 = -5$

$x = 6/4 = 3/2$ $\qquad 4x = -4$

$\underline{x = 3/2}$ $\qquad \underline{x = -1}$

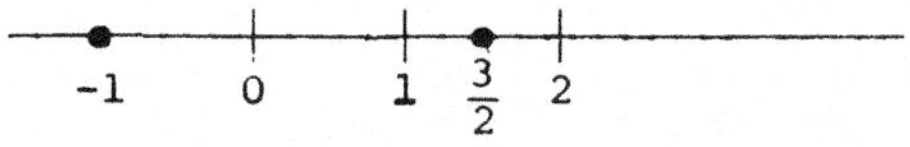

20. $|3x - 2| \geq 6$ $\qquad -(3x - 2) \geq 6$

$3x - 2 \geq 6$ $\qquad 3x - 2 \leq -6$

$3x \geq 8$ $\qquad 3x \leq -4$

$x \geq 8/3$ $\qquad x \leq -4/3$

Solution: $\left(-\infty, -\dfrac{4}{3}\right], \left[\dfrac{8}{3}, \infty\right)$

23. $x^2 - 2x + 6 = 0$

$x^2 - 2x + 1 = -6 + 1$

$(x - 1)^2 = -5$

$x - 1 = \pm i\sqrt{5}$; $\qquad x = 1 \pm i\sqrt{5}$

26. $49x^2 - 9 = 0$

$49x^2 = 9$

$7x = \pm 3$ $\qquad x = \dfrac{3}{7}, -\dfrac{3}{7}$

29. $3r^2 = 2r + 5$

$3r^2 - 2r - 5 = 0$ $\qquad b^2 - 4ac = 4 + 60 = 64$

Therefore, there are two real rational roots.

32. $x + \sqrt{2x - 3} = 4$

$\sqrt{2x - 3} = 4 - x$

$2x - 3 = 16 - 8x + x^2$

$x^2 - 10x + 19 = 0$ $\qquad$ Therefore, $x = \dfrac{10 \pm \sqrt{24}}{2} = \underline{5 \pm \sqrt{6}}$

35. $\left(1 - \dfrac{2}{x}\right)^2 - 8\left(1 - \dfrac{2}{x}\right) + 15 = 0$

let $u = 1 - 2/x$ $\qquad$ so that $u^2 - 8u + 15 = 0$

$(u - 5)(u - 3) = 0$

$u = 5$ $\qquad u = 3$

$5 = 1 - \dfrac{2}{x}$ $\qquad 3 = 1 - \dfrac{2}{x}$

$x = -\dfrac{1}{2}$ $\qquad x = -1$

38. $x^2 + 4x - 5 \geq 0$

$(x + 5)(x - 1) \geq 0$ when $x \leq -5$ or $x \geq 1$

Interval notation: $(-\infty, -5]$ or $[1, \infty)$

Chapter 2 - A SHORT QUIZ

1. The exponent on the variable in a linear equation is ___?___

2. A solution to a linear equation is ___?___ ___?___

3. Which of the following are linear equations?

   (a) $2x - 7 = \frac{1}{y}$ (b) $4 - x = 7x$ (c) $5x^2 + x = 1$

4. Is $\frac{7}{3}$ a solution of $3x - 5 = 2$?

5. The sum of two consecutive odd integers is 76. Write an equation that will allow you to find the integers.

6. Solve $\frac{1}{R} = \frac{1}{a} + \frac{1}{b}$ for b.

7. If $a > b$ and $m < 0$, place the correct inequality symbol between am _?_ bm .

8. Graph the inequality $-3 \leq x < 4$ on a number line.

9. Write this interval as an inequality: $[-3, 5)$.

10. $\frac{-5}{3x - 7} < 0$ only when $3x - 7$ is ___?___.

11. For what values of x is $|3x - 2| = -1$?

12. Write an absolute value inequality expressing this graph:

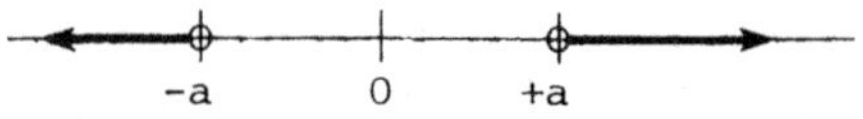

13. $ax^2 + bx + c = 0$ is called the general quadratic equation; the constants a, b, and c are real with what restrictions?

14. In the quadratic formula $b^2 - 4ac$ is called the ___?___ and when it is negative we know that the roots are ___?___.

15. Under what special condition does a quadratic equation have two identical roots?

16. Write a quadratic equation having 1 and -7 as roots.

17. Solve by completing the square:

    $3x^2 + 2x + 1 = 0$

18. Two integers differ by 5 and their product is 150. What are they?

19. A man can do a job in a hours and his son can do it in b hours. How long will it take them together?

20. Given $x^2 + 4x - 21 < 0$, what would be your first step in finding the solution?

21. Given $(x - 4)(x + 2)(x + 5) > 0$. Using a sign analysis diagram, determine the solution.

22. Under what condition of $3x + 5$ will the following be true: $\frac{3x + 5}{x - 7} = 0$?

Quiz Answers Chapter 2

1. 1 (one)

2. a number value of the variable that makes the statement (equation) true.

3. b

4. Yes

5. $n + n + 2 = 76$

6. $b = \frac{Ra}{a - b}$

7. $<$

8. 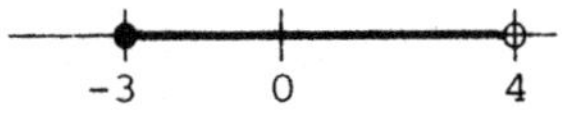

9. $-3 \le x < 5$

10. positive

11. Ø

12. $|x| > a$

13. $a \neq 0$

14. discriminant, complex

15. Discriminant = 0 .

16. $(x - 1)(x + 7) = 0$ <u>or</u> $x^2 + 6x - 7 = 0$

17. $3x^2 + 2x + 1 = 0$

$$x^2 + \frac{2}{3}x + \frac{1}{9} = -1 + \frac{1}{9}$$

$$(x + \frac{1}{3})^2 = -\frac{8}{9}$$

$$x + \frac{1}{3} = \frac{\pm 2i\sqrt{2}}{3}$$

$$x = \frac{-1 \pm 2i\sqrt{2}}{3}$$

18. 10, 15 also -10, -15

19. $\frac{ab}{a + b}$ hours

20. Factor the left-hand side.

```
21.  x - 4 ---------------------------0+++++++
     x + 2 ----------0++++++++++++++++++++++++
     x + 5 --0++++++++++++++++++++++++++++++++
```

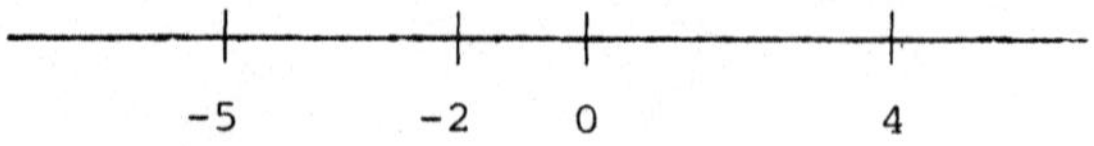

Answer: $-5 < x < -2$; and $x > 4$

22. only if $3x + 5 = 0$

CHAPTER 3

Functions

Section 3.1 Rectangular Coordinate Systems

BASIC CONCEPTS you should learn from this unit.

1. The rectangular coordinate system consists of two perpendicular number lines, a horizontal one, called the x-axis, and a vertical one, called the y-axis.

2. Each point in the coordinate plane can be located by two coordinates (x, y), each of which is a real number.

3. The Distance Formula $d = \sqrt{(x_2 - x_1)^2 + (y_2 - y_1)^2}$ provides an easy way to obtain the distance between two points in the plane.

4. The graph of an equation in two variables is a line in the plane whose points are found by constructing a table of x and y values from the equation and plotting those pairs of coordinate values.

5. Symmetry prevails

   A. with respect to the y-axis if $-x$ replacing x leaves the equation unchanged;

   B. with respect to the x-axis if $-y$ replacing y leaves the equation unchanged;

   C. with respect to the origin if both of the foregoing replacements leave the equation unchanged.

WHAT YOU SHOULD KNOW HOW TO DO upon completion of this unit.

1. Plot points on a set of coordinate axes.

2. Find the distance between two points in the plane.

3. Determine whether three points are collinear.

4. Sketch graphs of equations in two variables.

5. Determine the symmetry of the graph of an equation without graphing.

SOLUTIONS to Exercise Set 3.1

2.

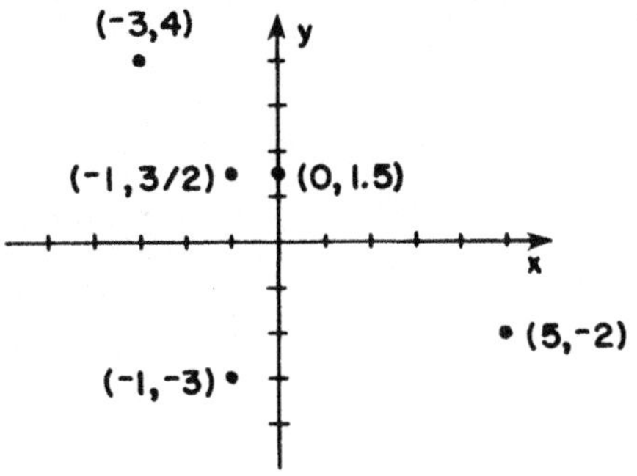

5. $P_1(-1, -5)$, $P_2(-5, -1)$

$$d = \sqrt{(-4)^2 + (4)^2} = \sqrt{32} = 4\sqrt{2}$$

8. $P_1(-\frac{1}{2}, 3)$, $P_2(-1, -\frac{3}{4})$

$$d = \sqrt{\left(-\frac{1}{2}\right)^2 + \left(-\frac{15}{4}\right)^2} = \sqrt{\frac{1}{4} + \frac{225}{16}} = \frac{\sqrt{229}}{4}$$

11. $R\left(-1, \frac{1}{2}\right)$, $S\left(-\frac{3}{2}, 1\right)$, $T(2, -1)$

$$\overline{RS} = \sqrt{\frac{1^2}{2} + \left(-\frac{1}{2}\right)^2} = \sqrt{\frac{1}{2}} = \frac{\sqrt{2}}{2}$$

$$\overline{RT} = \sqrt{\frac{3}{2}^2 + 3^2} = \sqrt{\frac{9}{4} + 9} = \frac{\sqrt{45}}{2}$$

$$\overline{ST} = \sqrt{2^2 + \left(\frac{7}{2}\right)^2} = \sqrt{4 + \frac{49}{4}} = \frac{\sqrt{65}}{2}$$

Thus $\overline{RS}$ is the shortest side.

14. Let $A(2, -3)$, $B(-1, -1)$, $C(3, 4)$

Then

$$\overline{AB}^2 = 2^2 + 3^2 = 13$$

$$\overline{AC}^2 = 7^2 + 1^2 = 50 \qquad 13 + 41 \neq 50$$

$$\overline{BC}^2 = 5^2 + 4^2 = 41$$

Therefore, ABC is <u>not</u> a right triangle.

17. $A(-1, 2)$, $B(1, 1)$, $C(5, -1)$

$$\overline{AB} = \sqrt{1^2 + 2^2} = \sqrt{5} \qquad \overline{AB} + \overline{BC} = \overline{AC}$$

$$\overline{AC} = \sqrt{3^2 + 6^2} = 3\sqrt{5} \qquad \text{yes, collinear}$$

$$\overline{BC} = \sqrt{2^2 + 4^2} = 2\sqrt{5}$$

20. $A(-1, 5)$, $B(1, 1)$, $C(-2, -8)$

$$\overline{AB} = \sqrt{6^2 + 2^2} = 2\sqrt{10}$$

$$\overline{AC} = \sqrt{3^2 + 1^2} = \sqrt{10} \qquad \underline{\text{not}} \text{ collinear}$$

$$\overline{BC} = \sqrt{9^2 + 3^2} = \sqrt{100} > 10$$

23. A(9, 2), B(11, 6), C(3, 5), D(1, 1)

$\overline{AB} = \sqrt{4^2 + 2^2} = 2\sqrt{5}$

$\overline{CD} = \sqrt{4^2 + 2^2} = 2\sqrt{5}$ equal

$\overline{AD} = \sqrt{1^2 + 8^2} = \sqrt{65}$

$\overline{BC} = \sqrt{1^2 + 8^2} = \sqrt{65}$ equal

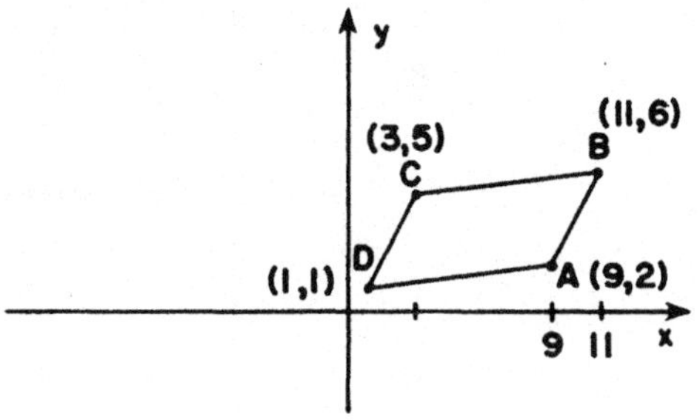

Yes, the points are the vertices of a parallelogram.

26. Plotting points to see locations:

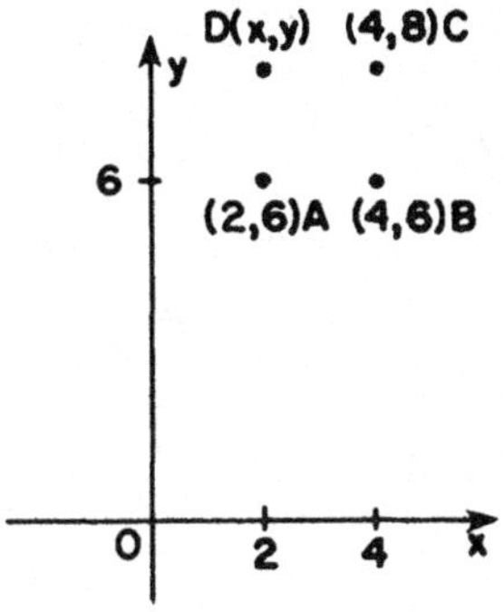

It would seem that (x, y) should be (2, 8). Verify by the distance formula.

29. $y = \sqrt{x}$

x	0	1	4	9	16
y	0	1	2	3	4

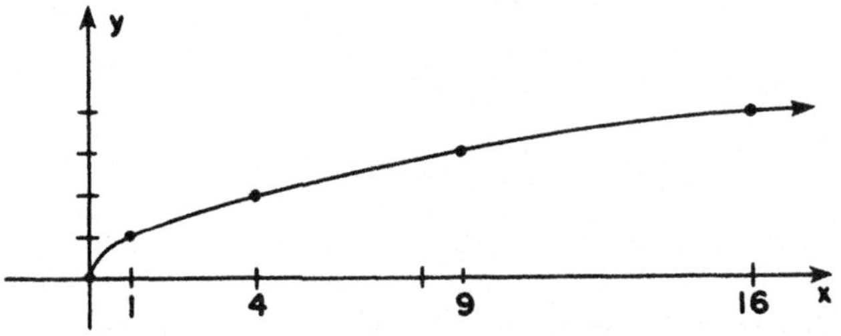

32. $y = 2 - |x|$

x	0	2	-2	3	-3	5	-5
y	2	0	0	-1	-1	-3	-3

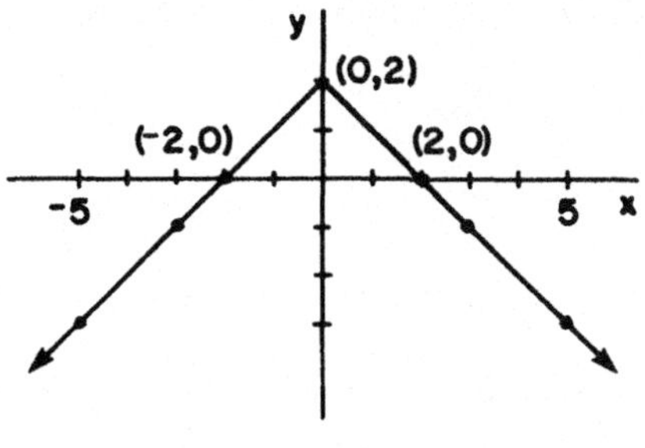

35. $y = x^3 + 1$

Substitution of $(-x)$ for x or $(-y)$ for y reveals no symmetry with either axis.

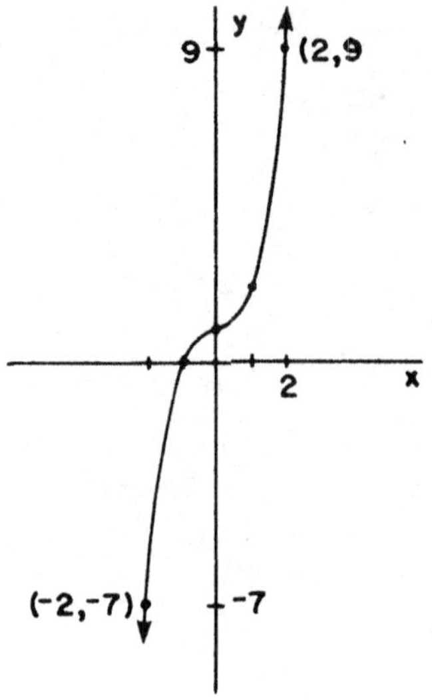

38. $y = 3x$

Substitution of $-x$ <u>and</u> $-y$ leaves the equation unchanged. Therefore, there is symmetry with the origin.

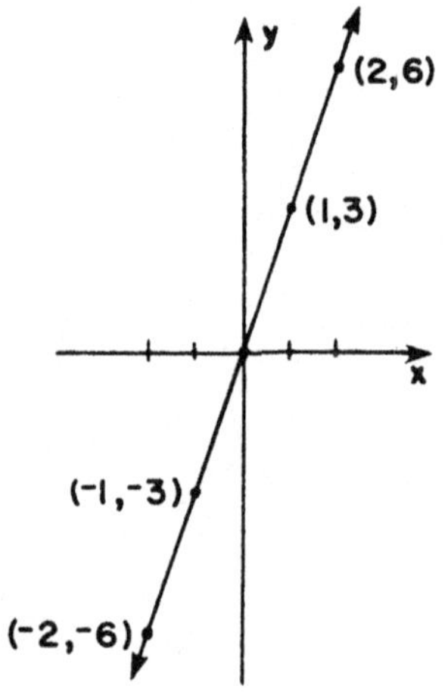

41. $y^2 = x - 4$

$-x$ for x CHANGES equation

$-y$ for y DOES NOT CHANGE equation

Therefore, there is only x-axis symmetry.

44. $y = (x - 2)^2$

Substituting $-x$ for x changes the equation, and substituting $-y$ for y changes the equation. Both substitutions at once also change the equation. Therefore, there is no symmetry with the axes or origin.

47. $y^2x + 2x^2 = 4x^2y$

Substituting $-x$ for x changes the first term to negative;

substituting $-y$ for y changes the last term to negative.

Both substitutions at once yield $-y^2x + 2x^2 = -4x^2y$, which is not the same as the original equation.

Therefore, there is no symmetry.

52. $4x^2 + 9y^2 = 36$: Since both x and y are squared and since $(-x)^2 = x^2$ and $(-y)^2 = y^2$, we have symmetry with respect to the x-axis, the y-axis, and the origin.

Section 3.2 Functions and Function Notation

BASIC CONCEPTS you should learn from this unit.

1. Definition of a function as a rule that assigns exactly one y-value to each x-value.
2. Definition of Domain as the set of all x-values.
3. Definition of Range as the set of all y-values.
4. If a vertical line intersects the graph of an equation at only one point, the equation represents a function.
5. Restrictions on the Domain are often required:
   A. x may not cause a denominator to be zero.
   B. x may not cause an even root (square root, fourth root, etc.) to be negative.

   We are dealing here only with Real numbers.
6. Domain restrictions will sometimes but not always cause restrictions in range.
7. The function notation $f(x)$ is read "f of x."

   In $f(x) = 3x + 7$, the $f(x)$ is simply a convenient way of denoting the rule.

   $f(2)$ in this case is $3 \cdot 2 + 7 = 13$ and is called the Function Value when $x = 2$.

WHAT YOU SHOULD KNOW HOW TO DO upon completion of this unit.

1. Define function, Domain, and Range.
2. Determine the Domain and Range of a given function.
3. Determine if a given equation is or is not a function.
4. Evaluate a function, that is, find $f(x)$ if given an x value.

SOLUTIONS to Exercise Set 3.2

2. $y = x^2 + x \qquad -2 \leq x \leq 1$

x	y
-2	2
-1	0
0	0
1	2
-1/2	-1/4

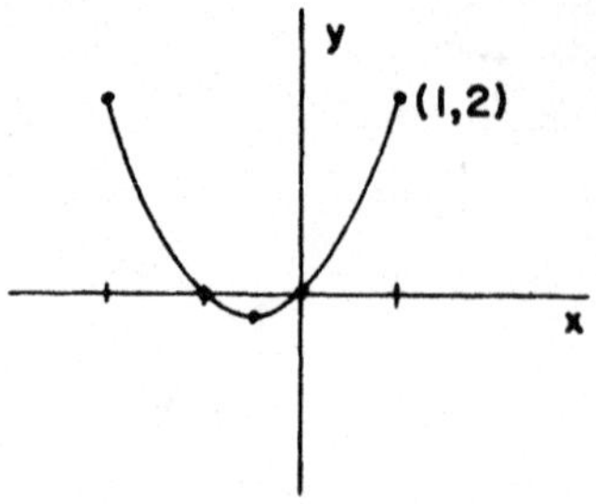

the Range is: $-\frac{1}{4} \leq y \leq 2$

5. $y = \sqrt{x - 1}$

x	y
1	0
2	1
5	2

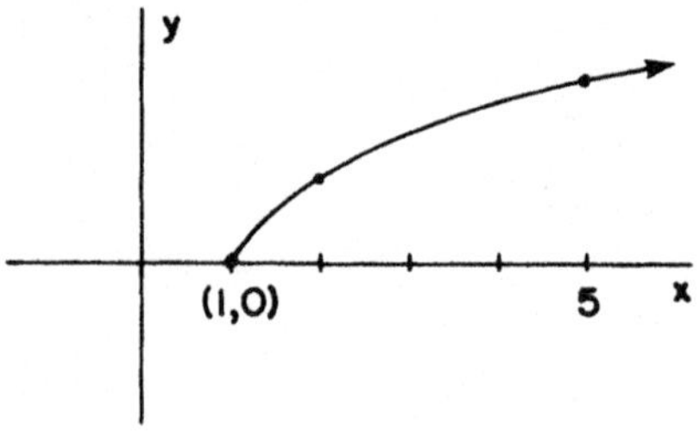

the Range is: $y \geq 0$

8. $f(x) = \sqrt{5 - x}$ the Domain is: $x \leq 5$

($x > 5$ causes a negative radicand.)

11. $f(x) = \dfrac{\sqrt{x - 1}}{x - 2}$ the Domain is: $x \geq 1$ so that $x - 1 \geq 0$

but $x \neq 2$ (zero denominator)

14. $f(x = x^2$ in order that $f(x) = 2$ we need to choose $x = \pm\sqrt{2}$.

17. $f(x) = 2x^2 + 5$

$f(0) > 2 \cdot 0 + 5 = \underline{5}$

20. $f(3x) = 2(3x)^2 + 5 = \underline{18x^2 + 5}$

23. $g(-3) = (-3)^2 + 2(-3) = 3$

26. $g(-x) = (-x)^2 + 2(-x) = x^2 - 2x$

29. Use a calculator $\dfrac{(-2.73)^2 + 1}{3(-2.73) - 1} = \dfrac{8.4529}{-7.19} \approx -1.176$

32. $f(-x) = \dfrac{(-x)^2 + 1}{3(-x) - 1} = \dfrac{x^2 + 1}{-3x - 1}$

35. $r(-8.27) = \dfrac{-8.27 - 2}{(-8.27)^2 + 2(-8.27) - 3} = \dfrac{-10.27}{48.8529} = -0.210$

38. $2r(a) = 2 \cdot \dfrac{a - 2}{a^2 + 2a - 3} = \dfrac{2a - 4}{a^2 + 2a - 3}$

41. $I = prt$ Here $P = x$ dollars, $r = 7\%$, $t = 4$

$I(x) = 0.07 \cdot 4 \cdot x$ or $I(x) = 0.28x$.

44. Since the perimeter, P, of a square is 4 times one side and the area is a side squared, if the area is A, then one side is $\sqrt{A}$ and $P = 4\sqrt{A}$.

Section 3.3 Graphs of Functions

BASIC CONCEPTS you should learn from this unit.

1. The graph of a function is the graph of $y = f(x)$. For example, if $f(x) = -2x + 7$ we would graph $y = -2x + 7$.

2. There are ways to describe functions other than by equations, for example, as a table of paired values.

3. A function is increasing if $f(x_2) > f(x_1)$ when $x_2 > x_1$ and decreasing if $f(x_2) < f(x_1)$ when $x_2 > x_1$.

4. A constant function will have the same value for all allowable x.

5. All polynomials are functions.

WHAT YOU SHOULD KNOW HOW TO DO upon completion of this unit.

1. Graph a given function.

2. Determine intervals when a function is increasing, decreasing, or constant.

3. Given a relationship between two variables in words, write a function statement of the variables.

SOLUTIONS to Exercise Set 3.3

2. $f(x) = 3 - 2x$

decreasing on all x

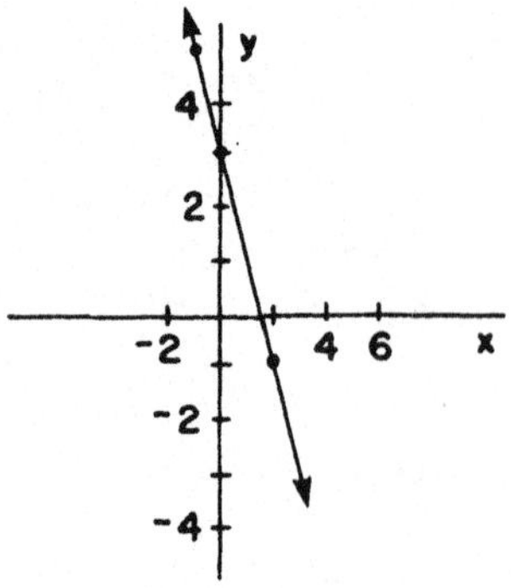

5. $f(x) = 9 - x^2$

increasing: $x < 0$

decreasing: $x > 0$

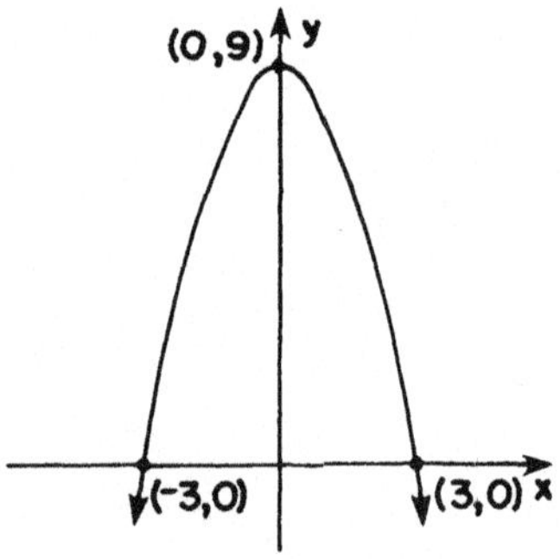

8. $f(x) = |1 - x|$

increasing: $x > 1$

decreasing: $x < 1$

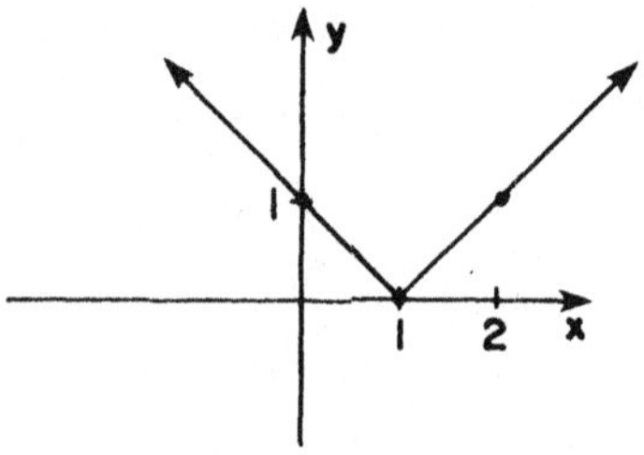

11. $f(x) = \begin{cases} x, & x < 2 \\ 2, & x \geq 2 \end{cases}$

increasing: $x < 2$

never decreasing

constant: $x > 2$

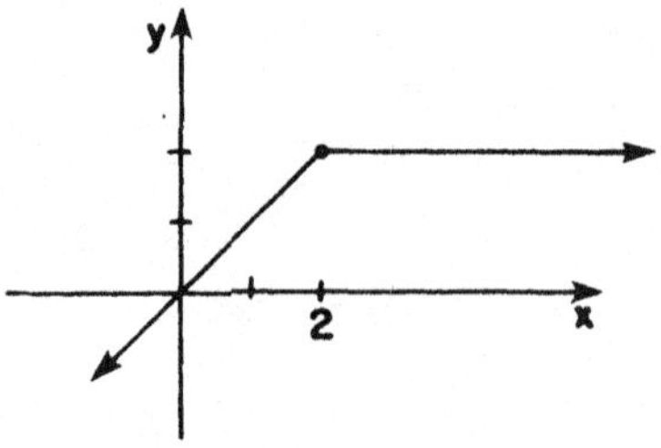

14. $f(x) = \begin{cases} 2 & \text{if } x \text{ is an integer} \\ -1 & \text{if } x \text{ is NOT an integer} \end{cases}$

$f(x)$ always a constant (either 2 or -1)

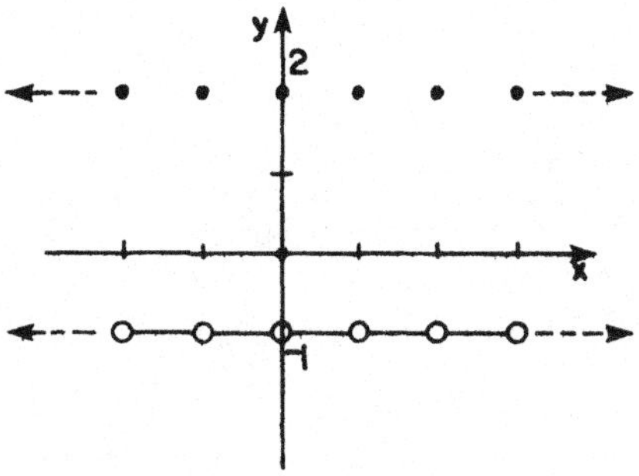

17. $f(x) = x^2$

$g(x) = 2x^2$

$h(x) = \frac{1}{2}x^2$

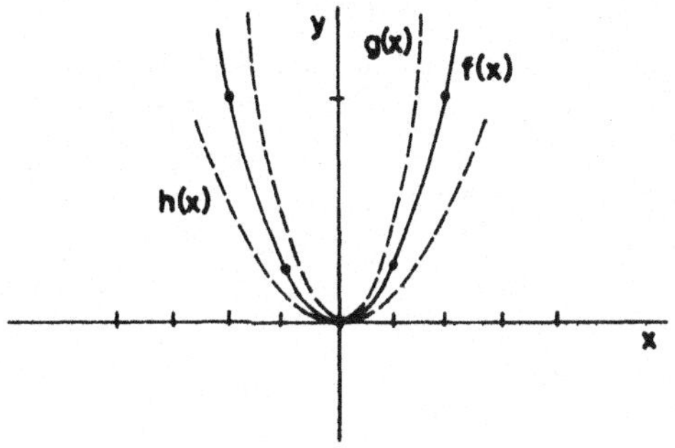

20. $f(x) = x^2 - 2$

$g(x) = 2 - x^2$

x	f(x)	x	g(x)
0	-2	0	2
2	2	2	-2
-2	2	-2	-2

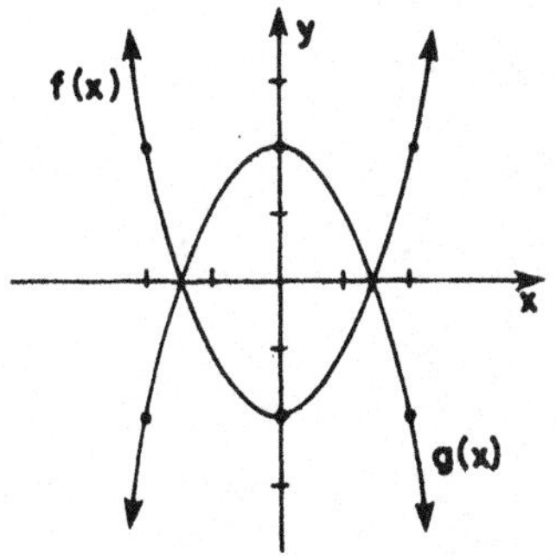

23. $f(x) = x^3$

$g(x) = -x^3$

Use table of values.

x	f(x)	x	g(x)
0	0	0	0
1	1	1	-1
-1	-1	-1	1

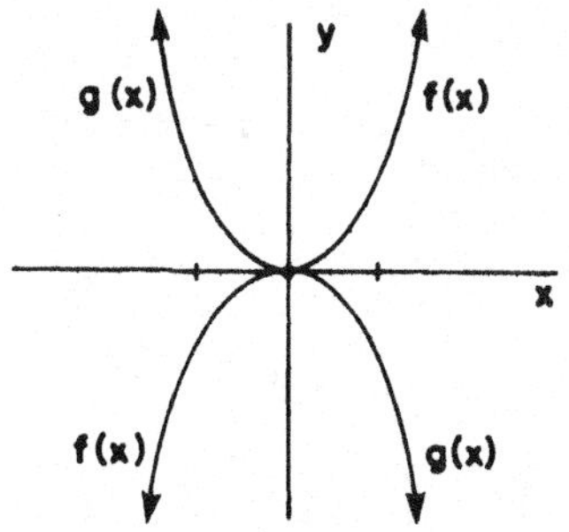

26. The Function will have three parts, corresponding to the three salary groups.
Dues = f(s) with s = exmployees salary.

$$f(s) = \begin{cases} 60 & \text{if} \quad s < 8000 \\ 60 + 0.01s & \text{if} \quad 8000 \leq s < 15000 \\ 130 + 0.02s & \text{if} \quad s \geq 15000 \end{cases}$$

29. m = number of miles

(a) C(m) = cost function

C(m) = \$14 + 8¢ per mile

Therefore, $C(m) = 14 + 0.08m$

(b) DOMAIN = allowable values of m, $m > 0$

(c) $C(100) = 14 + 0.08(100) = \$22$

Section 3.4 Linear Functions

BASIC CONCEPTS you should learn from this unit.

1. The slope of a line is defined as the ratio

$$\frac{y_2 - y_1}{x_2 - x_1}$$

for any two points $P_1(x_1, y_1)$ and $P_2(x_2, y_2)$ on the line.

2. The point-slope form of the straight line is $y - y_1 = m(x - x_1)$ where m is the slope and (x_1, y_1) are the coordinates of a point on the line.

3. The point (0, b) on a straight line is called the y-intercept, since it is where the line crosses the y-axis.

4. $y = mx + b$ is the slope-intercept form of the straight line.

5. The general first degree equation is $Ax + By + C = D$ and solving for y yields $-\frac{A}{B}x - \frac{C}{B}$, from which we can read the slope $m = -\frac{A}{B}$ and the y-intercept $b = -\frac{C}{B}$

6. A horizontal line has slope = 0.
A vertical lines has an undefined slope.

7. Parallel Lines have the same slope.
Perpendicular Lines have slopes that are negative reciprocals, $m_1 = \frac{2}{3}$, $m_2 = -\frac{3}{2}$.
Then line 1 would be perpendicular to line 2.

WHAT YOU SHOULD BE ABLE TO DO upon completion of this unit.

1. Determine the slope of the line passing through two known points. Positive slope implies increasing function.

2. Write the equation of a line if given the slope and one point on the line, or if given two points on the line.

3. Write the equation of a line if given the slope and y-intercept of the line.

4. From the equation of a line find the slope and y-intercept.

5. From their equations determine if two lines are parallel, perpendicular, or neither.

6. If given a word relationship between two variables, write a linear equation expressing the relationship.

SOLUTIONS to Exercise Set 3.4

2. $m = \dfrac{-3 - 3}{-1 - 2} = \dfrac{-6}{-3} = 2$ increasing

5. $m = \dfrac{1 - 2}{\frac{3}{2} - \frac{1}{2}} = \dfrac{-1}{1} = -1$ decreasing

8. $m\,\overline{AB} = \dfrac{4 - 2}{3 + 3} = \dfrac{2}{6} = \dfrac{1}{3}$

$m\,\overline{BC} = \dfrac{-2 - 4}{5 - 3} = \dfrac{-6}{2} = -3$

$m\,\overline{CD} = \dfrac{-4 + 2}{-1 - 5} = \dfrac{-2}{-6} = \dfrac{1}{3}$

$m\,\overline{DA} = \dfrac{2 + 4}{-3 + 1} = \dfrac{6}{-2} = -3$

OPPOSITE SIDES Do have the same slopes, so we have a parallelogram.

11. Point (0, 0) tells us the intercept is 0, and using $y = mx + b$, we have $y = 3x$.

14. $P_1(-3, 5)$; $P_2(1, 7)$ $m = \dfrac{7 - 5}{1 + 3} = \dfrac{2}{4} = \dfrac{1}{2}$

Using $y - y_1 = m(x - x_1)$ we obtain

$y - 5 = +\frac{1}{2}(x + 3)$ or $y = +\frac{1}{2}x + 13$

17. $P_1\left(-\frac{1}{2}, -1\right)$; $P_2\left(\frac{1}{2}, -1\right)$ $m = \dfrac{-1 + 1}{\frac{1}{2} + \frac{1}{2}} = 0$ Horizontal line. Equation: $y = -1$

20. $m = -3$, $b = -3$ Use $y = mx + b$ to get

$y = -3x - 3$

23. $m = \frac{1}{3}$, $b = -5$ similar process: $y = \frac{1}{3}x - 5$

26. $2x - 5y + 3 = 0$ First solve for y.

$-5y = -2x - 3$

$y = \frac{2}{5}x + \frac{3}{5}$

From which $m = \frac{2}{5}$, $b = \frac{3}{5}$

29. $3x + 4y + 2 = 0$ $\quad m = -\frac{A}{B} = -\frac{3}{4}$

$b = -\frac{C}{B} = -\frac{1}{2}$

32. (a) Horizontal Line is of the form $y = k$; in this case $y = -2$.

(b) Vertical Line is of the form $x = k$; $x = -5$.

35. Similar to Exercise 32: (a) $y = -9$; (b) $x = 9$.

38. (a) Rearrange to read $-5x + 2y + 4 = 0$. Then the slope $= -\frac{A}{B} = \frac{5}{2}$ and this would also be the slope of all lines parallel to the given line.

(b) Perpendicular Lines have their slope = the negative reciprocal, or $-\frac{2}{5}$.

41. (1, 3); $y = -3x + 2$ $\quad m = -3$

(a) Required line: $y - 3 = -3(x - 1)$ or $\underline{y = -3x + 6}$

(b) Perpendicular line: $y - 3 = \frac{1}{3}(x - 1)$ or $\underline{y = \frac{1}{3}x + \frac{8}{3}}$

44. (-1, -3) $\quad 3y + 4x - 5 = 0$ $\quad m = -\frac{4}{3}$

(a) Parallel line: $y + 3 = -\frac{4}{3}(x + 1)$, $4x + \underline{3y + 13 = 0}$

(b) Perpendicular line: $y + 3 = \frac{3}{4}(x + 1)$, $\underline{-3x + 4y + 9 = 0}$

47. (a) Essentially we have two points or pairs of values $P_1(10, 13.50)$ and $P_2(12, 15.90)$

$$m = \frac{15.90 - 13.50}{12 - 10} = \frac{2.40}{2} = 1.2$$

Now use $y - y_1 = m(x - x_1)$

$y - 13.50 = 1.2(x - 10)$

$y - 13.50 = 1.2x - 12$ or $\underline{y = 1.2x + 1.5}$

(b) Using this, if $y = 22$ we have

$22 = 1.2x + 1.5$ or $1.2x = 20.5$ and

$x = \$17.08$ approximately.

50. Given $cx - 5y + 8 = 0$ and that the x-intercept is 4 means that the line passes through (4, 0), which implies that $C \cdot 4 - 5 \cdot 0 + 8 = 0$ or $4C = -8$ and $C = -2$.

53. Given $f(x) = ax + b$ and $a > 0$, for $x_2 > x_1$, $ax_2 > ax_1$ and likewise regardless of b, $ax_2 + b > ax_1 + b$, which now means that $f(x_2) > f(x_1)$ whenever $x_2 > x_1$ and thus when $a > 0$ we have $f(x) = ax + b$ increasing.

A similar line of thought will prove the second part of the problem.

56. (a) $\angle CAQ = \angle BQC$ Both are complements of $\angle CQA$.

(b) $\Delta ACQ \sim \Delta BCQ$ Both are right triangles with corresponding angles equal.

(c) $\frac{\overline{CQ}}{\overline{AC}} = \frac{\overline{CB}}{\overline{CQ}}$ Corresponding parts of similar triangles are proportional.

(d) $m_1 = \frac{\overline{CQ}}{\overline{AC}}$; $m_2 = -\frac{\overline{CQ}}{\overline{CB}}$ Definition of slope $= \frac{y_2 - y_1}{x_2 - x_1}$

(e) $m_2 = -\frac{1}{m_1}$ Steps (c) and (d).

Section 3.5 Direct and Inverse Variation

BASIC CONCEPTS you should learn from this unit.

1. If two variables vary DIRECTLY, the RATIO of the variables is constant. For example, $x/y = k$.
2. If two variables vary INVERSELY, the PRODUCT of the variables is constant. For example, $xy = k$.
3. If a variable varies JOINTLY as two or more other variables, it varies DIRECTLY as the PRODUCT of the other variables. For example, M varies jointly as P and q means

$$\frac{M}{P \cdot q} \longleftarrow \text{Product of P and q}$$

WHAT YOU SHOULD BE ABLE TO DO upon completion of this unit.

1. Write a relationship of Direct Variation, of Inverse Variation, of Joint Variation.
2. From appropriate data, determine a constant of variation.
3. Knowing the constant of variation and values for all but one of the variables, find the value of that variable.

SOLUTIONS for Exercise Set 3.5

2. (a) Inverse variation implies: $x \cdot y = k$, and for any x, y pair we see that $k = \underline{6}$.

(b) $y = \frac{6}{x}$ or $xy = 6$

(c)

x	...	12	15	18	24	60
y	...	$\frac{1}{2}$	$\frac{2}{5}$	$\frac{1}{3}$	$\frac{1}{4}$	$\frac{1}{10}$

5. (a) $\frac{s}{t^2} = k$ and $\frac{10}{100} = \frac{1}{10} = k$

(b) $s = kt^2$ therefore, $s = \frac{1}{10} \cdot 5^2 = 2.5$

8. (a) $V \cdot P^2 = k$: $\frac{2}{3} \cdot 6^2 = k$; $k = 24$

(b) $V = \frac{k}{P^2}$; $V = \frac{24}{8^2} = \frac{24}{64} = \frac{3}{8}$

11. (a) $\frac{Ms^2}{r^2} = k$; $\frac{4 \cdot 2^2}{4^2} = k = 1$

(b) If $r = 6$, $s = 5$; $\frac{M \cdot 25}{36} = 1$; $M = \frac{36}{25}$

14. (a) $\frac{A \cdot d^3}{b^2 \cdot c^2} = k$; $\frac{18 \cdot 8}{4^2 \cdot 3^2} = \frac{18 \cdot 8}{16 \cdot 9} = 1 = k$

(b) $A = \frac{kb^2c^2}{d^3} = \frac{1 \cdot 9^2 \cdot 4^2}{3^3} = \frac{16}{3}$

17. $R \cdot A = k$; $20 \cdot 8 = k = 160$

$R = \frac{K}{A} = \frac{160}{12} = \frac{40}{3}$ ohms

20. w = weight of body

s = distance

and $w \cdot s^2 = k$; $400 \cdot 4000^2 - k = 6{,}400{,}000{,}000$

$$w = \frac{k}{s^2} = \frac{64 \times 10^8}{(10^3)^2} = \frac{64 \times 10^8}{10^6} = 64 \times 10^2 = 6400 \text{ lb}$$

(1000 miles from the center would be 3000 miles below the surface.)

23. Let C = illumination

I = intensity

s = distance from source

$\frac{C \cdot s^2}{I} = k$ We have $\frac{50 \cdot 2^2}{400} = k$; $k = \frac{200}{400} \cdot \frac{1}{2}$

$$C = \frac{KI}{s^2} = \frac{\frac{1}{2} \cdot 3840}{4^2} = \frac{1920}{4^2} = 120 \text{ candlepower}$$

Section 3.6 Combining Functions; Inverse Functions

BASIC CONCEPTS you should learn from this unit.

1. Two or more FUNCTIONS can be combined by addition, subtraction, multiplication, and division with due care not to divide by zero.

2. A COMPOSITE FUNCTION is essentially a function of a function. In other words, we can apply one function to a variable and another function to the result. We often use the phrase "f of g of x" and write $(f \circ g)(x)$.

3. A ONE-TO-ONE FUNCTION is one in which for each x there is exactly one y <u>and</u> for each y there is exactly one x .

   Test: If a HORIZONTAL line intersects the graph of a function exactly once, then the function is one-to-one.

4. The inverse of a function is found by interchanging the roles of x and y and then solving for y .

   The inverse of a one-to-one function is also a function.

   $f^{-1}(x)$ is the notation for an inverse function.

5. Since y's become x's and vice versa in inverses, the inverse of a function is a reflection about the line $y = x$.

WHAT YOU SHOULD KNOW HOW TO DO upon completion of this unit.

1. Apply the four arithmetic operations to two or more functions.
2. Find Composite Functions.
3. Recognize one-to-one functions.
4. Use the horizontal line test.
5. Find the inverse of a function.
6. Show that $f\left[f^{-1}(x)\right] = x$.

SOLUTIONS to Exercise Set 3.6

2. $f(x) + g(x) = x^2 + 1 + x - 2$

   Therefore $(f + g)(2) = 2^2 + 1 + 2 - 2 = 5$

5. $f \cdot g(x) = f(x) \cdot g(x) = (x^2 + 1)(x - 2)$

   Therefore, $f \cdot g(2) = (2^2 + 1)(2 - 2) = 0$

8. $\frac{f}{g}(x) = \frac{f(x)}{g(x)} = \frac{x^2 + 1}{x - 2}$

   Therefore, $\left(\frac{f}{g}\right)(-2) = \frac{(-2)^2 + 1}{-2 - 2} = \frac{5}{-4}$

11. $f(x) = 2x + 1 \qquad g(x) = 2x^2 + x$

    $f \circ g = f[g(x)] = f(2x^2 + x) = 2(2x^2 + x) + 1 = 4x^2 + 2x + 1$

14. $(g \circ f)(3) = g[f(3)] = g(7) = 98 + 7 = 105$

17. $(g \circ f)(x - 1) = g[f(x - 1)] = g[2(x - 1) + 1] = g[2x - 1] = 2(2x - 1)^2 + (2x - 1)$

    $= 8x^2 - 6x + 1$

20. $f(x) = x^2 + 4 \qquad g(x) = \sqrt{x + 2}$

    $g[f(x)] = g(x^2 + 4) = \sqrt{x^2 + 4 + 2} = \sqrt{x^2 + 6}$

23. See solution 20. The domain of $(g \circ f)(x)$ is <u>all reals</u>.

26. $f(x) = \sqrt{x + 1}$, $g(x) = x + 2$

(a) $(f \circ g)(x) = \sqrt{x + 2 + 1} = \sqrt{x + 3}$

(b) $(g \circ f)(x) = \sqrt{x + 1} + 2$

29. $f(x) = 2x + 4$, $g(x) = \frac{1}{2}x - 2$

$(f \circ g)(x) = 2\left(\frac{1}{2}x - 2\right) + 4 = x$

and

$(g \circ f)(x) = \frac{1}{2}(2x + 4) - 2 = x$

32. $f(x) = x^3$, $g(x) = \sqrt[3]{x}$

$(f \circ g)(x) = \left(\sqrt[3]{x}\right)^3 = x$

$(g \circ f)(x) = \sqrt[3]{x^3} = x$

35. $f(x) = 2x + 3$

$y = 2x + 3$

$x = 2y + 3$

$\frac{x - 3}{2} = y$

$f^{-1}(y) = \frac{x - 3}{2}$

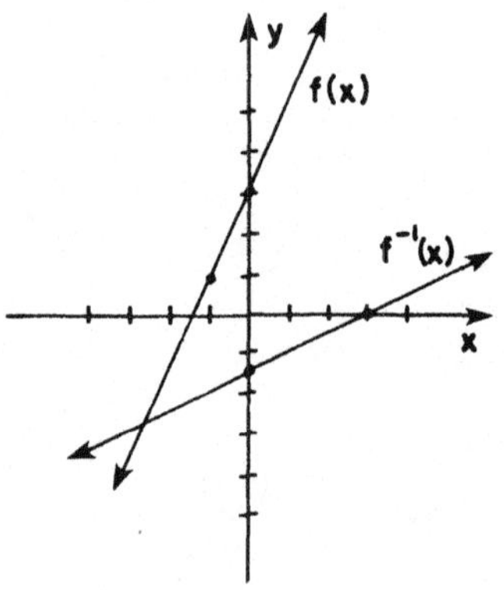

38. $f(x) = \frac{1}{2}x + 1$

$y = \frac{1}{2}x + 1$

Solving for x we have $x = 2y - 2$

Switch x and y. $f^{-1}(x) = y = 2x - 2$

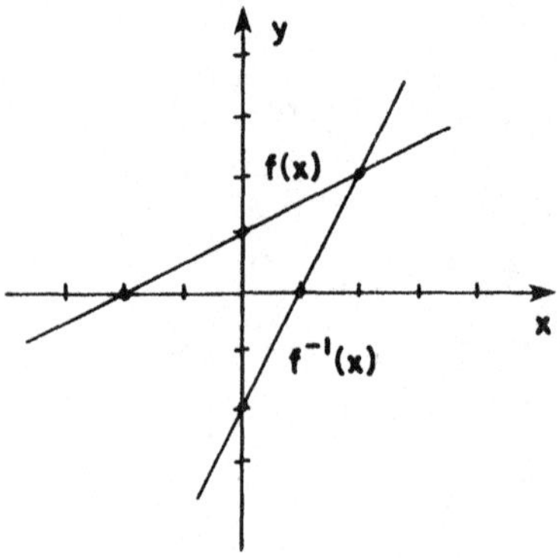

41. $f(x) = x^3 + 1$

$y = x^3 + 1$

$x = \sqrt[3]{y - 1}$

$f^{-1}(x) = y = \sqrt[3]{x - 1}$

x	0	1	2	9	-7
4	-1	0	1	2	-2

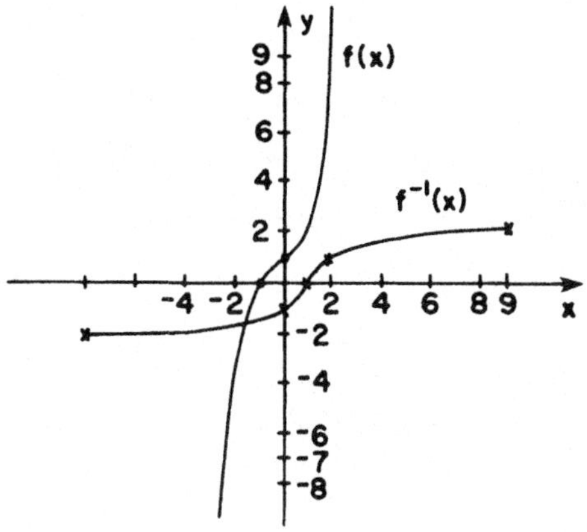

44. $f(x) = 3 - 5x$

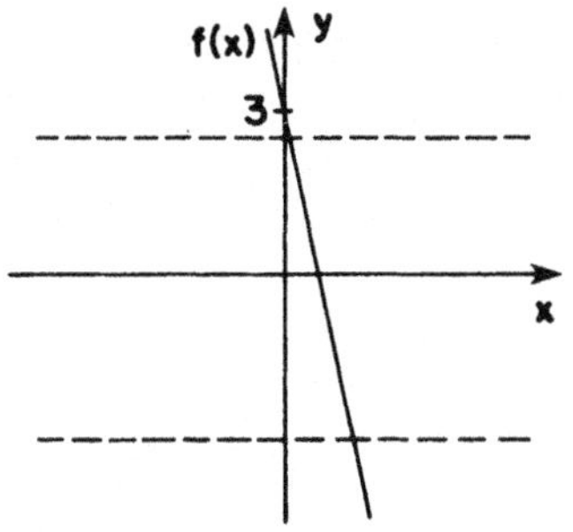

Yes, one-to-one.

47. $f(x) = -x^3 + 1$

x	0	1	2	-2
y	1	0	7	10

Yes, one-to-one

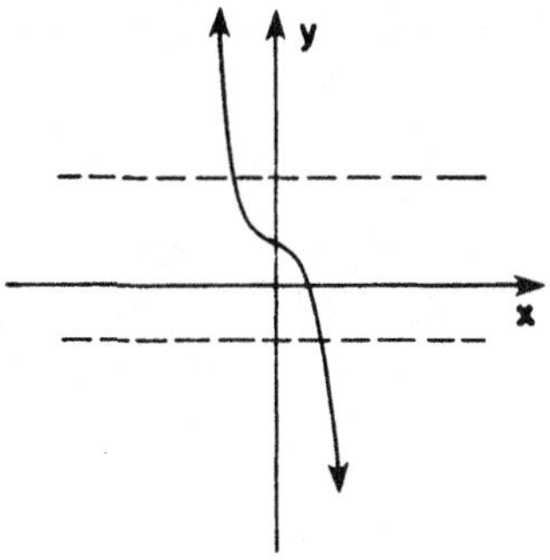

50. $f(x) = \begin{cases} x^2 - 4x + 4 & x \leq 2 \\ x & x > 2 \end{cases}$

x	0	1	2	4
y	4	1	0	4

no, not one-to-one

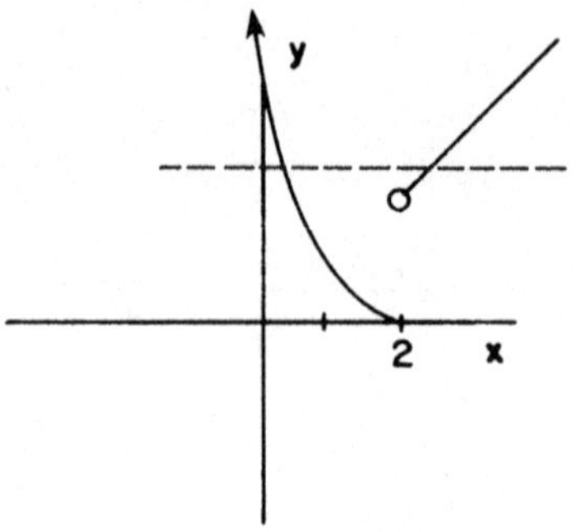

53. $f(x) = ax + b$

$y = ax + b$

$x = \frac{y - b}{a}$

$y = \frac{x - b}{a} = f^{-1}(x)$

SOLUTIONS to Review Exercises for Chapter 3.

2. A(3, -4), B(-2, -6), C(-1, 2)

$\overline{AB} = \sqrt{(-2)^2 + 5^2} = \sqrt{29}$

$\overline{AC} = \sqrt{6^2 + 4^2} = \sqrt{52}$

$\overline{BC} = \sqrt{8^2 + 1^2} = \sqrt{65}$ Therefore, BC is the longest side.

5. $y^2 = 1 - x^3$ $-x$ for x means that $y^2 = 1 + x^3$ and changes the equation.
$-y$ for y implies $y^2 = 1 - x^3$.
There is no change; therefore there is x-axis symmetry only.

8. The vertical line test fails $\frac{1}{m}$ there is no function.

11. $f(x) = \sqrt{x - 1} = 15$ $15^2 + 1 = 226$

Thus if $\underline{x = 226}$, then $f(x) = \sqrt{226 - 1} = 15$.

14. $f(x) = x^2 - x$; $f(y - 1) = (y - 1)^2 - (y - 1) = \underline{y^2 - 3y + 2}$

17. $f(x) = \begin{cases} x - 1 & \text{for } x \leq -1 \\ x^2 & \text{for } -1 < x \leq 2 \\ -2 & \text{for } x > 2 \end{cases}$

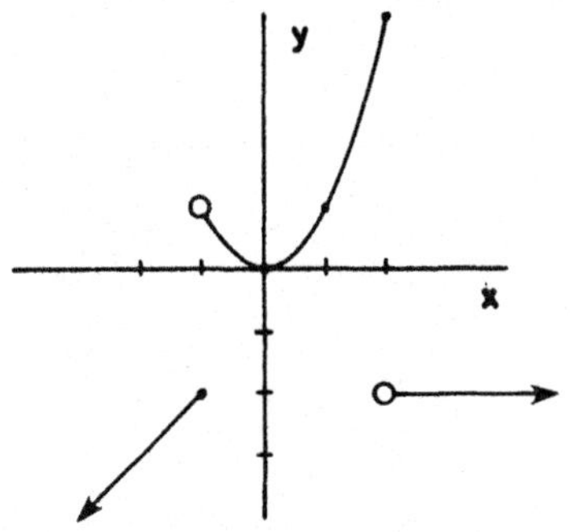

$f(x) \uparrow$ for $(-\infty, -1]$ and $(0, 2]$

$f(x) \downarrow$ for $(-1, 0)$

$f(x)$ constant for $x > 2$

20. $A(-4, -6)$, $B(-1, 3)$

$$m_{AB} = \frac{3 + 6}{-1 + 4} = \frac{9}{3} = 3$$

23. Slope = 0, so $y = 0x + 3$ or $y = 3$.

26. Direct Variation implies that $R/q = k$. Therefore,

$20/5 = k = 4$. $R = kq = 4 \cdot 40 = \underline{160}$.

29. $f(x) = x + 1$ $g(x) = x^2 - 1$

$f(x) + g(x) = x + 1 + x^2 - 1 = x^2 + x$

32. $f(x) = x + 1$ $g(x) = x^2 - 1$

$\left(\frac{f}{g}\right) x = \frac{x + 1}{x^2 - 1}$ DOMAIN: All Reals except ±1, which cause denominator to be zero.

35. $f(x) = \sqrt{x} - 2$ $g(x) = x^2$

$(f \circ g)(x) = \sqrt{x^2} - 2$

38. $(g \circ f)(2) = (\sqrt{2} - 2)^2 = 2 - 4\sqrt{2} + 4 = 6 - 4\sqrt{2}$

Chapter 3 - A SHORT QUIZ

1. The Rectangular Coordinate System consists of two ___?___ Number Lines, one called the ___?___, and the other called the ___?___.
2. A point with abscissa of -2 would be in the ___?___ or ___?___ quadrant.
3. State the Distance Formula.
4. GRAPH the equation $y = |x - 2|$.
5. What kind of Symmetry does the graph of $y = x^2 + 1$ have?
6. If a rule assigns exactly one y-value to each x-value, the RULE is called a ___?___.

7. What is the DOMAIN of $f(x) = \dfrac{x}{\sqrt{x - 1}}$?

8. Which of these are functions?

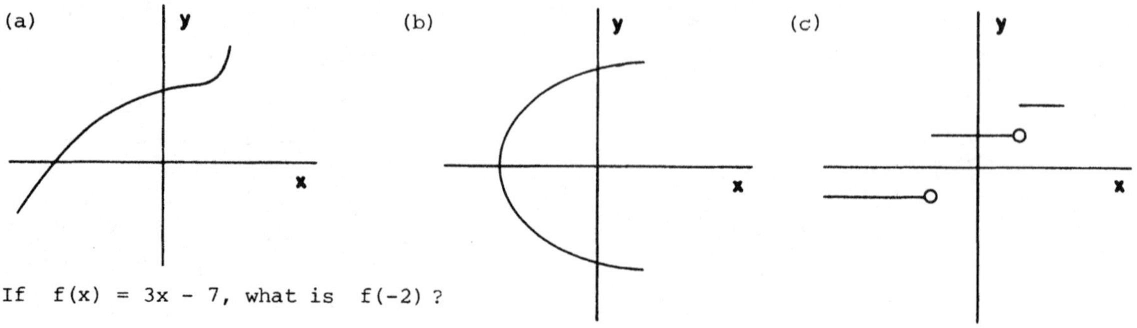

9. If $f(x) = 3x - 7$, what is $f(-2)$?

10. In Question 9 what is $f(a + h)$?

11. Given that $f(x_1) > f(x_2)$ whenever $x_1 < x_2$, is $f(x)$ increasing or decreasing?

12. By graphing $f(x) = x^2 - 2x + 1$ determine the interval where $f(x)$ is decreasing.

13. Is $f(x) = 7$ increasing, decreasing, or neither?

14. A function of the form $f(x) = mx + b$ is a ___?___ function.

15. Refer to Question 14: m represents the ___?___ of the line.

16. Generally, Slope is defined as equal to ___?___

17. Find the slope of a line passing through the points $P_1(3, -2)$; $P_2(-4, -7)$.

18. The graph of a decreasing linear function would have a ___?___ slope.

19. Write the point-slope form of the straight line equation.

20. If $2x - 5y - 7 = 0$, what is the slope? What is the y-intercept?

21. What kind of slope does a vertical line have?

22. If two lines are parallel, their slopes would be ___?___ .

23. Given that $x \cdot y = k$ we know that x and y vary ___?___ (directly or inversely).

24. Write an equation showing that x varies directly as the square of y and inversely as the square root of z .

In Questions 25 - 30 use the following: Given $f(x) = x^2 - 1$ and $g(x) = 2x$.

25. Find $(f + g)(x)$.

26. Find $\left(\frac{g}{f}\right)(x)$.

27. Find $(f \circ g)(x)$.

28. Find $f^{-1}(x)$.

29. Is $f^{-1}(x)$ a FUNCTION?

30. Find $g^{-1}(x)$.

Quiz Answers Chapter 3

1. perpendicular, x-axis, y-axis

2. second, third

3. $d = \sqrt{(x_2 - x_1)^2 + (y_2 - y_1)^2}$

4. $y = |x - 2|$

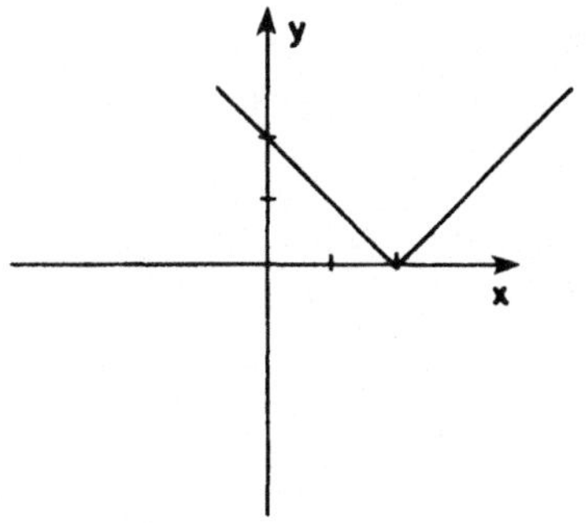

5. y-axis symmetry

6. function

7. D: $x > 1$

8. a, c

9. -13

10. $3a + 3h - 7$

11. decreasing

12. $(-\infty, 1)$

13. neither

14. linear

15. slope

16. $m = \dfrac{y_2 - y_1}{x_2 - x_1}$

17. $m = \frac{5}{7}$

18. negative

19. $y - y_1 = m(x - x_1)$

20. $\frac{2}{5}$, $-\frac{7}{5}$

21. undefined

22. equal

23. inversely

24. $\dfrac{x\sqrt{z}}{y^2} = k$

25. $x^2 + 2x - 1$

26. $\dfrac{2x}{x^2 - 1}$

27. $(2x)^2 - 1$

28. $f^{-1}(x) = \pm\sqrt{x + 1}$

29. no

30. $g^{-1}(x) = \frac{x}{2}$

CHAPTER 4

Exponential and Logarithmic Functions

Section 4.1 Exponential Functions

BASIC CONCEPTS you should learn from this unit.

1. Functions of the form $f(x) = a^x$ are called Exponential Functions. The variable, x, appears as an exponent on some base a.

   But there is a restriction: The base, a, must be positive $(a > 0)$ and not equal to one $(a \neq 1)$.

2. Some characteristics of exponential functions are

   A. $f(x) = a^x$ is increasing if $a > 1$;

   $f(x) = a^x$ is decreasing if $0 < a < 1$.

   B. $f(x) = a^{-x}$ is decreasing if $a > 1$.

   $f(x) = a^{-x}$ is increasing if $0 < a < 1$.

   C. If $a^x = a^y$, then $x = y$ and if $a^x = b^x$ then $a = b$.

3. Three applied examples are discussed in the text:

   A. The Growth Model: $Q(t) = q_0e^{kt}$. Here time is the usual variable and since the base, e, equals approximately 2.71828, which is greater than 1, the function is increasing (growth).

   B. The Decay Model: $Q(t) = q_0e^{-kt}$. This is of the form $f(x) = a^{-x}$ (see concept 2B) and so is decreasing, since the base, again e, is greater than 1.

   C. The Compound Interest Formula

      i. $S = P(1 + i)^n$ Here the base $(1 + i)$ is always greater than 1, so we have an increasing function.

      ii. $S = Pe^{rt}$ for continuous compounding.

WHAT YOU SHOULD KNOW HOW TO DO upon completion of this unit.

1. Graph exponential functions with various bases and exponents.

2. Solve simple exponential equations such as $2^x = 2^3$ and $(a + 1)^x = (2a - 1)^x$.

3. Use the growth model, decay model, and the compound interest formulas to solve applied problems.

SOLUTIONS to Exercise set 4.1

1. $f(x) = 4^x$

   $f(0) = 1;\ f(2) = 16$

   $f(-2) = 1/16$

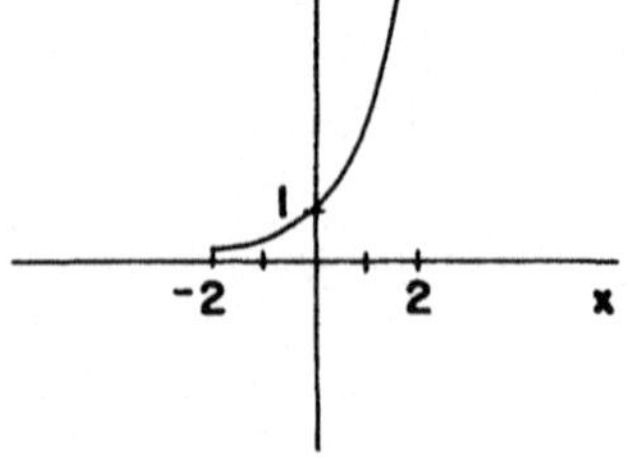

4. $f(x) = 10^{-x}$

$f(0) = 1; \quad f(1) = \frac{1}{10}$

$f(-1) = 10; \; f(-2) = 100$

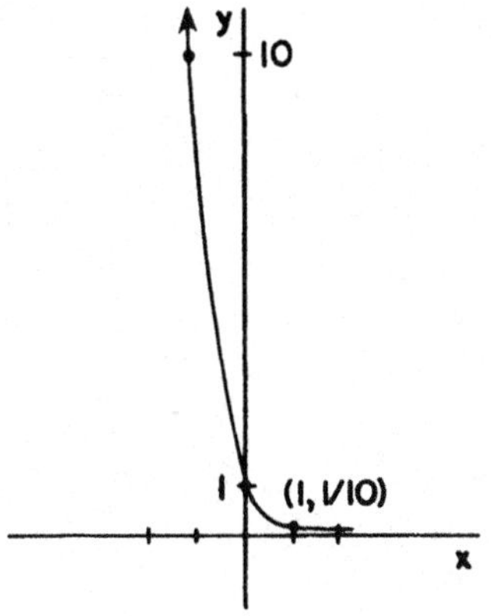

6. $f(x) = 2^{x-1}$

$f(0) = 2^{-1} = \frac{1}{2}$

$f(1) = 2^{0} = 1$

$f(2) = 2^{1} = 2$

$f(-1) = 2^{-2} = \frac{1}{4}$

Note that this graph of $f(x) = x^{x-1}$ is just the graph of $f(x) = 2^{x}$ shifted to the right one unit point for point.

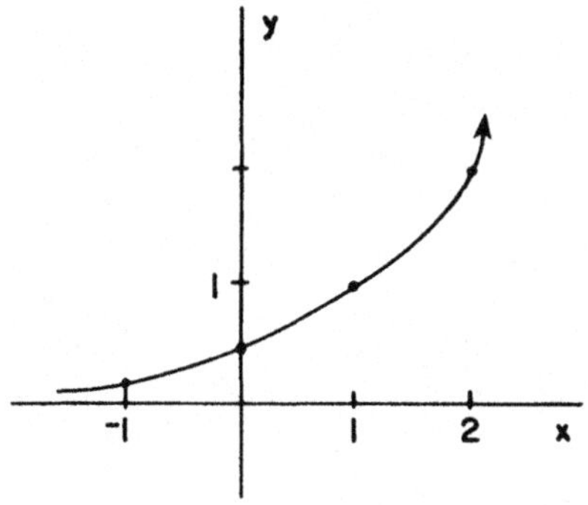

7. $f(x) = 2^{|x|}$

Here the exponent on 2 is always non negative.

$f(0) = 1; \quad f(1) = 2$

$f(-1) = 2; \quad f(2) = 4$

$f(-2) = 4$

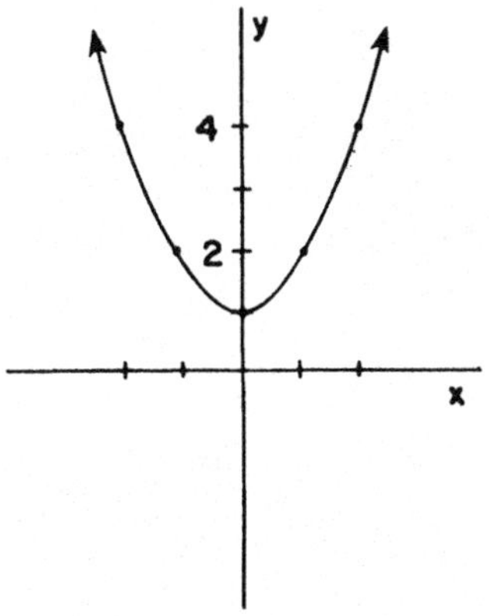

10. $f(x) = 3^{-2x}$

The exponent is opposite in sign to x, so $f(x)$ is decreasing.

$f(0) = 1$ $f(-1) = 3^2$

$f(1) = 1/3^2$ $f(-2) = 3^4$

$f(2) = 1/3^4$

The graph is decreasing very rapidly when $x < 0$ and decreases very slowly after $x > 0$.

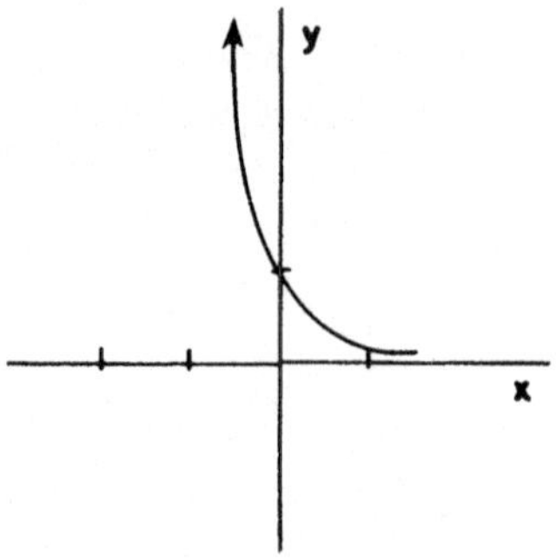

15. $3^x = 9^{x-2}$

$3^x = (3^2)^{x-2} = 3^{2x-4}$ using the $(x^m)^m = x^{mm}$ property of exponents

Thus $x = 2x - 4$

and $x = 4$

Check: If $3^4 = 9^{4-2}$, then $3^4 = 9^2$

$81 = 81$

19. $e^{x-1} = e^3$

$x - 1 = 3$

$x = 4$

20. $e^{x-1} = 1$

$x - 1 = 0$ Because $e^0 = 1$

Thus $x = 1$

23. $(a + 1)^x = (2a)^x$

$a + 1 = 2a$

$1 = a$

25. a. The basic formula to use here is the exponential growth model $Q(t) = q_0 e^{kt}$. Thus we see from $Q(t) = 200e^{0.25t}$ that $q_0 = 200$; this is the initial number of bacteria. Also, $k = 0.25$.

b. $t = 20$, so $Q(t) 200 \cdot e^{(0.25)(20)} = 200e^5 = 29683$

c.

t	1	4	8	10
Q	257	544	1478	2436

In the basic formula $Q(t) = 200e^{(0.25)5}$, let $t = 1, 4, 8, 10$ successively to obtain the Q values in the table.

29. Using the decay formula, $Q(t) = q_0 e^{-kt}$ with $q_0 = 1000$, we obtain, when the rate of decay is 4%, $k = 0.04$, and $t = 10$. Thus $Q(t) = 1000_e^{(0.04) \cdot 10} = 1492$ grams.

32. We use the compound interest formula $S = P(1 + i)^n$, in which n = number of conversion periods and $i = r/k$, r being the annual interest rate, and k is the number of conversion periods per year.

For Investment A $r = 8\%$ and $k = 4$, and so

$S = P(1 + 0.02)^4 = P(1.0824)$

For Investment B $r = 9\%$ and $k = 1$, and so

$S = P(1.09)^1 = P(1.09)$

35. Using the continuous compounding formula $S = Pe^{rt}$ with $P = 5000$, $r = 0.09$, and $t = 5$, we have

$S = 5000e^{(0.09)(5)} = 5000e^{0.45} = 5000(1.5683)$

therefore

$S \approx \$7842$

39. INVESTMENT A:

$S = 1000(1.04)^{20}$

$S = 2191.12$

INVESTMENT B:

$S = 1000e^{0.08 \cdot 10} = 1000e^{0.8}$

$S = 2225.54$

The approximate difference is thus \$34.42.

Section 4.2 Logarithmic Functions

BASIC CONCEPTS you should learn from this unit.

1. The logarithmic function $y = \log_a x$ is the inverse of the exponential function $y = a^x$.
2. The domain of the logarithmic function is restricted: $x > 0$.
3. The base, a, of the logarithmic function is restricted: $a > 0$ and $a \neq 1$. (Note that this is exactly the same restriction the base, a, has in the exponential function $y = a^x$.)
4. The statement of the logarithmic function, $y = \log_a x$ if and only if $x = a^y$, tells us that if we raise a to the power y we get x; in other words, y is an exponent.
5. $\ln x = \log_e x$ is the usual way to write the base e logarithmic function.
6. $a^{\log_a x} = x$ and $\log_a a^x = x$ are two simple identities to be learned.

WHAT YOU SHOULD KNOW HOW TO DO upon completion of this unit.

1. Change back and forth from the exponential form to the logarithmic form; for example,

$$3 = \log_2 8 \text{ if and only if } 8 = 2^3$$

2. Solve elementary log equations; for example,

$$\log_2(x - 1) = \log_2 10$$

means

$$x - 1 = 10$$

so

$$x = 11$$

3. Use the identities and concepts above to evaluate expressions; for example,

$$2^{\log_2 7} = 7$$

$$\log^{0.001} = -3 \qquad (10^{-3} = 0.001)$$

4. Graph simple log functions.

SOLUTIONS to Exercise Set 4.2

In many of the exercises we use the basic concept that the <u>logarithm</u> of a number to some base <u>is</u> the <u>exponent</u> we would use on the base to obtain the number (see Basic Concept 4). Thus·

1. $\log_2 4 = 2$ means $2^2 = 4$

3. $\log_9 \frac{1}{81} = -2$ means $9^{-2} = \frac{1}{81}$

5. $\ln 20.09 = 3$ means $e^3 = 20.09$

9. $\ln 1 = 0$, $e^0 = 1$

13. $25 = 5^2$ implies $\log_5 25 = 2$

17. $\frac{1}{8} = 2^{-3}$ implies $\log_2 \frac{1}{8} = -3$

21. $1 = e^0$ implies $\log_e 1 = 0$ or $\ln 1 = 0$

25. $\frac{1}{3} = 27^{-1/3}$ implies $\log_{27} \frac{1}{3} = -\frac{1}{3}$

29. $\log_{25} x = \frac{1}{2}$ implies $25^{1/2} = x$, so $x = 5$, $x = \sqrt{25}$. Recall that $x > 0$.

33. $\ln x = \frac{1}{2}$, $e^{1/2} = x$, $x = \sqrt{e} \approx 1.649$

37. $\log_x \frac{1}{8} = \frac{1}{3}$, $x^{1/3} = \frac{1}{8}$, then cubing both sides gives $x = \frac{1}{512}$

41. $\log_{x+1} 24 = \log_3 24$ implies that $x + 1 = 3$, $x = 2$

45. $3^{\log_3 6} = 6$

49. $\log_5 5^3 = 3$

53. $\log_7 49 = \log_7 7^2 = 2$

57. $\ln 1 = 0$, $(e^0 = 1)$

61. $\log 10{,}000 = 4$, $(10^4 = 10{,}000)$

65. $f(x) = \log_4 x$

x	1	$\frac{1}{4}$	4	16
y	0	-1	1	2

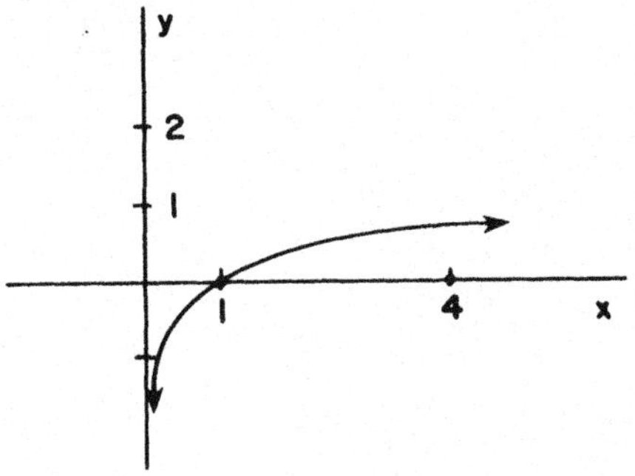

69. $f(x) = \ln \frac{x}{2}$ Here the base is e, so the function values (y-values) will be powers of e, yielding x/2 .

$\ln \frac{x}{2} =$

x	2	1	4	6	8
y	0	-0.693	0.693	1.098	1.386

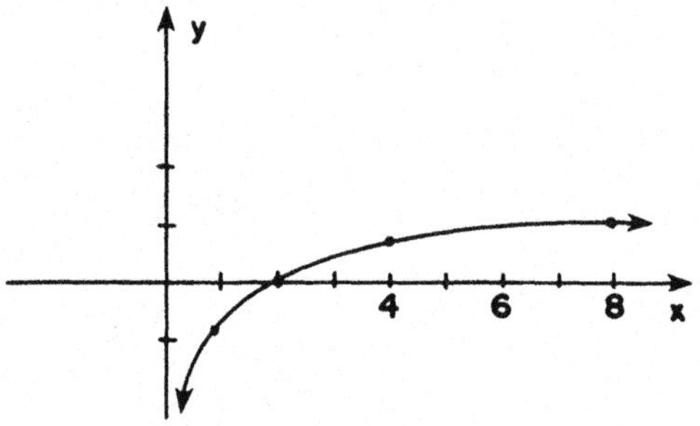

Section 4.3 Properties of Logarithms

BASIC CONCEPTS you should learn from this unit.

1. There are three properties of logarithms and they are virtually the same as the properties of exponents (which you already know). This is not too surprising, since logarithms really are exponents.

   A. $\log_a(x \cdot y) = \log_a x + \log_a y$ In words this equation says that
   The Log of a product equals the sum of the logs of the factors.

   B. $\log_a\left(\frac{x}{y}\right) = \log_a x - \log_a y$ In words this equation states that
   The Log of a Quotient equals the Log of the DIVIDEND MINUS the Log of the Divisor.

   C. $\log_a x^m = m \log_a x$ In words this equation says that
   The Log of a number raised to a power equals the exponent multiplied by the Log of the number.

That's all! But a couple of warnings should be mentioned:

A. Remember that $\log_a (x + y) \neq \log_a x + \log_a y$.

B. Remember that $\log_a (x - y) \neq \dfrac{\log x}{\log y}$.

If you have not yet memorized the three properties listed above, you should do so now! Solving problems will be much easier when you have learned them.

WHAT YOU SHOULD KNOW HOW TO DO upon completion of this unit.

1. Change logarithmic expression into simpler logarithmic form; for example,

$$\ln 2x^3 = \ln 2 + 3 \ln x$$

2. Write expressions containing several logarithms as a single logarithm; for example,

$$3 \log_a x - 2 \log_a z = \log_a \frac{x^3}{z^2}$$

3. Use the properties of logarithms to find the logarithms of products, quotients, roots, etc; for example, if $\log 2 = 0.30$ and $\log 3 = 0.47$, then

$\log 12 = \log 2 \cdot 2 \cdot 3 = \log 2 + \log 2 + \log 3$ and so

$\log 12 = 0.30 + 0.30 + 0.47 = 1.07$

SOLUTIONS to Exercise Set 4.3

1. $\log_{10} (120 \times 36) = \log_{10} 120 + \log_{10} 36$, using the Product Rule (Property 1.A).

3. $\mathrm{Log}_3 3^4$ — Parentheses are not really necessary, are they?

$= 4 \log_3 3$ — Power Rule (Property 1.C)

But

$\log_3 3 = 1$

So

$4 \log_3 3 = 4$

7. $\log_a \left(\dfrac{x}{yz}\right) = \log_a x - [\log_a y + \log_a z]$

$= \log_a x - \log_a y - \log_a z$

11. $\log_a (x^2 y^3) = \log_a x^2 + \log_a y^3$

product rule

$= 2 \log_a x - 3 \log_a y$

power rule

13. $\log_a \sqrt{xy} = \log_a (xy)^{1/2} = \log_a x^{1/2} y^{1/2}$ by properties of Radicals and Exponents

then:

$\log_a x^{1/2} + \log_a y^{1/2}$

(by the product property), and finally

$\frac{1}{2} \log_a x + \frac{1}{2} \log_a y$

by the Power Rule

17. $\ln(\sqrt{x}\,\sqrt[3]{y}) = \ln x^{1/2} + \ln y^{1/3} = \frac{1}{2} \ln x + \frac{1}{3} \ln y$

21. $\log 6 = \log (2 \cdot 3)$ Think of 6 as a product.

so $= \log 2 + \log 3$

$= 0.30 + 0.47 = 0.77$

24. $\log \sqrt{5} = \log 5^{1/2} = \frac{1}{2} \log 5 = \frac{1}{2}(0.70) = 0.35$

27. $\log \frac{15}{2} = \log \frac{3 \cdot 5}{2} = \log 3 + \log 5 - \log 2$

$= 0.47 + 0.70 - 0.30$

$= 0.87$

29. $\log \sqrt{7.5} = \log(7.5)^{1/2} = \frac{1}{2} \log \frac{15}{2}$

$= \frac{1}{2}(0.87)$ (from Exercise 27)

$\log \sqrt{7.5} = 0.435$

33. $\frac{1}{3} \ln x + \frac{1}{3} \ln y = \ln x^{1/3} + \ln y^{1/3} = \ln x^{1/3} \cdot y^{1/3} = \ln(xy)^{1/3}$ <u>or</u> $\ln \sqrt[3]{xy}$

35. $\frac{1}{3} \log_a x + 2 \log_a y - \frac{3}{2} \log_a z = \log_a x^{1/3} + \log_a y^2 - \log_a z^{3/2} = \log_a \frac{x^{1/3} y^2}{z^{3/2}}$

37. $\frac{1}{2}(\log_a x + \log_a y) = \frac{1}{2} \log_a x + \frac{1}{2} \log_a y = \log_a x^{1/2} + \log_a y^{1/2} = \log_a x^{1/2} y^{1/2}$

$= \log_a (xy)^{1/2}$ or $\log_a \sqrt{xy}$

41. $\frac{1}{2} \log_a (x - 1) - 2 \log_a (x + 1) = \log_a \frac{(x - 1)^{1/2}}{(x + 1)^2}$ or $\log_a \frac{\sqrt{x - 1}}{(x + 1)^2}$

43. $3 \log_a x - 2 \log_a (x - 1) + \frac{1}{2} \log \sqrt[3]{x + 1} = \log_a x^3 - \log_a (x - 1)^2 + \log \left[(x + 1)^{1/3}\right]^{1/2}$

$= \log \frac{x^3 \cdot (x + 1)^{1/6}}{(x - 1)^2}$

BASIC CONCEPTS you should learn from this unit.

1. The logarithm of a number has two parts, a mantissa, which is always positive, and a characteristic, which is any integer, positive, negative, or zero.

2. Common logs are BASE 10 logs.
   Natural logs are BASE e logs.

3. Log Tables supply only the mantissa. You must determine the characteristic.

4. Computation with logs utilizes the laws of logarithms and requires a knowledge of the use of log tables.

WHAT YOU SHOULD KNOW HOW TO DO upon completion of this unit.

1. Change the form of a number to Scientific Notation.

2. Given a number, find its common logarithm and/or natural logarithm.

3. Given the logarithm of a number, find the number.

4. Do various computations using three-digit numbers by means of logarithms.

SOLUTIONS to Exercise Set 4.4

1. $2725 = 2.725 \times 10^3$

4. $0.000914 = 9.14 \times 10^{-4}$

7. $296.2 = 2.962 \times 10^2$

10. $\ln 3.2 = 1.1632$

11. log 37.5: characteristic = 1
    mantissa = 0.5740
    answer = 1.5740

14.
$$\begin{aligned} \ln 60 &= \ln(5 \cdot 12) \\ &= \ln 5 + \ln 12 \\ &= 1.6094 + \ln 3 + \ln 4 \\ &= 1.6094 + 1.0986 + 1.3863 \\ &= \underline{4.0943} \end{aligned}$$

17. log 48,200 = characteristic + mantissa
    4 + 0.6830 = 4.6830

20. $\log 0.00532 = -3 + 0.7259 = \underline{0.7259 - 3}$

23. $\ln x = 1.0647$. Then from Table II, $x = 2.90$.

26. $\log x = 0.1903 - 2$. Then from the table, $x = 1.55 \times 10^{-2}$ or $x = 0.0155$.

29. $\log x = 0.7832 - 4$. Then from the table, $x = 6.07 \times 10^{-4}$ or $x = 0.000607$.

32. $\log x = -3,9004$. Convert first to a <u>positive</u> mantissa:

$$\begin{array}{r} +4.0000 \\ -3.9004 \\ \hline +0.0996 \end{array}$$

So -3.9004 is equivalent to $0.0996 - 4$

This implies that $x = 1.26 \times 10^{-4}$, approximately, since 0.0996 is not in the table.

35. $N = \dfrac{679}{321}$

$$\begin{aligned} \log 679 &= 2.8319 \\ -\log 321 &= \underline{2.5065} \\ \log N &= 0.3254 \\ N &= 2.12 \times 10^{0} = 2.12, \text{ approximately.} \end{aligned}$$

38. $N = (42.3)^3(71.2)$

$$\begin{aligned} \log N &= 3\log(42.3) + \log 71.2 \\ &= 3(1.6263) + 1.8525 \\ &= 4.8789 + 1.8525 \\ &= 6.7314 \end{aligned}$$

and $N = 5.39 \times 10^{6} = 539,000$ approximately.

41. $N = \dfrac{\sqrt{7870}}{(46.3)^4}$

$$\begin{aligned} \log N &= \frac{1}{2}\log 7870 - 4(\log 46.3) \\ &= \frac{1}{2}(3.8960) - 4(1.6656) \\ &= 1.9480 - 6.6624 = -4.7144 \\ &= 0.2856 - 5 \end{aligned}$$

Then $N = 1.93 \times 10^{-5}$ or 0.0000193

44. $T = 2\pi\sqrt{\dfrac{L}{G}}$ Given $L = 4.72$, $g = 32.2$, $\pi = 3.14$; so

$$T = 2(3.14)\sqrt{\frac{4.72}{32.2}} = 6.28\sqrt{\frac{4.72}{32.2}}$$

$$\begin{aligned} \frac{1}{2}(\text{Log of radicand}) &= \frac{1}{2}(\log 4.72 - \log 32.2) \\ &= \frac{1}{2}(0.6739 - 1.5079) = \\ &= \frac{1}{2}(-0.8340) = -0.4170, \text{ or} \\ &= 0.5830 - 1 \end{aligned}$$

Now add $\log 6.28 = \underline{0.7980}$

so $\log T = 1.3810 - 1 = 0.3810$ (from table)

or $\underline{T = 2.404}$

47. The compound interest formula is

$$A = P(1 + r)^n = P\left(1 + \frac{r}{k}\right)^n$$

In this problem, $n = 10$, $k = 2$, $r = 7.8\%$; so

$$A = 10{,}000\left(1 + \frac{0.078}{2}\right)^{10} = 10{,}000(1.039)^{10}$$

$$\log A = \log 10{,}000 + 10 \log 1.039$$

$$\log A = 4 + 0.1662 = 4.1662$$

Thus $A = \$14660$ Approximately

50. The formula is $A = \sqrt{s(s - a)(s - b)(s - c)}$ with

$s = \frac{1}{2}(a + b + c) = \frac{1}{2}(12.86 + 13.72 + 20.3) = 23.44$; $A = \sqrt{23.44(10.58)(9.72)(3.14)}$.

Now using logs we have

$$\log A = \frac{1}{2}(\log 23.44 + \log 10.58 + \log 9.72 + \log 3.14)$$

$$= \frac{1}{2}(1.3692 + 1.0253 + 0.9877 + .4969)$$

$\log A = 1.9396$

and

$A = 87.0'$ approximately

Section 4.5 Exponential and Logarithmic Equations

BASIC CONCEPTS you should learn from this unit.

1. An Exponential equation is one in which the variable appears in an exponent (for example, $2^{2x-1} = 56$). To solve equations of this type we usually take the Logarithm of both sides and then apply the laws of logarithms and rules for solving equations.

2. A Logarithmic equation is one in which we find the log of an expression containing the variable; for example, $\log x + \log(x + 21) = 2$.

   Such equations are normally solved by converting the logarithmic expression to one logarithm, then changing to exponential form.

WHAT YOU SHOULD KNOW HOW TO DO upon completion of this unit.

1. Recognize and solve EXPONENTIAL EQUATIONS.

2. Recognize and solve LOGARITHMIC EQUATIONS.

SOLUTIONS to Exercise Set 4.5

2. $2^x = 24$; $\log 2^x = \log 24$; $x \log 2 = \log 24$

   Therefore, $x = \log 24/\log 2 = 1.3802/0.3010 = 4.585$

5. $3^{2x} = 46$ implies that $\log 3^{2x} = \log 46$ or $2x \log 3 = \log 46$

   Therefore, $2x = \log 46/\log 3 = 1.6628/0.4771 = 3.4852$; $x = 1.743$

8. $3^{3x-2} = 23.1$ implies that $(3x - 2)\log 3 = \log 23.1$ implies that
$3x - 2 = \log 23.1/\log 3 = 2.8580$: $3x = 4.8580$
$\underline{x = 1.619}$

11. $2^{-x} = 15$ implies that $-x(\log 2) = \log 15$ or $-x = 3.907$
$x = -3.907$

14. $3^{-3x+2} = 2^{-x}$ implies that $(-3x + 2)\log 3 = -x(\log 2)$ or
$(-3x + 2)(0.4771) = -0.3010x$; $-1.4313x + 0.9542 = -0.3010x$
$0.9542 = 1.1303x$; $x = 0.844$

17. $e^{2x+3} = 20$ implies that $(2x + 3)\ln e = \ln 20$ (natural log here because of e) implies that $2x + 3 = \ln 20 = 2.9957$
$2x = -0.0043$; $\underline{x = -0.0022}$ approximately

20. $\log x - \log 3 = 2$; $\log \frac{x}{3} = 2$; $\frac{x}{3} = 10^2 = 100$,
so $x = 300$.

23. $\log x + \log(x - 3) = 1$; $\log x(x - 3) = 1$; $x^2 - 3x = 10$
$x^2 - 3x - 10 = 0$ $(x - 5)(x + 2) = 0$; $x = 5$, $x = -2$
Discard the negative root; so $x = 5$.

26. $\log(7x - 2) - \log(x - 2) = 1$; $\log\left(\frac{7x - 2}{x - 2}\right) = 1$ implies that
$\frac{7x - 2}{x - 2} = 10$; $7x - 2 = 10x - 20$; $\underline{x = 6}$

29. $\log_2(x + 4) = 3 - \log_2(x - 2)$ implies that $\log_2(x + 4) + \log_2(x - 2) = 3$
$\log_2(x + 4)(x - 2) = 3$ implies that $x^2 + 2x - 8 = 8$; $x^2 + 2x - 16 = 0$
$x = 3.123$ approximately

32. $50{,}000 = 20{,}000\, e^{0.05t}$ implies that $2.5 = e^{0.05t}$
$\ln 2.5 = 0.05t \ln e = 0.05t$
$0.9163 = 0.05t$ so $t = \underline{18.33}$ years

35. Use $A = P\left(1 + \frac{r}{k}\right)^n$. Let $P = 1$ and $A = 2$.
Since we wish the amount to double,
$2 = \left(1 + \frac{.08}{2}\right)^n$; $2 = (1.04)^n$; $\log_2 = n \log 1.04$
$n = 0.3010/0.0170 = 17.7 = 18$. NOW n is the number of compounding periods and we are compounding semiannually, so it would take 9 years.

38. $430 = 500e^{3k}$ $\frac{430}{500} = e^{3k}$; $\ln 430 - \ln 500 = 3k$
$3k = -0.1508$; $\underline{k = 0.05}$ = 5% rate of decay

SOLUTIONS to Review Exercises

2. $2^{2x} = 8^{x-1}$ Note that $8 = 2^3$ so $2^{2x} = (2^3)^{x-1}$.
$2^{2x} = 2^{3x-3}$ and $2x = 3x - 3$; $x = 3$

5. $27 = 9^{3/2}$ implies that $\log_9 27 = \frac{3}{2}$

6. $8 = 64^{1/2}$

10. $\log_5 \frac{1}{125} = x - 1$ implies that $\frac{1}{125} = 5^{x-1}$ and since $5^{-3} = \frac{1}{125}$
$5^{-3} = 5^{x-1}$ and $-3 = x - 1$; $\underline{x = -2}$

13. $\log_3 3^5 = 5$

16. $e^{\ln 3} = 3$

19. $\log_a \frac{x(2 - x)^2}{(y + 1)^{1/2}} = \log_a x + 2 \log_a (2 - x) - \frac{1}{2} \log_a (y + 1)$

22. $14 = 2 \cdot 7$ so $\log 14 = \log 2 + \log 7$
$= 0.30 + 0.85$
$\log 14 = 1.15$

25. $0.7 = 7 \times 10^{-1}$, so $\log 0.7 = 0.85 - 1$

28. $\ln 3x + 2(\ln y - \frac{1}{2} \ln x) = \ln 3x + 2 \ln y - \ln x = \ln \frac{3xy^2}{x} = \ln 3y^2$

31. $0.098 = 9.8 \times 10^{-2}$

34. $N = (0.765)(32.4)^2$; $\log N = \log 0.765 + 2 \log 32.4$
$= 0.8837 - 1 + 3.0211 = 2.9048$
$N = 803$ approximately

37. The Exponential Decay Model is $Q(t) = Q_0 e^{-kt}$ (see Section 4.1). If $Q_0 = 1$, we will then want $Q(t) = \frac{1}{2}$. $\frac{1}{2} = e^{-0.06t}$; $\ln \frac{1}{2} = \ln e^{-0.06t}$;
$\ln 1 - \ln 2 = -0.06t$; $\ln 2 = 0.06t$ implies that $t = 11.5$ hr.

40. $\log(2x - 1) = 2 + \log(x - 2)$; $\log(2x - 1) - \log(x - 2) = 2$;
$\log \frac{2x - 1}{x - 2} = 2$; $\frac{2x - 1}{x - 2} = 100$ implies that $2x - 1 = 100x - 200$
$x = \frac{199}{98}$

Chapter 4 - A SHORT QUIZ

1. Functions of the form $f(x) = a^x$ are called ____?____ functions.
2. For functions like $f(x) = a^x$ what restriction is placed on a ?
3. The graph of the function $f(x) = a^x$ will be ____?____ (increasing or decreasing) if $a > 1$.
4. Graph the function $f(x) = 3^{-x}$.
5. If $b^5 = z^5$, what do we know about b and z ?
6. Solve for x : $2^x = 8^{x+2}$
7. If the number of bacteria in a culture is described by $Q(t) = 80e^{0.3t}$, and t is measured in hours, (a) what is the initial number of bacteria in the culture? (b) Find the number of bacteria after 25 hours.
8. The inverse of the exponential function $y = 2^x$ is ____?____ .
9. Write $\log_{3b} 6 = \frac{1}{2}$ in exponential form.
10. Solve for x : $\log_3 x = -4$.
11. Graph $y = \log_3 x$.
12. Solve for x : $\log_2 x^2 = \log_2 9$.
13. Complete these statements of the properties of logarithms.
    a. $\log_a (x \cdot y) =$ ____?____
    b. $\log_a \left(\frac{x}{y}\right) =$ ____?____
    c. $\log_a x^n =$ ____?____
14. Write $\ln(4xyz)$ in terms of simpler logs.
15. Write $\frac{2}{3}(4 \ln x - 3 \ln y)$ as a single logarithm.
16. Approximate $N = \dfrac{\sqrt[4]{47.9}}{(3.12)^3}$ using logarithms.
17. Solve for x : $5^{2x-3} = 19$
18. Solve for x : $\log(x + 197) - \log(x - 1) = 2$

Quiz Answers Chapter 4

1. Exponential
2. $a > 0, \quad a \neq 1$
3. increasing

4.

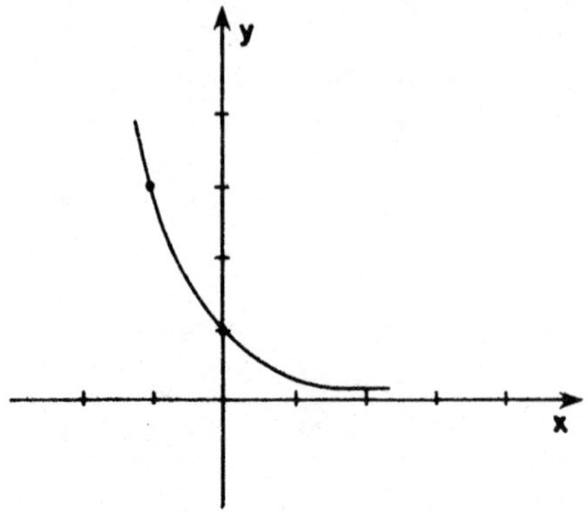

5. $b = z$

6. $x = -3$

7. a. 80

   b. 144, 644

8. $y = \log_2 x$

9. $36^{1/2} = 6$

10. $x = 1/81$

11. $y = \log_3 x$

x	y
1	0
3	1
$\frac{1}{3}$	-1

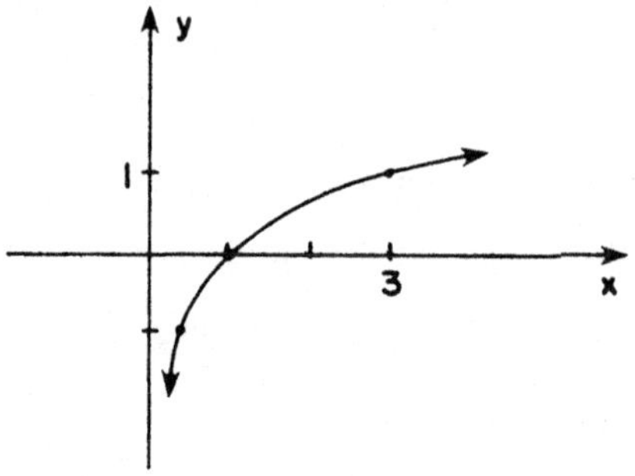

12. $x = \pm 3$

13. a. $\log_a x + \log_a y$

    b. $\log_a x - \log_a y$

    c. $n \log_a x$

14. $\ln 4 + \ln x + \ln y + \ln z$

15. $\ln \frac{x^{8/3}}{y^2}$

16. 0.0866

17. $x = 2.415$

18. $x = 3$

CHAPTER 5

Trigonometry: The Circular Functions

Section 5.1 The Wrapping Function

BASIC CONCEPTS you should learn from this unit.

1. The Wrapping Function, W, assigns to each real number a point on the unit circle (a circle of radins 1 centered at the origin). Thus if t is a real number, $W(t) = P(x, y)$ is on the circle; for example,

$$W(0) = (1, 0)$$

Think of a number line that is bent around the circle with the positive numbers starting at (1, 0) and going counterclockwise and the negative numbers going clockwise.

2. Each point determined by $W(t)$ will thus have rectangular coordinates (x, y).

3. Since 2π = the circumference of the unit circle, real numbers $2\pi \leq t < 4\pi$ will repeat the points of real numbers $0 \leq t < 2\pi$... . In essence this tells us that $W(t + 2\pi) = W(t)$ or $W(t + 6\pi) = W(t)$. A similar statement can be made for the negative numbers.

4. The coordinates of the point determined by $W(t)$ can be found, but not always easily. For some values of t, however, we can find these coordinates readily [for example, $W\left(\frac{\pi}{4}\right) = \left(\frac{\sqrt{2}}{2}, \frac{\sqrt{2}}{2}\right)$] by means of algebra and plane geometry.

WHAT YOU SHOULD KNOW HOW TO DO upon completion of this unit.

1. Replace any real number s with another real number t, such that $0 \leq t < 2\pi$ and $W(s) = W(t)$.

2. Plot the location of $W(t)$ on the unit circle.

3. Find the rectangular coordinates of $W(t)$ for special values of t, for example, $W\left(\frac{\pi}{4} + n\pi\right)$ where n is any integer; $W\left(\frac{\pi}{6} + n\pi\right)$; etc.

SOLUTIONS to Exercise Set 5.1

2. $s = \frac{13\pi}{2} = \frac{\pi}{2} + 6\pi = \frac{\pi}{2} + 3(2\pi)$

   Thus $t = \frac{\pi}{2}$; $W(s) = W\left(\frac{\pi}{2}\right)$

5. $s = -\frac{21}{2}\pi$ (Recall that $t = s + 2n\pi$ for this purpose) so that

   $t = -\frac{21}{2}\pi + 12\pi = \frac{3\pi}{4}$

   and $0 \leq t < 2\pi$ as required.

8. $s = \frac{11\pi}{2}$ and $t = \frac{11}{2}\pi - 4\pi = \frac{3\pi}{2}$

11. $s = \frac{27\pi}{5}$; $t = \frac{27\pi}{5} - 4\pi = \frac{7\pi}{5}$

14.

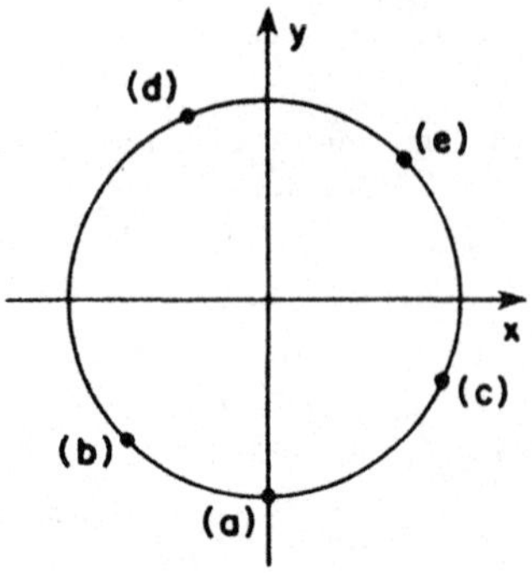

17. $W(5\pi) = W(\pi) = (-1, 0)$

20. $W\left(-\frac{3}{2}\pi\right) = W\left(\frac{\pi}{2}\right) = (0, 1)$

23. $W\left(\frac{4}{3}\pi\right) = \left(-\frac{1}{2}, -\frac{\sqrt{3}}{2}\right)$

26. $W\left(-\frac{19}{3}\pi\right) = W\left(-\frac{\pi}{3}\ n\right) = W\left(\frac{5}{3}\pi\right) = \left(\frac{1}{2}, -\frac{\sqrt{3}}{2}\right)$

29. $W\left(-\frac{5}{6}\pi\right) = W\left(\frac{7}{6}\pi\right) = \left(-\frac{\sqrt{3}}{2}, -\frac{1}{2}\right)$

32. $W\left(\frac{25}{3}\pi\right) = W\left(8\frac{1}{3}\pi\right) = W\left(\frac{1}{3}\pi\right) = \left(\frac{1}{2}, \frac{\sqrt{3}}{2}\right)$

35. $t = \frac{3}{4}\pi$ or $t = -\frac{7}{4}\pi$

38. $W(t) = \left(-\frac{1}{2}, -\frac{\sqrt{3}}{2}\right)$ so $t = \frac{4}{3}\pi$ or $-\frac{2}{3}\pi$

41. $W(t) = \left(\frac{3}{5}, \frac{4}{5}\right)$ Looking at the circle, we have

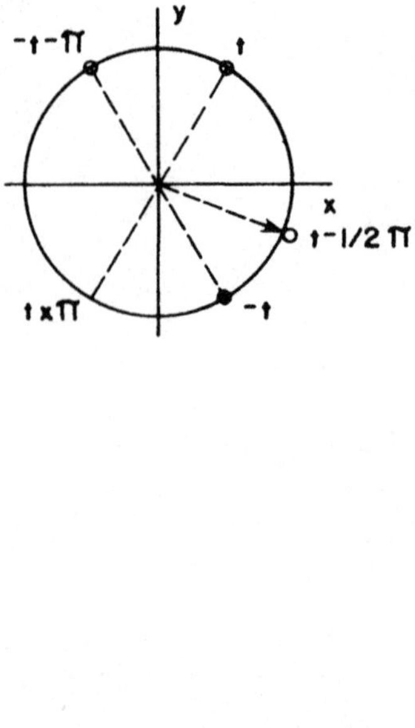

and so

$$W(t + \pi) = \left(-\frac{3}{5}, \frac{4}{5}\right)$$
$$W\left(t - \frac{\pi}{2}\right) = \left(\frac{4}{5}, -\frac{3}{5}\right)$$
$$W(-t) = \left(\frac{3}{5}, -\frac{4}{5}\right)$$
$$W(-t - \pi) = \left(-\frac{3}{5}, \frac{4}{5}\right)$$

44. If (a, b) is on the unit circle then $a^2 + b^2 = 1$.

To show that (b, a) is on the circle, note that $b^2 + a^2 = a^2 + b^2 = 1$.

For (b, -a), note that $b^2 + (-a)^2 = a^2 + b^2 = 1$ and likewise for (-b, a) and (-b, -a). The sum of the squares of the coordinates in each case equals $a^2 + b^2 = 1$, and so the point lies on the circle.

Section 5.2 The Sine, Cosine, and Tangent Functions

BASIC CONCEPTS you should learn from this unit.

1. The Definitions: if t is real and W(t) = (x, y), then

   $\sin t = y$

   $\cos t = x$

   $\tan t = \frac{y}{x} \quad x \neq 0$

   The domain of tan t is restricted, so that $x \neq 0$, that is,

   D: tan t = all reals except $\left(\frac{\pi}{2} + n\pi\right)$, where n is an integer.

2. The SIGNS of these functions in each quadrant are important and can be easily recalled from this diagram:

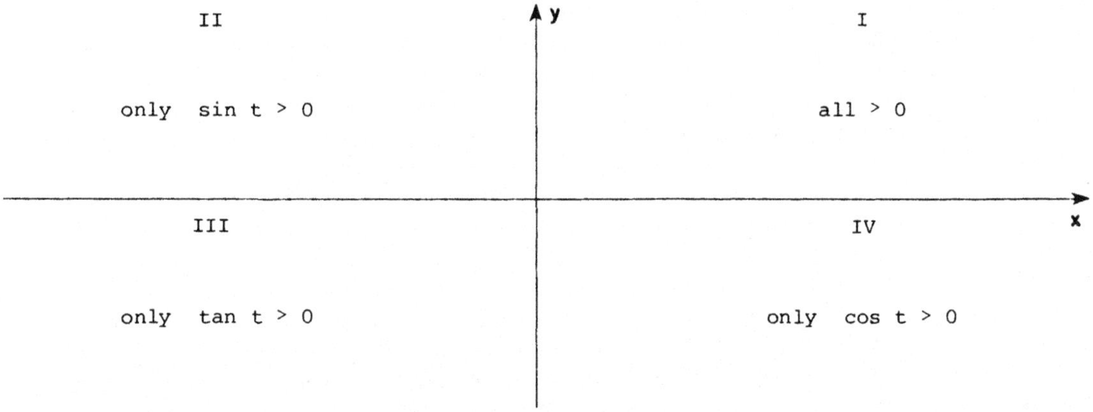

3. The BASIC SYMMETRY OF THE CIRCLE makes it easy to find sin(-t), cos(-t), and tan(-t):

   $\sin(-t) = -\sin t$

   $\cos(-t) = \cos t$

   $\tan(-t) = -\tan t$

4. If f(-x) = f(x), f is said to be even.

   If f(-x) = -f(x), f is said to be odd

5. An identity is an equation that is true for all values of the variable. Certain basic identities are very useful and should be learned. Recall that $\sin t = y$, $\cos t = x$, $\tan t = y/x$, and $x^2 + y^2 = 1$ for the unit circle.

We then have

$$\sin^2 t + \cos^2 t = 1$$

$$\tan t = \frac{\sin t}{\cos t}$$

for all allowable t.

WHAT YOU SHOULD KNOW HOW TO DO upon completion of this unit.

1. Define $\sin t$, $\cos t$, and $\tan t$ in terms of x and y.

2. State the quadrant in which each function is + or − .

3. Determine $\sin(-t)$, $\cos(-t)$, $\tan(-t)$ using symmetry.

4. Use the BASIC IDENTITIES to transform expressions; for example,

$$\tan t \sin t + \cos t = \frac{\sin^2 t}{\cos t} + \cos t = \frac{\sin^2 t + \cos^2 t}{\cos t} = \frac{1}{\cos t}$$

SOLUTIONS to Exercise set 5.2

2. $t = \frac{3\pi}{4}$; hence $W(t) = \left(-\frac{\sqrt{2}}{2}, \frac{\sqrt{2}}{2}\right)$

and then $\sin t = \frac{\sqrt{2}}{2}$, $\cos t = \frac{-\sqrt{2}}{2}$, $\tan t = -1$.

5. $t = \frac{7}{4}\pi$; $W(t) = \left(\frac{\sqrt{2}}{2}, \frac{-\sqrt{2}}{2}\right)$

so $\sin t = -\frac{\sqrt{2}}{2}$, $\cos t = \frac{\sqrt{2}}{2}$, $\tan t = -1$

8. $t = \frac{5\pi}{6}$; $W(t) = \left(\frac{-\sqrt{3}}{2}, \frac{1}{2}\right)$

and $\sin t = \frac{1}{2}$, $\cos t = -\frac{\sqrt{3}}{2}$, $\tan t = -\frac{1}{\sqrt{3}}$

11. $t = \frac{2}{3}\pi$; $W(t) = \left(-\frac{1}{2}, \frac{\sqrt{3}}{2}\right)$

$\sin t = \frac{\sqrt{3}}{2}$, $\cos t = \frac{1}{2}$, $\tan t = -\sqrt{3}$

14. $\sin t < 0$ and $\tan t > 0$ implies that $W(t)$ is in the quadrant III because $\sin t < 0$ in quadrants II and IV and $\tan t > 0$ in quadrants III and IV.

17. Given $\sin t < 0$ and $\cos t < 0$ implies that $W(t)$ is in quadrant III because
$\sin t < 0$ in quadrants III and IV
$\cos t < 0$ in quadrants II and III.

20. $t = \pi/2$

23. $t = \frac{\pi}{4}, \frac{3}{4}\pi$

26. $t = \frac{7\pi}{6}, \frac{11\pi}{6}$

29. $t = \frac{\pi}{3}, \ \frac{2}{3}\pi$

32. $t = \frac{3\pi}{4}, \ \frac{5}{4}\pi$

35. $t = \frac{\pi}{6}, \ \frac{5}{6}\pi$

38. Given $\tan t = -\frac{3}{4}$ and $W(t)$ in quadrant II, $\tan t = \frac{3}{-4}$, if $W(t)$ is in quadrant II, therefore, $\cos t = -4/5$.

41. $\cos t = \frac{4}{5}$ and $\sin t < 0$ means that $W(t)$ is in quadrant IV.
Also, $\sin^2 t + \cos^2 t = 1$, so $\sin^2 t + \frac{16}{25} = 1$ $\quad \sin^2 t = \frac{9}{25}$ $\quad \sin t = -\frac{3}{5}$

44. $\tan t = -\frac{5}{12}$ and $\sin t > 0$ means that $W(t)$ is in the Quadrant II,
so $\tan t = 5/-12$ and $\sin t = 5/13$.

47. $\frac{1 - \sin^2 t}{\sin t} = \frac{\cos^2 t}{\sin t} = \frac{\cos t}{\sin t} \cdot \cos t = \frac{1}{\tan t} \cdot \cos t = \underline{\frac{\cos t}{\tan t}}$

50. $\frac{1 - \cos^2 t}{\sin t} = \frac{\sin^2 t}{\sin t} = \sin t$

53. $(\sin t - \cos t)^2 = \sin^2 t - 2 \sin t \cos t + \cos^2 t$ expanding the binomial
$= (\sin^2 t + \cos^2 t) - 2 \sin t \cos t$
$= 1 - 2 \sin t \cos t$

Section 5.3 Graphs of Sine, Cosine, and Tangent

BASIC CONCEPTS you should learn from this unit.

1. A periodic function is one in which the function values repeat at a constant interval. The wrapping function $W(t)$ repeats at intervals of 2π (each "wrap" around the circle). The interval referred to is called the period of the function. Symbolically, we could say that if $f(x + c) = f(x)$ for some real number c, then $f(x)$ is periodic and c is its period.

2. The period of $\sin t$ and $\cos t$ is 2π.

3. Letting $y = \sin t$ and the vertical axis still be the y-axis but the horizontal axis now be the t-axis, we can graph $\sin t$, $\cos t$, and $\tan t$ on this new rectangular system.

4. Both $\sin t$ and $\cos t$ show their periodic or (wave) nature by repeating y values at 2π intervals. For example,

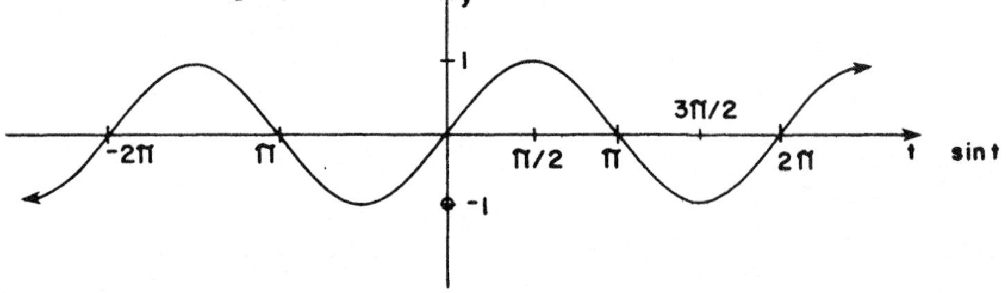

5. The tangent function is also periodic, but its period is π! Further, as t approaches $\frac{1}{2}\pi + n\pi$, n an integer, then $\tan t$ approaches either $+\infty$ if t approaches from the left or $-\infty$ if t approaches from right.

6. Composites of the trigonometric functions, for example, $(1 - \cos t)$, $\sin t + \cos t$, etc., can be graphed by adding the separate parts. For instance, if $y = \sin t + \cos t$ for $t = \frac{\pi}{4}$, then

$$y = \sin \frac{\pi}{4} + \cos \frac{\pi}{4} = \frac{\sqrt{2}}{2} + \frac{\sqrt{2}}{2} = \sqrt{2}.$$

WHAT YOU SHOULD KNOW HOW TO DO upon completion of this unit.

1. Define periodic function.
2. State the period of $\sin t$, $\cos t$, and $\tan t$.
3. Graph simple composites of the three trigonometric functions.
4. Determine Domain and Range of simple composites.

SOLUTIONS TO Exercise Set 5.3

2. $f(t) = -1 + \sin t$ Subtract 1 from each $\sin t$ value.

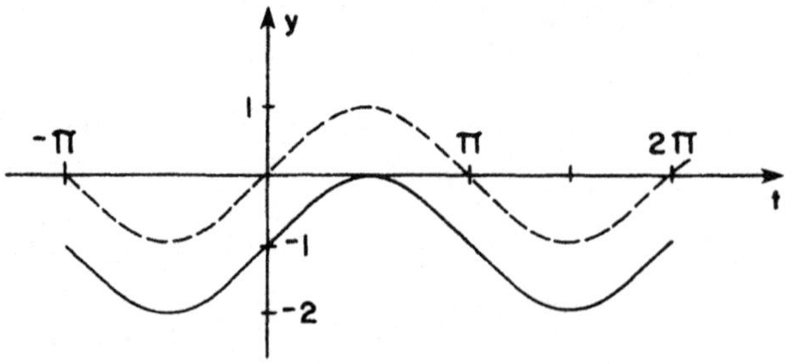

5. $f(t) = \sin t + \frac{1}{2} \cos t$ Plot each part and add.

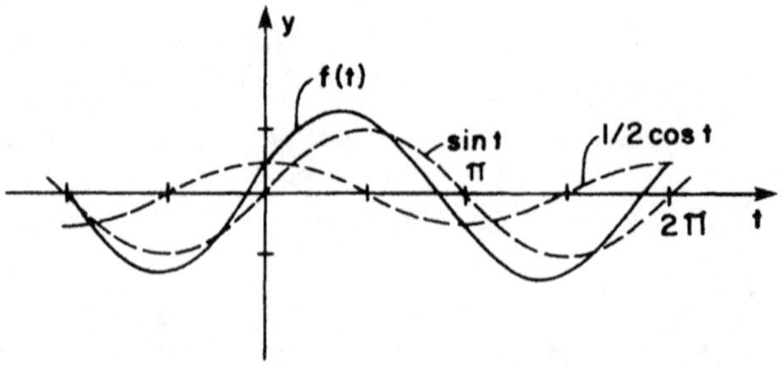

8. $f(t) = \sin(-5) + \cos t$ Plot each part noting that $\sin(-t) = -\sin t$ AND THEN ADD THE PARTS.

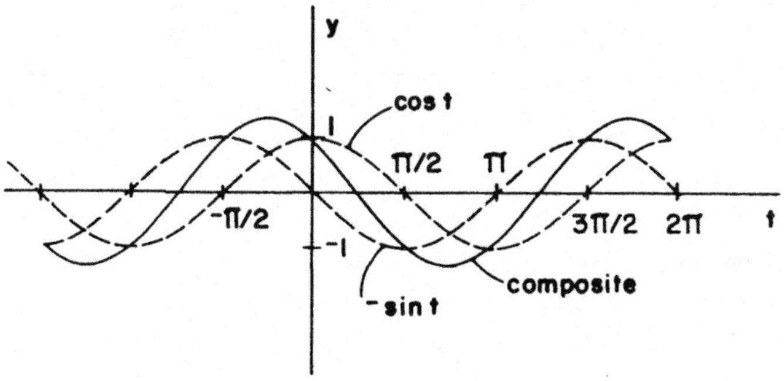

11. Prove that the period of $\sin t$ is 2π.

Assume a period = c; thus

$\sin(t + c) = \sin t \quad 0 < c < 2\pi$ for all t

If $t = 0$, we have

$\sin c = \sin 0 = 0$

and since $0 < c < 2\pi$,

$c = \pi$

(the only value where $\sin t = 0$).

However, $\sin(t + \pi)$ for $t = \pi/2$ yields

$\sin\left(\frac{\pi}{2} + \pi\right) \neq \sin\frac{\pi}{2}$

leaving 2π as our only choice for c.

(Note that $\sin(t + c) = \sin t$ and $t = 0$ gives $\sin c = 0$; also $\sin(t + 2\pi) = \sin t$ does hold at $t = \pi/2$.)

14. Graph of $\sin t$:

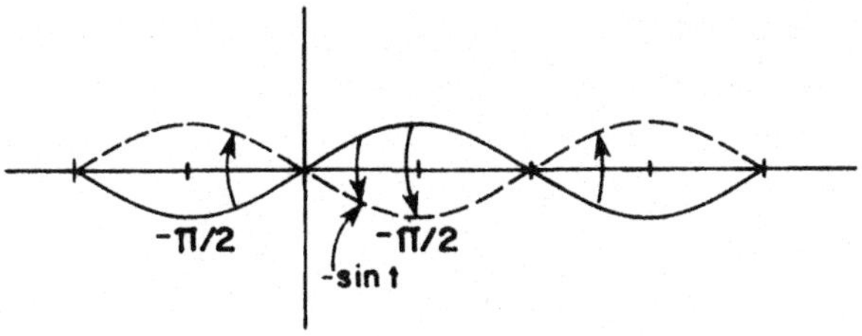

Notice for example, that $\sin - \pi/2 = -1$. Also, $-\sin \pi/2 = -1$. Other points could likewise be verified.

BASIC CONCEPTS you should learn from this unit.

1. VALUES of sin t, cos t, and tan t for $\frac{\pi}{2} < t < 2\pi$ can be found from the usual tables ($0 < t \leq \pi/2$) by using a reference number t' instead of t and appending the proper sign according to whether t is in the second, third, or fourth quadrant.

2. t' (the reference number) is found from

   $t' = \pi - t$ if t is in quadrant II;

   $t' = t - \pi$ if t is in quadrant III;

   $t' = 2\pi - t$ if t is in quadrant IV.

3. The following approximations for sin t and cos t can be used to obtain values accurate to three decimal places. (You need not memorize these expressions.)

$$\sin t \approx t - \frac{t^3}{6} + \frac{t^5}{120} - \frac{t^7}{5040}$$

$$\cos t \approx 1 - \frac{t^2}{2} + \frac{t^4}{24} - \frac{t^6}{720}$$

4. Hand-held calculators also have built-in trigonometric functions, which makes finding values easy.

WHAT YOU SHOULD KNOW HOW TO DO upon completion of this unit.

1. USE A TABLE to find sin t, cos t, and tan t for any real number t by first using a reference number t'.

2. Use a calculator to find sin t, cos t, and tan t .

3. Use the polynomial approximations given in basic concept 3 to find sin t and cos t .

SOLUTIONS to Exercise Set 5.4

2. sin 0.48: 0.48 is in quadrant I so we can read the solution directly from the table: sin 0.48 = 0.4617.

5. tan 3.44: $3.44 > \pi$

   $3.44 - 3.14 = 0.30 = t'$ tan 0.30 = 0.3089

   tan t in quadrant III is also greater than 0, so

   tan 3.44 = 0.3089

8. tan 6.05: $2\pi = 6.28$ Thus 6.05 is in quadrant IV.

   $6.28 - 6.05 = 0.23 = t'$

   tan 0.23 is approximately equal to 0.2339 Use negative value because t is in quadrant IV. tan 6.05 = -0.2339

11. $\tan(-3.27)$ Since $\pi = 3.14$, we are in the second quadrant and we use $t' = -(t + \pi) = 0.13$ to find the values $\tan 0.13 = 0.1317$. Use negative, however: $\tan(-3.27) = -0.1317$

14. $t = 1.10$

$$\cos t \simeq 1 - \frac{1.21}{2} + \frac{1.4641}{24} - \frac{1.772}{720}$$
$$\simeq 1 - 0.605 + 0.0610 - 0.0025$$
$$\simeq 0.454$$

$\cos t = 0.4536$ with a calculator

17. $\tan 0.1 = (\sin 0.1)/(\cos 0.1) = \left(0.1 - \frac{0.1^3}{6} + \frac{0.1^5}{120} - \frac{0.1^7}{5040}\right) \div \left(1 - \frac{0.1^2}{2} + \frac{0.1^4}{24} - \frac{0.1^6}{720}\right)$

$= (0.1 - 0.00017 + 0.00000 - 0.00000) \div (1 - 0.005 + 0.0000 - 0.0000)$

$= (0.09983) \div (0.995) = 0.1003$

$\tan 0.1 = 0.1003$ with the calculator

20. $\sin 1.32 = 0.9687$

23. $\tan 2.77 = -0.3878$

26. $\cos 7.08 = 0.6990$

29. $\sin 2.62 = 0.4969$

32. The $\cos t$ approximation is

$$\cos t = 1 - \frac{t^2}{2} + \frac{t^4}{24} - \frac{t^6}{720}$$

Since all powers of t are even powers, the substitution of $-t$ for t causes no change. Therefore, $\cos(-t) = \cos t$ and $\cos t$ is even!

Section 5.5 GRAPHS: Amplitude, Period, and Phase Shift

BASIC CONCEPTS you should learn from this unit.

1. The AMPLITUDE of a periodic function is one half the difference between the maximum and minimum values.

2. For $A \sin x$ and $A \cos x$ the amplitude $= |A|$.

3. The period of functions like $y = \sin Bx$ and $\cos Bx$ can be deduced if we know that the period of $\sin x$ is 2π. When

$Bx = 0 \qquad Bx = 2\pi$

$x = 0 \qquad x = \frac{2\pi}{B}$

so the period of $y = \sin Bx$ is $2\pi/B$.

4. Functions such as $y = A\sin(Bx + C)$ have a further characteristic known as PHASE SHIFT; it is determined by

$$Bx + C = 0 \qquad Bx + C = 2\pi$$
$$x = -\frac{C}{B} \qquad x = \frac{2\pi}{B} - \frac{C}{B}$$

If $-\frac{C}{B} > 0$, the shift is to the right $\frac{C}{B}$ units.

If $-\frac{C}{B} < 0$, the shift is to the left $\frac{C}{B}$ units.

5. All of the foregoing are useful in graphing functions such as $y = A\sin(Bx + C)$ and $y = A\cos(Bx + C)$.

WHAT YOU SHOULD KNOW HOW TO DO upon completion of this unit.

1. Determine the amplitude, period, and phase shift of a periodic function.
2. Sketch the graph of periodic functions of the form $y = A\sin(Bx + C)$ and $y = A\cos(Bx + C)$.

SOLUTIONS to Exercise Set 5.5

2. $f(x) = \frac{1}{4}\cos x$ amplitude $= \frac{1}{4}$ period $= 2\pi$

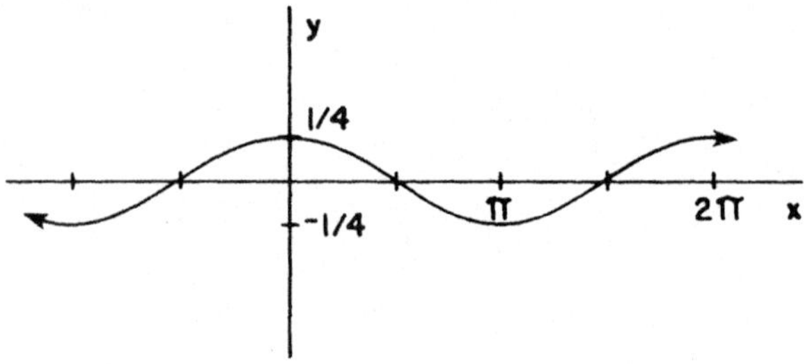

5. $f(x) = -2\sin 4x$ amplitude $= 2$ period $= \frac{2\pi}{4} = \frac{\pi}{2}$

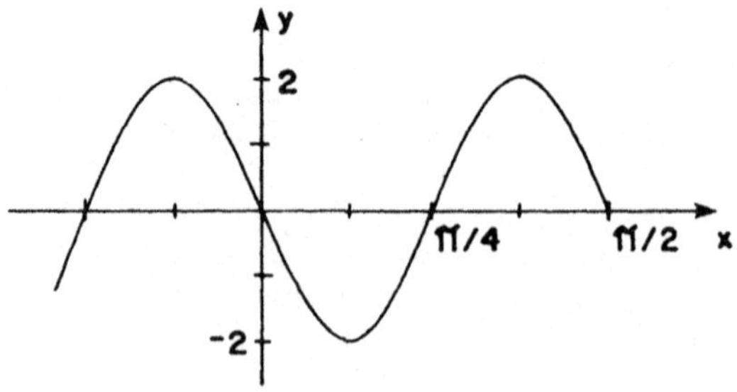

8. $f(x) = 4 \sin 4x$ amplitude = 4 period $= \frac{2\pi}{4} = \frac{\pi}{2}$

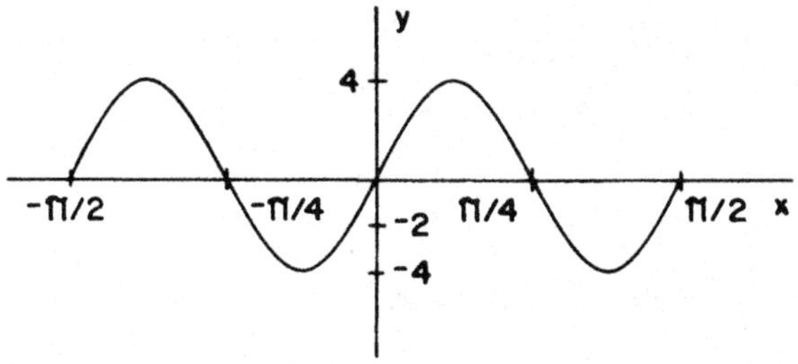

11. $f(x) = -3 \cos 3x$ amplitude = 3 period $= \frac{2\pi}{3}$

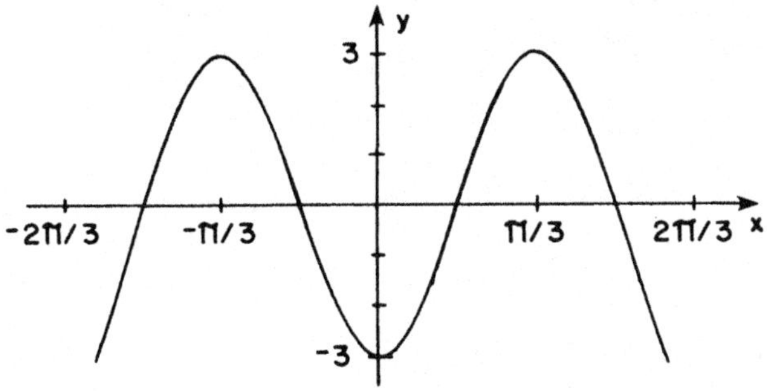

14. $f(x) = \frac{1}{2} \cos\left(x + \frac{\pi}{2}\right)$ amplitude $= \frac{1}{2}$ period = 2π

phase shift $= -\frac{\pi}{2}$ (left $\frac{\pi}{2}$)

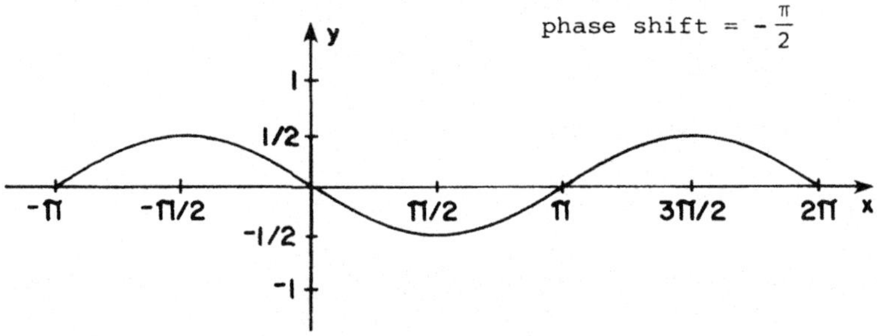

17. $f(x) = \frac{1}{3}\sin\left(3x + \frac{3\pi}{4}\right)$ amplitude $= \frac{1}{3}$ period $= \frac{2\pi}{3}$

phase shift $= \frac{-C}{B} = -\frac{\pi}{4}$ (left $\frac{\pi}{4}$)

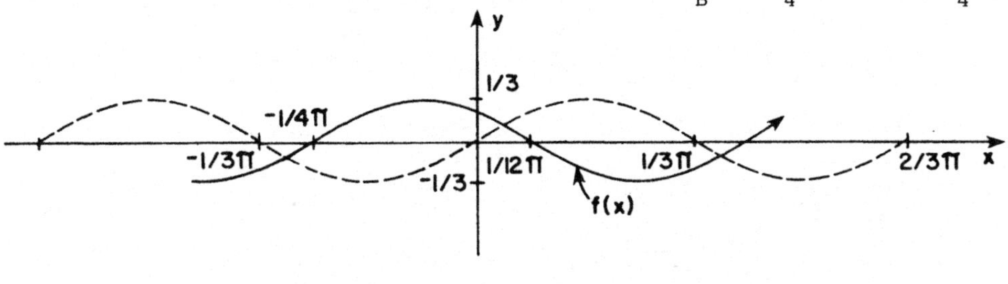

20. $f(x) = 6\sin\left(\frac{x}{2} + \frac{\pi}{2}\right)$ amplitude = 6 period = 4π

phase shift = $-\frac{C}{B} = -\pi$

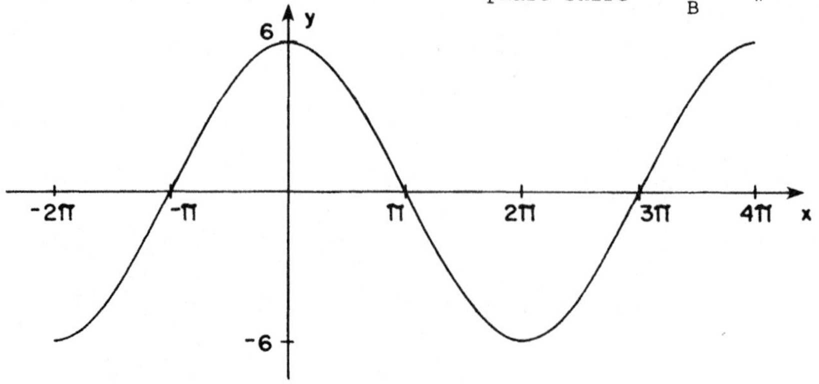

23. $y = 3\cos\left(-\frac{x}{3} + \frac{2\pi}{3}\right) = 3\cos\left[-\left(\frac{x}{3} - \frac{2\pi}{3}\right)\right] = 3\cos\left(\frac{x}{3} - \frac{2\pi}{3}\right)$

amplitude = 3 period = $\frac{2\pi}{3}$ phase shift = $+2\pi$

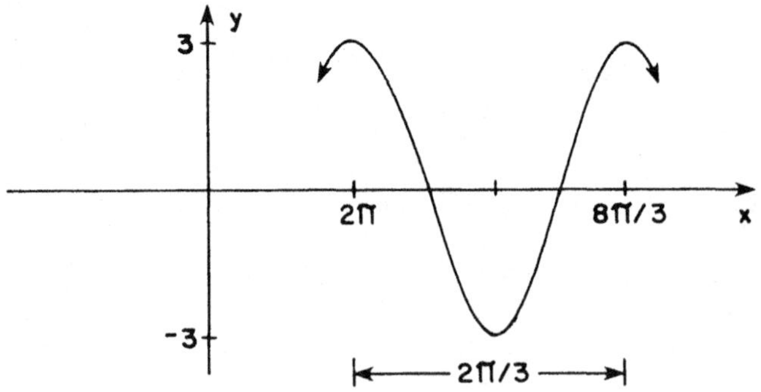

Section 5.6 Secant, Cosecant, and Cotangent

BASIC CONCEPTS you should learn from this unit.

1. $\sec t = \frac{1}{\cos t}$, $\csc t = \frac{1}{\sin t}$ and $\cot t = \frac{1}{\tan t}$

2. Values of these three functions can be found from the appropriate sine, cosine, or tangent. For example,

$$\csc\frac{\pi}{6} = \frac{1}{\sin\frac{\pi}{6}} = \frac{1}{\frac{1}{2}} = 2$$

3. The signs of $\csc t$, $\sec t$, and $\cot t$ are the same as those of $\sin t$, $\cos t$, and $\tan t$, respectively.

4. These functions can also be readily graphed from use of the sine, cosine, and tangent values.

WHAT YOU SHOULD KNOW HOW TO DO upon completion of this unit.

1. Determine values of the secant, cosecant, and cotangent.

2. Determine the quadrant of $W(t)$ if given appropriate information about the values of the functions of t; for example, $\sec t < 0$, $\sin t > 0$ implies that $W(t)$ is in quadrant II.

3. Determine value of t, $0 \leq t < 2\pi$, if given appropriate information on values of the functions of t. For example, $\tan t = \sqrt{3}$, $\csc t < 0$ implies that $W(t)$ is in quadrant III. $\tan t$ is positive and $\csc t$ is negative, thus $t = 4/3\,\pi$.

SOLUTIONS to Exercise Set 5.6

2. $t = \frac{\pi}{6}$

$\sin t = 1/2$, therefore, $\csc t = 2$

$\cos t = \sqrt{3}/2$, therefore, $\sec t = 2/\sqrt{3}$

$\tan t = 1/\sqrt{3}$, therefore, $\cot t = \sqrt{3}$

5. $t = \frac{5\pi}{6}$

$\sin t = 1/2$, therefore, $\csc t = 2$

$\cos t = -\sqrt{3}/2$, therefore, $\sec = -2/\sqrt{3}$

$\tan t = -\sqrt{3}/3$, therefore, $\cot t = -\sqrt{3}$

8. $t = \frac{7}{4}\pi$

$\sin t = -\sqrt{2}/2$, therefore, $\csc t = -2/\sqrt{2}$

$\cos t = \sqrt{2}/2$, therefore, $\sec t = 2\sqrt{2}$

$\tan t = -1$, therefore, $\cot t = -1$

11. $\frac{5\pi}{4}$

$\sin t = -\sqrt{2}/2$, therefore, $\csc t = -2/\sqrt{2}$

$\cos t = -\sqrt{2}/2$, therefore, $\sec t = -2/\sqrt{2}$

$\tan t = 1$, therefore, $\cot t = 1$

14. If $\sec t = -1$, $\cos t = -1$; thus, $t = \pi$

17. If $\cot t = 1$, $\tan t = 1$; thus, $t = \frac{\pi}{4}, \frac{5\pi}{4}$

20. If $\cot t = \frac{\sqrt{3}}{3}$, $\tan t = \sqrt{3}$; thus, $t = \frac{\pi}{3}, \frac{4}{3}\pi$

23. If $\cot t = -\sqrt{3}$, $\tan t = -\frac{\sqrt{3}}{3}$; thus, $t = \frac{5}{6}\pi, \frac{11}{6}\pi$

26. $\tan t < 0$ in quadrants II and IV

$\sec t < 0$ in quadrants II and III, therefore, $W(t)$ is in quadrant II

29. $\sec t < 0$ in quadrants II and III

$\cot t > 0$ in quadrants I and III, therefore, $W(t)$ is in quadrant III

32. $\csc t < 0$ in quadrants III and IV

$\cot t > 0$ in quadrants I and III, therefore, $W(t)$ is in quadrant III

35. $\sec t = -2$ $\qquad$ $\csc t > 0$

therefore $\cos t = -\frac{1}{2}$ and

$W(t)$ is in quadrant II or III $\qquad$ $W(t)$ is in quadrant I or II

Finally, then $W(t)$ is in quadrant II, and $t = \frac{5\pi}{6}$

38. $\sec t = \sqrt{2}$ $\qquad$ $\cot t > 0$

$W(t)$ is in quadrants I or IV $\qquad$ $W(t)$ is in quadrants I or III

therefore, $W(t)$ is in quadrant I and $t = \pi/4$

41. $\cot 3.37$ $(3.37 > \pi)$ Reference number $= 3.37 - 3.14 = 0.23$

$\cot 0.23 = 4.275$

44. $\csc 2.48$ $\quad 3.14 - 2.48 = 0.66 < \pi/2$, so $W(t)$ is in quadrant II

$\csc 0.66 = \underline{1.631} = \csc 2.48$

Section 5.7 The Inverse Trigonometric Functions

BASIC CONCEPTS you should learn from this unit.

1. Sin t, cos t, and tan t have <u>inverses</u> just as other functions do, but certain restrictions are necessary. For example,

   Inverse sine (arcsin or $\sin^{-1}$) is defined by

   $\sin^{-1} y = x$ if and only if $\sin x = y$ and $-\frac{\pi}{2} \le x \le \frac{\pi}{2}$.

   Inverse cosine (arccos or $\cos^{-1}$) is defined by

   $\cos^{-1} y = x$ if and only if $\cos x = y$ and $0 \le x \le \pi$.

   Inverse tangent (arctan or $\tan^{-1}$) is defined by

   $\tan^{-1} y = x$ if and only if $\tan x = y$ and $-\frac{\pi}{2} < x < \frac{\pi}{2}$.

2. The restrictions on x above are necessary in order to make the sin, cos, and tan functions one-to-one over some interval.

3. These <u>same</u> restrictions cause corresponding restrictions in the <u>range</u> of the <u>inverse</u> functions.

4. $\sin^{-1}$, $\cos^{-1}$, and $\tan^{-1}$, of course, have graphs which are reflections over the line $y = x$ just as with graphs of other inverses.

WHAT YOU SHOULD KNOW HOW TO DO upon completion of this unit.

1. Evaluate the three inverse functions for special values, for example, $\sin^{-1}\left(-\frac{\sqrt{2}}{2}\right)$.

2. Evaluate the three trigonometric functions for values given indirectly as $\cos\left(\arcsin -\frac{1}{2}\right)$.

3. Solve trigonometric equations using the inverse function as a means of expressing the result.

SOLUTIONS to Exercise Set 5.7

2. $\arccos\left(\frac{\sqrt{3}}{2}\right) = \frac{\pi}{6}$ the only value in the restricted range $0 \le y \le \pi$

5. $\arcsin\left(-\frac{\sqrt{2}}{2}\right) = -\frac{\pi}{4}$ Recall the restriction $-\frac{\pi}{2} < y < \frac{\pi}{2}$.

8. $\tan^{-1}\left(\frac{\sqrt{3}}{3}\right) = \frac{\pi}{6}$ in the restricted interval

11. $\cos^{-1}(0) = \frac{\pi}{2}$

14. $\arcsin\left(\frac{\sqrt{2}}{2}\right) = \frac{\pi}{4}$

17. $\cos^{-1}\left(-\frac{1}{2}\right) = \frac{2}{3}\pi$ The restriction is $0 \le y \le \pi$.

20. $\arctan(1.398) = 0.9512$ approximately

23. $\arcsin(0.9636) = 1.3003$

26. $\cos\left(\arcsin -\frac{1}{2}\right)$ We seek the real number t that has a $\sin = -\frac{1}{2}$, then the $\cos t \cdot t = -\frac{\pi}{6}$; $\cos t = -\frac{\sqrt{3}}{2}$.

29. $\cos^{-1}\left(\sin \frac{9\pi}{4}\right)$ First find $\sin \frac{9\pi}{4} = \sin \frac{\pi}{4} = \frac{\sqrt{2}}{2}$; we have $\cos^{-1}\left(\frac{\sqrt{2}}{2}\right) = \frac{\pi}{4}$.

32. $\sin^{-1}\left(\cos \frac{\pi}{6}\right)$ $\cos \frac{\pi}{6} = \frac{\sqrt{3}}{2}$ and $\sin^{-1}\left(\frac{\sqrt{3}}{2}\right) = \frac{\pi}{3}$

35. $12\cos^2 x - \cos x - 1 = 0$

$(3\cos x - 1)(4\cos x + 1) = 0$

$\cos x = \frac{1}{3}$, $\cos x = -\frac{1}{4}$

$x = \arccos\left(\frac{1}{3}\right)$ or $x = \arccos\left(-\frac{1}{4}\right)$

38. $3\cos^2 x - 7\cos x - 6 = 0$

$(3\cos x + 2)(\cos x - 3) = 0$

$\cos x = -\frac{2}{3}$, $\cos x = 3$ $\cos x$ never $= 3$

which means $x = \arccos\left(-\frac{2}{3}\right)$

SOLUTIONS to Review Exercises Chapter 5

2. If $t = -\frac{15}{2}\pi$ $W(t) = W\left(-\frac{15}{2}\pi + \frac{12}{2}\pi\right) = W\left(-\frac{3}{2}\pi\right) = W\left(\frac{1}{2}\pi\right)$ $t' = \frac{\pi}{2}$.

5. $t = \frac{16}{5}\pi$, $W(t) = W\left(\frac{16}{5}\pi - \frac{10}{5}\pi\right) = W\left(\frac{6}{5}\pi\right)$; $t = \frac{6}{5}\pi$.

8. $W\left(\frac{5}{6}\pi\right) = \left(-\frac{\sqrt{3}}{2}, \frac{1}{2}\right)$

11. Given $W(t) = \left(\frac{4}{5}, -\frac{3}{5}\right)$ (that is, quadrant IV).

Then $W(t - \pi)$ would be in quadrant II, by symmetry. Therefore, $W(t - \pi) = \left(-\frac{4}{5}, \frac{3}{5}\right)$.

14. $W\left(t - \frac{\pi}{2}\right)$ would be in quadrant III and $W\left(t - \frac{\pi}{2}\right) = \left(-\frac{3}{5}, -\frac{4}{5}\right)$.

17. $\sec\left(-\frac{5}{4}\pi\right) = \dfrac{1}{\cos\left(-\frac{5}{4}\pi\right)} = \dfrac{1}{-\frac{\sqrt{2}}{2}} = -\dfrac{2}{\sqrt{2}}$

20. Given $\sin t = -\frac{\sqrt{2}}{2}$ and $W(t)$ in quadrant III, $t = \frac{5}{4}\pi$.

23. $\sec t = -2$ and $W(t)$ in quadrant II implies $\cos t = -\frac{1}{2}$; thus $t = \frac{2}{3}\pi$.

26. $\sin t = \frac{12}{13}$ and $\cos t < 0$. To find $\tan t$, we use

$$\sin^2 t + \cos^2 t = 1$$

$$\cos^2 t = 1 - \sin^2 t$$

$$\cos^2 t = 1 - \frac{144}{169} = \frac{25}{169}; \quad \cos t = -\frac{5}{13}$$

$$\tan t = \left(\frac{12}{13}\right) \div \left(-\frac{5}{13}\right) = -\frac{12}{5}$$

29. $\dfrac{\sin t}{\cos^2 t} = \dfrac{\sin t}{\cos t} \cdot \dfrac{1}{\cos t} = \tan t \sec t$

32. $f(x) = 1 - \sin x = -(\sin x - 1)$

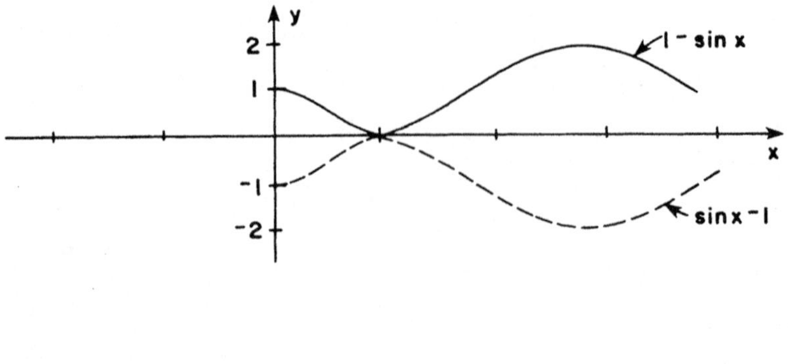

35. $f(x) = 4\sin\left(-x + \frac{\pi}{2}\right)$ Recall that $\sin(-x) = -\sin x$.

We have $f(x) = -4\sin\left(x - \frac{\pi}{2}\right)$.

Amplitude 4 ; period $= \frac{2\pi}{B} = 2\pi$; Phase Shift $= -\frac{C}{B} = +\frac{\pi}{2}$

38. $\tan(\cos^{-1} 1)$. First find $\cos^{-1} 1 = 0$ by the $0 \leq y < 2\pi$ restriction.

Now $\tan 0 = 0$

Chapter 5 - A SHORT QUIZ

1. The wrapping function assigns to each real number a ___?___ on the unit circle.
2. Since the circumference of a circle = 2π, and if we let t be any real number, $W(t) =$ ___?___ , W being the wrapping function.
3. Each point on the unit circle has two coordinates (x, y) such that __?__ + __?__ = 1 .
4. If $t = \frac{9}{4}\pi$, find a t' such that $W(t) = W(t')$.
5. What is $W(\pi)$?
6. If t is a real number and $W(t) = (x, y)$, how do we define $\sin t$ and $\cos t$?
7. In which quadrants is $\tan t < 0$?
8. Draw a unit circle and show by symmetry that $\sin(-t) = -\sin t$. (For this question use $W(t)$ in quadrant II.)
9. Using the basic identities, transform $\frac{1 - \cos^2 t}{\cos^2 t}$ into $\tan^2 t$.
10. If $\sin t = -\frac{3}{5}$ and if $\tan t < 0$, find $\cos t$.
11. Because the function values of $f(t) = \sin t$ repeat after 2π, again after 4π, etc., $\sin t$ is called a ___?___ function.
12. Composite trigonometric functions such as $f(t) = \sin t - \cos t$ can be graphed by ___?___ . (Give a detailed explanation.)
13. For what values of t is $\tan t$ undefined?
14. What is the period of $\sin t$? of $\tan t$?
15. Describe in words and using the graph of $\sin t$ the location of $\sin t - 2$.
16. Values of $\sin t$ for $t > \pi/2$ can be found by using a reference number t' and using the appropriate sign. If t is in quadrant III, then $t' =$ ___?___ .
17. If $\cos 1.2 = 0.3624$, what is $\cos(-1.2)$?
18. Given $f(x) = A\sin x$, what is the amplitude of $f(x)$?
19. In words, what is meant by amplitude?

20. Given $f(x) = A \sin(Bx + C)$ what is the period of $f(x)$?

21. Given $f(x) = 2 \cos(3x - 2\pi)$, find the (a) amplitude, (b) period, (c) phase shift.

22. Graph on the same axes (a) $f(x) = \sin x$, (b) $f(x) = 2 \sin x$, and (c) $f(x) = 3 \sin x$.

23. How is the secant t function defined?

24. What is the domain of sec t ?

25. In which quadrant(s) is $\sec t < 0$?

26. What is the range of sin t ? of tan t ?

27. Since sin t is not one-to-one, what restriction was made in order to define arcsin t (the inverse of sin t) ?

28. Find $\arccos \frac{1}{2}$.

29. Find $\tan^{-1}\sqrt{3}$.

30. Use $\cos^{-1} x$ to express the solutions of this equation exactly:

$$7 \cos^2 x - 4 = 0$$

Quiz Answers Chapter 5

1. point
2. $W(t + 2\pi)$
3. x^2, y^2
4. $t' = \pi/4$
5. $W(\pi) = (-1, 0)$
6. $\sin t = y$
   $\cos t = x$
7. II and IV
8.

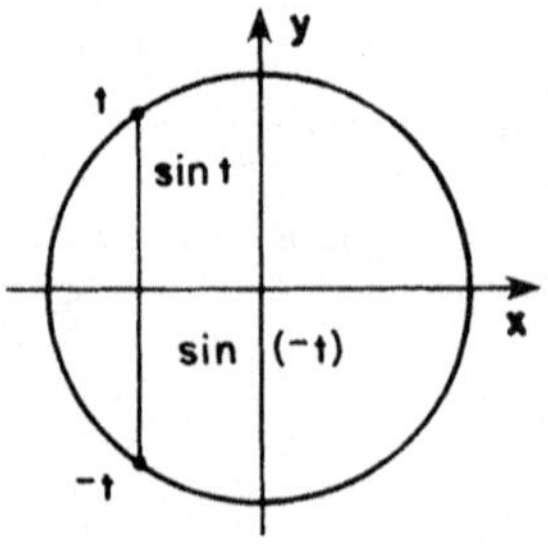

9. Since $\sin^2 t + \cos^2 t = 1$, $\sin^2 t = 1 - \cos^2 t$;

thus $\dfrac{1 - \cos^2 t}{\cos^2 t} = \dfrac{\sin^2 t}{\cos^2 t} = \tan^2 t$

10. $\cos t = \frac{4}{5}$

11. periodic

12. Finding values of $\sin t$ and of $\cos t$ and subtracting.

13. $0 + n\pi$ (n an integer)

14. 2π, π

15. $\sin t - 2$ is located point for point two units below the graph of $\sin t$.

16. $t - \pi$

17. 0.3624

18. A

19. One half the difference of the maximum and minimum values.

20. $\frac{2\pi}{B}$

21. (a) 2, (b) $\frac{2\pi}{3}$, (c) $\frac{2\pi}{3}$

22.

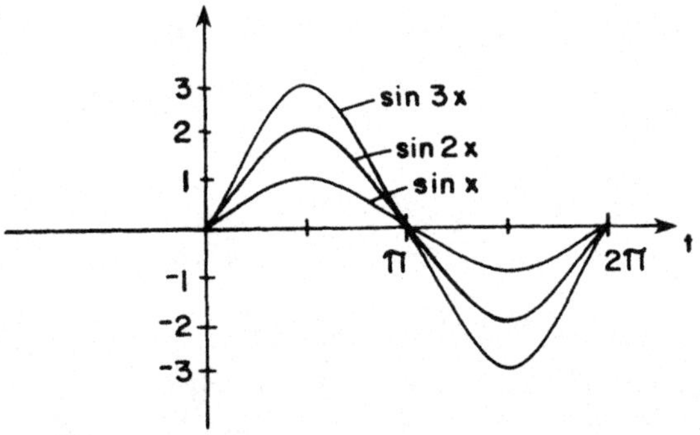

23. $\sec t = \dfrac{1}{\cos t}$

24. Domain of $\sec t$: all reals except $t = 0 + 2n\pi$, n an integer

25. II and III

26. Range of $\sin t$ is $[-1, 1]$; range $\tan t$ is $(-\infty, +\infty)$

27. $-\frac{\pi}{2} \le t \le \frac{\pi}{2}$

28. $\pi/3$

29. $\pi/3$

30. $x = \arccos 4/7, \arccos(-4/7)$

CHAPTER 6

Angles and Triangles

Section 6.1 Angles and Their Measurement

BASIC CONCEPTS you should learn from this unit.

1. An angle is the result of rotating a half line about its end point; and is said to be positive if the rotation is counterclockwise, and negative if the rotation is clockwise.

2. Angles are measured in two basic ways:

   A. In degrees, a degree being $\frac{1}{360}$ of a complete revolution; and

   B. In radians, a radian being $\frac{1}{2\pi}$ of a complete revolution. This comes from the fact that the circumference of a unit circle is 2π and so we see that radian measure is a measure of the arc an angle intercepts.

3. 2π radians = 360°

   π radians = 180°

4. The formula for converting radian measure to degree measure and vice versa is

$$\frac{\text{radian measure of angle}}{\pi \text{ radians}} = \frac{\text{degree measure of angle}}{180^\circ}$$

5. The reference angle, θ', of an angle, θ, is the acute angle formed by the terminal side of θ and the x-axis.

6. If an angle, θ, subtends an arc of length s in a circle of radius r, then the radian measure of $\theta = s/r$

WHAT YOU SHOULD KNOW HOW TO DO upon completion of this unit.

1. Define positive and negative angles.

2. Determine the quadrant of a given angle.

3. Convert between radian and degree measure.

4. Given any angle in degrees or radians, determine the reference angle of the given angle.

5. If given any two of the three variables in the formula $\theta = s/r$, (θ in radians, s and r in the same unit of length) be able to find the third variable.

SOLUTIONS to Exercise Set 6.1

2. $\theta = 182^\circ$ Since $180^\circ < \theta < 270^\circ$, θ is in quadrant II.

5. $\theta = 141^\circ$; $90^\circ < \theta < 180^\circ$. θ in quadrant II

8. $\theta = 555^\circ$ First subtract 360° to find an angle less than 360° that is coterminal with θ. This is 195° between 180° and 270° and is in quadrant III. Thus θ, too, is in quadrant III.

11. $\theta = -195^{\circ}$ $-180^{\circ} > \theta > -270^{\circ}$, so θ is in quadrant II. To see this, imagine that we are rotating a half line clockwise and that between 0° and -90° the angle formed by this rotation is in quadrant IV, between -90° and -180° it is in quadrant III, and so on.

14. $\theta = -3\pi/5$ $-\frac{\pi}{2} > \theta > -\pi$, so θ is in quadrant III.

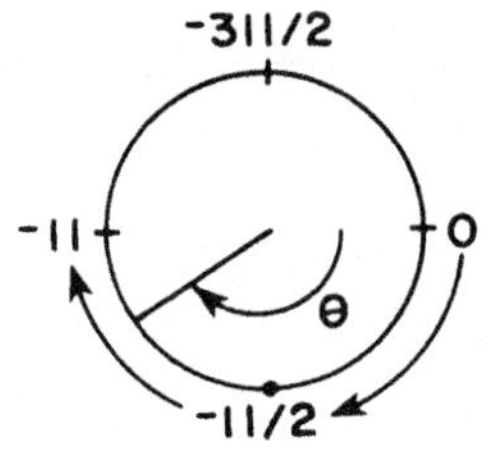

17. $\theta = \frac{13\pi}{3} = 4\pi + \frac{\pi}{3}$ The 4π portion is simply two complete revolutions and can be disregarded in this case; θ is coterminal with $\frac{1}{3}\pi$.
Now $0 < \theta < \frac{\pi}{2}$, so θ is in quadrant I.

20. 200°. Let t = the radian measure of the angle; then $t/\pi = 200/180$ and $t = 200\pi/180 = 10\pi/9$.

23. 75°; $\frac{t}{\pi} = \frac{75}{180}$, $t = \frac{75\pi}{180} = \frac{5\pi}{12}$

26. -570° is coterminal with -210°

$$\frac{t}{\pi} = \frac{-210}{180}, \quad t = -\frac{7\pi}{6}$$

29. 120°; $\frac{t}{\pi} = \frac{120}{180}$, $t = \frac{2\pi}{3}$

32. 196.54°; $\frac{t}{\pi} = \frac{196.54}{180} = 1.09$, hence $t = 1.09\pi$ approximately

35. $\frac{3\pi}{2}$; Using the same formula with θ as the degree measure required, we obtain

$$\frac{3\pi/2}{\pi} + \frac{\theta}{180}; \quad \frac{3}{2}\cdot 180 = \theta; \quad \theta = \underline{270^{\circ}}$$

38. $-\frac{7\pi}{12}$; $-\frac{7/12\pi}{\pi} = \frac{\theta}{180}$; $\theta = -\frac{7}{12}\cdot 180 = -105^{\circ}$

41. $\frac{5\pi}{2}$; $\frac{5\pi/2}{\pi} = \frac{\theta}{180}$; $\theta = \frac{5}{2}\cdot 180 = 450^{\circ}$

44. $\frac{9\pi}{2}$; $\frac{9\pi/2}{\pi} = \frac{\theta}{180}$; $\theta = \frac{9}{2}\cdot 180 = 810^{\circ}$

47. $30^{\circ}, 390^{\circ}$ are coterminal. (Subtracting 360° from 390° gives 30°.)

50. 120°, $\frac{14\pi}{3}$ 120° converts to $2\pi/3$ radians and subtracting 4π (two revolutions) from $14\pi/3$ also results in $2\pi/3$. Hence, the given angles are coterminal.

53. 130° The best way here is to subtract 130° from $180^{\circ} = 50^{\circ}$ as the reference angle.

56. 25° is its own reference angle.

59. $\frac{12\pi}{5} = 2\pi + \frac{2\pi}{5}$ and since $\frac{2\pi}{5} < \frac{\pi}{2}$, it is acute, so $\frac{2\pi}{5}$ is the reference angle.

62. $-\frac{2\pi}{3}$

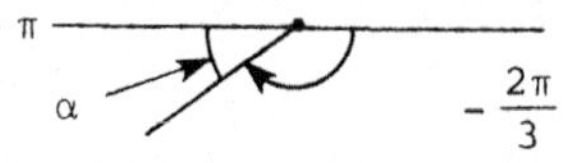

If α is the reference angle we see that $\alpha = \frac{\pi}{3}$

65. Use the formula $\theta = s/r$. Here $\theta = 4/7$ radians or (using $\pi = 3.14$)
$\frac{4/7}{\pi} = \frac{\theta}{180}$ and $\theta = \frac{4.180}{7\pi} = \frac{720^{\circ}}{7\pi} \approx 32.75^{\circ} \approx 32^{\circ}45'$

68. Convert 45° to $\pi/4$ radians. Then using $\pi = 3.14$ $\pi/4 = s/150$ and $s = 37.5\pi$ cm $= 117.75$ cm .

Section 6.2 Trigonometric Functions of Angles

BASIC CONCEPTS you should learn from this unit.

1. A trigonometric function of a specified angle measured in radians is the same as that trigonometric function of that same angle measured in degrees. Which is to say that the value of the trigonometric function depends on the angle itself rather than on the method of measurement. For example,

$$\sin 135^{\circ} = \sin \frac{3\pi}{4}$$

since 135° and $\frac{3\pi}{4}$ radians measure the same angle.

2. The Reference Angle concept allows us to use only acute angles when finding the trigonometric values of any angle. (Section 6.1 explains how to determine the reference angle.)

WHAT YOU SHOULD KNOW HOW TO DO upon completion of this unit.

1. Find the value of any trigonometric function for any angle measured in degrees (nearest 10') as well as in radians.

2. Find the reference angle of any angle measured in degrees.

SOLUTIONS to Exercise Set 6.2

2. 300°. Reference angle $= 360^{\circ} - 300^{\circ} = 60^{\circ}$. The cosine and secant will be positive, the others negative for a fourth quadrant angle. Thus

$\sin 300^{\circ} = -\sqrt{3}/2$, $\cos 300^{\circ} = 1/2$, $\tan = -\sqrt{3}$

$\csc 300^{\circ} = -2\sqrt{3}/3$, $\sec = 2$, $\cot = -\sqrt{3}/3$

5. 325° Reference angle $= 360^{\circ} - 315^{\circ} = 45^{\circ}$. The cosine and secant will be positive, the others negative.

$\sin 325^{\circ} = -\sqrt{2}/2$, $\cos 325^{\circ} = \sqrt{2}/2$, $\tan 325^{\circ} = -1$

$\csc 325^{\circ} = -\sqrt{2}$, $\sec 325^{\circ} = \sqrt{2}$, $\cot 325^{\circ} = -1$

8. -180° Reference angle is zero but the cosine and secant will be negative

$\sin 180^\circ = 0$, $\cos 180^\circ = -1$, $\tan 180^\circ = 0$

$\csc 180^\circ =$ undefined, $\sec 180^\circ = -1$, $\cot 180^\circ =$ undefined

11. $\sin 14^\circ 30' = 0.2504$

14. $\csc 260^\circ 50' = 2.2153$

17. $\text{Sin } 246^\circ 10' = -\sin 66^\circ 10' = 0.9147$

20. $\tan(-143^\circ 30') = \tan(-143^\circ 30' + 180^\circ) = 36^\circ 30' = 0.7400$

Section 6.3 Right Triangle Trigonometry

BASIC CONCEPTS you should learn from this unit.

1. The trigonometric functions of acute angles can be defined in terms of the sides of a right triangle as follows. Given θ as an acute angle in a right triangle, we have

$$\sin\theta = \frac{\text{side opposite } \theta}{\text{hypotenuse}} \qquad \cos\theta = \frac{\text{side adjacent to } \theta}{\text{hypotenuse}}$$

$$\tan\theta = \frac{\text{side opposite } \theta}{\text{side adjacent to } \theta} \qquad \csc\theta = \frac{\text{hypotenuse}}{\text{side opposite } \theta}$$

$$\sec\theta = \frac{\text{hypotenuse}}{\text{side adjacent to } \theta} \qquad \cot\theta = \frac{\text{side adjacent to } \theta}{\text{side opposite } \theta}$$

2. The foregoing definitions plus the Pythagorean Theorem allow us to find unknown angles and sides of right triangles if we know either: any two sides, or an acute angle and one side.

3. Right triangle trigonometry can be used to solve a great many practical problems in areas as diverse as carpentry and navigation.

WHAT YOU SHOULD KNOW HOW TO DO upon completion of this unit.

1. Given the sides and angles of a right triangle, find the trigonometric functions of the acute angles using the relationships presented in basic concept 1.

2. "Solve" a right triangle for missing parts.

3. Use right triangle trigonometry to find solutions to certain practical problems.

SOLUTIONS to Exercise Set 6.3

2. $\sin\theta = 5/13$, $\csc\theta = 13/5$

$\cos\theta = 12/13$, $\sec\theta = 13/12$

$\tan\theta = 5/12$, $\cot\theta = 12/5$

5. $\sin\theta = 2/\sqrt{5} = 2\sqrt{5}/5$, $\csc = \sqrt{5}/2$

$\cos\theta = \sqrt{5}/5$, $\sec = \sqrt{5}$

$\tan\theta = 2$, $\cot = 1/2$

8. $\sin\theta = \sqrt{x^2 - 1}/x$, $\csc\theta = x\sqrt{x^2 - 1}/(x^2 - 1)$

$\cos\theta = 1/x$, $\sec\theta = x$

$\tan\theta = \sqrt{x^2 - 1}$, $\cot\theta = \sqrt{x^2 - 1}/(x^2 - 1)$

11. P(-1, -1) Using a reference triangle and reference angle θ', we obtain

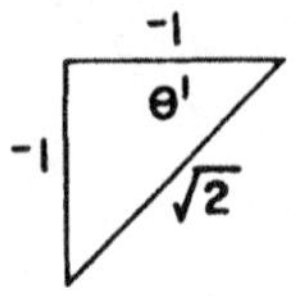

$\sin\theta = -\sqrt{2}/2$ $\csc\theta = -\sqrt{2}$

$\cos\theta = -\sqrt{2}/2$ $\sec\theta = -\sqrt{2}$

$\tan\theta = 1$ $\cot\theta = 1$

14. By a reference triangle,

$\sin\theta = 5/13$ $\csc\theta = 13/5$

$\cos\theta = 12/13$ $\sec\theta = 13/12$

$\tan\theta = 5/12$ $\cot\theta = 12/5$

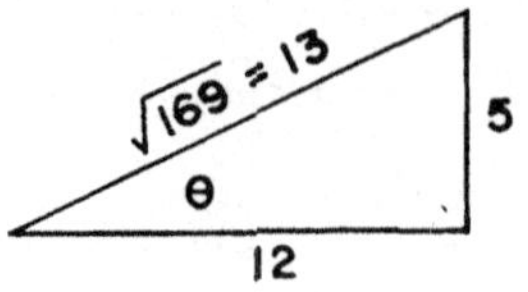

17. By a reference triangle,

$\sin\theta = -5/13$ $\csc\theta = -13/5$

$\cos\theta = -12/13$ $\sec\theta = -13/12$

$\tan\theta = 5/12$ $\cot\theta = 12/5$

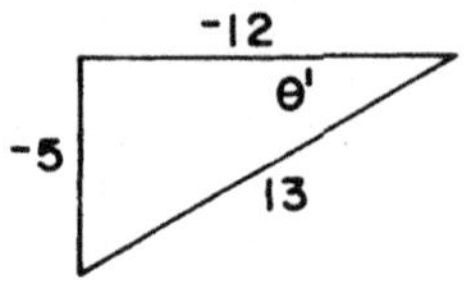

20. By a reference triangle,

$\sin\theta = -\sqrt{5}/5$ $\csc\theta = -\sqrt{5}$

$\cos\theta = -2\sqrt{5}/5$ $\cos\theta = -\sqrt{5}/2$

$\tan\theta = 1/2$ $\cot\theta = 2$

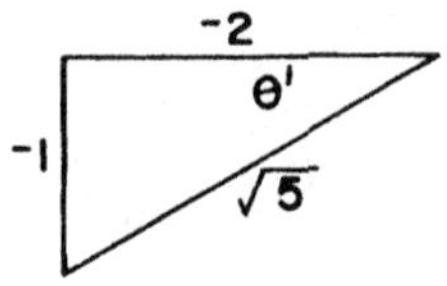

23. From the drawing, we see that

$\cot\theta = h/6.5$, and so $h = 6.5 \cot\theta$.

26. Here $\sec\theta = h/2.8$, so $h = 2.8 \sec\theta$.

29. $b = 40$, $\beta = 40^\circ$, find c.

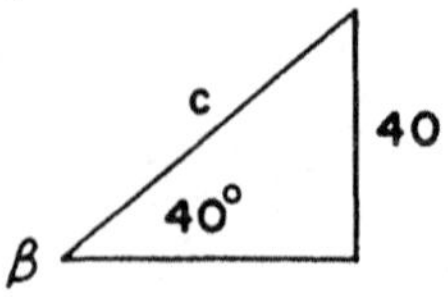

$\csc\beta = \frac{c}{40}$;

$$c = 40 \csc 40^\circ = 40(1.556) = 62.2$$

32. $b = 60$, $\alpha = 53^\circ$; find c.

$c/60 = \sec 53^\circ$; $c = 60 \sec 53^\circ$

$c = 60(1.6616) = 99.7$

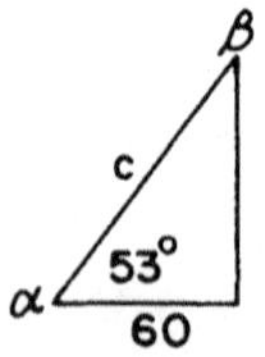

35. $\tan\left(\sin^{-1} - \frac{5}{13}\right)$. We use the restricted domain for $\sin^{-1}\theta$, as in sec 5.7. That is, $-90^\circ \leq \theta \leq 90^\circ$ $(\pi/2 = 90^\circ)$. This leads us to

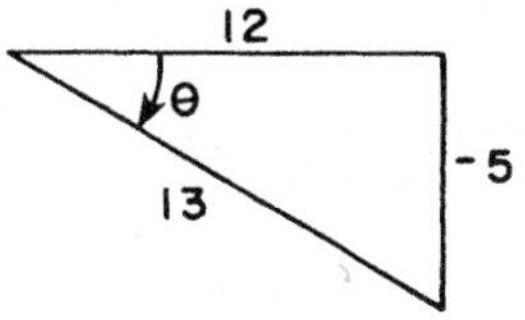

and so $\tan\theta$ is $-5/12$.

38. cos(arcsin-2/3). By reasoning similar to that in Solution 35 we have

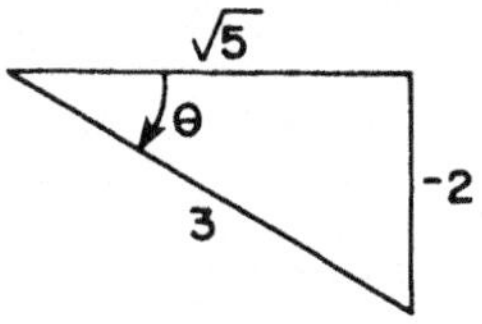

$\sin\theta = -\frac{2}{3}$

and $\cos\theta = \frac{\sqrt{5}}{3}$

41. From the sketch

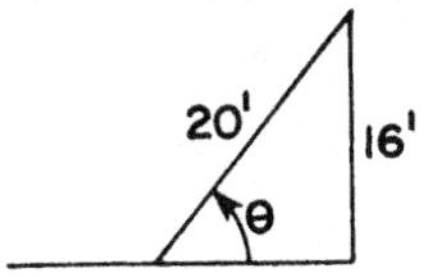

we see that

$\sin\theta = \frac{16}{20} = \frac{4}{5} = 0.8$

and from a table or by using an electronic calculator we find that $\theta = 53.1^\circ = 53^\circ 10'$ (rounded to the nearest 10').

44. B is the location of the boat and

A is the top of the rig

$\angle B = 16^\circ$ and $d/120 = \cot 16^\circ$

$d = 120 \cot 16^\circ = 418.5 \approx 419'$

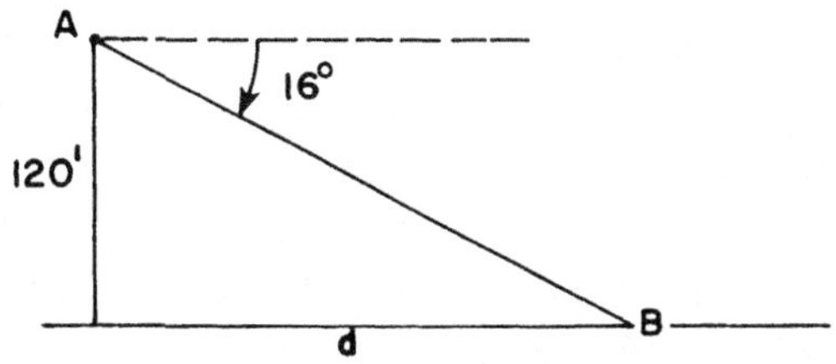

47. The observer is at point A and the rocket at its apogee is at point B;

$d/425 = \tan 66^\circ 20' = 2.2817$

$d = 425(2.2817) \approx 970'$

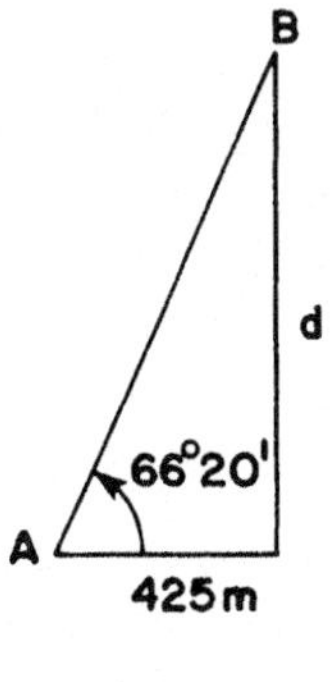

50. From the sketch we see that

$\cos\theta = 13/15 = 0.8667$ and

$\theta = 29^{\circ}55'$, ϕ also equals $29^{\circ}55'$

and the apex angle $\beta = 180^{\circ} - 59^{\circ}50'$

$= \underline{120^{\circ}10'}$

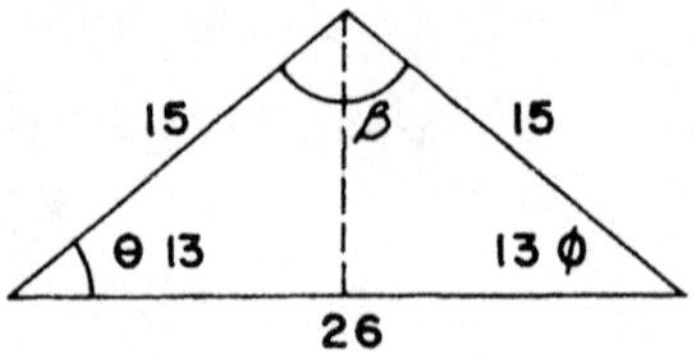

53. Let the height of building A be h ft

The distance, d, between the buildings

can be found from $d/110 = \cot 29$;

$d = 110(1.8040) = 198'$. Now

$h/198 = \tan 15^{\circ} = 0.2679$. So finally

$h = 198(0.2679) = 53'$.

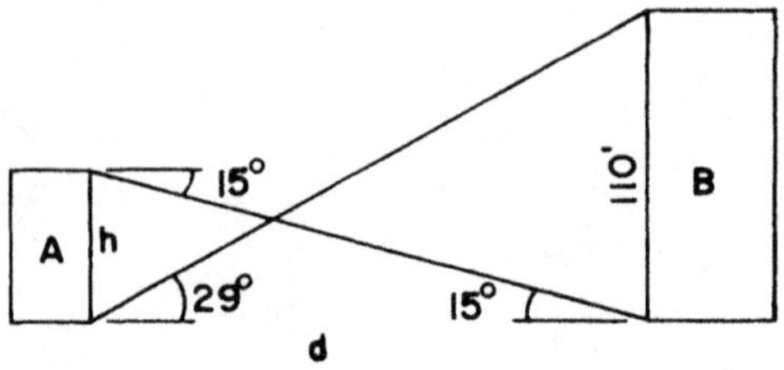

Section 6.4 Law of Cosines

BASIC CONCEPTS you should learn from this unit.

1. Given an oblique triangle with angles α, β, and λ, and sides a, b, c, opposite them, there is a relationship among the sides and angles called the Law of Cosines:

$$a^2 = b^2 + c^2 - 2bc\cos\alpha$$
$$b^2 = a^2 + c^2 - 2ac\cos\beta$$
$$c^2 = a^2 + b^2 - 2ab\cos\lambda$$

2. Oblique triangles can be solved by the Law of Cosines if either

   A. All three sides are known; or

   B. Two sides and the angle between those sides are known.

WHAT YOU SHOULD KNOW HOW TO DO upon completion of this unit.

1. Solve triangles for missing parts if given the information specified in basic concept 2B.

2. Solve, using the Law of Cosines, practical verbal problems with appropriate known data.

SOLUTIONS to Exercise Set 6.4

2. $a = 5$, $b = 12$, $c = 15$; find λ.

Using $c^2 = a^2 + b^2 - 2ab\cos\lambda$, we have $15^2 = 5^2 + 12^2 - 2\cdot 5\cdot 12\cos\lambda$,

which leads to

$\cos\alpha = \dfrac{225 - 25 - 144}{-120} = \dfrac{56}{-120} = -0.4667$ and $\lambda = 117^{\circ}50'$.

5. $a = 10$, $b = 12$, $\lambda = 108^{o}$. Using $c^2 = a^2 + b^2 - 2ab \cos \lambda$ gives

$c^2 = 100 + 144 - 240(-0.3090)$

$c^2 = 244 + 74.16 = 318.16$

$\underline{c \simeq 17.8}$

8. $a = 11$, $c = 15$, $\beta = 33^{o}$, find λ.

β is the angle between sides a and c, and

$b^2 = a^2 + c^2 = 2ac \cos \beta$

allows us to find b.

$b^2 = 121 + 225 - 2 \cdot 11 \cdot 15 \cdot (0.8387) = 69.2$; $b = 8.3$.

Now to find λ we use $c^2 = a^2 + b^2 - 2ab \cos \lambda$

$$225 = 121 + 69.2 - 2(11)(8.3) \cos \lambda$$

$$\cos \lambda = \frac{225 - 121 - 69.2}{-182.6} = -0.1906$$

$\lambda = 101^{o}$

14. Draw the diagram in which C is the location of the ship; then a, the distance required, would be found from

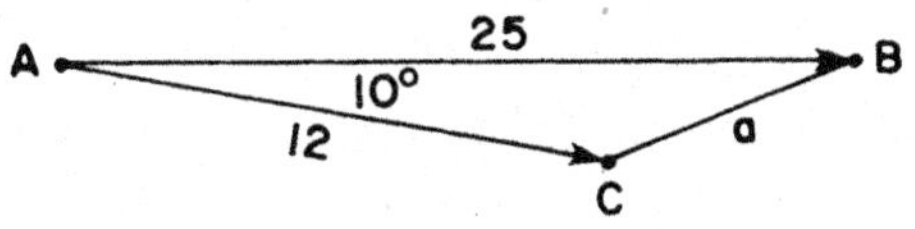

$a^2 = 12^2 + 25^2 - 2 \cdot 25 \cdot 12(\cos 10^{o})$

$a^2 = 144 + 625 - 600(0.9848) = 178.12$

$a = 13.3$ m

17. λ is the angle between the known sides a and b, so

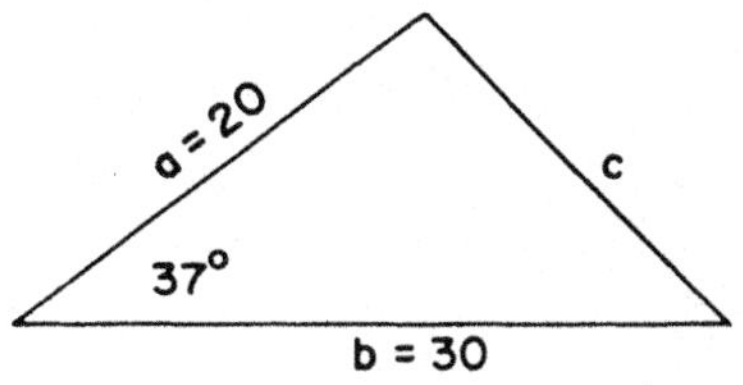

$c^2 = 20^2 + 30^2 - 2 \cdot 20 \cdot 30 \cdot \cos 37^{o}$

will determine the third side:

$c^2 = 400 + 900 - 958.4 = 391.6$

$c = 18.5$

$P = 20 + 30 + 18.5 = 68.5$

20. (a) Given the three forms of the Law of Cosines:

$$a^2 = b^2 + c^2 - 2bc \cos \alpha$$

$$b^2 = a^2 + c^2 - 2ac \cos \beta$$

$$\underline{c^2 = a^2 + b^2 - 2ab \cos \lambda}$$

Adding them yields

$a^2 + b^2 + c^2 = 2a^2 + 2b^2 + 2c^2 - 2bc \cos \alpha - 2ac \cos \beta - 2ab \cos \lambda$

Multiplying by (-1) and rearranging gives

$a^2 + b^2 + c^2 = 2(bc \cos \alpha + ac \cos \beta + ab \cos \lambda)$

(b) Having proved part (a), we can use it to easily prove statement (b). Simply divide the conclusion of (a) by $2abc$ and statement (b) is immediate.

BASIC CONCEPTS you should learn from this unit.

1. Given an oblique triangle with sides a, b, c and angles α, β, λ, respectively, opposite them, there is a relationship called the Law of Sines as follows:

$$\frac{a}{\sin \alpha} = \frac{b}{\sin \beta} = \frac{c}{\sin \lambda}$$

2. This Law of Sines may be used to "solve" triangles if
   - A. Any side and two angles are known, or
   - B. Two sides and the angle opposite one of these sides is known.

3. If two sides and the angle opposite one of them are known, several possibilities exist:
   - A. No triangle is possible;
   - B. Exactly one right triangle is possible;
   - C. Two triangles are possible;
   - D. Exactly one acute triangle is possible.

WHAT YOU SHOULD KNOW HOW TO DO upon completion of this unit.

1. State the Law of Sines.
2. Recognize and explain when the Law of Sines is applicable.
3. Explain what is meant by "the Ambiguous Case."
4. Use the Law of Sines to "solve" triangles.

SOLUTIONS to Exercise Set 6.5

2. $\alpha = 74^{\circ}$, $\lambda = 36^{\circ}$, $c = 6.8$

   since

$$\frac{a}{\sin \alpha} = \frac{c}{\sin \lambda}$$

   we have

$$\frac{a}{0.9613} = \frac{6.8}{0.5879}$$

   and

$$a = \frac{6.8(0.9613)}{0.5879} = \underline{11.1}$$

5. $\alpha = 42^\circ 20'$, $\lambda = 78^\circ 40'$, $b = 20$; find a.

Angle $\beta = 180^\circ - (42^\circ 20' + 78^\circ 40') = 59^\circ$; then

$$\frac{20}{\sin 59} = \frac{a}{\sin 42^\circ 20'}$$

and

$$a = \frac{20(\sin 42^\circ 20')}{\sin 59^\circ}$$

$$a = \frac{13.468}{0.8572} = 15.7$$

8. $\beta = 32^\circ$, $b = 20$, $c = 14$

Since $b/\sin\beta = c/\sin\lambda$, we have

$$\frac{20}{0.5299} = \frac{14}{\sin\lambda}$$

and then

$$\sin\lambda = \frac{14(0.5299)}{20} = 0.3709$$

and so $\underline{\lambda = 21^\circ\ 46'}$.

$$\alpha = 180^\circ - 32^\circ - 21^\circ 46' = 126^\circ 14'$$

11. $\lambda = 45^\circ$, $b = 7$, $c = 6$; find a.

Since $c/\sin\lambda = b/\sin\beta$, we have $6/\sin\lambda = 7/\sin\beta$ and

$$\frac{6}{0.7071} = \frac{7}{\sin\beta}$$

$$\sin\beta = \frac{7(0.7071)}{6}$$

$$\sin\beta = 0.8250$$

$$\beta = 55^\circ 40' \quad \text{or} \quad \beta = 124^\circ 20'$$

This is the ambiguous case: we have two possible β values and so two possible values for the angle α, $79^\circ 20'$ and $10^\circ 40'$. Now

$$\frac{a}{\sin 79^\circ 20'} = \frac{6}{\sin 45}$$

from which we have

$$a = \frac{6 \cdot (0.9827)}{0.7071} = \underline{8.3}$$

for a possible value of a, <u>or</u>

$$\frac{a}{\sin 10^\circ 40'} = \frac{6}{\sin 45}$$

$$a = \frac{6(0.1851)}{0.7011} = \underline{1.6}$$

for the second possible value of a.

14. Drawing a sketch may help:

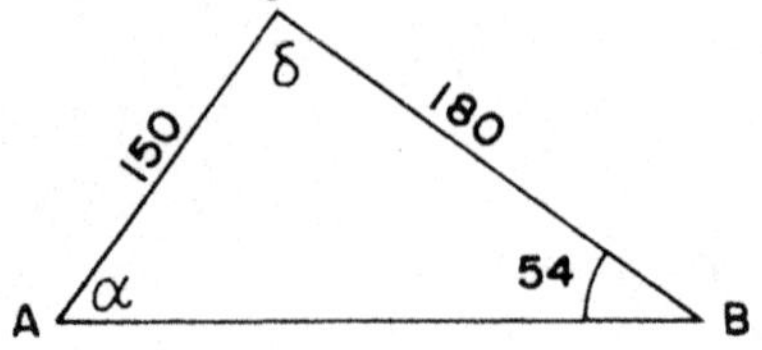

$\angle ABC = 54^\circ$

$$\frac{180}{\sin\alpha} = \frac{150}{\sin 54}$$

$$\sin\alpha = \frac{180(0.8090)}{150} = 0.9708$$

$$\alpha = 76^\circ 10'$$

and now

$$\angle\lambda = 180^\circ - 54^\circ - 76^\circ 10' = 49^\circ 50'$$

Side c is the length of the tunnel and

$$\frac{c}{\sin 49^\circ 50'} = \frac{150}{\sin 54^\circ}$$

$$c = \frac{150(0.7642)}{0.8090}$$

$$c = 141.7 \text{ meters}$$

17.

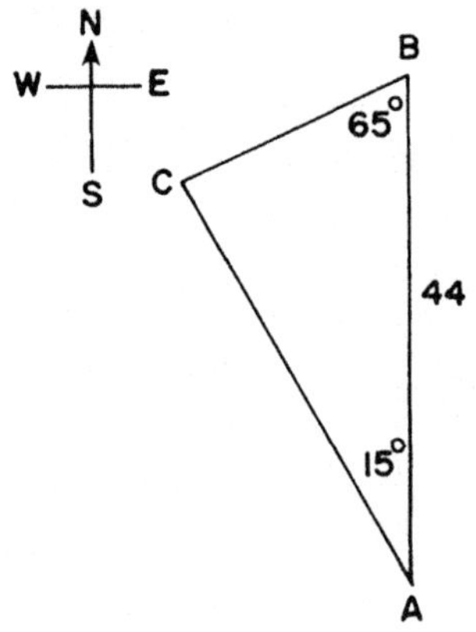

Let A be the ship's position at 2 P.M., B its position at 4 P.M., and C the light.

Angle C = 180 - 65 - 15 = 100°

$$\frac{b}{\sin 65} = \frac{44}{\sin 100}$$

$$b = \frac{44(0.9063)}{0.9848}$$

$$b = 40.5 \text{ miles}$$

20.

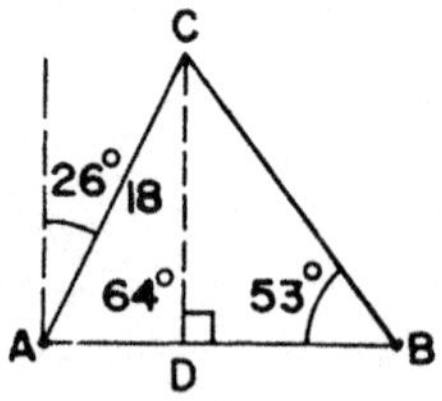

Let B be the island's location, A the ship at 5 P.M., and C the ship at 6 P.M.

Angle A = $90^\circ - 26^\circ = 64^\circ$

Angle BCD = 37°

and then

Angle B = 53°

$$\frac{\overline{BC}}{\sin 64^\circ} = \frac{18}{\sin 53^\circ} \qquad \overline{BC} = \frac{18(0.8988)}{(0.7986)} = 20.3 \text{ miles}$$

23. We are given the lengths of sides a and b of a triangle and angle α (which is opposite a).

(a) If $b \sin \alpha > a$, prove that there is no triangle.

Proof: We know that $a/\sin \alpha = b/\sin \beta$, which can be rearranged to $\sin \beta = b \sin \alpha/a$. But if $b \sin \alpha > a$, then $\sin \beta > 1$, which is impossible. Hence no triangle is possible.

(b) If $b \sin \alpha = a$, prove that there is a right triangle.

Proof: Using $\sin \beta = b \sin \alpha/a$ from part (a) we see that if $b \sin \alpha = a$, then $\sin \beta = 1$, and β is a right angle, so that we have a right triangle.

(c) If $b \sin \alpha < a < b$, prove that there are two triangles with the given parts.

Proof: Consider the diagram below.

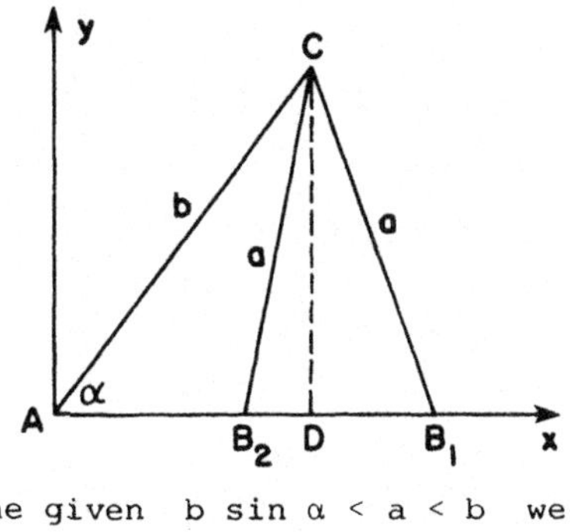

The distance $\overline{CD}$ from C to the x-axis is $b \sin \alpha$.

With the given $b \sin \alpha < a < b$ we are assured that side a will reach the x-axis, and further that there are two possible locations for B (as indicated); hence two triangles are possible.

(d) If $b \leq a$, prove that there is one acute triangle.

Proof: Consider the diagram below.

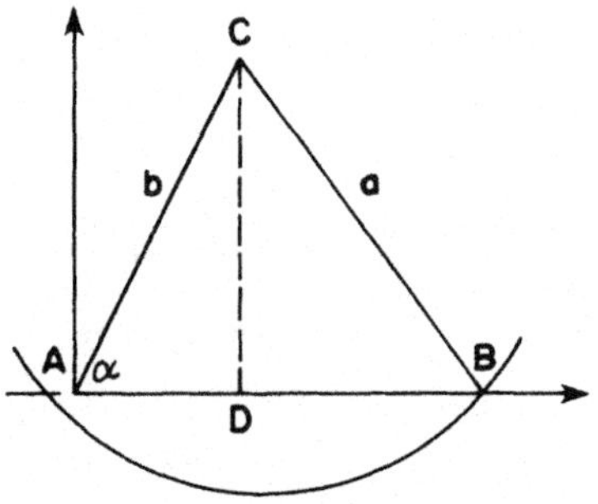

Again we have $\overline{CD} = b \sin \alpha$, but now $b \leq a$, so side a can only intersect the positive x-axis at <u>one</u> point, B, forming a triangle that includes angle α. We know that α is a part of the triangle so there is only one triangle.

SOLUTIONS to Review Exercises for Chapter 6

2. $\frac{3\pi/2}{\pi} = \frac{\text{degree measure}}{180}$, which leads to $180 \cdot 3/2 = 270^{\circ}$ as the required degree measure.

5. Are 100° and $5\pi/9$ coterminal? Coverting 100° to radians, we have $100\pi/180 = 5\pi/9$ and thus the given angles are the same angle and so are coterminal.

8. 310° is in quadrant IV and the reference angle is $360^{\circ} - 310^{\circ} = 50^{\circ}$.

11. Using the formula $\theta = s/r$, we obtain $\theta = 14/10 = 7/5$ as the radian measure of θ .

14. We are given $a = 3$, $c = 5$, and $\lambda = 90^{\circ}$. By the Pythagorean Theorem we have $a^2 + b^2 = c^2$ or $9 + b^2 = 25$ and $b^2 = 16$, so $b = 4$. Now $\tan \beta = 4/5$.

17. Given $P(\sqrt{2}, -\sqrt{2})$, we note that P is in quadrant IV. Further, $\cot \theta = x/y = \sqrt{2}/-\sqrt{2} = -1$.

20. Given $a = 50$, $b = 60$, and $\lambda = 90^{\circ}$. From the diagram we can conclude that $\tan \alpha = 50/60$.

$\tan \alpha = 0.8333$

$\alpha = 39^{\circ}50'$

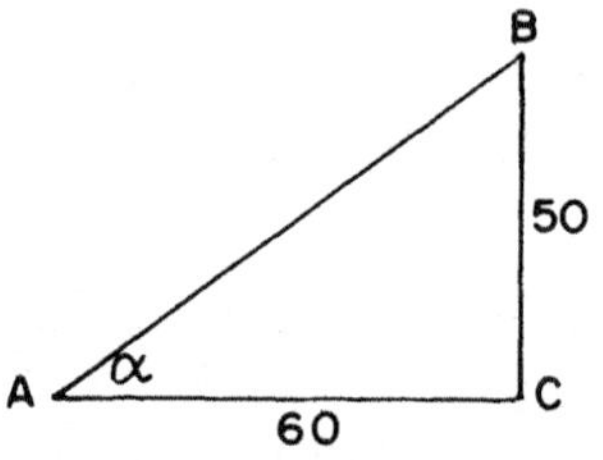

23. The diagram illustrates the conditions of the problem.

$\sin 65^{\circ} = \frac{h}{6}$

$h = 6 \sin 65^{\circ}$

$h = 6(0.9063)$

$h = 5.4$ meters as the required height

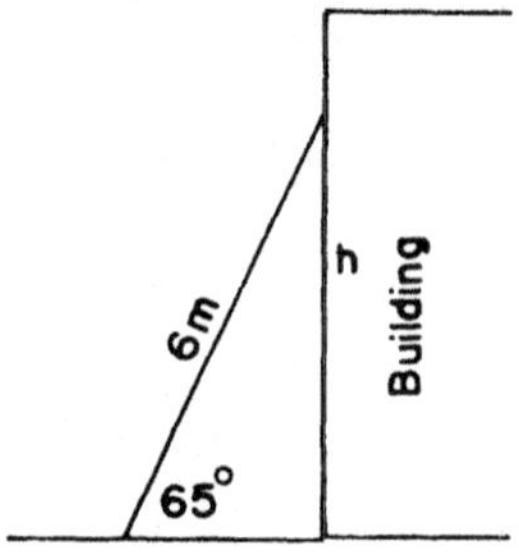

26. Consider the diagram:

α is the smaller angle formed by the diagonal and the sides. Also, we note that

$$\tan \alpha = \frac{16}{22} = 0.7273$$

and thus

$$\alpha = 36^{\circ}$$

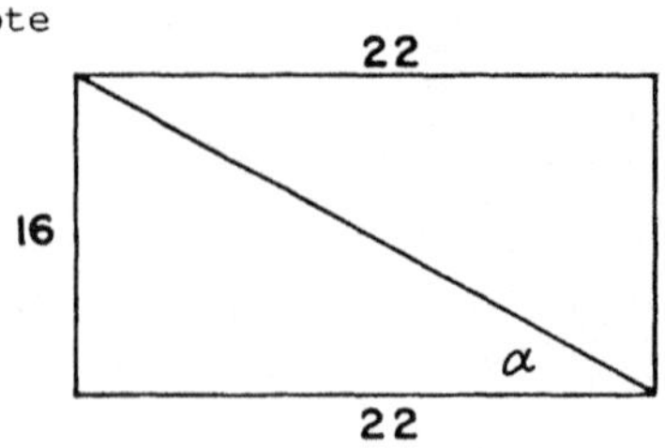

1. Angles are measured in two basic ways, or by two different units of measurement. What are these ways? ___?___ and ___?___

2. A reference angle θ' of an angle θ is an acute angle formed by the terminal side of θ and the ___?___ axis.

3. Find the reference angle for $\theta = 198^{\circ}$.

4. If $\theta = 11\pi/6$ radians, what is the reference angle?

5. Convert $\pi/3$ radians to degrees.

6. Convert 120° to radians.

7. What is the arc length subtended on a circle of radius 4 inches by a central angle of $\pi/6$ radians?

8. True or False: $\sin 120^{\circ} = \sin(2\pi/3)$.

9. True or False: $\cos 120^{\circ} = -\cos 60^{\circ}$.

10. Given the following right triangle, define $\sin\alpha$, $\cos\alpha$, and $\tan\alpha$ in terms of the sides.

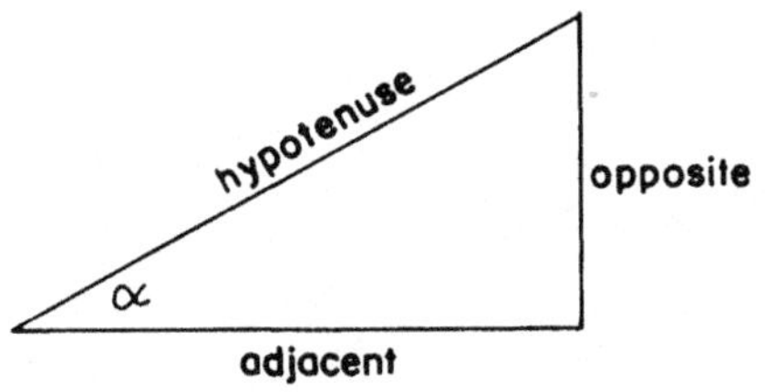

11. In the following triangle, find $\cot\theta$.

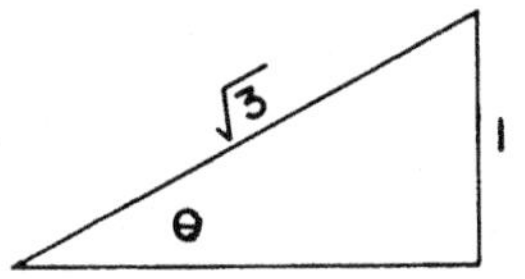

12. Find $\cos\theta$ if the point $P(-12, -5)$ lies on the terminal side of θ.

13. Find $\sin\left(\arctan \frac{3}{4}\right)$.

14. If a right triangle has the parts indicated in the following sketch, find $\tan\beta$ and side a.

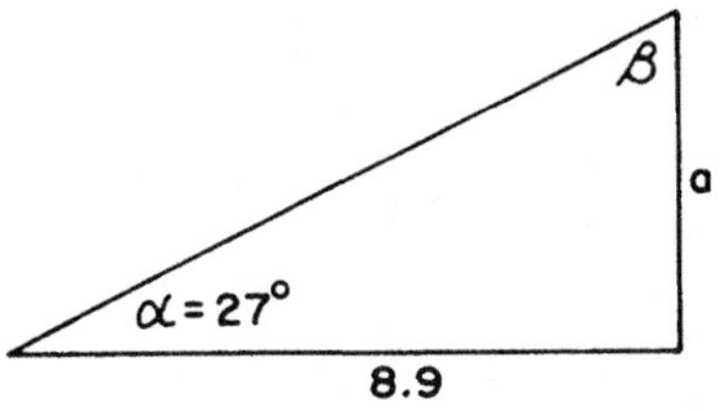

15. A person standing 40 feet from the base of a flagpole determines the angle of elevation of the top of the pole to be $39^\circ 50'$; how tall is the flagpole?

16. Which law would be used, the Law of Sines or the Law of Cosines, to solve a triangle if the three sides were known?

17. If the sides of a triangle are $a = 16$, $b = 10$, and $c = 17$, find the angle α (the angle opposite side a).

18. Using the following information, draw a diagram indicating the known angles and sides: At 2 P.M. a ship sailing S25°E at 12 miles per hour sights a lighthouse bearing N75°E, and again at 4 P.M. the same lighthouse bears N30°E.

19. Draw a sketch illustrating the ambiguous case of the Law of Sines in which there are two possible triangles.

20. Use the Law of Sines to find side a of a triangle if the angle opposite a, α, is 32°, $\beta = 47^\circ$, and $c = 18$.

Quiz Answers Chapter 6

1. degrees, radians
2. x
3. 18°
4. $\frac{\pi}{6}$
5. 60°
6. $\frac{2\pi}{3}$
7. $\frac{2\pi}{3}$
8. T
9. T
10. $\sin \alpha = \frac{\text{opposite}}{\text{hypotenuse}}$

    $\cos \alpha = \frac{\text{adjacent}}{\text{hypotenuse}}$

    $\tan \alpha = \frac{\text{opposite}}{\text{adjacent}}$
11. $\cot \theta = \sqrt{2}$
12. $\cos \theta = -\frac{12}{13}$
13. $\frac{3}{5}$
14. $\tan \beta = 1.9626$

    $a = 4.5$
15. 33.4 ft
16. Law of Cosines
17. 67°
18.

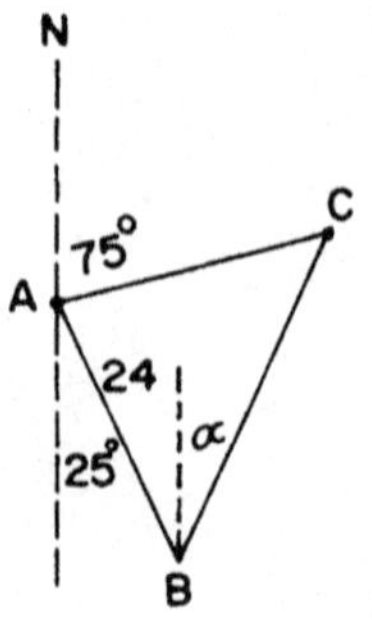

C is the lighthouse;

A is the ship at 2 P.M.;

B is the ship at 4 P.M.

$\alpha = 30^\circ$

19.

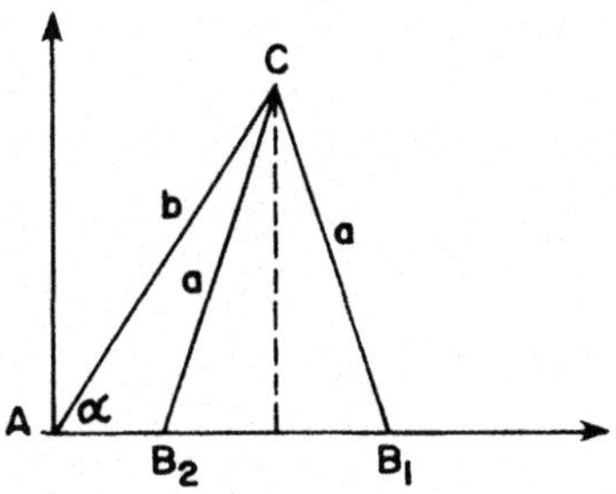

If we know α , a , and b , with $a < b$, then we have two possible triangles: AB_1C and AB_2C .

20. $a = 9.7$

CHAPTER 7

Analytic Trigonometry

Section 7.1 Trigonometric Identities

BASIC CONCEPTS you should learn from this unit.

1. A trigonometric identity is true for all values of the variable, whereas a trigonometric equation is true for only some values of the variable.

2. There are several fundamental trigonometric identities:

$$\tan t = \frac{\sin t}{\cos t} \qquad \cot t = \frac{1}{\tan t}$$

$$\cot t = \frac{\cos t}{\sin t} \qquad \sin^2 t + \cos^2 t = 1$$

$$\csc t = \frac{1}{\sin t} \qquad \tan^2 t + 1 = \sec^2 t$$

$$\sec t = \frac{1}{\cos t} \qquad \cot^2 t + 1 = \csc^2 t$$

3. Trigonometric identities are useful for two reasons:

   A. They allow us to simplify expressions;

   B. They allow us to verify other identities.

WHAT YOU SHOULD KNOW HOW TO DO upon completion of this unit.

1. Distinguish between a trigonometric identity and an equation.

2. State at least eight fundamental trigonometric identities.

3. Use the fundamental identities to verify other identities by transforming one side into the other, or by transforming both sides into the same expression.

SOLUTIONS to Exercise Set 7.1

2. $\cot x \sec x = \csc x$

$$\frac{\cancel{\cos x}}{\sin x} \cdot \frac{1}{\cancel{\cos x}} =$$

$$\frac{1}{\sin x} =$$

$$\csc x = \csc x$$

5. $\sin \alpha \sec \alpha = \tan \alpha$

$$\sin \alpha \frac{1}{\cos \alpha} =$$

$$\frac{\sin \alpha}{\cos \alpha} =$$

$$\tan \alpha = \tan \alpha$$

8. $1 - 2 \sin^2 t = 2 \cos^2 t - 1$

$$1 - \sin^2 t - \sin^2 t =$$

$$\cos^2 t - (1 - \cos^2 t) =$$

$$\cos^2 t - 1 + \cos^2 t =$$

$$2 \cos^2 t - 1 = 2 \cos^2 t - 1$$

11. $\dfrac{\sin u}{\csc u} + \dfrac{\cos u}{\sec u} = 1$

$\dfrac{\sin u}{\dfrac{1}{\sin u}} + \dfrac{\cos u}{\dfrac{1}{\cos u}} =$

$\sin^2 + \cos^2 = 1$ basic identity

14. $\sin^4 x + 2\sin^2 x\cos^2 x + \cos^4 x = 1$ factoring, we obtain

$(\sin^2 x + \cos^2 x)^2 =$

$1^2 = 1$

17. $\dfrac{\sec w \sin w}{\tan w + \cot w} = \sin^2 w$

$\dfrac{\dfrac{1}{\cos w}\cdot\sin w}{\dfrac{\sin w}{\cos w} + \dfrac{\cos w}{\sin w}} =$ multiply by $\cos w/\cos w$:

$\dfrac{\sin w}{\sin w + \dfrac{\cos^2 w}{\sin w}} =$ multiply by $\sin w/\sin w$:

$\dfrac{\sin^2 w}{\sin^2 w + \cos^2 w} =$

$\sin^2 w = \sin^2 w$

20. $\dfrac{1 + \tan^2 u}{\csc^2 u} = \tan^2 u$

$\dfrac{\sec^2 u}{\csc^2 u} =$

$\dfrac{\dfrac{1}{\cos^2 u}}{\dfrac{1}{\sin^2 u}} =$

$\dfrac{\sin^2 u}{\cos^2 u} =$

$\tan^2 u = \tan^2 u$

23. $$\frac{\sin^2 \alpha}{1 + \cos \alpha} = 1 - \cos \alpha$$

$$\frac{1 - \cos^2 \alpha}{1 + \cos \alpha} =$$

$$\frac{\cancel{(1 + \cos \alpha)}(1 - \cos \alpha)}{\cancel{(1 + \cos \alpha)}} =$$

$$1 - \cos \alpha = 1 - \cos \alpha$$

26. $$\frac{\sin \beta}{1 + \cos \beta} + \frac{1 + \cos \beta}{\sin \beta} = 2 \csc \beta$$

$$\frac{\sin \beta}{(1 + \cos \beta)} \frac{(1 - \cos \beta)}{(1 - \cos \beta)} + \frac{(1 + \cos \beta)}{\sin \beta} \cdot \frac{(\sin \beta)}{(\sin \beta)} =$$

$$\frac{\sin \beta - \sin \beta \cos \beta}{1 - \cos^2 \beta} + \frac{\sin \beta + \sin \beta \cos \beta}{\sin^2 \beta} =$$

$$\frac{\sin \beta - \sin \beta \cos \beta + \sin \beta + \sin \beta \cos \beta}{\sin^2 \beta} =$$

$$\frac{2 \sin \beta}{\sin^2 \beta} =$$

$$\frac{2}{\sin \beta} = 2 \csc \beta$$

29. $$\frac{\cot y}{1 + \cot^2 y} = \sin y \cos y$$

$$\frac{\frac{\cos y}{\sin y}}{1 + \frac{\cos^2 y}{\sin^2 y}} \cdot \frac{(\sin^2 y)}{(\sin^2 y)} =$$

$$\frac{\sin y \cos y}{\sin^2 y + \cos^2 y} =$$

$$\sin y \cos y = \sin y \cos y$$

32. $\sin(-\theta)\sec(-\theta) = -\tan \theta$

$$\frac{\sin(-\theta)}{\cos(-\theta)} =$$

$$\frac{-\sin \theta}{\cos \theta} =$$

Recall that $\sin(-\theta) = -\sin \theta$ and $\cos(-\theta) = \cos \theta$

$$-\tan \theta = -\tan \theta$$

35. $$\frac{1 + \tan x}{1 + \cot x} = \frac{\sec x}{\csc x}$$

$$\frac{(1 + \tan x)}{(1 + \cot x)} \cdot \frac{\sin x \cos x}{\sin x \cos x} =$$

$$\frac{\sin x \cos x + \sin^2 x}{\sin x \cos x + \cos^2 x} =$$

$$\frac{\sin x \cancel{(\cos x + \sin x)}}{\cos x \cancel{(\sin x + \cos x)}} =$$

$$\frac{\sin x}{\cos x} =$$

$$\frac{\sec x}{\csc x} = \frac{\sec x}{\csc x}$$

38. $$2 \csc^2\theta - \csc^4\theta = 1 - \cot^4\theta$$

$$\csc^2\theta(2 - \csc^2\theta) =$$

$$\csc^2\theta[2 - (1 + \cot^2\theta)] =$$

$$(1 + \cot^2\theta)(1 - \cot^2\theta) =$$

$$1 - \cot^4\theta = 1 - \cot^4\theta$$

41. $$\frac{\sec \lambda - \csc \lambda}{\sec \lambda + \csc \lambda} = \frac{\tan \lambda - 1}{\tan \lambda + 1}$$

$$\frac{\left(\frac{1}{\cos \lambda} - \frac{1}{\sin \lambda}\right)}{\left(\frac{1}{\cos \lambda} + \frac{1}{\sin \lambda}\right)} \cdot \frac{(\sin \lambda)}{(\sin \lambda)} =$$

$$\frac{\frac{\sin \lambda}{\cos \lambda} - 1}{\frac{\sin \lambda}{\cos \lambda} + 1} =$$

$$\frac{\tan \lambda - 1}{\tan \lambda + 1} = \frac{\tan \lambda - 1}{\tan \lambda + 1}$$

44. $$\cos^4 u - \sin^4 u = \cos^2 u - \sin^2 u$$

$$(\cos^2 u + \sin^2 u)(\cos^2 u - \sin^2 u) =$$

$$1(\cos^2 u - \sin^2 u) =$$

$$\cos^2 u - \sin^2 u = \cos^2 u - \sin^2 u$$

47. $\sin x = \sqrt{1 - \cos^2 x}$ is not an identity when $x = -\pi/6$ since $\sin(-\frac{\pi}{6}) = -\frac{1}{2}$ whereas $\sqrt{1 - \cos^2(-\frac{\pi}{6})} = +\frac{1}{2}$

50. $\sin\theta + \cos\theta = \sec\theta + \csc\theta$ is not an identity when $\theta = 0^\circ$ since $\csc\theta = \frac{1}{\sin\theta}$ and $\sin 0^\circ = 0$, causing $\csc 0^\circ$ to be undefined, whereas the left-hand side is defined when $\theta = 0^\circ$.

Section 7.2 The Addition Formulas

BASIC CONCEPTS you should learn from this unit.

1. There are six important identities called the Addition Formulas.
   A. $\sin(s + t) = \sin s \cos t + \cos s \sin t$
   B. $\sin(s - t) = \sin s \cos t - \cos s \sin t$
   C. $\cos(s + t) = \cos s \cos t - \sin s \sin t$
   D. $\cos(s - t) = \cos s \cos t + \sin s \sin t$
   E. $\tan(s + t) = \dfrac{\tan s + \tan t}{1 - \tan s \tan t}$
   F. $\tan(s - t) = \dfrac{\tan s - \tan t}{1 + \tan s \tan t}$

2. Two angles whose sum is 90° $(\pi/2)$ are called complementary angles.

3. The sine of an acute angle, α, is equal to the cosine of its complement. That is

$$\sin\alpha = \cos(90^\circ - \alpha)$$

   similarly

$$\cos\alpha = \sin(90^\circ - \alpha)$$
$$\tan\alpha = \cot(90^\circ - \alpha)$$

4. Functions satisfying the identities in basic concept 3 are called cofunctions. The secant and cosecant are another pair of cofunctions.

WHAT YOU SHOULD KNOW HOW TO DO upon completion of this unit.

1. State the six basic addition formulas.
2. Use the addition formulas to solve problems like those found in the exercise set in the text. [Example: Find $\sin(\pi/6 + \pi/4)$.]
3. Use the addition formulas to help in verifying identities.
4. Express a trigonometric value, such as $\sin 40^\circ$, or $\tan \pi/3$, in terms of cofunctions.

SOLUTIONS to Exercise Set 7.2

2. To show that $\sin(s + t) = \sin s + \sin t$ is <u>not</u> an identity choose, for example $s = 30^\circ$, $t = 30^\circ$; then

$\sin(60^\circ) = 0.8660$

But

$\sin 30^\circ + \sin 30^\circ = 0.5 + 0.5 = 1.0$

5. To show that $\tan(s + t) = \tan s + \tan t$ is <u>not</u> an identity choose $s = 30^\circ$, $t = 30^\circ$; then

$$\tan 60^\circ = 1.7321$$

But

$\tan 30^\circ + \tan 30^\circ = 0.5774 + 0.5774 = 1.1548$

(There are of course other correct answers for Exercises 2 and 5.)

8. Find $\sin(\pi/6 - \pi/4)$

The addition formula gives

$$\sin(\pi/6 - \pi/4) = \sin \pi/6 \cos \pi/4 - \cos \pi/6 \sin \pi/4$$
$$= \frac{1}{2} \cdot \frac{\sqrt{2}}{2} - \frac{\sqrt{3}}{2} \cdot \frac{\sqrt{2}}{2} = \frac{\sqrt{2} - \sqrt{6}}{4}$$

11. $\cos(30^\circ + 180^\circ) = \cos 30^\circ \cos 180^\circ - \sin 30^\circ \sin 180^\circ$
$$= \left(\frac{\sqrt{3}}{2}\right)(-1) - \frac{1}{2}(0) = -\frac{\sqrt{3}}{2}$$

14. $\sin(270^\circ - 45^\circ) = \sin 270^\circ \cos 45^\circ - \cos 270^\circ \sin 45^\circ$
$$= (-1)\left(\frac{\sqrt{2}}{2}\right) - (0)\left(\frac{\sqrt{2}}{2}\right)$$
$$= -\frac{\sqrt{2}}{2}$$

17. $\cos 7\pi/12 = \cos(\pi/4 + \pi/3) = \cos \pi/4 \cos \pi/3 - \sin \pi/4 \sin \pi/3$
$$= \frac{\sqrt{2}}{2} \cdot \frac{1}{2} - \frac{\sqrt{2}}{2} \cdot \frac{\sqrt{3}}{2} = \frac{\sqrt{2} - \sqrt{6}}{4}$$

20. $\cos\left(\frac{5\pi}{6}\right) = \cos\left(\pi - \frac{\pi}{6}\right) = \cos \pi \cos \frac{\pi}{6} + \sin \pi \sin \frac{\pi}{6}$
$$= (-1)\left(\frac{\sqrt{3}}{2}\right) + (0)\left(\frac{1}{2}\right) = -\frac{\sqrt{3}}{2}$$

23. $\sin 47^\circ = \cos 43^\circ$ cosine of the complementary angle

26. $\tan 84^\circ = \cot 6^\circ$ cofunction of the complementary angle

29. $\sin t = -3/5$, $W(t)$ is in quadrant III. Find $\sin(\pi/2 - t)$. Note that $\cos t$ will be $(-4/5)$.

$$\sin(\pi/2 - t) = \sin \pi/2 \cos t - \cos \pi/2 \sin t$$
$$= (1)(-4/5) - (0)(-3/5) = -4/5$$

32. $\sec \theta = 5/3$, θ is in quadrant I. From $\sec \theta = 5/3$ we know that $\cos \theta = 3/5$; then we find that $\sin \theta = 4/5$.

$$\sin(\theta + \pi/6) = \sin \theta \cos \pi/6 + \cos \theta \sin \pi/6$$
$$= \left(\frac{4}{5}\right)\left(\frac{\sqrt{3}}{2}\right) + \left(\frac{3}{5}\right)\left(\frac{1}{2}\right) = \frac{4\sqrt{3} + 3}{10}$$

35. $\sin s = 3/5$, $\cos t = -12/13$, $W(s)$ is in quadrant II and $W(t)$ in quadrant III.

First, from the information given, we deduce that

$$\cos s = -4/5 \quad \text{and} \quad \sin t = -5/13$$

then

$$\sin(s + t) = \sin s \cos t + \cos s \sin t$$
$$= \left(\frac{3}{5}\right)\left(-\frac{12}{13}\right) + \left(-\frac{4}{5}\right)\left(-\frac{5}{13}\right)$$
$$= -\frac{36}{65} + \frac{20}{65} = -\frac{16}{65}$$

38. $\sec \alpha = 5/3$, $\cot \beta = 15/8$, α is in quadrant IV and β in quadrant III.

First, we need $\tan \alpha$ and $\tan \beta$. To get $\tan \alpha$, we note that if $\sec \alpha = 5/3$, then $\cos \alpha = 3/5$, and if $\sin \alpha = -4/5$, then $\tan \alpha = -4/3$. Also, if $\cot \beta = 15/8$, then $\tan \beta = 8/15$.

Now

$$\tan(\alpha - \beta) = \frac{\tan \alpha - \tan \beta}{1 + \tan \alpha \tan \beta}$$

$$= \frac{-4/3 - 8/15}{1 + 4/3 \cdot 8/15} = \frac{-28/15}{77/45} = \underline{-84/77}$$

41. Prove that $\tan 2\alpha = \dfrac{2 \tan \alpha}{1 - \tan^2 \alpha}$

First write $\tan 2\alpha = \tan(\alpha + \alpha)$; then

$$\tan(\alpha + \alpha) = \frac{\tan \alpha + \tan \alpha}{1 - \tan \alpha \tan \alpha}$$
$$= \frac{2 \tan \alpha}{1 - \tan^2 \alpha} = \frac{2 \tan \alpha}{1 - \tan^2 \alpha}$$

44. $$\frac{\sin(s+t)}{\sin(s-t)} = \frac{\tan s + \tan t}{\tan s - \tan t}$$

$$\frac{\sin s \cos t + \cos s \sin t}{\sin s \cos t - \cos s \sin t} =$$

$$\frac{\frac{\sin s \cos t}{\cos s \cos t} + \frac{\cos s \sin t}{\cos s \cos t}}{\frac{\sin s \cos t}{\cos s \cos t} - \frac{\cos s \sin t}{\cos s \cos t}} =$$

We divide the numerator and the denominator by $\cos s \cos t$.

$$\frac{\frac{\sin s}{\cos s} + \frac{\sin t}{\cos t}}{\frac{\sin s}{\cos s} - \frac{\sin t}{\cos t}} =$$

$$\frac{\tan s + \tan t}{\tan s - \tan t} = \frac{\tan s + \tan t}{\tan s - \tan t}$$

47. $$\tan(x + \pi/4) = \frac{1 + \tan x}{1 - \tan x}$$

$$\frac{\tan x + \tan(\pi/4)}{1 - \tan x \tan(\pi/4)} =$$

$$\frac{\tan x + 1}{1 - \tan x \cdot (1)} = \frac{1 + \tan x}{1 - \tan x}$$

50. $$\cot(u + v) = \frac{\cot u \cot v - 1}{\cot u + \cot v}$$

$$\frac{1}{\tan(u+v)} =$$

$$\frac{1}{\frac{\tan u + \tan v}{1 - \tan u \tan v}} =$$

$$\frac{1 - \tan u \tan v}{\tan u - \tan v} =$$

$$\frac{\frac{1}{\tan u \tan v} - 1}{\frac{1}{\tan v} + \frac{1}{\tan u}} =$$

We divide the numerator and the denominator by $\tan u \tan v$.

$$\frac{\cot u \cot v - 1}{\cot v + \cot u} = \frac{\cot u \cot v - 1}{\cot u + \cot v}$$

53. $$\frac{\sin(x+h) - \sin x}{h} = \sin x\left(\frac{\cos h - 1}{h}\right) + \cos x\left(\frac{\sin h}{h}\right)$$

$$\frac{\sin x \cos h + \cos x \sin h - \sin x}{h} =$$

$$\frac{\sin x(\cos h - 1)}{h} + \frac{\cos x \sin h}{h} = \sin x\left(\frac{\cos h - 1}{h}\right) + \cos x\left(\frac{\sin h}{h}\right)$$

BASIC CONCEPTS you should learn from this unit.

1. There are three fundamental identities called the double-angle formulas:

   A. $\sin 2t = 2 \sin t \cos t$

   B. $\cos 2t = \cos^2 t - \sin^2 t$

   C. $\tan 2t = \dfrac{2 \tan t}{1 - \tan^2 t}$

2. Alternate forms of $\cos 2t$ are

   $\cos 2t = 1 - 2 \sin^2 t$

   $\cos 2t = 2 \cos^2 t - 1$

3. There are three fundamental identities called the half-angle formulas:

   $\sin \frac{t}{2} = \pm\sqrt{\frac{1 - \cos t}{2}}$

   $\cos \frac{t}{2} = \pm\sqrt{\frac{1 + \cos t}{2}}$

   $\tan \frac{t}{2} = \pm\sqrt{\frac{1 - \cos t}{1 + \cos t}}$

   You should also know that the sign to be used in the half-angle formulas depends on the quadrant of $W(t/2)$.

WHAT YOU SHOULD KNOW HOW TO DO upon completion of this unit.

1. State the double-angle formulas.

2. State the half-angle formulas.

3. Use these six formulas for problems similar to the following:

   A. If $\cos t = -4/5$ and $W(t)$ is in quadrant II, find $\sin 2t$.

   B. Express $\sin 3t$ in terms of $\sin t$ and $\cos t$.

   C. If $\cos 77^\circ = 0.22$, find $\cos 154^\circ$.

   D. Find exact values of $\sin 15^\circ$, $\cos 75^\circ$, and similar trigonometric values.

   E. If $\cos \alpha = 3/5$ and α is in quadrant I, find $\sin \alpha/2$.

4. Use the double- and half-angle formulas to assist in verifying identities.

SOLUTIONS to Exercise Set 7.3

2. $\cos x = -5/13$ and $W(x)$ is in quadrant III; find $\sin 2x$.

   use of $\sin^2 + \cos^2 = 1$ yields $\sin x = -12/13$; then

   $\sin 2x = 2 \sin x \cos x$

   $= 2(-12/13)(-5/13) = 120/169$

5. $\csc t = -17/8$ and $W(t)$ is in quadrant IV; find $\tan 2t$. We will need $\tan 2t$. First, though, since $\csc t = -17/8$, $\sin t = -8/17$. Then using $\sin^2 t + \cos^2 t = 1$ we find that $\cos t = +15/17$ and then $\tan t = -8/15$. Now

$$\tan 2t = \frac{2 \tan t}{1 - \tan^2 t} = \frac{-16/15}{1 - 64/225}$$

$$= -\frac{240}{161}$$

8. $\sec 5x = -15/12$ and $W(5x)$ is in quadrant III; find $\tan 10x$.

Since $\sec 5x = -15/12$, $\cos 5x = -12/15$

and again use of $\sin^2 5x + \cos^2 5x = 1$ yields

$\sin 5x = -9/15$

$10x = 2 \cdot 5x$

so we use the double-angle formula

$$\tan 10x = \frac{2 \tan 5x}{1 - \tan^2 5x}$$

$$= \frac{2 \cdot (9/12)}{1 - 81/144} \qquad (\tan 5x = 9/12)$$

$$= 24/7$$

11. $\sin 42^\circ = 0.67$; find $\cos 84^\circ$.

$\cos 2t = 1 - 2 \sin^2 t$,

$\cos 84^\circ = 1 - 2 \sin^2 42^\circ = 1 - 2 \cdot (0.67)^2 = 0.1022$

14. Find $\cos 75^\circ$ using a half-angle formula:

75° is half of 150° and $\cos 150^\circ$ is $-\sqrt{3}/2$ (reference angle of 30°). Then

$$\cos 75^\circ = +\sqrt{\frac{1 + (-\ 3/2)}{2}} = +\sqrt{\frac{2 - \sqrt{3}}{4}}$$

$$= \frac{+\sqrt{2 - \sqrt{3}}}{2}$$

17. $\csc 165^\circ = 1/\sin 165^\circ$ and $\sin 165^\circ = +\sqrt{(1 - \cos 330^\circ)/2}$ and since $\cos 330^\circ = \cos 30^\circ = \sqrt{3}/2$, we have

$$\sin 165^\circ = +\sqrt{\frac{1 - \sqrt{3}/2}{2}} = \frac{\sqrt{2 - \sqrt{3}}}{2}$$

and then

$$\csc 165^\circ = \frac{2}{\sqrt{2 - \sqrt{3}}}$$

20. $\sin \frac{\theta}{2} = \sqrt{\frac{1 - \cos \theta}{2}}$. Then $\sin \frac{\theta}{2} = \sqrt{\frac{1 - 3/5}{2}}$, $\sin \theta/2 = \sqrt{2/10} = \sqrt{1/5}$
The sine is positive because θ is in quadrant I and thus $\theta/2$ is also.

23. If $\cot \beta = 3/4$, then $\tan \beta = 4/3$, and $\cos \beta = 3/5$.

$$\tan \beta/2 = \sqrt{\frac{1 - \cos \beta}{1 + \cos \beta}}$$

$$\tan \beta/2 = -\sqrt{\frac{1 - 3/5}{1 + 3/5}} = -\frac{1}{2}$$

26. If $\sec 6\alpha = -13/12$, then $\cos 6\alpha = -12/13$, and

$$\sin 3\alpha = +\sqrt{\frac{1 + 12/13}{2}} = +\sqrt{\frac{25}{26}}$$

The sign is plus because 3α is in quadrant II.

29. $\tan 2y = \frac{2 \cot y}{\csc^2 y - 2}$. We multiply the right-hand side by $\sin^2 y/\sin^2 y$.

$$= \frac{2 \frac{\cos y}{\sin^2 y} \cdot \sin^2 y}{\left(\frac{1}{\sin^2 y} - 2\right)\sin^2 y}$$

$$= \frac{2 \sin y \cos y}{1 - 2 \sin^2 y}$$

$$\tan 2y = \frac{\sin 2y}{\cos 2y} = \tan 2y$$

32.
$$\cos 4\beta = 1 - 8 \sin^2\beta \cos^2\beta$$

$$\cos(2\beta + 2\beta) =$$

$$\cos^2 2b - \sin^2 2b =$$

$$1 - \sin^2 2b - \sin^2 2b =$$

$$1 - 2 \sin^2 2b =$$

$$1 - 2(2 \sin b \cos b)^2 =$$

$$1 - 8 \sin^2 b \cos^2 b = 1 - 8 \sin^2 b \cos^2 b$$

35. $\sin \frac{t}{2} \cos \frac{t}{2} = \frac{\sin t}{2}$

$$= \frac{2 \sin \frac{t}{2} \cos \frac{t}{2}}{2}$$ by a double-angle formula

$$\sin \frac{t}{2} \cos \frac{t}{2} = \sin \frac{t}{2} \cos \frac{t}{2}$$

38. $\dfrac{1 - \cos 2\beta}{1 + \cos 2\beta} = \tan^2\beta$

Using the half-angle formula and thinking of β as half of 2β we have

$$\tan \beta = \sqrt{\frac{1 - \cos 2\beta}{1 + \cos 2\beta}}$$

Now operating on the right-hand side we have

$$= \tan \beta \cdot \tan \beta$$

$$= \sqrt{\frac{1 - \cos 2\beta}{1 + \cos 2\beta}} \cdot \sqrt{\frac{1 - \cos 2\beta}{1 + \cos 2\beta}}$$

$$= \frac{1 - \cos 2\beta}{1 + \cos 2\beta} \quad \text{as required}$$

41. $\dfrac{2 \tan \alpha}{1 + \tan \alpha} = \sin 2\alpha$

The left-hand side = $\tan 2\alpha$ by the double-angle formula for the tangent function. But $\tan 2\alpha \neq \sin 2\alpha$. Therefore we do <u>not</u> have an identity.

44. $$\cos 2t + \cot 2t = \cot 2t(\sin t + \cos t)^2$$

$$\cot 2t(\sin 2t + 1) =$$

$$\cot 2t(2 \sin t \cos + \sin^2 t + \cos^2 t) =$$

$$\cot 2t(\sin^2 t + 2 \sin t \cos t + \cos^2 t) =$$

$$\cot 2t(\sin t + \cos t)^2 = \cot 2t(\sin t + \cos t)^2$$

Section 7.4 The Product-Sum Formulas

BASIC CONCEPTS you should learn from this unit.

1. There are eight formulas known as the <u>product-sum formulas</u>. Four of them express a product as a sum:

A. $\sin s \cos t = \dfrac{\sin(s + t) + \sin(s - t)}{2}$

B. $\cos s \sin t = \dfrac{\sin(s + t) - \sin(s - t)}{2}$

C. $\cos s \cos t = \dfrac{\cos(s + t) + \cos(s - t)}{2}$

D. $\sin s \sin t = \dfrac{\cos(s - t) - \cos(s + t)}{2}$

The other four express a sum as a product:

E. $\sin s + \sin t = 2 \sin\left(\dfrac{s + t}{2}\right)\cos\left(\dfrac{s - t}{2}\right)$

F. $\sin s - \sin t = 2 \cos\left(\dfrac{s + t}{2}\right)\cos\left(\dfrac{s - t}{2}\right)$

G. $\cos s + \cos t = 2 \cos\left(\dfrac{s + t}{2}\right)\cos\left(\dfrac{s - t}{2}\right)$

H. $\cos s - \cos t = -2 \sin\left(\dfrac{s + t}{2}\right)\sin\left(\dfrac{s - t}{2}\right)$

DO NOT MEMORIZE THESE FORMULAS!

WHAT YOU SHOULD KNOW HOW TO DO upon completion of this unit.

1. Use the product-sum formulas to
   A. Express a product as a sum or difference;
   B. Express a sum or difference as a product.
2. Use the product-sum formulas to evaluate a sum.
3. Use the product-sum formulas to verify identities.

SOLUTIONS to Exercise Set 7.4

2. $-3 \cos 6x \sin 2x = -3\left[\dfrac{\sin(6x + 2x) - \sin(6x - 2x)}{2}\right]$

$= -3/2[\sin 8x - \sin 4x]$

5. $-2 \cos 2\theta \cos 5\theta = -2\left[\dfrac{\cos(2\theta + 5\theta) + \cos(2\theta - 5\theta)}{2}\right]$

$= -[\cos 7\theta + \cos(-3\theta)^2]$

$= -[\cos 7\theta + \cos 3\theta]$

8. $-\sin 2u \cos 4u = -\left[\dfrac{\sin(2u + 4u) + \sin(2u - 4u)}{2}\right]$

$= -1/2[\sin 6u + \sin(-2u)]$

$= -1/2(\sin 6u - \sin 2u)$

11. $\sin 120^\circ \cos 60^\circ = \dfrac{\sin(120 + 60) + \sin(120 - 60)}{2}$

$= 1/2(\sin 180 + \sin 60)$

$= 1/2(0 + \sqrt{3}/2) = \underline{\sqrt{3}/4}$

14. $\cos 8t - \cos 2t = -2 \sin \dfrac{8t + 2t}{2} \sin \dfrac{8t - 2t}{2}$

$= -2 \sin 5t \sin 3t$

17. $\sin(\alpha + \beta) + \sin(\alpha - \beta) = 2 \sin \dfrac{(\alpha + \beta) + (\alpha - \beta)}{2} \cos \dfrac{(\alpha + \beta) - (\alpha - \beta)}{2}$

$= 2 \sin \alpha \cos \beta$

20. $\cos 5\theta + \cos 3\theta = 2 \cos \dfrac{5\theta + 3\theta}{2} \cos \dfrac{5\theta - 3\theta}{2}$

$= 2 \cos 4\theta \cos \theta$

23. $\cos \dfrac{3\pi}{4} - \cos \dfrac{\pi}{4} = -2 \sin \dfrac{\left(\frac{3\pi}{4} + \frac{\pi}{4}\right)}{2} \sin \dfrac{\left(\frac{3\pi}{4} - \frac{\pi}{4}\right)}{2}$

$= -2 \sin \dfrac{\pi}{2} \sin \dfrac{\pi}{4}$

$= -2 \cdot 1 \cdot \dfrac{\sqrt{2}}{2} = -\sqrt{2}$

26. $$\cos 70^\circ - \cos 10^\circ = -\sin 40^\circ$$

$$-2 \sin \frac{70^\circ + 10^\circ}{2} \sin \frac{70^\circ - 10^\circ}{2} =$$

$$-2 \sin 40^\circ \sin 30^\circ =$$

$$-2 \sin 40^\circ \cdot \frac{1}{2} =$$

$$-\sin 40^\circ = -\sin 40^\circ$$

29. $$\frac{\sin t - \sin s}{\cos t - \cos s} = -\cot \frac{s + t}{2}$$

$$\frac{2 \cos \frac{t + s}{2} \sin \frac{t - s}{2}}{-2 \sin \frac{t + s}{2} \sin \frac{t - s}{2}} =$$

$$-\frac{\cos \frac{t + s}{2}}{\sin \frac{t + s}{2}} =$$

$$\cot \frac{s + t}{2} = -\cot \frac{s + t}{2}$$

32. $$2 \sin\left(\theta + \frac{\pi}{4}\right) \sin\left(\pi - \frac{\pi}{4}\right) = -\cos 2\theta$$

$$2\left[\frac{\cos\left[\left(\theta + \frac{\pi}{4}\right) - \left(\theta - \frac{\pi}{4}\right)\right] - \cos\left[\left(\theta + \frac{\pi}{4}\right) + \left(\theta - \frac{\pi}{4}\right)\right]}{2}\right] =$$

$$\cos \frac{\pi}{2} - \cos(2\theta) =$$

$$0 - \cos 2\theta = -\cos 2\theta$$

35. $$\sin ax \cos bx = \frac{1}{2}\left[\sin(ax + bx) + \sin(ax - bx)\right]$$

Section 7.5 Trigonometric Equations

BASIC CONCEPTS you should learn from this unit.

1. A trigonometric equation is an equation involving a trigonometric function.
2. Due to the periodic nature of the trigonometric functions, if a trigonometric equation has a solution, it will have an infinite number of solutions. This is generally resolved by seeking the solutions only for values of the variable between 0 and 2π.
3. Trigonometric equations are solved by using methods learned earlier for solving algebraic equations. Factoring and the use of a substitute variable are helpful.
4. Solutions can be expressed in radians or in degrees.
5. Use of the known trigonometric identities is also helpful in solving the equations.

WHAT YOU SHOULD KNOW HOW TO DO upon completion of this unit.

1. Solve trigonometric equations using the methods of this chapter.

2. Express all the solutions to a trigonometric equation as well as those between 0 and 2π .

3. Express solutions in degrees or radians.

SOLUTIONS to Exercise Set 7.5

2. $2 \cos \theta + 1 = 0$

$$2 \cos \theta = -1$$

$$\cos \theta = -\frac{1}{2}$$

$$\theta = \frac{2\pi}{3}, \frac{4\pi}{3}; \ 120^\circ, \ 240^\circ$$

5. $4 \cos^2 \alpha = 3$

$$\cos^2 \alpha = 3/4$$

$$\cos \alpha = \pm\sqrt{3}/2$$

$$\alpha = \pi/6, \ 5\pi/6, \ 7\pi/6, \ 11\pi/6$$

$$= 30^\circ, \ 150^\circ, \ 210^\circ, \ 330^\circ$$

8. $2 \cos^2 \alpha - 1 = 0$

$$2 \cos^2 \alpha = 1$$

$$\cos^2 \alpha = 1/2$$

$$\cos \alpha = \pm\sqrt{2}/2$$

$$\alpha = \pi/4, \ 3\pi/4, \ 5\pi/4, \ 7\pi/4$$

$$= 45^\circ, \ 135^\circ, \ 225^\circ, \ 315^\circ$$

11. $2 \cos^2 \theta - 3 \cos \theta + 1 = 0$

$$(2 \cos \theta - 1)(\cos \theta - 1) = 0$$

$2 \cos \theta - 1 = 0$ or $\cos \theta - 1 = 0$

$2 \cos \theta = 1$ $\quad$ $\cos \theta = 1$

$$\cos \theta = 1/2$$

$$\theta = \pi/3, \ 5\pi/3, \ 0$$

$$= 60^\circ, \ 300^\circ, \ 0^\circ$$

14. $\tan 3\beta = -\sqrt{3}$

$3\beta = 2\pi/3\ ,\ \ 5\pi/3$

Then

$\beta = 2\pi/9\ ,\ \ 5\pi/9\ ,\ \ 8\pi/9\ ,\ \ 11\pi/9\ ,\ \ 14\pi/9\ ,\ \ 17\pi/9$

$= 40^\circ\ ,\ \ 100^\circ\ ,\ \ 160^\circ\ ,\ \ 220^\circ\ ,\ \ 280^\circ\ ,\ \ 340^\circ$

17. $2\cos^2\theta - 1 = \sin\theta$. Since $\cos^2\theta = 1 - \sin^2\theta$, we can write

$2(1 - \sin^2\theta) - 1 = \sin\theta$

Then

$2 - 2\sin^2\theta - 1 = \sin\theta$

and

$-2\sin^2\theta - \sin\theta + 1 = 0$

$2\sin^2\theta + \sin\theta - 1 = 0$

$(2\sin\theta - 1)(\sin\theta + 1) = 0$

$2\sin\theta = 1$ $\qquad\qquad$ $\sin\theta = -1$

$\sin\theta = 1/2$

$\theta = \pi/6\ ,\ \ 5\pi/6\ ,\ \ 3\pi/2$

$= 30^\circ\ ,\ \ 150^\circ\ ,\ \ 270^\circ$

20. $2\cos^2\theta\tan\theta - \tan\theta = 0$

$\tan\theta(2\cos^2\theta - 1) = 0$

$\tan\theta = 0$ $\quad$ or $\quad$ $2\cos^2\theta - 1 = 0$

$\cos^2\theta = +1/2$

$\cos\theta = \pm\sqrt{2}/2$

$\theta = 0\ ,\ \ \pi$ $\qquad\qquad$ $\theta = \pi/4\ ,\ \ 3\pi/4\ ,\ \ 5\pi/4\ ,\ \ 7\pi/4$

$\theta = 0^\circ\ ,\ \ 180^\circ$ $\qquad\qquad$ $\theta = 45^\circ\ ,\ \ 135^\circ\ ,\ \ 225^\circ\ ,\ \ 315^\circ$

23. $3\cot^2\theta - 1 = 0$

$3\cot^2\theta = 1$

$\cot^2\theta = 1/3$

$\cot\theta = \pm\sqrt{3}/3$

$\theta = \pi/3 + n\pi\ ,\ \ 2\pi/3 + n\pi$ $\quad$ n being an integer

$n\pi$ is needed for all solutions.

26. $\tan 3x - 1 = 0$

$\tan 3x = 1$

$3x = \pi/4 + n\pi$

so

$x = \pi/12 + n\pi/3$

29. $4\cos^2 2t - 3 = 0$

$\cos^2 2t = 3/4$

$\cos 2t = \pm\sqrt{3}/2$

$2t = \pi/6 + n\pi$ and $5\pi/6 + n\pi$

so

$t = \pi/12 + n\pi/2$, and $5\pi/12 + n\pi/2$

32. $\sin 2t + 3\cos t = 0$ Use $\sin 2t = 2\sin t\cos t$.

$2\sin t\cos t + 3\cos t = 0$

$\cos t(2\sin t + 3) = 0$

$\cos t = 0$ $2\sin t + 3 = 0$

$t = \pi/2 + n\pi$ $\sin t = -3/2$, reject since $|\sin t| > 1$

35. $\tan^2 x - \tan x = 0$

$\tan x(\tan x - 1) = 0$ $\tan x - 1 = 0$

$\tan x = 0$ $\tan x = 1$

$x = 0 + n\pi$ $x = \pi/4 + n\pi$

38. $2\cos^2 x - 5\cos x - 3 = 0$

$(2\cos x + 1)(\cos x - 3) = 0$

$2\cos x + 1 = 0$ $\cos x - 3 = 0$

$\cos x = -1/2$ $\cos x = 3$

yields $x = \emptyset$

$x = 2\pi/3 + 2n\pi$, and $4\pi/3 + 2n\pi$

41. $3\tan^2 u + 5\tan u + 1 = 0$ Let $x = \tan u$.

$3x^2 + 5x + 1 = 0$

$x = \dfrac{-5 \pm \sqrt{25 - 12}}{6}$ by the quadratic formula

$x = \dfrac{-5 \pm \sqrt{13}}{6} = 1.4342$ or -0.2324

$\tan u = -1.4342$ $\tan u = -0.2324$

$u = 5.32$ radians $u = 6.05$ radians

BASIC CONCEPTS you should learn from this unit.

1. Complex numbers of the usual form, $a + bi$, can be located as points on the complex plane using an x-axis and a y-axis for orientation. The y-axis is called the imaginary axis; the x-axis is called the real axis.

2. The absolute value of a complex number is the distance from the origin to the point.

3. For a complex number, $a + bi$

   $a = r\cos\theta$ and $b = r\sin\theta$

   a, b, r, and θ defined as in this diagram:

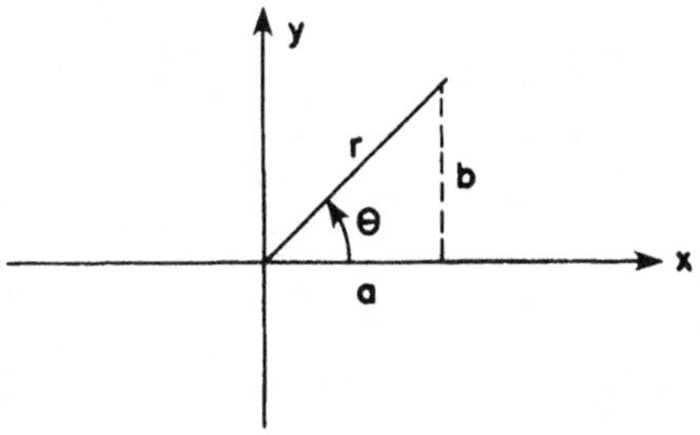

   Then $a + bi = r\cos\theta + (r\sin\theta)i$

   The right-hand side is called the polar form of the complex number.

4. In the polar form r is called the modulus and θ the argument:

$$r = \sqrt{a^2 + b^2} = |a + bi|$$

5. Multiplication and division of numbers in the polar form can be done by these formulas:

   A. $r_1(\cos\theta_1 + i\sin\theta_1)\cdot r_2(\cos\theta_2 + i\sin\theta_2) = r_1r_2\left[\cos(\theta_1 + \theta_2) + 2\sin(\theta_1 + \theta_2)\right]$

   B. $\dfrac{r_1(\cos\theta_1 + i\sin\theta_2)}{r_2(\cos\theta_1 + i\sin\theta_2)} = \dfrac{r_1}{r_2}\left[\cos(\theta_1 - \theta_2) + i\sin(\theta_1 - \theta_2)\right]$

6. DeMoivre's Theorem, which states that

$$\left[r(\cos\theta + i\sin\theta)\right]^n = r^n(\cos n\theta + i\sin n\theta)$$

   enables us to find the nth power of a complex number.

7. The formula

$$\sqrt[n]{r}\left[\cos\left(\frac{\theta + 2\pi k}{n}\right) + i\sin\left(\frac{\theta + 2\pi k}{n}\right)\right], \qquad k = 0,1,2,\ldots,n - 1$$

   enables us to find the nth roots of a complex number: $r(\cos\theta + i\sin\theta)$.

WHAT YOU SHOULD KNOW HOW TO DO upon completion of this unit.

1. Locate complex numbers on the plane.
2. Find the absolute value of a complex number.
3. Change the form of a complex number from polar form to rectangular form and vice versa.
4. Multiply two complex numbers in polar form.
5. Divide two complex numbers in polar form.
6. Use DeMoivre's Theorem to find the nth power of a complex number.
7. Use the formula in basic concept 7 to find the nth roots of a complex number.

SOLUTIONS to Exercise Set 7.6

2. $|-7 + 6i| = \sqrt{(-7)^2 + 6^2} = \sqrt{49 + 36} = \sqrt{85}$

5. $|-6 - 2i| = \sqrt{(-6)^2 + (-2)^2} = \sqrt{36 + 4} = \sqrt{40} = 2\sqrt{10}$

8. $2 + 2i$ $\quad a = 2 = r\cos\theta, \quad b = 2 = r\sin\theta$

$$r = \sqrt{a^2 + b^2} = 2\sqrt{2}, \quad \sin\theta = \frac{\sqrt{2}}{2} = \cos\theta$$

Thus in the polar form $2 + 2i = 2\sqrt{2}\left(\cos\frac{\pi}{4} + i\sin\frac{\pi}{4}\right)$

11. $-1 + i$ $\quad a = -1 = r\cos\theta$

$b = +1 = r\sin\theta$

$r = \sqrt{1 + 1} = \sqrt{2}$

Thus

$\sin\theta = \sqrt{2}/2$
$\cos\theta = -\sqrt{2}/2$ $\quad \theta = 135^\circ$

and $-1 + i = \sqrt{2}(\cos 135^\circ + i\sin 135^\circ)$

14. $3i = 0 + 3i$ $\quad a = 0 = r\cos\theta$

$b = 3 = r\sin\theta$

$r = \sqrt{0 + 9} = 3$

Then

$\sin\theta = 1, \quad \cos\theta = 0, \quad \theta = \pi/2$

and $3i = 3(\cos\pi/2 + i\sin\pi/2)$.

17. $\sqrt{2}(\cos 135^\circ + i\sin 135^\circ) = \sqrt{2}\left(-\frac{\sqrt{2}}{2} + i\frac{\sqrt{2}}{2}\right)$

$= -1 + i$ See solution 11.

20. $4(\cos 240^{\circ} + i \sin 240^{\circ}) = 4\left[-\frac{1}{2} + i\left(-\frac{\sqrt{3}}{2}\right)\right]$

$= -2 - 2i\sqrt{3}$

23. Using the product formula we have

$$2(\cos 10^{\circ} + i \sin 10^{\circ}) \cdot (\cos 320^{\circ} + i \sin 320^{\circ}) =$$

$$2(\cos 330^{\circ} + i \sin 330^{\circ}) =$$

$$2\left(\frac{\sqrt{3}}{2} + i \cdot -\frac{1}{2}\right) = \sqrt{3} - i$$

26. $\underline{-\sqrt{3} + i}$ $a = -\sqrt{3} = r \cos \theta$

$b = 1 = r \sin \theta$

$r = \sqrt{3 + 1} = 2$

Then

$\sin \theta = 1/2$, $\cos \theta = -\sqrt{3}/2$, $\theta = 5\pi/6$

and $-\sqrt{3} + i = 2(\cos 5\pi/6 + i \sin 5\pi/6)$.

-2 $a = -2 = r \cos \theta$

$b = 0 = r \sin \theta$

$r = \sqrt{4} = 2$

Then

$\sin \theta = 0$, $\cos \theta = -1$, $\theta = \pi$

and $-2 = 2(\cos \pi + i \sin \pi)$.

The product then is

$$2\left(\cos \frac{5\pi}{6} + i \sin \frac{5\pi}{6}\right) \cdot 2\left(\cos \pi + i \sin \pi\right) = 4\left(\cos \frac{11\pi}{6} + i \sin \frac{11\pi}{6}\right) = 4\left[\frac{\sqrt{3}}{2} + i\left(-\frac{1}{2}\right)\right]$$

$= 2\sqrt{3} - 2i$ in $a + bi$ form

29. For the number 5, $a = 5 = r \cos \theta$, $b = 0 = r \sin \theta$, $r = 5$, $\sin \theta = 0$, $\cos \theta = 1$

Thus $\theta = \pi/2$ and $5 = \underline{5(\cos \pi/2 + i \sin \pi/2)}$

For the number $-2 - 2i$, $a = -2 = r \cos \theta$, $b = -2 = r \sin \theta$, $r = 2\sqrt{2}$, $\sin \theta = -\sqrt{2}/2$,

and $\cos \theta = -\sqrt{2}/2$, $\theta = 5\pi/4$

and $-2 - 2i = \underline{2\sqrt{2}(\cos 5\pi/4 + i \sin 5\pi/4)}$.

The product then becomes

$$5\left(\cos \frac{\pi}{2} + i \sin \frac{\pi}{2}\right) \cdot 2\sqrt{2}\left(\cos \frac{5}{4}\pi + i \sin \frac{5\pi}{4}\right) = 10\sqrt{2}\left(\cos \frac{5\pi}{4} + i \sin \frac{5\pi}{4}\right)$$

$$= 10\sqrt{2}\left[-\frac{\sqrt{2}}{2} + i\left(\frac{-\sqrt{2}}{2}\right)\right]$$

$$= -10 - 10i$$

32. $(\sqrt{3} = i)^{10} = \left[2\left(\cos \frac{11\pi}{6} + i \sin \frac{11\pi}{6}\right)\right]^{10}$

$= 2^{10}\left(\cos 10 \cdot \frac{11\pi}{6} + i \sin 10 \cdot \frac{11\pi}{6}\right)$

$= 2^{10}\left(\cos \frac{55\pi}{3} + i \sin \frac{55\pi}{3}\right)$

$= 2^{10}\left(\cos \frac{\pi}{3} + i \sin \frac{\pi}{3}\right)$

$= 2^{10}\left(\frac{1}{2} + i \frac{\sqrt{3}}{2}\right)$

$= 512 + 512i\sqrt{3}$

35. $(-1 - i)^7 = \sqrt{2}\left(\cos \frac{5\pi}{4} + i \sin \frac{5\pi}{4}\right)^7$

$= (\sqrt{2})^7\left(\cos 7 \cdot \frac{5\pi}{4} + i \sin 7 \cdot \frac{5\pi}{4}\right)$

$= (\sqrt{2})^7\left(\cos \frac{35\pi}{4} + i \sin \frac{35\pi}{4}\right)$

$= (\sqrt{2})^7\left(\cos \frac{3\pi}{4} + i \sin \frac{3\pi}{4}\right)$

$= (\sqrt{2})^7\left(-\frac{\sqrt{2}}{2} + i \frac{\sqrt{2}}{2}\right) = -2^3 + 2^3 i$

$= -8 + 8i$

38. We want to find the square roots of -25. For -25, $r = 25$ and $\theta = \pi$; also, $n = 2$. Applying the formula yields

$$\sqrt{25}\left[\cos \frac{(\theta + 2\pi k)}{n} + i \sin \frac{\theta + 2\pi k}{n}\right], \quad k = 0, 1$$

which yields

$$5\left(\cos \frac{\pi}{2} + i \sin \frac{\pi}{2}\right) \quad \text{for } k = 0$$

and

$$5\left(\cos \frac{3}{2}\pi + i \sin \frac{3\pi}{2}\right) \quad \text{for } k = 1$$

41. $x^3 + 8 = 0$ is equivalent to $x^3 = -8$ and $x = \sqrt[3]{-8}$. We seek the three cube roots of -8. In the polar form $-8 = 8(\cos \pi + i \sin \pi)$. Now applying the formula

The roots are

$$\sqrt[3]{8}\left(\cos \frac{\pi + 2\pi k}{3} + i \sin \frac{\pi + 2\pi k}{3}\right), \quad k = 0, 1, 2$$

First root $= 2\left(\cos \frac{\pi}{3} + i \sin \frac{\pi}{3}\right) = \underline{1 + i\sqrt{3}}$

Second root $= 2\left(\cos \frac{3\pi}{3} + i \sin \frac{3\pi}{3}\right) = \underline{-2}$

Third root $= 2\left(\cos \frac{5\pi}{3} + i \sin \frac{5\pi}{3}\right)$

$= 2\left[\frac{1}{2} + i\left(-\frac{\sqrt{3}}{2}\right)\right] = 1 - \sqrt{3}$

44. $x^4 + 16 = 0$ or $x^4 = -16$ $x = \sqrt[4]{-16}$

$-16 = 16(\cos\pi + i\sin\pi)$ and the roots are $\sqrt[4]{16}\left(\cos\frac{\pi + 2\pi k}{4} + i\sin\frac{\pi + 2\pi k}{4}\right)$, $k = 0,1,2$

First root $= 2\left(\cos\frac{\pi}{4} + i\sin\frac{\pi}{4}\right) = 2\left(\frac{\sqrt{2}}{2} + i\frac{\sqrt{2}}{2}\right) = \sqrt{2} + i\sqrt{2}$

Second root $= 2\left(\cos\frac{3\pi}{4} + i\sin\frac{3\pi}{4}\right) = 2\left(-\frac{\sqrt{2}}{2} + i\frac{\sqrt{2}}{2}\right) = -\sqrt{2} + i\sqrt{2}$

Third root $= 2\left(\cos\frac{5\pi}{4} + i\sin\frac{5\pi}{4}\right) = 2\left(-\frac{\sqrt{2}}{2} - i\frac{\sqrt{2}}{2}\right) = -\sqrt{2} - i\sqrt{2}$

Fourth root $= 2\left(\cos\frac{7\pi}{4} + i\sin\frac{7\pi}{4}\right) = 2\left(\frac{\sqrt{2}}{2} - i\frac{\sqrt{2}}{2}\right) = \sqrt{2} - i\sqrt{2}$

SOLUTIONS to Review Exercises Chapter 7

2. $$\frac{\cos^2 x}{1 - \sin x} = 1 + \sin x$$

$$\frac{1 - \sin^2 x}{1 - \sin x} =$$

$$\frac{(1 - \sin x)(1 + \sin x)}{(1 - \sin x)} = \qquad \text{factoring}$$

(the factor $(1 - \sin x)$ is struck out in numerator and denominator)

$$1 + \sin x = 1 - \sin x$$

5. $\cos(45^\circ + 90^\circ) = \cos 45^\circ \cos 90^\circ - \cos 45^\circ \cos 90^\circ = \frac{\sqrt{2}}{2}\cdot 0 - \frac{\sqrt{2}}{2}\cdot 1 = -\frac{\sqrt{2}}{2}$

11. $\tan\frac{2\pi}{7} = \cot\left(\frac{\pi}{2} - \frac{2\pi}{7}\right) = \cot\left(\frac{7}{14}\pi - \frac{4\pi}{14}\right) = \cot\frac{3}{14}\pi$

14. $\sin t = -3/5$, $W(t)$ is in quadrant III

Then $\cos t = -4/5$ and $\tan t = 3/4$; but $\tan t = \tan(t + \pi)$ because the period of the tangent is π. Thus $\tan(t + \pi) = 3/4$.

17. If $\csc u = -5/4$ and $W(u)$ is in quadrant IV, then $\sin u = -4/5$ and $\cos u = 3/5$.
Since $\cos 2u = \cos^2 u - \sin^2 u$, $\cos 2u = 9/25 - 16/25 = -7/25$.

20. If $\sin\theta = 0.5$ and θ is in quadrant II, then $\cos\theta = -\sqrt{3}/2$. Since $\sin 2\theta = 2\sin\theta\cos\theta$, $\sin 2\theta = 2\cdot 1/2(-\sqrt{3}/2) = -\sqrt{3}/2$.

23. If $\cot t = -4/3$ and $W(t)$ is in quadrant IV, then $\tan t = -3/4$ and so $\cos t = +4/5$.
Then by the half-angle formula

$$\tan\frac{t}{2} = \sqrt{\frac{1 - 4/5}{1 + 4/5}} = -\frac{1}{3} \qquad W\left(\frac{t}{2}\right) \text{ is in quadrant II.}$$

26. $\frac{\pi}{8} = \frac{1}{2} \cdot \frac{\pi}{4}$ $\quad \sin \frac{t}{2} = \sqrt{\frac{1 - \cos t}{2}}$ with $t = \frac{\pi}{4}$. Thus $\sin \frac{\pi}{8} = \sqrt{\frac{1 - \sqrt{2}/2}{2}} = \frac{\sqrt{2 - \sqrt{2}}}{2}$

29.
$$\frac{1}{2} \sin 2y = \frac{\sin y}{\sec y}$$
$$\frac{1}{2} \cdot 2 \sin y \cos y =$$
$$\sin y \cdot \frac{1}{\sec y} = \frac{\sin y}{\sec y}$$

32. By the product formula, $\cos 3x - \cos x = -2 \sin \frac{4x}{2} \sin \frac{2x}{2} = -2 \sin 2x \sin x$

35.
$$2 \cos^2 \alpha - 1 = 0$$
$$2 \cos^2 \alpha = 1$$
$$\cos^2 \alpha = \frac{1}{2}$$
$$\cos \alpha = \pm \frac{\sqrt{2}}{2} \qquad \alpha = \frac{\pi}{4}, \frac{3\pi}{4}, \frac{5\pi}{4}, \frac{7\pi}{4}$$

38.
$$\cos^2 \alpha - 2 \cos \alpha = 0$$
$$\cos \alpha (\cos \alpha - 2) = 0$$
$$\cos \alpha = 0 \quad \text{or} \quad \cos \alpha = 2 \qquad \text{This yields } \emptyset .$$
$$\alpha = 90^\circ + n \cdot 180^\circ$$

41. $|2 - i| = \sqrt{2^2 + (-1)^2} = \sqrt{5}$

44. For $-3 + 3i$
$$a = -3 = r \cos \theta$$
$$b = 3 = r \cos \theta$$
$$r = 3\sqrt{2}$$
Thus $\cos \theta = -\sqrt{2}/2$, $\sin \theta = \sqrt{2}/2$, and $\theta = 135^\circ$;
$$-3 + 3i = 3\sqrt{2}(\sin 135^\circ + i \sin 135^\circ)$$

47. -2
$$a = -2 = r \cos \theta$$
$$b = 0 = r \sin \theta$$
$$r = 2$$
and $\cos \theta = -1$, $\sin \theta = 0$, and thus $\theta = 180^\circ$;
$$-2 = 2(\cos 180^\circ + i \sin 180^\circ)$$

50. $2(\cos 210^\circ + i \sin 210^\circ) \cdot (\cos 240^\circ + i \sin 240^\circ) = 2(\cos 450^\circ + i \sin 450^\circ)$
$$= 2(\cos 90^\circ + i \sin 90^\circ)$$

53. First convert -9 to the trigonometric form, $9(\cos \pi + i \sin \pi)$. Then use the root formula to obtain $\sqrt{9}\left[\cos\left(\frac{180 + 2\pi k}{2}\right) + i \sin\left(\frac{180 + 2k\pi}{2}\right)\right]$, $k = 0, 1$

First root $= 3(\cos 90^\circ + i \sin 90^\circ)$

Second root $= 3(\cos 270^\circ + i \sin 270^\circ)$

Chapter 7 - A SHORT QUIZ

1. A trigonometric ___?___ is true for only some values of the variable, whereas a trigonometric ___?___ is true for all values of the variable.

2. In terms of the same functions, give two alternate forms of $\sin^2 x + \cos^2 x = 1$.

3. True or False: $\tan x = \frac{\sec x}{\csc x}$.

4. Eight fundamental identities are given in the text; list any five of them.

5. In what ways are identities useful in the study of trigonometry?

6. Complete this identity:

   $\sin(s + t) = \sin s \cos t$ ___?___

7. Complete this identity:

   $\tan(s - t) =$ (___?___)/1 + tan s tan t

8. If the sum of two angles is 90°, the angles are called ___?___ .

9. The $\sin \alpha =$ ___?___ $(90^\circ - \alpha)$.

10. Write the right-hand side of this identity in two different ways:

    $\cos 2t = \cos^2 t - \sin^2 t$

11. If $\cos t = -4/5$ and is in quadrant II, what is $\sin t$?

12. If $\cos t = -4/5$ and is in quadrant II, what is $\sin t/2$?

13. If $\cos 67^\circ = 0.3907$, what is $\cos 134^\circ$? Do not use tables or the trigonometric functions on your calculator.

14. The periodic nature of the trigonometric functions assures us that if there is a solution to a trigonometric equation, then there are ___?___ solutions.

15. If a and b are real numbers, and if $a \cdot b = 0$, then what can be deduced about a and b?

16. If $\tan x = 1$, find <u>all</u> the values of x.

17. If $4 \sin^2 2t = 3$, then

    $\sin^2 2t =$ ___?___

    and

    $\sin 2t =$ ___?___

18. Joe claims that $(\cos\theta - 2)(2\cos\theta + 3) = 0$ has no solution. Is he correct? Why or why not?

19. If $\cos x/2 = 0$ with $[0, 2\pi)$ and we let $t = x/2$, express the restriction on t using interval notation.

20. If the absolute value of a number, real or complex, is its distance from the origin, explain why $|a + bi| = \sqrt{a^2 + b^2}$. Use a diagram if you wish.

21. Given the following diagram, express a and b in terms of a trigonometric function of θ.

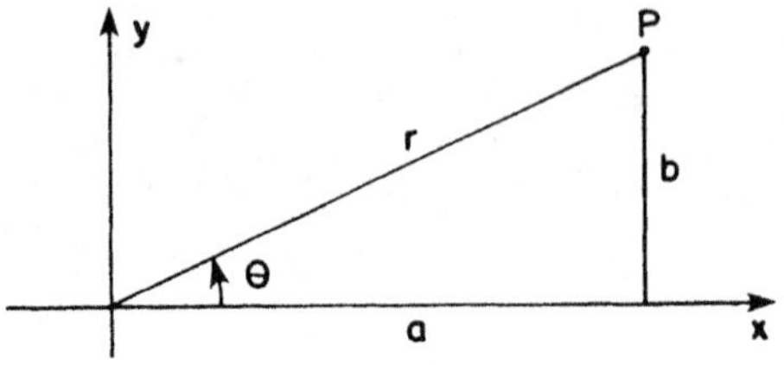

22. The author states that one of the fourth roots of unity is i. Show how this assertion could be verified.

23. Draw the axes of the complex plane labeling the imaginary axis and the real axis.

24. Another name for the trigonometric form of a complex number is the ___?___ form.

25. If $s^n(\cos n\phi + i \sin n\phi) = r(\cos\theta + i\sin\theta)$, then they are represented by the same point and their moduli are equal. That is, $s^n = r$, or we could write $s =$ ___?___

Quiz Answers Chapter 7

1. equation, identity

2. (a) $\sin^2 x = 1 - \cos^2 x$; (b) $\cos^2 x = 1 - \sin^2 x$

3. True

4. Check your answer in your text or on the first page of Chapter 7 in this study guide.

5. Answers will vary. See your text.

6. $+ \cos s \sin t$

7. $\tan s - \tan t$

8. complementary angles

9. $\cos\alpha$

10. (a) $1 - 2\sin^2 t$; (b) $2\cos^2 t - 1$

11. 3/5

12. $3\sqrt{2}/10$

13. -0.6947

14. infinitely many

15. $a = 0$, $b = 0$, or both a and $b = 0$

16. $\pi/4 + n\pi$

17. $3/4$, $\pm\sqrt{3}/2$

18. yes, $\cos\theta > 0$

19. $[0, \pi)$

20. 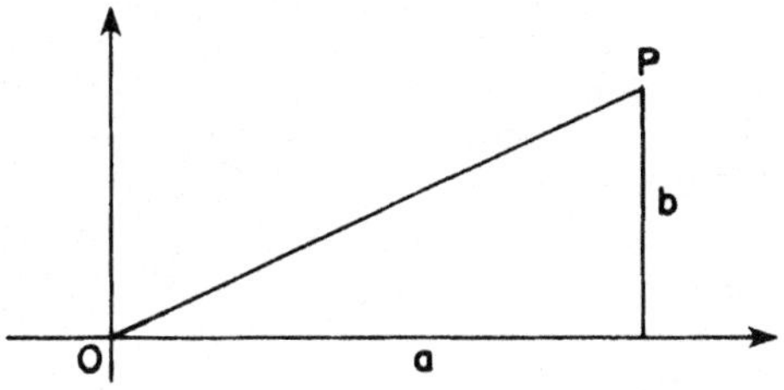

By the Pythagorean Theorem, $\overline{OP} = \sqrt{a^2 + b^2}$.

21. $b = r\sin\theta$, $a = r\cos\theta$

22. $i \cdot i \cdot i \cdot i$ should equal 1.

$(-1)\cdot(-1) = 1$

23. 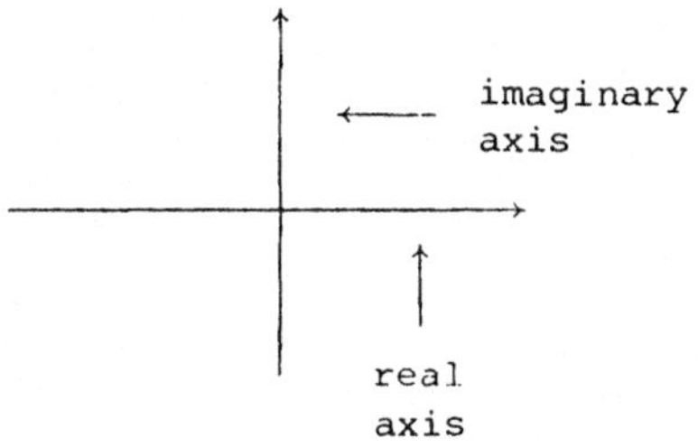

24. polar

25. $\sqrt[n]{r}$

CHAPTER 8

Analytic Geometry: The Conic Sections

Section 8.1 Analytic Geometry

BASIC CONCEPTS you should learn from this unit.

1. Analytic geometry enables us to combine the ideas of algebra with, or relate the ideas of algebra to, the ideas of geometry.

2. The distance formula is

$$d = \sqrt{(x_2 - x_1)^2 + (y_2 - y_1)^2}$$

3. The midpoint formula

$$x = \frac{x_1 + x_2}{2} ; \quad y = \frac{y_1 + y_2}{2}$$

yields the (x, y) coordinates of the midpoint of a line segment joining two points $P_1(x_1, y_1)$ and $P_2(x_2, y_2)$.

WHAT YOU SHOULD KNOW HOW TO DO upon completion of this unit.

1. State the distance formula and the midpoint formula.

2. Use the two formulas to find the distance between two points and the midpoint of a line segment.

3. Use the formulas to assist in the proof of theorems from plane geometry.

SOLUTIONS to Exercise Set 8.1

2. Given (1, 1) and (-2, 5) , the coordinates of the midpoint would be

$$x = \frac{1 + (-2)}{2} = -\frac{1}{2}, \quad \text{and} \quad y = \frac{1 + 5}{2} = 3$$

The midpoint is $(-\frac{1}{2}, 3)$.

5. In similar fashion for the line segment joining (-2, 1) and (-5, -3) we have

$$x = \frac{-2 - 5}{2} = -\frac{7}{2} \quad \text{and} \quad y = \frac{1 - 3}{2} = -1$$

The midpoint is $(-\frac{7}{2}, -1)$.

8. In similar fashion for the points (1, -3), (3, 2) we obtain

$$x = \frac{1 + 3}{2} = 2, \quad y = \frac{-3 + 2}{2} = -\frac{1}{2}$$

The midpoint is $(2, -\frac{1}{2})$.

11. In similar fashion for (1, -1) and (-1, 1) we obtain

$$x = \frac{1 - 1}{2} = 0, \quad y = \frac{-1 + 1}{2} = 0$$

and the midpoint is (0, 0).

14. First, position the rectangle as in this diagram, labeling the midpoints M, P, Q, R, as shown. Since M, P, Q, and R are midpoints of the sides, we can find their coordinates by the midpoint formula.

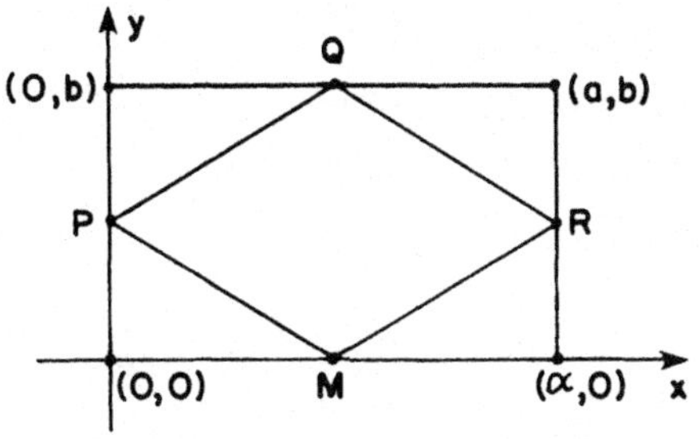

Thus for M

$$x = \frac{0 + a}{2}, \quad y = \frac{0 + 0}{2}; \quad M\left(\frac{a}{2}, 0\right)$$

for P

$$x = \frac{0 + 0}{2}, \quad y = \frac{0 + b}{2}; \quad P\left(0, \frac{b}{2}\right)$$

for Q

$$x = \frac{0 + a}{2}, \quad y = \frac{b + b}{2}; \quad Q\left(\frac{a}{2}, b\right)$$

and for R

$$x = \frac{a + a}{2}, \quad y = \frac{0 + b}{2}; \quad R\left(a, \frac{b}{2}\right)$$

Now

$$\overline{MP} = \sqrt{\left(\frac{a}{2} - 0\right)^2 + \left(0 - \frac{b}{2}\right)^2} = \frac{\sqrt{a^2 + b^2}}{2}$$

$$\overline{PQ} = \sqrt{\left(0 - \frac{a}{2}\right)^2 + \left(\frac{b}{2} - b\right)^2} = \frac{\sqrt{a^2 + b^2}}{2}$$

$$\overline{QR} = \sqrt{\left(\frac{a}{2} - a\right)^2 + \left(b - \frac{b}{2}\right)^2} = \frac{\sqrt{a^2 + b^2}}{2}$$

$$\overline{RM} = \sqrt{\left(a - \frac{a}{2}\right)^2 + \left(\frac{b}{2} - 0\right)^2} = \frac{\sqrt{a^2 + b^2}}{2}$$

Noting that all four line segments are equal, we do have a rhombus formed by MPQR .

17. Place the rectangle as shown in this diagram and show that $\overline{AC} = \overline{BD}$ by the distance formula.

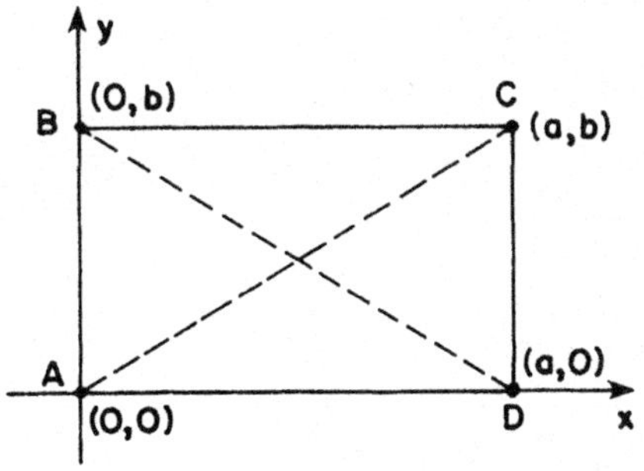

$$\overline{AC} = \sqrt{(0 - a)^2 + (0 - b)^2} = \sqrt{a^2 + b^2}$$

$$\overline{BD} = \sqrt{(0 - a)^2 + (b - 0)^2} = \sqrt{a^2 + b^2}$$

Thus the diagonals are equal.

BASIC CONCEPTS you should learn from this unit.

1. The term conic section comes from the fact that when a plane is passed through a cone at various angles, four different geometric figures are determined: a circle, a parabola, an ellipse, and a hyperbola.

2. Definition of a circle as the set of all points in a plane equidistant from a fixed point. The fixed point is called the center and the given distance the radius.

3. The standard form of the equation of a circle is $(x - h)^2 + (y - k)^2 = r^2$, with center (h, k) and radius r.

WHAT YOU SHOULD KNOW HOW TO DO upon completion of this unit.

1. Define a circle.

2. Write the general equation of a circle in standard form.

3. Write the specific equation of a circle with a given radius and given center.

4. Rewrite into standard form the equation of a circle given in general form (the general form being $Ax^2 + Ay^2 + Dx + Ey + F = 0$, $A \neq 0$), by completing the square method.

5. Determine the center and radius of a circle from its equation in standard form or in general form.

SOLUTIONS to Exercise Set. 8.2

2. From the standard form of the equation of the circle we have, for $(h, k) = (-3, 0)$ and $r = 3$,

$$[x - (-3)]^2 + (y - 0)^2 = 3^2 \quad \text{or} \quad x^2 + 6x + 9 + y^2 = 9$$

or finally

$$x^2 + 6x + y^2 = 0$$

5. Similarly, for $(h, k) = (0, 0)$, $r = 3$, we obtain

$$(x - 0)^2 + (y - 0)^2 = 3^2 \quad \text{or} \quad x^2 + y^2 = 9$$

8. Also, for $(h, k) = (2, 2)$, $r = 2$, we obtain

$$(x - 2)^2 + (y - 2)^2 = 4$$

11. Given the standard form of the equation of $(x - h)^2 + (y - k)^2 = r^2$ we can read directly from the problem, $(x - 2)^2 + (y + 2)^2 = 4$, that $h = 2$, $k = -2$, $r = 2$. Thus the center is $(2, -2)$ and the radius is 2.

14. Given $x^2 + (y - 2)^2 = 4$. To see this one, think of x^2 as $(x - 0)^2$. Then $h = 0$, $k = 2$, and $r = 2$; the center is $(0, 2)$, the radius 2.

17. Rearrange the given equation to read $x^2 + 4x + 4 + y^2 - 8y = 0$. Next, complete the square on the y variable, obtaining this equation:

$$x^2 + 4x + 4 + y^2 - 8y + 16 = 16$$

Then rewrite as

$$(x + 2)^2 + (y - 4)^2 = 4^2$$

The center is $(-2, 4)$, the radius is 4.

20. Dividing the given equation by 2 and rearranging, we obtain

$$x^2 + 4x + y^2 - 6y = 4$$

Now complete the square:

$$x^2 + 4x + 4 + y^2 - 6y + 9 = 4 + 4 + 9$$

and so

$$(x + 2)^2 + (y - 3)^2 = 17$$

The center is $(-2, 3)$; the radius $\sqrt{17}$.

23. Dividing by 3 and rearranging yields $x^2 - 4x + y^2 + 6y = 5$, and completing the square gives

$$x^2 - 4x + 4 + y^2 + 6y + 9 = -5 + 4 + 9$$

Then $(x - 2)^2 + (y + 3)^2 = 8$; the center is $(2, -3)$, the radius $\sqrt{8} = 2\sqrt{2}$.

26. Rearranging the given equation yields $x^2 - 4x + y^2 - 6y = 13$ and completing the square gives us

$$x^2 - 4x + 4 + y^2 - 6y + 9 = 13 + 4 + 9$$

or $(x - 2)^2 + (y - 3)^2 = 26$. This is thus a circle.

31. Divide by 2 and rearrange to get $x^2 - 3x + y^2 - 2y = 1$; completing the square gives

$$x^2 - 3x + \frac{9}{4} + y^2 - 2y + 1 = 1 + \frac{9}{4} + 1$$

and then

$$\left(x - \frac{3}{2}\right)^2 + (y - 1)^2 = \frac{17}{4}$$

which is a circle.

34. Rearrange to read $x^2 + x + y^2 + y = 0$; then

$$x^2 + x + \frac{1}{4} + y^2 + y + \frac{1}{4} = \frac{1}{4} + \frac{1}{4}$$

and thus

$$\left(x + \frac{1}{2}\right)^2 + \left(y + \frac{1}{2}\right)^2 = \frac{1}{2}$$

which is a circle.

Section 8.3 The Parabola

BASIC CONCEPTS you should learn from this unit.

1. Definition of a parabola as the set of all points equidistant from a point and a line, the given point being called the focus and the line being called the directrix.

2. The axis of the parabola is the line through the focus and perpendicular to the directrix.

3. The vertex of the parabola is the point where the axis intersects the parabola.

4. If the distance from the vertex to the focus of a parabola is p units and the vertex is at the origin, then the equation of the parabola is

   A. $x^2 = 4py$ and the graph has y-axis symmetry, or

   B. $y^2 = 4px$ and the graph has x-axis symmetry.

   For simplicity we let $a = 4p$; then the equations become $x^2 = ax$ and $y^2 = ay$.

5. If $a > 0$ in the equations in concept 4, the graph opens upward for $x^2 = ay$ and to the right for $y^2 = ax$. When $a < 0$, the directions of opening are downward and to the left, respectively.

6. The standard forms of the equations of the parabola generally are

   A. $(x - h)^2 = a(y - k)$
   in which case the vertex is (h, k), axis $x = h$, and the direction is upward if $a > 0$, downward if $a < 0$; and

   B. $(y - k)^2 = a(x - h)$
   in which case the vertex is (h, k), axis $y = k$, and the direction is to the right if $a > 0$, to the left if $a < 0$.

7. The method of completing the square can be used to write the equation of a parabola in its standard form.

WHAT YOU SHOULD KNOW HOW TO DO upon completion of this unit.

1. Define the terms parabola, vertex, focus, axis, and directrix.

2. Rewrite the equation of a parabola into the standard form.

3. From the standard form of the parabola equation, determine the vertex, axis, direction, focus, and directrix.

4. Graph a parabola from its equation.

SOLUTIONS to Exercise Set 8.3

2. $x^2 = -4y$

a < 0; the direction is downward;

the axis is the y-axis.

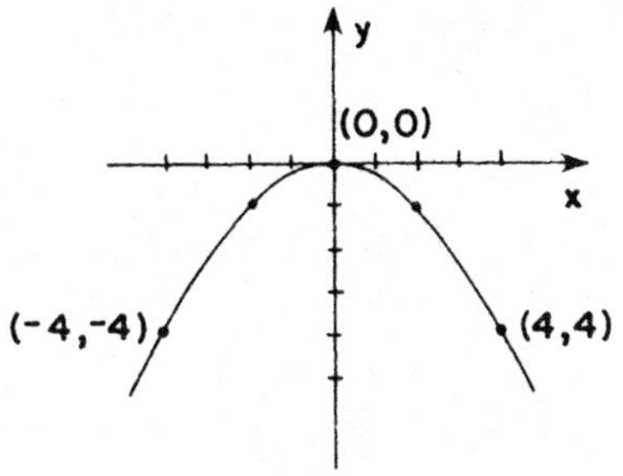

5. $x^2 + 5y = 0$

$x^2 = -5y$

A < 0; the direction is downward;

the axis is the y-axis.

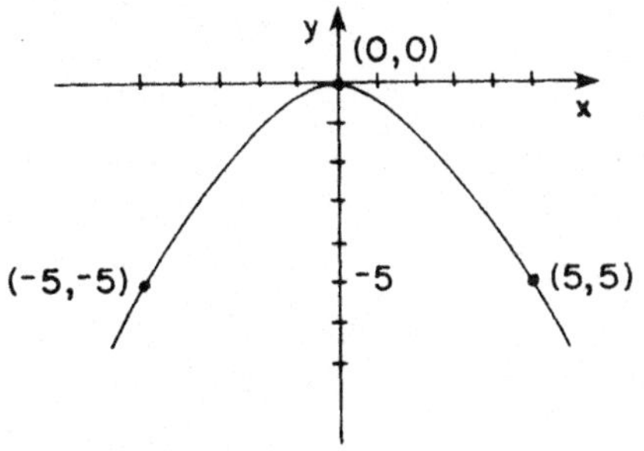

8. $(y - 2)^2 = -2(x + 1)$

The vertex is at (-1, 2);

a < 0; the direction is to the left;

the axis y = 2 .

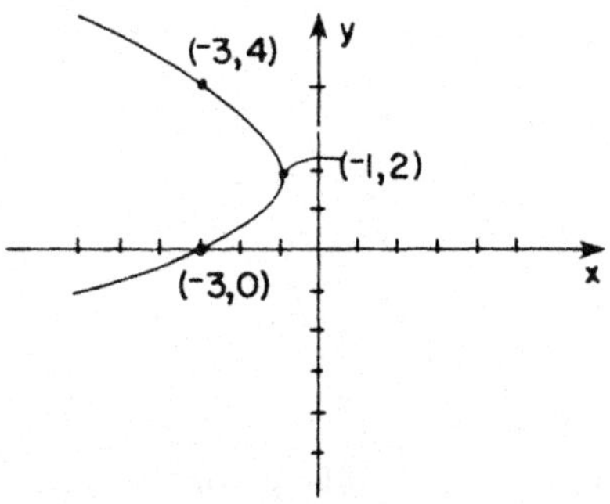

11. $y^2 = -2(x + 1)$

The vertex is at (-1, 0);

the axis is the x-axis;

a < 0, so the direction is to the left.

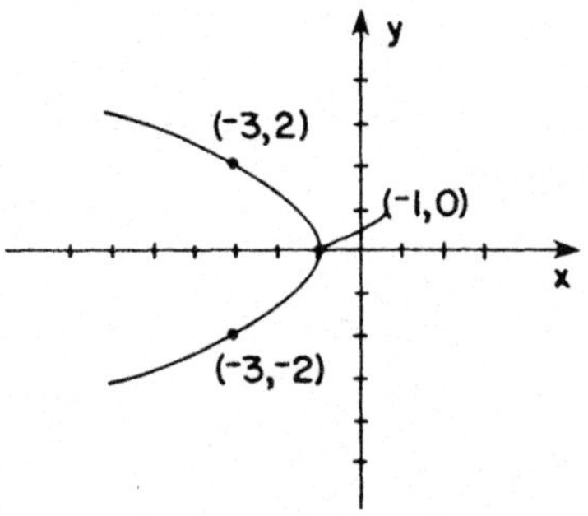

14. $x^2 + 4x + 2y - 2 = 0$. Rearranging and completing the square, we have $x^2 + 4x + 4 = -2y + 2 + 4$; $(x + 2)^2 = -2(y - 3)$. From this we determine the vertex to be $(-2, 3)$, the axis to be $x = -2$, and the direction to be downward.

17. $x^2 - x + 3y + 1 = 0$. Rearranging and completing the square yields $x^2 - x + 1/4 = -3y - 1 + 1/4$; $(x - 1/2)^2 = -3(y + 1/4)$. The vertex is at $(1/2, -1/4)$, the axis $x = 1/2$, and the direction is down.

20. $x^2 + 2x - 5y - 19 = 0$. Rearranging and completing the square yields $x^2 + 2x + 1 = 5y + 20$ or $(x + 1)^2 = 5(y + 4)$. The vertex is at $(-1, -4)$, the axis $x = -1$, and the direction is up.

23. $y^2 + 6y + \frac{1}{2}x + 7 = 0$. Rearranging and completing the square yields $y^2 + 6y + 9 = -\frac{1}{2}x - 7 + 9$ or $(y + 3)^2 = -\frac{1}{2}(x - 4)$. The vertex is at $(4, -3)$, the axis $y = -3$, and the direction is to the left.

26. $y^2 - 6y + 2x + 17 = 0$. Rearranging and completing the square yields $y^2 - 6y + 9 = -2x - 17 + 9$ or $(y - 3)^2 = -2(x + 4)$. The vertex is at $(-4, 3)$, the axis $y = 3$; the graph opens to the left.

Section 8.4 The Ellipse and Hyperbola

BASIC CONCEPTS you should learn from this unit.

1. Definition of an <u>ellipse</u> as the set of all points the sum of whose distances from two fixed points is a constant, the two points being called the <u>foci</u>.

2. The standard form of the equation of an ellipse that is centered at the origin is

$$\frac{x^2}{a^2} + \frac{y^2}{b^2} = 1$$

with $\pm a$ being the x-intercepts and $\pm b$ being the y-intercepts.

3. Definition of a <u>hyperbola</u> as the set of all points the difference of whose distances from two fixed points is a constant, the two points being called the <u>foci</u>.

4. The standard forms of the equation of a hyperbola that is centered at the origin are

A. $\frac{x^2}{a^2} - \frac{y^2}{b^2} = 1$

(In this case the foci are on the x-axis and x-intercepts are $\pm a$);

B. $\frac{y^2}{a^2} - \frac{x^2}{b^2} = 1$

(In this case the foci are on the y-axis and y-intercepts are $\pm a$).

5. The asymptotes of a hyperbola of the form in concept 4A are lines with equations

$$y = \pm\frac{b}{a}x$$

The asymptotes of a hyperbola of the form in concept 4B are lines with equations

$$y = \pm\frac{a}{b}x$$

WHAT YOU SHOULD KNOW HOW TO DO upon completion of this unit.

1. Recognize the equation of an ellipse in its standard form.
2. Graph an ellipse from its equation.
3. Rewrite the equation of an ellipse into the standard form.
4. Recognize the equation of a hyperbola in the standard form.
5. Graph a hyperbola from its equation, using intercepts.
6. Write the equations of the asymptotes of a hyperbola and use them in graphing the curve.
7. Rewrite the equation of a hyperbola into the standard form.

SOLUTIONS to Exercise Set 8.4

2. $\frac{x^2}{4} + \frac{y^2}{16} = 1$ (2, 0), (-2, 0) are the x-intercepts and (0, 4), (0, -4) the y-intercepts.

The graph is

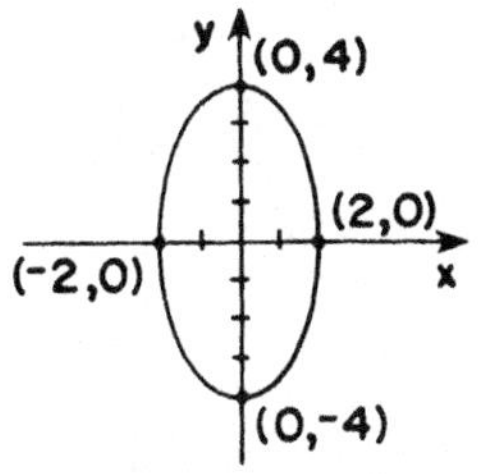

5. $\frac{x^2}{16} + \frac{y^2}{25} = 1.$ (4, 0), (-4, 0) are the x-intercepts and (0, 5), (0, -5) the y-intercepts;

the graph is

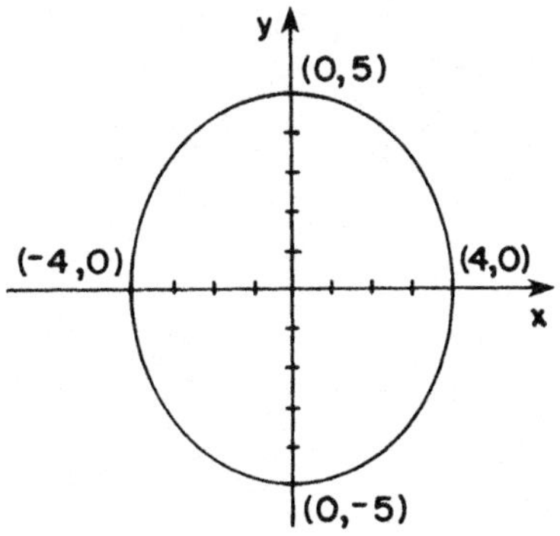

8. $16x^2 + 9y^2 = 144$. Dividing by 144 yields the standard form

$$\frac{x^2}{9} + \frac{y^2}{16} = 1,$$

and from this we see the x-intercepts, (3, 0), (-3, 0), and the y-intercepts, (0, 4), (0, -4).

11. $4x^2 + 16y^2 = 4$. Dividing by 4 gives us the standard form

$$\frac{x^2}{1} + \frac{y^2}{\frac{1}{4}} = 1$$

The x-intercepts are (1, 0), (0, -1); the y-intercepts are $\left(0, \frac{1}{2}\right)$, $\left(0, -\frac{1}{2}\right)$.

14. $5x^2 + 6y^2 = 50$. Dividing by 50 and rewriting $\frac{6}{50}$ as $\frac{25}{3}$ in the denominator yields the standard form,

$$\frac{x^2}{10} + \frac{y^2}{\frac{25}{3}} = 1$$

and thus the x-intercepts are $(\sqrt{10}, 0)$, $(-\sqrt{10}, 0)$ and the y-intercepts are $\left(0, \frac{5\sqrt{3}}{3}\right)$, $\left(0, \frac{-5\sqrt{3}}{3}\right)$.

17. $\frac{x^2}{25} - \frac{y^2}{16} = -1$ rewritten in standard form is

$$\frac{y^2}{16} - \frac{x^2}{25} = 1$$

and thus the intercepts are (0, 4) and (0, -4). The graph is thus

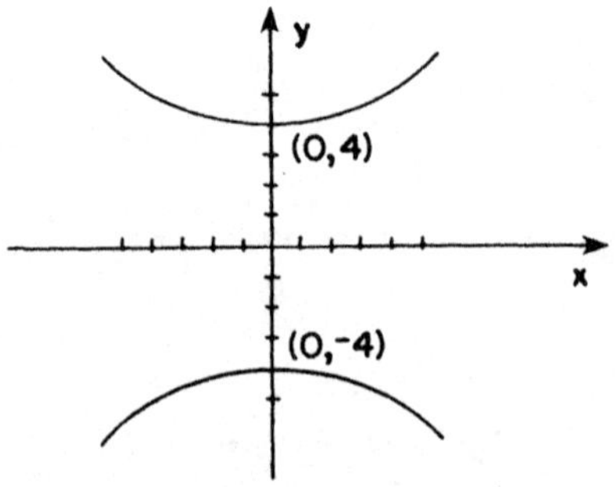

20. $$\frac{y^2}{49} - \frac{x^2}{25} = 1$$

is already in standard form and the intercepts are (0, 7), (0, -7). The graph is

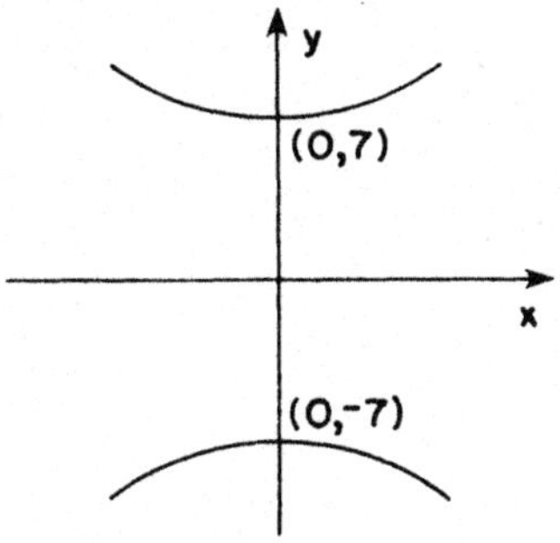

23. $16x^2 - y^2 = 64$. Dividing by 64 yields

$$\frac{x^2}{4} - \frac{y^2}{64} = 1$$

The intercepts are (2, 0), (-2, 0).

26. $2x^2 - 3y^2 = 6$. Dividing by 6 yields

$$\frac{x^2}{3} - \frac{y^2}{2} = 1$$

The intercepts are $(\sqrt{3}, 0)$, $(-\sqrt{3}, 0)$.

29. $16x^2 - 9y^2 = 144$. First divide by 144 to put the equation in standard form:

$$\frac{x^2}{9} - \frac{y^2}{16} = 1$$

with the $x^2/9$ term positive, the asymptotes are

$$y = \pm\frac{b}{a}x \quad \text{or} \quad y = \frac{4}{3}x \quad \text{and} \quad y = -\frac{4}{3}x$$

and the graph is

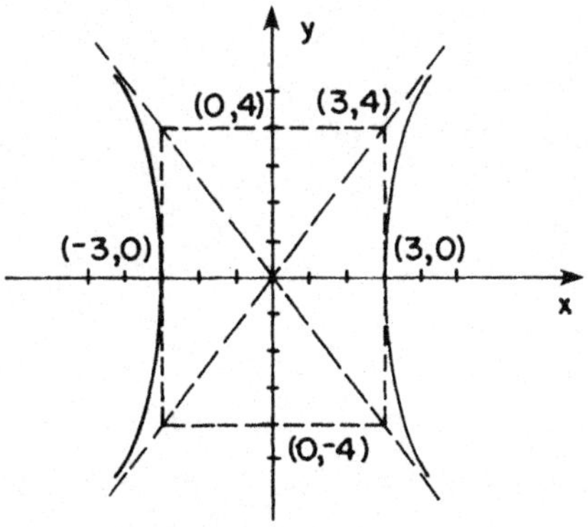

32. $25x^2 - 9y^2 = 225$ Divide by 225

$$\frac{x^2}{9} - \frac{y^2}{25} = 1$$

The asymptotes will be

$$y = \pm\frac{b}{a}x \quad \text{or} \quad y = \pm\frac{5}{3}x$$

The graph is

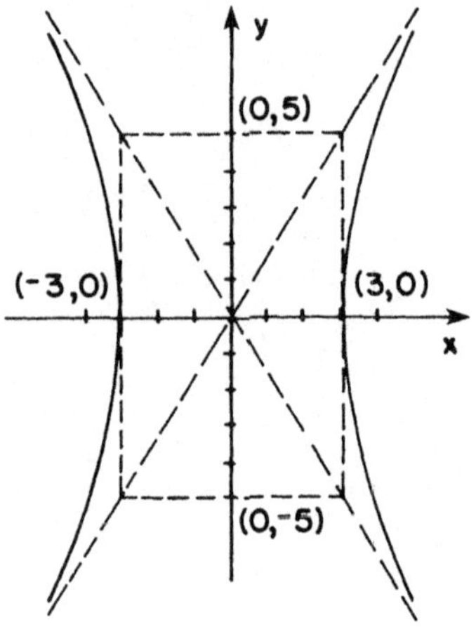

35. Referring to the figure

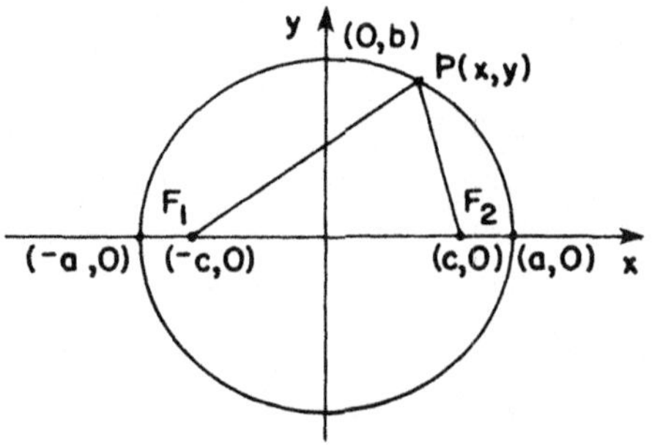

we see that $F_1P + PF_2 = 2a$. By the distance formula this relation can be written as

$$\sqrt{(x + c)^2 + y^2} + \sqrt{(x - c)^2 + y^2} = 2a$$

or

$$\sqrt{(x + c)^2 + y^2} = 2a - \sqrt{(x - c)^2 + y^2}$$

Then, squaring, we obtain

$$x^2 + 2xc + c^2 + y^2 = 4a^2 - 4a\sqrt{(x - c)^2 + y^2} + x^2 - 2xc + c^2 + y^2$$

or $4xc - 4a^2 = -4a\sqrt{(x - c)^2 + y^2}$. Dividing by 4 gives $xc - a^2 = -a\sqrt{(x - c)^2 + y^2}$,

squaring again, we get

$$x^2c^2 - 2a^2xc + a^4 = a^2x^2 - 2a^2xc + a^2c^2 + a^2y^2$$

35. (continued)

or $x^2(a^2 - c^2) + a^2y^2 = a^2(a^2 - c^2)$. Dividing by $a^2(a^2 - c^2)$ yields

$$\frac{x^2}{a^2} + \frac{y^2}{a^2 - c^2} = 1$$

Now substitute $b^2 = a^2 - c^2$ to get the standard form:

$$\frac{x^2}{a^2} - \frac{y^2}{b^2} = 1$$

Section 8.5 Identifying the Conic Sections

BASIC CONCEPTS you should learn from this unit.

1. There are two basic forms of the equation of a conic section:

   A. The general equation

   $$Ax^2 + Cy^2 + Dx + Ey + F = 0$$

   B. The standard form equation, which is obtained from the foregoing by completing the square.

2. The various conic sections can be identified from the general equation as follows.

   A. Either A or $C = 0$ (not both): Parabola.

   B. $A = C$: Circle.

   C. $A \neq C$ and $AC > 0$: Ellipse.

   D. $A \neq C$ and $AC < 0$: Hyperbola.

3. When the equation is in standard form much valuable information about the conic sections is seen by inspection: the location of the center or vertex; the location of the intercepts; the axis; the direction of opening; and the asymptotes. (See Table 2 in the text for details.)

WHAT YOU SHOULD KNOW HOW TO DO upon completion of this unit.

1. Identify a conic section as a circle, parabola, ellipse, or hyperbola by analyzing the general equation.

SOLUTIONS to Exercise Set 8.5

2. $-x^2 + 4y^2 + 2x - 3y + 5 = 0$. Here $A \neq C$ and $AC = -5 < 0$, so the conic section is a hyperbola.

5. $36x^2 - y^2 + x - y + 2 = 0$. Here, too, $A \neq C$ and $AC < 0$, so the conic section is a hyperbola.

8. $2y^2 - 3x + y + 4 = 0$. $A = 0$, so the conic section is a parabola.

11. $4x^2 + 9y^2 - x + 2 = 0$. $A \neq C$ and $AC > 0$. The conic section might be an ellipse, but completing the square reveals that

$$\left(x^2 - \frac{1}{4}x + \frac{1}{64}\right) + \frac{9}{4}y^2 = -\frac{1}{2} + \frac{1}{64} = -\frac{31}{64}$$

and since $-\frac{31}{64}$ is nonpositive, we have no graph.

14. $x^2 + y^2 + 6x - 2y + 10 = 0$. $A = C$; we might have a circle. Check by completing the square:

$$(x^2 + 6x + 9) + (y^2 - 2y + 1) = -10 + 9 + 1 = 0$$

So we have a circle with zero radius or <u>a point</u>. The point is the center: $(-3, 1)$.

SOLUTIONS to Review Exercises Chapter 8

2. Using the midpoint formula,

$$x_m = \frac{-2 - 3}{2} = -\frac{5}{2} \quad \text{and} \quad y_m = \frac{0 + 5}{2} = \frac{5}{2}$$

The midpoint is $\left(-\frac{5}{2}, \frac{5}{2}\right)$.

5. $P_1P_2 = \sqrt{(3 - 2)^2 + (4 + 1)^2} = \sqrt{26}$, $\quad P_2P_3 = \sqrt{(-1 - 3)^2 + (1 - 4)^2} = 5$

$P_3P_4 = \sqrt{(-2 + 1)^2 + (-4 - 1)^2} = \sqrt{26}$, $\quad P_4P_1 = \sqrt{(-2 - 2)^2 + (-4 + 1)^2} = 5$

Since opposite sides are equal, the quadrilateral is a parallelogram.

8. $y^2 = x^2 - 9$. Substituting $-x$ for x and $-y$ for y leaves the equation unchanged, so we have symmetry with both axes and the origin.

11. $3x^2 - 2x - 4y^2 = 6$. Replacing x by $-x$ changes the $-2x$ term to $+2x$, so there is no symmetry with the y-axis. But replacing $-y$ for y does not change the equation, so there is symmetry with the x-axis.

14. The standard form of the circle equation is $(x - h)^2 + (y - k)^2 = r^2$. Substituting the given values yields $(x + 3)^2 + (y + 3)^2 = 4$.

17. $x^2 + y^2 + 4x - 6y = -10$. Completing the square gives

$$x^2 + 4x + 4 + y^2 - 6y + 9 = -10 + 4 + 9$$

or

$$(x + 2)^2 + (y - 3)^2 = 3$$

from which we read the center to be $(-2, 3)$ and the radius $= \sqrt{3}$.

20. $x^2 + y^2 - 2x - 2y = -8$. Completing the square gives

$$x^2 - 2x + 1 + y^2 - 2y + 1 = -8 + 1 + 1$$

or

$$(x - 1)^2 + (y - 1)^2 = -6$$

which is impossible.

23. $y^2 + 3x + 9 = 0$ in standard form is $y^2 = 3(x + 3)$. The vertex thus is $(-3, 0)$, the axis is $y = 0$, and since $a = +3$, the parabola opens to the right.

26. $x^2 + 4x + 2y + 5 = 0$. Completing the square and rearranging gives $x^2 + 4x + 4 = -2y - 5 + 4$ or $(x + 2)^2 = -2(y + 1/2)$. Thus the vertex is $(-2, -1/2)$, the axis is $x = -2$, and $a = -2$, which is less than 0, so the direction of the opening is downward.

29. $9x^2 - 4y^2 = 36$ in standard form is

$$\frac{x^2}{4} - \frac{y^2}{9} = 1$$

This hyperbola intercepts the x-axis at $(\pm 2, 0)$.

32. $9x^2 - 16y^2 = 144$ in standard form is

$$\frac{x^2}{16} - \frac{y^2}{9} = 1$$

This hyperbola intercepts the x-axis at $(\pm 4, 0)$.

35. $4x^2 - 4y^2 = 1$. Rearrange to

$$\frac{x^2}{1/4} - \frac{y^2}{1/4} = 1$$

to see that $a^2 = (1/2)^2$ and $b^2 = (1/2)^2$. Intercepts are found by setting $y = 0$; then $x = \pm 1/2$, and the intercepts are at $(\pm 1/2, 0)$. The asymptotes are $y = \pm b/a\,x$ or $y = \pm x$.

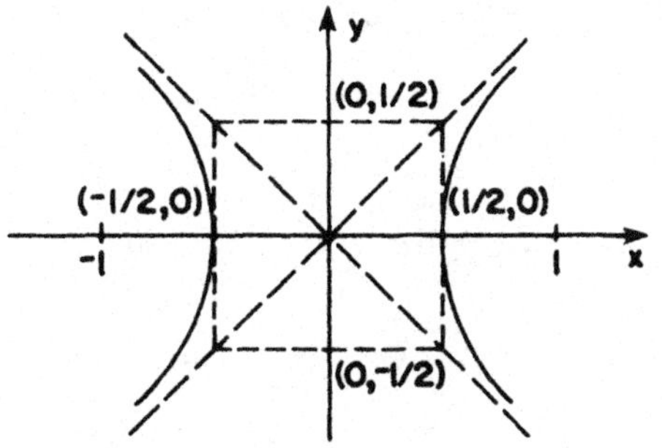

38. $6x^2 - 7y^2 - 5x + 6y = 0$. Using the criteria in Section 8.5, we see that $A = 6$ and $C = -7$, so $A \neq C$ and $AC < 0$; hence the conic section is a hyperbola.

Chapter 8 - A SHORT QUIZ

1. State the midpoint formula.

2. $(x - h)^2 + (y - k)^2 = r^2$ is the standard form of which conic section?

3. In Question 2, what are h, k, and r ?

4. Which conic section is represented by

$$\frac{x^2}{a^2} - \frac{y^2}{b^2} = 1 \quad ?$$

5. In Question 4, which axis is intersected?

6. $(y - 2)^2 = -(x - 3)$ is the equation of a ____?____ whose axis is parallel to which axis? (x-axis or y-axis)

7. Write the equation of a hyperbola in standard form that intercepts the y-axis at $(0, \pm 2)$.

8. Write an equation of a parabola that is symmetric to the y-axis and opens downward with its vertex at $(0, 1)$.

9. Which axis is intersected by $x^2 = 4(y - 2)$?

10. What is the direction of opening of the graph of the equation in Question 9?

11. What are the x- and y-intercepts of

$$\frac{x^2}{2} + \frac{y^2}{9} = 1 \quad ?$$

12. What type of symmetry is present in the graph of Question 11?

13. What is the relationship between a and b in the standard form of the ellipse equation if the ellipse is centered at the origin and has its longer axis along the y-axis?

14. True or False: The focus of a parabola lies on the directrix.

15. True or False: All points on a parabola are equidistant from the focus and the directrix.

16. True or False: The axis of a parabola is a line through the vertex and parallel to the directrix.

17. Write the equation of the directrix for this parabola: $y^2 = -4(x - 1)$.

18. In the standard form of the equation of a parabola, $(x - h)^2 = a(y - k)$, what do h and k represent?

19. Where does a hyperbola intersect its asymptotes?

20. Write the equation of a conic section such that the graph will be a single point.

Quiz Answers Chapter 8

1. $x_m = \dfrac{x_1 + x_2}{2}$; $y_m = \dfrac{y_1 + y_2}{2}$

2. Circle

3. (h, k) are the coordinates of the center, r is the radius.

4. Hyperbola

5. x-axis

6. Parabola; x-axis

7. $\frac{y^2}{4} - \frac{x}{7} = 1$

8. $x^2 = (y - 1)$

9. y-axis

10. upward

11. The x-intercepts are $(\pm\sqrt{2}, 0)$; the y-intercepts are $(0, \pm 3)$.

12. Both axes and origin

13. $b > a$

14. False

15. True

16. False

17. $x = 2$

18. Coordinates of the vertex

19. It does not intersect the asymptotes.

20. $(x - 2)^2 + y^2 = 0$ (Any circle with radius = 0 will do.)

CHAPTER 9

Systems of Equations and Inequalities

Section 9.1 Systems of Linear Equations

BASIC CONCEPTS you should learn from this unit.

1. A system of linear equations is a pair of linear equations with two unknowns.

2. Solving a system of equations means finding values of the unknowns that satisfy both equations.

3. A system that has a solution is said to be consistent. A system with no solution is inconsistent.

4. A system can be solved by graphing, in which case the solution is the coordinates of the point where the two lines intersect.

5. There are three possible results when two linear equations are graphed:

   A. Two different lines intersect once;

   B. Two parallel lines are obtained;

   C. Both equations graph onto the same line.

6. The substitution method of solving a system yields an exact solution, whereas graphing is approximate.

7. If the substitution method results in a mathematical inconsistency, the system has no solution.

8. If the method results in a mathematical identity, the system has many solutions.

WHAT YOU SHOULD KNOW HOW TO DO upon completion of this unit.

1. Define and recognize a system of linear equations.

2. Solve a system of two linear equations by graphing, and algebraically by substitution.

3. Define the terms "consistent system" and "inconsistent system."

4. Illustrate graphically a consistent system and an inconsistent system.

SOLUTIONS to Exercise Set 9.1

2. (1) $x - y = 1$

   (2) $x + y = 5$

(1)

x	y
1	0
3	2

(2)

x	y
1	4
4	1

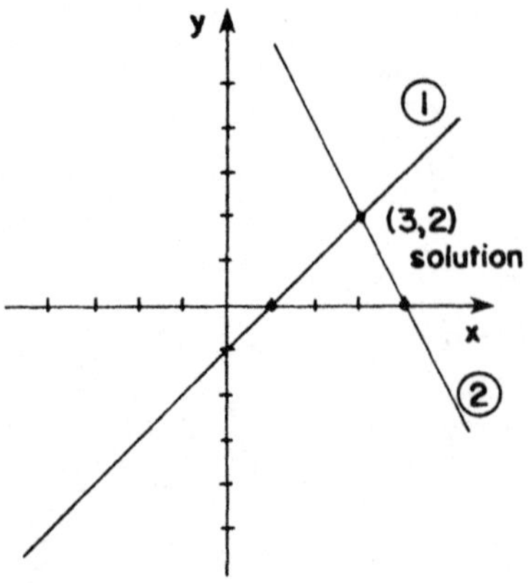

5. (1) $x + 3y = -2$

(2) $3x - 5y = 8$

(1)

x	y
-2	0
1	-1

(2)

x	y
1	-1
-4	-4

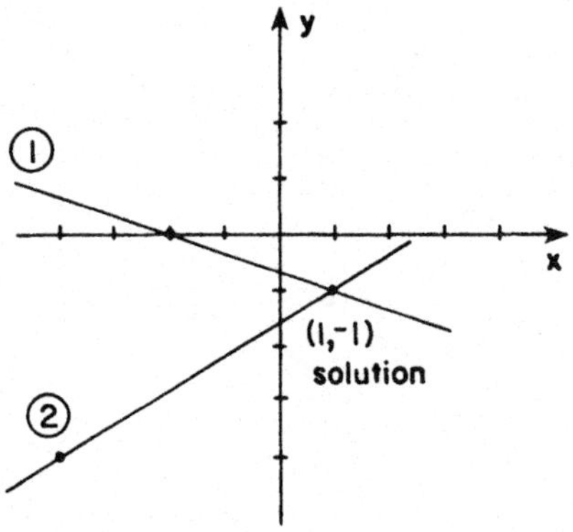

8. (1) $3x + 2y = 10$

(2) $-9x - 6y = 8$

(1)

x	y
0	5
4	-1

(2)

x	y
0	-4/3
-2	5/3

Lines appear parallel.

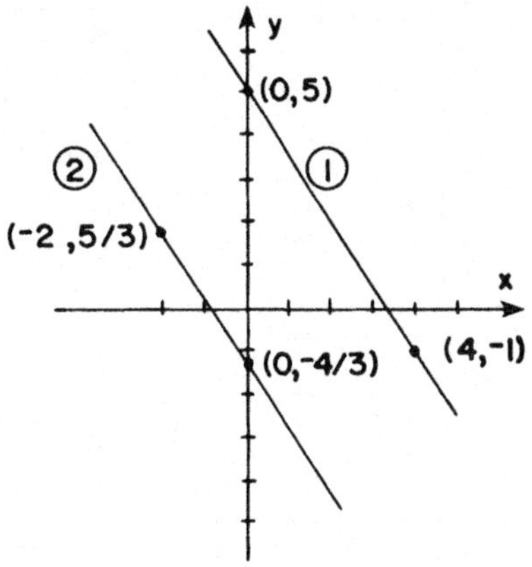

11. (1) $x + 3y = 3$

(2) $x - 5y = -1$

From (1) $x = -3y + 3$

Now substitute in (2), obtaining

$$(-3y + 3) - 5y = -1$$
$$-8y = -4$$
$$y = 1/2$$

$$x = -3y + 3$$
$$x = -3\left(\frac{1}{2}\right) + 3$$
$$x = 3/2$$

14. (1) $3x - y = -9$

(2) $2x - y = -7$

From (1) $y = 3x + 9$

Now substituting in (2) we get

$$2x - (3x + 9) = -7$$
$$-x - 9 = -7$$
$$x = -2$$

$$y = 3(-2) + 9$$
$$y = 3$$

17. (1) $3x - y = 18$ From (1) $y = 3x - 18$

(2) $\frac{3}{2}x - \frac{1}{2}y = 9$ and substituting in (2) yields

$$\frac{3}{2}x - \frac{1}{2}(3x - 18) = 9$$

$$3x - (3x - 18) = 18$$

$0 = 0$ an identity, thus any solution of (1) is a solution of (2) .

Section 9.2 Solving by Elimination

BASIC CONCEPTS you should learn from this unit.

1. The method of solving a system of two linear equations by elimination consists of manipulating the coefficients of the variables so that when the two equations are added one variable vanishes, and the resulting equation has only one variable and is easily solved.

2. If the method results in $0x + 0y = C$, $c \neq 0$, there is an inconsistent system. (no solution).

3. If the method results in $0x + 0y = 0$, then the two equations were essentially the same and there are many solutions.

WHAT YOU SHOULD KNOW HOW TO DO upon completion of this unit.

1. Solve a pair of linear equations by the method of elmination.

2. Detect an inconsistent system by the result of this method.

3. Determine that the two equations were essentially the same by the result of this method.

SOLUTIONS to Exercise Set 9.2

2.
$$x - 2y = 8$$
$$2x + y = 1$$

Multiply the second equation by 2, obtaining a new system:

$$x - 2y = 8$$
$$4x + 2y = 2$$

Adding gives $5x = 10$ and $x = 2$. Now substituting 2 for x in the first equation gives $2 - 2y = 8$ and $y = -3$.

5. $$x + 2y = 6$$
$$2x + 4y = 12$$

Multiply the first equation by -2 to obtain a new system, then add

$$\begin{array}{r} -2x - 4y = -12 \\ \underline{2x + 4y = 12} \\ 0x + 0x = 0 \end{array}$$

which is always true. Thus there are many solutions.

8. $$x + 2y = 0$$
$$5x - y = 22$$

Multiply the second equation by 2, obtaining a new system. Then add.

$$\begin{array}{r} x + 2y = 0 \\ \underline{10x - 2y = 44} \\ 11x \qquad = 44 \end{array}$$

Hence $x = 4$ and substituting 4 for x in the first equation gives $y = -2$.

11. $$x + 2y = 1$$
$$5x + 2y = 13$$

Multiply the first equation by -1, obtaining a new system. Then add.

$$\begin{array}{r} -x - 2y = -1 \\ \underline{5x + 2y = 13} \\ 4x \qquad = 12 \end{array} ; \quad x = 3 .$$

Substituting gives $y = -1$.

14. $$2x - 3y = 8$$
$$4x - 6y = 16$$

Multiply the first equation by -2, obtaining a new system. Then add.

$$\begin{array}{r} -4x + 6y = -16 \\ \underline{4x - 6y = 16} \\ 0x + 0y = 0 \end{array}$$

which is an identity. So there are many solutions.

17. Let x = number of pounds of nuts
y = number of pounds of raisins

$x + y = 2$ (We want 2 lb of mix.)

$2.10x + .90y = 2(1.62)$ value of the nuts

plus the value of the raisins equals the value of the mix. Multiply the first equation by -90 and the second equation by 100; we obtain

$$\begin{array}{rl} -90x - 90y &= -180 \\ 210x + 90y &= 324 \\ \hline 120x &= 144 \end{array}$$

Add these equations.

$$x = 144/120 = 6/5 \text{ lb}$$

and substituting gives $6/5 + y = 2$, $y = 4/5$ lb. 6/5 lb of nuts and 4/5 lb of raisins.

20. Let x = the amount invested at 12%, and
y = the amount invested at 10%

$x + y = 6000$ Sum of the two investments = \$6000.

$0.12x + 0.10y = 620$ Sum of the interests = \$620

Multiply the first equation by -10 and the second by 100 to obtain

$$\begin{array}{rl} -10x - 10y &= -60000 \\ 12x + 10y &= 62000 \\ \hline 2x &= 2000 \end{array}$$

$$x = 1000$$

and clearly $y = \$5000$.

Section 9.3 Applications

BASIC CONCEPTS YOU SHOULD LEARN from this unit.

1. Word problems can often be set up with two equations with two unknowns and then solved.
2. The BREAK-EVEN point in a business is when the number of items sold produces income or revenue exactly equal to their cost.
3. Revenue and costs are often depicted by algebraic equations.
4. In an economic system the equilibrium point is the solution of the demand equation and the supply equation. The ability tc solve a system of equations, then, assists us in determining the equilibrium point.

WHAT YOU SHOULD KNOW HOW TO DO upon completion of this unit.

1. Set up and solve word problems using two unknowns and forming two equations.
2. Determine the break-even point in a manufacturing and selling type problem.
3. Determine the equilibrium point in a supply and demand type problem.

SOLUTIONS to Exercise Set 9.3

2. Let x = the number of ten-cent stamps
y = the number of twenty-cent stamps

Then:

$$x + y = 27$$
$$10x + 20y = 300$$

Multiply the first equation by -10 to obtain

$$\begin{aligned} -10x - 10y &= -270 \\ \underline{10x + 20y} &= \underline{300} \\ 10y &= 30 \\ y &= 3 \end{aligned}$$

Add these equations.

and so $x = 24$.

Check: $3 + 24 = 27$

and $10(24) + 20(3) = 300$

5. Let A = the amount invested in Type A
B = the amount invested in Type B

Then $A + B = 6000$ Sum of the investments = \$6000.

$.08A + .10B = 520$ Sum of the dividends = \$ 520.

$$\begin{aligned} -8A - 8B &= -48000 && \text{Multiplying by } -8 \\ \underline{8A + 10B} &= \underline{52000} && \text{Multiplying by } 100 \\ 2B &= 4000 \\ B &= \$\,2000 \end{aligned}$$

Then $A = \$4000$.

8. Let a = the number of units of grain A
b = the number of units of grain B

$2a + 3b = 18$ fat equation

$80a + 60b = 480$ calories equation

$$\begin{aligned} 2a + 3b &= 18 \\ \underline{-4a - 3b} &= \underline{-24} \\ -2a &= -6 \end{aligned}$$

$$\begin{aligned} 6 + 3b &= 18 \\ 3b &= 12 \\ b &= 4 \end{aligned}$$

$a = 3$ units of grain A

$b = 4$ units of grain B

11. Let x = the speed of the bike in still air
y = the speed of the wind

Using the formula $d = rt$ we get

$4(x - y) = 45$ and

$3(x + y) = 45$

or $4x - 4y = 45$ or $12x - 12y = 135$

$3x + 3y = 45$ $\quad$ $12x + 12y = 180$

$24x = 315$

and $x = \frac{315}{24} = \frac{105}{8}$ mph

and since

$3x + 3y = 45$ or $x + y = 15$, $y = 15 - \frac{105}{8} = \frac{15}{8}$

14. Let t = the tens digit
u = the units digit

Then $3t + u = 14$

$t + 2u = 18$

will be the equations, or

$-6t - 2u = -28$

$t + 2u = 18$

$-5t = -10$

and $t = 2$, and from the first equation $6 + u = 14$ and $u = 8$. Thus the number is 28.

17. Let a = the amount invested in model A
b = the amount invested in model B

Then $a + b = 18000$

$.12a + .18b = (.16)(18000)$

are the equations; or

$a + b = 18000$ or $-2a - 2b = -36000$

$2a + 3b = 48000$ $\quad$ $2a + 3b = 48000$

$b = \$12000$

Then the first equation yields $a = \$\ 6000$

20. Graphing each equation yields

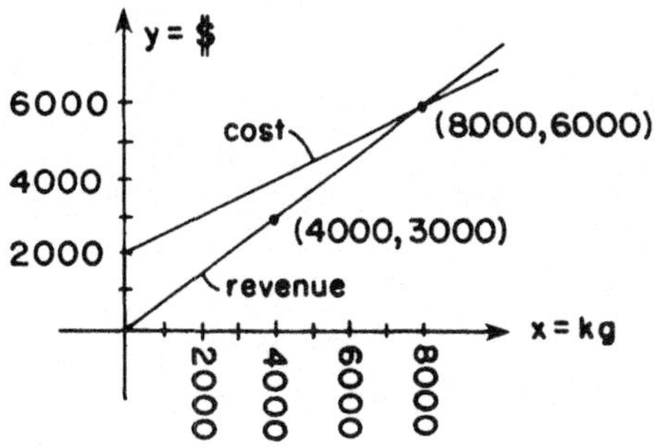

(a) Break-even point occurs at 8000 kg.

(b) Revenue for 8000 kg: $R = .75(8000) = \$6000$

23. a. Equilibrium occurs when $S = D$ or in this case $2p + 10 = -p + 22$. Solving this for p yields $p = 4$.

b. The number sold at this price can be found from either equation: $s = 2p + 10 = 18$.

Section 9.4 Systems of Linear Equations in Three Unknowns

BASIC CONCEPTS you should learn from this unit.

1. Three linear equations in three unknowns can be solved by the substitution method and by the elimination method.

2. Gaussian elimination is another method of solving systems of linear equations. The method consists of applying these three operations.

   A. Interchange any two equations.

   B. Multiply an equation by a nonzero constant.

   C. Replace an equation by itself plus a constant times another equation.

3. If Gaussian elimination results in

   A. $0x + 0y + 0z = C,\ C \neq 0$, the system is inconsistent;

   B. $0x + 0y + 0z = 0$, the system is consistent and has many solutions.

WHAT YOU SHOULD KNOW HOW TO DO upon completion of this unit.

1. Find the solutions (if any) to a system of three linear equations by Gaussian elimination.

2. If a system has infinitely many solutions, state the general solution.

3. Recognize when a system has no solution.

SOLUTIONS to Exercise Set 9.4

2. $2x + 3y + 4z = -12$
$x - 2y + z = -5$
$3x + y + 2z = 1$ Interchange the first and second equations, obtaining

$x - 2y + z = -5$
$2x + 3y + 4z = -12$ Add -2 times the first equation to the second equation, and
$3x + y + 2z = 1$ add -3 times the first equation to the third equation, obtaining

$x - 2y + z = -5$
$7y + 2z = -2$ Add -1 times the second equation to the third equation,
$7y - z = 16$ obtaining,

$x - 2y + z = -5$
$7y + 2z = -2$
$-3z = 18$

Now from the third equation $z = -6$, and back substitution yields $x = \frac{27}{7}$, $y = \frac{10}{7}$

5. $x + y + z = 2$
$x - y + 2z = 3$ Add -1 times the first equation to the second equation, and
$3x + 5y + 2z = 6$ -3 times the first equation to the third equation, obtaining

$x + y + z = 2$
$-2y + z = 1$
$2y - z = 0$ Add the second equation to the third equation, obtaining

$x + y + z = 2$
$-2y + z = 1$
$0y + 0z = 1$ an inconsistency: so there is no solution.

8. $4x + 2y - z = 5$
$3x + 3y + 6z = 1$ Add -1 times the second equation to the first equation,
$5x + y - 8z = 8$ obtaining

$x - y - 7z = 4$
$3x + 3y + 6z = 1$ Add -3 times the first equation to the second equation, and
$5x + y - 8z = 8$ -5 times the first equation to the third equation, obtaining

$x - y - 7z = 4$
$6y + 27z = -11$
$6y + 27z = -12$ Add -1 times the second equation to the third equation, obtaining

$x - y - 7z = 4$
$6y + 27z = -11$
$0y + 0z = -1$ an inconsistency: so there is no solution.

11. $2x + y + 3z = 8$

$-x + y + z = 10$

$x + y + z = 12$ Interchange the first and third equations, obtaining

$x + y + z = 12$

$-x + y + z = 10$ Add the first equation to the second equation, and add -2 times

$2x + y + 3z = 8$ the first equation to the third equation, obtaining

$x + y + z = 12$

$2y + 2z = 22$

$-y + z = -16$ Interchange the second and third equations, obtaining

$x + y + z = 12$

$-y + z = -16$ Add 2 times the second equation to the third equation,

$2y + 2z = 22$ obtaining

$x + y + z = 12$

$-y + z = -16$

$4z = -10$

From this $z = -5/2$ and back substitution yields $y = 27/2$, $x = 1$.

14. $2x - y + z = 2$

$3x + y + 2z = 3$

$x + y - z = -1$ Interchange the first and third equations, obtaining

$x + y - z = -1$

$3x + y + 2z = 3$ Add -3 times the first equation to the second equation, and add

$2x - y + z = 2$ -2 times the first equation to the third equation, obtaining

$x + y - z = -1$

$-2y + 5z = 6$ Add -1 times the third equation to the second equation,

$-3y + 3z = 4$ obtaining

$x + y - z = -1$

$y + 2z = 2$ Add 3 times the second equation to the third equation,

$-3y + 3z = 4$ obtaining

$x + y - z = -1$

$y + 2z = 2$

$9z = 10$

$z = 10/9$, $y = -2/9$, $x = 1/3$ by substitution

17. $x - 2y + z = -5$

$2x + z = -10$

$y - z = 15$

Add -2 times the first equation to the second equation, obtaining

$x - 2y + z = -5$

$4y - z = 0$

$y - z = 15$

Interchange the second and third equations, obtaining

$x - 2y + z = -5$

$y - z = 15$

$4y - z = 0$

Add -4 times the second equation to the third equation, obtaining

$x - 2y + z = -5$

$y - z = 15$

$3z = -60$

From this $\underline{z = -20}$ and $\underline{x = 5}$, $\underline{y = -5}$ by back substituting.

20. Let x = number of chairs, y = number of coffee tables, and z = number of dining tables. Then

$10x + 12y + 15z = 780$ sanding time equation

$6x + 8y + 12z = 540$ staining time equation

$12x + 12y + 18z = 900$ varnishing time equation

Divide the second equation by 2 and the third equation by 6 to simplify

$10x + 12y + 15z = 780$

$3x + 4y + 6z = 270$

$2x + 2y + 3z = 150$

Add -1 times the third equation to the second equation, obtaining

$10x + 12y + 15z = 780$

$x + 2y + 3z = 120$

$2x + 2y + 3z = 150$

Interchange the first and second equations, obtaining

$x + 2y + 3z = 120$

$10x + 12y + 15z = 780$

$2x + 2y + 3z = 150$

Add -10 times the first equation to the second equation, and -2 times the first equation to the third, obtaining

$x + 2y + 3z = 120$

$-8y - 15z = -420$

$-2y - 3z = -90$

Interchange the second and third equations. Multiply the second equation by -1, obtaining

$x + 2y + 3z = 120$

$2y + 3z = 90$

$-8y - 15z = -420$

Add 4 times the second equation to the third equation, obtaining

20. Continued

$x + 2y + 3z = 120$ From this $z = 20$ dining tables

$2y + 3z = 90$ $y = 15$ coffee tables

$-3z = -60$ $x = 30$ chairs

x and y obtained by back substitution.

Section 9.5 Systems Involving Non-Linear Equations

BASIC CONCEPTS you should learn from this unit.

1. A system of equations containing either one linear equation and one quadratic equation, or two quadratic equations, can be solved by the substitution method or the elimination method.

2. Considering the graph of a system helps in understanding how many solutions there are.

WHAT YOU SHOULD KNOW HOW TO DO upon completion of this unit.

1. Solve a system of two equations containing either one or two second degree equations.

2. Graph systems of equations containing second degree equations.

3. Translate word problems into systems of equations.

SOLUTIONS to Exercise Set 9.5

2. $x^2 + 4y^2 = 32$ Solve the second equation for $x = -2y$ and substitute into the
$x + 2y = 0$ first equation.

$(-2y)^2 + 4y^2 = 32$; $4y^2 + 4y^2 = 32$, $8y^2 = 32$, $y^2 = 4$

and: $y = +2$; $y = -2$ Substituting these in the second equation gives the
$x = -4$; $x = 4$ matching x values.

Graph: an ellipse and a line. There are two solutions.

5. $x^2 - 2x + y^2 = 3$ Solve the second equation for $y = 4 - 2x$ and substitute
$2x + y = 4$ this in the first equation.

$x^2 - 2x + (4 - 2x)^2 = 3$ or $x^2 - 2x + 16 - 16x + 4x^2 = 3$;

or $5x^2 - 18x + 13 = 0$; factoring gives $(5x - 13)(x - 1) = 0$

and so $x = 13/5$, $x = 1$ Substituting these in the second equation gives the
$y = -6/5$, $y = 2$ y values.

Two solutions.

8. $x^2 - y^2 = 3$

$x^2 + y^2 = 5$ Add the first equation to the second equation, obtaining

$x^2 - y^2 = 3$

$2x^2 = 8$

so $x = +2$ $x = -2$

$y = +1$ $y = +1$

Substituting $x^2 = 4$ in the first equation gives the y values. Also

$x = +2$ $x = -2$

$y = -1$ $y = -1$

Hence, there are four solutions.

11. $2y^2 - x^2 = -1$

$4y^2 + x^2 = 25$ Add the first equation to the second equation, obtaining

$2y^2 - x^2 = -1$

$6y^2 = 24$

so $y = +2$ $y = -2$

$x = 3$ $x = 3$

Substituting these values in the first equation gives the x values.

$y = 2$ $y = -2$

$x = -3$ $x = -3$

Hence, there are four solutions.

14. $16y^2 + 5x^2 = 26$ 4 times the first equation and 5 times the second equation

$25y^2 - 4x^2 = 17$ yields

$64y^2 + 20x^2 = 104$

$125y^2 - 20x^2 = 85$ Now add the first equation to the second equation, obtaining

$64y^2 + 20x^2 = 104$

$189y^2 = 189$

so $y^2 = 1$ and

$y = 1$ $y = -1$ Substituting these values in the first equation gives the

$x = \sqrt{2}$ $x = \sqrt{2}$ x values.

$y = 1$ $y = -1$

$x = -\sqrt{2}$ $x = -\sqrt{2}$

17. $x^2 - 3xy - 2y^2 - 2 = 0$

$x - y - 2 = 0$

Solving the second equation for y gives $y = x - 2$. Substituting this into the first equation gives:

$$x^2 - 3x(x - 2) - 2(x - 2)^2 - 2 = 0$$

which simplifies to

$$4x^2 - 14x + 10 = 0 \quad \text{or} \quad 2x^2 - 7x + 5 = 0$$

and factoring yields

$$(2x - 5)(x - 1) = 0$$

so $x = 5/2$, $x = 1$. Substituting in the second equation yields the y values $y = 1/2$, $y = -1$.

20. Let x = the width and y = the length. Then $2x + 2y$ = the perimeter and xy = the area, so the two equations become

$$\begin{aligned} 2x + 2y &= 22 \\ xy &= 30 \end{aligned} \quad \text{or} \quad \begin{aligned} x + y &= 11 \\ xy &= 30 \end{aligned}$$

and from the first equation we get $x = 11 - y$ and substituting in the second we get $y(11 - y) = 30$ or $11y - y^2 = 30$ or $y^2 - 11y + 30 = 0$; $(y - 5)(y - 6) = 0$, $y = 6$, or $y = 5$, and hence $x = 5$, $x = 6$.

One dimension is 5 ft, the other is 6 ft.

Section 9.6 Systems of Linear Inequalities in Two Variables

BASIC CONCEPTS you should learn from this unit.

1. The graph of a linear equation in two variables divides the plane into two half planes.
2. The graph of a linear inequality is simply one of the two half planes referred to above and may or may not include the line separating the half planes.
3. The solution to a system of linear inequalities is the set of all points whose coordinates satisfy all the inequalities.
4. The graph of a system of inequalities is the space or area occupied by the satisfying points, usually indicated by shading.
5. Certain word problems are best solved by setting up a system of inequalities and then solving the system.

WHAT YOU SHOULD KNOW HOW TO DO upon completion of this unit.

1. Graph a single inequality, indicating the solution.
2. Graph a system of inequalities and indicate the solution to the system.
3. Solve word problems that lead to a system of linear inequalities.

SOLUTIONS to Exercise Set 9.6

2. $y \geq x + 3$

Graph the line $y = x + 3$;
then choose a test point and
shade the appropriate half plane.
Using $(0, 0)$ as a test point
yields a false statement.

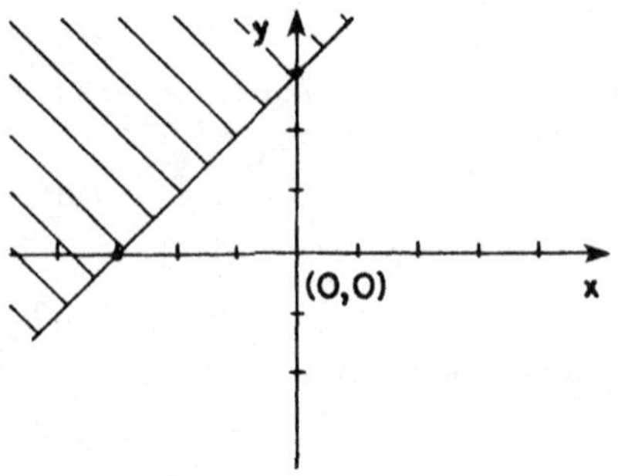

5. $y \leq x - x$

Graph $y = 4 - x$;
choose the origin as a test point.

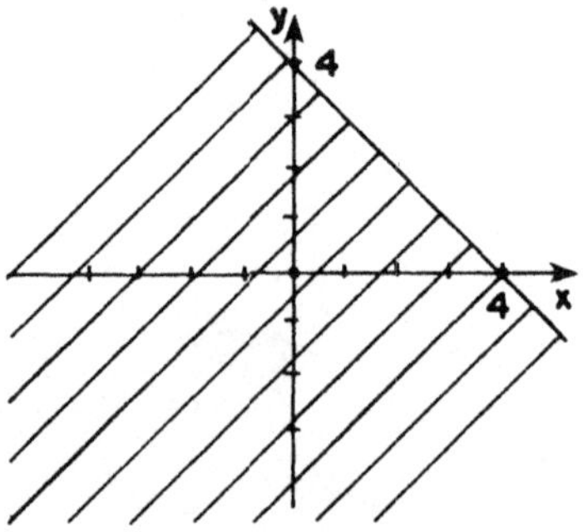

8. $y \leq 2x$

Graph $y = 2x$;
choose $(1, 1)$ as a test point.

$1 \leq 2$ which is a true statement.

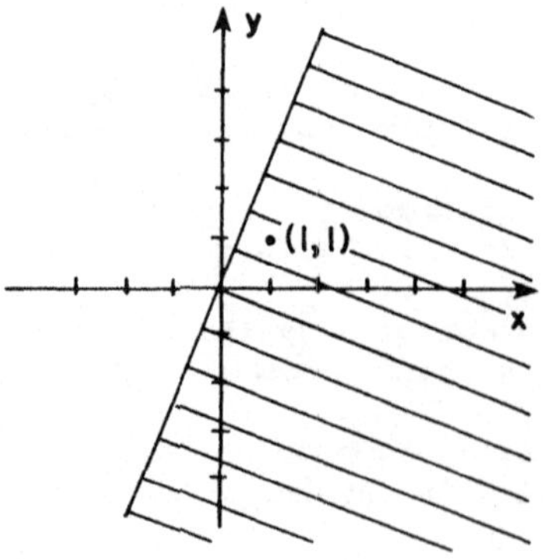

11. $x \leq 4$

Graph $x = 4$;

choose $(0, 0)$ as a test point.

$0 \leq 4$ which is a true statement.

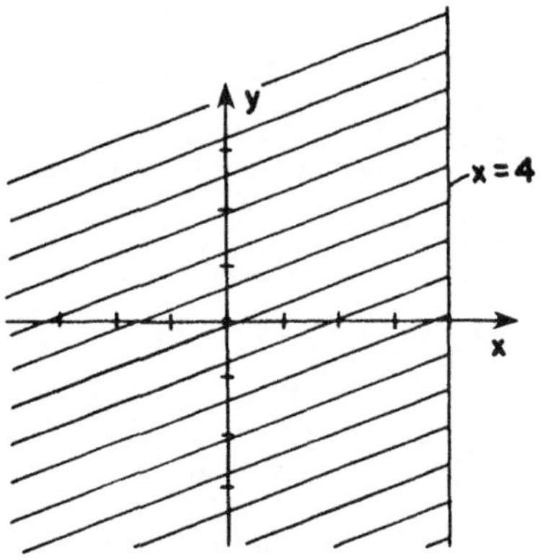

14. $5y \leq 25$

$y \leq 5$

Graph $y = 5$;

choose $(0, 0)$ as a test point.

$0 \leq 5$ which is a true statement.

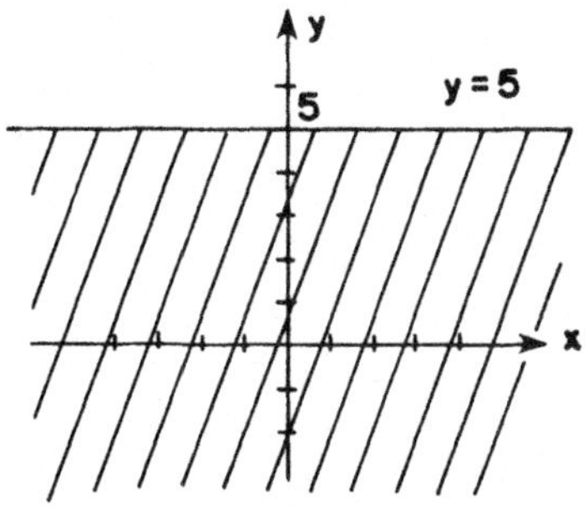

17. $-2 \leq x \leq 3$

Graph $-2 = x$
and $x = 3$.

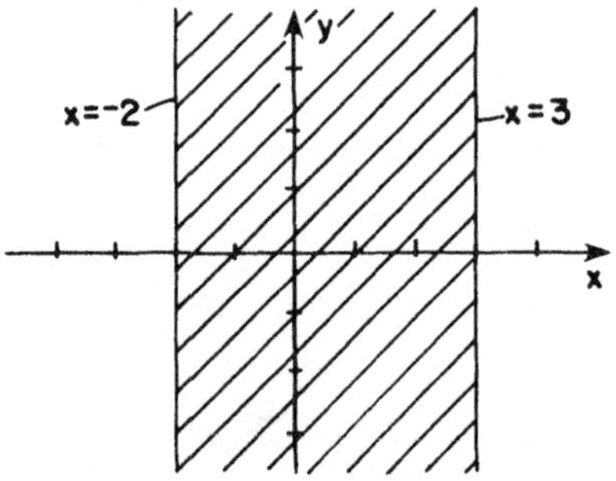

20. Since x = ounces of cheese, $100x$ = number of calories provided by the cheese. Likewise, $110y$ and $80z$ represent the calories in the bread and apples. Then we have

$$100x + 110y + 80z \leq 1500$$

as an appropriate inequality.

23. $3x - y \geq -7$

$3x + y \leq -2$

Graph the lines

$3x - y = -7$ and

$3x + y = -2$.

Try the test points

(0, 0) fails in the second inequality

(0, 6) fails in the second inequality

(-4, 0) fails in the first inequality

(-2, -2) satisfies

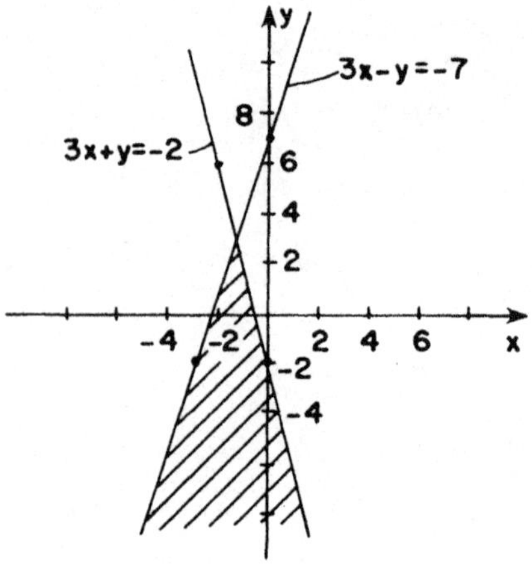

26. $2x - y \geq -3$

$x + y \leq 5$

$x \geq 1$

Graph the lines

$2x - y = -3$

$x + y = 5$

$x = 1$

Try test points to determine the solution.

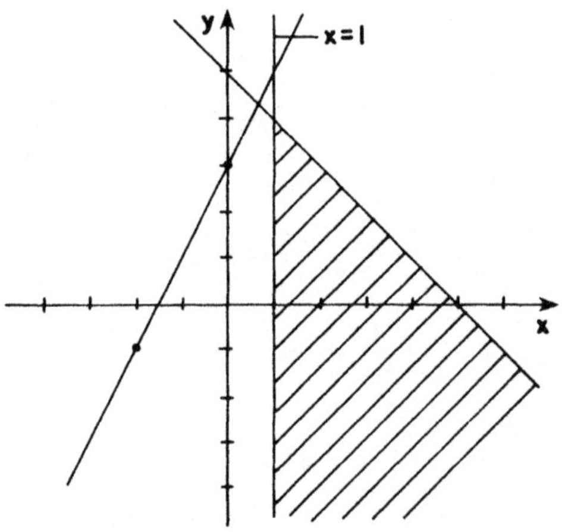

29. $3x + y \leq 6$

$x - 2y \leq -1$

$x \geq 2$

Graph the lines

$3x + y = 6$

$x - 2y = -1$

$x = 2$

Trying test points in all the areas is one way, but noting that satisfying points have to be below the first line, above the second and to the right of the third shows there is no solution.

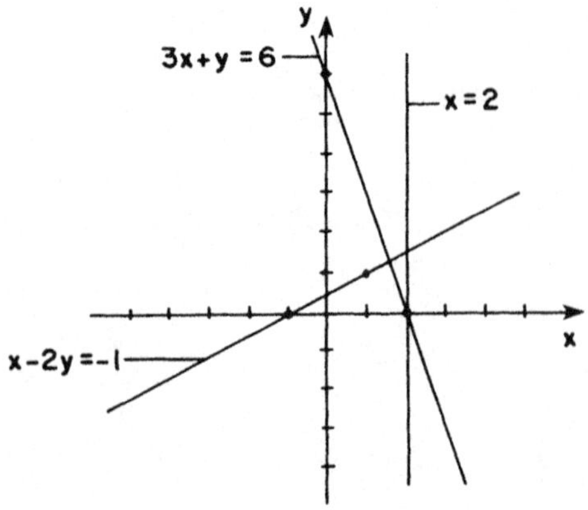

32. $2x + 3y \geq 18$

$x + 3y \geq 12$

$4x + 3y \geq 24$

$x \geq 0$ (y-axis)

$y \geq 0$ (x-axis)

Graphing the lines as before and noting the direction solution points must be from each line yields the solution.

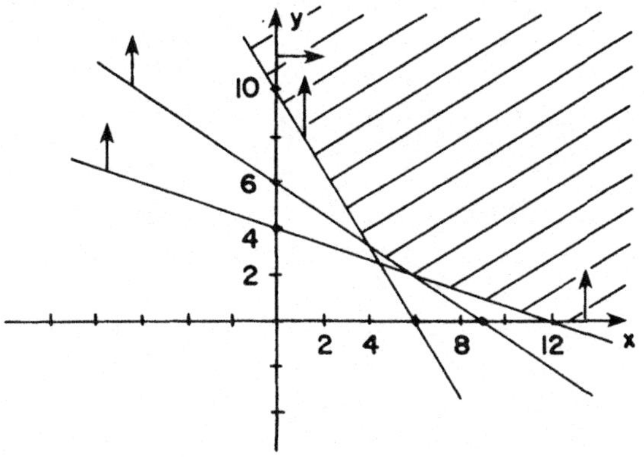

35. Let x = amount in common stocks
y = amount in preferred stocks

Then inequalities can be written:

(a) $x \leq 50{,}000$

(b) $y \leq 35{,}000$

(c) $x + y \leq 60{,}000$

(d) $x \leq 2y$

Also $x \geq 0$

$y \geq 0$

The shaded area represents possible combinations.

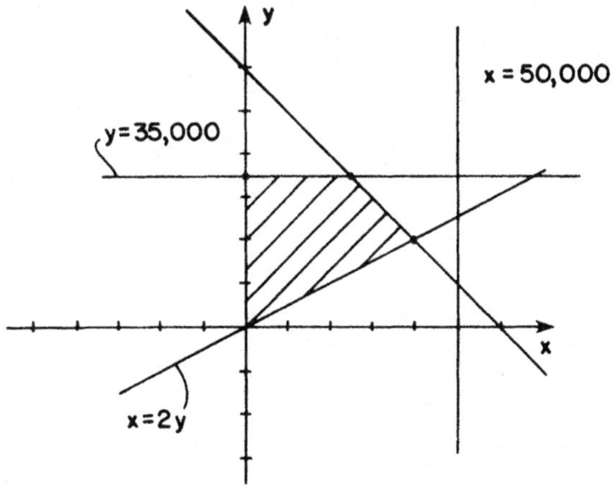

SOLUTIONS to Review Exercises Chapter 9

2. $4x - y = 4$

$2x + y = -1$

Graph each line on the same axes.

Estimate the intersection as $\left(\frac{1}{2}, -2\right)$.

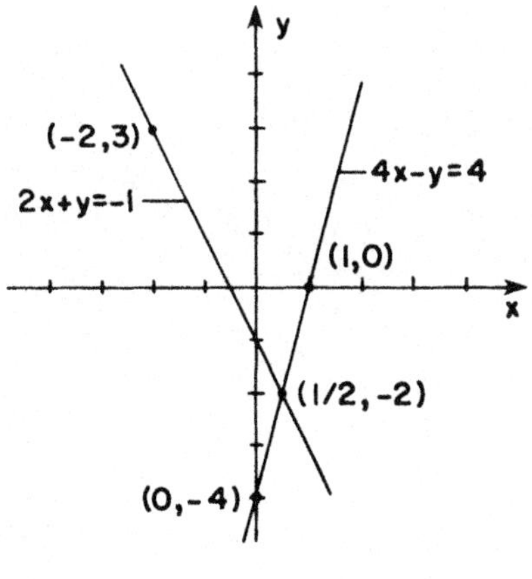

5. $2x + y = 0$

$x - 3y = \frac{7}{4}$

Solving the first equation for y yields $y = -2x$; substitute this for y in the second equation.

$$x - 3(-2x) = \frac{7}{4}; \quad x + 6x = \frac{7}{4}; \quad 7x = \frac{7}{4}, \quad x = \frac{1}{4}$$

and then $y = -\frac{1}{2}$.

8. $2x + 3y = -7$

$-x + 2y = 7$

Solving the second equation for x yields $x = 2y - 7$; substitute this in the first equation.

$$2(2y - 7) + 3y = -7; \quad 4y - 14 + 3y = -7; \quad y = 1, \quad x = -5$$

11. $-3x + y = -13$

$2x - 3y = 11$

Multiply the first equation by 3; then add.

$$\begin{array}{rcl} -9x + 3y & = & -39 \\ 2x - 3y & = & 11 \\ \hline -7x & = & -28 \end{array}$$

and so $x = 4$, $y = -1$ by substituting 4 for x in either equation.

14. $\frac{1}{3}x + \frac{1}{2}y = -1$

$\frac{1}{2}x - \frac{1}{4}y = \frac{5}{2}$

Multiply the second equation by 2, then add.

$$\begin{array}{rcl} \frac{1}{3}x + \frac{1}{2}y & = & -1 \\ x - \frac{1}{2}y & = & 5 \\ \hline \frac{4}{3}x & = & 4 \end{array}$$

Then $x = 3$, $y = -4$.

17. x = cost of hamburger $\qquad 5x + 4y = 22$

y = cost of steak $\qquad 3x + 7y = 28.15$

Multiply the first equation by -3 and the second equation by 5; then add.

$$\begin{array}{rcl} -15x - 12y & = & -66 \\ 15x + 35y & = & 140.75 \\ \hline 23y & = & 74.75 \end{array}$$

$$y = 3.25, \quad x = \$1.80$$

20. $C = 4025 + 9x$

$R = 16x$

Since the break-even point occurs when costs equal revenue, we simply set the right-hand sides of the equations equal to each other:

$$16x = 4025 + 9x$$

and thus $x = 575$ hours and the total revenue at $x = 575$ is $16 \cdot 575 = \$9,200$.

23. $5x - y + 2z = 10$

$-2x + 3y - z = -7$

$3x + 2z = 7$ Add the third equation to the second equation, obtaining

$5x - y + 2z = 10$

$x + 3y + z = 0$

$3x + 2z = 7$ Interchange the first and second equations, obtaining

$x + 3y + z = 0$

$5x - y + 2z = 10$ Add -5 times the first equation to the second equation, and

$3x + 2z = 7$ add -3 times the first equation to the third equation, obtaining

$x + 3y + z = 0$

$-16y - 3z = 10$ Add -9 times the second equation to 16 times the third

$-9y - z = 7$ equation, obtaining

$x + 3y + z = 0$

$-16y - 3z = 10$

$11z = 22$

From this $z = 2$, then $y = -1$, $x = 1$ by back substituting.

26. $x + 2y = 0$

$\underline{-x + 4y = 5}$ Add the equations.

$6y = 5$; $y = 5/6$, then $x = -5/3$.

29. $x^2 + y^2 = 25$

$3x + y = 15$

Solving the second equation for $y = -3x + 15$ and substituting in the first equation yields:

$x^2 + (-3x + 15)^2 = 25$; $x^2 + 9x^2 - 90x + 225 - 25 = 0$;

$10x^2 - 90x + 200 = 0$; $x^2 - 9x + 20 = 0$; $(x - 5)(x - 4) = 0$

$x = 5$ $x = 4$

$x = 0$ $y = 3$

32. $x^2 + y^2 = 9$

$y = x^2 + 3$

From the first $x^2 = 9 - y^2$; substitute this in the second equation to get

$y = 9 - y^2 + 3$ or $y^2 + y - 12 = 0$

and by factoring

$(y + 4)(y - 3) = 0$; $y = -4$ and $y = 3$.

Substituting -4 in the second equation fails to yield real x values. But $y = 3$ yields $x = 0$ for a good solution.

35. $x^2 + y^2 = 16$

$x^2 + y = 1$

Estimate: $x = 2.1$

$y = -3.8$

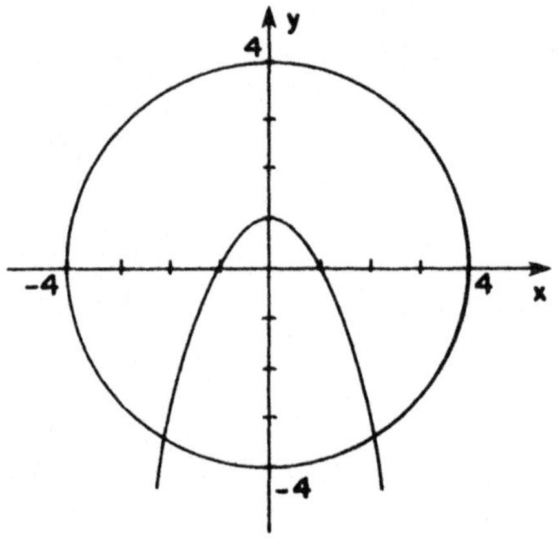

38. $2x + 3y \leq 2$

$x - y \geq 1$

Graph the lines

$2x + 3y = 2$

$x - y = 1$

Note where, in relation to the lines, the solution points must lie (below both lines).

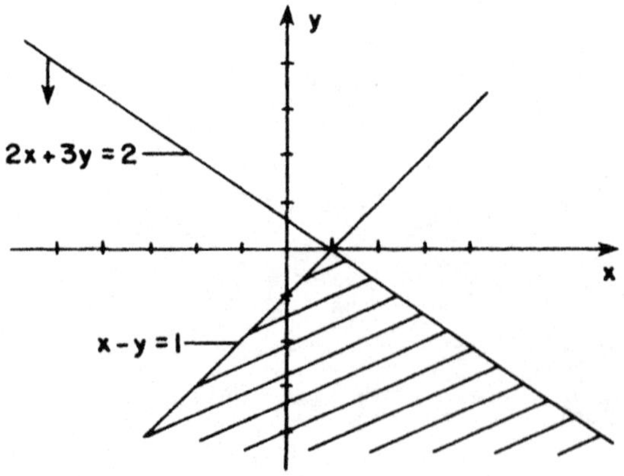

41. $2x + y \leq 4$

$2x - y \leq 3$

$y \geq 0$ (x-axis)

$x \geq 0$ (y-axis)

Graph the lines

$2x + y = 4$

$2x - y = 3$

Note that the solution is below the first line and above the second line, and above the x-axis and to the right of the y-axis.

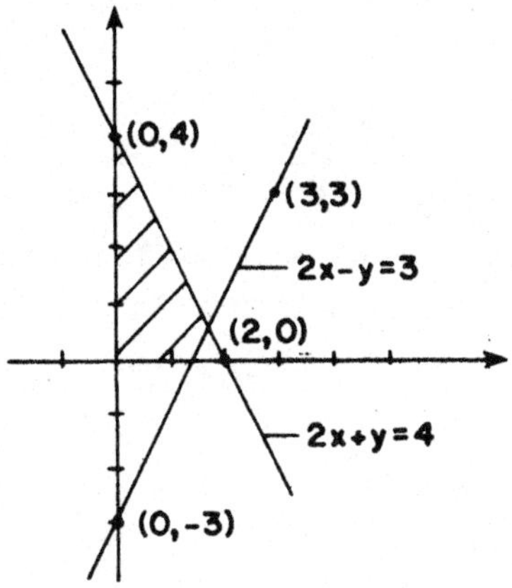

Chapter 9 - A SHORT QUIZ

1. What is meant when we say an (x, y) pair of numbers is a solution to a system of equations?

2. If a system of equations does not have a solution, the system is said to be ___?___ .

3. If a system of equations has one or more solutions, it is a ___?___ system.

4. Is (-2, 4) a solution to the system

$x - 2y = -10$

$2x + 3y = 8$?

Show work to justify your answer.

5. Draw sketches to illustrate all the possibilities for a system of two linear equations.

6. In solving the system in question 4 by substitution the first step might be to solve the first equation for x, getting $x = 2y - 10$. What is the next step?

7. If a system of two linear equations has exactly one solution, where on the graph of the system does the solution appear?

8. If the two equations of a system are really just different forms of the same equation, how many solutions are there?

9. If in solving a system by elimination $0x + 0y = 0$ is obtained, what is signified?

10. To solve the system

$2x - 3y = 5$

$5x + 9y = 4$

by elimination of the y variable, what is your first step?

11. Sometimes $0x + 0y = k$ $(k \neq 0)$ is obtained. Example: $0 = 7$. Since this is ___?___ true the system is ___?___ .

12. In a business situation of manufacturing and selling a product the <u>equilibrium price</u> is the price of the product at which ___?___ and ___?___ are equal.

13. In solving a system of three linear equations by Gaussian elimination the operations allowed are (a) ___?___ any two equations; (b) ___?___ an equation by a nonzero constant; and (c) ___?___ an equation by ___?___ .

14. When the Gaussian elimination is completed the value of one variable is evident. How are the values of the other variables found?

15. In this system what is the maximum number of solutions you could expect? Why?

$$x^2 + y^2 = 10$$
$$x + y = -3$$

16. If a system of equations consists of an equation of an ellipse and an equation of a parabola, and if the system has three solutions, sketch a graph of this situation.

17. The solution to a linear inequality is generally a ___?___ .

18. Is the solution to $y < 3x - 1$ above or below the line $y = 3x - 1$?

19. Does the test point (3, 0) satisfy this system?

$$2x - 3y \le 2$$
$$x + y \le 6$$

20. Try to graph the solution set to this system.

$$y \ge x^2$$
$$x + y \le 3$$

Quiz Answers Chapter 9

1. When substituted into each equation a true statement results.

2. inconsistent

3. consistent

4. <u>yes</u> $-2 - 8 = -10$
   $-4 + 12 = 8$

5.

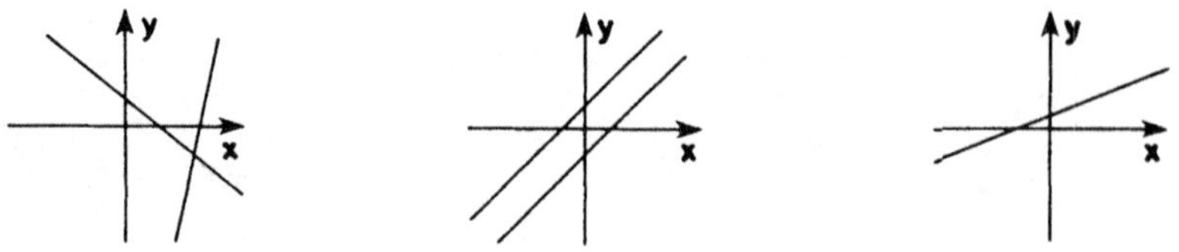

one solution — no solution — same line

6. Substitute $(2y - 10)$ for x in the second equation.

7. at the intersection of the lines

8. infinitely many

9. infinitely many solutions

10. Multiply the first equation by 3 .

11. never; inconsistent

12. supply, demand

13. (a) interchange; (b) multiply; (c) replace, the sum of itself and a nonzero multiple of another equation.

14. by back substituting

15. Two. The system is a line and a circle.

16.

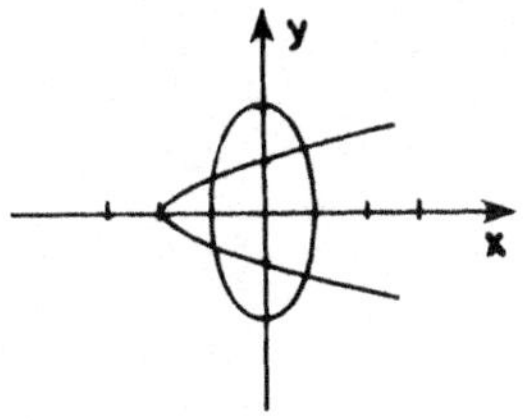

17. a half plane

18. below

19. no $(6^6 - 0 \leq 2$ is false)

20.

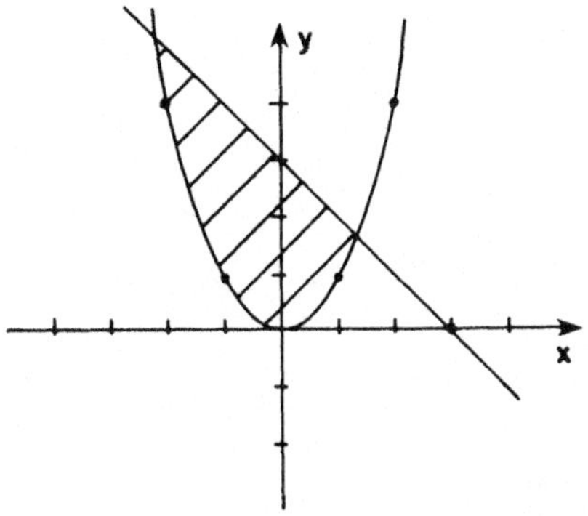

Try some test points.

CHAPTER 10

Matrices and Determinants

Section 10.1 Matrices and Linear Systems

BASIC CONCEPTS you should learn from this unit.

1. Definition of a matrix as a rectangular array of numbers.

2. By conventional agreement, an $m \times n$ matrix has m rows and n columns.

3. A square matrix has as many columns as rows.

4. Double subscripting is a convenient way to identify elements in a matrix. a_{ij} is the element in the ith row and jth column.

5. From a system of linear equations we obtain a coefficient matrix using the coefficients of the variables as entries, and an augmented matrix, which includes the right-hand sides of the equations.

6. The methods of Gaussian elimination - now called elementary row operations - can be applied to an augmented matrix of a linear system of equations to obtain the solution.

7. Obtaining pivot elements (a_{11}, a_{22}, etc.) equal to 1 is helpful.

8. The Gauss-Jordan variation simply has zeros above and below the pivot elements that are 1. The solution then appears in the right column.

WHAT YOU SHOULD KNOW HOW TO DO upon completion of this unit.

1. Define "matrix."

2. Write the coefficient matrix and the augmented matrix for a system of linear equations.

3. Identify elements of a matrix by double subscripting.

4. Solve a system of linear equations by applying row operations to the augmented matrix of the system.

5. Solve a system of linear equations by Gauss-Jordan elimination.

SOLUTIONS to Exercise Set 10.1

2. 1×4 (one row, four columns)

5. 3×3

8. $b_{13} = 8$, $b_{21} = 4$, $b_{33} = -6$, $b_{42} = 9$

11. (a) $\begin{bmatrix} \frac{1}{2} & 1 & 1 \\ 2 & -1 & -4 \\ 4 & 2 & -3 \end{bmatrix}$ (b) $\begin{bmatrix} \frac{1}{2} & 1 & 1 & 4 \\ 2 & -1 & -4 & 6 \\ 4 & 2 & -3 & 8 \end{bmatrix}$

14. $$\begin{aligned} 4x \quad\quad &= 2 \\ -7x + 8y &= 3 \end{aligned}$$

17. $$\begin{bmatrix} 1 & 2 & 0 & 3 \\ 0 & 1 & -2 & 4 \\ 0 & 0 & 1 & 2 \end{bmatrix}$$ From the third row we simply read that $\underline{z = 2}$.

Now back substitute in row 2: $y - 2(2) = 4$, so that $\underline{y = 8}$; and finally substitute the z and y values in row 1: $x + 2(8) + 0 = 3$; so $\underline{x = -13}$.

20. $$\begin{bmatrix} 1 & -4 & 2 & -4 \\ 0 & 1 & 3 & -2 \\ 0 & 0 & 1 & 5 \end{bmatrix}$$ From the last row we get $z = 5$ and back substitution gives $y = -17$ and $z = -82$.

23. The augmented matrix:

$$\begin{bmatrix} 1 & 1 & 1 & 4 \\ 2 & -1 & 2 & 11 \\ 1 & 2 & 1 & 3 \end{bmatrix}, \begin{bmatrix} 1 & 1 & 1 & 4 \\ 0 & -3 & 0 & 3 \\ 0 & 1 & 0 & -1 \end{bmatrix}, \begin{bmatrix} 1 & 1 & 1 & 4 \\ 0 & 1 & 0 & -1 \\ 0 & 1 & 0 & -1 \end{bmatrix},$$

$$\begin{bmatrix} 1 & 1 & 1 & 4 \\ 0 & 1 & 0 & -1 \\ 0 & 0 & 0 & 0 \end{bmatrix}$$

which indicates many solutions with $y = -1$. We may choose z at random . One solution is $(2, -1, 3)$; another is $(0, -1, 5)$.

26. $$\begin{bmatrix} 2 & 1 & -1 & -2 \\ -2 & -2 & 3 & 2 \\ 3 & 1 & -1 & -4 \end{bmatrix}, \begin{bmatrix} 2 & 1 & -1 & -2 \\ -2 & -2 & 3 & 2 \\ 1 & -1 & 2 & -2 \end{bmatrix}, \begin{bmatrix} 1 & -1 & 2 & -2 \\ -2 & -2 & 3 & 2 \\ 0 & -1 & 2 & 0 \end{bmatrix},$$

$$\begin{bmatrix} 1 & -1 & 2 & -2 \\ 0 & -4 & 7 & -2 \\ 0 & +1 & -2 & 0 \end{bmatrix}, \begin{bmatrix} 1 & -1 & 2 & -2 \\ 0 & 1 & -2 & 0 \\ 0 & -4 & 7 & -2 \end{bmatrix}, \begin{bmatrix} 1 & -1 & 2 & -2 \\ 0 & 1 & -2 & 0 \\ 0 & 0 & -1 & -2 \end{bmatrix}$$

And $z = 2$; Substituting in the second row gives
$y = 4$; and these two values in row 1 give
$x = -2$.

29. $\begin{bmatrix} 1 & 1 & -1 & 2 & 0 \\ 2 & 1 & 0 & -1 & -2 \\ 3 & 0 & 2 & 0 & -3 \\ -1 & 2 & 0 & 3 & 1 \end{bmatrix}$, $\begin{bmatrix} 1 & 1 & -1 & 2 & 0 \\ 0 & -1 & 2 & -5 & -2 \\ 0 & -3 & 5 & -6 & -3 \\ 0 & 3 & -1 & 5 & 1 \end{bmatrix}$, $\begin{bmatrix} 1 & 1 & -1 & 2 & 0 \\ 0 & 1 & -2 & 5 & 2 \\ 0 & -3 & 5 & -6 & -3 \\ 0 & 3 & -1 & 5 & 1 \end{bmatrix}$,

$\begin{bmatrix} 1 & 1 & -1 & 2 & 0 \\ 0 & 1 & -2 & 5 & 2 \\ 0 & 0 & -1 & 9 & 3 \\ 0 & 0 & 5 & -10 & -5 \end{bmatrix}$, $\begin{bmatrix} 1 & 1 & -1 & 2 & 0 \\ 0 & 1 & -2 & 5 & 2 \\ 0 & 0 & 1 & -9 & -3 \\ 0 & 0 & 1 & -2 & -1 \end{bmatrix}$, $\begin{bmatrix} 1 & 1 & -1 & 2 & 0 \\ 0 & 1 & -2 & 5 & 2 \\ 0 & 0 & 1 & 9 & -3 \\ 0 & 0 & 0 & 7 & 2 \end{bmatrix}$

From the last row, now we get $W = 2/7$

and this back substituted yields $z = -3/7$

and these two values in the second row yield $y = -2/7$

and these three put in the first row yield $x = -5/7$

32. $\begin{bmatrix} 2 & 1 & -1 \\ 3 & -1 & -7 \end{bmatrix}$, $\begin{bmatrix} 2 & 1 & -1 \\ 1 & -2 & -6 \end{bmatrix}$, $\begin{bmatrix} 1 & -2 & -6 \\ 2 & 1 & -1 \end{bmatrix}$,

$\begin{bmatrix} 1 & -2 & -6 \\ 0 & 5 & 11 \end{bmatrix}$, $\begin{bmatrix} 1 & -2 & -6 \\ 0 & 1 & \frac{11}{5} \end{bmatrix}$, $\begin{bmatrix} 1 & 0 & -\frac{8}{5} \\ 0 & 1 & \frac{11}{5} \end{bmatrix}$,

which gives directly $x = -8/5$, $y = 11/5$.

35. $\begin{bmatrix} 2 & 1 & -1 & 9 \\ 1 & -2 & 2 & -3 \\ 3 & 3 & 4 & 11 \end{bmatrix}$, $\begin{bmatrix} 1 & -2 & 2 & -3 \\ 2 & 1 & -1 & 9 \\ 3 & 3 & 4 & 11 \end{bmatrix}$, $\begin{bmatrix} 1 & -2 & 2 & -3 \\ 0 & 5 & -5 & 15 \\ 0 & 9 & -2 & 20 \end{bmatrix}$,

$\begin{bmatrix} 1 & -2 & 2 & -3 \\ 0 & 1 & -1 & 3 \\ 0 & 9 & -2 & 20 \end{bmatrix}$, $\begin{bmatrix} 1 & -2 & 2 & -3 \\ 0 & 1 & -1 & 3 \\ 0 & 0 & 7 & -7 \end{bmatrix}$, $\begin{bmatrix} 1 & -2 & 2 & -3 \\ 0 & 1 & -1 & 3 \\ 0 & 0 & 1 & -1 \end{bmatrix}$,

$\begin{bmatrix} 1 & -2 & 2 & -3 \\ 0 & 1 & 0 & 2 \\ 0 & 0 & 1 & -1 \end{bmatrix}$, $\begin{bmatrix} 1 & -2 & 0 & -1 \\ 0 & 1 & 0 & 2 \\ 0 & 0 & 1 & -1 \end{bmatrix}$, $\begin{bmatrix} 1 & 0 & 0 & 3 \\ 0 & 1 & 0 & 2 \\ 0 & 0 & 1 & -1 \end{bmatrix}$,

from which now we read $x = 3$, $y = 2$, $z = -1$.

38. $\begin{bmatrix} 4 & 1 & -1 & -1 \\ 1 & -1 & 2 & 3 \\ -1 & 2 & -1 & 0 \end{bmatrix}$, $\begin{bmatrix} 1 & -1 & 2 & 3 \\ 4 & 1 & -1 & -1 \\ -1 & 2 & -1 & 0 \end{bmatrix}$, $\begin{bmatrix} 1 & -1 & 2 & 3 \\ 0 & 5 & -9 & -13 \\ 0 & 1 & 1 & 3 \end{bmatrix}$,

$\begin{bmatrix} 1 & -1 & 2 & 3 \\ 0 & 1 & 1 & 3 \\ 0 & 5 & -9 & -13 \end{bmatrix}$, $\begin{bmatrix} 1 & -1 & 2 & 3 \\ 0 & 1 & 1 & 3 \\ 0 & 0 & -14 & -28 \end{bmatrix}$, $\begin{bmatrix} 1 & -1 & 2 & 3 \\ 0 & 1 & 1 & 3 \\ 0 & 0 & 1 & 2 \end{bmatrix}$,

$\begin{bmatrix} 1 & -1 & 2 & 3 \\ 0 & 1 & 0 & 1 \\ 0 & 0 & 1 & 2 \end{bmatrix}$, $\begin{bmatrix} 1 & -1 & 0 & -1 \\ 0 & 1 & 0 & 1 \\ 0 & 0 & 1 & 2 \end{bmatrix}$, $\begin{bmatrix} 1 & 0 & 0 & 0 \\ 0 & 1 & 0 & 1 \\ 0 & 0 & 1 & 2 \end{bmatrix}$

so the solution is $x = 0$, $y = 1$, $z = 2$.

Section 10.2 Matrix Operations and Applications

BASIC CONCEPTS you should learn from this unit.

1. Definition of equal matrices: Two matrices are equal if they have the same dimension and their corresponding elements are equal.
2. Matrices are added by adding corresponding elements.
3. Definition of scalar multiplication: Multiply each element of the matrix by the scalar.
4. Definition of matrix multiplication.
5. The product of an $m \times n$ matrix times an $n \times r$ matrix is a matrix of dimension $m \times r$.
6. Matrix multiplication is not commutative.
7. Matrix multiplication is associative.

WHAT YOU SHOULD KNOW HOW TO DO upon completion of this unit.

1. Define equal matrices.
2. Add two matrices.
3. Multiply a matrix by a scalar.
4. Recognize when two matrices can and cannot be multiplied.
5. Multiply two matrices.
6. Write a matrix equation for a linear system of equations.

SOLUTIONS to Exercise Set 10.2

2. Since corresponding elements must be equal

$a = 5$, $b = -6$ (since $a + b = -1$), $c = 3$ (since $2c = 6$),

$d = -7$ (since $c - d = 10$)

5.
$$2A = 2\begin{bmatrix} 2 & 3 & 1 \\ -3 & 4 & 1 \end{bmatrix} = \begin{bmatrix} 4 & 6 & 2 \\ -6 & 8 & 2 \end{bmatrix} \quad \text{and}$$

$$3G = 3\begin{bmatrix} -2 & 4 & 2 \\ 1 & 0 & 3 \end{bmatrix} = \begin{bmatrix} -6 & 12 & 6 \\ 3 & 0 & 9 \end{bmatrix}, \quad \text{then}$$

adding corresponding elements yields the sum

$$\begin{bmatrix} -2 & 18 & 8 \\ -3 & 8 & 11 \end{bmatrix}$$

8.
$$2B = 2\begin{bmatrix} 2 & -1 \\ 3 & 2 \\ 4 & 1 \end{bmatrix} = \begin{bmatrix} 4 & -2 \\ 6 & 4 \\ 8 & 2 \end{bmatrix}, \quad \text{a } 3 \times 2 \text{ matrix}$$

and since D is a 2×2 matrix, $2B - D$ is not possible.

11. The product CB will have dimension 3×2, which cannot be added to D, a 2×2 matrix.

14.
$$\begin{bmatrix} 2 & 3 & 1 \\ -3 & 4 & 1 \end{bmatrix} \cdot \begin{bmatrix} 1 & 2 & 3 \\ 4 & -1 & 2 \\ 3 & 2 & 5 \end{bmatrix} = \begin{bmatrix} 17 & 3 & 17 \\ 16 & -8 & 4 \end{bmatrix} = AC$$

$$2 \cdot \begin{bmatrix} -3 & 2 \\ 4 & 1 \end{bmatrix} \cdot \begin{bmatrix} -2 & 4 & 2 \\ 1 & 0 & 3 \end{bmatrix} = \begin{bmatrix} -6 & 4 \\ 8 & 2 \end{bmatrix} \cdot \begin{bmatrix} -2 & 4 & 2 \\ 1 & 0 & 3 \end{bmatrix} = \begin{bmatrix} 16 & -24 & 0 \\ -14 & 32 & 22 \end{bmatrix} = 2DC$$

Adding corresponding elements yields

$$\begin{bmatrix} 33 & -21 & 17 \\ 2 & 24 & 26 \end{bmatrix}$$

17.
$$2GE = 2\begin{bmatrix} -2 & 4 & 2 \\ 1 & 0 & 3 \end{bmatrix} \cdot \begin{bmatrix} 1 & -3 & 2 \\ 3 & 2 & 4 \\ 1 & 1 & 2 \end{bmatrix}$$

$$= \begin{bmatrix} -4 & 8 & 4 \\ 2 & 0 & 6 \end{bmatrix} \cdot \begin{bmatrix} 1 & -3 & 2 \\ 3 & 2 & 4 \\ 1 & 1 & 2 \end{bmatrix} = \begin{bmatrix} 24 & 32 & 32 \\ 8 & 0 & 16 \end{bmatrix}$$

and

$$3A = \begin{bmatrix} 6 & 9 & 3 \\ -9 & 12 & 3 \end{bmatrix}$$

Now subtracting $2GE - 3A$ yields

$$\begin{bmatrix} 18 & 23 & 29 \\ 17 & -12 & 13 \end{bmatrix}$$

20.
$$AB = \begin{bmatrix} 1 & 2 \\ 3 & 2 \end{bmatrix} \cdot \begin{bmatrix} 2 & -1 \\ -3 & 4 \end{bmatrix} = \begin{bmatrix} -4 & 7 \\ 0 & 5 \end{bmatrix}$$

and

$$BA = \begin{bmatrix} 2 & -1 \\ -3 & 4 \end{bmatrix} \cdot \begin{bmatrix} 1 & 2 \\ 3 & 2 \end{bmatrix} = \begin{bmatrix} -1 & 2 \\ 9 & 2 \end{bmatrix}$$

clearly $\neq AB$.

23. Applying the rules for multiplying, we find

$$AI = \begin{bmatrix} (a_{11} + 0 + 0) & (0 + a_{12} + 0) & (0 + 0 + a_{13}) \\ (a_{21} + 0 + 0) & (0 + a_{22} + 0) & (0 + 0 + a_{23}) \\ (a_{31} + 0 + 0) & (0 + a_{32} + 0) & (0 + 0 + a_{33}) \end{bmatrix} = \begin{bmatrix} a_{11} & a_{12} & a_{13} \\ a_{21} & a_{22} & a_{23} \\ a_{31} & a_{32} & a_{33} \end{bmatrix} = A$$

Also

$$IA = \begin{bmatrix} (a_{11} + 0 + 0) & (a_{12} + 0 + 0) & (a_{13} + 0 + 0) \\ (0 + a_{21} + 0) & (0 + a_{22} + 0) & (0 + a_{23} + 0) \\ (0 + 0 + a_{31}) & (0 + 0 + a_{32}) & (0 + 0 + a_{33}) \end{bmatrix} = \begin{bmatrix} a_{11} & a_{12} & a_{13} \\ a_{21} & a_{22} & a_{23} \\ a_{31} & a_{32} & a_{33} \end{bmatrix} = A$$

26. Let A be the coefficient matrix of the system, X be a column matrix $\begin{bmatrix} x \\ y \end{bmatrix}$, and B be a column matrix $\begin{bmatrix} 6 \\ -2 \end{bmatrix}$. Then AX = B is

$$\begin{bmatrix} 7 & -2 \\ -2 & 3 \end{bmatrix} \cdot \begin{bmatrix} x \\ y \end{bmatrix} = \begin{bmatrix} 6 \\ -2 \end{bmatrix}$$

29. Following a procedure similar to that shown in solution 26, we get

$$AX = B: \quad \begin{bmatrix} 3 & -1 & 4 \\ 2 & 2 & \frac{3}{4} \\ 1 & -\frac{1}{4} & 1 \end{bmatrix} \cdot \begin{bmatrix} x \\ y \\ z \end{bmatrix} = \begin{bmatrix} 5 \\ -1 \\ \frac{1}{2} \end{bmatrix}$$

32. Finding AX and setting it equal to B, we have

$$\begin{bmatrix} 1 & 7 & -2 \\ 3 & 6 & 1 \\ -4 & 2 & 0 \end{bmatrix} \cdot \begin{bmatrix} x \\ y \\ z \end{bmatrix} = \begin{bmatrix} 3 \\ -3 \\ 2 \end{bmatrix}$$

or finally

$$\begin{aligned} x + 7y - 2z &= 3 \\ 3x + 6y + z &= -3 \\ -4x + 2y &= 2 \end{aligned}$$

as the system.

35.

$$AI_n = \begin{bmatrix} a_{11} & a_{12} & \cdots & a_{1n} \\ a_{21} & a_{22} & \cdots & a_{2n} \\ \vdots & & & \vdots \\ a_{n1} & & \cdots & a_{nn} \end{bmatrix} \cdot \begin{bmatrix} 1 & 0 & \cdots & 0 \\ 0 & 1 & \cdots & 0 \\ \vdots & & & \vdots \\ 0 & & \cdots & 1 \end{bmatrix} = \begin{bmatrix} a_{11} & a_{12} & \cdots & a_{1n} \\ \vdots & & & \vdots \\ a_{n1} & & \cdots & a_{nn} \end{bmatrix}$$

To see this notice that the element in the first row, first column of the product is really the sum of $a_{11} \cdot 1 + (n - 1)$ zeros or simply a_{11}. Also the element in the first row, second column of the product is $0 + a_{12} + (n - 2)$ zeros. Similar arguments can be made for each element in the product matrix, which turns out to be A itself.

$$I_nA = \begin{bmatrix} 1 & 0 & \cdots & 0 \\ 0 & 1 & \cdots & 0 \\ \vdots & & & \vdots \\ 0 & & \cdots & 1 \end{bmatrix} \cdot \begin{bmatrix} a_{11} & a_{12} & \cdots & a_{1n} \\ a_{21} & a_{22} & \cdots & a_{2n} \\ \vdots & & & \vdots \\ a_{n1} & & \cdots & a_{nn} \end{bmatrix} = \begin{bmatrix} a_{11} & a_{12} & \cdots & a_{1n} \\ a_{21} & & \cdots & \vdots \\ a_{n1} & & \cdots & a_{nn} \end{bmatrix}$$

which is also equal to A, by a similar discussion. Now since $AI_n = A$ and $IA_n = A$, $AI_n = I_nA$, as required.

BASIC CONCEPTS you should learn from this unit.

1. A square $n \times n$ matrix A is invertible if an $n \times n$ matrix B can be found such that $AB = BA = I_n$, I_n being the identity matrix of order n. Then B is the <u>inverse</u> of A.

2. The inverse of matrix A, that is, A^{-1}, can be found by considering the matrix

$$\left[\begin{array}{c|c} A & I_n \end{array}\right]$$

and by row operations converting A into I_n. Then I_n (the right portion) will become A^{-1}.

3. Not all $n \times n$ matrices are invertible <u>and</u> the process of concept 2 will reveal which are or are not.

4. If $AX = B$ is a linear system, A being the coefficient matrix and if the inverse A^{-1}, of A exists, then the system has exactly one solution

$$X = A^{-1}B$$

5. Use of the inverse of a matrix to solve systems is especially useful if we have several systems with the same coefficient matrix.

WHAT YOU SHOULD KNOW HOW TO DO upon completion of this unit.

1. Verify that a matrix is the inverse of another matrix.

2. Find the inverse of a matrix if it has an inverse.

3. Determine that a matrix does not have an inverse.

4. Solve a system of linear equations using the inverse of the coefficient matrix.

SOLUTIONS to Exercise Set 10.3

2. $$AB = \begin{bmatrix} 3 & -1 \\ -2 & 2 \end{bmatrix} \cdot \begin{bmatrix} 1/2 & 1/4 \\ 1/2 & 3/4 \end{bmatrix} = \begin{bmatrix} 3/2 - 1/2 & 3/4 - 3/4 \\ -1 + 1 & -1/2 + 3/2 \end{bmatrix} = \begin{bmatrix} 1 & 0 \\ 0 & 1 \end{bmatrix}$$

   Since $AB = I$, yes, B <u>is</u> the inverse of A.

5. First form the augmented matrix

$$\left[\begin{array}{cc|cc} -1 & 5 & 1 & 0 \\ 2 & -4 & 0 & 1 \end{array}\right]$$

   and now transform the left-hand 2×2 matrix into I_2 by row operations;

$$\left[\begin{array}{cc|cc} 1 & -5 & -1 & 0 \\ 0 & 6 & 2 & 1 \end{array}\right]; \quad \left[\begin{array}{cc|cc} 1 & -5 & -1 & 0 \\ 0 & 1 & 1/3 & 1/6 \end{array}\right]; \quad \left[\begin{array}{cc|cc} 1 & 0 & 2/3 & 5/6 \\ 0 & 1 & 1/3 & 1/6 \end{array}\right]$$

   Now the right-hand 2×2 matrix is the inverse of A.

8. Form the augmented matrix as

$$\left[\begin{array}{ccc|ccc} 2 & 1 & 0 & 1 & 0 & 0 \\ 1 & 1 & 0 & 0 & 1 & 0 \\ 1 & 1 & 1 & 0 & 0 & 1 \end{array}\right]$$

Now perform row operations to transform the left-hand 3×3 into I_3.

$$\left[\begin{array}{ccc|ccc} 1 & 1 & 1 & 0 & 0 & 1 \\ 1 & 1 & 0 & 0 & 1 & 0 \\ 2 & 1 & 0 & 1 & 0 & 0 \end{array}\right], \left[\begin{array}{ccc|ccc} 1 & 1 & 1 & 0 & 0 & 1 \\ 0 & 0 & -1 & 0 & 1 & -1 \\ 0 & -1 & -2 & 1 & 0 & -2 \end{array}\right], \left[\begin{array}{ccc|ccc} 1 & 1 & 1 & 0 & 0 & 1 \\ 0 & 1 & 2 & -1 & 0 & 2 \\ 0 & 0 & 1 & 0 & -1 & 1 \end{array}\right],$$

$$\left[\begin{array}{ccc|ccc} 1 & 1 & 1 & 0 & 0 & 1 \\ 0 & 1 & 0 & -1 & 2 & 0 \\ 0 & 0 & 1 & 0 & -1 & 1 \end{array}\right], \left[\begin{array}{ccc|ccc} 1 & 1 & 0 & 0 & 1 & 0 \\ 0 & 1 & 0 & -1 & 2 & 0 \\ 0 & 0 & 1 & 0 & -1 & 1 \end{array}\right], \left[\begin{array}{ccc|ccc} 1 & 0 & 0 & 1 & -1 & 0 \\ 0 & 1 & 0 & -1 & 2 & 0 \\ 0 & 0 & 1 & 0 & -1 & 1 \end{array}\right]$$

Now the right-hand 3×3 matrix is the inverse of

$$\begin{bmatrix} 2 & 1 & 0 \\ 1 & 1 & 0 \\ 1 & 1 & 1 \end{bmatrix}$$

11. Form the augmented matrix

$$\left[\begin{array}{cc|cc} 1 & 3 & 1 & 0 \\ -1 & 4 & 0 & 1 \end{array}\right]$$

and transform the left-hand 2×2 matrix into I_2

$$\left[\begin{array}{cc|cc} 1 & 3 & 1 & 0 \\ 0 & 7 & 1 & 1 \end{array}\right], \left[\begin{array}{cc|cc} 1 & 3 & 1 & 0 \\ 0 & 1 & \frac{1}{7} & \frac{1}{7} \end{array}\right], \left[\begin{array}{cc|cc} 1 & 0 & \frac{4}{7} & -\frac{3}{7} \\ 0 & 1 & \frac{1}{7} & \frac{1}{7} \end{array}\right]$$

and the right-hand 2×2 matrix is the inverse required.

14. $$\left[\begin{array}{ccc|ccc} 8 & 7 & -1 & 1 & 0 & 0 \\ -5 & -5 & 1 & 0 & 1 & 0 \\ -4 & -4 & 1 & 0 & 0 & 1 \end{array}\right], \left[\begin{array}{ccc|ccc} 8 & 7 & -1 & 1 & 0 & 0 \\ -5 & -5 & 1 & 0 & 1 & 0 \\ 1 & 1 & 0 & 0 & -1 & 1 \end{array}\right], \left[\begin{array}{ccc|ccc} 1 & 1 & 0 & 0 & -1 & 1 \\ -5 & -5 & 1 & 0 & 1 & 0 \\ 8 & 7 & -1 & 1 & 0 & 0 \end{array}\right],$$

$$\left[\begin{array}{ccc|ccc} 1 & 1 & 0 & 0 & -1 & 1 \\ 0 & 0 & 1 & 0 & -4 & 5 \\ 0 & -1 & -1 & 1 & 8 & -8 \end{array}\right], \left[\begin{array}{ccc|ccc} 1 & 1 & 0 & 0 & -1 & 1 \\ 0 & 1 & 1 & -1 & -8 & 8 \\ 0 & 0 & 1 & 0 & -4 & 5 \end{array}\right], \left[\begin{array}{ccc|ccc} 1 & 1 & 0 & 0 & -1 & 1 \\ 0 & 1 & 0 & -1 & -4 & 3 \\ 0 & 0 & 1 & 0 & -4 & 5 \end{array}\right]$$

$$\left[\begin{array}{ccc|ccc} 1 & 0 & 0 & 1 & 3 & -2 \\ 0 & 1 & 0 & -1 & -4 & 3 \\ 0 & 0 & 1 & 0 & -4 & 5 \end{array}\right]$$

and the right-hand 3×3 matrix is the inverse required.

17. $$\begin{bmatrix} 1 & 0 & -1 & 1 & 0 & 0 \\ 2 & 1 & 0 & 0 & 1 & 0 \\ 0 & 1 & 1 & 0 & 0 & 1 \end{bmatrix}, \begin{bmatrix} 1 & 0 & -1 & 1 & 0 & 0 \\ 0 & 1 & 2 & -2 & 1 & 0 \\ 0 & 1 & 1 & 0 & 0 & 1 \end{bmatrix}, \begin{bmatrix} 1 & 0 & -1 & 1 & 0 & 0 \\ 0 & 1 & 2 & -2 & 1 & 0 \\ 0 & 0 & -1 & 2 & -1 & 1 \end{bmatrix}$$

$$\begin{bmatrix} 1 & 0 & -1 & 1 & 0 & 0 \\ 0 & 1 & 2 & -2 & 1 & 0 \\ 0 & 0 & 1 & -2 & 1 & -1 \end{bmatrix} \quad \left[\begin{array}{ccc|ccc} 1 & 0 & 0 & -1 & 1 & -1 \\ 0 & 1 & 0 & 2 & -1 & 2 \\ 0 & 0 & 1 & -2 & 1 & -1 \end{array}\right]$$

and the right-hand 3×3 matrix is the inverse required.

20. The coefficient matrix is

$$\begin{bmatrix} 2 & -3 \\ 3 & 1 \end{bmatrix}$$

We find its inverse:

$$\left[\begin{array}{cc|cc} 2 & -3 & 1 & 0 \\ 3 & 1 & 0 & 1 \end{array}\right], \begin{bmatrix} 2 & -3 & 1 & 0 \\ 1 & 4 & -1 & 1 \end{bmatrix}, \begin{bmatrix} 1 & 4 & -1 & 1 \\ 2 & -3 & 1 & 0 \end{bmatrix}, \begin{bmatrix} 1 & 4 & -1 & 1 \\ 0 & -11 & 3 & -2 \end{bmatrix},$$

$$\begin{bmatrix} 1 & 4 & -1 & 1 \\ 0 & 1 & -\frac{3}{11} & \frac{2}{11} \end{bmatrix}, \left[\begin{array}{cc|cc} 1 & 0 & \frac{1}{11} & \frac{3}{11} \\ 0 & 1 & -\frac{3}{11} & \frac{2}{11} \end{array}\right]$$

Now find the product of the inverse times the constant-term column matrix:

$$\begin{bmatrix} \frac{1}{11} & \frac{3}{11} \\ -\frac{3}{11} & \frac{2}{11} \end{bmatrix} \cdot \begin{bmatrix} -5 \\ -13 \end{bmatrix} = \begin{bmatrix} -4 \\ -1 \end{bmatrix}$$

Thus $x = -4$ and $y = -1$.

23. Find the inverse of the coefficient matrix

$$\left[\begin{array}{ccc|ccc} 2 & -1 & 3 & 1 & 0 & 0 \\ 3 & -1 & 1 & 0 & 1 & 0 \\ 1 & 1 & 1 & 0 & 0 & 1 \end{array}\right], \left[\begin{array}{ccc|ccc} 1 & 1 & 1 & 0 & 0 & 1 \\ 3 & -1 & 1 & 0 & 1 & 0 \\ 2 & -1 & 3 & 1 & 0 & 0 \end{array}\right], \left[\begin{array}{ccc|ccc} 1 & 1 & 1 & 0 & 0 & 1 \\ 0 & -4 & -2 & 0 & 1 & -3 \\ 0 & -3 & 1 & 1 & 0 & -2 \end{array}\right],$$

$$\left[\begin{array}{ccc|ccc} 1 & 1 & 1 & 0 & 0 & 1 \\ 0 & 1 & 3 & 1 & -1 & 1 \\ 0 & 0 & 10 & 4 & -3 & 1 \end{array}\right], \quad \left[\begin{array}{ccc|ccc} 1 & 1 & 1 & 0 & 0 & 0 \\ 0 & 1 & 3 & 1 & -1 & 1 \\ 0 & 0 & 1 & 4/10 & -3/10 & 1/10 \end{array}\right],$$

$$\left[\begin{array}{ccc|ccc} 1 & 1 & 1 & 0 & 0 & 1 \\ 0 & 1 & 0 & -2/10 & -1/10 & 7/10 \\ 0 & 0 & 1 & 4/10 & -3/10 & 1/10 \end{array}\right], \left[\begin{array}{ccc|ccc} 1 & 1 & 0 & -4/10 & 3/10 & 9/10 \\ 0 & 1 & 0 & -2/10 & -1/10 & 7/10 \\ 0 & 0 & 1 & 4/10 & -3/10 & 1/10 \end{array}\right],$$

$$\left[\begin{array}{ccc|ccc} 1 & 0 & 0 & -2/10 & 4/10 & 2/10 \\ 0 & 1 & 0 & -2/10 & -1/10 & 7/10 \\ 0 & 0 & 1 & 4/10 & -3/10 & 1/10 \end{array}\right]$$

Then

$$\begin{bmatrix} -\frac{2}{10} & \frac{4}{10} & \frac{2}{10} \\ -\frac{2}{10} & -\frac{1}{10} & \frac{7}{10} \\ \frac{4}{10} & -\frac{3}{10} & \frac{1}{10} \end{bmatrix} \cdot \begin{bmatrix} -11 \\ -5 \\ -1 \end{bmatrix} = \begin{bmatrix} 0 \\ 2 \\ 3 \end{bmatrix}$$

or

$$x = 0$$
$$y = 2$$
$$z = 3$$

26. The coefficient matrix is $\begin{bmatrix} 2 & 1 \\ 3 & -1 \end{bmatrix}$. We find its inverse:

$$\left[\begin{array}{cc|cc} 2 & 1 & 1 & 0 \\ 3 & -1 & 0 & 1 \end{array}\right], \left[\begin{array}{cc|cc} 2 & 1 & 1 & 0 \\ 1 & -2 & -1 & 1 \end{array}\right], \left[\begin{array}{cc|cc} 1 & -2 & -1 & 1 \\ 2 & 1 & 1 & 0 \end{array}\right], \left[\begin{array}{cc|cc} 1 & -2 & -1 & 1 \\ 0 & 5 & 3 & -2 \end{array}\right],$$

$$\left[\begin{array}{cc|cc} 1 & -2 & -1 & 1 \\ 0 & 1 & 3/5 & -2/5 \end{array}\right], \left[\begin{array}{cc|cc} 1 & 0 & 1/5 & 1/5 \\ 0 & 1 & 3/5 & -2/5 \end{array}\right]$$

and multiply

26. continued

$$\begin{bmatrix} 1/5 & 1/5 \\ 3/5 & -2/5 \end{bmatrix} \cdot \begin{bmatrix} -1 \\ -7 \end{bmatrix} = \begin{bmatrix} -8/5 \\ 11/5 \end{bmatrix}$$

so

$$x = -8/5$$
$$y = 11/5$$

29. We find the inverse of the coefficient matrix:

$$\begin{bmatrix} 2 & 1 & -1 & 1 & 0 & 0 \\ 1 & -2 & 2 & 0 & 1 & 0 \\ 3 & 3 & 4 & 0 & 0 & 1 \end{bmatrix}, \begin{bmatrix} 1 & -2 & 2 & 0 & 1 & 0 \\ 2 & 1 & -1 & 1 & 0 & 0 \\ 3 & 3 & 4 & 0 & 0 & 1 \end{bmatrix}, \begin{bmatrix} 1 & -2 & 2 & 0 & 1 & 0 \\ 0 & 5 & -5 & 1 & -2 & 0 \\ 0 & 9 & -2 & 0 & -3 & 1 \end{bmatrix},$$

$$\begin{bmatrix} 1 & -2 & 2 & 0 & 1 & 0 \\ 0 & 1 & -1 & 1/5 & -2/5 & 0 \\ 0 & 9 & -2 & 0 & -3 & 1 \end{bmatrix}, \begin{bmatrix} 1 & -2 & 2 & 0 & 1 & 0 \\ 0 & 1 & -1 & 1/5 & -2/5 & 0 \\ 0 & 0 & 7 & -9/5 & 3/5 & 1 \end{bmatrix},$$

$$\begin{bmatrix} 1 & -2 & 2 & 0 & 1 & 0 \\ 0 & 1 & -1 & 1/5 & -2/5 & 0 \\ 0 & 0 & 1 & -9/35 & 3/35 & 5/35 \end{bmatrix}, \begin{bmatrix} 1 & -2 & 2 & 0 & 1 & 0 \\ 0 & 1 & 0 & -2/35 & -11/35 & 5/35 \\ 0 & 0 & 1 & -9/35 & 3/35 & 5/35 \end{bmatrix},$$

$$\begin{bmatrix} 1 & -2 & 0 & 18/35 & 29/35 & -10/35 \\ 0 & 1 & 0 & -2/35 & -11/35 & 5/35 \\ 0 & 0 & 1 & -9/35 & 3/35 & 5/35 \end{bmatrix}, \begin{bmatrix} 1 & 0 & 0 & 14/35 & 7/35 & 0 \\ 0 & 1 & 0 & -2/35 & -11/35 & 5/35 \\ 0 & 0 & 1 & -9/35 & 3/35 & 5/35 \end{bmatrix}$$

Multiplying the inverse by $\begin{bmatrix} 9 \\ -3 \\ 11 \end{bmatrix}$ yields $\begin{bmatrix} 3 \\ 2 \\ -1 \end{bmatrix}$

or

$$x = 3$$
$$y = 2$$
$$z = -1$$

32. We find the inverse of the coefficient matrix:

$$\begin{bmatrix} 4 & 1 & -1 & 1 & 0 & 0 \\ 1 & -1 & 2 & 0 & 1 & 0 \\ -1 & 2 & -1 & 0 & 0 & 1 \end{bmatrix}, \begin{bmatrix} 1 & -1 & 2 & 0 & 1 & 0 \\ 4 & 1 & -1 & 1 & 0 & 1 \\ -1 & 2 & -1 & 0 & 0 & 1 \end{bmatrix}, \begin{bmatrix} 1 & -1 & 2 & 0 & 1 & 0 \\ 0 & 5 & -9 & 1 & -4 & 1 \\ 0 & 1 & 1 & 0 & 1 & 1 \end{bmatrix},$$

32. continued

$$\begin{bmatrix} 1 & -1 & 2 & 0 & 1 & 0 \\ 0 & 1 & 1 & 0 & 1 & 1 \\ 0 & 0 & -14 & 1 & -9 & -4 \end{bmatrix}, \begin{bmatrix} 1 & -1 & 2 & 0 & 1 & 0 \\ 0 & 1 & 1 & 0 & 1 & 1 \\ 0 & 0 & 1 & -1/14 & 9/14 & 4/14 \end{bmatrix},$$

$$\begin{bmatrix} 1 & -1 & 0 & 2/14 & -4/14 & -8/14 \\ 0 & 1 & 0 & 1/14 & 5/14 & 9/14 \\ 0 & 0 & 1 & -1/14 & 9/14 & 4/14 \end{bmatrix}, \begin{bmatrix} 1 & 0 & 0 & 3/14 & 1/14 & 1/14 \\ 0 & 1 & 0 & 1/14 & 5/14 & 9/14 \\ 0 & 0 & 1 & -1/14 & 9/14 & 4/14 \end{bmatrix}$$

and multiplying the inverse by $\begin{bmatrix} -1 \\ 3 \\ 0 \end{bmatrix}$ yields $\begin{bmatrix} 0 \\ 1 \\ 2 \end{bmatrix}$

or

$$x = 0$$
$$y = 1$$
$$z = 2$$

35. $X = A^{-1}B_1$ or

$$\begin{bmatrix} 3 & -2 & 4 \\ 2 & -1 & 0 \\ 0 & 4 & 1 \end{bmatrix} \cdot \begin{bmatrix} 1 \\ -1 \\ 5 \end{bmatrix} = \begin{bmatrix} 25 \\ 3 \\ 1 \end{bmatrix}$$

and $X = A^{-1} \cdot B_2$ or

$$\begin{bmatrix} 3 & -2 & 4 \\ 2 & -1 & 0 \\ 0 & 4 & 1 \end{bmatrix} \cdot \begin{bmatrix} 4 \\ 3 \\ -2 \end{bmatrix} = \begin{bmatrix} -2 \\ 5 \\ 10 \end{bmatrix}$$

38. Using x and y as the amounts invested at 10% and 15%, respectively, make three systems:

$$\begin{aligned} x + y &= 3{,}000 \\ 10x + 15y &= 360{,}000 \end{aligned} \; ; \quad \begin{aligned} x + y &= 3{,}000 \\ 10x + 15y &= 400{,}000 \end{aligned} \; ; \quad \begin{aligned} x + y &= 3{,}000 \\ 10x + 15y &= 5{,}000 \end{aligned}$$

Find the inverse of the coefficient matrix:

$$\begin{bmatrix} 1 & 1 & 1 & 0 \\ 10 & 15 & 0 & 1 \end{bmatrix}, \begin{bmatrix} 1 & 1 & 1 & 0 \\ 0 & 5 & -10 & 1 \end{bmatrix}, \begin{bmatrix} 1 & 1 & 1 & 0 \\ 0 & 1 & -2 & 1/5 \end{bmatrix}, \left[\begin{array}{cc|cc} 1 & 0 & 3 & -1/5 \\ 0 & 1 & -2 & 1/5 \end{array}\right]$$

Multiply the inverse times each constant-term column matrix:

(a) $\begin{bmatrix} 3 & -1/5 \\ -2 & 1/5 \end{bmatrix} \cdot \begin{bmatrix} 30{,}000 \\ 360{,}000 \end{bmatrix} = \begin{bmatrix} \$18{,}000 \\ \$12{,}000 \end{bmatrix} \begin{matrix} = x \\ = y \end{matrix}$

38. continued

(b) $\begin{bmatrix} 3 & -1/5 \\ -2 & 1/5 \end{bmatrix} \cdot \begin{bmatrix} 30,000 \\ 400,000 \end{bmatrix} = \begin{bmatrix} \$10,000 \\ \$20,000 \end{bmatrix} \begin{matrix} = x \\ = y \end{matrix}$

(c) $\begin{bmatrix} 3 & -1/5 \\ -2 & 1/5 \end{bmatrix} \cdot \begin{bmatrix} 30,000 \\ 500,000 \end{bmatrix} = \begin{bmatrix} -10,000 \\ 40,000 \end{bmatrix}$

which is not possible!

Section 10.4 Determinants and Cramer's Rule

BASIC CONCEPTS you should learn from this unit.

1. Definition of a determinant as a number associated with a square matrix.

2. The determinant of a 2×2 matrix is

$$\begin{vmatrix} a_{11} & a_{12} \\ a_{21} & a_{22} \end{vmatrix} = a_{11}a_{22} - a_{21}a_{12}$$

3. The minor of an element a_{ij} in a matrix is the matrix remaining after the row and column in which a_{ij} is located (that is, row i and column j) are deleted.

4. The cofactor of an element a_{ij} is the product of the minor of a_{ij} times $(-1)^{i+j}$.

5. A determinant of order higher than 2×2 is the sum of all the products found by multiplying each element in a row or a column by its cofactor. This process is often called evaluating a determinant or expanding a determinant.

6. Cramer's rule is a method of solving systems of linear equations. Here it is applied to a system of two equations, but the rule extends to three or more equations:

$$x = \frac{|D_1|}{|D|}, \quad y = \frac{|D_2|}{|D|}$$

with $|D|$ defined as the determinant of the coefficients; $|D_1|$ is $|D|$ with its first column replaced with the column of the right-hand sides of the equations (that is, the column of constant terms); $|D_2|$ is $|D|$ with its second column replaced with the column of right-hand sides of the equations.

WHAT YOU SHOULD KNOW HOW TO DO upon completion of this unit.

1. Define determinant, minor, cofactor.

2. Find determinants of the second, third, and fourth order (that is, evaluate these kinds of determinants).

3. Solve systems of linear equations by means of determinants.

SOLUTIONS to Exercise Set 10.4

2. $\begin{vmatrix} 3 & 4 \\ -1 & 2 \end{vmatrix} = (3)\cdot(2) - (-1)\cdot(4) = 6 + 4 = 10$

5. $0\cdot 3 = 1\cdot 0 = 0$

8. Computing the minors:

(a) a_{12} : $4\cdot 0 + 5\cdot 3 = 15$

(b) a_{22} : $3\cdot 0 - 5\cdot 2 = -10$

(c) a_{23} : $3\cdot(-2) - (5)\cdot(-1) = -1$

(d) a_{32} : $3\cdot(-3) - 4\cdot 2 = -17$

11. $4\cdot 8 - 5\cdot(-8) + 2\cdot(-10) = 52$

14. $-1\cdot(21 - 7) + 2\cdot(21 - 14) = 0$

using the first column.

17. Using the third row we have

$$2\begin{vmatrix} 3 & 2 & 0 \\ -1 & -3 & 0 \\ 4 & 1 & 3 \end{vmatrix} - 2\begin{vmatrix} 3 & 2 & 1 \\ -1 & -3 & -1 \\ 4 & 1 & 3 \end{vmatrix} = 2(-27 + 6) - 2(-24 + 5 + 4) = 2(-21) - 2(-15) = -12$$

20. Using the first column we have

$$1\begin{vmatrix} -1 & 4 & -1 \\ 3 & 1 & -4 \\ 2 & 0 & 2 \end{vmatrix} - 2\begin{vmatrix} 1 & 0 & 1 \\ -1 & 4 & -1 \\ 2 & 0 & 2 \end{vmatrix} = 1(-2 - 24 - 30) - 2(8 - 8) = -56$$

23.

$$x = \frac{|D_1|}{|D|} = \frac{\begin{vmatrix} 9 & 1 & -1 \\ -3 & -2 & 2 \\ 11 & 3 & 4 \end{vmatrix}}{\begin{vmatrix} 2 & 1 & -1 \\ 1 & -2 & 2 \\ 3 & 3 & 4 \end{vmatrix}} = \frac{-105}{-35} = 3$$

$$y = \frac{|D_2|}{|D|} = \frac{\begin{vmatrix} 2 & 9 & -1 \\ 1 & -3 & 2 \\ 3 & 11 & 4 \end{vmatrix}}{-35} = \frac{-70}{-35} = 2$$

23. continued

$$z = \frac{|D_3|}{|D|} = \frac{\begin{vmatrix} 2 & 1 & 9 \\ 1 & -2 & -3 \\ 3 & 3 & 11 \end{vmatrix}}{-35} = \frac{35}{-35} = -1$$

26.

$$x = \frac{|D_1|}{|D|} = \frac{\begin{vmatrix} -1 & 1 & -1 \\ 3 & -1 & 2 \\ 0 & 2 & -1 \end{vmatrix}}{\begin{vmatrix} 4 & 1 & -1 \\ 1 & -1 & 2 \\ -1 & 2 & -1 \end{vmatrix}} = \frac{-1(1 - 4) - 3(-1 + 2)}{4(1 - 4) - 1(-1 + 2) - 1(2 - 1)} = \frac{0}{-14} = 0$$

$$y = \frac{|D_2|}{|D|} = \frac{\begin{vmatrix} 4 & -1 & -1 \\ 1 & 3 & 2 \\ -1 & 0 & -1 \end{vmatrix}}{|D|} = \frac{4(-3) - 1(1) - 1(-2 + 3)}{-14} = \frac{-14}{-14} = 1$$

$$z = \frac{|D_3|}{|D|} = \frac{\begin{vmatrix} 4 & 1 & -1 \\ 1 & -1 & 3 \\ -1 & 2 & 0 \end{vmatrix}}{D} = \frac{4(-6) - 1(2) - 1(3 - 1)}{-14} = \frac{-28}{-14} = 2$$

29. $$\begin{vmatrix} a_1 + b_1 & a_2 + b_2 \\ c & d \end{vmatrix} = d(a_1 + b_1) - c(a_2 + b_2) = a_1d + b_1d - a_2c - b_2c$$

$$\begin{vmatrix} a_1 & a_2 \\ c & d \end{vmatrix} = a_1d - a_2c ; \qquad \begin{vmatrix} b_1 & b_2 \\ c & d \end{vmatrix} = b_1d = b_2c$$

Then adding the last two results we obtain $a_1d + b_1d - a_2c - b_2c$, the same as the first result.

32. If

$$A = \begin{bmatrix} a_{11} & a_{12} \cdots & a_{1n} \\ \vdots & & \vdots \\ a_{n1} & \cdots & a_{nn} \end{bmatrix}, \quad B = \begin{bmatrix} ka_{11} & ka_{12} & \cdots & ka_{1n} \\ a_{21} & a_{22} & \cdots & a_{2n} \\ \vdots & & & \vdots \\ a_{n1} & & \cdots & a_{nn} \end{bmatrix},$$

32. continued

then

$$|B| = Ka_{11}\left|\begin{matrix}\text{cofactor}\\ \text{of } Ka_{11}\end{matrix}\right| - Ka_{12}\left|\begin{matrix}\text{cofactor}\\ \text{of } Ka_{12}\end{matrix}\right| + \ldots Ka_{1n}\left|\begin{matrix}\text{cofactor}\\ \text{of } Ka_{1n}\end{matrix}\right|$$

$$|B| = K\left[a_{11}\left|\begin{matrix}\text{cofactor}\\ \text{of } Ka_{11}\end{matrix}\right| - a_{12}\left|\begin{matrix}\text{cofactor}\\ \text{of } Ka_{12}\end{matrix}\right| + \ldots a_{1n}\left|\begin{matrix}\text{cofactor}\\ \text{of } Ka_{1n}\end{matrix}\right|\right]$$

But this is precisely $K|A|$, since the cofactor of Ka_{ij} = cofactor of a_{ij} and we have proved that $|B| = K|A|$.

SOLUTIONS to Review Exercises Chapter 10

2. For matrix A , $a_{24} = -1$.

5. $\begin{bmatrix} 3 & -7 \\ 1 & 4 \end{bmatrix}$

↑ (second column) column of y coefficients

↑ (first column) column of x coefficients

8. $$\begin{bmatrix} -2 & 4 & 5 & 0 \\ 6 & -9 & 4 & 0 \\ 3 & 2 & -1 & 0 \end{bmatrix}$$

implies the system

$$-2x + 4y + 5z = 0$$
$$6x - 9y + 4z = 0$$
$$3x + 2y - z = 0$$

11. The matrix $\begin{bmatrix} 1 & -4 & 2 & -18 \\ 0 & 1 & -2 & 5 \\ 0 & 0 & 1 & -1 \end{bmatrix}$

shows that $z = -1$; then the second row is

$y - 2(-1) = 5$ or $y = 3$

and the first row is

$x - 4(3) + 2(-1) = -18$ or $x = -4$

14. The augmented matrix of the system is

$$\begin{bmatrix} 3 & -1 & -17 \\ 2 & 3 & -4 \end{bmatrix}, \begin{bmatrix} 1 & -4 & -13 \\ 2 & 3 & -4 \end{bmatrix}, \begin{bmatrix} 1 & -4 & -13 \\ 0 & 11 & 22 \end{bmatrix}, \begin{bmatrix} 1 & -4 & -13 \\ 0 & 1 & 2 \end{bmatrix},$$

so that $y = 2$. Back substituting yields $x - 4(2) = -13$, $x = -5$

17. By the definition of matrix equality $2x = -6$, and so $x = -3$.

20. $$B - A = \begin{bmatrix} 2 & -1 \\ 3 & 2 \end{bmatrix} - \begin{bmatrix} -1 & 5 \\ 4 & -3 \end{bmatrix} = \begin{bmatrix} 3 & -6 \\ -1 & 5 \end{bmatrix}$$

23. $$CD = \begin{bmatrix} -1 & 0 \\ 0 & 4 \\ 2 & -2 \end{bmatrix} \cdot \begin{bmatrix} 1 & 3 & 4 \\ -1 & 0 & -6 \end{bmatrix} = \begin{bmatrix} -1 & -3 & -4 \\ -4 & 0 & -24 \\ 4 & 6 & 20 \end{bmatrix}$$

26. $$CB = \begin{bmatrix} -1 & 0 \\ 0 & 4 \\ 2 & -2 \end{bmatrix} \cdot \begin{bmatrix} -1 & 5 \\ 4 & -3 \end{bmatrix} = \begin{bmatrix} 1+0 & -5+0 \\ 0+16 & 0-12 \\ -2-8 & 10+6 \end{bmatrix} = \begin{bmatrix} 1 & -5 \\ 16 & -12 \\ -10 & 16 \end{bmatrix}$$

29. $$\begin{bmatrix} -2 & 3 & 1 & 0 \\ 1 & 4 & 0 & 1 \end{bmatrix}, \begin{bmatrix} 1 & 4 & 0 & 1 \\ -2 & 3 & 1 & 0 \end{bmatrix}, \begin{bmatrix} 1 & 4 & 0 & 1 \\ 0 & 11 & 1 & 2 \end{bmatrix}, \begin{bmatrix} 1 & 4 & 0 & 1 \\ 0 & 1 & 1/11 & 2/11 \end{bmatrix},$$

$$\begin{bmatrix} 1 & 0 & -4/11 & 3/11 \\ 0 & 1 & 1/11 & 2/11 \end{bmatrix}$$

with the right 2×2 being the inverse required.

32. $$\begin{bmatrix} 1 & 2 & -2 & 1 & 0 & 0 \\ 3 & -1 & 0 & 0 & 1 & 0 \\ 0 & 1 & 4 & 0 & 0 & 1 \end{bmatrix}, \begin{bmatrix} 1 & 2 & -2 & 1 & 0 & 0 \\ 0 & -7 & 6 & -3 & 1 & 0 \\ 0 & 1 & 4 & 0 & 0 & 1 \end{bmatrix}, \begin{bmatrix} 1 & 2 & -2 & 1 & 0 & 0 \\ 0 & 1 & 4 & 0 & 0 & 1 \\ 0 & -7 & 6 & -3 & 1 & 0 \end{bmatrix},$$

$$\begin{bmatrix} 1 & 2 & -2 & 1 & 0 & 0 \\ 0 & 1 & 4 & 0 & 0 & 1 \\ 0 & 0 & 34 & -3 & 1 & 7 \end{bmatrix}, \begin{bmatrix} 1 & 2 & -2 & 1 & 0 & 0 \\ 0 & 1 & 4 & 0 & 0 & 1 \\ 0 & 0 & 1 & -3/34 & 1/34 & 7/34 \end{bmatrix},$$

$$\begin{bmatrix} 1 & 2 & 0 & 28/34 & 2/34 & 14/34 \\ 0 & 1 & 0 & 12/34 & -4/34 & 6/34 \\ 0 & 0 & 1 & -3/34 & 1/34 & 7/34 \end{bmatrix}, \begin{bmatrix} 1 & 0 & 0 & 4/34 & 10/34 & 2/34 \\ 0 & 1 & 0 & 12/34 & -4/34 & 6/34 \\ 0 & 0 & 1 & -3/34 & 1/34 & 7/34 \end{bmatrix}$$

32. continued

The right 3×3 is the inverse of the coefficient matrix, and multiplying it times the column matrix $\begin{bmatrix} -4 \\ -2 \\ 1 \end{bmatrix}$ yields $x = -1$, $y = -1$, $z = 1/2$

35. $\begin{vmatrix} 2 & -1 \\ 6 & -3 \end{vmatrix} = (2)(-3) - (6)(-1) = 0$

38. $\begin{vmatrix} 1 & 2 & -1 \\ 0 & 3 & 4 \\ 0 & 0 & -1 \end{vmatrix} = 1\begin{vmatrix} 3 & 4 \\ 0 & -1 \end{vmatrix} + 0 + 0 = -3$

using the first column.

41. $x + 2y = 2$

$2x - 7y = 48$

$$x = \frac{|D_1|}{|D|} = \frac{\begin{vmatrix} 2 & 2 \\ 48 & -7 \end{vmatrix}}{\begin{vmatrix} 1 & 2 \\ 2 & -7 \end{vmatrix}} = \frac{-110}{-11} = 10$$

$$y = \frac{\begin{vmatrix} 1 & 2 \\ 2 & 48 \end{vmatrix}}{-11} = \frac{44}{-11} = -4$$

44.

$$x = \frac{\begin{vmatrix} -5 & 3 & 1 \\ -3 & 2 & 2 \\ -2 & 1 & -2 \end{vmatrix}}{\begin{vmatrix} 2 & 3 & 1 \\ 0 & 2 & 2 \\ 4 & 1 & -2 \end{vmatrix}} = \frac{-5(-6) + 3(-7) - 2(4)}{2(-6) + 4(4)} = \frac{1}{4}$$

$$y = \frac{\begin{vmatrix} 2 & -5 & 1 \\ 0 & -3 & 2 \\ 4 & -2 & -2 \end{vmatrix}}{4} = \frac{2(10) + 4(-7)}{4} = -2$$

$$z = \frac{\begin{vmatrix} 2 & 3 & -5 \\ 0 & 2 & -3 \\ 4 & 1 & -2 \end{vmatrix}}{4} = \frac{2(-1) + 4(1)}{4} = \frac{1}{2}$$

1. A matrix is a ____?____ .

2. If a matrix has dimension $m \times n$, this means it has ____?____ columns.

3. Write the 3×3 identity matrix.

4. If N is a square matrix, what is $N \cdot N^{-1}$?

5. Two matrices are equal only if ____?____ .

6. In general, can a 3×4 matrix be added to any other 3×4 matrix?

7. If M and N are matrices of the same dimension, can they be multiplied if they are not square?

8. If P and Q are square matrices, does $P \cdot Q = Q \cdot P$?

9. The augmented matrix of a system of linear equations contains the coefficient matrix plus the ____?____ as the last column.

10. In transforming $\begin{bmatrix} 2 & 1 \\ -3 & 7 \end{bmatrix}$ into $\begin{bmatrix} 2 & 1 \\ 1 & 9 \end{bmatrix}$ what operation was done?

11. The product AB of two matrices will have as many rows as ____?____ and as many columns as ____?____ .

12. If A is a 2×3 matrix and B is a 3×5 matrix, is AB defined? Is BA defined?

13. A determinant is a ____?____ associated with a ____?____ matrix.

14. If the two rows of matrix $\begin{bmatrix} 2 & 3 \\ -1 & 5 \end{bmatrix}$ are interchanged, what change occurs to the determinant?

15. In the system $\begin{matrix} ax + by = c \\ dx + ey = f \end{matrix}$, show a solution for x in terms of determinants.

16. To multiply a matrix by a scalar, we must multiply ____?____ of the matrix by the scalar.

17. What is a scalar?

18. Into what sort of matrix does the Gauss-Jordan method transform the coefficient matrix?

19. In question 18 where does the solution to the system appear?

20. To find the inverse of matrix A (order n) we first form the matrix $\left[A \mid I_n \right]$ and then by ____?____ on this entire matrix transform A into ____?____ . A^{-1} then appears on the right half.

Quiz Answers Chapter 10

1. rectangular array of numbers.

2. n

3. $\begin{bmatrix} 1 & 0 & 0 \\ 0 & 1 & 0 \\ 0 & 0 & 1 \end{bmatrix}$

4. N

5. corresponding elements are equal.

6. yes

7. no

8. no

9. constant terms

10. The second row was replaced by the sum of itself and 2 times the first row.

11. A, B

12. AB is defined; BA is not.

13. number, square

14. The sign of the determinant is changed.

15. $x = \dfrac{\begin{vmatrix} c & b \\ f & e \end{vmatrix}}{\begin{vmatrix} a & b \\ d & e \end{vmatrix}}$

16. each element

17. simply a number

18. Identity matrix of the same order

19. in the right column

20. row operations, I_n

CHAPTER 11

Roots of Polynomials

Section 11.1 Polynomial Division and Synthetic Division

BASIC CONCEPTS you should learn from this unit.

1. Dividing a polynomial by another polynomial is similar to long division in arithmetic in that there is a quotient which may or may not contain a remainder.

2. Symbolically,

$$\frac{P(x)}{D(x)} = Q(x) + \frac{R(x)}{Dx}$$

   or more neatly, $P(x) = Q(x)\cdot D(x) + R(x)$ (the degree of $R(x)$ < the degree of $P(x)$).

3. Synthetic division is the name given to a "shortcut" process of long division of polynomials.

WHAT YOU SHOULD KNOW HOW TO DO upon completion of this unit.

1. Divide one polynomial by another and write the answer in the form

$$\frac{P(x)}{Dx} = Q(x) + \frac{R(x)}{D(x)}$$

2. Divide a polynomial by a polynomial of the form $(x - r)$ using synthetic division.

SOLUTIONS to Exercise Set 11.1

2.

$$\begin{array}{r|l}
 & x + 1 \\
\hline
x + 2 & x^2 + 3x + 3 \\
 & \underline{x^2 + 2x} \\
 & \quad x + 3 \\
 & \quad \underline{x + 2} \\
 & \quad\quad 1
\end{array}$$

Solution:

$$x + 1 + \frac{1}{x + 2}$$

5.

$$\begin{array}{r|l}
 & 3x^3 - 9x^2 + 25x - 75 \\
\hline
x + 3 & 3x^4 \qquad - 2x^2 + 1 \\
 & \underline{3x^4 + 9x^3} \\
 & -9x^3 - 2x^2 \\
 & \underline{-9x^3 - 27x^2} \\
 & \qquad +25x^2 \\
 & \qquad \underline{+25x^2 + 75x} \\
 & \qquad\qquad -75x \\
 & \qquad\qquad \underline{-75x - 225} \\
 & \qquad\qquad\qquad +226
\end{array}$$

The solution then is:

$$3x^3 - 9x^2 + 25x - 75 + \frac{226}{x + 3}$$

8.

$$\begin{array}{r} 3x + 3 \\ x^2 - x \overline{)\,3x^3 \qquad - 2x - 1} \\ \underline{3x^3 - 3x^2} \qquad\qquad \\ -3x^2 \qquad\quad \\ \underline{3x^2 - 3x} \quad \\ +x - 1 \end{array}$$

Solution:

$$3x + 3 + \frac{x - 1}{x^2 - 1}$$

11.

-2	1	-1	-6	5
		-2	+6	0
	1	-3	0	5

$Q(x) = x^2 - 3x + \frac{5}{x + 2}$

or $Q(x) = x^2 - 3x + R(x) = \frac{5}{x + 2}$

14.

-3	1	+0	+0	+0	-81
		-3	+9	-27	+81
	1	-3	+9	-27	+0

$Q(x) = x^3 - 3x^2 + 9x - 27$

$R(x) = 0$

17.

-2	1	+0	+0	+0	+0	+32
		-2	+4	-8	+16	-32
	1	-2	+4	-8	+16	+0

$Q(x) = x^4 - 2x^3 + 4x^2 - 8x + 16$

$R(x) = 0$

20.

-3	8	+4	-1	-5
		-24	+60	-177
	8	-20	+59	-182

$Q(x) = 8x^2 - 20x + 59$

$R(x) = \frac{-182}{x + 3}$

Section 11.2 The Remainder and Factor Theorems

BASIC CONCEPTS you should learn from this unit.

1. Remainder Theorem: If a polynomial $P(x)$ is divided by $(x - r)$, then the remainder is $P(r)$.

2. The Remainder Theorem coupled with synthetic division is an easy way to obtain (x, y) pairs for a polynomial.

3. Factor Theorem: A polynomial $P(x)$ has a factor $(x - r)$ if and only if $P(r) = 0$.

WHAT YOU SHOULD KNOW HOW TO DO upon completion of this unit.

1. For a given polynomial $P(x)$, find $P(r)$ by using the Remainder Theorem.

2. Determine by the Factor Theorem (that is, by synthetic division) whether $x - r$ is a factor of a polynomial $P(x)$.

3. Use the Remainder Theorem to assist in plotting points in graphing a polynomial.

SOLUTIONS to Exercise Set 11.2

2. $P(x) = x^4 - 3x^2 - 5x$

$P(r) = 3$

-1	1	+0	-3	-5	+0
		-1	+1	+2	+3
	1	-1	-2	-3	+3

5. $P(x) = x^6 - 3x^4 + 2x^3 + 4$

$P(r) = 0$

-1	1	+0	-3	+2	+0	+0	+4
		-1	+1	+2	-4	+4	-4
	1	-1	-2	+4	-4	+4	0

8.

-2	2	+1	+0	-5
		-4	+6	-12
	2	-3	+6	-17

R = -17

11.

-2	1	+0	+0	+0	+0	-30
		-2	+4	-8	+16	-32
	1	-2	+4	-8	+16	-62

R = -62

14. Using the Remainder Theorem we obtain

r	3 + 5 + 1 + 5	-2
1	3 8 + 9 + 14	12 = P(1)
0	...	-2 = P(0)
-1	3 + 2 - 1 + 4	-6 = P(-1)
-2	3 - 1 + 3 - 1	0 = P(-2)
-3	3 - 4 + 13 - 34	66 = P(-3)

x	y
1	12
0	-2
-1	-6
-2	0
-3	66

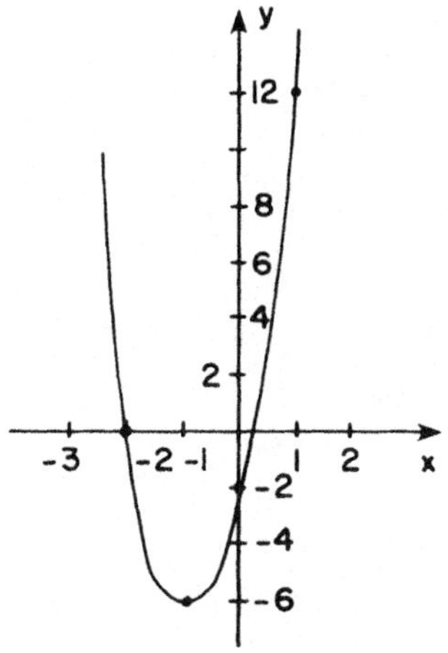

17.

r	1 - 10 + 0 + 0 + 1	
10	1 0 0 0 + 1	P(10)
6	1 - 4 - 24 - 144 - 863	P(6)
1	1 - 10 - 10 - 10 - 9	P(1)
0	... + 1	P(0)
-1	1 - 11 + 11 - 11 + 12	P(-1)
-4	1 - 6 + 24 - 96 + 396	P(-4)

x	y
10	1
6	-863
1	-9
0	+1
-1	+12
-4	+396

(continued on page 204)

17. continued

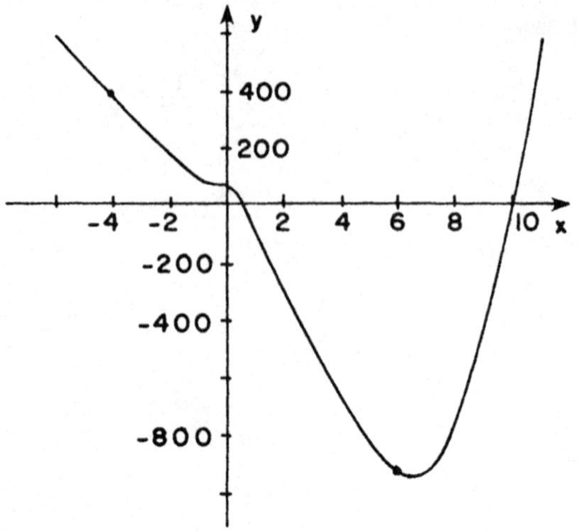

20.

1	1	+4	-3	+1
		1	+5	+2
	1	+5	+2	+3

R = 3, so x - 1 is <u>not</u> a factor.

23.

-3	1	+0	+0	+27
		-3	+9	-27
	1	-3	+9	0

R = 0, so x + 3 is a factor.

26.

3	1	+0	+0	+27
		+3	+9	+27
	1	+3	+9	+54

R = 54, so x - 3 is <u>not</u> a factor

29. We synthetically divide and determine K so that the remainder = 0 .

2	1	-3	+K	-1
		+2	-2	2(K-2)
	1	+1	+(K-2)	

$$2(K - 2) = 1$$
$$2K - 4 = 1$$
$$2K = 5$$
$$\underline{K = 5/2}$$

32.

r	2 +	0 + 3		+ 0	2
		$2r + 2r^2$		$2r^3+3r$	$2r^4+3r^2$
	2	$+2r + 2r^3+3$		$+ 2r^3+3r$	0

For the sum of $2 + (2r^4 + 3r^2)$ to equal 0 we have the equation

$$2r^4 + 3r^2 = -2$$

which is impossible, since $r^4 > 0$ and $r^2 > 0$ for all real r .

BASIC CONCEPTS you should learn from this unit.

1. The conjugate of a complex number $a + bi$ is $a - bi$.

2. The product of two conjugate complex numbers is a real number; for example, $(a + bi)(a - bi) = a^2 + b^2$.

3. The quotient of two complex numbers is found by multiplying the dividend and the divisor by the conjugate of the divisor.

4. The reciprocal of a complex number $(a + bi)$ is found by division:
$$\frac{1}{a + bi} \cdot \frac{(a - bi)}{(a - bi)} = \frac{a - bi}{a^2 + b^2} \text{ or } \frac{a}{a^2 + b^2} - \frac{bi}{a^2 + b^2}$$

5. Five basic properties of complex numbers are listed here, using the symbols Z and $\overline{Z}$ to represent a complex number and its conjugate, respectively.
   A. $\overline{Z} = \overline{W}$ if and only if $Z = W$.
   B. $\overline{Z} = Z$ if and only if Z is real .
   C. $\overline{Z + W} = \overline{Z} + \overline{W}$.
   D. $\overline{Z \cdot W} = \overline{Z} \cdot \overline{W}$.
   E. $\overline{Z^n} = \overline{Z}^n$, n a positive integer.

6. Factor Theorem: r is a root of $P(x)$ if and only if $x - r$ is a factor of $P(x)$.

7. Fundamental Theorem of Algebra: Every polynomial $P(x)$ of degree ≥ 1 has at least one complex root.

8. Linear Factor Theorem: A polynomial $P(x)$ of degree $n \geq 1$ can be written as the product of n linear factors such that $P(x) = a(x - r_1)(x - r_2) \ldots (x - r_n)$.

9. Number of Roots Theorem: If $P(x)$ is a polynomial of degree $n \geq 1$, then $P(x)$ has exactly n roots among the complex numbers.

10. The same number may occur as a root of a polynomial more than once, say K times. It then is said to have multiplicity K . For example, in
$$P(x) = x^3 + 3x^2 + 3x + 1 = (x + 1)^3$$
the number -1 is a root of multiplicity 3 .

11. The "depressed" equation $Q(x) = 0$, taken from
$$P(x) = (x - r)Q(x)$$
can be used to advantage in finding the roots of $P(x)$.

12. Conjugate Root Theorem: If $P(x)$ is a polynomial of degree $n \geq 1$ with real coefficients and if $a + bi$ $(b \neq 0)$ is a root of $P(x)$, then $a - bi$ is also a root. (In common language: Complex roots occur in conjugate pairs.)

WHAT YOU SHOULD KNOW HOW TO DO upon completion of this unit.

1. Multiply a complex number by its conjugate.

2. Divide two complex numbers.

3. Find the reciprocal of a complex number.

4. Write a polynomial $P(x)$ if given its roots.

5. Find the roots of a polynomial if given some of the roots by using the Factor Theorem, Conjugate Root Theorem, and Number of Roots Theorem.

SOLUTIONS to Exercise Set 11.3

2. $(3 + i)(3 - i) = 9 - i^2 = 9 + 1 = 10$

5. $(-4 - 2i)(-4 + 2i) = 16 - 4i^2 = 16 + 4 = 20$

8. $\frac{(1 + 3i)}{(2 - 5i)} \cdot \frac{(2 + 5i)}{(2 + 5i)} = \frac{2 + 11i - 13i^2}{4 + 25} = \frac{-13 + 11i}{29} = \frac{-13}{29} + \frac{11i}{29}$

11. $\frac{(3 - 2i)}{(2 - i)} \cdot \frac{(2 + i)}{(2 + i)} = \frac{6 - i - 2i^2}{4 + 1} = \frac{8 - i}{5} = \frac{8}{5} - \frac{i}{5}$

14. $\frac{(5 - 2i)}{-3i} \cdot \frac{+3i}{+3i} = \frac{15i - 6i^2}{-9i^2} = \frac{+6 + 15i}{9} = \frac{2}{3} + \frac{5i}{3}$

17. $\frac{1}{(4 + 3i)} \cdot \frac{4 - 3i}{(4 - 3i)} = \frac{4 - 3i}{16 + 9} = \frac{4}{25} - \frac{3i}{25}$

20. $\frac{1}{-7i} \cdot \frac{7i}{7i} = \frac{7i}{49} = \frac{i}{7}$

23. Let $Z = a + bi$, then $\overline{Z} = a - bi$;
Let $W = c + di$, then $\overline{W} = c - di$.

$Z \cdot W = (a + bi)(c + di) = ac - bd + (ad + bc)i$

and so $\overline{Z \cdot W} = ac - bd - (ad + bc)i$

Also $\overline{Z} \cdot \overline{W} = (a - bi)(c - di) = ac - bd - adi - bci = (ac - bd) - (ad + bc)i$

and we have proved that $\overline{Z \cdot W} = \overline{Z} \cdot \overline{W}$.

26. $P(x)$ must equal

$$(x - 5)(x + 5)(x - 1)(x + 1) = (x^2 - 25)(x^2 - 1)$$

or $P(x) = x^4 - 26x^2 + 25$ (using the Factor Theorem and the Linear Factor Theorem).

29. $P(x) = (x - 4)[x - (1 - \sqrt{3})][x - (1 + \sqrt{3})] = (x - 4)(x^2 - 2x - 2)$

$= x^3 - 6x^2 + 6x + 8$

32. $P(x) = (x - 3)^2(x + 2)(x - 2) = (x^2 - 6x + 9)(x^2 - 4)$

$= x^4 - 4x^2 - 6x^3 + 24x + 9x^2 - 36 = x^4 - 6x^3 + 5x^2 + 24x - 36$

Using synthetic division to find $P(4)$, we obtain

```
4 | 1  -6  +5  +24  -36
       4  -8  -12  +48
    ---------------------
    1  -2  -3  +12  +12
```

Since $P(4) = 12$ as required, the number $a = 1$ and $P(x) = x^4 - 6x^3 + 5x^2 + 24x - 36$.

35. Given $(x - 3)(x + 1)(x - 2) = 0$ the roots can be read directly as $3, -1, 2$.

38. $(x^2 - x)(x^2 - 2x + 5) = 0$. From $x^2 - x = 0$, we get $x(x - 1) = 0$ so $x = 0$, $x = 1$ and from $x^2 - 2x + 5 = 0$ we get $x = \dfrac{2 \pm \sqrt{4 - 20}}{2} = 1 \pm 2i$. So the roots of the equation are $0, 1, 1+2i, 1-2i$.

41. $(x - 5)^3(x + 5)^2 = 0$ directly tells us that 5 is a root of multiplicity 3 and -5 is a root of multiplicity 2.

44. $(x - 1)^2(x + 4)$ will do it; or

$(x^2 - 2x + 1)(x + 4) = x^3 + 4x^2 - 2x^2 - 8x + x + 4$

$= x^3 + 2x^2 - 7x + 4$

47. Using synthetic division is one way.

```
-1 | 1  +0  -3  -2
        -1  +1  +2
     ---------------
     1  -1  -2   0
```

So the depressed equation is

$x^2 - x - 2 = 0$

which factors to $(x - 2)(x + 1) = 0$, so that $x = 2$, $x = -1$ are the other two roots.

50. By synthetic division

```
-1 | 1  -2  -7  -4
        -1  +3  +4
     ---------------
     1  -3  -4   0
```

we get the depressed equation $x^2 - 3x - 4 = 0$. The left-hand side factors:

$(x - 4)(x + 1) = 0$

so the other roots are $x = 4$, $x = -1$.

53. If $1 + 3i$, -2 are the <u>only</u> roots, then $[x - (1 + 3i)][x - (-2)] = 0$ is the polynomial equation, so

$$P(x) = x^2 - (1 + 3i)x + 2x - 2 - 6i$$

or

$$P(x) = x^2 + (1 - 3i)x - (2 + 6i)$$

56. As in solution 53 we can write

$$P(x) = (x + 2)(x - 3)(x - 1 - 2i) = x^3 - (2 + 2i)x^2 - (5 - 2i)x + 6 + 12i$$

59. If $(3 - i)$ is a root, then so is $(3 + i)$. Multiplying $[x - (3 - i)] \cdot [x - (3 + i)] = x^2 - 6x + 10$, which must divide the given polynomial. Doing this we obtain

$$\begin{array}{r|l} & x - 1 \\ \hline x^2 - 6x + 10 & x^3 - 7x^2 + 16x - 10 \\ & \underline{x^3 - 6x^2 + 10x} \\ & \quad -x^2 + 6x - 10 \\ & \quad \underline{-x^2 + 6x - 10} \\ & \qquad\qquad\qquad 0 \end{array}$$

Now, the required product is $(x^2 - 6x + 10)(x - 1)$.

62. If $(-2 + i)$ is a root, then so is $(-2 - i)$ and multiplying $[x - (-2 + i)] \cdot [x - (-2 - i)]$ yields $x^2 + 4x + 5$. Dividing this into the given polynomial, we obtain

$$\begin{array}{r|l} & x^2 - x - 6 \\ \hline x^2 + 4x + 5 & x^4 + 3x^3 - 5x^2 - 29x - 30 \\ & \underline{x^4 + 4x^3 + 5x^2} \\ & \quad -x^3 - 10x^2 - 29x \\ & \quad \underline{-x^3 - 4x^2 - 5x} \\ & \qquad -6x^2 - 24x - 30 \\ & \qquad \underline{-6x^2 - 24x - 30} \end{array}$$

so that

$$(x^2 + 4x + 5)(x^2 - x - 6) \quad \text{or} \quad (x^2 + 4x + 5)(x - 3)(x + 2)$$

is the required product.

65. $x - a - bi = P(x)$; or $x - (a + bi) = P(x)$

BASIC CONCEPTS you should learn from this unit.

1. Descartes' Rule of Signs:

   A. The number of positive roots of $P(x)$ is either equal to the number of variations in sign of $P(x)$ or less than that number by an even number.

   B. The number of negative roots of $P(x)$ is either equal to the number of variations in sign of $P(-x)$ of less than that number by an even number.

2. The Rational Root Theorem: If the coefficients of a polynomial

   $P(x) = a_n x^n + a_{n-1} x^{n-1} + \ldots + a_0$, $a_n \neq 0$, and if p/q is a rational root in lowest terms, then p divides a_0 and q dividies a_n.

3. A rational root may be a multiple root.

WHAT YOU SHOULD KNOW HOW TO DO upon completion of this unit.

1. Describe the nature of the roots of a polynomial using Descartes' Rule of Signs.

2. Find all the rational roots of a polynomial equation using Descartes' Rule of Signs, the Rational Root Theorem, and the Conjugate Root Theorem.

SOLUTIONS to Exercise Set 11.4

2. $2x^6 + 5x^5 + x^3 - 6 = 0$; one change of sign, so one positive root.

   $P(-x) = 2x^6 - 5x^5 - x^3 - 6$; one change of sign, so one negative root.

   one positive, one negative, four complex roots

5. $x^5 - 4x^3 + 7x - 4 = 0$; three changes of sign, so there are either three positive roots or one positive root.

   $P(-x) = -x^5 + 4x^3 - 7x - 4$; two changes of sign, so there are two or no negative roots. Four possibilities exist: (1) three positive, two negative roots; (2) three positive, zero negative, and two complex roots; (3) one positive, two negative, and two complex roots; (4) one positive and four complex roots.

8. $x^5 + 6x^4 - x^3 - 2x - 3 = 0$; one change of sign, so there is exactly one positive root.

   $P(-x) = -x^5 + 6x^4 + x^3 + 2x - 3 = 0$; two changes of sign, so there are two negative roots or none. The possibilities are one positive, two negative, and two complex roots; or one positive, zero negative, and four complex roots.

11. $x^8 + 7x^3 + 3x - 5 = 0$; one change of sign, so there is one positive root.

    $P(-x) = x^8 - 7x^3 - 3x - 5 = 0$; one change of sign, so there is one negative root. The answer is one positive, one negative, and six complex roots.

14. $3x^3 - x^2 - 3x + 1 = 0$; Since $a_n = 3$ and $a_0 = 1$, the possible rational roots are ± 1, $\pm 1/3$.

	3	-1	-3	+1
1/3	3	0	-3	0

This synthetic division shows that 1/3 is a root, and the depressed equation $3x^2 - 3 = 0$ has +1 and -1 for roots. Thus the roots are 1/3 , +1 , -1 .

17. $5x^6 - x^5 - 5x^4 + 6x - x^2 - 5x + 1 = 0$. The possibilities are ± 1 and $\pm 1/5$. Using synthetic division, we obtain

	5	-1	-5	+6	-1	-5	+1	
1	5	4	-1	+5	+4	-1	0	1 is a root
-1	5	-1	0	5	-1	0		-1 is a root
1/5	5	0	0	5	0			1/5 is a root
-1	5	-5	+5	0				-1 is a root

From $5x^2 - 5x + 5 = 0$ we see that the roots of this depressed equation are complex, so we have located all the rational roots.

20. $6x^4 + 2x^3 + 7x^2 + x + 2 = 0$; there are no changes of sign, so there are <u>no</u> positive roots.

$P(-x) = 6x^4 - 2x^3 + 7x^2 - x + 2 = 0$; there are four changes of sign, so there are <u>four</u>, <u>two</u>, or <u>no</u> negative roots. Possibilities are -1 , -2 , -1/2 , -1/3 , -2/3 , -1/6 . Using synthetic division, we obtain

	6	+2	+7	+1	+2
-1	6	-4	11	-10	12
-2	6	-10	+27	-53	108
-1/2	6	-1	15/2	-11/4	27/8
-1/3	6	0	7	-4/3	22/9
-1/6	6	+1	+41/6	-5/36	437/216
-2/3	6	-2	25/3	-41/9	136/27

Since no zero remainders occurred, we conclude that there are <u>no</u> rational roots.

23. $4x^4 + x^3 + x^2 + x - 3 = 0$; one positive root, so there are three or one negative roots. Possibilities are ± 1, ± 3, $\pm 1/2$, $\pm 3/2$, $\pm 1/4$, $\pm 3/4$.

	4	+1	+1	+1	-3	
1	4	5	6	7	4	
1/2	4	3	5/2	9/4	-15/8	
3/4	4	4	4	4	0	3/4 is a root
-1	4	0	4	0		-1 is a root

23. continued

Then the depressed equation $4x^2 + 4 = 0$ gives the two remaining roots: $4x^2 = -4$, $x^2 = -1$, $x = \pm i$. The roots are $3/4$, -1, $\pm i$.

26. $12x^4 - 8x^3 - 75x^2 - 16x - 5 = 0$; there is one positive root and three or one negative roots. Possible roots are ± 1, ± 5, $\pm 1/2$, $\pm 1/3$, $\pm 1/4$, $\pm 1/6$, $\pm 1/12$, $\pm 5/2$, $\pm 5/3$, $\pm 5/4$, $\pm 5/6$, $\pm 5/12$.

Testing these by synthetic division reveals no rational roots. Thus the roots must be irrational or complex.

29. $2x^4 - x^3 - 28x^2 + 30x - 8 = 0$. Possible rational roots are ± 1, ± 2, ± 4, ± 8, $\pm 1/2$.

	2	-1	-28	+30	-8	
1/2	2	0	-28	16	0	1/2 is a root
-4	2	-8	4	0		-4 is a root

The depressed equation $2x^2 - 8x + 4 = 0$ or $x^2 - 4x + 2 = 0$ gives

$$x = \frac{4 \pm \sqrt{8}}{2} = 2 \pm \sqrt{2}$$

Roots are $\frac{1}{2}$, -4, $2 \pm \sqrt{2}$.

32.

	1	-4	-k	+6k	+9	
3	1	-1	-3-k	-9+3k	0	if $3(-9 + 3k) = -9$, $k = 2$

so if $k = 2$, the equation becomes

$$x^4 - 4x^3 - 2x^2 + 12x + 9$$

	1	-4	-2	+12	+9	
3	1	-1	-5	-3	0	3 is a root
-3	1	-7	19	-45	+144	no root
9	1	5	43	384	+	no root
-9	1	-13	115	-	+	no root
-1	1	-5	+3	+9	0	-1 is a root
1	1	-3	-5	+7	+	no root

38. Given that p is a prime, we let $x = \sqrt{p}$, or $x^2 = p$, or $x^2 - p = 0$. Now since p is prime the only possible rational roots to this equation are ± 1 and $\pm p$. Testing these quickly reveals that none are roots. Hence the two roots must be irrational and so p is irrational.

SOLUTIONS to Review Exercises Chapter 11

2.

	1	-3	+0	+2	-5
-2	1	-5	10	-18	31

R = 31

$Q(x) = x^3 - 5x^2 + 10x - 18$

5.

	2	+4	+3	+5	-2
-2	2	0	3	-1	0

Since R = 0, x + 2 is a factor of P(x).

8. $\frac{2 + i}{-5i} \cdot \frac{5i}{5i} = \frac{10i + 5i^2}{-10i^2} = \frac{-1}{2} + i$

11. $\frac{1}{-4i} \cdot \frac{4i}{4i} = \frac{4i}{16} = \frac{i}{4}$

14. $P(x) = (x - 3)(x - \sqrt{-3})(x + \sqrt{-3}) = x^3 - 3x^2 + 3x - 9$

17. $P(x) = (x - i)(x - i)(x + i)(x + i) = (x^2 - i^2)(x^2 - i^2)$
$= (x^2 + 1)(x^2 + 1) = x^4 - 2x^2 + 1$

20.

	1	-2	-9	+4
4	1	2	-1	0

so 4 is indeed a root. The depressed equation $x^2 + 2x - 1 = 0$ yields

$$x = \frac{-2 \pm \sqrt{4 + 4}}{2} = -1 \pm \sqrt{2}$$

as the other two roots.

23. $x^5 - x^4 + 3x^3 - 4x^2 + x - 5$ has five changes of sign, so there are at most five positive roots.

$-x^5 - x^4 - 3x^3 - 4x^2 - x - 5$ has no changes of sign, so there are no negative roots.

26. For $6x^3 - 5x^2 - 33x - 18 = 0$ the possible rational roots are ±1, ±2, ±3, ±6, ±18, ±1/2, ±3/2, ±9/2, ±1/3, ±2/3, ±1/6. Some synthetic division

	6	-5	-33	-18
3	6	13	6	0

reveals that 3 is a root. Then the depressed equation $6x^2 + 13x + 6 = 0$ yields -3/2 and -2/3 as the other two roots.

29. $6x^3 + 15x^2 - x - 10 = 0$. Trying synthetic division with the possible roots will yield

$$\begin{array}{r|rrrr} & 6 & +15 & -1 & -10 \\ \hline -1 & 6 & 9 & -10 & 0 \end{array}$$

so -1 is a root. Now use the depressed equation $6x^2 + 9x - 10 = 0$ to obtain

$x = \dfrac{-9 \pm \sqrt{81 + 240}}{12}$ or $\dfrac{-9 \pm \sqrt{321}}{12}$ as the two remaining roots.

Chapter 11 - A SHORT QUIZ

1. How many roots does a polynomial of degree n have?

2. In the following, where $P(x)$, $D(x)$, $Q(x)$, and $R(x)$ are polynomials, the degree of $R(x)$ will be ___?___ than the degree of $P(x)$.

$$\frac{P(x)}{D(x)} = Q(x) + \frac{R(x)}{D(x)}$$

3. In the expression in Problem 2, if $R(x) = 0$, we know that $D(x)$ is a ___?___ of $P(x)$.

4. In the expression

$$\frac{P(x)}{D(x)} = Q(x) + \frac{R(x)}{D(x)}$$

if $D(x) = x - r$, then $R(x)/D(x)$ is the value of $P(x)$ at $x =$ ___?___.

5. If $a + bi$, a complex number with $b \neq 0$, is a root of a polynomial $P(x)$, then ___?___ is also a root of $P(x)$.

6. Multiply $(a + bi)(a - bi)$.

7. Write a polynomial $P(x)$ in factored form if its roots are known to be $6, -3, 4$.

8. If $P(x) = 3x^2 - 2x + 5$, use synthetic division to find $P(-2)$.

9. If $P(x) = 4x^3 + 3x^2 + x + 2$, how many positive roots are there? How many negative roots are possible?

10. Determine the roots of $(x-2)^4(x + 5)^3 = 0$.

11. Synthetic division of $(x^3 - x^2 - 2x + 8) \div (x + 2)$ reveals

$$\begin{array}{r|rrrr} & 1 & -1 & -2 & +8 \\ \hline -2 & 1 & -3 & 4 & 0 \end{array}$$

showing that -2 is a root of the third degree polynomial. Use the depressed equation to find the other roots.

12. Does a polynomial always have a root in the complex numbers?

13. Complex numbers are usually expressed in the form $a + bi$. What kind of numbers are a and b? What does i represent?

14. Write the equation with real coefficients of degree two that has $-7i$ as one of its roots.

15. If $x - r$ is a factor of $P(x)$, then ____?____ is a ____?____ of the equation $P(x) = 0$.

Quiz Answers Chapter 11

1. n

2. less than

3. factor

4. r

5. $a - bi$

6. $a^2 + b^2$

7. $P(x) = (x - 6)(x + 3)(x - 4)$

8. 21

9. no positive roots; three or one negative roots

10. 2 is a root of multiplicity 4;
    -5 is a root of multiplicity 3.

11. $\frac{3}{2} \pm \frac{i\sqrt{7}}{2}$

12. yes

13. a and b are real numbers;
    i represents $\sqrt{-1}$.

14. $x^2 + 49 = 0$

15. r, solution or root

CHAPTER 12

Topics in Algebra

Section 12.1 Arithmetic Progressions

BASIC CONCEPTS you should learn from this unit.

1. An infinite sequence is a function whose domain is the set of natural numbers.

2. A finite sequence is a sequence that terminates.

3. An arithmetic sequence (sometimes called an arithmetic progression) is a sequence such that each term differs from its predecessor by a fixed number, called the common difference.

4. Symbolically, the nth term in an arithmetic sequence can be written by:

   A. a recursive formula, $a_n = a_{n-1} + d$;

   B. the formula $a_n = a_1 + (n - 1)d$, which relates a_n to a_1.

5. Adding the terms of a sequence results in a series. Thus, adding the terms of an arithmetic sequence would yield an arithmetic series. In symbols,

$$S_n = a_1 + a_2 + \dots + a_{n-1} + a_n$$

6. Two formulas for the sum of an arithmetic series are

   A. $S_n = \frac{n}{2}(a_1 + a_n)$;

   B. $S_n = \frac{n}{2}[2a_1 + (n - 1)d]$.

WHAT YOU SHOULD KNOW HOW TO DO upon completion of this unit.

1. Define arithmetic sequence and arithmetic series.

2. Find a_n if given a_1 and d of an arithmetic sequence.

3. Find d if given a_n and a_1 of an arithmetic sequence.

4. Find a_n if given a_{n-1} and d of an arithmetic sequence.

5. Recognize an arithmetic sequence or progression.

6. Find the sum of n terms of an arithmetic series if given sufficient information.

SOLUTIONS to Exercise Set 12.1

2. Recall that the domain is the set of natural numbers; then if $a_n = 2n + 1$, the first four terms are $2\cdot1 + 1$, $2\cdot2 + 1$, $2\cdot3 + 1$, $2\cdot4 + 1$, or $3, 5, 7, 9$.

5. Using $n = 1, 2, 3, 4$ successively, we have $\frac{1}{3}, \frac{4}{5}, \frac{9}{7}, \frac{16}{9}$ as the first four terms.

8. $\frac{8}{2}, \frac{11}{2}, \frac{14}{2}, \frac{17}{2}$ is an arithmetic progression with $d = \frac{3}{2}$.

11. $0, \frac{1}{4}, \frac{2}{4}, \frac{3}{4}$ is an arithmetic sequence with $d = \frac{1}{4}$.

14. 12, 8, 4, 1 is <u>not</u> an arithmetic sequence, since there is no common difference.

17. $a_1 = 3$; $a_2 = 3 + \left(-\frac{1}{2}\right)$; $a_3 = 3 + 2\cdot\left(-\frac{1}{2}\right)$, and $a_4 = 3 + 3\left(-\frac{1}{2}\right)$. Thus we have $3,\ \frac{5}{2},\ 2,\ \frac{3}{2}$.

20. Here we have $6,\quad 6 + \frac{5}{2},\quad 6 + 2\cdot\left(\frac{5}{2}\right),\quad 6 + 3\left(\frac{5}{2}\right)$, or $6,\ \frac{17}{2},\ 11,\ \frac{27}{2}$.

23. $a_{12} = a_1 + (12 - 1)d$, so $a_{12} = 14 + 11(-2) = -8$.

26. If $a_1 = \frac{1}{2}$ and $a_{12} = 6$, we can write $6 = \frac{1}{2} + 11\cdot d$ and find $d = \frac{1}{2}$; then $a_{30} = \frac{1}{2} + 29\cdot\left(\frac{1}{2}\right) = \frac{30}{2}$ or 15.

29. If $a_1 = -\frac{1}{4}$ and $a_{41} = 10$, we can write $10 = -\frac{1}{4} + 40d$ and find $d = \frac{41}{160}$; then $a_{22} = -\frac{1}{4} + 21\cdot\left(\frac{41}{160}\right) = \frac{821}{160}$.

32. $S_n = \frac{n}{2}[2a_1 + (n - 1)d]$, so in our case $S_{24} = \frac{24}{2}\left[2(-4) + 23\cdot\frac{1}{2}\right] = 42$.

35. Here $S_{40} = \frac{40}{2}[2(82) + 39\cdot(-2)] = 1720$.

38. In this instance $a_1 = 44$; $d = -3$, $S_n = 340$, and n is unknown. Thus we can write

$340 = \frac{n}{2}[2\cdot 44 + (n - 1)(-3)]$ and solve for n: $340 = \frac{n}{2}(91 - 3n)$; $680 = 91n - 3n^2$;

and then $3n^2 - 91n + 680 = 0$ and $(3n - 40)(n - 17) = 0$ and thus $n = 17$.

41. $a_1 = \frac{1}{2}$, $a_n = \frac{7}{4}$, $S_n = \frac{27}{4}$. We use the formulas to make two equations:

(1) $\frac{27}{4} = \frac{n}{2}[1 + (n - 1)d]$

(2) $\frac{7}{4} = \frac{1}{2} + (n - 1)d$

We solve for d in (2):

$$d = \frac{5}{4(n - 1)}$$

and substitute this result in (1):

$$\frac{27}{4} = \frac{n}{2} + \frac{n}{2}\cdot\cancel{(n-1)}\,\frac{5}{4\cancel{(n-1)}}$$

$$\frac{27}{4} = \frac{n}{2} + \frac{5n}{8}$$

$$54 = 4n + 5n$$

$$n = 6$$

44. Since $a_3 = 9$ and $a_6 = 18$, the common difference $d = (18 - 9)/3 = 3$. Subtracting 2d or 6 from a_3, that is, $9 - 6 = 3$, we get $a_1 = 3$. Now $S_{12} = \frac{12}{2}[2 \cdot 3 + (11)(3)] = 234$.

Section 12.2 Geometric Progressions

BASIC CONCEPTS you should learn from this unit.

1. A geometric sequence or progression is a sequence such that each term is obtained by multiplying the previous term by a fixed number. In symbols, $a_n = r \cdot a_{n-1}$ for $n > 1$; r is called the common ratio.
2. The nth term of a geometric sequence or progression is given by $a_n = a_1 r^{n-1}$
3. Geometric means are the terms between the first and last terms.
4. A geometric series is the sum of the terms in a geometric progression.
5. $S_n = \frac{a_1(1 - r^n)}{1 - r}$ is the formula for the sum of a geometric series.
6. If $|r| < 1$, then r^n gets closer and closer to zero as n gets larger, and thus the sum of an infinite geometric series with $|r| < 1$ becomes

$$S_n = \frac{a_1}{1 - r}$$

WHAT YOU SHOULD KNOW HOW TO DO upon completion of this unit.

1. Define geometric sequence and geometric series.
2. Find a_n if given a_1 and r of a geometric sequence.
3. Find r if given several terms of a geometric sequence.
4. Insert geometric means between two terms of a geometric sequence.
5. Recognize a geometric sequence.
6. Find the sum of n terms of a geometric series.
7. Find the sum of an infinite geometric series that has $|r| < 1$.

SOLUTIONS to Exercise Set 12.2

2. $(-3) \cdot (-4) = 12$; $(-3) \cdot 12 = -36$; $(-3)(-36) = 108$; so yes, this is a geometric sequence; $r = -3$.
5. Yes, it is geometric with $r = 0.2$.

8. $a_1 = -4$, $a_2 = (-4) \cdot 2 = -8$, and so on, yielding $-4, -8, -16, -32, \ldots$.

11. $-3, -6, -12, -24$ are found by multiplying a term's predecessor by 2, the common ratio.

14. $a_n = a_1 \cdot r^{n-1}$, so $a_6 = 18 \cdot \left(-\frac{1}{2}\right)^5 = 18\left(-\frac{1}{32}\right) = -\frac{9}{16}$.

17. $a_5 = a_1 \cdot r^4$, so $\frac{1}{27} = 3 \cdot r^4$ and then $r^4 = \frac{1}{81}$ and $r = \frac{1}{3}$ or $-\frac{1}{3}$. Then

$a_7 = 3 \cdot \left(\frac{1}{3}\right)^6 = \frac{1}{243}$.

20. $a_7 = a_4 \cdot r^3$, so $1 = \frac{1}{4} \cdot r^3$, $r^3 = 4$; $r = \sqrt[3]{4}$.

23. $a_n = a_1 r^{n-1}$, so $32 = \frac{1}{2} r^{n-1}$, and $64 = r^{n-1}$. But $r = 2$, and thus $2^6 = 2^{n-1}$ or $n = 7$.

26. We would write -3, a_2, a_3, 192, and find r from $192 = -3r^3$, $r^3 = -64$, and finally $r = -4$. Using this value to find a_2, we get $a_2 = -3 \cdot (-4) = 12$, and then $12 \cdot (-4) = -48$ for a_3.

29. $$S_n = \frac{a_1(1 - r^n)}{1 - r}$$

so with $a_1 = 3$ and $r = \frac{1}{3}$ we have

$$S_7 = \frac{3[1 - (1/3)^7]}{1 - 1/3} = \frac{9[1 - (1/3^7)]}{2} = \frac{9 - (1/3^5)}{2} = \frac{2186}{486} = \frac{1093}{243}.$$

32. $$S_6 = \frac{2\left(1 - \frac{2}{3}\right)^6}{1 - \frac{2}{3}} = 6\left(1 - \frac{64}{729}\right) = 6\left(\frac{665}{729}\right) = \frac{1330}{243}$$

35. $a_4 = a_1 \cdot r^3$, so $-\frac{54}{8} = 2r^3$, $r^3 = -\frac{27}{8}$, and $r = -\frac{3}{2}$. Now

$$S_5 = \frac{2[1 - (-3/2)^5]}{1 - (-3/2)} = \frac{4}{5}\left[\left(1 + \frac{243}{32}\right)\right] = \frac{4}{5}\left(\frac{275}{32}\right) = \frac{25}{8}$$

38. Here $a_1 = 20000$, $r = 1.05$, and we seek a_{11}.

$a_{10} = 20000(1.05)^{10} = 32577.9 = 32578$ people

41. $a_1 = 1$, $r = \frac{1}{2}$, $S = \frac{1}{1 - \frac{1}{2}} = 2$.

44. $a_1 = \frac{1}{2}$, $r = -\frac{1}{2}$, $S = \frac{0.5}{1 + .5} = \frac{0.5}{1.5} = \frac{1}{3}$.

47. $a_1 = 0.5$, $r = 0.5$, $s = \frac{0.5}{1 - 0.5} = 1$.

50. $3.6\overline{6} = 3 + 0.6 + 0.06 + 0.006 + \ldots$.
If we choose $0.6 = a_1$ and $r = 0.1$, the sum from 0.6 on is $\frac{0.6}{1 - 0.1} = \frac{2}{3}$ and the final sum is $3\frac{2}{3}$ or $\frac{11}{3}$.

53. Writing $0.325\overline{325}$ as $0.325 + 0.000325 + \ldots$, we have $a_1 = 0.325$ and $r = 0.001$.
Then the sum becomes $\frac{0.325}{1 - 0.001} = \frac{325}{999}$.

Section 12.3 Mathematical Induction

BASIC CONCEPTS you should learn from this unit.

1. The principle of mathematical induction is a method of proving certain statements involving the natural numbers.

2. Mathematical induction is defined as follows: If a statement involving the natural numbers, n,

   A. is true when $n = 1$, and

   B. whenever the statement is true for some natural number k, it is also true for the natural number $k + 1$, then

   the statement is true for all n.

WHAT YOU SHOULD KNOW HOW TO DO upon completion of this unit.

1. Define mathematical induction.

2. Prove certain types of statements by mathematical induction.

SOLUTIONS to Exercise Set 12.3

2. To prove $1^2 + 3^2 + 5^2 \ldots + (2n - 1)^2 = \frac{n(2n + 1)(2n - 1)}{3}$

Step 1. Evaluating the formula when $n = 1$ yields

$$\frac{1(2\cdot 1 + 1)(2\cdot 1 - 1)}{3} = 1$$

which does equal 1^2, which verifies the formula when $n = 1$.

Step 2. For $n = k$ we have

$$1^2 + 3^2 + \ldots + (2k - 1)^2 = \frac{k(2k + 1)(2k - 1)}{3}$$

Adding the square of the next consecutive odd integer to both sides gives

(continued on page 220)

2. continued

$$1^2 + 3^2 + \ldots + (2k - 1)^2 + [2(k + 1) - 1]^2 = \frac{k(2k + 1)(2k - 1)}{3} + [2(k + 1) - 1]^2$$

$$= \frac{k(2k + 1)(2k - 1)}{3} + (2k + 1)^2$$

$$= \frac{k(2k + 1)(2k - 1) + 3(2k + 1)^2}{3}$$

$$= \frac{(2k + 1)[k(2k - 1) + 3(2k + 1)]}{3}$$

$$= \frac{(2k + 1)(2k^2 + 5k + 3)}{3}$$

$$= \frac{(2k + 1)(k + 1)(2k + 3)}{3}$$

$$= \frac{(k + 1)[2(k + 1) + 1][2(k + 1) - 1]}{3}$$

and the statement is proved.

5. Prove $5 + 10 + \ldots + 5n = \frac{5n(n + 1)}{2}$

Step 1. Evaluating the formula when $n = 1$ yields

$$\frac{5 \cdot 1(1 + 1)}{2} = 5$$

as it should to verify the formula.

Step 2. For $n = k$ we have

$$5 + 10 + \ldots + 5k = \frac{5k(k + 1)}{2}$$

Adding the next multiple of 5 gives

$$5 + 10 + \ldots + 5k + 5(k + 1) = \frac{5k(k + 1)}{2} + 5(k + 1)$$

$$= \frac{5k(k + 1)}{2} + \frac{10(k + 1)}{2}$$

$$= \frac{5(k + 1)(k + 2)}{2}$$

$$= \frac{5(k + 1)[(k + 1) + 1]}{2}$$

and the statement is proved.

8. Prove that $1^3 + 2^3 + 3^3 + \ldots + n^3 = \frac{n^2(n+1)^2}{4}$.

Step 1. Evaluating at $n = 1$ yields

$$\frac{1^2(1+1)^2}{4} = 1$$

as it should to verify the formula, since $1^3 = 1$.

Step 2. For $n = k$ we have

$$1^3 + 2^3 + 3^3 + \ldots + k^3 = \frac{k^2(k+1)^2}{4}$$

adding the next consecutive cube yields

$$\begin{aligned} 1^3 + 2^3 + \ldots + k^3 + (k+1)^3 &= \frac{k^2(k+1)^2}{4} + (k+1)^3 \\ &= \frac{k^2(k+1)^2 + 4(k+1)^3}{4} \\ &= \frac{(k+1)^2(k^2 + 4k + 4)}{4} \\ &= \frac{(k+1)^2(k+2)^2}{4} \end{aligned}$$

and the statement is proved.

11. Prove that $a_n = a_1 + (n-1)d$ for an arithmetic progression.

Step 1. Evaluating when $n = 1$ yields $a_1 + (1-1)d = a_1$ or the first term of the progression, as it should.

Step 2. For $n = k$ we have $a_k = a_1 + (k-1)d$. We know that adding d to a_k will produce a_{k+1} for an arithmetic progression, so we add d to both sides:

$$\begin{aligned} a_k + d &= a_1 + (k-1)d + d \\ &= a_1 + kd \\ a_{k+1} &= a_1 + [(k+1) - 1]d \end{aligned}$$

as required to prove the statement.

14. Prove that $a + ar + ar^2 + \ldots + ar^{n-1} = \frac{a(1-r^n)}{1-r}$.

Step 1. Evaluating when $n = 1$ yields

$$\frac{a(1-r)}{1-r} = a$$

as it should to verify the statement.

(continued on page 222)

14. continued

Step 2. For $n = k$ we have

$$a + ar + \ldots + ar^{k-1} = \frac{a(1 - r^k)}{1 - r}$$

Adding the next term to both sides yields

$$\begin{aligned} a + ar + \ldots + ar^{k-1} + ar^k &= \frac{a(1 - r^k)}{1 - r} + ar^k \\ &= \frac{a(1 - r^k) + ar^k(1 - r)}{1 - r} \\ &= \frac{a(1 - r^k + r^k - r^{k+1})}{1 - r} \\ &= \frac{a(1 - r^{k+1})}{1 - r} \end{aligned}$$

and the statement is proved.

Section 12.4 The Binomial Theorem

BASIC CONCEPTS you should learn from this unit.

1. The expansion of $(a + b)^n$ where n is any natural number is simply the result of carrying out the multiplication.
2. Pascal's triangle is one way of determining the coefficients of the terms of the expansion.
3. Another way to find the coefficients is the following: In the preceding term, multiply the coefficient by the exponent of a and then divide by one more than the exponent of b. This works, of course, for all terms except the first.
4. Employing the binomial theorem is another way, although this is essentially the same as concept 3.
5. The sum of the exponents in any term of the binomial expansion equals n.
6. The product of the first n natural numbers is n factorial, which in symbols is written $n!$.
7. $0! = 1$ by definition.

WHAT YOU SHOULD KNOW HOW TO DO upon completion of this unit.

1. Defininе expansion of a binomial.
2. Define $n!$ (n factorial).
3. Expand a binomial using the binomial theorem.
4. Find a specific term in a binomial expansion.

SOLUTIONS to Exercise Set 12.4

2. $$(2a - 3b)^6 = (2a)^6 - 6(2a)^5(3b) + \frac{30(2a)^4(3b)^2}{2!} - \frac{120(2a)^3(3b)^3}{3!} + \frac{360(2a)^2(3b)^4}{4!} - \frac{720(2a)(3b)^5}{5!} + (3b)^6$$

$$= 64a^6 - 576a^5b + 2160a^4b^2 - 4320a^3b^3 + 4860a^2b^4 - 2916ab^5 + 729b^6$$

5. $$(2 - xy)^5 = 2^5 - 5\cdot 2^4xy + \frac{20\cdot 2^3(xy)^2}{2!} - \frac{60\cdot 2^2(xy)^3}{3!} + \frac{120\cdot 2(xy)^4}{4!} - \frac{120(xy)^5}{5!}$$

$$= 32 - 80xy + 80x^2y^2 - 40x^3y^3 + 10x^4y^4 - x^5y^5$$

8. $$(x - y)^7 = x^7 - 7x^6y + \frac{42x^5y^2}{2!} - \frac{5\cdot 42x^4y^3}{3!} + \frac{4(210)x^3y^4}{4!} - \frac{3(840)x^2y^5}{5!} + \frac{2\cdot(2520)xy^6}{6!} - \frac{5040}{7!}y^7$$

$$= x^7 - 7x^6y + 21x^5y^2 - 35x^4y^3 + 35x^3y^4 - 21x^2y^5 + 7xy^6 - y^7$$

11. $$\left(\frac{1}{3}x + 2\right)^3 = \left(\frac{x}{3}\right)^3 + 3\left(\frac{x}{2}\right)^2\cdot 2 + \frac{6(x/3)\cdot 2^2}{2!} + 2^3$$

$$= \frac{x^3}{27} + \frac{2x^2}{3} + 4x + 8$$

14. The first four terms of $(x - 3)^{12}$ are

$$x^{12} - 12x^{11}\cdot 3 + \frac{11\cdot 12x^{10}\cdot 3^2}{2!} - \frac{10\cdot 132x^9\cdot 3^3}{3!} = x^{12} - 36x^{11} + 594x^{10} - 5940x^9$$

17. The first four terms of $(2x - 3y)^{14}$ are

$$(2x)^{14} - 14(2x)^{13}(3y) + \frac{13\cdot 14(2x)^{12}(3y)^2}{2!} - \frac{12\cdot 13\cdot 14(2x)^{11}(3y)^3}{3!}$$

$$= 16{,}384x - 344{,}064x^{13}y + 3{,}354{,}624x^{12}y^2 - 20{,}127{,}744x^{11}y^3$$

20. The first four terms of $\left(x - \frac{1}{y}\right)^{15}$ are

$$x^{15} - 15x^{14}\cdot\frac{1}{y} + \frac{14\cdot 15x^{13}}{2!}\cdot\frac{1}{y^2} - \frac{13\cdot 14\cdot 15x^{12}}{3!}\cdot\frac{1}{y^3} = x^{15} - \frac{15x^{14}}{y} + \frac{105x^{13}}{y^2} - \frac{455x^{12}}{y^3}$$

23. $$\frac{12!}{11!} = \frac{12\cdot 11\cdot 10\cdot\ \ldots\ \cdot 2\cdot 1}{11\cdot 10\cdot\ \ldots\ \cdot 2\cdot 1} = 12$$

26. $$\frac{7!}{9!} = \frac{7\cdot 6\cdot 5\cdot\ \ldots\ \cdot 2\cdot 1}{9\cdot 8\cdot 7\cdot 6\cdot\ \ldots\ \cdot 2\cdot 1} = \frac{1}{72}$$

29. $\frac{6!}{3!} = 6 \cdot 5 \cdot 4 = 120$

32. $\frac{(n + 1)!}{(n - 1)!} = \frac{(n + 1)(n)(n - 1) \cdot \ldots \cdot 3 \cdot 2 \cdot 1}{(n - 1)(n - 2) \cdot \ldots \cdot 3 \cdot 2 \cdot 1} = (n + 1)(n) = n^2 + n$

35. The fifth term of $(x/2 - y)^{12}$ is the term with y^4 and $(x/2)^8$ and the coefficient of $\frac{12!}{4!8!}$. Finally, then, we have $\frac{495}{256} x^8 y^4$.

38. The next to last term of $(a + 4b)^5$ contains a and $(4b)^4$ and a coefficient of $\frac{5!}{4!1!}$ or $5a(256b^4) = 1280ab^4$.

41. We want the term containing $(3x)^4$ and $(4y)^3$ and a coefficient of $\frac{7!}{3!4!}$. Our term, then, is $35(81x^4)(64y^3) = 181{,}440x^4y^3$.

44. Since $(x^2)^4 = x^8$ we seek the term containing $(x^2)^4$ and $\left(\frac{1}{y}\right)^4$ with a coefficient of $\frac{8!}{4!4!}$ or $\frac{70x^8}{y^4}$.

47. $$(1 + 0.3)^6 = 1^6 + 6 \cdot 1^5(0.3) + \frac{30 \cdot 1^4(0.3)^2}{2!} + \frac{120 \cdot 1^3(0.3)^3}{3!} + \frac{360 \cdot 1^2(0.3)^4}{4!} + \frac{720 \cdot 1 \cdot (0.3)^5}{5!} + \frac{720 \cdot (0.3)^6}{6!}$$

$$= 1 + 1.8 + 1.35 + 0.54 + 0.1215 + 0.01458 + 0.000729$$

$= 4.8268$ rounded to four decimal places.

Section 12.5 Counting: Permutations and Combinations

BASIC CONCEPTS you should learn from this unit.

1. The fundamental counting principle is: If one event can occur in m ways and a second event can happen in n ways, then both events can happen in $m \cdot n$ ways.

2. A permutation is an arrangement that can be made using all or some of the elements of a set of objects without repetition.

3. The number of permuations of n objects taken r at a time is

$$P(n, r) = \frac{n!}{(n - r)!}$$

4. If we are interested in the number of ways of selecting objects and the order in which the objects are selected is not important, we use the word combinations to describe the situation.

5. The number of combinations of n objects taken r at a time is

$$C(n, r) = \frac{n!}{r!(n - r)!}$$

WHAT YOU SHOULD KNOW HOW TO DO upon completion of this unit.

1. Determine whether a problem is a permutation problem or a combination problem, or whether it involves both principles.

2. Calculate the number of permutations of n objects taken r at a time.

3. Calculate the number of combinations of n objects taken r at a time.

SOLUTIONS to Exercise Set 12.5

2. There are five letters in study, so the answer is $5!$ or 120 .

5. Each time the coin is tossed it has two possible ways to come up, and since there are eight events, or tosses, the answer is $2^8 = 256$.

8. $P(6, 6) = \frac{6!}{(6 - 6)!} = \frac{6!}{0!} = 6! = 720$.

11. $P(8, 3) = \frac{8!}{(8 - 3)!} = \frac{8!}{5!} = 8 \cdot 7 \cdot 6 = 336$.

14. $P(8, 4) = \frac{8!}{4!} = 8 \cdot 7 \cdot 6 \cdot 5 = 1680$.

17. $P(3, 1) = \frac{3!}{2!} = 3$.

20. a. Since there are ten people and no restrictions are imposed, we have simply $P(10, 10) = 10! = 3{,}628{,}800$.

    b. The men and women would have to occupy alternate seats according to the condition; that is, M W M W M W M W M W or the reverse of this. We thus have $2 \cdot 5 \cdot 5 \cdot 4 \cdot 4 \cdot 3 \cdot 3 \cdot 2 \cdot 2 \cdot 1 \cdot 1$ as the solution, or we could say $2 \cdot 5! \cdot 5! = 28{,}800$.

23. Since the e's and the d's are not distinguishable, we would have

$$\frac{6!}{3!2!} = 60$$

26. Using the letters of MISSISSIPPI four at a time:

    There are two possibilities involving four of the same letter; each can occur in one way for a total of two ways.

    There are eight possibilities involving three of one letter and one of another letter; each can occur in $\frac{4!}{3!} = 4$ ways, for a total of 32 ways.

    There are three possibilities involving two of one letter and two of another letter; each can occur in $\frac{4!}{2!2!} = 6$ ways, for a total of 18 ways .

26. continued

There are two possibilities involving one of one letter, one of another, and two of a third letter; each can occur in $\frac{4!}{2!} = 12$ ways, for a total of 24 ways.

Adding these, we have $2 + 32 + 18 + 24 = \underline{76}$ ways

29. $C(9, 3) = \frac{9!}{3!6!} = \frac{9\cdot8\cdot7}{3\cdot2} = 84$.

32. $C(7, 1) = \frac{7!}{1!6!} = 7$.

35. $C(n, n-1) = \frac{n!}{(n - 1)!\cdot[n - (n - 1)]!} = n$.

38. The faculty members can be chosen in $C(8, 2)$ or 28 ways, and the students in $C(10, 3)$ or 120 ways; $28\cdot120 = 3360$.

41. a. $26\cdot25\cdot24 = 15,600$.

b. $26\cdot26\cdot26 = 17,576$.

44. The basic counting principle is used here, yielding $3\cdot5\cdot2 = 30$ cars.

47. The four prisoners can be selected in $C(12, 4)$ or 495 ways, and the three officers in $C(10, 3)$ or 120 ways. The final answer, then, is $495\cdot120 = 59,400$.

50. a. Seven are defective and we can select four of these seven in $C(7, 4) = 35$ ways.

b. Two of the seven defectives can be selected in $C(7, 2) = 21$ ways; two of the five good transistors can be selected in $C(5, 2) = 10$ ways. Finally, there are $10\cdot21 = 210$ ways to select two good and two defective.

c. Five are good, so we have $C(5, 4) = 5$ ways to select four of these five.

d. $C(7, 3)\cdot C(5, 1) = 35\cdot5 = 175$ ways.

Section 12.6 Probability

BASIC CONCEPTS you should learn from this unit.

1. The probability of an event occurring is defined as

$$\frac{\text{number of successful outcomes}}{\text{total number of outcomes}}$$

2. A probability of 1 means that the event is certain to occur, and a probability of 0 means that the event cannot occur.

3. If p is the probability that an event will occur, then $1 - p$ is the probability that the event will not occur.

4. If two independent events have probabilities of p_1 and p_2, then the probability that both events will occur is p_1p_2.

WHAT YOU SHOULD KNOW HOW TO DO upon completion of this unit.

1. Define probability.

2. Find the probability that an event will not occur if the probability that it will occur is known.

3. Compute the probability of an event happening using the principles of combinations, permutations, and other methods of counting.

4. Compute the probability that two or more events will happen if the individual probabilities are known.

SOLUTIONS to Exercise Set 12.6

2. The probability of a 5 not showing up is $\frac{5}{6}$ for each die. For both dice not to show a 5 is then $\frac{5}{6}\cdot\frac{5}{6} = \frac{25}{36}$. Now the probability of at least one 5 showing up is $1 - \frac{25}{36}$ or $\frac{11}{36}$.

5. a. Similar to Problem 2: $1 - \frac{25}{36} = \frac{11}{36}$.

   b. There are five ways a sum of eight can occur, so the probability of a sum of eight is $\frac{5}{36}$.

   c. The probability that a 3 or a 4 will not appear means that a 1, 2, 5, or 6 will occur. So for each die the probability we seek is $\frac{2}{3}$ and for both dice $\frac{2}{3}\cdot\frac{2}{3} = \frac{4}{9}$.

8. a. The women can be selected in $C(18, 3) = 816$ ways. The total number of ways to select the committee is $C(30, 3) = 4060$ ways. Thus we have $\frac{816}{4060}$ or $\frac{204}{1015}$.

   b. This is equivalent to having an all-male committee, which is $\frac{C(12, 3)}{C(30, 3)}$ by reasoning similar to that used for part (a). This is $\frac{220}{4060}$ or $\frac{11}{203}$.

   c. The woman can be chosen in $C(18, 1) = 18$ ways and the two men in $C(12, 2) = 66$ ways; both events can occur in $18 \cdot 66 = 1188$ ways. So the probability of occurrence is $\frac{1188}{4060} = \frac{297}{1015}$.

   d. This is equivalent to the committee not having all men; this is $1 - \frac{11}{203} = \frac{192}{203}$. (Refer to part (b) above.)

11. Since there are 13 hearts to start with, then 12, then 11, then 10, the probability we seek is

$$\frac{13}{52}\cdot\frac{12}{51}\cdot\frac{11}{50}\cdot\frac{10}{49} = \frac{17{,}160}{6{,}497{,}400} = \frac{11}{4165}$$

14. a. This would be $1 - 0.6 = 0.4$.

b. $(0.4)(0.4) = 0.16$.

17. That 2% are defective implies that 98% are good, and the probability of selecting one good one is 0.98.

a. $(0.98)^4 = 0.922$ approximately .

b. $(0.2)^4 = 1.6 \times 10^{-7}$.

SOLUTIONS to Review Exercises Chapter 12

2. Letting $n = 1, 2, 3$ in order, we have

$$\frac{1-1}{1+1}, \frac{8-1}{2+1}, \frac{27-1}{3+1} \text{ or } 0, \frac{7}{3}, \frac{13}{2}.$$

5. $a_{21} = -2 + (20) \cdot 2 = 38$.

8. Since $a_{23} = a_1 + 22d$, we have $-15 = -4 + 22d$ and thus $d = -\frac{1}{2}$. Now $a_{26} = -4 + (25)(-\frac{1}{2}) = -\frac{33}{2}$.

11. For 2, -6, 18, -54 the common ratio is -3 .

14. Using $a_n = a_1 r^{n-1}$, we have $a_1 = \underline{-2}$; $a_2 = -2(-1) = \underline{2}$; $a_3 = -2(-1)^2 = \underline{-2}$; $a_4 = -2(-1)^3 = \underline{2}$.

17. $\frac{1}{72} = 3 \cdot r^3$ so that $r^3 = \frac{1}{216}$ and $r = \frac{1}{6}$; hence the two means required are

$$3 \cdot \frac{1}{6} = \frac{1}{2} \quad \text{and} \quad 3 \cdot \frac{1}{36} = \frac{1}{12}$$

Finally we have $3, \frac{1}{2}, \frac{1}{12}, \frac{1}{72}$.

20. $S = \frac{a}{1-r} = \frac{5}{1 - \frac{1}{2}} = 10$.

23. $$(2x - y)^4 = (2x)^4 - 4(2x)^3(y) + \frac{12(2x)^2y^2}{2} - \frac{24(2x)y^3}{3!} + y^4$$
$$= 16x^4 - 32x^3y + 24x^2y^2 - 8xy^3 + y^4$$

26. $6! = 6 \cdot 5 \cdot 4 \cdot 3 \cdot 2 \cdot 1 = 720$

29. $P(6, 3) = \frac{6!}{3!} = 120$.

32. $C(n, n-1) = \frac{n!}{(n-1)!(n-n+1)!} = n$.

35. This is a combination problem since the order of selection is not important. Thus we have $C(10, 6) = \frac{10!}{6!4!} + \frac{10 \cdot 9 \cdot 8 \cdot 7}{4 \cdot 3 \cdot 2 \cdot 1} = 210$.

38. The two white pens can be selected in $C(4, 2) = 6$ ways and the total number of ways in which two pens can be selected from twelve is $C(7, 2) = 21$ ways. $\frac{6}{21} = \frac{2}{7}$ is the probability we seek.

41. There are six husband-wife teams and there are $C(6, 2)$, or 15, ways to choose two of these teams. Second, from twelve people there are $C(12, 4) = 495$ ways to choose four. So there is a probability of $\frac{15}{495} = \frac{1}{33}$ of selecting two husband-wife teams.

Chapter 12 - A SHORT QUIZ

1. An infinite sequence is a function whose domain is the set of all ___?___ .

2. A sequence that terminates is called ___?___ .

3. The fifth term of $a_n = \frac{2n}{n + 1}$ is found by ___?___ .

4. A sequence whose successive terms differ from each other by a fixed number is called a(an) ___?___ sequence.

5. A recursive formula such as $a_n = a_{n-1} + d$ allows us to find any term in a sequence if we know the ___?___ term.

6. An arithmetic <u>series</u> is simply the ___?___ of an arithmetic sequence.

7. The number $0.33\overline{3}$ could be written as the infinite series $0.3 + 0.03 + 0.003 + \ldots$. In this case the series is ___?___ with a common ratio of ___?___ .

8. You already know that $0.33\overline{3} = \frac{1}{3}$. Verify this by finding the sum of the infinite series in problem 7 .

9. If we expand $(x + y)^4$, we get $x^4 + 4x^3y + 6x^2y^2 + 4xy^3 + y^4$. Find now the following <u>combinations</u>:

   $C(4, 1)$; $C(4, 2)$; $C(4, 3)$; $C(4, 4)$.

   What do you observe?

10. What do you suspect $C(4, 0)$ is? Test your guess by the combinations formula.

11. Solve for x : $x \cdot C(6, 4) = P(6, 4)$.

12. Selecting five committee members from a group of eight people is a combination problem because ___?___ .

13. In many states auto license plates consist of three letters followed by three numerals usually with repetitions allowed. By what principle would the number of possible plates be determined?

14. If your name had three letters, what is the probability that you would get a license plate with your name?

15. Phone numbers in this country consist of a three-digit area code, with the first digit non-zero, and then seven more digits. Repetitions are allowed. Theoretically, how many phone numbers are possible?

16. Probability is defined as the ____?____ divided by the ____?____ .

17. If the probability of an event occurring is x , what is the probability of it not occurring?

18. If a coin is tossed 25 times and comes up heads each time, what is the probability it will come up heads on the twenty-sixth toss?

19. What is the probability that a coin would come up heads 25 times in a row?

20. Empirical probability is probability obtained from ____?____ rather than from theory.

Quiz Answers Chapter 12

1. Natural numbers

2. Finite

3. Substituting 5 for n in $\frac{2n}{n+1}$ and then simplifying

4. Arithmetic

5. Preceding

6. Sum

7. Geometric, 0.1

8. $S = \frac{0.3}{1 - 0.1} = \frac{0.3}{0.9} = \frac{1}{3}$.

9. $C(4, 1) = 4$; $C(4, 2) = 6$; $C(4, 3) = 4$; $C(4, 4) = 1$. These are the coefficients of the expansion $(x + y)^4$.

10. 1 $\quad C(4, 0) = \frac{4!}{0!4!} = 1$.

11. $x = 2$

12. order of selection is not important.

13. counting principle

14. $\frac{1 \cdot 10^3}{26^3 10^3} = \frac{1}{26^3}$.

15. $9 \cdot 10^9 = 9{,}000{,}000{,}000$.

16. Number of successful outcomes, ... , total number of outcomes

17. $1 - x$

18. 1/2

19. $\left(\frac{1}{2}\right)^{25}$

20. Experience

A B C D E F 1 G 2 H 3 I 4 J 5

Printed in Dunstable, United Kingdom

91343011R00134

Ingo Sommer

DIE STADT DER 500 000

Ingo Sommer

Die Stadt der 500 000

NS-Stadtplanung und Architektur in Wilhelmshaven

vieweg

Schutzumschlag

oben: Entwurf Raumordnungsplan Dezember 1938, BAW
unten: Vor Wilhelmshavens Stadtmodellen 01. 04. 1939, v. l. Hallbauer, Franzius, Hitler, Müller, Himmler, SKF

Für Druckkostenzuschüsse ist zu danken:

Wohnungsbaugesellschaft Jade, Wilhelmshaven
Wilhelmshavener Spar- und Baugesellschaft
Bauverein Rüstringen, Wilhelmshaven
Bremische Gesellschaft für Stadterneuerung, Stadtentwicklung und Wohnungsbau
Gerd Möller-Stiftung, Wilhelmshaven
Oldenburgische Landschaft

Ursprünglich erschienen bei Friedr. Vieweg & Sohn Verlagsgesellschaft mbH, Brauschweig/Wiesbaden, 1993.

Einbandentwurf: Ingo Sommer
Gesamtherstellung: Lengericher Handelsdruckerei, Lengerich
Gedruckt auf säurefreiem Papier

ISBN 978-3-528-08851-4 ISBN 978-3-322-89736-7 (eBook)
DOI 10.1007/978-3-322-89736-7

Den Häftlingen und Zwangsarbeitern, die unter
der NS-Diktatur die Stadt aufbauen mußten

Inhalt

Im Textteil der Arbeit sind Personen, Parteien, Ränge, Institutionen, Gesellschaften, Firmen, Architekturschulen, Behörden, Gesetze, Verordnungen, Programme und Pläne *kursiv* geschrieben.

Anmerkungen sind nach Teil 1–8 geordnet und ab Seite 338 zusammengefaßt.

Hinweise auf Abbildungen befinden sich außerhalb des Textfeldes.

1. Einführung *1*

2. Wilhelmshavens Architekturgeschichte und Baudienststellen *9*

2.1. Berliner Dienststellen und der Aufbau Wilhelmshavens bis zum Ersten Weltkrieg *9*
2.2. Bauen für die Republik und Arbeitsbeschaffung *15*
2.3. Baudienststellen *23*

3. Wohnsiedlungen bis 1945 *41*

3.1. Gartenstadt als Leitlinie *41*
3.1.1. Wilhelmshaven *45*
3.1.2. Rüstringen *45*
3.2. Wohnungsbaugesellschaften in Wilhelmshaven/Rüstringen *47*
3.2.1. Kriegerheimstätten *51*
3.3. Erste Kleinsiedlungen von 1933 *53*
3.4. Früher NS-Siedlungsbau bis 1935 *58*
3.4.1. Adolf-Hitler-Siedlung *62*
3.4.2. Kleinsiedlungen in Himmelreich *64*
3.4.3. Wohnkolonie Kniphausersiel *66*
3.5. Gebaute Gemeinschaftssiedlungen ab 1935 *66*
3.5.1. Von der Wohnsiedlung Nord-Ost zum Inselviertel *68*
3.5.2. Arbeiter-Wohnstadt Voslapp *74*
3.5.3. Siedlung Cäciliengroden *82*
3.5.4. Fedderwardergroden, die Neue Stadt *84*
3.5.5. Marinesiedlung Altengroden *96*
3.5.6. Familienlager Neuengroden *103*
3.5.7. Werftarbeiter-Stadt Sande-Neufeld *109*
3.6. Altstadt-Wohngebiete ab 1935 *113*
3.6.1. Högers Neu-Siebethsburg *115*

3.6.2. Hansaviertel *122*
3.6.3. Jadeviertel *127*
3.7. Siedlungsutopien bis zum Schluß *129*
3.7.1. Siedlung Burg Kniphausen *130*
3.7.2. Schilldeich *132*
3.7.3. Feders Langewerth *133*
3.7.4. Nördliche Stadtparkkolonie *136*
3.7.5. Siedlung um die Wurt Brederwarden *137*
3.7.6. Reichsheimstätten in Schortens *142*

4. Militärplanungen im Raum Wilhelmshaven bis 1945 *145*

4.1. Militärische Interessen und Stadterweiterung *145*
4.2. Marineeigener Hochbau *148*
4.2.1. Marinegerätelager Roffhausen *150*
4.2.2. Marinelazarett Sanderbusch *152*
4.2.3. U-Bootstützpunkt *156*
4.2.4. Kasernenanlagen *157*
4.2.5. Bauten für die Wehrertüchtigung *161*
4.2.6. Technische Marinebauten *163*
4.3. Nordwerft und IV. Einfahrt *168*
4.4. Bauten für die Flieger *174*

5. Monumentalplanungen des „III. Reiches" in Wilhelmshaven *179*

5.1. Verwaltungsforum *183*
5.2. NS-Kulturforum *188*
5.3. Marineforum *191*
5.4. Aufmarschplätze und Achsen *194*

6. Pläne und Bauten für die Zeit „danach" *199*

6.1. Die eingebunkerte Stadt *199*
6.2. Erste Nachkriegsplanungen *206*
6.3. Arbeitsstab Speer in Wilhelmshaven *209*
6.4. Nachkriegsbauen *214*

7. Einzelaspekte des Bauprogramms *221*

7.1. Wehrhafter Städtebau im Spiegel von Wirtschafts- und Bebauungsplänen *221*
7.2. Leitbild der gegliederten und durchgrünten Stadt *227*
7.3. Gründekoration *236*
7.4. Ausstattung und Einrichtung *244*
7.5. Bauten für die Führung *252*
7.6. Bauten der Technik *259*

7.7. Bauten für die Gemeinschaft *263*
7.7.1. Schulen *264*
7.7.2. HJ-Heime *275*
7.7.3. NSDAP-Gemeinschaftshäuser *278*
7.7.4. Gewerbebauten *283*
7.7.5. Andere öffentliche Bauten *290*
7.8. Bauarbeiter und Bauarbeiten *296*
7.9. Kunst am Bau *306*
7.10. Zensur und Widerstände *313*

8. Schlußbemerkungen *327*

9. Anhang *332*

9.1. Bibliographie *332*
9.2. Anmerkungen *338*
9.3. Abbildungsnachweis *354*
9.4. Archivmaterialien *356*
9.5. Kurzbiographien *357*
9.6. Bildteil *381*
9.7. Tabelle *431*
9.8. Personenverzeichnis *433*
9.9. Abkürzungen *438*

1. Einführung

„Ruhe da! Nicht einsteigen, bevor der Zug hält!" befahl der führende Deckoffizier. Und gleich darauf, ehe noch die einlaufende Eisenbahn zum Stehen gebracht war, hing die blaue Reihe von Matrosen und Maaten laut schreiend an den Coupétüren. Mich, sowie vier andere Seeleute, lud ein gönnerischer Zufall in ein Frauenabteil zweiter Klasse ... Ich ließ ein Fenster herab, um mir noch einmal das Bild der leidigen Kasernenstadt einzuprägen. Nach einiger Zeit bat mich die Dame mit weicher, gütevoller Stimme, das Fenster der kalten Zugluft wegen zu schließen[1].

Joachim Ringelnatz, 1915

Die kurze Architekturgeschichte Wilhelmshavens enthält Epochen äußerst unterschiedlicher gesellschaftlicher Ziele, politischer Abläufe und kultureller Ereignisse. Sie umfaßt nicht einmal 140 Kalenderjahre. In diese Zeit fallen die Gründung der *Preußischen Flotte*, des *Norddeutschen Bundes*, des *Deutschen Reiches*, der Kolonien, zahlreicher neuer Städte, des „III. Reiches" und mehrerer deutscher Republiken. In diese Zeit fallen aber auch industrielle Revolution, Architekturrevolution, Erster und Zweiter Weltkrieg sowie der Untergang des Kaiserreiches, des *Deutschen Reiches* und *Preußens*. Die letzten deutschen Stadtgründungen Mannheim, Erlangen und Karlsruhe waren um 1700 begonnen worden. Die Residenzstädte konnten für die drei Stadtgründungen des 19. Jahrhunderts, Bremerhaven (1827), Ludwigshafen (1842) und Wilhelmshaven (1853), nicht Vorbild sein[2].

Wilhelmshaven wurde, abhängig von den jeweiligen Zielen um den Kriegshafen, dreimal gegründet. Die Aufrüstung der *Preußischen Marine* leitete die für das wilhelminische Stadtbild einflußreiche 1. Stadtgründungsphase ein. Sie endete mit den Anfängen der Moderne und den Arbeiten der Städtebaureformer *Martin Wagner* und *Willy Hahn*. In der 2. Stadtgründungsphase 1933 – 45 wurde die Stadt für nationalsozialistische Kriegspolitik erweitert. Nach größenwahnsinnigen Planungen sollte Wilhelmshaven bis zu 500.000[3] Menschen aufnehmen. Gerade 18.000 Wohnungen wurden noch gebaut. Nach Zerstörung, Demontage und Verlust der *Marine* wurde die auf 50.000 Einwohner geschrumpfte Stadt erst langsam wieder aufgebaut. Aufrüstung, Jadevertiefung, Ölhafenbau und einsetzende Wiederindustrialisierung führten in den 60er Jahren zur 3. Stadtgründungsphase mit einem starken Wirtschaftsaufschwung.

Diese Arbeit befaßt sich mit der Nahtstelle zwischen dem Bauen der 20er und 30er Jahre und der Architektur des NS-Regimes am Beispiel Wilhelmshaven und dem
129 umgebenden *Wohnsiedlungsgebiet*. Sie dokumentiert die architektonischen und städtebaulichen Entwicklungslinien, die Einflüsse der Dienststellen und die Rahmenbedingungen vor, während und nach der NS-Herrschaft. Weder zu Beginn des totalitären „III. Reiches" noch nach dessen Ende, wurden stadtplanerische oder baukünstlerische Entwicklungen abgeschnitten. Am Beispiel dieser Stadt und dieser Region lassen sich personelle und stilistische Kontinuität belegen, lassen sich Gewichte architektonischer und städtebaulicher Leitbilder ablesen. Deswegen umfaßt die Arbeit auch die Zeit „davor", aber auch den Übergang zur Zeit „danach". Diese Arbeit ist eine regional-lokale Architekturgeschichte. Es kommt darauf an, das Besondere herauszuarbeiten. Die Unterschiede und Beziehungsgeflechte sollen in den Zusammenhang gestellt werden.

Von Bedeutung für das Erscheinungsbild Wilhelmshavens waren von Anfang an die Berliner Dienststellen. Wer waren die Planer? Wie waren die Organisationsformen? Gab es Architekturschulen, Lehrer oder Grundsätze? Hatten die bis 1937 zersplitterten Gemeinwesen an der Jade Einfluß auf die städtebauliche Entwicklung? Wurde das Bauen nach 1933 anders? Spiegelt sich an den Bauten totalitäre Architektur? Ist an der Geometrie der Formen die diktatorische Staatsform erkennbar?

Wilhelmshavens Architekturgeschichte bis 1945 ist, da die Stadt wenig aus sich selbst heraus gewachsen ist, anhand der äußeren Einflüsse und der wechselnden Impulse nachzuzuzeichnen. Die regionale Eingrenzung ermöglicht es, Verflechtungen von Personen und Beziehungen von Institutionen zu benennen, die bisher wenig bekannt waren, und über die kaum geschrieben wurde. Verfasser wesentlicher Literatur zur Baugeschichte des NS-Systems wie *Joachim Petsch, Reinhard Merker, Roswitha Mattausch, Klaus Backes, Wolfgang Schäche, Ute Peltz-Dreckmann, Anna Teut, Dieter Bartetzko, Hartmut Frank, Tilman Harlander, Gerhard Fehl, Christian Schneider*, gingen davon aus, daß Wolfsburg und Salzgitter besonders herausragende Beispiele für NS-Siedlungspolitik seien. Erst nachdem der Verfasser 1985 *Niels Gutschow* und *Werner Durth*[4] auf Wilhelmshaven hinwies, und Material zur Verfügung stellte, haben Stadterweiterung und Neugestaltung dieses Kriegshafens in der Literatur Berücksichtigung gefunden.

Wohl keiner der an Wilhelmshavens Aufbau beteiligten Architekten, Planer, Unternehmer, Handwerker, Hafenbauer oder Wohnungswirtschaftler hat einen Hauch von Baugeschichte gespürt, als er zwischen 1933 und 1945 antrat, die rigorosen Befehle zu
1 vollziehen und die vorgeschriebenen Bauaufgaben zu erfüllen. Das NS-System machte sich die Fachleute unterschiedlichster Disziplinen als Erfüllungsgehilfen, Technokraten und Vollzugsbeamte zum Ausbau des Kriegshafens und der Marinerüstung nutzbar. Es holte sie aus allen möglichen Regionen des Reiches und der besiegten Länder in die Stadt oder verpflichtete sie wider Willen zum Dienst auf den Großbaustellen. Kaum jemand der Bauschaffenden konnte eigene Vorstellungen einbringen oder sich gar widersetzen. Architektengenerationen, die durch *Werkbund, Internationalen Stil, Bauhaus*, funktionalen Städtebau, *Stuttgarter-Schule* oder *Hamburger-Schule* geformt waren, interpretieren heute

als Zeitzeugen befragt – die Wilhelmshavener Aufbaujahre nicht als sonderlich aufregend.

Diese Arbeit dürfte, möglicherweise mehr als vergleichbare, auf persönlichen Kontakten des Verfassers mit damals an Planung und Bau Beteiligten beruhen. Immer wieder wechselten bei den Befragten Verwunderung über das Interesse an ausgerechnet dieser Epoche der Stadtgeschichte mit bereitwillig sachlicher Auskunftsfreude.

Die meisten Planunterlagen und Entwurfszeichnungen wurden, da sie der „Geheimhaltung" unterlagen, 1945 vernichtet. Das gesamte *Wilhelmshavener Stadtarchiv* verbrannte. Die Verflechtung der Wilhelmshavener Stadtplanung mit der Wehrmacht ließ es manchen Baumeister ratsam erscheinen, Skizzen und Entwürfe, Pläne und Modelle, Fotos und Berichte, nicht über das Ende des „III. Reiches" aufzuheben. Insofern ist die Quellensituation dürftig, das Archivmaterial dünn gesät, fast nichts aufgearbeitet. Den knappen Aufzeichnungen einiger weniger, den vereinzelten privaten Sammlungen, den lückenhaften Zeitungsjahrgängen und den protokollierten Erinnerungen von Beteiligten, kommt daher bei der Zusammenfügung von Wilhelmshavens Architekturgeschichte besondere Bedeutung zu.

Für den Verfasser als Angehöriger der nächstfolgenden Architektengeneration waren Gespräche mit denjenigen, die 1933 – 45 Baupolitik gemacht haben, von großer Bedeutung. Tatsächliche oder drohende Arbeitslosigkeit am Anfang der 30er Jahre – dann die Chance, mitzuarbeiten am Aufbau einer Stadt für fast 1/2 Million Menschen! Immer wieder drängt sich für den Auskunftheischenden die Frage auf: Wie hätte man selbst gehandelt? Hätte man sich von den großdimensionierten Bauaufgaben einnehmen lassen? Schließlich waren die Baudienststellen keine Parteigliederungen, die Architektenbüros keine Mordinstrumente. Das Ziel, eine Stadt für 500.000 Menschen zu vollenden, rückte in greifbare Nähe. Es kam anders. Die wenigsten der Planer und Baubeamten von 1933 – 45 ahnten, daß sie selbst die in Schutt und Asche gelegte Stadt wieder aufbauen mußten. Die vermeintliche national-„sozialistische" Revolution war der Apokalypse sehr nahe gekommen und hatte ganze Stadtquartiere sowie Hafenanlagen als Zeugen der Baukultur ausgelöscht. Daß die betroffenen Baumeisterjahrgänge die Kraft hatten, einerseits ihre begonnene Stadterweiterung abzuschließen, andererseits Trümmer wegzuräumen, ist wohl nur aus der Alternativlosigkeit der Nachkriegszeit zu erklären. Viereinhalb Jahrzehnte sind verstrichen, ohne daß systematisch baugeschichtliches Material gesichert wurde, ohne daß die Stadtgeschichtsschreibung diese 2. Stadtgründungsphase Wilhelmshavens verarbeitet hat. Obwohl sich im „III. Reich" die neu bebaute Stadtfläche mehr als verdoppelt hat, die Einwohnerzahl sprunghaft anstieg,
das *Wohnsiedlungsgebiet* weit in das Umland reichte, großangelegte Militärplanungen *36*
begannen, gigantische Monumentalentwürfe gezeichnet wurden, war bisher Desinteresse an dieser Stadtentwicklungsphase festzustellen.

Die 1957 herausgegebene, knapp 1.500-seitige offizielle „Stadtchronik" widmet der Baugeschichte 1933 – 45 nur wenige Seiten. Auch nachfolgende Veröffentlichungen verdrängten Stadtplanung und Architektur des NS-Systems.

Die kritische Auseinandersetzung mit nationalsozialistischer Stadterweiterung und Neugestaltung ist jetzt, 50 Jahre „danach", aktueller geworden. Die Notwendigkeit dazu jedoch wird nicht überall gesehen, sie wird in einzelnen Fällen sogar politisch bekämpft. Es überwiegt jedoch die Bereitschaft, in einer Stadt anderes zu sehen als die Ansammlung von Häusern und Fabriken. Das zunehmende Interesse an der Historie vor Ort, das stärker werdende kulturelle Gefühl in der Lokalpolitik, das sich entwickelnde geschichtliche Denken in den Behörden und das Einbeziehen baugeschichtlicher Aspekte in Entscheidungen, ist erheblich gewachsen.

Das Gesellschaftsbild von Architekten und Stadtplanern[5] ist grundsätzlich weit angelegt. Immer zeichnet es die Umrisse von Kultur und Politik der jeweiligen Regime und Systeme. Nationalsozialistische Baukultur legte sich wie ein Netz über die Städte und selektierte den Stilpluralismus der Weimarer Republik. Modernes Bauen hatte nur als Form ausnahmsweise Bestand. Die dazugehörige fortschrittliche Geisteshaltung hatte ab 1933 keine Chance mehr. Das Gewicht lag auf traditionalistischen Bauformen und totalitären Stadtgestaltungszielen. Der Widerspruch zwischen großstadtfeindlicher Kulturpolitik und radikaler Neugestaltung löst sich am Beispiel Wilhelmshaven so schnell nicht auf. Nebeneinander und zeitgleich entstanden moderne technische Zweckgebäude für Rüstung und Militär, Wohnsiedlungsgebiete in reduziertem Historismus und Planungen für maßstablose NS-Einschüchterungsarchitektur. Diese unterschiedlichen Ziele fügen sich nicht zu einer einheitlichen NS-Architekturideologie.

In diesem Zusammenhang ergibt sich die Frage nach der Gleichschaltung der bauenden Verwaltungen, Architekten und Wohnungsbaugesellschaften vor Ort. Die Kontrolle der Baukultur fand über die Ideologie hinausgehend am Entwurf, am Detail und am Objekt statt. Dem Verfasser ist es wichtig, nachzuvollziehen, wie Zensur und bauamtliche „Beratung" die Architekten und Planer zu Erfüllungsgehilfen machten, zu überprüfen, ob Türgewände, Dachneigungen, Gesimsformen und Sprossenteilungen einen offiziellen oder regionalspezifischen NS-Stilkatalog ergaben. In Wilhelmshaven bauten unterschiedlichste Ressorts an der Stadt für 500.000 Menschen. *Landesplanungsgemeinschaft, Marine*, Oldenburger Regierungsdienststellen, *Luftwaffe, Reichsarbeitsministerium, Reichssiedlungsämter, Deutsche Arbeitsfront*, kommunale Dienststellen und der *Arbeitsstab Speer*, hatten Gemeinsamkeiten, aber auch widersprüchliche Zielsetzungen. Die NS-Machthaber legten für Wilhelmshaven ein anmaßend riesiges Bauprogramm auf. Die Einwohnerzahl sollte verfünffacht und die Stadt Waffenschmiede des Reiches werden, erheblich größer als Wolfsburg oder Salzgitter. Die Erweiterung der Kriegshafenstadt wurde, begleitet von großem Propagandaaufwand, in kürzester Zeit nach knappsten Termin-
1 plänen begonnen, aber nicht vollendet. Eine uniforme Architekturtheorie läßt sich unter der Propagandadecke nur schwer erkennen. Diese Arbeit will dazu beitragen, am regionalen Beispiel die Widersprüche zwischen uneinheitlicher Architekturideologie, maßloser Propaganda und tatsächlichen Bauabläufen offenzulegen.

*

Der Gedanke, Planungen und Projekte für Wilhelmshaven von 1933 – 45 zusammenzutragen, entstand beim Verfasser, nachdem er 1984 für eine der vielen Wilhelmshavener Gemeinschaftssiedlungen[6] die Architekturgeschichte aufgeschrieben hatte. Das Material wurde mehr, damals Beteiligte meldeten sich, und Hinweise leiteten das Augenmerk auf Stadtteile mit ähnlicher Baugeschichte. Bald wurde klar, daß die Planungen für den Raum Wilhelmshaven weit größer und noch wahnsinniger waren als erwartet und bisher bekannt. Vor 50 Jahren erlebte Wilhelmshaven eine Planungseuphorie und Betriebsamkeit, die heutige Vorstellungskraft sprengt. Im Laufe der Zeit kam es zu verschiedenen Zwischenberichten und Veröffentlichungen des Verfassers. Auch wurde Material anderen Autoren zur Verfügung gestellt[7]. Dies war vorteilhaft, denn die Rückläufe und Hinweise führten zu neuen Quellen und anderen Sichtweisen. Es wurde offenbar, daß sich Institutionen und Bewohner einer durch nationalsozialistische Baukunst stark geprägten Stadt erst mit dem Abstand vieler Jahre den baukulturellen Zielen und städtebaulichen Leitlinien des „III. Reiches" stellen.

Stadterweiterung und Neugestaltung für eine Stadt von 500.000 Menschen können höchst unterschiedlich beschrieben werden. Die Breite des Untersuchungsgebietes erfordert Eingrenzungen. Planungen und Projekte, Utopien und Visionen dieser Jahre zusammenzutragen, ist eine Gratwanderung zwischen genauer Vollständigkeit und großzügiger Auswahl. Da sich diese Arbeit detailliert mit Bestand und Entwicklung von Architektur und Stadtplanung im Dienste des NS-Systems befaßt, müssen aus Platzgründen bestimmte Aspekte ausgeklammert werden. Vieles kann nur exemplarisch beschrieben werden. Immerhin waren für Wilhelmshaven allein 500 unterschiedliche Wohnungstypen entworfen worden, und für mehr als 30 Siedlungsgebiete wurden *Bebauungspläne* in oftmals vielfältigen Varianten aufgestellt. Wiederaufbauplanungen setzten bereits 1940 ein, überschnitten sich mit *Wirtschaftsplänen* und *Bebauungsplänen*. Die Eingrenzung bereitet Schwierigkeiten. Mit Recht wird der Leser erwarten, daß wesentliche gebaute Beispiele, die einzelnen Gemeinschaftssiedlungen, die wichtigsten Planungsabschnitte untersucht und dargestellt werden. Erst in der Bearbeitung stellte sich heraus, wie viele bisher unbekannte Projekte und Zielplanungen, Entwürfe und Planungen dies waren. Andererseits mußte entschieden werden, welche Beispiele auszuwählen, welche Spur weiterzuverfolgen war. Viele Funde ergaben sich zufällig, manche Hinweise hätten möglicherweise noch zu bemerkenswerten Erkenntnissen geführt. Wünschenswerte Vergleiche mit ähnlichen Planungen andernorts konnten nur ausnahmsweise angestellt werden.

Die intensive Einarbeitung in den sperrigen Themenbereich und das lange Studium der technischen Unterlagen führte manchmal zu einer nüchternen und abgehobenen Betrachtungsweise. Immer bestand die Gefahr, die faschistischen Ziele des NS-Systems auszublenden. Die Bewältigung des umfangreichen Materials erlaubte es nur ausnahmsweise, die militärischen Zielsetzungen und politischen Absichten des „III. Reiches"

einzuarbeiten. Die Architekturgeschichte dieser Jahre reiht sich trotzdem in die wahnsinnigen Untaten des NS-Systems ein, obwohl dies auf den ersten Blick nicht erkennbar ist. Wilhelmshavens NS-Architekturgeschichte hat auch zu tun mit menschenverachtender Zwangsarbeit, brutaler Zerstörung, übersteigerter Monumentalität und hemmungsloser Verschwendungssucht. Die Ziegel, aus denen Wilhelmshaven neu gestaltet wurde, kamen auch aus Neuengamme. Dort mußten Häftlinge im Klinkerwerk Ziegel brennen – 50.000 kamen um. Eine von 70 Außenstellen des *Konzentrationslagers Neuengamme* wurde von der *KMW* in Wilhelmshaven aufgebaut. Sie war Teil eines
155 großen Arbeitslagers, das wie viele andere dem Aufbau der Stadt diente. *KZ*-Häftlinge wurden – soweit bisher bekannt – für Abriß und Enttrümmerungsarbeiten, zum Bombenräumen und zur Baureifmachung eingesetzt.

★

Diese Arbeit beruht vor allem auf Quellen vor Ort: Bauakten der Behörden, Materialien von Baugesellschaften, Archive der *Stadt*, Sammlungen von Privaten, Protokolle von Sitzungen, alte Zeitungsausgaben, Beschlüsse politischer Gremien, ausgewertete Schriftwechsel, Plankammern von Baubehörden, Gespräche mit Betroffenen. Nur wenige der damals Beteiligten leben noch. Ihr Erinnerungsvermögen schwindet. Eile war geboten. Die überörtlichen Archive wurden gezielt vor allem zu Kontrollzwecken und zum Ausfüllen bestimmter Lücken benutzt. Sie erwiesen sich, bezogen auf Planmaterial, Fotos und Entwürfe, nicht als ergiebig. Verwirrend sind für den Leser möglicherweise die häufig wechselnden Straßennamen, die geänderten Gemeindegliederungen, die unterschiedlichen Behördenbezeichnungen, die militärischen Ränge und die hierarchischen Beamtentitel. Die 12-jährige NS-Herrschaft brachte unglaubliche Organisationsformen, absurde Aufgabenstellungen und chaotischste Arbeitsweisen hervor. Häufig wechselnde Bezeichnungen und Schreibweisen verwirren zusätzlich.

Nur ein Teil dessen, was für Wilhelmshaven und das umgebende *Wohnsiedlungsgebiet*
1 geplant war, wurde auch gebaut. Viele Gebäude, vor allem Bauten für die Gemeinschaft, Sozial-, Verkehrs- und Monumentalbauten blieben Papier. Bei der Auswahl von Abbildungen wurde darauf geachtet, bisher wenig bekannte Fotos und Pläne zu präsentieren. Der Verfasser hat sich bemüht, zeitgenössische Originalzeichnungen in verkleinerter Form zu verwenden und auf Nachzeichnungen zu verzichten. In wenigen Fällen mußten verblichene Pläne aufgearbeitet, sperriger Umfang beschnitten werden.

Obwohl Bauten immer Gemeinschaftsleistungen sind, unterschiedliche Architekten und wechselnde Mitarbeiter daran arbeiten, hat der Verfasser versucht, die entwerfenden Hauptpersonen, denen die geistige Urheberschaft zuzusprechen ist, zu nennen. Die hierarchischen Unterschriftsregelungen bei Baubehörden ermöglichten nicht immer, den wirklichen Planer herauszufinden. Bei Abbildungen beziehen sich die Jahreszahlen in der

Regel auf das Bau- bzw. Planungsjahr. Auch spätere Fotos tragen gelegentlich frühere Jahreszahlen.

Auch die typologische Einordnung von Bauten des „III. Reiches" war nicht einfach, denn Projekte hatten wechselnde Nutzungen und sich verändernde Inhalte. Ein Arbeiterlager für Kriegsgefangene sollte später Wohnstadt werden, ein Werftlehrlingsheim Kaserne. Marinedienstgebäude sollten sich zu einem *Marineforum* entwickeln.

Mehr als heute wurden im „III. Reich" Architektur und Stadtplanung zusammen gesehen, wenn auch die Spezialisierung von Studienschwerpunkten und Ämterzuordnung längst eingesetzt hatte. Stadtplanung wurde wissenschaftlicher und rechtlicher, Architektur rationeller und technischer. Wegen der Verflechtungen von ästhetischen, technischen, sozialen und politischen Ansprüchen hat sich der Verfasser für die zusammenhängende Darstellung von abgrenzbaren Siedlungsgebieten, von Militär und Monumentalplanungen entschieden.

Die wesentliche Zielsetzung der Arbeit liegt darin, Planungs- und Entscheidungsabläufe für Bauten und Projekte zu rekonstruieren, Entwurfsleitlinien nachzuspüren und die gebauten Ergebnisse exemplarisch zu präsentieren. Da Ziele schnell wechselten, da Planungen „über den Haufen geworden" wurden, Bauten niemals fertig wurden, erschienen tabellarische Auswertungen und vereinfachende Übersichten nicht möglich. Angewandte Untersuchungsmethodik und Ähnlichkeit der Planungsprozesse führen zu Wiederholungen und Überschneidungen in der Darstellung.

Es müßte den Rahmen sprengen, alle im Raum Wilhelmshaven vorgekommenen Bautypen für sich aufzuführen, zu untersuchen, zu beschreiben, abzuleiten und einzuordnen. Alle Betrachtungsweisen haben immer mit lokal begrenzbaren Großsiedlungen, konkreten Militärbauten oder erträumten Monumentalplanungen zu tun. Im Sinne einer Zusammenfassung werden daher in Teil 7 ausgewählte Bautypen, festgestellte Planungsideologien, tatsächliche Entwurfspraxis und baukulturelle Einzelaspekte querschnittartig näher betrachtet. Sicher lassen sich die vom Verfasser gewählten Einzelaspekte vermehren, erweitern, umgruppieren und ausbauen. Die Eingrenzung läßt spätere Vertiefungsarbeiten zu. Sie liefert Ansatzpunkte, weiter zu forschen, gezielter zu suchen, intensiver zu detaillieren.

Zu danken ist den Mitarbeiterinnen und Mitarbeitern von Archiven und Plankammern, die dem Verfasser geholfen haben, den Ämtern und Behörden-Chefs, die Einblick in Akten gewährten. Die Arbeit wurde 1990 als Dissertation von der Universität Oldenburg angenommen. Der Verfasser hat sie insbesondere im Abbildungsteil auf den neusten Stand gebracht und überarbeitet. Er bedankt sich bei den Gutachtern *Prof. Dr. Peter Springer* (Universität Oldenburg) und *Prof. Dr. Dieter Kimpel* (Universität Stuttgart). Ganz besonderer Dank gebührt *Helmuth Baur*, ohne dessen umfassende Detailkenntnisse, ohne dessen unermüdliche Gesprächsbereitschaft die Arbeit nicht entstanden wäre. Als damals beteiligter Planer gab er dem Verfasser vielfältige Hinweise und kenntnisreiche Erläuterungen der früheren Planungsziele. Ihm gelten Hochachtung und Dank.

2. Wilhelmshavens Architekturgeschichte und Baudienststellen

Es ist das einzige militärische Etablissement, welches das deutsche Reich an der Nordsee besitzt, und wird noch immer mit rastlosem Eifer ausgebaut, so daß es in kurzer Zeit eine Festung ersten Ranges darstellen muß. Seine Lage im Grunde des Jadebusens schützt es vor einem Bombardement von der Seeseite. Außer den Batterien, die dasselbe bis zur Mündung der Bucht beschützen, bietet eben die Bucht selbst ihm eine natürliche Verteidigung, da diese nach Entfernung der Seezeichen für jede feindliche Flotte so gut wie unerreichbar ist. Das Fahrwasser in dieselbe hinein ist außerordentlich gewunden, die Strömung sehr schnell, und wenn es Kanonenboote wagen sollten, sich einen Weg zu erzwingen, so wären sie aus kurzer Entfernung einem verheerenden Feuer schwerer Batterien ausgesetzt, die alles in Grund und Boden schießen, was den Torpedos etwa entgangen sein sollte. Vorläufig hat der Hafen nur einen Zugang; binnen zwei Jahren wird er jedoch noch einen zweiten besitzen, an dem man Tag und Nacht arbeitet und der ihm einen weiteren Vorzug gewähren wird[1].

Jules Verne/Paul Verne, 1881

2.1. Berliner Dienststellen und der Aufbau Wilhelmshavens bis zum Ersten Weltkrieg

Alle wesentlichen Entscheidungen zum Aufbau des „einzigen militärischen Etablissement, welches das deutsche Reich an der Nordsee besitzt", fielen in Berlin. Keineswegs *2*
aber waren nur Militärs beteiligt. Schon 1822–23 unternahm der junge Baumeister *Gotthilf Hagen* (1796–1884)[2] auf Empfehlung des weitgereisten Geographen *Alexander von Humboldt* (1769–1859), der sich intensiv mit Physik und Klima der Meere befaßt hatte, eine Studienreise von Berlin aus an die Nordsee. *Gotthilf Hagen* hatte sich besonders den architektonischen Werken des Vorklassizisten und Gründers der Berliner Bauakademie *David Gilly* (1748–1808) gewidmet[3]. 1822 hatte *Hagen* seine Baumeisterprüfung bei *Karl Friedrich Schinkel* (1781–1841) in Berlin abgelegt, die einzige Staatsprüfung im Baufach, die damals möglich war. Mit *Schinkel* verband ihn später eine enge Freundschaft[4]. Der preußische *Minister für Handel und Gewerbe* hatte *Hagens* Reise auf Fürsprache *Alexander von Humboldts* bezuschußt. 1826 veröffentlichte *Hagen* die

Ergebnisse seiner Reise[5], die ihn an alle wesentlichen europäischen Küsten bis nach Venedig geführt hatte. Er bemerkte anerkennend die Leistungen der „hydraulischen Architektur" an der oldenburgischen Nordseeküste und hielt in Skizzen die Gerätschaften zum Deichbau fest. Nirgendwo würden mehr „die Deiche zurückgelegt", und an sehr vielen Stellen finde ein starkes Anwachsen des Landes statt. Er bemerkte, daß die Küste des Butjadinger Landes nun als sicher gelten kann und studierte den „bedeutenden Sielbau bei Fedderwarden und die Eindeichung des Wapeler Grodens"[6]. Beeindruckt registrierte er den starken „Strom, der bei jeder Fluth und jeder Ebbe in den weiten Busen der Jahde herein- und hinausgeht". *Hagen*, der sich als „Wasserbaukünstler" oder auch „Hydrotekt" bezeichnete, war 1832–49 Lehrer für Wasserbau an der Berliner Bauakademie und der Artillerie- und Ingenieurschule Berlin – beide Vorläufer der erst 1879 gegründeten Königlich-Technischen Hochschule zu Berlin[7]. Das Militär bediente sich für die Ausbildung von Ingenieuroffizieren an der 1816 gegründeten Artillerie- und Ingenieurschule[8] – seit 1822 im von *Georg Heinrich Bürde* entworfenen Neubau Unter den Linden etabliert – gerne der Fachkunde des 1831 zum *Oberbaurat* in der *Preußischen Oberbaudeputation* ernannten *Gotthilf Hagen*. Dieser hatte, obwohl mit Militärbauten beauftragt, seinen Lebensschwerpunkt in der bürgerlichen Kultur- und Kunstszene Berlins. Er war befreundet mit dem Architekten *Peter Christian Wilhelm Beuth* (1781–1853), dem Direktor des 1827 aus der Gewerbeschule gebildeten Preußischen Gewerbeinstituts, in dem die bürgerliche Baukunst zuhause war. Der Architekt *Martin Gropius* (1824–80), der Bildhauer und Maler *Christian Daniel Rauch* (1777–1857), *Alexander von Humboldt* - auf dessen Betreiben *Hagen* 1842 in die *Akademie der Wissenschaften* Berlin berufen wurde – und *Karl Friedrich Schinkel* gehörten zu seinem kunstsinnigen Freundeskreis[9]. Auch *Friedrich August Stüler* (1800–65), *Johann Heinrich Strack* (1805–80) und *Friedrich Hitzig* (1811–81) gehörten als Architekten zu dem Freundeskreis. 1848 schloß sich der *geheime Oberbaurat* an der Bauakademie, *Gotthilf Hagen*, mit seinen „Fähnlein der Bauakademiker" dem neugebildeten Künstlerkorps an, um die „geweihten Räume *Schinkels*" zu verteidigen[10]. Auf Empfehlung des Kurators der Artillerie- und Ingenieurschule Berlin, *Prinz Adalbert von Preußen* (1811–73), wurde *Hagen* 1848 von dem *Reichsverweser* des *Deutschen Bundes, Erzherzog Johann*, als sachverständiger Delegierter *Preußens* für Wasserbaukunst in die Frankfurter Nationalversammlung berufen. Eigentlich wegen eines *Gesetzes für die deutschen Ströme* nach Frankfurt geholt, wurde *Hagen* bald von dem Vorsitzenden der *Technischen Marinekommission* der Paulskirche, *Prinz Adalbert von Preußen*, dem Neffen des *Königs Friedrich Wilhelm III.* für seine Pläne benutzt[11]. *Adalbert* hatte schon 1835–36 einen genauen *Flotten-Gründungsplan* aufgestellt und 1848 in einer *Denkschrift* zur Errichtung einer Kriegsmarine für den gesamten *Deutschen Bund* aufgerufen. *Adalberts* Feststellung, daß militärische Linienschiffe „wie es scheint auch in die Jahde bis zur Höhe von Heppens" einlaufen können, war zweifellos das Wissen des Zivilisten *Gotthilf Hagen*[12].

1850 löste *Friedrich Wilhelm IV.* (1795–1861) die *Oberbaudeputation*, die das preußische Bauwesen zusammengehalten hatte, auf. Nachfolgeorganisation wurde die *Bauabteilung*

im *Handelsministerium*. Das *Handelsministerium* in der Wilhelmstraße 79–80 wurde 1854
von *Friedrich August Stüler* erbaut[13]. Hier wurde *Hagen* am 21.01.1850 *Vortragender Rat*.
Unter *Hagen* arbeitete dort auch von 1866–76 der gebürtige Wittmunder Wasserbauer
Ludwig Franzius (1832–1903)[14]. Das gesamte Staatsbauwesen dirigierte dort *Stüler*, der
Vertrauter und Freund des baulustigen Königs war[15]. Erstmalig benötigten die Berliner
Dienststellen für ihre ehrgeizigen Flottenpläne im Jahre 1853 auch die Mitarbeit eines
Landesherzogs[16]. *Großherzog Nikolaus Friedrich Peter von Oldenburg* (1827–1900), trat im
Jahde-Vertrag Preußen Land für das „Königlich-Preußische Jadegebiet" ab[17].

Erstmalig benötigten die Berliner Dienststellen aber auch eine eigene *Königliche* 2
Admiralität, die 1854 auf Betreiben des seit dem 01.03.1849 die *Preußische Marine*
befehligenden *Adalbert von Preußen* und *Friedrich Wilhelm IV.* (1840–61 *König*) gegründet
wurde. Sie wurde 1861 nach Amtsantritt von *Wilhelm I.* (*König* von 1861–88, ab 1871
Kaiser) in *Preußische Admiralität* umbenannt. Ab 1869 hieß sie *Marineministerium*. Nach
Reichsgründung und Kaiserkrönung im Jahre 1871 wurde sie in *Kaiserliche Admiralität*
umbenannt[18].

Der „Wasserbaukünstler" *Gotthilf Hagen* mußte mit Gründung der *Königlichen*
Admiralität 1854 deren *Dezernat für Land- und Wasserbau* mit übernehmen[19]. Ab
September 1855 schließlich wurde er Vorsitzender der *Kommission für den Bau eines*
Kriegshafens an der Jade, der *Hafenbaukommission*. In dieser Zeit entstanden in Berlin und in
den Zeichenbüros auf der Baustelle an der Jade *Hagens* Baupläne für Wilhelmshaven, die 6
der architekturbesessene *König Friedrich Wilhelm IV.* am 25.06.1856 genehmigte[20].

Die Ursprünge der rechtwinkeligen Blockstruktur Wilhelmshavens, die klassizisti- 9
schen Parallelen zur von *Schinkel* ausgestalteten Berliner Straße Unter den Linden, gingen
nicht auf die Militärs zurück, sondern auf die Königlich-preußische Architektur aus dem
Umfeld der Berliner Bauakademie[21]. *Schinkels* 1834 entworfenes Ziegelgebäude der
Bauakademie am Werderschen Markt in Berlin war Vorbild auch für die fiskalischen 10
Bauten an der Jade[22]. Dieses Hauptwerk *Schinkels* (es beherbergte nicht nur die
Bauakademie, sondern auch die *Oberbaudeputation* und *Schinkels* Dienstwohnung) war
von sachlicher Schönheit. Neu waren die vorgefertigte Bauornamentik und die
Kappengewölbe auf gerastertem Stützensystem, die wiederum große Fensteröffnungen
erlaubten. Die Ziegelsichtbauweise der Berliner Bauakademie, aber auch die norddeut-
sche Backsteingotik und die Ziegelbauten, die er in Norditalien 1823 kennengelernt hatte,
haben Wilhelmshavens ersten Baumeister *Gotthilf Hagen* bewogen, möglichst von Stuck
und Putz als Ersatz für weither zu transportierenden Naturstein abzusehen[23].

Nachfolger von *Gotthilf Hagen*, der die militärische Dienstverpflichtung nach
Wilhelmshaven nur ungern absolvierte[24], der trotzdem aber Stadtgrundriß und
Grundlagen der Architektur maßgeblich bestimmte, wurde 1856 *Baurat Pfeffer*[25]. In
Wilhelmshaven waren es dann die *geheimen Marinebauräte* bzw. *Admiralitätsräte Göker*
(1856–73), *König* (1873–74), *Rechtern* (1876–91), *Bieske* (1892–95), *Brennecke*
(1895–1904), *Moeller* (1904–08), *Rollmann* (1908–13) und *Behrendt* (1913–19)[26], die vor

Ort die Berliner Befehle baulich durchzusetzen hatten. Sie leiteten jeweils das *Hafen-* bzw.
10 *Strombauressort der Marinewerft*, das für Wilhelmshaven wesentliche Bedeutung hatte.

Alle Fäden für den Aufbau Wilhelmshavens liefen in Berlin im Gebäude der *Admiralität*
2 am Leipziger Platz 12–13 zusammen[27]. Hier, in der ab 1734 aufgebauten Friedrichstadt-Erweiterung[28], lag wenige Schritte entfernt Leipziger-/Ecke Wilhelmstraße das 1846 nach Entwürfen von *Wilhelm Drewitz* und *Friedrich August Stüler* umgebaute *Kriegsministerium. Wilhelm Drewitz* (1806–88) und *August Ferdinand Fleischinger* (1804–85), beeinflußten in der *Bauabteilung des Kriegsministeriums* fast alle Bauten der preußischen Militärverwaltung, den Kasernenbau und die Dienststellen in den Provinzen[29].

Neben der obersten staatlichen Baubehörde, dem *Handelsministerium*, dem *Kriegsministerium* und der *Admiralität* wurden für Architektur und Städtebau Wilhelmshavens noch weitere Baumeister der *Berliner-Schule* entscheidend. *Johann Heinrich Friedrich Adler* (1827–1908), wie alle bisher genannten Architekten Absolvent der Berliner Bauakademie, und später auch dort Lehrer, baute 1868–70 zusammen mit *Friedrich Hitzig* das
11 Marinestationsgebäude und 1869–72 die neugotische Elisabethkirche[30]. *Kaiser Wilhelm II.* stiftete eine weitere Kirche (nachdem die zuvor gebaute der Marinegemeinde vorbehalten blieb), die nach einer Unzahl von Vorentwürfen 1901 gleichfalls nach Plänen von *Adler* an der Ecke Peter-/Adalbertstraße entstand. *Adler*, später *Professor* für Baugeschichte und Leiter der Ausgrabungen in Olympia, hat viele Kirchbauten in Preußen realisiert. Dies lag nah, denn er war 1877–99 *Vortragender Rat für Kirchenbau* im *Ministerium für öffentliche Arbeiten*, das ab 1878 oberste staatliche Baubehörde wurde. 1868 legte er einen spektakulären Entwurf für den Berliner Dom vor, der für den Bau in die engste Wahl kam[31].

Zu den Architekten, die von Berlin nach Wilhelmshaven geschickt wurden, gehörten auch *Alfred Breslauer* (1866–1953) und *Paul Salinger*. Die Vielbeschäftigten bauten in Berlin zumeist klassizistisch oder neobarock (Villa Borsig, Haus Andreae usw.)[32]. In Wilhelmshaven bauten sie das Seemannshaus Bismarck-/Kirchstraße. Es wurde 1903 durch *Kaiser Wilhelm II., Prinz Heinrich* und *Tirpitz eingeweiht*[33].

Bedienstete der Berliner Militärbehörden, insbesondere der *Admiralität* bzw. des *Marineministeriums*, ab 1889 dann des *Oberkommandos der Marine*, des *Reichsmarineamtes* oder des *Marinekabinetts* ließen sich nur ungern zum Aufbau Wilhelmshavens an die Jade versetzen[34]. Der machtbesessene *Alfred von Tirpitz* (1849–1930), 1877–85 in der
2 *Admiralität*, ab 1895 *Chef des Stabes*, 1898 *Staatssekretär* im *Reichsmarineamt* am Leipziger Platz und *Staatsminister*, machte mit dem „marinesüchtigen jungen *Kaiser Wilhelm II.*" Berlin zur Marinestadt[35]. Kompromißlos trieb er seine maritimen Visionen voran und drängte Deutschland in einen übersteigerten Flottenenthusiasmus. Stärker noch als das Marinefieber 1848 den ganzen *Deutschen Bund* ergriff, steigerte sich die Reichshauptstadt Berlin in einen wahrhaften Marinerausch. Die Schiffe lagen zwar in den befestigten Kriegshäfen Wilhelmshaven und Kiel, Militär- und Bauplanung jedoch entstanden in Berlin und bestimmten dort kulturelles Umfeld und Alltagsleben.

Das nach 1848 die Architektur beherrschende preußische Verständnis von bürgerlichen Freiheiten, die bewußt ruhige Ordnung königlich-preußischer Architektur wurde zugunsten von aufgewärmten Schmuckformen preisgegeben. Stuckfiguren und überladene Bauten bestimmten nicht nur das Stadtbild in Berlin, sondern auch den Aufbau der Exklave Wilhelmshaven.

Kaiser Wilhelm II. versprach den Deutschen eine schillernde Zukunft. Gerne zog er, nicht nur bei Besuchen in Wilhelmshaven, seine Marineuniform an, sondern posierte auch in Berlin zum Ärger des Heeres als *Admiral*[36]. Eigenhändig entwarf oder änderte er mit *Tirpitz* seine *Flottengesetze* (1898, 1900, 1906, 1908 und 1912) und skizzierte Marinedenkmäler für Wilhelmshaven.

Es war ganz im Sinne wilhelminischer Repräsentation[37], daß, als die preußische
Admiralität am Leipziger Platz nicht mehr ausreichte, unter *Tirpitz* ein Architektenwett- *2*
bewerb für ein neues *Reichsmarineamt* (mit *Marinekabinett* und *Admiralstab*) durchgeführt
wurde. An der Königin-Augusta-Straße, dem späteren Tirpitzufer (heute Reichpietsch-
Ufer 72–76)/Ecke Stauffenbergstraße wurde 1911–14 von *Reinhardt* und *Süßenguth* die
protzige Marinezentrale mit 800 Diensträumen gebaut[38]. Die 91 m lange und 20 m hohe
graue Muschelkalkfassade wurde klassisch streng mit barocken Formelementen sichtba- *4*
res Zeichen der Macht und wilhelminischen Geistes. Der feudale *Kaiser* stellte sich an die
Spitze bürgerlich-wirtschaftlicher Interessen – entsprechend war die Zustimmung für
seine ostentative Architektur. Das neue Gebäude am Landwehrkanal beherbergte nach
dem niederschmetternden Ausgang des Ersten Weltkrieges wieder die *Admiralität*,
danach den *Chef der Marineleitung* und das *Reichswehr-Ministerium*, später in der NS-Zeit
dann das *OKM*, zeitweise auch das *Marineverwaltungsamt* und die jeweiligen *Bauabtei-
lungen*. 1938 wurde, da auch die *Oberkommandos* von Wehrmacht und Heer unterzu-
bringen waren, an der damaligen Bendlerstraße 11–13 (heute Stauffenbergstraße) ein
Erweiterungsbau in moderner Stahlbetonskelettbauweise erforderlich. Dieser „Bendler-
block“ spielte eine wesentliche Rolle in der Widerstandsbewegung und bei den
Ereignissen des 20.07.1944.

Die Berliner Marinedienststellen waren in ihrer Architekturauffassung vor dem Ersten
Weltkrieg noch den wilhelminischen Ornamenten verbunden. Hektischer Aufbau der
Marine und rapide Expansion preußischer Stadtgründungen ermöglichten kaum
Architekturqualität oder gar Neuerungen. Das *Baudezernat des Reichsmarineamtes* oder der
jeweiligen *Admiralität* gab die Kasernengrundrisse, wie 1911 bei den Wilhelmshavener
Mühlenwegkasernen, für die Matrosendivision (auf oldenburgischem Gebiet) vor, und *13*
Regierungsbaumeister Willibald Strempel vom *Marine-Garnison-Bauamt Wilhelmshaven*
zeichnete die renaissanceartige Fassade mit Klinkern und Sandsteinornamenten. Wilhel-
minische Militärbauten an der Jade waren neben den vielen Hallen und Werkstätten der *9*
Werft vor allem:

1865 Wohnhaus *Oberwerftdirektor*
1867 *Marine-Kommandantur*, Kronprinzenstraße 20
1868 Kaserne Moltkestraße, südlich Ostfriesenstraße (Stralsunder Seeartillerie)

11 1868–70 Marinestationsgebäude, an der Stirnseite der Adalbertstraße
1869–71 *Marinelazarett*, Roon-/Kaiserstraße
1870 Werftwasserturm
1870 Stadtkaserne, zwischen Königstraße und Roonstraße
1873 *Stationsgericht* mit Arrestanstalt, Gökerstraße
1873–77 Werftkaserne, nördlich Ostfriesenstraße (2. Werftdivision)
1873–77 Hafenkaserne, nördlich Ostfriesenstraße (2. Matrosendivision)
1875 *Marine-Observatorium*, Altendeichsweg
10 1876 Werfttor 1
1878 Marine-Waschanstalt, Kanalstraße
1878 Marine-Friedhof Gökerstraße
1885–88 1.000-Mann-Kaserne (später Jachmann-Kaserne), zwischen Moltke- und Ostfriesenstraße (2. Seebataillon und 2. Matrosen-Artillerie-Abteilung)
1885 Sanitätsdepot
1885 Marinesportanlage
1887 *Werftkrankenhaus* Roonstraße
1891 Deckoffiziersschule, Jachmannstraße
1892 Signalstation I. Einfahrt
1899 Graf-Spee-Kaserne, zwischen Moltke- und Ostfriesenstraße
16 1902 Werftspeisehaus
1902 Marine-Offiziers-Kasino, Hollmann-/Gökerstraße
1903 Seemannshaus, Bismarck-/Kirchstraße
12 1904–07 Kaserne Gökerstraße (2. Seebataillon, später 2. Marineartillerieabteilung)
14 1905 *Marineintendantur*, Peter-/Wallstraße
1905 Torpedokaserne, Göker-/Roon-/Königstraße
1907 Zweites Marinegefängnis, Mühlenweg
1909 Admiralswohnungen, Viktoria-/Wall-/Peter-/ Parkstraße
1909 Zweite Torpedokaserne, Roon-/König-/ Friedrichstraße
1910 *UTO-Werft,* Westhafen
13 1912–13 Mühlenwegkasernen (II. Matrosendivision)

Viele der aufgeführten Bauten repräsentierten profane *Neugotik*[39] oder zinnenbewehrten *Normännischen Stil*[40], vergleichbar den Berliner Kasernenbauten des *Kriegsministeriums*
11 unter *Fleischinger* und *Drewitz*. Das Stationsgebäude am Adalbertplatz von *Adler* und *Hitzig* war vergleichbar den Schloßbauten von *Hitzig* (1811–81)[41]. Die Bauten der Garnison, die Kasernen, Depots, Sportanlagen, Kasinos, Friedhöfe und Kommandanturgebäude, wurden ab 1880 von der *Kaiserlichen Garnison-Bauverwaltung*, ab 1890 vom *Marine-Garnison-Bauamt* entworfen. Die Bauten der *Werft*, die Brücken, Signalstationen, Werkstätten überließen die Berliner Dienststellen dem *Hafenbauressort* der *Marinewerft*. Als Ministerialinstanz und Geldgeber jedoch beeinflußten sie die Architektur fortwährend.

Rüstringen bzw. die Vorläufergemeinden der 1911 gegründeten Stadt verfügte über nur wenige Militärbauten (Kasernen Mühlenweg und Gökerstraße, Marinegefängnis 12, 13
usw.), allerdings über oldenburgische Staatsbehörden, die nicht unmittelbar den Berliner Baudienststellen unterstanden.

Der *Übersichtsplan der marinefiskalischen Gebäude und Anlagen bei Wilhelmshaven und* 9
Umgegend um 1891 vom *Reichsmarineamt* bearbeitet, dokumentierte 54 Gebäude der Garnison und 77 Gebäude, die der *Werft* unterstanden. Er war eine Zwischenbilanz für die ehrgeizigen Berliner Baubehörden, in der die Stadt und ihre Menschen, private Wohnungen, Plätze und Straßen kaum vorkamen. Fast schien es, als wollte man in Berlin das Wachsen eines Gemeinwesens nicht zur Kenntnis nehmen. Schließlich sollte *Preußen* nach dem *Jahde-Vertrag* von 1853 weder einen Handelshafen, noch eine Handelsstadt anlegen und die Ansiedlung von Handwerkern und Gewerbetreibenden verhindern helfen.

Für Wilhelmshaven und die umliegenden Gemeinden waren die Jahre bis zum Ersten Weltkrieg ereignisreich. Das „II. Reich" hielt sich bereit, neben England seine Rolle als Großmacht zu spielen. Im 2. *Flottengesetz* war der weitere Marineausbau vorprogrammiert und die Kolonie Tsingtau wurde der *Marine* unterstellt. Mit *Wilhelm II.* hatte Deutschland vermeintliche weltpolitische Bedeutung erlangt. Die Zukunft sollte nicht mehr dem konservativen Adel gehören, sondern den Technikern und Ingenieuren. Der *Alldeutsche Verband* (1891), der *Deutsche Flottenverein* (1898), der *Deutsche Werkbund* (1907) setzten auf nationale Zukunft und Wirtschaftskraft. Die Aufbruchstimmung der Jahrhundertwende führte auch zu künstlerischen und architektonischen Umwälzungen. Dekorationen und Ornamente bekamen einen anderen Wert. Große Linien, technische Konstruktionen und klare Geometrie wurden wichtiger. *Jugendstil* und *Werkbundarchitektur* wurden auch in Wilhelmshaven/Rüstringen bedeutend, gefördert von den *Stadtbauräten Paul Wolf, Martin Wagner* und *Hermann Zopff.*

2.2. Bauen für die Republik und Arbeitsbeschaffung

Von 1900 bis 1911 war die *Marine* zu imposanter Größe gewachsen und verschlang zuletzt über die Hälfte des Militäretats[42]. Kurz vor dem Ersten Weltkrieg verlor *Tirpitz* den ehemals großen Einfluß. Als die Marinezentrale an der Königin-Augusta-Straße in 4
Berlin fertiggestellt war, und der Krieg begann, begab sich *Wilhelm II.* nicht dorthin oder gar nach Wilhelmshaven, sondern zum *Generalstab. Tirpitz* war isoliert, gewann aber 1917 noch einmal Einfluß auf den U-Boot-Krieg. Die Unzufriedenheit über seine Marinepolitik und der verlorene Krieg führten zu den Matrosenaufständen, zur Revolution, zum Ende des Kaiserreiches und auf den Weg zur Republik.

Das schubweise Wachsen der *Marine* hatte auf die Entwicklung Wilhelmshavens und das 1911 aus den oldenburgischen Nachbargemeinden gegründete Rüstringen katastrophalen Einfluß. Kurz vor dem Ersten Weltkrieg, losgelöst von wilhelminisch-kaiserlicher

Architektur, gingen das preußische Wilhelmshaven und das oldenburgische Rüstringen eigene Wege. Besorgt um das einheitliche Stadtbild hatte Wilhelmshaven 1876 den zivilen Vermessungsrevisor *Schnedler* und Heppens/Rüstringen den privaten Geometer *Aschenbeck* mit *Bebauungsplänen* beauftragt. In Ermangelung ortsansässiger Baumeister griffen die *Gemeinden* oft auf auswärtige Architekten aus Berlin, Bremen, Oldenburg, Hannover und anderen Städten zurück. Nachdem die Berliner Dienststellen 1906 den renommierten *Professor* der Berliner Technischen Hochschule, *Theodor Goecke* (1850–1919) empfahlen, kam es 1908 zu einem ersten *Bebauungsplan*[43]. *Goeckes* Stadterweiterungsplanung mit unregelmäßigen Baublocks ging auf die Planungen für Berlins südliche Vororte zurück. Seine Absichten scheiterten nicht zuletzt am Widerstand des ersten Rüstringer *Stadtbaurates Martin Wagner* (1885–1957). Er bekam nach Architekturstudium und Praxis bei *Hermann Muthesius*[44] und im *Bauamt* Berlin-Weißensee von 1911 bis 1914 großen Einfluß auf die städtebauliche Entwicklung der Jadestädte.

Wagner, sozialdemokratischer Reformer, Mitglied der *Deutschen Gartenstadtgesellschaft* und des *Deutschen Werkbundes*, hat mit Siedlungsgebieten, Erwerbshäusern, Verkehrsplanungen, städtebaulichen Wettbewerben, Ausstellungen und fortschrittlicher Grundstückspolitik wesentliche Spuren in Wilhelmshaven/Rüstringen hinterlassen[45]. Er legte sich offen mit dem bei den Berliner Dienststellen hochangesehenen *Professor Bruno Möhring* (1863–1929) an, der vom *Reichsamt des Inneren* mit Entwürfen für eine Arbeitersiedlung beauftragt war[46]. Er holte die Wegbereiter der Moderne, *Leberecht Migge, Hans Poelzig, Fritz Bräuning, Hans Bernoulli* für Wettbewerbsentwürfe und den Berliner *Stadtbaurat Ludwig Hoffmann* als Vorsitzenden eines Preisgerichtes nach Rüstringen[47]. Die Einschaltung von *Peter Behrens* scheiterte am Widerstand der Rüstringer Ratsherren. Wegen der Bahnhofsneuplanung setzte er sich 1911–14 mit dem Dresdener *Professor Cornelius Gurlitt* auseinander. Nicht zu unterschätzen war auch *Wagners* Einfluß auf Kunst und Kultur der Jadestädte. Vergeblich kämpfte er gegen das Ressortdenken der *Marine*, gegen die Konkurrenzsituation der Kommunen und die Dominanz der Berliner Militärdienststellen. 1914 ging er enttäuscht nach Berlin zurück und wurde dort der einflußreichste Baupolitiker der Weimarer Republik – heftig bekämpft von Konservativen und Nationalsozialisten.

Die Berliner Dienststellen konnten nach dem verlorenen Ersten Weltkrieg nicht verwinden, daß ausgerechnet von „ihrem" Wilhelmshaven die Matrosenaufstände ausgegangen waren. Kiel und Wilhelmshaven verloren in Berlin an Bedeutung. Das neue Parlament der Weimarer Republik hatte mit der Marine nicht viel im Sinn und der *Versailler Friedensvertrag* von 1919 reduzierte die *Reichsmarine* auf unbedeutende Größe[48].

4 Im Gebäude der *Marineleitung* residierte nach 1918 die einflußlose *Admiralität* mit *Vizeadmiral von Trotha*. 1920 wurde sie umbenannt in *Chef der Marineleitung*. Auch das *Reichswehr-Ministerium* residierte dort.

Die städtebauliche Entwicklung in Wilhelmshaven/Rüstringen kam wegen des Verlustes der Kriegsflotte und der damit verbundenen Arbeitsplätze zum Stillstand. Wilhelshaven wurde zwar 1919 kreisfreie Stadt, stand aber vor dem wirtschaftlichen Ruin, da kaum Werftarbeiter benötigt wurden, um die wenigen, dem *Reich* verbliebenen, Schiffe zu unterhalten.

★

Der Erste Weltkrieg war für die Wilhelmshavener/Rüstringer Baudienststellen eine Erholungspause. Stadt und Umland blieben von Schäden verschont. *Willy Hahn* schrieb „Vom Aufbau einer neuen Stadt", von der Muße der Kriegszeit, der Belebung gemeindlicher Bautätigkeit, der Steuerung der *Arbeitsbeschaffungsmaßnahmen* und der „Zeit der Architektenwettbewerbe". Schwerpunkte waren im Ersten Weltkrieg die „Ausführung derjenigen Arbeiten, die zur Herrichtung von Verteidigungsbauten an den wichtigsten Plätzen der Grenzlande in Angriff genommen sind". Personelle Lücken in den *Bauämtern* und Wohnungsbaugesellschaften schloß man wie auch andernorts durch Einschaltung freier Architekten, die sowieso unter ausbleibenden Privataufträgen zu leiden hatten.

Ziel der weitvorausdenkenden technischen Baubeamten war es, Vorbereitungen für die Zukunft zu treffen, Planungen für die Zeit nach dem Kriege voranzutreiben. So war es auch 25 Jahre später, als Wiederaufbauplanungen bereits ab 1940 für die Zeit nach dem „Endsieg" vorbereitet wurden. *Martin Wagner* stellte 1915 eine regelmäßig nach den Kriegen aufkommende Wohnungsnot fest[49]. Im *Preußischen Abgeordnetenhaus* wie auch im *Reichstag* wurden für den städtischen Wohnungsbau nach dem Kriege siedlungspolitische Beschlüsse gefaßt. Bereits in seiner Rüstringer Zeit hatte *Martin Wagner* den sich ab 1912 ausbreitenden Kleinwohnungsbedarf festgestellt[50]. Er entwickelte in seiner Berliner Zeit dann Grundlagen zur Wohnungsbaufinanzierung, befaßte sich mit demographischen Abhängigkeiten und arbeitete darauf hin, Mittel für das Kleinwohnungswesen aus denjenigen Quellen schöpfen zu müssen, „die von den Spargroschen des kleinen Mannes gespeist werden". Sparkassen, Versicherungsgesellschaften, Berufsgenossenschaften und Hypothekenbanken sollten gezwungen werden, nicht mehr Millionenobjekte zu beleihen, sondern das Geld in Wohnungsbaugenossenschaften zu deponieren.

Städtebauliche Gesamtplanungen kamen wegen der fehlenden wirtschaftlichen Perspektiven für Wilhelmshaven/Rüstringen zum Erliegen. *Willy Hahns Rüstringer Bauklassenplan von 1917*[51] und *Hermann Zopffs Bebauungsplan für Wilhelmshaven von* 1918[52] blieben die letzten Visionen bis zu den 1936 einsetzenden großangelegten NS-Planungen. Von den Jahren der Weimarer Republik[53] waren nur die Jahre 1924–30 stabil. Die Jahre nach dem Ersten Weltkrieg ermöglichten im ganzen Reich keine umfangreiche Bautätigkeit. Die 1911–37 nebeneinander bestehenden Städte Wilhelmshaven und Rüstringen befaßten sich mit den Folgen des Krieges und Notmaßnahmen, mit

Grünplanung und Selbstversorgung, mit nie gebauten Siedlungsprojekten und Visionen, mit mühsamer Suche nach wirtschaftlichem Auskommen und *Arbeitsbeschaffungsmaßnahmen*. Diese Aufgaben erfüllten allein die *Bauverwaltungen* der *Städte* unter ihren *Stadtbauräten Hermann Zopff* (Wilhelmshaven), *Willy Hahn* (Rüstringen) und seinem Nachfolger *Carl Haefner*[54]. Erst 1924 kam die Bautätigkeit wieder in Gang und Architekten bekamen Aufträge.

In Berlin (Hufeisensiedlung, Siemensstadt, Onkel Toms Hütte), Frankfurt (Praunheim, Römerstadt, Westhausen, Höhenblick, Bruchfeldstraße), Stuttgart (Weißenhofsiedlung), Dessau (Törten) und Köln (Kalkerfeld), konnten fortschrittliche Planer ihre Vorstellungen vom *Neuen Bauen* durchsetzen. Selbst im provinziellen Celle (Georgsgarten, Italienischer Garten), in Magdeburg (Groß-Diesdorfer Straße) und Kiel (Arbeitsamt), konnten *Bauhaus*-Architekten oder Verfechter *Internationalen Stils* bauen. In Wilhelmshaven/Rüstringen entstanden zu dieser Zeit städtebaulich nicht minder moderne Projekte, wie die Komplettierung von Siebethsburg, Admiral-Klatt-Straße,
27 Anfänge des Inselviertels, Südstrandanlagen, Bauhof Oststraße, Kriegsopfersiedlung Neuengroden, Karl-Hinrichs-Stift, Parkanlagen Rüstringen, Altenheim Lindenhof, Verlagshaus Brune, Marinesiedlung Athenstraße. Abkehr vom Innenhof, Räume mit einem Höchstmaß an Licht und Luft, intensive Durchgrünung, charakterisierten den neuen Wohnungsbau schon früh auch in Wilhelmshaven/Rüstringen. Die Wilhelmshavener Projekte waren städtebaulich modern, in ihrem Äußeren jedoch oft traditionell und beeinflußt durch *Hermann Zopff, Fritz Höger, Ernst Peters, Franz Iwersen* und *Richard Freygang*.

Auch das 1922 als Nachfolgebehörde des *Marine-Garnison-Bauamtes* gegründete *Marinestandortbauamt* – zeitweise als *Marinebauamt* bezeichnet – baute städtebaulich modern, äußerlich jedoch traditionell. In der Lilienburgstraße entwarf *Willibald Strempel* 1925–27 vier Wohnblöcke, dreigeschossige Ziegel-Miethäuser für die marineeigne *Gemeinnützige Gesellschaft zur Schaffung von Wohngelegenheiten für Reichsangehörige*. In der Athenstraße baute er eine kleine Marinesiedlung in Putzbauweise. *Otto Lückemeyer*, der Leiter des *Marinestandortbauamtes*, entwarf 1925 das *Offizier*-Wohnhaus Parkstraße 21 und zusammen mit *Willibald Strempel* die südliche Randbebauung des Inselviertels, Bismarckstraße 1–29.

Zwischen den Kriegen entstanden vereinzelt auch äußerlich moderne Bauten in
18 Wilhelmshaven/Rüstringen: Kaufhaus *Karstadt* von dem *Olbrich*-Schüler *Philipp Schäfer*
(1921–24), Rüstringer Rathaus von *Fritz Höger* (1928–29), Verlagshaus *Brune* (1929),
15, 19 Kaufhaus *EPEGE* von *Robert Fuchs* (1930), *Konsum*-Bürogebäude von *Ernst Peters* (1931)
usw. *Gustav* und *Willi Lübbers*, beide von den Nationalsozialisten ab 1937 mit
umfangreichen Bauaufträgen bedacht, entwarfen davor modern und im Zeitgeist der
20 Weimarer Republik: *AOK*-Gebäude (1929–31), *Kuhlmann*-Gewerbebau (1936), Ver-
21 waltungsgebäude *Hermann Möller* (1934) usw. *Höger, Fuchs, Minke, Peters* und *Freygang*
planten und bauten noch bis 1936 für private und genossenschaftliche Bauherren sachlich,
funktional oder *Bauhaus*-orientiert. Viele fortschrittliche Projekte jedoch wie die

Kriegerheimstättensiedlung Neuengroden von *Willy Hahn* (1919), die Bahnhofsplatz-
bebauung von *Hermann Zopff* (1918)[55], der Lagunen- und Sportpark Neuengroden von
Leberecht Migge (1918–20), das Projekt Leo von *Richard Freygang* (1929), die Nordstrand- 17
anlagen von *Werner Haschenburger* (1929–30) usw. blieben Papier.

★

Die 1920 gegründete *NSDAP* konnte wegen der schlechten wirtschaftlichen Verhältnisse und der politischen Situation im *Deutschen Reich* nach dem Ersten Weltkrieg von einer unbedeutenden antikommunistischen und antisozialdemokratischen Sekte zur mitgliederstarken Partei werden. Dem wirtschaftlichen Wachstum der Vorkriegsjahre[56] folgten wirtschaftliche Depression und Arbeitslosigkeit. Verschärft stellte sich dieser Abstieg in Wilhelmshaven/ Rüstringen dar. Die Umstellung von Kriegsproduktion auf Friedenswirtschaft, die der Industrie kaum glückte, konnte *Marine* und *Werft* erst recht nicht gelingen. Den ehemaligen Soldaten und entlassenen Werftarbeitern glückte es nicht immer, im bürgerlichen Leben Fuß zu fassen, und für die parlamentarische Demokratie von Weimar waren nur wenige zu gewinnen. Die *NSDAP* bekam ihre ersten Mitglieder aus der Jugend, die eine Zukunftsaussicht haben wollte, und aus der nationalistischen Mittelschicht. Die Furcht vor sozialem Abstieg trieb der *NSDAP* ganze Bevölkerungsgruppen in die Arme[57], Angestellte, Selbständige, Handwerker, Kleinunternehmer, und nach der Weltwirtschaftskrise ab 1932 dann auch Teile der Arbeiterschaft. Die *NSDAP* wurde vom Mittelstand beherrscht – ihm ging es um eine traditionell geordnete Gesellschaft, die ihren vermeintlichen Besitzstand gegen Kommunisten und Sozialdemokraten wehrhaft verteidigt[58]. Äußerst geschickt band die *NSDAP* auch Studenten, Beamte und Bildungsbürger an sich, denn Furcht vor einer modernen Industriegesellschaft erstreckte sich durch alle gesellschaftlichen Gruppierungen.

In Wilhelmshaven/Rüstringen waren die Verhältnisse anders – der ungünstige Standort hatte eine wirkungsvolle Industrialisierung verhindert, trotzdem aber hatte der Flottenaufbau Konjunktur in die Region gebracht, auch ohne Unternehmer und private Kapitalgeber. In Wilhelmshaven/Rüstringen gab es immer nur das Staatsunternehmen *Marine*. Der Aufbau Wilhelmshavens hat wenig mit Industrie zu tun. Die Grundlage für eine Industrialisierung, der bürgerliche Geldadel, kam in der Stadtentwicklung nicht vor. Die beiden Städte waren nach dem Verlust der Kriegsflotte 1919 wirtschaftlich am Ende, und um die wenigen verbliebenen Schiffe zu unterhalten und zu ersetzen, bedurfte es nur einer kleinen *Werft*. Eine Friedenswirtschaft konnte sich kaum etablieren. *Deutsche Werke*[59], Fremdenverkehr und *WRIHALA* versuchten mühselig wirtschaftliche Wiederbelebung.

Die 1928 gegründete Ortsgruppe Wilhelmshaven/Rüstringen der *NSDAP* konnte seit den Reichstagswahlen vom 20.05.1928 beträchtlichen Stimmenzuwachs verzeichnen, jedoch anfangs kaum in die *SPD*-Wählerschaft einbrechen. Sie gewann bürgerliche

Kräfte, Handwerker, Selbständige, ehemalige Marineangehörige usw. Sie konnte ihre Macht nur ausdehnen, indem sie neue Bevölkerungsgruppen für sich einnahm.

- Gezielte Siedlungspolitik und neue Wohnungsprojekte sollten nicht mehr nur die „Minderbemittelten" der Weimarer Republik, sondern die von *Hitler* gewünschten Bevölkerungsgruppen erreichen.
- Die Baukonjunktur wiederum sollte Arbeitsplätze schaffen und das Verhältnis zu Kapital und Industrie verbessern.
- Militär, Hafen und Werftplanungen sollten die Wirtschaft kurzfristig beleben und mittelfristig auslasten.

In Wilhelmshaven/Rüstringen war die Ausgangslage für eine „neue" Politik einfacher, da die Errichtung von industriellen Produktionsstätten wie in Wolfsburg (*Stadt des KdF-Wagens*) oder Salzgitter (*Stadt der Hermann-Göring-Werke*) wegfiel. Die Arbeitsplätze bei *Marine* und *Werft* konnten ohne Umwege schnell reaktiviert werden. Die
126 großangelegten Erweiterungen von *Werft* und Hafen, von Stadt und Infrastruktur, waren schon frühzeitig eingeplant. Als die Nationalsozialisten am 30.01.1933 die Macht übernahmen, war der Höhepunkt der Krise erreicht, und nur sofortige öffentliche Baunachfrage konnte den wirtschaftlichen Kollaps verhindern. Zu Anfang waren die Arbeitgeber völlig gegen *Arbeitsbeschaffungsmaßnahmen* eingestellt.

Hitler sprach am 23.05.1932 in einer Propagandaversammlung auf dem Rüstringer Schützenplatz geschickt die Ängste bürgerlicher Schichten vor „entsetzlicher Verelendung" an, und drohte, daß unter Marxisten nicht 6, sondern 40 Millionen Arbeitslose zu beklagen wären und Deutschland von einem furchtbaren Blutbad heimgesucht werde. Deutschland überwinde seine politischen Nöte, wenn es außenpolitisch erstarke[60]. Für Wilhelmshaven/Rüstringen hieß außenpolitische Stärke: Marineaufrüstung. Große Teile der Bevölkerung erwarteten die Stärkung von deutscher Wehrhoheit und deutschem Wehrwillen, „und damit auch des kommenden Aufschwungs unseres Vaterlandes und seiner Hafen-Doppelstadt Wilhelmshaven-Rüstringen"[61]. 8.637 Arbeitslose zum Jahresanfang, Senkung der Rüstringer Wohlfahrtssätze (10 bis 12 %) und großangelegte *NSDAP*-Wahlkundgebungen verunsicherten auch Teile der sozialdemokratischen Arbeiterschaft.

Eine Belebung der Bauproduktion trat erst langsam ab 1934 ein, trotz:

- 1. *Vierjahresplan vom* 01.02.1933 (nationaler Wiederaufbau)
- *Kraftfahrzeugsteuergesetz vom* 11.04.1933
- 1. *Gesetz zur Verminderung der Arbeitslosigkeit vom* 01.06.1933 (*Reinhard-Plan*)
- 600 *Millionen-Programm* des *Reichskommissars für Arbeitsbeschaffung (Gereke-Programm)* 1933[62]
- *Gesetz über Steuererleichterungen vom* 15.07.1933
- 2. *Gesetz zur Verminderung der Arbeitslosigkeit vom* 21.09.1933 (Autobahnbau)
- 500 *Millionen-Programm* 1934 (2. *Reinhard-Programm*)[63]

Arbeitsbeschaffungsmaßnahmen waren die Leitlinien aller Planungen in Wilhelmshaven/Rüstringen vor und nach 1933. Bereits 1918 bewilligte Rüstringens Stadtrat 500.000 Mark für Notstandsarbeiten[64] zum Ausbau des Stadtparks nach den Planungen von *Migge*. Auch Wilhelmshaven bekam entsprechende Haushaltsmittel für Notstandsmaßnahmen. Die *Oldenburgische Regierung* wurde initiativ und begann 1927 mit der Eindeichung des Voslapper Grodens für 12 Mio. Mark. Der Landflughafen Mariensiel (1927)[65], der Seeflughafen am Fliegerdeich (1928), der Südstrand (1928), das Rüstringer
Rathaus (1928–29), die Grünanlagen vor dem Rüstringer Rathaus (1929), die 112
Kurparkerweiterung bis zur Mozartstraße (1932), waren einige der wirtschaftsbelebenden *Arbeitsbeschaffungsmaßnahmen* der Weimarer Republik und galten als „produktive Erwerbslosenfürsorge". Wilhelmshaven und Rüstringen unterstützen auch den 1932 eingeführten *FAD*, der 1935 Pflicht wurde. Junge Menschen wurden in die Arbeitsdienstlager nach Dangast, Cloppenburg[66], Rüstersiel oder Wangerooge geschickt, Deichbau und Landwirtschaft, Wegebau und Moorkultivierung standen im Vordergrund. 1933 gab Rüstringen 1,1 Mio. Mark und 1934 561.000 Mark für *Arbeitsbeschaffungsmaßnahmen* aus. Wilhelmshaven sah 1934 400.000 Mark vor. Nach der Machtübernahme der *NSDAP* standen Straßenneubau und Straßenunterhaltung im Vordergrund. 1933–36 wurde die Ebkeriege ausgebaut, die Fortifikationsstraßen wurden geschottert,
die Herbartstraße zwischen Mühlenweg und Hegelstraße wurde gepflastert und die 135
Straßen der Adolf-Hitler-Siedlung wurden neu angelegt. Später kamen die Erschließung 26
des Geländes zwischen Mühlenweg und Kirchreihe, der Durchbruch der Peterstraße, der 23
Ausbau des Schwarzen Weges, das Umlegen der Bismarckstraße und die Neuanlage der Störtebekerstraße hinzu. Auch die Siedlungen Coldewei-Himmelreich und die Adolf-
Hitler-Siedlung gehörten zu den *Arbeitsbeschaffungsmaßnahmen.* 24

In Jever, seit dem 01.06.1936 zum *Wohnsiedlungsgebiet Jever-Rüstringen* gehörig, 129
stilisierte *Bürgermeister Johann Folkerts* Kläranlage, Kanalisation, Gräben, Hotelumbauten,
Kabelarbeiten, Friedhofserweiterung und sogar den gesamten Fliegerhorstbau Upjever 104
zu *Arbeitsbeschaffungsmaßnahmen.*

In Wilhelmshaven/Rüstringen setzte unmittelbar nach der Machtergreifung vom 30.01.1933 die Vorbereitung eines Baubooms ein, der seinesgleichen sucht. Noch bevor die *Oberbürgermeister Friedrich Paffrath* am 18.03.1933 und *Emil Bartelt* am 08.07.1933 aus ihren Ämtern verjagt wurden, besichtigte der am 16.06.1932 gekürte *NSDAP-Ministerpräsident Carl Röver*, Oldenburg, die Hafenanlagen[67], sicher auch, um die Voraussetzungen für die Marineaufrüstung zu prüfen.

Der größte zivile Hafennutzer, die *WRIHALA*, wurde dann auch 1934 auf Druck des *Reiches* liquidiert, um Platz für Baumaßnahmen und Wiederaufrüstung zu schaffen. Im Februar 1933 wurde mit dem Ausbau der Ebkeriege als *Arbeitsbeschaffungsmaßnahme* begonnen. Die Ebkeriege war wichtige Straßenverbindung ins Binnenland und Voraussetzung zum Erreichen der Ziegelei Mariensiel und Voraussetzung für den Bau der Arbeiterlager. Die *Gemeinden* des Jadegebietes forderten[68] auch den vollspurigen Ausbau der Eisenbahn Sande-Aurich als *Arbeitsbeschaffungsmaßnahme.* Am 08.06.1933 wurde das Lager des *FAD* in Rüstersiel gerichtet. Mehrere *SS*-Heime, der neue *Vorwärts*-Sportplatz, ein Feldpolizeiheim, die Sparkasse am Park und eine Freilichtbühne am Rosenhügel waren erste Baumaßnahmen, die von den Nationalsozialisten gefördert wurden.

Als *Hitler* zusammen mit *von Blomberg* und *Raeder* am 12.12.1933 den heimkehrenden Kreuzer Köln in der III. Einfahrt begrüßte, wurde er von Tausenden umjubelt – er hatte sich längst zur Marineaufrüstung entschlossen. Die *NSDAP*-Machthaber sprangen auf fahrende Züge und hefteten sich Baumaßnahmen auf ihre Fahnen, die noch in der Weimarer Republik geplant oder begonnen waren. Das vor der Machtübernahme begonnene Siel in Berne wurde nach seiner Fertigstellung Adolf-Hitler-Siel[69] getauft. Die Wilhelmshavener Siedlung südlich Sedan, nach einem alten *Bebauungsplan* entstanden
26 und ab 1934 gebaut, bekam gleichfalls den Namen *Hitlers*. Die propagandistische
24 Ausschlachtung von Richtfesten, Grundsteinlegunge und Stapelläufen hatte sich längst
156 ausgezahlt. Bei den Reichstagswahlen vom 12.11.1933 erhielt die Einheitsliste der *NSDAP* 92 % der Stimmen. Am 12.06.1934 konnten die regionalen Zeitungen stolz den Rückgang der Arbeitslosigkeit in den Jadestädten melden[70].

Männer wurden in Lohn und Brot gebracht. Frauen sollten durch Ehestandshilfen zurück an den Herd – aber nur „nationalgesinnte, gesunde, gut beleumundete Deutsche"[71]. Vom Berufsleben waren Frauen für lange Dauer ausgeschlossen. „Doppelverdienertum" wurde scharf bekämpft. In Wilhelmshaven/Rüstringen startete das „III. Reich" massive *Arbeitsbeschaffungsmaßnahmen*, die meisten liefen auf den Instanzenweg zu den Behörden und zur *Marine*. Aber auch die Bürger wurden zur Arbeitsbeschaffung angehalten. Der Präsident der *Deutschen Haus- und Grundbesitzer, Obersturmbannführer Tribius* hielt am 23.11.1934 die Haus-und Grundbesitzer[72] in Wilhelmshaven/Rüstringen an, mehr zu investieren. Er kündigte Hauszinssteuersenkung und sogar Zwangsmitgliedschaft an. Die Hausbesitzer wurden von der *Stadt Rüstringen* aufgefordert, Hauswarte und Putzfrauen als *Arbeitsbeschaffungsmaßnahme* einzustellen, und auf den üblichen „Treppendienst der Hausfrauen zu verzichten".

Die Arbeitsmarktanstöße nutzte die *NSDAP* geschickt für sich aus. Schon zum „Tag der Arbeit" am 01.05.1934 machte *Reichsstatthalter Carl Röver* vor dem Rüstringer Rathaus klar: „Als der Führer *Adolf Hitler* an die Regierung kam, war es seine erste
153 Aufgabe, durch eine großangelegte Arbeitsschlacht die Arbeitslosigkeit herabzudrücken. Das ist glänzend gelungen. Mehr als drei Millionen sind wieder in den Arbeitsprozeß eingegliedert worden". Selbst in Gottesdiensten wurde dankbar die Überwindung des Klassenkampfes vermerkt, der „das Volk an den Rand des Ruins brachte"[73]. Die auch in Wilhelmshaven/Rüstringen langsam erkennbare bessere Wirtschaftslage wurde umgebogen zum Kampf gegen „Nörgler, Böswillige, Narren und Kritikaster", die nicht erkennen wollen, daß die „Kaufkraft gestiegen ist", daß „das Geld rollt ... Wer das bisher Geschaffene nicht einsieht, ist böswillig", so tönte der *NSDAP-Oberbürgermeister* von Rüstringen[74].

2.3. Baudienststellen

Das für den Aufbau Wilhelmshaven wichtige staatliche Bauwesen in *Preußen* ging auf die 1799 gegründete Bauakademie Berlin und die 1808 installierte *Oberbaudeputation* zurück. *Oberlandesbaudirektor Schinkel* prägte als oberster preußischer Baubeamter und Lehrer Generationen von Architekten und technischen Beamten, die auch für Wilhelmshaven wichtig wurden. *Friedrich Wilhelm IV.* löste die *Oberbaudeputation* kurz nach *Schinkels* Tod auf. Das Kollegium wurde in das *Handelsministerium* übernommen. 1878 wurde der preußische Staatshochbau dem *Ministerium für öffentliche Arbeiten,* 1920–45 dem *Preußischen Finanzministerium* unterstellt.

Für Wilhelmshavens nicht-militärische Bauten waren ab 1873 die *Kreisbauinspektoren* der *Königlichen Bauverwaltung* der *Landdrostei Aurich,* später die *Kreisbauinspektionen* und ab 1911 die *Königlichen Hochbauämter* beim *Amt* bzw. *Landkreis Wittmund* zuständig [75]. Wilhelmshaven war am 01.04.1873 in die *Preußische Provinz Hannover* und damit in das *Königreich Preußen* einbezogen worden. Die *Provinzialbauverwaltung Hannover* hat wahrscheinlich wenig Einfluß auf Wilhelmshaven genommen.

Der 1853 zwischen dem *Königreich Preußen* und *Oldenburg* abgeschlossene *Jahde-Vertrag* war Fundament der Entwicklung Wilhelmshavens und der umliegenden Gemeinden. *Preußen* hatte früh sein Augenmerk auf die abseits der großen Ereignisse liegende Nordseeküste gerichtet. Nach den Vorarbeiten von *Prinz Adalbert von Preußen* und *Gotthilf Hagen* führte der Aufbau der *Preußischen Marine* Berliner Unterhändler in diese Region, und am 25.09.1852 den architektonisch ambitionierten *König Friedrich Wilhelm IV.* selbst. Am 18.08.1853[76] ratifizierte er den *Jahde-Vertrag* und *Prinz Adalbert* übergab das Terrain am 23.11.1954 mit viel Berliner Prominenz.

Die Architekturgeschichte der jungen Stadt war eng verflochten mit den örtlichen Hafen- und Garnisonbaudienststellen, deren Zuordnung und Organisation häufig
wechselten. Durch *Kabinettsordre vom* 11.02.1854 wurde der *Admiralität* in Berlin am 2
Leipziger Platz die Leitung des Projektes Wilhelmshaven übertragen[77]. Am 18.07.1855 wurde eine *Hafenbaukommission* eingesetzt – die erste Behörde an der Jade überhaupt[78]. Vorsitzender war *Gotthilf Hagen.* Hochbauangelegenheiten von *Werft* und *Marine* und der Stadtbau wurden 1855–73 von der *Hafenbaukommission* „miterledigt". Da jedoch die *Admiralität* noch nicht über eine eigene *Bauverwaltung* verfügte, war das *Handelsministerium* als oberste preußische Hochbauverwaltung und Nachfolgebehörde der *Oberbaudeputation* Ministerialinstanz für den Aufbau Wilhelmshavens. *Hagen* war im November 1854 mit dem *Chef des Stabes* an die Jade gekommen und zeigte sich unzufrieden über den Fortgang der Arbeiten vor Ort[79]. Die *Hafenbaukommission* bestand formal bis zum 01.04.1889[80], hatte aber nur bis 1872 Einfluß auf Architektur und Planung. Der Leitung
der *Hafenbaukommission* unterstand auch das *Hafenbauressort* der *Werft*[81]. Die *Marinein-* 10
tendantur für Wilhelmshaven war bis 1871 in Berlin bei der *Admiralität* angesiedelt. Es 2
kann kein Zweifel bestehen, daß man den Wilhelmshavener Dienststellen nicht allzuviel
Können zutraute, ganz sicher nicht den Entwurf von Hafenanlagen und Stadtgrundriß. 6

Das *Baudezernat der Admiralität* war verpflichtet, in Bauangelegenheiten die übrigen Ministerien zu beteiligen, und erkämpfte sich sehr zum Ärger des *Kriegsministeriums* zunehmend Selbständigkeit[82].

Mit der Auflösung des *Admiralitätskommissariats Oldenburg* (1873) gingen Hochbauangelegenheiten an die 1872 neu gegründete *Kaiserliche Intendantur* der *Marinestation der Nordsee* in Wilhelmshaven über. Die *Hafenbaukommission* der *Werft* verlor ihren Einfluß auf den Hochbau. Die *Stationsintendantur* entwickelte sich zur Mittelinstanz und baute eine eigene *Bauabteilung* auf, deren Leiter lange Jahre *Intendantur- und Baurat Valentin Bugge*[83], später *Baurat Ernst Zimmermann* war[84]. Die Ämterbezeichnungen der Marinebaudienststellen wechselten häufiger. Immer waren jedoch die Bauabteilungen eng mit
14 der örtlichen *Marineintendantur* verflochten. Es bildete sich um 1880 bei der *Kaiserlichen Marine-Garnison-Verwaltung* eine eigene *Garnison-Bauverwaltung* als Ortsbaudienststelle. *Bugge, Schubert, Häußler, Balfanz, Fleinert, Müller,* waren u.a. Leiter vor dem Ersten Weltkrieg. Dieses ab 1890 so bezeichnete *Marine-Garnison-Bauamt Wilhelmshaven,* zeitweise aufgeteilt in *Garnison-Bauämter I* und *II,* existierte bis zum Ende des Ersten Weltkrieges.

Die *Verfassung des Deutschen Reiches vom* 16.04.1871 sah nur Wilhelmshaven und Kiel als befestigte Kriegshäfen der *Kaiserlichen Marine* vor. Erst jetzt wurde vor Ort die
11 *Marinestation der Nordsee* mit ihrer *Kaiserlichen Intendantur Wilhelmshaven* für Bauangelegenheiten zuständig[85].

Das Marinebauwesen in Berlin unterstand ab 1871 dem *Baudezernat der Kaiserlichen Admiralität* und ab 1889 dem *Reichsmarineamt* als Ministerialinstanz. Darunter war die Anfang der 70er Jahre gegründete *Stations-* später *Marineintendantur Wilhelmshaven* Aufsichtsbehörde. Sie war zuständig für alle Marinedienststellen westlich der Elbe. Die *Marineintendantur Kiel* war zuständig für die Bereiche östlich der Elbe.

14 Die kleine *Bauabteilung der Intendantur Wilhelmshaven* hatte reine Aufsichts-, Prüf- und Genehmigungsfunktionen als Mittelinstanz. Die konkrete Bautätigkeit lag bei der *Kaiserlichen Garnison-Bauverwaltung.* Als Ortsbaudienststelle löste sie sich von der *Intendantur,* wurde zunehmend eigenständiger und bis 1918 allein für Hochbauten zuständig[86]. Die *Garnisonbauordnung vom* 03.04.1909 regelt den Marinehochbau.

10 Hafen- und Hochbau der *Werft* verblieben bei der *Werft* bzw. ab 1935 bei der *HNBD* und wurden erst 1943 wieder mit dem Hochbau zum *Marineoberbauamt Wilhelmshaven*
11 vereinigt. Die *Marinestation der Nordsee* verfügte noch lange Jahre über eine eigene *Königlich Preußische Festungs-Bau-Direktion* und zeitweise über die *Kaiserliche Neubauverwaltung.*

Nach dem verlorenen Krieg wurden die Baudienststellen von *Heer* und *Marine* aufgelöst und dem *Reichsschatzministerium* unterstellt. Nur *Post, Verkehrsministerium* und Eisenbahn behielten eigene Bauverwaltungen.

★

Nach dem Ersten Weltkrieg wurden in Wilhelmshaven alle marinebezogenen *Bauämter* und das *Hafenbauressort* aufgelöst. 1919 wurden *Reichsvermögensämter* und ab 1922 verschiedene *Reichsbauämter* eingerichtet. Leiter der Reichsvermögensämter waren u.a. *Beck, Klett, Hermeking*. Sie wurden später zu einem *Reichsbauamt* zusammengefaßt und der *Reichsbauververwaltung* unterstellt. Das *Reichsbauamt Wilhelmshaven* hatte die Liegenschaften der *Marine* zu verwalten, zu verwerten und umzubauen. Es war Nachlaßverwalter der Bauten und Hafenanlagen der wilhelminischen *Marine*. Das Personal wurde übernommen. Das *Reichsbauamt Wilhelmshaven* existierte noch bis 1949, allerdings ohne militärische Aufgaben. Die Aufgaben des 1922 aufgelösten *Reichsvermögensamtes* Wilhelmshaven gingen an das *Finanzamt*. Das *Reichsbauamt Wilhelmshaven* hatte noch mit dem Bau von *Finanzamt*, Reichswasserstraßen und Bauten der *NSDAP* zu tun. Leiter war über viele Jahre *Paul Berg*.

Ab 1922 hatte die *Reichsmarine* der Weimarer Republik wieder eine eigene Bauverwaltung, das *Marinestandortbauamt*, zeitweise auch als *Marinebauamt* bezeichnet. Es wurde im wesentlichen von *Otto Lückemeyer* aufgebaut, der den Marinehochbau im Raum Wilhelmshaven maßgeblich beeinflußte. Durch das *Wehrgesetz vom* 23.03.1921 war die *Reichswehr* als Berufsarmee geschaffen worden und mit ihr die aus freiwilligen Soldaten und Militärbeamten bestehende *Reichsmarine*. Nachdem der NS-Staat 1935 wieder die Wehrhoheit an sich gerissen hatte und die *Marine* offen in *Kriegsmarine* umbenannt wurde, konnte sich das *Marinestandortbauamt Wilhelmshaven* zur wichtigsten Wilhelmshavener Baubehörde entwickeln. Der Zuständigkeitsbereich reichte weit in das Binnenland, bis an den Bodensee. Leiter waren *Wilhelm Naud, Otto Lückemeyer* (bis 1941), *Ludwig Bellwinkel* (1941–43), *Alfred Hucke* (ab 1943).

In Berlin etablierte sich 1935 das *OKM* im Gebäude Tirpitzufer/Bendlerblock. Wie *4*
auch vorher die *Reichsmarine* und die *Kaiserliche Marine*, verfügte es über eine eigene *Hochbauabteilung* innerhalb des *Marineverwaltungsamtes*[87]. Ministerialinstanz war das *Marineverwaltungsamt des OKM*. Leiter waren während des „III. Reiches" *Ministerialdirektor Theodor Schreiber* und 1939–44 *Hans Benda*. *Abteilungsleiter Hochbau* waren die *Ministerialdirigenten Friedrich Schaller* und ab 1939 *Ernst Fitzler*. Für Wohnungsbau war der Jurist *Oberintendanturrat Walter Tornow* 1938–44 zuständig. Die *Hochbauabteilung* unterstand 1934–45 *Ministerialdirigent Walter Kelm*, der einem kleinen Stab von Architekten und *Bauräten* vorstand, zu denen *Rambacher, Kussin, Bellwinkel, Block, Geisenhainer* und *Zenker* gehörten. Viele von ihnen waren mit dem Baugeschehen in Wilhelmshaven eng verbunden. Daneben gab es außerhalb des *Marineverwaltungsamtes* eine *Hafenbauabteilung*, die *Alfred Eckhardt, Hafenbaudirektor* bei der Wilhelmshavener *Werft* 1919–33, leitete. Mittelinstanz war während des „III. Reiches" wie bisher die *Bauabteilung der Marineintendantur*. Die *Marineintendantur* hatte entscheidenden Einfluß auf *14*
die Finanzierung des Wohnungsbaues. Zuständig war der Jurist *Karl Hasso von Widekind*. Er bewirtschaftete alle Reichsdarlehen an Wohnungsbaugesellschaften und mischte sich in Gestaltung und Architektur ein. Ortsbaudienststellen der *Kriegsmarine* blieben die *Marinestandortbauämter* bis 1943. *Hitlers* zielgerichtete Rüstungspolitik beanspruchte das

OKM zusehends mehr und ließ kaum Raum für Verwaltungs- und Bautätigkeiten. Das Hochbauwesen wurde 1943 aus dem *Marineverwaltungsamt des OKM* ausgegliedert, ebenso das Hafen- und Strombauwesen aus dem *Hauptamt-Kriegsschiffbau.* Hieraus entstand in Berlin als neue Ministerialinstanz das *Amt Bauwesen* des *Allgemeinen Marinehauptamtes des OKM.* Leiter war bis zur Eingliederung in die *OT* auch der Wilhelmshavener *Alfred Eckhardt.*

Für Hafen- und Hochbau der *Werften* waren in Kiel und Wilhelmshaven[88] *Hafenbaudirektoren* eingesetzt. Nicht selten wechselten sie zwischen den beiden Kriegshäfen von und nach Berlin oder in ausländische Kriegshäfen.

Ab 1908 wurden die Ressorts in Hafen und Strombau aufgeteilt, 1922 wieder zusammengeführt, 1933 wieder getrennt. Während der Weimarer Republik waren die *Hafenbauressorts* zeitweise den *Reichsbauämtern* zugeordnet. *Wilhelm Krüger* und *Georg*
101 *Frede* bereiteten die Hafenerweiterungen, die Planung der *Nordwerft* und die IV. Einfahrt im *Hafen- und Strombauressort* der *Marinewerft* vor. *Alfred Eckhardt*, ab 1933 im *OKM*, war als Wilhelmshavener bestens mit den Gegebenheiten vertraut, und galt als Motor des Kriegshafenausbaus auf ministerieller Ebene. Als Ortsbaudienststelle für den Hafenneubau und *Hitlers* größenwahnsinnige Ausbaupläne wurde 1935 die *HNBD* unter *Carl Tiburtius* (ab 1939 *Ernst Beck*) gegründet. Sie entwickelte sich mit ihren zwei *Marinehafenbauämtern* zu einer der größten Behörden in Wilhelmshaven und befehligte ein Heer von Bauarbeitern und Firmen. Sie baute eigene Arbeiterlager und bereitete die
10 Umsiedlung ganzer Stadtteile vor[89]. Die *KMW* behielt eine ambitionierte *Hochbauabteilung.* Sie unterstand den *Hafenbaudirektoren Pein* und *Frede.* Unter *Drenckhan, Trittler* und *Hacke* entwarf *Hans Klaucke* außerordentlich modern.

Anfang 1943[90] löste das *OKM* sämtliche Hoch- und Tiefbaudienststellen der *Marine* auf. Die Nationalsozialisten wollten eine straffere Führung. Kiel und Wilhelmshaven erhielten je eine *Marineoberbaudirektion* als Mittelinstanz für Hoch- und Tiefbau als Gruppe der jeweiligen *Marineoberkommandos.* In Kiel wurde 1943 der langjährige Leiter des *Marinestandortbauamtes Wilhelmshaven, Otto Lückemeyer*, Chef der *Marineoberbaudirektion.* In Wilhelmshaven fiel diese Aufgabe an *Georg Frede*, der bis dahin das *Hafenbauressort der KMW* geleitet hatte. Ortsbaudienststelle für Hoch- und Tiefbau wurde ab 01.01.1043 das neu eingerichtete *Marineoberbauamt* unter der Leitung des Hafenbauers *Walter Triebel.* Sein Vertreter wurde der Hochbauer *Alfred Hacke.* Er hatte die Funktion des alten *Marinestandortbauamtes.* Die Umstrukturierung[91] belegte die veränderten Zielsetzungen der Nationalsozialisten. Architektur wurde keine eigene Rolle mehr zugemessen. Das Marinebauwesen wurde auch unter militärischen Aspekten in die eroberten und besetzten Gebiete ausgedehnt. 42 *Marinebauämter*, 10 *Marineoberbauämter* und 4 *Marinebaudirektionen* vom Mittelmeer bis zum Polarkreis zeugten von *Hitlers* kriegerischen Zielen[92].

GBI Albert Speer hatte bereits in seinen ersten Neugestaltungsplänen für Berlin nördlich
5 der Versammlungshalle am Königplatz ein neues *OKM* geplant. Am „Großen Bassin" hatten *Paul Bonatz* und *Kurt Dübbers* bis 1943 verschiedene, nie realisierte Planungen für

die oberste Zentrale der *Kriegsmarine* vorgesehen. Am 25.07.1944 wurde das *Amt Bauwesen,* Ministerialinstanz im Berliner *Allgemeinen Marinehauptamt des OKM*, aufgelöst, und alle nachgeordneten Baudienststellen der *Kriegsmarine Rüstungsminister Albert Speer* und seiner *OT* unterstellt[93]. Jetzt war *Speer* nicht nur *GBI* und für den Wiederaufbau bombenzerstörter Städte zuständig, sondern auch „Chef des Marinebauwesens" für die Endphase des Krieges. In der Hektik des untergehenden „III. Reiches" wurde die neue Organisation nur noch in Ansätzen durchgesetzt. Es bildeten sich *OT-Bauleitungen* für *Heer, Marine* und *Luftwaffe*, teils blieb aber auch alles beim alten. In Wilhelmshaven wurde die neue Organisation noch durchgesetzt und eine *Marine-Betriebsabteilung* als Mittelinstanz gebildet[94].

⋆

Neben den dominierenden Baudienststellen der *Marine* war noch die der *Luftwaffe* für Wilhelmshaven wichtig. Luftwaffenbaudienststellen konnten nicht wie die *Marine* auf gewachsene Strukturen zurückgreifen. 1918 war die *Luftwaffe* völlig aufgelöst worden, und erst 1934 begannen das *Luftflottenkommando* in Braunschweig und das *VI. Luftgaukommando See* in Kiel[95] mit planerischen Vorarbeiten. *Hermann Göring* wurde am 01.05.1935 *Oberbefehlshaber der Luftwaffe.*

Für die Bauten der Flieger war die *Bauabteilung* des *Reichsluftfahrtministeriums* Ministerialinstanz[96]. Mittelinstanz war die Bauabteilung der jeweiligen *Luftgaukommandos*, erst Kiel, später Hamburg. In Wilhelmshaven[97], auf dem Gelände der ehemaligen Siebethsburg, etablierte sich 1938 die Ortsbaudienststelle, die zu Anfang als *Luftwaffenkommando See, Bauverwaltungsbezirk Wilhelmshaven*, bezeichnet wurde. Erst 1939 wurde sie in *Luftwaffenbauamt Wilhelmshaven* umbenannt. Langjähriger Leiter war *Baurat Schäfer*, der das Landhaus des Gutsbesitzers *Georg Janßen* für seinen *Bauverwaltungsbezirk* umbaute, und 1938 auch noch um Anbauten, Messebaracken, Schlafbaracken und Dienstwohnungen ausweitete. Dem *Luftwaffenbauamt Wilhelmshaven* unterstanden verschiedene, dezentral untergebrachte, häufig wechselnde Dienststellen als *Bauleitungen der Luftwaffe:* Bremerhaven, Upjever, Delmenhorst, Nordenham, Wangerooge, Langeoog, Mariensiel, Helgoland, Norderney, Fliegerdeich, Wilhelmshaven usw. Neben den *Bauräten Schäfer, Weving, Schütt, Jaunisch, Breitschwert* und *Steinbrecher* waren für Hochbauplanungen *Heinrich von Griesbach, Arthur Nevermann, Walter Sommer, Heinrich Pauls, Josef Vosshans, Adalbert Jahn* und *Karl Higgen* dort beschäftigt. Auch *Klaus Rasch*, ab 1939 Leiter der *Hochbauabteilung der Stadt Wilhelmshaven*, arbeitete dort 1936. Die Architekten der *Luftwaffe* hatten einen außerordentlich großen Arbeitsbereich und entsprechend „großartig" war auch ihr Selbstverständnis. Uniformen wurden in der Regel nicht getragen. Auch *Walter Hunzinger*, Wilhelmshavens *Stadtbaudirektor* 1937–39, kam aus der *Bauverwaltung der Luftwaffe*. Sein Lebensstil wurde ihm allerdings 1939 zum Verhängnis, und er wurde von der *Stadt* entlassen.

Josef Vosshans war vom *Hafenbauressort der KMW* zum *Luftwaffenbauamt* gekommen. Unter *von Griesbach* arbeitete er in den *Bauleitungen der Luftwaffe:* Seefliegerhorst Fliegerdeich/Wilhelmshaven (1937–39), Bremerhaven (1940) Helgoland (1941) und später in Norwegen[98]. Architekt *Breitschwert* plante Flughafen und Fliegerkasernen auf
126 der Insel Langeoog. Auch die Planungen für eine riesige Luftwerft und Land- sowie
127 Wasserflugplatz im Heppenser bzw. Rüstringer Groden ressortierten beim Wilhelms-
128 havener *Luftwaffenbauamt.*

104 Die Architekten des *Luftwaffenbauamtes Wilhelmshaven* waren durchaus ambitionierte
105 Planer mit ästhetisch hohen Ansprüchen an Unterkunftsgebäude und Flugzeughallen. So zeugen noch heute vorhandene Kasernen und Fliegerunterkünfte an Wilhelmshavens Fliegerdeich, in Upjever, Langeoog und Wangerooge[99] von einer einheitlichen Linie. Die Wilhelmshavener Architekten des *Luftwaffenbauamtes* beteiligten sich im Kriege sogar an überregionalen Architektenwettbewerben[100]. Besonders beliebt waren Bauleitungen auf den Nordseeinseln. *Vosshans* berichtet noch heute vom erlebnisreichen Bau der Flugplatzanlagen und Gebäude auf der Düne von Helgoland. Beton wurde aus den herausgesprengten roten Felssteinen der U-Bootbunker gemischt.

Im Mai 1944 wurde das gesamte Bauwesen der *Luftwaffe*, wie auch das von *Heer* und *Marine*, der *OT* von *Rüstungsminister Albert Speer* unterstellt. Vehementer Förderer der Luftwaffenbauten für Wilhelmshaven war der gebürtige Wilhelmshavener *Erhard Milch* (1892–1972), *Staatssekretär im Reichsluftfahrtministerium* von 1933–44 und *Generalinspekteur der Luftwaffe* 1938–45.

★

Die Oldenburger Dienststellen wollten anfangs die Inbesitznahme des Jadegebietes durch *Preußen* nach bürgerlichem Recht abwickeln. Verwaltungsbehörde war 1854–73 das *Admiralitätskommissariat* in Oldenburg. Es war Mittler zwischen der *Berliner Admiralität* und den oldenburgischen Behörden. Die *Hafenbaukommission* an der Jade jedoch unterstand unmittelbar der *Berliner Admiralität.* Das Herzogtum Oldenburg war ebenso marinebegeistert wie Berlin. Der *Landtag* stimmte am 19.01.1851 dem *Jahde-Vertrag* trotz Bedenken zu. 1873 erhielt *Wilhelmshaven* eine Kommunalverfassung und erste Zivilbebehörden, denn die *Regierung Oldenburg* sah sich immer weniger in der Lage, Wilhelmshaven sachliche und personelle Hilfe zu gewähren. 1873 wurde *Wilhelmshaven* dem *Amt* (seit 1885 *Kreis) Wittmund* unterstellt. Von dort aus bzw. dem nach Wilhelmshaven entsandten *Landrätlichen Hilfsbeamten* wurden Baugenehmigungen ausgestellt, öffentliche Bauten entworfen, Verkehrsanlagen geplant. Die *Gemeinden* an der Jade hielten sich noch keine eigenen Baubeamten oder gar *Stadtbauräte.* Nach 20 Jahren militärischer Verwaltung entstanden bürgerliches Selbstverständnis und Baukultur erst langsam. Nach der Reichsgründung entwickelte sich wirtschaftliches Leben an der *Jade*, infolge der Zugehörigkeit zu den Staaten *Preußen* und *Oldenburg* sehr uneinheitlich und nach verschiedensten kommunalrechtlichen Prinzipien. *Wilhelmshaven* befand sich noch 1873 in größerer Abhängigkeit als eine Landgemeinde. Durch die *Hafenbaukommission* waren die wesentlichen Infrastrukturmaßnahmen gebaut worden. Die Architekturauffassungen der Königlichen Zivilverwaltung der preußischen Behörden und der kommunalen Dienststellen gingen weit auseinander. Fachlich geschultes Personal fehlte. Erst 1884 stellte Wilhelmshaven einen ersten *Bauinspektor* ein. Die meisten Bürger waren noch wenig mit Wilhelmshaven verbunden. Viele Zivilpersonen kamen, um schnell Geld zu verdienen, zogen oft wieder fort. Militärs und Beamte waren meist in Wilhelmshaven, und nicht im Umland ansässig, wurden oft versetzt und nur schwer an der Jade heimisch. Werftarbeiter wohnten zumeist in den umliegenden Gemeinden Bant, Heppens und Neuende. Erst mit *Paul Wolf* bekam Wilhelmshaven 1907 einen architektonisch geschulten *Stadtbaurat*. Er holte auswärtige Architekten in die Stadt, entwarf selbst, installierte eine Gestaltungssatzung und machte sich um die hygienischen Verhältnisse verdient. Ihm folgte 1910–37 *Hermann Zopff*. In Rüstringen folgten auf *Martin Wagner* ab 1914 *Willy Hahn* und ab 1921 *Carl Haefner*. Unter *Walter Hunzinger* und seinem Nachfolger ab 1939, *Walter Temp*, war die städtische *Bauverwaltung* so organisiert (Stand 1940)[101]:

3.	*Bauverwaltung (Temp)*
30.	*Bauverwaltungsamt (Ober-Bloibaum)*
31.	*Stadtplanung, Garten- und Friedhofswesen (Temp,* bis 39 *Hallbauer)*
31. a	*Stadterweiterungsamt (Baur,* bis 39 *Bartning)*
31. b	*Garten- u. Friedhofsamt (Stein,* bis 39 *Fresenius)*
31. c	*Städtebauamt (Baumann)*
32.	*Baupolizei (Neumann)*
32. a	*Baupolizei (Maas)*
32. b	*Baupflege (Koch,* davor *Siemer)*
32. c	*Wohnungsamt (Barlage)*
32. d	*Schätzungsamt (Pront)*
33.	*Hochbau (Rasch)*
33. a	*Bauunterhaltungsamt (Eilts)*
33. b	*Neubauamt (Rasch)*
33. c	*Luftschutzamt (Link)*
34.	*Tiefbau (Lehn)*
34. a	*Straßenbauamt (Schumacher)*
34. b	*Straßenreinigungs- u. Kanalplanungsamt (Bohländer)*
34. c	*Kanalbauamt (Aichinger)*
34. d	*Vermessungsamt (Kämmerer)*
34. e	*Bodenuntersuchungsstelle (Haarnagel)*

Die *Bauverwaltung der Stadt Wilhelmshaven* wehrte sich immer wieder gegen die planerische Bevormundung aus Berlin. *Oberbürgermeister Müller* machte häufig Vorschläge, wie schneller und wirkungsvoller zu arbeiten wäre[102]. In regelmäßigen Abständen sollten „gemeinschaftliche Besprechungen" mit der Aufsichtsbehörde, dem *MDF*, und dem Auftraggeber, dem *OKM*, stattfinden. Wilhelmshaven vermißte „klare Richtlinien" und „befruchtende Hilfestellung". Trotz personeller Aufstockung der *Bauverwaltung*[103] und eigens hinter dem Rathaus aufgestellter Bürobaracken blieb die planerische Arbeit vor Ort offenbar unbefriedigend. „Es ist eine auf die Dauer für die *Stadt* unmögliche Arbeitsdisposition, wenn in einem Augenblick von den Wehrmachtsstellen geltend gemacht wird, daß Tausende und Abertausende von Wohnungen bis zu einem bestimmten Termin für die Wehrmacht sichergestellt werden müssen, und seitens der *Stadt* alles dazu von ihr zu Leistende termingerecht geleistet werden soll, aber auf der anderen Seite immer wieder erkennbar wird, daß bei den verantwortlichen Reichsstellen die für die Erfüllung dieser Forderungen notwendigen Voraussetzungen noch nicht geschaffen sind"[104].

Die städtische *Bauverwaltung* lehnte vorsorglich alle Verantwortung wegen der zögerlichen planerischen Entscheidungen in Oldenburg und Berlin ab. Die Konflikte wurden auch der Öffentlichkeit nicht vorenthalten.

Der *Erlaß* des preußischen *Ministerpräsidenten Göring*, „daß sich niemand bei Fragen der Errichtung von Siedlungen kleinlich anstellen solle"[105], blieb ohne nachhaltige Wirkung. Daran änderten auch die Verflechtungen der städtischen Spitzenbeamten mit der *NSDAP*, die Verbindungen zur *Gauleitung*, die Mitgliedschaft in *LOB* und *Ahlhorner Architektenkreis*, nur wenig.

★

Wichtige zivile Baubehörde, Mittler zwischen Wilhelmshaven und Berlin, war der
3 *MDF, Julius Pauly*, in Oldenburg. Unter den *Ministerialräten Hans Ostendorf* und *Wilhelm Ostendorf* war der *Dezernent* im *MDF, Fritz Karl Schneider*, 1935–45 umfassend für die Aufsicht über Stadtplanung, Hochbau und Baupolizei zuständig. Bedeutsam für die Siedlungstätigkeit im Land Oldenburg waren das 1920 gegründete *Siedlungsamt*, das in den dreißiger Jahren gegründete *Oldenburgische Domänenamt* und das *Reichsheimstättenamt Gau Weser-Ems*. Die Ämter wurden zur Beschaffung von Bauland und zu siedlungspolitischen Zwecken eingeschaltet. Der *MDF* verstand seine Planungsaufsicht umfassend.
117 *Fritz Karl Schneider* griff selber in *Wirtschaftspläne* und *Bebauungspläne* ein, änderte nicht
37, 30 nur, sondern legte von sich aus für viele Gemeinschaftssiedlungen eigene Entwürfe vor. Er war Teilnehmer aller wichtigen Besprechungen in Wilhelmshaven und Berlin. Wilhelmshavens Planungen wurden in Oldenburg vorrangig bearbeitet. Sie bekamen beim Posteingang den Stempel: „Aufbau Wilhelmshaven! Sofort verlegen!"[106]. Der *MDF* wurde von der städtischen *Bauverwaltung* häufig eingeschaltet, wenn die Marinedienststellen die *Stadt* zu umgehen versuchten. Nach der *Verordnung über die baupolizeiliche Behandlung von öffentlichen Bauten vom* 20.11.1938 bekam die städtische *Bauverwaltung* die Marineplanungen nur nachrichtlich. Oftmals erfuhr die *Stadt* erst durch den *MDF* und auf umständlichen Wegen von den Planungsabsichten oder

Bauanträgen der *Marine*. Der *MDF* hatte auch die Finanzhilfeanträge der *Stadt Wilhelmshaven* zu prüfen und nach Berlin weiterzuleiten. Zuständig waren *Josef Wohlschläger* und *Kurt Siedenburg*. Man schaltete hierzu das *Landesbauamt Oldenburg* ein[107]. Die langsame Bearbeitung wurde *Hitler* am 01.04.1939 in Wilhelmshaven vorgetragen. Er versprach, „die Fragen zu lösen"[108]. *MDF Pauly* verfügte bereits 5 Tage später „schnellere Verfahrensgänge"[109], da er fürchtete, „daß es einen großen Skandal gibt, und daß die Schuldigen zu fassen versucht werden".

Kurt Siedenburg wurde zum *Landesbeauftragten für den Neuaufbau Wilhelmshavens* ernannt[110]. *Gauleiter Röver* wurde *Reichskommissar für die Erledigung von Beihilfeanträgen*. Nach dem *Hitler*-Besuch wurde auch die *Marine* gegenüber der städtischen *Bauverwaltung* mitteilsamer.

★

Einflußreichste, zivile Ministerialbehörde für die Stadterweiterung Wilhelmshavens war der *RAM*. *Arbeitsminister Franz Seldte* unterstand u.a. das gesamte Siedlungs- und Wohnungswesen von 1935–45. Die anfängliche Zusage der Nationalsozialisten, „jedem Arbeiter sein Heim", mußte scheitern. Die Wirtschaft verlangte Subventionsabbau, und so ressortierte der „unumgängliche Rest an öffentlicher Wohnungsfürsorge"[111] im *RAM*. Städtebau und Planungswesen lagen bei der *Abteilung IV c*. 1935 wurde *Reichssiedlungskommissar Gottfried Feder* entmachtet. Kleinsiedlungen wurden bedeutungsloser und Werksiedlungen, Wohnungen für Soldaten und Rüstungsarbeiter rückten als Reichsaufgabe in den Vordergrund. Nach dem *Gesetz zur Förderung des Wohnungsbaus vom* 30.03.1935 wurden Kleinwohnungen, Mietwohnungen und Volkswohnungen vom *RAM* durch Darlehen an Gemeinden und gemeinnützige Wohnungsbaugesellschaft gefördert. *Abteilungsdirigent Scholtz, Ministerialrat Schmidt, Joachim Fischer-Dieskau, Helmut Döscher, Max Büge, Karl Pfeiffer*[112], waren für Gemeinschaftssiedlungen u.a. in Wolfsburg, Salzgitter und Wilhelmshaven zuständig. Der *RAM* griff massiv in die Architektur ein, erließ die *Verordnung über die Baugestaltung vom* 10.11.1936,
fertigte Alternativen zu den *Bebauungsplänen* von Voslapp und Fedderwardergroden, z.B. *30*
den *Europahausplan vom* 12.12.1938. Eine Sonderrolle spielte die *Prüfstelle des RAM für* *37*
Gemeinschaftssiedlungen und städtebauliche Planungen. Die dort angestellten Architekten und *Assessoren Friedrich Nicolaus, Gertraude Herde, Alexander Herde* u.a., wurden von der *Deutschen Bau- und Bodenbank* bezahlt. Dort wurden *Bebauungspläne* und Finanzierungsanträge geprüft, Gutachten eingeholt, Korrekturen eingearbeitet[113]. Diese Sonderdienststelle des *RAM* ging ab 1940 in der Dienststelle des *Reichskommissars für den sozialen*
Wohnungsbau, ab 1942 *Reichswohnungskommissar*, auf. *Robert Ley* als Leiter hatte den *154*
RAM weitgehend entmachtet. Der *Reichswohnungskommissar* blieb jedoch für Wilhelmshavens Gemeinschaftssiedlungen bedeutungslos.

★

Die am 26.06.1935 gegründete, von *Hanns Kerrl* geleitete und *Hitler* persönlich unterstellte *Reichsstelle für Raumordnung*[114] hatte die *Landesplanungsgemeinschaften* aufzubauen und zu überwachen. Alle Planungsvorhaben in den Gauen waren der *Reichsstelle* zu melden. Diese ließ sich ausführlich berichten, griff bei Konfliktfällen ein und versuchte auszugleichen. Die *Reichsstelle* war mit Hoheitsaufgaben ausgestattet und gebildet worden, weil sich die ressortverhafteten Ministerialinstanzen in Berlin nicht absprachen, und ein starker Koordinierungsbedarf entstanden war. Bezogen auf das *Reich* wurde Raumordnung als Reichsplanung, bezogen auf das *Land* als Landesplanung bezeichnet. Die egoistischen Interessen und Grundstücksansprüche sollten ausgeglichen werden. Die *Reichsstelle für Raumordnung* legitimierte sich so: „*Adolf Hitler* erst rief das deutsche Volk zur Besinnung auf, bewies die richtige Einsicht in die Fehler der Vergangenheit, und stellte für die Staatslenkung den Grundsatz auf, nur noch einem Ziel zu folgen: der Sicherung des Bestandes der Nation, ihres Volkstums und ihrer Rasse“[115]. Zuarbeit für die Raumordnung lieferte die im Dezember 1935 gegründete *Reichsarbeitsgemeinschaft für Raumforschung Berlin*, die von *Gottfried Feder* geleitet wurde. Die *LOB* ließ sich von der *Arbeitsgemeinschaft für Raumforschung* der Universität Göttingen unter *Professor Arthur Schürmann* beraten. Er war gleichzeitig im Beirat der *LOB*.

Die *Reichsstelle für Raumordnung* und die für Wilhelmshaven zuständigen Dezernenten
129 *Siemer* und *Schmitz* wurden im *Wohnsiedlungsgebiet Wilhelmshaven-Jeverland* nur punktuell tätig:

- Zusammen mit der *LOB* organisierte sie 1936 Arbeitstagungen zu Siedlungsfragen und Moorkolonisation im Gau Weser- Ems[116].
- 1937 befaßte sie sich mit dem 1. *Raumordnungsplan* der *LOB* von *Georg Linde*.
- Massiv schaltete sie sich am 01.06.1938 ein, als *Oberregierungsrat Siemer* von der *Reichsstelle* 50 Vertreter aller militärischen und zivilen Baudienststellen zu koordinieren versuchte und *Teilwirtschaftspläne* genehmigte[117].
- Anfang 1939 lud sie zu einer Besprechung des von der *LOB* im Auftrag des *MDF* entworfenen 2. *Raumordnungsplans* für den Raum Wilhelmshaven nach Berlin[118].
- *Gerhard Isenberg* von der *Reichsstelle* legte 1942 eine Untersuchung zu den Einwohnerzielzahlen Wilhelmshavens vor[119].
- Im Juli 1942 schaltete *Wilhelm Wortmann* von der *LOB* die *Reichsstelle* ein, als er gegen
116 die Planungen von *Walther Schulz* in Wilhelmshaven von 1941–42 Stellung bezog[120].

Sowohl die *Reichsstelle*, als auch die *Landesplanungsgemeinschaften* hatten koordinierende und hierarchische Funktion. *Hitler* zielte darauf, den Ländern Planungsbefugnisse im Bereich öffentlicher Bauaufgaben zu entziehen, lockerte diese Haltung jedoch ab 1940 wieder. Nach dem *Gesetz über die Regelung des Grundbedarfs der öffentlichen Hand vom* 29.03.1935 sollte die *Reichsstelle* darüber wachen, „daß der deutsche Raum in einer den Notwendigkeiten von Volk und Staat entsprechenden Weise gestaltet wird“. Vor allem ging es um die Koordination des Landerwerbs. Durch die Eingemeindung vom

01.06.1938 war Wilhelmshavens Stadtgebiet so erweitert worden, daß überhaupt erst stadtplanerisch zusammenhängend gedacht werden konnte. Es wäre unweigerlich zu Stillständen gekommen, wenn nicht durch die Eingemeindung Austauschland verfügbar gemacht worden wäre.

Das *Reichsernährungsministerium* und die *Landesbauernschaft Weser-Ems* bestanden darauf, daß alle militärischen Dienststellen ihren bisher „geheimen" Flächenbedarf offenlegen sollten, um nicht Erbhofbauern im Gelände anzusiedeln, das dann doch wieder für Wehrmachtszwecke in Anspruch genommen werden muß[121]. Die *RUGES Berlin* kündigte an, sich im Rathaus in einem Büro zu etablieren, um die Umsiedlung vorzubereiten[122]. Mit dem Eingreifen der *Reichsstelle für Raumordnung* vom 01.06.1938 schienen die Zuständigkeitsstreitereien und Kompetenzunklarheiten erst einmal beendet. Der *Reichsfinanzminister* und der *Reichsinnenminister* hatten den *RAM* „als Prüfungs- und Kontrollinstanz für die Planungsarbeiten und die wirtschaftliche Durchführung des Gesamtprogramms bestimmt"[123].

Die *Reichsstelle für Raumordnung* hatte für Stadterweiterung und Neugestaltung Wilhelmshavens nur verwaltungsinterne und theoretische Bedeutung. Sie erreichte niemals das Gewicht der Dienststellen von *GBI Speer*. Allgemein wurde festgestellt, daß die *Reichsstelle* versagt habe[124]. Insbesondere vermochte sie nicht, die Interessen von *Marine*, *Luftwaffe* und *Stadt* in Wilhelmshaven auszugleichen. Die von *Hitler* gewünschte Zentralisierung der Reichsplanung in Berlin setzte sich nicht durch.

*

In der Weimarer Republik waren erste *Landesplanungsverbände* im Ruhrgebiet, in Oberschlesien und in Hamburg, entstanden. In Oldenburg hatte das *Siedlungsamt* seit 1920 ähnliche Aufgaben, abgestimmt auf Kleinsiedlungspolitik, Erwerbslosensiedlungen, *Landansässigkeitsprogramme* und Neulandgewinnung. Zur „Ordnung des deutschen Raumes" wurde im Juni 1935 die *Reichsstelle für Raumordnung* in Berlin gegründet. Zur Durchführung der hoheitlichen Reichsplanung war das „III. Reich" in 23 Planungsräume aufgeteilt, in Preußen waren dies Provinzen, in außerpreußischen Ländern Reichsstatthalterbezirke[125]. In den Planungsräumen waren 1936 jeweils unter Vorsitz des *Reichsstatthalters Landesplanungsgemeinschaften* gebildet worden. Die Aufstellung der *LOB* sollte bis zum 31.07.1936 vollzogen sein. Im Gau Weser-Ems koordinierte *Landesplaner Georg Linde* 1936–38 die Geschäfte der *LOB* in Oldenburg, anfangs als *Reichsbeauftragter* bzw. *Landesplanungsreferent beim Reichsstatthalter* in Oldenburg. 1936–45 war *Wilhelm Wortmann* zuständiger *Landesplaner* mit Dienstsitz in Bremen. Die *LOB* hatte ihren Hauptsitz beim *Reichsstatthalter Röver* in Oldenburg und unterstand ihm direkt. Bremen war tief gekränkt, daß der *Gauleiter* und *Reichsstatthalter* im kleineren Oldenburg residierte. Erst nach *Rövers* Tod verlegte sein Nachfolger, *Paul Wegener*, den Dienstsitz des *Reichsstatthalters* nach Bremen ins „Haus des Reiches"[126]. Ab 1938, nachdem *Georg Linde* zum Kriegshafenbau nach Frankreich abkommandiert wurde, war *Wilhelm*

Wortmann allein zuständiger *Landesplaner* für den Raum Wilhelmshaven und umfassend
129 einflußreicher *Planungsreferent* im Gau Weser-Ems. Mit seinen Mitarbeitern *Karl
Dillschneider* und *Rudolf Schuster* bekam er großen Einfluß auf das gesamte Baugeschehen
im Nordwesten des Reiches. Gleichzeitig leitete er den *Ahlhorner Architektenkreis* und den
Arbeitskreis Stadt- und Landesplanung des NSBDT.

Carl Röver, Gauleiter seit 1928, *Reichsstatthalter* seit 1933, *Reichsverteidigungskommissar*
seit 1939 und *Gauwohnungskommissar* ab 1940, benannte im November 1936 *Reichsmi-
nister Kerrl* die knapp 30 Mitglieder des *Beirates der LOB*[127]. Alle wesentlichen
Parteigliederungen, Oldenburger Ministerien, Bremer Senatsdienststellen, wirtschaftli-
che Interessenvertreter, landwirtschaftliche Kammern und Träger öffentlicher Belange,
waren vertreten. Aus Wilhelmshaven/Rüstringen waren dies *Oberbürgermeister Gustav
3 Nutzhorn* und *Komm. Admiral Otto Schultze.* Aus Oldenburg waren *Ministerpräsident
Georg Joel* und *Ministerialrat Wilhelm Ostendorf* vom *MDF*, vertreten. 1937 rückten
Oberbürgermeister Wilhelm Müller und der *Komm. Admiral Alfred Saalwächter* nach.

Die *LOB* sollte nicht selbst planen, sie sollte politische Leitlinien, bevölkerungspoliti-
sche Ziele und wehrpolitische Belange durchsetzen. Sie war eng mit der *NSDAP* und
ihren Gliederungen verflochten. Die Gebietskörperschaften, die Städte, Gemeinden und
Kreise blieben für ihre Fachplanungen, *Wirtschaftspläne, Flächennutzungspläne, Bebauungs-
pläne* und *Fluchtlinienpläne*, zuständig. Da viele Städte planerisch ungeübt waren, wurden
anfangs *Bezirksstellen der Landesplanungsgemeinschaften* eingerichtet. In Wilhelmshaven
leitete *Wilhelm Hallbauer* die *Bezirksstelle der LOB* von Oktober 1936 bis Dezember 1937.
Die Gemeinden hatten im Rahmen ihrer Selbstverwaltung Stadtplanung zu betreiben.
Raum-, Reichs- und Landesplanung waren nicht ihre Sache. Insbesondere den größeren
Städten, die sich erweiterten, wurde aufgegeben, die Pläne mit den kleinen Nachbar-
gemeinden abzustimmen oder diese als Auftragsangelegenheit mit aufzustellen. Dies
129 wurde für das *Wohnsiedlungsgebiet Wilhelmshaven-Jeverland* vielfach bedeutsam.

Aufsichtsbehörde für Wilhelmshavens Stadtplanung jedoch war nicht die *LOB*,
3 sondern der *MDF.* Der Sonderstatus von Wilhelmshaven und die Kriegswichtigkeit der
Rüstungsindustrie, die Rivalitäten zwischen *Marine* und *Luftwaffe, RAM* und *MDF*,
führten bei fast allen *Wirtschaftsplänen* und *Bebauungsplänen* zu Konflikten, und
erforderten die Einschaltung von *LOB* und Berliner Ministerialinstanzen.

Das „III. Reich" hatte anfangs den Eindruck von Stadtfeindlichkeit und Kleinsied-
lungsromantik erweckt. Die *LOB* jedoch führte die bei *Carl Röver* beliebte „völkische"
Gesinnung so gut es ging auf methodisches Planungsverständnis und rationale
25 Wirtschaftlichkeit zurück. Längst hatte *Hitler* erkannt, daß agrarromantische Siedlungs-
gebiete aus Grundstücksmangel nicht zu realisieren waren. Die *Reichsstelle für Raumord-
nung* achtete bei der Besetzung der *Beiräte der Landesplanungsgemeinschaften* auf die
Beteiligung von *Gauwirtschaftsberatern*, Banken, *Wirtschaftsministerien*, Unternehmern,
Finanzämtern, Wehrwirtschafts-, Handwerks- und Handelskammern. Carl Röver versuchte,
den *Beirat der LOB* zu übergehen, und die von der *Reichsstelle für Raumordnung*
vorgesehene Gemeinschaftsarbeit der Planungsträger zu unterlaufen. In der Sitzung vom

11.04.1938[128] schlug er vor, nur dann zusammenzukommen, wenn man auch etwas vorzutragen habe. Hemdsärmelig appellierte er an die „Kameradschaft", und man solle sich „gegen großzügige Planung" nicht sperren. Während andernorts *Landesplanungsgemeinschaften* längst wieder wissenschaftlich arbeiteten, versuchte *Röver*, der *Rosenberg* und *Darré* eng verbunden war, die *LOB* „völkisch" und bodenverbunden zu orientieren. Er proklamierte länger als andere das „landschaftlich gebundene" Bauen[129], machte die „landschaftseigene Normung" zur Maßgabe des Wohnungsbaus[130], und den *Ahlhorner Architektenkreis* zum Ideologiezentrum rassenpolitischen Wohnungsbaus. *Gau-Organisationsleiter Heinrich Walkenhorst, DAF-Gauobmann Bruno Dieckelmann, Landesbauernführer Jaques Groeneveldt, Gaubeauftragter für das Volkstum Kurt Thiele* und *Oberstarbeitsführer zur Loye* sicherten den Einfluß der *NSDAP* im *Beirat der LOB.*

Die *LOB* schulte die Planer der Gebietskörperschaften und die Mitarbeiter der *Stadterweiterungsämter*[131]. Bezogen auf Wilhelmshaven kam man zusammen wegen: der Lage des Bahnhofs, der Eingemeindung der Gebietskörperschaften, des Deichbaues, der Reichsstraßenplanung, der Zusammenlegung zu Groß-Wilhelmshaven, der Fremdenverkehrsangelegenheiten und vor allem wegen der neuen Gemeinschaftssiedlungen. Hauptaufgabe der *LOB* war es, die Reichsplanung, bezogen auf das Land Oldenburg, als Landesplanung durchzusetzen. Für den Raum Wilhelmshaven erstellte sie:

- 1. *Raumordnungsplan vom August* 1937 mit Erklärungen von *Georg Linde* vom 16.12.1937[132].
- 2. *Raumordnungsplan vom Dezember* 1938 für den Raum Wilhelmshaven[133].

1940 wurden die *Landesplaner* gleichzeitig *Generalreferenten für Raumordnung* und damit noch enger an den *Gauleiter* gerückt[134]. Er sollte mit seinem *Generalreferat für Raumordnung* die Funktion der Städte in seinem Gau bestimmen. *Hitler* wollte offensichtlich die Stellung der *Gauleiter* stärken, die Planung dezentralisieren, und übertrug ihnen umfangreiche Befugnisse und zusätzliche Funktionen, wie z.B. ab 1940 die des *Gauwohnungskommissars* oder die des *Beauftragten für die Neugestaltung.* Unübersehbar jedoch wurden Zuständigkeitschaos, Behördenwillkür und Machtkämpfe.

★

Die *NSDAP* hatte auf Betreiben ihres architekturbesessenen *„Führers"* ihre eigene parteiinterne *Bauverwaltung* aufgebaut. Zu nennen sind *Reichsschatzmeister Franz Xaver Schwarz*, der die Parteibauten finanzierte[135] und das *Hauptamt für Technik* als oberste *Bauverwaltung der NSDAP* in München. Die *NSDAP*-Zentrale hatte beispielsweise mit den Wilhelmshavener Gemeinschaftshäusern zu tun. *Siedlungsbeauftragter der NSDAP* war *Johann Wilhelm Ludowici* im Stabe *Heß.*

Auch die *DAF* hatte ihre eigenen Baudienststellen. *Julius Schulte-Frohlinde* leitete die 1936 aus der Planungsabteilung des *Reichsheimstättenamtes der DAF* und dem *DAF-Amt „Schönheit der Arbeit"* zusammengeschlossene *Bauabteilung der DAF.* Gleichzeitig war er

Leiter des *Arbeitskreises Baugestaltung im NSBDT* Berlin. Das *Amt „Schönheit der Arbeit"* wurde am 30.01.1934 gegründet und von *Albert Speer* formell geleitet. Es befaßte sich mit Gestaltung der Arbeitsplätze, Hygiene in Werkhallen und Gestaltung von Arbeitsgeräten. *Julius Schulte-Frohlinde* baute Schulungsburgen, *KDF*-Erholungsstätten und *DAF*-Bauten, wirkte bei den Mustersiedlungen mit und wurde auch von der *Stadt Wilhelmshaven* zur Beratung eingeschaltet[136]. Mustersiedlungen der *DAF* in Wilhelmshaven jedoch kamen offiziell nicht in Betracht, obgleich der *RAM* Fedderwardergroden
38 beispielhaft herausstellte und in der Fachpresse verbreitete. Die *Bauabteilung der DAF*, in der 1939–42 auch der spätere *Gutschow*-Mitarbeiter *Friedrich Lübbers* arbeitete, war mit dem *Arbeitsstab Speer* eng verflochten.

Mit *Fritz Todt*[137] hatte die *NSDAP* einen Baumanager mit schnell wachsender Machtfülle in ihren Reihen: 1932 leitete er den *Kampfbund deutscher Architekten und Ingenieure (KDAI)* ab 1933 war er *Generalinspektor für das deutsche Straßenwesen, Leiter des Hauptamtes für Technik der NSDAP* und *Reichswalter des NSBDT*. Der *NSBDT* hatte alle technisch-wissenschaftlichen Verbände und Vereine unter Führung des *Hauptamtes für Technik* zusammengefaßt. *Todt* wurde 1938 *Generalbevollmächtigter für die Regelung der Bauwirtschaft* und 1940 *Reichsminister für Bewaffnung und Munition*, bis *Albert Speer* 1942 seine Funktionen übernahm. Im Land Oldenburg wurden, wie auch in anderen Ländern, *Gaubeauftragte für die Regelung der Bauwirtschaft* von Fritz Todt eingesetzt. In Wilhelmshaven war 1940–45 *Stadtbaudirektor Walter Temp Gebietsbeauftragter*, Vertreter war *Egon Forster*. Die *Gebietsbeauftragten* hatten weitgehende Befugnisse, konnten Arbeiter disponieren und über die Kriegswichtigkeit von Projekten mit entscheiden.

Entsprechend dem *Hauptamt für Technik* in der *NSDAP*-Zentrale München war im Gau Weser-Ems[138] ein *Gauamt für Technik* aufgebaut worden. Es wurde bis 1942 von *Friedrich Gerking*, danach von *Dr. Hopf*, geleitet. Beide waren gleichzeitig *Gauwalter* des *NSBDT*. In Wilhelmshaven war *Fritz Seetzen Kreisobmann* des *NSBDT*, gleichzeitig Leiter des *Kreisamts für Technik der NSDAP*. Die verwirrenden, häufig wechselnden Bezeichnungen, Organisationsformen und Leitungsfunktionen unter *Carl Röver* bis 1942 und *Paul Wegener* 1942–45, waren in ihren Verflechtungen undurchdringlich. Hoheitsfunktionen und Parteifunktionen waren miteinander verflochten und führten zu unglaublicher Machtfülle und Verfilzung von Partei und Staat. Die Apparate von *Wehrmacht* und Wirtschaft, *NSDAP* und Ministerien waren vernetzt. Entscheidungsprozesse, die mit Architektur und Bauen zu tun hatten, lassen sich in der *Gauleitung* kaum rekonstruieren. Allenfalls sind besondere „völkische" Schwerpunkte, besonders kulturelle Vorlieben oder besondere Machtzusammenballungen zu erkennen. *Carl Röver* hing den „völkisch"-nationalen Tendenzen, den reaktionär-bodenverhafteten Ideologien lange an. Der *Gaukulturwalter Ernst Schulze (*gleichzeitig *Gaupropagandaleiter, Leiter der Reichspropagandastelle, Gaugeschäftsführer* und seit 1943 *Gauinspekteur*), der *Gauobmann der NS-Kulturgemeinde Johann Köllmann*, der *Gaubeauftragte für Volkstum Kurt Thiele*, der *Landesleiter Weser-Ems der RDBK Hans Fricke* und der *Leiter des Wohnungsbauwesens im*

Gau Weser-Ems Bruno Dieckelmann, stärkten die rückschrittliche Architekturauffassung von *Röver*.

Die *Nordische Kunsthochschule Bremen*[139] unterstand den Kulturfunktionären der *Gauleitung* und sollte die traditionelle Leitlinie stärken. *Gau-Heimstättenamt* und *LOB*, die mehr politisch-wirtschaftliche Funktionen ausüben sollten, wurden gleichfalls Instrumente zur Durchsetzung der baukulturellen Ziele. Ungezählte weitere Funktionen und Beziehungen sollten sich auf Stadterweiterung und Neugestaltung von Wilhelmshaven auswirken: z.B. der *Gaubeauftragte für Lagerbetreuung* und die *Referentin für Heimstättenwesen* in der *NS-Frauenschaft*.

Das *Reichsheimstättenamt Gau Weser-Ems*, eines von 33 sogenannten *Gau-Heimstättenämtern*, war als Amt der *DAF*, und somit direkt der Oldenburger *Gauleitung* unterstellt. *Carl Röver* hatte 1939 den *Gauobmann der DAF Bruno Dieckelmann*, mit der Leitung des Wohnungsbauwesens im Gau Weser-Ems beauftragt, bevor er 1940, wie auch sein Nachfolger *Paul Wegener* ab 1942, *Gauwohnungskommissar* wurde. Geschäftsführer des *Gau-Heimstättenamtes* waren u.a. *Adolf Köhne*[140] und *Johannes Richter*. Arbeitsausschüsse und Arbeitsgemeinschaften für Gestaltung achteten auf die Parteilinie und förderten bestimmte Architekten, u.a.: *Heuer, Lübbers, Bürgin, Hörold*. Sie sorgten aber auch dafür, daß andere Architekten wie *Höger, Iwersen, Minke, Freygang, Peters* und *Fuchs* von Aufträgen ferngehalten wurden. Das *Reichsheimstättenamt* war 1934 in Berlin als *Heimstättenamt der DAF* entstanden. Leiter war viele Jahre *Paul Steinhauser*. Die *NSDAP* sollte den Wohnungsbau „führen"[141]. Baugestaltung und Planung sollten auf weltanschauliche Grundlage ausgerichtet werden. Nicht die verhaßten „Kaninchenställe" der *17*
Weimarer Republik, sondern Wohnungsbau nach bevölkerungs- und kulturpolitischen Erfordernissen war erwünscht.

Das *Reichsheimstättenamt der DAF* gründete 1940 eine eigene Wohnungsbaugesellschaft, die *Neue Heimat*, die eigene „Führertypen" als genormte Wohnungsgrundrisse entwickelte[142]. Geschäftsführer der *Neuen Heimat* im Gau Weser-Ems war *Wilhelm Küster*. 1940 gab es im Alt-Reich 3.490 Gemeinnützige Wohnungsbauunternehmen, davon 2.826 Genossenschaften, in denen verdeckt noch Sozialdemokraten und Gewerkschafter Unterschlupf gefunden hatten. Die konsequente Anbindung des Wohnungsbaus an die *Gauleiter* ab 1940 durch die Gründung der *Neuen Heimat* war für Wilhelmshavens Gemeinschaftssiedlungen nicht mehr bedeutsam. In Wilhelmshaven kam die *Neue Heimat* nicht mehr an Projekte. Die Anzahl der kleinen Wohnungsbaugesellschaften jedoch, die sich aus der Weimarer Republik herübergerettet hatten, wurde drastisch reduziert. Am 04.06.1941 hatte der *Oldenburgische Innenminister* verfügt, daß nur noch die *Wohnungsbaugesellschaft Jade*, die *Wilhelmshavener Spar- und Baugesellschaft* und der *Bauverein Rüstringen* als gemeinnützige Wohnungsbauunternehmen anerkannt wurden. Er beabsichtigte, auch diese noch zusammenzulegen, möglicherweise in die *Neue Heimat* zu überführen[143].

Das *Reichsheimstättenamt Gau Weser-Ems der DAF* war auch eng mit dem *Deutschen Siedlerbund (DSB)* und der *LOB* verflochten. Mitarbeiter waren langjährig erprobte

Mitglieder der *NSDAP*[144]. Hier wurden Architekten ausgewählt, hier wurden Siedlungsgedanken ideologisiert, hier wurden Kleinsiedler beraten, hier wurden Referenten beschafft. So war das *Gau-Heimstättenamt der DAF* zuständig für Eignungsscheine für Kleinsiedler in der Siedlung Sanderbusch, und setzte sich mit der *NS-Volkswohlfahrt* für „erbgesunde" und kinderreiche Familien ein. *Heinz Wetzel*, dem ländlichen Bauen verpflichteter Städtebauprofessor der *Stuttgarter-Schule* und *Hauptabteilungsleiter im Reichsheimstättenamt*, hielt verschiedentlich Vorträge im Gau Weser-Ems[145].

In Sonderfällen plante das *Gau-Heimstättenamt* selber, so u.a. in Voslapp 1937,
67 Schortens[146], Papenburg usw. *Johannes Richter* war eng mit *Georg Linde* von der *LOB* verbunden. Man legte gemeinsam Standorte für Wohngebiete fest, bot kleinen Gemeinden und Ämtern Planungshilfe an, um den „Gemeindeaufbau gesund zu erhalten und zur gesellschaftlichen Ordnung zu zwingen".

Bedeutsam für die NS-Planungsideologie im nordwestdeutschen Raum war der *Arbeitskreis Stadt- und Landesplanung des NSBDT Gau Weser-Ems*, den *Wilhelm Wortmann*[147] leitete. *Friedrich Heuer, Richard Homann, Hermann Gildemeister, Heinz Däke, Eduard Scotland, Gustav Wolf, Walter Temp* und *Fritz Karl Schneider* gehörten dazu. Man traf sich, man hatte reichsweite Verbindungen und arbeitete im *Ahlhorner Architektenkreis* zusammen.

In Wilhelmshaven ließen sich die *Kreisleiter der NSDAP, Josef Mainzer* (1933–35), *Ernst Meyer* (1935–43) und *Bernhard Horstmann* (1943–45) vom *Beauftragten der NSDAP für Bauangelegenheiten*, Architekt *Willi Lübbers*, beraten.

Baukulturellen Einfluß hatte aber auch der *Kreisobmann der NS-Kulturgemeinde, Hermann Ahner*, und der 1934 gegründete *Ring Volkstum und Heimat.*

Für Bauten der *NSDAP* und ihrer Organisationen war das *Reichsbauamt Wilhelmshaven* unter *Paul Berg* zuständig[148]. Oftmals jedoch wurden *Stadtverwaltung* oder Wohnungsbaugesellschaften eingeschaltet. Im August 1940 veranstaltete die *Stadtverwaltung* zusammen mit der *NSDAP* einen Architektenwettbewerb für Gemeinschaftshäuser in Fedderwardergroden[149]. *Ministerpräsident Joel* hatte vom *Oberbürgermeister* Planungshilfe verlangt. Die Entwürfe der *NSDAP*-Gemeinschaftshäuser mußten im „Braunen Haus" in München zur Genehmigung vorgelegt werden. Die *NSDAP* hatte totalen Einfluß auf die Baupolitik der *Stadt. Willi Lübbers* war Stadtrat im *Magistrat.* Der größte Teil der *NSDAP-Ortsgruppenleiter* war im Rat vertreten[150]. Wenn eine Beschwerde einging, schrieb der *Kreisleiter* dem *Stadtbaudirektor.* Bei allen wichtigen Gesprächen mit den Berliner Ministerialbehörden war der *Kreisleiter* dabei. Die *NSDAP* als Hoheitsträger erwartete, daß *Stadt* und *Marine* sie über alles informieren, bis hin zu den neuen Namen von Straßen und Gastwirtschaften[151].

Der totalitäre Anspruch der *NSDAP* reichte bis in die Personalpolitik der *Bauverwaltung.* Der *RAM* versuchte, sich gleichfalls einzumischen. Wilhelmshavens *Stadtbaudirektor Walter Hunzinger* wurde im Februar 1939 entlassen. Im März beklagte sich der *RAM:* „Der *Oberbürgermeister Müller* und sein *Mag. Oberbaurat Hallbauer*, sind den großen Aufgaben, die der Ausbau der *Kriegsmarine*, und damit der Stadt Wilhelmshaven, mit sich

bringt, in keiner Weise gewachsen. Sie haben infolge ihrer Schwerfälligkeit mehrfach versagt“[152]. Der passive Widerstand der *Stadt* müsse „gebrochen werden“. Man *112*
befürchtete, daß *Müller* und *Röver* den *Hitler*-Besuch am 01.04.1939 zu Angriffen gegen den *RAM* nutzen würden mit dem Ziel, dem *RAM* städtebauliche Befugnisse zu entziehen. *Max Büge* vom *RAM* und auch das *OKM* wollten, daß die Stelle des neuen *Stadtbaudirektors* ausgeschrieben wird, und daß die Bewerber „jederzeit rückhaltlos für den nationalsozialistischen Staat eintreten“[153].

„Deutschblütige Abstammung“ war gefordert, einschließlich der Ehefrau. Empfohlen wurden von den Berliner Dienststellen und von der *LOB: Hans Högg* (Hildesheim), *Hans-Bernhard Reichow* (Stettin) und *Helmut Döscher* (Berlin). Die *Stadt Wilhelmshaven* ging auf die Vorschläge und Empfehlungen nicht ein. Sie beugte sich der *NSDAP* und stellte auf Vorschlag von *Kreisleiter Ernst Meyer* und nach Entscheidung von *Gauleiter Röver* den Bauingenieur *Walter Temp* ein. Er hatte sich beim Bau der IV. Einfahrt als Organisator der *HNBD* hervorgetan. Im Mai 1939 trat der Nicht-Architekt seinen Dienst an.

★

Die Architekten, die Wilhelmshaven aufbauen sollten, waren „Kinder“ des Kaiserreiches. Sie waren meistens national und unpolitisch erzogen, wurden in der Weimarer Republik Studenten und waren häufig in der Jugendbewegung engagiert. Nur wenige waren politisch oder gar sozialistisch orientiert. Viele wurden Architekten, weil das vom Vater so gewollt war. Als Studienorte wählte man nicht etwa das *Bauhaus*, sondern München, Stuttgart, Karlsruhe, Berlin, Dresden und später auch die NS-orientierte *Hochschule für Baukunst Weimar*[154].

In Dresden lehrten u.a. *Fritz Schumacher* (1899–1909), *Cornelius Gurlitt* (1893–1920), *Heinrich Tessenow*[155] (1920–25), *Hans Poelzig* (1916–20) und *Wilhelm Kreis*[156] (1904–08). Es studierten dort *Hermann Zopff* und *Wilhelm Wortmann*.

In München lehrten u.a. *Theodor Fischer* (1908–38) und *German Bestelmeyer* (1922–42). Es studierten dort *Gottfried Feder, Karl Pfeiffer, Benno Schachner, Albert Speer, Fritz Karl Schneider, Klemens Pitz, Gustav Wolf, Carl Tiburtius.*

In Stuttgart lehrten u.a. *Theodor Fischer* (1902–08), *Paul Bonatz* (1908–42), *Paul Schmitthenner* (1918–49) und *Heinz Wetzel* (1922–). Es studierten dort *Julius Schulte-Frohlinde, Konstanty Gutschow, Karl Maas, Karl Franzius, Kurt Geisenhainer, Willi Lübbers, Friedrich Lübbers, Friedrich Heuer.*

In Berlin lehrten u.a. *Theodor Goecke* (1902–19), *Hermann Jansen* (1919–35), *Heinrich Tessenow* (1926–42), *Hans Poelzig* (1923–35), *Bruno Taut* (1930–32), *Gottfried Feder* (1936–41) und *German Bestelmeyer (1919–22)*. Es studierten dort *Otto Lückemeyer, Alexander Herde, Karl Dillschneider, Ewald Zenker, Albert Speer, Ferdinand Schütt, Georg Linde.*

In Karlsruhe lehrten u.a. *Roman Heiligenthal* (1927–49), *Otto Ernst Schweizer* (1930–60) und *Hermann Billing (1906–37)*. Es studierten dort *Helmuth Baur, Hans Bartning, Rudolf Bürgin, Horst Linde, Werner Haschenburger.*

An der *Hochschule für Baukunst Weimar* lehrten u.a. *Paul Schultze-Naumburg* (1930–40), *Fritz Norkauer* und *Otto Bartning* (1926–30). *Egon Forster* studierte an dieser „nationalsozialistischen Kunsterziehungsanstalt".

Für den im „III. Reich" in Wilhelmshaven sich entwickelnden besonderen *Marinestil* schien insbesondere die *Stuttgarter-Schule* bedeutsam gewesen zu sein. Hier fanden die Architektengenerationen ihre Idole, die nicht nur Entwurf lehrten, sondern auch Weltanschauung. Noch heute sprechen die befragten Architekten voller Bewunderung von *Schmitthenner* und *Bonatz*, die angeblich Kunst und Technik, Tradition und Moderne, bruchlos an die Studenten vermitteln konnten. An der Berliner TH war es vor allem *Heinrich Tessenow*, der 1926 aus Stuttgart kommend die Architekturstudenten durch eine einfache und schlichte Gestaltung an sich band. Kaum einer der Architekten, die Wilhelmshaven aufbauten, studierte bei dem modernen *Hans Poelzig*. Keiner stand den *Ring-Architekten Peter Behrens, Hugo Häring, Erich Mendelsohn, Ludwig Mies van der Rohe, Hans Poelzig, Martin Wagner, Max* oder *Bruno Taut* nahe. Der Wohnungsbau der Weimarer Republik, vor allem die kompromißlos moderne Stuttgarter Weißenhofsiedlung (1927), polarisierte die Architekturauffassungen. 1928 bildete sich nach dem Rückzug von *Heinrich Tessenow* die Gruppe der konservativen *Block*-Architekten mit *Paul Schmitthenner, German Bestelmeyer, Paul Bonatz* und *Paul Schultze-Naumburg*. Die *Stuttgarter-Schule*, vor allem *Paul Schmitthenner*, sollte für Wilhelmshaven größte Bedeutung bekommen. *Schmitthenner* wußte mit der selbstbewußten Bescheidenheit seiner oft biedermeierlich wirkenden Bauten die tiefe Sehnsucht nach überschaubaren Lebensordnungen einfühlsam anzusprechen[157]. Aus dieser traditionalistischen Linie entwickelten sich regionale Stilarten, die *Schmitthenner* z.B. mit der Siedlung *Staaken* schon vorgegeben hatte.

Moderne Architektur wurde als bedrohlich empfunden. Die sozialistisch orientierten *Stadtbauräte Ernst May* und *Martin Wagner* waren als Vorbilder völlig ausgeschlossen – so bestätigten es alle befragten Architekten. Trotzdem konnten sich auch funktionale Städtebauauffassungen in Wilhelmshaven durchsetzen. *Hans Bartning* und *Helmuth Baur* waren ihrem Lehrer *Otto Ernst Schweizer* verpflichtet. Sie konzipierten moderne Siedlungsentwürfe ohne NS-Pathos und ohne gigantische Achsen. Auch die Architekten
93 der *KMW* um *Hans Klaucke* verzichteten fast gänzlich auf übersteigerte Maßstäbe und
96 repräsentativen Habitus und waren einer *Bauhaus*-orientierten Industriearchitektur
97 verbunden.

3. Wohnsiedlungen bis 1945

Bei der Ausführung der provisorischen Anlage ist in jeder Beziehung darauf Rücksicht zu nehmen, daß es späterhin angemessen erscheinen kann, dieselbe in eine definitive zu verwandeln. Es ist daher für die innerhalb der Enceinte einer solchen definitiven Anlage notwendig werdenden Bauten als: Kommandantur, Kaserne, Werkstätten, Magazine, Kirchen und Lazarett, das Alignement festzustellen. Die außerhalb jener Enceinte auszuführenden Privatbauten dürfen aber sowohl in bezug auf die Richtung der Straßen als auf Höhe und Material der Häuser nur unter Beachtung der für Bauten im engeren Rayon einer Festung geltenden Grundsätze ausgeführt werden. Ferner ist bei Bestimmung der für die Privatbauten auszuweisenden Plätze darauf Rücksicht zu nehmen, daß Raum für alle im Kriegsfall etwa zu errichtenden passagèren Befestigungen gelassen wird[1].

Friedrich Wilhelm IV., 1856

3.1. Gartenstadt als Leitlinie

Soziale Mißverhältnisse des Arbeitslebens traten in England, dem Vorreiter der Industrialisierung, eher zutage als in Deutschland. Auch die Versuche, Wohnungselend in Industriestädten[2] zu vermeiden, gingen auf England zurück.

Typisch englische Arbeiterviertel, wie sie *Karl Friedrich Schinkel*[3] auf seiner Englandreise 1826 skizzierte, elende Wohnquartiere, die *Gotthilf Hagen* 1822/23 in Paris[4] und 1852 in London sah, wurden in Preußen nicht kopiert. Sie blieben auch den neugegründeten Städten wie Wilhelmshaven erspart. Dafür entstanden aus dem Bautyp der preußischen Kaserne Mietshäuser, eben Miets-„Kasernen". Neue Städte wurden durch *Fluchtlinienpläne* und *Baugesetze* geordnet, durch Baumeister ersonnen und von Architekten geplant. In England glaubte man, Arbeiterhäuser seien keine Architektenaufgabe. Mit *Ebenezer Howard* und *Frederic Osborn* begann eine starke Bewegung, die bessere Städte propagierte und das Wohnen reformieren wollte. Die Gartenstadt, so hieß das Produkt der Reform, sollte autark sein und etwa 30.000 Einwohner umfassen. Kreisförmig angelegt und von Grüngürteln durchzogen sollten ländliches Wohnen und zentrales Einkaufen, stadtferne Kultur und gesunde Erholung, autonome Produktion und eigene Landwirtschaft miteinander existieren können. 1903 realisierte *Howard* zusammen mit *Raymond Unwin* die Gartenstadt Letchworth und ab 1919 zusammen mit *Louis de Soissons* die Gartenstadt Welwyn, beide nahe London gelegen. Die Gartenstadtidee zielte auf Kooperation, Selbständigkeit und gemeinsames Wirtschaften. Ihr theoretisches Fundament lehnte sich an die Konzepte utopischer Sozialisten wie *Robert Owen* und *Charles Fourier*. Formal ging das ideale Stadtmodell auf konzentrische Straßengrundrisse zurück, wie sie bereits in der Renaissance entwickelt wurden. Größerwerden der Städte und damit das vermeintlich zunehmende Wohnungselend sollten verhindert werden. Die Gartenstadt mit eigenem Haus und einem Stück Land sollte in England die Antwort sein auf die Massenquartiere von Manchester und London, in Deutschland auf die Mietskasernenübel in Berlin.

Zügig wurde die Gartenstadtidee nach Deutschland transportiert. 1902 gründete sich die *Deutsche Gartenstadtgesellschaft*[5], deren langjähriger Vorsitzender und späterer Ehrenvorsitzender Rüstringens *Oberbürgermeister Emil Lueken*[6] war. Die mit dem Aufbau Wilhelmshavens eng verflochtenen Architekten *Martin Wagner, Willy Hahn, Leberecht Migge, Hans Bernoulli* und *Theodor Goecke*, arbeiteten in der *Deutschen Gartengesellschaft* zusammen. Weitere Mitglieder waren: *Bernhard Kampffmeyer, Hans Kampffmeyer, Rudolf Eberstadt, Peter Behrens, Richard Riemerschmid, Franz Oppenheimer, Bruno Taut, Heinrich Tessenow* und der Lehrer von *Martin Wagner, Hermann Muthesius*. Die *Deutsche Gartenstadtgesellschaft* arbeitete eng mit dem *Bund deutscher Bodenreformer,* dem *Bund deutscher Mietervereine,* dem *Deutschen Werkbund*[7] und der 1904 von *Paul Schultze-Naumburg* gegründeten Heimatschutzbewegungen zusammen. Bemerkenswerte Gartenstädte entstanden in Deutschland[8]:

1909	Magarethenhöhe, Essen (*Georg Metzendorf*)
1909–13	Siedlung Hellerau, Dresden (*Richard Riemerschmidt/Hermann Muthesius/Heinrich Tessenow/Theodor Fischer*)
1910	Gartenstadt Frohnau, Berlin (*Joseph Brix/Felix Genzmer*)
1912	Gartenstadt Zehlendorf, Berlin (*Paul Mebes*)
1913	Siedlung Falkenberg, Grünau/Berlin (*Hans Bernoulli/Bruno Taut*)
1913	Siedlung Reform, Magdeburg (*Bruno Taut*)
1914–17	Gartenstadt Staaken, Berlin (*Paul Schmitthenner*)

Viele dieser Gartenstadt-Projekte zogen sich in ihrer Realisierung in die 20er und 30er Jahre. Andere Projekte wiederum wurden planerisch frühzeitig begonnen, aber erst später realisiert, wie:

1918	Siedlung Kleefeld, Hannover (*Paul Wolf*)
1919	Siedlung Laatzen, Hannover (*Paul Wolf*)
1919	Siedlung Lindenhof, Berlin-Schöneberg (*Martin Wagner/Leberecht Migge/Bruno Taut*)
1920	Siedlung Neu-Tempelhof, Berlin-Tempelhof (*Fritz Bräuning*)
1920	Siedlung Erdsegen, Bremen (*Wohnungsbauamt*)
1922	Siedlung Hammer, Kiel (*Willy Hahn*)

1924	Siedlung Freie Scholle, Berlin-Waidmannslust,
1924	Ceciliengärten, Berlin-Schöneberg (*Paul Wolf/Paul Mebes/Paul Emmerich*)
1926	Siedlungen Itterfeld, Bickendorf und Nordfriedhof, Köln (*Wilhelm Riphahn*)
1927	Westfalen-Siedlung, Bremen (*Wohnungsbauamt*)

Die Berliner Siedlung Lindenhof beruht auf den Rüstringer Erfahrungen *Martin Wagners* und dem Selbstversorgungskonzept von *Leberecht Migge*[9]. Die Mieter bildeten eine Genossenschaft zur Selbstverwaltung und schlossen sich von der Außenwelt ab. Kino, Altenheim, Restaurant, Läden, alles war vorhanden, was eine Gartenstadt ausmachen sollte. Tatsächlich jedoch war Lindenhof nicht autark, sondern ein Experiment zwischen Vorkriegsreformismus[10] und Wohnungsbau der 20er Jahre. *Wagner,* der in Rüstringen noch vehement das Kleinhaus durchsetzte, mißfiel zunehmend die antiurbane Strömung der *Gartenstadtgesellschaft.* Er schlug alsbald ihre Umbenennung vor in: *Gesellschaft zur Förderung der Stadtwirtschaft und Stadtkultur.* Viele der Gartenstädte haben sich tatsächlich zu Garten„vor"städten entwickelt, denn eigene Industrie und Arbeitsplätze hatten die wenigsten. Fast immer aber hatten die deutschen Gartenstädte Bezüge zu einem Arbeitgeber und zur nahen Großstadt. Hellerau war abhängig von der Fabrik der *Deutschen Werkstätten für Handwerkskunst Dresden,* Staaken von den *Munitionswerken Spandau,* die Siedlung Reform Magdeburg von den Magdeburger *Krupp-Werken,* die Margarethenhöhe in Essen gleichfalls von *Krupp.* Im nordwestdeutschen Raum hing die Arbeiterkolonie Koppenstraße (1911) in Varel von den *Hansa-Automobilwerken* ab, die Arbeitersiedlung Düperstraße (1910) in Delmenhorst von den *Linoleumwerken Schlüssel.* Die Arbeiterkolonie Nordenham war eine Werksiedlung der *Metallwerke Unterweser.*

Auch die späteren englischen Gartenstädte verdienen diesen Namen eigentlich nicht mehr. *Cadbury, Lever* usw. waren die Industriebetriebe, für die diese Siedlungen entstanden. *Howard* hatte die Anziehungskraft des Landes überschätzt, es handelte sich „leider nicht darum, wo die Leute hingehen, sondern darum, wo die Industrie hingeht"[11].

Die *Deutsche Gartenstadtgesellschaft* von 1902, die erste außerhalb Englands, war von vornherein realistischer orientiert als die britische *Garden-City-Association.* Theoretisch war sie zwar idealistisch, praktisch aber verwirklichten ihre führenden Architekten Garten „vor"städte, Villenkolonien und Werksiedlungen. Ähnlich war auch der *Deutsche Werkbund* von 1907 realistischer als die englische *Arts and Crafts*-Bewegung. Die deutschen Gartenstadt-Enthusiasten waren nicht wie die Engländer großstadtfeindlich, sondern zumeist akademisch gebildete Großbürger, die mit ihren gesellschaftspolitischen Vorstellungen genossenschaftlich orientiert waren. Der liberale Rüstringer *Oberbürgermeister Emil Lueken* war Schüler des *Werkbund*-Gründers *Friedrich Naumann*[12]. *Lueken,* wie dem linksliberalen *Naumann* ging es darum, die Arbeiterklasse durch soziale und kulturelle Formen in die bürgerliche Gesellschaft zu integrieren[13]. Es besteht kein Zweifel, daß *Lueken*, 28-jährig 1907 zum *Bürgermeister* von Heppens gewählt, von Anfang an Gartenstadtpolitik durchsetzen wollte. Er wurde *Bürgermeister* und *Oberbürgermeister* von Rüstringen, ging 1920 als *Oberbürgermeister* nach Kiel und wurde dort 1933 von den *Nationalsozialisten* aus dem Amt gejagt. Er holte die *Gartenstadtgesellschafts*-Mitglieder *Martin Wagner*[14] (1911) und *Willy Hahn* (1914) in seine Verwaltung und bezog *Hans Bernoulli* sowie *Leberecht Migge* in städtische Planungen ein[15]. Viele der zu seiner Amtszeit und danach in Rüstringen beschäftigten Architekten waren *Werkbund*-Mitglieder, wie: *Hermann Dieter* (Rüstringen), *Heinz Stoffregen* (Bremen), *August Abbehusen* (Bremen), *Otto Blendermann* (Bremen), *Adolf Rauchheld* (Oldenburg), *Hans Poelzig* (Breslau), *Fritz Höger* (Hamburg). Die Architekten des *Werkbundes* spielten in der Gartenstadtbewegung führende Rollen.

In Wilhelmshaven und den umliegenden Gemeinden hatten Gewerkschaften, Arbeitervereine und Sozialdemokraten wenig offiziellen Einfluß auf die reformerischen Bestrebungen des *Rüstringer Magistrats*. Dem Sozialdemokraten *Paul Hug* war noch 1902 die Bestallung als Vorstand des *Amtsverbandes Rüstringen* und 1904 die Zustimmung als *Gemeindevorsteher* von Bant versagt worden, obwohl er seit 1886 Beigeordneter von Bant und seit 1899 oldenburgischer Landtagsabgeordneter war. Trotzdem arrangierte sich die *SPD* mit den bürgerlichen Kräften. *Paul Hug* wurde 1911 für das neugeschaffene *Rüstringer Wohlfahrtsamt* und für die *Unterstützungskommission* des von *Emil Lueken* geleiteten *Hilfsvereins* zuständig[16] und bekam so Einfluß auf den Wohnungsbau.

In Berlin und in anderen deutschen Großstädten wurde die Gartenstadt-Ideologie[17] nach dem Ersten Weltkrieg durch den Großsiedlungsbau der 20er Jahre verdrängt. *Martin Wagner* hat die Integration der durchgrünten Siedlungen in den Massenwohnungsbau der Weimarer Republik mit den von ihm gegründeten gemeinwirtschaftlichen Wohnungsbaugesellschaften bewerkstelligt[18]. Die Hufeisensiedlung Britz (1925/27), die Großsiedlung Zehlendorf (1926), die Weiße Stadt (1929) und Siemensstadt (1929) belegten diesen Wandel in Berlin[19]. *Bruno Taut, Ernst May* und *Walter Gropius* folgten mit anderen Großsiedlungen. Es mußte auf Dauer scheitern, das ländliche Leben in Großstädten zu verwirklichen.

Anders in Wilhelmshaven/Rüstringen. Hier hatten in der Weimarer Republik Großsiedlungen keine Chance, und die Spuren der Gartenstädte ließen sich länger nachzeichnen. Sie gingen bruchlos über den Ersten Weltkrieg hinaus bis in das „III.
15, 17 Reich“. *Internationaler Stil* und *Bauhaus* kamen regional kaum zum Durchbruch. Begierig hatten *Emil Lueken, Martin Wagner, Willy Hahn* und *Leberecht Migge* die Gartenstadtgedanken aufgenommen und umgesetzt. Nach dem Weggang von *Lueken* und *Hahn* 1920/21 endete auch *Migges* Tätigkeit in Rüstringen. Ein gestalterisch denkender *Stadtbaurat* wurde erst 1937 wieder eingestellt. 1929–37 behalf man sich mit *Fritz Höger* als städtebaulicher Berater von Rüstringen. Die kommunalen Baubeamten *Max Neumann, Heinrich Siemer* und *Carl Haefner* blieben bei den traditionellen Wohnungsbauleitlinien.

3.1.1. Wilhelmshaven

Im konservativ-preußischen[20] Wilhelmshaven wurde 1906–33 die Baupolitik von *Oberbürgermeister Emil Bartelt, Paul Wolf* (1907–10) und *Hermann Zopff* (1910–37) bestimmt – soweit die *Marine* dies überhaupt zuließ. Ansätze von Gartenstadtgedanken fanden sich in Wilhelmshaven im Viertel für *Admiräle* und *Stabsoffiziere*, das 1909 vom *Marine-Garnison-Bauamt* im innerstädtischen Geviert Viktoria-/Park-/ Peter- Virchowstraße angelegt wurde. Die 6 jugendstil-orientierten Doppelhäuser[21] von *Marine-Intendantur- u. Baurat Friedrich Schubert* waren jedoch mehr Landhausarchitektur, wie sie *Hermann Muthesius*[22] 1903 aus England importiert hatte. Klinker, unsymmetrische Dachelemente, englische Schornsteine, Fachwerkgiebel, „kleinkarierte" Fensteraufteilung, Natursteinfenstergewände usw. waren dort vorherrschende Architekturelemente[23]. Diese Wohnkolonie für privilegierte Marinebedienstete war weder genossenschaftlich noch eigenständig. Sie war eine Werksiedlung der *Marine*. *Paul Wolf*, der später in Berlin, Hannover und Dresden traditionelle Siedlungen und Garten „vor"städte plante, hatte als Baubeamter in Wilhelmshaven bestenfalls Einfluß auf den *Bebauungsplan*.

Die Pläne für das spätere Wilhelmshavener Inselviertel nördlich der Werft gingen auf die 27
Vorkriegszeit zurück. *Hermann Zopff* legte 1918[24] eine Siedlungsplanung für das Inselviertel vor, die Gartenstadtelemente aufwies. Geschlossene Bebauung mit Toren schloß das geplante Viertel nach Süden und Osten stadtmauerartig ab. Reihenhäuser mit Hausgärten sowie eine östlich davon liegende Landhausbebauung für *Offiziere* und „sonstige Bürger" sollten gesunde und schöne „Heimstätten" werden. Die vorgesehene soziale Trennung der Schichten schlug einen Bogen zu den Stadtmodellen von *Theodor Fritsch*, die er 1896 zwei Jahre vor *Ebenezer Howard* veröffentlicht hatte[25]. *Fritsch* wollte kein Proletariat unterbringen, sondern „unliebsame Elemente" fernhalten. Körperliche und moralische Gesundheit sowie strenge hierarchische Gliederung waren Ziele seines konservativen Gartenstadtmodells. Im Inselviertel sollten die Arbeiter in den der Werft zugewandten Zeilen wohnen, die anderen Berufsgruppen abgestuft und geschützt dahinter. Die im Inselviertel strahlenförmig von Plätzen ausgehenden Straßen fanden sich bei *Howard* wie bei *Fritsch*. Die immerhin 73 ha große Siedlungsplanung von 1918 ließ keine Ansätze modernen Bauens erahnen, denn die Großsiedlungen in den Metropolen entstanden erst ab 1925. Wilhelmshavens/Rüstringens wirtschaftliche Situation erlaubte eine Realisierung des Inselviertels erst Schritt für Schritt nach *Zopffs* Plänen zwischen 1922 und 1938. Unterschiedlichste Architekten und Baugenossenschaften waren daran beteiligt.

3.1.2. Rüstringen

Im zumeist sozialdemokratisch regierten Rüstringen fiel die Gartenstadtbewegung auf erheblich fruchtbareren Boden[26] – allerdings lag hier der wesentliche Wohnungsbedarf für den Aufbau der neuen Stadt. Von 1880 bis 1910 stieg die Einwohnerzahl in Wilhelmshaven von 9.408 auf 19.865, in Bant/Heppens/Neuende von 8.606 auf 47.592. Das *Amt Rüstringen* ließ 1906–08 von dem *Gartenstadtgesellschafts*-Mitglied *Theodor Goecke,* Berlin, einen *Bebauungsplan* erarbeiten[27]. Die von *Goecke* 1906 vorgelegten Skizzen[28] für die Kleinwohnungskolonie Rüstringer Brücke ließen gewollt unregelmäßige Straßen und Platzsituationen erkennen, die auf Gemeinschaftseinrichtungen, Grünmittelpunkte und Mietergärten schließen ließen. Eine endgültige Bewertung von *Goeckes* Plänen fällt wegen der fehlenden Erläuterungsberichte schwer. Möglicherweise sollte eine riesige Garten„vor"stadt bis an die Umfangstraße herangeführt werden. Die Planung der Kleinwohnungskolonie Rüstringer Brücke wurde noch mit *Bruno Möhring, Martin Wagner* und *Willy Hahn* fortgesetzt, später eingestellt und im „III. Reich" noch einmal aufgewärmt, aber nicht realisiert.

Konkretere Gartenstadtbezüge waren bei den 1909 von *Emil Lueken* begonnenen und 1912 durch *Martin Wagner* überarbeiteten Villenviertelplanungen zu erkennen, beim Projekt Klaus-Groth-Straße (1913) und auch bei den *Bebauungsplänen* für Siebethsburg von *Paul Hakenholz* und *Paul Brandes.* Vorgärten, Gemeinschaftseinrichtungen, Mietergärten und Grünplätze verdeutlichten die Hinwendung zu ländlichem Leben in der Stadt und zur deutschen Variante der Gartenstadt. In Siebethsburg führten die fröhliche Naturverbundenheit und der unverbrauchte Idealismus des 1903 gegründeten *Bauvereins Rüstringen* zu einer starken Annäherung an das Gartenstadtideal von Hellerau (1909–13). Die Architektur von *Hakenholz* und *Brandes* war 1903–13 noch an englischer Landhausarchitektur und ihren Berliner Fortsetzungen orientiert.

Die geschlossensten Garten„vor"städte in Rüstringen waren die Stadtparkkolonie von *Martin Wagner* (1912–14) und die Kaiserin-Auguste-Viktoria-Siedlung von *Siegmar Brune* (1913–14). Als *Martin Wagner* 1911 erster Baubeamter der aus Bant, Heppens und Neuende gegründeten *Stadt Rüstringen* wurde, kamen für ihn wie für *Emil Lueken* Mietskasernen zur Bekämpfung der Wohnungsnot nicht in Betracht. Die kleine Garten„vor"stadt wurde zu günstigen Kosten von der *Stadt* erstellt und billig an einfache Bevölkerungsschichten weitergegeben. Umschlossen von ländlicher Umgebung stand die Stadtparkkolonie nicht für „Leben auf dem Lande", sondern für „Leben vom Lande"[29], denn die Familien sollten „auf ihrem Grund und Boden ihre gesamte Grünnahrung und auch Tierproduktion selber erzeugen". *Wagner* und *Lueken* griffen trotz der zahlreich bestehenden Wohnungsbaugesellschaften und trotz vehementer Proteste der Hausbesitzer planwirtschaftlich in die Siedlungspolitik ein, denn der private Kapitalmangel vor dem Ersten Weltkrieg ließ kaum andere Möglichkeiten. Mit der Stadtparkkolonie Rüstringen nahm *Wagner* Ziele des *Preußischen Wohnungsgesetzes von* 1918, des *Reichssiedlungsgesetzes von* 1919 und des *Reichsheimstättengesetzes von* 1920 vorweg. Er vertrat seine Ziele radikal im *Deutschen Werkbund* sowie in der *Deutschen Gartenstadtgesellschaft* und publizierte zum Ärger bürgerlicher Kreise lebhaft, so daß die Siedlung vielbesuchtes Anschauungsobjekt für liberalsozialistische Wohnungspolitik wurde. Bei aller Einfachheit waren die 100 Kleinhäuser am Stadtpark keineswegs

Behelfsheime. Die dekorarmen grauen Putzbauten mit geneigten Dächern biederten sich nicht der nordwestdeutschen Region an. *Wagner* hatte wenig Verständnis für die teure Ziegelbauweise.

Die Zielsetzung der Stadtparkkolonie fand sich bei der zur gleichen Zeit entstandenen Kaiserin-Auguste-Viktoria-Siedlung nicht wieder. Sie wurde im Auftrage der *Marineintendantur* für *Unteroffiziere* in der Nähe der Rüstringer Kasernenanlagen am Mühlenweg vom *Marine-Garnison-Bauamt* 1913–14 gebaut. Die *Marine* hatte sich seit 1870 völlig aus dem Wohnungsbau zurückgezogen, mußte aber nach dem Bau der Kasernenanlagen Gökerstraße (1904–07) und Mühlenweg (1912–13)[30] dringend Wohnungen schaffen. *12*
Für die Familien der *Unteroffiziere* wurde anfangs noch ein Wohnblock im reichseinheitlichen Stil der Mühlenweg-Kasernen gebaut[31]. Kurzfristig wurden sogar Wohnba- *13*
racken[32] für *Unteroffiziere* aufgestellt. Es muß Aufsehen erregt haben, daß ein begonnenes traditionell preußisches Wohnkonzept zugunsten einer damals beträchtlich modernen Kleinhausbebauung aufgegeben wurde. Einiges spricht dafür, daß die Rüstringer Gartenstadtenthusiasten diese Wende herbeigeführt haben, denn eine planerische Mitwirkung des *Magistrats* der *Stadt Rüstringen* war bei den fiskalischen Bauten zwingend. Geplant waren 1912 sechs U-förmig abgewinkelte Reihen von Kleinwohnungen[33]. Nur vier davon mit 70 Wohnungen wurden dann auch gebaut. Diese Werksiedlung der *Marine* war keine selbständige Gartenstadt. Wesentliche Elemente fehlten. Nebenerwerb für *Unteroffiziere* war undenkbar, Gärten durften bestenfalls der Erholung dienen. Genossenschaftliches Gedankengut war bei der *Marine* nicht gefragt. Selbstverwaltung hielt der Fiskus für überflüssig, denn dafür gab es die Garnisonbehörden. Formal war diese kleine Garten„vor"stadt nicht mit den schlichten Bauten der nordwestdeutschen Region vergleichbar. Da die Gartenstadt Staaken später gebaut wurde, dürften die Vorbilder für die Kaiserin-Auguste-Viktoria-Siedlung aus Dresden-Hellerau (1909–13) und England[34] importiert worden sein.

Die U-förmigen Reihen waren von *Siegmar Brune* äußerst durchdacht geplant worden. Ver- und Entsorgungsleitungen wurden kostengünstig zur Mitte gelegt. Die Höfe waren windgeschützt und wegen der niedrigen Traufhöhen hell und durchgrünt. Edelputz, Dachlandschaften, Gauben, Blendgiebel, und Jugendstilelemente arrangierten sich zu einer „englischen Gartenstadt", wie sie in Norddeutschland sonst nicht zu finden war.

3.2. Wohnungsbaugesellschaften in Wilhelmshaven/Rüstringen

Wohnungsbedarf wurde vor der Industrialisierung höchst praktisch gedeckt. Arbeiter bekamen ein Entgelt und für ihre Treue Wohnung, Kleidung und Nahrung. Im agrarischen Sektor hat sich diese Grundlage für Arbeitsverhältnisse lange gehalten. Kaufleute und Handwerksmeister beschafften ihrem Personal Wohnungen und, wenn nötig, bauten sie diese neu[35]. Hauseigentum mehrte das Ansehen des Betriebes und die

Reputation des Meisters oder Handelsherrn. Da das Eigentum am Grund vom Eigentum am Bauwerk rechtlich getrennt war, hatte der Grundbesitzer keine Möglichkeit, mit dem Boden zu spekulieren. Nach der gesetzlichen Zusammenlegung von Boden und Bauwerk zog sich der Staat immer mehr aus dem Wohnungswesen zurück und beschränkte sich auf das Ordnen des Bauens.

Mit der Aufhebung des Freizügigkeitsverbotes war auch ungehemmter Zuzug nach Wilhelmshaven möglich – soweit überhaupt Wohnungen zur Verfügung standen. Betriebe und *Werft* siedelten sich nicht nach und nach in einer vorhandenen Stadt an,
6 sondern wurden von Anfang an im neuen Stadtgrundriß bedacht. Umgekehrt
9 bestimmten *Werft* und *Marine* Städtebau, Wohnungswesen, Stadtentwicklung und Architektur. *Werft*, Hafen, *Marine* und die wenigen privaten Unternehmer verlangten Wohnungen in unmittelbarer Nähe der Arbeitsplätze. Neben einer wenig geordneten Wohnungsbautätigkeit um die Baubuden der Kommissionshäuser (1857) und in
6 Neu-Heppens (1860) wurde südlich der *Werft* Wohnungsbau nach dem *Bebauungsplan* von *Gotthilf Hagen* von 1856 betrieben. Zum kleineren Teil im Bautyp der „Bremer Häuser“[36], zum größten Teil im Stil der Berliner Mietskasernen – als „Bauen auf Spekulationen“[37]. Der Arbeitskräftebedarf der *Marine* und die vorherrschende Landflucht trafen sich günstig. So begannen Gründerzeitfassaden von Mietskasernen Wilhelmshavens Stadtbild zu bestimmen, unabhängig von sozialer Struktur des jeweiligen Stadtquartiers, unabhängig von sozialer Stellung der Bewohner. Haus- und Grundbesitzer waren wenig interessiert, kleine Häuser zu bauen – das aufgewandte Kapital lag zu lange fest. *Marine* und *Werft* bauten oder ließen bauen: für ihre *Offiziere* am Adalbertplatz (ab 1868), für ihre *Unterbeamten* an der Rheinstraße (1872), für ihre Arbeiter durch *Adolph de Cousser* in Bant (ab 1873), für ihre *Deckoffiziere* an der Kaiserstraße (1880). Die Wilhelmshavener *Werft* benutzte das Wohnrecht auch zur politischen Disziplinierung ihrer Arbeiter. Sie entließ nicht nur 1874 72 Schiffszimmerer wegen Mitgliedschaft in sozialistischen Organisationen sondern kündigte auch 90 Arbeitern die Werftwohnungen[38]. Erst mit den Wohnungsbaugenossenschaften (ab 1893) wurde der Wohnungsbau sozial und kurz vor dem Ersten Weltkrieg und danach sozialreformerisch.

Wilhelmshaven/Rüstringen steht exemplarisch für die Entwicklung des Wohnungs- und Siedlungswesens in Deutschland seit der Reichsgründung. Mindestens 30 Wohnungsbaugesellschaften haben als Genossenschaften unabhängig voneinander bestanden. Dazu kamen noch private Gesellschaften, Hauseigentümer, Geldinstitute[39] und Spekulanten. Die außergewöhnlich große Anzahl von Wohnungsbaugesellschaften im Jadegebiet war Indiz einer lebhaften Stadtentwicklung. Im gesamten Gebiet von Rheinland/Westfalen z.B. gab es 1900 insgesamt 385 Wohnungsbaugenossenschaften[40].

Auch in den Gemeinden an der Jade wurden die Wohnungsbaugenossenschaften nur unwesentlich aus den Geschäftsanteilen ihrer Mitglieder finanziert. Für die gewaltige Entwicklung der Wohnungsbaugenossenschaften war das aufkommende Sozialversicherungssystem verantwortlich, das sein Vermögen sicher und für gemeinnützige Zwecke anlegen wollte. Stärker als in anderen Städten hatte die regionale genossenschaftliche Entwicklung mit dem dominierenden Arbeitgeber zu tun – der *Marine*. Üblicherweise waren bei Dienst- oder Werkswohnungen Arbeits- und Mietvertrag miteinander gekoppelt. Wer das Werk – in Wilhelmshaven die *Werft* oder die *Marine* – verließ, mußte auch die Wohnung verlassen, es sei denn, er wurde Invalide oder Rentner.

Es existierten die unterschiedlichsten Rechtsformen bei den Wohnungsbaugesellschaften. Seit dem *Genossenschaftsgesetz vom* 01.05.1889 trat die Gemeinnützigkeit in den Vordergrund, und das bedeutete: Beschränkung der Dividende, Verzicht auf privaten Erwerb, keine Spekulation und kein Gewinn.

Arbeitereigenheime und Kleinwohnungskolonien, wie sie im Ruhrgebiet und anderen Städten entstanden, waren für die Stammbelegschaft eines Werkes die gewollte Verklammerung mit dem Arbeitgeber und die gewünschte Bindung an den Arbeitsplatz. Die häufig versetzten Militärs und Beamten jedoch waren in Wilhelmshaven/Rüstringen wie z.B. auch in Berlin höchst eigentumsunwillig. Auch aus diesem Grunde konnten sich Eigenheime und Einfamilienhäuser kaum ausbreiten. Schon in Berlin erfreuten sich die ersten Wohnungsbaugesellschaften der Befürwortung von *Friedrich Wilhelm IV*[41]. Berliner Mietskasernen wurden oft kopiert – allerdings: weniger Geschosse, weniger Hinterhöfe, weniger Stuckdekoration. Reine Arbeitersiedlungen wie in Bant blieben die Ausnahme und wurden gefürchtet als „Brutstätten" von Sozialdemokraten und Gewerkschaftern. Bauvereine, Baugesellschaften und Baugenossenschaften schienen besser geeignet, die soziale Kontrolle ihrer Mitglieder und Anteilseigner auszuüben. Dank der Einlagen durften sich die Mitglieder als Eigentümer fühlen und, falls sie *Werft* oder *Marine* verließen, in der Regel sogar bleiben, es sei denn, die *Marine* hatte sich Belegungsrechte erkauft.

Einzige marineunabhängige Werksiedlungen im Jadegebiet vor dem Ersten Weltkrieg waren die Ziegeleisiedlungen wie z.B. die in Himmelreich nördlich der Gemeindegrenzen[42]. Guter Lehmboden und der Bauboom in Wilhelmshaven hatte kurz nach 1900 die Ziegelei *Kleyhauer* in Himmelreich auf 50 ha Grund und Boden entstehen lassen. Der wohlhabende Ziegler und Bauunternehmer *Gerhard Hinrich Diedrich Kleyhauer* war auch vielfacher Hausbesitzer. Er erschloß nach einem *Siedlungsplan* und mit Hilfe der *Bodenkreditanstalt Oldenburg* die Ziegeleisiedlung Himmelreich. Es waren dies vor allem Häuser für seine Zieglermeister und für die Ziegeleiarbeiter, die als Saisonkräfte seßhaft gemacht werden sollten. Häuser für Bauhandwerker, Händler und Kaufleute kamen hinzu. Das Erschließungssystem war einfach, es folgte der Landstraße von Coldewei nach Rüstersiel und dem rechtwinkeligen Abbau der Lehmvorkommen. Das Wegegeviert hieß Ton-, Ziegel, Stein- und Klinkerstraße. Die sparsamen nichttypisierten Zieglerhäuser zeichneten sich durch einfachste Wohnungsgrundrisse, Satteldächer und Stallanbauten aus. Entwurfsverfasser waren nicht Architekten, sondern beauftragte Bauunternehmer. Die Grundfläche der Häuser war selten über 50 qm und die Dächer wurden erst nachträglich mit dem Wachsen der Familien ausgebaut. In Erwartung einer Nachkriegsbaukonjunktur erwarb die *Stadt Rüstringen* 1917–19 Ziegelei und Gelände und gründete kommunale Ziegelwerke. Die Arbeitersiedlung wuchs weiter, bis die Entwicklung 1931 zum Stillstand kam und die Ziegelei abgerissen wurde. Auf dem städtischen Gelände entstanden ab 1933 neue Kleinsiedlungen. Ab 1939 waren im Rahmen der Stadterweiterung noch weitere 120 Kleinsiedlungen geplant.

Wie industrielle Arbeitgeber ihre Belegschaft durch Wohnungen mit dem Werk verflochten und Arbeitsfrieden wollten, so handelten in Wilhelmshaven auch *Marine* und *Werft*. Wie anders sollten die zuströmenden fremden Arbeiter mit ihren abweichenden kulturellen und religiösen Hintergründen sonst untergebracht werden? Arbeiter- und Militär-Wohnungsbau waren sich ähnlich, wie sich auch Führungsaufbau und gesellschaftliche Gliederung ähnelten. Dies spiegelte sich in Architektur und Städtebau. Die Banter Werftarbeiterhäuser (ab 1873) zeigten als eingeschossige Doppel- und Reihenhäuser starke Parallelen mit der Siedlung Eisenheim/Oberhausen (1844), mit der Werksiedlung *Klävemann*, Donnerschweer Straße/Oldenburg (1872/73), mit den Werkswohnungen der Nordwolle/Delmenhorst (1884) oder mit den „Jutehäusern" in Delmenhorst (1890)[43].

Wilhelmshavener Wohnungsbaugesellschaften wurden hochoffiziell begünstigt:

- *Landesrat Liebrecht*[44] förderte 1893 maßgeblich die Gründung der *Wilhelmshavener Spar- und Baugesellschaft*. Immer wieder brachte er die „Dinge in Fluß"[45].
57 - *Staatssekretär des Innnern Graf Posadowsky* aus Berlin förderte 1903 die Gründung des *Bauvereins Rüstringen*[46].
- *Admiral Reinhard Scheer* erschien 1918 selbst auf der Gründungsversammlung der *Kriegerheimstätten- und Kleinwohnungsbaugesellschaft* und zeichnete 40.000 Mark.
- *Willy Hahn* wurde gar erster Geschäftsführer des *Eigenhandbauvereins, Reichsbund Wilhelmshaven-Rüstringen*, der von 1920 bis 1923 existierte.

Es ist festzustellen, daß die genossenschaftlich aufgebauten Wohnungsbaugesellschaften, vor allem die später „groß"gewordenen, im Staatsinteresse gestärkt wurden. Keine ihrer Siedlungen wurde eine reine Arbeiterkolonie wie z.B. die in Bant. Die Belegungsgepflogenheiten in Verbindung mit der eigentumsähnlichen Selbstbeteiligung brachten die gewünschte innere Stabilisierung. Die Vorteile von Arbeitereigenheimen, Werkswohnungen, Kleinwohnungskolonien und Mietwohnungen waren politisch miteinander verbunden. Auch das spätere Volkswohnungskonzept des „III. Reiches" benutzte geschickt diese Erkenntnisse und mußte sich keinesfalls auf Kleinsiedlungsideologie beschränken.

★

Eines besonderen staatlichen Wohlwollens erfreuten sich drei Wohnungsbaugesellschaften, die öffentlich gefördert und finanziert wurden, und die nicht als Genossenschaften bezeichnet werden konnten. Es waren dies:

- *Gemeinnützige Gesellschaft zur Schaffung von Wohngelegenheiten für Reichsangehörige*, Berlin[47], die 1923 gegründet wurde und später in *Gemeinnützige Deutsche Wohnungsbaugesellschaft* umbenannt wurde. Schwerpunkt dieser Gesellschaft war der Bau von Wohnungen für Angehörige der *Marine* und von Privathäusern für höhere *Offiziere*. Die Gesellschaft baute 1926 in der Lilienburgstraße und plante ab 1939 die Siedlung Altengroden. Es kam dann noch zu Nachkriegswohnungen am Adalbertplatz. *43–45*
- *Gemeinnützige Aktiengesellschaft für Angestelltenheimstätten* (*GAGFAH*), Berlin. Diese Gesellschaft, 1917 gegründet, plante 1932 in Wilhelmshaven Wohnungen[48].
- *Wohnungsbaugesellschaft Jade*, die 1937 von den Gebietskörperschaften gegründet *30*
  wurde und die an Wilhelmshavens Stadterweiterung im „III. Reich" beteiligt war. *34*
  Die *Wohnungsbaugesellschaft Jade* ist heute Wilhelmshavens größte Wohnungsbauge- *37*
  sellschaft. *53*

Es gelang der Baupolitik des „III. Reiches", nicht nur Wohnungsbaugesellschaften ohne Selbstverwaltungsorgane, sondern auch genossenschaftliche Wohnungsbaugesellschaften mit Selbstverwaltungsorganen für ihre Wohnungsbauziele einzuspannen. Durch Gleichschaltung und Einschleusung von Nationalsozialisten in Organe und Geschäftsführungen wurden Genossenschaften zu Instrumenten der großangelegten Stadterweiterungspolitik. Derartige Kraftanstrengungen waren bei privaten Wohnungsbaugesellschaften kaum erforderlich. Eine der wenigen privaten Gesellschaftsneugründungen während des „III. Reiches" war die *Wilhelmshavener-Rüstringer Wohnungsbaugesellschaft*, die ab 1934, von Wilhelmshavener Bauunternehmern gegründet, Wohnungen mit Reichsbaudarlehen erstellte. Im Villenviertel, an der Gökerstraße und in der Innenstadt baute diese Gesellschaft, bei der die Wilhelmshavener Baufirma *Hermann Möller* federführend war, Wohnungen für gut situierte Bevölkerungsgruppen, für *Offiziere* und öffentliche Bedienstete. Nach dem Kriege war die Gesellschaft an vielen Nachkriegsbauten beteiligt.

3.2.1. Kriegerheimstätten

Bereits im Ersten Weltkrieg veränderte die Kriegerheimstättenbewegung die Wohnungsdiskussion. Der *Hauptausschuß für Kriegerheimstätten*, der *Bund deutscher Bodenreformer* und der *Kriegerheimstättenverein* propagierten[49] Wohnungen für mittellose Kriegsbeschädigte, Kriegerwitwen und Kriegshinterbliebene. 1916 beschloß der *Rüstringer Stadtrat* die Bildung eines *Ausschusses für Kriegsbeschädigtenfürsorge* unter *Paul Hug*. 1919 gründete sich die *Freie Arbeitsgemeinschaft für Kriegersiedlungen Dresden*. Das Verbandsorgan war „Der Siedler". Die vom preußischen *Minister des Innern* geförderte Bewegung wurde gestützt durch den:

- *Kriegerdankbund*, in 10 Gaue und 87 Ortsgruppen gegliedert (4.100 Mitglieder)

- *Reichsbund der Kriegsbeschädigten, Kriegsteilnehmer und Kriegshinterbliebenen* (470.000 Mitglieder)
- *Reichsverband Deutscher Kriegsopfer* (400.000 Mitglieder)
- *Verband der Kriegsbeschädigten des Deutschen Reichskriegerbundes Kyffhäuser* (400.000 Mitglieder)
- *Bund erblindeter Krieger* (3.000 Mitglieder)[50].

Schon vor dem *Reichssiedlungsgesetz vom* 11.08.1919 hatte der Rüstringer Kaufmann *Karl Hinrichs* 100.000 Mark für eine Kriegerheimstättenanlage gestiftet. 1917 wurde die Gründung der *Kriegerheimstätten- und Kleinwohnungsbaugesellschaft* vorbereitet und 1918 vollzogen. Das Gründungskapital von 300.000 Mark kam vom *Kriegerheimstätten-Verein Wilhelmshaven-Rüstringen*, vom *Land Oldenburg*, von der *Stadt Rüstringen*, vom *Wohlfahrtsverein der Kaiserlichen Werft* sowie von *Admiral Scheer*[51]. Nach dem *Bürgschaftsgesetz vom* 16.06.1914 wollte die neugegründete Gesellschaft Tilgungsdarlehen vom *Reich* bekommen. Später und andernorts wurden auch Kriegerheimstätten auf der Grundlage des *Kapitalabfindungsgesetzes vom Juli* 1916 und des *Reichssiedlungsgesetzes von* 1919 finanziert.

Willy Hahns qualitätvolle Entwürfe von 1919 für die Kriegerheimstättensiedlung Rüstringen blieben Utopie und wurden Opfer der Inflation. Das Projekt ging auf in dem städtischen Bauvorhaben „Kleinwohnungen und Altenheim der Stadt Rüstringen"[52]. Der 1920 gegründete *Eigenhandbauverein, Reichsbund Wilhelmshaven-Rüstringen*, baute nach *Hahns* Planung 1921 16 Zweifamilienhäuser für Kriegsopfer als Teil einer Kleinwohnungskolonie in Neuengroden. 1922 wurden im Leiteweg Doppelhäuser durch den *Eigenhandbauverein* hochgemauert. Etwas später entstanden Kriegerheimstätten in der Stadt Oldenburg am Wittingsbrok/Brunsbrok durch eine dortige *Kriegerheimstätte-Baugesellschaft* (1923–27)[53].

Ideologisch hatten die Kriegerheimstätten keinen Bezug zur Gartenstadtbewegung, wenn auch in formaler Hinsicht die in Deutschland gebauten Garten„vor"städte sich in nichts von den Kriegerheimstätten unterschieden. Die Idee der Gartenstadt mußte mit dem Ersten Weltkrieg als beendet angesehen werden. Trotzdem wurde oftmals in der Nachkriegsnot auf alte Gartenstadtplanungen zurückgegriffen. Nach der Revolution von 1918 war die Ansiedlung von ehemaligen Kriegern, die „für ihr Land gekämpft hatten", ein Stück Wiedergutmachung und ein Schritt zu innenpolitischer Stabilität. Die große Anzahl von 808.000 Kriegsbeschädigten, 371.000 Witwen und 621.000 Waisen aus dem Ersten Weltkrieg ließ die Aufgabe „Kriegerheimstättenbau" als sehr groß erscheinen[54]. Tatsächlich jedoch kam die Bewegung schnell ins Stocken und wurde schließlich bedeutungslos. Zu einer eigenen architektonischen Form der Kriegerheimstätte kam es nicht.

3.3. Erste Kleinsiedlungen vor 1933

Das *Preußische Wohnungsgesetz vom* 28.03.1918 rückte erstmals die staatliche Aufgabe, Wohnungen zu schaffen, politisch in den Mittelpunkt. Bis dahin hatte sich das *Deutsche Reich* auf Wohnungsfürsorge zugunsten seiner Soldaten, Beamten und Angestellten beschränkt, in Wilhelmshaven/Rüstringen als Arbeitgeber der Werftarbeiter und *9*
Marineangehörigen. In die Zeit der fiskalischen Wohnungsfürsorge fielen z.B.:

1865–72 Häuser für Stationschef, *Oberwerftdirektor*, Adjutanten, *Stabsoffiziere*, „12 Männer" usw. Adalbertstraße
1869–71 Sielhäuser Hunte-/Weserstraße
1870–73 Beamten- und Arbeiterwohnhäuser an der Ostfriesenstraße (heutige Flensburger Straße)
1870 8 eingeschossige Doppelhäuser der *Hafenbaukommission*
1872–73 *Unterbeamten*-Wohnhäuser Rhein-/Marienstraße
1873–80 Werftarbeitersiedlung Bant
1890 4 *Deckoffiziers*-Wohnhäuser Weserstraße
1909–10 *Admirals*- und Stabswohnungen Viktoria-/Parkstraße
1913–14 *Unteroffiziers*-Familienwohnungen Mühlenweg (Kaiserin-Auguste-Viktoria-Siedlung)

Die Wohnungsknappheit nach dem Ersten Weltkrieg hatte ihre Ursache im Abflauen der privaten Bautätigkeit. Zins-, Lohn- und Baustoffkostenerhöhungen sowie die Konkurrenz der Wohnungsbaugenossenschaften machten Kleinwohnungsbau wieder attraktiv. Die Hausbesitzervereine kämpften gegen die Siedlung Siebethsburg und mobilisierten *Reichstagsabgeordnete*. Architekt *Julius Pollet*, später Gründer des *Kunstvereins* und vielfacher Hausbesitzer, verglich Siebethsburg gar mit Grunewald-Villen[55]. Der national-liberale Abgeordnete *Johannes Semler* forderte im *Reichstag*, die *Marine* solle die Genossenschaften nicht mehr derartig unterstützen. Die Bitte wurde von *Alfred von Tirpitz* zurückgewiesen[56].

Die aufkommende Kleinsiedlungsbewegung wurzelte in der Wohnungsnot des ausgehenden Kaiserreiches und den Maßnahmen zur Behebung des Wohnungsmangels durch die Weimarer Republik. Wohnungsfürsorge-Gesetze bildeten die Grundlage für Heimstätten:

- *Preußisches Wohnungsgesetz vom* 28.03.1918
- *Verordnung über das Erbbaurecht vom* 15.01.1919
- *Reichssiedlungsgesetz vom* 11.08.1919
- *Reichsheimstättengesetz vom* 11.05.1920

Für die praktische Umsetzung des Wohnungsbaus in den Gemeinden galt erstmalig, daß auch soziale, sittliche und künstlerische Gesichtspunkte die Grundstücksausnutzung bestimmen sollten[57]. Öffentliche Mittel, in der Regel aus der Hauszinssteuer, sollte es nur für „gesunde Wohnungen" geben. Das *Preußische Wohnungsgesetz* sah auch staatliche Beteiligung an gemeinnützigen Wohnungsbaugesellschaften vor, die Kleinwohnungsbau betreiben. *Preußen* wollte als zusätzliche Gesellschafter die Gebietskörperschaften, die großen Arbeitgeber und die *Landesversicherungsanstalten*[58].

Ziele des Wohnungsbaus in der Weimarer Republik waren: Behebung der dringendsten Wohnungsnot[59], Ausschaltung von Gewinn und Spekulation, Bauen im ländlichen Umland, aufgelockerte eingeschossige Siedlungen und kostensparende Bauweise. Mit geringstem Aufwand sollte die Bevölkerung in typisierten Heimstätten angesiedelt werden. Die angeblichen Vorteile des Landlebens wurden ebenso gepriesen wie später in der NS-Wohnungsbaupropaganda[60].Tatsächlich blieb der Wohnungsbau dieser Jahre weit hinter den Gartenstadtideen zurück, denn autarke Städte wurden kaum geschaffen. Der Wohnungsbau blieb in Wilhelmshaven/Rüstringen formal überaus traditionell.
17, 21 Moderner *Internationaler Stil* hatte nur ausnahmsweise eine Chance. Auch technisch und hygienisch blieb das gesunde Landleben eine Phrase. Es ging dem Staat immer um die billige Wohnungsproduktion und den politisch gemäßigten Heimstättenbesitzer.

*

Besonders in den Städten führten Nachkriegsarbeitslosigkeit und Wohnungsnot zu Eigeninitiative der jüngeren und aktiven Bevölkerungsschichten. Die *Stadtverwaltungen* von Wilhelmshaven und Rüstringen versuchten, die Selbsthilfe zu kanalisieren. Unabhängig von den etablierten großen Baugenossenschaften[61] schossen in Wilhelmshaven und Umgebung Bauvereine und Genossenschaften wie Pilze aus dem Boden:

- Als Abspaltung des *Beamten-Spar- und Bauvereins* bestand der *Verein für ländliche Bauten in Wilhelmshaven* 1903 – 07. Er wollte durch das *Reichsamt des Innern* finanziert werden[62].
- Die *Lebensborn, Gartenbaukolonie Bant*, wollte zwischen 1902 und 1918 Familien auf dem Land ansiedeln. Die Genossenschaft hatte 32 Mitglieder und war mit dem *Konsumverein* verflochten. Sie baute, gefördert von *Gemeindevorsteher Karl Jahn*, erste Siedlungen in Schortens und löste sich dann wegen interner Streitereien auf. Die finanziellen Erwartungen basierten auf Zusagen der *Rheinischen Betriebsgenossenschaft* und der *Landesversicherungsanstalt* Hannover/Oldenburg[63].
- 1904 gründete sich der *Bauverein Selbsthilfe*, der von der *Gemeinde Bant* unterstützt wurde. Er baute u.a. 36 Wohnungen an der Lindenstraße und schloß sich 1941 mit der *Wilhelmshavener Spar- und Baugesellschaft* zusammen[64].
- Gegenstück zum *Bauverein Rüstringen* war 1907–08 der *Bauverein Siebethsburg*, der Wohnraum für Werftfeuerwehrleute schaffen sollte.
- Die *Stadt Heppens* förderte den *Gemeinnützigen Bauverein*. Er baute 1911–12 an der Fritz-Reuter-Straße 90 Wohnungen.

- Für Angestellte der *Werft* wurde die *Baugenossenschaft Daheim* 1913 gegründet. Sie erstellte 1923 an der Fortifikationsstraße 12 Häuser.
- Der *Bauverein Schaar*, 1916 gegründet, baute für seine Mitglieder Erwerbshäuser beim Schützenhof Schaar.
- Wohnungen für Minderbemittelte wollte die *Eigenheimbaugenossenschaft des evangelischsozialen Vereins Wilhelmshaven-Rüstringen* erstellen, die 1917 gegründet wurde.
- Der *Bauverein Eigenheim* baute ab ca. 1917 für Beamte Eigenheime im Villenviertel aus Krediten gegen Hypothekenlebensversicherung.
- In der nördlichen Herbartstraße realisierte die 1921 gegründete *Reichsheimstättenbaugesellschaft Rüstringen* 20 Reihenhäuser für ihre Genossen im Erbbaurecht.
- Die *Gemeinnützige Bau- und Siedlungsgenossenschaft der Jadestädte* wurde 1924 gegründet. Sie baute Reihenhäuser am Altendeichsweg und wurde 1941 mit der *Wilhelms-* *28*
  havener Spar- und Baugesellschaft zusammengeschlossen.
- 1926 gründete sich ein *Gemeinnütziger Heimstättenverein des Mietervereins der Jadestädte.*
- Ende der zwanziger Jahre gründete sich die *Gemeinnützige Bau- und Siedlungsgenossenschaft „Leo"*, benannt nach einem Projekt in der Gökerstraße. Sie war am Bau der *17*
  Adolf-Hitler-Siedlung 1936 und an Bauten im Leiteweg 1922–26 beteiligt. *26*
- Aus 34 Genossen bestand die 1927 gegründete *Baugesellschaft Mieterfreunde.* Auch sie wurde mit der *Wilhelmshavener Spar- und Baugesellschaft* zusammengeschlossen[65].
- 1926 etablierte sich der *Gemeinnützige Bau- und Siedlungsverein Wilhelmshaven*, er brachte es auf einen Bestand von 30 Wohnungen[66].

Die Kleinsiedlungsbewegung war bis in die 20er Jahre Selbsthilfeorganisation und soziale Bewegung. Sie war die Antwort auf Arbeitslosigkeit[67] und Kurzarbeit – keinesfalls war sie gezielte Siedlungspolitik.

*

Im oldenburgischen Rüstringen waren aus der Not der Nachkriegsjahre heraus mannigfaltige Splittersiedlungen, Laubenkolonien, Kleingartensiedlungen und Behelfsheime entstanden. *Willy Hahn,* Rüstringens *Stadtbaurat* 1914–21, mußte sich wegen der wirtschaftlichen Not durch Wegfall der Existenzgrundlage *Marine* mit Sonderaufgaben befassen. Neben Notstandsarbeiten plante er:

1914–21	Kleinwohnungskolonie Rüstringer Brücke	
1915	Kleinhäuser südlich des Stadtteils Sedan, später Adolf-Hitler-Siedlung	*26*
1919	Kriegerheimstättensiedlung Neuengroden	
1919	Kolonie Neuengroden	
1919	Schrebergartenkolonie Neuengroden	

Im Vordergrund stand, die „wirtschaftliche Kraft des Einzelnen durch die Möglichkeit der eigenen Heranziehung von Früchten auf Pachtgelände zu stärken“[68]. Entscheidend veränderte Rüstringen seine Kleingartenpolitik, denn bisher wurde nur späteres Bauland als Kleingartengelände genutzt, und die Pächter etablierten sich dort nur vorübergehend. *Willy Hahn* plante noch im Ersten Weltkrieg Schrebergärten auf Gelände, das „infolge von grundbuchlichen Beschränkungen zugunsten der Kaiserlichen Fortifikation niemals einer Bebauung zugeführt werden“ konnte. *Hahn* entwarf seine Gartenkolonien in Übereinstimmung mit innenpolitischen Zielen. Gute Wohn- und Erholungsmöglichkeiten, Gefühl des Zuhauseseins auf eigener Scholle, fröhliches und gesundes Gartenleben, Seßhaftmachung der Bevölkerung propagierte auch die *Stadt Rüstringen*. Durch Hecken eingehegte Gärten, verschließbare Türchen, schöne Spielplätze, praktische Wasserstellen, beispielhafte Musterlauben und städtische Beratung waren 1915 Planungsinhalte für das kleingartenbewegte Rüstringen[69]. Die Wortwahl von *Hahn* fand sich später ähnlich aber mit anderem ideologischen Hintergrund in der Kleinsiedlungsideologie des „III. Reiches“ wieder.

Das Heimstättenwesen wurde, obwohl erheblich eher praktiziert, erst durch das *Reichsheimstättengesetz von* 1920 geregelt. Vorläufer war u.a. der Kleinwohnungsbau, wie er von *Martin Wagner* in seiner Rüstringer Zeit 1911–14 praktiziert wurde. „Das Bauen kleiner Wohnungen ist bekanntlich unrentabel. Ein Unternehmergewinn kann leicht aus dem Bau größerer Wohnungen herausgeschlagen werden. Der Kleinwohnungsbau ist dadurch ungebührlich vernachlässigt worden, und die gute bürgerliche Wohnung ist in der Hand nicht gut beratener Unternehmer stark gesunken“ So plädierte *Wagner* in seiner Eröffnungsrede zur *Rüstringer Architektur- und Gartenbauausstellung* 1911 für Kleinwohnungen[70]. Er realisierte ab 1912 in der Rüstringer Stadtparkkolonie 50 Kleinhäuser[71]. 600 bis 1.000 qm Gartenland, sparsame Seriengrundrisse, Gelegenheit zur Viehhaltung, geringe Baukosten, Wiederkaufsrecht des Ausgebers (der *Stadt Rüstringen*) und die Randlage waren vorweggenommene typische Kleinsiedlungsbestandteile der 30er Jahre. Weitere Kleinwohnungskolonien folgten an Thomas-Mann-Straße und Akazienstraße und sollten folgen an Rüstringer Brücke und Klaus-Groth-Straße. Unter *Willy Hahn, Wagners* Amtsnachfolger 1914–21, wurden noch weitere Kleinwohnungskolonien vorbereitet.

Im preußischen Wilhelmshaven plante 1918 *Hermann Zopff, Stadtbaurat* 1910–37 eine
27 73 ha große Wohnsiedlung – das heutige Inselviertel. Der westliche Teil war für Kleinhäuser und Reihenhäuser mit Gärten für Kriegsbeschädigte, Werftarbeiter, *Offiziere* und Beamte geplant[72]. Von dieser Kleinhaussiedlung wurden vor 1933 im wesentlichen der Altendeichs- und der Leiteweg realisiert. Miethäuser wurden an der Bismarckstraße gebaut. Bauen für Minderbemittelte als ursprüngliche Kleinsiedlungsidee war hier nur in Ansätzen erkennbar.

Als Initiatoren bzw. Ausgeber der Heimstätten kamen nicht mehr nur die Kommunen in Betracht, sondern *Domänenamt,* Bauvereine, Baugenossenschaften, *Siedlungsamt* usw.

Sie waren Träger der Heimstättenbewegung zwischen den Kriegen und auch danach.

Martin Wagner orientierte sich nach seinem Weggang von Rüstringen bald wieder am großstädtischen Kleinwohnungsbau ohne Gartenbau und ohne Selbstversorgung. Beeindruckt hatte ihn die ab 1910 unter kommunaler Regie geplante Bebauung Neu-Tempelhof, von *Bruno Möhring* 1912–13 begonnen und vom Tempelhofer *Stadtbaurat Fritz Bräuning* 1920–28 fortgesetzt[73]. In Rüstringen gingen anfangs *Wagners* Hausentwürfe noch nicht über 2 Geschosse hinaus – sie wurden aber zunehmend städtischer und entwickelten sich zu Reihenhäusern[74]. Das Interesse an der eigenen Scholle wurde vor und nach dem Ersten Weltkrieg vielfach zu hoch angesetzt. *Wagner* bevorzugte dann ab 1915 als *Abteilungsleiter* im *Zweckverband Groß-Berlin* wieder das kleinere Mehrfamilienhaus, das der Wohnweise entgegenarbeite, „die die Erhaltung der Volksgesundheit, der Sittlichkeit und der Wehrkraft des Deutschen Volkes zu unterbinden droht“[75]. *Wagner* schlug sich bald danach auf die Seite moderner Architekten und wurde Wegbereiter des preiswerten Massenwohnungsbaus für Arbeiter und Angestellte in der Weimarer Republik. In Wilhelmshaven/ Rüstringen blieb der preiswerte Wohnungsbau formal traditionell. Nur die *Bauhütte für das Jadegebiet* und
einige dem *Konsumverein* nahestehende Baugenossenschaften planten gemeinwirtschaft- *19, 20*
liche Projekte und moderne Formen. *17*

Das *Reichsheimstättengesetz von* 1920 versprach jeder deutschen Familie Wohnheimstätten, bestehend aus Einfamilienhäusern mit Nutzgärten und verpflichtete die *Gemeinden*, das notwendige Land für Heimstätten, Nutzgärten und Siedlungszwecke zu beschaffen. Politischen Ziele und Definition der Kleinsiedlung ergaben sich jedoch erst aus der *Notverordnung des Reichspräsidenten zur Sicherung von Wirtschaft und Finanzen und zur Bekämpfung politischer Ausschreitungen vom 06.10.1931*[76].

„Um die Seßhaftmachung der Bevölkerung auf dem Lande zu fördern, um die Erwerbslosigkeit zu vermindern und Erwerbslosen den Lebensunterhalt zu erleichtern, werden die landwirtschaftliche Siedlung sowie die Kleinsiedlung in der Umgebung von Städten und größeren Industriegemeinden (vorstädtische Kleinsiedlung) und die Bereitstellung von Kleingärten für Erwerbslose ... gefördert“. Es ging dem Staat nach dem Ersten Weltkrieg darum, den zurückkehrenden Soldaten, den Kriegerwitwen, Erwerbslosen und Kurzarbeitern Eigentum und Selbstversorgung zu schaffen. Durch Reichsdarlehen von 2.500 Mark wurde jede Siedlungsstelle gefördert – dafür hatten sich alle Siedler einer Betreuung zu unterwerfen, „damit sie aus Gartennutzung und Kleintierhaltung solche Erträge zu erzielen vermochten, die ihnen die notwendigen und angestrebten Erleichterungen ihres Lebensunterhaltes brachten“[77].

Erst gegen Ende der Weimarer Republik und nach der Inflation 1924 belebte sich der Wohnungsbau in Wilhelmshaven/Rüstringen. Der *Bauverein Rüstringen* baute in
Siebethsburg, das Inselviertel wurde begonnen, Mietwohngebäude entstanden in *27*
Bismarck- und Werftstraße.

3.4. Früher NS-Siedlungsbau bis 1935

Es war ein genialischer politischer Schachzug der NS-Baupolitiker, die Heimstätten- und Siedlungsbewegung der Weimarer Republik im „III. Reich" fortzuentwickeln. Seit Jahrhunderten waren Arbeiter vom Eigentum an Haus und Grund ausgeschlossen. Erst die Kleinsiedlungen ermöglichten die Überwindung „des wahrscheinlich schockierenden Wortes Proletariat"[78]. Die heutigen Kleinsiedler wehren sich heftig dagegen, mit „angedichteten Gedanken des sogenannten Blut und Boden des 3. Reiches" in Verbindung gebracht zu werden. Sie beziehen sich[79] auf den Sozialreformer *Victor Aimé Huber* (1800–69) (den Mitbegründer der *Berliner Gemeinnützigen Baugesellschaft*, 1848), auf den protestantischen Geistlichen *Friedrich von Bodelschwingh* (1831–1910), den Politiker *Friedrich Naumann* (1860–1919), den Bodenreformer *Adolf Damaschke* (1865–1935), den Arzt und Gründer der Kleingärtnerbewegung *Daniel Schreber* (1808–61) und den englischen „Gartenstadterfinder" *Ebenezer Howard* (1850–1928).

Im „III. Reich" wurde die Zielsetzung der Weimarer Republik, Wohnungen für Minderbemittelte und Erwerbslose zu schaffen, völlig verändert. Neue Siedlungen wurden vor allem dort gebaut, wo es für die Rüstung wichtig war[80], und das wurde vom *Reichsheimstättenamt, NSDAP* und *RAM* überwacht. Am 08.11.1935 wurde der *Deutsche Siedlerbund (DSB)* alleinige Organisation der Klein- und Eigenheimsiedler. Ihm oblagen Betreuung, Beratung, Schulung sowie Anleitung der Siedler. Für Wilhelmshaven kontrollierten dies der *Deutsche Siedlerbund e.V., Gaugruppe Weser-Ems* und das *Reichsheimstättenamt Gau Weser-Ems* in Oldenburg. Mobile Siedlerschulen, Zeitungen, Siedlerobleute, Kurse, Kreisgruppenlehrwarte, Siedlerlehrhöfe, Siedlerfrauenberaterinnen, Lehrbriefe und Kalender sollten die „Wohltaten" des neuen Regimes ins rechte Licht rücken. Die Geschäftsgrundlage auch für die Wilhelmshavener Siedlungen war klar: „Der Führer verlangt von seinem deutschen Arbeiter Höchstleistungen. Er hilft ihm aber auch, sein Leben so zu gestalten, wie es eines deutschen Menschen würdig ist, er will ihm ein Heim geben, in dem er sich glücklich und wohl fühlt"[81]. Die vom *Reich* organisierten *Siedler-Gaugruppen* hatten 1935 37.000 und im Jahre 1944 250.000 Mitglieder. Bis 1945 entstanden rund 75.000 Kleinsiedlungen in Deutschland. Andere Quellen nennen 180.000 Siedlerstellen – allerdings einschließlich der besetzten Gebiete[82].

Der Wohnungsbau des „III. Reiches" basierte zu Anfang fast ausschließlich auf Kleinsiedlungen. Das *Reichssiedlungsgesetz* der Weimarer Republik schuf die Grundlage hierzu durch Besiedlung von Grund und Boden des öffentlichen Besitzes, von Staatsdomänen oder von neu gewonnenem Land. Seit 1933 griff man verstärkt auf das *Reichssiedlungsgesetz* zurück. Im Bereich Weser-Ems konnten große Flächen Moor- und Marschenlandschaft durch das *Siedlungsamt Oldenburg* urbar gemacht und eingedeicht
1 werden. Die Stadterweiterung Wilhelmshavens und die Ausweisung von Siedlungsge-
127 bieten auf bestimmten Flächen hatte auch finanztechnische Hintergründe:

26 ○ Auf städtischem (Rüstringer) Grund entstand die Adolf-Hitler-Siedlung
24 (1934–36).

- Voslapp entstand ab 1938 auf 1928 neu eingedeichtem Land im Fedderwarder Bau- 30
  und Andelgroden. 32
- Cäciliengroden wurde ab 1938 auf einer Fläche hochgemauert, die zwar 1844 34
  eingedeicht, aber erst 1912–18 aufgespült wurde. 35, 36
- Die Siedlung Sanderbusch wurde ab 1935 auf dem Gelände des *Domänenamtes* gebaut. 25
  Die Anfänge lagen schon vor 1933.
- In Himmelreich vergrößerte sich ab 1933 eine Kleinsiedlung auf dem Gelände der ehemalig kommunalen Rüstringer Ziegelei.
- Die Wohnkolonie Kniphausersiel wurde 1929 mit 34 Siedlerstellen im neueingedeich- 22
  ten Domänengelände begonnen und im „III. Reich" weiter ausgebaut.

Das *Reichsheimstättengesetz vom* 10.05.1920 ermächtigte Gebietskörperschaften oder gemeinnützige Unternehmen, „Eigentum auszugeben", das allerdings an Bedingungen geknüpft war. Heimstätter konnten weder verkaufen, aufteilen oder vermieten, noch durften sie „Mißwirtschaft" betreiben"[83]. Die Baupläne mußten vom Ausgeber der Heimstätte genehmigt werden. Das *Reichsheimstättenamt Gau Weser-Ems*[84] Oldenburg stellte Eignungsscheine für Kleinsiedler aus, „daß gegen seine und seiner Familie Ansetzung als Kleinsiedler seitens des *Gau-Heimstättenamtes* in politischer und charakterlicher, gesundheitlicher und erbbiologischer Hinsicht und gegen die siedlerische
Befähigung Bedenken nicht erhoben werden"[85]. Wer in Voslapp Siedler werden wollte, 33
mußte seinen Antrag auf dem „Dienstwege" einreichen. Die Anträge wurden auch dem *Wehrkreisobmann* vorgelegt. Soziale Kontrolle funktionierte perfekt. Der Ansiedlungsablauf wurde überwacht und führte notfalls auch zur Aberkennung der Siedlereigenschaften nach einer Probezeit.

Die einstmals politisch unabhängige Kleinsiedlungsbewegung wurde durch das „III. Reich" von der Erwerbslosensiedlung zur Arbeitersiedlung uminterpretiert. Der gleichgeschaltete *Siedlerbund* übernahm ab 1935 die Aufgabe, seßhaftgewordene Arbeiter und Soldaten an Landleben und Eigentum zu gewöhnen. Wohl keine Wohnform ist so belächelt und bekämpft worden wie die Kleinsiedlung – das Häuschen mit Garten. Es wurde verantwortlich gemacht für: Verödung der Innenstädte, politische Entmündigung, lange Verkehrswege, zersiedelte Landschaft, überalterte Geschmacksmuster[86]. Es kann kein Zweifel bestehen, daß die Kleinsiedlungen der Weimarer Republik nicht auf Architektur und Städtebau zielten, sondern auf Sozialpolitik. Erwerbslose sollten durch Gartennutzung und Tierhaltung ihr Einkommen verbessern und öffentliche Kassen entlasten. Die Heimstätten sollten der Bodenspekulation entzogen werden. Da *Hitlers* Aufrüstungs- und Baupolitik die Arbeitslosenzahl zwangsläufig reduzierte, wurden Kleinsiedlungen ab 1933 für Kurzarbeiter[87] und später für Stammarbeiter gebaut. Ausbaustandard und Wohnungsgröße, Architekturanspruch und Dekoration stiegen – mit ihnen die Baukosten. Die Häuser in der Siedlung Sanderbusch 1935/37 waren spartanisch und einfach und kosteten 3.000 bis 5.000 Mark. Der Kleinsiedlungstyp 3 in der
Adolf-Hitler-Siedlung kostete 1935 6.000 Mark. Eine Kleinsiedlerstelle in Himmelreich 24

22 war 1934 mit 3.700 Mark kalkuliert. Siedlerstellen im Kniphausersieler Waagegroden
32 waren 1927 mit 4.800 bis 10.000 Mark kalkuliert worden. Voslapps Kleinsiedlungen
35 kosteten 1938 zwischen 7.500 und 9.500 Mark. Cäciliengrodener Siedlungen waren 1938
mit 9.000 Mark angesetzt. Bei den Kleinsiedlungen Voslapp und Cäciliengroden spielte
die Selbsthilfe kaum noch eine Rolle und wäre auch nicht im Sinne der *Marine*
gewesen.

★

Auch hautnahe Wohnungsnot war noch nach 1933 Motiv für Kleinsiedlungsbau. Die
Marinewerft drängte die *Magistrate* von Wilhelmshaven und Rüstringen im Oktober
1933[88] „im Rahmen der ihnen obliegenden Fürsorgemaßnahmen und gegebenenfalls
eines sich auf die Herstellung von Stadtrandsiedlungen und Kleinwohnungen erstrek-
kenden *Arbeitsbeschaffungsprogrammes* nach Möglichkeit für Abhilfe sorgen zu wollen".
Stadtbaurat Hermann Zopff mußte ablehnen, da Gartenland von 1.000 bis 1.500 qm für
52 Kleinsiedlungen in dem beengten Wilhelmshaven nicht zur Verfügung stand. Rüstrin-
gen verwies am 29.12.1933 auf die im Bau befindliche Stadtrandsiedlung Himmelreich
26 und die Adolf-Hitler-Siedlung für kinderreiche Familien. *Oberbürgermeister Gustav
Nutzhorn* deutete bereits an, daß künftig nur mit öffentlichen Mitteln der Wohnungsbau
gesteigert werden könne und er verwies auch auf die notwendige Unterstützung der
DAF.

Der *NSDAP*-Programmatiker *Gottfried Feder*, seit 1931 Vorsitzender des *Wirtschaftsrates der NSDAP* und seit 1934 *Reichssiedlungskommissar*, wurde schon im Februar 1933 nach Rüstringen geholt[89], auch, um die konzeptionelle Linie des künftigen Wohnungsbaus an der Jade abzustecken. Am 12.12.1933 besuchte *Adolf Hitler* die *Marine* in Wilhelmshaven zusammen mit dem *Chef der Marineleitung, Erich Raeder*, dem *Chef der Reichskanzlei, Hans Heinrich Lammers, Reichswehrminister General Werner von Blomberg*, um die Entwicklung von *Marine* und *Werft* zu besprechen. Die Machthaber des „III. Reiches" waren außerordentlich an Wilhelmshaven interessiert. *Minister Joseph Goebbels* führte am 27.01.1934 Gespräche im Offiziersheim und man kann annehmen, daß ihm die propagandistische Ausschlachtung der Aufrüstung in einer anschließenden Massenveranstaltung gut gelang. Bis in den Krieg hinein kam er immer wieder zu Großkundgebungen nach Wilhelmshaven. Am 01.03.1934 waren der *Reichsminister ohne Geschäftsbereich Hans Frank*, und vom 17. bis zum 19.10.1934 der *SS-Reichsführer Heinrich Himmler* in Wilhelmshaven/Rüstringen. Am 12.11.1934 kam der zum *Preußischen Staatsrat* avancierte national-sozialistische *Prinz August Wilhelm von Preußen* nach Wilhelmshaven. Er galt als monarchistischer Stimmenfänger. Am 01.07.1934 sollte die Hauptversammlung des *Deutschen Flottenvereins* im Rüstringer Rathaus den Interessen der *Marine* in Wilhelmshaven Geltung verschaffen. Auch 1935 gab sich, wie in den folgenden Jahren, die NS-Prominenz die „Türklinke in die Hand": *Reichsminister Rudolf Heß, Reichsbischof Ludwig Müller, Reichsinnenminister Wilhelm Frick, DAF-Leiter Robert Ley, Stabschef der SA* und *Preußischer Staatsrat Viktor Lutze, Reichswehrminister Werner von Blomberg* und der *Oberbefehlshaber der Kriegsmarine, Erich Raeder.* Propagandamärsche der Parteigliederungen, Flottenparaden, Auftritte von Sport- und Kulturprominenz sollten den Bewohnern, den Arbeitern und Soldaten der seit 1935 sogenannten „Stadt der Kriegsmarine" Hoffnung für die Zukunft machen.

Die Förderung des Siedlungswesens blieb auch nach dem

- *Gesetz über die Aufschliessung von Wohnsiedlungsgebieten (Wohnsiedlungsgesetz) vom* 22.09.1933
- *Gesetz über einstweilige Maßnahmen zur Ordnung des deutschen Siedlungswesens vom* 03.07.1934[90]

wichtige Aufgabe für das „III. Reich", um Wohnungen für das „schaffende Volk" zu bauen. Neue Dienststellen zur Durchsetzung der Wohnungspolitik wurden im „III. Reich" geschaffen[91].

Ab 1934 konzentrierten sich alle ministeriellen Kompetenzen der Wohnungsfürsorge, des Kleinsiedlungswesens und der Planung im *Reich* und in *Preußen* auf den *RAM.* Dies symbolisierte die enge Verbindung zwischen Arbeit und Wohnung, Arbeitsbeschaffung und Siedlungsbau. Die Ablösung der Gewerkschaften durch die *DAF* am 02.05.1933, die Ernennung von *Fritz Todt* zum *Generalinspektor für das deutsche Straßenwesen* am 28.06.1933, das *Nationale Aufbauprogramm* des *BDA* vom 10.04.1933, der Baubeginn der Reichsautobahnen am 23.09.1933, das *Gesetz über die Neubildung deutschen Bauerntums vom* 14.07.1933, der 1. *Vierjahresplan vom* 01.02.1933 setzten deutliche Zeichen für die Bauoffensive der *NSDAP.*

Städte und Gebietskörperschaften, Wohnungsbaugesellschaften und Genossenschaften, unterstützt von nationalsozialistischen Siedlungsprogrammatikern, setzten das vorstädtische Kleinsiedlungsprogramm der Weimarer Republik bis etwa 1935 bruchlos fort. Architekturideologische Festlegungen der *NSDAP* waren in Propaganda und parteioffiziösen Verlautbarungen noch dünn gesät und sehr allgemein gehalten[92]. Auch die Architekturkritik an Flachbau und *Neuer Sachlichkeit* war noch sehr verhalten. Die *Notverordnung vom* 06.10.1931 hatte auch im „III. Reich" Gültigkeit und die begonnenen Kleinsiedlungen wurden fortgeführt, wie z.B. die Siedlungen Berlin-Marienfelde, München-Am Hart[93], Nürnberg-Falkennest, Schweinfurt-Gartenstadt, Berlin-Falkensee und Essen-Eigene Tat. Formal waren noch keine nationalsozialistischen Besonderheiten zu erkennen. Verputzte oder verklinkerte Außenflächen, Satteldächer, knappe Grundrisse, Stallanbauten, Doppelhausformen usw. herrschten vor. Immer noch waren sie Notstandssiedlungen für Arbeitslose, Kinderreiche oder Kurzarbeiter. Städtebaulich wurden die genannten Siedlungen an vorhandene Straßen gelegt oder in enger Verbindung zu diesen angeordnet. Der Erschließungsaufwand sollte billig sein. Mit Gartenstädten oder Garten„vor"städten, mit Siedlungsbau der 20er Jahre oder den späteren Gemeinschaftssiedlungen für das *Wohnsiedlungsgebiet Wilhelmshaven-Jeverland* *129*
hatten die frühen Siedlungen bis 1935 wenig gemein.

3.4.1. Adolf-Hitler-Siedlung

Die Adolf-Hitler-Siedlung in Rüstringen war herausragendes Beispiel für den Wandel der Zielsetzungen im Kleinsiedlungswesen. Wie z.B. die 25 Häuser der „Anti-Weißenhof"-Mustersiedlung Kochenhof in Stuttgart 1933 unter der Leitung von *Paul Schmitthenner* auf einem Gelände entstanden, das bereits in der Weimarer Republik „beplant" wurde, so hatte auch die Adolf-Hitler-Siedlung ältere Ursprünge. Bereits *Theodor Goeckes Bebauungsplan* von 1908[94], *Martin Wagners Bebauungsplan* von 1913[95] und *Willy Hahns Bebauungsplan* von 1915 – verfaßt von *Heinrich Siemer* – legten die Grundzüge der Planung fest. Südlich des Stadtteils Sedan sollte eine Stadtrandkolonie mit 150 „Bremer Häusern"[96] entstehen. Der qualitätvolle *Siemer*-Entwurf wollte zwei- und dreistöckige Reihenhäuser in geschlossener Bebauung bzw. in Gruppen zusammengefaßt. *Siemer* hatte sich 1915 bereits von den freistehenden niedrigen Kleinsiedlungshäusern der Stadtparkkolonie von 1912 gelöst und wie *Wagner* und *Hahn* dem großstädtischen Haustyp zugewandt. Schmale hohe Einfamilienhäuser mit repräsentativen Fassaden und Vorgärten sollten entstehen. Anstatt des rechtwinkligen *Goecke*-Rasters von 1908 sah *Siemer* 1915 viertelkreisgefächerte Grundstückszuschnitte, platzartige Erweiterungen und öffentliche Grünanlagen vor. Sie zeugten von der städtebaulichen fortschrittlichen Auffassung der Rüstringer *Bauverwaltung* und von den Gartenstadtvorbildern. Die abgestuften Gebäudehöhen gingen auf *Wagners* 1913 bearbeitete *Zonenbauordnung* zurück.

1933 wurde die Planung für die Adolf-Hitler-Siedlung auf Initiative des neuen Rüstringer *Oberbürgermeisters Gustav Nutzhorn* wieder aufgenommen. Während 1932 noch diskutiert wurde, das Projekt der gewerkschaftseigenen *Bauhütte* zu übertragen, beschloß die *Baukommission* am 14.08.1933 den *Bebauungsplan* zu überholen. Unter *Magistratsbaurat Max Neumann* mußte Architekt *Heinrich Siemer* sein Gesamtkonzept von 1915 vereinfachen. Nur noch knapp 100 Kleinsiedlungshäuser sollten freistehend oder als Doppelhäuser in der damaligen Horst-Wessel-, der Dietrich-Eckart- und der Hans-Schemm-Straße entstehen. Die Grundstückszuschnitte blieben weiterhin viertelkreisgefächert. Öffentliche Grünflächen und Schmuckplätze fielen dem Rotstift zum Opfer und die abwechslungsreich verspringenden Baufluchtlinien wurden begradigt. Alles mußte vereinfacht und verbilligt werden. Am 18.04.1934 schrieb die *Stadt Rüstringen* einen Ideenwettbewerb für die Gebäudetypen aus, den wiederum *Heinrich Siemer* gewann.

Die unterschiedlichen Doppelhaustypen für kinderreiche Familien hatten durchschnittlichen Standard. Der Typ 2 in der heutigen Wilhelm-Krüger-Straße hatte im Erdgeschoß einen knappen Flur, Wohnküche und Wohnzimmer, eine steile Treppe ins Dachgeschoß und unter dem 55 Grad geneigten Satteldach Schlafzimmer für Eltern, Knaben und Mädchen. Der Haustyp sollte 7.600 Mark kosten. Alle Haustypen in der Adolf-Hitler-Siedlung hatten Stallgebäude, entweder im Anbau oder zwischen den Häusern – zur Kette ergänzt. Im Stall befanden sich Waschküche und Abort, Bäder fehlten. Die knappen Grundstücksgrößen von 250 bis 300 qm waren nicht auf

Nebenerwerb gerichtet, denn die Parzellierung entsprach noch den großstädtischen „Bremer Häusern".

Im Laufe der Realisierung wurden noch weitere Architekten für die Planung der Adolf-Hitler-Siedlung hinzugezogen. So plante 1935 *Robert Fuchs* in der Jahnstraße mit stilistischen Versatzstücken der 20er Jahre für verschiedene Bauherren. Er mußte sich jedoch von *Magistratsbaurat Max Neumann* und *Heinrich Siemer* Eingriffe in seine Planung gefallenlassen. Der Dachüberstand sollte vergrößert –, Giebelfenster mußten anders aufgeteilt werden. Von Architekt *Ernst Peters* stammen Entwürfe für freistehende Doppelhäuser mit Stallanbauten, die er 1936 für die heutige Fritz-Frerichs-Straße entwarf. Bauherr war die *Gemeinnützige Bau- und Siedlungsgenossenschaft der Jadestädte*. Die *Architekturabteilung des Westdeutschen Arbeiterverbandes* Köln entwarf für die *Gemeinnützige Bau- und Siedlungsgenossenschaft „Leo"* 1936 Doppelhäuser mit zwischenliegenden Ställen. Die Grundrisse zeigten die gleiche Typologie: Wohnküche und Wohnstube im Erdgeschoß, Schlafzimmer unter dem Dach, Waschküche und Klosett im Stall.

Die Wandlung der Architektur in der Adolf-Hitler-Siedlung von 1934 bis 1936 zeigte *24*
die anfängliche Richtungslosigkeit des Siedlungsbaues. Formale Anleihen an die 20er und 30er Jahre mit Ziegelornamentik und Fensterzusammenfassungen wurden erst schrittweise verdrängt. Langsam setzte sich in der Adolf-Hitler-Siedlung der Detailkatalog der heimatverbundenen Kleinsiedlung durch, wie er für den NS-Siedlungsbau typisch wurde: gleichmäßige Kreuzsprossen, steile Satteldächer, große Dachüberstände, symmetrische Fensteraufteilungen, hölzerne Klappläden, runde Regentonnen und bäuerliche Gärten. Die Häuser wurden zunehmend schmuckloser und schlichter. Bezieht man die städtebaulichen Entwürfe von *Willy Hahn* und *Heinrich Siemer* von 1915 in die Betrachtung ein, so dokumentierte die Adolf-Hitler-Siedlung städtebaulichen Rückschritt. Die privaten Vorgärten waren nicht mehr Teil des Straßenraumes, Hecken und Tore, Obstbäume und Eigentümergarten sollten ländliche Anwesen darstellen, obwohl die Grundstücksflächen für Selbstversorgung völlig ungeeignet waren.

★

Für Wilhelmshaven/Rüstringen hatte die Adolf-Hitler-Siedlung demonstrative Bedeutung. Sie wurde zur Mustersiedlung stilisiert wie: Stuttgart-Kochenhof (1933), München-Ramersdorf (1934), Düsseldorf-Schlageterstadt (1937), Braunschweig-Mascherode (1936–39). Die Adolf-Hitler-Siedlung war geeignet, die Skepsis sowohl der Arbeiter als auch der Unternehmerschaft gegenüber *Arbeitsbeschaffungsmaßnahmen* abzubauen und die Wirksamkeit öffentlicher Investititonsanstöße zu zeigen. Die *Stadtverwaltung Rüstringen* zog sich bald aus dem Projekt zurück und überließ Wohnungsbaugesellschaften, privaten Bauherren und mittelständischen Unternehmern die Adolf-Hitler-Siedlung. Oft wiederholte propagandistische Darstellungen sollten die Sorge der Bevölkerung zerstreuen, daß hier eine Primitivsiedlung entstehen könne. Nicht mehr Erwerbslose wie bei früheren Kleinsiedlungen sollten billig untergebracht

werden, sondern Arbeiter mit vielen Kindern[97] sollten mit einem Häuschen belohnt und für das NS-System gewonnen werden. Ab Februar 1935 wurden nur noch Vollbeschäftigte als Siedler zugelassen und damit wurde das soziale Programm der Weimarer
24 Republik endgültig beendet. Kleinsiedlungen waren zu Arbeitersiedlungen geworden[98]. Rüstringens *Oberbürgermeister Gustav Nutzhorn* schlachtete die Adolf-Hitler-Siedlung rücksichtslos für Propagandazwecke aus. Am 21.03.1934, noch vor dem Architektenwettbewerb[99], vollführte *Nutzhorn* den „1. Spatenstich“ für die Siedlung. Der Wilhelmshavener NS-*Oberbürgermeister Carl Renken* tat am gleichen Tag den „1.
27 Spatenstich“ im Inselviertel an der Saarbrücker Straße. Das war der Beginn der
153 alljährlichen *Arbeitsschlacht* in den Jadestädten – begleitet von Festakten im Rathaus bzw. im Neuen Schauspielhaus[100]. Es gehörte zur Dramaturgie des „Erfolges“, die verschiedenen Richtfeste von unterschiedlichen Bauabschnitten jedesmal im NS-Flaggenschmuck zu feiern. Bauarbeiter standen aufgereiht im Dachstuhl und *Gustav Nutzhorn* posierte pressewirksam als „Schöpfer der Adolf-Hitler-Siedlung im Kreise des erbgesunden Nachwuchses“[101]. Bei der Richtfeier an langen weißgedeckten Tischen grüßte von der Bühne das Bildnis des „Führers“, umgeben von dunklen Lebensbäumen. Deutscher Fleiß und deutsche Arbeit wurden gelobt. Von den 170 Bewerbern um ein Häuschen in der Adolf-Hitler-Siedlung waren von *Nutzhorn* selbst streng nach rassischen, erbgesunden und nationalsozialistischen Grundsätzen 50 ausgesucht worden.

3.4.2. Kleinsiedlungen in Himmelreich

Die Kleinsiedlungen Himmelreich und Coldewei waren Rüstringer Wohngebiete, die
1 sich außerhalb der Stadtgrenzen entwickelten. Erste Siedlungshäuser entstanden dort vor dem Ersten Weltkrieg als Ziegeleiarbeiter-Wohnhäuser. Nachdem die *Stadt Rüstringen* in der Ära *Lueken/Hahn* die Ziegelei[102] mit 50 ha aufgekauft hatte und sie 1932 stillegen mußte[103], entstanden ab 1933 in Himmelreich weitere Kleinsiedlungen als *Arbeitsbeschaffungsmaßnahmen.* Das *Hochbauamt Rüstringen* entwarf einfache geputzte Siedlungshäuschen an der Klinkerstraße nach Coldewei nach dem *Reichssiedlungsgesetz von* 1919. 10 Doppelhäuser wurden im Mai 1934 auf dem Gebiet der *Gemeinde Kniphausen* fertiggestellt[104]. Sie waren für Kriegsbeschädigte und Kinderreiche vorgesehen. Bis zu 1.500 qm große Grundstücksflächen erlaubten Nebenerwerb, zumal Gelände hinzugepachtet werden konnte. Streng wurden die Siedler von der *Stadt Rüstringen* durch Träger-Siedler-Verträge an ihre Pflichten gebunden. Die Vereinbarungen sahen eine monatliche Miete von 6,80 Mark im Jahre 1934 und ab 1935 10,40 Mark vor. Das dann einsetzende Reichsdarlehen von 2.750 Mark und ein Zinsnachlaß für kinderreiche Siedler ergaben nach 3 Jahren eine Belastung von 218,82 Mark/Jahr. Die Siedler waren also in den ersten Jahren Mieter. Sie konnten mit einer Kündigungsfrist von einem Monat für den Schluß des Kalendermonats gekündigt werden u.a., „wenn der Siedler sich als offensichtlich ungeeignet erwiesen hat“[105]. Der *Stadtmagistrat Rüstringen* bezeichnete in

einer zusätzlichen Protokollnotiz die Mietzeit als eine „Bewährungsfrist". Die Siedlerfamilie „soll während dieser Zeit beweisen, ob das Vertrauen, welches durch die Zuteilung der Stelle in sie gesetzt ist, auch gerechtfertigt war. Der Herr *Reichswirtschaftsminister* verlangt, daß Siedlerfamilien, die sich innerhalb dieser Zeit nicht bewährt haben, unnachsichtig von den Stellen entfernt werden"[106].

Rüstringens *Oberbürgermeister Gustav Nutzhorn* tönte anläßlich einer Einweihung im Mai 1934: „Vor der Zeit des Regierungsantritts *Adolf Hitlers* stand auf diesem Boden eine brachliegende Ziegelei ... doch jetzt im Dritten Reich wächst neues Leben aus den Ruinen. Eine neue Kolonie ist hier auf steinigem Boden entstanden, um der Bevölkerungspolitik des Führers zu dienen. Schwer und hart wie der Bau des Dritten Reiches wird auch der Aufbau dieser neuen Siedlung sein, die von nun ab eigene Scholle sein soll für viele Familien. Mit Schweiß wird der Boden gedüngt werden müssen, aber er wird einst Frucht tragen. Der frische Wind, der hier weht, wird keine internationalen Gedanken aufkommen lassen. Hier soll der Mensch wieder den tiefen reinen Segen der Verbundenheit zwischen Blut und Boden kennen und erleben lernen"[107]. Die NS-Machthaber gaben sich den Anschein, als hätten sie die Leistung vollbracht. Tatsächlich jedoch gingen die Ursprünge der Siedlung noch auf die Dienstzeit des Rüstringer *Oberbürgermeisters Friedrich Paffrath* zurück.

★

In eine spätere Phase der Kleinsiedlungsbewegung gehören die Projekte

- Kleinsiedlung Coldewei mit 53 Kleinsiedlerstellen
- Kleinsiedlung Himmelreich mit 66 Kleinsiedlungen

die *Hans Bartning* im März 1939 für das *Stadterweiterungsamt Wilhelmshaven* südlich und nördlich der Klinkerstraße, östlich und westlich der später aufgegebenen Eisenbahntrasse plante. Diese Kleinsiedlung südwestlich von Fedderwardergroden wurde trotz ihrer geringen Größe in den *Wirtschaftsplan Wilhelmshaven-Friesland vom Mai* 1940 aufgenommen. Das Projekt ging möglicherweise auf die Rüstringer Absichten von 1933 zurück, denn das knappe Erschließungssystem, der Verzicht auf jeglichen städtebaulichen Habitus und der Straßendorf-Charakter waren 1939 für Wilhelmshaven hoffnungslos überholt. Sie paßten wenig zum monströs angelegten Stadtausbau. Tatsächlich entstanden in Himmelreich im Hinblick auf *Bartnings* Kleinsiedlungsplanung noch vor dem Zweiten Weltkriege einige Heimstätten – der größte Teil jedoch wurde im Nachkriegswohnungsbau realisiert.

3.4.3. Wohnkolonie Kniphausersiel

Ähnlich der Adolf-Hitler-Siedlung, den Siedlungen Himmelreich, Sanderbusch und Schortens gingen auch die Ursprünge der Wohnkolonie Kniphausersiel auf die
1 Kleinsiedlungsbewegung der Weimarer Republik zurück. Nördlich des zu Rüstringen gehörenden Ortsteils Rüstersiel ließ das *Land Oldenburg* im Rahmen von *Arbeitsbeschaffungsmaßnahmen* am Westufer des Jadebusen 1928 Flächen eindeichen[108]. Hierzu gehörten nicht nur die Flächen, auf denen Voslapp gebaut wurde, sondern auch die gut 40 ha Waagegroden. Nach dem *Reichssiedlungsgesetz vom* 11.08.1919 sollten Erwerbslose auf dem nach der Eindeichung domäneneigenen Areal angesiedelt werden. 100 Bauplätze mit je 1.500 qm großen Nutzgärten waren nach dem *Bebauungsplan* der *Gemeinde Kniphausersiel*, die zu Fedderwarden gehörte, 1929 vorgesehen[109]. Träger der Kleinsiedlungsmaßnahme war wie bei der Siedlung Sanderbusch und weiteren Projekten im Oldenburger Land das *Siedlungsamt Oldenburg*, das Kleinsiedlungspolitik, Erwerbslosenansiedlungen und Landansässigkeitsprogramme in den 20er Jahren durchzusetzen hatte. *Georg Schipper* genossenschaftlich-orientierter *Ökonomierat* betrieb seit 1920 beim *Siedlungsamt Oldenburg* und ab 1932 als Leiter des *Domänenamtes Oldenburg* die Ansiedlung von Neubauern und Ansiedlern in Marsch und Watt. Er sammelte Interessenten[110] für Kleinsiedlungen, ließ kostenlos Typenentwürfe herausgeben, überwachte die Umsetzung, ließ Baugeräte zur Verfügung stellen, beschaffte die Finanzierung, kontrollierte die architektonische Gestaltung und organisierte die Selbsthilfe. 1929 begannen 34 Kleinsiedler nach 5 Typenentwürfen des *Siedlungsamtes* in Selbsthilfe, teils auch mit Unternehmerhilfe die Bauarbeiten. Im gleichen Jahr wurden die ersten Kleinsiedlerstellen bezogen.

Sofort nach der Machtübernahme der Nationalsozialisten betrieb das *Reichsheimstättenamt Gau Weser-Ems*, die Ansiedlungspolitik der Weimarer Republik weiter. Der
127 *Hallbauer*-Plan *Die Stadt der Dreihunderttausend vom* 15.03.1937 enthielt eine Wohnkolonie
22 Rüstersiel/Kniphausersiel von ca. 500 Siedlungshäusern. Tatsächlich wuchs die Wohnkolonie Kniphausersiel noch bis 1937 auf die jetzige Größe. Danach jedoch mußte ein
30 weiterer Ausbau zugunsten von Voslapp aufgegeben werden. Das *Schutzbereichsamt der Festungskommandantur* ließ wegen der geplanten benachbarten Flughäfen keinen weiteren Ausbau mehr zu. Das Schulprojekt Kniphausersiel wurde bereits 1936 nicht weiterverfolgt.

3.5. Gebaute Gemeinschaftssiedlungen ab 1935

Mietwohnungsbau kam nach 1933 nur sehr zögernd in Gang. Am 27.12.1933 hatte der *Bauverein Rüstringen* bei der *DAF* Baudarlehen für Arbeiterwohnungen beantragt[111]. Die *Marinewerft* unterstützte den Antrag mit dem Hinweis, daß 80 bis 100 Werftarbeiter noch in Notwohnungen untergebracht seien. Das *Kommando der Marinestation der Nordsee* forderte vom *Chef der Marineleitung* verzinsliche Baugelder für Wilhelmshaven/Rüstringen. Der Wohnungsbedarf sei von 218 im November 1932 auf 450 gestiegen. Die Wohnungsforderungen der *Marinewerft* gingen von verhältnismäßig geringem Bedarf aus. 1934 war das Ausmaß der Aufrüstung für die betroffenen Gebietskörperschaften noch nicht vorstellbar.

Im Juli 1934 schaltete sich die *Stadtverwaltung Rüstringen* ein und versprach, Grund und Boden für ein Projekt des *Bauvereins Rüstringen* von 109 Wohnungen in Erbpacht herzugeben. Am 01.07. stellten die Gemeinderäte von Wilhelmshaven für die Errichtung von städtischen Beamtenwohnungen in der Mozartstraße eigenes Gelände zur Verfügung.

Wohnungsbau als *Arbeitsbeschaffungsmaßnahme* wurde ab 1935 bedeutungsloser, da

- o die Arbeitslosigkeit rüstungsbedingt auch im Jadegebiet zurückging
- o „arbeitsunwillige Personen" von jeder Unterstützung ausgeschlossen wurden
- o Arbeitsdienst Pflicht wurde
- o die allgemeine Wehrpflicht wieder eingeführt wurde.

Arbeits- und Militärdienst funktionierten als Puffer für hunderttausende junger Menschen, die keine Gelegenheit bekamen, in ihren erlernten Beruf „einzusteigen". Verstärkt wurden auch junge Frauen beim *Marinebekleidungsamt*, der Marinewaschanstalt, in Telefonzentralen und in *Artillerie-Depots* der *KMW* angestellt.

Im *Wohnsiedlungsgebiet Wilhelmshaven-Jeverland* führte der Rüstungsboom zu einer lange nicht dagewesenen Wirtschaftskonjunktur. Nach dem 1. *Vierjahresplan* vom 01.02.1933 (nationaler Wiederaufbau) bereitete *Hitler* den 2. *Vierjahresplan* (Rohstoffunabhängigkeit) für 1936 vor. Wehrmacht und Industrie sollten in 4 Jahren kriegsfähig gemacht werden. Die Kriegsausrichtung der Wirtschaft führte zu einseitiger Steigerung der Rüstungsproduktion. Rüstungsbetriebe und deren Arbeitskräfte standen ab 1935/36 im Vordergrund der nationalsozialistischen Politik. Alle Wilhelmshavener Wohnungsbauten waren kriegswichtig, denn sie dienten mittelbar der Wehrhaftmachung. Der 2. *36*
Vierjahresplan betonte den Wohnungsbau für Rüstungsarbeiter und hob die Kontingentierung für Wehrmachtsbauten auf. Er behinderte aber auch den sonstigen Wohnungsbau bis zum *Verbot aller nicht kriegswichtigen Neubauten vom* 15.11.1939 und zum generellen „Bauverbot" vom Januar 1943[112].

Der *RAM* behielt sich ab 1939 die Anerkennung von Siedlungsprojekten als „Gemeinschaftssiedlungen" vor[113]. Um in den Genuß von Förderungen zu kommen, mußten die Siedlungen dem *Erlaß des RAM* vom 17.12.1938 entsprechen. Für das *Wohnsiedlungsgebiet Wilhelmshaven-Jeverland* brachten die Gemeinschaftssiedlungen vor *129*
allem eine Zunahme des Mietwohnungsbaues. Volkswohnungen wurden fester Bestand- *31*
teil der Gemeinschaftssiedlungen. Der Anteil von Kleinsiedlungen ging drastisch zurück[114]. Selbsthilfe wäre zu Lasten der Arbeitskraft gegangen. Propagandistisch jedoch blieb es bei den Idealen von Seßhaftmachung, Verwurzelung, Kinderreichtum, Heimatverbundenheit usw. Die Rüstungsarbeiter wurden zu Stammarbeitern stilisiert,

die als Gegenstück zu Erbhofbauern in einer „Stammarbeiterrolle" eingetragen werden sollten[115]. Durch die Ausweisungen zu *Wohnsiedlungsgebieten* ab 1936 wurde die Wohnungsproduktion zentralisiert und gestrafft. Alsbald reduzierte sich die Anzahl der Baugesellschaften auf den *Bauverein Rüstringen*, die *Wilhelmshavener Spar- und Baugesellschaft*, die *Gemeinnützige Gesellschaft zur Schaffung von Wohngelegenheiten für Reichsange-*
1 *hörige* und die *Wohnungsbaugesellschaft Jade*. Die meisten der realisierten Gemeinschaftssiedlungen ab 1935 lagen außerhalb des Altstadtgebietes. Auch Inselviertel, Altengroden und Neuengroden waren, obwohl innerhalb der „Maadelinie" gelegen, nicht innerstädtisch konzipiert.

3.5.1. Von der Wohnsiedlung Nord-Ost zum Inselviertel

Das „III. Reich" versuchte, sich nicht nur der Kleinsiedlungsbewegung zu bemächtigen, sondern auch des ganz alltäglichen genossenschaftlichen Wohnungsbaues. Die älteste gemeinnützige Wilhelmshavener Baugesellschaft, die *Wilhelmshavener Spar- und Baugesellschaft* war seit 1893 auf Mietwohnungsbau für ihre Mitglieder ausgerichtet. Ihre anfänglichen Pläne, Erwerbshäuser in eigener Regie zu bauen, zerschlugen sich. Die Gesellschaft baute dann vor der Jahrhundertwende auf preußischem Gebiet u. a. im Bereich der östlichen Bismarck- und Hinterstraße Arbeiterwohnhäuser.

1 Auf dem Gelände des späteren Inselviertels, nördlich der Bismarckstraße bis hin zur
9 Landesgrenze, lagerte u.a. die *Kaiserliche Werft* Baumaterial. Hinter der Landesgrenze begann die oldenburgische Gemeinde Heppens. Nach Osten hin lief in Fortsetzung der Jachmannstraße der einem Deich folgende Alte Deichsweg mit einigen Landstellen. Östlich davon riegelten seit den 80er Jahren bis nach dem Zweiten Weltkrieg militärische Anlagen das von Gleisen durchzogene Gelände bis zum Deich ab: Exerzierplätze, *Marine-Observatorium, Marine-Laboratorium*, Geschoßhöfe, Wagenschuppen, Entladestellen, Lagerhäuser, Schießstände und umfangreiche Kasernenanlagen. Das für Lagerzwekke und Landwirtschaft genutzte Areal von 350 m Breite und 1 km Länge war die einzige noch für eine Besiedlung im beengten Wilhelmhaven in Betracht kommende Fläche. Jedoch, eine Wohnbebauung wurde von der *Marine* bis zum Ende des Ersten Weltkrieges verhindert. Schon 1901 hatte die *Werft* einen *Bebauungsplan* für das Gebiet des heutigen Inselviertels verhindert. *Theodor Goecke* mußte in seinem *Bebauungsplan von* 1908 diese Fläche aussparen. 1918 konnte er dann endlich in seiner Monatszeitschrift „Der Städtebau" *Zopffs Bebauungsplan* für dieses Gelände vorstellen.

Wilhelmshavens *Stadtbaurat Hermann Zopff* hatte unmittelbar zum Ende des Krieges eine Planung für das Inselviertel, wie es später wegen seiner nach Inseln benannten Straßen genannt wurde, vorgelegt. Die *Marine* war bedeutungslos geworden und erhob
27 keinerlei Ansprüche mehr auf das Gelände. *Zopffs Bebauungsplan von* 1918 umfaßte jetzt das gesamte Dreieck nördlich der Werftanlagen zwischen Landesgrenze und Jadebusen. Auf 73 ha Bodenfläche sollten die „Wohnsiedlung Nord-Ost" und große Kur- und

Freizeitanlagen entstehen. Eingang zu diesem neuen Siedlungsgebiet[116] sollte der
Schnittpunkt Bismarckstraße/Jachmannstraße werden. Darüber hinaus waren Torbauten
geplant. Sie entstanden auch, so auf der Kreuzung Bismarckstraße/Siedlerweg und
Helgolandstraße/ Leiteweg. Die Betonung der Eingänge in *Zopffs* Planung von 1918,
aber auch die platzartigen Erweiterungen waren typische Merkmale der Gartenstadtge-
danken. Geschickt hatte *Zopff* rechtwinklige Baublöcke für das Inselviertel vermieden.
Leicht gebogene, geschlossene Baulinien führten auf den Schnittpunkt und „Stadtteil-
Eingang". Alle Höfe sollten ruhig gelegen sein und – obwohl Mietergärten – privat
bleiben. Mehrere unterschiedlich geformte Grünplätze sollten Unterzentren werden,
wenige betonte Hauptstraße Alleen. *Zopffs* Inselviertel-Konzept war 1918 städtebaulich
äußerst modern, und es wurde mit Recht in der Fachpresse herausgestellt. Der 27
Bebauungsplan von 1918 war erheblich kompromißloser und großstädtischer als die
laufenden Siedlungsprogramme im sozialdemokratischen Rüstringen. Moderner auch
als die Kriegerheimstätten und die von *Hakenholz* und *Brandes* entworfene Siedlung 57
Siebethsburg. Während man dort noch an „Landhausvillen für Arbeiter" orientiert war,
dachte *Zopff* schon in Stadtteilen, allerdings sozial getrennt für die verschiedenen
gesellschaftlichen Schichten[117]. Die Zeit der Weimarer Republik wurde auch in
Wilhelmshaven eine Zeit der Träume und Utopien. *Zopffs* Plan ging über das spätere
Inselviertel weit hinaus bis zum Deich und sollte Platz bieten für Villen, privilegierte
Wohngebiete, Schulen, Grünflächen, ein Sportstadion, Kuranlagen und Strandhäuser[118].
Nur ein Teil des *Zopff*-Planes, das Inselviertel, wurde von verschiedenen Wohnungs-
baugesellschaften ab 1922 realisiert. Die Bilanzwerte der Wohnungsbaugesellschaften
wurden nach der Inflation auf Goldmark umgestellt, und viele Genossen erhöhten ihre
Geschäftsanteile[119]. Trotz heraufziehender Weltwirtschaftskrise begann eine schubweise
Bautätigkeit im Inselviertel. *Zopffs Bebauungsplan von* 1918 blieb Leitlinie auch während 27
des „III. Reiches".

★

1922 begannen die Bauarbeiten im Ostteil des Leiteweges[120] mit Zweifamilien-Doppelhäusern des *Eigenhandbauverein Reichsbund Wilhelmshaven-Rüstringen.* Der Verein war 1920 unter Einfluß von *Willy Hahn* von Kriegsbeschädigten gegründet worden und plante auch große Projekte in Neuengroden. Die zweigeschossigen überaus sparsamen Gebäude im Leiteweg hatten Nutzgärten und zwischengefügte Stallanbauten. Zwei Typen wurden gebaut:

- Typ I mit Walmdach und Eingang durch den großen Stallanbau.
- Typ II mit Satteldach und Eingang durch das Wohnhaus. Der Stall als Zwischenbau war niedrig und klein.

Die Konzeption orientierte sich noch an vorstädtischen Kleinsiedlungen und dem *Reichsheimstättengesetz vom* 10.05.1920. Die beengten Grundstücksverhältnisse im Leiteweg erlaubten nur sehr kleine Nutzgärten. Man ging davon aus, daß noch mehr dieser Doppelhaustypen im Halligenweg und in der Saarbrücker Straße gebaut werden sollten[121].

Im Westteil des Leiteweges folgten 1926 zweigeschossige Reihenhäuser nach einem Entwurf von *Richard Freygang* für private Bauherren[122]. 1927 kamen noch zweigeschossige Reihenhäuser für „*Ibeling* und Genossen" nach einem Entwurf von *Ernst Peters* hinzu. Im Gegensatz zum Ostteil des Leiteweges waren die westlich gelegenen Ziegelrohbauten von *Freygang* und *Peters* mit ihren durchgehenden steilen Mansard- bzw. Walmdächern städtisch angelegt und lebhaft mit Ziegelornamenten dekoriert. Es dominieren Dreiecksformen bei Gauben, Treppenhausabschlüssen, Dachhäuschen und Eingangsdekorationen.

Die Realisierung des Inselviertels wurde 1924–33 am Siedlerweg und am Alten Deichsweg/Wangeroogestraße/ Norderneystraße/Borkumstraße[123] durch die *Gemeinnützige Bau- und Siedlungsgenossenschaft der Jadestädte* fortgesetzt. Es entstanden niedrige, spitzgiebelige typisierte Reihenhäuser in Ziegelrohbauweise nach Plänen von *Ernst Peters.* Die heimatschutzorientierten pfannengedeckten Klinkerhäuser mit Sprossenfenstern und (geplanten) Holzklappläden waren, obwohl von kleinstädtischem Habitus, keine Kleinsiedlungen, sondern Genossenschaftswohnungsbauten der 20er Jahre.

1925 begann im Inselviertel der Geschoßwohnungsbau für Reichsbedienstete an der Bismarckstraße. Die dreigeschossigen Putzbauten mit Ziegeldächern riegelten das gesamte Inselviertel nach Süden hin ab, unterbrochen nur durch das Torhaus[124] und zwei weitere Fußgängerdurchbrüche. Die Fassaden wurden an Fenstern und Eingängen abwechslungsreich gestaltet. Stuck, aber auch Ziegelumrandungen, stilisierte Spitzgiebel und flache Attikateile wechselten sich ab. Zurückspringende Bauteile, vorspringende spitze Treppenhäuser und Loggien unterbrachen den über 300 m langen Riegel. Fassaden und Dachformen lagen im traditionellen Zeitgeschmack der 20er Jahre. Für das *Marinestandortbauamt* und die Architekten *Willibald Strempel* und *Otto Lückemeyer*[125] kamen modernes Äußeres oder gar Flachdächer nicht in Betracht. Der Riegel an der Nahtstelle zwischen dem neuen Inselviertel und der Altstadt bewältigte geschickt den architektonischen Übergang. Immerhin lag ja der Platz mit der katholischen Garnisonkirche, der Petri-Kirche[126] gegenüber und das Seemannshaus[127] am westlichen Ende des Blocks. Hinter dem „Tor" zum Inselviertel folgten dann die niedrigen Reihenhäuser.

Das 1929 herausgegebene „Stadtpanorama Wilhelmshaven-Rüstringen" von *J. Ruep* und auch andere zeitgenössische Stadtpläne zeigten den Baublock Bismarckstraße/Leiteweg und Alter Deichsweg/Wangeroogestraße fertig erstellt. Ein kurz darauf herausgegebener Stadtplan enthielt bereits das gesamte Straßenraster mit den festgelegten Straßenbezeichnungen. 1930 „stand" die Planung.

Erst ab 1930 kaufte die *Wilhelmshavener Spar- und Baugesellschaft* Gelände im Inselviertel. Im südlichen Halligenweg[128] begann 1931 Architekt *Ernst Peters* mit zweigeschossigen Klinker-Zeilen, in denen er Satteldächer mit kubischen Treppenhaus-
29 türmen kombinierte. Er setzte sie ab 1934 in der Spiekeroogstraße und im nördlichen

Halligenweg fort. Die Typenhauszeilen bekamen über den zentral gelegenen Treppen-
häusern entweder ein Walmdach oder eine flache vorgeblendete leicht abgetreppte
Giebelattika mit Schlußstein. Zu Anfang waren die Fenster in den Klinkerfassaden noch
zusätzlich mit Klinkern umrahmt. Charakteristisch waren bei *Ernst Peters* die an den
Treppenhäusern vorspringenden, nach oben strebenden Fensterbänder. Zwischen den
schmalen Treppenhausfenstern sah *Peters* kreuzförmige Ziegelornamente oder Werkstein
vor. Ansonsten waren die genossenschaftlichen Mietwohnungszeilen schmucklos. Flache *29*
Betondächer überdeckten die Hauseingänge. Zunehmend wich *Peters* von *Zopffs* *27*
städtebaulichen Planungen ab. Die geschlossene Zeilenbebauung von bis zu 250 m Länge *28*
wurde aus Luftschutzgründen unterbrochen.

Die Stadt Wilhelmshaven selber hatte sich 1929 entschlossen[129], im Grundstücksgeviert
Halligen-/ Siedlerweg südlich der Saarbrücker Straße das Altenheim Lindenhof mit 110
Wohnungen nach Plänen von *Hermann Zopff* zu bauen. Der 1931 zum größten Teil
fertiggewordene Ziegelbau[130] war unsymmetrisch, nicht exakt rechtwinklig, und dem
strahlenförmigen Straßensystem angepaßt. *Zopff* selber hielt sich als Architekt an seine
1918 vorgegebene lange geschlossene Zeilen-Bebauung, so daß beim Lindenhof *27*
Flügellängen von über 100 m entstanden. *28*

1936 bereits[131] war bis auf Baltrumstraße und Saarbrücker Straße (Ostteil), Teile des Siedlerweges und den Südteil der Helgolandstraße das Inselviertel vollständig konzipiert, im Bau oder fertiggestellt. Es folgten dann noch in der Baltrumstraße ab 1936 geputzte Mietwohngebäude des *Beamten-Spar- und Bauverein*, die *Franz Iwersen* in traditioneller Formsprache entwarf. 1937 folgten Unteroffiziershäuser in der Helgoland-/Wangeroogestraße von *Willi Lübbers* für die *Wilhelmshavener-Rüstringer Wohnungsbaugesellschaft.*

Der Geschoßwohnungsbau der *Wilhelmshavener Spar- und Baugesellschaft* und ihres
Architekten *Ernst Peters* wurde 1936 wegen des zunehmenden Wohnungsbedarfs der
Marine noch einmal verstärkt. Er überschritt die westlich des Inselviertels liegende
Landesgrenze und erstreckte sich auf das Stadtgebiet des oldenburgischen *Rüstringen* im
Winkel zwischen Heppenser- und Saarbrücker Straße. Diese Fläche war von *Hermann* *28*
Zopff noch nicht „verplant". So konnte *Ernst Peters* im September 1936 einen
Bebauungsplan für das Gebiet um Pellworm-, Amrum- und Sylter Weg vorlegen[132]. Eine
ausschließlich lineare Randbebauung war wegen der großen Grundstückstiefe nicht
möglich. Erstmals mußten durch Wohnstraßen zusammengefaßte Nachbarschaften
gebildet werden. Die Architektur der 5 Gebäudetypen näherte sich den durch das „III.
Reich" veränderten Geschmacksmustern. Die steilen, hohen Ziegeldächer wurden
konsequent abgewalmt. Dachaufbauten wurden verkürzt und alles konzentrierte sich auf
die stark hervorgehobenen Haupttreppenhäuser mit dominierenden Dächern. Die
Zeilenbebauung der Weimarer Republik war endgültig aufgegeben worden.

Den Abschluß des Inselviertels bildete 1937 die Zeile Helgolandstraße 1–14. Die *Wilhelmshavener Spar- und Baugesellschaft* setzte mit Entwürfen von *Ernst Peters* die Randbebauung der Bismarckstraße von 1925–27 in die Helgolandstraße fort. Strikt nach

Zopffs Vorgaben von 1918 wurde das Torhaus Leiteweg 8–10 in die Zeile eingefügt – auch hier wieder zentriert auf das Walmdach. Der ältere Tordurchgang Siedlerweg von 1927 wirkte mit seinen Abtreppungen demgegenüber leicht und verspielt. Zur Saarbrücker Straße hin wurde die 1937 begonnene geschlossene Zeile Helgolandstraße 1–14 dann zugunsten von einzelstehenden traditionellen Mietwohnblocks aufgegeben. Lange Zeilenbebauungen entsprachen nicht den geschmacklichen und wohnungspolitischen Zielsetzungen der Nationalsozialisten.

1937 wurden alle Bautätigkeiten für das Inselviertel eingestellt, denn seit 1936 plante die
128 *HNBD* mit *Arnold Agatz Nordwerft* und IV. Hafeneinfahrt. Die Planung *Wirtschaftsgebiet*
Wilhelmshaven vom Juni 1937 sah östlich der Gökerstraße bis hin zum Jadebusen
ausschließlich Wehrmachtsanlagen vor. Der große Wohnungsbedarf für die Aufrüstung
erlaubte jedoch vorerst noch keinen Abriß der Wohngebiete, obwohl im *Wirtschaftsplan*
129 *Wilhelmshaven-Friesland vom Mai* 1940 auch das gerade erst fertiggewordene Inselviertel
als „aufzugebendes Wohngebiet" gekennzeichnet war. Nach der Planung von *Arnold*
Agatz vom 19.07.1938 sollte die *Nordwerft* bis an die Verlängerung der Jachmannstraße
101 nach Norden herangeführt werden. Die größenwahnsinnige Werftplanung des „III.
Reiches" wurde mit Ausnahme von Teilen der IV. Einfahrt, einigen Docks und
Werkstätten, nicht mehr fertiggestellt.

Die Bewohner des neuen Inselviertels ahnten nicht, daß ihre 1600 Wohnungen schon
wieder auf der Abrißliste standen, denn die *Wirtschaftspläne* waren „streng geheim", vor
allem, was ihre militärischen Zielsetzungen anging. 60 Bewohner des Alten Deichsweges
46 sollten schon 1939 in das Familienlager Neuengroden umgesiedelt werden. Zur
Jahreshauptversammlung der *Wilhelmshavener Spar- und Baugesellschaft* am 01.04.1938
wurde den Baugenossen freudig die bevorstehende Fertigstellung des Inselviertels
mitgeteilt.

★

Die Qualität des Inselviertels war für Wilhelmshavens Architekturgeschichte bedeutsam. Sie bewies, daß noch weit bis in das „III. Reich" hinein ein modernes städtebauliches Konzept verfolgt wurde. Es ging auf 1918 zurück und zeigte Ähnlichkeiten mit den überlangen Zeilen von Mehrfamilienhäusern, die *Ernst May*[133], *Martin Wagner*[134], *Otto Bartning*[135] und *Bruno Taut* häufig planten. *Zopff* hatte ausgiebige Studienreisen nach England, Frankreich, Holland, Spanien und Italien unternommen und kannte die internationale Entwicklung gut. Seine Beiträge waren in Fachzeitschriften gefragt. Sicher hatte er englische Gartenstädte gesehen[136] und die Stadterweiterung von Amsterdam Spaarndammerbuurt[137] kennengelernt. Sicher kannte er den *Berlage*-Plan von 1916 zur Erweiterung von Amsterdam[138]. Auch die Vorschläge von *Bruno Möhring* zum
Schöneberger Südgelände von 1912[139] dürften *Hermann Zopff* bekannt gewesen sein. Mit
27 dem *Bebauungsplan* Inselviertel legte *Zopff* einen städtebaulichen Entwurf vor, der die
Entwicklung vom gründerzeitlichen Mietshausblock zur Wohnzeile fortsetzte. Für

Wilhelmshaven/Rüstringen war das insofern aufsehenerregend, als alle vergleichbaren neuen Wohngebiete zu dieser Zeit auf Einzelhäusern, Landhausvillen oder Kleinsiedlungen beruhten.

Von den Zeitabläufen her hatten die Nationalsozialisten kaum eine Chance, das moderne Konzept von *Hermann Zopff* nennenswert zu beeinflussen. Aufmarschplätze, *HJ*-Heime, monumentale Achsen, *NSDAP*-Gemeinschaftshäuser und Ortsgruppen als Siedlungszellen blieben dem Inselviertel erspart. Das Vorgefundene konnte nicht mehr umformuliert werden. *Zopffs* Konzept ließ sich nicht umlenken, und die wenigen nationalsozialisten Vorbilder und Mustersiedlungen kamen für dieses bereits begonnene Wohngebiet zu spät. *Stadtbaurat Zopff* blieb bis 1937 im Amt und konnte seine Pläne noch zielbewußt durchsetzen. Allerdings wurde sein Konzept der geschlossenen Zeilen schrittweise aufgegeben zugunsten von Abschnitten und Einzelhäusern, obwohl alle Gebäudetypen für lange Reihenhausbebauungen geeignet waren. So machten die Zeilen
von *Ernst Peters* in der Tat einen künstlich „abgeschnittenen" Eindruck. Ab 1936 in den *28*
Baugebieten von Saarbrücker- und Heppenser Straße, Pellworm-, Amrum- und Sylter Weg[140] wurden die hohen abgeschnittenen Spitzgiebel abgewalmt. So paßten sie besser in den NS-Wohnungsbau-Formenkatalog. Einzige „reine" NS-Wohnungsbauten im Inselviertel blieben die wenigen Unteroffizierswohnhäuser, die *Willi Lübbers* 1937 in der Helgoland- und Wangeroogstraße baute. Dort wurde dann auch *Zopffs* Konzept völlig umgekehrt zugunsten von freistehenden Zweispännerbauten im Habitus von Einfamilienhäusern für die Führung.

Die Nationalsozialisten erkannten unmittelbar nach der Machtübernahme, daß das längst begonnene Inselviertel für ihre Politik nützlich sein könnte. Am 21.03.1934[141]
begann die *Arbeitsschlacht* mit einer Kundgebung im neuen Schauspielhaus und *153*
anschließendem „1.Spatenstich" durch *NSDAP-Oberbürgermeister Carl Renken* an der Saarbrücker Straße. Die *Wilhelmshavener Spar- und Baugesellschaft* vermeldete in ihrer Chronik stolz, daß sie dem Versuch zur Gleichschaltung getrotzt habe und „verhängnisvolle Eingriffe" abwehren konnte. In der NS-Baupropaganda las sich das so: „Die *Wilhelmshavener Spar- und Baugesellschaft* kann stolz darauf sein, an diesem großen Werk mitarbeiten zu können, und wir wissen es, daß alle ihre Mitglieder, die auch hier nicht Worte hören, sondern Taten sehen, sich am 10. April geschlossen hinter den Führer stellen, der auch ihnen geholfen hat"[142].

In der Wangeroogestraße ließen die Nationalsozialisten nachträglich einen „Grundstein" einmauern „Erbaut unter *Adolf Hitler"*. Auch Richtfeste entbehrten nicht der nationalsozialistischen Dekoration.

Für den heutigen Betrachter ist das Inselviertel ein städtebauliches Konzept der Weimarer Republik, das ein Vierteljahrhundert Gültigkeit besaß und dessen sich die Nationalsozialisten nicht bemächtigen konnten.

3.5.2. Arbeiter-Wohnstadt Voslapp

1 Das Voslapper Wohnungsbauprogramm war auf dem *Reichssiedlungsgesetz von* 1919 aufgebaut. Neu gewonnenes Land wie der 1928 im Rahmen von *Arbeitsbeschaffungsmaßnahmen*[143] eingedeichte Groden am Westufer des Jadebusen zwischen Rüstersiel und der Bauerschaft Voslapp sollten vorrangig besiedelt werden, da das Land im Besitz des
127 *Domänenamtes Oldenburg* war. Im Plan *Die Stadt der Dreihundertausend vom* 15.03.1937 erschien das Kreissymbol für die neue Siedlung erstmals. Der südöstlich geplante neue Rüstersieler Groden sollte Wasserflughafen werden und durch eine Waldanpflanzung zwischen Voslapp und Rüstersiel geschützt werden. Um die Nordstadt Fedderwardergroden für 25.000 Menschen waren die Siedlungsgebiete Breddewarden, Voslapp und Rüstersiel als Satelliten vorgesehen. Erst in der Planung *Wilhelmshaven Wirtschaftsplan vom* 03.08.1939 erschien dann noch weiter nördlich gelegen die Siedlung Utters. Der „geheime" *LOB*-Plan *Wirtschaftsgebiet Wilhelmshaven und Umgegend vom September* 1937 zeigte die Arbeiter-Wohnstadt Voslapp annähernd parzellenscharf als „weiträumiges" Kleinsiedlungsgebiet zwischen den Deichen nördlich der neuen „Maadelinie" – allerdings noch mit einem südlichen „Anhängsel".

Die *Bauverwaltung* der *Stadt Wilhelmshaven* unter *Walter Hunzinger* wurde erst ab 01.06.1938 für die Planung von Voslapp zuständig. *Wilhelm Hallbauer* jedoch hatte als Leiter des *Stadterweiterungsamtes* zusätzlich Kompetenzen als Geschäftsstellenleiter der *LOB*. Mehrere Siedlungskerne von 500/600 Wohnungen sollten „zu geschlossenen Ortsteilen etwa um Voslapp, Fedderwardergroden, Accum und Sande" zusammengefaßt werden, so schrieb *Hallbauer* am 06.08.1937[144]. *Georg Linde* von der *LOB* stand in der Tradition von *Gottfried Feder* und der Bindung an die eigene Scholle. Er propagierte vehement außerhalb der „Maadelinie" das „Eigenheim und die Siedlung" als ideale Wohnform[145]. 140 ha Land stünden im neuen Groden nördlich Kniphausersiel für eine Besiedelung zur Verfügung.

Aus der Sicht der anfangs für die Arbeiter-Wohnstadt Voslapp zuständigen, 1933 gebildeten, *Gemeinde Kniphausen* bzw. des *Amtes Friesland* hat sich Voslapp eher zufällig
22 entwickelt. *Amtshauptmann Hermann Ott* wollte für die Wohnkolonie Kniphausersiel eine neue Schule und begründete dies 1936[146] damit, daß die *Marine* ja 200 Wohnungen in Kniphausersiel bauen wolle. Unaufhaltsam stieg die Zahl. Im März 1937 schon waren es 500 Einfamilienhäuser, die in der Kolonie Kniphausersiel von der *Jade* für die *Marine* gebaut werden sollten. Am 15.03.1937 vermerkte der *Amtshauptmann*, daß die *Bauabteilung des OKM (Ministerialdirektor Schreiber* und *Oberregierungsrat Köhler)* anläßlich einer Sitzung bei der *Marineintendantur* von 1.000 Wohnungen gesprochen habe, hierin seien aber 500 für Kniphausersiel vorgesehene Einfamilienhäuser nicht enthalten. Der Bau der 500 Häuser der Wohnkolonie Kniphausersiel war in den Planungen *Die Stadt der*
127 *Dreihunderttausend vom* 15.03.1937 letztmalig dargestellt. Der *RAM* erklärte, Kniphausersiel werde vorerst zurückgstellt[147]. Grund war die „geheime" Luftwaffenplanung im Rüstersieler Groden. Am 26.07.1937 wurde *Amtshauptmann Ott*[148] in den Geschäftsräu-

men der *Jade* in Wilhelmshaven ein erster Plan des Architekten *Johannes Richter* vom *Reichsheimstättenamt Gau Weser-Ems*, Oldenburg für die Arbeiter-Wohnstadt Voslapp gezeigt. Die ganze Planung sei „nunmehr weiter nördlich gerückt". Sie sollte jetzt von „der *Jade* weiterbearbeitet" werden. Mitte 1937 stand fest, daß die *Marine* schnellstens mit Voslapp beginnen wollte. Das *OKM* wollte am 29.07.1937 für „840 Werkwohnungen für Marineangehörige in dem nördlichen Teil des Fedderwarder Baugrodens ... sofortige Erteilung der Bauerlaubnis"[149]. Die *Jade* verhandelte mit dem *Domänenamt* wegen der Grodenfläche. Die Kosten der Infrastruktur, der Straßen, elektrischen Leitungen und Klärgruben wurden von der *Jade* „mit in die Baukosten einbezogen." Die Wohnungsanzahl schwankte fortwährend um etwa 1.000.

Folgende Bebauungspläne sind für die Arbeiter-Wohnstadt Voslapp angefertigt worden:

- *Bebauungsplan Voslapp* des *Reichsheimstättenamtes Gau Weser-Ems*, Oldenburg, Juli 1937 (Unterschrift *Johannes Richter*)
- *Plan Gemeinschaftssiedlung-Voslapp des MDF vom November* 1937 aufgrund der 30
  Gemeinschaftsskizze vom 27.10.1937 (Unterschrift *Fritz Karl Schneider*).
- *Lageplan Voslapp* der *Wohnungsbaugesellschaft Jade* vom 02.03.1938
- *Lageplan mit Aufteilungsplan Voslapp, Wohnungsbaugesellschaft Jade*, Juni 1938
- *Fluchtlinienplan Nr.* 160, *Voslapp*, festgestellt Dezember 1938, förmlich festgestellt 07.01.1939
- *Siedlung Voslapp, Lageplan mit Typenverteilung*, ergänzt im Dezember 1941 durch die *Wohnungsbaugesellschaft Jade*

Mitte 1937 wurden erste Bodenuntersuchungen für Voslapp ausgewertet[150]. Im darauffolgenden November hatte das noch zuständige *Amt Friesland* die Kosten für die Arbeiter-Wohnstadt Voslapp an das *OKM* gegeben. Wilhelmshavens *Oberbürgermeister* befürchtete deshalb eine Benachteiligung anderer Siedlungen im Wilhelmshavener Stadtgebiet[151]. Die unproblematische Bereitstellung des Baugrundes durch das *Domänenamt Oldenburg* führte zur entscheidenden Bevorzugung des Projektes Voslapp, „das Zeugnis davon ablegen soll, wie die Wohnungsnot bekämpft und die Sozialpolitik des nationalsozialistischen Deutschlands in die Tat umgesetzt wird"[152].

Der erste *Bebauungsplan Voslapp* des Oldenburger Architekten *Richter* vom *Reichsheimstättenamt Gau Weser-Ems*, wurden vom *RAM (Karl Pfeiffer)*[153] und *OKM (Walter Kelm)* im Oktober 1937 überarbeitet und den Typenplänen von *Willi Lübbers* angepaßt.
Den *Plan Gemeinschaftssiedlung-Voslapp* zeichnete *Fritz Karl Schneider* vom *MDF* im 30
November 1937. *Friedrich Eisenbarth*, Bremen, entwarf die Gartenarchitektur. Die *Stadt* 33
Wilhelmshaven war erst nach der Eingemeindung ab 01.06.1938 beteiligt, *Oberbürgermeister Müller* jedoch seit 1937 als Aufsichtsratsvorsitzender der *Jade*.

Schneiders Plan Gemeinschaftssiedlung-Voslapp orientierte sich an den vorhandenen 30
Deichlinien. Die 2 km schnurgerade Hauptstraße[154] verlief auf dem seit 1761 bestehenden Kajedeich[155]. Die Begrenzung zum Jadebusen war der Voslapper Seedeich von 1928, und

zur geplanten Siedlung Fedderwardergroden schloß der Kniphauser Deich von 1718 ab. Die 4 Erschließungsstraßen paßten sich den gebogenen Deichlinien an. 7 Querstraßen verbanden die Erschließungsstraßen miteinander. In den so gebildeten Blöcken verteilte *Willi Lübbers* 7 Haustypen. *Schneiders* Planung *Gemeinschaftssiedlung-Voslapp* sah 4 Nachbarschaften und kleine platzartige Straßenerweiterungen vor.

Im *Lageplan Voslapp* vom 02.03.1938 tauchten erstmals auch 5 Volkswohnungsblöcke in eingeschossiger Reihenbauweise als 8. Wohnungstyp auf mit zusammen 46 Wohnungen[156]. Die Planungen für die Bauten der Gemeinschaft[157] „standen" 1937: 2 Schulen[158], *NSDAP*-Gemeinschaftshaus, *HJ*-Heim, Feuerwehr, Rathaus und Bürgermeisterhaus. Die 400 m lange Achse zwischen Marktplatz und Gemeinschaftshaus war als monumentale Aufmarschallee angelegt. Der *Plan Gemeinschaftssiedlung-Voslapp* des *MDF* wurde Anfang 1938 in der Partei[159] als Abkehr vom Mietskasernenbau der Weimarer Republik propagiert. Die *NSDAP* gebe dem Arbeiter ein Haus, das sein Eigentum werde, wenn er erbgesund und charakterlich geeignet sei. Zur Übergabe der Kleinsiedlungen an die Bewohner kam es nicht, wie vorgesehen, nach dreijähriger Probezeit, sondern erst ab 1955.

1 Die abseitige Lage der für 5.000 Einwohner vorgesehenen Arbeiter-Wohnstadt Voslapp außerhalb der Wilhelmshavener Stadtgrenzen wurde, wie auch bei der Siedlung Cäciliengroden, geschickt verschleiert: „Voslapp liegt nicht außerhalb der Welt" beteuerte die Parteipropaganda[160], obwohl es immerhin 10 km bis zu den Arbeitsplätzen bei der *KMW* waren. Die einzige leistungsfähige Straße kam von Rüstersiel und der einzurichtende Omnibusverkehr sollte mit einem „Gummibahnhof" am Rathaus verbinden. Das 10 km lange Straßennetz der Arbeiter-Wohnstadt Voslapp sollte mit Radwegen ausgestattet werden. Die seit 1911 marineeigene Werftbahn[161] wurde ab 1935 vom *Hafenbauressort* der *KMW* erheblich erweitert. Die Bahnhöfe für Fedderwardergroden (Hörn) und Voslapp (Schönengroden) entstanden erst 1941[162]. Zu einer Verlängerung der Werftbahn bis Breddewarden und Utters, zu der im *Wirtschaftsplan* 1939 vorgesehenen Ringschnellbahn kam es nicht mehr. Noch im Juni 1938 war die Verkehrsanbindung völlig unklar. „Entfernungen werden überbrückt, und in absehbarer Zeit wird eine Verkehrsregelung getroffen werden, damit Voslapp häufig und bequem mit Wilhelmshavens Stadtmitte verbunden ist"[163].

*

Die Entwürfe für Voslapps Kleinsiedlerstellen reichte Chefarchitekt *Willi Lübbers* von der *Jade* am 12.03.1938 als Bauantrag bei dem bis zum 01.06.1938 zuständigen *Amt Friesland* ein. Die Gebäudeentwürfe datierten vom 20.01.1938. Anfang 1938 lag auch die Genehmigung der Berliner Dienststellen zur Finanzierung über 6,5 Mio. Mark vor, nachdem *Willi Lübbers* verschiedentlich beim *OKM* vorstellig geworden war. Keinesfalls nur die Kleinsiedlungsideologie begrenzte die Geschoßzahl der Entwürfe für die Voslapper Wohnbauten. Anfang 1936[164] hatte das *OKM* die *Marineindendantur Wil-*

helmshaven angewiesen, niedrige aufgelockerte Bauweise zu bevorzugen. Dies war für die Wilhelmshavener Bodenverhältnisse günstig und im *Schutzbereichsplan* vorgegeben.

Früh müssen sich skeptische Stimmen erhoben haben, die das Ausmaß der Kleinsiedlung kritisch beurteilten. Man fürchtete, daß die Arbeiter der *KMW* möglicherweise für Kleinsiedlungen völlig ungeeignet seien. In allen Einzelheiten wurden wiederholt Finanzierung, Ausbaustandard und Ausstattung vorgestellt und in Parteipresse, Druckschriften, Ausstellungen und Musterwohnungen[165] dargelegt. *Lübbers* wehrte sich gegen die Annahme, die Häuser seien zu klein und zu primitiv, wie man es von den Erwerbslosen- und Kriegsopfer-Siedlungen der Weimarer Republik kannte. Schließlich war das Mauerwerk doch „massiv", ein Keller vorhanden, das Dachgeschoß zum Ausbau geeignet und die 600 bis 900 qm Land zur Eigenversorgung ausreichend[166].

Das Modell der Arbeiter-Wohnstadt Voslapp und einzelne Hausentwürfe wurden im Februar 1938 von der *KMW* ausgestellt[167]. „Es werden Häuser, die im Stil und im Baumaterial, in Form und Anlage unserer Landschaft in jeder Weise angepaßt sind. Die Fronten und auch die Dachform der einstöckigen Häuser sind sehr gefällig". Die monatliche Last für die künftigen Kleinsiedler sollte bis zum Eigentumsübertrag 30 bis 34 Mark betragen.

Im Januar 1938 begannen die Aufschließungs-, Vermessungs- und Erdarbeiten. Hunderte von Arbeitern waren in Lagern untergebracht[168]. Im März waren bereits „600
Bauleute am Werk"[169] in schwerer *Arbeitsschlacht.* Im Taktverfahren karrten 100 LKW *153*
Abbruchmaterial aus der IV. Einfahrt und Baustoffe nach Norden über Wilhelmshavens Stadtgrenze hinaus. *Willi Lübbers*, seit jeher mehr organisatorisch als baukünstlerisch interessiert, konnte erstmals seine Fähigkeiten unter Beweis stellen: Er kaufte in friesischen Ziegeleien Steine auf, erwarb auf der Leipziger Messe Herde und schaffte schlesische Handwerker heran. Zerfahrene Landstraßen wurden vom *Amt Friesland* gesperrt, kurz danach auf Druck der *NSDAP* jedoch wieder freigegeben. *Raeder* besuchte am 06.04.1938 die Baustelle und ließ sich den Fortgang der Planung von *Lübbers* erläutern. Im März 1938 wurden in Voslapp, der „Stadt für schaffende Volksgenossen"[170] erste Richtfeste mit Hakenkreuzfahnen zelebriert. *Friedrich Eisenbarth* begann noch im April,
während der laufenden Bauarbeiten Gärten anzulegen. Im Juni standen Musterwohnun- *33*
gen[171] zur Verfügung, und im Juli 1938 waren die ersten Häuser bezogen.

Der *Schmitthenner*-Schüler *Willi Lübbers* hatte Anfang 1938 eine ungewohnt große Entwurfsaufgabe zu lösen, die bisherige Vorstellungen weit überstieg: Eine Siedlung für 5.000 Menschen! Er gab vor, obwohl an den *Bebauungsplänen* nicht beteiligt, sich an der um 1500 gebauten Augsburger Fuggerei orientiert zu haben[172]. Auch die *Schmitthenner*-Siedlung Berlin-Staaken als Garten„vor"stadt mit kleinstädtisch-romantisierenden Reihenhäusern errichtet, dürfte kaum eine Rolle gespielt haben. Vergleiche boten sich mit den englischen Gartenstädten kaum an, eher schon mit den Stuttgarter Kleinsiedlungen Steinhaldenfeld und Sommerrain[173] mit Bremen-Grolland oder den Mustersiedlungen Mascherode und Ramersdorf. Gartenstadtelemente waren bei Voslapp zwar formal zu finden, jedoch politisch ins Gegenteil verkehrt. Es entstand nicht die autarke,

reformerische Stadt, sondern die marineabhängige Werksiedlung. Sie war aus Kosten-,
1 Grunderwerbs- und Luftschutzgründen weit aus der Stadt herausgerückt. Bezüge zu Dresden-Hellerau, zu den Wohnsiedlungen der Weimarer Republik oder den fortschrittlichen Rüstringer Siedlungen von *Wagner, Hahn* und *Migge* verboten sich. Voslapp war als Arbeitersiedlung der *Kriegsmarine* geplant. Soziale Ideen und bodenreformerische Ansätze fehlten völlig. Das NS-System ersetzte sie durch rassistische Phrasen, „völkische" Propaganda und dogmatische Ideologie. Wenn auch bei den Fassaden der Gewerbebauten um den Voslapper Marktplatz Idylle und Romantik nicht zu übersehen
32 waren, fielen Kleinsiedlungstypen und Volkswohnungen jedoch durch Schlichtheit, Einfachheit, Beschränkung und geringsten Standard ab. Der von Materialien und Formen ausgehende Eindruck solider Handwerkstradition täuschte. Rationelle Serienproduktion und wirtschaftlicher Baubetrieb standen für *Willi Lübbers* im Vordergrund. 668 Wohnungen waren als Doppelhäuser in den Typen D, E und F geplant. Nur 102, die Typen B und C, waren als freistehende Einfamilienhäuser vorgesehen. 5 Reihenhaus-
31 zeilen, Typ G, und 46 Volkswohnungen kamen etwas später hinzu. Alle Kleinsiedlungshäuser waren ausnahmslos aus Ziegeln gemauert. Die 55 Grad steilen Satteldächer waren mit 70 Grad steilen Krüppelwalmen versehen – eine Referenz an die Tradition ostfriesischer Bauernhäuser. Der 183 mal vorkommende Doppelhaus-Typ F mit insgesamt 366 Wohnungen z.B. war mit einer Wohnfläche von 47 qm nur minimal ausgestattet. Im Erdgeschoß: Windfang, Wohnküche (15 qm), Zimmer (13 qm), Waschküche (7 qm) und Außenabort. Eine steile Holztreppe führte nach oben. Dort befanden sich Trockenboden und 2 Kammern von je 11 qm, davon eine zum Selbstausbau vorgesehen. Kellerräume und Stall (im Garten) sollten die erforderliche Fläche für Kleinsiedlertätigkeit schaffen. In dem knapp bemessenen Siedlunghäuschen wohnten nicht selten Werftarbeiterfamilien mit 5 und mehr Kindern. Wenn auch die NS-Propaganda unablässig den Werftarbeitern einhämmerte, daß sie diese Wohltaten *Hitler* zu verdanken hätten, machte sich doch bald die Erkenntnis breit, daß die Siedlung weder fortschrittlich noch neu war, sondern sparsam und einfach.

Selbst der 43 mal gebaute freistehende Wohnungstyp B für Kinderreiche hatte nur 67 qm Wohnfläche – davon waren noch 2 Dachkammern für späteren Selbstausbau abzuziehen. Der technische und sanitäre Standard der Siedlungshäuser lag weit unter dem, was im städtischen Wohnungsbau und in den Großsiedlungen der Weimarer Republik längst üblich gewesen war. Bauschäden stellten sich ein, und die Freude über die eigenen vier Wände war oft nur kurz. Für die im Juli 1938 einziehenden
135 Erstbewohner waren die Verhältnisse miserabel und standen im krassen Gegensatz zu den
136 Wohnbedingungen für Führungspersonal. Fachlehrwarte, Mustereinrichtungen und Lehrküchen sollten die Neubewohner zum Kleinsiedler umerziehen. *Siedlerbund* und Parteidienststellen reglementierten Gartenarbeit und Tierzucht. Die Kritik wurde massiver, und selbst die Parteipresse[174] textete negative Schlagzeilen: „Die Verhältnisse in Voslapp gehen nicht so weiter!".

*

Mietwohnungen waren schon im November 1937 im *Plan Gemeinschaftssiedlung-* 30
Voslapp von *Schneider* vorgesehen. Anfangs sollten es zweigeschossige Reihenhäuser mit eigenen Gärten sein. Für den Anfang sollten zwei Familien darin wohnen, später sollte eine Familie das Haus kaufen können. Im März 1938 jedoch wurde das Konzept zu Gunsten von Volkswohnungen aufgegeben. Dies belegte nicht eine plötzliche Ablehnung des Kleinsiedlungsbaues oder der Agrarromantik. Es war schon 1935 durch *DAF-Leiter Robert Ley* klargerückt worden, daß es „Volksgenossen" gibt, die „keinen Sinn für die Bodenbewirtschaftung" haben und die „helle und schöne Mietwohnungen" bevorzugen[175]. Gartenbewirtschaftung, so fürchtete die *NSDAP* alsbald, halte die „Volksgenossen" von der Parteiarbeit ab. Bereits im *Gesetz zur Förderung des Wohnungsbaus vom 30.03.1935* hatte der *RAM* neben dem Kleinsiedlungsbau auch Mietwohnungsbau mit 35 Mio. Mark zur Förderung vorgesehen. Die Wohnungen sollten einfach sein und waren für ungelernte Arbeiter bestimmt[176]. Die Gesamtherstellungskosten einer
Volkswohnung waren auf 3.000 Mark begrenzt. Die Volkswohnungen sollten möglichst 31
nicht in geschlossenen Gemeinschaftssiedlungen oder großen Anlagen entstehen, sondern mit Kleinsiedlungen „durchmischt" werden. Die nationalsozialistischen Baupolitiker fürchteten die Rückkehr von Familienfeindlichkeit, Elendsquartieren und Mietwucher, die man zu bekämpfen vorgab. Tatsächlich wurde der Geschoß-Mietwohnungsbau in Wilhelmshaven alsbald zur Regel.

Gottfried Feders Versprechen, „jedem deutschen Menschen sein Häuschen im Grünen"[177] und die Ideologie der Gaumustersiedlungen waren nicht durchzuhalten. Das zuvor scharf bekämpfte Mietwohnhaus war für die staatlichen Rüstungsarbeiter von Wilhelmshaven sehr wohl geeignet, denn die Verwurzelung mit der Scholle erschien nur für die kleine Zahl der Stammarbeiter wichtig. Für das Heer der ungelernten Arbeiter erkannte man jedoch bereits damals die Notwendigkeit der „Mobilität"[178]. Es war klar, daß nach dem Zweiten Weltkrieg Wilhelmshaven nicht mehr so viel Rüstungsarbeiter benötigen würde.

Willi Lübbers, trat öffentlich gegen Mietwohnungen in Voslapp an: „Mietwohnung bleibt Mietwohnung. Jedes Deutschen Sinn steht nach eigenem Haus und Herd"[179]. *Lübbers* entwarf die zu 40 Zeilen addierten mehrgeschossigen Wohnblöcke für Voslapp
nicht mehr. *Jade*-Architekt *Willy Schemm*, zeichnete Anfang 1939 die Volkswohnungen, 31
nachdem die Kleinsiedlungen Voslapp bezogen waren. Der Grund für die verzögerte Planung der Volkswohnungen lag möglicherweise in ideologischen Auseinandersetzungen, möglicherweise auch in dem Architektenwechsel begründet.

Schemms Mietwohnungsplanungen sah vor[180]

o	Typ I	46 qm	84 Wohnungen
o	Typ I.V.	44 qm	84 Wohnungen
o	Typ VIII	59 qm	148 Wohnungen
o	Typ VIIIa	48 qm	76 Wohnungen
			392 Wohnungen

In 4 Teilabschnitten begann im Juni 1939 der Bau der sehr einfach ausgestatteten Volkswohnungen in Voslapp. Im September 1939 waren die Teilabschnitte 1 und 2 im Bau[181]. *Walter Temp*[182], obwohl Bauingenieur und erst seit 2 Monaten *Stadtbaudirektor* von Wilhelmshaven, griff im Juli 1939 „beratend" in *Schemms* Planung ein. „Ergebenst" teilte die *Jade* den Vollzug der gewünschten Änderungen mit. Aber auch das *Schutzbereichsamt* und die *Abteilung XXIV. der Marineintendantur* griffen ein und ließen im September 1939 Lattenmodelle der steilen Satteldächer anfertigen. *Karl Franzius* entschied, daß südlich und westlich des Marktplatzes Voslapp wegen der benachbarten Batterie behelfsmäßig Flachdächer von 1,50 m Höhe vorzusehen seien. Einige Satteldächer mußten sogar wieder abgebrochen werden. Sofort nach Kriegsende 1946 beantragte die *Jade* bei der *Militärregierung* die Herstellung der 1939 vorgesehenen Steildächer. Wegen der anhaltenden Wohnungsnot im Nachkriegs-Wilhelmshaven wurden Wohnungen in die Dachräume gebaut.

31 Der am häufigsten in Voslapp vorkommende Mietwohnungstyp VIII zeigte in *Schemms* Urentwurf vom Januar 1939 48 Grad steile Satteldächer, die an den Giebeln 65 Grad steil abgewalmt werden sollten. Ziegelbau, Kreuzsprossenfenster, Pfannendach, Fenstersymmetrie, Dachhäuschen, zentrierte Hauseingänge und schlanke Treppenhauslichtbänder spiegelten den knappen Detailkatalog des Massenwohnungsbaus wider. Der Zwang zur Rationalisierung beschränkte die Variationsbreite und uniformierte die Fassaden. Die Rückfronten waren völlig gleichförmig mit einem einzigen Fenstertyp vorgesehen: kein Ornament, keine Dekoration, kein Balkon, kein Blumenfenster. Nur Eingangstüren bekamen schließlich Werksteinumrahmungen. Nach weiteren Eingriffen der *Stadt* entfiel die Abwalmung zugunsten eines durch Klinkerornamente gegliederten Giebels. Bei dem sparsamen Zweispännertyp VIII war der Grundriß typisiert. Im Keller: Hauskeller, Waschküche und Luftschutzräume. Die 59 qm der Wohnungen teilten sich so auf: Wohnküche 15 qm, davon abgetrennt die Kochküche von 5 qm, Elternschlafzimmer 13 qm, 2 Kinderzimmer von 9 qm. Bad und WC waren auf engstem Raum (3 qm) zusammengedrängt. Aus Platzmangel war nur eine Sitzbadewanne möglich, Zentralheizung fehlte. Der Typ I.V. von 44 qm, der 84 mal in Voslapp vorkam, war ein noch sparsamerer Zweispänner. Er hatte nur 1 Kinderzimmer und die Kochküche war zu einer offenen Nische geschrumpft. Der Typ I nördlich der Geniusbankstraße war anfangs als abgewalmte Zeilenbauweise vorgesehen. Auf Anregung des *MDF*[183] griff *Stadtbaudirektor Temp* auch hier 1941 zensierend ein, strich die Abwalmungen, veränderte die

Fensteraufteilungen und betonte die Türeingänge durch Vordächer und aufgedoppelte Türen.

Zwei Blöcke von 90 m Länge sollten den Aufmarschplatz am *NSDAP*-Gemein- 31
schaftshaus an der Südseite begrenzen. Dieser Typ VIIIa wurde von *Willy Schemm* auch im Februar 1939 entworfen und im März von der *Baupolizei* genehmigt. Auch hier griff *Temp* ein, so daß am 04.11.1941 die Genehmigung ausgesprochen wurde. Es entfielen wiederum die Abwalmungen, Dachhäuschen und Eingänge wurden verändert. Die knapp 400 Volkswohnungen wurden erst ab 1941 bezugsfertig.

★

Die Arbeiter-Wohnstadt Voslapp hatte für die *NSDAP* besondere propagandistische Bedeutung, da hier „erstmalig in Wilhelmshaven und in dieser Geschlossenheit auch erstmalig im ganzen Reich, unseren schaffenden deutschen Menschen eine Heimat ..."[184] gegeben werde. Obwohl erst ab 01.06.1938 zu Wilhelmshaven gehörig, schlachtete *Kreisleiter Ernst Meyer* das Projekt für seine Partei rücksichtslos aus. Die NS-Klischees von „Deutschlands Zukunft", „glücklichen Kindern" und „froher Jugend" sowie die Versprechungen von *HJ*-Heimen, *NSV*-Kindergärten, Konzertabenden und Theatervorführungen waren in einer Werbebroschüre im Juni 1938 zusammengefaßt. Die Arbeiter hatten Stellen bei der *KMW* bekommen, und Wohnungen standen in Aussicht. Dies sollte „dem Führer gedankt werden". Das Leben mit Eigentum versprach innenpolitische Stabilität „Ruhe und Ordnung". „Die Volksgenossen aber, die dort eine Heimat finden und denen die Häuser, genau wie in Voslapp, in wenigen Jahren als Eigentum übergeben werden, haben das erreicht, was das Ziel so vieler ist. Sie bekommen ein eigenes Heim, das ihnen ihr alles sein wird und das sie dermaleinst ihren Kindern übergeben können. Ran ans Werk, das geschaffen werden soll für bodenständige deutsche Menschen, für unsere deutschen Arbeiter"[185]. So wurde Bewußtsein eingeschläfert. Langanhaltendes Presseecho begleitete die Propagandawelle, die immer wieder die Hausbauten vorstellte, Grundrisse erläuterte, Modelle abbildete. In Ausstellungen wurde die Wirkung noch verstärkt „der Arbeiter erhält nicht nur eine Wohnung, er erhält ein Haus, das nach 3 Jahren in sein Eigentum übergeht, das ihm gehören wird, ohne daß er über Kapital zu verfügen braucht"[186]. Daß deutschen Arbeitern Häuser „geschenkt" werden, um sie bodenständig zu machen, wurde nicht verheimlicht. Eigentum sei, wie auch der Garten, eine indirekte Lohnsteigerung. Schlimmer noch waren rassistische Begründungen für Einfamilienhausbau wie diese: „Man gibt dem Arbeiter, ohne daß er über eigene Mittel verfügen braucht, ein Haus, das bei Bewährung sein Eigentum wird. Maßgebend dafür ist nur, daß er erbgesund und charakterlich geeignet ist"[187]. Dankbare Bewohner wurden zitiert: „Wir bleiben für immer hier und halten fest an der Scholle, die uns übergeben ist"[188]. Keine Frage war für die *NSDAP*-Propaganda, wem das alles zu danken sei: *Hitler* ließ sich als der wahre Baumeister feiern[189]. 112

3.5.3. Siedlung Cäciliengroden

Cäciliengroden, eine Siedlung für 360 Familien von Arbeitern der *KMW*, entstand ab
1 1938 außerhalb Wilhelmshavens in der Gemeinde Oestringen auf einer Fläche, die 1844 eingedeicht, aber erst 1912–18 aufgespült wurde[190]. Auf der Basis des *Reichssiedlungsgesetzes von* 1919 war vorrangig, neugewonnenes Land zu bebauen. Der Standort Cäciliengroden hatte, wie auch der von Voslapp, mit günstigem Grunderwerb zu tun. Wie in Voslapp hatten die neuen Bewohner Anwartschaft auf eine Kleinsiedlung nach den *Bestimmungen über die Förderung der Kleinsiedlungen vom* 14.09.1937 und nach den Bewilligungsbescheiden des *MDF*. In Voslapp wie in Cäciliengroden wurden zusätzlich auch Volkswohnungen gebaut. Wie in Voslapp, war der *Deutsche Siedlerbund e.V., Gaugruppe Weser-Ems*, zuständig und die Anfang 1937 gegründete *Wohnungsbaugesellschaft Jade* mit Wilhelmshavens *Oberbürgermeister Müller* als Aufsichtsratsvorsitzendem Träger der Kleinsiedlungen. Auch in Cäciliengroden betrug die Probezeit für Kleinsiedler 3 Jahre. Nach dieser Zeit sollten die Siedlerstellen „zu Eigentum als Reichsheimstätte übertragen"[191] werden. Wie in der Arbeiter-Wohnstadt Voslapp, die einen zeitlichen Vorlauf von etwa 3 Monaten hatte, war auch in Cäciliengroden ein Architekt alleinverantwortlich für Wohnungsbau und Infrastruktur. Als Gartenarchitekt war für beide Siedlungen bis 1939 *Friedrich Eisenbarth* aus Bremen eingeschaltet.

127 Erstmalig zeigte der Plan *Die Stadt der Dreihunderttausend vom* 15.03.1937 von *Wilhelm Hallbauer* Cäciliengroden als Kreissymbol an der Stelle des Flughafens Mariensiel. Weiter südlich sollten eine Südstadt Sande für 25.000 Einwohner, westlich davon die Siedlungen Mitteldeich und Sanderbusch entstehen. Der Plan *Wirtschaftsgebiet Wilhelmshaven und Umgegend vom September* 1937 zeigt das Siedlungsgebiet genau in den späteren Umrissen südlich des Flughafens Mariensiel.

Der *Wirtschaftsplan Wilhelmshaven vom* 03.08.1939 von *Wilhelm Hallbauer* wies Cäciliengroden, das zu diesem Zeitpunkt bereits fertiggestellt war, als Teil des Wohngebietes Wilhelmshaven Süd aus, allerdings mit einer westlichen Erweiterung bis zum Industriegebiet zwischen Neudeich und Bahnlinie. Zu dieser Erweiterung kam es nicht mehr.

Der von dem Bremer *Schmitthenner*-Schüler *Friedrich Heuer* Anfang 1938[192] entworfene
34 *Bebauungsplan Cäciliengroden* bewegte sich ganz in der Tradition der Kleinsiedlungsbewegung und Gaumustersiedlungen. *Heuer*, der den Architektenauftrag nach Empfehlung durch *Willi Lübbers (Jade)*, *Wilhelm Wortmann (LOB)* und *Fritz Karl Schneider (MDF)* bekam, hatte sich 1936 mit der Siedlung Bremen-Grolland hervorgetan und Ende 1937 den Auftrag für Cäciliengroden bekommen. Der willkürlich wirkende Zuschnitt und die
1 abseitige Lage des 30 ha großen Areals, das die *Jade* am 01.04.1938 vom *Domänenamt Oldenburg* gekauft hatte, ergab sich durch die Parzellierung des neuaufgespülten Geländes[193]. Das Straßen- und Erschließungsgerüst war durch *Heuers* Städtebaulehrer *Heinz Wetzel* geprägt: regelmäßiger, rechtwinkliger und konzentrierter. *Heuer*, der später in nationalsozialistischen Ortsgruppensiedlungen dachte, interpretierte Cäcilien-

grodens Hauptstraße als Nord-Süd-Achse (500 m lang zwischen Schule und Marktplatz). *139*
10 Nachbarschaften zweigten von den 4 durchlaufenden Straßen ab. Jede der
Nachbarschaften bekam einen Platz von 20 x 20 oder 15 x 15 m. Drumherum standen 8 *35*
bis 12 Siedlungshäuser, frei oder als Doppelhäuser. Gartenarchitekt *Friedrich Eisenbarth*[194]
sah eine Dorflinde in der Mitte vor. Die Siedlergärten von 700 qm sollten dem
Nebenerwerb dienen, darum wurde für jede Kleinsiedlerstelle ein Stallgebäude errichtet.
Nur die Nord-Süd-Achse sollte öffentliches Grün in Form einer Allee bekommen. Die
Straßenprofile waren dörflich. Zwischen Entwässerungsgräben und Rasenbermen
wölbte sich die ländliche Klinkerfahrbahn. Fußwege wurden, wenn überhaupt, nur an
einer Seite der Straße angelegt. *35*

Cäciliengroden zeigte Merkmale nationalsozialistischer Baupolitik, denn es ging nicht
darum, erwerbslosen oder heimgekehrten Kriegern eine neue Existenz auf dem Land zu
schaffen, sondern eine Werksiedlung für die *KMW* aufzubauen. Hierzu gehörten auch
der *Norddeutsche Eisenbau*, die *Gießerei der KMW* und die Betriebe des benachbart *94*
geplanten Industriegebietes Sande. Die Problematik der abseits gelegenen Siedlung war
schon 1937 bekannt. *Wilhelm Hallbauer* war skeptisch[195], ob sich eigenes politisches und
gesellschaftliches Leben entwickeln würde. Wesentliche Teile der öffentlichen Aufgaben
würden wohl von Wilhelmshaven übernommen werden, weil die Arbeiter „durch ihre
tägliche Fühlung mit der Werft" dem kulturellen und wirtschaftlichen Leben in
Wilhelmshaven verbunden seien.

Die *NSDAP*-Propaganda[196] verniedlichte im Februar 1938 die abseitige Lage und *1*
meinte, alles sei doch besser als in Großstädten. Zum Glück haben wir „so viel Grund und
Boden, daß wir nicht auf den Bau von Mietwohnungen angewiesen sind, sondern solche
Häuser bauen können, zu denen noch ein Stück Land gehört, das den Menschen wieder
zur Heimat wird und sie bodenständig macht".

Kurz nach dieser ersten propagandistischen Vorstellung beantragte die *Jade* 500.000
Mark aus Reichsmitteln für Cäciliengroden. Dem *MDF* waren die Baukosten zu hoch.
Am 19.03.1938 bewilligte die *Deutsche Bau- und Bodenbank Berlin* 512.000 Mark und am
14.02.1939 das *OKM* knapp 1 Mio. Mark. Die Gesamtkosten betrugen 3,4 Mio Mark.
Die Baufirma *Herdejürgen und Harmsen* aus Einswarden begann im April 1938 mit knapp
200 Arbeitern die Siedlung. Im September 1938 wurden die ersten der geplanten 256
Kleinsiedlungen und 104 Volkswohnungen bezogen. 1.800 Einwohner sollten nach dem
Heuer-Bebauungsplan in den 360 Wohnungen von Cäciliengroden wohnen. Bauten der *34*
Gemeinschaft waren nicht, wie in späteren *Neuen Städten*, städtebaulich monumental
zusammengefaßt, sondern über die Siedlung verstreut[197]. Immerhin aber war die
Infrastruktur systematisch bedacht und im Gegensatz zu den Kleinsiedlungen Sander-
busch, Himmelreich, Kniphausersiel, Adolf-Hitler-Siedlung usw. war Cäciliengroden *1*
vergleichsweise üppig ausgerüstet. Nur Schule, Turnhalle/Appellhalle und einige Läden *139*
wurden weit nach Bezug der Siedlung 1940 noch hastig fertiggestellt. Die letzten der 360
Wohnungen waren Anfang 1939 bezogen. Die Erstbewohner waren aus allen Regionen *35*
des Reiches entweder dienstverpflichtet oder für die *Kriegsmarine* angeworben worden.

Am 09.01.1940[198] präsentierten *Amtshauptmann Hermann Ott* und *Chefintendant Nobe*
36 *Großadmiral Erich Raeder* das „fertige" Cäciliengroden.

Die 6 Haustypen in Cäciliengroden wiesen starke Ähnlichkeit mit *Heuers* Entwürfen für Bremen-Grolland auf. Die roten Ziegeldächer (55 Grad) waren allesamt wie auch dort extrem abgewalmt (80 Grad). Kunstvoll waren Gauben in die steilen Dächer eingearbeitet worden. Auch die aufgehenden Wände waren Ziegelbau. Kaum ein Haus hatte mehr als 60 qm Grundfläche. Dies entsprach einer Nutzfläche von knapp 40 qm. Die Grundrisse waren äußerst sparsam aufgeteilt. Im Erdgeschoß beim Typ 1 z.B.: zentraler Eingang, enger Flur, rechts Abstellraum, links Waschküche mit Toilette, Wohnküche (17 qm), dahinter Elternschlafzimmer (12 qm), Kinderzimmer (8 qm), Keller, Bad und Luftschutzraum[199] fehlten. Das Dach war nicht ausgebaut. Zu Reihen zusammengefaßte Volkswohnungstypen waren zumeist an der Nord-Süd-Achse gele-
135 gen und ähnlich knapp bemessen. Die großzügigen Bauten für Wilhelmshavens
136 Führung[200] die zeitgleich im Villenviertel und anderswo entstanden, bewiesen, daß die „Volksgenossen" des „III. Reiches" keinesfalls gleich behandelt wurden. Auch das
35 Äußere der Cäciliengrodener Siedlungshäuschen mit seinen dörflichen Attributen (Holzklappläden, Regentonnen, Rankgerüste usw.) war weit entfernt vom *Marinestil* der Offizierswohnungen in der Stadt. Für die Arbeiter in Cäciliengroden wurde auch an der Qualität gespart. Dies belegen auch umfangreiche Schäden, die sich schon 1940 zeigten. Der Bombenkrieg tat ein übriges. Kriegsgefangene mußten die Schäden noch im Krieg beseitigen. Die 256 Kleinsiedlungen wurden ab 1956[201] den Siedlern übertragen. Ab 1974 privatisierte die *Wohnungsbaugesellschaft Jade* auch die restlichen 104 Volkswohnungen. Die Infrastruktur (Kirche, Jugendheim, Dorfgemeinschaftshaus usw.) wurde erst nach dem Krieg Zug um Zug ergänzt.

3.5.4. Fedderwardergroden, die Neue Stadt

1 Fedderwardergroden entsprach wohl als erste Großsiedlung des *Wohnsiedlungsgebietes Wilhelmshaven-Jeverland* annähernd den nationalsozialistischen Vorstellungen von der
126 *Neuen Stadt.* Im *Neuordnungsplan vom* 20.11.1936 war nur eine neue Siedlung nördlich Rüstersiel angedeutet[202]. Diese Siedlung, mehrere Kilometer nördlich der Altstadt von
127 Wilhelmshaven, war dann im Plan *Die Stadt der Dreihunderttausend vom* 15.03.1937 als Nordstadt Fedderwardergroden für 25.000 Einwohner vorgesehen[203]. Daneben waren eine Südstadt *Sande* mit 25.000 Einwohnern – und eine Weststadt Langewerth mit 35.000 Einwohnern – geplant. Weitere 12 Siedlungsgebiete und die Kernstadt mit 140.000
128 Einwohnern ergänzten das Programm 1937. Auch der nächstfolgende Plan *Wirtschaftsgebiet Wilhelmshaven vom Juni* 1937 zeigte Fedderwardergroden noch systematisch als kreisrund ausgewiesenes Siedlungsgebiet. Die späteren *Wirtschaftspläne*[204] stellten die künftigen Stadtteilumrisse genauer dar. Die Einwohnerzahl sollte später auf 16.000 reduziert werden.

Im „III. Reich“ stand Wohnungsbau keinesfalls im Mittelpunkt – anders in
Wilhelmshaven. Hier mußten im Rüstungsinteresse schnell und zahlreich Marinewoh-
nungen und Werftarbeiter-Quartiere gebaut werden. Kleinsiedlungsprogramme mit *33*
dem ideologischen Ballast von Blut und Boden, Heimat und Erde, konnten den *35*
hochgesteckten Zielen für Wilhelmshaven auf Dauer nicht gerecht werden. Deswegen sah *Georg Linde* von der *LOB* schon am 27.08.1937 ein, daß „ein gewisser Prozentsatz anderer Wohnungen in Mietshäusern zu errichten“ sei[205]. Es sollten „keineswegs nur einförmige Arbeitersiedlungen mit nur wenigen Haustypen werden, wie sie leider auch heute noch immer wieder entstehen“. Von vornherein sollten die neuen Wilhelmshavener Siedlungen „charaktervoll“ und „Gebilde mit eigener Form und eigenem Formenausdruck verschiedener Art“ sein[206]. Möglichst viele gestaltende, bewährte Kräfte mit den verschiedensten Handschriften für die einzelnen Bauaufgaben[207] sollten herangezogen werden.

Für die *Neue Stadt* Fedderwardergroden lagen, als 1938 mit der Planung begonnen wurde, kaum Muster vor. Die *Jade*, als Bauherr von Fedderwardergroden, mußte ihre Richtlinien für die städtebaulich Planung von außen beziehen. Die *Stadt Wilhelmshaven* wurde offiziell erst am 1.6.1938 für Fedderwardergroden zuständig, als die Fläche eingemeindet wurde. *Oberbürgermeister Müller* jedoch war ab März 1937 als Aufsichts-
ratsvorsitzender der *Jade* für Fedderwardergroden zuständig. Die *Bebauungspläne* von *1*
Cäciliengroden, Voslapp, Hansaviertel usw. waren noch von den beauftragten Architekten zusammen mit den zuständigen *Bauverwaltungen* entwickelt worden. Bei Fedderwardergroden übernahmen wegen des Volumens von fast 25.000 Einwohnern die Berliner Baudienststellen, insbesondere das *RAM*, die „Planungsführung“. Der seit November 1937 neue *Stadtbaudirektor Walter Hunzinger* stand, obwohl angeblich[208] sehr erfahren, dem großen Wohnungsbauprogramm von 1938 fassungslos gegenüber[209]. *Oberbürgermeister Müller* hatte zur Vorbereitung der Planung für Fedderwardergroden Ende Februar 1938 in Berlin Gespräche mit *Ministerialrat Georg Heilmann* und Oberregierungsrat *Max Büge* vom *RAM* geführt[210]. Man verwies darauf, daß Wilhelmshaven nicht in das *Gesetz zur Neugestaltung deutscher Städte vom* 04.10.1937 einbezogen werde, da „dem Reich u.U. erhebliche Kosten“ entstünden. Wilhelmshavens Wunsch, *Neugestaltungsstadt* zu werden, war gescheitert. Zur Realisierung von Fedderwardergroden sollte Wilhelmshaven sich auf das *Landbeschaffungsgesetz* und die *Oldenburger Baugesetze* beschränken. Die *Stadt* jedoch versprach sich lebhafteren Mittelfluß, wenn sie als *Neugestaltungsstadt* eingestuft würde. Ersatzweise bot *Müller Julius Schulte-Frohlinde* vom Architekturbüro der *DAF* an, Fedderwardergroden zur Gaumustersiedlung zu machen. Diese Pläne scheiterten jedoch am Widerstand des *OKM*, das ein Einmischen der *DAF* als Erschwernis ansah.

Erst am 01.06.1938[211] rief eine zunehmende Desorientierung die Berliner *Reichsstelle für Raumordnung* auf den Plan. Es wurde klar, daß Groß-Wilhelmshaven ein Sonderfall war und die Behandlung der Bauprogramme als „Staatsgeheimnis“ eine planmäßige Landbeschaffung und Umsiedlung unmöglich machen würde. Gleichzeitig legte die

Reichsstelle für Raumordnung fest, daß der *RAM Wirtschaftsplan* und *Bebauungspläne* überprüfe und verbindlich feststelle. Das Reich sollte die erforderlichen Baumittel vorschießen und die *RUGES* das Land beschaffen. Der für Bauangelegenheiten
3 zuständige *MDF*[212] hatte Anfang 1938 versucht, die „in Aussicht genommenen Untersuchungen der Planung um Wilhelmshaven" mit Ausnahme von Voslapp und Cäciliengroden an freiberufliche Architekten zu vergeben, aber niemanden gefunden, der „für diese Aufgabe Zeit" fand[213]. Die *Bebauungspläne* für Fedderwardergroden
62 wurden eher zufällig begonnen, nachdem die Planungen Burg Kniphausen und die
63 Planung Schilldeich zu scheitern drohten.

Am 23.07.1938 teilte das *OKM* der *Marineintendantur Wilhelmshaven* mit, daß die Planung Fedderwardergroden, wie auch die Planung Schilldeich, im *RAM* demnächst fertiggestellt sei und die *Stadt Wilhelmshaven* den Bedarf an öffentlichen Gebäuden und einen Zeitplan für Erschließungsarbeiten für diese Stadtteile hergeben möge. Am 05.08.1938 wurde der Vorentwurf für Fedderwardergroden in einer Besprechung im *RAM* unter *Karl Pfeiffer* „vorläufig festgelegt".

Der erste bekannte *Siedlungsentwurf Fedderwardergroden* datiert vom 15.07.1938. Verfasser war *Fritz Karl Schneider* vom *MDF*. Am 26.08.1938 wurde Wilhelmshavens Gemeinderäten in „vertraulicher" Sitzung ein Überblick über die Gesamtplanung gegeben. Der *Oberbürgermeister* trug dort auch vor, daß er mit dem Mittelfluß höchst unzufrieden sei. Er beklagte öffentlich: „Das *Reich* gibt seine Mittel auf einem sehr umständlichen Wege, wir müssen diese Mittel auf dem Instanzenwege anfordern, das *Ministerium* verlangt sorgfältige Unterlagen und Nachweise der Notwendigkeit"[214]. Auch *Müllers Stadtbaudirektor Walter Hunzinger* beklagte in der Gemeinderatssitzung am 31.10.1938, „daß Wilhelmshaven ein ganz neues Problem sei, wie Salzgitter und Fallersleben, und von einer Größe, die die meisten gar nicht ahnen"[215].

Friedrich Nicolaus[216] *Oberregierungsrat* im *RAM*, gab am 23.11.1938 den angestrebten Zeitplan für Fedderwardergroden bekannt:

- Sofort — Baubeginn in den nicht kanalisierten Teilen
- 30.11.38 — Abschluß des Planes M. 1:1.000 durch die *Stadt*
- 31.01.39 — Pläne des *Tiefbauamtes* für Straßen, Plätze, Kanalisation, Vermessung
- Febr. 39 — Ausschreibung der Bauarbeiten
- 01.03.39 — Baubeginn Fedderwardergroden

Die *Stadt* sollte auch unverzüglich „Beihilfeanträge für Aufschließung und Schulen auf dem Dienstwege" stellen, Kosten für ein Arbeiterlager ermitteln und Bereitstellung von Arbeitskräften beantragen[217].

Am 16.12.1938 trug *Wilhelm Hallbauer*, Leiter des Wilhelmshavener *Stadterweiterungsamtes*, den *Bebauungsplan Fedderwardergroden*[218] im *OKM* im Beisein des *RAM* vor. Er wurde heftig kritisiert, da er von Planung und Modell vom 18.10.1938 abgewichen sei. Unmittelbar danach entstand im *RAM*, *Europahaus*, Berlin, Saarlandstraße 96, vom 19. bis 22.12.1938 der endgültige *Bebauungsplan Fedderwardergroden*. Dieser Plan vom

22.12.1938 ist eine Gemeinschaftsleistung aller beteiligten Baubehörden. Mutmaßungen über den einzelnen Architekten oder den besonderen Stadtplaner sind nicht zu belegen[219].

Der *RAM* holte die formelle Zustimmung des *OKM* zu diesem *Europahausplan* ein. Die Gebäudetypen von *Gustav Bork* lagen Ende 1938 dem *OKM* zur Auswahl vor. Der *RAM* drängte auf strikte Einhaltung des *Europahausplanes* vom 22.12.1938 und der gesetzten Termine. *Oberbürgermeister Müller* beklagte, daß der Berliner *Europahausplan* „eine städtebaulich gewiß erfreuliche, freihändig gezeichnete Skizze“ sei[220], aber Typenfestlegungen fehlten und der Plan ja noch exakt übertragen werden müsse. Erst dann könne man ihn den Aufsichtsinstanzen zur Feststellung im förmlichen Verfahren vorlegen. Noch im Januar 1939 verkleinerte das Wilhelmshavener *Stadterweiterungsamt* unter *Wilhelm Hallbauer* den unhandlichen *Europahausplan* vom Maßstab 1 : 1.000 in einen *Bebauungsplan* Maßstab 1 : 2.500 – ohne jegliche Änderungen. Für die *Stadt Wilhelmshaven* war enttäuschend, daß sie aus der „eigentlichen Planung“ für Fedderwardergroden, „welche nach der *Deutschen Gemeindeordnung* zu den Hoheitsaufgaben der *Stadt* gehört“, herausgedrängt wurde[221]. Diese Methode entbinde die Stadt „von einer ernsthafteren Verantwortung für die Gestaltung ihrer neuen Ortsteile weitgehend“.

Folgende *Bebauungspläne* markierten wesentliche Planungsstadien für Fedderwardergroden:

- *Siedlungsentwurf Fedderwardergroden vom* 15.07.1938 „Deckblatt 2“ vom *MDF* (Unterschrift *Fritz Karl Schneider*),
- *Europahausplan vom* 22.12.1938, Gemeinschaftsplanung von *RAM, OKM, Stadt Wilhelmshaven, Finanzministerium Oldenburg* (Unterschrift *Karl Pfeiffer*),
- *Bebauungsplan des Stadterweiterungsamtes vom Januar* 1939 (Unterschrift *Wilhelm Hallbauer*), exakte Verkleinerung des *Europahausplanes*
- *Modell Fedderwardergroden* zum *Hitler*-Besuch vom 01.04.1939 *38*
- *Fluchtlinienplan Nr.* 161 *Fedderwardergroden* fertiggestellt 05.04.1939, förmlich festgestellt 29.4.1939
- *Stadtplanentwurf* des *RAM*, (*Pfeiffer*) von 1940 mit Typenverteilung von *Hilgenstock* und *Bork*
- Überarbeiteter *Lageplan Fedderwardergroden* des *Stadterweiterungsamtes* von 1940 *37*
- Stadt Wilhelmshaven, Ortsteil Fedderwardergroden, *Übersichtsplan* 1943, *Stadtvermessungsamt*
- *Baustellenplan vom August* 1941 mit Typeneintragungen.

Dazu markieren die verschiedenen *Wirtschaftspläne* die Einbindung in die umgebenden Stadtteile.

Der *Schneider*-Plan vom 15.07.1938 zeigte den fast 200 m langen Aufmarschplatz mit einer gleichfalls 200 m langen Abknickung nach Westen hin. Dieser zentrale Ortskern für immerhin fast 16.000 Bewohner durchzog die Planung bis zu den Ansätzen der Realisierung. Der Platz sollte von zweigeschossigen Gewerbetrakten und Dienstleistungsbetrieben umschlossen sein. Auch die nationalsozialistischen Gemeinschaftshäuser im Schnittpunkt der Straßenachsen[222], im Süden[223] und im Norden[224] blieben Bestandteil aller weiteren Planungen. *Schneiders* Plan sah die 5 Schulen[225] an den auch später beibehaltenen Standorten vor. Die jeweils benachbarten 5 Sportplätze reduzierten sich mit dem *Europhausplan* auf 3 nördlich der Posener Straße. *Schneider* hatte die Schulen an den Stadtteilrand plaziert und die sich ergebenden Straßenschnittpunkte für kleine Platzgestaltungen nutzen wollen. Diese Platzentwürfe waren wenig monumental, sie hatten Dorfplatzcharakter. Auch der „Grüne Ring" als Gerüst der Grünplanung von *Heinrich Wiepking-Jürgensmann* blieb wegen seiner Funktion für das Entwässerungssystem immer Planungsbestandteil. Noch im November 1941 trug *Wiepking* seine Grünplanungen im *RAM* vor[226].

Die Eingriffe des *RAM* durch den *Europahausplan* vom 22.12.1938 lassen sich vereinfacht so beschreiben:

- o Fast alle rechtwinklig zur Straße stehenden Gebäudezeilen und Hausblocks wurden durch straßenbegleitende, traufständige Wohnhäuser ersetzt.
- o Die Anzahl der Einfamilienhäuser in Nachbarschaften oder gefangenen Straßen wurde durch den *RAM* stark reduziert[227].
- o Freistehende Einfamilienhäuser wurden innerhalb des „Grünen Ringes" nicht mehr
41 geduldet und nur für Führungspersonal am Ortsteilrand vorgesehen.
- *144* o Die Nord-Süd-Achse war mittig und in Blickrichtung begrenzt durch *NSDAP*-
146 Gemeinschaftshäuser im Süden und im Norden. Ende 1939 mußte die nördliche
66 Planung zugunsten einer Straßenverbindung nach Breddewarden aufgegeben werden, und es kam zu umfangreichen Umplanungen des nördlichen Stadtausganges[228].
- o Der *RAM* füllte das Baugebiet Fedderwardergroden bis zur westlich tangierenden Eisenbahnlinie auf.

Andere Planungsbestandteile entfielen erst später. Der von *Schneider* vorgesehene südliche Friedhof[229] erschien zu klein und lag unglücklich tief. Er wurde 1941 als großer *Zentralfriedhof Wilhelmshaven-Nord* nördlich Fedderwardergrodens vorgesehen[230]. Auch Garagenbauten gingen erst nach der Propagierung des *KDF-Wagens* und nach Erlaß der *Reichsgaragenordnung* in die Planung ein. Die Dauer der Bebauungsplanung für Fedderwardergroden war von 1938 bis zum Baubeginn relativ lang. Sie belegte ein Gegen- und Nebeneinander der Planungsbehörden, in dem die *Stadt Wilhelmshaven* erst langsam ihre Rolle finden mußte. In den Planungszeitraum fallen auch die Bemühungen, *Neugestaltungsstadt* zu werden, die Erweiterung des Stadtgebietes, die Kriegsvorbereitungen und die Ablösung des *Stadtbaudirektors Walter Hunzinger*. Fedderwardergroden war die erste Großsiedlung für Wilhelmshavens *Stadtverwaltung*. Für die *Bebauungspläne* von
34, 30 Cäciliengroden (1937) und Voslapp (1937) war die Wilhelmshavener Baubehörde nicht
25, 26 zuständig gewesen. Die Kleinsiedlungen Sanderbusch, Adolf-Hitler-Siedlung, Himmel-
67, 22 reich, Schortens, Kniphausersiel sowie die innerstädtischen Baugebiete Neu-Siebeths-

burg, Insel-, Jade- und Hansaviertel waren als Beispiele für die *Neue Stadt* wenig 57, 28
geeignet. 53

Im Februar 1938 besuchten daher *Oberbürgermeister Wilhelm Müller, Stadtbaudirektor Walter Hunzinger* und der Leiter des *Stadterweiterungsamtes, Wilhelm Hallbauer* die Mustersiedlung der *DAF* in Braunschweig-Mascherode sowie die Siedlung München-Ramersdorf und „eine Siedlung in Dessau“[231].

Das Wilhelmshavener *Stadterweiterungsamt* mußte sich an die Vormachtsstellung des *RAM* erst langsam gewöhnen. Die *LOB* war bei der Planungskoordination wenig hilfreich. Funktionsabläufe, Planungszuständigkeiten und Entscheidungsprozesse für die *Neue Stadt* waren den zeitgleich ablaufenden Planungen für die *Stadt des KDF-Wagens* (Wolfsburg) und die *Stadt der Hermann-Göring-Werke* (Salzgitter) sehr ähnlich und durch Gesetze und Richtlinien festgeschrieben. Es war kein Zufall, daß sich *RAM* und *OKM* bei Fedderwardergroden ab 15.12.1938 des Baumanagers *Fritz Hilgenstock* bediente, der mit dem Aufbau der *Reichswerke Hermann Göring* in der *Stadt der Hermann-Göring-Werke* Erfahrung gesammelt hatte[232].

★

Fedderwardergrodens Gebäudeplanung war vollständig und von Anfang an in der Hand der am 16.01.1937 gegründeten *Wohnungsbaugesellschaft Jade*. Die ausschließlich für den Aufbau der Stadt gegründete Gesellschaft hatte mit den innerstädtischen Wohngebieten Hansa- und Jadeviertel sowie den Kleinsiedlungen Cäciliengroden und Voslapp schon insgesamt 5.200 Wohnungen fertiggestellt bzw. begonnen, als ihr die *Bebauungs-*
pläne für Fedderwardergroden zur Ausführung übergeben wurden. Die organisatorische 37
Führung des gesamten Wilhelmshavener Wohnungsbauprogramms war eindeutig der 1
Jade zugemessen worden, denn sie habe sich bewährt, so schrieb *Wilhelm Hallbauer*, der Leiter der *Landesplanungsstelle* und des *Stadterweiterungsamtes* am 06.08.1937 dem *OKM*. Tatsächlich wurden die älteren einheimischen gemeinnützigen Wohnungsbaugesellschaften ab 1937 kaum mehr mit größeren Projekten für das Wohnsiedlungsgebiet
Wilhelmshaven-Jeverland beauftragt. 129

Bei der Gebäudeplanung waren, wie auch in *Wirtschafts-* und *Bebauungsplänen*, in erster Linie Luftschutz und Kosten zu beachten. So sollte das Baugebiet, da Wilhelmshaven „besonders luftgefährdet“[233] war, nicht vielgeschossig und eng gebaut werden, sondern „aufgelockert in 1- und 2-Geschoßbauten mit trennenden Grüngürteln, also weiträumig angelegt werden“[234]. „Der Wohnungsbau wird in Geschoßbau durchgeführt werden müssen. Hierbei ist im wesentlichen an die 2-geschossige Bauweise gedacht“, stellte *Wilhelm Hallbauer* am 10.12.1936 fest. In sich sollte bei der Gebäudeplanung wiederum berücksichtigt werden, was auch in Raum- und Stadtplanung Maßgabe war: Brandabschnitte, große Gebäudeabstände, Grünschneisen, Luftschutzräume usw.

Bebauungsplanung und Gebäudeplanung für die *Neue Stadt* Fedderwardergroden waren sorgfältig getrennt. Das Büro von *Fritz Hilgenstock*, Bauberatung und -organisation bekam den Auftrag, Fedderwardergroden zu bauen. *Hilgenstocks* Rolle ist wohl am ehesten mit der eines heutigen Generalplaners oder Generalübernehmers zu vergleichen. Er schaltete für die *Jade* selbständig Architekten ein, wie er für Salzgitter 1937 *Herbert Rimpl* unter Vertrag genommen hatte. *Gustav Bork* erinnert sich, daß er ein Kino für die *Stadt der Hermann-Göring-Werke* bauen sollte und hierbei *Hilgenstock* kennenlernte. Da beide gut miteinander auskamen, trug *Fritz Hilgenstock* dem *Bruno Paul*-Schüler *Gustav Bork* auch Fedderwardergroden an. Kleinere Wohnungsbauaufträge erhielten später auch noch *Hans Fricke*[235] und *Rudolf Bürgin*[236] aus Oldenburg sowie *Ludwig Hörold*[237]. *Hilgenstock*[238] war, als er von der *Wohnungsbaugesellschaft Jade* am 15.12.1938 den Auftrag als Generalplaner bekam, ein etablierter Bauingenieur, weniger Architekt. Er hatte an der TH Hannover studiert und sich 1928 in Berlin mit einem Bauleitungsbüro und etlichen Zweigstellen selbständig gemacht. Sicher haben die Bauabteilungen von *RAM* und *OKM* in Berlin ein fähiges Büro gesucht, das einer derart neuen Aufgabe gewachsen war. Die Auswahl wird nicht groß gewesen sein[239] Wie sich *Fritz Hilgenstocks*[240] Sohn erinnerte, habe sein Vater u.a. den Auftrag bekommen, weil er auf eigene Kosten Bodenerkundungen vorgenommen und erkannt habe, daß man bei der Verwendung leichter Schwemmsteine auf Pfahlgründung verzichten könne. Diese Begebenheit wirft ein Schlaglicht auf *Hilgenstocks* wirtschaftliches Denken. Sie war gleichzeitig Ursache für die Putzfassaden von Fedderwardergroden. Aus Kostengründen und Klinkerknappheit wurde auf den heimischen Ziegel verzichtet.

Das Büro von *Fritz Hilgenstock* wurde Ende 1938 in den Strandhotels an Wilhelmshavens Südstrand untergebracht, denn erst 1939 baute die *Stadt* Baracken. *Hilgenstock* überließ die Arbeit in Wilhelmshaven fast ganz seinem Schwager *Joseph Veeser*[241]. *Hilgenstock* kam nur zu Besprechungen mit Vertragspartnern und Politikern. Bei der *Jade* koordinierte der technische Geschäftsführer *Adolf Braunschweig* 1939–40 die Arbeiten.

Der Berliner *Europahausplan vom* 22.12.1938 reichte keinesfalls für eine Festlegung der Gebäudetypen aus, und *Oberbürgermeister Müller* bemängelte, daß *RAM* und *OKM* bisher keine Gebäude ausgewählt hätten[242]. Er stellte fest, daß nach den Berliner Plänen der Bau von Straßen und Leitungen jedoch möglich sei. Hart ging der *NSDAP-Oberbürgermeister* die Berliner Dienststellen *RAM* und *OKM* an und beklagte, daß die „Mittelvorlage des Reiches im Beihilfeverfahren“ nicht funktioniere, da nur 10 Mio. Mark für das gesamte Reichsgebiet für die Finanzierung der Gemeinschaftseinrichtungen zur Verfügung ständen. Vorsorglich wies *Müller* alle Schuld von sich, wenn das Bauprogramm für Fedderwardergroden nicht termingerecht ablaufe. In der ersten Ratssitzung 1939[243] verkündete *Müller* erstmals öffentlich den Bau von Fedderwardergroden und ließ sich 1,3 Mio. Mark für städtische Arbeiterbaracken in Fedderwardergroden bewilligen. Alles sollte besser werden als in Voslapp. Schulen und Geschäfte sollten gleich miterrichtet werden.

14 Am 02.03.1939 versuchte die *Marineintendantur* als Mittelinstanz des „ideellen“ Aufftraggebers *OKM* erstmalig, in einer großen Besprechung den Bauablauf zu koordinieren[244] und die Aufgaben zu verteilen. Die *Stadt* als Träger von 20 km Straßen und 44 km Entwässerung schloß gleichfalls mit dem Büro *Hilgenstock* einen Vertrag[245], weil das eigene Personal die Aufgaben nicht hätte bewältigen können. Wegen des außergewöhnlichen Baustoffbedarfes für Fedderwardergroden schaltete die *Marineintendantur* in einem Schnellbrief[246] das *OKM* ein, da wegen des bevorstehenden Krieges nicht mehr frei über Baustoffe verfügt werden konnte. *Friedrich Nicolaus, Oberregierungsrat* im *RAM*, rief am 23.03.1939 *OKM, Stadt Wilhelmshaven, Marineintendantur, Wohnungsbaugesellschaft Jade* und das Büro *Hilgenstock* nach Berlin, um letzte Bauvorbereitungen für Fedderwardergroden zu besprechen[247].

- Oberingenieur *Paulsmeier*[248] und *Magistrats-Oberbaurat Helmut Döscher*[249] schlugen als Gutachter vor, den Grundwasserstand für Fedderwardergroden durch Grabenvertiefung um 1 m abzusenken[250] sowie Straßen und Häuser auf 50 cm Sandaufschüttung zu 39
legen.
- Ein Sofortprogramm sollte den Totenweg zur Aufnahme des erwartenden Bauverkehrs nach Fedderwardergroden verlängern. Eine eigene Schiffsumschlaganlage am Banter Hafen und eine Lastwagenfahrbereitschaft von 85 LKW sollte aus dem Boden gestampft werden.
- Ziegelbauweise wurde wegen der schwierigen Materialbeschaffung gestrichen, obwohl sie „bodenständig und auch billig in der Unterhaltung" sei. Statt dessen sollten die Häuser „guten Putz" erhalten, keinesfalls nur einen Zementputz mit Anstrich.

Gustav Bork als hauptbeteiligter[251] Architekt legte in der gleichen Besprechung am 23.03.1939 dem *RAM* erstmals in seinem Atelier in Hannover entstandene Entwürfe und Modelle vor, die „im engsten Kreise geprüft wurden". Die architektonischen Leitlinien legte *Karl Pfeiffer* fest: „Die Lebhaftigkeit, die die Straßen und Häuserzüge zweifellos 38
haben sollen, darf nicht aufgrund willkürlicher Abänderungen der Elemente der Fassaden zustande kommen. Diese sollen vielmehr möglichst einheitlich sein und in getreuer Anlehnung an das Althergebrachte geformt werden. Die Konstruktion der Fenster und Türen ist durch alte Beispiele gegeben. Die Verhältnisse der Fenster der verschiedenen Häuser müssen aufeinander abgestimmt werden. Unterschiede in den Fassaden sollen auf Unterschieden in den Grundrissen beruhen. Daher ist es nötig, verschiedene Typen zu entwickeln."

Es muß angenommen werden, daß die Gebäudeplanung im Juni 1939 keinesfalls fertig war, denn die Parteipresse ging mit keinem Wort darauf ein, als sie das Projekt und die anlaufende Baustelle erstmals groß herausstellte[252]. Die Ver- und Entsorgungsleitungen[253] lagen im Juni 1939 zum größten Teil bereits in der Erde, und der Straßenbau begann gerade. Die Arbeiten für die Wohnungsbauten waren im April 1939 an acht Baufirmen 39
vergeben worden, kamen aber nicht in Gang. Erst Anfang 1940 lag für Fedderwardergroden der abgeschlossene Plan der Gebäudetypen von *Gustav Bork* mit Benummerung der fast 70 Wohnungstypen, Gewerbebauten, Sonderbauten, Trafostationen, Pumpwerke usw. vor. Architekt *Gustav Bork* hatte sich, wie er sich heute erinnert, im Stil der

Gegend anzupassen versucht. Er sei eigentlich immer den Idealen der 20er Jahre verhaftet gewesen.

Am 21.02.1939 trafen die ersten *KDF-Wagen* in Wilhelmshaven ein. *Hitler* hatte die „fast naive Automobil-Begeisterung" geschickt ausgenutzt[254], das Reichsautobahnnetz weiter ausgebaut und die *Reichsgaragenordnung* erlassen. Während der Berliner *Europahausplan* für Fedderwardergroden Garagen noch nicht vorsah, wies *Wilhelm Hallbauers* Plan vom Januar 1939 Garagen aus. Schon bei der Eröffnung der *24. Automobil-Ausstellung* in Berlin am 08.03.1934[255] hatte *Hitler* jedem deutschen Arbeiter ein Auto versprochen. In Wilhelmshavens Siedlungsplanung waren bisher nur Garagen in Wohngebieten für Führungskräfte, u.a. im Villenviertel und bei Gewerbebauten, vorgesehen. Nun mußte auch Fedderwardergroden noch eilig Garagen für *KDF-Wagen* bekommen und in den Gewerbetrakten wurden Tankstellen berücksichtigt.

Auch im März 1940, ein halbes Jahr nach Kriegsbeginn, waren offenbar noch keine Hochbauten fertiggestellt, denn die Stromversorgung, ob Freileitung oder Verkabelung, war noch völlig offen[256]. *Friedrich Nicolaus* vom *RAM* bekam am 19.12.1939 einen ausführlichen Bericht von *Fritz Hilgenstock* über die zunehmenden Schwierigkeiten[257]. Die ersten Wohnhäuser sollten wegen der Dringlichkeit gemäß *OKM/RAM-Erlass vom* 12.01.1940 Freileitungen bekommen.

Erst im Juli 1940 wurden die ersten Wohnungen in Kulmer- und Weichselstraße bezugsfertig, und Ende 1940 waren 369 Wohnungen fertiggestellt. Das gesamte Bauprogramm für Fedderwardergroden hatte sich gegenüber den Absichten von November 1938 enorm verzögert. In einer planerischen Zwischenbilanz resümierte *Stadtbaudirektor Walter Temp* im Oktober 1940[258], daß der *Lageplan* „jetzt in seinen Grundzügen" feststehe: 4.100 Wohnungen für Fedderwardergroden.

Die größte Angst der Nationalsozialisten war, daß sich politische Unruhe ergeben würde, weil Schulen, Arztpraxen, Postämter, Läden und Gemeinschaftsbauten nicht fertig werden könnten. Im Oktober 1940, wenige Wochen nach Fertigstellung der ersten
138, 142 Wohnungen, wurde auch die Schule A gerichtet. Die Schule B war begonnen worden,
141 und für weitere Schulen im Norden des Stadtteils waren die Entwürfe in Arbeit. Allein an
146 der Danziger Straße[259] waren drei Gemeinschaftshäuser geplant, im Norden, am
111, 144 Aufmarschplatz, dem „Langen Markt", im Achsenkreuz und im Süden.

39 Die Einstellung der Fedderwardergrodener Bauarbeiten kam nicht in Betracht, obwohl nichtkriegsgewichtige Bauten im Reich am 15.11.1939 stillgelegt wurden[260]. Der „*NSDAP*-Frontbrief" vermeldete Anfang 1941 stolz: „In allen Stadtteilen im Norden der Stadt wird weitergebaut". Es muß um die Stillegung von Fedderwardergroden jedoch Konflikte gegeben haben, denn der *RAM* verfügte im Februar 1942[261] die
142 Einstellung der Bauarbeiten bei der Schule B. Die ersten Bewohner mußten wegen der verzögernd entstehenden Gemeinschaftseinrichtungen auf die Zeit nach dem Krieg vertröstet werden.

Die Bilanz des Fedderwardergrodener Wohnungsbaues war[262]:

1939 100 Wohnungen gerichtet, 2.400 Wohnungen im Bau

1940	369 Wohnungen bezogen
1941	1.031 Wohnungen bezogen, 2.629 Wohnungen im Bau
1942	1.428 Wohnungen bezogen, 900 Notwohnungen in teilfertigen Gebäuden bezogen
1943	2.302 Wohnungen bezogen, 962 Wohnungen im Bau, 426 Bauten stillgelegt
1945	2.988 Wohnungen bezogen
1947	3.078 Wohnungen bezogen
1948	3.105 Wohnungen bezogen

Die Betrachtung der annähernd 70, fast ausnahmslos verputzten Wohngebäudetypen der *39–42*
Architekten *Bork, Fricke, Bürgin* und *Hörold* läßt erkennen, daß die NS-Mustersiedlungen
und Ausstellungen Ramersdorf (1934), Schlageterstadt (1937) Mascherode (1936–39)
und Frankfurt (1938)[263] für Fedderwardergroden geringe Bedeutung hatten. Obwohl
landwirtschaftliche Kleinsiedlerstellen keine Rolle mehr spielten, wurden ihre ländlichen
Formen zum Dogma. Die Gemeinschaftssiedlungen des „III. Reiches" wurden den
angeblichen Klassensiedlungen der Weimarer Republik gegenübergestellt[264]. Gemein-
schaft sollte sich ausdrücken durch Nachbarschaften an gefangenen Sackgassen, Plätzen, *37*
Straßenversätzen usw. Zwar ließen sich die in den Mustersiedlungen empfohlenen
dörflichen Dekorationen wie Brunnen, Linde, Sitzbank usw. nicht unmittelbar nach
Fedderwardergroden übernehmen, aber den „Volksgenossen" sollte „Gemütlichkeit" *111*
geboten werden.

Das kleine Häuschen als Lockmittel für die Arbeiter der *KMW* konnte sich in
Fedderwardergroden kaum auswirken, denn nur wenige Haustypen waren freistehende
Einfamilienhäuser am Rande der *Neuen Stadt.* Sie waren dem Führungspersonal
vorbehalten[265] und hatten eine Wohnfläche von bis zu 70 qm. Fensterläden, Obstspalier, *41*
Regentonnen, Sprossenfenster, Eingangspergola, Dachhäuschen, Giebelschlußsteine und
überhohe steile Satteldächer spiegelten, wie im Kleinsiedlungsbau, altertümliche Idylle
vor. Die knappe Grundstücksgröße war jedoch nicht für Nebenerwerb geeignet.

Fast alle Zeitzeugen bestätigen heute, daß Fedderwardergroden eine Großstadtsiedlung
in heimatverbundener Verkleidung werden sollte. Das war auch die Strategie von *RAM* *40*
und *OKM*, denn:

- 16.000 Menschen in Kleinsiedlungshäuschen unterzubringen, hätte ein Vielfaches an Fläche und Kosten benötigt.
- Landwirtschaftlicher Nebenerwerb hätte die neuen Bewohner von ihrer „kriegswichtigen" Arbeit und der gewünschten Parteiarbeit abgehalten.

Die Wurzeln des *Heimatschutzstils* kamen nicht aus Wilhelmshaven. Die Bauernhöfe[266] in Marsch und Geest waren kaum Vorbilder, auch nicht die Gründerzeitbauten der Innenstadt, die als Dokumente von Ausbeutung und Klassenherrschaft galten. Bei der äußeren Gestaltung der Neubauten blieb, wie auch in Wolfsburg, Salzgitter und anderswo nur übrig, die bodenständige Idylle der Mustersiedlungen fortzuentwickeln. Aus Rationalisierungsgründen mußten die erdgeschossigen Häuschen jedoch zu Doppel-, Ketten- und Reihenhäusern addiert werden. Innerhalb des „Grünen Ringes“ von Fedderwardergroden konnten die Gebäudetypen auch zu Gechossen gestapelt werden. Das Äußere der Wohngebäude sollte weder langweilig noch gekünstelt sein[267]. Die Forderung nach differierenden Fassaden führte zu den fast 70 Gebäudetypen.

Ein- wie auch zweigeschossige Wohnhaustypen wurden als Reihenhäuser zerlegt in „Endhaus links“ und „Endhaus rechts“ und „Mittelhaus rechts“ und „Mittelstücke“. Wie üblich, hatte Architekt *Gustav Bork*[268] in Fedderwardergroden einfache Kuben, Satteldächer, Hauseingänge, Blendgiebel, Dachhäuschen und Gauben symmetrisch angeordnet. Höhepunkt der Dekoration war jeweils der Hauseingang. Er diente der Unterscheidung und Auffindung: rechtwinkelig, mit Rundbogen, mit Stichbogen und durch Werksteinumrandungen oder Schlußsteine. Oftmals faßte *Bork* die Haustüren auch mit darüberliegenden Fenstern des Treppenhauses zu Vorsprüngen zusammen und krönte sie mit Ziergiebeln und Dekoration. Holzklappläden und Blumenfenster, sogar Fachwerkvorbauten, wurden zur Abwechselung eingefügt. Nach dem Typenplan für Fedderwardergroden[269] konnte die *Jade* unterschiedlichste Wohnhäuser baukostenartig zusammensetzen.

Die Theorie zur äußeren Gestaltung von Fedderwardergroden kam von den Gegnern der Moderne: *Paul Bonatz, Paul Schultze-Naumburg, Werner Lindner, Erich Böckler, Heinrich Tessenow* und *Paul Schmitthenner.*

⋆

Die Vorbereitungen für Fedderwardergroden wurden erst zögernd öffentlich. Zum
112 propagandistisch breit ausgewalzten *Hitler*-Besuch am 01.04.1939 präsentierte man der
109 *Reichsregierung* im Zimmer des *Oberbürgermeisters* im Wilhelmshavener Rathaus die
108, 38 Stadtmodelle[270] u.a. das Modell Fedderwardergroden. Das Interesse des „verhinderten Architekten“ *Hitler* galt jedoch den Neugestaltungs- und Monumentalplanungen. Obwohl er der *Stadt* die Lösung der Wohnungsfrage und 7.000 neue Wohnungen pro Jahr versprach, konnte er nur hoffen, „daß in den nächsten Jahren es gelingen wird, die Fragen zu lösen, die am brennendsten sind“[271]. *Hitler* hat nie die neuen Wohngebiete oder deren Baustellen besucht[272].

Die Wilhelmshavener Ratsherren jedoch nahmen lebhaft Anteil am Aufbau von Fedderwardergroden und zeigten sich gerne auf den Baustellen. Am 17.06.1939 schrieb die Parteizeitung[273] erstmals breitangelegt über die *Neue Stadt* und daß eine gewaltige
153 *Arbeitsschlacht* geschlagen werde. *Bebauungsplan* und Baustellenfotos wurden abgebildet
30 und die technischen Einzelheiten beschrieben. Beim Bau von Voslapp hatte es große Verärgerung wegen der zu spät begonnenen Gemeinschaftseinrichtungen gegeben. In Fedderwardergroden nun wurde die Infrastruktur ausführlich beschrieben und verspro-

chen, daß „in den nächsten Tagen der Bau der Schule im Südosten“ beginnt. Viele 138
Voraussagen der Nationalsozialisten blieben Utopie. Eine Eisenbahnlinie, die „an den Südwestrand von Fedderwardergroden grenzt“[274], wurde nie gebaut, denn aus Luftschutzgründen mußte der Bahnhof weitab von jeder Siedlung südlich Breddewarden 129
geplant werden. Auch er wurde niemals gebaut. Leeres Versprechen blieben auch eine Straßenbahn und die Ringschnellbahn. Die Besprechung im *RAM* vom 23.03.1939 zeigte, daß der öffentliche Personennahverkehr keinesfalls gelöst war. Der 1941 gebaute Bahnhof Hörn der Werftbahn sollte neben der Straßenbahn und neben Omnibusverbindungen „im wesentlichen den Berufsstoßverkehr aufnehmen“[275].

Fedderwardergroden spielte auch überregional eine propagandistisch große Rolle. Die Gemeinschaftssiedlung wurde in Fachzeitschriften und Aufsätzen von den Planungsbeteiligten als Mustersiedlung dargestellt[276]. Die*Neue Stadt*, so schrieb *Friedrich Nicolaus* vom *RAM* 1941, solle „der Wehrhaftmachung des ganzen großdeutschen Lebensraumes dienen“. Die Werkszeitung der *KMW* überbot sich in Superlativen. Man zählte Steine und Dachziegel, maß Kalk und Sand: 80 Mio. Steine, 11 Mio. Dachziegel, 300.000 cbm Füllsand, 4.000 Bauarbeiter, 50 km Schmutzwasserleitung usw. Die Propaganda stilisierte die Fehlplanung, daß nämlich erst 1 Jahr nach Kriegsbeginn die ersten Wohnungen bewohnt waren, daß erst 1942 das erste Kindertagesheim der *KMW* eingerichtet war, zum „Fortschritt des Wohnungsbaues“[277]. Das Ergebnis Fedderwardergroden blieb hinter den hochgeschraubten Erwartungen zurück: Halbfertige Wohnungen, verpatzte Fertigstellungstermine, mangelnde Versorgung.

Stadtbaudirektor Walter Temp und *Fritz Hilgenstock* informierten Ratsherren und Beigeordnete in Fedderwarden über die Fortschritte[278]. Die Sorge um Gemeinschaftseinrichtungen und das Fiakso von Voslapp hatte ihre Spuren bei den *NSDAP*-Ratsherren hinterlassen. Politische Unruhe unter den Werftarbeitern und schlechte Wohnverhältnisse für die Familien würden die „Moral der Rüstungsarbeiter“ untergraben. Die Werkszeitung der *KMW* drückte gern Reportagen mit gerade erst eingezogenen Wohnungsinhabern und gab das Ergebnis so wieder: „In meinem Leben war ich noch nie so glücklich“[279]. Die Wirklichkeit sah jedoch völlig anders aus. Obdachlose aus der Innenstadt mußten in halbfertigen Fedderwardergrodener Bauten untergebracht werden, und 1942 wurden 900 Notwohnungen hergerichtet. Die *Marineintendantur* hatte Extramittel für Obdachlosenunterbringung bereitstellen müssen[280]. Um die neuen Bewohner von Fedderwardergroden nicht unnötig zu verängstigen und Negativ-Propaganda zu verhindern, ließ die *Stadt* über Nacht die an 6 Neubauten auf Anordnung der *Marine* angebrachten Tarnanstriche entfernen[281]. Das außerordentliche Interesse der Berliner Ministerien an den Wohnverhältnissen für Rüstungsarbeiter wurde durch *Robert Ley*, dokumentiert. Er ließ sich Besuche in Fedderwardergroden nicht nehmen.
Am 09.06.1941 hielt er vor Bauarbeitern eine Rede. Sicher informierte sich *Ley* auch bei 154
seinen vorhergehenden Besuchen am 01.04. und 24.11.1939 über den Fortgang der Planung. Auch der oldenburgische *Ministerpräsident Georg Joel* kam nach Fedderwardergroden und hielt am 14.11.1939 eine Propagandarede im Arbeiterlager.

3.5.5. Marinesiedlung Altengroden[282]

126 Noch im *Neuordnungsplan vom* 20.11.1936 sollte die Bauerschaft Altengroden völlig
durch eine IV. Einfahrt in Rüstersiel und eine überdimensionale Großwerft[283] verdrängt
127 werden. Altengroden war erstmals im *Hallbauer*-Plan *Die Stadt der Dreihunderttausend vom*
15.03.1937 kreissymbolisch dargestellt. *Hitler* hatte bei seinen Wilhelmshaven-Besuchen
1936/37 dem Marineausbau höchste Priorität eingeräumt[284]. *Oberbürgermeister Wilhelm*
Müller setzte Mitte 1937 rigoros den Baubeginn für Altengroden auf den 01.03.1938 fest
[285]. Sein *Stadterweiterungsamt* hatte sich gut vorbereitet. Die „geheimen" Pläne *Wirt-*
128 *schaftsgebiet Wilhelmshaven vom Juni* 1937 und *Wirtschaftsgebiet Wilhelmshaven und*
Umgegend vom September 1937 zeigten ein „neues Wohngebiet in geschlossener
Bauweise". Gemeint waren Wohnflächen in Geschoßhäusern, keine Kleinsiedlungsflä-
chen. Dieses Wohngebiet für 7.000 bis 9.000 Wohnungen sollte sich südlich der
„Maadelinie" bis zum Rüstringer Stadtpark, westlich bis annähernd Fort Schaar und
östlich bis zur Fortifikationsstraße erstrecken – in der Mitte durchschnitten von der 1873
gelegten Fortifikationsstraße mit Feldbahn[286].

Mitte 1937 ergaben sich für das Projekt Altengroden beträchtliche Grunderwerbspro-
bleme, denn das von *Hallbauer* vorgesehene Baugelände war nur teilweise in öffentlicher
Hand. Umfangreicher Besitz gehörte zu Erbhöfen oder zur Stadt. 10 größere Hofstellen
und etliche Arbeiterhäuser lagen im Gebiet der geplanten Marinesiedlung Altengroden.
Die Erbhöfe sollten zu den außerhalb der Stadt gelegenen staatlichen Domänen
Hohewerther Grashaus und Schilldeich umgesetzt werden. Es gab dort angeblich
genügend Austauschland – auch „Land von Juden oder von Kirchen." Es kam zu
62 Konflikten, da gleichzeitig das *Amt Friesland* an der Siedlung Burg Kniphausen und an
63 Schilldeich plante. Dieser Konflikt wurde durch die Entscheidung der *LOB* zugunsten
von Altengroden entschieden. Die Siedlungen Schilldeich und Burg Kniphausen wurden
aufgegeben, wodurch wiederum Fedderwardergroden höchste Priorität bekam.

Es entwickelte sich ein starker Druck auf die Altengrodener Planung, da die *Marine* erkannt hatte, daß das Kleinsiedlungsprogramm mit seinen komplizierten Bestimmungen und der langatmige Weg über das *Amt Friesland*, das *Reichsheimstättenamt Gau Weser-Ems* und den *Deutschen Siedlerbund e.V. Gaugruppe Weser-Ems* eine Verzögerung für die Ansiedlung von Rüstungsarbeitern gebracht hätte. Man wollte im militärischen Interesse städtische Bebauung und möglichst schnell die Baulücken bis zur „Maadelinie" auffüllen, verkannte aber die Grunderwerbsprobleme[287].

Anfang 1938 wurde klar, daß Altengroden sich verzögern würde, verkleinert werden
30, 34 mußte und im übrigen Voslapp und Cäciliengroden wegen der leichteren Verfügbarkeit
des Baugrundes eher begonnen werden konnten[288]. Weitere Schwierigkeiten ergaben
sich aus den Landschaftsschutzaktivitäten von *Georg Heinrich Harms* und *Heinrich*
Oldewage[289]. Der Deichzug Altengrodener Weg und die Hofwurten mußten in die
Entwürfe einbezogen werden.

Die Planung für Altengroden hatte bis Mitte 1938 wegen der Umsiedlungsschwierigkeiten geruht, sie wurde erst wieder aufgenommen, als die Eingemeindung vom 01.06.1938 der *Stadt Wilhelmshaven* Planungshoheit auch für die Flächen des Austauschgeländes nördlich der Maade zubilligte. Der *RAM* vermerkte am 23.11.1938[290], daß zur Planung Altengroden mehrere Skizzen, und zwar je eine von der *Stadt Wilhelmshaven*[291], von *Karl Franzius* und von *Friedrich Heuer* vorliegen. Die Skizzen wurden den Beteiligten, dem *OKM* und dem *RAM* zugeleitet. *Karl Pfeiffer* vom *RAM* vertröstete *Oberbürgermeister Müller* am 16.12.1938, daß für Altengroden noch umfangreiche Abänderungen erforderlich würden und der ganze Plan skizzenhaft neu zu bearbeiten sei. Zusammen mit zwei *Assessoren* des *OKM* solle dies in Gemeinschaftsarbeit geschehen.

Ewald Zenker[292] von der *Bauabteilung des OKM* hat den *Bebauungsplan Altengroden* im Oktober 1938 gezeichnet, möglicherweise unter Einbeziehung der Skizzen von
Hallbauer, Franzius und *Heuer.* Wie bei Fedderwardergroden kam es zu verschiedenen 37
gemeinschaftlichen Erörterungen in Berlin. In jedem Fall hatte jedoch die *Bauabteilung des OKM* als Ministerialinstanz die Federführung. Der *RAM* war nicht mehr zuständig, da Altengroden keine Arbeitersiedlung werden sollte, sondern eine Marinesiedlung für gehobene Ansprüche für *Offiziere, Unteroffiziere* und Werftbeamte.

Zenker, Schüler von *Heinrich Tessenow* und *Hermann Jansen* arbeitete 1938–42 in der *Hochbauabteilung des OKM* unter *Walter Kelm.* Hier arbeiteten auch *Bellwinkel* und *Geisenhainer.* Im *Marineverwaltungsamt des OKM* ressortierten nur die grundsätzlichen Hochbauangelegenheiten der *Marine.* Die konkreten Planungen der örtlichen Baudienststellen mußten mit Ausnahme der *HNBD* über die Mittelinstanzen, die *Marineintendanturen,* vorgelegt werden. Die Stellungnahmen der verschiedenen zu beteiligenden Instanzen wurden nicht als Schriftsätze, sondern als Skizzen abgeliefert – nicht selten mußte die oberste, die Ministerialinstanz, die Skizzen zu einer „Gemeinschaftsskizze" zusammenfassen[293]. *Ewald Zenker* erinnerte sich[294], daß weder die städtischen *Bauämter* noch die Mittelinstanzen, noch die Ministerialinstanzen auf derart große Projekte vorbereitet waren. „Man mußte improvisieren, denn die militärische Führung hatte ehrgeizige Aufrüstungspläne" und „die Spitze der *Bauverwaltung* mußte sozusagen den Kopf hinhalten". *Marinestandortbauämter, Hafenbauressorts* und *HNBD* hatten als Ortsbaudienststellen genügend mit Kasernenneubau, Werften und Schleusen zu tun. Da auch die *Stadterweiterungsämter* personell schwach besetzt waren, plante man oftmals gleich in Berlin bei den Ministerien.

Die Marinesiedlung Altengroden war für den Admiralssohn *Zenker* erste Planungs- 1
aufgabe, als er im Oktober 1938 zur *Bauabteilung des OKM* kam. Er entwickelte den
Bebauungsplan Altengroden vom 14.02.1939 auf der Grundlage von *Hallbauers Wirtschafts-* 43
plänen und den *Landschaftsschutzkarten*, den Entwässerungsradien und Schutzbereichen. 128
Zenker orientierte sich stark an *Karl Pfeiffer* vom *RAM*, der „*Schmitthenner* sehr nahestand". Man tauschte sich mit dem *RAM* aus, ging gemeinsam auf Dienstreisen in die Küstenstädte und führte sich gegenseitig Beispiele vor. Es kann kein Zweifel bestehen,

daß sich die Entwurfsleitlinien für die Marinestädte in Berlin im Spannungsfeld zwischen
4 *OKM* und *RAM* entwickelten. *Ewald Zenker*, in Wilhelmshaven, aber auch in Berlin
aufgewachsen, weitgereist und gebildet, kannte natürlich die Bauten der Weimarer Republik, *Internationalen Stil* und *Bauhaus*. Die Entwürfe von *Bruno Taut, Hugo Häring, Martin Wagner* und *Ernst May* waren ihm vertraut, jedoch „nicht nach meinem Geschmack“[295].

Wesentliche Planungsschritte für Altengroden waren:

- *Skizzen* von *Wilhelm Hallbauer, Friedrich Heuer* und *Karl Franzius*, Mitte 1938
- *Ausschnitt aus der städtischen Landschaftsschutzkarte vom* 07.01.1939 (Unterschrift *Walter Hunzinger*) mit eingetragener Bebauungsplanung Altengroden-Nord vom 01.02.1939 (Unterschrift *Ewald Zenker*)

43 - *Bebauungsplan Altengroden vom* 14.02.1939 (Unterschrift *Walter Kelm, Ludwig Bellwinkel, Ewald Zenker*), basierend auf *Zenkers* Plan vom Oktober 1938
- *Nord-Abschnitt Altengroden vom* 29.02.1940 der *Gemeinnützigen Gesellschaft zur Schaffung von Wohngelegenheiten für Reichsangehörige*, Berlin (Unterschrift *Wilhelm Czermak* und *Friedrich Mellin*), eingetragene Wohnungstypen und Gewerbeplanungen, Stand 01.07.1940
- *Altengroden Bebauungsplan* der *Marineintendantur Wilhelmshaven* vom 04.10.1940 gemäß Plan *OKM* vom Juli 1940 (Unterschrift *Karl Franzius*) mit Wohnungstypen
- *Bebauungsplan Altengroden der Abteilung XXIV. der Marineintendantur Wilhelmshaven* vom 31.03.1942 (Unterschrift *Karl Franzius*)

43 Grundlegend und aussagekräftig war der *Bebauungsplan Altengroden vom* 14.02.1939. *Ewald Zenker* charakterisiert den Plan heute so[296]: „Bei der Planung für Altengroden habe ich angestrebt, die Struktur der Bebauung aus den örtlichen Gegebenheiten heraus zu entwickeln. Weiterhin wollte ich die Bewohner soweit wie irgend möglich vor dem Verkehrslärm auf der Hauptdurchgangsstraße schützen. Ein sinnvoller Wechsel zwischen ruhigen Nebenstraßen und Hauptstraßen mit unterschiedlichen Platzanlagen sollte der Bauanlage Vielseitigkeit und Lebendigkeit geben, was durch eine hierauf abgestimmte Einzelplanung der Häuser erst richtig zur Geltung hätte gebracht werden müssen. Die von mir angestrebte vielfache Durchgrünung des ganzen Wohngebietes war die Grundlage, für die sich später anschließenden Grünplanungen von *Wiepking-Jürgensmann*. Ich konnte meine Bearbeitung des Planes ungestört und ohne Beeinflussung von seiten der oberen Baubeamten zügig zuende führen. Er fand die Zustimmung von *Ministerialdirigent Kelm* und auch der obersten Marineleitung“.

43 *Ewald Zenker* reduzierte in seinem *Bebauungsplan Altengroden vom* 14.02.1939 die
128 Marinesiedlung gegenüber *Hallbauers Wirtschaftsplänen* von 1937 um die Hälfte auf 3.500
Wohnungen und ließ die Siedlung am Totenweg enden. Westlich davon wurde die Nördliche Stadtparkkolonie vom *Stadterweiterungsamt* 1939 weiterbearbeitet. Zwischen
98 Fort Schaar und dem verlängerten Totenweg wurden Baracken für Arbeiterlager und

militärische Zwecke aufgestellt und im Sinne der Schutzbereiche freigehalten. Südlicher Abschluß sollte 1938–39 das aus der Kernstadt auszulagernde städtische 400-Betten-Krankenhaus werden[297]. Für den Personenverkehr war ein östlich liegender Bahnhof der Ringschnellbahn gedacht.

Zenkers Bebauungsplan Altengroden wurde der *Stadt*, dem *MDF*, der *Marineintendantur* 43
und dem *Marinestandortbauamt* am 03.02.1939 erstmalige von *Ludwig Bellwinkel* als abgeschlossene Planung präsentiert[298]. Sogleich nach *Zenkers* Planfertigstellung wurde der Gartenarchitekt *Heinrich Wiepking-Jürgensmann* von Berlin aus beauftragt. Für die Detailplanung stand *Wilhelm Hübotter* zur Verfügung. In den Grünzügen verstreut sollten die 3 Bauernhöfe erhalten bleiben. 2 davon sollten zu *HJ*-Heimen umgenutzt werden[299]. Die den Deichzug von 1450 begleitenden Arbeiterhäuser hielt *Zenker* nicht für erhaltenswert – sie störten das Bild der Marinesiedlung für gehobene Ansprüche. Als *Tessenow*-Schüler war *Ewald Zenker* architekturtheoretisch auf die Ideale Bürgerlichkeit, Einfachheit, Ordnungsliebe, Fleiß, Sauberkeit, Ausdauer und Gemeinschaft ausgerichtet. Musische Vergleiche boten sich an: „keine Nebentöne, sondern einfache volle Töne, etwa wie das Marschlied oder wie der Choral"[300].

Es war für *Zenker* klar, daß der Altengrodener *Bebauungsplan*, wenn er schon 43
zweigeschossige Mietshäuer enthalten mußte, auf gemütliche Hausgärten nicht verzichten sollte. Er mußte es hinnehmen, daß die Reihenhaustypen alsbald aus der Planung gestrichen wurden. Trotzdem sollte die Marinesiedlung Altengroden qualitätvoller sein als die zeitgleich und nur wenig größer geplante *Neue Stadt* Fedderwardergroden. Städtebaulich war der *Zenker*-Plan sorgsam durchgearbeitet. Verkehrsstraßen, Erschließungsstraßen, Wohnstraßen und Wohnwege waren funktional differenziert abgestuft. Kaum eine Straße lief ins Leere, und überlegt wurden Einzelgebäude in die Blickwinkel gerückt. An den Erschließungsstraßen schirmten lange Zeilen die innenliegenden Grünflächen ab. Spiel- und Sportplätze, Badesee[301] und Ausflugsrestaurant, Garagenhöfe und ausgedehnte Fußwege sollten Altengroden zum gehobenen Anspruch einer *Offiziers*- und Beamtensiedlung machen. Altengroden wurde in der Propaganda[302] weniger aufdringlich präsentiert als die Arbeitersiedlungen. Im 2. Kriegsjahr, als Altengroden-Nord schon mitten im Bau stand, novellierte das *OKM* seinen *Bebauungsplan*, und *Karl Franzius* von der *Abteilung XXIV.* brachte den Plan am 04.10.1940 als *Altengroden Bebauungsplan* zu Papier. Die Bebauung wurde verdichtet und vereinfacht, die Typen anders verteilt. Auch für den Teil Altengroden-Süd kam es zu Vereinfachungen. Freistehende Zweifamilienhäuser wurden zu Zeilen kombiniert und die Gewerbeplanung wurde detailliert. *Karl Franzius* hatte sich als Architekt bei der *Marineintendantur* mit den Ladenbauten von Voslapp und Fedderwardergroden befaßt, und so arbeitete er eine perfekte Ladenverteilung in diesen *Altengroden Bebauungsplan* ein. Zentral und dezentral
sollten 15 Gewerbebauten in Altengroden entstehen. Realisiert wurde nur ab 1941 die 149
Ladenstraße 1 nach Plänen von *Erich Kühn*. *Franzius* ergänzte das Programm noch um 44
eine Oberschule in Altengroden-Süd und ungewöhnlicherweise um eine Kirche. Die Kirche wäre erst nach dem Kriege möglich gewesen – sie lag auf einer Werftbahnab-

zweigung, die noch lange gebraucht werden sollte. Die Novellierung von 1940 teilte im Südteil von Altengroden noch Flächen für gehobenes Luftwaffenpersonal und für frei verfügbare Einfamilienhäuser ab.

Franzius hatte als Architekt der Mittelinstanz ein umfassendes Planungsverständnis, und so legte er noch den *Bebauungsplan Altengroden vom* 31.03.1942 als Fortschreibung vor. Diese Planung war eine deutliche Abkehr von den *Tessenow*-Idealen *Zenkers*, dessen *OKM-Hochbauabteilung* sich im 3. Kriegsjahr kaum noch in Wilhelmshaven einmischen konnte. Nach der Überarbeitung von *Franzius* traten überdeutlich die Stadtteilzentren von Altengroden-Süd hervor. Die Plätze wurden pathetischer und für Aufmärsche geeignet gemacht. Das Hauptzentrum[303] weitete sich bedeutsam aus, und vor Kopf wurde das von Ladenzeilen und öffentlichen Gebäuden gesäumte *NSDAP*-Gemeinschaftshaus gerückt. Auch an Kultur wurde gedacht: Altengroden sollte ein Kino erhalten.

Zenker sah die Überarbeitung[304] kritisch. Er war der Auffassung, daß die Vielseitigkeit
43 und Lebendigkeit seines *Bebauungsplanes* durch abgestimmte Einzelplanungen nicht zum
Ausdruck kam.

★

1 Stärker als die anderen Gemeinschaftssiedlungen des *Wohnsiedlungsgebietes Wilhelms-*
129 *haven-Jeverland* nach 1935 war Altengroden mit der *Marine* verbunden. Im Gegensatz zu
34, 37 Cäciliengroden, Fedderwardergroden und Voslapp sollten in Altengroden die Mittel-
30 schicht, Marinebeamte, *Offiziere* und *Unteroffiziere*, wohnen. Die propagandistisch
vorgegebene „Volksgemeinschaft“ des „III. Reiches“ entwickelte sich in der Realität zu
einer Trennung der Stammarbeiter von den Angestellten, der Beamten von den
14 *Offizieren*, des Führungspersonals von den ungelernten Arbeitern usw. Die *Marineinten-*
dantur hatte ab 1935 den Wohnungsbau in Wilhelmshaven stark durch Reichsbaudar-
4 lehen gefördert. Obwohl *Marineintendanturrat Köhler* vom *OKM* das anstehende
Riesenbauprogramm durch Private abwickeln lassen wollte, bediente sich die *Marine* bei Altengroden ihrer eigenen Berliner Wohnungsbaugesellschaft. Schon 1936 hatte der *Oberbefehlshaber der Kriegsmarine, Erich Raeder*, die Mittelinstanzen in Kiel und Wilhelmshaven, angewiesen, „durch Inanspruchnahme von privaten Kapitalien für die Wohnungsbauten der *Marine* namhaft Mittel“ für Rüstungszwecke einzusparen[305]. Notgedrungen mußte man daher von der unmittelbaren „Wohnungsfürsorge“ der *Marine* auf das „Wohnungsbesetzungsrecht“ bei gemeinnützigen[306] Wohnungsbaugesellschaften auszuweichen. Knappe Fertigstellungstermine, Auslastung der Bauwirtschaft und personelle Verflechtungen[307] führten zur Einschaltung der marineeigenen *Gemeinnützigen Gesellschaft zur Schaffung von Wohngelegenheiten für Reichsangehörige* Berlin. Das Gesellschaftskapital des Reichsfiskus war durch *Ministerialrat Walter Kelm* und *Oberintendanturrat Walter Tornow* vom *OKM* und *Admiralstabsrichter Joachim Rudolphi* im Aufsichtsrat mehrheitlich vertreten.

Die Wohnungsbaugesellschaft[308] war 1923 mitten in der Inflation gegründet worden, um in den Marinestädten an Nord- und Ostsee den Wohnungsmangel zu beheben. Bereits 1925–27 hatte die Gesellschaft in Rüstringen[309] 50 Wohnungen für Bedienstete der *Reichsmarine* gebaut, ebenso in Bremerhaven, Cuxhaven, Kiel, Swinemünde, Berlin usw. Die Berliner Gesellschaft setzte gerne die Berliner Architekten *Mellin, Kühn, Renner, Kreft* und *Rahn* ein. Die *Hochbauabteilung des OKM* hatte großen Einfluß auf die *Gemeinnützige Gesellschaft zur Schaffung von Wohngelegenheiten für Reichsangehörige* – nur so ist es möglich gewesen, daß die gewährten Reichsbaudarlehen für Altengroden ungesichert blieben, denn auch das Baugelände war vom *Reich* erworben worden.

Der groß angelegte *Hitler*-Besuch am 01.04.1939 in Wilhelmshaven hatte starken 112
Zeitdruck erzeugt. *Gauleiter Carl Röver* gab den Druck weiter an *Oberbürgermeister Wilhelm Müller.* Am 04.05.1939 trat der neue *Stadtbaudirektor Walter Temp* seinen Dienst bei der *Stadt* an und versprach „ein Tempo anzuschlagen, das man früher nicht kannte"[310]. Hastig ließ er sofort nach seinem Dienstantritt die Erschließungsarbeiten beginnen. Das Nebeneinander von Hochbau und Straßenbauarbeiten führte zu starken Unzuträglichkeiten.

⋆

In Berlin zeichneten zur gleichen Zeit *Friedrich Mellin* und *Erich Kühn* von der *Gemeinnützigen Gesellschaft zur Schaffung von Wohngelegenheiten für Reichsangehörige* an den Hausentwürfen. Das Wilhelmshavener *Marinestandortbauamt* bildete anfangs die 44, 45
Postadresse der Gesellschaft vor Ort. *Otto Lückemeyer* unterschrieb für die *G.m.b.H.* Schon bald nach dem überhasteten „1. Spatenstich" vom 31.05.1939 dirigierte *Bauamtmann Alfred Eckermann* vom *Marinestandortbauamt* aus einer Bauleitungsbaracke an der Umfangstraße das Bauvorhaben. Das *Marinestandortbauamt* hatte in erstaunlicher personeller Verflechtung die örtliche Bauleitung für die Berliner Wohnungsbaugesellschaft. Im Juni 1939[311] reichte die *Gemeinnützige Gesellschaft zur Schaffung von Wohngelegenheiten für Reichsangehörige* die 13 Haustypen für Altengroden-Nord bei der städtischen *Baupolizeiabteilung* ein. Dem *MDF* wurde das Bauvorhaben nur mitgeteilt. Auch bei diesem „kriegswichtigen" Wohnungsbauprojekt wurde die *Bauverwaltung der Stadt* zensierend[312] tätig, strich „nach hiesiger Auffassung zu große Öffnungen" und kritisierte „untragbare Gartenansichten".

Der Bauabschnitt Ia, Altengroden-Nord, sah insgesamt 841 Wohnungen vor, davon 9 Typen mit 409 geräumigen Wohnungen für Beamte und *Offiziere* sowie 432 engere Wohnungen für *Unteroffiziere.* Die soziale Aufteilung sollte sich in den Wohnungstypen und in der städtebaulichen Anordnung spiegeln. *Unteroffiziers*-Wohnungstypen E 2, F, G und H waren an den großen durchlaufenden Straßen gelegen. Später kam noch Typ S hinzu. Beamte und *Offiziere* sollten in Wohnstraßen, Sackgassen und an den Wohnwegen angesiedelt werden. Die besonders geräumigen Reihenhaustypen J, L und M für *Offiziere* entfielen 1940. Sie sollten z.B. an der Nordseite der Klinkeburg als zweigeschossige Zeilenbebauung mit großzügigen 120 qm Wohnfläche und 4 1/2 Zimmern gebaut werden: Herrenzimmer, Wohnzimmer, Kochküche, Terrasse, Gäste-WC. Oben: 2 Kinderzimmer, Elternschlafzimmer und Bad. Im Keller Luftschutzraum und Zentralheizung. Der Standard mit Parkett, Warmwasseranlage, Elektroherd sowie einem geräumigen Hausgarten sollte sich von den *Unteroffiziers*-Wohnungstypen deutlich abheben. Ähnlich auch der Typ O[313], der als Zweifamilienhaus mit 3 1/2 Zimmern und 90 qm Wohnfläche gedacht war.

Typ G für Unteroffiziere wurde als freistehender Zweispänner mit einem besonders gestalteten Giebel in die Straßenachse gerückt[314], aber auch zu langen Zeilen kombiniert. Dieser Typ war mit 2 1/2 Zimmern und knapp 60 qm Wohnfläche erheblich schlichter ausgestattet: Wohnraum 11 qm, Kinderzimmer 7 qm, Kochnische 3,9 qm. Der Standard mit Einzelöfen und Kohleherd entsprach dem, was man *Unteroffizieren* zubilligte.

Der am häufigsten vorhandene, 1940 „nachgeschobene" Typ S war von der städtischen *Bauverwaltung* wegen seiner breiten Fenster kritisiert worden. *Friedrich Mellin* und *Erich Kühn* reichten eine Änderungszeichnung nach, bei der die Fenster durch Erker und Balkon kaschiert wurden – eine Bauweise, die auch in der Siedlung Berlin-Schlachtensee von *Erich Kühn* zu finden war[315].

44 Das Äußere der ab Herbst 1939 gebauten und ab April 1941 bezogenen Altengrodener Wohnhäuser ist grundsätzlich Ziegelsichtmauerwerk mit steilen Satteldächern. Der nicht
45 realisierte Reihenhaustyp J kam den kleinstädtischen Idealen von *Heinrich Tessenow* und *Ewald Zenker* noch am nächsten. Altengrodens Wohnungstypen wurden grundsätzlich einzeln entworfen. Es wurden keine *Kriegseinheitstypen* verwendet, wie sie *Ernst Neufert* propagierte[316]. Auch *Reichstypen für den Wohnungsbau*[317] nach *Johannes Jacob* oder *Landschaftsraumtypen* des *Reichsheimstättenamtes*[318] blieben Altengroden erspart. Besonders charakteristisch sind für Altengroden die 45 Grad-Giebel, die glatten Flächen und die kleinen Maßstäbe, wie sie von *Tessenow*-Schülern oft angewandt wurden. Gerne wurden auch die Giebel mit stilisierten Schlußsteinen, mit Ortgangziegelornamenten und Erkern versehen. Die Vorbilder für Altengroden waren in Berlin zu finden, bei den Siedlungen Fischtalgrund[319] Lichterfelde und Schlachtensee.

*

39 Im Gegensatz zu dem gut vorbereiteten Bauablauf von Fedderwardergroden kam die Marinesiedlung Altengroden nur mühsam voran. Obwohl zeitgleich begonnen, konnten in Fedderwardergroden die ersten Bewohner 1940 einziehen, die Altengrodener Bewohner erst ab April 1941[320]. Es hätte nicht passieren dürfen, daß Tiefbaumaßnahmen und Hochbauten in den gleichen Straßen begonnen wurden. Noch 1941[321] wurde darüber gestritten, wer die Gas- und Stromversorgung durchführen sollte. *Alfred Eckermann*, im Gegensatz zum routinierten Büro von *Fritz Hilgenstock* keinesfalls in der Abwicklung von Großsiedlungen erfahren, mußte mit vielen Unzuträglichkeiten kämpfen – die Hauptverwaltung der Wohnungsbaugesellschaft war weit weg, in Berlin.

Aus Berlin kam auch der Auftragnehmer für Altengroden, die *Arbeitseinsatzgesellschaft im Reichsstand des Deutschen Handwerks*. Zweigstelle war die *Baugesellschaft des Niedersächsischen Handwerks m.b.H.*

Otto Lückemeyer hatte „alles brauchbare Ziegelmaterial zwischen Niederrhein und Unterelbe für Altengroden angekauft[322]. Baumaterialien wurden auf einem eigens angelegten Güterbahnhof angeliefert. Dieser Baustellenbahnhof sollte nach den Planungen nach Kriegsende zugunsten der Siedlung Altengroden-Süd wieder abgebaut werden. Bauteile[323], Badewannen, Fliesen usw. wurden nach dem System des Kreuzerbaus in den fertiggestellten Kellern verschlossen, bis sie gebraucht wurden[324]. Zu Kriegsbeginn waren
700 Bauarbeiter in Altengroden tätig sie wohnten in Baracken westlich des Baugebietes. *98*
Durch Bomben obdachlos gewordene Wilhelmshavener Familien mußten in halbfertige
Wohnungen[325] und in Arbeiterbaracken verlegt werden. Anstatt der von *Hans Fricke* *45*
1940 geplanten Schule kam es im Kriege noch zu einer Schulbaracke an der Klinkeburg. *145*
Viele Schüler mußten in die Schule Rüstersiel. Dort befand sich auch der einzige größere Bunker. Nur 564 Wohnungen von geplanten 3.500 Wohnungen wurden bis 1945 in Altengroden fertiggestellt. Altengroden-Nord war demnach gerade bis zur Hälfte bezugsfertig oder notdürftig bewohnbar gemacht. Für Altengroden-Süd waren 1942 bereits einige Entwässerungsleitungen verlegt, ein Pumpwerk errichtet und der Straßenbau vorbereitet. Dann wurde auch dieser Neubau stillgelegt. In Altengroden-Nord wurde weiter gemauert, unterbrochen von Bomben und Fliegerangriffen. Die aufgelockerte luftschutzgerechte Planung hatte Altengroden größere Zerstörungen erspart. Auch nach Kriegsende wurde noch weiter gebaut. Der erste, von den Allierten eingesetzte Rat stellte Mittel zur Verfügung, um 44 im Rohbau „steckengebliebene" Wohnungen zu vollenden.

Helmuth Baur plante 1945 eine Behelfsheimsiedlung in Altengroden-Süd, die Utopie blieb. Derweil eigneten sich die Neubewohner von Altengroden-Nord unter primitivsten Bedingungen ihre Siedlung an. Die groß angelegte Grünplanung von *Heinrich Wiepking-Jürgensmann* war zu Ackerfläche und zur Fläche für Selbstversorgung und Kleintierzucht geworden. *Notgemeinschaft* und *Bürgerverein* halfen,wo es nötig war, aus der halbfertigen Marinesiedlung eine bewohnbare Nachkriegssiedlung zu machen. Die Gebäudeentwürfe von *Friedrich Mellin* und *Erich Kühn* von 1939 wurden noch bis weit in die 50er Jahre unverändert für Wiederaufbauten genutzt.

3.5.6. Familienlager Neuengroden

Das Neuengrodener Bauernland wurde von häufig wechselnden Landstellen bewirtschaftet[326]. 1865–68 entwickelte der preußische Fiskus plötzlich Interesse an den
ländlichen Liegenschaften, denn nach dem *Stoltz-Plan* sollte das Fort Rüstersiel in *8*
Neuengroden auf dem Gelände der späteren Siedlung von 1939 entstehen. Das Fort Rüstersiel rückte jedoch weiter nach Norden und wurde erst 1876–80 gebaut.

Fortifikationsstraße und Munitionsbahn verbanden das neue Fort Rüstersiel mit den älteren Heppenser Fortanlagen. An der Ausfallstraße etablierten sich private Bauherren, Händler, Kaufleute und Handwerker, die u.a. von den passierenden Soldaten lebten[327]. Kurz nach der Jahrhundertwende wurde sogar eine respektable zweigeschossige Schule an der Fortifikationsstraße erbaut. *Theodor Goeckes Bebauungsplan von* 1908 und *Willy Hahns Bauklassenplan von* 1917[328] sahen für das Gelände westlich der Fortifikationsstraße in Neuengroden offene Bauweise vor. Mit der Aufrüstungspolitik des „III. Reiches" wurde das Baugebiet wieder „entdeckt". Nach den ersten *Wirtschaftsplänen* von *Wilhelm Hallbauer* sollte zwischen Rüstringer Stadtpark und Fortifikationsstraße noch nicht
126 gesiedelt werden. Der *Neuordnungsplan vom* 20.11.1936 sah im Gebiet Altengroden/Stadtpark eine neue großangelegte Großwerft vor und östlich der Fortifikationsstraße im Gebiete des Heppenser Grodens einen Flughafen. Dazwischen auf der Fläche der ab 1939
1 hastig hochgemauerten Siedlung Neuengroden wäre eine Siedlung aus Schutzbereich-
127 gründen[329] nicht möglich gewesen. Die Planung *Die Stadt der Dreihunderttausend vom* 15.03.1937 sah östlich des Rüstringer Stadtparks keine Wohnbebauung mehr vor. Die Großwerft war mittlerweile zu Gunsten der *Nordwerft* entfallen, dafür war der neue Stadtteil Altengroden vorgesehen. Östlich, im aufzuspülenden Rüstersieler Groden, waren ein Wasser- und ein Landflughafen geplant.

128 *Wilhelm Hallbauers* Plan *Wirtschaftsgebiet Wilhelmshaven vom Juni* 1937 enthielt die neue „Maadelinie" und südlich davon im zusammenhängenden Rüstersieler/Heppenser Groden Wasser- und Landflughafen. Westlich einer neuen Haupt- Ausfallstraße sollte bis zur Heete gebaut werden können. Östlich der neuen Hauptausfallstraße begannen Sicherheitsstreifen und Wehrmachtsgelände. Das alte Neuengroden hätte aus militärischen Gründen abgerissen werden müssen.

Im nächstfolgenden *Hallbauer*-Plan *Wirtschaftsgebiet Wilhelmshaven und Umgegend vom September* 1937 war Neuengroden als Wohngebiet in „geschlossener Bauweise (Wohnflächen)" in den gleichen Umrissen wie im *Wirtschaftsplan Wilhelmshaven vom* 03.08.1939 erkennbar. Planerisch muß es im „III. Reich" um Neuengroden längere Zeit Unklarheiten gegeben haben, denn die *Luftwaffe* hielt die Flughafenplanung „geheim".

Die Zweckbestimmung des anfangs mit 570 Wohnungen geplanten Familienlagers Neuengroden unterschied sich grundlegend von anderen Siedlungsgebieten. Nach Meinung des *Stadterweiterungsamtes* und des *MDF* sollte es ein normaler innerstädtischer Stadtteil werden, ähnlich wie um Krähenbusch und Siebethsburg. Erst als die mit
9 Familien belegte Hafenkaserne[330] an der Ostfriesenstraße zum 01.07.1939 geräumt
28 werden sollte[331], und Teile des Inselviertels abgerissen werden sollten, um den Docks der
101 *Nordwerft* Platz zu machen, wurden Ersatzwohnungen in „Familienbaracken" erforderlich. Die *HNBD* war als Bauherr der *Nordwerft* für dieses Umsiedlungsprogramm zuständig. Sie bediente sich keiner der sonst für die Stadterweiterung tätigen Wohnungsbaugesellschaften, sondern beauftragte von sich aus und direkt die etablierte Wilhelmshavener Baufirma *Möller* als Generalunternehmer. *Möller* hatte außerordentlich große

und zahlreiche Wehrmachtsaufträge abzuwickeln. Über die Konditionen dieses riesigen
Planungs- und Bauauftrages ist nichts bekannt. Wegen der Eile fand die Vergabe
möglicherweise freihändig ohne Wettbewerb oder Ausschreibung statt. Die direkte
Beauftragung durch die dem *OKM* unmittelbar unterstehende *HNBD* unter *Carl
Tiburtius* führte dazu, daß weder *RAM* noch *MDF* noch die *Stadtverwaltung* Einfluß
hatten. Der bewußt unterlaufene Wettbewerb zwischen Planern, und der zurückge-
drängte Einfluß der planenden Behörden in Wilhelmshaven, Oldenburg und Berlin
wirkte sich auf Städtebau und Architektur negativ aus. Begründet wurde dies von
Zeitzeugen mit dem angeblichen Behelfscharakter der Siedlung. In den Bauunterlagen
finden sich nur wenige Belege darüber, daß die Siedlung ein Familienlager, gar eine
Behelfssiedlung werden sollte. Das *OKM* habe mündlich verlauten lassen, daß „das ganze
Lager ... nur vorübergehend bestehen soll, d.h., in einigen Jahren wieder beseitigt
wird“[332]. Möglicherweise war die eng benachbarte Flugplatzplanung der *Luftwaffe* so *128*
„geheim“, daß selbst die Baudienststellen in Wilhelmshaven und Berlin nicht vollständig
informiert wurden. Tatsächlich wären mehrgeschossige Mietwohngebäude mit hohen
Dächern in der Nähe der Einflugschneisen kaum genehmigungsfähig gewesen. Auch aus
Luftschutzgründen war ein Sicherheitsabstand geboten.

Der einfache Gebäudestandard der dann gebauten Wohnhäuser wies zwar auf ein
mögliches Provisorium hin, nicht jedoch der *Bebauungsplan*. Gegen den Behelfscharakter
sprach auch, daß im *Wirtschaftsplan Wilhelmshaven-Friesland vom Mai* 1940 das *129*
Siedlungsgebiet Neuengroden in der geplanten Form festgesetzt wurde. Demontable
Behelfsbauten wurden nicht in *Wirtschaftspläne* aufgenommen. Auch die Erweiterungs-
pläne von 1941 nach Norden hin bis zur verlängerten Prinz-Heinrich-Straße sprachen
gegen die Vorläufigkeit der Siedlung. Auch enthielt die letztgültige Planung vom *46*
03.03.1939 neben 570 Wohnungen eine komplette Infrastruktur mit: 16-klassiger Schule,
HJ-Heim, *KDF*-Autohalle, Ledigenheim, Spielplatz, *NSV*-Kinderheim, *NSDAP*-
Gemeinschaftshaus, Wirtschaft, Aufmarschplatz, Ärztehaus, Ortsteilverwaltung, Pump-
werk, Rektorwohnhäusern, Garagen, Bedürfnisanstalt, Trafos und 2 großen Ladentrak-
ten.

Die ersten *Bebauungspläne*, entworfen von *Theodor Schwarting* aus dem Baubüro des *46*
Generalunternehmers *Möller*, wurden noch mit dem bis Februar 1939 amtierenden
Stadtbaudirektor Walter Hunzinger besprochen. Der Leiter des *Stadterweiterungsamtes,
Wilhelm Hallbauer*, wurde offensichtlich übergangen, denn *Werner Hoffmann* von der
HNBD als Bauherr besprach die Angelegenheit direkt mit *Hunzinger*. Dieser war im
Baurecht wenig geschult und hatte noch kurz vor seiner Entlassung der *HNBD* geraten,
nicht erst lange auf den „juristischen Formelkram der Bauerlaubnis“ zu warten, sondern
sofort „loszubauen“[333]. *Hafenbaudirektor Carl Tiburtius* und der einflußreiche Generalun-
ternehmer *Friedrich Möller* bezogen sich auf diese Zusagen und das planerische
Einverständnis von *Walter Hunzinger*. Sie drohten, sofort nach Berlin zu fahren und sich
beim *OKM* zu beschweren, wenn die Stadt Schwierigkeiten mache. *Ministerialrat* *4*
Ostendorf vom gleichfalls übergangenen *MDF, Wilhelm Hallbauer* und *Hans Bartning* *3*

beharrten auf dem Recht der *Stadt*, über Gestaltung der Straßen und Gemeinschaftsbauten zu entscheiden. *Oberbaurat Hans Gerdes* von der *HNBD* versuchte, die offensichtlich von allen Architekten kritisierte *Möller*-Planung seriös darzustellen. Längst waren vollendete Tatsachen geschaffen worden. Am 01.02.1939 unterschrieben *Hallbauer* und *Bartning* vom *Stadterweiterungsamt* und *Hoffmann* von der *HNBD Schwartings Bebauungsplan*, allerdings mit dem Hinweis, daß die Änderungen des *Stadterweiterungsamtes* vom 24.01.1939 berücksichtigt werden sollten. Die am 31.01.1939 von der *Stadt* verfügte Stillegung der Baustelle wurde, nachdem sich *Friedrich Möller* heftig beschwert hatte, von *Bürgermeister Ulrich Balfanz* wieder aufgehoben. Losgelöst von dem Machtkampf mit *Friedrich Möller* befaßte sich *Wilhelm Hallbauer* vom *Stadterweiterungs-*
52 *amt* mit Neuengroden und legte in dem Plan *Wilhelmshaven städtebauliche Neuplanung vom* 29.03.1939 eine qualitätvolle Planung für Neuengroden bis zur Heete[334] vor.

112 Dieser Plan wurde *Hitler* bei seinem Wilhelmshaven-Besuch am 01.04.1939 präsentiert.

Folgende *Bebauungsplan*-Schritte lassen sich rekonstruieren:

- *Familienlager Neuengroden. Der Lageplan vom* 01.02.1939, Firma *Hermann Möller* (Unterschrift *Theodor Schwarting*), Plan Nr. 3049/100 mit Genehmigungsvermerk der *HNBD, Werner Hoffmann*, vom 01.02.1939 und Änderungshinweis des *Stadterweiterungsamtes* (Unterschrift *Wilhelm Hallbauer* und *Hans Bartning*)
- 46 *Familienlager Neuengroden. Der Lageplan vom* 03.03.1939, Firma *Hermann Möller* (Unterschrift *Theodor Schwarting*), Plan Nr. 3049/100 E
- *Familienlager Neuengroden* gleicher Plan wie vor, mit Haustypenbezeichnungen vom 03.03.1939, Plan Nr. 3049/110 B
- *Katasterplan* der *HNBD* vom 02.11.1940 mit Regenwasserabflußleitungen und eingestrichelten Wohngebäuden des 1. Bauabschnittes
- *Lageplan* des Nordteiles. (Deckblatt zu Plan Nr. 3049/500) vom 29.10.1941, Plan Nr. 3049/556
- *Lageplan* der *Stadt Wilhelmshaven* vom 01.08.1942 (Unterschrift *Egon Forster*). Der Plan zeigte einen Ausschnitt mit 102 Ausweichwohnungen für die *Stadt Wilhelmshaven*

Die genauere Betrachtung der *Bebauungspläne* zeigt bei dem 1. Plan von *Theodor Schwarting* überwiegend straßenbegleitende 30–50 m lange Wohnblocks der Giebeltypen E und F. Nur an der Freiligrathstraße, der Allmersstraße und an einigen Wohnwegen sollten die 15–18 m langen Wohnhaustypen A und B stehen.

46 Erst nach den Eingriffen des *Stadterweiterungsamtes* wurde die Hauffstraße zur Freiligrathstraße durchgelegt. Damit wurde die Siedlung „durchlässiger". Die 18 sehr provisorisch wirkenden Wohnwege wurden zugunsten von umfahrbaren kleinen Wohnblocks aufgegeben. Vermehrt wurden auch Garagen, sogar eine Autohalle von 150 x 30 m für *KDF-Wagen* vorgesehen. Das Ortsteilzentrum um den 100 x 50 m großen Aufmarschplatz[335], den Tsingtau-Platz, blieb mit *NSDAP*-Gemeinschaftshaus,
148 Verwaltung, Ärztehaus, Schule und Ladentrakt nahezu unverändert.

Mit dem *Lageplan* vom 29.10.1941 sollte sich die Siedlung Neuengroden nach Norden ausdehnen. Die *KDF*-Autohalle entfiel zugunsten eines Ledigenheimes. Der städtebauliche Anspruch der geplanten Siedlung Neuengroden war niedrig. *Theodor Schwarting* wollte wirtschaftliche Erschließung und rationellen Baubetrieb für die Firma *Möller*, die die Siedlung mit Planung zu einem Festpreis erstellen sollte. Grünplanung und Gärten, Kunst und Dekoration wurden erst gar nicht bedacht. Auch die sonst häufig zu findenden NS-typischen Bekenntnisse der Planer und die Berichte in der Lokalpresse fehlten. Bezüge zur Gartenstadt oder auch zum *RAM*-propagierten „wehrhaften Städtebau" kamen bei Neuengroden nicht vor. Nur die „bauberatenden" Eingriffe der Wilhelmshavener *Stadtverwaltung* und des *MDF* ließen Spuren der Architekturansprüche des „III. Reiches" erahnen.

★

Die Wohnungstypen in der Siedlung Neuengroden waren äußerst unterschiedlich 47
dimensioniert. So war der Typ G[336] als Zweispänner mit Garderobe, Gäste-WC, Diele, Küche, Kammer, Schlafzimmer, Herrenzimmer, Wohnzimmer, Eßzimmer und Balkon und einer Wohnfläche von über 100 qm üppig dimensioniert. Eßzimmer und Wohnzimmer waren durch Schiebetüren getrennt und unter dem Dach befanden sich noch Fremden- und „Mädchen"-Zimmer. Zentralheizung und Luftschutzkeller gehörten ebenso zur Ausstattung wie Waschküche und Fahrradraum. Das *Baupflegeamt* der *Stadt* unter *Max Neumann* „beriet" den Architekten *Theodor Schwarting*, setzte jedoch Veränderungen der Straßenansicht nur in Ansätzen durch. Vergrößerung von zu kleinen Fenstern, Schaffung eines zusammenhängenden großen Fensters im Schlafzimmer (!), Werksteinumrahmung mit Kreisbogen und Schlußstein[337] sollten die Straßenfassaden verbessern.

Typ E beispielsweise war ein äußerst sparsamer Zweispännergrundriß mit Wohnküche, Kochnische, Bad/WC, Kammer, Schlafzimmer und 50 qm Wohnfläche. Da auch im Dach noch je eine kleine Wohnung untergebracht werden mußte, wurden Schleppgauben auf die steile Dachfläche gesetzt. Dieser sparsame Wohnungstyp mußte mit Einzelöfen beheizt werden.

Im Familienlager Neuengroden sollten sehr schnell unterschiedlichsten militärischen Rängen Ersatzwohnungen zur Verfügung gestellt werden. Die durchgehend zweigeschossigen Gebäudetypen teils mit Sattel-, teils mit Walmdächern, sollten schnell und billig hochgemauert werden. Klinkersteine wären zu teuer geworden und waren knapp.
Qualitätvolle Architektur konnte sich nicht durchsetzen. Generalunternehmer *Friedrich* 47
Möller hätte durchaus Architekten wie *Erich zu Putlitz* oder *Willi Lübbers* einsetzen können, mit denen er in anderen Ortsteilen der Stadt zusammenarbeitete. Er setzte aus wirtschaftlichen Gründen jedoch sein eigenes Baubüro mit *Theodor Schwarting* ein. Die in Wilhelmshaven beliebten Architekturdetails des *Marinestils* und die von der städtischen

Baupflege gewünschten Dekorationen wurden in Neuengroden dem Zeit- und Kostendruck geopfert.

Die *Stadtverwaltung Wilhelmshaven* hat verschiedentlich versucht, auf die Gebäudeplanung von Neuengroden Einfluß zu nehmen. Am 31.01.1939 begann Generalunternehmer *Möller* mit 60 Arbeitern auf der Baustelle. Postwendend untersagte die *Baupolizei Friedrich Möller* die Bauarbeiten, obwohl sie als „von der *Marine* geplante Eigenbauvorhaben"[338] galten. Das städtische *Bauamt* wollte den Baubeginn nur genehmigen, „wenn die Vertreter des *Reiches* vorher ihre Bereitwilligkeit über die Zusatzbauten erklärt
148 haben". Mit dieser Bereitschaft, Schulen und Gemeinschaftsbauten herzustellen und auch einzurichten, hielt sich die *Marine* offensichtlich zurück. Außer einer Ladenzeile bekam die Siedlung Neuengroden dann auch keine Gemeinschaftsbauten mehr. *Bürgermeister Arthur Kellerhoff* schaltete am 10.02.39 *Ministerialrat Ostendorf* vom *MDF* ein, um die Zusicherung durchzusetzen, daß die Gemeinschaftseinrichtungen auch „tatsächlich durchgeführt würden". Es war längere Zeit strittig, ob die *HNBD* als Bauherr überhaupt einer baupolizeilichen Genehmigung für die Siedlung Neuengroden bedurfte. Schließlich reichte *Oberbaurat Hans Gerdes* von der *HNBD* die Pläne über die *Stadt* doch noch beim *MDF* ein[339].

Ostendorf und *Schneider* vom *MDF* waren mit *Möllers* Architekturleistung offensichtlich völlig unzufrieden und wünschten am 25.02.39 die Einschaltung von Privatarchitekten, „damit diese an städtebaulich hervorragenden Punkten des Stadtteils liegenden Sonderbauten auch ein entsprechendes Gesicht erhalten"[340]. Der ausschließlich wirtschaftlich orientierte *Möller* versprach, für eine bessere Gestaltung der Gemeinschaftseinrichtungen Privatarchitekten einzuschalten, wenn das *OKM* die Honorare bezahle. Man hielt *Friedrich Möller* am 02.03.1939 die Gewerbebauten am Markt in Voslapp als „gutes Beispiel" vor und übergab ihm die entsprechenden Ansichtszeichnungen. *Möller*
148 reichte am 04.04.1939 neue Pläne für die nicht mehr realisierten Läden am Aufmarschplatz ein. Er setzte zu Lasten jeglicher Architekturansprüche rigoros seine Vorstellungen von Baurationalisierung durch.

Die Vermutung, Neuengroden solle nach dem „Endsieg" wieder abgerissen und durch qualitätvollere Bebauung ersetzt werden[341], war, wenn überhaupt richtig, nur auf die Qualität der Wohnungsbauten zu beziehen. Architektur und Konstruktion der *Möller*-Wohnhäuser wurden von den Fachleuten einhellig kritisiert. So fand sich in der Propaganda, die sich ausführlich mit Voslapp, Fedderwardergroden, Cäciliengroden
46 usw. befaßte, kaum ein Foto. Die *Bebauungspläne* für Neuengroden enthielten jedoch alles, was ein Stadtteil benötigt. Auch die typischen Elemente des NS-Städtebaus, Nachbarschaften, Aufmarschplatz, Gemeinschaftshaus usw. waren geplant. Die Realität
47 der gebauten Objekte jedoch blieb weit hinter den Ansprüchen zurück. Die Häuser entsprachen nur oberflächlich den Gaumustersiedlungen und Vorstellungen des *RAM*. Nur mühsam konnten einige Walmdächer und Sprossen auf dem Wege der „Bauberatung" durchgesetzt werden. Viele Elemente des *Heimatschutzstils* wie Fensterläden, Schmuckgitter, Werksteinumrahmungen, (Führer-)Balkone fehlten.

*

Die Bauarbeiten für die Siedlung Neuengroden begannen nach Aufhebungen der Bausperre Anfang Februar 1939 an der heutigen Freiligrathstraße und der ersten Nachbarschaft zwischen Allmers-/Eichendorff-/Grothstraße[342]. Von den geplanten 140 Wohnhäusern wurden nur 78 gebaut. Dazu noch der Ladentrakt Hauffstraße, ein Pumpwerk und der Garagentrakt Drostestraße.

Die hastig begonnenen Bauarbeiten für Neuengroden mußten durch die *HNBD*
frühzeitig eingestellt werden. Der Auslöser für das Familienlager Neuengroden, die
Nordwerft, kam nur zögernd voran. Die Abrißarbeiten im Inselviertel wurden verschoben 101, 28
und auch später niemals durchgeführt.

Im November 1940[343] waren in Neuengroden 120 Wohnungen bewohnt, 40 weitere sollten bis Ende 1940 fertig werden. Der Rest von 340 Wohnungen stand im Bau. Noch 1942 wollte die *Bauverwaltung* der *Stadt* 13 Wohnhäuser der Typen E und C zwischen Storm-, Raabe- und Hauffstraße als Ausweichwohnungen für Bombengeschädigte vorsehen[344]. Auch hierzu kam es nicht mehr.

3.5.7. Werftarbeiter-Stadt Sande-Neufeld

In den frühen *Wirtschaftsplänen* von 1937 war Sanderbusch als Siedlungsfläche im 127
Zusammenhang mit der Südstadt Sande für 25.000 Einwohner ausgewiesen. Der Plan
Wirtschaftsgebiet Wilhelmshaven und Umgegend vom September 1937 zeigte das Siedlungs-
gebiet Sanderbusch parzellenscharf zusammen mit Sande-Neufeld als „Neue Wohnge-
biete in weiträumiger Bauweise (Kleinsiedlungsflächen)“. Die Siedlung Sanderbusch 25
zwischen der Bahnlinie Sande-Jever, dem Ems-Jade-Kanal und der projektierten
Reichsstraße 69 war schon während der Weimarer Republik vom *Siedlungsamt* bzw.
Domänenamt Oldenburg begonnen worden. Die Nationalsozialisten setzten das Kleinsied-
lungskonzept mit Unterstützung der *Marine* fort[345]. Der Plan *Wilhelmshaven Wirtschafts-* 1
plan vom 03.08.1939 von *Wilhelm Hallbauer* sah das ausschließlich eingeschossige
Wohngebiet Wilhelmshaven Süd vor, bestehend aus den Ortsteilen Sanderbusch, Sande,
Gödens und der gerade fertiggestellten, jedoch bereits zur Erweiterung vorgesehenen
Siedlung Cäciliengroden. Ein großer Teil des Baugeländes sollte von Juden aus 34
Neustadtgödens kommen[346], andere Teile von dem *Grafen von Wedel* und von
Kirchen.

Östlich der projektierten neuen Reichsstraße 69 sollte sich ein schmales, 5 km langes, 129
Industriegebiet von Idagroden nach Norden bis zum Ems-Jade-Kanal erstrecken,
durchzogen von Stammgleisen und Industriestraßen und optimal für Güterumschlag und
Rüstungsaufgaben erschlossen. Werftbahnhof und Ölumschlagsanlage waren bereits vor
dem Ersten Weltkrieg gebaut. *Gießerei der KMW*, *Norddeutscher Eisenbau* und Bahnhofs- 93
ausbau sollten die Fortsetzung dieses *Industrieumsiedlungsprogrammes* sein. Weitere 51

97 Arbeitsstätten lagen nicht weit entfernt. Die *Westwerft* mit „3.000 Mann“ ging 1937
wieder in Betrieb. Eine große Anzahl von *Artillerie-Depots* und Hallen zwischen
Ems-Jade-Kanal und Jadebusen sowie im Umfeld Wilhelmshavens lagerten Waffen und
Munition. Alle diese Betriebe hingen vom einzigen Arbeitgeber in Wilhelmshaven ab,
der *Marine*.

Landesplaner *Georg Linde* war die Zentralisierung des Wohnungsbaues in Wilhelms-
haven Süd zu voluminös[347], „kleine Gemeindekörper können oft ein besseres Eigenleben
auf nationalsozialistischer Grundlage führen als große“, und er verwies auf „alte
Wohngebiete für Marineangehörige ... seit Jahrzehnten“ in Jever, Heidmühle (Schor-
tens), Varel und der Friesischen Wehde. Diese Wohngebiete sollten nach dem
129 *Wirtschaftsplan Wilhelmshaven-Friesland vom Mai* 1940 noch ausgebaut und erweitert
werden.

157 Kurz vor dem *Hitler*-Auftritt in Wilhelmshaven am 01.04.1939 kam es zu hektischen
Besprechungen, und der *Minister des Innern* und der *MDF* verfügten[348], „daß auch das
1 Gebiet um Sande, an dessen Bebauung bisher nur für eine spätere Zeit gedacht war, schon
jetzt in den Ausbau Wilhelmshavens einbezogen werden muß“. Die *Stadt Wilhelmshaven*,
nicht der *Amtshauptmann* in Jever, sollte die „notwendige Ergänzung des *Wirtschaftsplanes*
... für den Raum Sande“ ausarbeiten und zur Prüfung vorlegen. *Landesplaner Wilhelm
Wortmann* sollte den Hauptplan danach dem *RAM* zur Genehmigung vorlegen. Man
besprach den notwendigen Bedarf von 8.000 Wohnungen für Sande unter strengster
„Geheimhaltung“ zusammen mit der *Marine* beim *MDF* am 22.05.1939 und am
31.05.1939 in Berlin beim *RAM*[349].

Der *RAM* hatte eine Ideenskizze für Sande aufgestellt, die auch „die Einbindung des
94 Werftarbeiterlagers in der Nähe des Bahnhofs Sande“ berücksichtigte. Der *Kreisplaner*
von Jever, *Gattnar*, machte noch Vorschläge[350], die *Fritz Karl Schneider* vom *MDF* und
Wilhelm Wortmann von der *LOB* verwarfen, weil Wilhelmshaven Süd angeblich zu nahe
an das Schloß Gödens gerückt sei. Die *Stadt Wilhelmshaven* bemühte sich, für den Raum
Sande/Gödens Planungshoheit zu bekommen. Das *Reichsheimstättenamt Gau Weser-Ems*
sollte entwerfen und eine *DAF-Bauträgergesellschaft* bauen[351]. Große Grundstücksteile
waren bereits von Juden „erworben“.

Das *Hafenbauressort der KMW* zog das Projekt an sich und projektierte in größter Eile
ab März Sande-Neufeld. Das von *Hallbauer* und *Gattnar* verfochtene, bereits in
Sanderbusch begonnene und fortgesetzte eingeschossige Kleinsiedlungskonzept wurde
bei Neufeld zu einem „massiv ausgeführten Arbeiterlager“[352] uminterpretiert. *Tuchsche-*
48 *rers* erste *Bebauungspläne* paßten sich in der Gebäudeanlage den mutmaßlichen,
rechtwinkligen, abgeziegelten Flächen an. Alle Gebäude sollten sich um den zentral
gelegten Platz mit Feuerlöschteich gruppieren. Der Platz sollte durch eine Ost-und eine
Westachse angebunden werden. Das Westtor sollte zum Sportplatz, zum Krankenrevier
und zu weiteren Wohnzeilen führen. Durch Abweichungen in der Abziegelung kam es
später zu einer keilförmigen Platzanlage.

Sande-Neufeld war Wohnlager für das Industriegebiet Sande und für die *Westwerft* – eine Stadt für sich. Es mußten dort Kriegsgefangene und Zwangsarbeiter aus von *Hitler* unterworfenen Ländern ihren traurigen Alltag zubringen. Sande-Neufeld wurde von der *KMW* bezeichnet als „Arbeiterunterkünftee-Salzengroden" Der *MDF* benutzte den Begriff „Arbeitslager". Der Fußweg zur *Westwerft* war nur 4 km. Weitere Barackenlager befanden sich in der Nähe: Mariensiel, Middelsfähr, Roffhausen.

Die Werkszeitung der *KMW* berichtete im Juli 1940 vom „fröhlichen" Richtfest des 4.
der 6 Blöcke von Sande-Neufeld. Blöcke, das waren die 6 Gemeinschaftshäuser für die *50*
„Gefolgschaft" der *KMW*, umgeben von jeweils 3 bis 4 je 60 m langen zweigeschossigen *49*
Wohnzeilen. Die *Marine* bezifferte die Anzahl der Bewohner auf „3.600 Mann", die im
„Gemeinschaftslager mit 6 Blocks" wohnen sollten. Die Werftarbeiter-Stadt muß jedoch
vor allem gegen Kriegsende erheblich dichter belegt gewesen sein. Insgesamt 24
Wohngebäude hatten, verteilt auf 2 Geschosse, jeweils 40 Zimmer an bis zu 60 m langen,
1,20 m breiten Fluren. Die 14–18 qm großen „Stuben" waren je mit 6–8 ausländischen
Arbeitern belegt. Der Ausbaustandard der *Hilgenstock*-Entwürfe blieb weit hinter den
Volkswohnungen des „III. Reiches" zurück. In den nicht ausgebauten Dachböden waren *31*
jeweils lange Massenwaschgelegenheiten für die 250 Bewohner einer Wohnzeile
installiert. Es fehlte jeglicher Wohn- oder Sanitärkomfort. „Schönheit der Arbeit" kam
nur in der Propaganda vor. Arbeitsschutz oder soziale Regelungen spiegelten sich nicht in
den Grundrissen – mit Ausnahme der 6 Gemeinschaftshäuer. Die Propagierung der *50*
„schönen, luftigen Wohnräume" und der „hygienisch einwandfreien Wasch- und
Badeanlagen" war angesichts der überbelegten Ausländer-Quartiere mehr als menschenverachtend. „Das Gemeinschaftslager Sande wird zweifellos ein Lager werden, wie wir es für unsere Arbeiter haben wollen"[353]. Tatsächlich sollten die ausländischen Arbeiter nur Zwischenbewohner sein, denn die Grundrisse des Büros *Fritz Hilgenstock* waren äußerst durchdacht so entworfen, „daß nur geringe Veränderungen notwendig sind, um diese Räume in schöne große 4-Zimmerwohnungen mit Bad und Nebenräumen zu verwandeln". Der Massenwohnungsbau von Sande-Neufeld sollte billigster Wohnraum für die in den unterworfenen Gebieten „selektierten" Arbeitsfähigen sein. Die Rüstungsprodukte der *KMW* sollten mit ihrer Hilfe billig hergestellt werden. *Hafenbaudirektor Georg Frede* brauchte sich bei seiner Richtfestrede am 11.07.1940 nicht zu verbiegen: Sande-Neufeld war keine Werksiedlung, niemand sollte dort heimisch werden. Er stellte statt dessen heraus, daß „trotz aller Schwierigkeiten" die 9 Mio. Mark teure Werftarbeiterstadt „in kürzester Zeit" von „13 Firmen mit etwa 300 Arbeitern"
hochgemauert wurde. Hauptbauunternehmer war, wie in Cäciliengroden und anderen *34*
Gemeinschaftssiedlungen die Firma *Herdejürgen und Harmsen.*

Der letztgültige *Bebauungsplan* von *Karl Tuchscherer* gruppierte *Hilgenstocks* Gebäude *48*
streng symmetrisch um den 350 m langen, jetzt trapezförmigen Platz. Der *MDF* empörte sich über die nun „keilförmige Hoffläche", und daß von den Gegebenheiten der abgeziegelten Fläche abgewichen wurde „Die Planung der Gebäudeanlage hätte wesentlich verbessert werden können, wenn der Beginn der Bauarbeiten nicht schon

erfolgt wäre"[354]. Vorbilder für Sande-Neufeld sucht man in den städtebaulichen
Leitlinien von *RAM, Reichsheimstättenamt, DAF* und in den Veröffentlichungen von
Feder, Nicolaus, Speer, Gutschow, Heuer und *Lindner* vergeblich. Die Lagerstadt hatte einen
Haupteingang mit Wache, und möglicherweise war sie (mit Post, Kläranlage,
Umspannstation, Bunker, Krankenrevier, Verwalterwohnungen, Heizzentrale und
einigen Läden) sogar umzäunt. Ansonsten fehlte alles, was eine *Neue Stadt* ausmacht:
Keine Gärten, keine Grünanlagen, keine Schulen, keine Verkehrsverbindungen. Der sich
48 öffnende Platz, anfangs als Teich geplant, diente der Befehlsausgabe an die Werftarbeiter,
nicht aber Aufmärschen und Propagandaveranstaltungen. Die hierzu nötigen Attribute,
wie Balkone, Tribünen und Fahnenkulissen, fehlten. Trotzdem entbehrte der *Tuchscherer-Bebauungsplan* nicht monumentaler Gesten. Die Aufweitung des Platzes zu einem
Riegel von 250 m Breite zentriert den Blick auf einen flachen Torbogen, flankiert von 2
Gemeinschaftshäusern.

Die Absicht, nach *Hitlers* „gewonnenem" Krieg in Sande-Neufeld Wohnungen
einzurichten, täuscht. Sozialer Wohnungsbau, wie ihn die Nationalsozialisten verstanden,
wäre Sande-Neufeld kaum geworden, denn das Lager war nicht „für die Gesamtheit der
Schaffenden" vorgesehen. Die „Volksgemeinschaft" war dort nicht präsent und wurde
auch nicht beschworen. Rationell[355] sollte die Siedlung durch das Büro *Hilgenstock*, das
42 sich beim Aufbau von Salzgitter und Fedderwardergroden bewährt hatte, fertiggestellt
143 werden, ohne Rücksicht auf individuelle Baugestaltung *Amtshauptmann Ott* bemängelte
1940, daß ja nach dem Kriege nicht mehr 3.600 Werftarbeiter sondern 300 Familien in
Sande-Neufeld wohnen würden und Schulen, *HJ*-Heim, *NSV*-Kindergarten vergessen
worden seien[356].

Einzige traditionelle Zugeständnisse der *Hilgenstock*-Entwürfe waren die 45 Grad
49 steilen Satteldächer der Wohnhäuser und die breitgelagerten Dachflächen der 6
50 Gemeinschaftshäuser. Symmetrische Eingänge, Werksteinumrahmungen, ausgearbeitete
Schlußsteine und hochstehende Fenster bei den Gemeinschafthäusern ließen NS-typische
Dekoration erahnen.

★

Nach Ende des Krieges und dem Ende der *KMW* waren noch 2.500 ausländische
Arbeiter in Sande-Neufeld[357]. Das „Menschenreservoir" wurde für die Rüstung nicht
mehr benötigt. Der *Verwaltungsbezirk Oldenburg, Abteilung I, Landesplanung*, und der
Landkreis Friesland entwickelten eine *Gewerbeplanung* für Sande-Neufeld[358]. *Fritz Karl*
3 *Schneider*, auch nach 1945 in der Oldenburger Planung, nahm starken Einfluß auf die
Gewerbeansiedlung, und so entstanden in den Gemeinschaftshäusern und Wohnzeilen
u.a. Tuchweberei, Gardinenherstellung, Wirkerei, Chemische Fabrik, Fahrzeugbaubet-
93 rieb, Sperrholzfabrik usw. Verschiedene Gebietskörperschaften, die *Gießerei Sande* und
51 der *Norddeutsche Eisenbau* sowie die *Reichsbahn* belegten die ehemaligen Werftarbeiter-
quartiere als Werkswohnungen. Der neue Eigentümer, das *Bundesvermögensamt*, ließ

Mitte der 50er Jahre die schon 1939 geplanten Umbauten der Massenquartiere zu Wohnungen durchführen. 304 zwei- bis sechsräumige Wohnungen zwischen 30 und 90 qm entstanden, Schornsteine wurden eingezogen, Flure abgemauert, Bäder eingerichtet, Dachböden ausgebaut. Am 17.09.1979 verkaufte der *Bund* die ehemalige Werftarbeiter-Stadt Sande-Neufeld für 4,9 Mio. Mark an die bisherigen Mieter. Eine selbstverwaltete Eigentümergemeinschaft organisierte die Grundstücksgeschäfte. Einzig die 6 Gemeinschaftshäuser zeugen noch von der nationalsozialistischen Vergangen- *50*
heit.

3.6. Altstadt-Wohngebiete ab 1935

Als im Oktober 1933 die Wilhelmshavener *Marinewerft* die *Magistrate* von Wilhelmshaven und Rüstringen aufforderte[359], mit Stadtrandsiedlungen und Kleinwohnungen für verbesserte Wohnverhältnisse zu sorgen, wurde noch in sehr kleinen Dimensionen und Abschnitten und nur innerhalb der alten Stadtgrenzen gedacht. Das mit 27.861 Einwohnern auf 700 ha dicht besiedelte Wilhelmshaven konnte nur *Hermann Zopffs Bebauungsplan* für das Inselviertel von 1918 aus den Schubladen ziehen. Hinzu kamen *27*
noch einige kleine Bebauungsmöglichkeiten um Kieler Straße, Prinz-Heinrich-Straße und Baudissinweg[360].

Rüstringen mit knapp 50.000 Einwohnern auf 2.500 ha Stadtfläche bot erheblich mehr Besiedlungsmöglichkeiten. Sehr schnell aktiviert wurden die Restbauflächen um:

- Kettenstraße/Oldebrüggestraße, u.a. Reihenhäuser von *Alexander Minke* (1936)[361]
- Bülow-/Brommy-/Oststraße der *Wilhelmshavener Spar- und Baugesellschaft*, 170 Wohnungen von *Ernst Peters* (1938)[362]
- Adolf-Hitler-Siedlung (1934–36) *26*
- Siedlung „Kaninchenhausen" der *Jadestädtischen Wohnungsbaugesellschaft* an der Peterstraße 131–169, 100 Wohnungen von *Robert Fuchs* (1933)
- Erweiterung des Villenviertels nach Süden bis zum Mühlenweg, u.a. Widukindstraße *23*
  (1935–38)[363] und Nordseite Mühlenweg von *Fritz Höger* (1935)
- Baugrundstücke für Private um Hamburger-, Lübecker-, Stettiner-, Stralsunder- und *136*
  Danziger Straße im nördlichen Hansaviertel (1933–34)[364].

Die *Bauverwaltung der Stadt Rüstringen* setzte *Max Neumann* und *Heinrich Siemer* ein, neue Baugebiete auszuweisen. Eigenständige nationalsozialistische Architekturmuster, gar monumentale Muster, waren noch nicht vorhanden. Die Stadterweiterung wurde traditionell begonnen durch Schließen von Baulücken, Bebauung vorhandener Straßentrassen, Nutzung festgestellter *Fluchtlinienpläne* und Auffüllen nichtbebauter Flächen. Der Rüstringer *Magistrat* konzentrierte sich bei der Erschließung neuer Wohngebiete dann auf die fiskalischen Flächen nördlich der Bismarckstraße und östlich der Siebethsburg sowie die großen Baulücken zwischen Mitscherlich- und Genossenschaftsstraße.

1935 dachte auch die *Marine* noch in überschaubaren Dimensionen. 254 Wohnungen[365] seien für angeworbene Facharbeiter der *Werft* nötig. Ende 1936 bereits, angesichts von voraussichtlich 18.000 Werftarbeitern, lagen die Zielzahlen bei 9.000 Wohnungen. Zusätzlich wurde noch mit Wohnungsbedarf für verheiratete *Unteroffiziere* und für Arbeiter der *Westwerft* und der IV. Einfahrt gerechnet – ca. 1.500 Wohnungen[366].
Wilhelm Hallbauer, Chefplaner des *Wohnsiedlungsgebietes Wilhelmshaven-Jeverland*, dachte
vor 1936 zweigleisig: Auffüllen der Altstadt und neue Baugebiete in einem Siedlungs-
126 gürtel um die Stadt. Sein *Neuordnungsplan für Wilhelmshaven vom* 20.11.1936 sah
entsprechend die „Stadt" und zwei vor der Stadt liegende Siedlungen vor. Der
127 *Hallbauer*-Plan *Die Stadt der Dreihunderttausend vom* 15.03.1937 wies dann eine Kernstadt
für 140.000 Einwohner und die vor der Stadt liegende Nordstadt, Weststadt und Südstadt
aus. Rüstringen verfügte 1937 über 14.000 Wohnungen. Die städtischen Planer mußten
über eine Verdopplung innerhalb ihrer Stadtgrenzen nachdenken[367]. Da dem Luftschutz
schon durch das *Reichsgesetz über die Aufschliessung von Wohnsiedlungsgebieten vom*
22.09.1933 ein hoher Stellenwert eingeräumt wurde, ging es angesichts der hohen
Luftgefährdung Wilhelmshavens von Anfang an um Trennstreifen, Gliederung,
Durchgrünung, Aussiedlungsflächen, militärische Schutzbezirke und Bausperrgebiete[368].

128 *Hallbauers* Plan *Wirtschaftsgebiet Wilhelmshaven vom Juni* 1937 spiegelte die Ergebnisse
langer Auseinandersetzungen um Luftgefährdung und Bebaubarkeit der Innenstadt.
126 Nach seinem *Neuordnungsplan vom* 20.11.1936 sollten *Marine* und *KMW* vom Süden der
Stadt zum Norden nach Rüstersiel/Altengroden ausgesiedelt werden. Das Wohngebiet
„Stadt" sollte sich bis an die Weserstraße und die dann nicht mehr benötigte Werft
ausdehnen. Die Binnenhäfen sollten ziviler Nutzung zugeführt werden. Der dann
127 folgende Plan *Die Stadt der Dreihunderttausend vom* 15.03.1937, ließ alles beim alten, und
aus Schutzbereichsgründen wurde die „Kernstadt" von einem „Luftschutzgürtel" so
eingeengt, daß südlich der Marktstraße und östlich der Virchowstraße keine Neubauten
mehr möglich gewesen wären. Rüstringens *Oberbürgermeister Gustav Nutzhorn* befürchtete[369] durch die damit verbundene Bausperre des *Schutzbereichsamtes der Festungskommandantur* Schadensersatzansprüche und Verfall der Altstadt und legte die Angelegenheit
Ministerpräsident *Georg Joel* vor. Eine Änderung trat erst mit der Planung *Wilhelmshaven-Wirtschaftsplan vom* 03.08.1939 ein, als der Hauptbahnhof nicht mehr an die
107 Kirchreihe, sondern nach Schaar verlegt werden sollte. Zu diesem Zeitpunkt jedoch
waren die großen Wohngebiete der Altstadt, Hansaviertel, Jadeviertel und Neu-Siebethsburg bereits fertiggestellt oder im Bau. Durch den Wegfall des Bahnhofs
Kirchreihe wurden weitere Wohngebiete nördlich davon möglich, die bereits in die
52 *Städtebauliche Neuplanung vom* 29.03.1939 übernommen waren. Dieser, *Hitler* am
112 01.04.1939 in Wilhelmshaven vorgelegte Plan, schrieb den Altstadt-Wohnungsbau fort
und integrierte die Neugestaltungsziele. Die Problematik Aussiedlung oder Abriß südlich
108 der Peterstraße und östlich des *Marineforums* wurde nur noch in „geheimen" Planungen
angesprochen. Die Altstadt-Wohngebiete blieben von monumentalarchitektonischen

Eingriffen nahezu unberührt. Weder *Marine-* noch *Verwaltungs-, Kultur-* oder *Sportforum* 108
hätten größere Abrisse von Wohnungen erforderlich gemacht. Nur die südliche 109
Bauflucht der Bismarckstraße sollte, um der künftigen Ost-West-Achse Platz zu machen, zurückgesetzt werden. Auch die Wiederaufbauplanungen von *Schulz, Schneider, Heuer, Franzius* und *Gutschow* schonten die Altstadtgebiete.

Große Bedeutung für die Altstadt-Wohngebiete hatten Bauten für die Führung[370]. Im
Hansaviertel hatte sich ab 1933 um Hamburger-, Lübecker-, Stettiner-, Stralsunder- und 136
Danziger Straße Mittelstand angesiedelt. Dies setzte sich im südlichen Villenviertel um 23
Widukind- und Herbartstraße fort. Zu weiteren Einfamilienhaus-Gebieten nördlich und 135
westlich von Neu-Siebethsburg und westlich der Genossenschaftsstraße kam es nicht 52
mehr.

Wilhelm Hallbauer stellte in seinem „geheimen" Erläuterungsbericht[371] vom August 1937 fest, daß das Altstadtgebiet mit den „noch zugestandenen 1.600 Wohnungen für Wehrmachtszwecke ausverkauft" sei. Die restliche Besiedlung des Stadtgebietes durch Werftarbeiter sei von den örtlichen Wehrmachtsstellen „seinerzeit übereinstimmend abgelehnt worden"[372], u.a. auch wegen der Luftgefährdung der „erbtüchtigen Masse". Das *OKM* drängte jedoch, die Altstadt zu besiedeln, „weil die Aufschließung des Siedlungsgürtels mehr Zeit erfordert". Schließlich wurde im August 1937 die Zielzahl für die Altstadt auf 3.500 Wohnungen erhöht[373].

Die Zusammenlegung von Wilhelmshaven und Rüstringen am 01.04.1937 zu Groß-Wilhelmshaven und der Beginn des „gewaltigen Aufbauwerkes" sollten sich nach dem Willen der Nationalsozialisten auf die Altstadt positiv auswirken: „Die großen Stadtviertel, die nördlich und südlich der Bismarckstraße vom *Bauverein Rüstringen* und von der *Wohnungsbaugesellschaft Jade* aufgebaut werden ..., runden den Stadtkern so völlig ab, daß wir bald nicht mehr das frühere unrühmlich erwähnte zerrissene Stadtbild haben, sondern ein geschlossenes Stadtgebilde"[374]. Der „Aufschwung", die „vereinte Kraft" und die „werdende Großstadt" sollten dem „Nationalsozialismus zu verdanken" sein.

3.6.1. Högers Neu-Siebethsburg

In dem Bestreben, Arbeiter an die *Werft* zu binden, wurde die *Marine* von jeher aus Berlin gestützt. 1903 erwarb *Staatssekretär Graf Posadowsky* vom *Reichsamt des Innern* für den 1903 gegründeten *Bauverein Rüstringen* 45 ha Land vom Neuender Bauern *Georg Janßen*. Ein Erbbaurechtsvertrag ermöglichte genossenschaftlichen Wohnungsbau mit finanzieller Hilfe der *Landesversicherungsanstalt*. Sie vermittelte auch gleich die Architekten *Paul Hakenholz*[375] und *Paul Brandes*, Hannover.

Der erste *Bebauungsplan*[376] von 1903 legte um den Burghügel offene Bebauung fest. Sie blieb maßgeblich für den *Goecke-Plan* von 1908 und für *Hahns Bauklassenplan* von 1917 und auch für die Ausweitung während des „III. Reiches". Schon vor dem Ersten Weltkrieg gehörte der liberale, spätere Rüstringer *Oberbürgermeister Emil Lueken*,

Vorstandsmitglied der 1902 gegründeten *Deutschen Gartenstadtgesellschaft*, zu den Förderern des *Bauvereins*. Die Gemeindeverwaltungen übernahmen Bürgschaften und setzten sich für die Selbstbestimmung der jungen Genossenschaft ein. Zusammen mit *Martin Wagner* führte *Lueken* den *Bauverein* aus einer tiefen Krise, in die er durch den Berliner Generalunternehmer *Boswau und Knauer* geraten war[377].
Martin Wagner bewahrte den Bauverein 1913 auch vor dem waghalsigen Versuch, „eine zweite Siedlung", die Kleinwohnungskolonie Rüstringer Brücke, nach den Plänen von *Bruno Möhring* zu beginnen. *Wagner* selber legte 1913 alternative Pläne vor, nach denen 1914 die Baugruben ausgehoben wurden, bis der Erste Weltkrieg ein Ende setzte[378].
1936 empfahl *Oberbürgermeister Nutzhorn*, die Kleinwohnungskolonie Rüstringer Brücke nach Plänen von 1913 zu realisieren – für Arbeiter der *Westwerft*[379]. Das *Schutzbereichsamt* lehnte ab.

Nach *Wagner* förderte *Willy Hahn*, Rüstringens Leiter der *Bauverwaltung* von 1914–21, den *Bauverein*. Erst 1926 belebte sich der Wohnungsbau wieder, und 1928 erhielten *Hakenholz* und *Brandes* den Auftrag, ihren *Bebauungsplan* von 1903 weiterzuentwickeln[380]. Der *Bauverein* hatte 1932 über 100 Mietshäuser mit fast 1000 Wohnungen in seiner Verwaltung. Ausreichende Gemeinschaftseinrichtungen standen zur Verfügung: das Geschäftshaus „Siebethsburger Heim" (1914), die Volksschule an der Edenburgstraße (1904), die Wasserturmschule (1911)[381] sowie die Luft- und Badeanstalt Siebethsburger Straße (1911).

14 Als die *Stadt Rüstringen* auf Forderung der *Marineintendantur* 1933 schnell größere
53 Wohnbauflächen zur Verfügung stellen sollte, hatte sie neben dem Hansaviertel vor allem
57 Neu-Siebethsburg anzubieten. Hier wie dort waren noch Straßenstücke unbebaut, hier wie dort lagerten *Bebauungspläne* in den Schubläden. Die *Rüstringer Bauverwaltung* hatte in der Weimarer Republik weitergedacht, Planungen fortgeschrieben und sich auf kommende Bauaufgaben vorbereitet. 1936 standen 16 Kilometer fertige Straßen zur Bebauung bereit[382].

Da sich *Fritz Höger* schon 1934 mit *Neu-Siebethsburg* befaßte und auf die Vorarbeiten der *Stadt* und des *Bauvereins Rüstringen* zurückgreifen konnte, kamen alle raum- und
126–129 stadtplanerischen Festlegungen in *Wirtschaftsplänen* zu spät sie konnten nur noch nachvollziehen. Der offizielle *Stadtplan von Rüstringen – Wilhelmshaven vom* 01.06.1936, zeigte die ersten Bauabschnitte von Neu-Siebethsburg fertiggestellt und das Straßenraster
128 bis zur Werftstraße komplett gezeichnet. Der Plan *Wirtschaftsgebiet Wilhelmshaven vom Juni* 1937 wies schon mit Eintragungen auf die zu enteignenden Flächen westlich des Erbpachtgeländes hin. Als Neu-Siebethsburg fertiggestellt war, zeigte der *Wirtschaftsplan vom* 03.08.1939 ein zweigeschossiges Altstadtwohngebiet, das bis zur heutigen Straße Am
109 Sportforum und im Westen bis an das *Kulturforum* reichen sollte. Nur ein Teil davon wurde vom *Bauverein Rüstringen* mit *Höger* als Neu-Siebethsburg gebaut,

- weil das Gelände weiter westlich nach dem *Gesetz über die Landbeschaffung für Zwecke der Wehrmacht* mühsam enteignet werden mußte[383]

o weil der *Bauverein Rüstringen* nicht mehr mit größeren Projekten betraut wurde
o weil *Höger* ab 1936 bei der *Gauleitung* unerwünscht war[384].

Gescheitert sind auch *Högers* Planungen von 1935 zur Besiedlung des Nordrandes des Burggeländes der Siebethsburg, die er begonnen hatte, um die Zeit bis zu den Enteignungsbeschlüssen zu überbrücken[385]. Die finanzkräftige *Luftwaffe* hat 1938 das Burggelände erworben und das *Luftwaffenbauamt* mit Beamtenwohnhäusern dort etabliert. Der *Bebauungsplan* Neu-Siebethsburg lag umfassend von der *Stadt* genehmigt 57
erst im Juni 1938 vor[386]. Auch er konnte nur nachvollziehen, was *Höger* mit dem *Bauverein* begonnen und zum Teil vollendet hatte.

Man hatte sich mit *Teilbebauungsplänen* beholfen:

o 1935 das Gelände Mühlenweg/Edo-Wiemken-Straße[387]
o 1935 das Gelände Bismarck-/Störtebeker Straße[388]
o 1937–1938 tom-Brok-Straße[389]
o 1937–1938 „Tausend-Wohnungsprojekt" Schaarreihe, westlich Totenweg (nicht 61
realisiert)[390]
o 1939 54 Wohnungen Arngaststraße (außerhalb Neu-Siebethsburg)

★

Die städtebaulichen Leitbilder *Högers* waren im Ansatz modern, obgleich er in seinen frühen Schaffensperioden in Einzelhäusern und wenig städtebaulich dachte. Sein Weltbild als Architekt war bestimmt von einer starken Aversion gegen das „wilde Bauen" und die „bösen Früchte" des Spekulantentums[391]. Hamburger Kaufmannshäuser, großstädtische Verwaltungsgebäude und repräsentative Gestaltungsaufgaben waren das Arbeitsgebiet des Baugewerkschul-Absolventen *Höger*, seit er sich 1907 in Hamburg selbständig gemacht hatte. Ausnahmsweise entwarf er auch Siedlungen:

o 1920–1921 Kriegerheimstättensiedlung in Sasel bei Hamburg[392]
o 1926 Wettbewerbsentwurf mit großen Wohnblöcken an der Hamburger Jarrestraße[393]
o 1931 Entwürfe für Erwerbslosen-Siedlerhäuser in Hamburg[394]
o 1932 Entwürfe für die *Parksiedlungs-GmbH* Hamburg[395]
o 1932–1933 Entwürfe für Stadtrandsiedlungshäuser in Hamburg[396]

Wohnungsbau in der Weimarer Republik hieß auch für *Höger:* billige Herstellungskosten und Verzicht auf Repräsentation. *Höger* konnte kaum eine seiner Siedlungen realisieren. Mit den Wohnungen von Neu-Siebethsburg hat er erstmalig Siedlungsbau in 57
großem Stil für 1600 Wohnungen praktiziert. Aufgeschlossen allem Neuen gegenüber knüpfte er städtebaulich an die Siedlungen der 20er Jahre. Er kannte die Berliner Siedlungen der Weimarer Republik: Die Weiße Stadt 1929–31 (*Otto Rudolf Salvisberg*), Siemensstadt 1929–31 (*Hans Scharoun* u.a.), Onkel Toms Hütte 1926–32 (*Bruno Taut*

u.a.). *Höger* ließ sich politisch und ideologisch nicht festlegen. Seine Architektur sollte alltäglich sein, kein revolutionäres Manifest, kein *Bauhausstil*, keine sozialistische Utopie, auch keine nationalsozialistische. *Högers* Bekenntnis zusammen mit dem *Bauverein* zielte auf idealistische Schönheit und bürgerliche Wohnkultur, auf zurückreichende Überlieferungen und naturverbundenes Umfeld. Er scheute sich nicht, 1937 zusammen mit dem *Bauverein* gegen „monumentale Achsenblicke" und „dumpfe Monotonie" anzutreten. Stattdessen wollte man „wo man auch geht und steht die sonnige Gartenstadt" mit „überall schönen Blicken", „wechselnden interessanten Bildern" und „liebliche malerische Wirkung"[397].

Die *Hamburger-Schule* wollte unspektakuläre Architektur, vertraute Materialien und Qualität. *Fritz Schumacher,* 1909–33 Leiter der *Hamburger Baubehörde,* suchte Harmonie für das Stadtbild, keine gesellschaftliche Auseinandersetzung[398]. Diese traditionelle und gar nicht internationale Haltung wurde nicht nur für *Höger* prägend, sondern für viele Architekten, die später mit Stadterweiterung und Neugestaltung Wilhelmshavens zu tun hatten, und die aus Hamburg kamen: *Wortmann, Gutschow, Hallbauer, Forster, Boyken, zu Putlitz, Schoch, Tuchscherer, Suhr* usw.

57 In Neu-Siebethsburg arbeitete *Höger* mit dem begonnenen Straßenraster und
59 bescheidenen Variationen. Er veränderte Straßenprofile, ergänzte Knickpunkte, grup-
60 pierte Höfe, entwickelte Nachbarschaften und schloß an Altstadtbaugebiete bruchlos an.
Höger war sein eigenes Vorbild. Er zitierte sich knapp und diszipliniert selbst: lebhafte
60 Backsteinfassaden, hanseatische Zweckmäßigkeit, verbindende Tordurchgänge und
kleinmaßstäbliche Bauerngärten. Mit städtebaulichen Bekenntnissen hielt er sich zurück. Die Eigenarten von Materialien, das Spiel von Licht und Farbe, die Maßordnung von Ziegelsteinen, die Tradition des Backsteinbaus und die Herstellung von Baustoffen interessierten ihn mehr[399]. *Högers* verschworener Freundeskreis interpretierte seine baukünstlerischen Leistungen mit religiöser Inbrunst. Trotz *Högers* städtebaulicher Enthaltsamkeit wurden in Neu-Siebethsburg städtebauliche Eigenarten erkennbar:

57 o Die Anordnung der Gebäude berücksichtigte mehr als üblich Licht- und Sonnenverhältnisse. Nordfronten wurden möglichst vermieden.

o Neu geplante Plätze sollten die Querstraßen beruhigen, optisch verkürzen und geometrisch verändern.

59 o Blickachsen wurden – wo eben möglich – nicht symmetrisch ausgerichtet.

60 o Nachbarschaften an Wigbold- und Edzardstraße sowie vor Kopf der Werftstraße standen für *Högers* Bemühen um kleine Dimensionen und dörfliche Übersichtlichkeit in der Großstadt Wilhelmshaven.

o Akzentuierte Stadteingänge bezogen sich auf englische Gartenstadtvorbilder[400].

Höger, dessen norddeutscher, dekorationsreicher Expressionismus sich von der verwandten *Berliner-Schule* um *Bruno Taut* und *Erich Mendelsohn* unterschied, neigte zwar zu monumentalen Gesten bei Einzelbauten, vermied aber in Neu-Siebethsburg Achsen und symmetrische Ausrichtung. Obwohl *Hakenholz* und *Brandes* die Blockrandbebau-

ung unterbrochen hatten, ging *Höger* noch einen Schritt weiter. Er plazierte Wohnhaus-
zeilen senkrecht zur Straße. Zur besseren Belichtung, Besonnung und Auflockerung des
städtischen Wohngebietes versetzte er die Zeilen zusätzlich. Insofern war Neu- 57
Siebethsburg ein Beispiel modernen Städtebaues und belegte auch die Entwicklung vom
„Block zur Zeile"[401]. Auf die Wirkung von straßenbegleitenden Hauswänden und die
Möglichkeiten von monumentalen Platzumfassungen verzichtete *Höger* im Gegensatz
zur gängigen NS-Architekturauffassung. Auch wenn seine Architektur ab 1936 von
NSDAP-Seite nicht mehr gelitten war, hinderte dies weder die Stadtplaner noch die
Wohnungsbaugesellschaft Jade daran, sich bei den folgenden Siedlungen gerne auf das bei
den Bürgern beliebte Neu-Siebethsburg zu beziehen und Gartenstadtgedanken zu
beschwören.

⋆

Der *Bauverein Rüstringen* war im Juli 1933 von der *NSDAP* gleichgeschaltet worden, weil angeblich der „Aufsichtsrat aber nicht die Gewähr bot", im „nationalsozialistischen Sinne" zu arbeiten"[402]. 1934 wurde *Harald Marquardsen* als Geschäftsführer aus der „freien Wirtschaft" geholt. Er beauftragte 1934 *Höger* mit Entwürfen für die Erweiterung von Neu-Siebethsburg, die durch „wehrpolitische Maßnahmen" erforderlich werde. Nachdem es „unserem Führer gelungen ist, ein starkes einiges Deutschland zu schaffen, sind die Voraussetzungen gegeben, und wollen wir freudig mitarbeiten an dem großen Aufbauwerk unseres Vaterlandes"[403].

1935–36 bebaute der *Bauverein* nach *Högers* Plänen den bisher unbebauten durch
Martin Wagners Planungen sich ergebenden Zwickel am Mühlenweg/Edo-Wiemken-
Straße. Es sollte mit den 122 Wohnungen in 7 Häuserblocks schnell gehen. Das Gelände
gehörte der *Stadt Rüstringen* und dem *Reichsbauamt Wilhelmshaven. Högers* Bauprogramm
von 1934 enthielt bereits Anschlüsse an die nördliche Platzwand eines städtischen
Verwaltungsforums um das Rathaus. Finanziert wurden dieser erste Bauabschnitt und alle
weiteren durch Reichsbaudarlehen des *OKM* Berlin. Noch 1935 entwarf *Höger* weitere
408 Wohnungen für den *Bauverein* im Gebiet Bismarck-/Störtebeker-/Siebethsburger-
/Werftstraße. Die zumeist dreigeschossigen versetzten Zeilen mit den Wohnungstypen A
bis D standen quer zur Bismarck- und Störtebekerstraße. Sie gingen über in eine 57
Randbebauung der Querstraßen[404]. Städtebaulicher Schwerpunkt und Tor zu Neu-
Siebethsburg wurden die großstädtischen Eckgebäude Siebethsburger Straße/Bismarck-
straße, die Haustypen E, F und G. Dieser zweite Bauabschnitt, gleichzeitig die
Durchführungsabschnitt I bis IV, wurde Mitte 1936 begonnen und war schon 1937
fertiggestellt. 90 ortsansässige Handwerksmeister[405] beschäftigten 1.000 „Mann Gefolg-
schaft" und nach dem „Wunsch unseres Führers sogar im Winterbau". In den schon
vorhandenen Gartenhäusern mußten bis zu 6 Bauarbeiter auf 10 qm Fläche leben.

Die weiteren Bauabschnitte wurden am 26.03.1936 auf einer Generalversammlung des *Bauvereins Rüstringen* beschlossen. Das gesamte damals „1.300 Wohnungs-Programm"

61 für Neu-Siebethsburg sollte dann noch um das „Tausend-Wohnungsprojekt" Schaar-
52 reihe für „Volksgenossen der hiesigen freien Berufe"[406] westlich von Neu-Siebethsburg ausgeweitet werden, denn „Selbstverständlich ist der Wunsch unseres Führers, daß jeder Arbeiter sein eigenes Haus haben soll, noch schöner, jedoch mußten wir uns mit den gegebenen Tatsachen abfinden, da auf dem zur Verfügung stehenden Fiskusgelände, welches wir ja nur in Erbbaupacht haben, eine große Anzahl von Wohnungen eingerichtet werden mußte"[407]. Ein Jahr später kündigte *Harald Marquardsen* noch einmal an, daß man dazu übergehen werde, Ein- und Zweifamilienhäuser zu errichten, „die nach einer gewissen Anzahlung jeder Genosse von uns erwerben kann"[408]. Man werde künftig kleine Eigenheime mit 200 bis 400 qm Gartenland bauen. *Höger* hatte auch hierfür erste Pläne vorbereitet, und man hängte sie im Juli 1937 bei der Generalverammlung aus. „Es kommen also leider für diese Häuser nur Genossen infrage, die einigermaßen gut verdienen oder durch äußerste Sparsamkeit dazu in der Lage sind, sich von ihrem Verdienst etwas zurückzulegen. Also derjenige, der gewillt ist, auf andere Freuden zu verzichten, und dessen Frau äußerst sparsam wirtschaften kann, wird dazu in der Lage sein, falls er nicht schon im Besitz eines ersparten oder ererbten Kapitals ist". Ausführlich wurde den Baugenossen von der Behaglichkeit und Wärme berichtet, die Eigenheime im
61 „Innern und Äußeren zeigen und bringen". Obwohl *Höger* 19 Typenentwürfe mit individuellen Grundrissen vorlegte, um „Bauherren einzufangen", sollte der *Bauverein* 1938 noch einen Architektenwettbewerb durchführen[409]. *Höger* war zu diesem Zeitpunkt von der *NSDAP* schon fallengelassen worden. Das Projekt „Schaarreihe" wurde auch vom *RAM* nicht gefördert[410].

565 Wohnungen hatte der *Bauverein* bis 1936 fertiggestellt, und er hatte damit einen großen Erfahrungsvorlauf im Massenwohnungsbau, bevor die Gemeinschaftsiedlungen
53 Hansa- und Jadeviertel überhaupt begonnen wurden.

153 Die vehemente Bautätigkeit wurde als NS-*Arbeitsschlacht* propagiert. Im April 1938 führte der Geschäftsführer *Generaladmiral Erich Raeder* über die Baustelle und stellte klar[411], daß der *Bauverein* „auf Gedeih und Verderb ... mit der *Marine* verbunden" sei, denn 90 % der Mitglieder seien Marineangehörige und das *OKM* wichtiger Geldgeber. Das änderte jedoch nichts daran, daß der 1903 gegründete *Bauverein Rüstringen* mehr und mehr gegenüber der 1937 gegründeten *Wohnungsbaugesellschaft Jade* ins Hintertreffen geriet, nicht zuletzt auch wegen *Fritz Höger*. *Marquardsen* beschwerte sich hierüber 1938 bei *Raeder*. *Oberbürgermeister Müller* gab seinen Widerstand gegen *Höger* und den *Bauverein* 1939 zwar kurzzeitig auf, delegierte jedoch seinen *Stadtkämmerer* in den
63 Aufsichtsrat. Trotzdem bekam der *Bauverein* weder das Projekt Schilldeich noch
43 Altengroden-Süd, noch die Bebauung des Burggeländes, noch das „Tausend-
61 Wohnungsprojekt".

Ende 1938 waren von den nun 1.600 Wohnungen, die in 14 Einzelabschnitte zerlegt
1 worden waren, 1.400 fertiggestellt[412]. 1.100 davon waren durch die Bediensteten der *Marine* belegt, 169 durch die *Marineintendantur* und 300 Wohnungen durch den *Bauverein* vergeben. 12 Mio. Mark hatte der *Bauverein* ausgegeben, davon waren 9,5 Mio. Mark

Reichsbaudarlehen aus den Wohnungsfürsorgemitteln des *OKM*, die von der *Marineintendantur* bewirtschaftet wurden. Erst 1949 wurde klar, daß der Finanzierungsplan 14
„hinfällig" geworden war. Der Geschäftsbericht sprach nun von 17 Mio. Mark Gesamtkosten.

Der *Bauverein* sorgte sich 1938 um weitere Projekte, da das „Tausend-Wohnungspro- 61
jekt" ausblieb, obwohl bereits 33 Mio. Mark Gesamtanleihe genehmigt worden waren. 3.000 cbm Holz und 8.000 Schwellen hatte der *Bauverein* gekauft – er gab sie mit Kriegsausbruch an die *Marine* ab. Noch 1941 hoffte der *Bauverein* auf das Projekt und bot sich den Parteidienststellen an. „Mehr als bisher wird sich die Partei in das Wohnungsbauprogramm einschalten ... denn nur durch eine feste Hand kann im Bauwesen endlich das zur Ausführung kommen, was einer Großmacht, wie unser Vaterland sie ist, würdig ist"[413]. 1939 mußte der *Bauverein* Lücken füllen und begann das Projekt Arngaststraße, das er 1941 fertigstellte. Nach 5 Jahren Mitarbeit an Erweiterung und Neugestaltung der Stadt wurde der *Bauverein* mit seinem Architekten *Höger* nicht mehr benötigt.

★

Högers zahlreiche und vielfältige Wohnungstypen für Neu-Siebethsburg wurden für Wilhelmshavens Stadterweiterung und Neugestaltung nicht beispielgebend. Zwar wurden Ziegel und steile Dächer für viele Gemeinschaftssiedlungen, die Bauten von *Marine* und *Luftwaffe* bedeutsam, jedoch legte die Backsteinornamentik andere Wurzeln frei. Im Vordergrund stand der entschiedene Künstler *Höger*[414], der sich im Spannungsfeld von traditionellem Handwerk, niederdeutscher Herkunft und Worpsweder Kultur sah. Nichts verband ihn mit den akademisch ausgebildeten Architekten, den jungen *Schmitthenner*- und *Tessenow*-Schülern, die in Wilhelmshaven bald tonangebend wurden. Neu-Siebethsburg wurde noch im „III. Reich" Schlußpunkt einer ausklingenden Architekturauffassung, die ohne Nachfolger und ohne Schüler blieb[415]. Danach wurden die Wilhelmshavener Wohnungsbauten vereinfacht und geglättet, rationalisiert und verbilligt. Die Lust am Backsteinornament, das Spiel mit Ziegelformaten wäre ab 1936 für die tonangebenden NS-Baupolitiker undenkbar gewesen. Kreuzsprossen, Schlußsteine, schmiedeeiserne Gitter und Werksteinumrahmungen waren in Neu-Siebethsburg nicht zu finden. *Höger* kümmerte sich um jedes Ornament, detaillierte es, paßte es an, nichts überließ er dem Zufall. Am Äußeren seiner Gebäude wollte er Licht, Schatten und Farbe, keine erdrückenden Wände, keine einschüchternden Dimensionen, keine militärische Strenge. Er modellierte selber Ziegelschmuck und Reliefs und zog Künstler 160
hinzu. 162

Die *Höger*-Verehrung des *Bauvereins* nahm kultische Formen an. Zur Generalversammlung 1937 wurde sein Bild „grün geschmückt"[416]. Die Zeitungen druckten Danktelegramme zu seinem 60. Geburtstag. Die Verehrung hielt an. 1947 setzte ihm der *Bauverein* zu Lebzeiten ein Denkmal am „Höger-Tor" tom-Brok-Straße. 60

★

57 Mit Bauten der Gemeinschaft war Neu-Siebethsburg knapp ausgestattet. Die Läden am Dorfangerplatz Edo-Wiemken-Straße/Edzardstraße wurden nicht mehr eingerichtet, da die Finanzierung fehlschlug[417]. *Höger* entwarf noch 1939 kostenlos für die *NSV* ein Heimgebäude in der Papingastraße. Die Bauleitung lag beim *Bauverein*[418]. In dem im November 1939 gerichteten Ziegelgebäude kamen der *NSV*-Kindergarten, eine Rettungsstelle des *SHD* und Räume für das *NSDAP-Amt für Volkswohlfahrt* unter. Es wurde am 15.10.1944 durch Bomben zerstört, nur der Keller blieb erhalten[419].

Ein *NSDAP*-Gemeinschaftshaus für Siebethsburg war nicht vorgesehen. Im April 1939 verwendete sich die *Marineintendantur* dafür, daß die Ortsgruppe der *NSDAP* eine Mietwohnung vom *Bauverein* für ihre Gemeinschaftsräume bekam[420]. Die Ortsgruppe zog dann in das *NSV*-Heim Papingastraße.

3.6.2. Hansaviertel

1 1936, zwei Jahre nach Neu-Siebethsburg, begannen für Hansa- und Jadeviertel die konkreten planerischen Vorarbeiten im Rüstringer *Stadtbauamt* unter der Leitung von *Max Neumann*, zu einem Zeitpunkt, als das ganze Ausmaß der Stadterweiterung und Neugestaltung noch unklar war, die Jadestädte noch getrennt verwaltet wurden. Im
136 Bereich des Hansaviertels waren Hamburger-, Lübecker-, Stettiner-, Danziger- und Stralsunder Straße 1933–34 mit Ein- und Zweifamilienhäusern für Angestellte und „verdiente Volksgenossen" besiedelt worden. Das städtebauliche Konzept des Hansaviertels (benannt nach den Straßennamen) ging auf ein vorhandenes Straßenraster der Jahrhundertwende zurück. Schon der *Bebauungsplan* von *Theodor Goecke*[421] von 1908 zeigte die vorhandene Hamburger Straße zwischen Bismarckstraße und Bremer Straße sowie Lübecker- und Stettiner Straße. Die privaten Wohnungsbauten 1933–34 waren den Nationalsozialisten höchst willkommen – sie schrieben sie auf ihre eigenen Fahnen. Auch im weiteren Gebiet des Hansaviertels waren Mitscherlichstraße und Werftstraße, Peterstraße und Bremer Straße, Lessingstraße und die Querstraßen vorhanden oder lückenhaft bebaut. *Willy Hahns Bauklassenplan*[422] für Rüstringen von 1917 schrieb *Goeckes Bebauungsplan* fort, und so waren Oldeooge-, Uhland-, Goethe- und Schillerstraße sowie auch Teile der späteren Adolf-Hitler-Siedlung[423] längst vorhanden, als *Max Neumann* 1936 daran ging, Grundbesitzverhältnisse und weitere Bebaubarkeit zu untersuchen. Die Erschließung des Geländes zwischen Bismarck- und Peterstraße stand seit langer Zeit fest und war schon 1910 in gedruckten Stadtplänen gestrichelt
53 eingezeichnet. Der *Bebauungsplan* des *Stadterweiterungsamtes* für das „1000-Wohnungs-Bauprogramm" von 1937 veränderte dieses Konzept nur unwesentlich. Die Paul-Hug-Straße wurde auf das 1928–29 gebaute Rüstringer Rathaus geleitet, und um den

Krähenbusch herum mußten wegen des Naturschutzes Änderungen vorgenommen worden.

Moderne Bestandteile der Planungen von *Theodor Goecke, Martin Wagner, Leberecht Migge* und *Willy Hahn* jedoch wurden für das Hansaviertel 1937 fallengelassen:

- Die „grüne" Parkstraße, die Paul-Hug-Straße, wurde als Adolf-Hitler-Straße nicht nach Westen durchgelegt.
- Die dreigeschossige geschlossene Bebauung für das Hansaviertel wich einer aufgelokkerten Zeilenbauweise.
- Platzgestaltungen als Orientierungspunkte entfielen.

Der Grund und Boden für das Hansaviertel gehörte zu je einem Drittel der *Stadt Rüstringen* sowie den Privatgrundbesitzern *Nathenstedt* und *Fink*. Westlich der Werftstraße gehörte der Grund und Boden weitgehend dem Bauern *Fink* allein.

Städtebaulich ging es *Max Neumann* darum, die vor dem Ersten Weltkrieg in das Hansaviertel gelegten Straßen fortzuführen und eine architektonische Anpassung an die wilhelminischen Bauten von Mitscherlich-, Peter- und Werftstraße herzustellen. Durch Vor- und Rücksprünge sollte der Übergang von geschlossener zu offener Bebauung festgeschrieben werden. Vereinzelte Altbauten galt es einzubinden, und gestalterisch mußte der Übergang von Stuckbauten zu den gewünschten Klinkerbauten bewerkstelligt werden.

Der *Fluchtlinienplan Nr.* 150[424] zwischen Mitscherlich-, Bismarck-, Werft- und Peterstraße war noch vor Amtsantritt des neuen *Stadtbaudirektors Walter Hunzinger* am 21.07 fertiggestellt und am 09.11.1937 förmlich festgestellt worden. Die Interpretation
der Lessingstraße zu einer Ost-West-Achse oder des Otto-Meentz-Platzes zu einer *56*
Nord-Süd-Achse vom Ehrenmal bis zum *HJ*-Heim fand erst unter Einfluß von *Walter Hunzinger* statt.

1936–37 gab es für *Max Neumann* kaum Vorbilder für NS-typische innerstädtische Wohnbebauungen[425]. Parteioffizielle Propaganda befaßte sich noch ausschließlich mit Kleinsiedlungen, Heimstätten, Landansiedlungen und Programmen für bäuerlich romantische Mustersiedlungen. Diese waren einseitig auf Heimatverbundenheit und Nebenerwerb ausgerichtet. Beim zwei- und dreigeschossigen Mietwohnungsbau
konnte man 1936–37 nur auf *Fritz Högers* 1934 begonnene Planungen von Neu- *57*
Siebethsburg und die Weimarer Republik zurückgreifen, in Wilhelmshaven auch auf
den Wohnungsbau des Inselviertels von *Hermann Zopff. Max Neumann* und *Heinrich* *27*
Siemer, die beide noch unter den *Stadtbauräten Martin Wagner* und *Willy Hahn* für eine Reform des Städtebaues in Rüstringen gekämpft hatten, stellten sich auf die neuen Verhältnisse ein. Vergessen waren die modernen Grünkonzepte von *Leberecht Migge*[426], aufgegeben wurden die sozialreformerischen Leitlinien von *Willy Hahn*. Statt dessen
beschränkte sich die *Bauverwaltung* in ihrem *Bebauungsplan Hansaviertel* auf die *53*
Fortschreibung der rechtwinkligen Blockstruktur und reicherte diese noch um die *9*

52 NS-typische Achsenphilosophie an. Die schüchterne Nazifizierung gelang im Hansaviertel nur bedingt.

Möglicherweise wurde die Wende durch *Wilhelm Hallbauer* verursacht, der bereits 1936 als Leiter der *Landesplanungsstelle Wilhelmshaven* großen Einfluß auf Stadterweiterung und Neugestaltung bekam.

Rüstringen ergriff die Chancen, endlich die liegengebliebenen Bebauungen und halbfertigen Erschließungen zu vollenden sowie städtebauliche Lücken zu schließen. Diese Kontinuität hatte wie so oft wirtschaftliche Ursachen. Es kam darauf an, Wohnungsbau schnell zu beginnen, unnötigen Erschließungsaufwand zu vermeiden und
140 die vorhandene Infrastruktur der Altstadt zu nutzen. Nur ein neuer Schulbau (Schule am
147 Krähenbusch 1939, *Egon Forster*)[427] und ein *HJ*-Heim (1938–39, *Heinz Däke*)[428] wurden erforderlich. An der Lessingstraße befand sich noch das Gebäude des 1911 gebauten Lyzeums, der Frl.-Marien-Schule[429]. An der Schillerstraße betrieb die katholische Kirche noch bis 1938 eine katholische Mädchenschule, die danach von der *KMW* zum Heim für Jungarbeiter der Werft um- und angebaut wurde. Weitere Schulen befanden sich in den umgebenden Altstadtgebieten. Ein Feuerwehrgebäude war an der Oldeoogestraße seit 1935 vorhanden. Das geplante *Verwaltungsforum* sollte als nördlichen Platzabschluß einen großen Gewerbetrakt von *Rudolf Bürgin* bekommen, der für die Versorgung des Hansaviertels ausreichend gewesen wäre. Perfekte Gewerbeplanung setzte sich erst 1939 durch.

53 Der Einfluß der *Wohnungsbaugesellschaft Jade* auf den *Bebauungsplan Hansaviertel* war gering. Das Straßenraster lag im wesentlichen fest, als die *Jade* Anfang 1937 gegründet wurde, und *Willi Lübbers*, vom 01.02.1937–10.10.1938 ihr Chefarchitekt, war wenig städtebaulich, dafür umso mehr aber bauwirtschaftlich interessiert. Im übrigen oblag es keinesfalls alleine der *Jade*, das Hansaviertel zu komplettieren. So bauten auch die *Wilhelmshavener Spar- und Baugesellschaft*[430] und private Bauherren weiterhin im Hansaviertel.

*

Im Gegensatz zu fast allen anderen Gemeinschaftssiedlungen[431] für Wilhelmshaven beauftragte die *Jade* für das Hansaviertel verschiedene Architekten mit den Gebäudeent-
56 würfen. Annähernd 25 verschiedene Wohnhaustypen wurden von *Ernst Boyken* und *Rudolf Bürgin* aus Oldenburg sowie *Richard Freygang, Robert Fuchs, Franz Iwersen* und *Otto Klatt*[432] aus Wilhelmshaven entworfen. *Willi Lübbers*, der das Entwurfsbüro der *Jade* aus einer Baubaracke an der Lessingstraße leitete, entwarf nicht nur die meisten
55 Gebäudetypen – er bedachte auch gleich seinen Vater *Gustav Lübbers*, mit dem er eine private Bürogemeinschaft hatte, mit Planungsaufträgen. Die Vielfalt der Entwürfe entsprach *Wilhelm Hallbauers* Forderung, „möglichst viele gestaltende, bewährte Kräfte mit den verschiedensten Handschriften" heranzuziehen[433]. Die freien Architekten *Freygang, Fuchs* und *Iwersen* machten sich bei den NS-Machthabern unbeliebt und

bekamen keine nennenswerten Planungsaufträge mehr. *Boyken* und *Bürgin* jedoch wurden auch bei weiteren Entwürfen in anderen Stadtteilen hinzugezogen.

Im April 1937 begann die *Jade* mit den Bauarbeiten für das Hansaviertel zwischen Oldeooge-, Peter-, Goethe- und Bremer Straße. *Willi Lübbers* hatte für diesen ersten Bauabschnitt selbst den zweigeschossigen Wohnungstyp I entworfen. Er wiederholte ihn später häufig. Der Wohnungstyp I war bescheiden: Wohnküche, Schlafzimmer, Stube, WC mit Dusche kombiniert, Einzelöfen – das alles auf knappstem Raum. Die Keller enthielten von Anfang an Schutzräume, da die Innenstadt besonders gefährdet war. Je 2 Zweispänner waren zu einem Doppelhaus addiert. Die Ziegelbauten mit den abgewalmten Dächern deuteten auf die *Schmitthenner-Schule* hin, die auch beispielgebend für den Wohnungsbau des *Marinestandortbauamtes* und des *Luftwaffenbauamtes* war. Kreuzsprossen, werksteinumumrahmte Eingangstüren und Treppenhausfenster mit Schlußsteinen ließen die heimatschutzbewußte Formsprache erkennen. Die Parteipropaganda[434] stellte die besondere Ausgestaltung der Hauseingänge heraus, lobte die „Weiträumigkeit der Anlage" mit „Luft, Licht und Sonne" und die technische Ausstattung. Dieser erste Bauabschnitt des Hansaviertels wurde bereits im Oktober 1937 bezogen. Die späteren Bauabschnitte des „1000-Wohnungs-Bauprogramms" begannen zeitversetzt zwischen Juli 1937 und Anfang 1938 mit weiteren 800 Wohnungen zwischen Rathaus und Werftstraße. Anschließend an das *Verwaltungsforum* um das Rathaus entwarf *Gustav Lübbers* u.a. Wohnungen für Kinderreiche an der Mitscherlichstraße, den Typ IV B. Die „Gegenfront" zum Rathaus sollte „ruhig und vornehm" wirken[435]. Tiefbraune Handstrichklinker und quaderartige Türumrahmungen, schmiedeeiserne Oberlichtgitter und eine Plastik „Mutter und Kind" korrespondierten zu der Monu- *161*
mentalität des *Verwaltungsforums*.

Gegenüber dem 1936 gebauten „Haus des Handwerks"[436] durfte *Gustav Lübbers* neun zu Doppelhäusern kombinierte dreigeschossige Wohngebäude planen, den Typ IV A. *55*
1935 hatte *Fritz Höger* diese Adolf-Hitler-Straße noch in geschlossener Bauweise als gebogene Randbebauung mit einem vorspringenden Eckgebäude entworfen. *Willi Lübbers*, eingeschworener *Höger*-Gegner und seit 1937 Mitglied im *NSDAP-Magistrat*, vereinfachte die Bogenform und begradigte sie, angeblich, um ein „besonderes Gepräge" zu schaffen[437].

In Anpassung an die uneinheitliche wilhelminische Bebauung plante *Gustav Lübbers* an der Werftstraße einen 120 m langen, dreigeschossigen Block, verputzt, durch halbrunde Treppenhaustürme und bullaugenähnliche Rundfenster gegliedert. Er mußte sich hier korrigierende Eingriffe durch *Wilhelm Hallbauer* gefallen lassen. Die südlich anschließenden Bebauung von *Franz Iwersen*, an der Peterstraße[438] dürfte kaum auf Beifall der NS-Baupolitiker gestoßen sein. Auch *Robert Fuchs* knüpfte mit seinen dreigeschossigen Wohnhaustypen an der Werft-/Lessingstraße und formreich gegliederten Ziegelornamenten an seine früheren Bauten der Weimarer Republik. Die Propaganda entschuldigte[439], daß Fassaden in der Werftstraße keine „Ziegelsteine erhalten, sondern verputzt werden“, um „Anschluß an die bereits vor Jahren entstandenen“ Bauten zu bekommen.

Im Gegensatz zu der vielfältigen und formenreichen Architektur der Werftstraße konnten die Bebauungen der beiden Oldenburger Architekten, *Rudolf Bürgin* an der
56 Schillerstraße[440], und *Ernst Boyken* am Otto-Meentz-Platz[441], des Beifalls der Propaganda
63, 62 sicher sein. *Bürgin*, Planer von Schilldeich, der Siedlung Burg Kniphausen und
146 verschiedener Partei- und Wohnungsbauten in Fedderwardergroden, bediente sich bei seinen zweigeschossigen Ziegelbauten frühzeitig des beliebter werdenden NS-Formenkataloges. Symmetrie, (Führer-)Balkone, Bogenverbindungen und Kreuzsprossen wiesen die künftige Richtung.

Der ehemalige *Höger*-Mitarbeiter *Ernst Boyken*, der am Wilhelmshavener Rathaus
56 (1928–29) mitgearbeitet hatte, empfahl sich mit der Bebauung Otto-Meentz-Platz für
66, 37 weitere Planungsarbeiten in Breddewarden und Fedderwardergroden. Die markanten, den Platz flankierenden Zeilen mit je 7 Eingängen, lassen heute die damaligen Absichten nicht mehr erkennen. Der *Kreisobmann der NS-Kulturgemeinde Hermann Ahner*, lobte: „Es öffnete sich uns gegenüber dem Banter Friedhof ein weiter Platz, eine Anlage, die später so ähnlich werden soll wie der Adalbertplatz, für die aber jetzt schon der Rahmen geschaffen ist durch die Häuserzeilen, die rechts und links den Platz ... einfassen und durch den Krähenbusch im Hintergrund eine besondere Note erhalten ... jeder, der vor diesem Platz steht, wird seine Freude haben an der einheitlichen Architektur und an der
56 geschlossenen Wirkung dieser Platzanlage“[442]. Der Otto-Meentz-Platz sollte kein Grünplatz sein. Er war für politische Zwecke gedacht, als Appellplatz für die Wilhelmshavener *HJ*. *Ernst Boykens* Bebauung sollte Kulisse werden für Befehlsausgabe und weihevolle Feiern am Ehrenmal auf dem Banter Friedhof. Die Dimensionen des langen Platzes wurden betont durch die 100 m langen Blocks, ausgerichtet auf das vor
147 Kopf liegende *HJ*-Heim.

Die Wohnungspolitik von Kleinhäusern und Heimstätten, die den Nationalsozialisten zugeschrieben wird, war für das Hansaviertel bedeutungslos. Obwohl die *Stadt*
52 *Wilhelmshaven* und die *Jade* 1937–39 weitgehend den Wohnungsbau in der Altstadt fortsetzten, das Vorgegebene übernahmen und den Städtebau nur geringfügig veränderten, fehlte es nicht an Versuchen, nationalsozialistisches Gedankengut aufzupfropfen. So beschrieb die *NSDAP*-Parteizeitung[443] kleinbürgerlich die „sauberen Gardinen vor den Fenstern“, die einen „recht freundlichen Eindruck“ machen und die Ligusterhecken und „Gruppen von Tannen und Zierbüschen“, die den Straßenraum abschließen.

112 In einer Sonderausgabe zum *Hitler*-Besuch vom 01.04.1939 präsentierte *Hermann Ahner* das Hansaviertel symbolträchtig: „Wenn wir links blickten in die Querstraßen, war alles im Schatten, eng und dunkel; blickten wir rechts, erweiterten sich die Straßen ins Helle: Dort war die Nahtstelle zwischen dem alten und dem neuen Wilhelmshaven, dort sahen wir schon durch den Gegensatz von Schatten und Licht, wie ganz anders das neue Bauen das Straßenbild gestaltet als früher ... Was auf diesem Geländeteil rechts der Peterstraße nun entstanden ist, kennzeichnet das neue Bauen im Dritten Reich“.

3.6.3. Jadeviertel

Die Ostgrenze des Jadeviertels, die Werftstraße, markierte eine städtebauliche Zäsur, die
bereits *Willy Hahn* 1917 als westliche Begrenzung der geschlossenen 3geschossigen
Bebauung vorgesehen hatte. Nach seinem *Bauklassenplan*[444] sollte auf der Fläche des
heutigen Jadeviertels eine sternförmige straßenbegleitende „offene Bauweise in Grup- 1
pen" auf einen Kreuzungspunkt führen. Tatsächlich aber blieb dieses Areal im Geviert
von Genossenschafts-, Bismarck-, Werft- und Peterstraße bis auf die Adolf-Hitler- 26
Siedlung (1934–36) und die Siedlung „Kaninchenhausen" der *Jadestädtischen Wohnungs-
baugesellschaft* (1933) landwirtschaftlich genutzt. Die 10 Wohnhäuser der *Jadestädtischen
Wohnungsbaugesellschaft* Peterstraße 131–169 waren unter *Oberbürgermeister Friedrich
Paffrath*[445] als *Arbeitsbeschaffungsmaßnahme* der Weimarer Republik begonnen worden
und nach der Machtergreifung durch die Nationalsozialisten fertiggestellt worden.
Architekt *Robert Fuchs* entwarf 1933 die schlichten Zeilen in moderner Formsprache mit
waagerechten Putzbändern, abgestuften Giebeln und sprossenlosen Fenstern. Durch die
zu enge Zeilenanordnung waren die 10 Gebäude etwas mißglückt. Die Nationalsozia-
listen diffamierten die Bauten von *Robert Fuchs* u. a. wegen der niedrigen Geschoßhöhen
als „Kaninchenhausen" oder „Fuchsbauten".

Der zunehmende Wohnungsbedarf des „III. Reiches" erforderte ab 1937 eine 53
Ausweitung des Hansaviertels nach Westen. Die *Marineintendantur* verlangte von der 14
Stadt am 01.09.1937 die Planung eines Wohngebietes mit 800 Wohnungen für
Unteroffiziere und weitere 800 Wohnungen für Marinearbeiter in „weniger intensiver
Mischung als bislang"[446]. Eine 1938 verlängerte Bremer Straße als Ost-West-Achse und 52
ein neu durchgelegter Banter Weg als Nord-Süd-Achse erschlossen das Gelände. Der
Bebauungsplan wurde 1937 vom *Stadterweiterungsamt* unter *Wilhelm Hallbauer* im
Einvernehmen mit der Jade erarbeitet. Wegen der Umsiedlung der Erbhöfe, der
Einsprüche des *Reichsnährstandes* und der Notwendigkeit, Flugabwehranlagen einzuar-
beiten, hatten sich die Verhandlungen hingezogen[447]. *Wilhelm Hallbauer* handelte sich
wegen des *Bebauungsplanes Jadeviertel* eine Rüge von *Oberbürgermeister Müller* ein. Dieser
wollte die Planung zurückstellen, bis der neue *Stadtbaudirektor Walter Hunzinger* am
01.11.1937 seinen Dienst antrat[448].

Möglicherweise hat *Hunzinger* noch Einfluß genommen, denn der *Fluchtlinienplan*
Nr. 152 wurde erst am 01.06.1938 förmlich festgestellt[449]. Auch war eine gegenüber dem
Hansaviertel zunehmende NS-typische Kosmetik in der Anlage von Straßen und Plätzen 53
erkennbar. 10 Platzanlagen schoben sich in die Wohnblöcke, nicht wie bei *Högers*
Neu-Siebethsburger Planung als abwechslungsreiche, schiefwinklige Knickpunkte, nicht
wie in Voslapp oder Cäciliengroden als Nachbarschaftsplätze, sondern streng geome-
trisch auf Gebäudemitten und Straßenachsen ausgerichtet. Westlich des 1938 durchge-
legten Banter Weges war das Jadeviertel auf das dominierende Achsenkreuz Tilsiter-/
Marienburger Straße zentriert. In Ermangelung repräsentativer Bauten plante *Hallbauer*
Schulen[450] vor Kopf von Lessingplatz und Ernst-Moritz-Arndt-Straße. Westlich der 132

Pommerschen Straße sollte 1941 noch eine Sportanlage entstehen. Ein *NSDAP*-Gemeinschaftshaus[451] mit Ehrenhof und Bunker als „Wehrburg“ war im Kreuzungspunkt Banter Weg/Bremer Straße geplant. Südlich davon beidseitig der Nord-Süd-Achse Banter Weg je 200 m lange Gewerbetrakte[452]. Die perfekte Planung für die
52 Gemeinschaftsbauten unterschied das Jadeviertel von den anderen Altstadt-Wohngebieten.

53 Am Rande des Jadeviertels südlich der Peterstraße realisierte die *Jade* noch 30 Wohnungen für Polizeibeamte, genannt „Schupohausen“, die nach Plänen von *Willi Lübbers* im Oktober/November 1938 fertiggestellt wurden. Einige Läden und eine Tankstelle plante *Lübbers* gleich mit.

53 Der *Bebauungsplan Jadeviertel*, 1938 zum Abschluß gekommen, sollte bereits 1939 um ein Einfamilienhaus-Siedlungsgebiet nach Westen bis zu den Arbeiterunterkünften Ebkeriege erweitert werden. *Ernst Peters* sollte dort Einfamilienhäuser für die *Wilhelmshavener Spar- und Baugesellschaft* entwerfen[453]. Die strenge Rechtwinkligkeit der Blockstrukturen und Plätze war geschickt an die vorhandene Altbausubstanz angepaßt worden. So stuften sich an den Eingangstoren Lessingstraße und Marienburger Straße die Gebäudetypen zu der gewollten Portalwirkung ab. Die Rückfronten der „häßlichen“ Altbauten[454] der Werftstraße wurden durch zwei durchgehende große Reihenbauten an der Kolberger Straße verblendet. Sie sollten die Gründerzeitbauten dreigeschossig abdecken, unterbrochen nur durch das Tor zum Lessingplatz. Dieser Kolonnadendurchbruch mit seinen Kopfbauten befand sich, ähnlich schon 1917 bei *Paul Schmitthenners* Siedlung Staaken und wurde seither vielfach kopiert[455]. Gegenstück zum Tor Lessingplatz wurde der Eingang zur Marienburger Straße.

★

53 Der Bau der knapp 900 Wohnungen des Jadeviertels war von Anfang an auf die im Januar 1937 gegründete gleichnamige Wohnungsbaugesellschaft zugeschnitten und wurde im Februar 1938 als Bauabschnitt „Westlich der Werftstraße“ in vier Unterabschnitten begonnen. Die ersten Wohnungen konnten Ende 1938, der Rest 1939, bezogen werden. Die städtische Kanalisation und der Straßenbau begannen bereits 1937 mit der
54 Durchlegung des Banter Weges und provisorischer Knüppeldämme. Die nur drei Typenentwürfe für das Jadeviertel zeichnete die Bauabteilung der *Jade* unter dem am 01.12.1937 nach Wilhelmshaven gekommenen Architekten *Willy Schemm*. Er leitete das Entwurfsbüro, in dem die Architekten *Penger, Bochert, Jaeger, Rohlfs* und *Siepermann* arbeiteten – alle keine NS-Parteigänger[456]. Technischer Geschäftsführer war 1938–40 *Adolf Braunschweig*. Wie von der *Marineintendantur* 1937 gefordert, reichte *Schemm* die Entwürfe im Februar 1938 getrennt nach Wohngebieten für *Unteroffiziere* (U 1, U 2 und U 3) und Werftarbeiter (A 1) ein. Für die offene Bebauung hatte *Schemm* fast ausnahmslos Doppelhäuser entworfen. Gerne erwähnte die Parteipropaganda noch 1938, daß man sich an „die neue Siebethsburg“ erinnert[457].

Vier Baufirmen[458] arbeiteten an vier Stellen im Taktverfahren. Jede der Firmen hatte
ihre eigene „Gefolgschaftsunterkunft“ gebaut. Noch während des Baues wurde die
Wohnungsanzahl auf 1.000 erhöht und der *Bebauungsplan* nach Norden hin um den 53
Teilabschnitt U 4 erweitert, nachdem 1938 eine kleine Siedlung für städtische Bedienstete 132
nach Plänen von *Egon Forster* zu den Akten gelegt worden war.

Schemm hatte seine Haustypen so dimensioniert und gegründet, daß Klinkerfassaden
möglich waren. Zum Teil konnte er sogar auf Rammpfähle verzichten. Am 22.06.1938
feierte man das erste Richtfest mit Hakenkreuzfahnen, üblicher NS-Propaganda und
Dank an den „Führer“: „800 Volksgenossen bauen Wohnungen für 1.000 Familien, die in
wenigen Monaten ihr eigenes Heim bekommen werden“[459].

Die Hausentwürfe, obwohl keinesfalls Volkswohnungen für ungelernte Arbeiter, 54
fielen durch außerordentliche Einfachheit auf. Der Typ 1 mit 46 qm Wohnfläche im
Unterabschnitt A 1 für Arbeiter war ein zweigeschossiger Zweispänner mit: Diele 4 qm,
Wohnküche 13 qm, Schlafraum 15 qm, Stube 11 qm, Bad 3 qm. Auch in den Dächern
war jeweils noch eine Wohnung neben dem Trockenboden ausgebaut. Es gab
Luftschutzkeller, aber ohne die entsprechenden Türen. Sie sollten eingespart werden.

In den dreigeschossigen, über 100 m langen, Zeilenbebauungen an der Kolberger
Straße kombinierte *Willy Schemm* drei Typen. Er staffelte sie vor und zurück. Die
überlangen Zeilen mußten durch über Dach führende Brandwände gegeneinander
abgeschottet werden. Geschickt stufte *Willy Schemm* die Giebel in drei Sprüngen ab. Die
langen Zeilen an der Kolberger Straße waren mit ihren durchgehenden Dachgauben,
schlanken Treppenhauslichtbändern und breitliegenden Wohnungsfenstern durchaus
großstädtisch und als Übergang in die offene Bebauung des Jadeviertels gemeint. Weiter
westlich wurden die Einzelbauten schmuckloser, schlichter und durchgängig zweige-
schossig. Breit gelagerte Fensterformate und breite Dachgauben ließen kaum NS-
typische Details erkennen. Nur die Umrahmungen von Eingangstüren und Treppen-
hausfenstern waren Zugeständnisse an die vom „III. Reich“ gewünschte Architektur. Die
ein Jahr später entworfenen Volkswohnungen für Voslapp waren eine Weiterentwick- 31
lung der Wohnungstypen des Jadeviertels und diesen sehr ähnlich. 54

3.7. Siedlungsutopien bis zum Schluß

Für das *Wohnsiedlungsgebiet Wilhelmshaven-Jeverland* wurden 1933–45 zehn Gemein- 1
schaftssiedlungen gebaut. Darüber hinaus noch zahlreiche Kleinsiedlungen, Baulücken-
auffüllungen und Wohngebiete für Private und Führungspersonal. Alles in allem waren
dies annähernd 20.000 neue Wohnungen. Weder für Wolfsburg noch Salzgitter wurden
im „III. Reich“ ähnlich viele Wohnungen produziert. Zahlreiche weitere Siedlungen für 129
das *Wohnsiedlungsgebiet Wilhelmshaven-Jeverland* blieben Utopie oder in den Anfängen
stecken. Einige davon sollen hier noch vorgestellt werden. Der Plan *Die Stadt der* 127
Dreihunderttausend vom 15.03.1937 sah neben der verbleibenden „Kernstadt“ 15 neue

einzelne Siedlungen außerhalb der „Maadelinie" vor, die sich zur Südstadt Sande,
Weststadt Langewerth und Nordstadt Fedderwardergroden gruppieren sollten. Es blieb
129 bei dieser Dreiteilung bis zum letztgültigen *Wirtschaftsplan* von 1940. In den folgenden
Jahren fand eine Konzentration auf weniger und größere Siedlungsgebiete statt.

Nur eine der Siedlungen, der Ortsteil Gödens, lag außerhalb des *Bausperrgebietes* des
129 *MDF* vom 07.07.1939[460]. Erst der *Wirtschaftsplan* 1940 kam der tatsächlichen Entwick-
lung nahe. Er unterschied in „mögliche Siedlungsflächen", z. B. Utters und Cleverns,
66 „geplante Siedlungsflächen", z. B. Breddewarden, und „fertiggestellte und fertiggeplante
37, 30 Siedlungen", z. B. Fedderwardergroden und Voslapp. Nur wenige der Siedlungsutopien
sind bis zu *Bebauungsplänen* gediehen. Von Utters, Accum, Moorsum oder Gödens sind keine *Bebauungspläne* bekannt. Vorgesehen war, 1939 *Erich zu Putlitz* und *Ernst Prinz* Fedderwarden, Accum und Moorsum entwerfen zu lassen[461]. Die Planung Utters scheiterte am Widerstand des *Landesbauernführers* und an Einsprüchen des *Reichsnährstandes*[462]. Das Siedlungsgebiet Sande/Gödens sollte direkt vom *Reichsheimstättenamt Gau Weser-Ems* geplant werden. Für andere Siedlungsgebiete wiederum wurden sogar schon Entwässerungsleitungen verlegt (Altengroden-Süd), Wege trassiert (Breddewarden) oder Arbeiterlager vorgehalten (Langewerth).

3.7.1. Siedlung Burg Kniphausen

Das Siedlungsgebiet um die Burg Kniphausen ergab sich aus den Grundbesitzverhält-
nissen. *Georg Linde* stellte im August 1937 fest, daß um die Burg Kniphausen 180 ha
Siedlungsfläche möglich seien[463]. Die *LOB* zielte anfangs noch auf Heimstätten und
Kleinsiedlungsflächen. Die Grundstücksgrößen für Heimstätten erforderten auch nach
14 Meinung der *Marineintendantur* die Besiedlung außerhalb Wilhelmshavens und Rüstrin-
gens[464]. Die *Marineintendantur* verhandelte von sich aus mit den Bürgermeistern der umliegenden Gemeinden. Das *Reichsheimstättenamt Gau Weser-Ems* hatte nach einer Umfrage bei 300 Werftarbeitern Interesse für die Errichtung eines Siedlungshauses festgestellt.

Das Gelände außerhalb der „Maadelinie", so die *LOB* 1937, sei „zum weitaus größten Teil im Besitz der öffentlichen Hand (Staat und Kirche) von Juden oder aber nicht erbhofgebunden (Erbengemeinschaften und *Fürst Kniphausen)* . . . Das Wesentliche wird immer sein, daß man so wenig wie möglich an Erbhöfe herangeht"[465]. Gleichzeitig sollten die Drostei Schilldeich und die Domäne Hohewerther Grashaus Austauschgelände für innerstädtische Erbhöfe werden, ein Widerspruch, der zu Problemen führte, als sich im März 1938 herausstellte, daß die 1.800 geplanten Wohnungen um Burg Kniphausen „ausfallen"[466]. Es ist nicht feststellbar, warum sich das Projekt Kniphausen zerschlug, möglicherweise wegen Unstimmigkeiten mit dem *Fürst Kniphausen,* möglicherweise, weil der staatliche Besitz um Schilldeich und Hohewerther Grashaus schneller zu realisieren schien.

Das Siedlungsgelände für die Siedlung Burg Kniphausen war parzellenscharf im Plan *Wirtschaftsgebiet Wilhelmshaven und Umgegend vom September* 1937 eingetreten. Es sollte „Neues Wohngebiet in weiträumiger Bauweise (Kleinsiedlungsflächen)" werden. Die *Marineintendantur* schrieb dem zuständigen *Bürgermeister* der 1933 neugebildeten Groß-gemeinde Kniphausen im September 1937, daß 1.800 Wohnungen für Marineangehörige „im Umkreis der Burg Kniphausen" geplant werden sollen und die Kosten für Gemeinschaftseinrichtungen, Straßenentwässerung usw. gemeldet werden sollen[467]. Die am 01.04.1937 zu Groß-Wilhelmshaven zusammengeschlossenen Jadestädte, vor allem der neue Wilhelmshavener *Oberbürgermeister Wilhelm Müller,* sahen die Planungsaktivitäten in den Nachbargemeinden skeptisch, fürchteten Zersplitterung der Verwaltung und Entwässerungsprobleme[468]. Tatsächlich betrieb Wilhelmshaven bereits vehement die Eingemeindung der Nachbargemeinden, die jedoch erst am 01.06.1938 zustande kam.

Das Projekt Siedlung Burg Kniphausen (1.800 Wohnungen) war insofern planerischer 62
Vorläufer der Siedlung Schilldeich (1.500 Wohnungen). Es hatte nur von Dezemer 1937 63
bis März 1938 Bestand. Die späteren *Wirtschaftspläne* sparten die Siedlungsfläche um die 129
Burg, die weiterhin im Besitz des *Fürsten Kniphausen* blieb, aus.

★

Rudolf Bürgins Bebauungsplan Burg Kniphausen wies, wie auch Schilldeich, überge- 62
wichtig kleinsiedlungsbezogene Entwurfsmerkmale auf. Die vorhandenen Hofstellen
sollten erhalten bleiben. Das 2 km lange und 1,5 km breite Siedlungsgebiet war größer als 63
Schilldeich, größer als Voslapp, und sollte von einem „Grünen Ring" durchzogen 30
werden, wie er kurz danach auch für Fedderwardergroden entworfen wurde. Die auf 37
1438 zurückgehende Burg Kniphausen selber wurde nicht angetastet. Der gegenüberliegende Mönkeberger Busch jedoch, gleichfalls dem *Fürsten Kniphausen* gehörend, sollte nach *Bürgins* Planung eine große Platzanlage werden, von mehrgeschossigen Bauten der Gemeinschaft umgeben. Hier waren Rathaus, *NSDAP*-Gemeinschaftshaus, *HJ*-Heime, Gewerbetrakte und Feierhalle geplant. Weitere Gemeinschaftsbauten sollten sich westlich der Burg und an der den „Grünen Ring" begleitenden Straße um Aufmarschplätze gruppieren. Das traditionelle Kleinsiedlungskonzept war auf Nebenerwerb aufgebaut. Mehr noch als bei der Nachfolgeplanung Schilldeich hatte *Bürgin* bei Kniphausen auf durchlaufende Achsen und monumentale Blickverbindungen verzichtet. Die Straßen waren häufig versetzt, geknickt und gebogen. Das sehr flächige und weitläufige Siedlungskonzept Burg Kniphausen für fast 10.000 Menschen, von denen nur ein Teil in Mietwohnungen an der Ringstraße untergebracht werden sollte, entsprach der reinen Lehre des *Reichsheimstättenamtes Gau Weser-Ems.* Die gewünschte Verwurzelung mit dem Grund und Boden wurde jedoch Zug um Zug unwichtiger und von der *Marine* nicht mehr gewünscht. Das völlig überdimensionierte Kleinsiedlungsgebiet hätte viele

Nachteile gebracht: fehlende Verkehrsanbindung an die Werft, extrem lange Wohnwege, hohe Erschließungskosten usw. Es war zum Scheitern verurteilt.

3.7.2. Schilldeich

Konkreter Planungsanstoß für Schilldeich war die Besprechung vom 02.03.1938 bei
3 *Staatsminister Pauly* in Oldenburg, als *Amtshauptmann Ott* vom *Amt Friesland* widerstrebend beauftragt wurde, die Planung für Schilldeich zu beginnen[469], da „als Ersatz für die
62 ausfallende Siedlung Kniphausen . . . das Gelände etwas weiter nach Nordosten gelegen" ausgewählt worden sei. *Amtshauptmann Ott* versprach, „in 14 Tagen einen *Bebauungsplan* vorzulegen, der dann noch dem *Oberbürgermeister* von Wilhelmshaven zugehen soll", denn Wilhelmshaven wurde ab 01.06.1938 durch Eingemeindung zuständig. Die *LOB* wehrte sich heftig gegen den „Ersatzstandort" Schilldeich, da die Staatsdomäne Hohewerther Grashaus und die Drostei Schilldeich im Einvernehmen mit dem *Reichsnährstand* und der *Landesbauernschaft* als landwirtschaftliches Austauschgelände für Altengrodener Erbhöfe vorgehalten werden sollten[470]. *Landesplaner Linde,* der bisher nicht eingeschaltet worden war, unterstellte dem *MDF,* daß er lieber sein eigenes Domänengelände und das Gelände der *Stiftungskasse Oldenburg* vermarkten wolle, als Land des *Fürsten Kniphausen* aufzuschließen. Im übrigen liege Schilldeich im Schutzbereich der *Festungskommandantur,* der die Wohnungsgebiete innerhalb der „Maadelinie" umgebe. Der Wilhelmshavener *Magistrat* unterstützte das Projekt Schilldeich und wollte die bevorstehende Eingemeindung nicht torpedieren.

Das Gebiet von Schilldeich, außerhalb der Stadtgrenzen und außerhalb der „Maadelinie" gelegen, war nach nationalsozialistischer Auffassung für eine Wohnsiedlung geeignet. Der Jude *Samuel de Taube* „verkaufte" 1933 die ehemalige Drostei Schilldeich an die *Stiftungskasse Oldenburg (Landesfürsorgeverband)*[471]. Die Staatsdomäne Hohewerther
127 Grashaus umfaßte 68 ha Land. Sie war bereits im März 1937 in den Plan *Die Stadt der Dreihunderttausend* als Siedlungsgebiet „Hohewerth" eingetragen.

Die Siedlung Schilldeich sollte durch die *Jade* realisiert werden. Der *Bauverein Rüstringen* beschwerte sich über diese Entscheidung bei *Generaladmiral Raeder*[472], der sich im April 1938 sehr zufrieden über Neu-Siebethsburg gezeigt hatte. Die *Marineintendantur* und der Leiter der *Bauabteilung des OKM, Theodor Schreiber,* sprachen sich für den *Bauverein Rüstringen* aus. *Oberbürgermeister Müller,* gleichzeitig Aufsichtsratsvorsitzender der Jade, war der Auffassung, alle Wohnbauten für Marineangehörige sollten über die *Jade* abgewickelt werden.

*

Rudolf Bürgin, Schüler von *Schweizer* und *Billing,* hatte sich 1937 mit Schulbauten in
62 Oldenburg selbständig gemacht. Der Auftrag für den *Bebauungsplan* Kniphausen im

Dezember 1937 war eine seiner ersten größeren Aufgaben. Es folgte im März der Auftrag
für den „Ersatzstandort" Schilldeich. *Bürgin* orientierte sich mit beiden Siedlungsprojek- *63*
ten, die unmittelbar aneinandergrenzten, stark an den Vorstellungen von *Reichsheimstät-
tenamt* und *Deutschem Siedlerbund.* Als *Bürgin* Kniphausen und Schilldeich im Auftrage des
Amtes Friesland entwarf, waren Neu-Siebethsburg und Hansaviertel zur Hälfte fertigge- *57, 53*
stellt. Die Pläne für Voslapp lagen bereits vor und Cäciliengrodens Planung war in der *30, 34*
Bearbeitung. Eine großstädtische, moderne Bebauung kam für Schilldeich nicht in
Betracht. Der staatliche Grundbesitz legte die Anwendung des *Reichssiedlungsgesetzes vom*
11.08.1919 nahe. So plante *Bürgin* 1.000 Kleinsiedlungshäuschen für Siedler mit
„Eignungsschein" auf je 600 bis 1.000 qm großen Grundstücken[473]. Zusätzliches
Pachtland für Nebenerwerb war am Rande von Schilldeich vorgesehen. Dazu sollten
noch rd. 500 Volkswohnungen an dem als Anger bezeichneten Aufmarschplatz und den
größeren Straßen entstehen. Mit 6.000 Bewohnern war Schilldeich immerhin größer als
Voslapp geplant und in siedlungstypologischer Hinsicht ähnlich. Ähnlich wie in Voslapp
ging *Bürgin* bei seiner Schilldeich-Planung von den vorhandenen Wegeverbindungen
aus. Er führte alle Straßen zum Anger, den er bald zum Forum uminterpretierte,
umgeben von Verwaltungsbauten, Parteibauten, Geschäften und Werkstätten. Das
„Forum" sollte mit 100 × 250 m zwar räumlich geschlossen sein, es war jedoch nicht
monumental gedacht. Drei Schulen, zwei Sportplätze, *HJ*-Heime und *NSV*-Häuser
sollten im Siedlungsgebiet Schilldeich verteilt sein. Alle Wohnstraßen waren auffällig in
Nord-Süd-Richtung geplant. „Ein von Ost nach West verlaufender Grünzug" sollte
Landschaft und Natur in die Siedlung ziehen und parkartig gestalten. Die Architektur der
Einzelgebäude war traditionell gedacht – keine Doppelhäuser, keine Reihenhäuser. Die
Miethäuser sollten traufständig die Hauptstraßen betonen.

3.7.3. Feders Langewerth

Nach dem Plan *Die Stadt der Dreihunderttausend* sollte Langewerth 1937 Kern der *127*
Weststadt mit 35.000 Einwohnern werden, somit die größte und wichtigste Siedlung
außerhalb der „Maadelinie". Die *Marine* bevorzugte Langewerth, da „die Arbeiterschaft
ihre Arbeitsstellen noch mit dem Fahrrad erreichen kann"[474]. Der *Wirtschaftsplan vom*
03.08.1939 zeigte das 4 km lange und 1,5 km breite langgestreckte Siedlungsgebiet, das
zur Hälfte auf Wilhelmshavener Stadtgebiet und zur anderen Hälfte in Friesland liegen
sollte. Die geplante, weitgehend 3-geschossige, nur am Rande 2-geschossige Bebauung
wies darauf hin, daß große Bevölkerungsmassen untergebracht werden sollten. Von
78.000, gar von der „Stadt der 100.000" war die Rede[475]. Der *Wirtschaftsplan
Wilhelmshaven-Friesland vom Mai* 1940 jedoch zeigte von der gesamten Weststadt nur *129*
noch die auf Wilhelmshavener Stadtgebiet liegende südöstlichste Hälfte von Langewerth
als „mögliche Wohnfläche".

62 Das *Amt Friesland* hatte noch bis März 1938 das Siedlungsprojekt um die Burg
Kniphausen betrieben, das sich mit Langewerth überschnitten hätte. Im November 1938
drängte *Friedrich Nicolaus* vom *RAM,* die Vorarbeiten für Langewerth „unverzüglich
aufzunehmen“[476]. Der südliche Teil auf Wilhelmshavener Gebiet sei „zunächst in Angriff
zu nehmen“. In der ersten Hälfte des Jahres 1939 schrieb die *Stadt Wilhelmshaven* einen
64 beschränkten Architektenwettbewerb aus. *Gottfried Feder*[477] (Berlin), *Dietrich Suhr* (Kiel),
65 *Erich zu Putlitz*[478] (Hamburg) und *Ernst Prinz* (Kiel) wurden eingeladen. Die Auswahl
war Programm. *Fritz Höger* beispielsweise, der in Wilhelmshaven baute und in der Stadt
außerordentlich beliebt war, bot nicht die Gewähr für eine Lösung der Aufgabe in
nationalsozialistischem Sinne. Auch hätte es möglicherweise nahegelegen, *Gustav Bork,*
Rudolf Bürgin oder *Friedrich Heuer* einzuladen, die zu der Zeit im Raum Wilhelmshaven
planten oder bauten. Auch die Wilhelmshavener Architekten *Iwersen, Freygang, Minke,*
Peters, Fuchs u. a. bekamen keine Chancen, möglicherweise wegen der mangelnden,
3, 4 städtebaulichen Erfahrungen. Das *Oldenburger Staatsministerium* und das *OKM* drängten,
Langewerth zügig zu beginnen[479]. Die *Marineintendantur* jedoch mußte dem *OKM*
mitteilen, daß die „mehr städtische Bauweise“ in Verbindung mit der Weststadt (Accum
und Fedderwarden) zeitaufwendiger sei[480], denn auch Pumpwerke und Druckrohrlei-
tungen müßten für die Weststadt gebaut werden. Die Landbeschaffung für Langewerth
66 durch die *RUGES* sei bis zum 01.11.1939 zugunsten von Breddewarden zurückgestellt.
Das *OKM* wollte Langewerth am 01.11.1939 beginnen. Nachdem am 09.08.1939 der
64 1. Preis des Architektenwettbewerbes nach persönlichem Vortrag *Gottfried Feder*
zuerkannt worden war, bekam die *Stadt* Weisung von Berliner NS-Dienststellen, „*Feder*
nicht herauszustellen“[481]. Als Gründungsmitglied der *NSDAP, Reichstagsmitglied* und
ehemaliger *Reichssiedlungskommissar* war *Feder* zwar noch hoch angesehen, jedoch von
Hitler bereits 1935 auf einen Lehrstuhl nach Berlin abgeschoben worden. Ende 1940 stellte
Stadtbaudirektor Walter Temp Langewerth als „größten Stadtteil, den das neue Wilhelms-
haven in den nächsten Jahren erhalten wird“ heraus, nannte *Feder* aber nicht, und blieb in
der Aussage unverbindlich[482]. Zum gleichen Zeitpunkt stellte *Friedrich Nicolaus* vom
RAM das ganze Projekt Langewerth wegen mangelnder Bodentragfähigkeit in Frage
und empfahl, die Planung der Großsiedlungen um Sande wieder aufzunehmen[483]. Da
143 Sande jedoch noch nicht eingemeindet war, lag die Kompetenz hierfür beim *Amt*
109 *Friesland.* Wegen des benachbarten *Kulturforums* müsse, so *Nicolaus,* auch *GBI Albert Speer*
angesprochen werden. *Speer* verzichtete jedoch darauf, einzugreifen[484]. Die Bedenken
wegen der Bodentragfähigkeit wurden 1941 auf die gesamte Weststadt mit ihren jetzt
123.000 geplanten Einwohnern einschl. Fedderwarden, Accum und Moorsum, ausge-
64 dehnt[485]. Für *Feders* Planung, die ein- bis viergeschossig (19.000 Wohnungen) vorgesehen
war, wurde eilig eine Reduzierung auf 1 bis 2 Geschosse (15.000 Wohnungen)
erwogen[486]. Der *RAM* empfahl im August 1941, wegen der Bodenverhältnisse und der
Luftgefährdung eine völlig neue Weststadt, die 1000 m westlich des neuen Hauptbahn-
hofs „mit einer lockeren Bebauung einsetzen“ sollte und sich in den Raum Accum
entwickeln müsse. Wie in Danzig/Langfuhr könne eine monumentale Grünachse, die die

Forumsachse aufnimmt, die Wohnstadt mit Groß-Wilhelmshaven verbinden[487]. *Feder*
war mittlerweile gestorben, sein Projekt hatte kaum noch Chancen zur Realisierung.
Langewerth wurde zugunsten von Breddewarden zurückgestellt. Die Architekten *66*
Erich zu Putlitz und *Ernst Prinz,* mit dem zweiten und dritten Preis im Wettbewerb *65*
Langewerth ausgezeichnet, sollten noch 1939 Fedderwarden, Accum und Moorsum
entwerfen.

★

Es liegt nahe, *Feders* Planung Langewerth zu seinem kurz vorher erschienenen *64*
Standardwerk „Die neue Stadt" in Beziehung zu setzen. *Helmuth Baur,* schon damals im
Wilhelmshavener *Stadterweiterungsamt* tätig, erinnert sich gut an das außerordentlich
große Wohnsiedlungsprojekt. Es war das größte geschlossene Wohngebiet im *Wohn-* *129*
siedlungsgebiet Wilhelmshaven-Jeverland, möglicherweise im ganzen Reich. *Erich zu Putlitz* *65*
hatte in seinem monumentalen Wettbewerbsbeitrag – so *Baur,* das Siedlungsgebiet
erheblich erweitert. Der Entwurf von *Ernst Prinz,* ausgerichtet auf eine Mittelachse, habe
gewisse Parallelitäten zu *Herbert Rimpls* Entwurf für die *Stadt der Hermann-Göring-*
Werke.

Eingegrenzt durch die in den *Wirtschaftsplänen* vorgegebene, westlich tangierende
Ringschnellbahn mit 4 Bahnhöfen, dem neuen Hauptbahnhof Schaar sowie eine neu
geplante Umgehungsstraße wollte *Feder* Langewerth durch eine innere Ringstraße *64*
erschließen. Die von Südost nach Nordwest verlaufende Mittelachse sollte ein Grünzug
werden. Weitere Grünzäsuren sollten Langewerth in 16 Zellen aufteilen, jeweils mit
eigenen Gemeinschaftseinrichtungen, Parteihäusern usw. *Feders* „Die neue Stadt", war
weitgehend von *Fritz Rechenberg* bearbeitet worden[488]. Der hatte durch seine Arbeit bei
der *Wohnungs-AG der Hermann-Göring-Werke* mit *Herbert Rimpl* Kontakt. Es ist jedoch
unklar, ob Rechenberg, der ab 1939 in Berlin selbständiger Architekt war, an
Langewerth mitgearbeitet hat. Die von *Rechenberg* vertretene Funktionsteilung der
Stadtstruktur jedenfalls traf für Langewerth nicht zu, es ging ausschließlich um
Wohnungen. Die Arbeitsplätze lagen 4 bis 5 km entfernt: die *Westwerft,* die alte *KMW*
und die im Bau befindliche *Nordwerft.* Zwar propagierten *Feder* und *Rechenberg* die *101*
Idealstadt für 20.000 Menschen, jedoch war klar, „Die Großstadt kann also nicht
überhaupt abgelehnt werden . . . Erst der Umbau des Reiches (auch der Großstädte) zu
organischen Zellengemeinschaften mit größtmöglicher Selbständigkeit wird uns allmählich zu einer neuen Gemeinschafts- und Lebensform hinführen"[489]. Auch 100.000
Einwohner ließen sich nach *Feders* und *Rechenbergs* Meinung zu einer Gemeinschaft
zusammenfassen, entscheidend sei die weitere Kern- und Zellenbildung. Die Zellentheorie, die später von *Gutschow* und *Heuer*[490] zur NS-Ortsgruppe von 8.000 Einwohnern
uminterpretiert wurde, sollte ein „repräsentatives Verwaltungsviertel" nicht verhindern.
320 × 170 m groß sollte der Aufmarschplatz im Zentrum von Langewerth werden,
umgeben von Gemeinschaftshalle, Gewerbebauten, Verwaltungsstellen und Parteihäu-

sern. *Oberbürgermeister Müller* und sein *Stadtbaudirektor Temp* attestierten in einer
Begutachtung[491] vom 10.08.1939 den Architekten „Fleiß und Liebe“. *Feders* Entwurf
64 wurde nach Magistratsbeschluß zur Weiterbearbeitung vorgeschlagen. *Feder* wurde ein
Büro in Wilhelmshaven bewilligt. Für die einzelnen Zellen sollten verschiedene
Bauträger mit verschiedenen Architekten unter *Feders* künstlerischer Oberleitung
116, 118 arbeiten. Die Planungen von *Schulz* und *Gutschow* beendeten die Vision einer Weststadt
Langewerth. Bei *Gutschow* sollte Langewerth 1944/45 nur noch 12.000 Einwohner in
isolierter Lage, abgerückt vom *Kulturforum* und Hauptbahnhof, bekommen.

Es entstanden 1939 ein Wähleramt und eine Ausweichvermittlung, im Kriege ein großes Frauenlager und ein Bunker als militärische Bauten in Langewerth. Langewerth ist heute wie vor 1933 eine bäuerliche Siedlung mit wenigen 100 Einwohnern.

3.7.4. Nördliche Stadtparkkolonie

Weiteres Beispiel einer nicht mehr realisierten Siedlung war die von *Hans Bartning* und
Helmuth Baur geplante Nördliche Stadtparkkolonie innerhalb der „Maadelinie“. Nach
128 dem *Wirtschaftsplan von* 1937 sollte sie als Wohngebiet mit städtischem Charakter Teil der
43 Marinesiedlung Altengroden werden[492]. *Ewald Zenkers Bebauungsplan Altengroden vom*
14.02.1939 jedoch endete am verlängerten Totenweg, möglicherweise wegen Grunder-
werbsschwierigkeiten, möglicherweise wegen der abtrennenden Verkehrswege. Die
ersten Skizzen für die Nördliche Stadtparkkolonie fertigte *Hans Bartning* Anfang 1939.
Dieser in Teilen durchaus unfertige Plan wurde hastig zusammengestellt und auf den
112 *Hitler*-Besuch vom 01.04.1939 zugeschnitten. Die Nördliche Stadtparkkolonie sollte
noch sehr provisorisch an *Martin Wagners* Stadtparkkolonie von 1912 angeschlossen
werden. Auch die Verkehrsanbindung an den verlängerten Totenweg und die
Umfangstraße war in diesem *Bartning*-Plan noch nicht geregelt. Der Aufbau dieses
kleinen, für annähernd 500 Wohnungen vorgesehenen Siedlungsgebietes bezog sich auf
eine durchlaufende Magistrale, die zur Umfangstraße führen sollte.

Als *Hans Bartning* im August 1939 zum Kriegsdienst eingezogen wurde, setzte der am 24.07.1939 zur *Stadt Wilhelmshaven* dienstverpflichtete *Helmuth Baur* die Planung fort. *Baur*, wie auch *Bartning*, Schüler von *Otto Ernst Schweizer*, war keinesfalls der heimatschützenden *Stuttgarter-Schule* zuzurechnen, sondern der funktionalen Stadtplanung, die in Karlsruhe gelehrt wurde. *Baurs* städtebaulicher Entwurf für die Nördliche Stadtparkkolonie vom August 1939 war modern und unprätentiös. Er bezog sich auf die umgebenden Verkehrswege und schloß an die Wohnstraßen und Plätze, die *Martin Wagner* mit seiner Rüstringer Stadtparkkolonie 1912 vorgegeben hatte. Diese pragmatische Planungsauffassung ließ die Eigenständigkeit der Disziplin Städtebau und die sytematischen Wege der Planung erkennen, mit denen *Baur* später Wilhelmshavens Stadtentwicklung beeinflußt hat. Der Entwurfsprozeß war methodisch, wenig gefühlsorientiert, nicht auf Monumentalität ausgerichtet und schon gar nicht auf vordergründige optische Wirkung ausgelegt. Wurten und Bachläufe, Hofstellen und Grünflächen wurden von *Helmuth Baur* detailliert in die Planung der Nördlichen Stadtparkkolonie einbezogen. Die funktionale Planungsauffassung war auch erkennbar an der abgewogenen Infrastruktur und der durchdachten Verkehrsplanung. *Helmuth Baur* verkürzte die Magistrale, knickte Hauptdurchgangsstraßen ab und arbeitete den Charakter von Wohnstraßen heraus.

Viele Straßen waren gebogen, abgeknickt und verkehrsberuhigt vorgesehen. Schallschützend sollten sich Kleingärten zwischen die Umfangstraße und die Nördliche Stadtparkkolonie schieben. Schule und Kinderheim hat *Baur* nicht zur städtebaulichen Wirkung mißbraucht. Spielplätze, keine Selbstverständlichkeit in Plänen jener Jahre, lagen abseits der Verkehrsstraßen. Bedarfsgerecht sollten Läden an vier Stellen des Wohngebietes verteilt sein. Auffällig war an *Baurs* moderner Planung das Fehlen eines NS-Gemeinschaftshauses und eines Aufmarschplatzes. Eine Ladenzeile und eine Gaststätte sollten zusammen mit dem zu einem *HJ*-Heim umzubauenden Bauernhof den Stadtteilmittelpunkt bilden. Fehlende Symmetrie und die Leichtigkeit des Entwurfs unterstrichen die moderne Wirkung. Bemerkenswert war auch die Durchmischung von unterschiedlichsten Haustypen: Mehrfamilienhäuser, Eigenheime und Reihenhäuser. Vom 12-Familienhaus über Reihenhäuser und Doppelhaustypen bis zu freistehenden 1-geschossigen Hausformen war ein durchdachtes Spektrum vorgesehen, das in dieser Qualität für Wilhelmshaven damals ungewöhnlich war.

Das etwa zeitgleich geplante und etwa gleich große Familienlager Neuengroden
beispielsweise war städtebaulich bieder, einfallslos und fast dilettantisch. 46

An der Stelle der Nördlichen Stadtparkkolonie hat die *Stadt Wilhelmshaven* ab 1957 das Reinhard-Nieter-Krankenhaus nach Entwürfen von *Benno Schachner* geplant und 1967 fertiggestellt.

3.7.5. Siedlung um die Wurt Breddewarden

Die Siedlung Breddewarden kam als Einzige der Realisierung sehr nahe, war in vielen
Details durchdacht, vorbereitet und gezeichnet. Sie fand sich schon im Plan *Die Stadt der*
Dreihunderttausend vom 15.03.1937 als Siedlungstrabant der Nordstadt, durchzog alle 127
Planungsstadien und war noch in verschiedenen Varianten in *Gutschows* Wiederaufbau- 118
planungen von 1945 enthalten. Der Plan *Wilhelmshaven Wirtschaftsplan vom* 03.08.1939
zeigte Breddewarden außerhalb der Wilhelmshavener Stadtgrenzen liegend als Ortsteil
des Wohngebietes Wilhelmshaven Nord, parzellenscharf dargestellt in weitgehend
zweigeschossiger Bauweise im Kern und eingeschossiger Bauweise am nördlichen,
östlichen und westlichen Rand. Die Fläche um die alte Wurt gehörte seit 1933 zur
Gemeinde Kniphausen und zum *Amt Friesland*. Der tragfähige Geestuntergrund und die 143
luftschutzgünstige Lage ließen Breddewarden ideal erscheinen. Ein schneller Grunder-
werb jedoch schien noch 1937 nicht möglich. Im *Wohnsiedlungsgebiet Wilhelmshaven-* 1

Jeverland hatten die neuen Ortsteile Vorrang, deren Flächen schnell zu erwerben und
62, 63 deren Aufschließung billig schien, die Siedlungen Burg Kniphausen, Schilldeich,
30, 34 Voslapp, Cäciliengroden usw. *Amtshauptmann Hermann Ott* ließ daher noch keinen
Bebauungsplan aufstellen.

Der konkrete Anstoß zur Aufstellung eines *Bebauungsplanes* für Breddewarden kam
erst, als sich *Fritz Nobe* von der *Marineintendantur* zum Jahreswechsel 1938/39 wegen der
66 angeblich schnelleren Erschließbarkeit dafür aussprach, Breddewarden der Siedlung
64 Langewerth vorzuziehen[493]. Breddewarden lasse sich an das 1939 zu bauende Pumpwerk
Fedderwardergroden anschließen, und die *RUGES* sei vom *OKM* beauftragt, das Land
für Breddewarden bis zum 01.07.1939 zu beschaffen. Der Kern der Nordstadt,
Fedderwardergroden, war am 01.06.1939 in die Zuständigkeit der *Stadt Wilhelmshaven*
gekommen. Da der in Arbeit befindliche *Wirtschaftsplan vom* 03.08.1939 einen
„selbständigen städtebaulichen Raum mit den vier Ortsteilen Fedderwardergroden,
Voslapp, Breddewarden und Utters"[494] auswies, fühlte sich *Oberbürgermeister Müller*
gefordert. 1939 sollten noch 7.500 Wohnungen für Arbeiter und Angestellte geschaffen
37 werden, und neben Fedderwardergroden erschien nach Wegfall von Langewerth[495]
Breddewarden schnell zu realisieren. *Marineintendantur* und *OKM* befürworteten Ende
1938, daß die *Stadt Wilhelmshaven* Breddewardens Planungen in die Hand und am
01.03.1939 einen *Bebauungsplan* vorlegen solle, auch wenn die Eingemeindung erst zum
01.04.1940 vorgesehen sei – dann nämlich, wenn der erste Bauabschnitt Breddewardens
bezugsfertig sein sollte[496]. Straßenbau und Kanalisation sollte die *Jade* oder ein „sonstiger
Bauträger als Unternehmer durchführen. *Wilhelmshaven* könne die Straßen nach
Eingemeindung übernehmen. Die noch zuständige *Gemeinde Kniphausen* sollte die später
zuständig werdende *Stadt Wilhelmshaven* mit der Erstellung der Gemeinschaftsbauten
beauftragen. Die *Marineintendantur* war offenbar überzeugt, daß das Projekt durch das
Amt Friesland nicht zügig gefördert werden könne und stritt sogar dafür, daß
Wilhelmshaven Baupolizeibefugnisse für das außerhalb der Stadt liegende Gebiet von
Breddewarden bekommen solle. Man setzte auf die besonders enge Verflechtung
zwischen der *Stadtverwaltung Wilhelmshaven* und der *Wohnungsbaugesellschaft Jade*[497].

Auch bauwirtschaftlich erschien Breddewarden „immer noch schneller möglich"[498] als
Langewerth, denn die Baubaracken für die Unterbringung der Bauarbeiter von
Fedderwardergroden standen 1939 und boten sich zur Weiternutzung für das nur durch
einen 500 m breiten Grüngürtel getrennte Projekt Breddewarden an. Um das Projekt zu
beschleunigen, und dem Argument des zu knappen Arbeiterlagers zu begegnen,
98 beantragte die *Marine* am 23.05.1939 beim *RAM* über den *MDF* Baracken für 2000
Mann. Das *Marinestandortbauamt* stellte die zerlegbaren Holzbuden auf[499].

Die *RUGES* hatte bereits am 16.12.1938 den Auftrag zum Grunderwerb von
Breddewarden bekommen. Um die Bauern nicht zu beunruhigen, sollte es zu keiner
3 weiteren Prioritätenumstellung mehr kommen[500]. Das *Oldenburgische Staatsministerium*
14 und die *LOB* waren skeptisch und bezweifelten die Auffassung von *Marineintendantur* und
4 *OKM*, daß Breddewarden noch 1939 begonnen werden könnte und empfahlen nach wie

vor, Langewerth zu beginnen, mit dem Baubeginn für Breddewarden jedoch bis zur vorgesehenen Eingemeindung am 01.04.1940 zu warten, da Voslapp noch nicht zur Zufriedenheit fertiggestellt war und der Baubeginn von Fedderwardergroden unmittelbar bevorstand[501]. Bauarbeiten in drei unmittelbar benachbarten Ortsteilen hätten organisatorisch höchste Anforderungen und einen unverhältnismäßig hohen Aufwand bedeutet. Die Oldenburger Behörden stimmten schließlich zu, wenigstens die Planung im Hinblick auf die erhoffte Eingemeindung durch die *Stadt Wilhelmshaven* schon 1939 zu beginnen. *64*

Hans Bartning, seit November 1938 beim *Stadterweiterungsamt Wilhelmshaven*, begann Anfang 1939. Folgende *Bebauungspläne* sind bekanntgeworden:

- *Erster Entwurf eines Bebauungsvorschlages für Breddewarden,* verfaßt von *Hans Bartning* am 24.02.1939 (Unterschrift *Wilhelm Hallbauer*, 24.02.1939)
- *Breddewarden, erster Vorschlag*, verfaßt von *Hans Bartning* am 24.02.1939 (Unterschrift *Wilhelm Hallbauer*, 24.02.1939)
- *Wilhelmshaven Nord, Ortsteil Breddewarden,* vom 24.02.1939 (Unterschrift *Hans Bartning, Wilhelm Hallbauer*)
- *Wilhelmshaven Nord, Ortsteil Breddewarden, vom* 24.02.1939, geändert nach Besprechung mit *Marineintendantur* und *Niedersächsischer Bauträgergesellschaft* am 28.04.1939 (Unterschrift *Hans Bartning, Wilhelm Hallbauer*)
- *Wilhelmshaven Breddewarden*, gezeichnet von *Friedrich Heuer*, undatiert, geprüft von *Hans Bartning* 30.07.1939 (Unterschrift *Wilhelm Hallbauer* 03.08.1939)
- *Wilhelmshaven Breddewarden Lageplan vom* 18.11.1939, von *Friedrich Heuer*
- Darstellung Breddewarden entsprechend der Planung *Heuer* im *Stadtplan Wilhelms-* *66*
  haven vom Juni 1940 (Unterschrift *Walter Temp, Helmuth Baur*)
- Darstellung Breddewarden entsprechend der Planung *Heuer* im Plan *Städtebauliche Planung mit Verkehrs- und Grünplanung für Wilhelmshaven, Altstadt und Nord, vom Mai* 1940, abgeschlossen am 01.06.1941 (Unterschrift *Helmuth Baur*).

Die ersten Skizzen für Breddewarden gingen auf *Hans Bartning* zurück[502] – *Hallbauer* hat bestenfalls am 28.02.1939 den Erläuterungsbericht dazu verfaßt, möglicherweise aber nur unterschrieben. Die *Stadt* hatte pünktlich zum 01.03.1939 dem drängenden *OKM Bebauungspläne* vorgelegt.

Die im Erläuterungsbericht[503] beschriebenen städtebaulichen Entwurfsprinzipien gingen von der Landschaft und „vier alten großen Höfen" aus, die die Dorfwurt krönen sollten. Es war klar, daß hier die beherrschenden Bauten der Gemeinschaft angeordnet werden sollten. Wasserzüge, Baumbestand, Grünflächen und wechselnde Blicke sollten in die Planung einbezogen werden. Methodisch und systematisch baute *Hans Bartning* seine Planung von den Verkehrsbeziehungen her auf. Vom Bahnhof Breddewarden der Ringschnellbahn aus sollte der für über 3.000 Wohnungen gedachte neue Ortsteil durch eine 800 m lange Nord-Süd-Achse erschlossen werden. Die Blickbeziehung vom Bahnhof zur „Stadtkrone" auf der Wurt bestimmte den *Ersten Entwurf eines Bebauungs-*

vorschlages für Breddewarden, der die Grundlage abgab für alle weiteren Stufen. *Bartnings* Vorschläge mündeten in den vorläufigen *Bebauungsplan vom* 24.02.1939. Der wortgewandte *Hallbauer* interpretierte die Wurt als „Burghügel", zu dessen Füßen sich der Marktplatz langsam organisch entwickeln sollte. Der Ort sollte „im Halbkreis um diesen Wurthügel" angeordnet werden, gegliedert durch vorhandene Hauptwasserzüge im Grün. Nach Vortrag bei *Nicolaus* und *Ostendorf* sowie nach Abstimmung mit *Marine* und Bauträger wurde in einem am 28.04.1939 geänderten Entwurf der Wurthügel vom Durchgangsverkehr befreit und die geschlossene Bebauung des Ortskernes in offene Reihen umgewandelt. *Nicolaus* wollte keinen „aufwendig städtischen" Ortskern. Tatsächlich jedoch war die Infrastruktur für die möglicherweise 15.000 Einwohner von Breddewarden durchaus städtisch und perfekt vorgesehen: Filmtheater, Verwaltungsstelle, Post, Großgaragen, Cafes, 5 Tankstellen, 130 Läden und Gewerbebetriebe. Die vorhandenen Hofstellen sollten als Kinderheim, Gastwirtschaft, *NSDAP*-Gemeinschaftshaus und *HJ*-Heime eingebunden werden. Insgesamt waren 4 Schulen, 3 Sportplätze und 5 *HJ*-Heime für die „völkische" Erziehung der Breddewarder Jugend eingeplant.

Die Straßenstruktur von Breddewarden war modern, verkehrsgerecht und wenig auf Symmetrie und Blickpunkte ausgerichtet. Die Durchgangsstraße wurde im Zentrum abgeknickt. Viele Tangentialstraßen waren gebogen, zahlreiche Nachbarschaften und Wohnstraße sollten ein ruhiges Wohnen in Breddewarden ermöglichen. Der *Bartning*-Plan wurde am 06.03.1939 vom *MDF* genehmigt mit der Auflage, die Änderungsvorschläge für den Ortskern von *Friedrich Nicolaus* einzuarbeiten. Die Planung wurde neben
112 anderen Ortsteilplanungen am 01.04.1939 *Hitler* im Rathaus präsentiert[504]. Danach wurde der Bremer Architekt *Friedrich Heuer* eingeschaltet.

★

Für die Durchführung der 3.164 Wohnungen um die Wurt Breddewarden wurde im Februar 1939 von *OKM* und *RAM* die *Niedersächsische Bauträgergesellschaft Hannover* beauftragt[505]. Technischer Geschäftsführer war *Reg.Baumeister a.D. Spröde*. Da die *Bauverwaltung* personell äußerst knapp besetzt war, beauftragte die *Niedersächsische*
66 *Bauträgergesellschaft* 1939 *Friedrich Heuer* mit der Weiterbearbeitung des *Bebauungsplanes* und der Entwicklung von Typenplänen[506].

Friedrich Heuer, der sich 1936 bei der *NSDAP* mit seiner Siedlung Grolland beliebt
34 gemacht hatte, entwarf 1937 die Siedlung Cäciliengroden und 1938 *Bebauungspläne* für Altengroden. Er wurde von der *LOB*, vom *Reichsheimstättenamt Gau Weser-Ems* und von *Gauleiter Carl Röver* gefördert. *Heuer* sollte seine Planung so einrichten, daß die Bauarbeiten am 1. Juli 1939 beginnen könnten. Anläßlich einer Besprechung in der *Marineintendantur* am 07.06.1939 genehmigten *Karl Pfeiffer, Fritz Karl Schneider, Hans Bartning* und *Karl Franzius Heuers* Pläne für Breddewarden[507].

66 *Heuers Bebauungsplan* veränderte nicht die Konzeption von *Bartning*. Neu war der „Grünkeil", der sich von Norden her an die Wurt Breddewarden mit seinen

Gemeinschaftsbauten heranschieben sollte. Dies führte zu einer deutlichen Zerteilung Breddewardens in zwei gleich große „Zellen", verbunden nur durch den Ortskern und durch die Gemeinschaftsbauten. *Bartning* bezeichnete diese Aufteilung als „Schmetterling"[508]. *Heuers* Aufteilung in zwei Hälften war jedoch auch die Vorwegnahme der „Ortsgruppe als Siedlungszelle". Zusammen mit *Konstanty Gutschow* propagierte *Heuer* – orientiert an Parteigliederungen – wenig später die Idealstadt mit 8.000 Menschen für eine Ortsgruppe[509].

Durch *Heuers* Überarbeitung sollte Breddewarden Plätze für Feiern und Aufmärsche bekommen und im Zentrum „geschlossen" werden. In Planerläuterungen wurde jetzt nur noch NS-konform von Nord-Süd-Achse geschrieben, Begriffe, die bei *Hans Bartning* fehlten. Sein funktionales Trabanten-Konzept konnte durch *Heuer* bestenfalls noch im Zusammenhang mit den Gemeinschaftsbauten und der Gestaltung der Achsen nazifiziert werden.

Die Grünplanung für Breddewarden wurde von *Friedrich Heuer* unter dem Einfluß von *Heinrich Wiepking-Jürgensmann* ausgeweitet. Oberingenieur *Paulsmeier*, der auch für Fedderwardergroden und Voslapp tätig war, machte ein Bodengutachten, wonach das Grundwasser durch vier Becken abgesenkt werden sollte[510]. Mit dem Aushub wollte man die Straßen aufhöhen. Also wurde geplant, die Bachläufe in die Becken zu lenken, zu Seen aufzustauen und durch grachtenähnliche Wasserläufe miteinander zu verbinden. Der See südlich der Dorfwurt sollte eine Badeanstalt bekommen. Im November 1941 stellte *Heinrich Wiepking-Jürgensmann* den Grünplan Breddewarden nach Absprache mit *Friedrich Nicolaus, Wilhelm Wortmann* und *Fritz Karl Schneider* als Beihilfeantrag an das *RAM*[511].

Im August 1939 waren bereits eine Baustraße für Breddewarden fertiggestellt, der *Fluchtlinienplan* beim Stadterweiterungsamt in Arbeit, Typenentwürfe im Maßstab 1 : 100 von *Friedrich Heuer* gezeichnet, Gemeinschaftseinrichtungen geplant, Ausschreibungen veröffentlicht. Im August 1939 sollte mit dem Westteil begonnen werden[512]. Der Bauträger hatte schon Baustoffe bestellt und den Hafen Hooksiel für deren Anlandung ausgebaggert. Der Kriegsbeginn verzögerte den Baubeginn. Möglicherweise hatte *Stadtbaudirektor Walter Temp* die Termine gegenüber den Aufsichtsbehörden auch geschönt, denn noch im Januar 1940 ging es beim *RAM* um Geländevermessung, Bestandspläne, Typenpläne usw.[513]. Im Juli 1940 stimmten *RAM* und *OKM* endlich offiziell dem *Bebauungsplan* Breddewarden zu mit der Einschränkung, die Teiche und *66*
Wasserflächen zu vermindern und die Badeanstalt zu streichen[514].

1941 griff die *Marineintendantur* noch einmal in Breddewarden ein, um die ins Stocken *14*
geratene Planung „flott zu machen"[515]. Die *Stadt Wilhelmshaven* erklärte sich sogar bereit, selbst Grunderwerb für Breddewarden zu betreiben, da die *RUGES* nicht vorankam[516]. Die *Stadt* kündigte in der Propaganda den Baubeginn für die 3.000 Wohnungen mit „im Verlaufe des Jahres 1941" an[517]. Nach wie vor wurde die Eingemeindung von *Amtshauptmann* und *Ministerpräsident* herausgezögert. Plötzlich bezweifelte auch die *Marineintendantur* die bevorstehende Eingemeindung und überhaupt die Planungszustän-

digkeit der *Stadt Wilhelmshaven*[518]. Mittlerweile schritten die Tiefbauarbeiten voran. Es sind nur wenige Gebäudeentwürfe für Breddewarden bekanntgeworden, so z.B. für die „Schule Breddewarden 1" vom August 1939 von *Ernst Boyken*. *Heuers* Wohnhausentwürfe trennten streng zwischen:

- 2.831 Wohnungen für Arbeiter
- 249 Wohnungen für Angestellte
- 84 Wohnungen für *Unteroffiziere*

15 Wohnungstypen waren für Arbeiter, 5 für Angestellte vorgesehen. Gestaltungsfragen hatten für Breddewarden selbstverständliche Bedeutung. Die *Marineintendantur* als Mittelinstanz mischte sich ein. Sorgsam wurde mit *MDF* und *RAM* abgestimmt, daß der zweigeschossige Kernbereich von Breddewarden geputzte Fassaden bekommen durfte – die eingeschossige Randbebauung jedoch war in Ziegelbau vorgesehen. Die Stadtlandschaft Breddewarden sollte zur Wurt und zu den Parteibauten hin orientiert sein. Landschaft, Bodenart, Topographie, Himmelsrichtung, Fernblick, Wasserläufe, waren individuell zu berücksichtigen. Man wollte die Bewohner gewinnen, ihre persönliche Liebe zu Volkstum und Heimat benutzen. Das wäre mit der Gemeinschaftssiedlung um die Wurt Breddewarden sicherlich hervorragend gelungen, wenn sie nicht Utopie geblieben wäre.

3.7.6. Reichsheimstätten in Schortens

Schortens, Teil der 1933 gebildeten *Gemeinde Oestringen* war, wie auch Heidmühle und Ostiem, schon vor 1900 beliebtes Wohngebiet für Arbeiter der *Werft*. Billige Grundstücke ermöglichten bereits vor dem Ersten Weltkrieg Ein- und Auspendeln über die 1870 gebaute Bahnlinie Sande-Jever. Privat wurden kleine Nebenerwerbsstellen auf Grundstücken hochgemauert. *Karl Jahn* (1877–1958), Hochbautechniker bei der *Marineintendantur*, zog die Werftarbeiter in die Gemeinde Schortens, deren Vorsteher er 1919–33 war[519]. Zwischen den Kriegen teilten die Werftarbeiter häufig ihre Grundstükke auf oder erwarben für ihre Kinder Baugelände dazu und vermehrten die ungeordnete Besiedlung, gefördert durch das *Reichsheimstättengesetz von* 1920 und das *Reichssiedlungsgesetz von* 1919. Auch Domänengelände erweiterte die Siedlungsmöglichkeiten in Schortens. Viele kleine Bauvereine und Genossenschaften aus Wilhelmshaven/Rüstringen versuchten, sich in der Weimarer Republik in Schortens zu etablieren. Der Nebenerwerb sicherte vielen arbeitslos gewordenen Werftarbeitern ein Überleben zwischen den Kriegen.

Nach 1933 erfuhr der Kleinsiedlungsbau eine Belebung – das Konzept jedoch wurde nicht geändert. Jetzt trat das *Reichsheimstättenamt Gau Weser-Ems* unter *Johannes Richter* auf den Plan, möglicherweise auch das *Domänenamt Oldenburg* unter *Georg Schipper*. Beide planten für Gemeinden, die nicht dazu in der Lage waren[520]. Wie auch in

Sanderbusch, Kniphausersiel usw. setzte sich der Kleinsiedlungsbau bruchlos in das „III. 25, 22
Reich" hinein fort.

Bereits im September 1937, im Plan *Wirtschaftsgebiet Wilhelmshaven und Umgegend*, waren in Schortens neue Wohngebiete in weiträumiger Bauweise vorgesehen. Der Erläuterungsbericht sagte aus, daß in Heidmühle – Schortens die Pläne für 700 bis 800
Siedlungen und Eigenheime bereits fertig vorlagen. 1940 wies der *Wirtschaftsplan* für 129
Schortens aus:

- Klosterneuland als „geplante Siedlungsfläche"
- Klosterfelde als „geplante Siedlungsfläche"
- Papenmoorland als „mögliche Siedlungsfläche"
- Bohlswarfe als „mögliche Siedlungsfläche"

Die durch Grünstreifen luftschutzgerecht getrennten Siedlungsteile rundeten sich zu einer fächerförmigen Flächendisposition um den alten Ortskern von Heidmühle. *Bebauungspläne* für Klosterfelde, Papenmoorland und Bohlswarfe sind nicht bekanntgeworden. Die gesamte Neugestaltung Schortens jedoch, von *Johannes Richter* entworfen, war Gegenstand einer Sonderveröffentlichung in der Zeitschrift „Bauen Siedeln Wohnen" im März 1940. *Gauleiter Carl Röver* schrieb das Geleitwort und gab die Parole aus: „Die neu entstehenden Städte, Dörfer und Heimstätten im Gau Weser-Ems müssen in ihrer Grundlinie landschaftlich gebunden sein und der Wesensart unserer Bevölkerung Ausdruck verleihen". Der *Gauobmann der DAF*, gleichzeitig *Leiter des Wohnungsbauwesens im Gau Weser-Ems* und *Mitglied des Reichstages, Bruno Dieckelmann*, beschwor, „gesunde, kräftige Menschen, die nur in gesunden Wohnungen und schönen Heimen heranwachsen können ... Die Heimstätte der deutschen Familie, der Keimzelle und Erhalterin des ewigen Fortlebens unseres Volkes, muß mehr sein und mehr geben, als nur die Befriedigung eines nackten Lebens- und Wohnbedürfnisses".

Der 1940 veröffentlichte *Bebauungsplan Klosterneuland* zeigte südlich der späteren 67
Bundesstraße 210 ein Siedlungsgebiet mit 200 Kleinsiedlungsstellen auf großen Grundstücken sowie etwa die gleiche Anzahl von Mietwohnungen. An Gemeinschaftsbauten waren eine Schule, ein *NSDAP*-Gemeinschaftshaus, ein *HJ*-Heim, ein Gewerbetrakt und ein Aufmarschplatz vorgesehen. Die ein- und zweigeschossigen Gebäudetypen von *Johannes Richter* zeigten einfachen Standard, unten 2 Stuben, oben 2 Kammern. Das Äußere war schmucklos traditionell. Es hätte sich, wenn Schortens realisiert worden wäre, kaum von der im *Wohnsiedlungsgebiet Wilhelmshaven-Jeverland* üblichen Architektur unterschieden.

4. Militärplanungen im Raum Wilhelmshaven bis 1945

Ergriff uns die Flut, rasch schwemmt sie uns in den Jadebusen, fast bis in den Hafen, fast bis an den Kai. Kreidebleich stehen wir auf Deck, einundfünfzig Seemeilen auf einem Moldauschiffchen in stürmischer See – ach, arme „Lanna 8", Don Quichotte des Wassers, du siehst ramponiert aus! Morgen heißt es, alles Seewasser auspumpen, das Salz abschlagen, die Rohre reinigen und Süßwasser nehmen. Rings um uns der riesige Kriegshafen Wilhelmshaven. Jung ist das Gesicht dieser Stadt, riesig ihre Figur, aber scharf tritt ihr hippokratischer Zug hervor. Ein versenkter Hulk, verödete Bassins. Die Stadt stirbt[1].

Egon Erwin Kisch, 1920

4.1. Militärische Interessen und Stadterweiterung

Wilhelmshavens Militärbauten, insbesondere die Planungen der NS-Zeit können nicht ausschließlich unter baukünstlerischen Gesichtspunkten betrachtet werden. Die Wertung von Bauproduktion, Architekturform, Propagandainhalt und Dekoration müssen die militärischen Ziele und politischen Funktionen einbeziehen. Die Grundlagen für größenwahnsinnige Stadterweiterung und totalitäre Neugestaltung waren militärisch. Die Berliner Marinedienststellen wollten die *Bauprogramme* und haben sie unnachsichtig vorangepeitscht. Die hier zu betrachtenden marineeigenen Militärplanungen mußten sich auf den zivilen Städtebau auswirken. Militärische Wurzeln von Stadtplanung und Architektur lassen sich am Beispiel Wilhelmshaven gut freilegen. Sie bestimmten die Stadt seit ihrer Gründung und durchzogen die Architekturgeschichte bis heute.

Gotthilf Hagen, seit 1854 Leiter des *Dezernates für Land- und Wasserbau der Königlichen Admiralität* und ab September 1855 *Vorsitzender der Kommission für den Bau eines Kriegshafens an der Jade*, hatte den englischen Ingenieur *Sir John Rendel* und *Heinrich Hübbe* eingeschaltet. Man führte einen Architektenwettbewerb durch. Am 13.04.1855 besichtigten *Hagen, Rendel, Hübbe* und *Kapitän Hyltén-Cavallius* das Jadegebiet. Beide Wettbewerbsteilnehmer befaßten sich neben den wassertechnischen Belangen mit repräsentativen Paradeplätzen, Werften, Militärquartieren, Depots, Werkstätten, Schießständen, Exerzierplätzen und Kasernen. Städtebaulich besonders aufwendig muß der

Entwurf *Rendel* gewesen sein. Man schätzte ihn auf knapp 30 Mio. Thaler. Schließlich reichte *Hagen* am 29.05.1856 der Admiralität selbst Entwurf und *Denkschrift* ein.

Im 2. Teil der *Hagen-Denkschrift* von 1856 legte *Generalmajor Fischer* seine Vorstellun-
6 gen über die militärische Befestigung der Stadt vor[2]. *Fischer* hatte mit einem Schriftsatz an *Prinz Adalbert* vom 25.10.1854 gefordert, daß Hochbauten mind. 3.000 Schritte von den Punkten entfernt sein sollten, bis zu denen feindliche Schiffe sich nähern konnten. Aus dem Grenzverlauf der Stadt ergab sich die Lage der Flügelbefestigungen und des Walls. Ein Geschützturm mit 45 Geschützen sollte die Hafeneinfahrt sichern. Hinter dem Deich sollten noch Schießschartenmauern angeordnet werden. Das ganze umschloß ein
6 Wassergraben. *Friedrich Wilhelm IV.* genehmigte am 25.06.1856 den *Hagen-Plan*, *Fischers* Befestigungsentwurf und die Stadtanlage. Strikt wies er darauf hin, daß das „Alignement" für die Militärbauten festzustellen sei; im übrigen Richtung der Straßen, Höhe und Material der Häuser unter Beachtung der für Bauten im engeren Rayon einer „Festung" geltenden Grundsätze auszuführen seien[3].

Die militärischen Forderungen ruhten bis 1861, als *Major Klotz* den Bahnhof und ganz Heppens in die Umwallung einbeziehen wollte[4]. Der *Klotz-Plan* erweiterte das Stadtgebiet und machte Platz für 4 geschlossene Forts. 1863 ergänzte *Major Klotz* seine
7 Planung um eine weitere *Denkschrift*.

Die traditionelle Umwallung Wilhelmshavens sollte dem Zivilschutz dienen. Wie anders wäre es zu erklären, daß *Moltke* am 25.02.1864 die Umwallung ablehnte[5] mit dem Hinweis, daß der Hafen nur als Flottenstützpunkt, zum Instandsetzen und Ausrüsten der Schiffe dienen sollte – eine bürgerliche Stadt sollte Wilhelmshaven nicht werden.
8 Festungen, wie von *Fischer, Hagen, Klotz* und danach noch von *Stoltz*, entworfen, wurden sinnlos, da Geschütze immer längere Distanzen überwinden konnten. Alsbald verzichtete man ganz auf einen Schutz des Hafens und verschob die gesamte städtebauliche Anlage 300 m nach Osten. Schwimmende Batterien und Kanonenboote, nicht Bauten, sollten den Schutz des Hafens übernehmen. Nach den Erfahrungen des Krieges von 1870/71 wurden alle Umwallungspläne fallengelassen und es kam 1876–80 noch zu 4 Fortanlagen[6]: Heppens, Mariensiel, Schaar und Rüstersiel. Die Forts wurden mit Fortifikationsstraßen und Feldbahnen verbunden. Nach dem Ersten Weltkrieg durften die Fortanlagen stehen bleiben, jedoch nicht erweitert oder dem waffentechnischen Fortschritt angepaßt werden. Sie hatten somit keine militärische Bedeutung mehr[7].

Die militärischen Interessen der *Reichsmarine* der Weimarer Republik[8] blieben für Architektur und Städtebau Wilhelmshavens belanglos. Der Flottenbau war fast zum Erliegen gekommen, und in welchem Maße die *Marine* überhaupt sinnvoll sei, war strittig. Die Auflagen des *Versailler Vertrages* machten die *Marine* wertlos und wurden als demütigend empfunden. In Wilhelmshaven behalf man sich mit *Arbeitsbeschaffungspro-*
10 *grammen*. Viele Nebenbetriebe der *Reichsmarine* mußten schließen, die *Marinewerft* befaßte sich mit Ersatzbauten, Zwischenaufträgen und Reparaturen. Zurückgehende Einwohnerzahlen, ansteigende Arbeitslosenquoten und geringe Bautätigkeit führten zu

einer Stagnation der Stadtentwicklung, die von den aufkommenden Nationalsozialisten geschickt ausgenutzt wurde.

★

Städtebau und Architektur wurden im „III. Reich" von den Berliner Ministerialinstanzen bestimmt. Konflikte der um Planungshoheit kämpfenden *Stadt Wilhelmshaven* mit den umliegenden Gebietskörperschaften und mit der *NSDAP* blieben nicht aus. Die Wilhelmshavener Militärplanungen bewegten sich im Spannungsfeld diskreter Verschwiegenheit und spektakulärer Propaganda. *112*

Freigebliebene weiße Flächen in den sowieso schon „geheimen" *Wirtschaftsplänen* *126-*
deuteten auf die Vertraulichkeit militärischer Planungen. Verkäufliche Stadtpläne *129*
amputierten ganze Stadtteile und Hafenbereiche, soweit sie militärisch genutzt waren. Jegliche Hinweise auf Marinedienststellen fehlten im „Pharus-Stadtplan" 1941 von Groß-Wilhelmshaven. Ein Kuriosum für eine hochgerüstete „Seefestung", deren Einwohner ausschließlich von der *Marine* abhingen.

Freie Architekten wurden wegen des „Geheimschutzes" für die marineeigenen Militärbauten kaum eingeschaltet[9], wohl aber für Wohnungs- und Gemeinschaftsbauten. Baukunst, eigentlich die „öffentlichste" der Künste, war bei den Militärbauten „geheim". Auch behördeneigene Architekten und Ingenieure der militärabhängigen Ortsbaudienststellen in Wilhelmshaven, *HNBD, Marinestandortbauamt, KMW, Reichsbauamt* und *Luftwaffenbauamt*[10], hatten oft nur Teilinformationen von den ihnen übertragenen Bauten.

Wohnungsbau, Monumentalplanungen und Bauten der Gemeinschaft wurden von
der *NSDAP* propagandistisch ausgeschlachtet. Militärplanungen, und mit ihnen alle *101*
Pläne, die sie zeigten, waren zwar „geheim", ließen sich jedoch in der „Waffenschmiede" Wilhelmshaven kaum ernsthaft geheimhalten. *Wirtschaftspläne* für das *Wohnsiedlungsgebiet Wilhelmshaven-Jeverland* und die dazugehörigen Erläuterungsberichte waren „Geheime Reichssache", „Geheim" oder „Geheime Kommandosache". Auf Geheimnisverrat standen drakonische Strafen. Die *KMW* beispielsweise warnte ihre Arbeiter, in Alkohollaune „Geheimnisse" auszuplaudern. Stadtplaner und Architekten der Gebietskörperschaften, die Wilhelmshaven erweitern und neugestallten sollten, tappten oft im Dunkeln, und kannten die Ziele der Berliner Militärbehörden nicht. Jedes *Bauamt* hatte seine eigenen Sachbearbeiter für „Geheim"-Angelegenheiten.

Die auf Stadtplanung und Architektur bezogenen militärischen Interessen wurden von
der *Marineintendantur* als Mittelinstanz wahrgenommen. Die *Abteilung XXIV.* vertrat die *14*
militärischen Interessen, gab sie an die Stadtplaner weiter und kontrollierte die Ergebnisse. Jede Veränderung vom Wohnungsbauzielzahlen hatte militärische Ursprün-
ge und wurde vom *OKM* bestimmt. Die Wilhelmshavener *Stadtverwaltung* wehrte sich *4*
heftig und oft gegen die unvermittelten Entscheidungen der Wehrmacht. Dies nützte ebensowenig wie gute parteipolitische Verbindungen. Die *Oberkommandos* von *Marine*

und *Luftwaffe*, die für den Raum Wilhelmshaven wichtig waren, zitierten die Dienststellen nach Berlin und instruierten mündlich. Vermerke über Besprechungen waren die Regel. Verfügungen, *Erlasse* und *Dienstanweisungen* kamen selten vor.

Genaue Zahlen über die Steigerung der Rüstungsausgaben, insbesondere der
1 Bauinvestitionen in Wilhelmshaven, liegen nicht vor. Als grober Anhalt der Steigerungsraten mögen hilfsweise die „sonstigen einmaligen Ausgaben" im *Marinehaushalt* 1932–39 dienen. Hierin sind weder Schiffbau noch fortdauernde Ausgaben enthalten, jedoch Bauinvestitionen für die gesamte *Reichs-* bzw. *Kriegsmarine*[11]:

1932	21,5 Mio. Mark
1933	110,2 Mio. Mark
1934	170,3 Mio. Mark
1935	208,5 Mio. Mark
1936	331,3 Mio. Mark
1937	536,7 Mio. Mark
1938	894,0 Mio. Mark
1939	1.256,1 Mio. Mark

Die Kosten für alle marineeigenen Bauten einschließlich Hafenbau ohne Wohnungsbau wurden für Wilhelmshaven in der Zeit von 1933–45 mit 350 Mio Mark angegeben[12]. Man kann davon ausgehen, daß die Rüstungsausgaben des „III. Reiches" 60 % der Reichsausgaben ausmachten. Entsprechend der rasant steigenden Rüstung stiegen die Ausgaben für Bauten, entsprechend nahmen Bedeutung und Umfang der Baudienststellen auch in Wilhelmshaven zu.

4.2. Marineeigener Hochbau

Die *Marine* bediente sich in Wilhelmshaven/Rüstringen für ihre eigenen Hochbauten neben der *KMW* und der *HNBD* der Ortsbaudienstellen – bis zum Ersten Weltkrieg des *Marine-Garnison-Bauamtes*, von 1922–43 des *Marinestandortbauamtes*. Architektur und Dekoration, Anlage und Habitus von Kasernen, Kommandanturen, Stationsgebäuden, Lazaretten, Kasinos und Werftanlagen in Wilhelmshaven gingen bis zum Ersten Weltkrieg auf die Leitlinien der Berliner Dienststellen, das *Preußische Handelsministerium*, das *Kriegsministerium* und die *Admiralität* zurück[13]. Historisierend, martialisch und, soweit zur Repräsentation herangezogen, auch prächtig geschmückt, wurden Militärbauten auf die Region abgestimmt. *Garnison-Bauordnungen* und *Dienstanweisungen* aus Berlin regelten das Allernotwendigste. Der alltägliche militärische Hochbau wurde von den populären Baumeistern kaum beachtet. Unterkünfte und Garnisonbauten, Festungsanlagen und Rüstungswerkstätten kamen in Architekturführern und Bildbänden, Reiseführern und Kunstgeschichten nicht vor. Die militärischen Baubehörden traten wenig nach außen, schlossen sich ab und zogen ihren Nachwuchs selbst heran.

In der Weimarer Republik entstanden im Raum Wilhelmshaven kaum Militärbauten.
1935 wurde die *Reichsmarine* in *Kriegsmarine* umbenannt. Ab 01.06.1935 hieß die bisherige
Marineleitung am Tirpitzufer in Berlin *Oberkommando der Kriegsmarine (OKM)*. Der *Chef* 4
der Marineleitung hieß nun *Oberbefehlshaber der Kriegsmarine*. Ab 09.11.1935 trug die
Marineflagge *Hitlers* Hakenkreuz. Erst nach den *Wehrgesetzen von* 1935 konnte die
Marine wieder offen eigene Gebäude errichten und stockte ihre Baudienststellen auf. Die
militärische „Festung" als städtebauliche Leitlinie war unwichtig geworden. Marine- 8
hochbauten, das waren jetzt Funktionsgebäude, Unterkünfte, Sozialbauten, Zweckgebäude, Industriebauten.

Militärhochbau galt als amusisch und wenig künstlerisch. Er zog junge, ehrgeizige Architekten kaum an. Nur angesichts drohender Arbeitslosigkeit konnten die Marinebaubehörden ausreichend Fachpersonal rekrutieren. In Wilhelmshaven war der Marinehochbau jedoch keinesfalls unambitioniert. Die Generation der Marinebaubeamten der ausgehenden 30er Jahre hatte bei *Schmitthenner, Bonatz, Tessenow, Wetzel* und *Bestelmeyer* studiert und entwickelte unter der Leitung von *Otto Lückemeyer* ihren eigenen für Wilhelmshaven typischen *Marinestil*. Eine Ausnahme bildeten die Werftbauten der *HNBD* und des *Hafenbauressorts* der *KMW*. Vor allem der dort beschäftigte *Hans Klaucke* stach durch außerordentlich moderne Entwürfe hervor.

Exemplarisch sollen einige militärische Bautypen im Raum Wilhelmshaven vorgestellt werden[14]. Viele Bauten und Projekte harren noch der Erforschung und Aufarbeitung:

Marinefachschule (1937) 52
Marinekleiderkammer
Marineverpflegungsamt Schortens (1936)
Artilleriezeugamt Tannenhausen
Munitionsanstalt Schweinebrück (1935)
Artillerie-Depots (heutiges Industriegelände West)
Luftverteidigungskommando Rosenhügel[15]
Funkempfangsstelle Totenweg (1935/*Lückemeyer*)
Flakschießplätze Dangast
Kasernen der *6. Schiffsstammabteilung* Varel (1939–42/*Thees*) 88
Umbau Offiziersheim (1938–39/*Thees*)[16] 133
Ergänzungsbauten bei der *Westwerft* (1935–45)[17] 97
Ergänzungsbauten der *KMW* (ab 1937)[18] 96
Marinekraftfahrabteilung Ebkeriege (1939–40)
Marineinstandsetzungsamt Ulmenstraße (1937/*Franzius*) 69
Erweiterung *Marine-Observatorium* (1937/*Franzius*)
Ausweichvermittlung Antonslust (1939/*Thees*)
Wähleramt der Marine Langewerth (1939/*Thees*)
Wasch- und Badehaus *Westwerft* (1940/*Klaucke*)

9 1935 wurden erst einmal die ehemals Kaiserlichen Bauten der *Reichsmarine* wieder in Besitz genommen. Danach folgten Um- und Neubauten für alle denkbaren Militärzwecke. Keine Statistik führt sie auf, keine Zusammenstellung nennt Quadratmeter, Kubikmeter, Kosten, Architekten. Der Verfasser mußte sich mit Zufallsfunden und dünnen Archiven zufriedengeben. Insofern sind die Beispiele militärischer Architektur lückenhaft und ergänzungsbebedürftig. Sie können Grundlage für weiterführende Arbeiten sein. Beträchtlich viele marineeigene Bauten, mit Ausnahme der *KMW*-
121 Werkhallen, haben den Krieg überdauert, wurden wieder aufgebaut. Viele werden heute wieder militärisch genutzt.

129 Die Architektur der marineeigenen Bauten im *Wohnsiedlungsgebiet Wilhelmshaven-Jeverland* war, soweit öffentlich zugänglich, traditionell und restaurativ. Es ging um Darstellung der Wehrkraft und Symbolisierung des Regimes. Die bauliche Hülle für soldatische Eigenschaften ließ Stilpluralismus kaum zu. Das „III. Reich" verordnete keinen neuen Stil, es nutzte die vorhandenen Richtungen und übernahm Bewährtes, wo es paßte. So reichte die architektonische Vielfalt von der strohgedeckten Sendestelle bis
103 zur modernen Werfthalle, bis zum funktionalen Hangar.

Die militärischen Baubehörden mußten dem nationalsozialistisch verordneten Kampf gegen modernes Bauen folgen – jedoch nicht bedingungslos. Auch im Raum
93 Wilhelmshaven baute das *Hafenbauressort der KMW*, plante die *HNBD* äußerlich
102 modern, wo es zweckmäßig schien. Wo monströse Architektur jedoch Menschen beeinflussen sollte, wurden dominierende Achsen, strenge Symmetrie, hohe Steildächer und regionale Baustoffe verwendet. Architektur wurde autoritär benutzt, wo Menschen autoritär behandelt werden sollten. Da für Marinebauten kaum freie Architekten eingeschaltet wurden, konnten die Baubehörden selbst Gestaltung kanalisieren und Richtungen herausarbeiten. Das zentralisierte Marinebauwesen, konnte die verwirrende Stilvielfalt der Weimarer Republik selektieren, bündeln und umlenken.

4.2.1. Marinegerätelager Roffhausen

1 Für die Standortauswahl des *Marinegerätelagers Roffhausen* waren Luftschutzgründe maßgeblich. Es wurde bewußt außerhalb der Stadtgrenzen von Wilhelmshaven innerhalb der Gemeinde Oestringen angesiedelt. Es lag günstig zu:

- den seit 1936 in den *Wirtschaftsplänen* verzeichneten neuen Reichsstraßen 69 und
126– 210
129 - dem geplanten neuen Ortsgüterbahnhof im Südwesten Wilhelmshavens
94 - dem Industriegebiet um den Werftbahnhof Sande und dem *Industrieumsiedlungsprogramm*
- den *Artillerie-Depots* zwischen Mariensiel und Westhafen
- der seit 1939 in den *Wirtschaftsplänen* festgesetzten, westlich des *Marinegerätelagers* nach Norden abzweigenden Hauptstraße[19]

o der geplanten Ringschnellbahn, die das Gelände des *Marinegerätelagers* unmittelbar berührt hätte.

Noch 1937 lag das Grundstück des *Marinegerätelagers Roffhausen* im Schnittpunkt der
umliegenden neugeplanten Siedlungen Roffhausen, Mariensiel, Sanderbusch und
Weststadt Langewerth. Die Lage im Verkehrsschwerpunkt und die noch nicht verplante
freie Landschaft führten zur aufgelockerten Bauweise des *Marinegerätelagers*[20]. Das
Marinestandortbauamt unter *Otto Lückemeyer* erhielt 1936 den Planungsauftrag. Für den
dort tätigen *Schmitthenner*-Schüler *Kurt Geisenhainer* kam nur eine ländlich-regionale
Bauform mit Ziegeln und abgewalmten Pfannendächern in Betracht. 6 Lagerhäuser je 72
30 x 18 m und 3 je 65 m lange Hallen sollten Bauernhofdimensionen vortäuschen.
Besonders auffällig waren die jeweils über 10 m hohen Walmdächer im Verhältnis zu 73
dem nur 3,40 m hohen erdgeschossigen Ziegelmauerwerk. Die Konstruktion war jedoch
alles andere als ländlich. Bis zu vier Geschosse waren unter den Pfannendächern in
Betonskelettbauweise und strengem 6 m Stützenraster verborgen. Ausbau- und Kon-
struktionsraster waren gegeneinander versetzt, so daß nach außen hin Betonflächen nicht
in Erscheinung traten. Das stabile Betonskelett war auch gleichzeitig Tragkonstruktion
für Hebezeuge und Kranbahnen. Die schrägen Betonflächen unter der Dachziegelhaut
hatten die Aufgabe, Brandbomben abrollen zu lassen. Geschützt wurden die gelagerten
Marinegeräte, nicht die Menschen. Wasch- und Umkleideräume für Arbeiter und
Soldaten befanden sich jeweils in den ungeschützten, in herkömmlicher Zimmermanns-
bauweise konstruierten Spitzböden. Es entsprach ganz dem *Schmitthenner*-Vorbild[21],
Dachhäuschen und Gauben auf den geneigten Ziegelflächen zu plazieren. Die Dachform
war jeweils zur Traufe hin geschwungen, wie es bei Sparren mit Aufschieblingen der Fall
war. Auch Holzsprossenfenster, auskragende Traufgesimse, aufgedoppelte Holztore,
geschmiedete Laternen, Segmentbögen und verzierte Eisengeländer gehörten zum
Detailkatalog des *Marinegerätelagers*. Die Verwandtschaft mit dem im gleichen Jahr von
Kurt Geisenhainer entworfenen *Marinelazarett Sanderbusch* war nicht zu verkennen. In 77
Sanderbusch jedoch fanden sich mehr Dekorationen, (Führer-)Balkone, Werksteinum-
rahmungen, Hoheitssymbole, Steinmetzarbeiten und Kunstgegenstände.

Nichts anderes hatte auch *Paul Schmitthenner* gewollt: „zeitlose Architektur", die man überall in Deutschland verwenden sollte, für viele Zwecke geeignet, in verschiedenem Gewand. In Norddeutschland eben als Ziegelbau[22]. Regionales Bauen wurde bewußt dem *Internationalen Stil* gegenübergestellt. Im „III. Reich" wurden Zweckbauten und Industriehallen durchaus in moderner Formsprache erstellt[23]. Das Wilhelmshavener *Marinestandortbauamt* jedoch hat unter *Otto Lückemeyer* niemals *Internationalen Stil*, Flachdächer oder *Bauhaus*-Bezüge erwogen. Es verschleierte die kriegerischen Militärbauten mit rückschrittlicher Harmlosigkeit.

★

Städtebaulich war das *Marinegerätelager Roffhausen* ausschließlich nach Funktionsab-
läufen organisiert: Keine Zufälligkeiten, keine spitzen Winkel, keine gärtnerischen
72 Anlagen. Der Lageplan könnte ebenso die Rationalität eines modernen Industriebetriebes
widerspiegeln. Der Haupteingang war zentral angelegt mit Wache, Pförtnerhaus und
Vorplatz. Für Aufmärsche war kein Platz. Gegenüber waren ab 1940 40 Baracken für
2.000 Menschen[24] aufgestellt worden. Sie waren an einem Achsenkreuz von Lagerstraßen
aufgereiht. Dieses Barackenlager Roffhausen mit zwei Exerzierplätzen, Feuerlöschtei-
chen, Sportplatz und Offizierswohnungen beherbergte die Arbeitskräfte des *Marinege-
rätelagers*.

74 Das *Marinegerätelager Roffhausen* wurde nach dem Zweiten Weltkrieg übergangslos zu
Industrieanlagen umgewandelt – es gab keine geeigneteren Industriehallen als diese.
Schon im August 1946[25] lief in den ehemaligen Hallen die Produktion von Schreibma-
schinen an. Architekt *Franz Iwersen* baute sie nur geringfügig um und integrierte sie im
Laufe der Jahre in eine moderne Fabrikanlage. Da die Planungen von Ringschnellbahn
und nach Norden führender Hauptstraße nicht mehr realisiert worden waren, stand der
Expansion eines modernen Industriekomplexes nichts mehr im Wege. 4.300 Mitarbeiter
schafften 1952 in der Schreibmaschinenfabrik, zum größten Teil ehemalige Arbeiter der
KMW. Die Gebäude des ehemaligen *Marinegerätelagers* erfüllten noch lange ihre
73 Funktion als Teil des großen Industriebetriebes *AEG/Olympia*. Im Äußeren sind sie
traditionell gestaltet, im Inneren waren sie schon 1937 moderne Industriebauten.

Für die durch die Schreibmaschinen-Produktion einsetzende Nachkriegs-Wirtschafts-
blüte in Roffhausen war es günstig, daß die gegenüberliegenden Arbeiterbaracken noch
verfügbar waren. Sie boten Familien und Flüchtlingen, Arbeitskräften und Werksange-
74 hörigen provisorischen Wohnraum. Ab 1950 wurden Baracken nach Plänen von *Franz
Iwersen* durch Doppelhäuser ersetzt. Aus Kostengründen behielt er Wegestruktur und
technische Erschließung bei. Im Straßenraster sind noch heute Teile des Lagergrundrisses
zu erahnen. Weitere Siedlungen und Infrastrukturbauten ließen das ehemalige *Marine-
gerätelager* zu einer ansehnlichen Industriegemeinde wachsen.

4.2.2. Marinelazarett Sanderbusch

1 Das *Marinelazarett Sanderbusch*, das größte seiner Art in Deutschland überhaupt, war
128 erstmals im Plan *Wirtschaftsgebiet Wilhelmshaven vom Juni* 1937 als „Wehrmachts-Anlage"
zwischen dem Ems-Jade-Kanal und der alten Reichsstraße 210 ausgewiesen. Im Zuge der
Oldenburgischen Verwaltungsreform wurden Sande und Sanderbusch[26] 1933 der neugebil-
deten Großgemeinde Oestringen zugeschlagen[27]. Um das auf 1550 zurückgehende Gut
Sanderbusch herum hatte *Wilhelm Hallbauer* parzellenscharf den Flächenbedarf für das
neue *Marinelazarett* markiert, jedoch keinesfalls aus eigener Machtvollkommenheit, denn
die Baupläne waren nach Vorgaben des *OKM* zu dieser Zeit bereits nahezu fertig.

In Wilhelmshaven/Rüstringen gab es in den 30er Jahren ein *Werftkrankenhaus* (gebaut ab 1884), ein *Marinelazarett* (gebaut ab 1869), ein *Städtisches Krankenhaus* (gebaut ab 1889) und ein *Katholisches Krankenhaus* (gebaut ab 1904). Die vier Krankenhäuser lagen in der 9
Wilhelmshavener Südstadt, nahe den Arbeitsstätten der Bevölkerung. Dort mußte Erste Hilfe geleistet werden, dort waren Arbeitsunfälle häufig. Das *Marinelazarett*[28] entwickelte sich vor dem Ersten Weltkrieg bis auf 560 Betten. Ein zweites *Marinelazarett* mit 500 Betten wurde noch am 13.05.1914 genehmigt und sollte nahe dem *Marinebekleidungsamt* in Heppens gebaut werden. Die begonnenen Arbeiten wurden mit Kriegsausbruch eingestellt[29]. Mit der 1935 wiedereingeführten Wehrpflicht und der Aufrüstungspolitik des „III. Reiches" wurde das *Marinelazarett* in Wilhelmshavens Südstadt endgültig zu eng. 1939 mußte wegen des Raummangels auch die *Marine-Sanitätsschule* nach Sanderbusch ausquartiert werden. Ähnlich waren die Verhältnisse beim gleichfalls der *Marine* unterstehenden *Werftkrankenhaus*[30].

Da Wilhelmshavens Südstadt nach den *Wirtschaftsplänen* von 1937 keine neuen 128
Baugenehmigungen bekommen, schließlich ganz „ausgesiedelt" werden sollte, mußten neue Krankenhausstandorte gefunden werden. Die Planung *Wilhelmshaven Wirtschaftsplan vom* 03.08.1939 legte dann als neue luftgeschützte Standorte außerhalb des Stadtgebietes fest:

- *Marinelazarett Sanderbusch* mit *Sanitätsschule* und „Gefolgschafts"-Siedlung auf dem 75
  Gelände des Gutes Sanderbusch mit eigener Schnellbahnstation und neuer Straßenverbindung nach Wilhelmshaven.
- *Werftkrankenhaus* Barkeler Busch. Die Planungsarbeiten waren weit fortgeschritten, wurden aber mit Beginn des Krieges eingestellt. Beteiligt war *Hermann Distel.*
- *Städtisches Krankenhaus* zwischen Burg Kniphausen und Fedderwarden. Hierfür haben 151
  die Architekten *Hermann Distel* und *Benno Schachner* 1939–42 Planungen begonnen[31].

Auch der *Wirtschaftsplan Wilhelmshaven-Friesland vom Mai* 1940 fixierte die drei neuen 129
Krankenhausstandorte.

Der heraufziehende Zweite Weltkrieg ließ jedoch keine andere Wahl mehr, als die zum Abriß vorgesehenen Krankenhäuser in Wilhelmshavens Südstadt vorerst zu erhalten, sie sogar noch um Krankenhausbunker und Hilfskrankenhäuser zu ergänzen[32]. Die beengte Lage in der Südstadt führte dazu, daß die Krankenhäuser häufig zerstört – und der Betrieb nur mühsam aufrechterhalten wurde. Medizinische Versorgung für Bauarbeiter und Werfthandwerker, Marinesoldaten und Zwangsverpflichtete war dringend erforderlich. Die geplanten drei neuen Krankenhäuser im Grünen, weit ab von der Stadt in „gesunder Gegend mit frischer ozonreicher Luft"[33], blieben bis auf das Lazarett in Sanderbusch Utopie[34].

⋆

Entwurf und Bau des neuen *Marinelazaretts Sanderbusch* wurden 1936 dem *Marinestandortbauamt* unter *Otto Lückemeyer* übertragen. Der letzte Besitzer des Gutes Sanderbusch, *Redmar Daun*, verkaufte seine Ländereien samt Gebäude 1936 dem Fiskus zum Bau des Lazaretts. Parkgelände und Gutsgebäude wurden sofort unter Naturschutz gestellt und geschickt von Architekt *Kurt Geisenhainer* in die Gesamtplanung einbezogen[35].

In einem *Marinelazarett* wurden grundsätzlich nur Marinesoldaten von Marineärzten behandelt. Sie sollten gesunden und möglichst schnell wieder für den Dienst auf Schiffen „bereitgestellt werden". Wachgebäude, Arrestzellen, Mannschaftsräume, *Sanitätsschule*, Stabsarztzimmer, Offizierskasino, Tennisplätze usw. gehörten zum Raumprogramm. Die militärisch-hierarchische Gliederung war in Bau und Einrichtung, Außenanlage und Standard, nachvollziehbar. Wie zivile Krankenhäuser damals selbstverständlich nach Klassen gegliedert waren, wurde auch das militärische Krankenhaus nach Rängen abgestuft. Mannschaften mußten mit sparsamerer medizinischer Ausstattung zufrieden sein und sich Spinde und Waschtische teilen. Höhere Ränge hatten Anspruch auf eigene Bäder und Sofas in ihren Krankenzimmern. Familien oder Frauen von Mannschaftsdienstgraden hatten grundsätzlich keinen Anspruch auf medizinische Versorgung im *Marinelazarett*. Die medizinische Versorgung der Familien oblag den Soldaten-Familienärzten in den neuen Siedlungen.

Beim *Marinelazarett Sanderbusch* führte eine 150 m lange gerade Vorfahrt zu den zwei symmetrisch angeordneten Wachhäusern und zum „Kasernentor". Ein gewaltiger
75 Vorplatz von 80 m im Quadrat war für Aufmärsche vorgesehen. Der Lazarettgrundriß war durch die im Krankenhausbauwesen stark genormten Funktionsabläufe bestimmt[36]. Der Schwerpunkt lag in den Bereichen Chirurgie, Orthopädie, Geschlechtskrankheiten usw. Trotz der *Preußischen Vorschriften für Anlage, Bau und Einrichtung von Krankenanstalten*
76 *von* 1920 gelang es *Kurt Geisenhainer*, eine seiner *Schmitthenner*-Ausbildung entsprechende
77 heimatschutzorientierte Architektur durchzusetzen. Monumentale Architektur oder sklavische Symmetrie, wie sie beispielsweise *Hermann Distel* 1941 für das Universitäts-Klinikum Berlin[37] an der Ost-West-Achse vorgesehen hatte, kam für *Geisenhainer* nicht in Betracht.

Er orientierte sich bei seiner Planung 1936 angesichts der großen Baumassen an den in Europa üblichen horizontalen Linien[38], obwohl es Hochhauskrankenhäuser in USA und Skandinavien gab. Das Lazarett sollte in die Landschaft passen und sich mit seinen
77 Nebengebäuden, der „Gefolgschafts"-Siedlung und der *Marine-Sanitätsschule*[39] um den alten Gutshof Sanderbusch legen. Drei Geschosse und fast 50 Grad steile Hohlpfannendächer, in den Hauptbautrakten abgewalmt[40], formten die kompakten Baukörper. Nebengebäude und vorspringende Baukörper, Blendgiebel und technische Gebäude erhielten Satteldächer mit Giebeln. Zahlreiche Ziergiebel unterbrachen die langen Baufluchten. Unter den Blendgiebelvorsprüngen waren Kommando- oder Gemein-
76 schaftsräume, Ausblicke oder (Führer-)Balkone angeordnet. Die Bauform war heimatschutzorientiert, aber nicht aus Tarnungsgründen ländlich. Sichtbare „Rote Kreuze" auf den Dachflächen wiesen jeden Flieger auf die Zweckbestimmung hin[41].

Die mit braunen Pfannen gedeckten steilen Dächer hatten Aufschieblinge, die die gewünschte, leicht geschwungene Form brachten. Dachhäuschen waren verbrettert und durch ein gleichfalls geschwungenes Dach abgedeckt. Diese Formsprache, der *Marinestil,* war auch bei anderen *Geisenhainer*-Bauten zu finden.

Alle Bauteile des *Marinelazarett Sanderbusch* waren Ziegelbauten. Vereinzelte Ziegelornamente waren unter Fenstern, Ortgängen und an Gebäudeecken zu finden. Die Fenstereinschnitte im Haupthaus waren rechtwinklig und immer mit Werkstein umrahmt. Repräsentativere Fenster jedoch waren durch Bögen überdeckt, mit Blumenfenstern betont oder durch Lisenen verbunden.

Außentüren hatten grundsätzlich gebogene Stürze und wulstige Werksteingewände. Nie fehlte die geschmiedete Laterne mittig über dem Eingang. Die übergroßen Haupteingänge repräsentierten als monumentale Versatzstücke den Staat, das Militär, die *Marine.*

Die schmückenden Tierreliefs von *Rudolf Grau* sind noch vorhanden[42]. Leichtmetalltüren lassen die ehemals ausgefachten Eichentüren nicht mehr erahnen. Die vor den Haupteingängen angeordneten breitgelagerten Freitreppen waren damals für Patienten sicher überaus ungeeignet. Beliebt waren bei *Kurt Geisenhainer* profilierte Balkenköpfe aus Beton unter Blumenfenstern und Balkonen. Sie hatten keine statische Aufgabe, denn auskragende Betonplatten waren längst gebräuchlich und in *DIN*-Normen übernommen[43].

Auffällig sind für den heutigen Betrachter die vielen erhaltenen (Führer-)Balkone. Der oberste Feldherr, auch wenn er, wie anzunehmen ist, niemals in Sanderbusch war, konnte auf diese Weise immer präsent sein.

Besonders heimatschutzorientiert waren die Holzfenster mit ihren kreuzförmigen 77
Aufteilungen. Repräsentative Räume waren mit Doppelfenstern und weiteren Sprossen ausgestattet. Selbst Kellerfenster und Heizraumlichtbänder erhielten diese simpel, strenge Aufteilung. Im Inneren des Krankenhauses fallen trotz der Nachkriegsveränderungen auch heute noch die rustikalen Holztreppengeländer, die Türen mit Rahmen und Füllung, die detaillierten Türdrücker und die repräsentativen Versammlungsräume auf. Das *Marinelazarett Sanderbusch* bietet eine Fülle von Beispielen, für die im „III. Reich" vom *Marinestandortbauamt* gerne verwendeten Details.

Das erste deutsche *Marinelazarett* nach 1935 wurde am 31. 08. 1938 in Stralsund mit 320 Betten eröffnet. Oldenburg hatte schon Anfang 1937 eine neues *Standortlazarett* bekommen. Die Bauarbeiten für das größte deutsche *Marinelazarett* in Sanderbusch begannen 1937. Das Modell wurde auf der 1. *Deutschen Architektur- und Kunsthandwerk-Ausstellung* Januar–März 1938 im *Haus der Deutschen Kunst* in München gezeigt[44]. Die *Marinelazarette* Wilhelmshaven, Sanderbusch und Varel unterstanden dem *Sanitätsamt* Wilhelmshaven der *Marine*[45].

Erst nach Kriegsbeginn war das *Marinelazarett Sanderbusch* fertig: Ein Krankenhaus – nur für verletzte Marinesoldaten, nur für den Krieg. Im April 1941 wurde es von

Admiralarzt Dr. Schulte-Ostrup „feierlich in Dienst gestellt". Zu der östlich liegenden Siedlung für die „Lazarett-Gefolgschaft" mit 30 Doppelhäusern kam es nicht mehr.

4.2.3. U-Bootstützpunkt

Nach dem *Flottenabkommen vom* 18.06.1935 und den *Wehrgesetzen* konnte *Hitler* offen den
Aufbau der zuvor verbotenen U-Boot-Flotte angehen. In Wilhelmshaven folgten der
1935 stationierten *U-Flottille Weddigen* die *U-Flottille Saltzwedel* und 1937 die *U-Flottille*
1 *Lohs*. Am Nordufer des 1935 zum *Scheer*-Hafen umbenannten Zwischenhafens, auf dem
Gelände ehemaliger Müllplätze und Lauben plante *Heinz-Günther Thees* vom *Marine-*
78 *standortbauamt* ab 1936 den neuen U-Bootstützpunkt[46]. Die Raumprogramme wurden
Thees und dem Leiter des *Marinestandortbauamtes Otto Lückemeyer* vom *OKM* übergeben.
Der *Bebauungsplan* zwischen heutiger Emsstraße und Banter See sollte sich mit zwei
Liegehäfen und weiteren Flottillenkasernen von der Banter Ruine bis zur *Westwerft*
ziehen. Gebaut wurde jedoch nur der östliche Teil mit einem Liegehafen. Die Anlage war
mit ihrer eigenen Heizzentrale selbständig, und versorgte nicht nur den Stützpunkt,
sondern auch die Liegeplätze der Boote. Zum ersten, 1936 durch die Firma *Möller*
begonnenen Bauabschnitt, gehörten Heizzentrale, Tauchturm, Unterkünfte, Exerzier-
platz, Elektrozentrale und Wirtschaftsgebäude. Die Mannschaften von 4 großen und 4
mittleren Booten sollten untergebracht werden. Er wurde am 13.05.1937 von *Raeder* und
Dönitz eingeweiht[47]. Zum zweiten Bauabschnitt (1937–38) gehörten weitere Kasernen,
Nebengebäude, Turnhalle, Gemeinschaftsgebäude, Sportplatz usw. Im Kriege wurden
79 noch mehrere Bunker betoniert. Die Bauten direkt am Hafenufer mußten wegen ihres
Gewichtes auf Pfählen gegründet werden. Die Kosten für die gesamte Anlage beliefen
sich auf annähernd 2 Mio. Mark.

78 *Heinz-Günther Thees* gruppierte die streng symmetrische Anlage, ausgehend vom
80 zweigeschossig abgewalmten Wirtschaftsgebäude, entlang einer zur Emsstraße hin
abschließenden „Kasernenmauer". Die ausschließende Wirkung der Anlage und ihre
hermetische Absicherung gegenüber dem Stadtgebiet sind aus dem Modell gut
erkennbar und wurden unterstrichen durch massive Bewachungsanlagen. Eine hierar-
chische Gliederung der Kasernen war nicht nötig, denn Kommando- bzw. Stabsgebäude
fehlten dem U-Bootsstützpunkt. Die Zweckbestimmung als Kaserne ist aus dem
78 langgestreckten Exerzierplatz, der strikten Zeilenbauweise, dem gleichförmigen Fassa-
denaufriß und der parallelen Gebäudereihung erkennbar. Unterkunftsgebäude waren an
79 der Mittelachse gruppiert, in „Reih' und Glied" ausgerichtet. Die Mannschaftsgebäude
waren typisiert und gleich, im *Marinestil* hochgemauert. Sie unterschieden sich kaum von
31 dem heimatverbundenen Äußeren der Volkswohnungen.

U-Bootpersonal symbolisierte für *Hitler* militärische Auslese und Kampfleistung[48]. Er
81 machte selbst dem Stützpunkt am 09.04.1940 seine Aufwartung. War der Bau des
militärischen Komplexes einerseits „geheim", so tönte doch der *Kreisobmann* der lokalen

NS-Kulturgemeinde[49] von der vorbildlichen architektonischen Durchbildung, den freundlichen Farben, den geschmiedeten Laternen und der geschmackvollen Einrichtung[50]. Die *80, 134* Gesamtanlagen sollte die Banter Ruine einbeziehen und durch ein „Grünes Tor" betretbar sein. Das Wort „Kasernentor" wurde nicht benutzt. „Hier kann der U-Bootsmann sich wahrlich wohlfühlen, wenn er nach langer Fahrt in seinen Heimathafen Wilhelmshaven zurückkehrt". Der Betonung der Heimat entsprach auch die Architektur mit ihren steilen Dächern, Kreuzsprossen und Ziegelflächen. Kameradschaft untereinander und Vorbildlichkeit für die Jugend wurden beschworen. Das Einlaufen der neuen U-Boote habe „unsere Jugend" mit „leuchtenden Augen" verfolgt. Aus dem folgenden mörderischen Krieg kehrten über 30 000 deutsche U-Bootfahrer nicht zurück. 1948 zogen wiederum junge Menschen in die Kaserne ein. Englische Militärbehörden betrieben dort von 1948–72 eine Schule mit Internat. Die Schüler wurden kasernengerecht vor der Deutschen Bevölkerung abgeschlossen. Seit 1972 nutzt die *Bundesmarine* den erhaltenen Baukomplex.

4.2.4. Kasernenanlagen

Die Unterbringung von Marinesoldaten war immer eine Besonderheit, da Schiffsbesatzungen nicht wie die Soldaten von *Heer* und *Luftwaffe* permanent anwesend waren, sondern in der Regel auf den Schiffen wohnten. Im Ersten Weltkrieg stieg die Zahl der Marinesoldaten in Wilhelmshaven durch Kriegsdienstpflichtige auf bis zu 80.000. Riesige *9, 12* Kasernenanlagen der *Kaiserlichen Marine* standen zur Unterbringung bereit[51]. In der *13* Weimarer Republik sank der Personalbestand der Marine auf 5.000 Soldaten in Wilhelmshaven[52]. Viele der nutzlos gewordenen Kasernen wurden durch die *Reichsvermögensverwaltung* als Wohnungen vermietet oder als Notquartiere genutzt. Neubauten entstanden nicht. Erst das „III. Reich" begann wieder mit dem Kasernenbau, nachdem ab 16.03.1935 die allgemeine Wehrpflicht wieder eingeführt worden war. Da viele der Kaiserlichen Kasernen reaktiviert werden konnten, bedurfte es nur ergänzender Kasernenbauten im Raum Wilhelmshaven.

Die Kasernenanlage in Sengwarden beispielsweise ergänzte die *Befehlstelle Marinekommando Nord* und war auf Funktionalität, Technik und Modernität gerichtet[53]. Diese neue Kommandozentrale stand im Mittelpunkt von *Hitlers* Kriegführung und war entsprechend gefährdet. *Heinz-Günther Thees* mußte ab 1937 in regelmäßigen Abständen nach Berlin, um sich dort die „geheimen" Programmvorgaben mündlich übermitteln zu lassen. *Raeder* selbst kümmerte sich um das Projekt und verfolgte die Baustelle persönlich[54].

Aus Tarnungsgründen wurde die Anlage weit aus der Stadt herausgerückt und im *1* ländlichen Raum des *Amtes Friesland* zwischen Sengwarden und Wehlens auf 25 Hektar enteignetem Weideland gebaut. Es mußte sichergestellt werden, daß neue Stadtteile oder *129* Eisenbahnlinien die hochempfindliche Nachrichtenübermittlungstechnik, die Funk- und

86 Ortungsgeräte nicht behindern. Aus Luftschutzgründen bekamen die eingeschossigen
87 oberirdischen Unterkunftsgebäude ländliche Krüppelwalmdächer und wurden um einen
Platz geordnet bzw. an zwei Straßen nachbarschaftlich gruppiert. Der überwiegende Teil der militärisch sensiblen, hochspezialisierten Befehlszentrale, befand sich eingebunkert unter der Erde. Fünf unterirdische Bunker aus 1,8 m dicken Betonwänden waren durch Gänge verbunden.

Unter Überwachung der „GKdos-Stelle“ des *Marinestandortbauamtes* mußten 1937–40 Architekten und Zeichner, jeder für sich, an abgeschlossenen Teilen entwerfen, um die Ergebnisse danach den höheren Rängen abzuliefern. Diese setzten die Planungen zu einem „Ganzen“ zusammen.

Architektonisch bewegte sich die *Befehlstelle Marinekommando Nord* im von *Otto Lückemeyer* geprägten *Marinestil.* Die hermetisch abgeriegelte Kasernenanlage bekam ein von Wachen flankiertes Kasernentor. Eine Hierarchie der Gebäude war auf Anhieb nicht erkennbar, jedoch wurden die wichtigen Befehlstellen, Kommandozentralen und Nebengebäude nach militärischen Gesichtspunkten und achsial geordnet. Abseits jeder pathetischen Geste sollten die Gebäude traditionelle Hülle für moderne militärische Führung werden. Die abgezirkelten Außenflächen wurden nicht der Landschaft angepaßt, sondern nach perfekten Bewachungsgesichtspunkten geordnet – geeignet zum Aufziehen von Wachen, zum Visieren von Zielen, zum Kontrollieren von Menschen.

152 Der Bauablauf, ab 1938 dirigiert durch den mit Militäraufträgen reich bedachten
Friedrich Möller und die Firma *Peter Büscher,* stand unter strengsten Sicherheitsvorschrif-
98 ten. Die Bauarbeiter waren in den Barackenlagern von Sengwarden und Fedderwarden
kaserniert. Die Baukosten von 12,5 Mio Mark wurden zum größten Teil unter der Erde verbaut.

Die Kasernenanlage wurde nach dem Kriege Auffanglager für 1.500 ausländische Arbeiter. Sie ist heute Sitz des *Befehlshabers der Seestreitkräfte der Nordsee* und des *Marineabschnittskommandos Nordsee.*

★

Heinz-Günther Thees war auch Architekt der umfangreichen, noch heute erhaltenen
88 Kasenenanlagen in Varel. Diese große Militäranlage wurde Musterbeispiel des im
Marinestandortbauamt Wilhelmshaven gepflegten *Marinestils.* 1939, für die 6. Schiffsstammabteilung zeitgleich mit den Sengwarder Kasernen geplant, zog jedoch am 23.06.1942 die 12. Schiffsstammabteilung ein. Holländische Bauarbeiter, im westlich gelegenen Barackenlager untergebracht, mußten die aufgereihten Ziegelblocks im feuchten Wiesengelände hochmauern. Nach dem Kriege war die Kasernenanlage Flüchtlingslager, bis sie 1961 wieder von der *Bundeswehr* in Beschlag genommen wurde.

★

1934 begann nach 16jähriger Unterbrechung wieder der Bau von Zerstörern.
Liegeplätze und Landeanlagen waren am Nordufer des 1935 zum *Hipper*-Hafen 128
umbenannten Großen Hafens an der Gazellebrücke. Die Besatzungen der neuen
Zerstörer benötigten, wenn sie ausgetauscht wurden, Kasernen mit Ausstattung für
Hygiene, Versorgung und Geselligkeit.

Heinz-Günther Thees vom *Marinestandortbauamt* Wilhelmshaven entwarf 1937 den
Komplex der *Bonte*-Kasernen und schloß die Bauarbeiten noch 1938 ab[55]. Wie fast alle 82
anderen an den Hafen grenzenden militärischen Anlagen waren auch die *Bonte*-Kasernen
sorgsam zur Stadt hin abgeschottet und in diesem Fall zur Mainstraße hin durch
Wachgebäude abgeschlossen. Das walmgedeckte Hauptgebäude war trotz fehlender
repräsentativer Funktionen im Äußeren traditionell und in der Materialwahl gediegen.
Mit seinem (Führer-)Balkon orientierte es sich zu den Kajenanlagen. Kommandanten
konnten auf den breiten Balkon treten, Befehle ausgeben und militärische Rituale
beobachten. Ein von Sandstein-Adler und Hoheitszeichen gekröntes, von Schmiedeeisen
flankiertes Eingangsportal von *Rudolf Grau* war der Prozentanteil Kunst am Bau. In der 83
Eingangshalle der *Bonte*-Kasernen richtete die Zerstörer-Flottille eine Heldengedenkstätte 85
für gefallene Marinesoldaten ein. Das traditionelle Äußere der Kasernen setzte sich nach
innen hin fort. Kommandantenzimmer und Mannschaftsstuben unterschieden sich 84
deutlich in Ausstattung und Fläche, Standard und Details. Auch die Nebengebäude,
teilweise eingeschossig, waren traditionell gestaltet, unterbrochen nur durch einen steilen
Giebel zur Kajenanlage. Der gesamte Eindruck der Kasernen an der Gazellebrücke war
heimatlich und solide. Den Soldaten sollte ein Gefühl von Schutz und Geborgenheit 82
vermittelt werden. Die Weltläufigkeit der Seefahrt, das Besondere des Bauens am
Wasser, fand nicht seinen architektonischen Ausdruck. Den Besatzungen der Zerstörer
sollten ein Zuhause und eine heile Welt vorgespiegelt werden. Traditionsbewußt
erhielten die Kasernen und die Anlegebrücken 1940, als der *Zerstörer-Kommodore Friedrich
Bonte* im Seekrieg umkam, seinen Namen. Die Kasernen wurden ab 1954 wieder von der
Marine genutzt, 1979 aber zugunsten einer Uferpromenade abgebrochen.

⋆

Unterkunftslager für Arbeiter wurden nicht unbedingt als Barackenlager geplant. Sie 98
wurden zunehmend massiver, aufwendiger und dauerhafter gebaut, oft für spätere
Weiternutzung als Wohnungen oder Kasernen vorgesehen. Die Arbeiterunterkünfte
Ebkeriege waren vom Bautyp als Kasernen gedacht. Schnell, billig, raumsparend und 99
verkehrsgünstig sollte die „Werftgefolgschaft" untergebracht werden. Die Gebäude
waren anfangs als Mehrzweckgebäude vorgesehen, die alle Nutzungsperpektiven
offenlassen sollten. Das Gelände an der Ebkeriege war bereits 1930 von der *Marine* 1
erworben worden[56] und sollte städtebaulich einen Puffer zwischen dem 1939 fertiggestellten
Jadeviertel bzw. dem geplanten Einfamilienhausgebiet der *Wilhelmshavener Spar-* 52
und Baugesellschaft[57] und dem anschließenden Industriegebiet bilden. Die Arbeiterunter-

künfte waren auf Dauer angelegt und sollten über eine Station der Ringschnellbahn mit den Arbeitsstätten der *KMW* verbunden werden. Der *Wirtschaftsplan Wilhelmshaven vom* 03.08.1939 wies das Gelände in „zweigeschossiger Bauweise" aus. Im angrenzenden Industriegebiet entstand die *Marinekraftfahrabteilung* Ebkeriege (1939–40), mit großen Barackenlagern, südlich grenzte der neue Ortsgüterbahnhof an.

Entscheidend für die vergleichsweise solide Bauweise und die durchdachte städtebauliche
Anordnung der Arbeiterunterkünfte Ebkeriege war das in der Nachbarschaft geplante
107 *Kulturforum*. Die Bismarckstraße als monumentale Ost-West-Achse sollte von öffentli-
99 chen Bauten gesäumt werden. Die diagonal die Kasernenanlage Ebkeriege durchziehende
109 Straße[58] sollte auf den Ausgangspunkt des *Kulturforums* den „Runden Platz" führen bzw.
als Hauptachse sogar das *Kulturforum* bis zum neuen Personenbahnhof durchziehen. Der Entwurf des *Hafenbauressorts der KMW* von *Karl Tuchscherer*[59] von 1939 sah 7 U-förmige, zweigeschossige Kasernenblocks, Wirtschaftsgebäude, Heizzentrale, Appellplätze und eine Sportanlage vor. Die „Soldaten der Arbeit"[60] sollten gut untergebracht sein und den „Kameraden im Soldatenrock" in nichts nachstehen. Der ideologische Hintergrund war klar: „Wehrkraft und Wirtschaft" sollten Höchstleistungen vollbringen.

Wegen der Rüstungsaufträge der *KMW* wurden immer mehr junge Menschen ab 14 Jahren aus allen Teilen des Reiches zur Werft geholt[61]. So wurden die Arbeiterunterkünfte zu einem „Werftjugendheim" umfunktioniert. „Heimleiter" *Heuer* wachte darüber, daß die Erziehungsgrundsätze der *HJ* eingehalten wurden[62]. Die Jugendlichen sollten „ganze Kerle" werden. Durch „Zucht, Ordnung und Sauberkeit" militärischen
Drill, Appelle und Musterungen sollten sie zu „deutschen Männern" gemacht werden.
153 Ernteeinsätze, *Arbeitsschlachten*, Wandertage und Sportwettkämpfe zielten auf national-
sozialistische „Betriebserziehung". „Betriebsjugendwalter" sorgten für die „politische Schulung der Jugendlichen" innerhalb der *KMW*[63]. *NSDAP*-Funktionäre, Werftleitung und Militärs besuchten das Lehrlingsheim regelmäßig, hielten Propagandareden und überzeugten sich von „Ordnung und Disziplin"[64]. Volksempfänger in den Stuben übertrugen „Führerreden". Vorträge warnten vor den Lastern der Zivilisation, vor Rauchen und Trunksucht. Die Lehrlinge der *KMW* wurden durch „Kameradschaftsabende", „schneidige Spielmannszüge", stimmungsvolle Feiern, Ausflüge, Elternabende und „Stiftfeste" belohnt[65]. Der große Gemeinschaftssaal im „Werftjugendheim" Ebkeriege entsprach den Dekorationsklischees der *NSDAP* mit sichtbarer Holz-Holzkonstruktion, Sonnenrädern, Bühne und „Führerbild"[66]. Der Gemeinschaftssaal wurde Forum für Fanfarenmusik, Wettsingen, Lehrlingsfreisprechungen und „Heldengedenkfeiern". 1941 wurden alle Werftlehrlinge zur „vormilitärischen Ausbildung" in
99 die Marine-*HJ* übernommen[67]. Die einfache Lagerarchitektur der U-förmigen schlichten
Putzbauten sollte die schnelle Unterbringung von immerhim 500 Werftlehrlingen ermöglichen. Jeder der Blöcke bekam seinen eigenen Appellplatz, jeder Block seinen eigenen „Heimwärter". Die *Marine* wollte „die jungen Kameraden nach Schluß der Arbeitszeit" keine Minute sich selbst überlassen[68] und vollständig unter den Einfluß der *HJ* bringen. Die soziale Kontrolle sollte perfekt funktionieren. Die Werkszeitung der

KMW berichtete abschreckend von „ausgekniffenen Lehrlingen“, denen zu Hause „das Fell versohlt“ wurde. Die kasernierte Unterbringung in Stuben mit 8 Betten behinderte jegliche jugendliche Individualität. Die symmetrisch angeordneten Appellplätze in den Höfen der Blocks wurden für weihevolle Feiern, martialische Aufmärsche[69], „zackige Spielmannszüge“ und uniformierte Musikspektakel genutzt. Der benachbarte Sportplatz diente den Sportappellen der *Betriebssportgemeinschaft* der *KMW*.

Die Arbeiterunterkünfte Ebkeriege wurden von 300 Arbeitern des Generalunternehmers *Möller* hochgemauert[70]. Ziegelsteine waren 1939–40 knapp geworden. Schon bei
Fedderwardergroden, Neuengroden und Sande-Neufeld reichte es nicht mehr für den 41, 47
heimatverbundenen Baustoff. Da *Friedrich Möller* den 8 Mio. Mark Auftrag direkt von 49
der *KMW* bekommen hatte, war er bei der Wahl der Baustoffe frei. *Karl Tuchscherer* und *Wilhelm Diersen* vom *Hafenbauressort der KMW* und *Karl Higgen* von der Firma *Möller* wehrten gestalterische Auflagen der *Stadt Wilhelmshaven* ab. Ebkeriege war „niedere Architektur“ eben nur eine „Arbeiterunterkunft“. Im November 1940 wurde die Anlage fertiggestellt[71]. Noch im Kriege wollte die *Stadtverwaltung* in Ebkeriege Obdachlose unterbringen[72]. Nach dem Kriege zogen neuaufgebaute Betriebe der Texttilverarbeitung in das ehemalige „Werftjugendheim“ ein. Ab 1956 nahm die neue *Bundesmarine* den Komplex in Beschlag.

Mehr Kaserne als *HJ*-Wohnheim für Jungarbeiter der *KMW* war auch der Anbau für 284 Betten an der 1938 von den Nationalsozialisten geschlossenen katholischen
Cäcilienschule an der Schillerstraße im Hansaviertel[73]. *Karl Tuchscherer* vom *Hafenbau-* 53
ressort der KMW entwarf den neuen zweigeschossigen Trakt in traditioneller Formsprache[74]. Er sollte nicht billig sein wie an der Ebkeriege sondern innerstädtisch-gediegen, nicht behelfsmäßig wie in Sande-Neufeld sondern bürgerlich-solide.

4.2.5. Bauten für die Wehrertüchtigung

Sportliche Betätigung spielte für die Nationalsozialisten aus Gründen der Wehrertüchtigung eine große Rolle[75]. Sport sollte den angeblich idealen Menschen bilden, seinen Körper und seinen Charakter. Entschlußkraft, Abhärtung und Erziehung zu „wehrhaften Männern“ standen im Vordergrund der Propaganda. Die gesamte Sportbewegung war im *Deutschen Reichsbund für Leibesübungen* und ab 1938 im *NS-Reichsbund für Leibesübungen* unter dem *Reichssportführer* organisiert.

In der Weimarer Republik hatte es im Reich und ebenso in Wilhelmshaven/Rüstringen umfangreiche moderne Sportanlagen gegeben. Die fortschrittliche Rüstringer Grünpolitik hatte nach Planungen von *Leberecht Migge* große freizeitorientierte Anlagen geschaffen. Rüstringen stand im Zentrum des Arbeitersports[76] und plante, wie auch Wilhelmshaven, weitere moderne Sportanlagen. Sofort nach 1933 wurden die sozialdemokratischen Sportvereine verboten und die Einrichtungen den Parteigliederungen zur Verfügung gestellt. Licht- und Luftbäder wurden geschlossen – Freikörperkultur vorerst verboten.

9 Die Marine hatte sich schon früh für sportliche Betätigung ihrer Soldaten eingesetzt und 1885 eine erste Marinesportanlage in Betrieb gesetzt. Schon seit der Jahrhundertwende stellte sie ihren Offizieren Tennisplätze zur Verfügung. Fechthallen, Schießstände und Reitplätze gehörten zur selbstverständlichen Ausstattung einer Garnison.

Die Siedlungsplanungen der NS-Zeit enthielten ausgiebige Sportplatzflächen, die *Gottfried Feder* mit 5 qm „je Kopf der Bevölkerung" veranschlagte[77]. Die *Bebauungspläne* der Gemeinschaftssiedlungen im *Wohnsiedlungsgebiet Wilhelmshaven-Jeverland* enthielten
140 ausreichende Sportplätze und Turnhallen, zumeist bei den Schulen[78]. Die Turnhallen waren gleichzeitig auch Appellhallen. Die Realisierung der Sportanlagen in den neuen Siedlungen und Stadtteilen blieb oft leeres Versprechen. Es kam deswegen zu heftigen Auseinandersetzungen im Gemeinderat von Wilhelmshaven: „Kein Theater, keine Strandhalle, sondern Sportplätze" forderte *NSDAP*-Ratsherr *Karl-Heinrich Hinz* in der Sitzung vom 28.03.1939, denn „die schönste Kulturstätte" sei ein Sportplatz[79]. In kaum
72, 48 einem Arbeiterlager fehlten Sportplätze, keine Kaserne wurde ohne Fußballfelder, ohne Turnhalle geplant. Für die dem Führungspersonal zuzurechnenden Ärzte waren beim
75, 151 *Marinelazarett Sanderbusch* und beim *Städtischen Krankenhaus* Tennisanlagen geplant.

1937 gründete die *DAF* ihre *Wehrkreis-Sportgemeinschaft*, die auf den Betriebssport gerichtet war. Im Arbeitsleben von *KMW* und *Marine* spielten *Wehrkreis-Sportwarte* eine große Rolle. Die von dort betreuten *Betriebssportgemeinschaften* der *KMW* und der *HNBD* berichteten regelmäßig in der Werkszeitung der *KMW*. Die *Standort-Sportgemeinschaften* waren wiederum dem *KDF-Kreissportamt* zugeordnet. *Wehrkreis-Fachwarte* waren gleichzeitig *Sportfachwarte* der *KMW*[80]. Sport und Wehrmacht waren perfekt verflochten. Als *Hitler* die *Olympischen Spiele* 1936 geschickt für seine Propaganda
109 nutzte, war klar, daß Sportbauten auch in die künftige NS-Baupolitik einbezogen werden mußten. In Wilhelmshaven standen die jährlichen „Skagerrak-Gedächtnisspiele" auf dem großen Exerzierplatz ganz im Zeichen der NS-Propaganda. Ab 1939 dann wurden „Betriebs-Kriegsfußballmeisterschaften" durchgeführt. Die Freizeit der „Gefolgschaft" war wehrsportlich verplant. Die „Wehrkampftage" der *Marine-SA* wurden fester Bestandteil des lokalen Sportgeschehens.

Anfang 1937 planten *Max Neumann* und *Heinrich Siemer* eine Zusammenfassung der
52 vorhandenen zwei Sportplätze zwischen Ebkeriege und Genossenschaftsstraße zu einer *SA*-Sportanlage, Freiflächen und Hallen für Reit- und Schießsport sollten hinzukommen[81].

101 Als die Bauarbeiten für IV. Einfahrt und *Nordwerft* begannen, mußte die Garnison-
90 Schwimmanstalt weichen. Sie wurde nach Planungen von *Heinz-Günther Thees* vom *Marinestandortbauamt* 1937–38 im Zusammenhang mit der Marinesportanlage an der Freiligrathstraße neu gebaut. Die sporttechnisch zukunftsweisende Anlage mit Sprungturm, 50-m-Bahn und Nebenbecken bekam traditionelle Umkleidegebäude mit Ziegeldächern im *Marinestil*.

Prunkstück des Marinesports wurde die werfteigene Sportanlage zwischen Göker- und Freiligrathstraße, „fast vor den Toren der neuen Großwerft“[82]. Die Verbindung zwischen Rüstungsarbeit und Sport war gewollt und wurde propagiert. Sportplätze sollten „unmittelbar neben Industriewerken“ liegen, Freude gleich neben der schaffenden Arbeit. Das Stadion, für die „Sportgefolgschaft“ der *KMW* gebaut, bot bis zu 10.000 Menschen Platz. Die Krönung der Wilhelmshavener Sportbauten allerdings – das *Sportforum* – benannt nach dem *Sportforum* der Berliner Olympiabauten[83], blieb *109*
Utopie.

Die Sportappelle der *KMW* lauteten wenig später: „gerade im Kriege sind Leibesübungen notwendig. Sie vermehren erstens die Arbeitskraft, zweitens die Lebensfreude und den Lebensmut. So stärken sie die Heimatfront“[84]. *Robert Ley* rief zu „Sommersporttagen der Betriebe“ auf: „Leibeserziehung ist Kampf für Deutschland! Alle marschieren!“ Der *Oberwerftdirektor* überreichte als Wanderpreise *Hitler*-Büsten[85].

4.2.6. Technische Marinebauten

Die notwendige Umsiedlung von marinebezogenen und werftabhängigen Einrichtungen aus Wilhelmshavens Altstadt in das umgebende *Wohnsiedlungsgebiet*, insbesondere *129*
nach Oestringen[86], wurde 1937 von *Wilhelm Hallbauer* mit Luftgefährdung begründet, denn Gas-Werk, Schlachthof, Lazarette, Versorgungsanlagen und Depots seien in der Südstadt „ungünstig aufgereiht“[87]. *Hallbauer* wollte nur „diejenigen Anlagen“ belassen, „welche durch das Wasser standortsbedingt sind“. Ein großes *Industrieumsiedlungsprogramm* sollte beginnen.

Völlig private Industriebetriebe waren 1937 im Raum Wilhelmshaven kaum vorhanden[88]. Der *Norddeutsche Eisenbau Wilhelmshaven* war bereits 1937 in moderne *94*
Hallen nach Sande ausgesiedelt. Außer einer Werkstatt für Feinmechanik, die in Wilhelmshaven etabliert war, rechnete man nicht mit weiteren größeren privaten Industriefirmen, die sich niederlassen würden. Der *Wirtschaftsplan Wilhelmshaven vom 03.08.1939* wies demzufolge auch nur das Gebiet um den Sander Bahnhof und das Gebiet um den neuen Wilhelmshavener Ortsgüterbahnhof als Industriegebiet aus. Der wirkliche Industriebetrieb Wilhelmshavens war die staatliche Rüstungsschmiede, die *KMW*. *10*

⋆

Das neue Industriegebiet um den Bahnhof Sande lag günstig, denn dort fand der *94*
Massengüterumschlag statt, dort wurde 1941 der neue Personenbahnhof gebaut, dort *51*
war Ausgangspunkt der Werftbahn[89], dort befand sich eine Ölumumschlaganlage der
Marine, und ganz in der Nähe war auch das *Marinelazarett Sanderbusch*[90] fertiggestellt *75*
worden. Cäciliengroden, Sande-Neufeld und Sanderbusch waren die dazugehörigen *34, 48, 25*
Wohngebiete. Weitere Arbeiterlager und Wohnsiedlungen waren fest eingeplant. In das

128 5 km lange Industriegebiet, das bereits im *Wirtschaftsplan Wilhelmshaven vom Juni* 1937
1 ausgewiesen waren, sollte auch die Gießerei der *KMW* ausgesiedelt werden. Die Gießerei gehörte zum *Maschinenbauressort der KMW* unter *Maschinenbaudirektor Otto Klette.* Sie lag beengt südlich des Bauhafens und hatte dort keine Entwicklungsmöglichkeit. Erste Standortüberlegungen Anfang 1938 zielten in Richtung Schortens[91]. Die Eisenbahnlinie Sande-Jever von 1870 sollte einen nördlichen Abzweig bekommen. Die Bauplanung für den Standort unmittelbar neben dem Wasserwerk Feldhausen lag fertig vor, als Mitte 1938 der neue Standort im Industriegebiet Sande akut wurde. Die Bauentwürfe wurden völlig unverändert in Zuordnung, Funktion und Architektur auf den neuen Standort südlich der Ölumschlaganlage Sande Neudeich übertragen[92]. Die Planung des *Hafenbauressorts der KMW* war Beispiel außerordentlich moderner Industriearchitektur. Beteiligt waren unter *Drenckhan*, die Architekten *Klaucke, Heeren* und *Brinkmann.*

93 Der Gießereientwurf von 1938 beruhte auf zwei je 140 m langen „Schiffen", die durch
94 einen Querriegel vor Kopf verbunden waren. Die „Schiffe" erhielten Gießereieinrichtungen für Eisen-, Stahl-, Leichtmetall- und Bronzeguß. Geschützt vor den lärm- und staubemittierenden Hallen entstand die Modelltischlerei. Dort wurde konstruiert, dort wurden die Holzformen für Schiffschrauben, Turbinen, Motorgehäuse der Kriegsschiffe modelliert. Die drei Fabrikhallen wurden 193–40 als Stahlbinderkonstruktionen mit flachen Dächern und Ziegelverkleidung entworfen[93]. Das Konstruktionsraster lag innen. Es markierte sich jedoch durch Trennung der Ziegelfelder und Lichtbänder. Glasdächer und großzügige Fensteröffnungen ergaben sich aus der Nutzung und der Funktion der Halle. Düstere Fabrikationsräume hätten den Anforderungen an den modernen Rüstungsbetrieb und dem Ideal von „Schönheit der Arbeit" nicht entsprochen. Ständische Ordnung und traditionelle Baukultur kamen nicht in Betracht. Außer fleißiger Arbeit und präziser Gründlichkeit sollten die modernen Fabrikhallen nichts darstellen. Da weitgehend mit Elektroöfen geschmolzen wurde, mußte 1940 ein Trafogebäude nachgerüstet werden. Dieses auffällig *Bauhaus*-orientierte Technikgebäude entwarf *Hans Klaucke*, der modernste Architekt der *KMW.*

Auch die Außenanlagen der *KMW*-Gießerei waren wohlgeordnet. Ein Feuerlösch-
94 teich sollte Brände verhüten. Direkte Gleisanschlüsse zum Werftbahnhof Sande ermöglichten rationellen Arbeitstakt und pünktliche Lieferung für die staatliche Rüstung.

Der heute noch gut erhaltene Industriekomplex wurde von Bomben verschont und der Demontage entzogen. Bereits im März 1946 befaßte sich die englische *Militärregierung* mit der ehemaligen *KMW*-Gießerei, fertigte Bauaufnahmen der Gebäude und plante Wiederaufnahme der Produktion. Ab 1947 wurde die Gießerei von einem privaten Konsortium weiterbetrieben[94].

*

Der städtische Schlachthof, 1906 am Handelshafen in der Südstadt gebaut, war
gleichfalls nicht erweiterbar und platzte wegen der Zunahme an Rüstungsarbeiten und *128*
Marinesoldaten aus allen Nähten. Auch er war Bestandteil des *Industrieumsiedlungsprogrammes*. Die Fleischproduktion hatte sich von 1932 nach 1938 auf rd. 5 Mio. kg
verdoppelt. Für 6–7 Mio. Mark sollte ein neuer Schlachthof mit Kühlhaus im
Industriegebiet Sande gebaut werden. Die *Stadt Wilhelmshaven* forderte von *MDF* und *94*
RAM Anfang 1939 Finanzierungshilfe[95]. Da *Amtshauptmann Hermann Ott* und *Kreisplaner D.J. Gattnar* sich gegen den Standort sperrten, fand am 15.05.1939 im Bahnhof Sande eine Besprechung unter der Leitung von *Landesplaner Wilhelm Wortmann* statt. Der *MDF* bestimmte dann am 17.05.1939 den Standort unmittelbar am Bahnhof zwischen der alten Reichsstraße 69 und der Eisenlinie Sande-Jever. Die gleichfalls umzusiedelnde Gasanstalt sollte in Neu-Marienhausen entstehen. Der *RAM* empfahl 1941, für die Schlachthof-Planung die Architekten *Walter Ferse* und *Franz Meurer*, Berlin. Die Planung wurde 1942 eingestellt.

★

Das *Hafenbauressort der KMW* hatte neben Unterkunftslagern und Freizeitbauten für Rüstungsarbeiter alle technischen Bauten für die einzelnen Ressorts zu erstellen. Bauten mit repräsentativem Anspruch oder kulturellem Ausdruck kamen kaum vor, wurden dann aber traditionell verkleidet. Bauten der *KMW* sollten für Arbeit, Leistung und Ordnung stehen. Unter den *Hafenbaudirektoren Richard Pein* (1936–39) und *Georg Frede* (1939–45) entwickelte sich eine moderne, funktionale Architekturauffassung, getragen
vor allem von dem *Bauhaus*-orientierten *Hans Klaucke*. Sein Auftraggeber, die *KMW*, *96*
wollte zweckdienliche, gesunde und lichte Arbeitshallen. *Klaucke* und seine Mitarbeiter *97*
Boysen, Borchers, Heeren, Brinkmann und *Queisner* entwarfen moderne Materialien, *93*
kubische Formen, tragende Skelettkonstruktionen. Natürlich entsprachen die außeror- *95*
dentlich modernen Entwürfe von *Hans Klaucke* nicht dem in Wilhelmshaven verbreiteten *Marinestil*. Für die Architekten des *Marinestandortbauamtes*, für die städtischen Baubeamten, waren die Zweckbauten der *KMW* „niedere Baukunst"[96].

Zu den wenigen technischen Bauten in traditionellem Habitus gehörten beispielsweise
die Neubauten für das *Navigationsressort der KMW*. Seit 1876 wurden nautische *91*
Instrumente und Seekarten für die der Werft zugeteilten Schiffe verwaltet, Kompasse gewartet, Karten fortgeschrieben, Deviationen bestimmt und magnetische Instrumente gerichtet. Am Südufer des 1935 zum *Scheer*-Hafen umbenannten Zwischenhafens[97] entstanden 1938–39 Anlegebrücken, dahinter, auf reichseigenem aufgespültem Gelände,
ein Bürogebäude für Ingenieure, Techniker und Personal des *Navigationsressorts*. Es *1*
wurde in traditioneller Formsprache mit Walmdach und Ziegelaußenhaut der Landschaft angepaßt. Ein Schiffskammergebäude kam im Juni 1939 dazu. Das flachgelagerte Gebäude mit gering geneigtem Walmdach sollte Werkstätten beinhalten. Wegen der großen Spannweiten ließ es sich nicht mit steilen Dachformen überdecken. 1939 bekam

91 das für den bevorstehenden Seekrieg besonders wichtige *Navigationsressort* sein Hauptgebäude, das Plangebäude. Ein Wachgebäude mit Garagen folgte 1942. Das Schiffskammergebäude drückte mit seinen flachgestreckten Lichtöffnungen technische Zweckbestimmung aus, denn die Arbeitsräume sollten optimal belichtet werden. Das Büro- und Plangebäude jedoch, 80 m lang mit Seitenflügeln, Vorfahrt und (Führer-) Balkonen drückt die hierarchische Gliederung aus. Die Anlagen wurden achsial, symmetrisch, durch Türme gekrönt. Der Flakturm mit seinem notwendigerweise flachen Dach wurde als mittelalterlich schießschartenbewehrter Turm getarnt. Das *Navigationsressort* wurde ab 1948 als wissenschaftliches Institut und danach wieder für die *Bundesmarine* genutzt.

Gleichfalls in traditioneller Umhüllung entwarf die *KMW* die Erweiterung des Bürogebäudes ihres *Hafenbauressorts* 1943 zwischen Großem Hafen und Kaiser-Straße[98]. Der von *Hans Klaucke* ausnahmsweise im *Marinestil* entworfene, fast monumentale, zweigeschossige Bürotrakt, sollte den benachbarten Hafenbauhof der *KMW* abschirmen. Der Sitz des *Hafenbauressorts* repräsentierte für jeden sichtbar die Staatsmacht. Der Vergleich mit den sonst modernen Entwürfen der *Hochbauabteilung der KMW* verdeutlicht die Uneinheitlichkeit der Stile: Sachliche Modernität für technische Gebäude – traditionelle Umhüllung für Repräsentationsgebäude.

Die meisten technischen Bauten der *KMW* entzogen sich heimatverbundener Bauweise. Werfthallen in Marinestandorten wurden modern gebaut. In Bremerhaven beispielsweise entwarf *Egon Eiermann* 1941 eine Schiffbauhalle im *Internationalen Stil*. Das
1 neue Torpedolagerhaus am Südufer des Verbindungshafens unmittelbar hinter dem Südstrand, gebaut, war Beispiel technisch modernen Industriebaus. Nach Abschluß des *Flottenabkommens vom* 18.06.1935 war der Bau größerer Torpedoboote wieder möglich. Die alten Torpedolager reichten nicht mehr aus.

Das 80 m lange neue Lagerhaus sollte die vierfache Nutzlast eines üblichen Lagergebäudes aushalten können. Aus Luftschutzgründen sollte unter dem Ziegel-Satteldach ein geneigtes Betondach die Torpedos schützen. *Klaucke* bediente sich bei diesem, der Öffentlichkeit unzugänglichen Bau, moderner Formsprache[99]. Die schlanken hohen Fenster sollten sorgfältiges Lagern ermöglichen, eine Vorbedingung für die Sortierung der empfindlichen Torpedowaffen. Die durch Kranbahnen gegebenen Stapelhöhen erzwangen hohe Geschosse und große Lichtöffnungen. Die Funktion bestimmte die moderne Umhüllung. Äußerliches „Architekturmachen" wäre für das Torpedolagerhaus sinnlos gewesen[100]. Es sollte nichts dargestellt, nichts ausgedrückt, nichts dekoriert werden. Die Ziegelhaut sollte nicht regionale Bezüge ausdrücken, sondern gegen Wind und Wetter schützen. Wenn nicht der Luftschutz zwecks Abweisung von Brandbomben ein geneigtes Dach gefordert hätte, wäre die Abdeckung
89 flach geblieben. Das von *Hans Klaucke* entworfene Torpedolagerhaus war technisch
modern, zweckmäßig konzipiert und funktionsgerecht. Weitere außerordentlich moder-
96, 97 ne Entwürfe wurden von *Hans Klaucke* für Bauhafen und *Westwerft* vorgelegt und zum
Teil realisiert[101]. *Klaucke* wurde auch von der *HNBD* für Gestaltungsfragen der IV.

Einfahrt eingeschaltet. Er plante auch für die von der *KMW* Wilhelmshaven betreuten 102
Standorte Achim und Emden[102] und entwarf 1941–42 die beinahe futuristische
Außengestaltung von Bunkern auf dem Gelände der *KMW*. Die Bunker bei der 114
E-Werkstatt Nord und beim Tor 1 bekamen das Aussehen konstruktivistischer 10
Betonkuben mit vorgeblendeten Türmen, auskragenden Aussichtsbalkonen, geglieder-
ten Attikadekorationen, moderne Gegenstücke zu den „Wehrbunkern" mit mittelalter- 113
licher Hülle.

Technische Bauten für die *Marine* waren auch die Leuchttürme, die das *Hafenbauressorts* für das ab 1939 von *Friedrich Markworth* geleitete *Strombauressort der KMW* architektonisch betreute. Sie dienten der Sicherung und dem Auffinden der Wilhelmshavener Einfahrten. Die Leuchttürme Arngast und Voslapp wurden vor dem Ersten Weltkrieg erbaut. Den Leuchtturm Mellum-Plate entwarf *Hans Klaucke* vom *Hafenbauressort* nach militärisch, funktionalen Vorgaben. *Karl Tillessen* und *Adolf Imken* vom *Strombauressort der KMW* und die Firma *Möller* realisierten Mellum-Plate 1939–42. Die übliche runde Form der Leuchttürme ergab sich aus Strömung und Windschlüpfrigkeit. Mellum-Plate jedoch sollte nicht nur navigatorisch genutzt werden, sondern auch militärisch. Der kantige rechteckige Aufbau beherbergte 30 Soldaten, die je 4 Geschütze bedienen sollten, und auf dem unteren Ring zu plazieren waren. Die Obere Plattform über den leuchtturm-typischen Laternen sollte riesige Scheinwerfer tragen. Militärische Absicht war es, in die Jade einfahrende feindliche U-Boote aufzuspüren und zu beschließen. Diesem Zweck folgte die architektonische Form.

Kubisch einfach und modern, ohne seefahrtstümelnde Attitüde entwarf *Klaucke* auch
das Leuchtfeuer Minsener Oog auf der Buhne A. *Karl Tillessen* und *Werner Martens* 95
realisierten das Leuchtfeuer 1942–43. Ein ähnlich kubischer Leuchtturm, 1943 für die Außenjade geplant und ein Funkfeuerturm 1940 für Mellum entworfen, blieben Utopie.

★

Technische Marinebauten des *Marinestandortbauamtes* wurden grundsätzlich traditionell im *Marinestil* verkleidet. Sie waren im Gegensatz zu den Bauten der *KMW* nicht im abgeschlossenen Werftgelände, sondern häufig in der Stadt oder der freien Landschaft
gelegen. *Marinegerätelager, Marineinstandsetzungsamt, Marine-Observatorium, Wähleramt* 73, 69
Langewerth und Ausweichvermittlung Antonslust, wurden der jeweiligen Umgebung angepaßt. Sattel-oder Walmdächer, Ziegelaußenwände und Kreuzsprossen waren die Regel. Ende 1935 entwarf das *Marinestandortbauamt* Wilhelmshaven am Mühlenweg in Rüstringen südlich der Garnison-Waschanstalt eine neue *Waffenmeisterei* für die Kasernenanlagen Gökerstraße und Mühlenweg[103]. Das noch erhaltene technische Militärgebäude wurde 1936 in traditioneller Ziegelbauweise hochgemauert. Die streng symmetrische Anlage bekam ein Walmdach, doppelt so hoch wie das zu überdeckende Dachgeschoß. Der typische Kasernengrundriß mit langen Fluren und aneinandergereih-

ten Diensträumen, von Architekt *Heinrich Aven* streng nach Zugehörigkeit zu den zwei Kasernenanlagen entworfen, ordnete die Funktionen. Lager, Büro, Handwaffenwerkstatt, Ausgaberäume, sogar die Hofeingänge kamen zweimal vor. *Aven* war noch erkennbar ungeübt in der Handhabung des aufkommenden *Marinestils*. Die zweiflügeligen Fenster entwarf er noch sprossenlos. Gebaut jedoch wurden sie 1936 dann im Kreuzsprossenraster des *Marinestils*.

4.3. Nordwerft und IV. Einfahrt

Entscheidend für dauerhafte Funktion und bequeme Anfahrbarkeit des Kriegshafens war
9 seit jeher das Jade-Fahrwasser[104]. Die Wilhelmshavener Hafeneinfahrten erforderten gute Fahrwasserströmungen, um ein Zulaufen mit Sand und Schlick zu vermeiden. Bereits 1892 begann das *Hafen- und Strombauressort der Kaiserlichen Werft* mit den Arbeiten für einen 5,8 km langen Leitdamm vom Leuchtturm Arngast bis vor die beiden engen Hafeneinfahrten. Das äußere rauhe Fahrwasser sollte durch „spinnen"förmige Leitwerke östlich von Wangerooge umgeformt werden. Die Bauarbeiten wurden vor dem Ersten Weltkriege nicht mehr ganz fertiggestellt, aber ab 1926 fortgesetzt und 1935 rechtzeitig zu *Hitlers* Wiederaufrüstung abgeschlossen. Leitdamm und „Spinne" wurden nach 1945 durch die Alliierten gesprengt und verloren weitgehend ihre Funktion.

Motor der militärisch gewünschten und hartnäckig verfolgten Jade-Fahrwasser-Korrektion war *Wilhelm Krüger*, 1933–36 *Hafenbaudirektor*, davor seit 1901 im *Strombauressort*. 1908 und 1926, jeweils vor den Weltkriegen, gründete er *Wasserbauversuchsanstalten*, um wissenschaftliche Grundlagen für seine Ingenieurbauwerke zu schaffen. Er hielt Tagungen ab, betrieb in Wilhelmshaven die Gründungen des *Senckenberg-Institutes* (1928) und der *Provinzialstelle für Marschen- und Wurtenforschung* (1938). *Werner Haarnagel*[105] von der *Provinzialstelle*, lieferte mit seinen Baugrunduntersuchungen die notwendigen bodenmechanischen Erkenntnisse für die Hafenerweiterung. *Wilhelm Krüger* konstruierte auch die notwendigen Navigationshilfen für den Kriegshafen: den Leuchtturm Arngast (1909–10) und den Leuchtturm Voslapp (1906–07). 1939–43 folgten weitere Leuchttürme.

Wenig mit Architektur, aber viel mit „Festungsbau" zu tun hatte die großangelegte Eindeichung von Heppenser und Rüstersieler Groden[106]. Sie war wichtigste Voraussetzung für die Planungen von *Marine* und *Luftwaffe* des „III. Reiches" in Wilhelmshaven.
126 Der *Neuordnungsplan vom* 20.11.1936, der Plan für die *Stadt der Dreihunderttausend vom*
127 15.03.1937, der Plan *Wirtschaftsgebiet Wilhelmshaven vom Juni* 1937 und der endgültige
128 *Wirtschaftsplan Wilhelmshaven-Friesland vom Mai* 1940 wiesen die riesigen einzudeichen-
129 den Grodenflächen von 2 km Breite und 6 km Länge aus[107]. Die Eindeichung wurde in 2 Abschnitten vorgenommen. Der Heppenser Groden wurde zwischen 1935 und 1938 begonnen und 1954 fertiggestellt (515 ha). 1000 Deichbauarbeiter täglich waren hier

bereits vor dem Krieg beschäftigt. Der Rüstersieler Groden wurde erst nach dem Kriege 1960–63 vollständig eingedeicht (505 ha).

★

Die militärischen Absichten im marinebegeisterten Berlin des 19. Jahrhunderts gingen
sehr schnell erheblich über die bescheidenen Bauplanungen für einen „kleinen Hafen mit
Depot" hinaus. Wilhelmshaven wurde schon nach den ersten Skizzen von 1856 mehr als
ein provisorischer Not- und Winterhafen in einer „Festung". Was nützten Kriegsschiffe, *6*
wenn man sie nicht warten und reparieren, ausrüsten und munitionieren konnte? Bevor
Gotthilf Hagen eingeschaltet wurde, mutete man dem Offizier *Christensen* die Zeichenarbeiten zu. Dies dokumentiert deutlich die militärische Interessenslage. Auch die Arbeiten des nächsten Planers *Wallbaum* scheiterten. Erst mit den Plänen des Zivilisten *Hagen* vom 25.06.1856 gelang der Durchbruch. 1871 nahm die *Königliche Werft* ihre Arbeit auf[108].

Bis 1945 wurden in der Wilhelmshavener *Marinewerft* 165 Schiffe neu gebaut.
Daneben wurde unterhalten und instandgesetzt, gerüstet und abgerüstet. 1880 war die
erste Erweiterung fällig, eine Torpedowerft an der I. Einfahrt. Ab 1900 entstanden *9*
aufgrund der *Flottengesetze* große Hellinge und neue Trockendocks. 1910 kamen noch die U-Boot- und Torpedobootwerft, die spätere *Westwerft*, als Erweiterung hinzu (*UTO-Werft*). Die *Reichsmarine* der Weimarer Republik schrumpfte und suchte sich zivile Betätigungsfelder.

Die 1935 eigens für den Bau der *Nordwerft* gegründete Sonderdienststelle *HNBD*[109] als
Teil der *Kriegsmarine*, wurde 1936 vom *OKM* beauftragt, die *KMW* zu erweitern. Das *10*
Flottenabkommen von 1935[110] ermöglichte *Hitler*, nun auch offiziell die *Marine* aufzurüsten. In Wilhelmshaven sollten Großschiffe gebaut werden, mit denen *Hitler* seinen Krieg gewinnen wollte. Dies bedeutete eine völlig andere und erheblich größere *KMW*, die durch die überalterte III. Einfahrt nicht mehr erreichbar gewesen wäre. Hauptliegehafen der neuen, bis 350 m langen und 140.000 t schweren Kriegsschiffe und Flugzeugträger sollte Wilhelmshaven werden. Der Wahn von uneinnehmbaren Schiffen trieb zu immer größeren Schlachtschiffen, zu immer größeren Kalibern[111]. Schlachtschiffe, die andere besiegen sollten, mußten zwangsläufig noch größer sein. Das Wettrüsten war eine Aufgabe der Ingenieure, nicht der Seeoffiziere. Der Eifer der Kriegsschiffbauer konzentrierte sich auf die *Nordwerft* von Wilhelmshaven. Die *HNBD* wurde für das *OKM* und *Hitlers* Kriegspläne wichtigste Dienststelle im hysterischen Rüstungswettlauf. *Carl Tiburtius* und ab 1939 *Ernst Beck* als Chefs der *HNBD*, wurden die wichtigsten Marinebeamten und mit allen Machtbefugnissen ausgestattet.

1936 hatte *Wilhelm Hallbauer* noch vorgeschlagen, die neue Großwerft landeinwärts in *126*
das Gebiet von Altengroden zu verlegen. Die IV. Einfahrt sollte in Rüstersiel liegen und hätte den gesamten Ortsteil dem Abriß preisgegeben. Eine weitere Schleuse und die begradigte schiffbare Maade sollten mit dem Ems-Jade-Kanal verbinden. Die Planung

scheiterte wegen der hohen Kosten und am Grunderwerb. Nach den Flächenanforde-
4 rungen der Einzelressorts entstand ein Programm für die *Nordwerft*, das vom *OKM*
genehmigt wurde[112]. Der erste Entwurf, noch weitgehend an den Entwürfen für die IV.
Einfahrt von 1917 orientiert, wurde von der *HNBD* nur geringfügig verändert und vom
OKM abgelehnt. Wie so oft in Wilhelmshavens Baugeschichte beauftragten die Berliner
Dienststellen einen Planer aus der Reichshauptstadt: *Arnold Agatz* hatte sich mit dem Bau
der Nordschleuse in Bremerhaven (1928–33) und dem Ausbau der Columbuskaje
qualifiziert[113]. Eigentliche Schlüsselfigur bei der *Nordwerft*-Planung war jedoch *Alfred
Eckhardt*[114]. Von 1918–33 war er *Hafenbaudirektor* und Leiter des kombinierten *Strom- und
Hafenbauressorts der Marinewerft Wilhelmshaven*. 1933 avancierte er zum Leiter des
Marinehafenbaus im *OKM*. Ab 1942 war er Chef des *Amtes Bauwesen* des *OKM*.
Zweifellos hat *Alfred Eckhardt Arnold Agatz* als Planer ab 1935 hinzugezogen, da er wußte,
daß die Wilhelmshavener *HNBD* planerische Hilfe benötigte.

Bereits im Ersten Weltkrieg hatte der *Admiralstab* festgestellt, daß Schiffe „der neuen
Größenordnung" nur unter großen Schwierigkeiten die drei vorhandenen engen
Hafeneinfahrten passieren konnten. Am 10.09.1917 legte das *Hafenbauressort* der
Kaiserlichen Werft dem *Reichsmarineamt* in Berlin einen ersten Entwurf der IV. Einfahrt für
Wilhelmshaven vor[115]. Für 85 Mio. Mark sollten 2 Schleusenkammern von gigantischen
Ausmaßen[116] einschl. der notwendigen Hafenerweiterung gebaut werden. Der Entwurf
stammte von *Hafenbaudirektor Hubert Behrendt*, dem Chef des *Hafenbauressorts der
Kaiserlichen Werft*. Mit Sicherheit kannten auch *Tiburtius, Eckhardt, Beck* und *Frede* die
Planung. Nach dem verlorenen Ersten Weltkrieg wurde die Planung eingestampft.

Die expansive Flottenbaupolitik der NS-Machthaber ließ die voluminösen Pläne 1935
wieder akut werden. *Behrendts* Planung wurde modernisiert, gestrafft und in die
großangelegte *Nordwerft* eingepaßt. Für technische Änderungen größeren Umfangs war
wenig Zeit[117].

101 Das Baugelände für die *Nordwerft* lag zum Teil im ab 1935 eingedeichten Heppenser
28 Groden, zum Teil jedoch im Wohngebiet Inselviertel. Der weitere Bau des Inselviertels
129 wurde abrupt eingestellt[118]. Der *Wirtschaftsplan von* 1940 sah das gesamte Inselviertel bis
zur Gökerstraße als „aufzugebendes Wohngebiet" vor. Die *HNBD* begann ab 1939 mit
46, 47 Hochdruck das Familienlager Neuengroden[119] als Ersatzwohnungsbau für die Bewohner
9 des östlichen Inselviertels und der gleichfalls abzureißenden Hafenkasernen an der
Ostfriesenstraße. Auch der für NS-Aufmärsche beliebte große Exerzierplatz mußte der
Nordwerft weichen, so daß schon ab 1938 neue „Maifelder" angelegt werden mußten.
Auch Batterien, Geschoßhöfe und das alte Fort Heppens, mußten abgerissen werden. Das
gesamte Baugelände war in der Hand des Fiskus, so daß Grunderwerb nicht erforderlich
wurde.

101 Der Nordhafen sollte als Großwerft nach Norden hin erweitert werden und durch ein
65 m breites Schiebetor mit Schiebebrücke hermetisch abgesperrt werden. Am Westufer
des neuen Hafenbeckens sollten ein Baudock und zwei Reparaturdocks entstehen, jedes
119 fast 400 m lang und 60 m breit[120]. Eine Zeitlang wurde sogar über eine V. Einfahrt am

Nordufer nachgedacht. Am Westufer der IV. Einfahrt sollten Flugzeugträger liegen. 30.000 Menschen sollten nach *Hitlers* Kriegsplänen ab 1943 auf der *Nordwerft* arbeiten, Großschiffe bauen und reparieren. Die Bauplanungen für *Nordwerft* und IV. Einfahrt waren „Geheime Reichssache", und wurden in vielen Einzelgesprächen zwischen der *HBND* und der *Bauabteilung des OKM* in Berlin bei *Alfred Eckhardt* festgelegt. Informationswege über Mittelinstanzen waren nicht vorgesehen. Die „Geheimhaltung" verbot ausgiebigen Schriftwechsel.

Die *HNBD* setzte sich zumeist aus Bauingenieuren und Hafenbauern zusammen. Sie war, obwohl *Carl Tiburtius* Hochbau studiert hatte, Architekturbelangen gegenüber wenig aufgeschlossen. Auch das eingeschaltete Planungsbüro *Agatz* hatte seinen Schwerpunkt nicht in Architektur und Städtebau, sondern im Hafenbau. Es kann angenommen werden, daß von 1936–37 auch *Fritz Höger* zur architektonischen Beratung der *HNBD* herangezogen wurde[121]. Bei der *HNBD* war *Werner Hoffmann* einer der wenigen Architekten. Er schaltete in Architekturangelegenheiten auch *Hans Klaucke* von der *KMW* ein.

Die Planungsmethodik der *Nordwerft*, bezogen auf Hochbauten, war außerordentlich
modern und flexibel. Die im *Bebauungsplan* dargestellten Blöcke beinhalteten die *101*
unterschiedlichen Ressorts für Ausrüstung, Artillerie, Schiffbau, Torpedotechnik, *92*
Maschinenbau und Nachrichtenmittel. Das gesetzte Zeitziel sollte durch „Simultanpla-
nung" erreicht werden. 47 bebaubare Blöcke, durch Straßen und Eisenbahnlinien
getrennt, waren in „Fünferreihen" angeordnet. Die Gliederung sollte Ver- und
Entsorgung, Verkehr und Transporte ermöglichen, aber auch Brandüberschlag verhin-
dern. 8 Eisenbahnstammgleise sollten zwischen den Hallen liegen. Innerhalb der
Baublöcke sollten unterschiedliche Hallen angeordnet werden können. Gleichzeitig *103*
waren einige Blocks im Bau, andere in der Planung, wieder andere in ihrer Nutzung noch
völlig offen oder für Erweiterungszwecke vorgesehen. Diese von traditionellen
Entwurfsmethoden stark abweichende Arbeitsweise führte zu strenger Rasterung der bis
180 x 180 m messenden Baublocks. Die „Offenheit" der Planung ermöglichte das *92*
Freilassen von Lagerflächen, aber auch die Integration z.B. des 1937 von *Karl Franzius* *101*
erweiterten *Marine-Observatoriums.*

Grundsätzlich waren alle Hallen in Stahlskelettbauweise mit bis zu 100 m überspan-
nenden abgestützten Fachwerkbindern vorgesehen. Sie sollten mit Ziegeln verblendet
werden. Die modernen Flachdachhallen hatten versetzt angeordnete 4 m hohe Ober- *103*
lichter. Größtes Bauwerk sollte die Schiffbauhalle III werden, 184 m lang, 109 m breit und 27 m hoch. Die neuen Hallen waren ungewohnt groß, weit größer als die Hallen der *Kaiserlichen Werft.* 360 Menschen sollten in der Schiffbauhalle arbeiten. Andere, sogar klimatisierte Hallen, sollten bis zu 900 Arbeiter aufnehmen können. Die Hochbauten wurden luftschutzgerecht entworfen. Alle Werkshallen bekamen eingebunkerte Keller. Wegen der Splitterwirkung von Bomben sollten die Fenster erst 2 bis 2 1/2 m über Terrain beginnen. Alle Lichtquellen konnten durch Spannungssenkung bei Luftangriffen verdunkelt werden.

92 Die Architektur war überaus funktional und durfte funktionsgerechte Hüllen zeigen. Lichtbänder, Flachdächer und Glaswände zeugten von moderner Industrieplanung und rationaler Arbeitsorganisation. Traditionelle Kostümierung der *Nordwerft* schied von vornherein aus. NS-Propaganda spielte nur eine untergeordnete Rolle. Die 30.000 Arbeiter der *Nordwerft* sollten sachliche, zweckmäßige, helle und saubere Arbeitsplätze bekommen. „Schönheit der Arbeit" in den staatlichen Rüstungsbetrieben hieß in diesem Falle auch Stahl und Glas. Das hochspezialisierte technische Personal hatte mit Elektronik und Optik, Navigation und Mechanik, Nachrichtenübermittlung und Maschinenbau zu tun. Rückwärts gewandte Architektur wäre unpassend, möglicherweise sogar schädlich gewesen.

101 Das nicht mehr gebaute Verwaltungsgebäude der *Nordwerft*, dem man sich über einen symmetrischen Vorplatz nähern sollte, wäre das größte militärische Verwaltungsgebäude der *Kriegsmarine* überhaupt geworden. Das monumentale, symmetrische Äußere sollte Tor für 30.000 Arbeiter werden, sollte auf Disziplin und Leistungswillen einstimmen. Ein großer Personenbahnhof der Schnellbahn war eigens vorgesehen, um Arbeiter aus den neuen Stadtteilen heranzuschaffen.

100 Die Architektur der IV. Einfahrt mit ihren zwei Schleusenkammern war sachlich, modern und ausschließlich in Betonbauweise vorgesehen. Die wenigen technischen Gebäude hatten keine repräsentativen Aufgaben. Nach anfänglich eigenen Entwürfen der
102 *HNBD* schaltete *Ernst Beck* noch *Hans Klaucke* von der *KMW* ein. Der legte einen außerordentlich modernen, qualitativ hervorragenden Entwurf für die Architektur der IV. Einfahrt vor. Die Eingangsbauwerke sollten nach dem Willen des *OKM* die Macht des „III. Reiches" symbolisierende, weit ausladende Hoheitszeichen bekommen.

★

Im März 1936 begannen die Bauarbeiten für *Nordwerft* und IV. Einfahrt mit
100 Grundwasserabsenkungen und Erdarbeiten[122]. 267 Mio. Mark waren veranschlagt. Umfangreiche Wohnlager für über 20.000 Bauarbeiter mußten gezimmert werden.

Die renommiertesten Baufirmen des Reiches waren beschäftigt. Bisher unbekannte Probleme des Bauablaufs und der Logistik mußten durch die zu enormer Größe angewachsene *HNBD* gelöst werden. Die Kriegswichtigkeit sicherte alle gewünschten Baustofflieferungen und die Zuweisung aller erforderlichen Bauarbeiter, Zwangsverpflichteter, „Erziehungsgefangener" und Häftlinge. Kaum war das erste Gebäude, der Schnürboden, fertiggestellt, wurde es als Lager für weitere Baustoffe in Anspruch genommen. Selbst im Bremer Hafen mußten Lagerflächen für die *HNBD* angemietet werden. Die meisten Massengüter wurden auf dem auch zu diesem Zweck erweiterten
94 Werftbahnhof Sande umgeladen. Seilbahnähnliche Kabel-Kranbahnen mit 450 m Spannweite sollten Baustoffe an jeden beliebigen Ort der Riesenbaustelle transportieren können. Ein besonderer Stromlieferungsvertrag sicherte 10 Mio. KWh Stromabnahme für die Bauarbeiten. Der nach modernen Gesichtspunkten aufgestellte Bauzeitenplan war

auf die Kiellegung der geplanten Schlachtschiffe abgestimmt. Oftmals griff die
Bauabteilung des *OKM* ändernd in die Planung ein. Während des Bauablaufs 1940 befahl
Alfred Eckhardt, das Baudock VII zu einem U-Bootsbunker[123] mit eigenem Werkstatt- 101
gebäude und 6 Dockzellen umzurüsten. Mit dem Umbau wurde noch begonnen. Die
riesigen Docks für den Kriegsschiffbau galten als einmalig in der Ingenieurbaukunst. Die
maschinelle Ausrüstung sollt ermöglichen, daß sie in 5 Stunden völlig leergepumpt
werden konnten. Das großdimensionierte Hafentor wurde ab 1936 begonnen, aber nicht
fristgerecht fertiggestellt. An der dazugehörigen Schiebebrücke, gleichfalls 1936 begon-
nen, wurde noch 1944 gearbeitet. Sie wurde nicht mehr fertig.

Carl Tiburtius und sein Nachfolger aber 1939, *Ernst Beck*, hatten beim Bau der
Schleusenkammern der IV. Einfahrt mit ungewohnten Dimensionen zu kämpfen. Diese 100
„größte Seeschleuse der Welt" galt bei den Militärs wegen der Bauzeit von 6 Jahren als
„außergewöhnliche technische Leistung"[124]. *Hitler* trieb die *Bauabteilung des OKM* zu
größter Eile, da ihm die Arbeiten zu langsam vorangingen. Sein *Z-Plan von* 1938 und die
Aufkündigung des *Flottenabkommens* am 28.04.1939, sollten die ungehemmte Marinerü-
stung ermöglichen. Am 01.04.1939 besuchten *Hitler* und Teile des Kabinetts die nur
schleppend vorankommende Baustelle der *Nordwerft*[125]. *Carl Tiburtius*, Chef der *HNBD*,
führte *Hitler* über die Baustelle. Kurz danach wurde er in „Pension" geschickt und von
Ernst Beck abgelöst. *Tiburtius*, der kein Parteimitglied war, hatte zunehmend Konflikte
mit dem NS-Regime bekommen. *Raeder* hat den für die Marinerüstung wichtigsten
Techniker vor Schlimmerem bewahrt. Intern sollte die Ablösung ohne großes Aufheben
erfolgen. Die Affäre um *Irina Tiburtius*, seine Ehefrau wurde zwar nie öffentlich, ihre
antifaschistische Einstellung jedoch muß wohl durchgesickert sein[126].

Zu Kriegsbeginn waren weder *Nordwerft* noch IV. Einfahrt fertig. Techniker und 101
Facharbeiter der Wilhelmshavener *KMW*, für den Bau der Großkampfschiffe vorgese- 10
hen, wurden zu den Werftbetrieben nach Frankreich, Belgien, Holland und Norwegen
versetzt, um dort Schiffe zu reparieren. Baubeamte wurden zu den in den unterworfenen
Gebieten neu eingerichteten *Marinebauämtern* versetzt.

Am 07.11.1942 „taufte" Großadmiral *Raeder* die IV. Einfahrt. Nur die Ostkammer der 102
„*Raeder*-Schleuse" war nach sechsjähriger Bauzeit fertig geworden. Engländer sprengten
die Schleuse am 20.01.1949. Von den Hochbauten der *Nordwerft* waren nur die
Hauptschiffbauhalle, die Schweißhalle und das Heizkraftwerk begonnen worden, und
nur das Gebäude des Schnürbodens wurde fertiggestellt. Da absehbar war, daß die
Nordwerft ihre Funktion niemals würde erfüllen können, mußte die *Marine* bestellte
Aggregate abgeben, u.a. nach Wolfsburg zur *KDF-Wagen* Produktion. Wilhelmshavens
militärabhängige Stadtentwicklung wird dadurch gekennzeichnet, daß bereits 1955 die
Vorüberlegungen für den Wiederaufbau der IV. Einfahrt begannen. 1964 wurden die
ersten Schiffe durchgeschleust. Der dazugehörige Marinestützpunkt mit seinen Kasernen,
Kommandos und Stäben wurde 1968 fertiggestellt.

4.4. Bauten für die Flieger

Die *Kaiserliche Marine* hatte bis 1918 eigene Seefliegerkräfte. An Wilhelmshavens Fliegerdeich, der 1906 bis 1908 aufgeschüttet wurde, richtete sie 1914 eine Ablaufbahn, Hangars, Werkstätten und Mannschaftsgebäude, ein. Die Flugzeugwerkstatt unterstand
10 der *Kaiserlichen Werft*[127]. Nur wenige Stadtpläne zeigten die 1918 dort befindlichen „geheimen" Luftwaffenanlagen. Auf dem Fliegerdeich befanden sich 2 Flugzeughallen und hinter dem Deich Reparaturwerkstätten, Lagerhäuser, Prüfstände und Gleisanschlüsse. Die gesamte *Luftwaffe* wurde 1918 aufgelöst. Nach dem Ersten Weltkrieg etablierten sich in den Luftwaffengebäuden Fischereibetriebe (1920), das Cafe Strandlust (1922), das Seebad Möve (1920), eine Fischauktionshalle (1920), das städtische Schlickbad (1931), der Seeflughafen des *Luftfahrtvereins Jade* (1928) und die Seeflugzeugwerft *Focke-Wulf* (1928). Schon während der Weimarer Republik versuchte die *Reichsmarine*, die Restriktionen des *Versailler Vertrages* zu umgehen und förderte verdeckt private Seefliegerei und Flughafenbau[128].

129 Der Landflughafen Mariensiel entstand 1927 mit Unterstützung der *Marine*. Beide Flughafenprojekte, der Seeflughafen Fliegerdeich und der Landflughafen Mariensiel wurden zwar von dem privaten *Luftfahrtverein Jade* getragen, sie waren jedoch von der *Reichsmarine* finanziert[129]. Nach *Hitlers* Machtübernahme kam es zu weiteren getarnten Seefliegerstaffeln. Schon ab 1934 beanspruchte die *Marine* wieder die Schuppen des Seeflughafens am Fliegerdeich[130]. Der *Reichsminister für Luftfahrt* und *Oberbefehlshaber der Luftwaffe, Hermann Göring*, unterstellte alsbald die gesamte Seefliegerei seiner *Luftwaffe*[131].

Das Reichsgebiet war ab 1935 bis 1939 in 4 *Luftflottenkommandos* eingeteilt. Wilhelmshaven gehörte zum *Luftflottenkommando 2 (Nord)* in Braunschweig. *Göring* hielt es für zwingend, die Küsten durch Jagdverbände zu schützen. Er hatte sich für ein *Heinkel*-Flugzeug als Jagdflugzeug entschieden. Die bauliche Infrastruktur, die See- und Landfliegerhorste, waren teils noch vorhanden, teils mußten sie neu geschaffen werden. Territorial war die *Luftwaffe* entsprechend den Wehrkreisen in *Luftgaukommandos* gegliedert. Das für den Raum Wilhelmshaven zuständige *Luftgaukommando See* war in Kiel, später in Hamburg, etabliert. *Luftwaffe* und Luftfahrt, aufs engste miteinander verflochten, mußten entsprechend ihrem vehementen Ausbau lebhafte Bautätigkeit befriedigen. So baute *Wilhelm Kreis* 1935–37 das *Luftgaukommando* in Dresden, *Ernst Sagebiel* 1935–36 das *Reichsluftfahrtministerium* und 1936–39 den Flughafen Tempelhof, *German Bestelmeyer* das *Luftgaukommando* München, *Hermann Brenner* die *Deutsche Versuchsanstalt für Luftfahrt*. Die gestalterischen Leitlinien für das *Luftwaffenbauamt Wilhelmshaven* und seine Projekte kamen aus Kiel. Das dortige *Luftgaukommando VI.* ließ sich als Mittelinstanz alle Entwürfe vorlegen, veränderte, griff ein, zeichnete auch völlig neu.

*

1934 hatte das *Luftflottenkommando* den Raum Jever als strategisch günstigen und
wichtigen Punkt für die Luftverteidigung der deutschen Bucht festgelegt. Der Staatsforst
Jever gehörte dem *Land Oldenburg* und versprach als Standort gute Tarnung. Im April 129
1935 konnte *Amtshauptmann Hermann Ott* vom *Amt Friesland* dem *Luftgaukommando See*
melden, daß alle Eigentümer verkaufsbereit seien. Große Waldbestände wurden im Juni
1935 abgeholzt. Seit April 1935 war *Baurat Ferdinand Schütt* Chef der *Bauleitung der
Luftwaffe*, Fliegerhorst Upjever. Organisatorisch gehörte diese Bauleitung ab 1939 zum
Luftwaffenbauamt Wilhelmshaven unter *Baurat Schäfer*.

1937 sollte der Fliegerhorst Upjever fertig sein, und so verfügte *Ferdinand Schütt* 156
schließlich über 4.000 Arbeiter und technisches Personal, die am Bau beteiligt waren und 130
in Baracken und Massenquartieren lebten. 1935 wurden Gleise und Straßenanschlüsse
sowie die wichtigsten militärischen Gebäude begonnen. Zur Bauleitung gehörten bis zu
60 Architekten und Ingenieure. Aus Hamburg von *Fritz Höger* kamen *Arthur Nevermann*,
und *Peter August Müller* – viele kamen aus der Umgebung: *Adalbert Jahn, Karl Higgen,
Otto van Reeken.*

Die Luftwaffenanlagen in Upjever bewegten sich in einem Stilpluralismus, der
moderne, aber auch auf die unmittelbare Umgebung bezogene regionale Bezüge
aufwies. Die Sendestelle im Barkeler Busch (1935) von *Adalbert Jahn* gezeichnet, war als
Bauernhaus mit Strohdach getarnt. Auf dem Fliegerhorst errichtete Gebäude wurden
schon damals als *Marinestil* bezeichnet. Selbstverständlich mußte die Werfthalle I (1936)
mit ihren notwendigerweise großen stützenfreien Spannweiten mit Stahlbindern und
schlanken Betonstützen konstruiert werden[132]. Den großen Hangartoren, Oberlichtern,
Hallenfenstern und flachgeneigten Dächern fehlten bis auf die Klinkeranbauten jegliche
regionale Zugeständnisse. Die 1937 gebaute Werfthalle II, die Werfthalle III und der 104
Tower wurden gleichfalls ohne Kompromisse modern und sachlich mit Flachdächern,
gering geneigten Pultdächern oder sogar gebogenen Dächern konstruiert. Nur kleine
Anbauten, Mannschaftsräume an den Hallen und Fliegerkasernen bekamen heimattü- 68
melnd versproßte Fenster.

Am 15.04.1936 trafen die ersten Fliegerstaffeln in Upjever ein, stürmisch von
NSDAP-Gliederungen und Bauleuten begrüßt. Zum 01.05.1936 feierte man die 156
Einweihung mit einem Parademarsch auf Jevers Marktplatz. „So ist ... 1936 für Jever eine
neue Zeit angebrochen. Möge sie nach dem Willen unseres Führers eine Zeit des Friedens
sein zum Glück von Heimat und Vaterland" jubelte die lokale Presse[133]. Die 450 *Offiziere*
und Mannschaften waren in Wohnblöcken, Ledigenheimen und Beamtenheimen
untergebracht. Krankenhaus und die Wohnsiedlung des Fliegerhorstes wurden 1936 105
fertig, das Feuerwehrgebäude 1937 und das Kasino 1938. Die sachlich gebauten Hallen
galten als Zweckbauten. Sie sollten „nur" Flugzeuge beherbergen. Die traditionell
gewandeten Personal- und Kommandogebäude jedoch sollten militärische Autorität und
Hierarchie ausdrücken. Die verwöhnte *Luftwaffe* bekam einen durchaus luxuriösen
baulichen Standard. Entsprechend ihrer militärischen Bedeutung wurden die verschie-
denen Gebäude angeordnet und gestaltet. Großlichtbilder der Fliegerhorstbauten

wurden auf der *1. Deutschen Architektur- und Kunsthandwerk-Ausstellung* in München gezeigt.

Die *Wohnungsbaugesellschaft Friesland* baute um 1938 mit dem Architekten *Hans Ahmels* in Jever Mietwohnungen für Personal der *Luftwaffe*. Mittelstand und Arbeiter siedelten sich 1935–39 in der Schützenhofsiedlung, hinter der Siedlung beim Dannhalm, am Englischen Weg, beim Tivoli und in Moorwarfen an[134]. Ab Juni 1936 baute die
143 *Oldenburgische Regierung* dem 1933 neugegründeten *Amt Friesland* sein Amtsgebäude – Vorbereitung auf die künftige Rolle im *Wohnsiedlungsgebiet Wilhelmshaven-Jeverland*.

Gleich benachbart, in Addernhausen, entwarf *Peter August Müller*[135] 7 Doppelhäuser für Arbeiter und Angestellte des Fliegerhorstes. Die 1935–36 von *Adalbert Jahn* ausgeführten Klinkerhäuschen mit Spitzdächern und Stallanbauten machten den Eindruck von Nebenerwerbssiedlungen und erinnerten an *Högers* Hamburger Stadtrand-Siedlungshäuser.

Die Funktionsgebäude auf dem Fliegerhorst Upjever haben durchgängig abgewalmte Dächer im *Marinestil*. Im Wohnungsbau jedoch dominierten Satteldächer ohne Aufschieblinge. Die Doppelhäuser oder zu Ketten addierten Doppelhäuser der Flieger-
105 horstsiedlungen Upjever erinnern an Siedlungen von landwirtschaftlichen Arbeitern in einer Heidekolonie. Über den Eingängen waren die Satteldächer zumeist abgeschleppt und durch profilierte Holzstiele abgespreizt. Ähnliche Wohngebäude wurden für Luftwaffenpersonal auf Wangerooge[136] und Langeoog gebaut.

Neben dem Fliegerhorst Upjever, aber zu seinem Flughafenbereich gehörig, wurden noch weitere militärische Flughäfen ausgebaut, neu gebaut oder umgebaut: Hage, Varel, Marx, Zwischenahn und Wittmundhafen. Auch Scheinflughäfen wurden gebaut. Der äußerste Flugabwehrring des deutschen Hoheitsgebietes umfaßte vor allem die Inseln. Wie im Ersten Weltkrieg hatten die Nordseeinseln große strategische Bedeutung. Tausende Arbeiter wurden ab 1938 beim Bau von Bunkern und Batterien, Landebahnen und Fliegerkasernen, Befehlsbunkern und Beobachtungsständen auf den Nordseeinseln beschäftigt.

*

126 Für den Aufbau Wilhelmshavens war bereits in *Wilhelm Hallbauers Neuordnungsplan vom* 20.11.1936 der spätere Heppenser Groden für die *Luftwaffe* reserviert. In seinen
127, 128 Plänen vom 15.03.1937 und vom Juni 1937 waren ein nördlicher tideunabhängiger Wasserflughafen und ein südlicher Landflughafen mit Luftwerft vorgesehen. Die Maade sollte anfangs noch in den nördlichen Wasserflughafen münden, später dann als neue „Maadelinie“[137] nördlich davon in den Jadebusen fließen. Die *Luftwaffe* hielt ihre Pläne äußerst „geheim“, und so waren die späteren *Wirtschaftspläne* ohne genauere Aussage. Am 22.04.1939 hatte das *Reichsluftfahrtministerium* „die Errichtung eines Flughafens
101 östlich Rüstersiel genehmigt“. *Nordwerft*, und nördlich davon die Luftwaffenanlagen, sollten im Rüstersieler- und Heppenser Groden entstehen. Die Erläuterungsberichte, die

an sich schon „Geheime Reichssache“ waren, verloren nur knappe Sätze über die Luftwerft und den neuen Wasserflugplatz. *Wilhelm Hallbauer* erinnerte sich 1953 „mit Grauen der Auseinandersetzungen *Marine – Luftwaffe* in Wilhelmshaven 1938/39“[138] und die für die *Luftwaffe* unzureichende Eindeichung.

Die späteren *Bebauungspläne* von *Walther Schulz* von 1941 bis 1942 sahen zwar anfangs *116*
noch eine eingedeichte Wasserfläche im Rüstersieler Groden vor, der neue Landflughafen jedoch sollte zwischen Langewerth und Accum entstehen. Für den groß angelegten neuen Flugplatz waren Startbahnen in verschiedenen Richtungen geplant. Vermutlich war er für militärische Nutzung nicht vorgesehen, denn die *Schulz*-Pläne waren Wiederaufbauplanungen für friedlichere Zeiten. Möglicherweise 1941 muß die *Luftwaffe* ihre Planungen für einen Wasserflughafen fallengelassen haben. Der *Schulz*-Plan vom Dezember 1941 jedenfalls sieht an der Stelle des Wasserflughafens ein eingedeichtes
Strandbad vor. Die Wiederaufbauplanungen 1944–45 von *Konstanty Gutschow* vom *118*
Arbeitsstab Speer sahen für Wilhelmshaven keine weiteren Flughäfen mehr vor. Der Wasserflughafen war endgültig zum Badesee geworden.

*

Heute noch sichtbare Zeugen der Seefliegerei sind in Wilhelmshaven die drei *1*
Kasernengebäude am Fliegerdeich und möglicherweise das Wirtschaftsgebäude des *Flughafens* Mariensiel[139]. Am Fliegerdeich plante der *Bauverwaltungsbezirk Wilhelmshaven* des *Luftfahrtkommandos See Wilhelmshaven* bereits ab 1935 eine neue Ablaufbahn und ab 1936 drei neue Fliegerkasernen. Die Entwürfe fertigte *Heinrich von Griesbach*. Werkplanung und Bauleitung lagen bei Architekt *Josef Vosshans*, der 1937 von der *KMW* zum *Bauverwaltungsbezirk Wilhelmshaven* gekommen war.

Das zentrale Hauptgebäude der Fliegerkasernen mit seinen Wirtschaftsräumen legte sich unmittelbar mit einer Terrasse an den vorhandenen Deich. Die vom „III. Reich“ privilegierten Flieger hatten Blick auf den Jadebusen und zum Süden ausgerichtete Zimmer. Zwei symmetrisch flankierende weitere Kasernengebäude hatten keine Verbindung zur Deichkrone. Das Äußere der Fliegerkasernen war dem in Wilhelmshaven üblichen *Marinestil* angepaßt: Ziegelbauweise, Walmdächer mit Aufschieblingen, Kreuzsprossen, werksteinumrandete Eingangstüren und Treppenhausfenster.

5. Monumentalplanungen des „III. Reiches" in Wilhelmshaven

Hitlers Politik der folgenden Monate ist ein einziges großangelegtes Manöver, um diese Wendung herbeizuführen und die Fronten Europas nach seinen taktischen Überlegungen neu zu formen. Admiral Canaris, der bei ihm war, als die Nachricht über die englische Garantie für Polen eintraf, hat Hitlers Ausruf überliefert: „Denen werde ich einen Teufelstrank brauen". Schon am folgenden Tag nutzte er den Stapellauf der „Tirpitz" in Wilhelmshaven zu einer Rede gegen die britische „Einkreisungspolitik", drohte den „Trabantenstaaten, deren einzige Aufgabe es (sei), gegen Deutschland angesetzt zu werden" und deutete die Kündigung des deutsch-englischen Flottenvertrages an ...[1].

Joachim C. Fest, 1976

Schon früh wollten Baumeister dauerhaft, unvergänglich, massiv und in kolossalen Dimensionen bauen. Bewußt strebten sie an, den Eindruck des Voluminösen zu steigern: durch überragende Größe, gliedernde Staffelung, herausgehobene Hauptbauwerke, zentralisierte Achsen und vergröbernde Details. Durch übertriebene Maßstäbe und zitierte Stile, durch gigantische Flächen und ausgesuchte Materialien, durch brutale Symmetrie und achsiale Anordnung sollten Untertanen eingeschüchtert, Besucher beeindruckt werden. Das „III. Reich" und *Hitlers* Architekturbesessenheit ließen ganze Baumeister-Generationen hoffen, obwohl die Bauwirtschaft durch Wehrmachtsbauten längst zufriedengestellt war und repräsentative Bauprojekte den Rüstungsetat zu schmälern drohten. Monumentale Bauten und das nationalsozialistische Stadtbild wurden vor allem in den Köpfen der Planer bewegt und in *Stadterweiterungsämtern* gezeichnet. In Wilhelmshaven gibt es noch heute Spuren von Monumentalplanungen, begonnene Bauten, ins Leere laufende Straßen, unmotivierte Freiflächen, symmetrische Blickachsen, massige Gebäude.

Die Neugestaltung deutscher Städte sollte in der Geschichte beispiellos sein und nach *Hitlers* Befehl bis 1950 vollendet sein. Eilfertig nannte Wilhelmshaven seinen *Wirtschaftsplan* vom 15./16.06.1939 „*Wilhelmshaven* 1950"[2]. Monumentalismus durchzog alle Planungssektoren, erreichte Wohnungsbau, Militärplanung, Gewerbebau, und Gemeinschaftsbauten. „Führerstädte", Gauhauptstädte und auch ganz „normale" Städte wetteiferten darum, nach dem *Gesetz zur Neugestaltung deutscher Städte vom* 04.10.1937 als *Neugestaltungsstadt* anerkannt zu werden. Die Kriegsmarinestadt Wilhelmshaven spielte bei Repräsentations- und Staatsbauten eine Sonderrolle. Sie hatte kaum Geschichte und

sollte aus mehreren Gemeinden ab 1937 zu Groß-Wilhelmshaven zusammenwachsen, darüber hinaus noch um Trabanten-Städte erweitert werden. Parallel dazu trieb *Hitler* die
101 Marinerüstung zu ungeheuerlichen Ausbauplänen.

Hitler befaßte sich schon immer mit Architektur und hatte feste Vorstellungen, wie nationalsozialistische Städte aussehen sollten[3]. Er verfolgte zusammen mit *Speer* die Neugestaltungspläne genau – auch die Wilhelmshavener Entwürfe. Er ließ sich über die Planungen berichten, hatte Modelle in seiner *Reichskanzlei* und ließ im Januar 1938 neben Kriegsschiff-Modellen Wilhelmshavener Entwürfe auf der *1. Deutschen Architektur- und Kunsthandwerk-Ausstellung* in München zeigen[4].

Anläßlich seines Besuchs am 01.04.1939 besprach *Hitler* die Neugestaltung Wilhelmshavens mit den städtischen Planern. Er ließ sich detailliert informieren, griff in die Planung ein, äußerte sich zwar wortkarg, aber sachkundig[5]. In seiner anschließenden Rathaus-
112 platzrede vor 100.000 „Volksgenossen" versprach er der Stadt die Lösung der
106 Wohnungsfrage und ein neues Theater. Wilhelmshavener Planer fuhren regelmäßig zu den Ministerien nach Berlin und den Parteidienststellen nach München, lieferten Modelle ab, besprachen Leitlinien von Stadterweiterungen und Neugestaltung und bekamen exklusive Informationen.

Walter Hunzinger und *Egon Forster*[6] wurden 1938 in München durch das am 18.07.1937 fertiggestellte *Haus der Deutschen Kunst (Troost)* geführt und bekamen Zugang zu dem „streng geheimgehaltenen" Neubau der Berliner *Reichskanzlei* (*Speer*), der am 03.01.1939 fertiggestellt wurde. Sie studierten die Modelle des am 30.01.1937 ernannten *GBI Speer* zur Neugestaltung Berlins, besahen sich die im Entstehen begriffene Ost-West-Achse (*Speer*), die am 20.04.1939 eingeweiht wurde. *Karl Franzius* von der *Marineintendantur*
5 wurde von *Speer* nach Berlin gebeten, als er sich am Wettbewerb *OKM* beteiligt hatte.
107, 108 *Speer* war über Wilhelmshavens Planungen, über *Kulturforum* und *Marineforum*, gut informiert[7]. *Willi Lübbers* zeigte seine Modelle von Voslapps *NSDAP*-Bauten *Hitler* am 01.04.1939, transportierte sie zu *Heß* nach München, anschließend in die *Reichskanzlei* nach Berlin[8]. *Wilhelm Hallbauer* und *Heinz Baumann* stellten *Speer* u.a. im August 1938 in Berlin die ersten Skizzen für Wilhelmshavens Neugestaltung vor[9].

Ende 1935 befaßte sich die *Stadt Rüstringen* mit Theater- und Hallenbadplänen[10]. Wilhelmshavens *Stadtbaurat Zopff* entwickelte seine Bahnhofsplanungen von 1918 fort und *Gauleiter Carl Röver* versprach am 01.04.1937, „Wilhelmshaven zur schönsten Stadt
128 Deutschlands zu machen"[11]. Für Groß-Wilhelmshaven, dessen innerstädtische Neugestaltung 1937 annähernd festlag, hätte, nach Auffassung von *Wilhelm Hallbauer*, das *Groß-Hamburg-Gesetz vom* 26.01.1937 Chancen der Neugestaltung beinhaltet[12]. Das „Machtwort des Führers"[13] verlor sich im Gestrüpp von Zuständigkeiten, fehlenden gesetzlichen Regelungen und widersprüchlichen Forderungen von *Marine* und *Luftwaffe*. Die *Stadt* bemühte sich verzweifelt, die von der *Marine* aus Luftschutzgründen betriebene
94 Aufgabe der Altstadt, *Industrieumsiedlungsprogramm* und Dezentralisierung der neuen Gemeinschaftssiedlungen planerisch aufzufangen und das Stadtbild zusammenzufassen um architektonische Ansprüche durchzusetzen.

Das *Gesetz zur Neugestaltung deutscher Städte vom* 04.10.1937 endlich erschien Groß-Wilhelmshaven die Finanzierung der Neugestaltung zu öffnen. Das *Reich* hatte bis dahin „alle Entscheidungen in Planung und Durchführung der Stadterweiterung" entsprechend „dem zufälligen damaligen Bedürfnis" in Anspruch genommen[14]. Entsprechend hatte *Reichsinnenminister Wilhelm Frick* am 06.07.1938 den *MDF* beschieden, daß Reichsmittel für Wilhelmshavens Stadterweiterung und Neugestaltung nur vorfinanziert werden ohne „Präjudizierung für die Übernahme bestimmter späterer Verpflichtungen". Die Finanzierung von Wilhelmshavens Neugestaltung war ungesichert und unsolide. *Hitler* hatte verboten, über Baukosten und Finanzierung überhaupt zu sprechen[15]. Im August 1939 legte er fest, daß die *Neugestaltungsstädte* mit Ausnahme der „Führerstädte" Berlin, Hamburg, Linz, München und Nürnberg ihre monumentalen Baumaßnahmen selbst finanzieren müßten. In Wilhelmshaven war man, wie auch in anderen Städten, der verhängnisvollen Auffassung, die Bauten der *Neugestaltungsstädte* würden vom *Reich* finanziert, denn „alles Geld müsse ja vom deutschen Steuerzahler aufgebracht werden" und fließe durch den „Topf des *Reichsfinanzministers*"[16].

Das *OKM* hatte *Müller* und *Hallbauer* am 19.11.1937 nach Berlin gerufen und gebeten, *4*
die Kosten für Stadterweiterung und Neugestaltung zusammenzustellen[17]. Die *Stadt* war auch bereits bei *Rudolf Heß* in München vorstellig geworden, um die Neugestaltungsprojekte mit der Parteizentrale abzustimmen. *Oberbürgermeister Müller* machte am 20.12.1937 eine Eingabe an *Reichsstatthalter* und *Gauleiter Carl Röver*, um zu erreichen, daß *Hitler* das *Gesetz zur Neugestaltung deutscher Städte vom* 04.10.1937 auch auf Wilhelmshaven anwende[18]. „Durch die Wehrmachtspolitik des Führers" stehe „die *Kriegsmarine* vor einer gewaltigen Entwicklung". Die gesamten städtebaulichen Ausgaben beliefen sich auf 480 Mio. Mark für „den Aufbau einer neuen Stadt", die einen „überwältigenden Eindruck der Baugesinnung des neuen Deutschland" abgeben solle. Nach „den großen Richtlinien des Führers für Städtebau" solle eine breite Achse von Süden nach Norden die Stadt durchziehen. Am Hafen sollte sie „mit künstlerisch vollendeten Landeanlagen für den Führer" beginnen, durch das „große *Marineforum*" mit „einer weiten Platzanlage für militärische Paraden" führen und als „Traditionsstraße der *Marine* zu dem Forum der Bewegung" führen. Dort sollten sich „das Haus der Partei und ihrer Gliederungen, ein Stadttheater, Kunsthalle, Museum und Volksbücherei" erheben, anschließend daran das *Sportforum*. Abschluß der Achse sollte in der Nähe des Stadtparks eine große Festhalle mit Aufmarschgelände sein, „gedacht als Krönung des Ganzen". Eine Ost-West-Achse sollte am neuen Bahnhof beginnen, vorbei am *Verwaltungsforum* Rathausplatz, und am neuen Werfttor enden.

Der *MDF* mochte den Antrag der *Stadt Wilhelmshaven* nicht befürworten und hielt ihn *3*
für zwecklos, denn im Gegensatz zu Berlin, München, Nürnberg und Hamburg werde Wilhelmshaven „doch immer eine Werft- und Marinestadt bleiben und vor allem wird auch die Festung bei dem Stadtgebilde zum Ausdruck kommen"[19]. Im März 1938 machte *Ministerialrat Rodenberg* von der oldenburgischen Vertretung beim *Reich Oberbürgermeister Müller* klar, daß Wilhelmshaven keine *Neugestaltungsstadt* werden könne[20]. Auch

Oberregierungsrat Max Büge und *Ministerialrat Georg Heilmann* lehnten ab. *Müller* hatte vergeblich versucht, nachzuweisen, daß Wilhelmshaven auch Monumentalbauten vorbereite: *Marinefachschule*, Volksbücherei, Theater und Handelslehranstalt. Vor allem das *OKM* war 1938 gegen eine *Neugestaltungsstadt* Wilhelmshaven. Es fürchtete, daß sich dies gegen die *Marine* auswirken werde und Monumentalbauten sowieso fehlten. Die Anträge der *Stadt Wilhelmshaven* konnten kaum erfolgreich sein:

o *Stadtbaudirektor Hunzinger* stellte öffentlich fest, Wilhelmshaven sei „keine zivile Stadt", sondern Hauptwaffenplatz der Nordsee[21].

o *Hitler* war an dem in Wilhelmshaven vorherrschenden *Heimatschutzstil*, den Gemeinschaftssiedlungen, Kasernen und Klinkerbauten nicht interessiert[22]. Seine faschistischen Monumentalbauten sollten mit Natursteinen aufgemauert werden.

o *Hitler* wollte Neugestaltung nicht durch mittelmäßige Baubeamte und ehrgeizige *Oberbürgermeister* abwickeln lassen, sondern durch Architekten und *Gauleiter*, die er zu *Beauftragten für die Neugestaltung* machte.

o Die *Reichskanzlei* wurde mit Anträgen auf Anerkennung als *Neugestaltungsstadt* überhäuft. *Speer*, drängte *Hitler*, keine weiteren *Neugestaltungsstädte* mehr zuzulassen.

101 o Der Bau von *Nordwerft* und IV. Einfahrt hatte kriegswichtige Bedeutung. Weitere Architekturaufgaben hätte die Bauwirtschaft nicht verkraftet.

o Die *Stadt Wilhelmshaven* hatte, nachdem *Walter Hunzinger* Anfang 1939 entlassen war, gegen den Rat von *RAM* und *OKM* den Nicht-Architekten *Walter Temp* zum *Stadtbaudirektor* gemacht.

Als das „Hamburger Fremdenblatt" am 28.07.1938 meldete, Wilhelmshaven sei „in die Reihe der reichsseitig auszubauenden Städte einbezogen worden", stellte *Müller* klar, daß dies „blühender Unsinn und große Irreführung der Öffentlichkeit" sei[23]. Die *Stadt* müsse leider „auf einem sehr umständlichen Weg" die Mittel für die Neugestaltung auf dem Instanzenwege anfordern.

Auch als nach zwischenzeitlichem Verbot aller nicht kriegswichtigen Neubauten vom 15.11.1939 und nachfolgenden „Eroberungserfolgen" zwischen Mitte 1940 und Mitte 1941 *Hitler* die Anzahl der *Neugestaltungsstädte* auf 37 ausgedehnt hatte[24] und eine wahre Planungseuphorie ausbrach, war der Sonderfall Wilhelmshaven nicht unter den offiziell anerkannten Städten. Enttäuscht nahmen die lokalen Parteifunktionäre zur Kenntnis, daß Oldenburg als „Gauhauptstadt des Altreichs" *Neugestaltungsstadt* und *Röver* zum *Beauftragten für die Neugestaltung* ernannt wurde[25]. *Röver* ließ das *Planungsamt* der *Stadt Oldenburg* und freie Architekten bis 1942 an einem *Gauforum* und einer Nord-Süd-Achse zeichnen[26]. Wilhelmshaven stand bei *GBI Albert Speer* nicht mehr als *Neugestaltungsstadt* zur Disposition. Enttäuschend war für Wilhelmshaven auch, daß Bremen, Kiel und Wolfsburg *Neugestaltungsstädte* wurden. Das norwegische Trondheim, seit 1940 besetzt, sollte gar zum größten deutschen Kriegshafen ausgebaut werden. *Hitler, Speer* und *Raeder* hatten dort am 21.06.1941 Standorte für die Neuanlage einer 250.000-Einwohner-Stadt mit riesigen Kriegsmarineanlagen festgelegt[27].

Ähnlich wie Salzgitter wurde Groß-Wilhelmshaven, wenn auch nicht der Status der *Neugestaltungsstadt*, so jedoch eine herausragende Sonderrolle zugestanden. Erst Anfang 1944 wurde Wilhelmshaven *Wiederaufbaustadt* und damit gleichzeitig *Neugestaltungsstadt.*

5.1. Verwaltungsforum

Vorläufer der Planungen zu einem *Verwaltungsforum* war der Architektenwettbewerb, *1*
den *Martin Wagner* 1913 für das Rathaus-Umfeld, die Rüstringer Zentralanlagen ausschrieb. In kaiserlicher Tradition wollte er in der 1911 gegründeten oldenburgischen Stadt Rüstringen „die Wehrmacht Deutschlands nach außen hin repräsentieren helfen"[28]. Die Zentralanlagen symbolisierten den Willen, die „zersplitterten Verwaltungen und Instanzen zu gemeinsamer Arbeit" zu verbinden.

Wilhelmshaven/Rüstringen hatte 85.000 Einwohner und war mit 10.000 Arbeitern der *Kaiserlichen Werft* und 30.000 Marinesoldaten völlig marineabhängig. Berlins *Stadtbaurat* *10*
Ludwig Hoffmann saß dem Preisgericht Zentralanlagen vor, dessen Teilnehmer u.a. *Hans Poelzig, Heinz Stoffregen, Hans Bernoulli, August Abbehusen* und *Otto Blendermann* waren. *Peter Behrens* sollte nicht teilnehmen, *Fritz Höger* konnte nicht teilnehmen. Die Pläne mit Rathaus, zwei Schulen, Sparkasse, Bibliothek, Feuerwache, Kino, Saalbau und Geschäftshäusern blieben kriegsbedingt liegen. 1915 nahm *Wagners* Nachfolger, *Stadtbaurat Willy Hahn*, die Planung der Zentralanlagen mit einem zweiten Wettbewerb wieder auf. Zu nennenswerten städtebaulichen oder programmatischen Veränderungen kam es nicht. Ein zur Bismarckstraße abgeschlossener Rathausplatz sollte ein Rathaus von *Fritz Bräuning* sowie östliche und westliche Randbebauungen bekommen. Verkehrsfluß und Verwaltungsfunktionen sollten die Planung bestimmen. *Fritz Höger*, von dem für Bauangelegenheiten zuständigen *Bürgermeister Arthur Kellerhoff* und *Reinhard Nieter* nach Rüstringen geholt, konnte das Rathaus 1928–29 in dem bekannten niederdeutschen Klinkerexpressionismus bauen. Rüstringen war mit *Höger* zufrieden und ernannte ihn nach Fertigstellung des Rathauses in Ermangelung eines gestaltungsbewußten *Stadtbaurates* zum städtebaulichen Berater.

Wie in Berlin die Neugestaltungspläne bereits vor dem *Gesetz zur Neugestaltung deutscher Städte vom* 04.10.1937 einsetzten, hatten auch die Pläne für ein *Verwaltungsforum* frühe Wurzeln. Die Rüstringer Planer dachten jedoch noch nicht in Platzwänden und Aufmarschkolonnen – sie standen in der Tradition der 20er Jahre. Bereits im Februar 1934 schickte Rüstringens *Bürgermeister Arthur Kellerhoff Fritz Höger* zu *Friedrich Carl*, dem *Präsidenten* des *Landesfinanzamtes Weser-Ems* nach Bremen, um die Bebauung der freien Plätze um das Rathaus zu besprechen. Im Februar 1935 berichteten die Zeitungen von Theater- und Hallenbadplänen am Rathaus. *Fritz Höger* skizzierte Vorentwürfe für ein

Theater, dem Rathaus gegenüber an der Nordseite der Bismarckstraße[29]. Entstehen sollten im Rathausumfeld: „Haus des Handwerks“, *Zollamt, Reichsbauamt, Finanzamt,* Handwerkeraltenheim, Gymnasium, *Polizeipräsidium* und etwas weiter entfernt das Marinestationsgebäude. Städtebauliches Ziel war, alle wichtigen jadestädtischen Marine-, Reichs- und städtischen Behörden „auf einen Quadratkilometer“ städtebaulich zu zentralisieren[30]. Als *Hitler* im Juni und Oktober 1936 den Kriegsmarinestandort Wilhelmshaven besuchte, traf er Entscheidungen zu den Militärplanungen von Garnison und *KMW*. Von Monumentalbauten im Sinne des *Gesetzes zur Neugestaltung deutscher Städte vom* 04.10.1937 war noch nicht die Rede, und an Wohnungsbau und Werksiedlungen war *Hitler* kaum interessiert.

*

Das Wilhelmshavener *Finanzamt* war in der Weimarer Republik im Gebäude der
14 *Marineintendantur* an der Peterstraße untergebracht und eng mit der *Marine* verflochten[31]. Die vehemente Rüstungspolitik erforderte mehr Büroflächen für die *Marineintendantur* als wichtigste örtliche Mittelinstanz. Zur bevorstehenden Zusammenlegung zu Groß-Wilhelmshaven sollten alle wichtigen Reichsbehörden repräsentativ im *Verwaltungsforum* untergebracht werden. Der Leiter des *Reichsbauamtes Paul Berg* und *Finanzpräsident Friedrich Carl* fürchteten, daß das Raumprogramm für ein neues *Finanzamt* an der Südwestecke des Rathausplatzes ein monumentales Volumen nicht ermögliche, zumal *Zollamt* und *Reichswasserwerk* entfielen[32]. *Oberbürgermeister Gustav Nutzhorn* empfahl, daß die *Stadt Rüstringen* auf eigene Rechnung die Geschoßhöhe des gegenüberliegenden 4-geschossigen Realgymnasiums aufnehmen solle, um das gewünschte symmetrische Bauvolumen zu erzielen. Das *Reichsfinanzministerium* stimmte dem Vorschlag von *Baurat Schikore* vom *Landesfinanzamt Weser-Ems* zu, die Geschosse zu erhöhen, aber keine Reserveflächen einzubauen. Die Mehrkosten bis zu 20.000 Mark sollte *Rüstringen* tragen und dazu noch das Grundstück kostenlos hergeben. *Fritz Höger* und die *Bauverwaltung* sollten noch Gegenvorschläge machen[33]. *Höger* war bereit, im Rahmen seines Vertrages als städtebaulicher Berater tätig zu werden, wie er auch beim benachbarten „Haus des Handwerks“ städtebaulich beriet. Im Mai 1935 lagen zwei Vorschläge von *Schikore*, einer von *Höger* und zwei vom *Bauamt* vor. Das *Rüstringer Bauamt* empfahl, die Nord-Ost-Ecke zu bebauen. Das *Landesfinanzamt*, das die Südwestecke des Rathausplatzes bevorzugte, übte Druck aus und drohte, in Wilhelmshaven, und nicht in Rüstringen zu bauen. *Gustav Nutzhorn* wies sein *Bauamt* in die Schranken und so wurde entschieden, daß *Schikore* und *Höger* gemeinsam weiterarbeiten sollten. Ende Mai 1935 hatte *Höger* abermals einen Entwurf für das *Finanzamt* vorgelegt[34], aber schon im Juni war klar, daß seine Planung nicht genehmigt werden würde[35].

Trotzdem aber wurde *Höger* von *Schikore* bei der Fassade hinzugezogen – sie sollte „ganz schlicht“ gestaltet werden. *Högers* Fassadenberatung zog sich bis in den Juli 1935 hin, so daß die *Stadt* verärgert war und erwog, *Höger* auch den versprochenen freihändigen Auftrag für das neue *Arbeitsamt* zu entziehen und statt dessen einen Wettbewerb nach den Richtlinien der *RDBK* durchzuführen[36]. Im September 1935 legte *Schikore* seine Planung bei der *Stadt Rüstringen* vor. Diese gab die Pläne weiter an *Höger* mit der Bitte um „freundliche Genehmigung". Im Oktober 1935 machte *Höger* noch Alternativvorschläge zu *Schikores* Planung und ein Modell[37].

Im Februar 1936 begann der Bau[38] des *Finanzamtes* nach den von *Paul Berg* unterschriebenen Plänen. Bereits am 26.09.1936 feierte man mit NS-typischer Dekoration das Richtfest für das eine halbe Million Mark teure Projekt. *Fritz Höger* und der bauleitende Architekt *Scheeb* vom *Reichsbauamt Wilhelmshaven* wurden von Parteigliederungen und Behörden für die „bodenständige Architektur" gelobt, die „sich in ihrer schlichten Klinkerfassade dem Rüstringer Rathaus und dem Gesamtbild des Rathausplatzes[39] anpaßt. Die Arbeitsplätze seien „schlicht, gesund und schön". Die NS-Propaganda bejubelte das „deutsche Baumaterial" und das „Wahrzeichen einer neuen Zeit"[40]. *Arthur Kellerhoff* nannte die künftigen Ziele, „aus dem Rüstringer Rathausplatz ein Forum zu machen, das viele wichtige Behörden umfasse". Es sollte „eindrucksvoll" sein, wie keine oldenburgische Stadt es habe. Am 08.10.1937 weihte *Finanzminister Lutz Graf Schwerin von Krosigk* das neue Gebäude, das nun *Finanzamt, Zollamt* und *Reichsbauamt* enthielt, mit nationalsozialistischer Propaganda ein[41]. Vier Tage vorher hatte *Hitler* das *Gesetz zur Neugestaltung deutscher Städte* erlassen und die Berliner *Speer*-Planungen bekamen große Aktualität. Sicher ist es dem *Finanzminister* nicht schwergefallen, auch das *Finanzamt* Wilhelmshaven verbal in die Reihe der Neugestaltungsmaßnahmen des „III. Reiches" einzuordnen.

★

Auch das *Arbeitsamt Wilhelmshaven* war ein vorweggenommener Schritt zur Neugestaltung von Groß-Wilhelmshaven. Regulierung des Arbeitskräftebedarfs war für die Rüstungs-und Bauprojekte von größter Bedeutung. Seit 1928 war das dem *Reich* unterstellte *Arbeitsamt* im alten Heppenser Rathaus untergebracht. Der außerordentlich hohe Arbeitskräftebedarf erforderte schon bald ein neues Verwaltungsgebäude. Seit dem *Gesetz zur Ordnung der nationalen Arbeit vom* 01.05.1934 waren „soziale Ehre", „Führergrundsatz" und Weisungsrecht des *RAM* in das Arbeitsrecht eingefügt. Die für Wilhelmshaven kriegswichtigen Berufe bedurften der Lenkung, der Dienstverpflichtung, der Ausbildung und der sozialen Absicherung. *DAF* und *Marine, KMW* und Wirtschaft waren höchst interessiert an der Zuweisung der erforderlichen Arbeitskräfte und führten gemeinsam jährliche *Arbeitsschlachten* durch. *153*

Das *Arbeitsamt Wilhelmshaven* sollte 1935 wegen der erwünschten monumentalen Wirkung zusammen mit dem *Finanzamt* gebaut werden. Die Absichten scheiterten, denn das *Arbeitsamt*, bis dahin Mieter in einem städtischen Gebäude, beanspruchte wiederum städtische Räume. Die *Stadt Rüstringen* entschied sich, für 200.000 Mark an der Nord-Ost-Ecke des Rathausplatzes von *Fritz Höger* ein städtisches Gebäude zur

Weitervermietung an das *Arbeitsamt* zu erstellen. Am 01.10.1936 wurde das *Arbeitsamt* gerichtet und am 14.04.1937 pünktlich zur Zusammenlegung der Städte zu Groß-Wilhelmshaven übergeben. *Regierungsrat Richard Held* dankte im Rahmen eines Betriebsappells dem „Führer" für die Beseitigung der Arbeitslosigkeit und den „Aufstieg der Wirtschaft"[42]. *Högers Arbeitsamt* wurde im Kriege vollständig zerstört und nach dem Kriege zu einer Druckerei aufgebaut.

*

Nach der Zusammenlegung der Städte zu Groß-Wilhelmshaven am 01.04.1937 arbeitete der neue *Oberbürgermeister Wilhelm Müller* zielstrebig auf eine *Neugestaltungsstadt* im nationalsozialistischen Sinne hin. Das Gauforum in Weimar von *Hermann Giesler* war bereits 1937 unter *Hitlers* besonderer Protektion fertiggeworden und die monumentalen Baumaßnahmen der „Führerstädte" Berlin, München und Nürnberg wurden laut propagiert. Die bisherigen *Stadtbauräte Carl Haefner* und *Hermann Zopff* wurden in Pension geschickt. *Fritz Höger*[43] bekam die Kündigung seines städtebaulichen Beratervertrages. Der neue *Stadtbaudirektor Walter Hunzinger* übernahm ab 01.11.1937 die Neugestaltung von Groß-Wilhelmshaven. Schon 1936 hatte sich die *LOB* gegen *Höger* ausgesprochen. Man versprach, ihn bei der „weiteren Ausgestaltung des Rathausplatzes und dessen nähere Umgebung" hinzuzuziehen. Tatsächlich kam er für die Monumentalplanungen nicht mehr in Betracht. *Wilhelm Müller*[44] kündigte „außerordentliche Planungen" an und wollte Bauten, „die sich vor der Geschichte sehen lassen" können. „Nichts ist stärker geeignet, dem Menschen die Größe seiner Zeit und der eigenen Verpflichtung eindringlicher zu machen, als der tägliche Anblick der großen Bauwerke". *Müller* wollte „steinerne Zeugen der Macht des Reiches" und „Denkmäler unserer Zeit" für Groß-Wilhelmshaven. Die Mitwirkung *Högers* am *Verwaltungsforum* 1934–37 stand in der Tradition seiner Rüstringer Beraterfunktion. *Höger* galt, da er durch seine Bürohäuser bekanntgeworden war, als der geeignete Architekt für das *Verwaltungsforum*. Beim „Haus des Handwerks" hatte er sich 1935 jedoch schon nicht mehr durchsetzen können und beim *Finanzamt* entwarf er gerade noch Fassadenteile. Sechs Klinkerlisenen am Eingang bezogen sich in gleicher Maßordnung auf sein 1928–29 gebautes Rathaus. *Höger* zitierte beim *Finanzamt*, das oberste Rathausgeschoß, die Sprossenteilungen und das Sockelgeschoß. Der Naturstein-Hoheitsadler an der Gebäudeecke mußte seine Flügel „anlegen", um *Högers* Lisenen nicht zu verdecken.

Beim *Arbeitsamt* war *Höger* 1936–37 nicht nur städtebaulicher Berater, sondern Architekt im umfassenden Sinne. Walmdach und Ziegelornamente bestimmten das traditionelle Äußere. Wie beim gleich hohen *Finanzamt* war das Sockelgeschoß durch herausgehobene Ziegelschichten strukturiert und das Obergeschoß staffelgeschoßartig abgesetzt.

*

Der Neubau des *Polizeipräsidiums*[45] sollte wichtigstes Bauwerk des *Verwaltungsforums* werden. *Himmler* wurde im Sommer 1936 Chef der Polizei im gesamten Reich. Wilhelmshaven war Waffenschmiede der *Kriegsmarine*, und da bis zu 42.000 Menschen im Dienste von *KMW, HNBD* und der Baufirmen arbeiteten, nahmen Bespitzelungen, Zwangsmaßnahmen und Sicherheitsaufgaben zu. Der Aufbau der *Gestapo* in großem Zuge ergab sich aus dem *Gesetz über die geheime Staatspolizei vom* 10.02.1936. Das *Polizeipräsidium* in der Roonstraße 57 konnte die ausufernde Polizeimacht nicht mehr unter ihrem Dach beherbergen. Der Neubau des *Polizeipräsidiums* im *Verwaltungsforum* war bereits 1936 fest eingeplant. Erst am 03.08.1938 verkaufte die *Stadt* dem *Reich* die Nord-West-Ecke des Rathausplatzes für den Bau des neuen *Polizeipräsidiums.* Erst jetzt sollte das *Verwaltungsforum* totalitär-monumentale Ausmaße annehmen, sollten benachbarte Blöcke abgerissen und neu gestaltet werden. Die Zentrale der Polizeigewalt sollte sich in Fortsetzung des *Finanzamtes* und abgewinkelt zur neuen Ost-West-Achse, der Bismarckstraße, als walmgedeckter Ziegelbau fortsetzen. Eine durchgehende Säulenhalle sollte die von *Höger* beim *Finanzamt* begonnene Ziegelgestaltung des Sockelgeschosses kaschieren. *Hitler* selber drängte darauf, die verbindenden Arkaden zu streichen[46], da sie die Belichtung der unteren Räume behindern. Der NS-Staat mußte noch vor dem beabsichtigten aber nie durchgeführten Bau des *Polizeipräsidiums* eine Baracke für die *Gestapo* am Rathausplatz aufstellen. Im Gebäude der *NSDAP-Kreisleitung*, heute *Robert-Koch-Haus*, ließ *Klaus Rasch* 1940 Räume für die Gestapo bauen.

★

Stadterweiterung und Neugestaltung, Selbstbewußtsein und Nationalstolz, Aufgabenerweiterung und Repräsentation ließen die *Kommunalverwaltung* Groß-Wilhelmshavens anwachsen. Eine Rathauserweiterung erschien *Oberbürgermeister Müller* unverzichtbar. *Fritz Höger* hatte zwei Flügelbauten an der Südseite des Rathauses entworfen[47]. So beantragte *Müller* am 13.02.1939 beim *MDF* eine Erweiterung. Er begründete dies mit den erheblich zugenommenen Aufgaben der *Bauverwaltung.* Modelle und Pläne seien nicht unterzubringen. Deswegen könnten auch die „großen neuen Siedlungen der Marine" nicht so schnell erledigt werden"[48]. Wenn „nach vielen Jahren die Aufbauarbeiten der neuen Stadt im wesentlichen abgeschlossen" seien, könnten die *Bauämter* „auf ihren normalen Umfang zurückgeführt werden". Die Räume könnten dann für die Verwaltung der erheblich größeren Stadt benötigt werden. Verschiedene Modelle der Rathauserweiterung wurden *Hitler* am 01.04.1939 präsentiert[49]. Die Heimatschutzeuphorie hatte sich der Wilhelmshavener Planer bemächtigt. Wie sonst wäre es zu erklären, daß *Hitler* sogar Rathaus-Seitenflügel mit Walmdächern gezeigt wurden. 500.000 Mark *108*
sollten die zwei Flügelanbauten am Rathaus kosten. Nach dem *Hitler*-Besuch sollte gar das Rathaus in das *Kulturforum* verlegt werden. Die *Bauverwaltung* wurde provisorisch in Baracken untergebracht. Es blieb bis heute bei der Utopie von einer Rathauserweiterung.

⋆

Kern des *Verwaltungsforums* sollte der Rathausplatz werden. Im Dezember 1937 legte *Oberbürgermeister Müller* den Rathausplatz als „dritten Schwerpunkt der städtebaulichen Gestaltung" fest[50]. Entwürfe und Bauabsichten im *Verwaltungsforum* ließen sich zur Begründung des Antrages *Neugestaltungsstadt* hervorragend verwenden. Die Planungen jedoch waren weder abgeschlossen noch im Sinne von *GBI Speer*.

52 Im Januar 1939 machte die *Stadt* einen zweiten Anlauf und präsentierte Entwürfe und
Modelle des *Verwaltungsforums* der Öffentlichkeit. „Nur durch starke architektonische
und geldliche Mittel" glaubte die *Bauverwaltung*, die „Vielheit" und Zerrissenheit der
Gestaltung Wilhelmshavens ausgleichen zu können, so tönte die Propaganda[51]. Die
Planung, die auf *Egon Forster* und *Wilhelm Hallbauer* zurückging, hatte nur Bedeutung für
112 den *Hitler*-Besuch am 01.04.1939.

Der Rathausplatz, von *Höger* 1927 als Wasserbecken von 150 x 40 m geplant, 1930
dann ersatzweise mit Rasen und Rosen bepflanzt, bekam zunehmende Bedeutung für
157 nationalsozialistische Feiern militärischer Aufmärsche und Massenkundgebungen. Maifeiern und Lichtdome sollten die Bevölkerung beeindrucken und für das System gewinnen. Ratsherr *Karl-Heinrich Hinz* beantragte im März 1939 die Umbenennung des Rathausplatzes in „Platz der *SA*"[52]. *Für Hitlers* Massenkundgebung am 01.04.1939 vernichtete man die Begrünung und füllte den Platz zum Aufmarschplatz auf. 85 x 200 m sollte der „Platz der *SA*" messen. Die geschlossenen Platzwände waren als monumentale Kulisse geplant. „Tore" zum Platz sollten an der Bismarckstraße ein Ehrenmal, gerahmt von zwei „Säulen der Bewegung"[53], und jeweils begrenzende Pylonen an der Adolf-Hitler-Straße werden. Jenseits der Bismarckstraße plante *Rudolf Bürgin* 1938 einen viergeschossigen Riegel als abschließende Platzwand zu *Högers* Neu-Siebethsburg. Die Sparkasse sollte auf dem spitzen Winkel an der Einmündung Mühlenweg in die Bismarckstraße ihre Hauptverwaltung etablieren, „wie das Chile-Haus" in Hamburg.

5.2. NS-Kulturforum

Erst 1936 entschlossen sich die getrennten Städte Wilhelmshaven und Rüstringen zu einer
Gesamtplanung, und beauftragten *Wilhelm Hallbauer*, der bis zum Dezember 1937 die
Bezirksstelle der LOB leitete und seit der Zusammenlegung zu Groß-Wilhelmshaven
gleichzeitig der *Stadtplanung* vorstand. *Hallbauer* wollte anfangs das Stadtzentrum in der
126 Altstadt belassen, eine Großwerft mit einer IV. Einfahrt nach Altengroden/Rüstersiel
legen und Hafenbecken sowie die alte *KMW* ziviler Nutzung zuführen. Der Plan „wurde verworfen, da die Marinestellen in ihrem Ressort-Egoismus unbelehrbar und so auf die Fortsetzung ihrer bisherigen Planungspolitik eingeschworen war, daß sie weder die wirtschaftlichen, noch die militärischen Vorteile" gelten ließen[54]. Die zivile Stadt wurde vom Wasser abgedrängt und durch „Rayonbeschränkung" bis zur Lebensunfähigkeit

eingeschränkt. Es blieb nur „die Verlagerung des Zentrums an die Maade“[55], und die
Verlegung des Bahnhofs an die Umfangstraße in Schaar, wie der Plan *Die Stadt der* 127
Dreihunderttausend vom 15.03.1937 es vorschlug.

Kern der erträumten *Neugestaltungsstadt* Groß-Wilhelmshaven sollte ab 1937 ein
Kulturforum als neuer Stadtmittelpunkt werden. Anders als bei *Verwaltungsforum* und 108
Marineforum setzen die Planungen für das *Kulturforum* erst nach *Hitlers Gesetz zur*
Neugestaltung deutscher Städte vom 04.10.1937 ein. Die Standorte wechselten. Im Juni 1937
war gegen den Willen des *OKM* im Plan *Wirtschaftsgebiet Wilhelmshaven* der luftanfällige 128
Kopfbahnhof Kirchreihe erwogen worden, so daß die Fläche für ein *Kulturforum* in
Schaar durch die Eisenbahnlinie Oldenburg-Wilhelmshaven abgeschnitten worden
wäre. *Wilhelmshaven* beantragte am 20.12.1937 erfolglos, *Neugestaltungsstadt* zu werden
und bekam deutliche Hinweise aus Berlin, daß es an geeigneten Planern und einer
Konzeption fehle[56].

Die *Marine* war an einem *Kulturforum* desinteressiert und beobachtete argwöhnisch die Entwürfe, der Stadt einen zivilen Mittelpunkt zu geben. Im März 1938 drängte *Oberbürgermeister Müller* in Berlin *Ministerialdirektor Schreiber* und *Oberregierungsrat Köhler* von der *Bauabteilung des OKM*, beim *Reichsverkehrsminister* den Baubeginn des neuen Hauptbahnhofs in Schaar anzuordnen[57]. Bis 1943 sollte der neue Bahnhof fertig sein. Auch *Fritz Todt, Generalinspektor für das deutsche Straßenwesen*, hatte die Finanzierung der das *Kulturforum* südlich tangierenden Reichsstraße 69 noch nicht gesichert. Beide Verkehrsbauten waren Vorbedingung für ein *Kulturforum*. Nach *Müllers* Antrag vom 20.12.1937 sollte am „neuen Bahnhof“ der *Neugestaltungsstadt* eine „breite Verkehrsstraße als West-Ost-Achse“ beginnen[58]. Die Standortdiskussion verwirrte die Berliner Ministerialbehörden.

Das *OKM* setzte sich gegen den Kopfbahnhof Kirchreihe durch und erreichte, daß der
Hauptbahnhof mit seinem umgebenden Forum 1938 nach Schaar zwischen Maade und 52
Ebkeriege zurückverlegt wurde[59]. Die Entwurfsarbeiten konnten erst 1938 beginnen, als
Heinz Baumann und *Hans Bartning* zur *Bauverwaltung* der *Stadt Wilhelmshaven* kamen.

Wilhelm Hallbauer und *Heinz Baumann* vom *Stadterweiterungsamt* legten erste Skizzen
für ein *Kulturforum* in Schaar im August 1938 und danach noch verschiedentlich bei *GBI*
Albert Speer vor[60]. Diese Skizzen sind nicht mehr auffindbar. Einen vorläufigen Abschluß
bildete der 1. *Fassung Kultur und Sportforum vom* 24.01.1939 von *Hallbauer, Bartning* und
Baumann[61]. Nach geringen Veränderungen wurde der Planungsstand in die *Städtebauliche* 52
Neuplanung Wilhelmshaven vom 29.03.1939 übernommen. Ein Modell wurde angefertigt
und *Hitler* am 01.04.1939 von *Müller* und *Hallbauer* im Rathaus präsentiert. *Hitler* stellte, 109
da *Speer* sich noch immer gegen eine Vermehrung der *Neugestaltungsstädte* aussprach, der
Stadt 1 Mio. Mark aus seinem Verfügungsfond für ein Theater zur Verfügung und
verlieh damit *Wilhelmshaven* einen Sonderstatus. Die Planungskosten mußte der *GBI*
übernehmen.

Die 2,5 km lange Ost-West-Achse sollte im Westen an einem Viertelkreis-Platz nach 52
Nord-West abknicken und nach einer 1 km langen, 60 m breiten, 2-spurigen Parade-

straße vor dem monumentalen 300 m breiten Hauptbahnhofsgebäude Schaar enden[62]. Kern des *Kulturforums* sollte der Aufmarschplatz von 150 x 200 m für 120.000 Menschen werden. Das Theater (*Benno von Arent/Werry Roth*)[63] im Osten, der Saalbau der *NSDAP* für 5.000 Menschen (*Heinz Baumann*) im Westen, das Ehrenmal für die Toten der „Bewegung" im Süden und die Versammlungshalle im Norden sollten den Aufmarschplatz einrahmen. Nach dem Vorbild der Berliner *Dietrich-Eckart Bühne (Werner March* 1934–36) sollte ein Freilichttheater für 20.000 „Volksgenossen" in der Süd-West-Ecke des *Kulturforums* entstehen. Weitere Bauten der Gemeinschaft wie Kunsthalle, Museum, Volksbücherei, Haus der Partei, *Reichspost,* Markthalle und *Luftgaukommando* sollten im *Kulturforum* durch Achsen und Plätze, Grünstreifen und Ehrenhaine, Säulengänge und Wasserflächen verbunden werden. Eine weitere nordöstlich abknickende Achse, flankiert von Hallenbad und *KDF*-Hotel sollte zum *Sportforum* und Sportstadion für 40.000 Zuschauer führen. *Hans Bartning* stellte sich locker verstreute Bauten, abwechslungsreiche Grünflächen und „klare, kantige Backsteinarchitektur auf weiten Rasenflächen zwischen einigen Baumgruppen" vor und fürchtete die „römische Starrheit" des *Hallbauer*-Plans[64]. Nachdem *Hallbauer* und *Bartning* mit Beginn des Zweiten Weltkrieges Wilhelmshaven verlassen hatten, brachten *Stadtbaudirektor Walter Temp* (ab 04.05.1939) und *Hochbauabteilungsleiter Klaus Rasch* (ab 16.06.1939) neue Ideen ein, „mit dem Ziel, das *Kulturforum* im Sinne des üblichen NS-Programms und entsprechender Architektur wie die *Kulturforen* damals in vielen deutschen Städten geplant wurden"[65]. Als *Helmuth Baur* gegen die weitere Nazifizierung des *Kulturforums* Bedenken äußerte, wurde *Heinz Baumann* kurzerhand zur *Hochbauabteilung* versetzt. Dort arbeitete er nach den Vorgaben von *Temp* und *Rasch* und in Absprachen mit *GBI Speer.*

Unter dem Eindruck des begonnenen Krieges und der Bausperre für alle nicht kriegswichtigen Bauten vom 15.11.1939 verkleinerte *Baumann* die 2. *Fassung* erheblich, rückte sie direkt an die Ost-West-Achse und ließ zwischen Hauptbahnhof und *Kulturforum* aus Luftschutzgründen einen Grüngürtel von 1 km Breite. Er straffte die gesamte Anlage, machte sie symmetrischer und zentrierte sie auf eine neue 1 km lange Nord-Süd-Achse *Kulturforum – Sportforum.*

Unter dem Eindruck der ersten Kriegserfolge ließ *Hitler* Mitte 1940 Neugestaltungs-
107 pläne wieder zu. *Temp* und *Baumann* entwickelten jetzt das *Kulturforum* an alter Stelle zu monumentaler Architektur. Die Maßstäbe dieser 3. *Fassung* wurden größer, die Achsen überdeutlich, der Stil neuklassizistisch. Das Theater[66] von *Benno von Arent* und *Werry Roth* rückte an die Südseite der Paradestraße. Der Aufmarschplatz wurde jetzt zu einem reinen *NSDAP*-Forum, eingerahmt von Versammlungshalle, Rathaus und Parteibauten. Konzerthalle und Bibliothek, Museum und Läden, Hotels und Gebäudeblocks rückten an die zum Hauptbahnhof führende breite Achse, um weiteren rechtwinklig geschlossenen Blöcken Platz zu machen. Geschlossene Platzwände, gigantische Dimensionen, verbindende Hallen, vorgeblendete Säulen, markante Ecktürme und runde Plätze deuteten auf die Einflüsse des *GBI* und der *NSDAP.* Alle Planungen mußten von den Parteidienststellen abgezeichnet werden. *Kreisleiter Ernst Meyer*[67] bejubelte die Planung: „Wenn der

Fremde in einem Jahrzehnt auf dem hohen Bahndamm nach Wilhelmshaven einfährt, werden ihn nicht mehr Reklamewand und Pappdach, sondern schöne Wohnviertel, ein neuer Bahnhof und die Krone der Stadt, die große Halle der Volksgemeinschaft mit ihrem Turm begrüßen. Das uns vom Führer geschenkte Theater, das an Größe und Schönheit alle bisher in Nordwest-Deutschland vorhandenen Theater übertreffen wird, soll nach seinem Willen sofort nach dem Sieg Wirklichkeit werden. Viele mögen es nicht glauben, aber es ist so. Dort, wo heute noch die Kühe weiden, in Schaar, wird der Mittelpunkt des neuen Wilhelmshaven sein, das sich bis Sengwarden einerseits, bis zum Barkeler Busch und Sande andererseits erstrecken wird."

Am 26.11.1940 stellte *Temp* das *Kulturforum* in seiner 3. *Fassung RAM, OKM, MDF,* *106*
NSDAP usw. vor[68]. *Speer* hatte über *Gauleiter Carl Röver* darauf gedrungen, daß „bei der *107*
Planung für die kulturellen Anlagen in Wilhelmshaven die städtebaulichen Grundsätze des Führers Beachtung finden". Der Standort des *Kulturforums* hatte „die Zustimmung des Führers gefunden ... Pläne und Modelle des Hauptplatzes liegen z.Z. *Prof. Speer* vor, der den Führer darüber unterrichten wird". Trotz Planungseuphorie mochte *Wilhelm Müller* keinen weiteren Antrag stellen, Wilhelmshaven zur *Neugestaltungsstadt* zu machen[69]. Die Wilhelmshaven zustehende Sonderrolle war ausreichend.

Die formalen Anleihen, derer sich *Baumann* bei der 3. *Fassung* des *Kulturforums* bediente, waren nicht zu übersehen – es war das Gauforum Weimar, das 1937 nach Plänen von *Hermann Giesler* fertiggestellt worden war. Platzwände und Höhen, Baukörper und Proportionen des Wilhelmshavener *Kulturforums* entsprachen exakt dem Weimarer Vorbild. Der Glockenturm, der in Weimar auf *Hitlers* Wunsch noch hinzugefügt war, wurde peinlich genau in die Wilhelmshavener Planung übernommen. Der *Schultze-Naumburg*-Schüler *Egon Forster* und *Klaus Rasch*, beide Weimar persönlich besonders verbunden, kannten die Planungen bestens. *Klaus Rasch* war sogar in seiner Ausbildung unter *Hermann Giesler* an den Weimarer Planungen beteiligt.

Das Wilhelmshavener *Kulturforum* blieb Utopie. Die Wiederaufbauplanung von *116,*
Schulz, Schneider, Heuer und *Gutschow* legten 1941–45 weitere größenwahnsinnige *117*
Visionen vor[70]. Auch sie blieben Papier. *118*

5.3. Marineforum

Das für Wilhelmshavener Maßstäbe gigantische Projekt *Marineforum* entwickelte sich aus *1*
den 1935 einsetzenden Militärplanungen und dem marineeigenen Hochbau bereits vor dem *Gesetz zur Neugestaltung deutscher Städte vom* 04.10.1937. Unmittelbar nach dem *Gesetz für den Aufbau der Wehrmacht vom* 16.03.1935 begannen Überlegungen für den
Ersatz des 65 Jahre alten Marinestationsgebäudes Adalbertplatz. Die neue *Kriegsmarine* *11*
sollte zentralistischer, rationeller und schlagkräftiger geführt werden. Das Marinestationsgebäude war ab 1936 als erste Stufe des *Marineforums* vorgesehen, als die
Dimensionen der monumentalen Nord-Süd-Achse noch nicht offenbar waren. *Otto* *52*

Lückemeyer führte möglicherweise auf Empfehlung von *Wilhelm Wortmann* einen Architektenwettbewerb des *Marinestandortbauamtes* durch und holte den 1933 von den Nationalsozialisten entlassenen Hamburger *Oberbaudirektor Fritz Schumacher*[71] als Gutachter nach Wilhelmshaven. Beteiligt am internen Wettbewerb waren u.a. *Karl Franzius, Kurt Geisenhainer, Heinz-Günther Thees* und *Karl Maas.* Die Entscheidungsprozesse, die zu dem Entwurf des Marinestationsgebäudes führten, lassen sich nicht mehr nachvollziehen.

70 Die Planung von 1936, die *Karl Franzius* und *Kurt Geisenhainer* zugeschrieben werden muß[72], orientierte sich an den verfügbaren Grundstücksverhältnissen und den Anforderungen des *Marinestandortbauamtes.* Erst nach der Zusammenlegung der Städte zu Groß-Wilhelmshaven, trieb *Wilhelm Hallbauer* die Monumentalplanungen für das
108 *Marineforum* zielstrebig voran. Neubauten, die der Planung von 1939 im Wege gestanden hätten, wurden noch 1937 genehmigt. *Fritz Höger* entwarf 1936 unmittelbar gegenüber dem Haupteingang des Marinestationsgebäudes die Privathäuser *Mücke* und *Mehrtens.* Auch sie hätten dem *Marineforum* weichen müssen. Ende 1937 beantragte Wilhelmshaven, *Neugestaltungsstadt* zu werden.

52, 71 Die Nord-Süd-Achse sollte das seit April 1937 fertige Marinestationsgebäude, die gegenüberliegende Schiffsmaschineninspektion und weitere Monumentalbauten zum
128 *Marineforum* zusammenzufassen. Im Plan *Wirtschaftsgebiet Wilhelmshaven von Juni* 1937 war die spätere 2,5 km lange Nord-Süd-Achse noch als Grünachse dargestellt, wie sie als „Parkstraße" von *Martin Wagner* und *Willy Hahn* freigehalten worden war. Die Wandlung von einer „Parkstraße" der Weimarer Republik zu einer Monumentalachse des „III. Reiches" bewies den 1937 auch in Wilhelmshaven einsetzenden übersteigerten städtebaulichen Maßstab.

Die Planung *Marineforum* erschien realistischer als *Verwaltungsforum* und *Kulturforum,* denn die Militärbauten genossen Vorrang, waren schon lange fest eingeplant, und entsprachen der Sonderrolle Wilhelmshavens. *Wilhelm Hallbauer* legte am 18.10.1937 pünktlich zum *Gesetz zur Neugestaltung deutscher Städte* einen *Bebauungsplan* vor und *Willi Lübbers* riet der *Stadt Wilhelmshaven,* für das *Marineforum* ohne öffentliches Aufsehen „Gelände zu erwerben, wo sie dieses nur bekommen könne"[73].

Im Dezember 1937 ließ sich das *OKM* die Pläne *Hallbauers* erläutern[74]. *Oberbürgermeister Müller* bestand zusätzlich auf persönlichem Vortrag bei *Raeder.* Im Januar 1938 wurden Modelle des *Marineforums* auf der *1. Deutschen Architektur- und Kunsthandwerk-Ausstellung* in München gezeigt. Die Weiterplanung zog sich in die Länge. Erst am 06.04.1938 konnte *Müller* die Neugestaltungspläne *Raeder* im Rathaus präsentieren[75]. Am 22.04.1938 gab *Müller* den Gemeinderäten von Wilhelmshaven die Pläne des *Marineforums* und der Nord-Süd-Achse bekannt. Vom „Heiligtum der deutschen Nation, dem Ehrenfriedhof" zum Marinestationsgebäude sollte die Achse reichen und „Ehrfurcht vor der Größe der steinernen Zeugen der Macht des Reiches" einflößen[76]. Die *Marine* hatte ihre eigenen Prioritäten, die Planung ruhte und *Hitler* stattete dem Kriegshafen 1938 nicht den erwarteten Besuch ab. Zum Jahreswechsel versicherte die Parteipropaganda, daß das

neue Jahr wohl mit dem *Marineforum* beginne[77]. *Hallbauer* und *Franzius* arbeiteten
fieberhaft an Neugestaltungsplänen und Modellen des *Marineforums*, um sie *Hitler* am 108
01.04.1939 zu präsentieren. Die erste Version von 1938 mußte verändert werden[78], und
am 17.03.1939 fand eine letzte Abstimmung mit dem *RAM*, den Oldenburger Behörden
und dem *OKM* im Reichshaus Unter den Linden statt[79]. Der *Hitler*-Besuch am 112
01.04.1939 brachte das *Marineforum* nicht nennenswert weiter – die Finanzierung blieb
unklar. Die *Marine* wurde auf den Krieg vorbereitet und mußte ihre Haushaltsmittel in
kriegswichtige Projekte umsteuern. Im August 1939 kündigte die *Stadt* an, den
Straßenbau nach Norden hin zu beginnen[80]. Das heutige Brommy-Grün markiert das nie
weitergeführte *Marineforum*.

★

Das *Marineforum* war in traditioneller Ziegel- und Walmdachverkleidung als 70
marineeigener Hochbau nahezu fertig, als die Monumentalplanungen einsetzten.
Marinestationsgebäude und Schiffsmaschineninspektion unterschieden sich nicht von den
sonstigen Bauten des *Marinestandortbauamtes*. Die der Nordseeküste angepaßte *Stuttgarter-Schule* bestimmte die Ästhetik: vereinfacht, massig, dekorarm. Nur Eingänge wurden
hervorgehoben.

Haupteingänge waren ebenso wie andere besondere Gebäudeteile durch Werkstein-Pilaster, Balkenköpfe und Friese von *Richard Emil Kuöhl* dekoriert. Eine Wandscheibe zur
Mozartstraße, anfangs für das Hoheitszeichen reserviert, bekam das *Kuöhl*-Relief 71
„Seefahrt ist not“. Natursteingefaßte Freitreppen führten vom Appellplatz zum erhöhten 70
Eingangspodest, das von Adler und Hoheitszeichen auf einem Pylonen bewacht und von
Flaggenmasten gesäumt wurde. Die Eingangsfront mit ihren drei (Führer-)Balkonen,
den drei hohen Fensterbändern, den drei schlanken Laternen und den drei martialisch
dekorierten Balkenköpfen blieben ein hilfloser Versuch, den in Wilhelmshaven
gebräuchlichen *Marinestil* mit neuklassizistischen Elementen zu verbinden. Alle weiteren
Details blieben traditionell: werksteinumrahmte Fenster, regelmäßige Kreuzsprossen,
unterteilte Einzelgauben und stilisierte Balkenköpfe. Zugeständnis an die innerstädtische
Lage blieb der von *Franzius* und *Geisenhainer* geplante Arkadengang vor der gegenüberliegenden Schiffsmaschineninspektion.

Das Marinestationsgebäude wurde am 01.12.1936 gerichtet und am 22.04.1937 von *Otto Lückemeyer*, dem *Oberbefehlshaber der Kriegsmarine Erich Raeder* und dem *Kommandierenden Admiral der Marinestation der Nordsee Otto Schultze*, übergeben[81]. Das alte Marinestationsgebäude am Adalbertplatz wurde „Haus des kommandierenden Admirals“ und dessen Wohnsitz. Nach dem Kriege wurde das neue Marinestationsgebäude von verschiedenen Behörden und von der *Bundesmarine* genutzt. Die Schiffsmaschineninspektion war nach 1945 „Handelshof“ und wurde danach Stammdienststelle der *Marine*.

*

Kern des Neugestaltungsprojektes *Marineforum* war der 90 x 120 m große Appellplatz
108 für „militärische Paraden“ vor dem Marinestationsgebäude. Bei den ersten Planungen sollten die geschlossenen Platzwände durch ein südliches „Siegestor“ unterbrochen werden. Die nördlich liegende, 1911 gebaute, jugendstilorientierte Willehad-Kirche als natürlicher Blickpunkt, sollte durch einen zentral ausgerichteten 4-säuligen „Ehrentempel“[82] verdeckt werden, der später entfiel. Die Schiffsmaschineninspektion, 1937 fertiggestellt, sollte, um dem Appellplatz „Raum“ zu schaffen, zum größten Teil wieder abgerissen bzw. zurückgebaut werden. Mit weiteren Abrissen und Rückbauten war nördlich und südlich des Appellplatzes vor dem Marinestationsgebäude zu rechnen. Die *Hitler* am 01.04.1939 präsentierten Modelle verbanden das *Marineforum* mit dem
52 *Verwaltungsforum* durch die geschlossene Randbebauung der Adolf-Hitler-Straße. Das „Parkstraßen-Konzept“ war zugunsten einer achsenbetonten Einschüchterungsarchitektur aufgegeben worden, nur an einigen Stellen eingeengt durch Tore, Überbrückungen, Arkaden und Blendwalme.

Das *Marineforum* sollte nicht nur eine Ansammlung von vorhandenen und zu ergänzenden Einzelmonumenten werden. Es entwickelte sich ab 1937 zu einem Nord-Süd-Achsen-Konzept, abgeleitet aus den Planungen der *Neugestaltungsstädte.* Der *Monumentalstil* wurde beim *Marineforum* mit dem *Heimatschutzstil* vermischt, ähnlich den Ordensburgen, dem *Gauforum Weimar* und dem *Verwaltungsforum.* Das *Kulturforum* Wilhelmshaven von 1939 war, da auf der „grünen Wiese“ geplant, neuklassizistisch ausgerichtet.

Das *Marineforum* sollte, gesäumt von Kandelabern, Flaggen und Pylonen, nach Süden
52 hin bis zu „künstlerisch vollendeten Landeanlagen für den Führer“ im Bereich der Deichbrücke führen. Im Norden sollte die Traditionsstraße der *Marine* an einer neu geplanten, gigantischen *Marinefachschule* vorbei, abknickend bis zum „Heldenfriedhof“, „der Erinnerungsstätte der Marine“, führen[83].

Hitler und *Speer* konnten für das *Marineforum* nie ganz begeistert werden. Es fehlte der umfassende Neugestaltungsplaner, der die nationalsozialistische Weltanschauung zum Ausdruck bringen konnte. Zuständigkeitschaos und Rivalitäten zwischen *Marine* und *Stadt, KMW* und *Luftwaffe* ließen das *Marineforum* für *Hitlers* Zwecke unbrauchbar werden. Marinetradtition war nicht Parteitradition. Andere *Neugestaltungsstädte* mit NS-Tradition hatten Vorrang. Wilhelmshaven spielte zwar eine Sonderrolle, gehörte aber nicht dazu.

5.4. Aufmarschplätze und Achsen

Neben den drei herausragenden Forumsprojekten und den Marineplanungen fehlte es nicht an Versuchen, das gesamte Stadtbild Wilhelmshavens zu militarisieren. Die erst langsam zu Groß-Wilhelmshaven zusammenwachsenden Stadtteile boten scheinbar gute Chancen zur Neugestaltung im nationalsozialistischen Sinne. Der Kriegshafen sollte bei festlichen Anlässen repräsentieren und Besucher durch gigantische Bauten beeindrucken. Umfassend sollte der totalitäre Führungsanspruch dargestellt werden.

Nach *Gottfried Feder*[84] bedurfte eine Idealstadt von 20.000 Einwohnern eines Aufmarschplatzes für „Veranstaltungen der Partei und der Jugend sowie der Gliederungen der Partei". Er empfahl 5.000 qm in den Idealmaßen von 60 x 80 m für 1.000–2.000 Mann, die in „Formationen aufmarschieren", gesäumt von Rathaus, Partei- und Dienstgebäuden. In jedem Fall sollte man „von einem Balkon den Redner sprechen" lassen. Im Kern der Stadt empfahl *Feder* 8.000–10.000 qm bzw. eine Fläche bis 100 x 180 m mit „repräsentativem Hintergrund". Jede der neuen Gemeinschaftssiedlungen sollte ab 1938 eine Monumentalplanung in kleinem Maßstab bekommen, nicht jedoch die älteren Planungen, die auf das Wohnsiedlungskonzept der Weimarer Republik zurückgingen.

Die Versuche, nachträglich Aufmarschplätze und Achsen in die Gemeinschaftssied- 1
lungen einzufügen, wurden grotesk, gerieten zu einem Taumel, der die Konzeptlosigkeit
des „III. Reiches" klar hervortreten ließ. Die Kombination der in Wilhelmshaven
vorherrschenden Heimatschutzbauten mit Versatzstücken von *Speers* Neugestaltungs-
euphorie konnte nicht glücken. *Friedrich Heuer* interpretierte harmlose Siedlungswege in
seiner abgelegenen Siedlung Cäciliengroden zu Nord-Süd-Achsen und den Dorfplatz 34
zum Aufmarschplatz. Die Marinesiedlung Altengroden, 1938 von dem *Tessenow*-Schüler 43
Ewald Zenker geplant, wurde noch 1942 von *Karl Franzius* mit mehreren Aufmarsch-
plätzen vor Schulen und Gemeinschaftsbauten nazifiziert. Das Familienlager Neuengro- 46
den bekam 1939 einen Tsingtau-Platz für Aufmärsche. In der Werftarbeiter-Stadt 148
Sande-Neufeld plante *Karl Tuchscherer* 1939, den im Mittelpunkt liegenden Teich zum
Platz für Befehlsausgabe für die Werftarbeiter um. Selbst der Anger der nie gebauten 48
Siedlung Schilldeich bekam einen aufmarschgerechten Grundriß. Die Arbeiter-Wohn- 63
stadt Voslapp mit der in Nord-Süd-Richtung durchlaufenden Hans-Zenker-Straße 30
bekam eine rechtwinklig abknickende 400 m lange Aufmarschachse zu den erhöht auf
dem Deich liegenden *NSDAP*- und *HJ*-Häusern. Der *MDF* drängte 1940 auf eine
Abmachung zwischen *Stadt* und *Jade* über die Herrichtung eines Aufmarschplatzes vor
den Voslapper Gewerbebauten[85]. Die *Stadt* war der Auffassung, daß Aufmarschplätze
von den Bauträgern zu finanzieren seien.

Ausführlich beschrieb die Parteipresse die monumentalen Bebauungsplanelemente der
Neuen Stadt Fedderwardergroden. Man mußte annehmen, daß Nord-Süd-Achse und 37
Ost-West-Achse schon immer Planungsbestandteil gewesen seien[86]. Tatsächlich jedoch
waren vorhandene Landstraßen in die Planung übernommen worden. Die Nord-
Süd-Achse in Fedderwardergroden folgte ähnlich Voslapp den vorhandenen Gräben und
den Liegenschaftsgrenzen. Die Achsen liefen nicht exakt in Himmelsrichtung. Die
angeblich monumentalen Straßenbreiten ergaben sich aus der freizuhaltenden Trasse für
eine Fedderwardergrodener Straßenbahn und aus Luftschutzvorschriften. Die Parteipres-

se schrieb: „Wo sich die beiden Hauptachsen der neuen Wohnstadt schneiden, wird ein
111 großer Platz entstehen, der für Aufmärsche oder andere Zwecke zur Verfügung steht, an
dem auch ein Verwaltungsgebäude und das erste Gemeinschaftshaus der Partei errichtet
werden soll“. Am Nord- und Südende der Achse sollten weitere Aufmarschplätze
entstehen.

Viele der Aufmarschplätze blieben in Ansätzen stecken, und kaum einer wurde je für Paraden genutzt. Viele der auf Blickachsen und Symmetrie ausgerichteten städtebaulichen Gesten sind heute nicht mehr erkennbar: Lessingplatz, Otto-Meentz-Platz, Marienburger- und Tilsiter Straße. Zweifel sind angebracht, ob die Planer Aufmarschplätze bewußt entwarfen oder ob nicht die NS-Propaganda die Interpretation besorgte. Die Propaganda schreckte nicht davor zurück, den das Jadeviertel durchschneidenden Banter Weg als Nord-Süd-Achse zu beschreiben.

Prunkstück der Aufmarschplätze sollte das Zentrum der 100.000 Einwohner-Stadt
64 Langewerth von *Gottfried Feder* werden. Es sollte 320 x 170 m messen und wäre wohl die
monumentalste städtebauliche Anlage geworden. Auf Achsen und neuklassizistischen
Habitus verzichtete *Feder* jedoch. Er wollte seine 16 Ortszellen durch eine Ringstraße
65 verbinden. Die Wettbewerbsbeiträge von *Erich zu Putlitz* und *Ernst Prinz* jedoch sahen
für Langewerth monumentale Mittelachsen und symmetrischen Stadtgrundriß vor.

*

Nach dem Vorbild Berlin[87] entwarf *Wilhelm Hallbauer* einen „Runden Platz“ am
52 Kreuzungspunkt Mühlenweg/Gökerstraße. Er sollte Verkehrsverteiler für die benach-
barten Kasernenkomplexe und den Eingang zur neuen *Nordwerft* werden. Eine
geschlossene Blockrandbebauung von 200 x 200 m sollte den kreisförmigen Platz mit
einem Durchmesser von 100 m aufnehmen, dekoriert wie der „Runde Platz“ in Berlin
mit Pylonen und Fontänen. Werftversammlungshaus, Verwaltungsgebäude, NS-
Parteizeitung und Geschäftshäuser sollten den „Runden Platz“ säumen.

Große Ähnlichkeit mit *Speers* Planungen sollte auch der Mitte 1940 von *Heinz*
107 *Baumann* als Eingang zum *Kulturforum* geplante „Runde Platz“ von 200 m Durchmesser
am westlichen Ende der Ost-West-Achse bekommen. Die Dekoration mit einer hoch
aufragenden Siegessäule und die kreisförmige Grüngestaltung entsprachen der am
19.04.1939 fertiggestellten Neugestaltung des großen Sterns in Berlin, der von *GBI Speer*
gleichfalls mit einem Durchmesser von 200 m entworfen worden war. Die geschlossenen
Platzwände sollten sich auf den „Runden Platz“ der Berliner Nord-Süd-Achse
beziehen.

*

52 Das von *Wilhelm Hallbauer* in die Altstadt gelegte, übersteigerte Achsenkreuz von
5 x 4 km orientierte sich – wie auch die Achsenplanungen der Gauhauptstadt Olden-

burg[88] – an den 1936 begonnenen gigantischen Neugestaltungsplänen der Reichshauptstadt Berlin. Vorhandene Straßen oder sonstwie freie Flächen, vorhandene Randbebauungen und gradlinige Blickachsen waren Voraussetzungen für Achsenplanungen. Die Achsen hatten mehrfache Funktionen, denn die versprochene Motorisierung mit *KDF-Wagen* für „alle Volksgenossen" stand unmittelbar bevor. In Wilhelmshaven sollte das Achsenkreuz auch der schnellen Heranführung der Arbeitermassen zu *Nordwerft, Westwerft* und *KMW* dienen. Wie auch in der *Stadt der Hermann-Göring-Werke*[89] und der *Stadt* des *KDF-Wagens*[90] war das Achsenkreuz der „*Stadt der 500.000*" zugleich städtische Infrastruktur für eine moderne funktionale Stadt.

Möglicherweise waren die Achsen zu sehr an dem problemlosen Grunderwerb orientiert, so daß sie bei *GBI Speer* wenig überzeugten. Die langen ungebrochenen Perspektiven, die markant gestalteten Knickpunkte waren in Wilhelmshaven nur schwer herstellbar. Die Ost-West-Achse Bismarckstraße hätte die Verlegung der Baufluchten nach sich gezogen. Die erwünschten Blickachsen und Stadtkronen kamen im geplanten
Stadtbild nicht zum Ausdruck. Die 2 km lange sogenannte „Temp-Achse", die zwischen *52*
Siebethsburg und *Sportforum* auf das *Kulturforum* zulaufen sollte, blieb 1939 vergeblicher *109*
Versuch einer anbaufreien, schnurgeraden, zusätzlichen Ost-West-Achse. Auch die
Achsenplanungen von *Schulz, Schneider, Heuer* (1941–42) und *Gutschow* (1944–45) *116, 117*
blieben kulissenhafte Visionen totalitärer Einschüchterungsarchitektur. *118*

In Groß-Wilhelmshaven hatte der kriegswichtige Wohnungsbau größte Bedeutung. Die Versuche, die „*Stadt der 500.000*" als bedeutendsten Reichskriegshafen und als Waffenschmiede der Nation architektonisch auszuschlachten, scheiterten an dem uneinheitlichen Stadtbild, an den dominierenden Marineinteressen, an dem Leitbild der gegliederten durchgrünten Stadt und an den wenig überzeugenden Gestaltungskonzepten.

6. Pläne und Bauten für die Zeit „danach"

Natürlich ging dann beim jedesmaligen Passiren irgend eines Kriegsschiffs der Spektakel von Neuem los, was Angesichts des Waldes von Masten dem Hafen einen ganz wundervollen Anstrich verlieh. Die Stadt dagegen bietet gar nicht dar, und ist erst etwa so weit vorgeschritten wie Wilhelmshafen am Jadebusen[1].

Kronprinz Friedrich Wilhelm, 1869

6.1. Die eingebunkerte Stadt

Wilhelmshavens und Rüstringens Gebäude wurden während des Ersten Weltkrieges von Bombenangriffen verschont, obwohl die kriegführenden Staaten bereits Flugzeuge und Luftschiffe besaßen. Flugzeuge waren zu Beginn des Ersten Weltkrieges zunächst nur für Erkundungs- und Beobachtungszwecke eingesetzt[2]. Luftschutz, wie er im Zweiten Weltkrieg verstanden wurde, war 1914–18 noch weitgehend unbekannt, obwohl seit 1906 theoretisch an der Bekämpfung von Luftfahrzeugen gearbeitet wurde. Aus finanziellen Gründen schien es völlig undenkbar, die Gebäude der ganzen Stadt und den Hafen luftschutzgerecht herzurichten. Im Verlaufe des Ersten Weltkrieges jedoch sammelte man Erfahrungen, rüstete Stockwerksdecken mit Betonabdeckungen aus, mauerte Fenster mit Steinen zu, verstärkte Dächer durch Holzstiele und nagelte Schutzwände aus Planken. Im Vordergrund der baulichen Maßnahmen standen Fabriken und Rüstungsbetriebe. Das *Preußische Kriegsministerium* erarbeitete 1915 erste Anleitungen, die *Richtlinien für den Eigenschutz von Fabrikanlagen gegen Angriffe aus der Luft.* Im Juli 1916 gab es weitere *Richtlinien* heraus, die sich im wesentlichen auf Verdunkelung und Tarnung bezogen. 1923 entwarf das *Wehrministerium* der Weimarer Republik eine erste Anleitung für den Reichsluftschutz. Es fürchtete, daß Kriege in Zukunft über die eigentlichen Kampfgebiete hinausgreifen würden, um „Volks- und Kriegswirtschaft zu brechen". Nebeneinander liefen internationale Konferenzen, um die Humanisierung des Krieges, aber auch militärische Diskussionen um Kriegstechnik und Sprengstoffentwicklung. 1930 führte der *Preußische Innenminister* eine erste große Luftschutzübung durch. Die Militärs mißtrauten dem völkerrechtlichen Luftschutz, da in den meisten Städten militärische Objekte mit Wohngebieten der Bevölkerung vermischt waren. *Hitlers Gesetz über die Aufschließung von Wohnsiedlungsgebieten (Wohnsiedlungsgesetz) vom* 22.09.1933 folgte das *Luftschutzgesetz vom* 26.06.1935. Es war schon auf den nächsten Krieg abgestellt. Vor Ausbruch des Zweiten Weltkrieges wurden eilig 10 *Durchführungs-Verordnungen* und 2 *Änderungs-Verordnungen zum Luftschutzgesetz* erlassen. Der *Oberbefehlshaber der Kriegsmarine* ordnete am 10.12.1935 an, daß die *Marineintendantur* bei Wohnungen von vornherein Luftschutzkeller vorsehen solle – aber nur, wenn die Wohnungen in der Nähe von gefährdeten Gebäuden liegen. Dies liege im Interesse der Landesverteidigung und werde „von der Industrie und von der Zivilbevölkerung gefordert“. Die *Marine* müsse mit gutem Beispiel vorangehen. In Wilhelmshaven als Luftschutzort I. Ordnung seien selbst abgelegene Neubauten luftgefährdet. Arbeitersiedlungshäuser von 1 1/2 Geschossen wurden alsbald aus der Anordnung ausgenommen[3].

In Wilhelmshavens Stadtbild wurden die Auswirkungen des Luftschutzes auf Funktion und Architektur gut ablesbar. Die innerstädtischen Wohnbauten erhielten ab 1937 – wenn schon keine kompletten Luftschutzkeller – so doch verstärkte Decken und Gasschleusen. Für die Neubewohner schrieb die *Landesgruppe Nordmark des Reichsluftschutzbundes* Preise für Luftschutzbereitschaft aus. 1937 wurde mit einer *Durchführungs-Verordnung* der Luftschutz bei Neu-, Um- und Erweiterungsbauten geregelt und wenige Tage vor Ausbruch des Zweiten Weltkrieges am 17.08.1939 wurde der behelfsmäßige Luftschutz auf bestehende Gebäude durch eine weitere *Verordnung* ausgedehnt. Die
1 *Verordnungen* kamen für Wilhelmshavens neue Gemeinschaftssiedlungen fast schon zu
52 spät. Schwerpunkt aller Luftschutzmaßnahmen wurde die dichtbesiedelte Innenstadt. Wegen der aufgelockerten Bauweise rechnete man von vornherein nur mit Zufallstreffern in den neuen Stadtteilen. In halbfertigen Schulen wurden auch Untergeschosse zu Rettungsstellen und Hilfskrankenhäusern umgebaut[4].

Das Groß-Wilhelmshavener *Bunkerbauprogramm* lief erst während des Krieges an, als im Auftrag und auf Kosten des *Reiches* die *Stadt Wilhelmshaven* mit einem eigenen *Luftschutzbauamt* tätig wurde. Für Nationalsozialisten war Luftschutz Teil der militärischen Reichsverteidigung. Die Gestaltung der Stadt war untrennbar mit diesen militärischen Anforderungen verflochten. Bunker sollten auch ästhetisch Verteidigungsbereitschaft und Wehrwillen symbolisieren. Heute werden die 40 übriggebliebenen Betonklötze als düster und bedrohlich empfunden.

Im Alltag des „III. Reiches“ hatten Organisation und Ideologie des Luftschutzes hohen Rang – alle Bereiche des öffentlichen und privaten Lebens waren durchdacht. Es gab den besonderen Hafenluftschutz und den Luftschutz für Wehrmachtsanlagen, der in Wilhelmshaven der *Marine* unterstand. *Reichsbahn* und *Reichspost*, Krankenhäuser und Industrieanlagen, *Waffen-SS* und Wasserstraßen, Reichsautobahnen und Baustellen hatten ihre jeweiligen, im Detail verschiedenen Zielsetzungen und *Bunkerbauprogramme*. Alle Baubehörden in der „*Stadt der 500.000*“ bauten für ihren Amtsbereich Bunker und Luftschutzräume. Als die *Royal Air Force* am 04.09.1939 den Luftkrieg im Westen begann, war Wilhelmshaven erstes Hauptziel, höchste Zeit also für das *Sofortprogramm*. Der *Bauverein Rüstringen* konnte in Siebethsburg 20.000 „Volksgenossen" pünktlich zum Kriegsbeginn Luftschutzkeller anbieten.

GBI Albert Speer hatte Ende September 1940 von *Hitler* den Auftrag bekommen, systematisch bombensichere Bunker zu planen[5]. Dieses *Sofortprogramm*[6] sollte im Reichsgebiet von der *Organisation Todt* (*OT*) abgewickelt werden. Wie immer in solchen Fällen ernannte das NS-System Sonderbeauftragte. *Fritz Todt*, Schöpfer der Reichsautobahnen und 1940–42 *Reichsminister für Bewaffnung und Munition*, galt als geschickter Organisator und Ingenieur. Er kontrollierte sämtliche Straßenbauvorhaben, den Bau des Westwalls und das praktische Baugeschehen des „III. Reiches", mit einer Einschränkung: Architektur und Städtebau waren nicht seine Sache.

Das *Sofortprogramm* wurde durch „Führerbefehl" *Gesetz*. Wilhelmshaven wurde wegen seiner Wichtigkeit für Krieg und Rüstung unverzüglich in das *Sofortprogramm* aufgenommen. Für innerstädtische Wohngebiete, aber auch für die Stadterweiterung, waren behelfsmäßige Luftschutzmaßnahmen zu treffen. Bei allen Neubauten, insbesondere bei den Bauten der Rüstungsindustrie, sollten von vornherein bombensichere Luftschutzräume einbetoniert werden. Die Keller aller öffentlichen und privaten Gebäude mußten auf Eignung als provisorische Luftschutzräume überprüft werden. Der *Reichsbauausschuß* legte am 30.10. und 11.11.1940 die technischen Einzelheiten fest, die in einer *Anweisung für den Bau bombensicherer Luftschutzräume, Fassung Nov. 1940*[7] niedergeschrieben wurden. Alle *Stadtbaudirektoren* wurden am 14.11.1939 nach Berlin beordert. Man instruierte sie im Hause des *Vereins Deutscher Ingenieure* und versorgte die *Bauverwaltungen* mit *Speers* Typenplänen.

Für Ausschachtungsarbeiten und Tiefbunker war keine Zeit. In Wilhelmshaven wären sie auch wegen des hohen Grundwasserstandes nicht möglich gewesen. So wurden mit Ausnahme weniger militärischer Befehlstellen vor allem Hochbunker geplant. Im ganzen deutschen Reich wurden über 3.000 Bunker für die Zivilbevölkerung aus dem Boden
gestampft mit Platz für über 1 Mio. Menschen. Hinzu kamen noch Krankenhäuser mit *151*
Operationsbunkern und die verschiedensten Bunker für Militär, Partei, Bahn, *Post* und alle wichtigen Lebensbereiche.

Die *Hochbauabteilung der Stadt Wilhelmshaven* baute mit ihrem *Luftschutzbauamt* insgesamt „10 Rundtürme und 24 Vierkantbunker, dazu 11 kleinere Bunker"[8], die allerdings nicht bombensicher waren. Das zivile *Sofortprogramm* für Wilhelmshaven belief
sich auf immerhin 34 Mio. Mark. Die *Kriegsmarine* baute nach dem Programm des *OKM* *114*
30 militärische Bunker in Werft, Hafen-, und Kasernennähe. Ein Teil der Wilhelmshavener Bunker wurde nach dem Kriege gesprengt, einzelne wurden auch ab 1950 von
einer eigens gegründeten Gesellschaft zu Wohnhäusern umgebaut. Immerhin 40 Bunker *115*
stehen noch heute, 34 gelten als verwendbar. 24 sind wieder hergerichtet[9].

Wilhelmshaven war total verbunkert. So hatte das ebenfalls luftgefährdete Braunschweig mit über 200.000 Einwohnern nur 24 Bunker[10], Wilhelmshaven hatte 80 Bunker. Nicht mitgerechnet private Bunker, Deckungsgräben, Kleintürme, Splittergräben und militärische Befehlstellen. Es gab darüber hinaus noch 64 behelfsmäßige Bunker und 7.000 ausgebaute Luftschutzkeller[11].

113 Das *Luftschutzbauprogramm* war durchaus ein städtebauliches und architektonisches
114 Thema. An der Standortauswahl wurde das *Stadterweiterungsamt* beteiligt. Wichtigstes Ziel war, bei lang andauerndem Aufenthalt die Benutzer ruhig zu halten. Auch Bequemlichkeit sollte eine Rolle spielen. *Speers* Grundrißgestaltungen gingen wohl darauf ein, daß es nicht zu eng wurde und die physischen Bedingungen wie Kühlung, Belüftung, Wasserversorgung und elektrische Ausstattung nicht vernachlässigt wurden. Geschickt waren die Grundrisse der verschiedenen Geschosse unterschiedlich gestaltet. Treppen befanden sich in Außenzonen und Aufenthaltsflächen in den besonders sicheren Innenzonen. Betondicken von Wänden und Decken staffelten sich je nach Fassungsvermögen. *Heinz Baumann*, im September 1939 vom *Stadterweiterungsamt* zur *Hochbauabteilung* versetzt, ließ sich sogar einen Bunkertypenentwurf patentieren. An allen vier Ecken waren Eingänge, die Geschosse waren versetzt – so sollte sich der Bunker besonders schnell füllen und entleeren lassen. *Ernst Neufert*, Mitarbeiter *Speers* und Autor des „Bombensicheren Luftschutz im Wohnungsbau“[12], hat in seiner „Bau-Entwurfslehre“ 1940 Luftschutz für Architekten veröffentlicht. Er propagierte den „wohnlichen Luftschutzraum“ und erfand auch Friedensnutzungen für Bunker. Hier traten erstmalig Eckhausbunker als Symbole des Verteidigungswillens auf.

Die städtebaulichen Anforderungen an Luftschutzbunker gingen von der „Ortsgruppe als Siedlungszelle“ aus. So hatten Rüstersiel, Voslapp, Neuengroden, Aldenburg und Mariensiel „ihren eigenen“ Bunker. Die zwei typengleichen Voslapper Rundbunker hatten je 7 Etagen und konnten zusammen über 2.000 Personen aufnehmen. Die Baukosten lagen bei über 1 Mio. Mark, die Bauzeit bei je 9 Monaten[13]. Für Altengroden-Nord, Fedderwardergroden, Cäciliengroden und Sande blieben die Ortsteil-Bunker im Bau stecken, schließlich waren die Trabanten auch nur teilweise fertiggeworden[14]. Bauhandwerker und Bewohner mußten in die nachträglich angelegten provisorischen Luftschutzkeller oder in die Bunker anderer Stadtteile flüchten.

Friedrich Heuer, Architekt mehrerer Gemeinschaftssiedlungen hatte 1940 eine Bunker-Systemplanung entworfen[15]. In neuen Siedlungsgebieten sollten unterirdische Gänge zu Luftschutzbauten führen, die als Wohnhäuser getarnt waren. Partei- und Verwaltungsforen waren Luftschutzbauten zugeordnet, die eine „ganze Ortsgruppe“ als „Wehr- und Wachtürme“ beherrschen sollten. Diese Verbindung von Architektur und Wehrbereitschaft blieb aberwitziger Anspruch.

Auch im totalen Krieg wurde darauf geachtet, daß sich Luftschutzbunker in städtebaulicher und baukünstlerischer Hinsicht in ihre Umgebung einfügten. Sie sollten Größe und Macht baulich stützen. Baurechtliche Vorschriften über Gebäudehöhen, Abstände usw. waren zu beachten. Ausnahmeanträge mußten der *Baupolizei* vorgelegt werden. Die meisten Luftschutzbunker standen frei, manche wurden auch in Baulücken eingefügt[16]. Entscheidend war, daß sie von der Bevölkerung schnell zu erreichen waren. „Bei der Gestaltung sollte der Wehrcharakter nach Möglichkeit zum Ausdruck kommen.

Das reine Betonbauwerk konnte so ausgebildet werden, daß es bei entsprechender Gestaltung architektonisch befriedigte"[17]. Tatsächlich fiel auf, daß einige Luftschutzbunker, auch wenn sie heute als deplaziert und unangenehm empfunden werden, in ihrer Höhe und Dachform mit den benachbarten Gebäuden korrespondierten. In freier Landschaft waren es häufig Rundtürme mit kegelförmigem Dach, in der Innenstadt fügten sie sich in die Blockrandbebauung ein und in traditionellen Wohnsiedlungsgebieten waren auch Satteldächer zu finden. Die beiden *Tessenow*-Schüler *Alexander Herde (RAM)*[18] und *Ewald Zenker (OKM)* befaßten sich in Berlin mit den gestalterischen Aufgaben, Bunker städtebaulich einzuordnen. Der Wehrcharakter von Bunkern wurde in mittelalterliche Burgen, Stadttore und Türme gekleidet. Aber auch ganze Wohnzeilen mit vorgeblendeten Gauben und Fenstern wurden als Bunker für Wohngebiete entworfen.

Bei Wilhelmshavens Monumentalplanungen waren auch die von *Ernst Neufert* 107
erdachten Eckhausbunker zu finden, die einen zusätzlichen massiven und wehrhaften Ausdruck bringen sollten. Sie symbolisierten angebliche Dauerhaftigkeit und Unvergänglichkeit des „III. Reiches". *Speer*-Mitarbeiter *Friedrich Tamms*, kommentierte seine Bunkerentwürfe in NS-typischen Phrasen als „wahren Ausdruck des Glaubens eines Volkes an seine unzerstörbare Lebenskraft"[19].

Auch ältere Siedlungen bekamen architektonisch angepaßte Bunker. Zwischen die nicht unterkellerten Werftarbeiterhäuser in der Siedlung Bant von 1873 beispielsweise wurden kleine Bunker gefügt. Mit Rücksichtnahme auf die alte Bebauung bekamen ab 1939 je 2 Werfthäuser ein flaches, verklinkertes Zwischenbauwerk, zurückgesetzt und unter Traufhöhe gehalten.

Der für 530.000 Mark 1942 in Wilhelmshaven am Banter Weg gebaute Bunker und der Bunker Heppenser Straße hatten traditionelle Satteldächer. Die bebaute Fläche entsprach der von Wohnblocks der Umgebung, so daß eine perfekte Tarnung gegeben war. Der Rundbunker an der Rhein-Ecke Gökerstaße von *Heinz-Günther Thees* war verklinkert worden. Luftschlitze und Eingangstüren bekamen Werksteingewände. Der alte *Jugendstil*-Rathausbrunnen aus rotem Sandstein wurde geschickt angebunden, so daß
sich eine angenehme Platzwirkung ergab. Für das Jadeviertel war 1937 ein *NSDAP*- 53
Gemeinschaftshaus Banter Weg/Bremer Straße geplant, das als Eckturm einen Rundbunker bekommen sollte. Bunker sollten Luftschutz zum alltäglichen Bestandteil des Städtebaues machen. Traditionelle Stilelemente wurden auf den Bunkerbau übertragen:
Profilierte Dachsimse, umlaufende Balkone, repräsentative Umrahmungen und verklin- 115
kerte Rundbögen. Sie sollten den Bunker als Fremdkörper in der Stadtgestaltung 113
aufgehen lassen. Landschaftsgebunden und heimatverpflichtet sollten fast alle Wilhelms-
havener Bunker spätestens nach dem Kriege eine Klinkerhaut bekommen. Die aus dem 114
Beton herausstehenden Maueranker sind an vielen Luftschutzbauten noch sichtbar.

*

Luftschutz für Rüstungsbetriebe, für *Werft* und Hafen, wollte, daß die Arbeiter nur so kurz wie möglich ihren angestammten Platz verlassen sollten. Die kriegswichtige Produktion sollte weitergehen. Während der Fliegerangriffe wurde der Notbelegschaft (zur Beobachtung von Maschinen und Meßständen) ein eigener splittersicherer Schutzstand angeboten, um jederzeit in Arbeitsvorgänge eingreifen zu können. Derartige Schutzstände waren in Treppenwinkeln, hinter Stützen und im Mauerwerk verborgen.

Nachrichtenanlagen, Schaltzentralen, Maschinenhäuser und Turbinenaggregate in *Werft* und Hafen wurden baulich voneinander getrennt, um wenigstens teilweise den Betrieb aufrechtzuerhalten. Versorgungsleitungen wurden, um gegen Trümmersplitter zu schützen, in Rohrkanäle gelegt. Bei evtl. Schäden konnte man jederzeit die Kanäle öffnen. Ein genaues System von Kennzeichnungen und Hinweisschildern ermöglichte jede Reparaturmaßnahme an Wasser-, Gas- und Elektroleitungen. Im Detail sollten Form und Konstruktion gegen Sprengbomben, Brandbomben, aber auch gegen Luftbeobachtung schützen. Die massive Bauweise hatte den Vorzug, daß rahmenartige Ausbildung und fachwerkähnliche Tragwerke große Steifheit aufwiesen. Dies sollte sich nach dem Kriege als zweckmäßig erweisen, denn die Konstruktionsart ermöglichte den
115 Umbau von Bunkern zu Nachkriegswohnungen. Die architektonische Form der Türme, Zeltdächer, Kegelhauben, Satteldächer hatte einen einfachen Grund: man rechnete anfangs noch mit Bomben, die verzögert zünden würden. Die spitze Bauform hätte dafür gesorgt, daß Spreng- und Brandbomben erst auf dem Boden zur Explosion gekommen wären. Schutz gegen Luftbeobachtung wirkte sich architektonisch als Tarnung aus. Auffallende Bauwerke und Einrichtungen, die den Blick aus der Luft anzogen, sollten vermieden werden. Wenn sie schon notwendig waren, sollten sie als „Scheinbauten" den Gegner täuschen[20]. Tarnung legte sich wie ein Schleier über alle besonderen und eigenwilligen Gestaltungsabsichten. Im Interesse des Luftschutzes wurden helle Betonflächen, z.B. die Reichsautobahnen, dunkel gefärbt, Glasdächer überstrichen und Verdunkelungsanlagen entworfen. In Fedderwardergroden begann das *Luftschutzbauamt* 1942 Tarnanstriche auf die gerade geweißten Putzfassaden zu bringen. Das „Unsichtbarmachen" ganzer Wohn- und Siedlungsgebiete, Industrie- und Verkehrsanlagen, Rüstungsbetriebe und Wasserflächen hatte eine besondere Luftschutzästhetik zur Folge. In Voslapp bekamen Volkswohnungsbauten 1939 auf Betreiben des *Schutzbereichsamtes* anfangs Flachdächer, um das Schußfeld der benachbarten Fliegerabwehr zu sichern. Erst 1946 wurden die hohen Steildächer komplettiert.

Der Wettbewerb „Alarm" des *NSBDT* von 1940 zur „Gestaltung und städtebaulichen Einordnung von Luftschutzbunkern" kam für die meisten Städte zu spät. Die ersten Preisträger des Wettbewerbs, *Alexander* und *Gertraude Herde* beide Architekten beim *RAM*, realisierten ihre Vorschläge nicht mehr. Die Entscheidungen des *Reichsluftfahrtministeriums* fielen zu Gunsten von Großbunkern. *Herdes* meinten, daß Bunker Stadtplanung in starkem Maße beeinflussen –, nicht jedoch das Stadtbild beherrschen sollten. Im Einklang mit dem *RAM* waren sie der Auffassung, daß man künftig keine

Siedlungen mehr bauen werde, sondern Städte mit geschlossener Bauweise im Kern und aufgelockerten Rändern. Eine Siedlung mit Hunderten von Einzelhäusern sei eben eine Siedlung und keine Stadt. Die äußere Gestaltung von Luftschutzbunkern könne den Wehrcharakter zum Ausdruck bringen. Bunker sollten „nicht als Fremdkörper wirken, sondern vielmehr in Harmonie, mit ihrer Umgebung die Kraft und Stärke des deutschen Volkes zum Ausdruck bringen"[21]. Entsprechend waren die Entwürfe als Söller, Stadttürme und Zeughäuser ausgebildet. Die Vorschläge wurden verworfen zugunsten von städtebaulich isolierten Großbunkern, die auch in Wilhelmshaven „Kriegsbaukunst" dokumentierten. Von Lübeck sind Bunker in Form alter Stadttürme und historischer Torgebäude bekannt geworden und noch erhalten.

★

Konsequent wurden im *Wohnsiedlungsbiet Wilhelmshaven-Jeverland* durch ein *Industrie-* *129*
umsiedlungsprogramm alle nicht wassergebundenen Rüstungsbetriebe verlagert, insbesondere in das Industriegebiet um den Werftbahnhof Sande. Siedlungs- und Industriegebiete *94*
bekamen sichere technische Versorgung. Angestrebt waren dezentrale- oder Mehrfachanlagen, die man durch Absperrvorrichtungen leicht umleiten konnte. Man nannte die Redundanz „Vermaschung". Eine Vielzahl gesicherter Anlagen, Transformatoren und Umspannwerke sollten die Energieversorgung vor Luftangriffen schützen. Große Umspannwerke gehörten der Vergangenheit an. Die neuen Schaltwerke und Trafosta- *137*
tionen wurden entsprechend kleinmaßstäblich, mit Ziegeldächern und in heimatverbundener Formsprache errichtet.

Auch in der Wasserver- und -entsorgung wurden kleine dezentrale Anlagen, Pumpstationen, Notbrunnen und Ringleitungen Grundlage eines planmäßigen Luftschutzes. Pumpwerke, Heizzentralen und Trafos in Siedlungsgebieten und Militäranlagen wurden als Wohnhäuser getarnt und nicht mehr, wie in den 20er Jahren, modern entworfen.

Das Pumpwerk-Ost wurde 1939 von der *HNBD* trotz außen sichtbaren Betonskeletts mit Ziegelaußenhaut, Kreuzsprossen und Walmdach traditionell kostümiert. Verstärkte Decken und überdeckte Zulaufsysteme mit 80 cm dicken Wänden sollten den Einsturz nach Bombenabwurf verhindern. Ein völlig eingebunkertes Pumpwerk für die gesamte Südstadt entstand 1941–44 an der Weser-/Jadestraße. In Wilhelmshaven standen *KMW*, *Westwerft* und *Nordwerft* im Mittelpunkt des militärischen Bunkerbaus. Hier herrschten kubische Bauformen vor. Auf repräsentativen Habitus wurde verzichtet. *Hans Klaucke*
entwarf 1941–42 den Befehlsbunker beim Werfttor 1 und den Bunker bei der *114*
E-Werkstatt der *KMW* in auffällig moderner Formsprache. Kolonnaden, Vordächer, Türme und Attiken unterstützten das futuristische Aussehen.

Da sich die Bauarbeiten der *Nordwerft* in den Krieg erstreckten, mußten auch *101*
Luftschutzbauten bedacht werden. In den Kellern der Werkhallen, des Schnürbodens
und des Heizkraftwerkes und den unterirdischen Betriebsräumen der Schleuse wurden *100*

Bunker eingerichtet. Allein einer der Bunker, der während der Bauarbeiten an der Ostfriesenstraße gebaut werden mußte, faßte 1.000 Personen[22]. Um Kriegsschiffbau auch während der Fliegerangriffe fortzusetzen, plante man von vornherein auch Verdunk-
103 lungsanlagen hinter senkrechter Verglasung für die Bauhallen ein. In Rüstungsbetrieben verdunkelte man auch durch Spannungsreduzierung.

Der militärische Bunkerbau hatte eigene, besondere Ziele, nicht die Menschen zu schützen, sondern das Kriegsgerät. Mit Kriegsbeginn verfolgte das *OKM* sinnlose
78 Planungen: im U-Bootstützpunkt Banter See einen U-Boot-Großbunker[23] zu bauen, die
101 IV. Einfahrt zu überdachen und das Dock VII der *Nordwerft* zu U-Boot-Montagebunkern umzubauen[24]. Während 1943–45 *KZ*-Häftlinge und Kriegsgefangene in Farge bei Bremen den U-Bootsbunker „Valentin"[25] betonierten, ebensolche in Brest, Lorient, St. Nazaire, La Pallice, Bordeaux, Bergen, Trondheim, Hamburg, Kiel und Bremen, gebaut wurden[26], blieb es in Wilhelmshaven bei den Planungen.

89 Die Torpedowerft hinter dem Südstrand verbarg unter dem Ziegeldach eine geneigte Betonkonstruktion, die vor Brandbomben schützen sollte. Fast alle, auch kleinen, militärischen Gebäude, wie z.B. die Ausweichvermittlung Antonslust, die ab 1935 gebaut wurden, bekamen ihre eigenen Luftschutzkeller, verstärkte Decken oder stählerne Fensterläden, die gegen Bombensplitter schützen sollten. Auch der Fliegerhorst Upjever
68 hatte sein eigenes, kriegsmäßiges Gesicht, allerdings ohne formalen Gestaltungswillen. Jagdbomber und Munition wurden durch eigens entworfene Bunker geschützt und durch Tarnstrich, so gut es ging, unsichtbar gemacht[27]. Militärische Bunker waren nur für die Funktion entworfen – der ideologische Überbau von Wehrhaftigkeit und Kraft durfte fehlen – er war sowieso dem Militär angehaftet.

NS-Propaganda unterstützte die Ziele des baulichen Luftschutzes auf ihre Weise – denn der verstärkte Bau von Bunkern machte mißtrauisch. Die Zeitungen berichteten von der angeblichen Sicherheit, von den meterdicken Wänden. Die „Hamburger Zeitung" schrieb über einen Bunker: „Er ist nun mein Haus, mein Schutz und Freund, geworden, der gewaltige Betonklotz inmitten der Häuserruinen. Was da auch noch kommen mag, er hält durch ..."[28]. *Hitler* sah die Bunker für Wehrmacht und *Marine*, für Rüstungsbetriebe und Werften, vor allem als Notwendigkeit, sein Kriegsmaterial zu schützen und den Nachschub zu sichern. Bei den kämpfenden Soldaten befürchtete er nachlassenden Mut und geringere Einsatzbereitschaft durch geschützte Stellungen[29]. *Speer* erinnerte sich, daß *Hitler* bis in die Nachtstunde hinein einzelne Bunkertypen für Verteidigungsanlagen entwarf.

6.2. Erste Nachkriegsplanungen

Planungen für die Zeit „danach" begannen für die *„Stadt der 500.000"* mit dem *Erlaß zur Vorbereitung des deutschen Wiederaufbaus nach dem Kriege vom* 15.11.1940, denn die Bemühungen, *Neugestaltungsstadt* nach dem *Gesetz zur Neugestaltung deutscher Städte vom*

04.10.1937 zu werden, waren gescheitert. Trotzdem hatte Wilhelmshaven einen Sonderstatus. Allenthalben bekamen die Neugestaltungsplanungen Auftrieb. Der 1940 neu eingesetzte *Reichskommissar für den sozialen Wohnungsbau* (ab 1942 *Reichswohnungskommissar*), *Robert Ley*, bereitete genormte und rationalisierte Wohnungsproduktion vor. Im Gau Weser-Ems lagen die Aufgaben bei *Gauwohnungskommissar, Gauleiter Carl Röver*[30]. Zusammen mit den Gauen Hannover und Braunschweig und der *Stadtschaft für Niedersachsen* schrieb *Röver* Anfang 1941 den Wettbewerb „Wohnen nach dem Kriege“[31] aus. Preisrichter waren u.a. *Karl Elkart, Gerd Offenberg, Hans Spiegel, Fritz Schumacher* und *Wilhelm Wortmann*. Der in Aussicht gestellte Massenwohnungsbau für die Nachkriegszeit – man nahm an ab 1942 – ließ auch in Wilhelmshaven unabhängig von den nur mühsam vorankommenden Gemeinschaftssiedlungen und Forumsplanungen wieder auf die Sonderrolle hoffen.

Rövers Wettbewerb, im Mai 1941 in Hannover ausgestellt, hatte für die von *Walther Schulz* 1941 und 1942 erarbeiteten Wilhelmshavener Nachkriegsplanungen Bedeutung. *116*
Schulz war im September 1941 von der *Stadt Göttingen* wegen Arbeitsmangel nach Wilhelmshaven abgeordnet worden. Hier ersetzte er den zum Kriegsdienst einberufenen *Helmuth Baur*. Die Maßgaben für die Nachkriegsplanungen *Schulz* waren klar: Auf der Basis des *Führererlasses vom* 15.11.1940 wollte die NSDAP den einheitlichen niederdeutschen Wohnungsbau. „Landschaftsgebundene Wünsche“ und „nordisch bestimmtes Lebensgefühl“ sollten herausgearbeitet werden. Auch wo die Kleinstadt sich zur Großstadt entwickelte, sollte am „ein- oder zweigeschossigen Wohnbau, der das Wohnen nicht vom Boden loslöst“, festgehalten werden. Detailliert wurden heimatverbundene Gestaltungsmerkmale vorgegeben[32].

*

In seiner knapp 9-monatigen Zeit als Leiter der *Stadtplanungsabteilung* fertigte *Walther Schulz* verschiedene *Bebauungspläne*[33], *Wirtschaftspläne* und Skizzen. Nach seinen Plänen verfestigte sich die Abriegelung der Stadt vom Wasser durch *Westwerft, KMW* und *Nordwerft*. Da sich die Flugplatzplanungen im Heppenser- und Rüstersieler Groden *128*
zerschlagen hatten, plante *Schulz* einen Handelshafen an der Mündung der begradigten Maade im Voslapper Groden. Rückgrat seiner Planung von 1941 sollte eine streng symmetrische, 6 km lange, Südwest-Nordost-Achse werden, vom nach Schaar zurückgelegten Hauptbahnhof bis zum „Runden Platz“ mit Freilichtbühne auf dem ehemaligen Fort Rüstersiel. Die Achse war nicht mehr auf die Altstadt bezogen, sondern auf den neuen Handelshafen im Voslapper Groden. Das neue *Kulturforum* sollte vom *Sportforum* durch einen „Großen See“ getrennt werden. Eine weitere Wasserfläche, vergleichbar *Speers* „Großem Bassin“[34] der Berliner Neugestaltungspläne von 1937, sollte dem Hauptbahnhof vorgelagert werden. Wie in Berlin der Nordbahnhof, sollte in Wilhelmshaven der neue Hauptbahnhof vor Kopf des 600 x 250 m „Großen Bassin“ liegen, umgeben von monumentaler Bebauung. Die Seen sollten zugleich luftschutzge-

rechte Feuerlöschteiche sein. Die neue „Schwerlinie" des *Wirtschaftsplans* 1941 für eine 300.000 Einwohner-Stadt Wilhelmshaven sollte gesäumt werden von Verwaltungs- und Kulturbauten, Geschäftszentrum und Rathaus, Kreisleitung und Parteibauten, Vergnügungsstätten und Hotels. Weitere Monumentalplanungen im Sinne der Neugestaltungspläne von 1937 waren nach wie vor das *Marineforum*, der „Runde Platz" und das *Verwaltungsforum*[35]. Die Öffnung der Stadt zum Meer bei Rüstersiel war gekennzeichnet durch große gestaltete Grünräume und eine korrigierte Maade, begleitet von Wehrsportgelände, Festwiesen, Friedhof, Botanischem Garten, Zeltlagerplatz und Zoo[36]. Die Bebauung im *Wohnsiedlungsgebiet Wilhelmshaven-Jeverland*, sollte zugunsten einer ortsgruppen-gegliederten konzentrierten Nord-West-Stadt reduziert werden[37]. Das Verkehrskonzept des *Wirtschaftsplanes* 1941 von *Walther Schulz* war funktional und modern. Die Achse sollte nach Südwest abknicken zu einem Kreisel, der mit weiteren „Runden Plätzen" den Verkehr der *KDF-Wagen* verteilen sollte. 10 Straßenbahnlinien sollten den öffentlichen Personennahverkehr übernehmen. Wilhelmshavens neuer ziviler Flugplatz sollte westlich des Hauptbahnhofs auf dem weniger tragfähigen Boden von Langewerth liegen.

116 Weitere *Bebauungspläne* von *Schulz* von 1942[38] nazifizierten die Achsenplanung weiter und orientierten die Bahnhofslage zu einer 4 km langen Nord-Süd-Achse. Die „Große Straße" sollte von einem monumentalen *Kulturforum* am neuen Südbahnhof Bismarckstraße zum „Runden Platz" mit Wasserbecken von 300 m Durchmesser südwestlich Fedderwardergroden führen, unterbrochen von einem weiteren „Runden Platz" westlich von Altengroden. Ein „Großes Bassin", grachtenähnliche Wasserläufe und ausgedehnte Grünzüge sollten die Stadt gliedern. *Konstanty Gutschow* vom *Arbeitsstab Speer* kritisierte 1945 die „Monumentalachse ohne Endziel" von *Schulz*[39].

Scharf kritisierten auch *OKM* und *LOB* die *Bebauungspläne* von *Schulz*[40]. Auch der *Reichsverkehrsminister* nahm am 10.03.1942 negativ Stellung. Die Entfernung des von *Schulz* neugeplanten Hauptbahnhofs zur Stadtmitte sei erheblich zu lang, und die Nordstadt habe „keinen sinnvollen Zusammenhang mit dem Forum". Das *Marineforum* habe keinen Bezug zu der von *Schulz* entworfenen „neuen Stadtmitte". Auch der *Schulz-Bebauungsplan* 1942 mit dem neuen Südbahnhof Bismarckstraße Bismarckstraße bringe keine Anbindung zur Altstadt. Die *Bebauungspläne* seien in Maßstab und Umfang übersteigert „und auf eine Mehr-Millionenstadt zugeschnitten". *Schulz* schaltete noch erfolglos den Leiter der *Akademie für Städtebau, Reichs- und Landesplanung Berlin, Reinhold Niemeyer*, ein.

⋆

Die von *Landesplaner Wortmann* frühzeitig erhobene Kritik an den *Schulz*-Plänen bewegte *Fritz Karl Schneider* vom *MDF* zu mehreren Alternativplanungen. Im Januar 1942 entwickelte er zwei Versionen, die sich nur in der Aufteilung der Gemeinschaftssiedlungen geringfügig unterschieden[41]. Die Planungen von *Schulz* erschienen *Schneider* zu aufwendig. Er selbst verzichtete auf lange Achsen, die für eine Randbebauung oder monumentale Gestaltung nicht in Betracht gekommen wären. Nach *Schneiders* Meinung sollten *Marineforum* und *Verwaltungsforum* beibehalten werden, und nördlich des wieder aktuellen Kopfbahnhofs Kirchreihe als *Kulturforum* fortgesetzt werden. *NSDAP-Kreisleitung,* Theater und Festhalle sollten sich um ein 500 x 900 m „Großes Bassin" nach Berliner Vorbild gruppieren. Westlich anschließend war das *Sportforum* geplant, nördlich des „Großen Bassin" lagen „Heldenfriedhof" und Stadtpark. Die weiteren Gemeinschaftssiedlungen sollten Kniphausersiel, Neuengroden-Ost, Breddewarden, Fedderwarden, Burg Kniphausen, Langewerth, Himmelreich und Utters sein. Die Altstadt sollte sparsam arrondiert werden.

Der *Bebauungsplan Stadtplanung Wilhelmshaven vom April* 1942, den *Schneider* zusam- *117*
men mit *Friedrich Heuer* vorlegte, konzentrierte die Planung noch dichter an der Altstadt, monumentalisierte das *Kulturforum* und dehnte es nach Westen um Aufmarschplätze und massierte Randbebauungen aus. Das *Sportforum* sollte an den zum Badesee uminterpretierten Wasserflugplatz im Rüstersieler Groden verlegt werden. *Stadtbaudirektor Walter Temp* mußte nach dem Weggang von *Schulz* im Mai 1942 die Planungen alleine vorantreiben. Er bediente sich verschiedener Gutachter: *Döscher* (Berlin), *Blum* (Hannover), *Wentzel* (Aachen), *Lademann* (Hamburg) und *Richard* (Berlin). Die *LOB* teilte bereits Mitte 1942 mit, daß z.Z. „die Weiterführung der Planungsarbeiten durch Anordnung des Führers überall eingestellt sei"[42]. Anfang 1943 kam es zu entsprechenden Erlassen, denn *Hitler* räumte der Rüstungsproduktion höchsten Rang ein[43]. *Temp* schrieb *Helmut Döscher* am 26.06.1943, daß „jede Planungsarbeit" weiter „nicht vertretbar" sei, da die Luftangriffe „umfassende Veränderungen in das Stadtbild getragen haben"[44].

6.3. Arbeitsstab Speer in Wilhelmshaven

Mit *Hitlers Erlaß über die Vorbereitung des Wiederaufbaues bombengeschädigter Städte vom* 11.10.1943 und der Einrichtung des *Arbeitsstabs zur Wiederaufbauplanung zerstörter Städte* durch *Albert Speer* im Dezember 1943, entwickelten sich auch im stark bombenzerstörten Wilhelmshaven wieder Gestaltungsziele. *Speers* Ämterhäufung als *Generalbauinspektor (GBI)* für die Reichshauptstadt (ab 03.01.1937) als *Generalbevollmächtigter für die Regelung der Bauwirtschaft*, Leiter der *OT* und *Rüstungsminister* (ab 08.02.1942) machte ihn einflußreicher als *Reichswohnungskommissar Robert Ley*, der als Leiter der *DAF* und *Reichsorganisationsleiter der NSDAP* Wilhelmshaven eng verbunden war. *Speer* bat *Gauleiter Paul Wegener* im Dezember 1943 um Benennung der Städte im Gau Weser-Ems, die *Wiederaufbaustädte* werden sollten[45]. Die *Gauleiter* – gleichzeitig *Gauwohnungskommissare* und *Beauftragte für die Neugestaltung* in ihren Gauen sollten *Speer* Mitte 1944 auch geeignete Architekten nennen. Der *Arbeitsstab* sollte nur beraten, übernahm aber die Kosten der zu beauftragenden Planer. *Speer* wollte „sich bereits im Kriege darüber schlüssig" werden, wie die *Wiederaufbaustädte* aussehen sollten. Er warb um Planungsverständnis bei den *Gauleitern*, denn die Zeit zur Vorbereitung werde unterschätzt. Er kündigte *Gauleiter Wegener* an, daß der „Führer" außerdem wünscht, daß die Planungen größerer Städte ihm zur Genehmigung vorgelegt werden. *Speer* rechnete auch mit „bestimmt zu erwartenden Änderungswünschen des Führers".

Wilhelmshaven lag in dem von *Speer* ausgegebenen Kriterienkatalog für *Wiederaufbaustädte*. Die zu erwartende Schrumpfung der staatlichen Rüstungsindustrie sollte Menschen und Maschinen auf neue Bauaufgaben umlenken. Insofern bestand ökonomischer Zwang, Groß-Wilhelmshaven zur *Wiederaufbaustadt* zu machen. Auch die Resignation der Bevölkerung über Stadtzerstörung und Wohnungsnot sprachen für Wilhelmshaven. *GBI Speer* mit seinem *Arbeitsstab* sollte die *Wiederaufbaustädte* und die eingeschalteten Architekten beraten. Der Rahmen „künftiger Gestaltung" sollte abgesteckt werden. *RAM, DAF* und *OKM* hatten im *Arbeitsstab Speer* keine Kompetenz mehr. Es ging *Speer* nicht um „irgendwelche baukünstlerische Ideen" oder um Stadtzentren, auch nicht um die in den *Neugestaltungsstädten* beliebten monumentalen Achsen, sondern um Sanierung und Straßengerippe, Verkehr und Wohnungsbau. 3,5 Mio. Wohnungen sollten im Reichsgebiet nach dem Kriege fehlen[46].

Unmittelbar nach *Hitlers Erlaß über die Vorbereitungen des Wiederaufbaues bombengeschädigter Städte vom* 11.10.1943 nahm *Stadtbaudirektor Walter Temp* Kontakt zu *Wilhelm Wortmann* von der *LOB* und *Fritz Karl Schneider* vom *MDF* auf. 250- bis 300.000 Einwohner sollte die *Wiederaufbaustadt* haben. Für den Bauingenieur *Temp*[47] waren Hafen und Werft, Ortsgüterbahnhof und Straßen, vor allem die „*Temp*-Achse" unverrückbar. Der Bahnhof Schaar sollte wegen der *KMW* wieder näher in die Stadt rücken. Die Ringschnellbahn sollte zweigleisig über *Westwerft, KMW, Nordwerft* bis Rüstersiel und Voslapp fahren, und ein Verkehrsflughafen sollte die erhoffte internationale Bedeutung Wilhelmshavens unterstreichen. Das neue Parteiforum sollte alle *NSDAP*-Gliederungen und eine Festhalle mit 30.000 Sitzplätzen enthalten. Das Stadtforum war für Stadttheater, Konzerthalle, Stadtbücherei und alle weiteren Verwaltungsgebäude vorgesehen. *Mari-*
1 *neforum*, Gemeinschaftssiedlungen und Freiflächen sollten weiter ausgebaut werden.

Am 17.01.1944 besprachen *Gauleiter Paul Wegener* und *Wilhelm Wortmann* die Wiederaufbauplanung mit den Bürgermeistern von Bremen, Emden und Wilhelmshaven. Am 25.02.1944 teilte *Albert Speer Gauleiter Paul Wegener* mit, daß Wilhelmshaven, Bremen und Emden *Wiederaufbaustädte* würden[48]. Im Entwurf zum *Führererlaß vom* 19.09.1944 standen 42 Städte auf der Liste des *Arbeitsstabs Speer*. Als Planer waren vorgesehen für Bremen *Cäsar Pinnau* und *Wilhelm Wortmann*, für Emden *Wilhelm Wortmann*, für Wilhelmshaven *Konstanty Gutschow*. Osnabrück kam noch später hinzu. *Hans-Hermann Klaje* vom *Arbeitsstab Speer* koordinierte die Wiederaufbauplanungen von Berlin aus[49]. Am 13.06.1944 hatten *Gutschow, Wortmann, Oberbürgermeister Müller* und *Stadtbaudirektor Temp* in Wilhelmshaven das erste Planungsgespräch. Groß-Wilhelmshaven hatte seinen Sonderstatus verloren und endlich den langersehnten Status als *Neugestaltungsstadt* bekommen, denn auf die *Wiederaufbaustädte* sollte das *Gesetz zur Neugestaltung deutscher Städte vom* 04.10.1937 Anwendung finden. Die von *Hitler* und

Dönitz diktierte Zielplanung für Wilhelmshaven lag bei 150.000 Einwohnern mit einer Erweiterungsmöglichkeit für weitere 50.000 Einwohner. *Stadtbaudirektor Temp* blieb bei seiner Zielzahl von 250.000.

Gutschow bearbeitete die Wiederaufbauplanung zusammen mit *Richard Zorn, Wilhelm Wucherpfennig, Jeanette Hoppe, Gustav Schleicher, Friedrich Lübbers*[50] und *Rudolf Hillebrecht* aus seinem Hamburger Büro und von den *Speer*-Büros Berlin und Wriezen aus. Als Gutachter zog er *Lademann, Max Karl Schwarz* und *Otto Blum* hinzu. In Wilhelmshaven wurde noch das Architekturbüro des ehemaligen technischen Geschäftsführers der *Wohnungsbaugesellschaft Jade, Adolf Braunschweig* für Zuarbeit eingeschaltet. Im Oktober 1944 übergab *Gutschow Oberbürgermeister Müller* und *Otto Lehn*, dem Vertreter des am 01.09.1944 zum Kriegsdienst eingezogenen *Walter Temp* die ersten Skizzen.

Für *Gutschows* Wiederaufbauplanungen galt die städtebauliche Leitlinie der gegliederten und durchgrünten Stadt. Im Gegensatz zu *Neufert, Rimpl* und *Tamms*, die die typisierte genormte „Notstadt" verfochten, orientierten sich *Gutschow, Hillebrecht, Schulte-Frohlinde, Reichow, Heuer* und *Wortmann* an landschaftsbezogenem Bauen und regionaler Architektur. Dieses hatte *Gauleiter Röver* mit seinem Wettbewerb „Wohnen nach dem Kriege" von 1941 festgelegt. Dies war auch Leitlinie des *Reichsheimstättenamtes Gau Weser-Ems*, des *Ahlhorner Architektenkreises* und aller *NSDAP*-Gliederungen in Weser-Ems.

★

Der Wiederaufbau nach dem „Endsieg" sah eine erheblich veränderte Stadt vor[51]. *118*
Systematisch und variantenreich hatte das vom *GBI* beauftragte Büro *Gutschow* alle Gegebenheiten vor Ort aufgearbeitet und in zahlreichen Alternativen untersucht. Die Werft sollte verkleinert und Teile davon nach Bremerhaven oder Brunsbüttelkoog verlegt werden. Nebenbetriebe der *KMW* würden auf die Geest umgesiedelt werden. Die friedensmäßige Zukunft Wilhelmshavens sollten höhere Kommando-Stellen, Marineausbildung und Badebetrieb sein. Trotzdem sollten die Marineanlagen von Mariensiel bis zur *Nordwerft* die Stadt vom Jadebusen abschneiden. Einzig der als *101*
Wasserflughafen vorgesehene See im Rüstersieler Groden sollte Badesee werden. „Das Stadtgesicht sollte gegliedert und durchgrünt", durch Schutzbereiche sortiert und von „kleinen Kanälen im Sinne von Grachten" durchzogen sein. Der Durchgangsbahnhof Schaar, viel zu weit von den Werfttoren gelegen, sollte durch die Kopfbahnmöglichkeiten Kirchreihe, Bismarckstraße und Südstadt ersetzt werden. Forumsanlagen, Aufmarschplätze und Monumentalanlagen im Anschluß an das begonnene *Marineforum* *108*
sollten den Bahnhof über eine 4 km lange Nord-Süd-Achse, an der Theater, *Sportforum* und Rathaus gelegen sein sollten, mit dem *NSDAP*-Kreishaus und der „Halle der 30.000" verbinden. Von hier aus sollte man einen Blick auf den 2,5 km lagen und 1,5 km breiten Badesee haben, „wo die Stadt ans Meer tritt". Eine Variante zeigte das *Verwaltungsforum* versetzt zum Kopfbahnhof Kirchreihe.

Systematisch trieb der *Arbeitsstab Speer* die Planungen weiter. *Gutschow* erarbeitete zweite, dritte und vierte Skizzen im Dezember 1944 und Januar 1945. Die letzten Skizzen waren vom März 1945 datiert, sie unterschieden sich wiederum durch die Lage der Kopfbahnhöfe und Durchgangshöfe und die darauf aufbauenden Achsen. Die zweiten Skizzen besprachen *Konstanty Gutschow, Friedrich Lübbers, Arthur Kellerhoff* und *Friedrich Heuer* bei *Gauleiter Paul Wegener* am 04.01.1945 in der Oldenburger *Gauleitung*. Die vierte und letzte Planung vom März 1945 griff Leitgedanken, „das Wasser zum charakteristisch gestaltenden Element in Wilhelmshavens Stadtantlitz" zu machen, wieder auf[52]. *Oberbaudirektor Plate* (Bremen) und *Oberregierungsrat Brahms* (Oldenburg) berieten *Gutschow* bei den Süßwasserläufen und Salzwassergrachten, die Wilhelmshaven durchziehen sollten. Die Grachten sollten die neuen Gemeinschaftssiedlungen umschließen und vor allen repräsentativen Bauten „Große Bassins" bilden. Sie sollten die Bauten der Gemeinschaft und der *NSDAP* „in ihrer Wirkung steigern". Am *Marineforum* war symbolträchtig „eine Anlegestelle für Barkassen geplant". Auch der Rüstringer Rathausplatz sollte, wie von *Höger* 1927 geplant, ein „Bassin" werden. 450 x 100 m sollte ein großes Wasserbecken vor dem neuen Hauptbahnhof messen. Die 1,7 km lange schnurgerade Gracht vom Hauptbahnhof Kirchreihe zum *NSDAP*-Kreishaus hatte Achsenfunktionen wie eine Paradestraße.

Je mehr Wilhelmshaven durch Bomben in Schutt und Asche sank, um so utopischer entwickelten sich die Wiederaufbaupläne des *Arbeitsstabs Speer*. Am 27.04.1945 übergab der *Arbeitsstab Speer Oberbürgermeister Müller* die letzten Wiederaufbaupläne – 9 Tage später wurde Wilhelmshaven an die Alliierten übergeben.

★

Der *Arbeitsstab Speer* schaltete Anfang 1945 den in Berlin bei der *OT* befindlichen *Marinebaurat Karl Franzius* ein[53]. Als das gesamte Marinebauwesen am 25.07.1944 der *OT* und somit *Rüstungsminister Albert Speer* unterstellt wurde, kam *Franzius* nach Berlin und zeichnete dort u.a. am 18.01.1945 einen Wiederaufbau-Vorschlag. Die utopische,
119 über 20 km lange Bandstadt, sollte sich mit gegliederten „ortsgruppengerechten Gemeinschaftssiedlungen" von der Altstadt über Inhausersiel, Crildumersiel, Hohenstiefersiel bis nach Horumersiel ziehen – begleitet von einer Schnellbahn, die über einen Damm bis Wangerooge führen sollte. Neues Stadtzentrum sollte Inhausersiel mit Hauptbahnhof und Monumentalanlagen auf dem Deich werden. Die *Franzius*-Planung von 1945 wies nördlich von *Nordwerft* und IV. Einfahrt weitere Werft- und Hafenbecken, Docks und Anlegestellen bis nach Voslapp aus. Dort sollte die V. Einfahrt liegen. Im zerstörten Berlin gab man Wilhelmshaven wohl kaum eine militärische Zukunft, und so war der Hafen möglicherweise für Handel und Fischerei, Schiffbau und Linienverkehr gedacht.

★

Nach *Hitlers* Unterwerfung der Nachbarländer setzte eine Nachkriegs- und Wiederaufbauplanung ein, die sich vom alltäglichen Elend des Nationalsozialiusmus weit entfernte. Architekten und Baubeamte bekamen eine übergroße Machtfülle. Symptomatisch dafür war die Karriere von *Albert Speer*, der es aufgrund der Architekturbesessenheit *Hitlers* innerhalb von 10 Jahren vom *Tessenow*-Assistenten an der TU Berlin zum übermächtigen *Rüstungsminister* brachte. Weitere Architekten und Ingenieure wie *Feder, Todt* und *Rosenberg* säumten *Hitlers* Weg. Nach dem „Sieg" sollte ein eigenes *Bauministerium* geschaffen werden. In den Mittelinstanzen sollte „ein verantwortlicher Baubeamter" den *Regierungspräsidenten/Reichsstatthaltern/Reichsverteidigungskommissaren* beigeordnet werden[54].

Im Gau Weser-Ems war man keinesfalls vom Baugeschehen in Berlin abgeschnitten. Der personelle Austausch der Baubeamten und die regelmäßigen Besprechungen des *Arbeitsstabs Speer* am Pariser Platz in Berlin und in der Außenstelle Wriezen funktionierten. Im Vorgriff auf künftige Organisationsformen und Empfehlung des Leiters der *Akademie für Städtebau, Reichs- und Landesplanung*, des *Landesplaners* von Berlin-Brandenburg, *Reinhold Niemeyer*, wurde *Gauleiter Paul Wegener* der *Ahlhorner Architektenkreis* beigeordnet. Man traf sich im „Haus des Reiches" in Bremen oder im *Blockhaus Ahlhorn*, das der *RAD* 1936 für die *NSDAP* gezimmert hatte.

Die Beratung des *Reichsstatthalters* sollte weit über Architektur hinausgehen. Der *Arbeitskreis Stadt- und Landesplanung des NSBDT*[55] wurde von *Wilhelm Wortmann* am 15.10.1943 gebildet[56] – 4 Tage nach dem *Erlaß über die Vorbereitungen des Wiederaufbaues bombengeschädigter Städte vom* 11.10.1943. Der *Arbeitskreis* hieß später *Gaugruppe Weser-Ems der Deutschen Akademie für Städtebau, Reichs- und Landesplanung, Arbeitskreis im NSBDT*. Die Gaugruppenleiter aus Schleswig Holstein (*Jensen*), Hannover-Ost (*Mangel*), Weser-Ems (*Wortmann*), Hannover (*Elkart*) und Hamburg (*Gutschow*) trafen sich regelmäßig und waren mit dem *Arbeitsstaab Speer* und untereinander eng verbunden. Architekten und Hochschullehrer, Planer und Landschaftsgestalter, richteten im *Ahlhorner Architektenkreis Stadtbauräte* und Baubeamte ideologisch aus. *Hermann Gildemeister* (Bremen) sprach zu „Grundformen der Stadtentwicklung", *Wilhelm Wortmann* zu „Grundlagen für die Gesundung und Neugestaltung des Raumes Bremen", *Eduard Scotland* zu „Baugestaltung", *Wolfgang Triebel* zu „Hochbau-Flachdachbau", *Friedrich Heuer* zu „Wohnungsbau vom rassenpolitischen Gesichtspunkt aus", *Richard Homann* über „Landschaftsraum", *Walter Wickop* über „Baumethoden im Wohnungsbau"[57].
Der *Ahlhorner Architektenkreis* galt als Theorie-Werkstatt einer stark „völkisch" und immer noch heimatschutzorientierten Architekturauffassung. Er tat sich durch rassenpolitisches Gedankengut hervor. Im Gegensatz zu der in Berlin durch *Neufert, Rimpl* und *Tamms* propagierten Bausystematik hielt der *Ahlhorner Architektenkreis* an *Rosenberg* und *Schultze-Naumburg* fest. Diese konservative Architekturauffassung blieb für Wilhelmshaven folgenlos, da Wohnungsbau bis 1945 nicht mehr stattfand. Viele Mitglieder des *Ahlhorner Architektenkreises* bekamen jedoch in erstaunlicher Kontinuität Einfluß auf das Nachkriegsbaugeschehen Wilhelmshavens.

6.4. Nachkriegsbauen

Nach Ende des Zweiten Weltkrieges in der „meistzerstörten Stadt im Lande Niedersachsen"[58] wurde die militärische Infrastruktur durch die Alliierten weitgehend unbrauchbar gemacht. Wilhelmshaven sollte niemals mehr Kriegsschiffe bauen oder beherbergen. Der Leitdamm im Jadebusen, die Leitwerke östlich Wangerooge, die IV., III. und II. Einfahrt wurden gesprengt und zum Teil mit Trümmerschutt verfüllt und eingedeicht[59]. Man dachte sogar an ein Verlanden des ganzen Jadebusens. Nur die I. Einfahrt blieb erhalten. Durch Sprengung der Kaianlagen und Dämme aus Trümmerschutt wurden die Hafenbecken 1949 unbrauchbar gemacht. So wie die Planer ab 1941, erkannten auch die Kommunalpolitiker ab 1945, daß die Militärhäfen möglicherweise künftig als Handelshäfen genutzt werden müssen und versuchten, Zerstörungen zu verhindern.

Der Heppenser Groden, Voraussetzung für *Nordwerft*, IV. Einfahrt und Flughäfen, war
101 nach dem Krieg tristes Überbleibsel militärischen Größenwahns. Er war versumpft, und große Flächen standen unter Wasser. Nur ein Teil war mit Sand aufgespült[60]. Gesprengte Eisenbahnzüge, abgebrannte Arbeiterbaracken und kahle Gebäudereste deuteten auf die militärische Vergangenheit. Das städtische *Amt für Wiederaufbau und Wirtschaftsförderung* erarbeitete 1945–46 eine *Trümmersatzung*, die alle Bewohner ab 16 Jahren zu „Handdiensten" verpflichtete. Nichtverpflichtete mußten sozialgestaffelt Ablösebeträge zahlen. *Wilhelm Aichinger*, Leiter des *Tiefbauamtes*, schätzte, daß eine halbe Mio. cbm Schutt beseitigt werden müßte. Es wurde erheblich mehr. Der *Exekutivausschuß* beschloß im November 1946 das *Ortsstatut über die Durchführung von Aufräumarbeiten in der Stadt Wilhelmshaven*. *Karl Higgen* vom Bauunternehmen *Möller*, leitete die Beseitigung der zivilen Trümmer[61]. Lorenzüge transportierten Trümmermengen in das ehem. *Nordwerft*-Gelände. Jetzt wurde *Hitlers* Rüstungs-Prunkstück Schuttabladeplatz der Nachkriegszeit. Der *Wiederaufbau-Ausschuß* stellte aus finanziellen Gründen Mitte 1948 die Trümmerbeseitigung ein. Der nun zuständige *Bauausschuß* vergab die Trümmerbeseitigung 1950 nach Ausschreibung an die Firma *Möller* – teilweise als *Arbeitsbeschaffungsmaßnahme*. Die *Stadt* konnte nach Beseitigung von 1 Mio. cbm Schutt ihre Wiederaufbauplanungen ernsthaft beginnen. Der Wille zur Neugestaltung von Grund auf wurde durch die Wilhelmshavener Haus- und Grundbesitzer empfindlich gedämpft. Gerichte verurteilten die *Stadtverwaltung* in „Trümmerprozessen", die im Rausch des Neubeginns von Privatgrundstücken entnommenen Steine und Materialien zu entschädigen[62].

Die umliegenden Gebietskörperschaften hatten sofort nach 1945 versucht, die Eingemeindungen von 1938 zu revidieren und Groß-Wilhelmshaven wieder zu verkleinern. Umgekehrt betrieb Wilhelmshaven die Eingemeindung des Industriegebie-
94 tes um den Bahnhof Sande. *Friedrich Paffrath* von den Nationalsozialisten verjagter *Oberbürgermeister*, kehrte am 01.07.1945 zurück und wurde *Oberstadtdirektor* von Wilhelmshaven. Der nichtbelastete, von den Nationalsozialisten verfolgte ehemalige Leiter des *Tiefbauamtes*, *Otto Lehn*, leitete von 1945–49 das *Baudezernat*[63]. Die für

Wilhelmshavens Stadterweiterung und Neugestaltung verantwortlichen Bau- und Verwaltungsbeamten wurden zum größten Teil 1945 entlassen, einige kehrten nach „Entnazifizierung" in den Dienst zurück. Nicht nur NS-Funktionäre und Beamte[64], auch Bauten und Straßenbezeichnungen, Anlegebrücken und Denkmäler, Kunst- und Platzgestaltung, wurden „entnazifiziert". *HNBD, Marinestandortbauamt, Luftwaffenbauamt, Hafenbauressort der KMW* bzw. die später gebildeten *Marineoberbaudirektion/Marineoberbauamt/OT Bauleitung Wilhelmshaven* wurden aufgelöst. Auffangbehörden wurden: die *Gruppe Strombau*, die *Dienststelle Land- und Wasserbau*, das *Seewasserstraßenamt*, das *Naval Dockyard*, das *Verwaltungsamt für Reichs- und Staatsvermögen*, die *Außenstelle des Wasserwirtschaftsamtes Varel,* ab 1949 das *Staatshochbaumt* und das *Wasser- und Schiffahrtsamt.* Viele der arbeitslosen Architekten und Baubeamten hatten sich selbständig gemacht oder kehrten in die Auffangbehörden zurück.

Die Hochbauten des größten staatlichen Rüstungsbetriebes, der *KMW*, wurden nach
der Demontage der Maschinen 1949 gesprengt und 1950–54 vom *Staatshochbauamt*
abgeräumt, mit Ausnahme von Schiffbauhalle, Lehrlingshochhaus und Werfttor 1. *10*
Gebäude der Torpedowerft, der *Westwerft* und der Gießerei der *KMW* waren längst von *89, 93*
Nachfolgebetrieben genutzt. Erhalten blieben auch das *Marinelazarett Sanderbusch,* das *76*
Marinegerätelager Roffhausen und viele technische Gebäude. *Wilhelm Aichinger* hatte für *73*
die Enttrümmerung von *KMW* und *Westwerft* 2,6 Mio. Mark geschätzt. Geschickt kombinierte er mit *Karl Tillesen, Friedemann Schwichow, Henry Picker* und dem Bauunternehmen *Freytag* die Abräumung der *KMW* mit dem Wirtschaftswegebau Heppenser Groden als *Arbeitsbeschaffungsmassnahme.* Nach den Plänen von *Max Karl Schwarz* von 1949 sollten im Heppenser Groden 52 Kleinsthöfe, 18 Gärtnerhöfe und 10 Gärtnereien angesiedelt werden[65]. Die landwirtschaftliche Nutzung verzögerte sich wegen des sterilen Seesandes. Die Besiedlung wurde hinausgeschoben, und 1955 begannen die Planungen für den Wiederaufbau der IV. Einfahrt.

Wilhelmshaven war wegen der republikweit höchsten Arbeitslosenziffern zum *Notstandsgebiet* erklärt worden[66]. *Henry Picker*, ehemals *Oberregierungsrat* im „Führerhauptquartier", fungierte als *Industriebeauftragter* für Wilhelmshaven in Bonn[67]. Nach den *Richtlinien der Bundesregierung für die Hilfe des Bundes an die finanzschwachen Länder im Rahmen des Arbeitsbeschaffungsprogramms vom März* 1950 wurde ein breites Programm
aufgelegt. 1949–50 wurde der 1939 für *Hitlers* Propaganda-Auftritt hastig aufgefüllte *112*
Rathausplatz nach der Planung von *Gerhard Graubner* mit Pflastersteinen aus der *123*
Westwerft befestigt, Symbol für die zivile Zukunft. Bis 1952 wurden Kanalisationen komplettiert, Straßen ausgebaut, Parkanlagen aufgeforstet, Friedhöfe erweitert und Meliorationsarbeiten begonnen. Auch andere Dienststellen schlossen sich der *Arbeitsbeschaffung* an: das *Finanzamt* wurde wiederhergestellt, die *Landeszentralbank* aufgebaut, das Maadesiel hochgemauert und der Neubau der *Post* begonnen. Wie auch in der Anfangsphase des „III. Reiches" mußte sich die neue Republik mit *Arbeitsbeschaffungsmaßnahmen* und massiven Stützungen behelfen, um die Wirtschaft anzukurbeln.

*

Bereits 1945 gründeten die Alliierten ein deutsch-englisches *Port Emergency Committee*, einen Wiederaufbau-Ausschuß für den Hafen[68]. Die *Stadt* sollte eine Planung zur betrieblichen Unterbringung der entlassenen Werftarbeiter vorlegen. Anfang 1946 dann sollte die ehemalige *KMW*, immer noch unter Leitung des *Oberwerftdirektors Paul-Willy Zieb*, geschlossen werden. Um das zu verhindern, legten mehrere *Denkschriften* Leitlinien
122 für eine friedensmäßige Zukunft Wilhelmshavens fest. Fischwirtschaft, Gemüseanbau, Torfgewinnung und Folgeindustrien sollten etabliert werden. Nachfolgeunternehmen und Flüchtlingsbetriebe fanden sich jedoch nur zögernd, und Wilhelmshaven stand vor der gleichen Situation wie zu Anfang der Weimarer Republik. *Otto Lehn*[69], Nachkriegs-*Stadtbaurat* von 1945–49 befaßte sich weiter mit den Wiederaufbauplänen des *Arbeitsstabs Speer*, die er ab September 1944 in Vertretung von *Walter Temp* betreut hatte. 1946 versuchte er, *Karl Franzius* mit einem *Stadtbebauungsplan* zu beauftragen, ein Jahr,
119 nachdem *Franzius* seine Bandstadt für den *Arbeitsstab Speer* entwickelt hatte. Der *Bauausschuß* steuerte jedoch auf den *Ideenwettbewerb* 1948 zu, den *Helmuth Baur* dann 1947 vorbereitete und auf das Stadtzentrum erweiterte.

Die Stadtplaner hatten mit der Feststellung von Fluchtlinien, mit *Bebauungsplänen* für Behelfsheimgebäude und mit der Anpassung von vorhandenen aus dem „III. Reich" übernommenen Planunterlagen zu tun[70]. Aufmarschplätze in den neuen Gemeinschaftssiedlungen wurden zu Kirchplätzen umgeplant, und ein *Verteilungsplan der Kirchengrundstücke* billigte nun den Religionsgemeinschaften Standorte zu. Kleingärten wurden ausgewiesen, Barackenstandorte gesucht (Nissenhuts) usw. Die von den Nationalsozialisten nur spärlich mit Bauten der Gemeinschaft bedachten Siedlungen mußten auf ihre Defizite (Sportplätze, Schulen, Jugendheime usw.) untersucht werden. Neuartige städtebauliche Überlegungen wurden diskutiert. So sollten im Gelände der *Westwerft* und
121 des ehemaligen Bauhafens der *KMW* Wohngebiet entstehen, um für alle Zeiten Rüstungsproduktion zu verhindern[71]. Andererseits ermöglichte die Ausweisung von
122 Wohngebieten die Zuteilung von Baumaterialien und Energieversorgungen durch Hannover. *Helmuth Baur* begann Ende 1945 mit Vorarbeiten für einen neuen *Wirtschaftsplan*, der im Oktober 1947 vom Rat als *Vorläufiger Wirtschaftsplan* verabschiedet wurde. Daneben bearbeitete das *Niedersächsische Amt für Landesplanung und Statistik* ab Anfang 1948 ein *Gutachten über den Wiederaufbau Wilhelmshavens*. Gleichzeitig betrieb *Stadtbaurat Otto Lehn* den *Ideenwettbewerb* 1948. *Kurt Brüning* vom *Niedersächsischen Amt für Landesplanung und Statistik*, hatte den ehemaligen *Landesplaner* der *LOB Wilhelm Wortmann* mit dem *Gutachten* beauftragt, zusammen mit dem ehemaligen Mitarbeiter des
122 *Arbeitsstabs Speer, Max Karl Schwarz*. Das *Gutachten* sollte Grundlagen für den Nachkriegs-*Flächennutzungsplan* liefern. Die Weiterbeschäftigung der Planer des „III. Reiches" und des *Arbeitsstabs Speer* wurde allenthalben in den Städten praktiziert und belegt, daß die „Stunde Null" nicht stattfand.

Wortmann empfahl in dem *Gutachten*, wie schon in seiner Zeit als NS-*Landesplaner* Wilhelmshaven zum Land hin aufzubauen[72]. Der Bahnhof sollte in die ehemalige *KMW* 10
nördlich der Friedrich-Ebert-Straße verlegt werden. Zu diesem Zeitpunkt betrieb die *Stadt Wilhelmshaven* noch die seit 1937 geplante Verlagerung des Bahnhofs an den Westrand der Stadt. „Die Binnenalster" um die ehemalige *KMW* sollte nach *Wortmann* neben dem Bahnhof auch Anlegestelle für Bäderdampfer und Promenadenufer werden. Die Gemeinschaftssiedlungen des „III. Reiches" sollten wiederhergestellt, aufgefüllt und 1
abgerundet werden[73]. Das innerstädtische „Bremer Haus" und der skandinavische Wohnungsbau sollten künftig Vorbilder für Wilhelmshaven werden. Alle Ansätze waren bescheiden, restaurativ und wenig revolutionär: Die Harmonie zwischen Stadt und Land sollte wieder hergestellt werden. Wilhelmshaven sollte – wie schon im „III. Reich" – gegliedert und durchgrünt werden. Die Abwendung von großstädtischen Idealen hin zu „reich gegliederter Stadtlandschaft" entsprach auch den Vorstellungen der Alliierten. *Wortmann* wollte in seinem erst 1950 fertiggestellten *Gutachten* wie *Fritz Schumacher* in Hamburg, 1945 „großstädtischen Geist und beschauliche Lebensverhältnisse" miteinander verbinden[74]. *Max Karl Schwarz*, ehemals Gutachter des *Arbeitsstabs Speer* und landschaftlicher Berater des Reichsautobahnbaus, erarbeitete die gärtnerischen Grundlagen für Wilhelmshaven. Gärtnereien und Kleinsiedlungen, Trümmerbegrünung und Landwirtschaften, Kleinsthöfe und Obstmostereien, Konservenfabriken und Unterglaskulturen sollten – in brachliegenden Grodenflächen und ehemaligen Schutzbereichen 122
angesiedelt – Wilhelmshaven wirtschaftlich gesunden. Die Bundesbehörden durchkreuzten *Wortmanns Gutachten* und richteten ihr Augenmerk alsbald auf eine militärische Zukunft Wilhelmshavens[75].

Der *Ideenwettbewerb* 1948, zu dem u.a. *Höger, Zinsser* und *Fiederling* eingeladen wurden, zu dem *Gustav Wolf* (Münster), *Fritz Karl Schneider* (Oldenburg), *Klaus-Georg Schöning* (Hannover), *Otto Lehn, Klaus Rasch, Heinz Baumann* und *Helmuth Baur* als Preisrichter herangezogen wurden, belegte die personelle Kontinuität der Nachkriegsarchitektur. Das nicht zu Ende gebaute *Verwaltungsforum* um das Rathaus sollte u.a. wegen des „Fehlens der langen Platzwände" planerisch überarbeitet werden. Man wollte „Neugestaltung", dieser Begriff von *GBI Speer* blieb aktuell. Viele der mit Wilhelmshavens Stadterweiterung und Neugestaltung verflochtenen Architekten nahmen neben den Eingeladenen teil: *Willi Lübbers, Karl Franzius, Rudolf Bürgin, Ernst Boyken, Egon Forster.* Preisträger *Gerhard Graubner* jedoch löste die „unangebrachte repräsentative Starrheit" 123
des *Verwaltungsforums* durch „ungezwungenen Ausdruck" auf und verkleinerte die monumentalen Ausmaße des Rathausplatzes[76]. *Graubner* wurde, weil ein fachlich geschulter *Stadtbaurat* fehlte, „städtebaulicher Berater des Rates der *Stadt Wilhelmshaven*" und bekam großen Einfluß auf Wilhelmshavens Nachkriegsplanung bis Mitte der 50er Jahre[77].

Revolutionäre städtebauliche Umwälzungen waren ernsthaft nicht zu erwarten, denn das Bau- und Planungsrecht blieb in seinen Grundzügen, in der Systematik und in seinem Behördenaufbau auch nach 1945 gleich. Erst 1973 bekam das 1946 neugebildete *Land*

Niedersachsen seine *Bauordnung*. Die 1952 von der *Stadt Wilhelmshaven* erlassene *Bauordnung* wurde u.a. aufgrund des § 2 der *Verordnung über Baugestaltung vom* 10.11.1936 erlassen. Das *Preußische Wohnungsgesetz vom* 28.03.1918 galt bis 1973. Erst das *Bundesbaugesetz vom* 23.06.1960 ersetzte die *Notverordnung des Reichspräsidenten vom* 06.10.1931. Erst 1960 wurden das *Gesetz über die Aufschließung von Wohnsiedlungsgebieten (Wohnsiedlungsgesetz) vom* 22.09.1933 und das *Gesetz zur Neugestaltung deutscher Städte vom* 04.10.1937 aufgehoben. Bis 1973 galt die nationalsozialistische *Verordnung über Baugestaltung vom* 10.11.1936.

Neben der *Stadt Wilhelmshaven* und dem *Niedersächsischen Amt für Landesplanung und Statistik* betrieben auch Private Stadtplanung. *Klaus Rasch*, erst 1953 wieder von der *Stadt* übernommen, plante für das *Dennoch-Programm* 1949 einen Innenstadtbebauung um den
14 Bahnhof und ab 1947 ein Kulturhaus im Gebäude der ehemaligen *Marineintendantur*. In
13 ehemaligen Kasernen, Arbeiterlagern, Krankenhäusern und Instituten sollte sich nach einer *Denkschrift vom* 26.11.1946 eine „Heim-Universität" nach amerikanischem Vorbild etablieren. Anfang 1947 legten *Fritz Boysen* und *Heinz Baumann* Planungen hierzu vor. Die Hochbauten waren traditionell orientiert. Ecktürme und *Marinestil*, Fensterumrahmungen und Sprossen, Zentrierung und Symmetrie, Säulenvorbauten und Walmdächer – alles blieb beim alten. Moderne Architektur kam nicht vor.

★

Nach 1945 stieg Wilhelmshavens Einwohnerzahl durch Rückkehr von Evakuierten und Kriegsheimkehrern sowie durch Flüchtlinge rapide an. Im September 1945 erließen die Alliierten ein Zuzugsverbot. Die von *Gutschow* und *Hillebrecht* Anfang 1944 festgestellte, sehr hohe Wohnungsbelegung von 6 Einwohnern pro Wohnung[78] wurde
12 durch die Umnutzung von Arbeiterbaracken und Kasernen langsam abgebaut. Elementare Bedürfnisse wie das „Dach über dem Kopf", „Ackerbau und Viehzucht" hatten hohen Rang. Schon im August 1945 wurde ein *Wohnungsausschuß* gebildet und Wohnungen nach dem *Kontrollratsgesetz Nr.* 18 vergeben. Viele Bauarbeiter waren in
1, 45 den nicht zu Ende gebauten Gemeinchaftssiedlungen des „III. Reiches" geblieben, lebten
98 in den Lagern, Baubuden und zimmerten für ein warmes Essen Schweineställe und Hühnergehege, Kaninchenboxen und Gartenzäune für die zurückgekommenen Familien. Selbsthilfe war so wichtig wie nach dem Ersten Weltkrieg. Erst 1948 wurden deutsche Bestimmungen zum *Kontrollratsgesetz* der Alliierten in Kraft gesetzt. Mittel für den Wohnungsbau sollte eine *Wohnraumsteuer* herbeischaffen, die ab 01.04.1949 erhoben wurde.

Die ersten größeren Wiederaufbauten orientierten sich an den noch vorliegenden Entwürfen und vorhandenen Plänen von *Bork, Schemm, zu Putlitz, Lübbers* usw. Eine Auseinandersetzung mit der Architektur des „III. Reiches" fand nicht statt. Veränderungen an Grundbesitzverhältnissen, Wirtschaftsstruktur oder Baurecht wurden weder entworfen noch durchgesetzt. Die ab 1950 neugebauten Schulen entstanden unter der

Leitung von *Egon Forster* ausnahmslos nach Mustern der Vorkriegsjahre, teilweise sogar
auf den vorhandenen Fundamenten. Die örtlichen Wohnungsbaugesellschaften, im
„III. Reich" gleichgeschaltet, wurden auf demokratische Grundlage gestellt, jedoch nur
zögernd unterstützt. Sie behielten ihre Gemeinschaftssiedlungen, Volkswohnungen und 1
Kleinsiedlungen fertig, halbfertig oder zerstört.

Der Wiederaufbau hatte die alten Muster[79]. Die *Wilhelmshavener Spar- und Baugesell-*
schaft beantragte 1950, Gebäude am Alten Deichsweg im Inselviertel nach „alter Art" 28
wieder herzustellen, da sie ausgebombt waren. Architekt *Ernst Peters* schrieb im
Erläuterungsbericht zum Wiederaufbau der Spiekeroogstraße: „Der Wiederaufbau
geschieht nach den früheren Unterlagen und Ausführungen. Änderungen werden nicht
vorgenommen". Auch der *Bauverein Rüstringen* orientierte sich bei den Wiederaufbauten
an „Wirtschaftlichkeitserwägungen" und mauerte die Baulücken nach *Fritz Högers* alten 57
Plänen hoch. *Högers* Entwürfe an der Arngaststraße wurden weiterentwickelt – nach
altem Vorbild. Nach *Höger* sicherte sich der *Bauverein Rüstringen* die Dienste des
ehemaligen *Höger*-Mitarbeiters *Hellmut Suhr.* Es entstanden nach dessen Plänen das
Flüchtlingssonderprogramm Siebethsburg (1950–52), die Bebauung am Berliner Platz
(1954) und die Wohnbebauung am Störtebeker-Platz (1950) aus den Vorgaben des
Ideenwettbewerbs 1948 von *Graubner.*

Die *Wohnungsbaugesellschaft Jade* verdrängte schnell die militärisch-städtebaulichen
Leitlinien des „III. Reiches". Die luftschutzgerechte Stadt des *RAM* wurde uminterpre-
tiert zu „neighbourhoods" und zur „idealen Wohnstadt"[80]. Konservative Leitlinien
hatten bei der *Jade* lange Bestand. Die Olympia-Siedlung von 1954 setzte das Jadeviertel
bruchlos fort. Das eigene Verwaltungsgebäude an der Lessingstraße baute man nach dem 124
traditionellen Entwurf von *Egon Forster* 1952–53. Mit seinem hohen Walmdach, den
Einzelgauben und Kreuzsprossen war es Relikt der *Hochschule für Baukunst*, Weimar.

Die *Wilhelmshavener-Rüstringer Wohnungsbaugesellschaft* baute gleichfalls nach alten
Mustern weiter. Am 11.05.1948 beantragte sie den Wiederaufbau der Wohnhäuser
Gökerstraße 140–142a „nach vorliegenden Zeichnungen von *Erich zu Putlitz"*. Für die
Gökerstraße 166–176 lagen gleichfalls noch die Entwürfe von *zu Putlitz* vor. *Karl-Heinz*
Kraak änderte sie 1951 um. Bedarfsgerecht waren jetzt keine Offizierswohnungen,
sondern 56 Kleinwohnungen zu entwerfen. Die Randbebauung wurde erstes *Sonder-*
programm des sozialen Wohnungsbaues und von der *Landestreuhandstelle Hannover* gefördert.
Im Äußeren hielt sich *Kraak* an Form und Material des *Putlitz*-Entwurfes. Im
innerstädtischen Bereich Viktoriastraße/Parkstraße hatte *zu Putlitz* kurz vor Kriegsaus-
bruch noch Entwürfe für Marinewohnungen im Auftrag der *Wilhelmshavener-Rüstringer*
Wohnungsbaugesellschaft fertiggestellt. Ein Teil der Gebäude wurde begonnen, im Kriege
aber zerstört. 1949 plante *Karl-Heinz Kraak* den Wiederaufbau. Die Grundrisse wurden
nachkriegsgerecht verändert. Die „Bauberatung" der *Stadt* forderte beim Wiederaufbau 120
NS-typische Details, Treppenhausvorbauten und (Führer-)Balkone[81]. Auch die von
Höger 1934–36 begonnenen Bauten der *Wilhelmshavener-Rüstringer Wohnungsbaugesell-*
schaft am Mühlenweg wurden schon bald nach dem Kriege komplettiert, denn das

23 Wohngebiet zwischen Widukindstraße und Mühlenweg war nicht fertiggeworden. Architekt *Heinz Folkerts* schlug 4 Geschosse vor. Der *Bauausschuß* berief sich auf die *Verordnung über die Regelung der Bebauung vom* 15.02.1936 und ließ nur 3 Geschosse zu[82].

Neben dem Mietwohnungsbau waren Kleinsiedlungen und Eigenheime eine schnelle Möglichkeit zur Wohnraumbeschaffung. Begonnene Kleinsiedlungsgebiete konnten
26 wiederaufgebaut und fortgesetzt werden. Die ehemalige Adolf-Hitler-Siedlung wurde von *Ernst Peters* und *Karl Tuchscherer* nach vorhandenen Plänen unverändert rekonstruiert. Der *Niedersächsische Fußball-Toto-Verband* baute 1950 in der Adolf-Hitler-Siedlung auf alten Fundamenten ein zerstörtes Haus nach Plänen von Architekt *Julius Kohlscheen* wieder auf, mit Satteldach und Stallanbau, Trockenboden und Kammer, und verloste es unter der Wilhelmshavener Bevölkerung. Eigenheimpolitik wurde zielstrebig angeheizt. Die Siedlung Himmelreich wurde bruchlos nach Plänen von *Hans Bartning* von 1939 fortgeführt und mit heimatschutzorientierten einfachen Satteldachhäusern und sparsamsten Grundrissen zu Ende gebaut. Stallanbauten waren überholt, großzügige Bäder und sprossenlose Fenster kündeten vom neuen Wirtschaftswunder.

52 Nicht benutzte Monumentalachsen der NS-Neugestaltungsplanung von 1937 waren nach dem Kriege wieder verfügbar. Sie wurden zu Grünachsen, Schrebergärten, Baugebieten oder öffentlichen Bedarfsflächen (Hallenbad, Krematorium) umgewidmet. Die Fortsetzung des *Marineforums* nach Norden wurde ab 1945 Grünanlage. Im nördlichsten Bereich wurde sie Wohngebiet für Einfamilienhäuser. *Senator Wilhelm Rehbein*, 1946 *CDU*-Mitbegründer in Wilhelmshaven, initiierte das *Kardinal Graf Galen-Siedlungswerk* und rief die Katholiken von Wilhelmshaven auf, ein „christliches Siedlungs-Gemeinschaftswerk" zu beginnen. *Rehbein* stellte die sozialpolitische Aufgabe in den Mittelpunkt seiner Siedlung, rief zu Spenden und freiwilliger Arbeitsleistung auf[83]. Zusammen mit Architekt *Alexander Minke* entstanden zwischen 1952 und 1955 70
125 kleinsiedlungsorientierte Klinkerhäuser mit Satteldächern.

Der Einfamilienhausgedanke wurde nach 1945 auf demokratische Grundlage gestellt und von dem Zwang überlagert, mit möglichst wenig öffentlichen Mitteln möglichst viel Wohnraum zu schaffen. Pläne von 1940 wurden noch bis weit in die 50er Jahre weiterbenutzt oder kopiert. Der interessierte Betrachter hat heute kaum eine Unterscheidungsmöglichkeit, da die verschiedenen Modernisierungswellen längst die Nachkriegsbauten erfaßt haben. Die Übernahme alter architektonischer und städtebaulicher Leitbilder hatte handfeste rechtliche und wirtschaftliche Hintergründe.

7. Einzelaspekte des Bauprogramms

Bis heute aber kann ich nur sagen, daß ich enttäuscht bin und daß die nationalsozialistische Bewegung alles das, was ich in meiner Vergangenheit geleistet habe, mit auf ihr Programm schreiben muß, wenn sie ihr eigenes Programm zur Durchführung bringen will. Ich habe meine Baugildenidee schon 1920 herausgebracht und einiges mehr als 100 Baubetriebe auf dieser Grundlage mit über 20.000 Bauarbeitern auf die Beine gestellt. Das war ein nationales und soziales Werk, wie es nationaler und sozialer nicht vom Deutschen Kampfbund ausgeführt werden kann[1]!

Martin Wagner, 1933

7.1. Wehrhafter Städtebau im Spiegel von Wirtschafts- und Bebauungsplänen

Der Standort der *Stadt des KDF-Wagens* (Wolfsburg) konnte völlig neu entworfen werden. Die *Stadt der Hermann Göring Werke* (Salzgitter) lag seit dem Ersten Weltkrieg wegen der Erzvorräte fest. Wilhelmshaven als Waffenschmiede und Kriegshafen hatte sich ab 1853 als eine der wenigen Stadtneugründungen des 19. Jahrhunderts etabliert. Das ländliche Umland schien ab 1935 einen Ausbau des Siedlungsgebietes bis auf 500.000 Menschen problemlos zu ermöglichen. Die raumplanerische Frage, wie und wo die „Festung" Wilhelmshaven erweitert und neugestaltet werden sollte, wurde 1937 noch vor dem ersten *Raumordnungsplan* von *Wilhelm Hallbauer* gestellt: „Die Verlagerung der neuen Stadt als Ganzes auf die wesentlich bessere Lebensbedingungen bietende Geest, für die eingehende Voruntersuchungen in Schortens/Heidmühle bei Jever von der *Landesplanungsstelle* gemacht sind, scheitert an der Trägerschaft dieser neuen Stadt mit allen dann dort notwendigen Verwaltungseinrichtungen, kulturellen und Versorgungsanlagen sowie Verkehrseinrichtungen"[2]. Die *LOB* hatte ernsthaft die Umsiedlung der ganzen Stadt erwogen.

Immer wieder drohte das Konzept für Wilhelmshavens Stadterweiterung auch wegen der schlechten Bodenverhältnisse auseinanderzulaufen. Nach einer Spitzenbesprechung am 26.11.1940 wurde festgestellt, „man dürfe nicht den Fehler wiederholen, auf solchem Boden eine ganze Stadt zu entwickeln"[3]. *Friedrich Nicolaus* vom *RAM* fragte 1941 „ob es richtig ist, den Wohnungsbau im Marschland durchzuführen. Bebauung an und für sich" sei „der Marsch wesensfremd"[4].

Landesplaner Wilhelm Wortmann schrieb am 27.01.1941 der *Reichsstelle für Raumordnung* nach Berlin, daß die „natürlichen Vorbedingungen für die Entwicklung einer Großstadt und eines Großhafens nicht günstig" seien. Er bemängelte das sich überstürzende Tempo, „das bei allen Maßnahmen vorgeschrieben wurde. Ein geklärter, endgültiger Gesamtplan für den militärischen Ausbau lag bei Kriegsbeginn noch nicht vor". Im Hinblick auf eine „gesunde Raumordnung" sei zu wünschen, „daß der Ausbau des Kriegshafens nicht den vor Kriegsausbruch angenommenen Umfang annimmt"[5]. Die Einwohner-Zielzahlen für Wilhelmshaven schwankten hektisch zwischen großspurigen 500.000 (1941) und bescheidenen 150.000 (1944).

Kriegswichtigkeit und Vorrangigkeit der Stadterweiterung und Neugestaltung Wilhelmshavens manifestierten sich in *Hitlers Erlaß vom* 27.01.1939, der besagte, daß der „Aufbau der Kriegsmarine allen anderen Aufgaben, einschließlich der Aufrüstung der beiden anderen Wehrmachtsteile und einschl. des Exports, vorgeht"[6]. Durch *Görings Ausführungserlaß vom 02.02.1930* als *Beauftragter für den Vierjahresplan* sollten „vor allem die erforderlichen Arbeitskräfte und Rohstoffe in vollem Umfang zugeteilt werden"[7].

Wilhelmshavens Aufbau wurde aber auch immer wieder und unvermittelt gestoppt, beispielsweise durch *Hitlers* Anordnung von 1942, die Planungsarbeiten „überall" einzustellen[8]. 1943 stoppte *Stadtbaudirektor Walter Temp* weitere Planungsarbeiten, „da in absehbarer Zeit nicht mit einer Beendigung des Luftkrieges" zu rechnen sei[9].

Die *Raumordnungspläne* vom Dezember 1937 und Dezember 1938[10] und die reichsplanerischen Ansätze der Wiederaufbauplanungen stellten den Standort Wilhelmshaven nie ernsthaft in Frage, zu sehr waren Garnison und Wohngebiet miteinander verflochten. Reichsplanung unterstand mit der *Reichsstelle für Raumordnung Hitler* persönlich. Seine verbrecherischen militärischen Befehle wurden dort raumplanerisch nachvollzogen. Abweichende Ziele wurden niemals formuliert. Auch die Frage, wie und ob Wilhelmshaven nach 1945 aufzubauen sei, ob der Jadebusen verlanden, ob die Stadt gar unter Wasser gesetzt werden sollte, wurde keine Frage der Raumordnung, sondern eine Entscheidung der Siegermächte, gekettet an Existenzfragen und Überlebenswillen.

*

Das *Gesetz über die Aufschließung von Wohnsiedlungsgebieten (Wohnsiedlungsgesetz) vom* 22.09.1933, war das erste von den Nationalsozialisten erlassene Baugesetz. Es gab dem
3 *Land Oldenburg* die Möglichkeit, Gebiete mit bevorstehendem Bevölkerungszugang zu *Wohnsiedlungsgebieten* zu erklären. Die für das *Wohnsiedlungsgebiet* aufzustellenden *Wirtschaftspläne* sollten Pläne auf lange Sicht[11] sein, sie sollten die bauliche Entwicklung

„an der Wurzel" kontrollieren, bevor sie sich in Investitionen verfestigt haben. Im Gau Weser-Ems wurden durch den *MDF* zu *Wohnsiedlungsgebieten* erklärt:[12]

1934 Bremen
1935 Zwischenahn
1936 Nordenham, Oldenburg, Lemwerder, Jever-Rüstringen
1937 Delmenhorst, Jever-Wilhelmshaven
1938 Wilhelmshaven – Jeverland

Die Grenzen des *Wirtschaftsgebietes Wilhelmshaven-Jeverland* von 1938 wurden im *Wirtschaftsplan Wilhelmshaven-Friesland vom Mai* 1940 festgesetzt. Durch das *Gesetz über* 129 *einstweilige Maßnahmen zur Ordnung des Deutschen Siedlungswesens vom* 03.07.1934, das *Heimatschutzgesetz vom* 17.09.1934 und die *Verordnung über Baugestaltung vom* 10.11.1936 wurde der restaurative Baustil festgelegt und liberaler Städtebau endgültig beendet.

Zwischen Oktober 1936 und Dezember 1937 leitete *Wilhelm Hallbauer* die *Bezirksstelle* bzw. *Landesplanungsstelle der LOB* in Wilhelmshaven[13]. Sie wurde auch als *Geschäftsstelle Jadestädte-Jeverland der LOB* bezeichnet. *Hallbauer* war dem neuen *Landesplaner Wilhelm Wortmann* von *Fritz Schumacher* empfohlen worden und im Oktober 1936 als *Beauftragter der LOB* für Wilhelmshaven/Rüstringen eingestellt worden. Er begann seine Arbeit mit Vorträgen im Rüstringer Rathaus zu „Stadterweiterung und Landesplanung im Bereich der neuen Stadt Wilhelmshaven"[14] oder zu „Stadtbaukunst einst und jetzt und das neue Wilhelmshaven". Unter verschiedensten Bezeichnungen und in unterschiedlichsten Funktionen legte er seine *Wirtschaftspläne* vor:

- *Neuordnungsplan vom* 20.11.1936 *(Bezirksstelle).* 126
- *Die Stadt der Dreihunderttausend vom* 15.03.1937 (*Bezirksplanungsstelle Jadestädte-* 127 *Jeverland*). 128
- *Wirtschaftsgebiet Wilhelmshaven vom Juni* 1937 (*Landesplanungsgemeinschaft Oldenburg-Bremen, Geschäftsstelle Jadestädte-Jeverland*).
- *Übersichtsplan der neuen Planung für Wilhelmshaven, Stand Juli* 1937 mit Erläuterungsbericht zum Übersichtsplan der neuen Planung für Wilhelmshaven vom 06.08.1937 (Leiter der *Landesplanungsstelle* und des *Stadterweiterungsamtes*)[15].
- *Wirtschaftsplan Wilhelmshaven und Umgegend, September* 1937.
- *Wilhelmshaven, städtebauliche Neuplanung vom* 29.03.1939 (*Stadt Wilhelmshaven Stadt-* 52 *erweiterungsamt*). Dieser Plan war weniger ein *Wirtschaftsplan*. Er war Neugestaltungsplan für die Altstadt, zusammengestellt für den *Hitler*-Besuch am 01.04.1939.
- *Wilhelmshaven* 1950 vom 15./16.06.1939 (*Stadt Wilhelmshaven Stadterweiterungsamt*)[16].
- *Wilhelmshaven Wirtschaftsplan vom* 03.08.1939 (*Stadt Wilhelmshaven/Stadterweiterungsamt*).

Es war Aufgabe der *Stadt*, *Wirtschaftspläne* aufzustellen. Die *LOB* jedoch griff verschiedentlich mit eigenen *Wirtschaftsplänen* in die Aufgaben der *Stadt* ein. *Landesplaner*

Georg Linde legte am 31.08.1937 einen *Plan für die Ausweisung neuer Wohngebiete im Rahmen des Wirtschaftsplanes*[17] vor, den er als Teil des *Wirtschaftsplanes* für das *Wohnsiedlungsgebiet Wilhelmshaven Jadegebiet* bezeichnete und der *Marine* am 27.08.1937 erläuterte. Im Januar 1940 schließlich fertigte die *LOB* einen kompletten *Wirtschaftsplan*
129 *Wilhelmshaven-Friesland,* der wegen des kurzzeitigen Kriegsdienstes von *Wilhelm Wortmann* im Mai 1940 von *Karl Dillschneider* unterschrieben wurde. Dieser *Wirtschaftsplan vom Mai* 1940 hatte den Umfang eines *Raumordnungsplanes* und basierte auf dem *LOB-Raumordnungsplan vom Dezember* 1938[18]. Möglicherweise griff die *LOB* in die *Wirtschaftspläne* ein, weil *Hallbauer* ab November 1939 bei der *Stadt* ausgeschieden war. Der Einfluß der *LOB* war seit Kriegsbeginn erheblich gestiegen. *LOB*-Vorsitzender, *Gauleiter Carl Röver* war *Reichsverteidigungskommissar* und ab 1940 *Gauwohnungskommissar* geworden. Zunehmend verwischten sich die Grenzen zwischen *Wirtschaftsplänen* und
116, 117 *Raumordnungsplänen.* Die Wiederaufbauplaner *Schulz, Schneider* und *Heuer* (1941–42)
118 und *Gutschow* (1944–45) fertigten die *Raumordnungspläne* für ihre *Wirtschaftspläne* selbst, möglicherweise, weil die *LOB* kriegsbedingt zu knappe Planungskapazität hatte.

Wirtschaftspläne, die außerhalb der *Wohnsiedlungsgebiete Flächennutzungspläne* hießen, waren städtebauliche Programme für die Zukunft. *Bebauungspläne* legten Aufbau und Gestaltung fest. Die entsprechenden dürren, gesetzlichen *Fluchtlinienpläne* wurden von NS-Baupolitikern für städtebauliche „Entartung" der Weimarer Republik verantwortlich gemacht[19]. Man wollte jetzt Gestaltung festlegen, landschaftsgebunden entwerfen und baukünstlerisch planen. Der *RAM* empfahl, daß Architekten gleichzeitig *Bebauungspläne* mit aufstellen sollten. Er stellte schon 1936 den Städten für städtebauliche Planungen und Wettbewerbe Haushaltsmittel zur Verfügung. Ausstellungen und „Muster guter Pläne" sollten über Städtebau aufklären[20].

126– Die Öffentlichkeit wurde über Wilhelmshavens „geheime" *Wirtschaftspläne* nur
129 zögernd informiert. Sie mußte sich mit Teilinformationen zufrieden geben. Spannungsreich schilderte die Presse[21], daß im Rathaus „in so manchem Zimmer der Männer, die sich mit dem Ausbau Wilhelmshavens zur Großstadt zu beschäftigen haben", Pläne, Modelle, Entwürfe und Skizzen zu finden seien. Ein Überblick über die Gesamtplanung fand in vertraulicher Gemeinderatssitzung erstmals am 26.08.1938 statt. Städtebauliche Wettbewerbe, wie vom *RAM* 1936 empfohlen, wurden, mit Ausnahme von Langewerth, nicht durchgeführt, wohl aber Wettbewerbe zu Einzelbauten.

★

Die Idee, Wilhelmshavens Stadtgebiet und damit die „Seefestung" durch Eindeichungen zu erweitern, ging auf die *LOB* zurück. Wie sonst hätte man den größenwahnsinnigen Flächenanforderungen der Wehrmacht gerecht werden können[22]? Um den Flächenbedarf der den Militärplanungen folgenden Wohngebiete zu sichern, hatte die *LOB* am 01.06.1938 umfangreiche Eingemeindungen durchgesetzt. Im Mittelpunkt der landesplanerischen Bemühungen stand die Verbesserung der Verkehrswege in die abgelegene „Festung“ Wilhelmshaven. Immer wieder versuchten die Planer, die im Reichsautobahn- und Eisenbahnbau seit 1933 bevorzugten Ost-West-Verbindungen nach Norden umzulenken. *Fritz Todt* hatte in seiner „Braunen Denkschrift“ die militär-strategische Vision entwickelt, daß 300.000 Soldaten in 100.000 beschlagnahmten Autos in zwei Nächten von der Ost- zur Westgrenze des Reiches transportiert werden sollten[23]. In dieser Wehrmachtsstrategie hatte die *Marine* keinen Platz. In Raum- und Stadtplanung jedoch waren neue Nord-Süd-Verbindungen verankert, um Soldaten und Arbeiter heranzuführen, Material und Kriegsgerät zu transportieren, Treibstoff und Energie nach Wilhelmshaven zu schaffen. Die Verkehrsverbindungen in den Norden jedoch blieben schlecht.

Die Entflechtung des äußerst luftgefährdeten Sander Verkehrsknotenpunktes hatte wichtigste Planungspriorität für die *LOB*. *Röver* und *Todt* hatten Ende November 1937 eine „eingehende Besprechung“ der *LOB* mit *GBI Speer*[24] zur Verkehrserschließung des Raumes Oldenburg-Wilhelmshaven initiiert. Aus militärischen Gründen sollte die neue Reichsautobahn Oldenburg östlich umgehen. Die *LOB* arbeitete 1937 an „der Weiterführung nach Ostfriesland und gegebenenfalls auch nach Wilhelmshaven“.
Entscheidend war die neue Reichsstraße 69, die im großen Bogen auf Wilhelmshavens *127*
Bismarckstraße führen sollte. Zur neuen, gleichfalls weit ausholenden Eisenbahnerschließung mit einem großangelegten Ortsgüterbahnhof, kam es nur in Ansätzen. Brückenpfeiler, Bahndamm, Stellwerke, Büros, Ladestraßen, Entwässerungen und ein Verwal-
tungsgebäude wurden allerdings noch bis 1941 gebaut[25]. Der Personenbahnhof beim *107*
Kulturforum blieb Absicht.

★

„Wehrhafter Städtebau“[26] wurde vor allem von *Friedrich Nicolaus* vom *RAM* propagiert, der Wilhelmshavens Stadtererweiterung und Neugestaltung stark beeinflußte. Die gegliederte und durchgrünte Stadt sollte nicht nur vor Luftangriffen schützen und Militärbauten sichern, sondern auch die Plazierung von Flugabwehrgeräten, Nachrichtenübermittlungsstellen, logistischen Einrichtungen, Militärbaracken, Schützengräben und mobilen Einheiten ermöglichen. *Nicolaus* verknüpfte die aktuellen militärischen Forderungen des *Schutzbereichsamtes der Festungskommandantur* mit historisch abgeleiteten germanischen und mittelalterlichen Wehrgedanken. Er entwickelte Linien von „germanischer Stadtbildung“ über den Städtebau des deutschen Ritterordens bis zu modernen „Festungsanlagen“. Städtebaulich geschulte Soldaten sollten mit militärisch erfahrenen Städtebauern zusammenarbeiten.

Die verschiedenen *Festungspläne* für das „Preußische Marineetablissement“[27] am *6, 7*
Jadebusen mit Wall und Graben, später auf Fortanlagen reduziert, war für *Nicolaus* *8*
ebenso Markstein auf dem „Wege zur deutschen Befreiung“, wie *Hitlers Wehrgesetze von* 1935 und der 2. *Vierjahresplan von* 1936. Er forderte für das „III. Reich“ die „zwangsläufige Ableitung aller städtebaulichen Entschließungen aus den Geboten der

Landesverteidigung". Jede Neugründung und Erweiterung einer Stadt habe ihre Ursache „in einer auf die Wehrbereitschaft von Volk und Reich gerichteten Maßnahme". Der Städtebau sollte mit den Augen des Soldaten gesehen werden. *Nicolaus* war es, der im *RAM* darauf achtete, daß *Bebauungspläne* Flächen für Batterien und Kriegseinrichtungen frei ließen[28], daß Bunker Verteidungskraft ausstrahlten, daß Architektur „Wehrbereitschaft" zeigte. Die NS-Baupolitiker konnten sich auf *Hitler* berufen, der in seiner Kulturrede 1937 die „höhere Mission" der Baukunst mit der „Kriegskunst" verglichen hatte. *Landesbaupfleger Gustav Wolf,* dem *Ahlhorner Architektenkreis* verbunden, und Preisrichter im *Ideenwettbewerb* 1948, belehrte 1940[29], daß Baupflege „nach dem Vorbilde unserer Wehrmacht alle Kampfmittel" für die „Erziehung zum guten Bauen mobilisieren" müsse. *Friedrich Heuer*, mit Wilhelmshavens Stadtplanung vielfach verflochten, referierte 1943 vor dem *Ahlhorner Architektenkreis* und forderte: „Auflockerung der Ballungen, um unseren Volksgenossen die biologischen gesunden Vorbedingungen zu geben durch Steigerung der Geburtenzahl, die Zahl unserer wehrtüchtigen Männer für die Zukunft bedeutend zu heben"[30].

Besonders die Monumentalplanungen des NS-Staates wurden mit Wehrhaftigkeit und Militärtradition begründet. Ost-West-Achsen und Nord-Süd-Achsen wurden auf die Hauptstraßen römischer Heerlager zurückgeführt[31], rechte Winkel und Symmetrie auf friderizianische Stadtgrundrisse, Appellplätze und Aufmarschplätze auf mittelalterliche Paradeplätze[32], „Runde Plätze" und Straßenkreuze auf barocke Residenzstädte. Angeblicher „Niedergang" und „Entartung" der Baukunst der Weimarer Republik hatten nach Auffassung der Nationalsozialisten ihre Ursache in nachlassender „Wehrbereitschaft."
64 *Gottfried Feder* war sich sicher, daß Verteidigungsfähigkeit und Festungseigenschaften die Schönheit des Stadtbildes garantierten.

Kriegsarchitektur und militärische Muster sollten auch in Wilhelmshavens Planungen für Stadterweiterung und Neugestaltung Eingang finden. Zutreffend bezeichneten die Baudienststellen[33] Wilhelmshaven als „Festung, in der jetzt klar umgrenzte Gebiete für die Stadterweiterung", andere für „Abwehrbatterien und sonstige Kampfanlagen" zur
1 Verfügung stehen. Die scharfe Umgrenzung der Gemeinschaftssiedlungen wurde die moderne „Stadtumwallung", bestimmt durch das *Schutzbereichsamt* und den *Verkehrsoffizier* der *Kommandantur.* Sie hatte durch alle Planungsstufen Bestand und symbolisierte die „Wehrbereitschaft" des „III. Reiches". Die wenigen Einfallstraßen wurden unmerklich „Stadttore" in den Befestigungsanlagen.

Wilhelmshavens Städtebau war immer durch militärische Zweckmäßigkeit beherrscht. In Vorbereitung des Zweiten Weltkrieges änderten sich die Ziele, wurden moderne Kampfeinrichtungen wichtig. Sie legten sich wie ein Schleier auf Gestaltung und Funktion der „Seefestung". Der architekturbesessene oberste „Kriegsherr" wies der „*Stadt der 500.000*" eine Sonderrolle zu. Seine Architekten gaben dem kriegerischen Zweck bauliche Form und wehrhaften Ausdruck.

7.2. Leitbild der gegliederten und durchgrünten Stadt

Wohl kaum eine Stadt spiegelte so konsequent den Zusammenhang zwischen drohendem Krieg und Stadtstruktur wie Wilhelmshaven. Unabhängig von den städtebaulichen Absichten berücksichtigten bereits die *Raumordnungspläne* den möglichen Luftkrieg und zu befürchtende Bombenabwürfe. Die Stadterweiterungspolitik zielte auf Rüstungseinrichtungen, Wirtschaftsbetriebe, Rohstoffproduktion und Wohnungsbau. Die Stadtneugründungen, vor allem Wolfsburg (90.000 Einwohner) und Salzgitter (130.000 Einwohner) sollten für sich eine ausreichend wirtschaftlich tragfähige Basis von Handel, Gewerbe und Industrie bekommen. Das erheblich größer geplante Wilhelmshaven hat von Anfang an nicht Wirtschaftsbetriebe haben sollen, sondern Rüstungsbetriebe. Damit war es ein Sonderfall. Nach 1848 gab es in Deutschland nur eine einzige militärisch bedingte Stadtneugründung: Wilhelmshaven[34]. Bremerhaven war 1827 wegen des Handelshafens, Ludwigshafen 1842 wegen des Industriehafens, gegründet worden.

Die Stadtgestalt galt, seit es Flugzeuge gab, nicht nur als „luftempfindlich", sondern wegen des Kriegshafens und der Rüstungsbetriebe als „stark luftgefährdet"[35]. Nach der Begriffsbestimmung des *Reichsluftfahrtministeriums* war Wilhelmshaven sogar „besonders luftgefährdet"[36]. Insofern hat bei allen Entwürfen die Trennung der neuen Großsied-
lungen durch Grüngürtel zentrale Bedeutung gehabt. Das *OKM* forderte Wilhelmsha- *4*
vens *Marineintendantur* im August 1938 auf, einen Plan nach der Verordnung des *14*
Reichsluftfahrtministeriums herzustellen, der die 500 m Sicherheitsgürtel zwischen Wohngebieten und luftgefährdeten Anlagen zeigt[37]. Zwar spielte die Vermeidung von Industriezusammenballung auch bei der Planung für Salzgitter[38] und andere Stadterweiterungen eine Rolle, doch Wilhelmshaven und sein *Wohnsiedlungsgebiet* standen besonders beispielhaft für das stadtplanerische Oberziel „Luftkrieg und Luftschutz".

Die Wilhelmshavener *Festungskommandantur* verfügte über ein eigenes *Schutzbereichsamt*, das dafür sorgte, daß Flak-Schußwinkel, Schutzbereichsgrenzen, Bauverbote und Bausperren eingehalten wurden. Erstaunlicherweise erinnerten sich viele befragte Architekten und Planer von damals nur schwer an die Luftschutzaspekte ihrer Entwürfe. Statt dessen behaupten sie, von Anfang an nur das Ziel der durchgrünten „Gartenstadt" vor Augen gehabt zu haben[39].

Das *Gesetz über die Aufschließung von Wohnsiedlungsgebieten (Wohnsiedlungsgesetz) vom* 22.09.1933[40], in seiner Zielsetzung ein weiter Vorläufer des heutigen *Baugesetzbuches*[41], räumte dem Luftschutz hohen Rang ein. Die *Wirtschaftspläne* sollten „eine Auflockerung der Großstädte anstreben"[42]. Auch nach dem *Erlaß des Reichsarbeitsministers vom* 18.11.1933 sollte nicht „allein der Gesichtspunkt der Erzielung einer städtebaulich möglichst günstigen Bebauung ausschlaggebend sein".

Auf der Jahrestagung der *Deutschen Akademie für Städtebau, Reichs- und Landesplanung* vom 16.02.1940 referierte der Architekt *Karl Otto, Oberbaurat* im *Reichsluftfahrtministerium* über Luftkrieg und Städtebau[43]. Bei der Planung neuer Stadtanlagen seien die

Erfahrungen des Luftkrieges ernst zu nehmen, und durch entsprechende städtebauliche
Maßnahmen seien Städte und Siedlungen so zu ordnen, „daß sie gegen Luftkriegswir-
kungen hart und unempfindlich werden. Eine richtige städtebauliche Gliederung und
Aufteilung des gesamten Stadtkörpers gewährleistet zugleich den größten Schutz für das
einzelne Haus und den einzelnen Betrieb gegen die Luftkriegsgefahr. Die sich aus der
Luftkriegsführung zwangsläufig ergebenden städtebaulichen Maßnahmen bilden somit
die Grundlage jedes luftschutzmäßigen Bauens. Die Erfolgsaussichten eines Luftangriffs
hängen ausschlaggebend von der Struktur einer baulichen Anlage ab. Die Angriffswir-
1 kungen sind um so geringer, je weiträumiger und aufgelockerter die Anlage gestaltet ist."
Karl Otto stellte die Forderung auf, die einzelnen Flächen des Stadtgefüges – Bauflächen,
Grünflächen, Verkehrsflächen – bereits so anzuordnen, daß mehrere Stadt-Zellen
38 entstehen, die durch breite Trennstreifen voneinander abgesetzt sind. Trennstreifen
sollten ein Netz über Stadterweiterungen und Neuplanungen legen und entweder aus
Grünflächen, Parkanlagen, Wasserläufen, Erholungs-, Sport-, Kleingartenflächen oder
42 aus Straßenzügen mit außergewöhnlicher Breite bestehen.

Die Auflockerung der Bauflächen und die Abgrenzung durch Trennstreifen hatte nur
ein Ziel: Brandabschnitte zu bilden und das Ausbreiten von Feuer zu verhindern. Diese
Trennstreifenideologie sollte sich auch auf die Verkehrsflächen auswirken, die möglichst
30 um die Ortsteile herumführen sollten. Bei den Wilhelmshavener Gemeinschaftssiedlun-
37, 43 gen Voslapp, Fedderwardergroden und Altengroden war dies nur noch teilweise
66 möglich, denn sie waren im Bau. Für die geplanten Stadtteile Utters, Breddewarden,
64, 67 Langewerth, Neustadtgödens, Schortens, Cleverns usw. jedoch wurde dieser Aspekt im
129 *Wirtschaftsplan Wilhelmshaven-Friesland vom Mai* 1940 streng beachtet.

Als höchst anfällig stufte man auch Großbahnhöfe ein, da sie besonders gefährdete
Angriffsziele sein würden. Die *Deutsche Akademie für Städtebau, Reichs- und Landesplanung*
empfahl mehrere kleinere Bahnhöfe. Tatsächlich wiesen die späteren *Wirtschaftspläne*
viele Bahnhöfe und kleinere Haltestellen aus. Vor allem sollten die Bahnhöfe von
Baugebieten durch möglichst breite Freiflächen, Plätze, Grünanlagen o. ä. getrennt
werden[44]. Der *Wirtschaftsplan Wilhelmshaven vom* 03.08.1939 wies außerdem noch die
Ringschnellbahn mit 20 kleinen Bahnhöfen aus.

Während die früheren Planungen auf den Schutz von Bahnhöfen wenig Rücksicht
nahmen, war sowohl im *Wirtschaftsplan Wilhelmshaven vom* 03.08.1939 als auch im
129 *Wirtschaftsplan Wilhelmshaven-Friesland vom Mai* 1940 keiner der neugeplanten Bahnhöfe
mehr im Wohngebiet gelegen. Im Durchschnitt lagen die größeren Bahnhöfe 500 m
107 vom Wohngebiet entfernt. Der vorgesehene Hauptbahnhof in Schaar[45] z. B. war zum
64 neugeplanten Stadtteil Langewerth durch einen breiten Grünstreifen und zur Altstadt
51 durch die großen Platzanlagen des *Kulturforums* isoliert. Auch der Personenbahnhof
94 Sande und der Werftbahnhof Sande lagen isoliert von Wohngebieten.

1 Besonders bei den neuen Gemeinschaftssiedlungen sorgte man sich darum, daß
wirtschaftliche Erwägungen eine zu hohe Ausnutzung des Bodens zur Folge haben
könnten. Pro Hektar Nettobauland hielt man maximal 150–180 Menschen für

möglich[46]. Diesen Zielen kamen auch die Wohnungsbauideologie des NS-Regimes, des *Reichsheimstättenamtes*, der *DAF*, des *RAM* und des *Amtes „Schönheit der Arbeit"*, entgegen. Für Wohngebiete hielt die *Deutsche Akademie für Städtebau, Reichs- und Landesplanung* grundsätzlich die offene Bauweise, bestenfalls die Zeilen- und Reihenhausbauweise für geeignet. Man hatte im Ersten Weltkrieg festgestellt, daß Sprengbomben vor allem in Altbaublöcken eine verheerende Wirkung haben. Deshalb hielt man bereits sehr früh Entkernen der Altbaublöcke und das Schaffen von zusätzlichen Brandabschnitten für richtig.

Daß den Standortfestlegungen für Anlagen der Landesverteidigung, für Kasernen, Häfen und Rüstungsbetriebe das besondere Augenmerk der Stadtplaner galt, verwunderte nicht. Diese Anlagen sollten möglichst in Gewerbegebiete, in das Vorfeld der Städte oder in unbebaute Ortslagen mit genügend großem Abstand von Wohngebieten. Ihre Umgebung sollte als Bauverbotsgebiet eingestuft werden. Die Trennung von bebauten Ortslagen mußte darüber hinaus auch die erwünschte und notwendige Tarnung ermöglichen.

Das Wilhelmshavener Militärbauprogramm 1935–45 berücksichtigte weitgehend *1*
und von vornherein den Luftschutz. Die *Befehlstelle Marinekommando Nord* wurde weit *86*
außerhalb, westlich Sengwarden vorgelagert. Der Fliegerhorst Upjever entstand in der *104*
ostfriesischen Landschaft südlich Jever in bewaldeter Umgebung. Der neue Befehlsstand des *Luftverteidigungskommandos* wurde in den Wilhelmshavener Stadtpark gelegt – er ähnelte einem Forsthaus, war aber von großen Baracken- und Bunkeranlagen umgeben. Bei Tannenhausen entstand das *Artilleriezeugamt*, bei Schweinebrück eine *Munitionsanstalt*, in den Dörfern Antonslust und Langewerth wurden Telefonvermittlungsstellen der
Marine und in Roffhausen das *Marinegerätelager* hochgemauert. Im Äußeren waren diese *72*
Bauten landwirtschaftlichen Gebäuden zum Verwechseln ähnlich. Aber auch strategisch
nicht ganz so empfindlichen Bauten, wie das *Navigationsressort der KMW* am *91*
Anton-Dohrn-Weg, sowie das *Marinelazarett Sanderbusch*, waren gleichfalls in Grün *75*
gebettet.

Die Situation der *KMW* in Alt-Wilhelmshaven mußte von vornherein unter *9*
Luftkriegsbedingungen als äußerst ungünstig angesehen werden. So sah dann der
Wirtschaftsplan Wilhelmshaven-Friesland vom Mai 1940 südlich der Ebertstraße und östlich *129*
der Gökerstraße/Freiligrathstraße keine Wohnbebauung mehr vor. Im großen Stil sollten
die Altbauten abgerissen werden, u.a. auch das gerade erst fertiggewordene Heppenser *28*
Inselviertel. Mit gebührendem Abstand von annähernd 1.000 m sollte östlich der
Gökerstraße die neue *Nordwerft* entstehen[47]. Die alten Werft- und Militärgebäude in der *101*
Südstadt waren nicht mehr haltbar, und fast alle Einrichtungen waren zur Umsiedlung in das *Wohnsiedlungsgebiet* und das Industriegebiet Sande vorgesehen. Auch der Seeflughafen am Südstrand lag strategisch äußerst ungünstig. Er sollte nördlich der *Nordwerft*,
wiederum durch Trennstreifen, von Fedderwardergroden und Voslapp abgegrenzt, neu *128*
entstehen.

*

In seinem Erläuterungsbericht[48] 1937 ging *Wilhelm Hallbauer*, detailliert auf die Luftgefährdung und deren Folgen ein, und stellt fest, daß die Luftwaffen „ganze bisherige Stadt- und Hafenanlagen unzweckmäßig und nicht länger tragbar erscheinen" lasse. Beim Abwurf einer Bombe am 5 1/2 km langen Südstrand würden auf Anhieb lebenswichtige Teile der „Festung" außer Betrieb gesetzt. Es sei lebenswichtige Aufgabe, eine weitgehende Dezentralisation von Wohnungsbau, militärischen Anlagen und Versorgungsanlagen herzustellen[49]. *Hallbauer* forderte Schutzgürtel von mindestens 500 m Breite zwischen den Explosionsbereichen der im Ernstfalle mit Munition ausgerüsteten Wehrmachtsanlagen, sowie den vordringlichen Kampfzielen für Flieger und den eigentlichen Wohngebieten, denn „Arbeitermassen auf der Werft lassen sich im Ernstfalle auf Sirenensignal im Heldenkeller unterbringen, Kinder in Kinderwagen und im Garten, Frauen beim Einholen und in der Waschküche, lassen sich nicht rechtzeitig aus gefährdeten Gebieten entfernen".

128 Der Plan *Wirtschaftsgebiet Wilhelmshaven vom Juni* 1937 berücksichtigte daher auch umfangreiche Aussiedlungsgebiete in Heppens, Neuengroden und der Südstadt, denn „Frauen und Kinder sind das wertvollste Zukunftskapital eines jeden Volkes. Die noch für Bauzwecke innerhalb des Stadtgebietes in Anspruch zu nehmenden Flächen müssen so aufgelockert und dünn besiedelt werden, daß Brand- und Gasgefahr möglichst vollständig ausgeschaltet werden"[50]. Konsequent wurden im Plan *Wirtschaftsgebiet Wilhelmshaven vom Juni* 1937 die neuen Wohngebiete außerhalb der Luftgefahrenzonen ausgewiesen, einstweilen noch als einfache, isoliert voneinander entworfene, symbolhafte Kreise. Es sollte vermieden werden, daß Millionen Aufwendungen für die Werftanlagen nicht vergeblich gewesen sein sollten „weil der Faktor Mensch, der sie bedienen soll, im entscheidenden Augenblick versagt; denn der Werftarbeiter wird nur dann in Ruhe bei seiner Arbeit bleiben, wenn er Frau und Kinder außerhalb der eigentlichen Gefahrenzone weiß"[51].

Neue Versorgungsanlagen für Stadt und „Festung" (Elektrizität, Gas, Wasser usw.) sollten in Ringsystemen derart erfolgen, daß Treffer an einer Stelle nicht die gesamte Versorgung lahmlegen, sondern das Ringsystem die Versorgung auch weiterhin sicherstellt[52]. Eine „bomben"sichere Gasversorgung von Bremen aus wurde noch 1942 verlegt. Als besonders gefährdet sah man die Kreuzung von Ems-Jade-Kanal, der alten Reichsbahnlinie und der Reichsstraße bei Mariensiel an. Die Militärs forderten eine Entflechtung dieses Punktes. Noch konsequenter beachteten die Wiederaufbauplaner Luftkriegsaspekte[53]. In der Denkschrift über den *Bebauungsplan* für Wilhelmshaven vom 24.11.1941 beschrieb *Walther Schulz* seine Planungen. Er sah vor, daß die noch weiter nördlich gelegene Bismarckstraße Grenze zwischen den südlich gelegenen luftempfindlichen Anlagen und den nördlich sich entwickelnden Wohngebieten sein solle. Längs der *KMW* sah er eine 500 m breite Schutzzone vor, und auch der neue Hauptbahnhof sollte in einer ebensolchen Grünzone liegen. *Schulz* propagierte die abseitige Lage des

Hauptbahnhofs in einer Grünzone, die mit einzelnen Gebäuden durchsetzt ist, „so daß sich dem Fremden als erstes über die tiefliegenden formal gehaltenen Gärten hinweg bis zum Dunst des fernen Sees ein Bild imponierender Weite ergibt, das sich erst beim Näherschreiten durch die Häuserblocks zu Räumen schließt. Dieser inmitten einer Stadt unerwartete Eindruck soll einmalig und für Wilhelmshaven charakteristisch werden“[54]. Nüchterne Luftschutzziele wurden romantisch verbrämt und in Grünideologie verpackt.

Seine Erläuterungen zum *Bebauungsplan* vom Dezember 1941[55] verschweigen Luftkriegsaspekte völlig. Sie stellten die freie Verbindung aller Grünanlagen unter sich und mit der umgebenden Landschaft heraus. Tatsächlich jedoch war es Hauptaufgabe von *Walther Schulz*, die Stadt unter Beachtung der Luftschutzforderungen weiterzuentwik- *116*
keln und den *Wirtschaftsplan* entsprechend zu verändern[56].

Die Entwürfe des *Arbeitsstabs Speer* von *Konstanty Gutschow*, 1944–45, berücksichtig- *118*
ten grundsätzlich und konsequent die Trennung der Stadtteile durch Grün, Verkehrsstraßen und Grachten. Allerdings sollte nach *Gutschows* Zielplanung die Entwicklung Wilhelmshaven anders verlaufen: Begrenzung der Werftgröße, Absonderung von werftzugehörigen Nebenbetrieben und Aussiedlung in die Geest. *Gutschow* riskierte lebhaften Pendlerverkehr zwischen Arbeitsstätten und Wohngebieten, und sein Ziel war es, wirtschaftspolitisch ausgleichende Kräfte zu entwickeln. Er wollte höhere Kommandostellen der *Kriegsmarine* für Wilhelmshaven, um ein Gegengewicht zu Arbeitern und Angestellten zu schaffen. Er wollte Beamte, Marineausbildung, Schulung und auch das „Nordseebad Wilhelmshaven“[57]. Die Planung *Gutschow* in der Endphase des „III. Reiches“ verlor in den Erläuterungsberichten kaum ein Wort zu Luftschutzaspekten, in den Entwürfen jedoch wurde peinlich genau darauf geachtet.

Die luftschutzbedingten baulichen Vorgaben wurden jedoch noch weit detaillierter. Man hatte ermittelt, daß enge Altstadtstraßen, durch Trümmer versperrt, Brandbekämpfung und Hilfeleistung erschwerten. Es fügte sich gut, daß Hauptverkehrsstraßen wegen des durch den *KDF-Wagen* erwarteten stärkeren Individualverkehrs schon ausreichend groß gewählt wurden. Genügende Breite war aber auch deswegen erforderlich, um den ungehinderten Verkehr auch bei einem etwaigen Einsturz der Gebäude zu gewährleisten. Die notwendige Mindestbreite zwischen beiderseitigen Straßenbebauungen sollte für Erschließungsstraßen aus der Summe der Höhen der beiderseitigen Bebauung zuzüglich einer Fahrbahnbreite (etwa 5,50 m) errechnet werden[58]. So war man sich sicher, daß der Verkehr auch bei einem Einsturz der Gebäude aufrechterhalten bleibt. Die notwendige Mindestbreite zwischen Häusern war und ist im übrigen mit den Abstandsvorschriften in den Bauordnungen[59] verankert. Grundsätzlich und im Regelfall muß ein Gebäude mindestens einen Abstand von der Gebäudegrenze haben, der der Gebäudehöhe entspricht. Auch in den Vorschriften des *Bundes* zum baulichen Zivilschutz sind entsprechende Abstandsvorschriften und Böschungswinkel von möglichen Trümmerhaufen verankert[60].

Daß die luftschutzbedingten städtebaulichen Grundsätze mit der anfangs romantischen, stadtfeindlichen Siedlungsideologie der Nationalsozialisten übereinstimmten, muß hier nicht näher ausgeführt werden[61]. Kleinsiedlungsprogramme, obwohl nicht von den Nationalsozialisten erfunden, sollten die Liebe zu Boden, zu Heimat und Landschaft, stärken. *Friedrich Heuer*, vielbeschäftigter Architekt im *Wohnsiedlungsgebiet Wilhelmshaven* und *Konstanty Gutschow*, von *GBI Albert Speer* 1944 beauftragter Wiederaufbauplaner für die Stadt, brachten 1940 die „Ortsgruppe als Siedlungszelle" heraus[62]. Dieses „Handbuch" für Stadterweiterung propagierte die „Stadtlandschaft", die den „Zielen des nationalsozialistischen Reiches dienen soll". Die aufgelockerte dezentrale Siedlungsweise hatte sich an die politische Neugliederung anzulehnen. Sie sollte Liebe zur Heimat fördern, volksnahe Verwaltung ermöglichen und „Siedlungsgebilde luftangriffssicher gliedern". In Anlehnung an die Ortsgruppen der *NSDAP* sollten 6 untergliederte Siedlungszellen die Gefahren des Bombenabwurfs mindern. Gegen Brandbombenabwurf sollten mit städtebaulichen Mitteln sogenannte Brandinseln geschaffen werden. Die Gliederung der Siedlungsgebiete sollte unabhängig vom Bunkerbau eine Hauptrolle spielen.

Im November 1940 hatte die *Fachgruppe Bauwesen* im *NSBDT* im Auftrage des *Reichsluftfahrtministeriums* den Wettbewerb „Alarm" ausgeschrieben. Es ging hierbei um städtebauliche Einordnung, Gestaltung, Funktion und Grundrisse von Luftschutzbunkern. Der Wettbewerb wurde von *Alexander* und *Gertraude Herde* gewonnen[63]. Beide waren in der *Prüfstelle des RAM für Gemeinschaftssiedlungen und städtebauliche Planungen* als Architekten beschäftigt. *Alexander Herde*, obwohl rechte Hand von *Friedrich Nicolaus*, bekam aus unerfindlichen Gründen jedoch keinen Auftrag zur Weiterbearbeitung. Er verarbeitete das Thema 1941 zu seiner Dissertation „Der Luftschutzbunker im Wohngebiet, seine Grundrißgestaltung und städtebauliche Einordnung". Im Gegensatz zu der ortsgruppenorientierten Planung von *Friedrich Heuer* hat *Alexander Herde* differenzierte Lösungsmöglichkeiten angeboten, obwohl er sich für eine strenge Ordnung des Luftschutzwesens im Städtebau aussprach: „Wie es im militärischen, das die Veranlassung zur Errichtung dieser Bauten gab, nichts zufälliges und willkürliches gibt, so müssen auch hier, im Grundriß, im Aufbau und in der städtebaulichen Anlage, Straffheit und Zwangsläufigkeit der Anordnung oberstes Gesetz sein". Es gab für die Planer im *RAM* keine Zweifel, daß die grundsätzlichen, städtebaulichen Forderungen an Luftschutz mit Straßenführung, Wohngebietsausdehnung und Funktionstrennungen zu tun haben mußten. Es sollten bereits bei der Aufstellung des städtebaulichen Vorentwurfes die Luftschutzzusammenhänge geklärt werden. Auf der Basis der *Anweisung zum Bau bombensicherer Luftschutzräume, Fassung Nov.* 1940, war es für die Bunkerplaner im *RAM* wichtig, den Schutzsuchenden keine langen Wege zuzumuten. Die Standorte von Bunkern sollten sich an einer Höchstentfernung Bunker/Wohnhaus von 200 m orientieren. *Alexander Herdes* Musterentwürfe, die weder Monumentalität noch Repräsentation ausstrahlten, zeigten integrierte Luftschutzbunker als Teile der Bebauung. Die Stadtplanung Wilhelmshavens wurde Jahrzehnte dadurch bestimmt, daß die *Marine*

große Flächen zu Schutzbezirken erklärt hatte, die von jeder Bebauung freizuhalten waren. Schutzbezirke waren vorgesehen für zukünftige Abwehrbatterien, Flaks und sonstige Kampfanlagen[64]. Ziel war, Schutzgürtel um das alte Stadtgebiet zu legen, und zusätzlich einen Schutzbezirk zwischen Kirchreihe und Rüstringer Stadtpark auszuweisen. Vor allem der letztere Schutzbezirk wurde für die Bürger als künftiges Erholungs-
zentrum propagiert. Spätere NS-Planungen sahen hier *Sportforum*, Stadien und 52
Kampfbahnen vor. Mit diesem militärischen Schutzbezirk im Zentrum der Stadt war die 109
Grünplanung aus der Ära *Wagner/Migge/Hahn* beendet. Sie hatte strahlenförmig nach außen und an die Jadeufer führende Grünverbindungen vorgesehen. Nun forderte die *Marine* die gesamte Abriegelung der Wasserseite. *Wilhelm Hallbauer* warnte davor, Parks und Sportanlagen, Versammlungsstätten und Grüngebiete zu vernachlässigen, denn „soll Wilhelmshaven nicht wieder geeigneter Nährboden für revolutionäre Störungen werden, ist diese Fürsorge für die Freizeitgestaltung der Werftarbeitermassen lebenswichtig“[65].

Ein Vergleich des Planes *Wirtschaftsgebiet Wilhelmshaven vom Juni* 1937 mit einer vom 128
Verfasser vorgenommenen Kartierung der Flak-Stellungen weist nach, daß die großen Abwehrkanonen des Zweiten Weltkrieges tatsächlich dann in diesen Schutzbezirken aufgestellt wurden. Die Stadtplanung im Jahre 1937 hatte längst den „totalen Krieg“ berücksichtigt.

In verschiedenen Erläuterungsberichten bedauerten die damaligen Planer immer wieder, daß der Zugang zum Wasser trotz der vielen vorgesehenen Grüngürtel und trotz
des zentralen Schutzbezirkes zwischen Kirchreihe und Rüstringer Stadtpark nicht mehr 52
möglich sein würde. Auch in den Erläuterungen der *LOB* zum *Raumordnungsplan Wilhelmshaven von* 1937 wurden die Kleinsiedlungen zur „besten Wohnform für den Arbeiter“ erklärt und die aus lufttechnischen Gründen weitgehende Auflockerung und Dezentralisation des Wohnungsbaus propagiert. Die großräumige Ausdehnung des Wohngebietes aus militärischen Gründen wurde immer wieder mit dem gesunden Leben auf nationalsozialistischer Grundlage, der wahren Verbindung von Stadt und Land, und den nationalsozialistischen Zielen begründet. Diese Ziele spielten in der NS-Architekturpropaganda die größte Rolle.

Daß hierbei die Luftgefährdung in den Hintergrund rückte, war nur zu verständlich. Immerhin schrieben die Planer: „Für die Zivilbevölkerung ist es bedauerlich, daß der Zugang zum Wasser nur sehr bedingt möglich ist. Die Notwendigkeiten der Landesverteidigung müsse in einer Festung aber unbedingt voranstehen. Bei den Einzelplanunen müssen künftig alle nur denkbaren Möglichkeiten geschaffen werden, um die Menschen in ihrer Freizeit auch an das Meer heranbringen zu können. Ohne gesunde Menschen sind schließlich auch die besten militärischen Anlagen und Betriebe ohne Wert“[66]. Die Planer versuchten zu beschwichtigen und klarzumachen, daß die Lage der Stadt in der Flachlandschaft der Marsch und die Abriegelung von ihrem natürlichen Erholungsgebiet, der See, durch reichliche Schaffung von öffentlichen Grünflächen und Grünzügen ausgeglichen werde. Sie versprachen, die Bearbeitung der *Bebauungspläne* so zu vertiefen, daß auch die Anlage eines Badestrandes mit den nötigen Nebenanlagen sichergestellt werden könne. Es blieb bei diesen Versprechungen. Weder wurden die Grüngürtel und Parks bepflanzt, noch gar ein neuer Badestrand angelegt.

★

Inwieweit hat sich das Leitbild der gegliederten und durchgrünten Stadt als Ergebnis rückwärtsgewandter Wohnungsbauideologie und luftkriegsgerechter Planung im Zweiten Weltkrieg bewährt? Grundsätzlich ist herauszuzustellen, daß die Architekten, Stadt-, Landes- und Raumplaner kaum Einblick in die detaillierte Militärplanungen hatten[67]. Militärische Infrastruktur war „Staatsgeheimnis", die Flächen, in denen Geschütze und Fliegerabwehrkanonen standen oder stehen sollten, waren militärische Schutzbezirke. Die Hoffnung von *Wilhelm Hallbauer*, klar ausgewiesene Flächen für jetzige und zukünftige Abwehrbatterien und sonstige Kampfanlagen zu bekommen, hatte sich offensichtlich zerschlagen. Trotzdem wurden die großen Flak-Stellungen in einem Ring um Wilhelmshaven gelegt. Dazu kamen noch kleinere Flak-Stellungen in der Innenstadt,
91 auf den Dächern vor öffentlichen Gebäuden und an exponierten Stellen.

Schwerpunkt der Bombardierungen war die Innenstadt[68]. Die neuen Großsiedlungen Voslapp, Fedderwardergroden, Altengroden, Neuengroden, Neu-Siebethsburg und das Jadeviertel wurden nur wenig bombardiert. Die neue lockere Siedlungsstruktur war als Bombenziel ungeeignet. Nach der *Harris*-Doktrin[69] war es vorrangiges Ziel der Alliierten, die Bevölkerung zu demoralisieren, den Glauben an das „III. Reich" zunichte zu machen und die Lebensgrundlagen zu zerstören. Dieses Ziel konnte durch Bombardierungen der Innenstadt erreicht werden. Der Aufwand, auch nur eines der großflächigen neuen Siedlungsgebiete „auszuradieren", wäre ungleich höher gewesen. Bombenabwürfe in der Innenstadt hatten ein Vielfaches an Wirkung. Brände breiteten sich rasch aus und zusammenbrechende Gebäude verschütteten gleichzeitig das innerstädtische Straßennetz. Die Schwierigkeiten mit Wilhelmshavens Siedlungsstruktur drückte ein Geheimvermerk des *US-Strategical bombing survey* von 1942 aus[70]: „The Residential Area is not particularly vulnerable to bombing because of its open character, the dominance of small houses and the low density of population." Auch die Einzelauswertung weiterer Bombenabwürfe nach eingesehenem Kartenmaterial
45 beweist, daß die alliierten Bomber die neuen Stadtteile kaum angriffen[71].

Bombenabwürfe in militärischen Schutzbezirken wurden von den zivilen Dienststellen nur teilweise registriert. Man kann aber davon ausgehen, daß diese Gebiete durch Flak verteidigt wurden und mittels bunkerähnlicher Bauweise, Vernebelungsanlagen[72] und Tarnungen besser geschützt waren. Zerstörungen an Militäranlagen wurden, soweit nötig, aus kriegstechnischen Gründen auch sofort beseitigt. Der „totale Krieg" sollte weitergehen.

Auch die Auswertung der Gebäudezerstörungen beweist, daß vorwiegend Straßenzüge mit geschlossener Blockbebauung in der Innenstadt bombardiert wurden und dort

Häuser in Schutt und Asche fielen. Die verschiedenen „Zerstörungskarten"[73] sparten die neuen Stadtteile aus, da dort kaum Zerstörungen festgestellt wurden.

Die Bilanz der Gebäudezerstörungen 1939–45 war bedrückend. Hundertmal wurde die Stadt aus der Luft angegegriffen. 11.045 Sprengbomben, 73.205 Brandbomben und 35 Minenbomben waren auf den zivilen Teil der Stadt gefallen[74]. *Friedrich Paffrath*, Nachkriegs-*Oberstadtdirektor* von 1945–55, vermutete in dem ersten Verwaltungsbericht nach dem Kriege, daß auf den militärischen Teil die gleiche Bombenanzahl gefallen sei. Die „Chronik der Stadt"[75] jedoch registrierte nur 41 Angriffe auf *KMW* und Hafen im Zweiten Weltkrieg. Rund 60 in Hafen und Werft liegende Wasserfahrzeuge wurden beschädigt, zerstört oder versenkt. Immerhin wurden 2.105 zivile Gebäude in Wilhelmshaven total zerstört und fast 13.000 teilweise – fast alle in der Innenstadt. In den letzten Kriegswochen verließ die Bevölkerung fluchtartig die Stadt oder wurde evakuiert. Kinder wurden aufs Land „verschickt". Am Tage der Kapitulation, am 06.05.1945, waren nur noch 54.000 Menschen in Wilhelmshaven[76].

Die gegliederte und durchgrünte Stadt und die getrennt voneinander geplanten 129
Gemeinschaftssiedlungen hatten sich im Luftkrieg „bewährt". Die aufgelockerte 1
Bauweise war gut zu propagieren, da sie den Zielen sowohl der *NSDAP* als auch den angeblichen Wünschen der Bewohner entsprach. Nach Kriegsende hatte der Luftschutz keine Bedeutung mehr. So konnten die Planer:

- die räumliche Verbindung der weit auseinander gezogenen Wohngebiete durch Gestaltung der Freiräume fordern[77]
- die Gesundung der Städte durch festumrissene, in sich geschlossene Einheiten am Stadtrand erhoffen[78]
- weiträumige Parklandschaften, Wander- und Radfahrwege entwerfen, die zum Wasser führen sollten[79]
- die zurückgelassene, luftschutzgerechte Stadt als „Stadtlandschaft" loben[80]

Im Gegensatz zum ehemaligen *Landesplaner Wilhelm Wortmann*, der sich auch nach 1945 noch gerne der Propagandaformel von der „Stadtlandschaft" bediente, räumten die ehemaligen leitenden Baubeamten, *Helmuth Baur*[81] und *Klaus Rasch*[82] die Wichtigkeit des städtebaulichen Luftschutzes ein. Hinzu kamen noch handfeste Vorteile für den Bauablauf. Nur die getrennt liegenden Stadtteile erlaubten überhaupt gleichzeitiges Bauen an mehreren Siedlungen und Militärobjekten. Die späteren Grün- und Schutzgürtel waren dringend erforderlich für das Heranschaffen von Baustoffen und Maschinen, Materialien und Bauteilen. Es schwante den Planern wohl bald, daß die
Anbindung der nördlichen Trabanten[83] und die Schaffung der nicht fertiggewordenen 1
Folgeeinrichtungen eine schwere Hypothek für Wilhelmshaven werden würde.

Von sich aus wäre es möglicherweise nicht zu den vorhersehbar teuren Trabantensiedlungen gekommen. Bis 1939 wurden in Wilhelmshaven die innerstädtischen
Stadtteile[84] komplettiert und ergänzt – teilweise nach älteren, vorhandenen *Bebauungs-* 52
plänen in der luftgefährdeten Innenstadt. Erst ab 1937 setzten sich die luftkriegsgerecht

abgesonderten isolierten Stadtteile mit geplanter eigener Infrastruktur durch. Der weitgehend nur vorgeschobene Gliederungs- und Durchgrünungsprozeß wurde, je näher der Krieg heranrückte, um so kompromißloser. Die Gartenstadtidee spielte nur eine Nebenrolle.

7.3. Gründekoration

126–129 Neben der raumplanerischen und städtebaulichen Funktion, der militärischen und luftschützenden Bedeutung des Grüns, gab es immer auch Gartenarchitektur. Erläuterungsberichte zu Wilhelmshavens ersten *Wirtschaftsplänen* gingen zwar ausführlich auf Schutzbezirke, Grüngürtel, Schußfelder und Abriegelungen ein, wenig jedoch auf das konkrete Umfeld von Wohnungen.

Paul Wolf, 1907–10 *Stadtbaurat* im preußischen Wilhelmshaven, hatte, wie zu Anfang auch *Martin Wagner*, nur ein begrenztes technisches Verhältnis zu Gärten. Mit geringem Aufwand sollten die Erdgeschoßwohnungen durch Vorgärten vom Straßenraum getrennt werden. Hausgärten sollten Aufenthalt im Freien bieten[85]. Die peinliche Ordnung der englischen Gärten wurde gelobt. Daß ein „Hausgarten für den einfachen Arbeiter" erzieherisch wirken kann, daß er sogar aus eigenem Antrieb und ohne Anleitung etwas anlegen kann, erschien *Paul Wolf* noch 1926 erwähnenswert. Der kämpferische *Leberecht Migge* hingegen dachte zu dieser Zeit bereits ökologisch und revolutionär. Er orientierte sich an sozialen Zielen, neuen Selbsthilfekonzepten und genossenschaftlichem Siedeln. Er verachtete Eigentum an Grund und Boden[86] und überwarf sich 1933 mit *Reichsernährungsminister Richard Darré* wegen der Blut- und Bodenideologie des „III. Reiches".

Konservative Architekten wollten erziehend wirken und Arbeitern mit genauen Anweisungen Gartenkultur „beibringen", wie es dem traditionellen Ideal entsprach. Häuser des Mittelstandes und vornehme Villen bekamen schmückende Gartenanlagen nach reichhaltigen Formenkatalogen und Musterbüchern von Gartenarchitekten. Belehrend stellte *Paul Schmitthenner*, strikter Gegner neuer Architektur, 1932 seinen eigenen Villengarten[87] als Beispiel heraus mit: „Steinhof, Vogelbrunnen, Wiesenhof, unterer Obstwiese, oberer Obstwiese, Sitzplatz, Stauden, Terrassen". Das „Ordnungschaffen" täuschte darüber hinweg, daß die *Stuttgarter-Schule* Häuser als Stück der Landschaft sehen wollte[88].

Nationalsozialistische Akzente setzte 1929 erstmals *Paul Schultze-Naumburg:* „Bauwerke als Landschaft haben ein ausgesprochen deutsches Gesicht"[89] und mahnend erinnerte der nationalsozialistische Kulturkämpfer daran, daß der Baum den Deutschen heilig war, und man sogleich nach dem Hausbau einen Baum pflanzen solle[90]. Wenn Gärten Ideale, Träume und Wunschbilder von Menschen spiegelten, dann sollte den Kleinsiedlern und „Volksgenossen" Großstadtfeindlichkeit und Kleinbürgerlichkeit aufgezwungen werden. Eine konservative Gartenideologie rückte die Verbundenheit mit der Scholle, die

Mutter als Siedlerfrau[91] (mit Schürze und Kind), den erbgesunden Nachwuchs und den von Kleinsiedlungsorganisationen gegängelten Untertan in den Mittelpunkt. In der neuen Arbeiter-Wohnstadt Voslapp mußten zahlreiche Besitzer zwangsweise ihre 30
Siedlung verlassen, weil sie „absolut kein Interesse für ihre Kleinsiedlerstelle besaßen", und die Gärten angeblich „vollkommen vernachlässigten"[92]. Mit einem „10.000 RM-*Programm*" wurden die nicht mehr dem traditionellen Gartenideal entsprechenden Freiflächen durch die *Marine* rekonstruiert und der offiziellen Ästhetik angepaßt[93]. Auch Gärten wurden geregelt, verordnet und kontrolliert. Individuelle Gartennutzung fand bei *Siedlerbund* und *NSDAP* keine Befürworter. Gärten als Stätten von Müßiggang und Erholung, Spiel und Sonnenbad, paßten nicht in das nationalsozialistische Konzept. Das Gartenideal im *Wohnsiedlungsgebiet Wilhelmshaven-Jeverland* war der traditionelle Bauerngarten mit schnurgeraden Hecken, symmetrischen Anpflanzungen und streng vorgegebener Aufteilung.

Die Leerformel von „Licht, Luft und Sonne"[94] in den Gärten legte sich kontrollierend über die Grünplanung, überwacht vom *Ortsgruppenleiter* der *NSDAP*, vom *Gemeinschaftsleiter* und vom *Fachlehrwart des Deutschen Siedlerbundes*. Soziale Kontrolle verbot, Hecken, Sträucher und Bäume „auf eigenes Risiko" zu pflanzen, oder gar Pflanzen „in die Siedlung hinzuschmuggeln", denn Hecken und Bäume mußten nach Straßen 32
einheitlich gepflanzt werden[95]. „Licht, Luft und Sonne" standen für großstadtfeindliche 33
Ideologie und dafür, daß Kinder nicht mehr in dunklen Höfen spielen mußten[96]. Die deutsche Mutter sollte ihr Kind in Sicherheit wissen, um „unbekümmert ihren häuslichen 161
Pflichten nachgehen zu können".

Die Pläne des Bremer *Gartenbauinspektors Friedrich Eisenbarth* für Voslapp und 30
Cäciliengroden belegten die Reglementierung der Gartenarchitektur. Selbst die Freiflä- 34
chen militärischer Bauten, soweit nicht für Aufmärsche erforderlich, wurden als Bauerngärten dekoriert. Die Ausweichvermittlung Antonslust bekam noch im Kriege Obstgärten, Ziergarten, Sitzgarten und strenge Beetaufteilung nach Bauernart.

Friedrich Nicolaus vom *RAM* sah die Notwendigkeit, Gartengestaltung frühzeitig in den Planungsablauf einzubeziehen, um „Natur- und Menschenwerk" durch Grün zu verbinden. Obwohl die Gartenplanungen von *Heinrich Wiepking-Jürgensmann* und *Wilhelm Hübotter* für die neuen Gemeinschaftssiedlungen auf der Gartenkultur der 20er Jahre beruhten, stülpte *Friedrich Nicolaus* NS-Ideologie darüber: „Wie das Blut die Adern des Körpers auffüllt, so soll das Grün die neue Stadt durchdringen. Je besser Stadt und Grün miteinander in Beziehung gesetzt sind, um so klangvoller wird die Musik eines Bebauungsplanes schon in der Zeichnung oder im Modell und viel mehr dann schließlich in der vollendeten Stadt auf den Menschen zu wirken vermögen"[97].

Auch in der Gartenarchitektur kämpften die maßgeblichen Städtebauer ab 1933 gegen Individualismus und angeblich „verworrenen Liberalismus". Öffentliche Grünanlagen sollten durch Symmetrie und Wiederholung Form und Aussage bekommen. Grün sollte auch Mittel nationalsozialistischer Inszenierungen werden: Hecken als Ersatz für aufmarschierte Reihen, Bäume als Ersatz für flatternde Fahnenreihen, Wege nach vorn

109 orientiert, als Ersatz für Paradestraßen. Groß-Wilhelmshavens *Kulturforum* sollte mit der
Gartenarchitektur eine nützliche Ergänzung bekommen. Stadtplätze seien Plätze mit
eigenen Gesetzen, gegen die man sich nicht ungestraft versündige – so der Fachbeauf-
tragte des *Deutschen Heimatbundes, Werner Lindner*[98]. Er sprach sich für das symmetrische
Hausbaumpaar mit Ruhebank und für aufgereihte Baumalleen aus. *Lindner* kämpfte
gegen „abwegige Art von Stadtgärtnerei"[99], „Bretzelwege" und gegen „englische
9 Landschaftsgärten". Wilhelmshavens Kurpark von 1870, Rüstringens Stadtpark von
1915–19, und die fortschrittliche Grünpolitik der Weimarer Republik hatten im „III.
Reich" keine Chance auf Fortsetzung. 1933–45 wurde kaum öffentliches Grün und
kaum ein Park angelegt. Es blieb bei widersprüchlicher Propaganda, die eine einheitliche
nationalsozialistische Grünideologie nicht erkennen ließ. Es blieb bei verbalen Leerfor-
meln[100] vom Bachlauf, der die Heimat gestaltet und die Menschen beglückt, von
Baumkronen, die zu einem glücklichen Rahmen verhelfen, von lichter Berankung, die
Fenster umspielt, von stets neu beglückenden Wundern, von Lauben, die zur
Gartengestaltung gehören. Fotos nach dem Schema: gut gestaltet/schlecht gestaltet,
lieferten das hier passende Dogma nationalsozialistischer Gartenarchitektur.

★

30, 34 Für die Siedlungen Voslapp und Cäciliengroden hatte die *Jade* den *Gartenbauinspektor*
Friedrich Eisenbarth beauftragt. Die eigene Scholle der Kleinsiedler sollte bewirtschaftet
und aufgeteilt werden. Da viele der erwarteten Rüstungsarbeiter aus großen Städten und
anderen Regionen kamen, bedurfte es der Anleitung. Schon zum Baubeginn von
Voslapp im Februar 1938 stellte die *KMW* Hausmodelle aus und propagierte den
33 bäuerlichen Kleinsiedlungsgarten. Zur Kanalisierung der gewünschten gärtnerischen
Nutzung wurden den neuen Bewohnern alle Gartengeräte, Schubkarren, Spaten, Harke,
Forke und Hacke, versprochen. Sechs Hühner sollten ebenso im Stall stehen wie
wahlweise ein Lamm, eine Ziege oder ein Ferkel[101]. Alles das war in den veranschlagten
Baukosten von 7.500 bis 9.500 Mark enthalten. „Die Gärtner sind gekommen und legen
... einen Mustergarten an, der für spätere Bepflanzung der anderen Grundstücke als
Vorbild dienen soll. Bei der Frühlingsstimmung, die in dieser Zeit über unserer
Landschaft liegt, denken wir schon ein Jahr weiter und malen uns aus, wie es dann in
Voslapp aussehen wird, wenn dann auch dort Bäume und Sträucher zu sprießen anfangen
und Blumen Gärten zieren werden. Dann werden die Volksgenossen, die dort draußen
eine Heimat erhalten, mit leuchtenden Augen den Frühling sehen"[102]. Nachdem dann die
ersten Bewohner eingezogen waren, wurde nicht etwa individuell gepflanzt: „Der
deutsche Mensch ist nicht mehr sich selbst überlassen, die Zeit des hemmungslosen
Liberalismus mit seiner scheinheiligen und falschen Freiheit ist gottlob überwunden, und
sie wird niemals wiederkehren"[103].

Architekt *Willi Lübbers* und der *Deutsche Siedlerbund Gaugruppe-Weser Ems*, instruierten
die neuen Bewohner Voslapps im Juli 1938 im Feierraum des Arbeiterlagers über die

Richtlinien für Gestaltung und Anlage der Gärten, daß das „Stückchen Land, das zu jeder Wohnung gehört, in Schuß kommt“[104]. Obwohl die Kleinsiedlungen als Eigentum der „Volksgenossen“ propagiert wurden[105], fand die Anpflanzung der Bäume und Sträucher gemeinschaftlich und unter Anleitung statt.

In Cäciliengroden beeilte man sich, im Herbst 1938 gleichfalls mit Hilfe von *Eisenbarth* *35*
Gartenanlagen herzurichten, um so den Bewohnern die Arbeit der Bestellung ihrer Gärten im kommenden Frühjahr zu erleichtern[106]. Architekt *Friedrich Heuer* begründete bei der Richtfeier den dörflichen Charakter und die „Grüneinhüllung“ des Vorortes: „Jeder soll sein eigenes Stück Land bekommen, um mit dem Boden zu verwurzeln, um auf seiner Heimatscholle festzustehen und sich dort täglich Kraft für sein Tagwerk zu holen“[107]. Es wurden nicht nur Häuser eingehüllt, sondern auch kritisches Bewußtsein. Grün sollte Kraftspender für die Rüstungsarbeit sein.

Ab August 1938 rief *Eisenbarth* die neuen Voslapper Siedler verschiedentlich zusammen und erläuterte seine Gartenarchitektur anhand von Plänen und Fotos. Voslapp solle „schön und landschaftsgebunden“ werden. Das Wichtigste aber war: „Die Siedler werden nicht nur in ihrem eigenen Interesse versuchen, möglichst viel aus dem ihnen zugeteilten Boden herauszuholen, sondern dazu auch verpflichtet, um unsere Volkswirtschaft von unnötiger Einfuhr zu entlasten“[108]. Für die Kleinsiedlungen stand nicht Schmuck im Vordergrund, sondern wirtschaftlicher Vorteil für die Bewohner – so tönte die Propaganda. Nicht die großen Grünanlagen der *Neugestaltungsstädte* waren das Vorbild für Kleinsiedlungen, sondern das niederdeutsche Dorf mit seinen Bauerngärten und niedrigen Obstbäumen. Hohe Straßenbäume und öffentliche Grünanlagen, Spielplätze und Parks, kamen für die Kleinsiedlungsgebiete nur ausnahmsweise in Betracht.

Kleinsiedlungen und Einfamilienhäuser waren für das „III. Reich“ besonders geeignet, traditionelle Architekturströmungen durchzusetzen. Gärten sollten angeblich heimatlose Großstädter an den Boden binden und vorindustrielles Leben spiegeln. Dem „entwurzelten“ Großstadtmenschen wurde der „verwurzelte“, heimatverbundene „Volksgenosse“ gegenübergestellt. Die Adolf-Hitler-Siedlung und die Siedlungen in Cäciliengroden, *24, 35*
Voslapp, Sanderbusch, Himmelreich, Kniphausersiel, bekamen noch heimatschutzorien- *32, 22*
tierte Einfamilienhäuser mit Gärten. Die nicht mehr realisierten Planungen für Schilldeich, Langewerth, Breddewarden, Altengroden-Süd, Utters, Accum, Fedderwarden, Moorsum, Burg Kniphausen, Gödens, Cleverns usw. bewiesen, daß Kleinsiedlungsideologie und Einfamilienhausbau im Siedlungsgürtel um Wilhelmshaven fortgesetzt werden sollte.

Auch zu dem großflächigen Kleingartengelände zwischen Hohewerth und Heiligengroden, das im *Wirtschaftsplan Wilhelmshaven-Friesland vom Mai* 1940 ausgewiesen *129*
wurde, kam es nicht mehr. Der große Zentralfriedhof Wilhelmshaven-Nord zwischen Fedderwardergroden und Utters[109] blieb, ebenso wie der Friedhof südlich Breddewarden, gleichfalls in der Planungsphase stecken.

★

Völlig anders und erheblich schwieriger gestaltete sich Gründekoration in den
Altstadtgebieten Wilhelmshavens. Anknüpfend an die 20er Jahre bemühte sich die *Jade*
bei der Schließung von Baulücken in der Werftstraße, von der Straßenflucht
zurückgesetzte Bebauungen und Vorgärten zu schaffen. Optisch mußte Gründekoration
dazu herhalten, den Übergang zwischen geschlossenen Randbebauungen und aufgelok-
kerten Zeilenbauweisen zu kaschieren. Bei den neuen Schmuckplätzen der Innenstadt
läßt sich eine völlige Abkehr von der Grünpolitik der Ära *Wagner/Migge/Hahn*
53 nachweisen. Der Otto-Meentz-Platz hatte sein Vorbild im 1868 von *Friedrich Hitzig* und
9, 11 *Johann Heinrich Friedrich Adler* konzipierten Adalbertplatz[110] mit Doppelstraße und
147 gleicher Breite. Der 200 m lange Platz zwischen dem 1939 gebauten *HJ*-Heim im
Norden und einem Ehrenmal an der Peterstraße sollte als Appellplatz für die Jugend
53 dienen[111]. Der Lessingplatz war gleichfalls als Appellplatz für die geplante Schule gedacht.
Ähnlich auch die Plätze westlich der Werftstraße.

Der *Bauverein Rüstringen* war als 1903 gegründete Baugenossenschaft lange den Idealen
von Nachbarschaft und Gartenstadt besonders verbunden. In Abkehr von der beengten
Großstadt der Gründerzeit wollten die Baugenossen grüne Straßenräume, offene
Bebauung, Vorgärten, Gemeinschaftseineinrichtungen und Nutzgärten. Die Planer bis
1928 (*Hakenholz* und *Brandes*) trafen auf feste Vorstellungen über Gartengestaltung und
Freiraumnutzung in Siebethsburg. Sie orientierten sich an den deutschen Gartenstadt-
theoretikern *Wilhelm Heinrich Riehl* und *Theodor Fritsch*[112]. Leben im Grün, Ruhe unter
Bäumen, Selbstversorgung im Garten, waren gesellschaftspolitische Ziele der „Elite der
Arbeiterschaft“ – der selbstbewußten, keineswegs großstadtfeindlichen, Siebethsburger
Baugenossen[113]. Stadt- und Landleben sollten miteinander verbunden werden. Von den
in Wilhelmshaven tätigen Architekten *Martin Wagner*[114], *Leberecht Migge*[115] und *Willy*
Hahn[116], waren fortschrittliche Grünkonzepte bekannt. Auch zu Anfang des „III.
57 Reiches“ verfolgte der *Bauverein Rüstringen* seine Ziele. Private Gärten zur Selbstver-
sorgung gruppierten sich um Gartenlauben[117]. Sträucher, Obstbäume und Hecken,
59 wurden vom *Bauverein* gleich mitgepflanzt. Straßenplätze bekamen gestaltete Vorgärten
und Großgrün. Freiflächen wurden bepflanzt, und mit Kunst und Brunnen deko-
riert.

Fritz Höger, der Architekt von Neu-Siebethsburg, orientierte sich keinesfalls bei der Grünplanung an symmetrischen Achsen[118] und Aufmarschplätzen, sondern an seiner heimatverbundenen Vorstellung von harmonischem Gefüge, positiven Körpern und behaglichen Gärten[119]. Wie Räume sollten die Gärten gestaltet sein, mit „Material, das Ihr euer Material nennen dürft, aus dem Stoff der Mutter Erde und der unglaublich reichhaltigen Pflanzenwelt“. Gartenarchitekten sollten auch Gärtner sein, wie ein Hausarchitekt auch Bauhandwerker sein sollte. Haus und Garten sollten zu einer Einheit verwachsen mit viel freier Natur und mit Tieren. *Fritz Höger* sorgte sich früh um den Zustand der vorhandenen Bäume an den Straßen. Er kämpfte um jeden alten Baum in

Marsch und Moor. Pappel, Weide, Erle, Esche, Birke und Eiche gehörten zu seinem gartenarchitektonischen Werkzeug. Der *Bauverein Rüstringen* bezog getreu seinen genossenschaftlichen Gedanken die Bedürfnisse der Gemeinschaft in seine Grünplanung ein.

Die gemeinschaftliche „Bleiche" hinter dem Haus, der Spielplatz mit Sandkasten für die kleinen Kindern, der Jugendspielplatz für die Heranwachsenden, wurde in den Kleinsiedlungsgebieten des „III. Reiches" jedoch eingespart, obwohl „besonders kinderreiche, erbgesunde Familien aufwachsen" sollten[120]. Die in Siebethsburg immer noch vorherrschenden fortschrittlichen Grünideen mögen daraus zu erklären sein, daß die Siedlung seit ihrem Bestehen sozialdemokratisch geprägt war. Hier waren die Anhänger der *Freiluftbewegung* und *Naturfreunde* zu Hause. Seit 1911 hatte der *Verein für Gesundheitspflege und Naturheilkunde Wilhelmshaven-Rüstringen* eine Licht-, Luft- und Badeanstalt betrieben[121]. Hier, wie auch im Licht- und Luftbad des *Arbeiter-Turnvereins Rüstringen* an der Genossenschaftsstraße und im Stadtpark-Luftbad[122] waren *Naturfreunde* und Freikörperkultur-Anhänger etabliert. Die sozialreformerische Natur- und Grünideologie war den Nationalsozialisten nicht geheuer, obwohl es gemeinsame Wurzeln gab. Das „III. Reich" fürchtete zu recht, daß sich in Arbeiterturnvereinen und *Naturfreunde*-Gruppen Widerstand formieren könnte. Das Licht- und Luftbad an der Genossenschaftsstraße wurde bald zum *SA*-Schießstand umgebaut.

Der seit 1933 von den Nationalsozialisten gleichgeschaltete *Bauverein* sorgte selbst für Gleichschaltung von Heckenschnitt und Unkraut-„jäten". „Unser ganzes Bauvorhaben zeigt eine gewisse Diszipliniertheit" und dieses sollte nach dem Geschäftsbericht 1938 bei den Gartenanlagen zum Ausdruck kommen. Die vom *Bauverein* angelegten Gärten durften ab 1933 nicht verändert werden – selbst große Sträucher durften nicht hinzugepflanzt werden, denn sonst entstünde ein „buntes Durcheinander".

Grün war in der Weimarer Republik in Wilhelmshaven, und vor allem in Rüstringen nicht Dekoration, sondern soziales Grün[123]. „Der durch aufreibende Büro- oder Werkstatttätigkeit entnervte Mensch muß einen Flecken freier Natur haben, wo er sich erholen, wo er neue Kraft zu neuem Schaffen sammeln kann. Unserer Jugend, den Kleinen und Allerkleinsten muß Gelegenheit zu frohem Spiel auf freier Fläche ... geboten werden"[124]. Grünflächen sollten nicht der Repräsentation dienen, sie sollten Volksparks sein. *Leberecht Migges* Plan von 1915[125] für Rüstringens Stadtpark, zeigte Anlagen für alle Bevölkerungsschichten: Sportwiesen, Tanzzelte, Luftbad, Tennisplätze, Schrebergärten, Kanäle, Bürgergärten, Spielflächen, Sommerwirtschaften und Turnplätze. Eine ähnlich fortschrittliche Konzeption brachte *Leberecht Migge* 1911 in seine Planungen für Oldenburgs Dobbengelände[126] und 1918 in den Rüstringer Sport- und Lagunenpark ein.

★

Gründekoration in Fedderwardergroden durch *Heinrich Wiepking-Jürgensmann*[127], den Gartengestalter des Tannenbergdenkmals (1927) und der Olympiabauten Berlin (1936), war weder an der Kleinsiedlungsideologie von Voslapp und Cäciliengroden noch an den fortschrittlichen Genossenschaftsgedanken orientiert. Kleinsiedlungen wurden den Nationalsozialisten zunehmend ein Ärgernis, da Gartenarbeit „Volksgenossen" keine Zeit mehr für Parteiarbeit ließ.

38 Fedderwardergroden wurde anfangs als „Grüne Stadt" bezeichnet[128]. Nicht das private, sondern das öffentliche Grün stand im Mittelpunkt. Brunnen und Plastiken sollten das Grün ergänzen. Ein innerer und ein äußerer „Grüner Ring" sollten vorhandene Wassergräben aufnehmen, und die alten Bauernhöfe berücksichtigen. Auch die Straßen vom Ring zum Stadtrand sollten Grünzüge werden. Vorgärten waren nicht privates, sondern öffentliches Grün. *Wiepking-Jürgensmann* lehnte strikt jegliche Zäune und Grenzbetonungen ab. Für *Karl Pfeiffer* vom *RAM*, dem das endgültige städtebauliche Konzept Fedderwardergroden zuzuschreiben ist, war es wichtig, die vorhandene Topographie, jeden Wasserlauf und jede Baumgruppe in die Planung einzuarbeiten[129].

Schon anläßlich der ersten propagandistischen Ausschlachtung von Fedderwardergroden am 17.06.1939, legte die *NSDAP*-Parteizeitung ihr Schwergewicht auf die Darstellung von Grünplanungen in den neuen Stadtteilen. Der ganze Ortsteil werde „am Rande durch einen Grünstreifen eingefaßt, der zudem die Bauernhöfe, die in ihrem Garten- und Baumbestand völlig erhalten bleiben sollen, untereinander verbindet". Die Aufgaben des damals schon sogenannten „Grünen Ringes" waren jedoch zuerst technischer Natur. Der von Grün begleitete Ringgraben sollte den Grundwasserstand für die Bauarbeiten senken und später das Regenwasser in die Maade leiten.

★

Natur-, Landschafts- und Vogelschutz spielten früh auch bei den Nationalsozialisten eine Rolle in der Gründekoration. Die Gebiete wurden aufgenommen und kartiert. Im Januar 1939 kamen 62 Wilhelmshavener Landschaftsteile unter Landschaftsschutz – natürlich nur dort, wo sie militärischen Absichten nicht hinderlich waren[130]. Bestandteil vieler Grünplanungen waren Hecken, die man gerne mit militärischen Attributen belegte. Der Wind war der „Feind", die „Macht" des Windes sollte „gebrochen" werden usw.[131]. Windschutzhecken dienten auch dem Vogelschutz. Stadterweiterung und Neugestaltung Wilhelmshavens führten zu erheblichen Bodenbewegungen und Eingriffen in die Natur. Alte Wehranlagen, Gräben, Wurten und Siedlungen wurden ausgegraben – für den Städtebauideologen *Friedrich Nicolaus* vom *RAM* Beleg eines wehrhaften Stadtbaus[132]. 1935 entstand das *Reichsnaturschutzgesetz*. Es spiegelte den starken Einfluß der *Heimatschutzbewegung* im „III. Reich". Die ehrenamtlichen Heimatgeologen, *Heinrich Schütte, Wilhelm Krüger* sowie die Naturschutzbeauftragten *Georg Heinrich Harms* und *Heinrich Oldewage* sicherten Bodenfunde[133], stellten 1939 die erste

Landschaftsschutzkarte auf und publizierten über Natur- und Vogelschutz. Die Bodenfundstellen wurden, wenn sie den Baubetrieb behinderten, jedoch rücksichtslos überbetoniert. So verschwanden durch die Bauarbeiten viele historische Wurten im Stadtgebiet[134]. Die 1938 in Wilhelmshaven neu eingerichtete *Provinzialstelle für Marschen und Wurtenforschung* diente auch als städtische *Bodenuntersuchungsstelle* sowohl der Vorbereitung von Bauprogrammen, als auch der historischen Küstenforschung[135]. Ab 1939 wurden geschützte Landschaftsteile zwar nicht in den Monumental- und Militärplanungen, aber immerhin bei den Siedlungsplanungen berücksichtigt. *Heinrich Wiepking-Jürgensmann* sparte in seinen *Landschafts- und Gestaltungsrahmenplänen* für die neuen Stadtteile Wurten aus, sah Gulfhäuser für Gemeinschaftszwecke vor und gruppierte Grün an Gräben und Bachläufen.

Die Durchführung der Grünplanung der öffentlichen Flächen lag bei der *Stadt Wilhelmshaven*, die der Wohnungsbauten übernahm der jeweilige Bauträger[136]. Ab 1939/40 lag die Gesamtbegutachtung allein bei *Heinrich Wiepking-Jürgensmann*, nachdem der städtische Leiter des *Gartenamtes Wolfgang Fresenius*, ein Schüler von *Wiepking*, eingezogen wurde[137]. *Wiepking* stellte seine Wilhelmshavener Grünplanungen sogar öffentlich vor. Am 04.07.1939 hielt er in der *Admiral-Scheer-Schule* eine Versammlung ab. *Wiepking-Jürgensmann* wurde im September 1941 von der *Stadt* gedrängt, seine Grünpläne fertigzustellen und zusätzlich für jede zehnte Familie in Volkswohnungen außer den Hausgärten noch Kleingärten auszuweisen. Die nachfolgenden Objektplanun-
gen lagen dann bei Gartenarchitekt *Wilhelm Hübotter* (Altengroden), *Wilhelm Heintz*[138] 43
(Breddewarden) usw. Auch Voslapp sollte grünplanerisch noch einmal überarbeitet 66, 30
werden, nachdem die *Jade* Gartenarchitekt *Friedrich Eisenbarth* 1939 „abserviert" hatte. Für Voslapp war der Bremer *Gartenbaudirektor Richard Homann* vorgesehen. Grünplaner des *Arbeitsstabs Speer* in Wilhelmshaven war *Max Karl Schwarz*, 1944–45. *Schwarz* wurde
auch 1949–50 noch als Gutachter eingeschaltet[139]. 122

Die ideologischen Wurzeln von *Heinrich Wiepking-Jürgensmann*[140] gingen auf *Wilhelm Heinrich Riehl* und die „Naturgeschichte des deutschen Volkes" zurück. *Riehl* wollte den Gegensatz zwischen Stadt und Land aufheben. *Wiepking* wollte „wieder ehrlich bäuerlich" denken, aber nicht ästhetisch: „Ästhetisch denkt kein Bauer". In bedrückender Propagandasprache beschwor er „urewige Volkstumskräfte" und „das älteste Ackervolk des Weltnordens".

Für die NS-Machthaber war Gründekoration wichtiger Bestandteil von Grundsatzerklärungen und Etatreden. *Oberbürgermeister Müller* bediente sich häufiger des *Martin-Wagner*-Begriffes von der „Grünen Stadt am Meer", verschleierte aber damit, daß die „militärischen Zweckaufgaben" die Stadt vom Wasser abdrängten. Grünkonzepte sollten den Verlust überspielen: ein Netz von Erholungswegen, Grünzüge als Promenaden, eigenes Gartenland für jede Wohnung[141]. Der Neubau der Strandhalle 1938 sollte 110
den Verlust von Grün und Wasser kompensieren. Gründekoration diente auch der 131
ideologischen Mystifizierung, und mußte für das Begehren zur Anwendung des *Gesetzes zur Neugestaltung Deutscher Städte vom* 04.10.1937 herhalten: „Die besondere Bedeutung

Wilhelmshavens als Reichskriegshafen an der Nordsee, wo sich das Heiligtum der Deutschen Nation, der Ehrenfriedhof, befindet, rechtfertigt wohl die Anwendung des Gesetzes"[142]. Wilhelmshaven sei Wallfahrtsort für Hunderttausende geworden.

Grün hatte nicht nur seine Begründung in Dekoration und Selbstversorgung. Die *Marineintendantur* wies die von ihr betreuten Wohnungsbaugesellschaften schon im Januar 1938 an, den vom „Führer und Reichskanzler geführten Kampf um die Ernährung des deutsches Volkes" zu unterstützen und Anbauflächen für Früchte in Mietergärten zu vergrößern[143]. Aus angeblich wehrwirtschaftlichen Gründen verpflichteten *RAM*[144] und *Marineintendantur*[145] die Wohnungsbaugesellschaften, Seidenraupenzucht (Fallschirmseide) und Angorakaninchenhaltungen (Wolle für die *Luftwaffe*) in den Mietergärten zu fördern.

7.4. Ausstattung und Einrichtung

Überwiegend orientierten sich Menschen an Kultur und Gewohnheiten nächsthöherer Schichten oder Klassen. Besonders im wilhelminischen Kaiserreich war dieses Verlangen ausgeprägt, als Großbürger so wohnen wollten wie Adelige, Handwerker wie Fabrikbesitzer, Arbeiter wie Beamte, Mannschaften wie *Offiziere*. Die Geschichte der Einrichtungsarchitektur ist eine lange Geschichte fortdauernder Imitationen. Zeitungen und Ausstellungen informierten über neueste Strömungen und verbreiteten moderne Ideen. Sozialer Status sollte an Kleidungsstücken, Eßgewohnheiten, technischen Geräten, Fahrzeugen und Wohnungseinrichtungen ablesbar sein. Der Begriff von Schönheit und guter Form war immer an die jeweiligen Verhältnisse und die aktuellen Zeitabläufe gebunden[146]. Wie die Architektur der Bauten, war auch die Wohnungsausstattung in Deutschland bis 1900 historisierend. Die Gründerzeitfassaden in Wilhelmshaven orientierten sich im Zeitalter *Wilhelms II.* an den Berliner Strömungen und an dem, was Fassadenkataloge boten. Schließlich importierten die zahlreichen Architekten aus der Reichshauptstadt weltstädtische Moden und Ansprüche auch in die Exklave Wilhelmshaven. Entsprechend wurden auch bürgerliche Vorlieben für Ausstattungen und Innenarchitektur transportiert. Kaiserliche Mariner und Werftbeamte kauften ihre Serienmöbel in Magazinen. Geplanter Innenausbau kam bestenfalls bei Versammlungsstätten oder im Schiffbau vor[147]. Die wenigen vorhandenen Fotos von Inneneinrichtungen der Offizierskasinos, Ballhäuser, Ausflugslokale und Strandhallen zeigen keine Unterschiede zu dem, was andernorts üblich war.

Regional hatte nach der Jahrhundertwende die *Oldenburger Landes-, Industrie- und Gewerbeausstellung* von 1905, an der *Peter Behrens* maßgeblich beteiligt war, Einfluß auf Alltagskultur, Warenerscheinungsbild und Publikumsgeschmack[148]. Mit *Martin Wagner* wurde 1911 ein führendes *Werkbund*-Mitglied[149] Baubeamter in Rüstringen. In dem von ihm mitgegründeten Wilhelmshavener *Kunstverein* organisierte er Ausstellungen zu Architektur und Gartenbau, Parkanlagen und Friedhöfen, Raumschmuck und Haushaltsgegenständen[150]. *Wagner* beschickte auch die Kölner *Werkbund-Ausstellung* von 1914 und stellte im *Bremen-Oldenburger Haus* seine allerdings mehr städtebaulichen Entwürfe aus[151]. In Vorträgen, Zeitungsartikeln und Ausstellungen forderte er im Sinne des *Werkbundes* die Gestaltung von Massenprodukten.

Wohnungen für Marineangehörige waren in der Weimarer Republik äußerst knapp und, falls überhaupt vorhanden, auch knapp bemessen. Noch 1933 beklagte das *Kommando der Marinestation* der *Nordsee* den Mangel an Wohnungen unter 50 Mark Miete für Soldaten und Beamte der unteren Gehaltsgruppe[152]. Viele Soldaten wohnten mit ihren Familien jahrelang in Wilhelmshavener/Rüstringer Kasernen, die zu Woh- *12*
nungen umgebaut waren. Die Kasernen wiederum sollten nach nationalsozialistischem Willen im Rahmen der Aufrüstung ab 1935 wieder ihrer militärischen Zweckbestimmung zugeführt werden.

Wagners Nachfolger, *Willy Hahn*, kritisierte offenherzig die „kapitalistische Wohnungsproduktion"[153] und erreichte einen Ratsbeschluß, mit dem er sein *Hochbauamt* um ein eigenes *Wohnungsamt* erweiterte. Die engagierten Beamten der sozialdemokratisch regierten Stadt Rüstringen inspizierten die Wohnungen, registrierten Belichtung, Belüftung, Heizung, Mietpreis und halfen Mißständen ab. *Hahn* entwarf, wie z.B. auch *Peter Behrens*[154] in Berlin, einfache Möbel für Kriegsgetraute. Die *Stadt* gab 60.000 Mark und der *Arbeitgeberverband für das Tischlergewerbe* übernahm die Ausführung von 45 Einrichtungen. Die sachlichen Mustermöbel wurden 1918 in der Kunsthalle ausgestellt. Der Preis der Küchenmöbel lag bei 200 Mark, die Schlafstubeneinrichtung kostete 1.000 Mark[155]. Die Möbel verkauften sich schlecht, möglicherweise wegen der Geldknappheit, möglicherweise wegen des schlichten Anspruchs, möglicherweise, weil sie nicht dem Publikumsgeschmack entsprachen.

★

Auffällig unterschied die *Marine* ab 1933 die hierarchische Gliederung von Wohnungen für Marineangehörige, für Arbeiter, für Angestellte und für Beamte. Wohnungsbedarf für Arbeiter wurde nach 1933 mit „gesundheitlich, sittlichen und innenpolitischen" Gründen belegt. „Drei- bis vierräumig müßten die Wohnungen schon sein[156].
Deutlich war die Abkehr von der sparsamen Kleinwohnungspolitik der Weimarer *17*
Republik. Sie war auch von den Wohnungsbauträgern gewollt. Für 1936 beispielsweise waren folgende Wohnungen vorgesehen:

- für *Offiziere* (über 80 Mark Miete): Göringstraße (heute Parkstraße)
- für *Unteroffiziere* (20 bis 55 Mark Miete): Mühlenweg, Totenweg, Zedeliusstraße, Helgolandstraße
- für Werftarbeiter: Halligenweg, Bismarckstraße, Adolf-Hitler-Siedlung, Dietrich-Eckart-Straße, Saarbrücker Straße, Spiekeroogstraße
- für Beamte und Angestellte: Bismarckstraße, Herbartstraße, Baltrumstraße[157].

Garagen für Autos – Wohlstandssymbole schon damals – (die ersten *KDF-Wagen* trafen am 21.02.39 in Wilhelmshaven ein) waren nur in Wohnblocks mit größeren Wohnungen – und vor allem für *Offiziere* oder Gewerbetreibende vorgesehen[158]. Ab 1939 allerdings wurden für jede 5. Wohnung Garagen vorgesehen. Für die *KDF-Wagen* sollte Fedderwardergroden 800 Garagen bekommen[159]. Eine Vermischung der militärischen Ränge in Wohngebieten war nicht vorgesehen. Der *Kommandierende Admiral* beschwerte sich selbst darüber, daß in Altengroden Arbeiter vorübergehend in *Unteroffiziers*-Wohnungen ziehen sollten[160].

Da bei den Wehrmachtswohnungen häufiger Mieterwechsel zu erwarten war, sollte die Ausstattung schon dauerhafter sein als allgemein üblich[161]. In der Tat wies das Wilhelmshavener Wohnungsbauprogramm einen beträchtlichen Standard auf. Dauerhafte Materialien, solide Baustoffe und bewährte Konstruktionen sollten zusätzlich ein Gefühl von Gediegenheit erzeugen. Die Angst vor großstädtischem Wohnungselend und die Sehnsucht der Marineangehörigen nach einem traulichen Zuhause und einer neuen Heimat wußten die Nationalsozialisten für ihre Ziele auszunutzen. Es bedurfte keines Beweises dafür, „daß die Arbeiter jemals außer unter dem Diktat der Not sich freiwillig für einen kulturellen Code der Kargheit entschieden hätten ... Einfach und sparsam lebte man als Arbeiter selbst bis zum Überdruß“[162]. Die vom Direktor des *Landesmuseums Oldenburg, Walter Müller-Wulckow*, herausgegebenen „Blauen Bücher“ mit ihren beliebten Geschmacksanleitungen[163] oder „Die Kunst im Dritten Reich“ wurden bestenfalls von Bildungsbürgern gelesen.

33, 41 Das Äußere der neuen Wohnsiedlungen und Heimstätten suggerierte dem neuangekommenen Werftarbeiter oder Marinesoldaten Schutz und baldiges Zuhause. Das Innere sollte Behaglichkeit und Wärme versprechen[164]. „Wer solch eine Wohnung sieht, wünscht seinen Möbelwagen schon vor der Tür“[165], appellierte die Werkszeitung der *KMW* an die Gefühle ihrer „Gefolgschaft“. Von vornherein war klar, daß *Bauhaus*-Möbel, internationales Design, Stahlrohrsessel und die funktionale Formsprache der 20er Jahre als Einrichtung nicht in Betracht kamen. Kunstgewerbe oder verschwenderischer Ausbau standen für Marineangehörige nicht zur Diskussion.

Nach dem Verbot der Gewerkschaften hatte deren „Nachfolger“, die *DAF*, 1934 das *Amt „Schönheit der Arbeit“* unter *Albert Speer* gegründet[166]. Es war anfangs nur für Betriebsstätten und Diensträume zuständig. Erst später – durch Zusammenschluß mit dem *Architekturbüro der DAF* und dem *Reichsheimstättenamt* – bekam es Einfluß auf den Wohnungsbau der *Hauptabteilung Schönheit des Wohnens*. Schon 1937–38 hatte das *Reichsheimstättenamt* Mustermöbelentwürfe herausgebracht[167], die sich an den 20er Jahren und dem *Werkbund* orientierten. In den zwei *Deutschen Architektur- und Kunsthandwerk-Ausstellungen* 1938 und Zeitungsartikeln wurden deutscher Wohnungsstil und Musterentwürfe für Tische, Stühle, Lampen und alle möglichen Ausstattungsdetails verbreitet[168].

Wilhelmshaven sollte für die Rüstung ausgebaut werden. Hierzu bedurfte es Wohnungen in ausreichender Anzahl und Größe. Ein reichhaltiges Marktangebot an

Möbeln sollte kurz vor dem Kriege die Wirtschaft und den Arbeitsmarkt nicht belasten und Rohstoffe schonen helfen. Die Stahlproduktion arbeitete für den Krieg. Selbst der *KDF-Wagen* kam nicht mehr auf's Fließband. Es fügte sich gut, daß schlichte Holzmöbel in das empfohlene, bodenständige Formgefühl der *NSDAP* paßten. Im Juni 1938, als in der neuen Großsiedlung Voslapp die ersten der 1.300 Wohnungen fertiggestellt wurden, 30
stattete die *Wohnungsbaugesellschaft Jade* zwei der Häuser mit Mustermöbeln der *DAF* aus[169]. Man wollte zeigen, wie die Wohnung am „praktischsten und zweckmäßigsten ist". Die Lokalzeitungen lobten, durch Fotos illustriert, bunte Gardinen, schöne Möbelformen, Läufer, Beleuchtungskörper, Beizfarben und schlichten Wandschmuck. Ausgiebig beschrieb der Wilhelmshavener Kurier am 21.06.1938 die Mustereinrichtungen, die durch Möbel aus Rüster, Gardinen, Kissen, Lampen, Tischdecken, den roten Kokosteppich und Keramikteller an der Wand „anheimelnd" wirke. Beim Musterschlafzimmer wurde herausgestellt, daß es von der *RDBK* als „besonders wertvoll" anerkannt worden sei. Tatsächlich aber müssen die Mustermöbel die knappen Räume der zwei Voslapper Siedlungshäuschen total ausgefüllt haben. Gar nicht zu dem handwerklichen Charakter der Voslapper Mustereinrichtungen paßte der Bakelit-Volksempfänger vor dem Fensterbrett. Rundfunk in jedes Haus zu legen, bedeutete Propagandamöglichkeit für die *NSDAP*. Die Zeitungen riefen dazu auf, regelmäßig und überall die Rundfunkreden *Hitlers* zu hören[170].

Im Juli 1937 hatte die *Wohnungsbaugesellschaft Jade* eine der 1.000 Wohnungen im Projekt „rund um den Krähenbusch probeweise installiert" und der Öffentlichkeit vorgestellt. Dusche, fließendes Warmwasser, kombinierter Gas-/Kohleherd und Anschlußmöglichkeiten für Volksempfänger belegten den Ausbaustandard[171].

Die wirtschaftliche Situation des „III. Reiches" ließ Kühlschrank und Elektrogeräte, die längst erfunden und im Ausland gebräuchlich waren, kaum zu. Gern wurden die Vorzüge der Bescheidenheit herausgestellt. Die Werkszeitung der *KMW* warnte, daß man auch zum Sklaven seiner Möbel werden könne. Es gehörte zur Strategie des „III. Reiches", die aus den Städten kommenden Frauen für das ländliche Siedlerleben umzuerziehen. Das *Deutsche Frauenwerk* stellte für die Wilhelmshavener Siedlerfrauen eine eigene „Siedlerfrauenberaterin" ein. Sie leitete an und beriet in Hauswirtschaft, Mütterdienst, Gartenbau und Einrichtungsfragen[172]. Im Juli 1939 propagierte der *Reichsstand des Deutschen Handwerks* in einer Wanderausstellung „Deutsche Werkstoffe im Handwerk"[173] in der Wilhelmshavener Kunsthalle – angeblich zeitlose Gestaltung und materialgerechte Verarbeitung von Holz. Schirmherr *Göring*, Beauftragter für den *Vierjahresplan*, hatte sich zum Ziel gesetzt, „neue Werkstoffe" zu zeigen – d.h., alles dafür zu tun, metallische Rohstoffe zu schonen.

Jährliche *Arbeitsschlachten, Handwerkertage, Gaukulturwochen, Frühjahrsoffensiven, Gau-* 153
tage der Technik usw. machten offenbar, daß es den Nationalsozialisten um Aufrüstung und Bauprogramme ging, erst danach um Schönheit des privaten Wohnens. Formgestalter und Innenarchitekten wurden durch das *Reichsheimstättenamt* verdächtigt, Wohnstil nur bei der Oberschicht durchzusetzen[174]. Die *NSDAP*-Propaganda tönte, daß es bei

der Einrichtung von Wohnungen nicht um Ästhetik oder Konsum gehe, sondern „jeder einzelne muß auf seinen höchsten Leistungswert gebracht werden". Wohnkultur sollte, so schrieb es immer wieder die Werkszeitung der *KMW*, gesundes, glückliches Familienleben schaffen. Durch Sprossenfenster fallendes Licht, bäuerliche Gardinenumrahmungen, rustikaler Bilderschmuck, einfache Holzmöbel, saubere Schondeckchen und das allgegenwärtige Bild des Diktators symbolisierten die Wohnungsgemütlichkeit des „III. Reiches".

39–42 Die *Kriegsmarine* bevormundete auch die Werftarbeiterfamilien der *Neuen Stadt* Fedderwardergroden[175]. Die Wohnungen seien – Schmuckkästchen, die Beheizung geschehe durch „gute Dauerbrandöfen" und die Häuser seien „so gestellt, daß Licht, Sonne und Luft von allen Seiten" herankomme. In einer Elektrolehrküche sollten die Hausfrauen die Bedienung ihres Elektroherdes lernen. Sanft, aber deutlich, wies die Werftarbeiterzeitung darauf hin, daß die Familien „Freude daran haben, diese Wohnungen zu pflegen. Sie wohnen in Musterbauten des neuen großen Wohnungsplanes".

In den lokalen Zeitungen stellte die Propaganda immer wieder die bürgerliche Behaglichkeit der Wohnungen heraus, wohl auch, um andere Unzulänglichkeiten zu vertuschen (schlechte Straßenverhältnisse, fehlende Beleuchtung, mangelnde Einkaufsmöglichkeiten, nicht fertiggewordene Schulen usw.) Die Entwurfsleitbilder des *Reichsheimstättenamtes* für die Inneneinrichtungen kompensierten Mängel der baulichen Hülle – zumindest eine Zeitlang – bis auch der Außenraum der Siedlungen ansehnlich sein würde. So wurden beispielsweise in Fedderwardergroden auch kleinste Mietwohnungen von der Werftarbeiterzeitung zu „Schmuckkästchen" erklärt[176]. Technische Errungenschaften wie Dauerbrandöfen, Ausguß und Duschen, wurden ausgiebig gelobt. „Die Gardine ist die Visitenkarte der Hausfrau" und Bettkissen sollten nicht in den Fenstern zur Straße gelüftet werden, sondern ggf. zum Garten[177]. Auch Einrichtung wurde zum Instrument sozialer Kontrolle mißbraucht.

Die Besuche von *Oberwerftdirektor* und *Großadmiral* in den neuen Wohnungen wurden propagandistisch breit ausgewalzt. Zeitungen brachten maßstabsgenaue Grundrisse aller Wohnungstypen und animierten ihre Leser und die künftigen Mieter, ihre Möbel in Pappe auszuschneiden und in den Grundrissen hin- und her zu schieben[178]. Dies war Wohnungsfürsorge, wie sie die *KMW* verstand, und die dem „Führer" zu danken sei.

★

Grundsätzlich anders wurden öffentliche und halböffentliche Gebäude, Werft- und Marine-, Partei- und Verwaltungsbauten ausgestattet. Muster taugten wenig. Sachliche Formen, historische Versatzstücke, heimattümelnde Traditionen, weltläufiger Dampferstil[179], selbst Kunststoffe, wurden eingesetzt, wo sie paßten.

130 In der Arbeiterkantine des Fliegerhorstes Upjever beispielsweise waren überlange, blankgescheuerte Holztische, in Viererreihen geordnet, Holzstuhl an Holzstuhl.

Aufgereihtes dickwulstiges Eßgeschirr war ausgerichtet auf das Bild des „Führers" und sein Hoheitssymbol. Die Kantine unterschied sich kaum von Feierräumen in Gemeinschaftslagern der *KMW*. Die Muster von symmetrischen Massenaufmärschen, von Berliner, Nürnberger und Münchner Parteibauten wiederholten sich beklemmend – diesmal im einzig möglichen Material, Holz. Bar und Kegelbahn für die Flieger – die Lieblingskinder des „III. Reiches" – waren vergleichbar feudal ausgestattet und an großbürgerlichen Einrichtungen orientiert.

Die Entwürfe zum Innenausbau der Dienstvilla von *Stadtbaudirektor Walter Hunzinger* zeigten den gediegenen Wohlstand der NS-Führungsclique. Detailliert waren vom Biedermeierzimmer bis zur Eckbank im Weinkeller, von der Eingangsdiele bis zum Damenzimmer, alle Ausstattungsqualitäten vorgegeben.

Das *NSV-Müttererholungsheim* bei Schledehausen wies einfache Formen und schlichte Gestaltung auf: überlange, nach vorn ausgerichtete Tische, Stühle und Lampen, diesmal geschmückt durch Blumen und Tischdecken – der angemessene hygienische Rahmen für eine Wohlfahrtseinrichtung.

Andere öffentliche Gebäude – wie das *HJ*-Heim am Krähenbusch – lassen noch heute *147*
den NS-typischen Einrichtungsstil erahnen. Architekt *Heinz Däke* entwarf eine zweigeschossige Eingangshalle, um Dekoration mit weit herunterhängenden Fahnen zu ermöglichen. Nur so kamen die Symbole zur Wirkung. Fahnenreihen sorgten für die gewünschte Emotionalisierung[180].

Das *SS*-Heim Kirchreihe und das *NSDAP* Schulungsheim Ruttelerfeld, ausgestattet mit offenem Kamin, Holzbalkendecken und bequemem Mobilar, spiegeln noch heute den feudalen Stil und die Ansprüche der nationalsozialistischen Führung.

Auch die 1938 von *Walter Hunzinger* entworfene Strandhalle belegte, daß Innenein- *110*
richtungen ausschließlich auf die Nutzer bezogen wurden: hier für *KDF*-Gruppenreisende und Passagiere von Schiffahrtslinien. Rustikale Umgebung wäre völlig deplaziert
und hätte lächerliche Wirkung erzeugt. Die opferschalenähnlichen Lampen ähnelten den *131*
Umgangsleuchten im Berliner Olympiastadion.

Fritz Högers Rüstringer (ab 1937 Groß-Wilhelmshavener) Rathaus von 1928/29 galt früher als Beispiel eines besonderen norddeutschen Klinker-Expressionismus. Es war von
den Nationalsozialisten nicht umzudeuten. Aufgepappte Adler, montierte Hoheitszei- *112*
chen, nationalsozialistischer Bilderschmuck und Fahnen konnten die grundsätzliche andere Architektursprache von *Fritz Höger* nicht völlig umdeuten[181].

Beispielhaft für „Schönheit der Arbeit" waren die Gemeinschaftseinrichtungen in der
Werftarbeiter-Stadt Sande-Neufeld[182]. Tausende Werftarbeiter teilten sich auf 6 große *50*
Feiersäle. Sichtbare Holzkonstruktionen, große Sonneräder, portalähnliche Podeste und nach vorn ausgerichtete Tischreihen hatten das Ziel, die Identifikation der hier unter unwürdigen Bedingungen wohnenden, zumeist kriegsgefangenen Werftarbeiter mit „ihrer Werft" zu stärken. Handwerkliche, ländliche Formen ließen sich gut mit dem betrieblichen Führerprinzip verbinden. „Gleichheit" sollte vorgespiegelt, aber keinesfalls
politisch eingeräumt werden. Die Wohnverhältnisse der Zwangsarbeiter von Sande- *49*

Neufeld waren miserabel. Überbelegte Zimmer und Massenwaschgelegenheiten auf den Trockenböden waren das Gegenteil von „Schönheit der Arbeit". Lange Flure und aufgereihte Massenquartiere waren so entworfen, daß sie nach dem Kriege durch Einziehen weniger Mauern Familienwohnungen ergeben sollten. Rüstung, Krieg, Sieg und anschließende friedliche Zeiten waren im Grundriß eingeplant.

★

Die Ausbau- und Einrichtungszeichnungen für das von *Kurt Geisenhainer* entworfene
75 *Marinelazarett Sanderbusch* von 1937 sind noch fast vollständig erhalten[183]. Die Krankenzimmer und Behandlungsräume für Mannschaften sollten seriell ausgestattet werden. Für die höheren militärischen Ränge des *Marinelazaretts* jedoch planten die Werkstätten für Möbelbau *Ernst Tietjen*[184] genaueste Einzelheiten in vielen Details:

- Die Geschwaderarzt-Zimmer bekamen Holzvertäfelungen, Einbaumöbel, Heizkörperverkleidung, Verdunklungen, Bilderschmuck, Lehnsessel, Bücherborde und eingebaute Waschtische. Bilder von der See, Lampen mit flachen Glasschalen und geblümte Stores waren der Rahmen für Repräsentation und ärztliches Standesbewußtsein.
- Der *Unteroffiziers*-Speiseraum war getrennt vom Mannschaftsspeiseraum. Geputzte Wände, rustikale Holzvertäfelungen und schmiedeeiserne Türbeschläge präsentierten einen völlig anderen Formenkatalog. Wappen mit Eichenlaub dekorierten die Wandvorsprünge. Unter den hölzernen Kassettendecken hingen an Ketten Wagenräder mit runden Lampenschalen. Buntkarierte Stores, Übergardinen, einfache Holzstühle und Tische signalisierten „Volksgemeinschaft" und „Gleichheit".
- Die Beratungszimmer der marineärztlichen Ränge waren feudal mit Clubsesseln, Ablagetischen, Stores, Gardinen, gestickten Wandteppichen und Landschaftsbildern ausgestattet.
- Der zentrale Versammlungsraum für das *Marinelazarett Sanderbusch* wurde unter das Dach gelegt, um die charakteristische, bei Gemeinschaftshäusern übliche, gewaltige Dachkonstruktion[185] sichtbar zu lassen. Auf einem Podest war ein nationalsozialistischer „Altar" geplant: eine Holzwand mit Hoheitssymbol, 3 Schwertern, aufgesetztem Eichenlaub und Palmendekor, darüber schwere Eichenbalken als Türsturz. Ein nationalsozialistischer Feierraum, wie er besser vom *Amt „Schönheit der Arbeit"* nicht hätte entworfen werden können. Die architektonische Ausrichtung sollte den feierlichen Charakter in Analogie zu einem Kirchenschiff unterstreichen.

Ärzten, *Offizieren,* Matrosen, Kapitänen, jeder Hierarchiestufe war eine besondere Einrichtung zugewiesen. Oben symbolisierte weltläufige repräsentative Sachlichkeit die Herrschaftsansprüche, unten überwogen bäuerliche Stilelemente für die lenkbare „Gefolgschaft". Auch Inneneinrichtung sollte dazu beitragen, die *Kriegsmarine* im militärischen Sinne nationalsozialistisch zu gliedern.

⋆

Noch klarer ließ sich das schichtenspezifische Denken und Handeln von politischer Führung und Marineführung bei den Wilhelmshavener Kasernenbauten ablesen. Die 1937 – 38 von *Heinz-Günther Thees* vom *Marinestandortbauamt* gebauten *Bonte*-Kasernen waren keine auf Dauer angelegten Truppenunterkünfte. Die Zerstörer-Besatzungen „wohnten" auf den Schiffen. Nur im Fall des Austausches der Besatzung benutzte man die *Bonte*-Kasernen als Wohnheim. Die Inneneinrichtung spiegelte das Prinzip von Befehl *82*
und Gehorsam. Das Kommandanten-Sitzungszimmer war mit seinem langen Arbeitstisch auf das Bildnis des *Großadmirals* ausgerichtet. Der Befehlshaber war somit „anwesend". Parkettfußboden, Holzbalkendecke, aufgereihte Lampenschirme, Holzverkleidungen und Polsterstühle sollten das gediegene Ambiente für die Militärführung schaffen. Die Mannschaftsräume waren einfacher eingerichtet mit bäuerlichen Attribu- *84*
ten, geflochtenen Stuhlsitzen, geblümten Tischdecken, bunten Vorhängen sowie riesigen geschmiedeten Lampenrädern. Die Deckenbalken waren nicht, wie bei den Kommandanten-Zimmern, aus Holz, sondern sie bestanden aus Beton und auch das Parkett war einfacher.

Ähnlich hatte *Heinz-Günther Thees* die Kasernen des U-Bootstützpunktes für die *78–81*
Flottille Saltzwedel am Zwischenhafen ab 1936 entworfen. Die Einrichtung der Mannschaftskantine war bäuerlich schlicht. Der Ausbaustandard für die U-Boot- *134*
Kommandantenzimmer war entsprechend feudaler. Die verordnete Schlichtheit des „III. Reiches" wurde nur scheinbar eingehalten.

Auch die Unterkunftsgebäude der *KMW* wurden entsprechend den Rängen eingerichtet. Das 1940 entworfene Wasch- und Badehaus für die *Westwerft* wurde mit unterschiedlichem Sanitärkomfort für Mannschaften, *Unteroffiziere* und *Offiziere* ausgestattet. *Unteroffiziere* hatten mehr Platz in den Umkleideräumen, *Offiziere* durften sich in geräumigen Badezellen von 5 qm reinigen.

⋆

Wilhelmshavens Offiziersheim[186], 1902 in historisierendem Stil als kaiserliches Marine-Offiziers-Kasino gebaut, entsprach mit seinem Schnörkel und Plüsch[187] keinesfalls mehr den Auffassungen der nationalsozialistischen Machthaber. *Heinz-Günther Thees* bekam 1938 den Auftrag zu einer völlig neuen Innengestaltung. Er zog *Clemens Müller-Schiedmayer* von den *Vereinigten Werkstätten* Bremen und den örtlichen Innenarchitekten *Hugo Adena* hinzu. Der großangelegte, zweigeschossige U-förmige Baukörper enthielt Fähnrichsmesse, U-Bootszimmer, Lesezimmer, Admiralszimmer, Damenzimmer, Empfangszimmer, Tanzsaal, Billardsalon usw. [188]. Mannschaften hatten selbstverständlich keinen Zutritt zu dem Offiziersheim – es sei denn, sie gehörten zum Küchen- oder Bedienungspersonal. Notgedrungen mußte ein schlichter Mannschaftsraum geschaffen werden. Er war wiederum bäuerlich eingerichtet mit blankgescheuerten

Tischen, ohne Decken in rustikalem Ambiente. Die Büste des Diktators vor Kopf „überwachte" den Raum – Aufforderung zu Ordnung und Sittsamkeit.

133 In den Gesellschaftsräumen des Offiziersheimes waren die Wände raumhoch
holzvertäfelt. Insgesamt war die Innenarchitektur dieser Räume keinesfalls bäuerlich. Sie entsprach mehr den Schiffsinterieurs, wie sie *Paul-Ludwig Troost* ab 1914 für den *Norddeutschen Lloyd* Bremen entworfen hatte. Die Repräsentanten des „III. Reiches", *Hitler* und *Göring,* griffen gleichfalls gerne bei ihren privaten Räumen auf *Troosts* Dampferstil zurück[189]. Auch die *Kriegsmarine* bediente sich gerne dieser Detailkataloge, um sich vom Pomp und Kitsch der Gründerzeit abzusetzen.

Im Offiziersheim Wilhelmshaven war alles handwerklich und künstlerisch durchgearbeitet. Jeder bezogene Clubsessel war von Hand gefertigt, jede stoffbezogene Bank war individuell aufgepolstert. Wertvolle Teppiche auf dem blankpolierten Parkett, ja sogar Auslegware, bewirkte gedämpfte Club-Atmosphäre. Die Damenzimmer waren mit zierlichen hellen Möbeln ausstaffiert. Spiel- und Billardsalons entsprachen dem, was international beliebt war. Hier konnten aufwendige Feste gefeiert werden, hier war man unter sich. Ölbilder von *Hitler* und *Hindenburg* sollten die holzvertäfelten Wände schmücken und die politische Führung repräsentieren.

7.5. Bauten für die Führung

Privater Einfamilienhausbau im *Wirtschaftsgebiet Wilhelmshaven-Jeverland* wurde durch
1 gewaltige Massenwohnungsbauprogramme überdeckt und war daher für Propagandazwecke ungeeignet. Tatsächlich jedoch benötigten die vielen *Offiziere*, Handwerksmeister, „Betriebsführer," Ingenieure, Angestellten und Beamten, die mit Stadterweiterung und Neugestaltung Wilhelmshavens zu tun hatten, ihre besonderen Wohngelegenheiten.
136 Zahlreiche Einfamilienhäuser im Hansa- und Villenviertel zeigten schon ab 1933 den
135 relativen Wohlstand, den das neue Regime versprach. Wohnungsbau für privilegierte Bevölkerungsgruppen wurde unauffälliger geschaffen als Wohnungsbau für die breiten Massen. Der hierarchische Bevölkerungsaufbau (Gefolgschaft-Führer) war im „III. Reich" allgegenwärtig. Trotzdem sollte sich jeder als „Volksgenosse" in der „Volksgemeinschaft" fühlen. Klassenunterschiede waren zwar offiziell abgeschafft, nicht aber der von den Nationalsozialisten gewünschte organische Bevölkerungsaufbau mit Rangunterschieden. Der Städtebau, wie ihn z.B. *Gottfried Feder* 1939 in seinem Standardwerk „Die neue Stadt"[190] propagierte, sollte sich nicht von der „leeren Form rein architektonischen Gestaltungswillens"[191] leiten lassen, sondern von der „inneren Struktur der sozialen Gemeinschaft". Das organische Gemeinschaftsleben einer Beamtenstadt, einer Industrie- und Gewerbestadt, einer Bäderstadt, einer *KDF-Stadt,* einer *Hermann-Göring-Stadt*[192], sollte nach *Feder* sehr unterschiedlich sein. Unerwartet und beachtlich hoch sei immer wieder der Anteil der Gewerbetreibenden und Selbständigen. In der architektonischen Praxis hatte *Feder* keine Gelegenheit mehr, seine „völkischen"

Vorstellungen von der Struktur einer sozialen Gemeinschaft zu bauen, denn sein Wettbewerbsentwurf für Wilhelmshaven Langewerth von 1939 mit 16 Ortsgruppen *64*
wurde nicht realisiert. Sein Modell der Gesellschaft orientierte sich an 8 Stufen „verschiedener Einrichtungen"[193], die aber kaum auf die „Wehrmachtsstadt" Wilhelmshaven anwendbar waren. *Feder* versuchte, seine Theorie einer Handels- und Zahlungsbilanz auch auf Städte mit behördlichen Einrichtungen und Wehrmachtsteilen anwendbar zu machen. Die Sonderrolle der Rüstungsschmiede Wilhelmshaven und ihr spektakulärer Ausbau auf 500.000 Menschen kam in *Feders* Theorien nicht vor.

Das „III. Reich" konnte die weitverbreiteten Träume vom luxuriösem Heim oder der feudalen Großstadtwohnung nicht ersticken. Uniformiertes Wohnen fand für die Privilegierten des NS-Staates nicht statt. Die Spitzen von Verwaltung[194], *KMW* und Militär lebten mit ihren Familien äußerst zurückgezogen und ruhig in Villenvierteln. Bildbände von *Paul Schmitthenner* oder *Paul Schultze-Naumburg*, auch Wohnzeitschriften hielten den Wunsch nach großbürgerlichem Leben, exklusivem Bauen und komfortablem Wohnen hoch.

Das Musterbuch der Architekten *Ernst Neuferts* „Bau-Entwurfslehre"[195], dokumentierte nach wie vor prachtvolle Villen-Grundrisse mit Beispielen von *Peter Behrens, Ludwig Hilberseimer* und sogar aus England und Amerika. Die zwei *Deutschen Architektur- und Kunsthandwerk-Ausstellungen*, München,[196] zeigten 1938 prachtvolle und großzügige öffentliche und private Bauten. Mancher gutgestellte Wilhelmshavener[197] schaute sich auf dem Weg in die Berge oder auf einer *KDF*-Reise Fotos und Modelle im *Haus der Deutschen Kunst* an. Das gesellschaftliche, abgeschlossene Leben der Mittel- und Oberschicht war nach wie vor lebendig. Die besondere hierarchische Gliederung von *Werft* und *Marine*, Behörden und Baudienststellen erzeugte eine große Anzahl von Führungsebenen und herausgehobenen Stellungen, die ihren jeweiligen eigenen Lebensstil in Offiziersheimen, im Jadeklub und in privaten Zirkeln pflegten[198].

Der Leiter des Wilhelmshavener *Stadterweiterungsamtes*, *Wilhelm Hallbauer*, faßte 1937 seine Zielvorstellungen so zusammen: „Wenn also irgendwo in Deutschland der Begriff Volksgemeinschaft und Volksverbundenheit nicht nur eine seelisch erwünschte Forderung darstellt, sondern eine hervorragende praktische Notwendigkeit und Selbstverständlichkeit, dann ist dies hier im neuen Wilhelmshaven der Fall"[199].

Landesplaner Georg Linde, ein überzeugter Nationalsozialist, ging in seinem Erläuterungsbericht vom 27.08.1937 fast ausschließlich auf bevölkerungspolitische Zielsetzungen des Wohnungsbaus ein:

„1. Das Eigenheim und die Siedlung sind die Idealform für den deutschen Arbeiter (Angestellten, Beamten). Sie sind auch am besten imstande, das Problem des Kindes zu lösen, soweit die Wohnung darauf Einfluß hat. Von dieser Lösung aber hängt die ganze Zukunft des Deutschen Volkes ab. Mietwohnungen werden aber auch in Zukunft ebenfalls in mehr oder weniger großem Umfang je nach Betriebsverhältnissen und der Geeignetheit der infrage kommenden Gefolgschaftsmitglieder nötig sein.

2. Klassensiedlungen und Klassenwohnviertel sind weitgehendst zu vermeiden. Für Wilhelmshaven und Umgegend ist die Durchführung aus verschiedenen Gründen besonders schwierig, aber keineswegs unmöglich"[200].

Gegenläufig zu den NS-Phrasen von der „Volksgemeinschaft" versuchte man in *Bebauungsplänen* für neue Stadtteile, Wohnungsbau für Privilegierte unauffällig mit
30 unterzubringen. Bei genauer Durchsicht der *Bebauungspläne* z.B. von Voslapp oder
43 Altengroden, waren das Haus des *Bürgermeisters*, private Baugrundstücke, Häuser für Schulrektoren und Lehrer auszumachen. Ärzte und Apotheker mußten ebenso in den neuen Stadtteilen an bevorzugter Stelle angesiedelt werden. Die Rektor-Wohnhäuser bei der Schule Voslapp, von *Willi Lübbers* 1939 für die *Stadtverwaltung* entworfen, waren mit 90 qm Wohnfläche gegenüber den benachbarten Kleinsiedlungshäusern opulent geplant. Im Erdgeschoß: Windfang, Diele, 2 Wohnzimmer (mit Schiebetür verbunden), WC. Im ausgebauten Dachgeschoß: 3 Schlafzimmer, Diele, Einbauschränke, Bad. Im Keller: Luftschutzräume, Waschküche und Zentralheizung. Der Ausbaustandard hob sich deutlich von den umliegenden Siedlungshäusern ab. Der Klinkerbau sollte isolierende Hohlschichten, im Eingangsbereich und in den Naßräumen Kunststeinbelag und in den beiden Wohnzimmern Parkettfußboden erhalten[201]. Die Außentür aus massiver Eiche sollte durch Werksteingewände umrahmt werden. Als Krönung der Rektorwohnungen waren zwei Blumenfenster vorgesehen.

Im Standardwerk für großstädtische Stadterweiterungen von 1940, der „Ortsgruppe als Siedlungszelle"[202] wurde der Gegensatz zwischen „reichen Villenvierteln" und „ausgesprochenen Arbeiterquartieren" oder „Plutokratenortsgruppen" und „Proletarier- oder Habenichtse-Ortsgruppen" als schädlich für eine „vertiefte Volks- und Lebensgemeinschaft" beschrieben. Die von *Friedrich Heuer* dort propagierte Ortsgruppeneinheit aus 6 Parteizellen sollten zwar von Mietwohnungen im Geschoßbau bis zum Einzelhaus reichen, sie sollten aber nicht unbedingt in der „kleinsten Zelle ... die verschiedenen Wohnarten nach Maßgabe des Bevölkerungsquerschnitts" berücksichtigen. Tatsächlich waren in der Muster-Ortsgruppe von 1.800 Einwohnern nur gut 70 freistehende Einfamilienhäuser vorgesehen. Nach *Heuers* Gewerbeplan waren viele dieser Einfamilienhäuser für Ärzte, Zahnärzte usw. vorgesehen. Die Einfamilienhäuser lagen jeweils am Bebauungsrand der „Zellen" ruhig und mit idealen Besonnungsverhältnissen. Oftmals wurden sie auch für sich abgesondert und an Plätzen zusammengefaßt.

Klaus Rasch, stellte 1939 in seiner Dissertation[203] fest, daß in Wilhelmshaven der subjektive Wohnungsbedarf, d.h. der „für zahlungswillige und zahlungsfähige" Wohnungssuchende, „von den Angehörigen der Reichsbetriebe und Behörden ausgesprochen hoch" sei und schnellste Behebung verlange.

Auch die Wohnungsbaugesellschaften konnten sich den höheren Ansprüchen nicht gänzlich verschließen. Der *Bauverein Rüstringen* plante das große Wohngebiet Schaar-
52 reihe als großstädtische Alternative zu Siedlungshäusern. Dieses „Tausend-Wohnungs-
61 projekt" mit eigenen Bauten der Gemeinschaft zwischen Siebethsburg und den

Monumentalbauten von *Kulturforum* und *Sportforum* sowie nördlich davon war für 109
Führungskräfte vorgesehen und auch für „die in der freien Wirtschaft tätigen Volksgenossen“[204]. Auch als das Riesen-Bauprogramm im Mai 1939 auf 450.000 Einwohner für die Stadt Wilhelmshaven aufgestockt wurde, behielt man den privaten Wohnungsbau im Auge. Der *Oberbürgermeister* versicherte Stadtrat *Willi Lübbers*, daß
man in Langewerth Flächen für Private vorsehen werde. 64

Auch die *Stadtverwaltung* griff verschiedentlich zugunsten ihrer eigenen „Gefolgschaft“
in den Wohnungsbau ein. 1938 entwarf *Egon Forster* eine kleine Siedlung für städtische 132
Angestellte im Jadeviertel in der Ernst-Moritz-Arndt-Straße. Weitere „Gefolgschafts“- 53
Siedlungen waren westlich Neuengroden vorgesehen.

Die *Marine* griff ab 1938 auch selbst in den Wohnungsbau für ihre *Offiziere* ein, ohne sich einer Wohnungsbaugesellschaft zu bedienen. Sie beantragte[205] z.B. beim städtischen *Bauamt* die Genehmigung für 8 Offizierswohnhäuser: Viktoriastr. 6, Marktstr. 1[206], Holtermannstr. 32/34 (Doppelhaus), Holtermannstr. 36, Schulstr. 51/53 (Doppelhaus) Hegelstr. 58. *Karl Franzius* entwarf zusammen mit *Kurt Geisenhainer* Wohnhäuser für höhere *Offiziere* und Stabsangehörige und typisierte sogar Grundrisse und Ansichten. Die Villen wirkten derart großzügig, daß sie nach dem Kriege sofort von der *Militärregierung* in Anspruch genommen wurden. Das Gebäude Hegelstr. 58 war nach der Freigabe bis 1982 Offiziersheim der *Bundesmarine*. Der Entwurf zum Einzelhaustyp a 3 hatte z.B. eine Bruttogeschoßfläche von 280 qm, verteilt auf 2 Geschosse. Hinzu kamen Keller (mit Luftschutzräumen), ausgebautes Dachgeschoß und Garage. Im Erdgeschoß öffneten sich hinter der großzügigen Diele Herrenzimmer (20 qm), Wohnzimmer (26 qm) und Eßzimmer (26 qm), die durch 3 Schiebetüren miteinander verbunden waren. Wintergarten, Küche, Kleiderablage und Terrasse ergänzten das repräsentative Erdgeschoß. Im Obergeschoß folgten Gästezimmer, 3 Schlafzimmer, Mädchenzimmer und Balkon. Über 3 m Geschoßhöhe, Eichen-Parkett, verflieste Bäder und elektrische Rufanlagen (für das Hauspersonal) gehörten zum Standard. Durch die 3 Schiebetüren ergab sich eine zusammenhängende Wohnfläche von fast 100 qm, ausreichend für die Empfänge der dort wohnenden *Offiziere*.

Die Villen waren Ziegelbauten und wurden von gewaltigen Ziegel-Walmdächern mit markanten Gesimsen gekrönt[207]. Zur Straße hin bekamen die Offizierswohnhäuser oftmals Blumenfenster mit profilierten Balkenköpfen. Die weißen Holzfenster mit Kreuzsprossen wurden häufig mit Werkstein umrahmt. Alle Offizierswohnhäuser bekamen auf Balkenköpfen ruhende Balkone mit schmiedeeisernen Geländern[208]. Sie waren Kopien der (Führer-)Balkone, wie sie bei öffentlichen Bauten im „III. Reich“ gängig waren. Zur seitlich liegenden werksteinumrahmten Eingangstür gehörte eine Freitreppe. Untergeordnete Fenster und Garteneingänge waren schmiedeeisern und „wehrhaft“ vergittert. Selbst die Garage war sorgsam durchdetailliert mit aufgedoppelter, zweiflügeliger Holztür, Ziegeldach und austragendem Gesims. Die Gestaltungsdetails waren so oder ähnlich bei allen marineeigenen Offizierswohnungen von *Karl Franzius* und *Kurt Geisenhainer* zu finden.

Wohnungen für *Offiziere*, Werft- und Wehrmachtsbeamte wurden auch offiziell anders gesehen als Wohnungen z.B. für Werftarbeiterfamilien. Sie spielten in der nationalsozialistischen Propaganda eine untergeordnete Rolle, da Führungsgruppen sowieso ein „besonders festes Abhängigverhältnis" hatten[209]. Wohnungsbau für Arbeiter jedoch wurde propagandistisch dick aufgetragen, schön gefärbt und mit Ideologie verbrämt. Erhaltung von Sitte, Moral und Kultur[210] bezog sich nicht auf die privilegierte Bevölkerungsgruppe Wilhelmshavens, sondern auf die Werftarbeiter, die eigentlich immer revolutionärer oder sozialistischer Umtriebe verdächtigt wurden.

★

Die *Wilhelmshavener-Rüstringer Wohnungsbaugesellschaft* hatte sich seit ihrer Gründung
1934 auf Wohnungen für Führungspersonal eingerichtet. Die Gründer, ortsansässige,
mittelständische Bauunternehmer und Handwerksbetriebe, hatten die Zeichen der Zeit
erkannt und finanzierten über Reichsbaudarlehen ihren schnell wachsenden Wohnungs-
bestand. Noch 1935 hatte die Gesellschaft *Fritz Höger* mit dem Entwurf von 8
52 Wohnblocks nördlich des Mühlenweges zwischen Bülow- und Schulstraße für
Unteroffiziere beauftragt. Nur 4 davon wurden von *Höger* gebaut. Dann wurde auf
58 NS-Wunsch die Zusammenarbeit mit *Höger* eingestellt. Rechtsanwalt *Meyer*[211], der erste
65 Geschäftsführer, hatte danach das große Hamburger Architekturbüro *Erich zu Putlitz* und
August Schoch mit Entwürfen für Offizierswohnhäuser beauftragt und im Juli 1938
entsprechendes Gelände an der Gökerstraße und im Villenviertel erworben[212]. *Zu
Putlitz*[213], bei dem auch *Egon Forster*[214], gearbeitet hatte, entwarf für die Gökerstraße
140–170, Leibniz-, Fichte- und Holtermannstraße großzügige Wohnungen für *Offiziere*
der *Kriegsmarine*. *Zu Putlitz* war einer der meistbeschäftigten Architekten des „III.
Reiches" und hatte 1936 im Auftrage *Baldur von Schirachs* die Akademie für deutsche
Jugendführung in Braunschweig gebaut.

Die Wohnungstypen am Rande des Villenviertels an der Gökerstraße waren vorwiegend Dreizimmerwohnungen: Wohnzimmer (20 qm) und Eßzimmer (13 qm), durch Schiebetüren verbunden, Schlafzimmer (17 qm). Sie waren erheblich komfortabler ausgebaut als die Wohnungen in den neuen Stadtteilen. Unter dem Dach gab es Kammern für das Personal, durch Klingelleitungen mit den Geschoßwohnungen verbunden. Die dreigeschossigen geputzten Randbebauungen mit ihren hohen Walmdächern waren schmucklos einfach und jeweils symmetrisch auf die Hauseingänge bezogen. Bei den dreiteiligen Fenstern hatte *zu Putlitz* auf Kreuzsprossen verzichtet, nicht jedoch auf die Werkstein-Umrahmung von Fenstern, Eingangstüren und Treppenhauslichtbändern. Ein Teil der Villenviertel-Randbebauung wurde im Kriege zerstört, jedoch bereits 1948 durch die *Wilhelmshavener-Rüstringer Wohnungsbaugesellschaft* wieder ausgebaut. Die nicht fertiggewordenen Teile wurden erst 1951 nach völlig anderen Grundrissen, aber in Anlehnung an die Architektur von *zu Putlitz* von *Karl-Heinz Kraak* wieder aufgebaut[215]. Noch bis weit in die 50er Jahre baute die *Wilhelmshavener-Rüstringer*

Wohnungsbaugesellschaft nach dem Muster von *Erich zu Putlitz*, nicht nur im Villenviertel,
sondern auch im Stadtzentrum. Im Innern des Villenviertels waren die Stabsoffiziers- 120
wohnhäuser von *zu Putlitz* noch erheblich komfortabler geplant als am Rande. In der
Holtermannstraße[216] hatten die Mietwohngebäude nur noch zwei Geschosse. Großzü-
gige Grundrisse, Garagen und Dachterrassen zwischen den symmetrisch angelegten
3 Blocks dokumentierten den herausragenden Standard.

Für *Unteroffiziere* ließ die *Wilhelmshavener-Rüstringer Wohnungsbaugesellschaft* im
Inselviertel von *Willi Lübbers* 1937 mehrere Vierfamilienhäuser bauen. Die zweigeschos- 28
sigen Zweispänner in Ziegelbauweise waren jeweils durch gewaltige gaubenlose
Walmdächer abgeschlossen. Selbstverständlich waren die vier Mietwohnungen für
Unteroffiziere mit 50 bis 60 qm Wohnfläche sparsam ausgelegt. *Willi Lübbers* verzichtete
auch bei diesen Häusern nicht auf den (Führer-) Balkon vor dem Treppenhausfenster. Die
unteren Führungsschichten mußten sich mit weniger Komfort und weniger Wohnfläche
zufriedengeben als *Offiziere*.

★

Nur unwesentlich großzügiger als die Wohnungen für *Unteroffiziere* im Inselviertel
waren die Reichspostwohnungen an der Bremer Straße, von *Oberpostbaurat Martini*
entworfen. Die gleichfalls zweigeschossigen Zweispänner hatten von vornherein
ausgebaute Dachgeschosse und mehrere Gauben in den Ziegel-Walmdächern. Im
südlichen Teil des Villenviertels, um Herbart-, Widukind-, Sachsen- und Schulstraße und 52
um das *SS*-Heim Kirchreihe[217] entstanden 1935–38 bis hin zum Mühlenweg noch 23
zahlreiche private Bauten für Führungspersonal. Viele der Einfamilienhäuser entwarf
Willi Lübbers[218], gleichzeitig Stadtrat im *NSDAP-Magistrat* und seit 01.02.1937 techni-
scher Direktor der *Wohnungsbaugesellschaft Jade*. Seine Entwürfe für Unternehmerhäuser 135
im südlichen Teil der Herbartstraße übertrafen den Standard der Offiziershäuser noch
erheblich. Gerne bediente sich *Lübbers* des gleichen Detailkataloges: heimatlicher
Ziegelbau, schmucklose Außenhaut, steile Walmdächer, rote oder braune Hohlpfannen-
deckung, vereinzelte Gauben, weiße Holzfenster, werksteinumrahmte Öffnungen,
regelmäßige Kreuzsprossen usw. Nicht selten führten zweiläufige Treppen oder
großzügige Freitreppen zur zentral und symmetrisch liegenden Eingangstür hinauf.

Andere Wohngebiete, wie z.B. das Hansaviertel um Stralsunder- und Danziger Straße, 53
wurden 1933–34 für Angestellte und verdiente „Volksgenossen“ gebaut, die in der freien
Wirtschaft tätig waren. Viele der kleinen Klinkerhäuschen wurden von den Architekten
Franz Iwersen und*Hanns Iwersen* in traditioneller Formsprache der 20er Jahre entworfen. 136
Typisch waren Ziegeldächer, Schleppgauben, eckige oder runde Erker, Klinkerstreifen
seitlich der Fenster und rautenförmige Giebelfenster. NS-typische Details haben hier
kaum Spuren hinterlassen.

Wo es erforderlich erschien, interpretierten Architekten bestehende Bauten in ihrem
Sinne um. Das Haus eines Bauunternehmers im Villenviertel z.B. war 1913 in gemäßigt

moderner *Werkbund*-Architektur entstanden[219], ohne historisierenden Schmuck, ohne traditionelle Ornamente. Die *Stadt Wilhelmshaven* baute es für ihren neuen *Stadtbaudirektor* zur Dienstvilla aus. *Walter Hunzinger* wurde in der Öffentlichkeit vorgestellt als ein Mann, „von dessen Wirken noch Jahrhunderte später Zeugnis ablegen werden“[220]. Das Haus sollte an die Architekturvorstellungen des neuen *Stadtbaudirektors* angepaßt werden. Die Umbauarchitektur[221] drückte Repräsentation, Amtsautorität und Personenkult aus. Aus einer unpathetischen einläufigen Eingangstreppe wurde nach dem Umbau eine überbreite, zweiläufige Treppe mit Muschelkalkstufen und schmiedeeisernem Geländer. Die Treppenpodestumwehrung sollte anfangs die persönlichen Initialen des *Stadtbaudirektors* erhalten. Anstatt der bescheidenen Eingangstür in der Fläche dominierte nach dem Umbau von 1938 eine wulstige Werksteinumrandung mit einem großen Schlußstein. Die Fassade sollte flankiert werden von einem schmiedeeisernen Leuchtenpaar und Holzklappläden. Die *Werkbund*-Architektur war nach dem Umbau nicht mehr zu erkennen. Der nationalsozialistische Amtsinhaber konnte einziehen. Entsprechend prachtvoll fiel der innere Umbau aus: Eichenparkett, Balkendecken, Biedermeierverzierungen, Kamin, Hausbar, getäfelte Wände, Blumenfenster.

43 Das Konzept der Marinesiedlung Altengroden hob sich schon zu Anfang aus den Arbeitersiedlungen heraus. Das *OKM* hatte die 3.500 Wohnungen exklusiv für Beamte der *KMW, Offiziere* und *Unteroffiziere* der *Marine* geplant. Es baute dort auch nicht die *Jade*, sondern eine marineeigene Wohnungsbaugesellschaft aus Berlin, die engstens mit
4 den Spitzen des *OKM* verflochten war. Die Haustypen für Beamte und *Offiziere* waren andere als die für *Unteroffiziere*. 1939 wurden sogar noch bestimmte Gebiete nur für das privilegierte Luftwaffenpersonal und für Eigenheime ausgewiesen.

Bauten für die Führung waren auch die Beamten-Zweifamilienwohnhäuser, die vom *Luftwaffenkommando See, Bauverwaltungsbezirk Wilhelmshaven*, an der Edo-Wiemken- und Siebethsburger Straße im Naturschutzgebiet des Burghügels von Siebethsburg gebaut wurden. Unter *Baurat Heinrich Pauls* arbeiten *Klaus Rasch* und *Josef Vosshans* 1938 an den Entwürfen. Die *Luftwaffe* hatte das Gelände 1938 von Privat erworben und damit *Högers* Pläne für eine Besiedlung des Burggeländes zunichte gemacht. Auffällig waren die Putzaußenflächen der zweigeschossigen Walmdachhäuser und das Holzfachwerk der Erker. Auch fehlten Kreuzsprossen. Ansonsten jedoch hielt sich das *Luftwaffenbauamt* an die in Wilhelmshaven vorherrschenden, bei den Marinebehörden gebräuchlichen Detailkataloge: Walmdächer, Einzelgauben, Blumenfenster, Balkenköpfe, Wintergärten mit Balkonen, geschmiedete Lampen und Geländer. Die Häuser für die Beamten der *Luftwaffe* wurden durch Bomben zerstört und 1951 auf Betreiben des *Stadtplanungsamtes* in unveränderter Art wieder aufgebaut.

Noch mitten im Kriege achtete die *Marineintendantur* sorgfältig darauf, die Wohnungen für Arbeiter, Angestellte, *Unteroffiziere* sowie *Offiziere* zu trennen. Nach der *Preisstoppverordnung* durften Arbeiterwohnungen 25 Mark je cbm, Offizierswohnungen jedoch 30 Mark je cbm kosten[222].

7.6. Bauten der Technik

Kaum ein technisches Bauwerk im *Wohnsiedlungsgebiet Wilhelmshaven-Jeverland* wurde 129
nicht durch die *Marine* beeinflußt[223]. Wichtige logistische Einrichtungen, wie beispiels-
weise die Wasserversorgung, mußten 1938 aus militärischen Gründen an die *Kriegsmarine*
übergeben werden. In Wilhelmshaven flossen die weitaus meisten Mittel in Bauten der
Rüstungs-, Hafen-, Waffen-, Versorgungs- und Verkehrstechnik. In der Bedeutungs-
hierarchie der nationalsozialistischen Architekturtheoretiker gehörten Versorgung und
Verkehr im Gegensatz zu Bahnhöfen und Postgebäuden nicht zu „Einrichtungen höherer
Ordnung“[224]. Bauten der Technik kamen in den Heimatschutz-Standarwerken von
Lindner und *Schultze-Naumburg* nicht vor. Die Bauten von Reichsdienststellen und Partei,
Regierung und Gliederungen hatten in Architektur und Stadtplanung des NS-Systems
größten propagandistischen Wert. Bahnhofsgebäude beispielsweise sollten auf die 51
Benutzer wirken, jedoch nicht die technischen Stellwerke. Tankstellen wurden
landschaftsgebunden entworfen, nicht jedoch Ölumschlaganlagen. Fußgängerbrücken
über einen Wassergraben durften kleinteilig aussehen, Eisenbahnbrücken mußten
schlechterdings weit gespannte Betonkonstruktionen zeigen. Bauten der Technik
rangierten hinter Monumental- und Kulturbauten, hinter Gemeinschafts- und Siedlungs-
bauten. Die NS-Kulturpolitik filterte aus den vorhandenen Baustilen die für sie
brauchbaren heraus und unterschied nach propagandistischer Brauchbarkeit. Wohnge-
bäude, Militär- und Monumentalbauten bekamen eine Verkleidung, die etwas ausdrük-
ken sollte. Technische Bauten in den Wohnsiedlungen, wie Transformatoren und
Pumpwerke, Tankstellen und Gewerbebetriebe, wurden gleichfalls traditionell verklei- 149
det. Technische Bauten, soweit sie der Öffentlichkeit unzugänglich waren oder wegen 97
ihrer Größe nicht anders zu bewältigen waren, durften Funktionen zeigen, Flachdächer
haben, kubische Formen darstellen. 103

Bei Bauten der Technik konnte auf Symbolträchtigkeit und Bedeutungsinhalt fast ganz verzichtet werden. *Hitler* wäre es nicht eingefallen, eine Fabrik im neuklassizistischen Repräsentationsstil zu bauen. *Speer* schrieb den angeblich von *Hitler* bevorzugten Stil der Parteipresse zu. Im Gegenteil habe er sich „für einen Industriebau in Glas und Stahl“ geradezu begeistern können.

Jeder Bauherr der Weimarer Republik konnte jeden beliebigen Baustil anwenden.
Alle möglichen Grundrißfunktionen konnten mit allen möglichen Architekturen
umhüllt werden. Ab 1935 wurden Bauten der Technik durchaus auch modern, 92–
zweckmäßig, rationell und funktional, aber nicht ideologisch modern gebaut. Die 97
Architekten waren nicht Neuerer, sondern Ingenieure und Konstrukteure im Dienste des
Staates. Nur in Einzelfällen waren sie dem *Bauhaus* verpflichtet. Bestimmte Bautypen
verlangten aus technischen und wirtschaftlichen Gründen modernes Äußeres.

★

Millionensummen in der Kriegsmarinestadt Wilhelmshaven wurden für Wasserbau-
ten und Kanalisationen, Uferbefestigungen und Deichbauten, Kläranlagen und Pump-
werke, veranschlagt. Die Hafenanlagen waren beispielhaft für das ganze Reich, für die
internationale Fachwelt. *Erich Raeder, Alfred Eckhardt, Arnold Agatz* und *Carl Tiburtius*
103 präsentierten der Hauptversammlung der *Hafenbautechnischen Gesellschaft* im Mai 1937
101 die begonnenen „Wunderwerke" von *Nordwerft* und IV. Einfahrt[225]. Die damals „größte
102 Doppel-Seeschleuse der Welt" ist noch heute Hauptbestandteil des Standardwerkes
„Erfahrungen mit Grundbauwerken" von *Arnold Agatz* und seinem damaligen
Mitarbeiter an den Wilhelmshavener Planungen, *Erich Lackner*[226].

Korrespondierend zu militärischem Größenwahn, architektonischen Utopien und städtebaulichen Projekten entwickelten Ingenieure noch weitergehende technische Visionen ungeahnten Ausmaßes für die *„Stadt der 500.000"*. Hierzu gehörte auch der von *Daniel Picker* forcierte Bau eines Viaduktes über den Jadebusen, der am 31.05.1939 in Berlin besprochen wurde[227]. Die Vision lebte nach dem Kriege wieder auf und entwickelte sich zu einem *Brückenbauverein* Ende der 50er Jahre und zu den Jadeuntertunnelungsplänen von *Stadtbaurat Ulrich Tappe*[228] im Jahre 1971. Ähnliche überspannte Pläne gab es zu neuen Kanalverbindungen in das Binnenland.

Technische Visionen waren auch Bestandteile der Planungen nach dem *Erlaß zur
Vorbereitung des deutschen Wiederaufbaus nach dem Kriege vom* 15.11.1940. So plante *Walther*
116 *Schulz* neben neuen Bahnhofsanlagen einen großen Verkehrsflughafen und einen
Handelshafen Voslapp mit einem 2 km langen Hafenbecken, 4 neuen Kaianlagen, Docks
und einem eigenen Verschiebebahnof parallel zum Geniusdeich. Der *Arbeitsstab Speer*
118 plante einen städtischen Hafen als Dock mit Umschlagsanlagen nördlich der *Nordwerft*.
Exakt nach dieser Planung wurde der Speicherpolder der Maade 1948 im Rüstersieler
Groden angelegt.

⋆

Die „Maadelinie" spielte bei den Konzeptionen für neue Gemeinschaftssiedlungen eine
126 große Rolle. Noch 1936 sollte sie als neuer Kanal über eine Schleuse und eine Großwerft
in Altengroden mit dem Jadebusen verbunden werden. Die Fachleute stritten 1937 heftig
über technische Lösungen für eine Kanalisation der Stadterweiterung. In jedem Fall sollte
die Maade als Vorfluter „begradigt, vertieft" und „ausgebaut" werden[229]. Pumpwerke in
Mariensiel und Rüstersiel sollten „bei Hochwasser die Wassermengen über den Deich
schaffen". Als Sachverständiger wurde *Helmut Döscher* vom *RAM* eingeschaltet.
127 Modellversuche simulierten den Maade-Ausbau. Anfang 1937 noch sollte die „Maade-
linie" zwischen Rüstersieler und Heppenser Groden, zwischen nördlichem Wasser- und
südlichem Landflughafen in die Jade eingeleitet werden. Die *Luftwaffe* bestand jedoch
darauf, daß ihre militärischen Anlagen nicht durchschnitten werden sollten, und so wies
128 der Plan *Wirtschaftsgebiet Wilhelmshaven vom Juni* 1937 dann die nach Nord-Ost gerade
durchlaufende neue „Maadelinie" aus, die südlich von Voslapp in den Jadebusen münden

sollte. Im Mai 1939 ordnete *Oberbürgermeister Müller* an, „die Maade-Korrektion im Beihilfeverfahren durchzuführen“[230]. Die sich ausweitende Stadterweiterung erforderte ein ständiges Vergrößern des Maade-Querschnittes. Der *RAM* erhob nach Rücksprache mit dem *Reichsernähungsminister* Ende 1940 „keine Forderungen wegen der Lage der Ausmündung der Maade bei Voslapp“[231]. 1940–44 dann wurde die Maade nördlich Altengroden für 600.000 Mark in der neuen Linienführung begradigt. Das *Kanalbauamt der Stadt Wilhelmshaven* unter *Wilhelm Aichinger* mußte *RAD* und Zwangsverpflichtete einsetzen, um die Maßnahme im Kriege abzuschließen. Zu der bereits in querenden Straßentrassierungen berücksichtigten nordöstlichen Weiterführung nach Voslapp kam es nicht mehr.

Die zur städtischen Entwässerung gehörigen Bauten der Technik wurden ausnahmslos traditionell gebaut und im *Heimatschutzstil* verkleidet: Pumpwerk Nord (1940), Pumpwerk Grodenfähre (1944), Pumpwerk Rosenhügel (1940), Pumpwerk Fedderwardergroden (1940), Pumpwerk Neuengroden (1939), Pumpwerk Altengroden (1943), Pumpwerk Ost (1939)[232]. Die Kläranlage Coldewei nördlich der Maade war allerdings äußerst modern konzipiert. Die 0,8 Mio. Mark teure Anlage für die Ortsteile Altengroden, Fedderwardergroden und Breddewarden war bausteinartig aufgebaut. Entsprechend der wachsenden Stadterweiterung sollte auch die Kläranlage wachsen. Die gesamten Entwässerungsmaßnahmen für Groß-Wilhelmshaven beliefen sich auf 8 Mio. Mark.

Umfangreiche Entwässerungsinvestitionen für den neuen Ortsgüterbahnhof Ebkeriege (1940–43) und für die nie begonnene Siedlung um die Wurt Breddewarden[233] *66*
(1940–44) blieben nutzlos.

★

Die neuen Straßenbauten für Stadterweiterung und Neugestaltung von Groß-Wilhelmshaven ressortierten bis auf einige von der *Marine* ab 1938 beschlagnahmte Straßen im Bereich von *KMW* und *Nordwerft* bei der städtischen *Tiefbauabteilung* unter *Otto Lehn.* Der Straßenbau unterstand dem Leiter des *Straßenbauamtes, Heinz Schumacher.* Für 10 Mio. Mark wurden die neuen Siedlungen mit Straßen erschlossen[234]. Fast die Hälfte davon floß 1939–44 in die Straßen der *Neuen Stadt* Fedderwardergroden. Die neue *42*
Reichsstraße 69, die im weiten Bogen von der Bismarckstraße nach Sande führen sollte, war im wesentlichen militärisch begründet und unterstand nicht der *Stadt Wilhelmshaven,* sondern der *Reichsstraßenverwaltung.*

Für die Nationalsozialisten rangierte der Architekt vor dem Straßenbauingenieur. Der Architekt sollte „zuerst seinen Gedanken genau und bestimmt ausdrücken“[235]. Er sollte „die eigentliche, lebenerfüllte Gestaltung der Straßen und Platzräume beeinflussen, Modelle bauen und die Wechselbeziehungen zwischen Straße und Haushöhen klären. Der Straßenbauer sollte erst danach die tiefbautechnischen Pläne machen. Die Straßen wurden in ihrer „Führung“ durchdacht, nach Blickwinkeln trassiert und auf Wirkung ausgelegt. Brücken bekamen, soweit eben möglich, traditionelle Formen und wurden aus heimischen Materialien gebaut.

★

Innerhalb der „Maadelinie“ hatten die *Gas- und Elektrizitätswerke Wilhelmshaven* genormte Transformatoren entwickelt, die auf einer Grundfläche von 5 x 6 m Trafo, Schaltraum, Ölgrube und oftmals auch einen Schaltraum für die Straßenbeleuchtung enthalten sollten. Die Ziegelbauten wurden dort placiert, wo sich Stadtteile entwickelten. Das Mauerwerk wurde an den Ecken durch wechselweise herausgehobene Ziegelschichten betont. Die Zeltdächer waren mit Pappe beklebt und die Türen zweckmäßig, ohne Umrandung, mit Oberlichtern ausgerüstet. Zug um Zug jedoch bekamen die Trafos das Äußere von Walmdachhäuschen, die heimatschutzorientierte Architektursprache des *Marinestils*. Trafos sollten sich der Umgebung der neuen Stadtteile anpassen und bekamen beispielsweise in Fedderwardergroden konsequent geputzte Außenwände. *Willi Lübbers*, einflußreichster und meistbeschäftigter Architekt des „III. Reiches“ in Wilhelmshaven bekam auch die großen Bauaufträge der *Gas- und Elektrizitätswerke*. Am Alten Banter Weg war das 1909 errichtete, im *Jugendstil* entworfene Umspannwerk I für die groß angelegte Stadterweiterung nicht mehr ausreichend[236]. *Lübbers* baute 1938 die Haupt-
137 umspannstelle in Heimatschutz-Formsprache. Der zweigeschossige abgewinkelte Ziegelrohbau mit hohem Walmdach erhielt zur Straße hin einen werksteinumrahmten Eingang mit schmiedeeisern vergittertem Oberlicht. Kreuzsprossenfenster und Dachgauben gaben dem voluminösen Gebäude den besonderen Anstrich. Nur dort, wo es unbedingt erforderlich war, in den Schalträumen und Umspannsälen, mußten überhohe Stahlfenster eingepaßt werden. Aber auch diese wurden mit Sprossen aufgeteilt. Die großen Montagetore bekamen noch Werksteinumrahmungen, Schlußsteine und Eckquader. Technik sollte kaschiert werden und nicht stören. Nicht umgehen ließ es sich, daß hinter der Hauptumspannstelle am Alten Banter Weg vier Transformatoren im freien Gelände aufgestellt wurden. In ähnlich traditioneller Architektur wie die Hauptumspannstelle, wurde auch das Hauptumspannwerk Roffhausen, nordwestlich von Sande, hochgemauert.

Für Wilhelmshaven war ein erheblich höherer Stromverbrauch zu erwarten, als es die 1920 gegründeten *Gas- und Elektrizitätswerke Wilhelmshaven-Rüstringen* angenommen hatten. Nach der Machtübernahme der Nationalsozialisten kam es zu heftigen Angriffen von *Oberbürgermeister Renken* auf die Mitgesellschafterin, die *Thüringer-Gasgesellschaft*. Erst 1935 gelang es *Wilhelm Keppler, Hitlers* Wirtschaftsberater, den Streit zu schlichten, denn der kriegswichtige Ausbau Wilhelmshavens stand unmittelbar bevor[237]. 1937 flammte der Streit wieder auf, als es um die Gas- und Stromversorgung der neuen Stadtteile ging. Der Strom für Groß-Wilhelmshaven wurde im wesentlich in Wiesmoor produziert – auch aus Torf –, zu dessen Gewinnung Zwangsarbeiter, Häftlinge, Kriegsgefangene und *RAD* eingesetzt wurden. Für den außergewöhnlich hohen

Strombedarf der Bauarbeiten an IV. Einfahrt und *Nordwerft* wurde sogar das Kraftwerk 100
Farge bei Bremen erweitert. Voslapp und Fedderwardergroden bekamen nach einem 30, 37
Sonderabkommen von 1937 ihren Strom nicht von den *Gas- und Elektrizitätswerken*, sondern vom*Landeselektrizitätsverband.* Gasversorgung kam nicht in Betracht.

Gasleitungen wurden von Luftschutz- und Militärplanungen stark beeinflußt. Die alten Gaswerke lagen nahe dem Zwischenhafen (1869) und am Rüstringer Rathaus (1896). Noch 1942 wurde eine 100 km-lange Gasleitung von Bremen nach Wilhelmshaven gelegt, um die Versorgungssicherheit zu erhöhen.

7.7. Bauten für die Gemeinschaft

Die neuen Gemeinschaftssiedlungen für das *Wohnsiedlungsgebiet Wilhelmshaven-Jeverland* 1
stellten in erster Linie die vom „III. Reich" gewünschten Bindung der Belegschaft an die Staatsbetriebe *Werft* und *Marine* dar. Arbeiter, aus allen Gegenden des Reiches, für die Aufrüstung nach Wilhelmshaven angeworben, abgeschoben, dienstversetzt oder zwangsverpflichtet, sollten in ihrem alltäglichen Leben halbwegs zufriedengestellt werden. Man sorgte sich um alles, was zu einer *Neuen Stadt* gehört: Schulen, *HJ*-Heime, *NSDAP*-Gemeinschaftshäuser, Verwaltungsgebäude, Sportanlagen, Gewerbebauten, *NSV*-Häuser, *Postämter,* Krankenhäuser, Feuerwehrgebäude, Rathäuser usw. Die Baupolitiker des „III. Reiches bezeichneten die gesamte hochbauliche Infrastruktur vereinfachend als Gemeinschaftseinrichtungen[238]. Bauten für die Gemeinschaft gingen im „III. Reich" weit über das hinaus, was als lebensnotwendige Infrastruktur in der Weimarer Republik üblich war. Ein perfektes Netz von Freizeitbauten sollte Einfluß und Kontrolle der *NSDAP* sichern, die „Volksgenossen" gängeln und die „Volksgemeinschaft" überwachen. Die NS-Diktatur wollte sich von der „verhaßten" Weimarer Demokratie, vom „Elend" der Großstädte abheben und versprach eine „glorreiche" Zukunft.

Wie auch bei den vergleichbaren Großsiedlungen Wolfsburg und Salzgitter blieb trotz des *Runderlasses des Reichsarbeitsministers betr. Finanzhilfe des Reiches zu den Aufschliessungsarbeiten und Gemeinschaftseinrichtungen in Gemeinschaftssiedlungen vom* 17. *Dezember* 1938 die Finanzierung von vielen Gemeinschaftsbauten ungeklärt, wie die *Marineintendantur* am 17. April 1939 einräumte.

Hitler löste seine Versprechungen nicht ein. Viele Bauten für die Gemeinschaft blieben Fragmente oder in den Anfängen stecken. Schlimmer noch, Gemeinschaftsleben, das war in Wilhelmshaven Leben in Baracken und Baubuden, Fehlen von Versorgung und Kultur.

Im Folgenden soll es um Schulen, *HJ*-Heime, *NSDAP*-Gemeinschaftshäuser, Gewerbebauten, *Postämter,* das „Haus des Handwerks", das *Städtische Krankenhaus* und die Strandhalle als Beispiele für öffentliche Bauten gehen. Die Auswahl erfolgte so nach der Fülle des vorliegenden Materials, aber auch weil gerade diese Gebäude der Symbolisierung von Macht und „Volksgemeinschaft" dienten. Besonders bei diesen Gebäudetypen versuchten die Architekten des „III. Reiches", Gebäudeformen historisch abzuleiten oder regional zu begründen. Die Planer der Gemeinschaftsbauten lehnten sich an *Schultze-Naumburg* und den *Heimatschutzstil* an.

Die *Stuttgarter-Schule, Schmitthenner* und *Tessenow* waren maßgeblich, nicht die moderne Architektur der Weimarer Republik, nicht die reformerischen Ideen des *Bauhauses*. Auch waren Parallelen zu Kasernenbauten und spätwilhelminische Dekoration unübersehbar. Bauten der Gemeinschaft sollten weder städtisch noch monumental sein. Sie sollten Dorf und Kleinstadt vorspiegeln. Darüber hinaus gab es für Wilhelmshaven aber auch noch Gemeinschaftsbauten im Zusammenhang mit den Monumental-
108, 109 planungen[239] von *Verwaltungsforum, Kulturforum, Marineforum* und *Sportforum*. Es ist festzustellen, daß Bauten der Gemeinschaft in den ersten Siedlungen des „III. Reiches" nicht vorkamen. Die auf die Weimarer Republik zurückgehenden Wilhelmshavener
28 Großsiedlungen (z.B. Inselviertel) wiesen nur ausnahmsweise Gemeinschaftsbauten auf. Umgekehrt ist festzustellen: Eine perfekt über das Siedlungsgebiet verteilte Planung von Gemeinschaftsbauten, wie sie sich ab Ende 1937 durchsetzte, war auch typisch für die Architektur des „III. Reiches". Mit fortschreitender Zeit wurden die Bauten der Gemeinschaft mehr und mehr zur Vision städtebaulicher Wirkung und monumentaler Machtdarstellung herangezogen. Sie sollten der Betonung von Achsen, dem Abschluß von Plätzen, der Krönung von Erhebungen und der Abgrenzung von Siedlungsrändern dienen. Nur ein Teil der Gemeinschaftsbauten für Wilhelmshaven wurde realisiert, vor allem Gewerbetrakte, Schulen und *HJ*-Heime usw. Unterlagen über *SS-*, *SA-* und *NSV*-Heime sowie militärische Gemeinschaftsbauten waren nur lückenhaft aufzutreiben[240].

Verwaltungsgebäude für Gebietskörperschaften, Ortsteile, Gemeinden, Städte oder
129 Landkreise innerhalb des *Wohnsiedlungsgebietes Wilhelmshaven-Jeverland* waren mannigfaltig vorgesehen, wurden aber nur ausnahmsweise realisiert. So baute der *MDF* für das *Amt Friesland* 1936–38 einen Verwaltungssitz in städtebaulich bevorzugter Lage als
143 monumentalen Platzabschluß. Architekt war *Josef Wohlschläger* vom *MDF*.

7.7.1. Schulen

Das „III. Reich" erkannte frühzeitig, daß seine „Revolution" scheitern mußte, wenn Indoktrination und „völkische" Erziehung vernachlässigt würden. Waren im NS-Parteiprogramm von 1920 noch unverbindliche Bekenntnisse zum „Ausbau unseres gesamten Volksschulbildungswesens"[241] zu finden, so rückten nach 1933 hohle Phrasen wie Erziehung zu „eigenem Volkstum", zu „Nationalstolz", zu „wehrpolitischer Ertüchtigung" und zu „Ehrfurcht vor der militärischen Tradition" in den Mittelpunkt. In seinem Buch „Mein Kampf" hatte *Hitler* seine rassistischen Erziehungsziele offengelegt. Nationalsozialistische Vererbungslehre Rassen- und Familienkunde wurden in den

Schulunterricht aufgenommen, um den Sinn für „Auslese" und „Führergedanken" zu stärken.

Seelisch-charakterliche Erziehung sollte im „III. Reich" Jugendliche zu den angeblich klassischen Idealen führen. Kleidung und Mode, Frisuren und Auftreten wurden durch NS-Organisationen reglementiert. Schulgebäude und Sporteinrichtungen, Kulturbauten und Jugendheime wurden in die nationalsozialistischen Ziele gepreßt. Erziehungseinrichtungen sollten hygienisch verbessert, Gebäude von Grund auf saniert und Schulstandorte an die Stadtränder verlegt werden[242]. Die sachlich-großstädtischen Schulen der Weimarer Republik[243] waren für die engstirnige „geistig-sittliche" Erziehung zu nationalsozialistischen Zielen wenig geeignet. *Martin Elsaesser,* fortschrittlicher *Stadtbaudirektor* im Frankfurt der Zwanziger Jahre, wollte Schulräume, „die der Erzieher braucht". Schulen sollten möglichst im Grünen liegen, vielleicht als Pavillon-Schulen am Rande der Stadt. Bewegliche Möblierung für alle möglichen Zwecke, Gruppenunterricht in Nebenräumen, Waschbecken und Regale in Klassen sollten „der Entwicklung der Kinder freien Raum gewähren". Die Ziele im Schulbau waren modern und zukunftsweisend. Der Maßstab des Kindes sollte im Bau festgehalten werden, „aber auch, wo die Massenschule gebaut werden muß, ist wichtig, sie nicht mehr als Monumentalbau mit städtebaulicher Absicht (Straßenabschluß, Blickpunkt) zu bauen, sondern als Reich des Kindes zu gestalten. Leichte und bewegliche Bauweise, Beziehung zu kindlichem Maßstab in der Bemessung und Aufteilung der Fenster und Türen wird die Wirklichkeit, die Heiterkeit und Zwangslosigkeit bringen, in der unsere Kinder aufwachsen und gedeihen sollen". Es sind in dieser Zeit allenthalben, vor allem in Großstädten, hervorragende Bildungsbauten im *Internationalen Stil* und in *Neuer Sachlichkeit* entstanden.

Nationalsozialistische Erziehungsgedanken richteten sich nicht auf humanistische, umfassende Bildung, sondern auf „rassische" Auslese nach körperlichen, charakterlichen, geistigen und „völkischen" Gesichtspunkten. Schulen sollten im „III. Reich" gebaut werden, um den „besonders gut Veranlagten der deutschen Jugend so zu erziehen, daß er fähig wird, später in gehobenen oder führenden Stellen unser politisches, kulturliches und wirtschaftliches Volksleben maßgebend mitzugestalten"[244]. Koedukation oder Einübung demokratischen Verhaltens galten als „liberalistisch und schädlich".

Der *Speer*-Mitarbeiter *Ernst Neufert* stellte noch 1940 in seiner „Bau-Entwurfslehre"[245] moderne Schulen in „Flachbauformen" sowie englische und amerikanische Schultypen vor. Von mehrgeschossigen Schulen riet er wegen der schwierigen Entleerung bei Luftangriffen ab. Der NS-Ideologe und Architekturprofessor *Gottfried Feder*[246] forderte an erster Stelle mehr Platz für körperliche Ertüchtigung, Turnhöfe und Turnhallen. Für Wilhelmshavens neue Stadtteile und das *Wohnsiedlungsgebiet* war die Schulbauplanung wegen der fehlenden Beispiele außerordentlich mühsam. *Gottfried Feder* empfahl für seine NS-Idealstadt von 20.000 Einwohnern:

- 5–6 Volksschulen mit je 8 Doppelklassen (bei „wachsendem Volk" 2 für späteren Ausbau)

○ 1 Mittelschule mit 6 Klassen

○ 2 höhere Schulen mit 9 Klassen oder 2 höhere Schulen ohne Mittelschule mit 27 Klassen.

127 Für den Raum Wilhelmshaven war die Siedlungsstruktur von Anfang an durchgrünt und gegliedert vorgesehen. Die Richtzahlen für Schulen in isolierten Stadtteilen unterschiedlichster Einwohnerstruktur machten Festlegungen und Verteilung von Bildungseinrichtungen für die Planer schwierig. Prognosen über Schülerzahlen waren damals wie auch heute fast unmöglich.

Im Gegensatz zum Schulbau der Weimarer Republik überhöhte das „III. Reich" die städtebauliche Bedeutung der Schulen. *Friedrich Nicolaus*[247] vom *RAM* forderte, daß die Schule „durch ihre Baumasse das Orts- und Straßenbild beherrschen" solle. Ihr Baukörper würde erst den „Maßstab geben für die Umgebung, den Platz und die Straße, in deren Blickpunkt sie etwa steht". Die Forderung, daß die Schule da sein sollte, wenn die Wohnungen der *Neuen Städte* bezogen werden, ließ sich nirgendwo einlösen. Provisorien und Schulbaracken wurden die Regel. Die Berliner Dienststellen schienen schon 1938 bemerkt zu haben, daß der Schulbau zu schwerfällig in Gang kam[248]. *Max Büge* vom *RAM* bemängelte, daß erst die Schulen Voslapp und Cäciliengroden Finanzierungshilfen bekommen hätten. Für die übrigen sieben Schulen sollte die Stadt Wilhelmshaven in „kürzester Frist" Entwürfe vorlegen. „Wir bitten daher, dem *RAM* Architekten namhaft zu machen, die schnell und vorbildlich arbeiten und die Gewähr bieten, brauchbare Entwürfe in kürzester Frist auszuarbeiten". Der *RAM* behielt sich vor, die Auswahl selbst zu treffen und geeignete Architekten heranzuziehen.

Schulen für Wilhelmshaven wurden nur dort gebaut, wo es unumgänglich war. Obwohl Massen von Jungarbeitern und Werftlehrlingen in den staatlichen Rüstungsbetrieben arbeiteten, kam es weder zum Bau einer für 4 Mio. Mark geplanten Gewerblichen – noch einer für 3 Mio. Mark geplanten Kaufmännischen Berufsschule.

★

30 Voslapps Schulbauplanung entwickelte sich noch sehr zufällig und unkoordiniert. 1936 rechneten der *Bürgermeister* von Kniphausersiel und der *Amtshauptmann* von Friesland[249] mit einer Zunahme der Schülerzahlen. Da die vorhandene Schule recht alt sei, verlangte *Amtshauptmann Ott* von *Bürgermeister Rastede* Bau- und Finanzierungspläne für eine neue Schule. Schließlich hatte auch das Nachbardorf Fedderwarden gerade seine Schule bekommen. Man wollte die *Marine* an den Baukosten beteiligen, denn diese wolle „in Kniphausersiel etwa 200 Familien ansiedeln"[250]. Das von Architekt *Ludwig Lehmann* am 14.10.1936 vorgelegte Projekt einer 8-klassigen Volksschule für die Wohnkolonie Kniphausersiel zwischen Flut- und Butjadinger Straße wurde zu den Akten gelegt, als sich herausstellte, daß anstatt der Wohnkolonie Kniphausersiel die Arbeiter-Wohnstadt Voslapp gebaut werden sollte.

Mitte 1937[251] ging man von 840 Wohnungen für Marineangehörige und zwei 8-klassigen Schulen in Voslapp aus. *Ministerialdirektor Theodor Schreiber* Chef der *Bauabteilung des OKM* wollte 250 Mark Schulbauzuschuß pro Wohnung leisten. Der zuständige *Oldenburgische Minister der Kirchen und Schulen* wies am 25.09.1937 bei 1.000 Wohnungen für Voslapp 1.200 schulpflichtige Kinder nach. Man hatte keine Erfahrungswerte, und *Regierungsrat Böckmann* setzte die bei „Neubauernsiedlungen" zu erwartenden Kinder von 0,98 auf 1,2 herauf[252]. Da keine Klasse über 50 Schüler umfassen sollte, waren bei 1.200 Schulkindern 24 Schulklassen zu erwarten. *Böckmann* rechnete damit, daß „nur eine ganz geringe Zahl von Schülern auf höhere und mittlere Schulen" gehen dürfte, so daß er drei 3-klassige Volksschulen empfahl. Die Werksiedlung Voslapp war eine Stadt für Arbeiter. Nichts deutet auf die Überwindung von Klassenschranken hin. Eine derartige Schule mit Luftschutzräumen werde ohne Grundstück und Inventar 175.000 Mark kosten. Eine Turnhalle würde noch einmal 50 bis 54.000 Mark kosten. Alles in allem rechnete man für Voslapp bei drei Schulen, einer Turnhalle, Dienstwohnungen, Grunderwerb und Einrichtung mit 919.000 Mark. *Amtshauptmann Hermann Ott* besprach die Problematik am 19.11.1937 im *OKM* und mit *Ministerialrat Augustin* vom *Reichsfinanzminister.* Das *Land Oldenburg* wollte sich mit 25 % an den Kosten der Schule beteiligen. Der *Plan Gemeinschaftssiedlung-Voslapp* des *MDF* enthielt eine 16-klassige und eine 8-klassige Schule. Da die Wohnungsanzahl für Voslapp noch einmal im Februar 1938 erhöht wurde, und zwar auf 1.235, kam es dann zur endgültigen Festlegung auf zwei 8-klassige und eine 16-klassige Schule, zwei Turnhallen, Appellhalle und einen Sportplatz für insgesamt 1,17 Mio. Mark[253]. Die zuständige Gemeinde Kniphausen beauftragte 1937 den *NSDAP*-Stadtrat und technischen Geschäftsführer der *Jade, Willi Lübbers,* mit der Planung aller Schulen in Voslapp. *Lübbers* legte seine ersten Pläne für eine 16-klassige Schule in der Tiarksstraße erst im Februar 1938, zwei Monate nach Baubeginn von Voslapp, vor. *Amtshauptmann Ott* fürchtete um die pünktliche Einschulung der Voslapper Kinder zum 1. Oktober 1938.

Fritz Karl Schneider[254] vom *MDF* war mit den Schulbauplänen von *Lübbers* keinesfalls zufrieden. Die Fenster waren ihm zu niedrig, und die „ganzen Pläne machen den Eindruck einer übereilten Entwurfsbearbeitung und sollten für die Ausführung doch wesentlich sorgfältiger durchgeführt werden". Auch die Ausbildung der Ansichten war für *Schneider* „recht bescheiden". Tatsächlich wurde in Voslapp nur die 16-klassige Volksschule an der Tiarksstraße fertig. Es kam wegen der späten Fertigstellung im November 1940 zu großen Unzuträglichkeiten, so daß an den zwei weiteren Schulstandorten an der Hunrichsstraße und der Tiefstraße Schulbaracken aufgestellt werden mußten. Die Entwurfspläne von *Lübbers* für die Schule Tiarksstraße vom 26.02.1938 zeigten eine kasernenähnliche, zweigeschossige Anlage mit 100 m langen Fluren und nach Süden ausgerichteten acht 54 qm großen Klassen ohne jegliche Fachräume. Die Entwurfsqualität blieb weit hinter dem zurück, was in der Weimarer Republik an Bildungsbauten andernorts üblich war. Die überlangen „Kasernen"-Flure und die umfangreichen Luftschutzeinrichtungen wiesen auf eine Militarisierung des nationalsozialistischen Schulwesens. Im November 1938 legte *Lübbers* neue Entwürfe vor. Die Schulklassen waren nun höher, die Fenster gestreckt, die Dachflächen klarer gegliedert, größer und steiler. Das Äußere der Schule wurde Detail für Detail nazifiziert. Es sollte „einen städtebaulich wichtigen Akzent“ bilden und „zusammen mit dem später zu errichtenden Gemeinschaftshaus von See her in der Hauptsache die Silhouette der Siedlung maßstäblich“ beeinflussen[255]. Holländische Handstrichklinker für die Verblendung, sparsame Verzierung in den Giebeln, Werksteinumrahmungen um die Portale, weit ausladende Freitreppen und Gesimse, schmiedeeiserne Ziergitter und Lampen, geschwungene Schornsteinhauben und aufgedoppelte Türen, vergitterte Oberlichter und plastische Hoheitssymbole waren die Kosmetik, mit der „völkische“ Atmosphäre geschaffen werden sollte. Die 16 Klassen mit Katheder-Podium, dreireihiger Bankmöblierung und fehlenden Fachräumen waren auch pädagogisch ein Rückschritt. Lehrerinnen und Lehrer erhielten getrennte Zimmer. Eine symmetrische Eingangshalle, die man von unten her über Freitreppe und Innentreppe betrat, sollte auf die „seelisch-charakterlichen“ Erziehungsziele einstimmen. Über der Eingangstreppe befand sich ein heute nicht mehr vorhandener (Führer-)Balkon, dahinter dann in Ober- und Dachgeschoß eine 30 m lange, 8 m breite und 7 m hohe Appellhalle. Sie sollte mit ihrem sichtbaren Sprengwerk bis in den Dachraum reichen. Diese Halle für Befehlsausgabe und nationalsozialistische Feiern bekundete die Abkehr des „III. Reiches“ von kindgerechter Erziehung und die Hinwendung zu menschenverachtenden Zielen. Die Appellhalle wurde die erste nationalsozialistische Feierhalle in Wilhelmshaven.

Zwei Rektorwohnhäuser, im März 1939 von *Willi Lübbers* geplant und im August 1940 genehmigt, wurden nicht mehr gebaut. An ihrer Stelle wurde die „Turnhalle für Leibesübungen“ dann 1964 errichtet. Die Luftschutzkeller der Schule wurden nach dem stärker werdenden Bombenkrieg zum Hilfskrankenhaus hergerichtet.

Als im Juli 1938 die Voslapper Siedlungshäuser fertig waren, mußten 120 Kinder aus den südlichen Abschnitten die alte Schule Kniphausersiel besuchen, 70 Kinder aus dem mittleren Teil kamen nach Coldewei und 30 bis 40 Kinder aus dem nördlichen Teil mußten nach Sengwarden. 230 Kinder konnten überhaupt nicht eingeschult werden. Im November 1938 wurden Baracken des Arbeiterlagers und ein umgesetzter Pulverschuppen der *Marine* für Schulzwecke herangezogen. Am 07. November 1940 konnte die Schule Tiarksstraße bezogen werden. Auf die Baracken jedoch konnte man nicht verzichten.

*

Friedrich Heuer entwarf nicht nur die Siedlung Cäciliengroden für die *Wohnungsbau-*
34 *gesellschaft Jade*, sondern auch planungstreu die dazugehörigen Gemeinschaftsbauten. 1938 waren vorgesehen: Schule, Turnhalle/Appellhalle, *HJ*-Heim, Sportplatz, Verwaltungsgebäude, *NSDAP*-Gemeinschaftshaus, Gaststätte, Verwaltungsgebäude und 22 Läden.

Es kam jedoch nur zum Bau einiger Läden, der 8-klassigen Schule und der Turnhalle/Appellhalle. Der *RAM*[256] stimmte am 17.09.1938 *Heuers* Schulplanung zu, allerdings waren ihm die Kosten zu hoch. Schließlich stellten die Berliner Behörden[257] dem *Land Oldenburg* 300.000 Mark Finanzbeihilfe zur Weitergabe an den *Amtshauptmann* in Jever und von dort an die *Gemeinde Oestringen* in Heidmühle zur Verfügung. Ende 1938, als große Teile der Siedlungen bezogen waren, begannen die Bauarbeiten. *Heuer* entwarf die Schule Cäciliengroden als langgestreckten niedrigen Bau mit je einer *139*
Abknickung am Ende. Diese Abknickungen sollten Verwaltungsräume und Turnhalle/Appellhalle aufnehmen. *Karl Pfeiffer* vom *RAM* und *Fritz Karl Schneider* vom *MDF* griffen in *Heuers* Planung ein. Der Luftschutzkeller entfiel, der Eingangsdachreiter bekam eine neue Form und bei der Turnhalle/Appellhalle wurde die Bühne gestrichen. Da absehbar war, daß es zu einem *NSDAP*-Gemeinschaftshaus nicht mehr kommen konnte, wurde angeordnet, daß die Halle gleichzeitig den Parteiveranstaltungen der neuen Ortschaft dienen sollte[258]. Hier wurden später Schulentlassene „verpflichtet", Ritterkreuzträger ausgezeichnet, Gefallene geehrt, Parteiversammlungen durchgeführt, Mutterkreuze verliehen und Impfungen veranstaltet[259]. In den benachbarten Schulklassen etablierten sich ein *NSV*- und ein Werftkindergarten.

In die Architektur der Schule wurde häufig und hastig eingegriffen. Aus militärischen Gründen und wegen des benachbarten Flughafens Mariensiel durften die Schulklassen nicht wie vorgeschrieben 3,50 m, sondern nur 3,30 m hoch sein. Diese militärische Anordnung drückte das Höhenprofil der sowieso schon eingeschossig geplanten Schule noch weiter herunter und verstärkt den flachen Eindruck des „Schulgehöftes".

Der Bau der Schule Cäciliengroden mußte „im militärischen und wehrpolitischen Interesse beschleunigt durchgeführt werden"[260]. *Heuer* und seine Mitarbeiter *Bollermann* und *Gummels* setzten im Dezember 1938 durch, daß die Bauarbeiten beschränkt ausgeschrieben und an die Firma *Herdejürgen und Harmsen* aus Einswarden vergeben wurden, obwohl diese nur wenige 100 Mark billiger war als der Bauunternehmer *Sievers* aus Sande. Die Einswardener Firma hatte die gesamte Siedlung Cäciliengroden hochgemauert und verfügte noch über die Baustelleneinrichtungen und Arbeiterlager.

Im August 1940 war die Schule Cäciliengroden fertig, Als der *Oberwerftdirektor* mit *Wehrkreisobmann* und *NSDAP*-Funktionären im Oktober Cäciliengroden besichtigte[261], konnte *Friedrich Heuer* eine Schule präsentieren, die mit Ziegelfassaden, Walmdächern, Eingangsturm und Sprossenfenstern ein ländlich-sittliches Gesicht zeigte. Tatsächlich jedoch war die Schule durch die nachträglichen Eingriffe etwas mißglückt. Die Traufen waren derart tiefgerückt, daß sie unmittelbar an die Fensteroberkante grenzten.

Die düster und schmucklos wirkende Schule wurde propagandistisch als „modern" dargestellt und wohl auch empfunden: getrennte Tische und Stühle anstatt Bänke, neuzeitliche Sanitäranlagen anstatt Plumpsklo, moderne Zentralheizung anstatt Kanonenöfen, einzelne Klassen anstatt Schichtunterricht. Das alles sollten die „glücklichen Bewohner" *Adolf Hitler* danken. Das NS-System jedoch forderte unnachsichtig

Gegenleistungen ein. 1941 verließ der erste Jahrgang die Schule Cäciliengroden vorfristig. Er wurde für die Rüstungsproduktion der *KMW* benötigt.

★

129 Die Schule in Cleverns südlich von Jever wurde im Vorgriff auf das dort geplante
Siedlungsgebiet 1938 von den Architekten *Hans Ahmels* und *Adalbert Jahn*[262], entworfen
und von der zuständigen *Gemeinde Oestringen* gebaut. Schon 1935–36 hatte das *VI.*
Luftgaukommando See bzw. der *Reichsluftfahrtminister* den benachbarten Fliegerhorst
105 Upjever gebaut[263]. Die dazugehörige Fliegerhorstsiedlung entstand ab 1936. Keinesfalls
wollte man die erforderliche Schule für die noch erheblich größer geplante Siedlung
Cleverns zu spät erstellen. Es ist nicht bekannt, wie sich die Planungen für Cleverns
entwickelten. Die Schule jedenfalls wurde bereits 1938 fertiggestellt. Der eingeschossige
Winkelbau mit Ziegelfassaden und Satteldachform bewegte sich im üblichen, simplen
Stilkatalog. *Amtshauptmann Hermann Ott* forderte am 09.03.1938 vom *Arbeitsamt*
Wilhelmshaven zusätzliche Arbeitskräfte für den Schulneubau Clevernsan, der „jetzt noch
viel dringender geworden sei, weil im Laufe dieses Sommers an der Betonstraße in
Rahrdum für die Luftwaffe 66 neue Unteroffizierswohnungen gebaut und die Kinder
dieser hier angesetzten Familien in die Schule Cleverns mit eingeschult werden
sollen".

★

66 Dem neuen Stadtteil Breddewarden[264], an dem *Hans Bartning* und *Friedrich Heuer* planten, wurden präzise die erforderlichen Gemeinschaftsbauten zugeteilt. Vier Schulen sollten entstehen. Die Schule Breddewarden 1, plante 1939 *Ernst Boyken* mit Sportplatz, Turnhalle/Appellhalle, Lehrerwohnhäusern, Schulgarten usw. Das Schulprojekt für diesen niemals gebauten Stadtteil war sehr weit gediehen, alle Pläne lagen in den Zeichnungsschränken. Der Klassentrakt sollte zweigeschossig werden. Die Flure waren im Gegensatz zu vielen anderen Schulentwürfen jener Jahre in der Mitte geplant (zweibündig). Die große Turnhalle/Appellhalle sollte, obwohl eingeschossig, fast so hoch wie der zweigeschossige Klassentrakt sein. Flache Bögen, natursteinumrahmte Fenster, zentraler Uhrturm, martialischer Flaggenschmuck usw. waren vorgesehene Dekorationsdetails. Da auch diese Schule Aufgaben als Festhalle zu übernehmen hatte, sollte sie besonders hervorgehobene Eingangsportale bekommen. Die Natursteingewände des überbetonten Eingangs sollten kreisförmig weit ausgerundet und mit einem markanten Schlußstein versehen werden. Die Türform symbolisierte den griechischen Buchstaben Omega. Die aufgedoppelte zweiflügelige Eichentür, flankierende Treppen und Wangenmauern markierten die sorgfältig komponierte Torwirkung. Die Planung von *Ernst Boyken* war verschiedentlich verändert worden. Sie wurde 1941 eingestellt. Wie weit die

Planungen der weiteren drei Breddewardener Schulen vorangetrieben wurden, ist unbekannt.

★

Die Gemeinschaftsbauten für das Familienlager Neuengroden wurden 1939 von Generalunternehmer *Möller* „mitgeplant". Ladenbauten, Schule und das Verwaltungsgebäude stießen auf herbe Entwurfskritik bei der *Stadt* und den Oldenburger Behörden. *Egon Forster* von der *Hochbauabteilung* entwarf im April 1939 eine völlig neue Schule als Änderungsvorschlag. Im August 1939 genehmigte der *Oldenburgische Minister der Kirchen und Schulen Forsters* Entwurf. Die Planungen zogen sich lange hin, und es kam zu Auseinandersetzungen, da die *Stadt* darauf beharrte, daß der Fiskus alle Kosten für die Schule, auch die der Einrichtungen, aus Reichsmitteln übernehmen sollte. *46*

1942 kritisierte der *Kreisschulrat* heftig *Forsters* Entwurf, da dieser in dem Z-förmigen Gebäude am geplanten Tsingtau-Platz in zwei Geschossen zwei achtklassige Schulen übereinander vorgesehen hatte. Er fürchtete, daß „zwei Befehlsstellen" die Schulordnung untergraben würden. Die Autorität würde halbiert, und auch die nationalsozialistischen Feiern kämen bei zwei Schulen nicht zur Wirkung. Die Schule Neuengroden wurde gleichfalls niemals gebaut.

★

Für die Altstadt-Wohngebiete Hansaviertel und Jadeviertel waren 1937 bereits drei Schulen vorgesehen. An der Adolf-Hitler-Straße sollte eine 18-klassige Volksschule entstehen[265], die *Egon Forster* 1939 plante, nachdem das Hansaviertel längst fertiggestellt war. Zwei weitere Schulen sollten vor Kopf des Lessingplatzes und vor Kopf der Ernst-Moritz-Arndt-Straße entstehen. Diese Schulen im Gebiet westlich der Werftstraße sollten die jeweils vorgelagerten Plätze beherrschen und mit städtebaulichen Mitteln der Ordnungsmacht des NS-Staates Bedeutung verschaffen. Nicht kindgerechte Erziehungsbauten sollten errichtet werden, sondern Schulen für Befehl und Gehorsam. Die letztgenannten zwei Schulen wurden nicht gebaut. Die Schule an der Adolf-Hitler-Straße jedoch – mitten im Hansaviertel – wurde auf Druck des dort angesiedelten Mittelstandes doch noch realisiert. Im Januar 1939[266] stellte man den Ratsherren die ersten Modelle der *Hochbauabteilung* von *Egon Forster* für die neue Schule vor. Er hatte die Pläne gezeichnet in „andersartiger architektonischer Haltung unserer Zeit, die auch bei Schulbauten keine falsche Pracht mehr vortäuschen will, sondern der inneren Haltung in der die Jugenderziehung heute ausgerichtet wird, auch äußerlich Ausdruck verleiht". Der Entwurf zeigte die gängige Z-förmige Anlage mit Klassentrakt in Ost-West-Lage, einen Querbau mit Haupteingang und Nebenräumen, die Turnhalle/Appellhalle und einen Wohnungsanbau. Sprossenfenster, Ziegelaußenhaut, Walmdach, Uhrturm und schmiedeeiserne Details spiegelten die traditionalistische architektonische Gestalt dieser Schule, *53* *132* *140* die 1939 geplant wurde und 1940 noch in Betrieb ging. Nach teilweiser Zerstörung wurde sie von *Egon Forster* nach dem Kriege wieder aufgebaut.

★

Fedderwardergroden sollte bereits in den frühesten Planungsstadien 5 Schulen
37 bekommen. Die zu den Schulstandorten gehörigen fünf Sportplätze wurden jedoch Zug um Zug auf drei vermindert. Folgende Schulen entstanden 1939–69 tatsächlich in Fedderwardergroden:

138 o Schule A (Schule Albrechtstraße) ab 1939, Architekt *Ludwig Hörold*
142 o Schule B (Schule Salzastraße) ab 1940, Architekt *Franz Iwersen*
o Schule C (Finkenburgschule) ab 1969, Architekt *Hochbauamt*
o Schule D (Herbartschule) ab 1952, Architekt *Egon Forster, Hochbauamt.* Die Planung von 1939 von *Rudolf Bürgin* wurde nicht realisiert.
o Schule E (Schule Nogatstraße) ab 1954, Architekt *Hochbauamt*
o Agnes-Miegel-Schule ab 1964 durch das *Hochbauamt.* Diese Mittelschule war anfangs bei der Schule D vorgesehen. Sie wurde aber erst 1964 an einem anderen Standort an der Warthestraße gebaut.

Fünf Schulen – das entsprach auch den Empfehlungen von *Gottfried Feder*[267], die er 1939 veröffentlichte.

Stadt und *NSDAP* sorgten sich darum, daß sich „die Zustände von Voslapp in diesem
112 neuen Ortsteil keinesfalls wiederholen"[268]. Nach dem *Hitler-* Besuch vom 01.04.1939 hatte man im *MDF* eine *Verwaltungsvorschrift* erarbeitet, wie das Antrags- und Bewilligungsverfahren zu beschleunigen sei[269].

138 Im Juni 1939 begannen die Bauarbeiten bei der Schule A. Ende November 1940 konnte sie eingerichtet werden und 1943 war sie soweit fertiggestellt, daß ein Hilfskrankenhaus im Keller eingerichtet werden konnte. Noch im März 1941 hatte *Stadtbaudirektor Walter Temp* dem *Magistrat* die Pläne für das an die Schule Albrechtstraße grenzende *HJ*-Heim vorgelegt. Es sollte vom *RAM* finanziert werden. *Ludwig Hörold* hatte den Gesamtkomplex um die Schule Albrechtstraße mit Lehrerwohnhäusern, Turnhalle/Appellhalle, *HJ*-Heim, Sportplatz, Gärten usw. geplant. Architektonischer Mittelpunkt des 110 m langen Schultraktes war der zentral angelegte Schuleingang mit halbrunder Freitreppe, stilisierten Eckquadern und betonter Tür. Streng symmetrisch rechts und links befanden sich Rundfenster, darüber ein schmiedeeisern umwehrter (Führer-)Balkon, auf den 3 Türen führten. Auch der Glocken-/Uhrturm fehlte nicht. Der zweigeschossige Klinkerbau mit Ziegelwalmdach wies unter dem Nebentrakt und zum Schulhof hin zahlreiche flache Bögen auf. Überlange Flure und verschiedene Ausstattungsdetails weisen auf *Hörolds* Vorbild für diesen Schulbau hin: den Kasernenbau. Erst 1947 wurde die Schule Albrechtstraße komplett fertiggestellt. Die Turnhalle/Appellhalle kam über die noch heute sichtbare Gründungsplatte nicht hinaus.

Die Schule B an der Salzastraße ging 1 Jahr später in Bau als die Schule A. Im Auftrage des *Hochbauamtes* hatten *Franz Iwersen* und sein Bruder *Hanns Iwersen* im Februar 1939 erste Entwürfe für eine 8-klassige Volksschule mit Turnhalle/Appellhalle, Sportplatz und 2 Lehrerwohnhäusern vorgelegt. Sie waren nicht geneigt, die Zensur von *Klaus Rasch* und *Egon Forster* in ihren Entwurf einzuarbeiten. Erst nachdem die *Baupolizeiabteilung* Details vorgab, zweiläufige Eingangstreppe und schräge Eisengeländer verboten hatte, legten die Brüder *Iwersen* einen neuen Entwurf vor. Im Mai 1940, noch während der Bauphase, hatte *Klaus Rasch Iwersens* bei einem neuen Entwurf „die Hand geführt". Helle Eckquader, wie *Hans Fricke* sie bei der Schule Altengroden vorgesehen hatte, mußten her, die Rundbögen wurden zu flachen Bögen uminterpretiert, einige (Führer-)Balkone kamen hinzu und das gesamte Eingangsportal wurde neu ausgerichtet. Eine breit gelagerte Treppe, Fahnenstangen rechts und links, nazifizierten den Entwurf. Mitte 1940 war das Kellergeschoß der Schule B fertiggestellt. Mitte 1940 war das Kellergeschoß der Schule B fertiggestellt. Mitten im Ausbau jedoch wurde der Bau bis auf den Luftschutzkeller stillgelegt. 1942 kaufte das *Hochbauamt* der Stadt von der Firma *Paul Horgass* aus Magdeburg eine 8-klassige Schulbaracke, die an der benachbarten Warthestraße aufgestellt wurde. 1946 beantragte die *Stadt* über das *Dezernat für Baulenkung* beim *Regierungspräsidenten Oldenburg* von den Alliierten die Zustimmung zur Fertigstellung der Schule B. *142* *145*

Rudolf Bürgin hatte Anfang 1939 nach einem Wettbewerb von der Stadt den Auftrag zur Planung des „Zentrum-Nord" von Fedderwardergroden bekommen. Die Pläne bis 1938 ließen die Nord-Süd-Achse nördlich mit der Schule D enden. Erst 1939 zum Wettbewerb wurde dann eine nie ganz realisierte Straßenverbindung nach Breddewarden vorgesehen und das Programm noch um ein *HJ*-Heim, eine Mittelschule, eine Sportanlage und ein *NSDAP*-Gemeinschaftshaus ausgeweitet. *Bürgins* Wettbewerbsplanungen zeigten einen geometrisch ausgerichteten Appellplatz, umgeben von *NSDAP*-Gemeinschaftshaus, Mittelschule, Schule D und angrenzendem *HJ*-Heim. Die Schule D war wie die Mittelschule von überlangen Klassenfluren durchzogen und mit den Klassenräumen zum Appellplatz orientiert. *Bürgin* verwendete den gängigen heimattümelnden Stilkatalog: steile Satteldächer, rohes Ziegelmauerwerk, steile Giebel, herausragende Schlußsteine (Führer-)Balkone, werksteinumwehrte Kreuzsprossenfenster, markante Eckquader, herausgehobene Ortgänge usw. Abwalmungen und Aufschieblinge an den Dächern sollten fehlen. Besonders dominierend ist in seinen Wettbewerbsplanungen von 1939 die reichhaltige Dekoration der Eingangsbereiche. Das Hauptportal der Schule zum Appellplatz hin sollte stichbogenüberwölbte Eingangstüren bekommen, durch Natursteinquader zusammengefaßt. *Bürgins* zahlreiche Werkzeichnungen lassen seine Vorliebe für Natursteingewände, Hoheitszeichen, Balkenköpfe, Schlußsteine und Schmiedeeisen erkennen. Keine Lampe, kein Türgriff, keine Kassettendecke, kein Geländer, keine Treppe war ihm unwichtig. Penibel und in großem Maßstab zeichnete er die Gestaltung vor. Kein Schulentwurf zeigt die nationalsozialistische Zielsetzung so offen wie die Schule D. *37* *146* *141*

Erziehung zum Gehorsam, Einschüchterung der Kinder, völkische Auslese und Drill
141 zur Ordnung lassen sich in den Fassadenaufrissen ablesen. Die Verwandtschaft zu
NS-Ordensburgen und zu nationalsozialistischen Erziehungsanstalten (*NAPOLA*)
146 drängt sich auf, insbesondere im Zusammenhang mit den benachbarten Bauten der
Gemeinschaft. Das *RAM*[270] mischte sich nocht mit verschiedenen Änderungsskizzen ein,
gleichfalls die städtische *Bauverwaltung* im Juli 1940. Bis auf den Gewerbetrakt
Kniprodestraße konnte *Rudolf Bürgin* nichts von seinem „Zentrum-Nord" realisieren.
Erst *Egon Forster* baute die Herbartschule 1951 an der Stelle des von *Rudolf Bürgin*
vorgesehenen *HJ*-Heimes. Die geplante Mittelschule wurde an der Warthestraße erst
1964 realisiert.

★

43 Die erste Planung der Marinesiedlung Altengroden von *Ewald Zenker* enthielt 1939
eine Schule in Altengroden-Nord und 2 Schulen in Altengroden-Süd. Anzahl und Lage
der Schulen wurden in späteren Planungsstadien mehrfach verändert. Nur die Planung
für die Schule I an der Ubbostraße in Altengroden-Nord wurde noch vom *Hochbauamt* in
die Wege geleitet. Der *Landesleiter Weser-Ems der Reichskammer der bildenden Künste, Hans*
Fricke aus Oldenburg, legte im August 1940 erste Entwürfe für die 16-klassige
145 Volksschule vor. *Fricke*, ein *Bauhaus*-Schüler, hatte früh seine Architekturauffassung über
Bord geworfen und war 1931 der *NSDAP* beigetreten. Er war einflußreicher
Architektenfunktionär und Berater von *Gauleiter Carl Röver* in Kulturangelegenheiten.
Die Planung passierte, wie nicht anders zu erwarten, ohne Kritik alle Dienststellen in
Oldenburg und Wilhelmshaven und wurde am 22.01.1941 vom Leiter der *Hochbauab-*
teilung, Klaus Rasch, genehmigt[271]. Man kannte sich, man hatte gleiche Architekturauf-
fassungen und man war seit langen Jahren *NSDAP*-Mitglied.

Die Baugrube wurde noch ausgehoben und die Pfähle in den Boden gerammt. 1942 jedoch mußte eine Schulbaracke an der Klinkeburg aufgestellt werden, denn kriegsbedingt konnte die Schule Altengroden nicht weitergebaut werden. Das *RAM* bewilligte Mittel für den 8-klassigen Barackenbau, in dem im Schichtunterricht bis 12 Klassen unterrichtet wurden[272].

Frickes Entwurf von 1940 zeigt eine großzügig beplante Landschaft am Rand der Marinesiedlung Altengroden zwischen Badeteich und Siedlungsrand. An alles war gedacht: normgerechter Sportplatz, Sportlerumkleidehaus, Turnhalle/Appellhalle, Lehrerwohnhäuser mit Garten, Wäschetrockenplatz, mehrere Schulhöfe, Rektorwohnhaus und ein Luftschutzkeller für 417 Menschen. Grundrißlich zeigt die Schule, wie auch andere Entwürfe der NS-Zeit, überlange Flure, kathederorientierte Klassenzimmer, jedoch wenig Fachräume. Im Äußeren ist der *Fricke*-Entwurf ein Musterbeispiel heimatschutzorientierter Architektur. Der zweigeschossige Klinkerbau sollte ein steiles Satteldach bekomme jedoch ohne die in Wihelmshaven beliebten ausgeprägten Dachaufschieblinge. Der Eingangstrakt sollte nicht abgewalmt werden, sondern durch eine strenge, hoch aufragende Giebelform mit 3 Stichbögen, eine breitgelagerte Eingangstreppe und 5 hohe Fenster dominieren. Die Natursteineckquader und Kreuzsprossen vermitteln einen herben Eindruck. Der lange Klassentrakt sollte zur Straße schlanke Klassenfenster und zum Schulhof 9 (Führer-)Balkone bekommen. Am Kopf des Lehrertraktes war dem Schulleiter ein besonders herausgehobener Balkon zugedacht, von dem er Befehle ausgeben und die Schulgemeinde übersehen konnte. Fenster, Türen, Balkone, Sockel und Treppen sollten mit Naturstein umrahmt werden bzw. aus Naturstein bestehen.

In erstaunlicher stilistischer Kontinuität setzten 1959 *Egon Forster* und *Walther Nehl* vom *Hochbauamt* die Planung von *Hans Fricke* fort[273], als es darum ging, auf den 150 bereits 1941 gerammten Pfählen weiterzubauen. 1961 wurde die Schule in Altengroden Nord eröffnet. Städtebaulich blieb es bei der Planung *Fricke*, im Detail jedoch paßte sich das *Hochbauamt* den Nachkriegsverhältnissen an und der im Grundriß Z-förmige Hauptbaukörper wurde „entnazifiziert". Der Haupteingang wurde nicht mehr zur Straße hin orientiert, er rückte zur Seite bzw. nach hinten. Die Abwalmungen im Nebentrakt wurden durch Giebel ersetzt. Eckquader, Natursteinsockel und Sprossenfenster entfielen. Anstatt der natursteinumrahmten Einzelfenster faßte man nunmehr die Fenster durch Betonkonstruktionen zu Bändern zusammen. Sonst blieb jedoch alles beim alten. Nach der Fertigstellung 1961 folgte noch ein 2. Bauabschnitt mit Lehrschwimmbecken und Turnhalle. Da jedoch Pfähle für diesen 2. Abschnitt nicht zur Verfügung standen und das Programm sich verändert hatte, wurde der *Fricke*-Entwurf bedeutungslos.

7.7.2. HJ-Heime

Die Organisation der *HJ* ging auf den *Jugendbund der NSDAP*[274] von 1922 zurück. Auf dem 1. Parteitag der wiedergegründeten *NSDAP* am 03. und 04.07.1926 in Weimar wurden die nationalsozialistischen Jugendbünde in *Hitlerjugend, Bund Deutscher Arbeiterjugend* umbenannt. Sie wurde die einzige offizielle Jugendorganisation der *NSDAP*. Schon 1932 hatte die *HJ* 108.000 Mitglieder und war in Norddeutschland stark vertreten. *Baldur von Schirach, Reichsjugendführer der NSDAP* und seit 1933 *Jugendführer des Deutschen Reiches,* trat an die Spitze der *HJ*, die 1936 6 Mio. Mitglieder zählte. Das „III. Reich" ließ im Erziehungswesen nur staatliches Schulwesen und Parteieinrichtungen gelten. Gewerkschaften, Kirchen, Sozialdemokraten, Kommunisten, Arbeiterwohlfahrt und Sportvereine hatten keinen Einfluß mehr auf die Erziehung. Schulpflicht bestand vom 6. bis zum 14. Lebensjahr. Mit 10 Jahren war es Pflicht, in das *Deutsche Jungvolk (DJV)* oder die *Deutschen Jungmädel (DJM)* einzutreten. Mit 14 Jahren erfolgte der Übertritt in die *HJ* und in den *BDM*. Mit 18 Jahren wurden *HJ* und *BDM*-Mitglieder in die *NSDAP* eingegliedert.

HJ-Heime zu bauen, war für die Nationalsozialisten seit 1934 wichtigstes Mittel, die Jugend für ihre Ziele einzunehmen. *Baldur von Schirach*, hatte 1937 dem Gebiet Nordsee der *HJ* 101 neue Heime versprochen, davon dem Land Oldenburg 37[275].

HJ-Heime für die 10- bis 12jährigen Jugendlichen waren flächendeckend vorgesehen. Sie waren die bauliche Hülle für angebliche „Volkstreue", „friedliebende" Erziehung und „völkische" Ausbildung. *Baldur von Schirach* proklamierte zu jedem Jahr neue Parolen. 1937 war das *Jahr der Heimbeschaffung*. Nach den Versprechungen von *Gauleiter Carl Röver* sollten die 37 *HJ*-Heime des Landes Oldenburg „in so würdiger Form entstehen, ... daß sie Zeugen seien unserer Zeit"[276]. Um eine bodenständige und in die norddeutsche Landschaft passende gleichartige Bauweise durchzusetzen, beauftragte die *Gebietsführung Nordsee der HJ* in Oldenburg, den Architekten *Heinz Däke* mit dem Entwurf aller 37 *HJ*-Heime im Lande Oldenburg. Im Jahre 1937, dem *Jahr der Heimbeschaffung* sind angeblich im *Deutschen Reich* 400 *HJ*-Heime[277] entstanden. Planungshinweise und Richtlinien wurden in „Werkheften für den Heimbau der Hitlerjugend" veröffentlicht. Reichsarchitekt der *HJ* war *Hanns Dustmann*[278]. *Ernst Neufert, Helmut Hentrich, Hanns Häuser* u. a. *GBI*-Architekten, entwarfen *Hitlerjugend*-Häuser im Reich, streng orientiert an den Maximen zur „soldatischen" Erziehung der jungen Generation. Der Reichsarchitekt der *HJ*, stufte die Architektur entsprechend der Region und der städtebaulichen Einordnung ab. Für Berlin entwarf er große Versammlungshallen im neuklassizistischen Stil, für Melle beispielsweise baute er 1936/37 ein *HJ*-Heim in Fachwerk. Auch die Baupläne für Wilhelmshaven mußten vom *Arbeitsausschuß für HJ-Heimbeschaffung* in Berlin genehmigt werden. Die Stadtteilplanungen für den Raum Wilhelmshaven ergaben 1940 ein Ziel von 30 *HJ*-Heimen. Tatsächlich realisiert wurden jedoch nur:

1938 — *HJ*-Heim für Jungarbeiter der Werft in der ehem. katholischen Cäcilienschule[279] mit Eßsälen, Gemeinschaftsräumen, Lesezimmern und Schlafsälen (Planung des Umbaus: *Hafenbauressort* der KMW)

147 1938–39 — *HJ*-Heim am Krähenbusch (als Ersatz für die Baracke der Rüstringer Jugendpflege) vor Kopf eines Appellplatzes[280] des heutigen Otto-Meentz-Platzes (Architekt: *Heinz Däke)*

1935–36 — Otto-Weddigen-Jugendherberge an der Freiligrathstraße. Sie unterstand gleichfalls der *HJ* und wurde fast nur von ihr genutzt[281]. Sie wurde von *Baldur von Schirach* am 30. 06. 1936 anläßlich des *HJ*-Gebietssporttages eingeweiht und 1938 noch erweitert (Architekt: *Heinz Däke)*

Die Parteipropaganda überschlug sich in Lobeshymnen: „Die Jugend des Führers kann in Herbergen wohnen, die der Jugend eines großen, starken Volkes würdig sind. Auch hieran sollte man denken, wenn am 10.04. das gesamte Volk antritt, um sich zum Führer zu bekennen."

Nach mehreren Vorentwürfen und Korrekturen des *Arbeitsausschusses für HJ-Heimbeschaffung* hatte *Heinz Däke* die Jugendherberge an der Freiligrathstraße zu einem Gebäude interpretiert, das Härte und Disziplin, Tugendhaftigkeit und Wehrbereitschaft symbolisieren sollte. Die Friedenstraße führt auf einen von der Stadt symmetrisch durch

Fahnen eingefaßten Vorplatz, der auch für Appelle benutzt wurde. In den vorlaufenden Entwürfen waren noch breitgelagerte Fensterformate enthalten. Die Ansicht zum Appellplatz wurde von Entwurfsschritt zu Entwurfsschritt „gestrafft". Die Sprossenfenster wurden schmal und im Erdgeschoß bekamen sie mittig darüber Reliefs. Der Haupteingang bekam eine wuchtige Terrakotta-Umrahmung, flankiert von Rundfenstern und Laternen, darüber ein „Wikingerschiff".

Am 23.03.1938 besuchte der *Reichsjugendführer* Wilhelmshaven und besprach das weitere Programm der *HJ*-Heime für die Stadt. Er äußerte sich befriedigt über die Aufbaupläne. *Schirach* war *Hitler* blind ergeben und organisierte im Reich gigantische Jugendaufmärsche. Mit dem groß-propagierten Heimprogramm wollte er die *HJ* noch weiter ausbauen.

Ein *Gesetz zur Förderung der Hitler-Jugend-Heimbeschaffung* trat am 30. 01. 1939 in Kraft. Den nationalistischen Schwärmereien und unechten Idealen entsprach die Architektur der *HJ*-Heime. Ausrüstung und Dekoration knüpfte an bündische Jugend und *147*
Wandervogelbewegung an oder wurden sogar von diesen übernommen. Alle Gebäude hatten Schulungs-, Gemeinschafts-, Gruppen- und Führerräume. Die Organisationsstruktur spiegelte sich in Appellfluren, Hoheitszeichen, Fahnenhallen und Kameradschaftsräumen. Hervorragend ließen sich die gestalterischen Ziele mit den landschaftsgebundenen Materialien Klinker und Dachziegel, mit den Bauformen Sprossenfenster und Walmdach verbinden. Sie symbolisierten Dauerhaftigkeit und Handwerkerleistung. Raumprogramm und Grundriß der *HJ*-Heime waren immer ähnlich aufgebaut[282]. Von der Eingangshalle aus sollte das Haus insgesamt zu übersehen sein. Vor dem *HJ*-Heim befand sich immer ein Appellplatz, am Krähenbusch ein Platz von 200 m Länge, der auf der gegenüberliegenden Seite der Peterstraße durch ein Ehrenmal abgeschlossen werden sollte. Im Inneren dann getrennte Räume für *Jungvolk, BDM, Jungmädel* und *HJ*. Im ganzen Reich war es so: Jungen rechts, Mädel links. Die zweigeschossige Halle sollte mit Fahnen dekoriert werden. Im Eingangsgiebel war das Zeichen für die *HJ* in Sandstein gemeißelt. 1939 wurde mit dem Bau des 75.000 Mark teuren *HJ*-Heimes Krähenbusch *147*
begonnen, es wurde jedoch niemals als *HJ*-Heim genutzt, denn nach Ausbruch des Krieges hatte der Wilhelmshavener *Polizeipräsident* das Gebäude zur Unterbringung von *Sicherheits- und Hilfskräften (SHD)* beschlagnahmt. *NSDAP-Kreisleiter Ernst Meyer* vertröstete *Oberbürgermeister Wilhelm Müller* am 18.05.1940 und versprach, das Gebäude schon bald nach dem Kriege als *HJ*-Heim bereitzustellen.

Auch *Gottfried Feder* hatte in seinem Standardwerk „Die neue Stadt" Musterraumprogramme für *HJ*-Heime und für Jugendherbergen vorgegeben[283]. „Führerzimmer", „Gefolgschaftsräume" und Massenlager verrieten in der Wortwahl die „völkischen" Ziele[284].

Die ersten Planungen für Voslapp von 1937 zeigten auf dem Deich zum Jadebusen ein *30*
HJ-Heim mit Aussichtsturm, das *Willi Lübbers* entworfen hatte. Es stand im Zusammenhang mit dem *NSDAP*-Gemeinschaftshaus vor Kopf der fast 400 m langen Aufmarschachse.

37 Für Fedderwardergroden[285] waren gleichfalls von Anfang an 3 *HJ*-Heime vorgesehen.
34, 66 Cäciliengroden sollte 1 *HJ*-Heim, Breddewarden 3 *HJ*-Heime bekommen. Auch für die
62–64 nicht gebauten Stadtteile Burg Kniphausen, Schilldeich, Utters, Langewerth usw. waren entsprechend den Richtzahlen des *Arbeitsausschusses für HJ-Heimbeschaffung* Häuser für die nationalsozialistische Jugend vorgesehen.

Beachtlich war, daß einige im Stadtgebiet vorhandene alte Bauernhöfe im Zusammenhang mit neuen Stadtteilen erhalten und zu *HJ*-Heimen umgebaut werden sollten.
46 Das Familienlager Neuengroden und die nördliche Stadtparkkolonie sollten je einen
43 *HJ*-Bauernhof, Altengroden sogar zwei *HJ*-Bauernhöfe bekommen. Es gab *HJ*-Heime[286], die durch Umnutzung alter Bauten die gewünschte traditionelle Ausstrahlung bekommen hatten. In Dinkelsbühl wurde das Kornhaus, in Hannover der Ballhof, in Neuburg ein Marstallgebäude, in Ingolstadt gar Teile der Stadtumwallung zum *HJ*-Heim umgebaut. Auch hier nutzten die Nationalsozialisten geschickt den romantischen Habitus. Jugendbewegung und Wandervogel, Freischar und Jugendherbergswerk hatten sich schon nach der Jahrhundertwende gerne in Burgen und Schlössern etabliert. Landschulheime bevorzugten Bauernhäuser oder Gutshöfe als naturnahe Mittelpunkte für ihre Reformziele.

Das größenwahnsinnige Ziel, die „*Stadt der 500.000*“ mit 30 *HJ*-Heimen flächendekkend zu überziehen, wurde durch den Kriegsausbruch beendet.

7.7.3. NSDAP-Gemeinschaftshäuser

Gemeinschaftshäuser waren in Städten und Wohngebieten, Betrieben und Werksiedlungen geplant. Sie hießen auch Gefolgschaftshäuser, Freizeithallen o. ä. und sollten der *NSDAP*-Parteiarbeit auf unterster Ebene dienen[287]. Sie sollten „Mittelpunkt der Zusammenfassung aller in der neuen Stadt lebenden Menschen sein“[288]. Wilhelmshavens Gemeinschaftshäuser galten als beispielhaft, und der *RAM* veröffentlichte sie als Muster. Städtebaulich sollten sie den „beherrschenden Abschluß des Hauptplatzes“ einer Siedlung bilden und im „leichtgebundenen Zusammenklang mit dem Ganzen“ ihren „wichtigen Auftrag erfüllen“. *NSDAP-Reichsorganisationsleiter Robert Ley* hatte für ihren Bau *Richtlinien* herausgegeben[289]. Gemeinschaftshäuser sollten den Feierraum der Dienststellen der Partei, eine kleine Gaststätte mit Küche, eine Gesundheitsstation, Kindergarten und Sportanlagen umfassen. Die städtebauliche und architektonische Verantwortung lag bei *GBI Albert Speer,*[290] der darauf Wert legte, daß die geplanten *NSDAP*-Gemeinschaftshäuser einen „beherrschenden Platz einnehmen“ und die erforderliche Verbindung zu *HJ*-Heimen, Aufmarschplätzen und Ehrenhöfen bekamen. *Speer* forderte, daß die Verwaltungsräume von Staat und Gemeinden in besonderen Gebäuden untergebracht sein sollten. Die *NSDAP* sollte für sich „deutlich erkennbar sein“. *Gauleiter Carl Röver* schrieb *Rudolf Heß* am 18.03.1940 zu den *NSDAP*-Gemeinschaftshäusern, „daß in den neuen Wilhelmshavener Siedlungsgebieten die Betreuung und

Bearbeitung der Volksgenossen durch die *NSDAP* intensiver sein muß als im übrigen Reich".

Gottfried Feder, der Planer von Langewerth, erklärte das Haus der *NSDAP* „zu den 64
wichtigsten Einrichtungen der zukünftigen Stadt"[291]. *Friedrich Heuer*, vielfach mit Wilhelmshavens Planung verbunden, propagierte die „Wiedergewinnung einer neuen Volksgemeinschaft" nach Zielen der *NSDAP*[292]. Er wollte die ideale Stadt nach Ortsgruppen gliedern. Je 8.000 Einwohner sollten eine Gemeinschaftshalle bekommen. In seinen Musterentwürfen vom Dezember 1940 bildet das Gemeinschaftshaus den Kopfbau einer 1.000 m langen Nord-Süd-Achse. In Seitentrakten sollten Parteigeschäftsstelle, Verwaltung, Gemeinderäume, *HJ* und *NSV* untergebracht werden. *Heuers*
Planung für die Ortsgruppengröße unterschreitende Siedlung Cäciliengroden von 1938 34
war stark ländlich orientiert. Die Gemeinschaftseinrichtungen sollten über die für 1.800
Einwohner geplante Siedlung verstreut sein: das *HJ*-Heim bei der Schule, das 139
NSDAP-Gemeinschaftshaus am Ende der Siedlung. Beide Häuser wurden nicht gebaut.

Auch die *Bartning*-Planung für Breddewarden von 1939 zeigt eine wenig monumentale *NSDAP*-Gemeinschaftshalle. Die im Süden der Siedlung liegende Wurt sollte landschaftsgebunden und locker mit Parteihäusern bebaut werden, aber auch vorhandene
Altbauten und Bauernhäuser einbeziehen. *Heuer* setzte die Planungen fort. 66

NSDAP-Gemeinschaftshäuser sollten Aufgaben übernehmen, die zuvor Gewerkschaftshäuser, Sportheime und Kirchen hatten. Die Nationalsozialisten hatten deren Struktur brutal zerschlagen. Dem Verfasser ist nur ein 1940 ausgewiesener Kirchenstandort für Wilhelmshaven bekannt[293]. Die privilegierte Lage der *NSDAP*-Gemeinschaftshäuser an Plätzen und Achsen versinnbildlichte die Verdrängung der Kirchen aus der
Stadt. Es war kein Zufall, daß viele Entwürfe für Gemeinschaftshäuser (Glocken-)Türme, 146
kreuzgang-entlehnte Grundrisse, Hallenschiffe und apsis-ähnliche Giebel aufwiesen.

*

Für Voslapp plante der einflußreiche *Willi Lübbers* noch vor seinem Abgang bei der 30
Jade im Oktober 1938 Bauten der Gemeinschaft. Rathaus, Feuerwehr und Bürgermeisterhaus sollten dörflich-bieder im Mittelpunkt stehen, allerdings durch eine 400 m lange Achse mit dem abseits gelegenen *NSDAP*-Gemeinschaftshaus vor dem Deich verbunden. Noch weiter abgesetzt sollte das *HJ*-Heim auf dem Deich angeordnet werden. Das die Achse abschließende über 100 m breite *NSDAP*-Gemeinschaftshaus mit seinem aufragenden Turm sollte bei festlichen Parteiveranstaltungen Ziel von Aufmärschen sein. Die *NSDAP* wollte repräsentieren und die „Volksgenossen" in der Arbeiter-Wohnstadt Voslapp durch Macht symbolisierende Bauten beeindrucken.

Willi Lübbers stellte seinen Entwurf des *NSDAP*-Gemeinschaftshauses am 01.04.1939 im Wilhelmshavener Rathaus *Adolf Hitler* vor. Dann transportierte er es eigenhändig nach München zu *Rudolf Heß*. Später fand es sich im Modellsaal des *GBI* in Berlin wieder[294]. Wiederholt mahnten die *NSDAP*-Funktionäre das Gemeinschaftshaus beim *RAM* in Berlin an. 1.200 Sitzplätze sollte der Feierraum für Parteiveranstaltungen haben, darüber hinaus eine Wirtschaft. Das Grundstück sollte von der *Jade* an die *Stadt* „übergehen". Die *Stadt* sollte das Gemeinschaftshaus bauen und es der *NSDAP* in Verwaltung geben[295]. Die Realisierung verzögerte sich immer mehr und so waren die Ratsherren froh, daß ihnen *Stadtbaudirektor Temp* im November 1940 die im Bau befindliche Appellhalle der Schule als Versammlungsraum präsentierte[296].

★

37 Für Fedderwardergroden waren ein Gemeinschaftshaus in der Mitte, eines im Norden
und eines im Süden vorgesehen[297]. Nach dem Wettbewerb von 1939 plante *Rudolf Bürgin*
146 das „Zentrum-Nord". Die Pläne des Gemeinschaftshauses zeigten eine kirchenschiffähnliche nationalsozialistische Feierhalle unter einem monströsen Satteldach. Ein markanter Turm sollte das „Zentrum Nord" überragen. Im Äußeren sollten rohe Ziegel, natursteinerne Eckquader, plastische Hoheitssymbole und flatternde Fahnen *NSDAP*-Architektur ausdrücken. Unter einer sichtbaren Holzbalkendecke mit Kassettenausfachung sollten sich die Fedderwardergrodener zu weihevollen Parteiveranstaltungen versammeln. Hohe Hallenfenster, Lampen als Opferschalen, vielfacher Fahnenschmuck und eine Heldengedenkstätte sollten den architektonischen Rahmen für Liturgien des NS-Regimes abgeben.

Das Gemeinschaftshaus „Fedderwardergroden-Mitte" sollte am nördlichen Kopf des
111 „Langen Marktes" entstehen, östlich davon das *HJ*-Heim. 200 m weiter südlich war ein großes quadratisches Verwaltungsgebäude mit innenliegendem Ehrenhof als Abschluß vorgesehen[298].

Für das „Zentrum-Süd" arbeiteten die *Abteilung XXIV. der Marineintendantur* und die *Bauabteilung des OKM* zeitgleich mit dem Wettbewerb[299] an Plänen für eine Freizeithalle. Anfangs dachte die *Kriegsmarine* an eine Freizeit-Baracke am Rande Fedderwardergrodens. Die Freizeithalle war als ein „Haus der Arbeit" für die Marine-„Gefolgschaft" gedacht, keinesfalls für die *NSDAP*. Die *KMW* wollte den Feierabend ihrer Arbeiter gestalten und Raum für Kultur und Frohsinn schaffen. Da sich die *Marineintendantur* mit den Plänen von *Karl Franzius* auf das „Zentrum-Süd" festgelegt hatte, kam es zu Konflikten mit der *NSDAP*, die dort ihr Gemeinschaftshaus errichten wollte. Die *Stadtverwaltung* stand hierbei auf Seiten der Partei und leistete Hilfe für den Wettbewerb. Auch *Gauleiter Carl Röver* wollte den Standort für die *NSDAP* sichern. Er schrieb noch
144 im März 1940 an *Heß* und wollte Unterstützung für sein Parteiprojekt. Nach den Planungen von *Karl Franzius* wurden die Rammarbeiten am 26. 09. 1940 begonnen.

Die *Marine* setzte sich möglicherweise deswegen durch, weil die Betreuung der 4.000
154 in Fedderwardergroden arbeitenden Bauhandwerker und das Wohlbefinden der Rüstungsarbeiter kriegswichtig waren. Verordnete Freizeit und Amüsement, Variete

und Film, Konzerte und Vorträge sollten über die miserablen Verhältnisse des zweiten Kriegsjahres hinwegtrösten.

Oberbürgermeister Müller[300] nahm die Entscheidung des *OKM* hin, da er damit rechnete, daß die *Marine* nach dem Kriege an Einfluß verliere und die *Stadt* dann ein fertiggestelltes *NSDAP*-Gemeinschaftshaus übernehmen könne. Die *Kreisleitung der NSDAP* war nach wie vor nicht einverstanden. Auch *Fritz Karl Schneider* vom *MDF* zeigte sich noch am 07. 01. 1941 überrascht, daß nunmehr die Freizeithalle für die *Marine* entstehe und der Bau begonnen war. Er kritisierte die nicht erfolgte Anzeige bei der höheren *Baupolizei* und empfahl wegen der geänderten Zweckbestimmung ein gemeinsames Gespräch mit der *NSDAP*. *Schneider* kritisierte das „vorgeblendete Fachwerk", das hier nicht ortsüblich sei. Auch der Chef der Wilhelmshavener *Baupolizeiabteilung, Max Neumann,* hatte „ernste Bedenken" gegen Fachwerk und schaltete das *Gewerbeaufsichtsamt Oldenburg* gegen die Fachwerkkonstruktion ein.

Architekt *Stuhlmann* von der *Abteilung XXIV.* der *Marineintendantur* teilte knapp mit, daß das Holz bereits gekauft sei. Die Bauarbeiten wurden dann doch Ende 1940 aus Baustoffmangel eingestellt. Die Umfassungsmauern standen noch bis weit in die Nachkriegsjahre. Der Standort wurde nach dem Kriege durch einen Supermarkt bebaut.

Der Kern des Entwurfs von *Karl Franzius* sollte die Halle sein – 250 m lang und 120 m *144*
breit, sollte sich der Fachwerkgiebel in die Blickrichtung der Nord-Süd-Achse schieben. *37*
Das gewaltige Satteldachgebäude mit einer Firsthöhe von 22 m sollte ein Theater für 1.400 Zuschauer abdecken. Das Äußere wollte bäuerlich-heimatverbundene Kulisse für ein modernes Veranstaltungsgebäude mit Klimaanlage, Vorführeinrichtungen, Bühnentechnik, Cafe usw. sein. Eine derartig wuchtige und voluminöse Gemeinschaftshalle aus Fachwerk zu planen, war weder historisch begründet, noch regional ableitbar, noch konstruktiv möglich. Das vorgespielte mittelalterliche Äußere der Halle und seiner flankierenden 2-geschossigen Bebauung sollte einen Aufmarschplatz, den Hochmeisterplatz, an zwei Seiten einfassen. Das Holzfachwerk sollte Bodenständigkeit symbolisieren und das fehlende Geborgenheitsgefühl transportieren. Die Architektur des Eingangsgiebels war trotz des verniedlichenden Fachwerks monströs und einschüchternd. Eine mittig ausgerichtete Treppe sollte zu der 17 m breiten Portalanlage führen. Darüber ein ebenso breiter Balkon zur Abnahme von Paraden. Noch weiter darüber sollte ein 5 m breiter Adler in der Spitze des Fachwerkgiebels den geometrischen Platz „überwachen". Die gewaltige Größe der Freizeithalle ließ die *Stadtverwaltung* um die Anziehungskraft ihres
geplanten innerstädtischen *Kulturforums* fürchten. *107*

★

Eine ähnliche Zweckbestimmung wie die Freizeithalle der Marine in Fedderwarder-
groden hatten die 6 Gemeinschaftshäuser in den Werftarbeiterquartieren von Sande- *48*
Neufeld. Am 11. 07. 1940 wurde beispielsweise das IV. Gemeinschaftshaus mit großem

Propagandaaufwand gerichtet. *Hafenbaudirektor Georg Frede* von der *KMW* lobte, „daß sich bei diesem Bau die ungebrochene Kraft der Heimat in schönster Weise bewährt
50 habe“[301]. Die 6 Freizeithallen in Sande-Neufeld waren „Gefolgschaftshäuser“ für Arbeiter der *KMW*, keine *NSDAP*-Parteihäuser. Sie sollten der Versorgung und Freizeitgestaltung, der „Freude und Wohlfahrt“ von jeweils 1.000 Werftarbeitern dienen. Mittelpunkt war jeweils neben den Wirtschaftsräumen der große Festraum mit Holzbalkendecke, Sonnenrädern und Bühne. Eine architektonische Zusammenfassung in riesigen Sälen sollte das Zusammengehörigkeitsgefühl der „Betriebsgemeinschaft“ herstellen. Die 6.000 Zwangsarbeiter, Kriegsgefangenen und Dienstverpflichteten der *KMW*, die in Sande-Neufeld zusammengepfercht waren, wurden nicht nur bei der Arbeit, sondern auch in der knapp bemessenen Freizeit bevormundet, gegängelt, doktriniert und überwacht.

★

52 Für die innerstädtischen Wohngebiete sollte am Kreuzungspunkt Banter Weg/Bremer
53 Straße ein *NSDAP*-Gemeinschaftshaus entstehen. Auf dem Gelände einer ehemaligen Wurt sollte nach Planungen des *Stadterweiterungsamtes* von 1937 ein kreuzgangähnliches Gebäudegeviert entstehen. Als Besonderheit war ein Rundbunker in das Gemeinschaftshaus einbezogen worden. Er sollte gestalterisch angeglichen werden. Den sich öffnenden Ehrenhof sollte man durch eine Säulenreihe betreten, die auf den Haupteingang ausgerichtet war. Dieses Projekt wurde ebenso nicht gebaut wie die *NSDAP*-Gemeinschaftshäuser in Altengroden, in Neuengroden und in den anderen Stadtteilen.

★

Die Wiederaufbauplanungen für Wilhelmshaven ab 1942 perfektionierten die Bauten der Gemeinschaft weiter zur Schreckensvision des totalen *NSDAP*-Staates. *Konstanty*
118 *Gutschow* vom *Arbeitsstab Speer* führt in seinen Planungen von 1944–45 nicht nur monumentale und zentrale Parteibauten (Volkshalle, Kreishaus usw.), sondern auch annähernd 40 *NSDAP*-Gemeinschaftshäuser auf[302]. Die soziale Kontrolle in den Stadtteilen sollte die „Volksgenossen“ völlig erfassen. Die Partei sollte alle Bereiche des öffentlichen Lebens lenken und überwachen. Niemals wären derartig viele Bauten der Gemeinschaft zu finanzieren gewesen, denn das „III. Reich“ stand wirtschaftlich auf tönernen Füßen. Die wahnwitzigen und menschenverachtenden Ziele, Wilhelmshaven in Ortsgruppen aufzuteilen, blieben Utopie. Schon bald war klar, daß keines der *NSDAP*-Gemeinschaftshäuser gebaut werden würde. Es blieb bei wenigen Freizeiträumen und Feierhallen für Werftbelegschaft und Bauarbeiter.

7.7.4. Gewerbebauten

Läden und Gewerbebauten waren während der Weimarer Republik fast ausschließlich der Privatinitiative überlassen. Die Nationalsozialisten betrachteten sich als Befreier des Handwerks von der „privatkapitalistisch geleiteten" Großindustrie. Nach dem „Führergrundsatz" aufgebaute Betriebe sollten wieder ihre angeblich volkswirtschaftlich notwendige Stellung bekommen. Die *NSDAP* biederte sich dem Mittelstand an. Gewerbebetriebe sollten eine wesentliche Funktion zur „Versorgung des Volksganzen" bekommen, aber auch steuernde Aufgaben „den Verbrauch nach Maßgabe der verfügbaren Güter zu lenken"[303]. Sowohl der Handel als auch das Handwerk waren jedoch keinesfalls frei, sondern eng an die *DAF* gekoppelt. Gewerbebauten in den neuen Siedlungen wurden als politisch überaus wichtig angesehen – wichtiger jedenfalls als die übrigen Bauten für die Gemeinschaft. *Gewerbestrukturplanung* wurde erstmals systematisch betrieben. Möglicherweise gab es ausländische Vorbilder.

Gottfried Feder, Kritiker der angeblich „entarteten Großstadt"[304] propagierte die Musterstadt für 20.000 Einwohner mit selbständigem Gewerbeleben. Seine Theorie der eigenen wirtschaftlichen Tragfähigkeit konnte wegen des fehlenden Umlandes und wegen der Marineabhängigkeit für Wilhelmshaven nicht greifen. Trotzdem hielten die Planer Siedlungen mit mangelnden gewerblichen Einrichtungen für „sozialpolitisch unerwünscht" und „wirtschaftspolitisch gefährdet"[305]. An Wilhelmshavener Siedlungen des „III. Reiches" sind noch heute die wirtschaftlichen Planungsziele ablesbar. Man wollte nicht die „krankhaft verbildeten Wasserköpfe", sondern die Illusion von „natürlich gewachsenen kleinen Mittelstädten". *Gottfried Feder* klassifizierte für eine 20.000 Einwohnerstadt die Gewerbebetriebe, die vorkommen müssen, – sollen oder – können und gab *Richtzahlen* heraus[306].

Grundrißbeispiele und Typenblätter gaben Anhalt zu Funktionsabläufen und Betriebsgrößen. Eine *Neue Stadt* von 20.000 Einwohnern sollte 140 Einzelhandelsbetriebe bekommen, allein 80 davon für Nahrungsverteilung. Diese Planzahlen waren der *Gewerbeplanung* für Fedderwardergroden zugrunde gelegt.

Die Schwerpunkte der planmäßigen NS-Gewerbeförderung lagen in den neuen Gemeinschaftssiedlungen. In Wilhelmshavens Innenstadt wurde kaum investiert. Die alten Geschäftszentren waren Rhein-, Markt- und Gökerstraße. Kaufhauskonzerne, *Konsum* und jüdische Geschäftsleute wurden bekämpft oder verjagt. Nur wenige Gewerbebauten in der Innenstadt wurden nach Bombenzerstörung noch im Kriege wieder aufgebaut.

Das kleinstädtische Gewerbeleben in den neuen Stadtteilen wurzelte in den Gartenstadtideen von *Ebenezer Howard* und in den konservativen Stadtmodellen von *Theodor Fritsch*. *Heinrich Tessenow*[307], der nicht müde wurde, Sauberkeit und Reinheit der Kleinstadt und des Handwerks zu loben, beeinflußte Architektengenerationen. Romantisch-verklärt verbreiteten Musterbücher des „III. Reiches" Beispiele kleinstädtischer Gewerbebauten aus der Vergangenheit und aus den nationalsozialistischen Mustersied-

149 lungen[308]. Kleinteilige Schaufenster, geschmiedete Reklameschilder, modellierte Zunftzeichen und in Stein gehauene Hoheitssymbole sollten Läden und Handwerksbetriebe schmücken.

Die rückwärts gewandten handwerklich-kleinstädtischen Traumbilder wichen 1940 einer nüchternen Betrachtung. Die „Ortsgruppe als Siedlungszelle" forderte die rationell geplante Ladenstraße nach funktionalen Gesichtspunkten. Es bestehe „kein Zweifel, daß der Einzelhandel und der Kleinbetrieb in Zukunft unter dem Zeichen der Einsparung menschlicher Arbeitskraft einer Rationalisierung bedarf"[309]. Die planmäßige Versorgung der Bevölkerung stand im Vordergrund; „völkisch"-romantische Kleinstadtvisionen gehörten bald der Vergangenheit an. *Hitler* „brauchte" Menschen für seinen Krieg. Handel und Gewerbe wurden planmäßig „durchkämmt", Menschen durch Maschinen ersetzt.

Anfang 1940 stellte die *Stadt* eine *Gewerbestrukturplanung* auf. Nach *Feder* entwarfen das *Stadterweiterungsamt* und *Heinz Baumann* Art und Zahl der Läden und Gewerbebauten für alle Neubaugebiete. Am 07.03.1940 wurden sie nach Absprache mit den anderen Beteiligten festgelegt. Ein *Kriegsbauprogramm* sollte die Realisierung wenigstens eines Teiles sichern.

Das Vorbild der „Ortsgruppe als Siedlungszelle" spiegelte sich in den *Gewerbeplänen*. Nach den Mustern von *Friedrich Heuer* sollten grundsätzlich Läden im Zusammenhang mit weiteren Gemeinschaftseinrichtungen im Zentrum der Ortsgruppe konzentriert werden. Sie waren geplant am Kreuzungspunkt von Achsen oder an Aufmarschplätzen. Die ideale Ortsgruppe mit ihren jeweils 8.000 Einwohnern sollte im Bedarfsfall zusätzlich kleine Läden in Eckgebäuden bekommen, aber immer orientiert zur Mitte, zur Achse, nicht dezentral.

148 Das äußere Erscheinungsbild der Handels- und Gewerbebauten war immer und ausschließlich traditionell. Nicht ein einziger Bau war modern oder sachlich, gar mit Flachdach oder außen sichtbarem Beton konstruiert. Trotzdem waren die Anforderungen an die Betriebsabläufe zeitgemäß. Große Gewerbehöfe ermöglichten rationelle Funktionsabläufe. Eigene Garagen für Lieferwagen oder *KDF-Wagen* gehörten zur selbstverständlichen Grundausstattung für die Betriebsführer. Bei den frühen Wilhelmshavener Wohnsiedlungen des „III. Reiches" wurde Gewerbeplanung mehr zufällig
28 betrieben. Das auf 1918 zurückgehende Inselviertel enthielt keine Gewerbebauten. Auch
26, 25 die Adolf-Hitler-Siedlung, die Kleinsiedlungen in Himmelreich und Sanderbusch, wurden noch nicht mit Gewerbebauten bedacht. 1937 war es strittig, ob die Werksiedlungen der *Marine* überhaupt „Geschäftshäuser" bekommen sollten. Erstmalig
30 bei der Arbeiter-Wohnstadt Voslapp kam es zu einer eigenständigen *Gewerbeplanung* nach Maßgaben der *Abteilung XXIV. der Marineintendantur* als Mittelinstanz.

Die *Kriegsmarine* selbst mußte viele Gewerbebauten in die Hand nehmen, da private Bauherren kaum zur Verfügung standen und die Wohnungsbaugesellschaften um ihre
1 Satzungstreue fürchteten. Es entstanden bis 1945 u. a.: in Fedderwardergroden 38 Gewerbebauten mit 60 Wohnungen, in Voslapp 15 Gewerbebauten mit 16 Wohnungen,

in Altengroden 7 Gewerbebauten mit 8 Wohnungen, in Cäciliengroden 6 Gewerbebebauten mit 6 Wohnungen, in Neuengroden 2 Gewerbebauten mit 4 Wohnungen[310]. In den zumeist reichseigenen Gewerbetrakten zahlten die „Betriebsführer" in der Regel Umsatzpacht. Sie sollten ihre Läden später kaufen können – deswegen waren sie in Reihenhausform entworfen. *Stadtbaudirektor Walter Temp* forderte die *Wohnungsbaugesellschaft Jade* am 25.10.1939 auf, die Trägerschaft für den „Trakt von Ladenbauten in Voslapp" zu übernehmen. Der Gewerbetrakt am Voslapper Markt sollte über Beihilfen des *RAM* finanziert werden. Das *Reich* sollte Eigentümer werden. Die Finanzierung kam jedoch ins Stocken, und am 11.05.1940 teilte die *Marineintendantur* den Gebietskörperschaften mit, daß alle Gewerbebauten in den neuen Siedlungen Voslapp, Cäciliengroden, Altengroden und Breddewarden nunmehr im Auftrag und für Rechnung der *Kriegsmarine* abgewickelt werden sollen. Die *Stadt* spielte nur noch die Rolle der *Baupolizei* und war für die Plätze zuständig.

In erster Linie hatten die Architekten bei den Ladentrakten die Grundrisse mit dem *Reichsstand des deutschen Handwerks* und dem *Landeshandwerksmeister Niedersachsen* sowie der *Wirtschaftsgruppe Einzelhandel* abzustimmen. Die Zeichnungen mußten von dort mit unterschrieben werden. In der *Bauabteilung des OKM* erarbeiteten die *Marinebauräte Bellwinkel* und *Block* eine nützliche Mappe mit Musterentwürfen für Ladenbauten in Marine-Siedlungen.

★

Eine erste systematische *Gewerbeplanung* wurde 1937 für Voslapp aufgestellt. Einzelne *30*
private Läden entstanden ab 1938 an der Flutstraße. Für die über 5.000 Einwohner waren erhebliche Versorgungsschwierigkeiten zu erwarten. Die ersten Gewerbebauten wie z.B. den des Schlachters *Luttmer* entwarf der Architekt der *Jade*, *Willi Lübbers*, gleich mit. Er verfiel auf die für Wilhelmshaven ungewöhnliche Fachwerkkonstruktion. Die *Baupolizei* bestand auf Eichenholz, denn man hatte mit den Siebethsburger Fachwerkgiebeln keine guten Erfahrungen gemacht. Fachwerk wurde jedoch im „III. Reich" zusehends beliebter. 1936–37 hatte *Hanns Dustmann* das *HJ*-Heim Melle in Fachwerk gebaut und auch der *Lübbers*- Lehrer *Paul Schmitthenner* war der Auffassung, daß Fachwerk „in allen deutschen Gauen" geeignet sei[311]. *Lübbers* wollte auch gleich die südlich benachbarten Gewerbebauten an der Flutstraße „als ein einheitliches Gefüge von 3–5 und 1-achsigen friesischen Fachwerkgiebeln" bauen. Er hielt das Fachwerk für eine „bodenständige Bauweise"[312]. Die romantisierend-modische Bauart wurde am 05.07.1938 von der *Baupolizei* versuchsweise genehmigt". Zu den benachbarten Gewerbebauten kam es nicht mehr. *Willi Lübbers* mußte im Oktober 1938 die *Wohnungsbaugesellschaft Jade* verlassen. Stattdessen beauftragte die *Jade* 1939 das Büro von *Fritz Hilgenstock* und *Gustav Bork*, die bisher nur in Fedderwardergroden Aufträge von der *Jade* zu erfüllen hatten, mit der Planung der Gewerbebauten in Voslapp. Seit 1937 zeigten die Voslapper *Bebauungspläne* den Marktplatz mit den umgebenden Gewerbebauten. Anfang 1939 legte *Gustav Bork*

seine Vorentwürfe der *Bauverwaltung* und dem *RAM* vor. Im Mai 1940 überarbeitete er seine Pläne und im nachfolgenden August gab *Karl Franzius* letzte Änderungsanweisungen[313], denn der Gewerbetrakt mit seinen 15 Läden sollte in Reichseigentum bleiben. Mit den Arbeiten am Gewerbetrakt wurde nach langen Auseinandersetzungen zwischen *Baupolizei* und *Marineintendantur* im Juli 1940 begonnen. Noch im Oktober 1940 kritisierte die städtische *Baupolizei* die Architektur des Gewerbetraktes, da sie nicht „bodenständig" sei[314]. Sie müsse überarbeitet werden.

Architektonisch und städtebaulich blieb der Marktplatz Voslapp unvollendet. Die gegenüberliegende Gewerbezeile von 100 m Länge wurde nicht mehr begonnen. Auch die am nördlichen Platzrand geplanten Gemeinschaftsbauten wurden nicht weiter verfolgt, denn 1941 wurde an deren Stelle ein Bunker für 1.320 Menschen hochbetoniert[315]. Die nördlich anschließenden Garagen und die vorgesehene Tankstelle blieben Planung. Der Ladentrakt von *Gustav Bork*, der im Kriege noch fertiggestellt wurde, ist in seinem hanseatisch strengen Habitus bei den heutigen Stadtteilbewohnern außerordentlich beliebt. Er ist gesuchte Kulisse für Märkte und Weihnachtsbasare, romantische Fassade einer Stadt, in der fehlende Tradition schmerzlich empfunden wird.

★

37 Für die Gemeinschaftssiedlung Fedderwardergroden waren zahlreiche zentrale und dezentrale Gewerbetrakte geplant. Der letztgültige *Bebauungsplan von* 1940 zeigte fast 180 Betriebe in verschiedenen Zeilen zusammengefaßt. Nicht alle wurden fertiggestellt, möglicherweise hatten die Machthaber des „III. Reiches" erkannt, daß die fast 600 „Gefolgschaftsmitglieder" für die Läden in den Kriegsjahren nicht zu „rekrutieren" waren. 70 Betriebe gehörten zum *Kriegsbauprogramm*.

Für die *Wohnungsbaugesellschaft Jade* entwarf Architekt *Ludwig Hörold* 1941 gegenüber
138 der gleichfalls von ihm geplanten Schule A nach *OKM*-Maßgabe eine Ladenzeile mit Frisörgeschäft, Schuhmacher, Schneider usw. 8 Betriebe sollten in 2 Trakten angesiedelt werden. Die Maßnahme hatte „Dringlichkeitsstufe I" und wurde mit allen Beteiligten, vor allem mit der *Abteilung XXIV. der Marineintendantur* abgestimmt. Ähnlich wie bei den Voslapper Gewerbebauten trat die *Marineintendantur* als Bauherr auf. Nach der *Verordnung über die baupolizeiliche Behandlung von öffentlichen Bauten vom* 20.11.1938 erteilte der *MDF* die Baugenehmigung. Der reichseigene Gewerbetrakt war von *Ludwig Hörold* gestalterisch an die Schule, das *HJ*-Heim und die Lehrerwohnungen angelehnt[316]. Die zweigeschossigen Gewerbebauten wurden mit hohen Satteldächern und traufständig zur Straße geplant. Kleine blumenfensterähnlich versproßte Schaufenster und viele den benachbarten Wohnbauten entnommene Details integrierten die Gewerbebauten unauffällig in das traditionelle Bild der Siedlung. Der städtischen *Baupolizei* mißfiel[317] die vorgesehene Verwendung von Edelputz an den Fassaden, da die gegenüberliegende Schule ein Ziegelbau war. Die *Bauverwaltung* setzte sich durch und die Gewerbebauten an der Albrechtstraße bekamen dann gleichfalls eine Ziegelaußenhaut.

Im Norden von Fedderwardergroden, an der Kniprodestraße, baute *Rudolf Bürgin* 37
1941 gegenüber seinem „Zentrum-Nord“[318] einen Gewerbetrakt. Er blieb einziges 146
Relikt seiner Planung. Auch *Bürgin* paßte sich an die traufständige zweigeschossige Satteldacharchitektur von Fedderwardergroden an. Kleine Schaufenster, hölzerne Klappläden und einzelne Dachhäuschen sollten an altertümliche Zunfthäuser erinnern.

Auch die von *Ernst Boyken* 1942 geplanten Geschäftshäuser an der Warthestraße ordneten sich im Äußeren völlig der Nachbarschaft unter. Einzige architektonische
Höhepunkte waren Fachwerkerker mit abgewalmten Dachhäuschen. Der „Lange 111
Markt“ in Fedderwardergroden sollte von rund 50 Läden gesäumt werden. Diesem utopischen, zentralen Aufmarschplatz mit *NSDAP*-Gemeinschaftshaus, *HJ*-Heim,
Gemeindehaus und *Post* gehörte die ganze Liebe der Stadtplaner. Für das Gemeindever- 150
waltungsgebäude wünschte der *RAM* 1941 monumentale 3-geschossige Bauweise oder ein hohes Obergeschoß. Zeugnis dieser Planung sind die Gewerbebauten an der Westseite des „Langen Marktes“ die erst 1960 gebaut wurden. Prunkstück der Fedderwardergrodener Gewerbeplanung war der fast 250 m lange zweigeteilte Ladentrakt an der Südseite der Posener Straße, die damals auch als Ost-West-Straße bezeichnet wurde. Er war der einzige zentrale Gewerbebau für Fedderwardergroden vor 1945. Das Büro *Fritz Hilgenstock* hatte am 15.12.1938 den Auftrag als Generalplaner für Fedderwardergroden von der *Wohnungsbaugesellschaft Jade* bekommen. Erst Mitte 1940, als die ersten Wohnungen von Fedderwardergroden schon bezogen waren, legte der für den Entwurf zuständige *Gustav Bork*, Hannover, Vorentwürfe für die Gewerbebauten an der Ost-West-Straße vor. Zu diesem Zeitpunkt hatte *Bork* die Planungen für die Gewerbebauten am Voslapper Markt bereits abgeschlossen. Der große Ladentrakt an der Südseite der Ost-West-Straße in Fedderwardergroden sollte eine Apotheke und rd. 20 mittelständische Gewerbebetriebe enthalten. An der Sortierung und Auswahl der Betriebe waren sowohl die *Marineintendantur* als auch der *Kreishandwerksmeister*, der *Kreiswirtschaftsberater*, der *Reichsstand des deutschen Handwerks* und die *Wirtschaftsgruppe Einzelhandel*, Berlin, maßgeblich beteiligt.

Schuhmacher, Fischgeschäft, Herrenschneider, Wäscherei, Bäcker, Fleischer, Putzmacherei, Schuhgeschäft, Frisör, Drogerie, Elektriker, Lebensmittelläden, Kohlenhändler, Schreibwarenladen und Leihbücherei sollten vertreten sein, keine Kaufhäuser, kein *Konsum*. Die hinter dem Ladentrakt liegenden Höfe sollten Zufahrten, Garagen, Autowerkstatt, Tankstelle usw. bekommen. Architektonische Glanzpunkte wurden die Eckgebäude (Apotheke) und ein mittlerer „Sonderbau“. Westlich abgesetzt sollte es dann noch zu einem zweite Gewerbetrakt mit einem Giebelgebäude kommen. Der Eckbau des ersten Gewerbetraktes, die Apotheke, ist heute noch in Fedderwardergroden ein optischer Merkpunkt. Er sollte mit dem Gasthaus „Zum Delphin“ ein nördliches Gegenstück in ähnlicher Formsprache bekommen. *Karl Franzius* entwarf die Pläne für den „Delphin“ 1942. Die 1940 von *Gustav Bork* vorgelegten und im Dezember 1940 von der *Abteilung XXIV. der Marineintendantur* genehmigten Pläne fanden nicht den Beifall der *städtischen Baupolizei* und der *Bauaufsicht* des *MDF*. Funktionale Mängel wurden festgestellt. Der *MDF* beschwerte sich am 05.12.1940[319], daß die „reichseigenen Gewerbebauten für die Ost-West-Straße" fast im Rohbau fertig seien[320] und immer noch nicht der örtlichen *Baupolizei* angezeigt wurden. Die *Marineintendantur* vernachlässige die Anzeigepflicht bei der *Stadt*. Nur Architekt *Friedrich Heuer* habe bisher seine Gewerbebauten in Cäciliengroden pünktlich vorgelegt. Auch in Voslapp sei ohne Anzeige mit dem Bau begonnen worden. Die Drohung wirkte. Im Dezember 1940, ein halbes Jahr nach Baubeginn, bekam die städtische *Bauverwaltung* die *Bork*-Pläne mit dem Genehmigungsvermerk der *Marineintendantur*. Tatsächlich war es zu Verzögerungen gekommen, weil auch *Karl Franzius* umfangreich in die Planung eingegriffen hatte.

Am 07.01.1941 wurden die *Gewerbeplanungen* der Ost-West-Straße beim *MDF* besprochen. Der Leiter der Wilhelmshavener *Baupolizei, Max Neumann*, kritisierte „nichtbodenständige Bauteile ... Zacken am Giebel ... überflüssige Attika-Aufbauten und obere Dacherker"[321]. Die Gestaltung war der Stadt „fremdartig". „Stuckschnecken am ‚Sonderbau' störten und die „WC-Fenster sind zu klein"[322]. Es kam Anfang 1941 sogar zu einer Stillegungsverfügung durch den *MDF*.

Streitigkeiten um die Architektur der Gewerbebauten an der Ost-West-Straße hinderten den *RAM* nicht daran, die Entwürfe 1941 als Beispiel „einer großen Platzwand", die „das Zusammenleben der Gemeinschaft auch äußerlich in die vom Staate gewollte feste Ordnung fügt"[323], zu veröffentlichen. Die zweigeschossigen Ladenbauten mit ihren mächtigen Satteldächern sollten den Eindruck einer mittelalterlichen Marktplatzbebauung machen. Eckgebäude und der herausgehobene „Sonderbau" sowie der abgesetzte zweite Gewerbetrakt sollten die lange Straßenfront auflockern. Eiserne Maueranker, geschmiedete Reklame, Ortgänge aus Naturstein, abgeschleppte Gauben, wuchtige Ladeneingänge, gebogene Tordurchfahrten, modellierte Zunftzeichen und kleine Sprossenteilungen symbolisieren die vorgespielte Tradition. Die Eingriffe von *Stadtverwaltung, MDF* und *Marineintendantur* zogen sich durch die gesamte Bauzeit. Noch 1942 hatten *Karl Franzius* und *Horst Linde*[324] von der *Marineintendantur* mit den Details zu tun.

Ein Gewerbetrakt in der Plauenstraße, den *Ernst Boyken* 1942 plante, erhielt im Kopfgebäude sogar ein Autohaus mit Tankstelle. Man rechnete damals wohl noch mit wenig Platzbedarf und so unterschied sich das Äußere des Autohauses nur unwesentlich von der „Scheibchen"-Architektur der Gewerbebauten. Lediglich ein scheunenartiger, unten offener Vorbau mit Walmdach deutet auf die technische Zweckbestimmung. Ein sprossenloses großes Schaufenster sollte den Blick auf ein (!) ausgestelltes Auto freigeben. *Gottfried Feder* ging damals nur von 14,5 *KDF-Wagen* für 1.000 Einwohner aus[325] . Er empfahl für eine Stadt der Größe Fedderwardergrodens 6 Tankstellen. Der Gewerbetrakt Plauenstraße wurde nicht mehr gebaut. Er sollte Teil eines 350 m langen Gewerbetraktes werden, der vor allem Handwerksbetriebe aufnehmen sollte. Durch einen kammartigen Grundriß sollten große Gewerbehöfe gebildet werden.

★

Für die Marinesiedlung Altengroden hatte *Ewald Zenker* von der *Bauabteilung des OKM* 1939 noch keine erkennbare besondere *Gewerbeplanung* vorgelegt. Erst der *Bebauungsplan von* 1940 sah u.a. 2 größere Ladenstraßen in Altengroden-Nord vor. Für den nicht realisierten Teil von Altengroden-Süd waren weitere dezentrale und zentrale Gewerbetrakte vorgesehen. Nur die „Ladenstraße 1" am Goedenser Weg wurde noch im Krieg gebaut. *Stadtbaurat a.D. Erich Kühn* aus Berlin entwarf im April 1940 Geschäftslokale für Bäckerei, Frisör, Lebensmittelgeschäft, Drogerie, Schneiderei und Fleischerei. Der Gewerbetrakt wurde durch die *Kriegsmarine* unter Mithilfe der *Marineintendantur* Wilhelmshaven als reichseigener Bau erstellt. Die städtische *Baupolizei*[326], der das Projekt angezeigt wurde, kritisierte den *Kühn*-Entwurf. Es kam zu einer Besprechung mit allen Beteiligten beim *MDF.* Die abgewalmten 2 Vorbauten waren der *Stadt* zu wuchtig und die Giebel sollten aus architektonischen Gründen nach einer *Änderungsskizze der Stadt vom* 05.12.1940 den benachbarten Wohnhäusern angepaßt werden. *43* *149*

Der Bau der Marinesiedlung Altengroden begann am 31.05.1939. Mit der 90 m langen Ladenstraße 1 am Goedenser Weg wurde erst 1941 begonnen. *Karl Franzius* von der *Abteilung XXIV. der Marineintendantur* griff noch nach Baubeginn in den Planungsablauf ein. *Erich Kühn* mußte es sich gefallen lassen, daß der nordöstliche der wuchtigen Vorbauten entfiel und der südwestliche einen verspielten Zwiebelturm bekam. *44*
Ansonsten wurden die *Kühn*-Pläne, die eine starke Ähnlichkeit mit den Voslapper Entwürfen von *Bork* aufwiesen, nicht mehr geändert. Ziegelbauweise, Einzelgauben und kleinversproßte, blumenfensterähnliche Schaufenster charakterisieren auch den Altengrodener Gewerbetrakt. Der Zugang zu den Läden wurde, wie auch in Voslapp, durch eine zusammenfassende Empore gebildet. Architekt *Kühn* dekorierte seine Ansichtszeichnungen 1940 mit Hakenkreuzfahnen, die den Ladentrakt-First überragten. Das Äußere der Altengrodener Gewerbebauten war ländlich, bieder und kleinstädtisch geplant – nichts Monumentales, nichts Einschüchterndes. Empore und NS-Fahnen blieben mißlungene Versuche der Nazifizierung.

Ab 1943 ging die Ladenstraße 1 in Betrieb. Die *Marineintendantur* warb „Betriebsführer" und „Gefolgschaften" an. Nicht selten mußten die „Betriebsführer" ihren Frauen die Leitung der Handwerksbetriebe überlassen, da sie selber für „kriegswichtige" Aufgaben eingesetzt wurden. Oftmals wurden die Gewerbebetriebe nur mühsam mit Hilfe von Dienstverpflichteten und Ausländern aufrechterhalten. Die Einzelhändler in den neuen Stadtteilen bekamen in den letzten Kriegsjahren Bedeutung als wichtige Versorgungseinrichtungen, die auch Naturalien umsetzten und Selbstangegebautes in Zahlung nahmen. Die Ladenstraße 2 an der Werdumer Straße wurde erst 1951 durch den *Konsum* gebaut.

★

53 Für das innerstädtische Jadeviertel entwarfen die Architekten *Franz* und *Hanns Iwersen* 1938 einen Gewerbetrakt. Den Anstoß gab ein Schlachter, der sich im Inselviertel aus „wehrpolitischen Gründen", wegen des drohenden Teilabrisses, nicht mehr ansiedeln durfte. Alsbald kamen weitere Ansiedlungswillige, ein Tischler, ein Bäcker usw. hinzu. Immerhin gelang es der *Stadtverwaltung*, die Gewerbetreibenden an einen einzigen Architekten zu binden und und ein einheitliches Äußeres durchzusetzen. Im November 1938 wurde schließlich von den Architekten *Iwersen* ein abgestimmter 100 m langer Gewerbetrakt entworfen, der in seiner nördlichen Hälfte noch heute steht. Er wurde 1939 fertiggestellt.

Die Grundstücke wurden den Gewerbetreibenden von der *Stadt* in Erbpacht überlassen. Nach dem *Fluchtlinienplan von* 1937, der beidseits der Nord-Süd-Achse Banter Weg je einen 200 m langen Ladentrakt vorsah, waren sie so großzügig bemessen, daß auch Schlachthaus, Tischlerwerkstatt und Backstube und, in den Obergeschossen, „Gefolgschaftswohnungen" möglich waren. Die *Marineintendantur* hatte bei diesen privaten Gewerbebetrieben keine Befugnisse und politisch war es höchst willkommen, dem Mittelstand private Investition zu ermöglichen. Der zweigeschossige Gewerbetrakt bewegte sich auf den ersten Blick mit Dachhäuschen, Kreuzsprossen und Hoheitszeichen (in Treppenhausfenstern) im gängigen Stilkatalog. Daß die Architekten ihre eigene Vorstellung einbrachten, ist an dem lebhaften Spiel der Materialien erkennbar. Den geputzten Außenflächen wurden in der Ladenfront Klinkersockel und strukturierte Türumrahmungen entgegengesetzt. Waagerecht durchgehende Ziegelbänder sollten die Werbung einfassen. Sogar breitgelagerte Fensterformate kommen vor.

An der Ecke Werftstraße/Peterstraße plante *Willi Lübbers* für den Automechaniker *Otto Fischer* im Juni 1938 eine Tankstelle. Auch dieser Gewerbebau bekam ein traditionelles Äußeres. Ähnlich der Tankstelle an der Reichsstraße 69 (südlich Sande) und an der Plauenstraße wurde der technische Zweckbau ländlich-bäuerlich verkleidet. Wie bei den Autobahnraststätten von *Friedrich Tamms*[327] krönte auch *Willi Lübbers* seine Tankstelle Peterstraße mit einem mächtigen Walmdach. *Hitler* strebte die motorisierte Volksgemeinschaft an. 1938 fuhren bereits 1,3 Mio. PKW im Reich und 1939 trafen die ersten *KDF-Wagen* in Wilhelmshaven ein.

7.7.5. Andere öffentliche Bauten

Die Hochbauten der *Post*[328] hatten mit Verkehr und Verwaltung, Dienstleistung und Gemeinschaft zu tun. Wilhelmshaven/Rüstringen hatte seit 1892 Fernsprechleitungen und ein *Fernsprechamt* neben dem Wilhelmshavener Rathaus. Die *Reichspost* war wichtiger Bauherr für den Aufbau der Stadt im „III. Reich". Vielfach wurden die mustergültigen Bauten der *Post* gelobt und daß „sie ihr möglichstes tut, um dem großen Publikum mit vollendet modernen Einrichtungen zu dienen". Herausgestellt wurde

auch, daß „in allen Räumen das Prinzip von der Schönheit der Arbeit" verwirklicht
werden soll".

Eine leistungsfähige Nachrichtenübermittlung war aus militärischen Gründen für
Wilhelmshaven besonders wichtig, und so plante die *Reichspostdirektion Oldenburg* 1939
unter *Postbaurat Friedrich Jochem* ein neues vergrößertes *Fernsprechamt*, das als kriegswich-
tig eingestuft wurde. Schon am 05.08.1938 hatte die *Post* ein Grundstück zwischen
Schiller- und Goethestraße von der *Stadt* erworben. Der zweigeschossige Funktionsbau
war keinesfalls sachlich und technisch geplant, obwohl die *Post* über hervorragende
Betriebsbauten der 20er Jahre verfügte[329]. Der Entwurf von *Jochem* zeigte heimatver-
bundenen Regionalismus mit gewaltigem Ziegelwalmdach, natursteinumrahmten
Öffnungen und kleinen quadratischen Sprossenteilungen. Er macht einen fast hanseati-
schen Eindruck. Die Werksteineckquader ließen die massive Bauweise scharf hervortre-
ten. Das Portal, gleichfalls in Naturstein gerahmt, sollte von symmetrisch angeordneten
schmiedeeisernen Lampen und Geländern dekoriert werden. Über dem Eingang mit
seiner Eichentür sollte der Adler mit Hoheitssymbol wachen. Der städtischen *Baupolizei*
waren die Planungen von *Friedrich Jochem* nicht monumental genug. Man wollte eine
Ost-West-Achse vom Lessingplatz bis zum *Fernsprechamt*[330] und keine Einmündung in *53*
den Posthof. 1940 versuchte auch das *Reichspostministerium Berlin* einen monumentaleren
Entwurf durchzusetzen. Die *Stadt Wilhelmshaven* und die Regierung Oldenburg jedoch
bevorzugten eine „strenge nordische Gestaltung der Fassaden"[331]. Es kam im „III. Reich"
nicht mehr zur Fertigstellung des *Fernsprechamtes* Schillerstraße. Stattdessen sah sich die
Reichspostdirektion Oldenburg 1940 gezwungen, eine Baracke als Vermittlungsstelle für
3.600 Anschlüsse aufzustellen. 1949 erst konnte die Post das *Fernsprechamt* vollenden –
sklavisch nach dem Entwurf von 1939. Nur das Portal wurde entnazifiziert. Adler und
Hoheitszeichen entfielen, auch die Granitfreitreppe wurde gestrichen. Die einstmals in
Kupferdeckung geplanten einzelnen Dachhäuschen wurden zu größeren Gauben
gusammengefaßt.

Für Fedderwardergroden war schon 1938 südlich des „Langen Marktes" ein großer *37*
Postkomplex mit Kraftfahrzeughalle, Postwohnungen, Hof und *Postamt* vorgesehen. Im
Juni 1940 legten *Postbaurat Jochem* und *Architekt West* die Planung vor. Das stark
rückwärts gewandte, biedermeierlich-romantische Äußere sollte durch steile Satteldä-
cher und geschlämmte Ziegelsteine[332] gekennzeichnet sein. Vorbauten, Erker, (Führer-) *150*
Balkon, Sockel, Gewände, Gesimse, Eckquader und Hoheitszeichen sollten aus Ober-
kirchener Sandstein modelliert werden. Eingangstür, Hoftore und Erker waren aus
Eichenholz vorgesehen. Die Bleiverglasung des Erkers und die kleinteiligen Kreuzspros-
sen, der gedrungene Arkadendurchgang und die schmiedeeisernen Details machten den
Eindruck eines mittelalterlichen Marktplatzes. Einem geschichtslosen Stadtteil sollte
durch die gediegene Selbstdarstellung einer Reichsbehörde Tradition verliehen werden.
Das Satteldach schien der *Stadt* zu hoch geplant. Sie verlangte ein Holzgerüst in der
vorgesehenen Größe und die Zustimmung der *Marineintendantur* bzw. des *Schutzbereichs-*

amtes. Der groß geplante Postbau blieb Planung. 1940 stellte die *Reichspost* ein
42 Notpostamt im freibleibenden Winkel der Gebäudeflügel auf.

Auch andere Stadtteile sollten ihre *Postämter* an zentraler Stelle bekommen. Im Voslapper Gewerbetrakt am Markt wurde es noch erstellt, nachdem das im *Bebauungsplan von* 1937 vorgesehene Postgebäude nördl. des Rathauses entfallen war. Für die nicht mehr gebauten Stadtteile waren jeweils *Postämter* in der Nähe der Ortsmittelpunkte, an den Aufmarschplätzen oder in den Gewerbetrakten vorgesehen. Nichts davon wurde gebaut.
123 Erst nach dem Kriege bekam Wilhelmshaven sein *Hauptpostamt* in den Resten des *Verwaltungsforums*. In den Ortsteilen behalf man sich lange mit Provisorien.

★

Die 1934 gegründete *Kreishandwerkerschaft Wilhelmshaven/ Rüstringen*[333] hatte ihr seit der Weimarer Republik bestehendes Sozialwerk, die *Handwerker-Baugenossenschaft der Jadestädte e.G.m.b.H.* 1935 mit dem Bau eines „Hauses des Handwerks" in der Paul-Hug-Straße 26/28 beauftragt. Das dreigeschossige Verwaltungsgebäude mit Handwerkeraltenwohnungen, Verwaltungs- und Gemeinschaftsräumen sollte knapp 100.000 Mark kosten. *Fritz Höger*, 1929–37 städtebaulicher Berater Rüstringens, hatte nach Fertigstellung des Rathauses 1929 großen Einfluß auf die Architektur der Stadt. Im August 1935 legte er einen bemerkenswerten städtebaulichen Entwurf für die Südseite der Paul-Hug-Straße vor – schräg gegenüber dem von ihm entworfenen *Finanzamt* am *Verwaltungsforum*. Eine leicht konkave Randbebauung sollte sich in Schiller- und Mitscherlichstraße fortsetzen. Das Eckgebäude Mitscherlich-/Paul-Hug-Straße war nach *Högers* Plänen von 1935 als dominierend vorspringender Bau vorgesehen. Im Scheitelpunkt der südlichen Randbebauung, auf der Hälfte der Paul-Hug-Straße, sollte das „Haus des Handwerks" als 3 1/2 geschossiges Torgebäude mit Satteldach und durchgehenden Gauben den Zugang zur Blockinnenfläche ermöglichen[334]. Am 03.09.1935 schickte *Höger* dem *Kreishandwerksmeister Wilhelm Völker,* Obermeister *Toni Rech* und *Magistratsbaurat Max Neumann* seine Pläne zu. Sie wichen offenbar von den ursprünglichen Plänen der *Handwerker-Baugenossenschaft* und ihres Architekten *Wilhelm Meichner* ab. *Höger* schrieb: „Die Architektur habe ich erheblich überholt und groß und schlicht und würdig gestaltet. Ich hoffe, daß Sie meine Vorschläge gut und schön finden und bejahen werden. Zur weiteren Mitarbeit und Hilfe bin ich gern bereit, und zwar unentgeltlich als Idealist und als der *Stadt* gegenüber verpflichtete, städtebauliche Berater".

Högers geplante Randbebauung integrierte das „Haus des Handwerks" und ließ eine einfache quadratische Toröffnung als Durchgang zum Hof frei. Symmetrisch rechts und links waren Treppenhäuser mit Ziegellisenen vorgesehen. Die Fensterformate waren waagerecht gelegt, ähnlich denen des benachbarten Rathauses. Im Rüstringer *Magistrat* war man mit *Högers* Plänen nicht zufrieden[335]. Man wollte einen gärtnerisch gestalteten Vorplatz und überhaupt das „Haus des Handwerks" 10 bis 15 m aus der Gebäudeflucht zurückgesetzt haben. Das Tor war den Rüstringer Beamten zu schmucklos und die Dachausbauten sollten verschwinden. Das „Haus des Handwerks" sollte nicht mit einem Mietshaus verwechselt werden können – obwohl es zu einem großen Teil aus Wohnungen bestand.

Die *Handwerker-Baugenossenschaft der Jadestädte* reichte im September 1935 den Entwurf ihres Architekten *Wilhelm Meichner* ein. *Högers* großstädtische Randbebauung wurde aufgegeben und das „Haus des Handwerks" als freistehendes Gebäude mit abgewalmtem Dach entworfen. Die Tordurchfahrt hatte einen 5 m breiten auf Klinkerlisenen ruhenden (Führer-)Balkon erhalten. Das schmiedeeiserne Gitter sollte an zentraler Stelle das Handwerkersymbol bekommen.

Der Grundstein wurde am 01.12.1935 gelegt[336]. Kurz vorher, am 29.10.1935 teilte *Kreishandwerksmeister Wilhelm Völker* dem Architekten *Fritz Höger* mit, daß der von ihm abgeänderte Entwurf vom August 1935 nur zum Teil angenommen wurde. Das *Bauamt* habe gewünscht, daß das „Haus des Handwerks" 15 m zurückgesetzt werde. *Höger* beschwerte sich postwendend[337] bei *Oberbürgermeister Nutzhorn*, zu dem er persönliche Kontakte unterhielt. Er stellte seine guten Beziehungen zu *Alfred Rosenberg* heraus und berichtete, daß ihn auch die Stadt Lübeck zum städtebaulichen Berater gemacht habe. *Rosenberg* liege „... außer Lübeck, auch der übrige Norden sehr am Herzen ... Er freute sich auch sehr zu dem Werden in Rüstringen". *Höger* wollte verhindern , „... daß nicht die Häßlichkeiten, die im ursprünglichen Projekt vorhanden waren, sich wieder eingeschlichen haben".

Möglicherweise erst nach Baubeginn mischte sich der *Höger*-Gegner *Willi Lübbers* ein und beeinflußte noch verschiedene Details. Das Dach bekam 7 Dachhäuschen und das schmiedeeiserne Gitter über der Tordurchfahrt bekam neben dem Handwerkersymbol auch das nationalsozialistische Hoheitszeichen. Ein weiterer (Führer-)Balkon über der hofseitigen Tordurchfahrt wurde von *Lübbers* zum „Ehrenhof des Handwerks" orientiert.

Das „Haus des Handwerks", am 13.06.1936 eingeweiht, versinnbildlichte das „neue Arbeitsethos" der nationalsozialistischen „Gemeinschaft". Es war gedacht als Gemeinschaftshaus für die Handwerker wurde aber propagiert als eine Weihestätte „deutschen Arbeitsgeistes".

★

Ende 1938 forderte *Oberbürgermeister Müller* die Bereitstellung von Zuschüssen für das neue *Städtische Krankenhaus*[338], da die veranschlagten Kosten von 8 Mio. Mark für 400 Betten nicht von der *Stadt* aufzubringen waren. Das alte Krankenhaus war 1889 gebaut worden. Sämtliche vier Krankenhäuser in Wilhelmshavens Südstadt mußten aus Luftschutzgründen, und da sie keine Entwicklungsmöglichkeit hatten, ausgesiedelt werden[339]. *Oberregierungsrat Engel* vom *Reichsinnenminister* forderte zunächst Entwürfe und Kostenermittlungen. Der *RAM* empfahl einen Architektenwettbewerb, „da die

Stadt über einen geeigneten *Baudirektor* nicht verfügt“[340]. Alternativ hatte der *RAM* im April 1939 den *Berliner Landesbaurat Richard Lang* eingeschaltet. *Lang* bekämpfte den von *Wilhelm Hallbauer* im März 1939 empfohlenen Standort zwischen Ehrenfriedhof und Altengroden-Süd. Wegen der Bodenverhältnisse und wegen des mangelnden Luftschutzes bevorzugte er eine periphere Lage. *Müller* versuchte zu vermitteln, und propagierte sogar zwei Krankenhäuser, da die Bettenzahl entsprechend den steigenden Einwohnerzahlen mittlerweile „auf 700 erhöht werden müsse“. Ein zusätzliches
64 *Städtisches Krankenhaus* sollte möglicherweise in *Gottfried Feders* 100.000 Einwohner-Stadt Langewerth geplant werden. Am 20.7.1939 schließlich beauftragte die Stadt mit Zustimmung des *RAM* den Hamburger Architekten *Hermann Distel*, der über gute Beziehungen zu *GBI Albert Speer* verfügte, und der vielen in Wilhelmshaven tätigen Hamburger Architekten gut bekannt war. *Distel* plante u.a. das Universitäts-Klinikum Berlin an der Ost-West-Achse in monumentaler Architektur im Zusammenhang mit der Hochschulstadt[341]. Ein renommierter Planer schien gerechtfertigt, denn die *Marine* hatte
75 mit ihrem soeben fertiggewordenen *Lazarett Sanderbusch* einen hohen Qualitätsmaßstab vorgegeben.

Es folgte ein langer Konflikt wegen des neuen Standortes zwischen Fedderwarden und der Burg Kniphausen, der außerhalb der Stadtgrenzen vom 01.06.1938 lag. Der neue Krankenhauskomplex lag sehr verkehrsungünstig und hätte keinen Ringschnellbahn-Anschluß bekommen. Er sollte durch eine Bus-Linie mit der Stadt verbunden werden. *Distel*, der *Werftkrankenhaus* und *Städtisches Krankenhaus* kombinieren wollte, konnte seine Idee nicht durchsetzen. Aus Luftschutzgründen wählte die *Marine* für ihr *Werftkrankenhaus* den Standort Barkeler Busch[342]. Die Planung für das *Städtische Krankenhaus* verzögerte sich möglicherweise, weil der renommierte *Distel* mit seinem fünfgeschossigen Entwurf auf Kritik stieß, weil die Bettenanzahl strittig war und weil die Planung für die freie Landschaft zu monumental erschien.

Der *RAM* hatte *Baurat a.D. Bischof*, Berlin, eingeschaltet, da er nicht über eigene Krankenhausarchitekten verfügte. *Bischof* forderte von *Distel* zusätzlich Alternativen an und riet, von 5 Geschossen auf 3 Geschosse und einen 4-geschossigen Haupttrakt abzustufen. Mittlerweile war nur noch von 450 Betten und zusätzlicher Erweiterungsmöglichkeit die Rede. Auch *Hochbauabteilungsleiter Klaus Rasch* und die Wilhelmshavener Chefärzte hatten Bedenken gegen die monumentale Architektur von *Distel* und schlugen 1941 vor, *Benno Schachner* aus Brünn hinzuzuziehen. Weil *Distel* verschiedentlich allein mit *RAM*-Gutachter *Bischof* in Berlin verhandelt hatte, ohne die *Stadt* hinzuzuziehen, kam es zu weiteren Auseinandersetzungen[343].

151 Die Planung *Distel* vom 01.10.1941 schließlich war eine von 6 Alternativen, die *Stadtbaudirektor Walter Temp* im Dezember 1941 dem *RAM* vortragen wollte. Die
75–77 Entwürfe waren entsprechend dem Vorbild des *Marinelazarettes Sanderbusch* jetzt der Landschaft angepaßt und dem *Marinestil* ähnlicher geworden. Das 10 ha große Krankenhausgelände sollte von einem „Runden Platz“ mit Obelisk durch ein Wachhaus, flankiert von zwei symmetrisch angeordneten Chefarztvillen, betreten werden. Die

Hauptachse sollte in einen U-förmigen großen Hof mit Behandlungsflügeln führen. Der Haupteingang, dekoriert durch Säulen und gekrönt von einem (Führer-) Balkon, sollte neuklassizistische Züge tragen. In traditioneller Walmdachform und in Ziegelrohbau war der leicht gebogene Bettentrakt mit den einzelnen Abteilungen geplant. Den Stationsflügeln sollte jeweils ein Bunker gegenüberstehen, der im Luftangriffsfall die Patienten aufnehmen sollte. Drei der Bunker waren mit Zinnenkranz und einer mit Zeltdach gekrönt – Versuche, die Bunker als Ritterburgen zu verkleiden[344]. In seinen Erläuterungen schrieb *Distel* am 07.10.1941: „Alle Baukörper sind in ihren Massen und in ihrer Dachform auf die einfachste Formel gebracht worden. Die Umrisse entsprechen der bodenständigen Bauweise ... Zur Belebung der roten Flächen werden Fensterumrahmungen, Gesimse und Pfeiler aus Natursteinen eingestreut“[345]. Geordnet wurden auch die kranken „Volksgenossen“. Geschlechtskranke und Tbc-Patienten sollten in einem Extragebäude „abgesondert“ werden. Entsprechend der hierarchischen Stellung des Personals bekamen Gärtner, Beamte, ledige Pfleger, Assistenzärzte, Schülerinnen und Schwestern eigene voneinander getrennte Wohnhäuser. Es gehörte zum Standesbewußtsein der Ärzte, Sport zu treiben, und also ordnete *Hermann Distel* gegenüber dem Gemeinschaftshaus für das Personal 2 Tennisplätze an.

Das *OKM* teilte dem *RAM* am 16.05.1942 mit, daß wegen des *Erlasses des Reichsmarschalls vom* 13.04.1942 die Krankenhaus-Planung eingestellt werden müsse, da „der Bau nicht kriegswichtig“ sei.

In erstaunlicher Kontinuität wurde *Distels* Mitplaner *Benno Schachner* 15 Jahre später von der *Stadt*[346] mit der Planung des dann tatsächlich realisierten *Städtischen Krankenhauses* an anderem Standort beauftragt.

★

Groß-Wilhelmshaven wollte der Öffentlichkeit, den auswärtigen Gästen und Seepassagieren am Südstrand ein „würdiges Gebäude“ präsentieren und sich gegen die Marineinteressen am Wasser festsetzen[347]. Die auf 1886 zurückgehende Strandhalle war 1926 durch einen reich gegliederten Ziegelbau von *Gustav Lübbers* ersetzt worden. *Walter Hunzinger*, ab 01.11.1937 *Stadtbaudirektor*, zog das Projekt an sich und entwarf den neuen
Halbrundbau in einer Formsprache, die der Großstadtarchitektur von *Erich Mendelsohn* *110*
und den Brüdern *Luckhardt* mehr entsprach, als dem in Wilhelmshaven vorherrschenden *Marinestil*[348]. *Hunzinger* paßte sich nicht der *Zopff*-Architektur des Südstrandes von 1928 an, sondern wollte mit monumentaler Geste und repräsentativem Pathos das Tor Wilhelmshavens zur Nordsee betonen. Der Besucher sollte vom Blick aufs Meer überwältigt sein. Die umlaufend hohen schlanken Fenster gaben ein gigantisches Panorama frei. Heimatverbundene Architektur hätte diesen Zweck nicht erfüllt.
Werksteinumrahmte Portale und Fenster sowie eine NS-übliche Innenarchitektur sollten *131*
für *KDF*- Reisende eine gemütliche Atmosphäre herstellen. Das der neuen Strandhalle hinderliche Seewasseraquarium von 1927 wurde 1938 als moderner Neubau westlich der

Strandhalle noch begonnen, jedoch erst 1950 fertiggestellt. Die „architektonisch groß ausgestaltete" Strandhalle abseits der Achsen- und Monumentalplanungen, wurde Ende 1938 fertiggestellt, mehrfach vom Kriege zerstört und wieder aufgebaut. *Hunzinger* setzte sich bei ihrem Bau über alle haushaltsrechtlichen Bestimmungen hinweg, zog sich den Ärger von Rat und Verwaltung zu, und wurde wenig später als *Stadtbaudirektor* entlassen.

7.8. Bauarbeiter und Bauarbeiten

Das irrwitzige Aufrüstungsprogramm des NS-Systems erforderte völlig neue, militäri-
1 sche Investitionen für Stützpunkte, Kasernen, Flughäfen und Wohnstätten. Die zu
101 erweiternden Rüstungsbauten führten vor allem zu dem Riesenprogramm *Nordwerft*. Es
sollte ab 1936 planmäßig realisiert werden, um die Schiffbaukapazitäten der privaten und marineeigenen Werften auszuweiten. An Arbeitskräften waren unterzubringen:

- Alter Werftarbeiterstamm von *KMW* und *Westwerft*
- Bauarbeiter für die IV. Hafeneinfahrt, für die geplante *Nordwerft* und alle weiteren Militär- und Werftbauten
- Neue Werftarbeiter für die *Nordwerft*
- Bauarbeiter für das *Wohnsiedlungsgebiet Wilhelmshaven-Jeverland*, Baustäbe, Bauleitungen, Baubehörden
- Mannschaften der Marinebetriebe, für die keine Kasernen zur Verfügung standen

Es begann ein Teufelskreis von Organisationsproblemen und Kriegsvorbereitungen. Womit sollten die Baudienststellen beginnen? Was war mit den notwendigen Gemeinschaftseineinrichtungen? Schließlich propagierte die *NSDAP* die perfekte städtebauliche Organisation des nationalsozialistischen Reiches, wollte die „Volksgemeinschaft" bilden, Parteibauten zur Verfügung stellen und die neuen Städte volksnah verwalten[349].

Hitlers Expansionspolitik erforderte gute Facharbeiter, Spezialisten und Technokraten an der „Heimatfront" und in den okkupierten Gebieten. Ohne Ansehen der Person wurden je nach Erfordernis benötigte Arbeitskräfte abkommandiert, vom Werftarbeiter bis zum Baubeamten. Ausschlaggebend war der Einsatzbedarf. Die Zuweisung von Arbeitskräften lag beim Wilhelmshavener *Arbeitsamt*, das eigens für die neuen Aufgaben 1936–37 am Rathausplatz neu erbaut wurde. Der *Generalbevollmächtigte für die Regelung der Bauwirtschaft, Fritz Todt*, wurde im Januar 1940 von *Erich Raeder* gebeten, für kriegswichtige Wilhelmshavener Bauten Arbeiter und Maschinen zuzuweisen. Die *OT, Dienststelle Wilhelmshaven* als Unterabteilung des *Generalinspektors für das Deutsche Straßenwesen, Abteilung Rüstungsbau, Außenstelle Hannover*, wurde zuständig. *Todt* hatte *Gaubeauftragte für die Regelung der Bauwirtschaft* eingesetzt, die gemeinsam mit den staatlichen *Landesarbeitsämtern* nach Dringlichkeitsstufen I–IV Arbeiter zuwiesen.

Gaubeauftragter war 1940–45 *Walter Temp. Karl Dönitz* überführte im Mai 1944 das gesamte Marinebauwesen zusammen mit den Baudienststellen von *Heer* und *Luftwaffe* in die *OT*.

Die folgenden Zahlen mögen einen Anhalt zur Wilhelmshavener Arbeitskräftesituation geben:[350]

o 11.12.1940	42.000 Menschen mittelbar und unmittelbar im Dienst der *Marine*
o 01.05.1942	38.501 Bedienstete auf den Werften einschl. der Nebenbetriebe
o 1.03.1945	31.000 Werftbedienstete

Allein am Ausbau der *Nordwerft* und am Bau der IV. Einfahrt waren beschäftigt:

o 1937	7.000 Menschen
o 1938	13.500 Menschen
o 1939	18.300 Menschen (davon 11.000 Dienstverpflichtete)
o 1940	21.200 Menschen (davon 8.000 Dienstverpflichtete)
o 1941	18.600 Menschen (davon 3.000 Dienstverpflichtete und 4.300 Kriegsgefangene)

Arbeiterlager für die *Nordwerft* mußten durch die *HNBD* gestellt werden. Das *101* *Hafenbauressort* der *KMW* stellte Baracken für Werftarbeiter auf, z.B. auf dem Schützenhof, am Alten Banter Weg. Für die wegen der Stadterweiterung nach Wilhelmshaven kommandierten Arbeiter war die *Stadtverwaltung*, aber auch das *Marinestandortbauamt* zuständig. Das *Marinestandortbauamt* erstellte u.a. die Lager Alten- *98* groden für 3.000 Arbeiter und Breddewarden für 2.000 Arbeiter[351]. Die *Marine* selbst stellte für ihre Mannschaften bei ihren Dienststellen kurzfristig, zumeist im Kriege, Unterkunftsbaracken auf, bei den Funkstellen, Vermittlungsstellen, Flak-Anlagen, Depots, Lazaretten und Kommandostellen. Fertigbaracken wurden aus allen Teilen des Reiches beschafft. Noch im Herbst 1944 wurde erwogen, 12.000 Häftlinge als Zwangsarbeiter für Rüstungs- und Bauarbeiten durch die *SS* in den Raum Wilhelmshaven schicken zu lassen. 1944–45 war ein Teil des Arbeiterlagers Banter Weg abgeteilt als *Konzentrationslager* (Außenstelle von Neuengamme). 1.100 Häftlinge waren dort gefangen, 236 kamen um. Es gibt Hinweise auf weitere Straflager. Zwangsarbeiter und *KZ*-Häftlinge wurden für besonders gefährliche Bauarbeiten eingesetzt. Sie mußten *155* abreißen, enttrümmern und Bomben beseitigen. An die 50 % der Arbeiter waren Ausländer.

Zwangsarbeiter wurden für alle vorkommenden Arbeiten eingesetzt, z.B. „Ostarbeiter" und Polen. *KZ*-Häftlinge mußten 1943 Aufräumungsarbeiten leisten. Ab 1944 *48* wurden sie auch in der Rüstungsproduktion der *Westwerft* eingesetzt. Baufirmen und *152* Wohnungsbaugesellschaften mußten ihre Arbeiter überwachen. Die *Marineintendantur* teilte im September 1941 mit, daß Bauarbeiter, die „rechtswidrig" ihre Arbeit in Deutschland verlassen", durch die *SS* in ein „Erziehungslager" eingereiht werden, wo sie mit „schwerer Arbeit beschäftigt werden". Bereits 1939 erwogen *KMW* und Polizei in Wilhelmshaven ein „Erziehungsgefängnis" einzurichten für nicht an ihre Arbeitsplätze zurückkehrende Dienstverpflichtete und ausländische Arbeiter.

⋆

Seit 1919 war in der *Reichsverfassung*[352] die Auffassung von Arbeit festgelegt: „Jeder Deutsche hat unbeschadet seiner persönlichen Freiheit die sittliche Pflicht, seine geistigen und körperlichen Kräfte so zu betätigen, wie es das Wohl der Gesamtheit erfordert". Ab 1933 stellten die Nationalsozialisten den Wert des „Arbeitsmenschen" für das Volk in den Vordergrund. Ohne Klassenkampf sollten alle Arbeiterschichten in die nationalsozialistische „Volksgemeinschaft" eingegliedert werden[353]. Das *Gesetz zur Ordnung der nationalen Arbeit von* 1934 wurde gesetzliche Grundlage einer mittelstandsfreundlichen Politik der Nationalsozialisten. Die propagandistisch auf Treue der „Gefolgschaft" zum Betriebs-„führer" aufgebaute „Betriebsgemeinschaft" sollte ausschließlich Volk und Staat nützen[354]. Unternehmerische Entscheidungen waren nach dem „Führergrundsatz" vorgesehen. *DAF*-„Vertrauens"männer der „Gefolgschaft" ersetzten während des „III.
153 Reiches" Betriebs- oder Personalrat. Sie blieben ohne Einfluß. *Arbeitsschlachten*, Leistungskämpfe der Betriebe und *Frühjahrs-Offensiven* sollten die Leistung steigern. Für die großangelegten Wilhelmshavener Baumaßnahmen war der Leiter des *Arbeitsamtes* als *Reichstreuhänder der Arbeit* des *RAM* zuständig. Er sollte für Arbeitsfrieden und Tarifordnung sorgen und konnte Löhne eigenverantwortlich festsetzen. Das gesamte nationalsozialistische Arbeitsrecht beruhte auf Militarisierung und der Leerformel der „sozialen Ehre". Als schlimmste Verstöße galten Verhetzung, Ehrkränkung, böswillige Ausnutzung der Arbeitskraft und Schwarzarbeit. „Der von der Regierung mit aller Kraft geführte Kampf gegen die Arbeitslosigkeit darf durch keine Schädlinge gestört werden", so lautete ein Merkblatt der Wilhelmshavener *Baupolizei* gegen Schwarzarbeit.

Die *Gewerbeordnung* gewährte Schutz des Betriebes, des Lebens und der Sittlichkeit. *Gewerbeaufsichtsämter* und das zeitweise von *Albert Speer* geleitete *Amt „Schönheit der Arbeit"* sowie die *KDF* hatten großen Einfluß. 1933 waren die Gewerkschaften verboten und durch die *DAF* ersetzt worden. Lohnverhandlungen wurden abgeschafft und sozialer Frieden hieß: Die Arbeiterschaft für die *NSDAP* gewinnen. Die Gleichschaltung der *NSDAP* erstreckte sich auch auf die Unternehmerschaft.

Leiter der *DAF* war von 1933–45 *Robert Ley*, er kontrollierte den Wohnungsbau in Wilhelmshaven und im ganzen Reich. Seit 1925 *Hitlers* engster Mitarbeiter, war der grobschlächtige Nationalsozialist trotz seiner exzentrischen Auftritte in Teilen der Bevölkerung beliebt. Die allmächtige *DAF* hatte schließlich 25 Mio. Zwangsmitglieder und nur einen einzigen Zweck: Das politische Bewußtsein der Arbeiter einzuschläfern und die Arbeitsleistung für die Rüstung zu steigern. *Ley* besuchte häufig Wilhelmshaven, um sich in Begleitung der *DAF*-Funktionäre vom Fortgang der Bauarbeiten zu
154 überzeugen. Augenzeugen[355] erinnerten sich, daß er am 09.06.1941 von Polizei und Presse

eskortiert, in einem Geländewagen zum Arbeiterlager Fedderwardergroden preschte und vor Bauarbeitern eine Rede hielt. Er verlas angetrunken das falsche (für den Auftritt bei der *Werft* vorgesehene) Manuskript.

NS-Prominenz veranstaltete sogar *NSDAP*-Kundgebungen in Arbeiterlagern[356].

Für Wilhelmshaven konnten zur Sicherung des Kräftebedarfs schon 1938 Dienstverpflichtungen ausgesprochen werden[357]. Um angesichts des Kriegsbeginns die Wohnungen für die *Marine* fertigzustellen, erließ der *MDF* im Juli 1939 eine Bausperre für die Nachbargemeinden. Am 15.11.1939 wurden reichseinheitlich die nichtkriegswichtigen Bauten verboten. Nicht so bei den kriegswichtigen Bauten in Wilhelmshaven. Die Dienstverpflichtungen oder Teildienstverpflichtungen betrafen alle Berufe und Fachgebiete, die für Stadterweiterung, Militärbauten und Werftausbau benötigt wurden. 1939 arbeiteten 11.000 Dienstverpflichtete in Wilhelmshaven.

Neben der Pflicht zum aktivem Dienst bei der Wehrmacht (2 Jahre) bestand seit 1932 der *FAD*. In Wilhelmshaven und Rüstringen gab es mehrere, auch temporäre, Arbeitsdienstlager[358]. Zwischen 1935 und 1945 war die Arbeitsdienstpflicht für Männer und Frauen vom 18. bis zum 25. Lebensjahr gesetzlich verankert[359]. *RAD Reichsleiter Konstantin Hierl, Reichsarbeiterführer* und *Staatssekretär* im *RAM* (ab 1943 *Reichsminister*) organisierte Deichbau, Landgewinnungsmaßnahmen, Moorkultivierung, Siedlungsvorbereitung und Wegebau. Er kam am 31.01.1936 nach Wilhelmshaven, um Arbeitseinsätze zu besprechen.

★

Für den Bau der *Nordwerft* und der IV. Einfahrt und die dort beschäftigten ca. 20.000 Arbeiter hat die *HNBD* mehrere Arbeiterlager fertiggestellt: 100

- Arbeiterlager Sudetenstraße. 5-zeilige zweigeschossige Baracken für 2.700 Menschen. Nachdem die erste Schleuse der IV. Einfahrt fertiggestellt war (1942) und die Arbeiten an der *Nordwerft* nur langsam vorangingen, waren dort auch Marinehelferinnen und *RAD* untergebracht. Nach dem Kriege wurden die Baracken zum Teil zerstört.
- Arbeiterlager Norderneystraße. 16 eingeschossige Baracken für 1.800 Menschen. Das Lager für wurde später auch noch Notunterkunft für Ausgebombte und Evakuierte. An der Borkumstraße entstanden zwei weitere Baracken für 140 Arbeiter.
- Arbeiterlager Rüstersiel (zwischen Kniphauser Deich und Flutstraße). 28 eingeschossige Baracken für 1.000 Menschen, Wirtschaftsbaracken usw. Ein Teil davon war bereits als *RAD*-Lager gebaut worden. Ab 1948 war das Arbeiterlager pädagogische Akademie und ab 1949 Hochschule Rüstersiel. Seit 1962 sind dort Dienststellen der *Bundesmarine* untergebracht.

Während die Arbeiterlager Sudetenstraße und Norderneystraße unmittelbar neben der *Nordwerft* lagen, war das Arbeiterlager Rüstersiel weiter abgelegen und durch die Werftbahn mit den Großbaustellen der *Nordwerft* verbunden. 101

Zur Erzielung einer guten Arbeitsleistung war es erforderlich, den Bauarbeitern ein gutes
130 Arbeitsklima zu schaffen. Nicht nur Feiern im Gemeinschaftsraum waren propagandis-
50 tisch wichtig. Auch die Brücke zur einheimischen Bevölkerung sollte geschlagen werden.
Die insgesamt über 40.000 Arbeiter wurden von den Einheimischen als Fremde angesehen. Kamaradschaftsaufrufe lauteten so: „Sie bringen dieses Opfer der Trennung für die Wehrhaftmachung und den wirtschaftlichen Aufschwung unseres Volkes, sie arbeiten daran ebenso wie die hier beheimateten Arbeiter und Handwerker ... sich mit ihnen verbunden fühlen als Arbeitskameraden, die am gleichen Werke schaffen, und als Volksgenossen im Dritten Reich“[360].

Der Wilhelmshavener *Polizeidirektor* befürchtete 1937 bei dem Arbeiterlager Sudetenstraße sittliche, polizeiliche und politische Probleme. Bauunternehmer *Friedrich Möller* wurde vom *Magistrat* zu Rate gezogen und konnte die Bedenken entkräften. Er habe „immer gute Erfahrungen“ mit Massenquartieren gemacht[361].

Die deutschen Bauarbeiter genossen gegenüber den Werftarbeitern finanzielle Privilegien. Die Werkszeitung der *KMW* Wilhelmshaven[362] mußte bei Stammwerftarbeitern häufig Mißgunst und Neid abbauen. Die von ihren Familien getrennt lebenden Bauarbeiter in den Lagern bekamen bis 1939 Trennungsentschädigung, Vollverpflegung (ohne Lebensmittelkarten), Schwerstarbeiterzulagen, freie Heimreisen und Unterhaltungsprogramme. „Der Nationalsozialismus hat die Betreuung des deutschen Menschen ganz bewußt auch auf das geistige Gebiet ausgedehnt. Wer fleißig arbeitet, hat auch Anspruch auf Ablenkung und Unterhaltung“.

Abseits von nach außen dargestellten bürgerlichen Moralvorstellungen billigte die *Stadt Wilhelmshaven* den ausländischen Arbeitern 3 Bordelle zu. Am Banter Weg, in Fedderwardergroden sowie zwischen Sande und Mariensiel baute die *Wohnungs- und Barackenbaugesellschaft mbH Berlin* auf Rat der *Kriminalpolizei* und der *DAF* Bordelle in Baracken[363].

In Wilhelmshaven wurden bis zu 40 Lager gezählt. Außerhalb der Gemeindegrenzen
72, 66 gab es weitere Lager und Quartiere, wie z.B. in Roffhausen, Breddewarden,
48 Sande-Neufeld, Fedderwarden, Sengwarden, Middelsfähr, und Mariensiel. Die Lager
99, 49 Sudetenstraße, Ebkeriege und Sande-Neufeld waren massiv und mehrgeschossig gebaut
und für eine spätere Weiternutzung für Militär- oder Wohnzwecke vorgesehen.

98 Die Barackenlagerplanung des *Marinestandortbauamtes* war rationalisiert und normiert.
Auf einem Quadrat von 200 x 200 m waren 12 zerlegbare Baracken um einen 60 x 60 m großen Hof angeordnet, exakt für „1.000 Mann“ ausreichend. Vor Kopf, ausgerichtet auf den Hof das Wirtschaftsgebäude. Die Quadranten wurden nach Erfordernis zusammengesetzt. Altengroden bekam 3, Breddewarden 2, die Marinekraftfahrzeugabteilung 1. Ein kleineres angefügtes Rechteck war der Sportplatz.

⋆

Die *Marinewerft* verfügte seit 1911 über eine eigene Werftbahn vom Verschiebebahn-
hof Sande aus. Ab 1935 baute das *Hafenbauressort* der *KMW* das Schienennetz auf 200 km
und 20 Haltestellen aus. Der Personenverkehr nach Voslapp wurde am 20.11.1941
aufgenommen. Die Bahn hieß ab 1942 Marinebahn, ab 1949 Vorortbahn. Im Zweiten
Weltkrieg wurden täglich bis zu 30.000 Menschen zu ihren Arbeitsplätzen auf
Kriegsschiffen, Baustellen und Werftanlagen befördert. Die Werftbahn übernahm auch
von Sande aus einen Großteil der Verteilung von Gütern[364] und Baustoffen, und sorgte 94
mit 600 Bediensteten für den Personen- und Baustellenverkehr. Die neuangelegten
Bahnhöfe und Bahnlinien berücksichtigten die Baustellen und Arbeitsplätze von 51
vornherein und waren wesentlichste Infrastrukturmaßnahme für rationellen Baubetrieb.
Auch Buslinien zu entlegenen Baustellen und zu den Lagern der Arbeiter mußten
eingerichtet werden.

Die Werftbahn erschloß die Baustellen Fedderwardergroden und Voslapp. Geschickterweise befanden sich die entsprechenden Gleistrassen im Schutzbereich zwischen den geplanten Ortsteilen. Eine Verlängerung der Trasse nach Nordwesten hin war zwischen den geplanten, aber niemals realisierten Stadtteilen Utters und Breddewarden als Ringschnellbahn vorgesehen. Auch die Baumaterialien für Altengroden wurden zum eigens angelegten Güterbahnhof zwischen Dodoweg und Kurt-Schumacher-Straße geliefert. Die geplanten Stadtteile Langewerth und die außerhalb der Gemeindegrenzen befindlichen Siedlungsgebiete um Schortens, Cleverns, Neustadtgödens und Cäciliengroden waren so vorgesehen, daß die Baustellen auf dem Schienenwege beliefert werden konnten. Alle im Wilhelmshavener Stadtgebiet verlegten zusätzlichen Kleinbahntrassen sollten nach Fertigstellung der Stadtteile wieder aufgenommen werden. Die großen, dann das Bild störenden Umschlagplätze für Baumaterialien sollten verschwinden. Tatsächlich jedoch bestanden die großen Umschlaganlagen für Baumaterialien auf Bahnhöfen und in Häfen oft bis Mitte der 60er Jahre und hatten Bedeutung für das Wirtschaftsleben der Nachkriegszeit. Die Gleiserschließung für die Bauarbeiten war
unabhängig von der tatsächlichen Eisenbahnplanung, die 1940 nach dem *Wirtschaftsplan* 129
Wilhelmshaven-Friesland ein anderes Konzept hatte. Für den Güterverkehr schließlich war
ein neuer Ortsgüterbahnhof an der Ladestraße geplant, mit Hochbauten, umfangreichen 52
Industrie- und Lagerflächen und entsprechenden Stammgleisen für die *Marine*. Der *Fluchtlinienplan* war 1941 fertiggestellt worden. Die *Reichsbahndirektion Münster* hatte Grunderwerbsprobleme. Im August 1941 mußte der Bau wegen Materialmangel eingestellt werden[365].

Viele Baumaterialien wurden per Schiff angeliefert. Der Mariensieler Hafen war zeitweise von Steinschiffen blockiert. Materialien mußten auch mit Pferdefuhrwerken nach Cäciliengroden, Sande-Neufeld, Sanderbusch und nach Roffhausen gebracht werden. Wilhelmshavener Baufirmen hatten eigene Löschvorrichtungen in den Häfen Mariensiel und Dykhausen gebaut. Holländische Transportschiffe brachten Schwemmsteine aus der Eifel und Ziegelsteine aus dem Emsland. Seegängige Schiffe löschten Baumaterialien im Handelshafen[366]. 1939 begann der Bauträger der niemals gebauten

Siedlung Breddewarden, die *Niedersächsische Bauträgergesellschaft*, den Hafen Hooksiel[367] auszubaggern. Von hieraus sollten Baustoffe nach Breddewarden, später auch Utters, über neue Straßen und Feldbahnen transportiert werden[368].

★

Das bauwirtschaftliche Steuerungsinstrumentarium der letzten 50 Jahre blieb erstaunlich kontinuierlich. Versuche, Bau- und Planungsprozesse systematisch und diszipliniert anzugehen, waren keine Erfindung der Nachkriegszeit. Bemühungen um Rationalisierung und Vorfertigung, Rasterung und Typengrundrisse gingen auf die 20er und 30er Jahre und davor zurück. Daß es eines *Bebauungsplanes, Fluchtlinienplanes, Wirtschaftsplanes, Flächennutzungsplanes* bedurfte, daß Erschließungsmaßnahmen zu veranschlagen waren, daß Ver- und Entsorgungseinrichtungen geplant werden mußten, daß öffentliche Einrichtungen von vornherein zu berücksichtigen waren, daß Wohngebäude nach den Bevölkerungsdaten zu konzipieren und zu finanzieren waren, daß Baumaßnahmen ordnungsgemäß ausgeschrieben werden mußten, wurde auch 1933–45 anerkannt und praktiziert. Völlig falsch wäre die Vorstellung, daß der baubesessene Diktator sowie seine maßgeblichen Architekten die bauwirtschaftlichen Abläufe und Regeln auf den Kopf
152 gestellt hätten. Es gab zwar Befehle zum Bau, aber beim Bau selbst herrschten keinesfalls Willkür und Chaos, sondern Ordnung und handwerkliche Präzision. Der herrische Bau„führer" war dem Vernehmen nach eher die Ausnahme. Man verstand sich als Techniker der Baustelle, nicht als „Einpeitscher". *Fritz Todt*, übermächtiger Baumanager der *NSDAP*, *Generalbevollmächtigter für die Regelung der Bauwirtschaft* ab 1938 und 1940–42 *Rüstungsminister* hatte übergreifenden Einfluß auf das gesamte Baugeschehen durch Lenkung der Materialbewirtschaftung.

Die *DIN* 276, noch heute maßgeblich für Kostenermittlungen von Hochbauten, geht bezeichnenderweise auf 1934 zurück. Die Bauutopien des NS-Staates standen sicher auf den Fundamenten von technischen Normen und Regeln der Baukunst. Deutscher Perfektionismus überlagerte die Bauwut des NS-Regimes. Baukosten mußten genau ermittelt werden, denn sie waren Grundlage der Zuschüsse, die vom *RAM* und *OKM* nach Wilhelmshaven flossen. Die Urangst der Architekten, wegen nichteingehaltener Kosten und Terminüberschreitung zur Verantwortung gezogen zu werden, ließ Unordnung und Hektik nicht zu. Zeitdruck, begleitende Propaganda, nie gekannte Dimensionen, völlig neue Aufgaben und die schwierigen Bodenverhältnisse in Wilhelmshaven, brachten genügend Unsicherheit in den Bauablauf, so daß niemand geneigt war, organisatorisch eingefahrene Schienen zu verlassen.

Den Kostenermittlungsmethoden für die Großsiedlungen in Wilhelmshaven kamen die einfachen Gebäudegrundrisse und Gebäudeformen zugute. Ohne moderne Abwinkelungen und Flachdächer, ohne Glasflächen und expressionistische Dachschwünge, ohne Klinkertexturen und Ziegelplastiken, ließen sich Risiken umgehen. Die technisch-wirtschaftlichen Rahmenbedingungen waren: Anerkannte Regeln der Technik niemals

verlassen, neue Konstruktionen nicht zuerst ausprobieren und kein unkalkulierbares Wagnis eingehen. Diese Rahmenbedingungen richteten sich auch gegen *Neues Bauen* 15
und *Internationalen Stil*. Das blieb auch Glaubensbekenntnis der Architekten, als es an den 17–21
Wiederaufbau nach dem Zweiten Weltkrieg ging. Völlig anders jedoch waren die bauwirtschaftlichen Abläufe bei Monumentalbauten und Wiederaufbauplanungen. Seriöse Kostenermittlungen und durchdachte organisatorische Vorbereitungen für *Kultur-, Marine-* und *Verwaltungsforum* Wilhelmshaven kamen kriegsbedingt nicht mehr 107, 108
zustande. Die Finanzierung war unsolide. Gemeinschafts-, Gesellschafts-, Kultur- und Schulbauten allerdings waren planerisch schon weit fortgeschritten, entworfen, ausgeschrieben, vergeben und finanziert.

Auch die Auslobung von Bauleistungen fand im „III. Reich" nicht willkürlich statt, sondern geregelt und nach Verdingungsordnung und Preisprüfung. Die Ideologie des „III. Reiches" war mittelstandsfreundlich. *Ministerialrat Thiele* vom *RAM* teilte dem *OKM* am 21.02.1936 mit: „Eine Beschränkung der Initiative der Privatwirtschaft sei in keiner Weise anzunehmen. Im Gegenteil sei es mehr als bisher möglich, das örtliche Handwerk zu beschäftigen ..."[369]. Eine Bevorzugung von bestimmten Auftragnehmern hätte sich niemand leisten können. Korruption wurde unnachsichtig verfolgt. So kam *Walter Hunzinger*, Wilhelmshavens *Stadtbaudirektor* von 1937–39, in den Verdacht von persönlicher Bereicherung. Von einem zum anderen Tag wurde er, obwohl altes Parteimitglied, vom Dienst suspendiert. Der Geschäftsführer der Wilhelmshavener Geschäftsstelle der *Baugesellschaft des Niedersächsischen Handwerks*, wurde im August 1941 wegen Unregelmäßigkeiten verhaftet. Verdachtsmomente gegen Baubeamte und Geschäftsführer wurden unnachsichtig verfolgt, aber auch wie bei *Otto Lehn* zur Disziplinierung eingesetzt.

Die *Verdingungsordnung für Bauleistungen*, die *VOB*, wurde bereits in der Weimarer Republik 1926 eingeführt. Das Verdingungsrecht begann schon 1830. Öffentliche Auftraggeber des „III. Reiches" achteten peinlich auf die gleichmäßige Behandlung des Handwerks. Die *NSDAP*-Dienststellen drängten 1934 die Gebietskörperschaften zu Beschlüssen, um „geachtete Handwerksmeister" zu beauftragen, die der *DAF* angehörten. Die „liberalistisch-jüdische Zeit" sei nun vorbei und die immer noch „Fernstehenden" sollten „doch nun endlich einmal zur Besinnung" kommen[370]. Bei großen Vorhaben, wie z.B. der *Nordwerft*, bediente man sich auch der auswärtigen Bauindustrie 101
mit ihrem besseren technischen Geschick oder stiftete mit Hilfe des *Generalbevollmächtigten für die Bauwirtschaft, Fritz Todt* und seines *Gaubeauftragten* leistungsfähige Firmenzusammenschlüsse. Notfalls wurden den Firmen Dienstverpflichtete und Gefangene „ausgeliehen", aber auch diese ordnungsgemäß abgerechnet und zu klaren Bedingungen. Sorgfältig wurde auch die Planung von der Ausführung getrennt. Pauschalvergabe war grundsätzlich verboten. Architekten, Statiker[371], Landmesser und Gartenarchitekten bekamen ordnungsgemäße Aufträge, die sogar noch vereinzelt nach dem Kriege abgerechnet wurden. Auch wurden Architektenwettbewerbe durchgeführt (Langewerth, Gemeinschaftshäuser). Auftragschancen als Architekten hatten nur die 64, 65

etablierten Mitglieder der *RDBK*. Eine Ausnahme bildete *Fritz Hilgenstock*, der als Nicht-Architekt „Generalplaner“ wurde.

★

Der Baubetrieb für die neuen Stadtteile wurde von den beauftragten Architektenbüros und Baubehörden aus jeweils dort aufgestellten Baubaracken aus benachbarten Altbauten oder aus bereits fertiggestellten Wohnungen dirigiert. In Fedderwardergroden z.B. etablierten sich die Baubeamten des *Marinestandortbauamtes* in den 6 zum Erhalt vorgesehenen Bauernhäusern, die später *NSV*-Heime werden sollten. Die Bauleitung der
101 *HNBD* für die *Nordwerft* residierte in der ehemaligen Hafenkaserne. Für Fedderwardergroden war im Auftrage der *Jade* der Baumanager *Fritz Hilgenstock* eingeschaltet. Das 220 ha große Baugelände war den Bauern von der *RUGES* abgekauft und an die *Jade* als Erbbaurecht weitergegeben worden. Bis zu 4.000 Bauarbeiter, Dienstverpflichtete und Gefangene, die auch aus Holland, Belgien und Polen kamen, mußten von Oberbauleiter
98 *Joseph Veeser* dirigiert werden. Im Juni 1939 waren Baracken für 700 Arbeiter aufgestellt. Die Baustoffe wurden mit der Werftbahn zum Bahnhof Hörn geliefert – eine Kleinbahn transportierte sie weiter. *Fritz Hilgenstock* drängte darauf, daß die Straßen in Fedderwardergroden vor Baubeginn vollständig erstellt wurden. So konnten die Baufahrzeuge und die mit Motorrädern beweglich gemachten Bauleiter zügig vorankommen. *Hilgenstock* hatte erkannt, daß halbfertige Straßen, wie z.B. in Voslapp und Cäciliengroden, den Baubetrieb hemmen, dauernde Reparaturen erfordern und letztlich den Betrieb hindern[372].

Die Bautätigkeit in Fedderwardergroden spielte sich vor allem während des Krieges ab. Die eingesetzten Baufirmen *Büscher* (Münster), *Hitzbleck* (Wuppertal), *Reh* (Berlin) u.a. wurden oft zur Beseitigung von Bombenschäden in die Innenstadt abkommandiert. Das Büro *Hilgenstock* mußte noch gegen Kriegsende für die *OT* Schützengräben in Ostfriesland planen. Immerhin 3.000 von 4.100 geplanten Wohnungen für Fedderwardergroden wurden 1945 durch das Büro *Hilgenstock* an die *Jade* übergeben. Von den knapp 1.300 Wohnungen in Voslapp wurden fast alle noch vor dem Kriege fertig.

Willi Lübbers[373], *Beauftragter der NSDAP für Bauangelegenheiten*, 1937–38 Chefarchitekt bei der *Jade* und gleichzeitig im *NSDAP-Magistrat* hat seine Bauleitung für Voslapp sehr selbstherrlich ausgeübt. Er kauft „auf eigene Kappe“ Material und warb 1.000 schlesische Bauhandwerker an. Der Architekt zögerte auch nicht, den gesamten Straßenbau mit durchzuführen. Der Unterbau wurde per LKW aus den Baugruben der *Nordwerft* herangeschafft. Rücksichtslos wurden die Landstraßen zerfahren und es kam zu großen Mißhelligkeiten und Schäden. Bis über 600 Arbeitskräfte[374] waren gleichzeitig in Voslapp tätig. Das Wilhelmshavener *Arbeitsamt* hatte die Personalorganisation zu leisten und die nötigen Facharbeiter an die Baufirmen zu vermitteln. Für Voslapp z.B. wurden allein 300 Maurer herangeschafft.

Auch für Cäciliengroden und Sande-Neufeld, die von der Baufirma *Herdejürgen und Harmsen* aus Einswarden gebaut wurden, hatte man die Handwerker baustellennah untergebracht. 100 Arbeiter in einer Baracke waren üblich. Ein nationalsozialistischer Gemeinschaftsraum fehlte in keinem Lager[375]. Das „Führerbild" symbolisierte Dankbarkeit für *Hitler,* der die Arbeit beschafft hatte. „Sie schaffen durch Ihrer Hände Arbeit Werte, die der Gemeinschaft des Volkes gehören ... Das sind die Ziele, die sich der Führer gesteckt hat, und bei deren Erreichung ihm das ganze deutsche Volk helfen wird, weil alle wissen, daß der Führer nur für das Volk arbeitet", so tönte die Propaganda. Die Bauarbeiter des Jadeviertels wohnten in Baracken am Banter Weg. Teile davon waren von 1944–45 *Konzentrationslager.* Für die Bauarbeiten am Hansaviertel, an denen vor allem die Firma *Kampen* beteiligt war, wurden Unterkunftsbaracken auf dem Lessingplatz plaziert.

34, 48 *130*

★

Wegen des beginnenden Krieges standen alle Bauarbeiten unter dem Diktat der *Vierjahrespläne*[376]. Dem *Arbeitsamt* Wilhelmshaven waren alle Bauarbeiten anzuzeigen mit genauer Angabe von Eisen- und Stahlmengen, unterschieden nach Moniereisen und anderen Bauteilen. Die *Vierte Anordnung zur Durchführung des Vierjahresplanes* sollte den Stahlverbrauch für das Bauen strikt einschränken, um die Rüstung zu verstärken. Für Bewehrung von Geschoßdecken und konstruktive Zwecke wurde Stahl in der Regel noch zugelassen. Es entwickelten sich die eisensparenden Baustahlgewebe. Bodenständige Satteldächer im Wohnungsbau, vor allem aber die hölzernen Hänge- und Sprengwerke der Gemeinschaftshallen und Parteihäuser kamen der Eisenkontingentierung entgegen. Türen, Fenster, Treppen und Einfriedigungen aus Stahl, Blech oder Flachdächer hatten keine Chance auf Genehmigung durch das *Arbeitsamt.* Sie paßten ohnehin nicht in das Architekturkonzept des „III. Reiches". Eine Ausnahme bildeten Hallen für Fliegerhorste (Upjever), für die *KMW,* für *Nordwerft* und *Westwerft* und für die Rüstungsbetriebe in Sande, die wegen ihrer Dimensionen Flachdächer, Eisentore und Stahlfachwerke bekommen durften. Auch die Betonherstellung war kontingentiert. Ab 1938 benötigten Bauunternehmer vom *MDF* ausgestellte „Zementschecks". Die Hektik des Baubetriebes, der unnachsichtige Zeitdruck und andauernde Personalmangel führten trotz traditioneller Bauabläufe zu konstruktiven Fehlern, Bauschäden und Fehlplanungen. So mußte 1938 die 1899 gebaute Petruskirche geschlossen werden, da sie wegen der Bauarbeiten an der benachbarten Nordwerft einzustürzen drohte. Beim Bau von Voslapp hatte *Willi Lübbers* große Bauschäden durch undichte Kellerwände in den Siedlungshäusern verursacht. Architekt *Friedrich Heuer* hatte mit dem Bautempo und den ungewohnten Bodenverhältnissen in Cäciliengroden große Schwierigkeiten. Am 16.10.1940 mußte *Marineintendanturrat Krüger* sich vor Ort umfangreiche Mängel, undichte Dächer und Versackungen vorhalten lassen[377]. *Amtsbaumeister Hanns Ladwig* mußte zahlreiche Mängel der Schule feststellen. Noch 1953 beklagte die *Gemeinde* den Pfusch beim Straßenbau, und

156 *97, 103* *93*

daß die Straßen 10 bis 15 cm zu tief angelegt seien. Lange war auch völlig unklar, ob man bei den neuen Stadtteilen erst die Straßen oder die Häuser bauen sollte. Nachdem in Voslapp große Schäden durch den Lastwagenverkehr zu beklagen waren, begann man dann in Fedderwardergroden, zuerst die Straßen zu bauen.

7.9. Kunst am Bau

Hitler war seit seiner frühesten Jugend der Kunst verfallen[378]. Hemmungslos benutzte er Architektur und bildende Kunst für politische Ziele. Er malte selbst und umgab sich mit Künstlern und Architekten. *Alfred Rosenberg*, Kulturideologe der Nationalsozialisten, hatte 1927 die *Nationalsozialistische Gesellschaft für deutsche Kunst* gegründet, die am 19.12.1928 in den *Kampfbund für deutsche Kultur (KDK)* überführt wurde. Der *KDK* nutzte geschickt die Angst der Bürger vor moderner Kunst, setzte sich für Heimatkunst und völkische Kulturideale ein. In Norddeutschland wurzelte der *KDK* in der 1904 gegründeten *Heimatschutzbewegung* um *Paul Schultze-Naumburg* und dem zeitgleich gegründeten *Verein für niedersächsisches Volkstum*. Am 18.07.1933 gründete *NSDAP-Kreiskulturwart Hermann Ahner* in Wilhelmshaven/Rüstringen den *KDK*, der jetzt offizielle Kulturorganisation der *NSDAP* geworden war. *Hitler* proklamierte auf dem Reichsparteitag 1933 die „rassische Kunst", legte in München den Grundstein zum *Haus der deutschen Kunst* und erließ das *Gesetz über die Errichtung einer Reichskulturkammer*. 1934 wurde der *KDK* auch in Wilhelmshaven/Rüstringen Unterorganisation des Ortsverbandes der *NS-Kulturgemeinde* von *Alfred Rosenberg*. Vorsitzender war wiederum *Kreisobmann Hermann Ahner*, gleichzeitig Leiter des *Kreiskulturamtes* der *NSDAP*. 1937 wurde die *NS- Kulturgemeinde* der *DAF*-Organisation *KDF* zugeschlagen. Die *NS-Kulturgemeinde*, Ortsverein Wilhelmshaven/Rüstringen, gründete u.a. den *Ring Volkstum und Heimat*, unter dem die Heimat- und Naturschutzvereine 1934 zwangsvereinigt wurden[379]. *KDK* (1933–34), *NS-Kulturgemeinde* (1934–37) und *KDF* (1937–45) trugen „völkisch"-nationalsozialistische Kultur nach Wilhelmshaven. Eine einheitliche Auffassung zur bildenden Kunst setzte sich nur zögernd durch. „Nordische" Kunstauffassung, martialische Marinemalerei und volksnahe Heimatkunst waren seit der Weimarer Republik in Wilhelmshaven bedeutsam.

Sofort nach 1933 begann die Dekoration öffentlicher Gebäude mit *Hitler*-Bildern, Hoheitszeichen und unsäglichem NS-Kitsch. Das *NSDAP-Kreiskulturamt* schaltete 1934 den *Verein der Kunstfreunde* gleich und diktierte die Wilhelmshavener Ausstellungspraxis[380]: „Volkstum im Ausland", „Die deutsche Front", „Mensch – Hygiene – Sport", „Die deutsche Marineschau", „Des Seemanns Feierabend", „Heimatbilder", „Kriegsmaler Engelhardt Kyffhäuser" usw.

Bauen war im „III. Reich" Teil nationalsozialistischer Kunstpolitik und Propaganda. *Joseph Goebbels*, der die gesamte NS-Kulturpolitik in der Hand hatte, trat am 27.01.34 in Wilhelmshaven auf. Verschiedene Male danach, zuletzt im März 1941, inszenierte er propagandistische Massenversammlungen in Wilhelmshaven. Die *Landesstelle Weser-Ems des Propagandaministeriums* und die *Hauptstelle Kultur der Gau-Propagandaleitung* führten vom 28.11.–06.12.1936 ihre *Gau-Kulturwoche* in Wilhelmshaven durch. *Carl Röver* präsentierte „großartige" Erfolgsbilanzen der „seit der Machtübernahme ausgeführten Bauten", der „angesiedelten Siedlerfamilien" und der „investierten Gelder"[381]. *Reichskulturwalter Franz Karl Moraller* kam zur *Gau-Kulturwoche"* nach Wilhelmshaven und redete „zu Kulturfragen[382]. *Rudolf Heß* kam im März 1935 und im November 1936 zusammen mit *Alfred Rosenberg* nach Wilhelmshaven. Der *Reichsinnenminister* und radikale *Bauhaus*-Gegner *Wilhelm Frick* im Mai 1935 und Juni 1937, der antisemitische *NSDAP*-Kulturpolitiker und „Stürmer"-Herausgeber *Julius Streicher* im März 1938.

Gau-Kulturpolitik fand ihr nationalsozialistisches Ideologiezentrum ab 1934 in der *Nordischen Kunsthochschule Bremen*, ab 1936 im *Blockhaus Ahlhorn* und ab 1935 in der *Niederdeutschen Kultstätte Stedingsehre*, die mit einer Gau-Schulungsburg in Bookholzberg am 13.07.1935 eröffnet wurde[383]. Der Blut- und Boden-Ideologe *Alfred Rosenberg* förderte den Ausbau der Kultstätte und stilisierte sie zum Gegenstück zu *Goebbels „Thingspielen"*. Seine verschiedenen Auftritte in *Stedingsehre* zogen *Röver* den Zorn von *Martin Bormann* zu[384]. Die Gau-Kunstpolitik wurde von *Gau-Propagandaleiter Ernst Schulze* und *Gau-Schulungsleiter Heinrich Buscher*[385] sowie dem *Gaustellenleiter Kultur, Hans Fricke,* bestimmt. *Fricke, Bauhaus*-Absolvent von 1925, war 1931 zur *NSDAP* konvertiert und hatte ab 1935 als *Landesleiter Weser-Ems der RDBK* unbeschränkten Einfluß auf die bildende Kunst in Nordwestdeutschland.

★

Kunst im Zusammenhang mit Bauwerken hatte in der Weimarer Republik wenig Bedeutung, da es an Wohnungsbau und Repräsentativ-Projekten mangelte. Als *Propagandaminister Joseph Goebbels* am 22.06.1934 den *Erlaß Aufträge an bildende Künstler und Kunsthandwerker bei Bauaufgaben der Staatshochbauverwaltung* herausgab, fanden viele arbeitslose Künstler endlich Möglichkeiten, „an den großen Aufgaben der Baukunst, dem künstlerischen Ausdruck des Volksganzen mit zu formen". Zielrichtung war neben nationalsozialistischer Kunst der Arbeitsbeschaffungseffekt. Die Prozentregelung für Kunst am Bau sollte im gesamten öffentlichen Bauwesen und bei den Wohnungsbaugesellschaften angewandt werden. *Goebbels* war faktisch 1933–45 „Kulturminister" des „III. Reiches", gleichzeitig Gründer der *RDBK* (1933), Organisator der Bücherverbrennungen in Berlin (1933), der Ausstellung *Entartete Kunst* im Münchner Hofgarten (1937) und der *Großen Deutschen Kunstausstellungen* im Münchner *Haus der Deutschen Kunst (1937–44)*.

Konsequent veranschlagten *Stadt Wilhelmshaven, Marinestandortbauamt, KMW, HNBD* und *Luftwaffenbauamt* bei den Kostenermittlungen nach *DIN* 276 Prozentanteile für bildnerischen Schmuck. Der *RAM* stellte 1940 für „guten, architektonischen oder bildnerischen Schmuck" Sonderdarlehen zur künstlerischen Ausgestaltung bereit[386].

Auch dem *RAM* ging es darum, „Künstlern zusätzliche Arbeit" zu schaffen. Es wurde „ein kunstpädagogisches Problem gegen ein sozialpolitisches eingetauscht"[387]. Den Anfang machten Büsten und Bilder, Adler und Fahnen, Hoheitszeichen und Pylonen an Bauten der Gemeinschaft. Die *NSDAP* war erleichtert, als der *Erlaß vom* 22.06.1934 die Abwälzung der Kosten auf Bauherren ermöglichte.

Der Rüstringer Maler und *Mackensen*-Schüler *Georg Harms* erfreute sich besonderer Vorliebe der Nationalsozialisten. Der heimatverbundene Künstler produzierte nicht nur Bilder, die er beispielsweise „Hinterm Pflug" und „Blut und Boden" nannte, er durfte auch *Hitler, Röver, Joel* u.a. porträtieren und in Wilhelmshavens repräsentativsten Gebäuden aufhängen. Am 20.06.1933 hing das erste *Hitler*-Bild im Wilhelmshavener Rathaus. Dankbar wurde vermerkt, daß im Zimmer des Rüstringer *Oberbürgermeisters Gustav Nutzhorn* „ein ganz hervorragendes Gemälde des Führers von einem neueren Meister des Pinsels" hängt[388]. Im Oktober 1938 wurde eine Büste des NS-Schriftstellers *Dietrich Eckart* in der im März nach ihm benannten Schule aufgestellt. Zur Eröffnung des *Marine- und Kolonialmuseums* im Mai 1935 wurde der Sitzungssaal des Rathauses Wilhelmshaven mit einer Statue von *Großadmiral Tirpitz* und einem Bild von *Admiral Scheer* dekoriert – beide Ehrenbürger der Stadt[389]. Am 19.12.1935 enthüllte *Oberbürgermeister Gustav Nutzhorn* ein Hoheitszeichen an der Sitztribüne des *Höger*-Rathauses. Es hatte sich herausgestellt, daß die expressionistische Architektur des Ratssaales keine freien Wände für Bilder ermöglichte. Ein Holzschnitzer aus Unterammergau hatte das Hoheitszeichen geschnitzt[390]. „Groß und breit wuchtet der Adler seine Fittiche vor dem Beschauer, in den Fängen den Kranz mit dem Hakenkreuz tragend, Zeichen des Kampfes, des Glaubens, der Hoffnung und des Sieges".

Hitler-Bilder wurden in den Werkhallen aufgehängt und pompös eingeweiht.
130 Kasernen, Offiziersheime, Kantinen, Lazarette, Arbeiterlager, Wahllokale, Mannschafts-
83 räume, *HJ*-Heime und das „Haus des Handwerks" bekamen „ihre" Hoheitszeichen.
Arbeitsplätze und Betriebssportgemeinschaften, Werkstätten und Lehrlingsheime wurden symbolisch von „Führer"-Bildnissen überwacht oder mit *Hitler*-Büsten dekoriert[391].
Zur Zusammenlegung der Städte zu Groß-Wilhelmshaven am 01.04.1937 grüßte die
„Büste des Führers" in einem Blumen-und Blatthain die Festgäste. Der „Führer" sollte
133 allenthalben präsent sein und der „Gefolgschaft" die Macht darstellen. Hoheitszeichen als
plastischer Schmuck wurden in Wilhelmshaven zumeist von *Rudolf Grau* entworfen. Er
war zentriert über den Eingängen des *Marinelazaretts Sanderbusch* (1937) und der
83, 95 *Bonte*-Kasernen (1937). Er fand sich am Leuchtfeuer Minsener Oog (1942–43), beim
147 *Fernsprechamt* Schillerstraße (1939), am *HJ*-Heim Krähenbusch (1938), an Kopfbauten der
102, 144 IV. Einfahrt (1940), am *Finanzamt* (1936). Jede Schule, jedes *NSDAP*-Gemeinschaftshaus,
jedes *HJ*-Heim bekam sein eigens entworfenes Hoheitszeichen. Der in allen Haltungen vorkommende Adler sollte Wachsamkeit, militärische Tugenden und Herrschaftsanspruch verkörpern. Reproduktionen durften nicht verwandt werden[392]. *Goebbels* hatte mit dem *Gesetz zum Schutz der nationalen Symbole vom* 19.05.1933 versucht, Massenkitsch zu verhindern. Dies gelang nur in Ansätzen.

*

Nach den *Olympischen Spielen* von 1936 entwickelte sich eine typische Formsprache der Bauplastik[393], deren Exponenten *Arno Breker* und *Joseph Thorak* wurden. Beide knüpften an die heroischen Männerakte des 19. Jahrhunderts. Die Formen wurden härter, die Dimensionen monumentaler, die Männer wehrhafter, die Züge ebener, die Haltung angespannter. In Wilhelmshaven sollten die großangelegten Forumsplanungen mit monumentalen Bauplastiken ausgestattet werden, mit Pylonen und Säulen, Standbildern und Reliefs. Es kam 1937 nur zu Ansätzen von Kunst am Bau in *Marineforum* und im *Verwaltungsforum*. An der Ostfassade des Marinestationsgebäudes wollten *Karl Franzius*
und *Kurt Geisenhainer* 1937 einen 4 m weit ausladenden Adler mit Hakenkreuz. Der 1937 *71*
beauftragte Hamburger Bildhauer *Richard Emil Kuöhl* produzierte das Relief „Seefahrt ist not“, verzierte Balkenköpfe mit Seetieren und kriegerischen Symbolen. Der Aufmarsch-
platz davor wurde von einem Pylonen mit Hoheitszeichen flankiert. *70*

Mit Seetieren und maritimen Symbolen dekorierte *Rudolf Grau* 1937 das *Marinelazarett* *76*
Sanderbusch.

Wilhelmshaven war für monumentale Bildhauerei, wie auch für den *Neu-Klassizismus* von *Albert Speer* wenig geeignet. Heimatkunst setzte sich auch in der Bauplastik durch.
Die Plastiken hießen „Wikingerschiff“, „Hein und Grete“, „Mutter und Kind“, *161*
„Schmied“, „Zimmermann“ usw. *Höger* entwarf für Neu-Siebethsburg 1936–39 Fische, *160, 162*
Koggen, Leuchttürme, Werftkräne, Ähren usw. als Ziegelreliefs. *Oehler* modellierte an Gebäudeecken Handwerkerfiguren, *Grau* stellte in den Freianlagen Harmonikaspieler
auf. *Högers* Platzentwurf vor Kopf der Werftstraße sollte mit einem Männerakt vor *60*
einem Wasserbassin dekoriert werden.

Die *Wohnungsbaugesellschaft Jade* ließ für die von *Goebbels* verordneten „Kunst am Bau Prozente“ in der *Neuen Stadt* Fedderwardergroden Zunftzeichen und Reliefs an Giebeln und über Hauseingängen anbringen. Knabenakte, von *Meh* 1941 entworfen, sollten einen Platz in Fedderwardergroden dekorieren[394]. Außerordentlich beliebt waren auch
Brunnen[395]. *Karl Franzius* zeichnete 1942 einen Brunnen für den „Langen Markt“ in *111*
Fedderwardergroden. Im Kriegsjahr 1940 schilderte die Werftarbeiterzeitung der *KMW* das Stadtbild von Fedderwardergroden: „Es soll eben nicht der Eindruck einer Reihe von Mietskasernen entstehen, sondern der einer Stadt mit weiträumigen Straßen, mit Höfen und Plätzen, teilweise auch durch Bildwerke geschmückt, mit lieblich idyllischen, lauschigen Winkeln“[396]. Die *Wilhelmshavener Spar- und Baugesellschaft* schreckte nicht davor zurück, im Inselviertel eine Tafel anzubringen: „Erbaut unter *Adolf Hitler*“. Für Voslapp fertigte *Scheuernstuhl* das Modell eines Brunnens vor dem Gewerbetrakt. Dies führte zur Verärgerung bei der *Stadt*, die wünschte, „daß derartiger Bildschmuck auf öffentliche Plätze gestellt und von der *Stadtverwaltung* in ihre Beihilfeanträge hereingenommen wird“[397]. Für Voslapps Schule sollte ein Hamburger Künstler die Appellhalle mit Fresken dekorieren[398]. Peinlich genau nach dem Vorbild von *Paul Schmitthenner*[399]

ließen *Kurt Geisenhainer* und *Egon Forster* Gitterwerke und Geländer mit Wappen
stilisierter Käfer ausstatten.

Plastik an Rüstungsbauten sollte nicht harmlos verniedlichen, nicht ländlich stimmen,
nicht heile Welt spiegeln, sondern Härte, Fleiß und Geschick ausdrücken. Die Arbeitswelt
wurde von der *DAF* ideologisch doktriniert. Das 1933 gegründete *Amt „Schönheit der
Arbeit"* der *DAF* überwachte nicht nur die architektonische Gestaltung und hygienische
10 Ausstattung, sondern beriet auch bei künstlerischer „Verschönerung". Die *KMW* wollte
Arbeitsmoral und Disziplin, Kameradschaft und Zusammengehörigkeit betonen.
Bauarbeiter, Soldaten[400] und Rüstungsarbeiter wurden im „III. Reich" in ihrer
Arbeitsumgebung naturalistisch dargestellt, zumeist in der Malerei. *Georg-Emil Baumann*
und *Diedrich Janssen-Jennelt* wurden zu Chronisten der Bauarbeiten in Wilhelmshaven.

Die Plastiken und Reliefs an den Werkstattgebäuden der *KMW* standen im Zeichen
der Kriegsproduktion, bei der die Menschen in den Hintergrund rückten. Militärische
159 Ausstrahlung sollte die Wehrkraft stärken und die „Soldaten der Arbeit" stolz machen.
158 Hammerschwingende Schmiede und muskelzeigende Körper standen schon damals im
153 anachronistischen Gegensatz zum modernen Technologiestandard der Arbeit auf der
KMW.

Wichtige Funktionen im „III. Reich" und davor hatten Gefallenen- und Kriegerdenkmäler. Daß sie Geschichte vergegenwärtigen sollten, war nicht neu. Jetzt aber ging es um die „Helden", die sich für die „Volksgemeinschaft" geopfert hatten, und nicht um das Leid der Menschen. Kampf um die nackte Existenz schien besonders geeignet zu plastischer Darstellung, zu Dekoration von gärtnerischen Anlagen und zu offizieller Kunstpropaganda. Kriegerdenkmäler waren in Wilhelmshaven/Rüstringen eine häufige Bauaufgabe. Der Ehrenfriedhof, nach Entwürfen von *Martin Wagner* und *Leberecht Migge* in Form eines „Grabgartens" angelegt, wurde im „III. Reich" zum „Heldenfriedhof" uminterpretiert. Er wurde Ruhestätte für 2.400 Gefallene der zwei Weltkriege. 1940 stellte *Wilhelm Kreis, Generalbaurat für die Gestaltung der deutschen Kriegerfriedhöfe*, dem *Marinestandortbauamt* Musterentwürfe für Grabkreuze zur Verfügung.

Zahlreiche Gefallenen-Ehrenmale sollten das Kriegsgeschehen im „III. Reich" heroisieren. Neben den vielen NS-Feiertagen sollte das Gedächtnis an Schlachten durch besondere Gedenktage[401] gefestigt werden: Heldengedenktag der Kriegsmarine 13.03., Tag der deutschen Wehrmacht 19.03., Tag der Skagerrakschlacht 30.05., Jahrestag des Weltkriegsbeginns 01.08., Tag der Seeschlacht bei Helgoland 28.08., Gedenktag der Toten der Bewegung 09.11., Tag der Helden der Falklandschlacht 08.12. Das Ritual der Totenehrung wurde zum „braunen Kult"[402]. Die Denkmäler wurden mit Hoheitszeichen und Opferschalen, Adlern und Pylonen dekoriert. Nicht stilles Gedenken über Ursachen und Leid des Krieges stand im Vordergrund, sondern die Architekturkulisse für Gedenktage und Aufmärsche. Riten wurden periodisch wiederholt und ergänzten sich mit Sonnenwendfeiern, Nationalfeiertag, Muttertag, zu einem ausgeklügelten NS-

Feierjahr. Gedenkstätten wurden unter *Hitler* aufgeblasene Heldendenkmale. Schlachten sollten verherrlicht und Fronterlebnisse geschönt werden.

Die *Schiffbauabteilung der KMW* beispielsweise hatte sich 1933 in ihrem Ressortgebäude ein Gefallenen-Ehrenmal entwerfen lassen. Auf „3 etwa 3 m hohen Schwertern aus Eichenholz“ waren die Namen der Weltkriegsgefallenen des Schiffbauressorts eingetragen. „Im Schmucke von Eichenkränzen und Blumen ist es ein mahnendes Bild, das in seiner stillen, Ehrfurcht gebietenden Weise jeden ... an die Taten des großen Krieges erinnert. Ein aus roten Blumen geflochtenes Hakenkreuz, das zu Füßen des Ehrenmals niedergelegt wurde, mahnt, der Toten nicht zu gedenken, sondern in ihrem Geiste zu handeln und für die Zukunft unseres Vaterlandes zu kämpfen, wie sie es getan“[403]. Ebenso
wie die Ressorts der *KMW*, hatten auch Kasernen häufig ihre „eigenen“ Gedenkstät- 85
ten.

Das am 12.05.1935 eröffnete *Wilhelmshavener Marine- und Kolonialmuseum* bekam an der Nordwand des Treppenhauses ein Ehrenmal, „Sie dienten und starben für Deutschland“, mit Zahlen der Toten, „welche angeben, wieviel deutsches Blut jene einst deutsche Erde, die uns Lüge, List und Tücke entriß, getrunken hat“[404]. Die von *Oberbürgermeister Wilhelm Müller* 1937 geplante Neugestaltung und Erweiterung des „Heldenfriedhofs“ zur „Erinnerungsstätte der Marine“[405] vor Kopf der Nord-Süd-Achse
scheiterte, wie auch das Ehrenmal im *Kulturforum* für die Toten der „Bewegung“. 107

Die Verflechtung des Totenkultes mit der Religion, wie beispielsweise in der Garnisonkirche, blieb die Ausnahme. Schon in den zwanziger Jahren hatten *Marinedekan Friedrich Ronneberger*[406] und der spätere NS-*Reichsbischof Ludwig Müller* eine Totengedenkstätte eingerichtet. Verschiedene Gedenktafeln kamen während des „III. Reiches“ hinzu. Als die 1869–72 von *Johann Friedrich Adler* gebaute Garnisonkirche im September 1942 durch Bomben zerstört wurde, baute *Karl Maas* vom *Marinestandortbauamt* das Hauptschiff der Kirche mit den Gedenkstätten wieder auf[407]. Großzügig hatte die *Marine*, unterstützt von *Reichsbischof Ludwig Müller*, erhebliche Mittel zur Verfügung gestellt. Die unverzügliche Wiederherstellung war für den Wehrwillen und die Verteidigungsmoral notwendig. Die künstlerische Darstellung des soldatischen Todes stand in der Tradition der Heldenehrung, die aus der Weimarer Republik übernommen wurde. Die *Marine* hatte ihre von der *NSDAP* unabhängige eigene Tradition, in der christliche Symbole Bedeutung behielten.

★

Die Nationalsozialisten dekorierten Aufmärsche und Festumzüge, Paraden und 112
Massenkundgebungen künstlerisch und architektonisch. Die Ausstaffierung von Propa- 157
gandaveranstaltungen war geplant und wurde wie ein Bauwerk entworfen. Die „Flucht in Mythos und Kult“[408] begann mit *Speers* Dekoration des Tempelhofer Feldes zum 1. Mai 1933[409], setzte sich fort in seinen Entwürfen für das Nürnberger Parteigelände ab 1934 und den Kulissen, die *Reichsbühnenbildner Benno von Arent*, der Architekt des

Wilhelmshavener Theaters, für den *Mussolini*-Besuch 1937 in Berlin entwarf. Billiger als
156 Parteibauten in Stein und Beton waren Dekorationen aus Pappe und Holz, waren Lichtdome und Himmelsbeleuchtungen[410]. Sie wurden wie effektvolle Bühnenbilder eingesetzt, um die massenpsychologische Wirkung von Feiern zu steigern. Die aufmarschierenden Parteigliederungen und Soldaten wurden Teil der Architektur, symmetrisch geordnet, blockweise zentriert. Auch Wilhelmshaven bekam seine NS-typische Dekorationen für Partei- und Militärveranstaltungen. Der ab 1869 erstellte
9 Wilhelmshavener Kurpark wurde beliebtes Objekt für die Selbstdarstellung der *NSDAP*. Der englische Landschaftsgarten bekam 1933 noch nach der Planung von *Hermann Zopff* eine Achse. Die historische Beziehung *Hindenburg – Hitler* sollte baulich Gestalt gewinnen. Der Parkeingang Bremer-/Virchowstraße wurde am 19.12.1933 Hindenburg-Tor getauft und mit einer Büste des Feldherrn und *Hitler*-Wegbereiters dekoriert. Der Parkeingang Göker-/Bismarckstraße bekam am gleichen Tag ein „Hitler-Tor"[411].

Unter Tannengrün-Dekoration mit Hoheitszeichen waren die Gäste versammelt, um die „Stätte der Schönheit und des Friedens einzuweihen ... *Hindenburg* und *Hitler* – sie sind unsere Ehrenbürger – diese beiden Namen sollen immerdar alle Männer und Frauen, alle Knaben und Mädchen, die hier ein- und ausgehen, zu Ehrfurcht und Dankbarkeit ermahnen". Die Architektur des Hitler-Tores war heimatschutzorientiert und weit entfernt von den später aufkommenden Parteidekorationen.

Die jährlichen Maifeiern standen im Mittelpunkt der nationalsozialistischen Aufmarschspektakel, die bis zum Bau der *Nordwerft* auf dem großen Exerzierplatz östlich des Inselviertels oder auf dem Bismarckplatz stattgefunden hatten. Der Westteil des Wilhelmshavener Kurparks wurde als Teil des *Marineforums* zum „Maifeld" stilisiert. Höhepunkt der Maifeiern waren jeweils die aus Berlin übertragenen *Hitler*-Reden. Die Inszenierungen hatten durchaus architektonische Ansprüche. Die Grundrisse der Aufmarschblöcke auf dem „Maifeld" wurden in den Zeitungen veröffentlicht[412]. Der Platz vor dem Friesenbrunnen war von Fahnen gesäumt und symmetrisch auf die Tribüne der Funktionäre ausgerichtet. Kriegsinvaliden, Arbeitsopfer und Veteranen erhielten Sitzplätze im Halbrund, dann die Aufmarschblöcke von *Kriegsmarine, RAD,* Polizei, *DAF, KMW*, Wehrmachtsbetrieben und „Formationen der Bewegung". Die Grünverbindung Rathaus – Kurpark von *Martin Wagner* wurde durch die Nationalso-
52 zialisten zur Aufmarschachse umgedeutet. Der Vorabend des 1. Mai wurde mit Parademärschen, Feuerwerk und Lichtwasserfällen vom Rathausturm gefeiert. Höhepunkt war 1938 ein Lichtdom. 250 m über dem Rathaus zerplatzten Bomben und „3,5 x 4,5 m große Hakenkreuzfahnen entrollen sich aus ihrem Innern und flattern langsam – viele von Fallschirmen gehalten – vom Licht der rund um den Rathausplatz verteilten Scheinwerfer der *Kriegsmarine* gesucht, zu Boden". Das Werfttor 1, das täglich die „Gefolgschaft" der *KMW* zur Arbeit empfing, symbolisierte für die Nationalsozialisten „Segen der Arbeit". Am 1. Mai 1934 wurde zwischen beiden Einfahrten ein „prachtvoll in Kupfer getriebenes Hoheitszeichen" angebracht[413]. NS-typische Dekora-

tionen für die zahlreichen Festanlässe fanden sich auch in Flugzeughangars und 156
Werfthallen, Werkstätten und Kantinen. 130

Hitler ließ für die Beisetzung der Gefallenen der „Deutschland" am 17.06.1937 eigens eine weihevolle Dekoration von *Karl Franzius* entwerfen. Pylonen und Sockel mit Opferschalen säumten den Weg zum „Heldenfriedhof", symmetrisch ausgerichtet und mit pseudoreligiösem Effekt. Am 01.04.1939 ließ *Hitler* seinen Wilhelmshaven-Besuch zu einer massenpsychologischen Inszenierung gestalten. *Högers* Rathausplatz wurde einpla-
niert und mit einer 100 m breiten Pappkulisse nördlich abgeschottet, durch Pylonen und 157
Ausfachungen, Hoheitszeichen und Fahnen gegliedert. Ein Wald von Fahnen und Girlanden, stilisierten Sockeln und Alleen, säumte die symmetrische Achse zwischen
Rathaustreppe und *Hitlers* Rednertribüne. *Högers* expressionistische Rathausfassade war 112
mit Parteifahnen nahezu verdeckt. Die damals eingeschraubten Befestigungshaken sitzen noch heute. Die streng ausgerichtete Dekoration war Kulisse für 100.000 Zuhörer und *Hitlers* wütende Ausfälle gegen England. In der Innenstadt wurden *Hitlers* Weg, die Taufkanzel der „Tirpitz" und der Eingang zur *KMW*, durch Flaggenpylonen, Kupferschalen und Pappstaffagen dekoriert.

7.10. Zensur und Widerstände

Die Biographien der mit der Erweiterung und Neugestaltung Wilhelmshavens verflochtenen Ingenieure und Architekten waren durch gemeinsame Wurzeln und ähnliche Rahmenbedingungen gekennzeichnet. Durchaus nicht alle wurden Nationalsozialisten, viele standen dem System ablehnend gegenüber. Andere bekamen nach Arbeitslosigkeit, Wartezeit, Studium oder gescheiterter Selbständigkeit endlich die Möglichkeit, in ihrem Beruf zu arbeiten. Manche ließen sich anfangs begeistern, gingen aber mehr und mehr auf Distanz zu dem verbrecherischen System. Andere waren frühzeitig der *NSDAP* beigetreten und bekamen Einfluß auf die Baupolitik Wilhelmshavens. Einige wenige verzichteten auf eine Berufskarriere, legten sich gar mit dem NS-System an oder gingen in Baubehörden oder Wohnungsbaugesellschaften auch nur einfach „in Deckung".

Beträchtliche Schwierigkeiten, Aufträge der neuen Machthaber zu bekommen, hatte
beispielsweise *Richard Freygang*. 1915 hatte er die Synagoge und auch Geschäftsbauten für 17
Juden und den *Konsum* entworfen. Auch *Franz Iwersen, Ernst Peters, Robert Fuchs* und 19
Alexander Minke konnten sich als Wilhelmshavener Architekten nur wenig Hoffnung auf 21
offizielle oder öffentliche Bauten machen. Einige hatten sich durch Bauten im *Bauhausstil* 15
unbeliebt gemacht. So blieben sie während der NS-Zeit auf Bauaufträge aus dem
Mittelstand angewiesen. Auch bei den frühen innerstädtischen Wohnungsbauprogram- 136
men fielen ab 1936 noch einige Aufträge für diese Architekten ab. Wenn es ihnen gelang, einen öffentlichen Auftrag zu ergattern, mußten sie sich häufig massive Zensur ihrer Entwürfe gefallen lassen.

Die im *Wohnsiedlungsgebiet Wilhelmshaven-Jeverland* tätigen Architekten, Planer und Baubeamten können nicht in „Gesinnungs-Schubladen“ sortiert werden. Die meisten waren von ihren Aufgaben besessen. Die Macht, der sie dienten, war austauschbar. Die Architekturauffassung war von ihrer jeweiligen Biographie beherrscht. Vielmehr als heute war man auf bestimmte Lehrer und Schulen ausgerichtet. Die Aufgabenstellungen waren in Wilhelmshaven oft gleich. Funktional entworfene Bautypen oder Stadtteile sollten traditionelles Aussehen bekommen. In Ausnahmefällen, bei technischen Bauten, durften Konstruktion und Zweck gezeigt werden. Die Form sollte der Propaganda folgen, nicht der Grundriß-Funktion[414]. Zu besonderer Blüte kam in Wilhelmshaven die *Stuttgarter-Schule*, abgestimmt jedoch auf die Region und die *Marine*. Sie hielt sich bis weit in die Nachkriegszeit. Vor allem in den zentralistisch geführten Baubehörden wurde den Beamten und angestellten Architekten die Auswahl des Stils abgenommen. Zur Architektenausbildung in der Weimarer Republik gehörte es, alle möglichen Stile zu beherrschen. Auch die der Moderne, auch die des *Bauhauses*. Man wollte jedoch nicht provinziell zugeschnürt sein, man kannte sich aus, man unternahm Studienreisen und sah sich um. Die vorhandene Stilvielfalt wurde jedoch bald eingeengt[415]. Das NS-Regime filterte die Stile heraus, derer es sich bedienen wollte. Dies gelang durch Zensur und „Bauberatung“, wenn nötig durch massiven Druck und Berufsverbot. Autoritäre Manipulation der Architektur kennzeichnete das Bauen des „III. Reiches“, nicht Stilrichtungen, Walmdächer, Kreuzsprossen, Werksteinumrahmungen allein.

★

Willy Schemm[416] beispielsweise war bis 1937 freier Architekt in Wuppertal gewesen. Da seine Frau rassisch verfolgt wurde, bekam er, obwohl Pflichtmitglied der *RDBK*, kaum noch Aufträge. Nach vielen Bewerbungen wurde er von der sich stark ausbreitenden *Wohnungsbaugesellschaft Jade*, deren Spitze damals fest im Griff der *NSDAP*
53 war, am 01.12.1937 angestellt. Er sollte das Jadeviertel westlich der Werftstraße entwerfen. Als *Willi Lübbers* seine Stellung als Chefarchitekt der *Jade* am 10.10.1938 verlor, war *Willy Schemm* Anwärter auf die technische Geschäftsführung. Es kam jedoch anders. Da er den Aufsichtsratsvorsitzenden, *Oberbürgermeister Wilhelm Müller*, den „nicht rassereinen“ Ehepartner gestanden hatte, zerschlug sich seine Berufung. *Schemm* fungierte verdeckt als *Jade*-Chefarchitekt. Technischer Geschäftsführer von 1938–40 wurde *Adolf Braunschweig*. Danach wurde der lauttönende Parteiredner *Erich Duvendack* in der Geschäftsführung nach vorne geschoben. *Schemm* war heilfroh, schon im November 1938 wegen der kriegswichtigen *Jade*-Bauten „uk“. gestellt zu werden. Die technische Abteilung der *Jade* hatte sich damals gegenüber der Verwaltungsbaracke in Unterwohnungen an der Oldeoogestraße etabliert. Hier arbeiteten unter *Schemm* die Architekten *Rohlfs, Penger, Bochert, Siepermann, Jaeger* u.a. Man war unter sich und froh, einen Schritt abseits zu stehen. Die Haltung dieser Architekten war durchaus NS-feindlich, ja antifaschistisch. Abends hörte man Radio London. Um den Nachstellungen der *Gestapo*

zu entgehen, zog *Schemm* mit seiner Familie 1941 aufs Land nach Ellens. Nach der Kapitulation 1945 wurde der NS-Propagandist *Duvendack* sofort entlassen. *Schemm* versuchte noch, ihm bei der „Entnazifizierung" zu helfen. 1945–59 war *Schemm* dann technischer Geschäftsführer und bekam großen Einfluß auf das Nachkriegsbaugeschehen in Wilhelmshaven.

★

Gartenbauinspektor *Friedrich Eisenbarth* aus Bremen, obwohl Mitglied der *RDBK* und damit nationalsozialistischer Baukultur und „anständiger Baugesinnung" verpflichtet, bekam ab 1938 beträchtliche Schwierigkeiten mit seinem Bauherrn, der *Wohnungsbaugesellschaft Jade* und der *Gaugruppe Weser-Ems* des *Deutschen Siedlerbundes*[417]. Als Gartenarchitekt von Voslapp und Cäciliengroden hatte er sich in den Einrichtungsfonds *30, 34*
des *Siedlerbundes* eingemischt. Als Schlichter wurde der *Landesleiter Weser-Ems* der *RDBK Hans Fricke*, eingeschaltet, der sich dafür aussprach, *Eisenbarth* die Planung zu belassen. Die Gaugruppe des *Siedlerbundes* bestritt die Kompetenz der *Landeskulturkammer*, da es nicht um Kunst gehe. *Eisenbarth* war Gartenarchitekt bei der *Stadt Bremen* gewesen und u.a. wegen „politischer Unzuverlässigkeit" entlassen worden. Dem *Gartenbaudirektor* von Bremen, und späteren Mitglied des *Ahlhorner Architektenkreises Richard Homann*, der *Eisenbarths* Wilhelmshavener Planungen übernehmen sollte, war es „absolut unverständlich, ... daß *Eisenbarth* heute Parteigenosse ist"[418]. *Eisenbarth* war auf Wunsch der *Bundesleitung Berlin des Deutschen Siedlerbundes* auch als *Kreisgruppenlehrwart des Siedlerbundes* entlassen worden. Nachdem er den *Siedlerbund* als „Laienverein" bezeichnet hatte, wurde er am 04.01.1939 auch von der *Jade* aus dem Vertrag entlassen, obwohl er entsprechende Vereinbarungen mit *Willi Lübbers* von der *Jade* und *Landesplaner Georg Linde* vorzuweisen hatte[419]. Der *DSB* trat gegenüber den Wilhelmshavener Wohnungsbaugesellschaften sehr selbstbewußt auf. Um die *Jade* gefügig zu machen, reichte es aus, mit *Rudolf Heß* zu drohen[420]. Es bedurfte nur des Hinweises auf politische Unzuverlässigkeit, und *Eisenbarth* war aus dem Vertrag entlassen.

★

Das Verhältnis der Machthaber des „III. Reiches" zur Architektur ist insbesondere am Wirken von *Fritz Höger* in Wilhelmshaven/Rüstringen festzumachen. Von Interesse ist, wie die Fäden zu *Höger* geknüpft wurden und wie sie unter NS-Einfluß wieder gelöst wurden.

Städtebau war nie *Högers* vordringliches Arbeitsfeld. Er befaßte sich mit besonderen Einzelbauwerken, mit niederdeutscher Kultur, mit heimischem Handwerk, mit natürlichen Werkstoffen und mit ausdrucksstarken Ziegeldekorationen. Der gelernte Zimmerer und Baugewerkschul-Absolvent stand früh in Opposition zu Hamburgs *Oberbaudirektor Fritz Schumacher*. Als Mitglied des *Deutschen Werkbundes* befaßte er sich mit dem Gedanken des Gartenstadt-Enthusiasten *Hermann Muthesius* (1861–1929) und des kulturkonservativen *Paul Schultze-Naumburg* (1869–1949). Besonders vertiefte er sich in die Thesen der 1904 gegründeten Heimatschutzbewegung[421]. *Höger* war befreundet mit dem Worpsweder Maler *Fritz Mackensen*, auch mit dem Hamburger Kunsthallenleiter *Alfred Lichtwark* und dem Sozialdemokraten *Paul Bröcker*.

Schon 1913 hatte *Martin Wagner Fritz Höger* zum Wettbewerb der Rüstringer Zentralanlagen eingeladen. Der vielbeschäftigte *Höger* lehnte ab, ebenso verweigerte er die Beteiligung an städtebaulichen Aufgaben unter *Willy Hahn* 1918[422]. 2 Jahre nach Fertigstellung des Chilehauses knüpften 1926 *Bürgermeister Arthur Kellerhoff* und *SPD-Stadtratsvorsitzender Reinhard Nieter* aus Rüstringen Verbindungen zu *Höger*, und beauftragten ihn mit der Rathausplanung. *Höger* baute nicht nur das Rüstringer Rathaus (1928–29), sondern auch Privatbauten für wohlhabende Bürger aus Wilhelmshaven und
58 Rüstringen, u.a. die Häuser *Meyer* und *Hoeland*, Baudissinweg 2 und 4 (1934–35), *Reif*, Bismarckstraße 92 (1935) sowie *Mücke* und *Mehrtens*, Viktoriastraße (1936). *Hoeland* kannte *Höger* aus Hamburg. Bauunternehmer *Friedrich Möller*, der 1934 die *Wilhelmshavener-Rüstringer Wohnungsbaugesellschaft* gegründet hatte, beauftragte *Höger* mit seinem ersten Miet-Wohnprojekt nördlich des Mühlenweges zwischen Bülow- und Schulstraße. 8 Wohnblocks mit je 12 Wohnungen sollten gebaut werden. Rechtsanwalt *Meyer* machte die Verträge für die Wohnungsbaugesellschaft und es verwundert nicht, daß *Hoeland* und *Meyer Höger* für ihre Privathäuser beauftragten.

Der 1933 von der *NSDAP* gleichgeschaltete *Bauverein Rüstringen* beauftragte 1934
57 *Höger* mit den Planungsaufgaben für Neu-Siebethsburg und beendete die 30-jährige Zusammenarbeit mit den Hannoveraner Architekten *Hakenholz* und *Brandes*. Der *Bauverein* schrieb im Geschäftsbericht 1934 lapidar, man habe den bekannten Architekten, „Herrn *Prof. Fritz Höger*", mit dem Entwurf der ersten 121 Wohnungen für Neu-Siebethsburg beauftragt, weil es nötig gewesen sei, die Bauten dem gegenüberliegenden Rathaus anzupassen. Gemütvolle Worte fand ein späterer Geschäftsbericht:[423] „Behaglichkeit und Wärme sollen diese Häuser im Inneren und Äußeren zeigen und bringen". *Harald Marquardsen*, seit dem 01.10.1934 Geschäftsführer des *Bauvereins Rüstringen*, war sicher, daß *Höger* dem NS-System genehm war. Schließlich waren die lokalen Verbindungen zu *Friedrich Möller* und *Arthur Kellerhoff*[424] nützlich. Auch plante *Höger* 1935 an den Projekten *Finanzamt* und *Arbeitsamt* sowie „Haus des Handwerks" in Rüstringen. Er war, da Rüstringen den Tiefbauingenieur *Carl Haefner* als *Stadtbaurat* hatte, von 1929–37 städtebaulicher Berater der *Stadtverwaltung*. Immerhin war der „Altmeister der Backsteinkunst" 1934 auch einflußreiches Verwaltungsratmitglied der *RDBK* und *Professor* der 1934 von den Nationalsozialisten gegründeten *Nordischen Kunsthochschule Bremen* geworden. Die *Nordische Kunsthochschule* unterstand *Carl Röver*. Fachlich zuständig war der *Landesleiter Weser-Ems der RDBK* und *Gaustellenleiter Kultur*, der Oldenburger Architekt *Hans Fricke*. *Höger*, der mit den „studierten Architekten" *Gottfried Feder* und *Alfred Rosenberg* bekannt war, hatte, wie viele Architekten, anfangs auf die *NSDAP* gesetzt und schon 1930 Kontakt zu *Hitler* und später auch *Göring* gesucht[425].

Als Gegner des *Bauhauses* und des *Internationalen Stils* mußte er empfänglich sein für die rückwärts gewandte „völkische" und heimatschutzorientierte Architekturauffassung der frühen *NSDAP*. „Volksverrottung" und „Gesinnungspest" würden, so lobte *Höger* noch 1933, durch deutschen Bau- und Lebensstil ersetzt[426]. *Höger*, Präsident der *Wirtschaftlichen Vereinigung deutscher Architekten*, hatte für seinen Verband sofort nach der Machtübernahme „völkische" Grundsätze, Arierklausel und Führerprinzip eingeführt[427]. Trotzdem machte sich *Höger* bald bei der *NSDAP* unbeliebt. Er hoffte auf „schöpferischen Aufbruch" – Aufträge jedoch gingen an *Albert Speer, Paul Ludwig Troost, Wilhelm Kreis* u.a. *Höger* konnte und wollte nicht neu-klassizistisch bauen. Die *LOB* teilte der *Stadtverwaltung Rüstringen* am 17.10.1936 mit, daß man die „Arbeiten des Arch. *Höger* grundsätzlich ablehne, weil dieser das erforderliche Verständnis für Siedlungsbauten und Planungsarbeiten vermissen" lasse. *Arthur Kellerhoff* schloß sich der *LOB* nicht an und verwies auf *Högers* „Monumentalbauten". 1936 verließ *Höger* zusammen mit*Fritz Mackensen* die *Nordische Kunsthochschule Bremen* wegen eines Konfliktes mit *Heß*[428]. Der Bruch mit dem NS-System war nicht mehr zu kitten. Im Oktober machte sich Geschäftsführer *Adolf Köhne* vom *Reichsheimstättenamt Gau Weser-Ems* mit *NSDAP*-Funktionären von Oldenburg nach Meppen auf, um zwei Kleinsiedlungen zu besichtigen, die *Höger* 1935 dort gebaut hatte[429]. „Die errichteten Anwesen" würden „zu Unrecht als Kleinsiedlungen bezeichnet". Sie seien eher „Fahrkartenhaus oder Warteraum", und vernichtend schließt die Kritik: „Die Siedleranwärter lehnen die Häuser ab". Damit war klar, daß *Höger* keines der im *Wirtschaftsgebiet Wilhelmshaven-Jeverland* *129*
anstehenden Großprojekte mehr übertragen werden durfte. Auch in Wilhelmshaven formierte sich Widerstand gegen ihn. *Willi Lübbers* drängte ihn aus der Rolle des städtebaulichen Beraters[430], die *Höger* seit 1929 für Rüstringen innegehabt hatte, kurz nachdem am 01.04.1937 die Städte zu Groß-Wilhelmshaven zusammengeschlossen wurden. Pünktlich zum 01.11.1937, als der neue *Stadtbaudirektor Walter Hunzinger* seinen Dienst antrat, bekam *Höger* seine Kündigung. Man machte ihm noch Hoffnungen auf einen Kino-Entwurf Mühlenweg/Bismarckstraße, eine Rathauserweiterung und die Teilnahme als Preisrichter an Schulwettbewerben für die neuen Stadtteile.

Nach dem Zerwürfnis mit dem NS-Regime sprach *Höger* 1937 mutig von der „fremden, kalten Gebärde" der offiziellen Staatsbaukunst. Sie führe zur Entfremdung vom „Volk und seiner Führung"[431]. Wie auch die bäuerliche bodenverbundene Kleinsiedlungsbewegung zunehmend unbedeutender wurde, hatte auch *Högers* niederdeutsche Architekturauffassung nicht länger Bestand im „III. Reich". Ähnlichkeit mit propagandistischen Vokabeln der NS-Kulturpolitiker änderte nicht daran, daß er mit seinen Entwürfen für die Reichsführerschule Hohen-Lychen (1935), die Nordische Halle Lübeck (1936), das Reichsehrenmal Beka (1937) und die Nordische Kultstätte (1943) bei den Nationalsozialisten nicht ankam. Es verwundert auch nicht, daß *Höger* mit Fürsprache der aus Hamburg gekommenen und mit der Baupolitik Wilhelmshavens eng verbundenen Planer *Wortmann, Hallbauer, Forster* und *Gutschow* nicht rechnen konnte. Im Gegensatz zu *Höger* wurde sein Widersacher *Fritz Schumacher*, der 1933 von den

Nationalsozialisten aus dem Amt gejagt wurde, noch 1936 vom *Marinestandortbauamt* als Gutachter für den internen Wettbewerb *Marineforum* nach Wilhelmshaven geholt und als Preisrichter für den Wettbewerb „Wohnen nach dem Kriege" 1941 eingeschaltet.

Bemerkenswert eng blieb die Beziehung von *Höger* zu *Gustav Nutzhorn, Harald Marquardsen* und *Arthur Kellerhoff. Nutzhorn* verschaffte *Höger* 1941 noch den Entwurfsauftrag für ein Präsidiumsgebäude, als er *Polizeipräsident* von Aussig geworden war. „Ihm gefiel *Högers* Klinkerbaustil, auch wenn der offizielle Baustil des Dritten Reiches ganz andere Wege ging"[432]. *Marquardsen* beauftragte *Höger* 1939 mit dem Projekt Arngaststraße und war auch sonst *Höger* persönlich bis zu dessen Tode 1949 freundschaftlich verbunden. *Kellerhoff*, der *Höger* 1926 nach Wilhelmshaven geholt hatte, und der von 1947–50 Kämmerer im Nachkriegs-Wilhelmshaven war, sorgte auch dafür, daß *Höger* zum *Ideenwettbewerb* 1948 eingeladen wurde.

Nachdem *Höger* von *NSDAP* und *Stadtverwaltung* 1937 abserviert worden war, wurde auch die Zusammenarbeit der *Wilhelmshavener-Rüstringer Wohnungsbaugesellschaft* mit *Höger* beendet. *Friedrich Möller* schaltete ab 1937 vor allem den Hamburger Architekten
65, 120 *Erich zu Putlitz* ein[433], der sich u.a. mit Parteibauten hervorgetan hatte. Der *Bauverein Rüstringen* beharrte auf *Höger*, wurde aber durch *Oberbürgermeister Müller*, gleichzeitig Aufsichtsratsvorsitzender der *Jade*, merklich von großen Bauaufgaben zurückgedrängt.
63, 43 Weder Schilldeich noch Altengroden-Süd, noch die Besiedlung des Siebethsburger
61 Burggeländes, noch das „Tausend-Wohnungsprojekt" Schaarreihe wurden dem *Bauverein* übertragen oder überhaupt realisiert. *Höger* war verbittert, zog sich mehr und mehr zurück – er traf nicht den Geschmack der *NSDAP*. Die abzuwickelnden, jedoch auslaufenden Aufträge des *Bauvereins Rüstringen* sicherten möglicherweise überhaupt sein Überleben während des „III. Reiches"[434].

Nach dem Zweiten Weltkriege bekam *Höger* noch kleine Aufträge des *Bauvereins*.
123 Über sein schlechtes Abschneiden beim *Ideenwettbewerb* 1948 war er verbittert. 1949 starb er. Als *Kurt Siedenburg*, Denkmalpfleger des *Verwaltungsbezirks Oldenburg*, 1964 Rathaus, *Finanzamt* und Rathausplatz unter Denkmalschutz stellen wollte, belebte sich letztmalig die „Anti-*Höger*-Front". *Egon Forster*, seit 1938 Baubeamter bei der *Stadt*, lehnte ab, da es nicht klar sei, ob *Högers* Werk „die Qualifizierung bereits verdient"[435].

★

Abseits von Berufsopportunismus und NS-Einfluß standen auch einige Hafenbaubeamte und angestellte Architekten dem System kritisch gegenüber. Zu nennen sind *Hafenbaudirektor Georg Frede*, Leiter des *Strom-* bzw. *Hafenbauressorts der KMW* ab 1936, *Hans Klaucke*, Architekt der *KMW* 1938–45, *Carl Tiburtius* und *Ernst Beck*, beide *Hafenbaudirektoren* und nacheinander Leiter der *HNBD*.

Die Affäre *Tiburtius*[436] wurde niemals öffentlich – zu wichtig war die Rolle, die *Carl Tiburtius* für Marinerüstung und *Hitlers* Kriegspläne spielte. Der studierte Hochbauer hatte schon 1908–11 im *Hafenbauressort der Kaiserlichen Werft* in Wilhelmshaven an den

Projekten der III. Einfahrt und der südlichen Hafenerweiterung gearbeitet. Während der Weimarer Republik dirigierte er den Kieler Hafenbau. Im kaiserlichen Berlin, nationalliberal erzogen, wurde er Burschenschafter und preußischer Baubeamter, aber niemals Nationalsozialist. Nach 1935 trug ihm sein alter Wilhelmshavener Kollege aus den Zeiten des *Kaiserlichen Hafenbauressorts, Alfred Eckhardt,* seit 1933 Leiter des Hafenbaus im *OKM*, die Leitung der neugegründeten *HNBD* Wilhelmshaven an. *Tiburtius*, damals
immerhin schon 61 Jahre alt, unterstand jetzt unmittelbar dem *OKM*. Er galt als absolute *4*
Fachautorität und Kenner des internationalen und lokalen Hafenbaus. Er gehörte zu den ganz wenigen erfahrenen Spezialisten, arbeitete in der *Hafenbautechnischen Gesellschaft*[437] und hatte überregionale Kontakte. Kein anderer als er kam infrage, als es um die Besetzung der für das „III. Reich" wichtigsten Hafenbaubehörde ging. Die *Nordwerft* sollte Arbeitsplatz für 30.000 Werftarbeiter werden. Hier sollten die größenwahnsinnigen Schlachtschiffe auf Kiel gelegt und ausgerüstet werden. Höchste Anforderungen an Präzision, Zuverlässigkeit und Zeitgenauigkeit wurden von den Baumeistern der
Nordwerft und IV. Einfahrt verlangt. *Carl Tiburtius* wurde unversehens zur Schlüsselfigur *101*
der NS-Rüstungspolitik. Nach der „Reichskristallnacht" vom 09.11.1938 hatte er die rassistische Politik der Nationalsozialisten im kleinen Kreis angegriffen. Dies reichte aus, um *Kreisleiter Ernst Meyer* und *Gauleiter Carl Röver* auf den Plan zu rufen. *Tiburtius* „setzte die Parteifunktionäre vor die Tür" und zeigte sich standhaft. Am 01.04.1939 noch führte
Tiburtius Hitler über die Baustelle von *Nordwerft* und IV. Einfahrt, als Reichsregierung *100*
und Gesamtkabinett den Kriegshafen besuchten.

Der Attentatsversuch seiner Frau, *Irina Tiburtius*, auf *Hitler* am 01.04.1939 wurde nie offiziell bestätigt[438]. Lokale Zeitzeugen bestätigten Gerüchte. Die vergleichsweise „harmlose" Reaktion auf die Affäre ist für heutige Betrachter kaum verständlich. Sie ist wohl nur zu erklären aus der überragenden Rolle, die *Tiburtius* für die Rüstungspolitik spielte.

Carl Tiburtius wurde sofort nach dem 01.04.1939 als Leiter der *HNBD* abgelöst. Sein Nachfolger wurde *Ernst Beck*[439]. *Raeder* konnte verhindern, daß *Carl Röver* die Familie *Tiburtius* in ein *KZ* abtransportieren ließ. Der hochspezialisierte und erfahrene Hafenbauer *Carl Tiburtius* wurde weiter gebraucht. *Raeder* schickte ihn zur Abwicklung großer Hafenbauprojekte 1941–42 nach Nordafrika und nach Odessa.

★

Das *Strom-* bzw. *Hafenbauressort* der *KMW* wurde ab 1936 von *Georg Frede*[440] geleitet, der das gesamte Werft-Bauwesen bis 1945 bestimmte. *Frede*, ein kulturell interessierter, kunstsinniger Baubeamter, und alles andere als Nationalsozialist, ließ seinen Ingenieuren und Architekten großzügige Entfaltungsmöglichkeiten. Dies galt vor allem für *Boysen, Queisner, Klaucke, Bauer* usw. Man beschäftigte Künstler, fertigte Reliefs aus dem
Arbeitsleben und hatte eine moderne Architekturauffassung, die durchaus auch *93–97*
Internationalen Stil, funktionales Äußeres, kubische Flachdachformen und *Bauhaus-*

Anklänge zuließ. Herausragendster Architekt der *KMW* war *Hans Klaucke*, der u.a. 1925–33 unter *Johannes Göderitz*[441] in Magdeburg gearbeitet hatte. Als *Bauhaus-Anhänger* hat er seine konsequent moderne Architekturauffassung nie versteckt.
114 Außerordentlich sachliche Entwürfe für Werkstätten, Brücken, Leuchttürme, Werft-
89 bahn-Stellwerke, Bunker, Lagerhäuser, Kraftwerke, Transformatoren, Gießereihallen, Schleusen-Betriebsgebäude waren stilistisch hervorragende Ergebnisse. *Klaucke* war ein begnadeter Zeichner und hinterließ eine ansehnliche Zahl von handgefertigten Perspektiven und Schaubildern. Er war krasser Gegner des NS-Regimes, und als ihm *Georg Frede* 1944 anbot, Marinebaubeamter zu werden, lehnte er ab mit den Worten: „In diesem Regime will ich kein Beamter werden"[442]. *Frede* denunzierte *Klaucke* nicht, schützte ihn weiter, ließ ihn arbeiten und hielt zu ihm.

Nach dem Zweiten Weltkrieg waren die Arbeitsplätze der Architekten beim *Hafenbauressort* verloren. Wie auch seine Kollegen *Tuchscherer* und *Boysen*, machte sich *Klaucke* in Wilhelmshaven selbständig. Bald trat er der *KPD* bei und bekam – wie nicht anders zu erwarten – keine Aufträge der öffentlichen Hand, vor allem nicht von der *Stadtverwaltung Wilhelmshaven*. Viele der alten Kreise hatten sich längst wiedergefunden. *Klaucke* jedoch wurde boykottiert. Glücklicherweise bekam er durch *Karl Tillessen* von den *Wasserwirtschaftsämtern* Aufträge für Schöpfwerke und Sielbauwerke, die er konsequent modern realisierte[443].

*

Auch bei der *Bauverwaltung* der *Stadt Wilhelmshaven* lief durchaus nicht alles so, wie es sich *Kreisleiter* und *Oberbürgermeister* vorstellten. Neben der Entlassung von *Stadtbaudirektor Walter Hunzinger* im Februar 1939 schlug der „Skandal" um *Magistratsbaurat Otto Lehn* hohe Wellen. Seit 1913, und nach der Zusammenlegung der Städte 1937 wiederum Leiter der Wilhelmshavener *Tiefbauabteilung*, war *Lehn* für die gesamte Erschließung, Ver- und Entsorgung der Großbauvorhaben unentbehrlich. *Lehn* war nebenamtlich Geschäftsführer der *Gas- und Elektrizitätswerke* und der *Entwässerungsgesellschaft*. Als Logenmitglied[444] war er kein Nationalsozialist und hatte der *SPD-Fraktion* des Rates nahegestanden.

Bereits 1932 wurden dem technischen Bürovorsteher des *Tiefbauamtes* fingierte Rechnungen vorgeworfen, die mit dem Selbstmord des Beschuldigten endeten[445]. *Oberbürgermeister Emil Bartelt*[446], gegen die *NSDAP* eingestellt, lehnte ein Dienststrafverfahren gegen *Lehn* wegen mangelnder Dienstaufsicht ab. Auch *Stadtbaurat Hermann Zopff* und der Auricher *Regierungspräsident Jan Berghaus* schützten *Lehn*. Nach der Machtübernahme erreichten *NSDAP-Oberbürgermeister Carl Renken, Senator Daniel Picker* und *Regierungspräsident Refardt* staatsanwaltliche Ermittlungen gegen *Lehn*. *Lehn* wurden von *Renken* die Dienstgeschäfte untersagt. Sie wurden 1933 an NS-Ortsgruppenleiter *Werner Haschenburger* übergeben. Obwohl sich Vorwürfe der persönlichen Bereicherung nicht erhärteten, wandte sich die *NSDAP* gegen eine Wiedereinstellung *Lehns*. Am 23.02.1934 teilte *Renken Lehn* mit, daß er ihn in den Ruhestand versetzen wolle. Er sei „in nationaler Beziehung als Beamter im nationalsozialistischen Staat ungeeignet" und „habe im Sommer 1933 die Hoheitszeichen der *NSDAP*" beim Umzug nicht gegrüßt. Aus Sparsamkeitsgründen lehnte der *Preußische Innenminister* am 23.03.1934 die Entlassung *Lehns* ab, auch *Staatsanwalt* und *Generalstaatsanwalt* stellten das Verfahren ein.

Die nationalsozialistischen Beigeordneten drohten auf Anstiftung von *Picker* ihre Ämter niederzulegen, falls *Lehn* in den Dienst zurückkehre. Auch bewegte er *Gauleiter Röver*, *Regierungspräsident Refardt* unter Druck zusetzen. Nach Freispruch auch der Dienststrafkammer, trat *Lehn* nach 2-jähriger Amtsenthebung seinen Dienst am 31.5.1935 wieder an. 3 Jahre hatte die *NSDAP* vergeblich versucht, *Lehn* loszuwerden. Nach dem Zweiten Weltkrieg wurde er bis 1949*Stadtbaurat.*

*

Markante Beispiele zensierender „Bauberatung" durchzogen die Jahre 1933–45. Die kommunalen Baubeamten in Wilhelmshaven/Rüstringen waren zumeist konservativ, im Einzelfall gemäßigt progressiv. Man las das „Zentralblatt der Bauverwaltung"[447], die „Deutsche Bauzeitung", die „Deutsche Bauhütte" oder ab 1937 auch „Die Kunst im Deutschen Reich". Die Bücher von *Schultze-Naumburg, Schmitthenner, Feder, Bonatz, Tessenow, Lindner, Böckler,* waren Mindestausstattungen der Verwaltungsbüchereien.

Die damals beteiligten Planer bestätigten einhellig, daß *Internationaler Stil,* und insbesondere die Weißenhof-Siedlung verpönt waren. Zensur reichte bis weit in die Nachkriegszeit. In der NS-Zeit übliche Deteils und beliebte Dekorationen finden sich als Korrekturhinweise und Deckblätter in Bauantragszeichnungen und behördlichen Baugenehmigungen.

Schon vor 1933 waren die *Bauverwaltungen* von Wilhelmshaven und Rüstringen architektonisch ausgeblutet. Nach dem Weggang von *Willy Hahn* verwaltete der Tiefbauingenieur *Carl Haefner* 1921–37 die Rüstringer Bauverwaltung, *Heinrich Siemer* und *Max Neumann* vertraten die architektonischen Positionen und zogen 1929–37 *Fritz Höger* hinzu. Gestalterisch denkende Einzelpersonen fehlten. Die Bautätigkeit war besonders im preußischen Wilhelmshaven so gering, daß sich architektonisch/ideologische Kämpfe kaum entwickeln konnten. Noch im Oktober 1933 bekam Architekt *Richard Freygang* von der *Baupolizei* Rüstringen ein außerordentlich modernes Zweifamilienhaus mit Eckfenstern und Flachdach in der Oldebrüggestraße genehmigt. *Fritz Höger*, von den Nationalsozialisten bis 1936 wohlgelitten, konnte mit den Häusern *Reif, Mücke* und *Mehrtens* noch gemäßigt moderne Architektur durchsetzen. Von 1933 bis zum Zusammenschluß zu Groß-Wilhelmshaven am 01.04.1937 blieben die *Bauverwaltungen* von Rüstringen und Wilhelmshaven personell im wesentlichen unter *Carl Haefner* und *Hermann Zopff* unverändert. Es gab keinen „roten *Stadtbaurat*", den die Nationalsozialisten hätten nach dem *Gesetz zur Wiederherstellung des Berufsbeamtentums vom* 07.04.1933 nach Hause schicken müssen. *BDA*[448] und *Deutscher Werkbund* wurden ab

1928 durch den von *Rosenberg* gegründeten *KDK* und den 1931 unter Einfluß von *Schultze-Naumburg* gegründeten *KDAI* attackiert. Mitte 1933 wurden *BDA* und *Deutscher Werkbund* dem *KDAI* unter Leitung von *Gottfried Feder* untergeordnet. Der *KDK* entfaltete in Wilhelmshaven vor allem nach dem Zusammenschluß zur *NS-Kulturgemeinde* (1934–37) große Aktivität.

1933 entstand unter *Propagandaminister Goebbels* die *Reichskulturkammer (RKK)*. Regional war in Oldenburg *Hans Fricke* von den Nationalsozialisten eingesetzt worden. Bald verlor der *KDAI* seinen Status als Berufsverband der Architekten, nachdem der *KDK* im Juni 1934 Unterorganisation der *NS-Kulturgemeinde* geworden war. Die *RDBK* wurde ab 1933 und ab 1935 verstärkt einzige Berufsorganisation der freien Architekten. Auch in Wilhelmshaven/Rüstringen mußten alle Architekten, wenn sie öffentliche Aufträge erhalten wollten[449], Mitglieder werden. Damit war ein wirksames Kontrollinstrument geschaffen, zumal sich nach 1933 die öffentlichen Aufträge massiv ausweiteten.

Das städtische *Baupflegeamt* der *Baupolizeiabteilung* bekam eine Schlüsselstellung im Baugenehmigungsverfahren. Zensiert und korrigiert wurden vor allem vorgelegte Entwürfe von Wohnungsbaugesellschaften und privaten Bauherren. Bauten von *HNBD, Luftwaffenbauamt, Marinestandortbauamt, KMW, Reichspost, Reichsbahn* und die städtischen Bauten selbst waren in der Regel baugenehmigungsfrei – sie wurden bestenfalls der *Baupolizei* angezeigt[450]. Die Baubehörden selbst übten Entwurfszensur innerhalb ihrer Hierarchien aus, aber auch gegenüber ihren beauftragten freien Architekten.

Die *RDBK* verfügte keinesfalls über einen Formkatalog zum Bauen im „III. Reich". Das am 01.10.1934 in Kraft getretene *Architektengesetz* verpflichtete die Baumeister zu „deutscher Kultur und Baugesinnung". In der *Verordnung über Baugestaltaltung vom* 10.11.1936 aufgrund des *Gesetzes über einstweilige Maßnahmen zur Ordnung des Deutschen Siedlungswesens vom* 03.07.1934 wurden „anständige" Baugesinnung und werkgerechte Durchbildung verlangt. Rechtsgrundlage für die „Bauberatung" waren oldenburgische Landesgesetze, wie das:

- *Gesetz für das Großherzogtums gegen die Verunstaltung von Ortschaften und landschaftlich hervorragenden Gegenden vom 11.01.1910 (Verunstaltungsgesetz)*
- *Gesetz für den Landesteil Oldenburg betreffend die Prüfung von Denkmälern und Bauten in künstlerischer Hinsicht vom* 29.06.1935[451].

Die oldenburgischen Gesetze erlaubten den Aufbau der „Bauberatung" in Rüstringen und ab 01.04.1937 in Groß-Wilhelmshaven.

24 In der Adolf-Hitler-Siedlung beispielsweise wurden 1936 Dachhäuschen gestrichen und Fensterfronten verändert. Ein 1935 vorgelegter Entwurf wurde im Dachprofil korrigiert mit dem Hinweis „Dachneigung wie in der A.-Hitler-Siedlung". 1935 wurden die von Architekt *Herenbrock* gezeichneten Klappläden in Holzklappläden mit Luftschlitzen verändert, Traufhöhen angepaßt und Schornsteinköpfe auf die Dachspitze zentriert.

Seit *Walter Hunzinger* 1937 *Stadtbaudirektor* geworden war, *Egon Forster* ab März 1938 und *Klaus Rasch* ab Juni 1939 zur städtischen *Bauverwaltung* gestoßen waren, wurde „Bauberatung" in großem Stil durchgesetzt. Die Stadtplaner *Wilhelm Hallbauer* und *Helmuth Baur* waren architektonisch weniger ambitioniert. *Hallbauers* Versuch, den Entwurf von *Gustav Lübbers* für die Randbebauung Werftstraße 111–123 sogar mit Hilfe von NS-Flaggen zu nazifizieren, mißlang.

Auch die *Abteilung XXIV. der Marineintendantur* griff zunehmend zensierend in Planungen ein. *Karl Franzius* hatte, bevor er 1939 zur *Marineintendantur* kam, ab 1935 beim *Marinestandortbauamt* gearbeitet und umfangreiche architektonische Erfahrung.

Egon Forster erinnerte sich[452], daß es mühevoll war, die freien Architekten an die gewünschten Architekturideen heranzuführen, und daß es vieler Korrekturen bedurfte. Das *Baupflegeamt* zeichnete auch komplette Bauvorlagen von freien Architekten neu. 1939 überarbeitete es eine Kleinsiedlerstelle in Himmelreich, zentrierte Türen, achtete auf Symmetrie und ergänzte Heimatschutzattribute: Regentonne, Schleppgaube, Holzsprossen, Verbretterungen.

Die Entwürfe für den Marktplatz Voslapp von *Gustav Bork* stießen beim *MDF* auf heftigen Widerstand wegen der „Ausbildung der Dachaufbauten, die auch mir in dieser Form durchaus nicht bodenständig erscheinen"[453]. *Klaus Rasch* schloß sich den Bedenken an – er drohte sogar mit Stillegung der Baustelle. *Borks* Entwürfe waren bereits Anfang 1939 beim *Stadterweiterungsamt* eingereicht worden und auch mit dem *RAM* in Berlin besprochen worden. Das hinderte jedoch keinen Baubeamten, das Äußere zu zensieren und mit Deckblättern zu korrigieren. Von Januar 1939 bis August 1940 versuchten alle beteiligten Dienststellen, den Bau zu beeinflussen, zu überarbeiten und ihren Stempel aufzudrücken. Erst einen Monat nach Baubeginn gab *Karl Franzius* die Planung frei mit der Auflage, die Blendgiebel umzuarbeiten und mit Walmdächern auszurüsten. Ebenso griff die *Bauverwaltung* bei den Fedderwardergrodener Gewerbetrakten ein. Zensur wurde als „Bauberatung" getarnt. Die monopolartige Stellung der öffentlichen Auftraggeber in Wilhelmshaven ließ freien Architekten keine andere Wahl.

Nicht immer jedoch hatte die *Stadt* Möglichkeiten, Gewerbebauten zu zu beeinflussen, denn die Bauanträge wurden nur angezeigt. Besorgt machten die städtischen Mitarbeiter 1941 Bleistifthinweise in den Plänen des Architekten *Ludwig Hörold* für den Gewerbetrakt Albrechtstraße. Der Sachbearbeiter versah die Festlegung „Edelputz" mit einem Fragezeichen, fragte nach Materialien und Fensterumrahmungen und malte hilflos in den Zeichnungen herum.

Auch der Ladentrakt Goedenser Weg in Altengroden, von *Erich Kühn*, Berlin, *149*
entworfen, wurde auf Verlangen der städtischen *Baupolizei* 1941 einer Überarbeitung
unterzogen. Einer der wuchtigen Vorbauten entfiel, und *Karl Franzius* von der
Marineintendantur entwarf eine verspielte Zwiebelturmform. Er schreckte nicht davor *44*

zurück, den Bau stillzulegen und den Zwiebelturm verschiedene Male abreißen und neu aufbauen zu lassen.

Auch das *Marinestandortbauamt* wurde mit Kleinigkeiten von der *Baupolizei* zurechtgewiesen. Die Garagen der Stabsoffiziershäuser im Villenviertel erhielten den zensierenden Zusatz „Ziegeldach", obwohl offensichtlich ein Ziegeldach gezeichnet worden war. Auch Entwürfe der etablierten *Hochbauabteilung* der *Reichspostdirektion* wurden von der Wilhelmshavener *Baupolizei* zensiert. Das *Fernsprechamt* an der Schillerstraße war der Stadt nicht „monumental" genug[454]. Man wünschte eine Achsenverbindung zum Lessingplatz.

Die *städtische Bauverwaltung* griff von Fall zu Fall auch in Grundrißplanungen ein. Architekt *Ludwig Hörold* mußte sich bei der Planung von Einfamilienhäusern für Fedderwardergroden 1941 im Glettkauer Weg harte Eingriffe von *Klaus Rasch* gefallen lassen. *Rasch* lehnte die Bauanträge von *Hörold* rundweg ab, „da die Grundrißgestaltung Mängel aufwies, die nicht ohne grundsätzliche Umarbeitung des Grundrisses zu beheben" seien[455]. Er schaltete sogar die *Marineintendantur* ein und ermahnte die *Jade*, „vor Einreichung des Antrages auf Baugenehmigung die Pläne mit der *Bauverwaltung* möglichst frühzeitig zu besprechen". *Rasch* strich Eingangstreppe und Podest aus dem Entwurf, zentrierte Giebel und Fenster, lenkte die Innentreppe um und machte einige Verbilligungsvorschläge. Insgesamt sollte der Einfamilienhaustyp einfacher und sparsamer werden. Das Büro *Fritz Hilgenstock*, das als Generalplaner die Baugenehmigung beantragt hatte, akzeptierte widerspruchslos die städtische „Bauberatung".

Zu großen Auseinandersetzungen kam es zwischen dem Generalunternehmer des
46 Familienlagers Neuengroden, *Friedrich Möller*, und dem *MDF*. *Friedrich Möller* hatte die gesamte Planung in seiner Baufirma durch *Theodor Schwarting* ausschließlich auf wirtschaftliche Ziele ausgerichtet[456]. Die Planung der Gemeinschaftsbauten und Läden sollte nach Meinung der *Stadt* „disziplinierter erfolgen". Schaufenster seien möglichst zu versprossen und die Umrahmungen sollten verbreitert werden[457]. Die Wilhelmshavener „Bauberatung" fertigte Deckblätter, und erst der *MDF* setzte die Änderungsvorschläge mit einer Verfügung an die *HNBD* durch[458]. Der *MDF* schien skeptisch zu sein und ordnete an, daß die *Stadt* die Einhaltung der Gestaltungsauflagen kontrollieren solle. Für
148 den Ladentrakt am Aufmarschplatz ließ *Möller* dann sogar durch einen Künstler eine
47 Perspektive zeichnen. Auch die verschiedenen Wohnhaustypen in Neuengroden riefen die Aufsichtsbehörden auf den Plan[459]. Runde Giebelfenster sollten entfallen, Dachgauben in einen Rhythmus gebracht werden, Fenster waren zu verändern. *Möllers* Entwurf für eine Neuengrodener Schule wurde von der *Stadt* völlig verworfen. *Egon Forster* zeichnete im April 1939 ein neues Projekt, das am 21.08.1939 vom *Minister der Kirchen und Schulen* genehmigt wurde. *Möller* war wenig geneigt, sich gestalterischen Auflagen unterzuzuordnen. Er ließ mitteilen, daß verschiedene Änderungsvorschläge zu spät gekommen seien und bestätigte knapp, daß er sich mit dem *Baupflegeamt* bereits geeinigt habe.

Auch technische Bauten blieben von zensierenden Eingriffen nicht verschont. Transformatorenhäuschen, die in ihrer äußeren Formsprache immer als sachlich nüchterne Serienbauten angelegt waren, und schon in den 20er Jahren mit Flachdächern ausgerüstet wurden, wandelten sich schrittweise zu Walmdachhäuschen. Schon 1935 war der städtischen *Baupolizei* der Trafo Ulmenstraße zu technisch. Das liegende rechteckige Oberlicht über der Eingangstür sollte halbrund werden. Der Korbbogen in der Ziegelwand sollte einen dominierenden Schlußstein bekommen[460].

*

Bespitzelung und Denunziation reichten im „III. Reich" bis in Baugenehmigungsverfahren. In der Kleinsiedlung Himmelreich beispielsweise hatte ein Siedler 1941 einen Stall mit Flachdach angebaut. Diesen Verstoß gegen nationalsozialistische „Baugesinnung" nahm *Kreisleiter Ernst Meyer* zum Anlaß einer Beschwerde beim städtischen *Bauamt*[461]. Der Siedler sei „auch heute noch ein ausgesprochener Gegner des Nationalsozialismus".

8. Schlußbemerkungen

In Schlicktown, die haben eine Ahnung[1].

Ernst Toller, 1930

Um Architektur und Stadtplanung für „*Die Stadt der 500.000*" zu verstehen, konnten die Bauten vom Verfasser nicht nur ideologieneutral künstlerisch bewertet werden. Historische, politische und gesellschaftliche Sichtweisen gehörten dazu. Architektur als öffentlichste der Künste ist ganz besonders Stilen und Moden, Zeitgeist und Individualität unterworfen. Um jedoch die Funktion zu erkennen, mußten in dieser Arbeit die Rahmenbedingungen des NS-Regimes bewußt gemacht werden. Hierzu gehören Ideologie und Propaganda des „III. Reiches", aber auch technische Zwänge wie Wasserbau, Bodenmechanik, Logistik, Baubetrieb, Verkehr. Die technologische Beschreibung darf nicht als Absicht mißverstanden werden, den Nationalsozialismus zu verharmlosen oder zu normalisieren. Auf den Fundamenten von Architektur und Stadtplanung begann das NS-System, einen mörderischen Rüstungswahn aufzubauen. Monströse Bauabsichten standen am Anfang des verbrecherischen Weges.

Viele laut propagierte städtebauliche Besonderheiten wurden als nationalsozialistische Neuerungen dargestellt, hatten aber handfeste technische Hintergründe: Die großzügige Durchgrünung hatte mit dem vorgeschriebenen Luftschutz zu tun, die feine Abstufung der Geschossigkeit mit Schußwinkeln von Kanonen, die weitläufigen Straßenbreiten mit Brandabschnitten, die Bachläufe mit Entwässerung, die verstreuten Teiche mit Löschwasser, die radikale Industrieumsiedlung mit Explosionsschutz. Straßenachsen sollten 52
nicht nur monumental sein, sondern auch dem flüssigen Erreichen der Arbeitsstätten dienen. Auch Baulandknappheit und überkommene Eientumsverhältnisse führten zu städtebaulichen Besonderheiten. Siedlungen entstanden dort, wo in der Weimarer Republik Land gewonnen oder erworben worden war. Eindeichungen ergänzten den gewaltigen Flächenbedarf für die Wehrmacht. Eigentumsverhältnisse in der Altstadt wurden nicht angestastet und Abrisse oder gewalttätige Eingriffe in Besitzverhältnisse wurden vermieden. Die weit ausladenden Planungen für das *Wirtschaftsgebiet Wilhelmshaven – Jeverland* mußten auch wegen des bäuerlichen Widerstandes schrittweise zurückgenommen werden. Die frühen Konzepte von 1936 hätten den landwirtschaftlichen Besitzstand erschüttert und waren daher zum Scheitern verurteilt. Eine riesige *Nordwerft* konnte nur auf neugewonnenen Grodenflächen geplant werden. Auch für 101
architektonische Besonderheiten gab es handfeste technische Hintergründe: Ziegelmate-

rial schien billig und ausreichend vorhanden, die begrenzte Geschoßzahl hatte mit
Tragfähigkeit des Bodens zu tun, offene Bebauung entsprach den Vorschriften über
42 „Trümmerkegel". Auf rationellen, gar vorgefertigten Massenwohnungsbau war die
regionale Bauwirtschaft nicht eingerichtet, und das Arbeitskräftereservoir ließ den
gleichzeitigen Beginn aller Gemeinschaftssiedlungen nicht zu. Gerade weil es für die
Stadterweiterung und Neugestaltung kaum Vorbilder gab, griffen die Planer auf die in
Wilhelmshaven gängigen Architekturmuster zurück. Der *Heimatschutzstil* hatte in der
Weimarer Republik in Wilhelmshaven/Rüstringen Tradition, er war Alltagsarchitek-
tur. Die zur See fahrende Bevölkerung hatte Behausungen bevorzugt, die das Gefühl
vom eigenen Nest, vom abgegrenzten Garten, vermittelten. Das Häuschen im Grünen
war die Gegenwelt zur Technik der Schiffe, zur Unrast der Seefahrt. Die Wohnungs-
baugesellschaften hatten sich diesem Bedürfnis angepaßt und moderne Bauformen
vermieden.

Baubehörden und Architekten entstammten überwiegend der *Stuttgarter-Schule.* Sie
paßten sich mit ihren Details den regionalen Verhältnissen an, so wie *Schmitthenner* es
empfohlen hatte. Dies war keinesfalls ein echter Regionalstil, sondern die Regionalisie-
rung der *Stuttgarter-Schule.* Das Ergebnis wurde ein besonderer, vom Verfasser so
genannter, *Marinestil,* der das Gesicht Wilhelmshaven über viele Jahrzehnte bis in die
Gegenwart bestimmte. In der Weimarer Republik hatten keinesfalls modernes Bauen,
Neue Sachlichkeit oder *Internationaler Stil* dominiert. Würde man dies voraussetzen, wäre
1933 tatsächlich das Ende moderner Architektur gewesen und 1945 die „Stunde Null"
geworden. Avantgardistische Strömungen, radikale Neuerungen oder sozialistische
Wohnungsbauexperimente spielten in der Architektur und Städtebaudiskussion eine
große Rolle, im Alltag jedoch wurde abseits der großen Städte traditionell gebaut.

Das Bauen nach 1933 wurde, wenn auch nicht messerscharf, von der Architektur der
Weimarer Republik abgetrennt, so doch Schritt für Schritt anders. Ästhetische und
regionale Traditionen der Baugenossenschaften und der Kleinsiedlungsbewegung
wurden in das „III. Reich" übernommen. Der ausschließliche Blick auf Stadtgrundriß
und Gebäudeäußeres kann das Bauen 1933–45 nicht als NS-Architektur entlarven. Die
109 zumeist nicht realisierten Monumentalbauten jedoch spiegeln etwas von den Absichten
108 des Diktators, der sich in Stein verewigen und seinen eingebildeten hohen Rang durch
restaurative Rückgriffe auf längst überholte Städtebautraditionen unterstreichen wollte.
Dazu reichte das vorhandene Gestaltungsinstrumentarium nicht, und ihm wurden Planer
und Baubeamte auf Anhieb nicht gerecht.

Es hat in der Architekturgeschichte nicht an Versuchen gefehlt, Entwürfe von Bauten,
Planungen von Städten, Ideen von Jahren für jedwede Region anwendbar zu machen.
Stadtplanung für Gartenstädte wurde in der Literatur verbreitet, Gründerzeitdekor
wurde nach Katalog bestellt, Architekturschulen wurden von *Professoren* verbreitet.
Auch die Nationalsozialisten hatten ihre Musterbücher. Die Erweiterung einer „See-
festung" allerdings kam darin nicht vor. „Blut und Boden"-Mustersiedlungen waren
52 nicht in den Kriegshafen übertragbar, und wie Monumentalachsen am Wasser enden

sollten, blieb offen. Von den unausgereiften Berliner Architekturleitlinien mußte zwangsläufig abgewichen werden.

Aus der regionalen Anpassung ergaben sich Fehlplanungen und Irrwege, Mißverständnisse und Kuriositäten, Konstruktionsmängel und Mißerfolge.

Viele Chancen und die Besonderheiten der Stadt blieben ungenutzt. Aus militärischen Gründen wurde Wilhelmshaven zum Meer hin abgeschottet. Bauen am Wasser kam *129* nicht vor. Wilhelmshaven wurde, das ist heute nur langsam korrigierbar, zur Binnenstadt am Meer.

★

So menschenfeindlich und totalitär das NS-System auch war, aus Symmetrie und Achsenordnung, Dekoration und Material allein läßt sich eine für Wilhelmshaven geschaffene, besonders „faschistische Architektur" nicht ableiten. Auch ein gewonnener Krieg, auch eine solide Finanzbasis, hätte Wilhelmshaven nicht zu einer nationalsozialistischen Musterstadt gemacht. Die Stadt wäre auch unzerstört ohne Rüstungsaufgaben kaum lebensfähig gewesen. Den noch im Kriege entworfenen ungeheuerlichen Wiederaufbauplanungen fehlte jegliches wirtschaftliche Fundament für eine zivile Stadtentwicklung. Die zahlreichen, aus Luftschutzgründen weit verstreut liegenden *1* Gemeinschaftssiedlungen, wären niemals zu einer Stadt zusammengewachsen. Der Aufwand für Infrastruktur und Bewirtschaftung hätte Groß-Wilhelmshaven, *„Die Stadt der 500.000"*, stranguliert. Das weiträumige *Wohnsiedlungsgebiet* stieß schon Ende der 30er Jahre auf heftigen Widerstand von Bauern und Grundbesitzern. Auch die Monumen- *126-* talplanungen konnten bestenfalls einen provinziellen Abklatsch der „Führerstädte" *129* liefern. Aus Mangel an Neugestaltungs-Naturstein und wegen der herben, klimatischen Gegebenheiten, hatte man sich in Wilhelmshaven längst dem *Marinestil* mit Ziegeln und geneigten Dächern zugewandt. Wilhelmshavens Planungen für Stadterweiterung und Neugestaltung, hektisch begonnen und niemals abgeschlossen, wurden auch exportiert. *Hitler* ließ „neue Städte" auch in Polen planen. Warschau bekam ebenso wie Lodz eine deutsche Bauverwaltung. *Wilhelm Hallbauer* und *Hans Bartning* wurden als Planer nach Polen exportiert. *Hitlers* Unterwerfungspolitik erforderte Kriegshäfen und Werften in den besetzten Ländern, und viele Architekten der Wilhelmshavener Marinebaubehörden bekamen spektakuläre Planungsaufgaben in Norwegen, Frankreich, Griechenland, sogar am Schwarzen Meer.

Auch in Wilhelmshaven wurden moderne Architektur, *Internationaler Stil, Bauhaus*, und funktionaler Städtebau bekämpft. *Neues Bauen* war für die propagandistischen Zwecke des „III. Reiches" unbrauchbar und wurde in der Bevölkerung mit Sozialismus, Demokratie, Technik und Internationalismus in Verbindung gebracht. Das *Architektengesetz von* 1934 erlaubte die Verordnung von „Deutscher Kultur und Baugesinnung" im Sinne der *NSDAP*. Viele Architekten bekamen keine Chance mehr. Andere verstanden die Wünsche der neuen Machthaber schnell und paßten sich den präsentierten Beispielen eilfertig an. Gerne wurde die Illusion eines eigenen nationalsozialistischen Stils geweckt. Tatsächlich jedoch wurde die Architektur der *Stuttgarter-Schule* auf die Küstenregion übertragen. Hieraus entstand der besondere *Marinestil*, der in Wilhelmshaven bald dominierte. Neu war nur die phrasenhafte Bedeutung, die man den Bauten zumaß: klare Ordnung, gemeinsames Erleben, nationale Werte, deutsches Wesen, „völkische" Gemeinschaft usw. Die Umhüllung der Bauten war traditionell, die Zuordnung landschaftsbezogen. Die Eile der schon 1934 im Raum Wilhelmshaven einsetzenden Rüstungsbauten ließ die Bildung neuer Stilformen nicht zu. Sie ermöglichte nur die Auswahl unter verfügbaren Stilen. In der Weimarer Republik hatte das Bürgertum den *Heimatschutzstil* mit steilen Dächern bevorzugt. Nur vereinzelt baute man Flachdächer, kubische Formen oder gar *Bauhausstil*. Wesentlichstes Kennzeichen des *Marinestils* wurden steiles Dach und Ziegelfassaden als Symbol für traditionelle Ordnung und nationale Auffassung. Die *Marine* „baute" auf Autorität, Befehl und Gehorsam. Nur so war sie angeblich zu führen, so sollte sie Kriege gewinnen. Die Architektur war ein Mittel unter vielen, der *„Stadt der 500.000"* ein Gesicht zu geben. Die neuen „Volksgenossen" konnten Propaganda ungelesen zur Seite legen, Volksempfänger abschalten. Vor der Architektur konnten sie die Augen nicht verschließen.

Unterschiedliche Umhüllungen von Bauten sollten den unterschiedlichen Zielsetzungen dienen. Wohnsiedlungen mußten schnell und wirtschaftlich hochgemauert werden. Die Ansprüche der *Marine* an ihre Werksiedlungen waren nicht hoch. Es waren mehr die ehrgeizigen Baubeamten, die an ihren mitgebrachten Architekturschulen festhielten. Militärplanungen, und in Wilhelmshaven vor allem Rüstungsbauten, sollten dem System von Befehl und Gehorsam, der „Schönheit der Arbeit", dienen, und durften im Einzelfall modern sein. Monumentalplanungen wurden mit höchsten ideologischen Anforderungen belegt.

★

Überraschend kurz war die Zeit, in der überhaupt die „Rüstungsschmiede des III. Reiches" geplant und gebaut wurde. Es waren vielleicht 5 bis 6 Jahre, denn die Wiederaufbauplanungen 1941–42 und 1944–45 galten nicht den Kriegs- und Rüstungszielen, sondern einer künftigen zivilen Stadtentwicklung. Das Repertoire für Architektur und Städtebau konnte entgegen den Ankündigungen der Nationalsozialisten nicht neugeschaffen werden. Die *NSDAP* und ihre Gliederungen versagten auf der ganzen Linie. Ohne städtische Mitarbeiter, ohne die Dienststellen der *Marine*, wären die Infrastrukturmängel weit schmerzhafter gewesen.

Während der Bearbeitung mußte der Verfasser eine Reihe vorschneller Urteile revidieren. Trotz verwirrender Zuständigkeiten und trotz chaotischer Behördenstrukturen war das Bauen des „III. Reiches" in Wilhelmshaven durch große Regelungsdichte gekennzeichnet und bürokratisch geordnet. Mehr als angenommen, wurden Entwürfe als Gemeinschaftsarbeiten angelegt. Kaum ein Plan ist allein einem Baumeister zuzuschreiben. Die lange Bearbeitung der *Bebauungspläne* und Entwürfe, der *Wirtschaftspläne* und Werkzeichnungen ließ den kreativen Architekten in den Hintergrund treten. Die Planer im NS-Regime hatten ein technokratisches Selbstverständis und entwickelten Fachbrüderschaften, die bis weit in die Nachkriegsjahre tragfähig blieben. Man behielt die städtebauliche Infrastruktur sorgfältig im Auge, dachte an Verkehrsnetze, trennte Funktionen, sortierte Nutzungen. Daß eine „Festung" nicht von sich aus schön sein konnte, muß den Architekten schnell klargeworden sein.

Es ist festzuhalten, daß durch die luftschutzgerechte Stadterweiterung Voraussetzungen geschaffen wurden für ein Zusammenwachsen der zersplitterten Siedlungen, für leistungsfähige Verkehrsbänder, für wirkungsvolle Innenstadtverbesserungen nach dem Kriege. Es gehört zur Ironie der Geschichte, daß die größenwahnsinnigen Planungen des „III. Reiches" Wilhelmshaven erst zu einer funktionierenden Stadt machten. Eindeichungen erlaubten in den Nachkriegsjahren Industrieflächen. Die IV. Einfahrt ermöglichte zivile Hafenwirtschaft. Freigehaltene Schutzbereiche ließen Platz für Infrastruktur und neue Zusammenhänge. Die zurückgelassenen ehemaligen Werftarbeiter lieferten das notwendige Facharbeiterpotential.

Die Planungen für den „Hauptwaffenplatz des Reiches" wurden schneller als gedacht von der Wirklichkeit überholt. Für heutige Bewohner haben die Gemeinschaftssiedlungen eine völlig andere Bedeutung als für die damaligen Bau-„Herren". Bürgerinnen und Bürger haben ihre Gehäuse in der zweiten oder dritten Generation in Besitz genommen, und wissen wenig über die kriegerische Zweckbestimmung. Die Mischung aus funktionierendem Städtebau und traditionellem Äußeren machen heute die Identifikation mit den ehemaligen Gemeinschaftssiedlungen leicht. In Wilhelmshaven wurden keine spektakulären Manifeste gebaut, keine bekannten Vorbilder kopiert, eher schon die Grundlagen für den sozialen Wohnungsbau der Nachkriegszeit vorweggenommen. Der begonnene Ausbau der Stadt, die Sonderrolle von Groß-Wilhelmshaven, hatte mit dem Ende der Nationalsozialisten seinen Abschluß gefunden. Das Bereithalten für Seekriege prägte fast ein Jahrhundert die Identität der Stadt. In neuer Bündnissituation spielt die Stadt eine andere Rolle mit ziviler Zukunft.

9. Anhang

9.1. Bibliographie

Die Bibliographie führt zitierte Quellen und Sekundärmaterialien auf. Darüber hinaus auch Literatur, die es dem Verfasser ermöglichte, sich mit dem Thema auseinanderzusetzen. Die Bibliographie kann nicht annähernd vollständig sein. Beim Erscheinungsort/ Verfasser/Herausgeber ist jeweils nur der erstgenannte aufgeführt.

Adalbert, Prinz von Preußen: Denkschrift über die Bildung einer deutschen Kriegsflotte, Potsdam 1848

Adè, H.: Der niederdeutsche Backsteinbaumeister Fritz Höger, in: Kunst und Volk, Sept. 1937

Agatz, A. u. a.: Erfahrungen mit Grundbauwerken, Berlin 1977

Ahner, H.: Die Neugestaltung Wilhelmshavens, in: WK 23.04. 1938

Ahner, H.: Hafenbauer gründeten Wilhelmshaven, Wilhelmshaven 1964

Ahner, H.: Neues Bauen in Wilhelmshaven, in: WK 01./02.04. 1939

Ahner, H.: Wie die Großstadt am Jadebusen entstand, in: Wilhelmshaven – Stadt- und Landschaft am Meer, Wilhelmshaven 1958

Ahner, H.: Wilhelmshavener Chronik, Artikelfolge WZ, 1968–1969

Akademie der Künste (Hg.): Martin Wagner 1885–1957, Wohnungsbau und Weltstadtplanung, Berlin 1985

Albers, G.: Entwicklungslinien im Städtebau, Ideen, Thesen, Aussagen 1875–1945, Düsseldorf 1975

Aljets, U.: Die Geschichte der Stedinger und das Dritte Reich, Wilhelmshaven 1987

Alting, H.: Zu Hause in Himmelreich/Coldewei, in: WZ 13.03. 1985

Alting, H.: Zu Hause in Roffhausen-Middelsfähr, in: WZ 10.10. 1985

Appelius, S. u. a.: Die braune Stadt am Meer, Wilhelmshavens Weg in die Diktatur, Hamburg 1985

Appelius, S.: Die Stunde Null, die keine war, Hamburg 1986

Architektenkammer der Freien Hansestadt Bremen u. a. (Hg.): Architektur in Bremen und Bremerhaven, Worpswede 1988

Architekten- und Ingenieurverein zu Bremen (Hg.): Bremen und seine Bauten, Bremen 1900

Arndt, K.: Neoklassizismus als Problem, in: Kunstchronik 1970

Arndt, K.: Der sture gepanzerte Irrtum, in: Die Zeit 14.07.1989

Asche, K.: Jugendstil in Oldenburg, Bd. I Oldenburg 1985, Bd. II Oldenburg 1988

Backes, K.: Hitler und die bildenden Künste, Köln 1988

Baedecker, K. (Hg.): Berlin-Tiergarten, Stadtführer, Freiburg 1984

Bahrdt, H. P.: Humaner Städtebau, Hamburg 1968

Bartetzko, D.: Zwischen Zucht und Ekstase, Berlin 1985

Bauhaus-Archiv (Hg.): Siedlungen der Zwanziger Jahre – heute, Ausstellungskatalog Berlin 1985

Baur, H.: Der Flächennutzungsplan Wilhelmshaven, Beilage zur Niederschrift über die Sitzung zur Überprüfung des Flächennutzungsplanes Wilhelmshaven, Wilhelmshaven 25.05.1951

Baur, H.: Die Affäre L. oder ein Lehrstück aus den dreißiger Jahren, Friedeburg 1988 (Typoskript)

Baur, H.: Erläuterungsbericht zum Flächennutzungsplan 1951 für die Stadt Wilhelmshaven, Wilhelmshaven 1951 (Typoskript)

Baur, H.: Städtebauliche Entwicklung Wilhelmshaven, Breddewarden, Wilhelmshaven 1987 (Typoskript)

Baur, H.: Städtebauliche Entwicklung Wilhemshaven, Kulturforum 1939/40, Wilhelmshaven 1985 (Typoskript)

Baur, H.: Städtebauliche Entwicklung Wilhelmshaven, Wettbewerb Langewerth, Wilhemshaven 1984/85 (Typoskript)

Bauverein Rüstringen (Hg.): Bauverein Rüstringen; Wilhemshaven ca. 1960

Bauverein Rüstringen (Hg.): 25 Jahre Bauverein Rüstringen, Wilhemshaven 1928

Bauverein Rüstringen (Hg.): Fünfzig Jahre 1983–1953, Wilhelmshaven 1953

Bauverein Rüstringen: Geschäftsberichte 1933–1945

Bauwelt (Hg.): Die große Speer-Feier des Leon Krier oder Klassik zum Völkermord, 31.07.1987

Bauwelt (Hg.): Die Legende von der „Stunde Null", Planungen 1940–1950, 28.12.1984

Bauwelt (Hg.): 30. Januar 1933, Zäsur in der Baugeschichte?, 28.01. 1983

Benze, R.: Erziehung im Großdeutschen Reich, Frankfurt 1939

Berndt, H.: Das Gesellschaftsbild bei Stadtplanern, Stuttgart 1968

Beseler, H. u. a.: Kriegsschicksale Deutscher Architektur, Bd. I, Nord, Neumünster 1988

Beuster, F.: Städtische Siedlungspolitik nach dem Kriege, Berlin 1915

Beutz, H.: Wilhemshaven, Einiges aus Vergangeheit und Gegenwart der Stadt Wilhemshaven, Wilhelmshaven 1950

Bibliographisches Institut: Meyers kleines Lexikon, Leipzig 1933

Blöcker, W. H.: Raumordnung, in: Raumforschung und Raumordnung, 1937/1

Blumenroth, U.: Deutsche Wohnungspolitik seit der Reichsgründung, Münster 1975

Börsch-Supan, E. u. a.: Berlin, Kunstdenkmäler und Museen, Reclams Kunstführer Deutschland, Bd. VII, Stuttgart 1977

Börsch-Supan, E.: Berliner Baukunst nach Schinkel 1840–1870, München 1977, 3 Bd.

Bohlen, E. u. a.: Wilhelmshaven auf alten Ansichtskarten, Wilhelmshaven ca. 1983

Bonatz, P.: Leben und Bauen, Stuttgatt 1950

Brenner, H.: Die Kunstpolitik des Nationalsozialismus, Hamburg 1963

Breyer, S. u. a.: Die deutsche Kriegsmarine 1935–1945, Bd. 3, Friedberg 1987

Brockhaus, F. A.: Der Volks-Brockhaus, Leipzig 1941

Brüggemann, R. u. a.: BBR, Versuch einer Rekonstruktion des Planungs- und Bauablaufes der Großraumsiedlung Berlin-Buckow-Rudow (Gropiusstadt), Berlin 1971 (Diplomarbeit)

Brüning, K. (Hg.): Gutachten über die Grundlagen für den Neuaufbau der Stadt Wilhelmshaven, Hannover 1950

Brune, W. (Hg.): Wilhelmshavener Heimatlexikon, 1. bis 3. Bd., Wilhelmshaven 1986–1987

Buddensieg, T. u. a. (Hg.): Die nützlichen Künste, Berlin 1981

Büge, M.: Finanzierungshilfe des Reiches für Gemeinschaftssiedlungen, Berlin 1939

Bültemann, M.: Architektur für das Dritte Reich, Berlin 1986 (Dissertation)

Bürgerverein Rüstersiel (Hg.): Himmelreich – Coldewei, Festschrift zur Fünfzigjährigen Gründungsfeier, Wilhemshaven 1951

Büsing, H.: Arbeitersport in Rüstringen und Wilhelmshaven, Wilhelmshaven 1985

Büsing, H. u. a.: ... das Volk vom Elend zu erretten, Revolution in Rüstringen und Wilhelmshaven, Wilhemshaven 1988

Büsing, H. u. a.: Einmal werden froh wir sagen: Heimat Du bist wieder mein! KZ in Wilhelmshaven-Rüstringen und Wilhelmshavener im KZ, Wilhelmshaven 1987
Büsing, H.: ... so viel unnennbare Leiden erduldet, zur Geschichte der Rüstringer und Wilhelmshavener Juden, Wilhelmshaven 1986
Bundesminister für Raumordnung, Bauwesen und Städtebau: Das Haus mit Garten, Forschungsauftrag, GEWOS, Bonn 1973
Clifford, D.: Geschichte der Gartenkunst, München 1966
Cramer, J. u. a.: Das Bremer Haus, Bremen 1982
Cramer, J.: Der rote November 1918, Revolution in Wilhelmshaven, Wilhelmshaven 1968
Credé, N.: Wege aus dem Chaos, Wilhelmshaven 1945–1949, Wilhelmshaven 1987
Daniel, A.: Der Bauplatz des Reiches, Dortmund 1987 (Magisterarbeit)
Dede, K.: An der Jade, Fischerhude 1978
Der Reichsstatthalter in Hamburg (Hg.): Die Ortsgruppe als Siedlungszelle, Hamburg 1940
Deutscher Siedlerbund e. V. (Hg.): Fünfzig Jahre Deutscher Siedlerbund 1935–1985, Köln 1985
Deutscher Werkbund: Bremen Oldenburg, Abteilung 1, Das Bremen-Oldenburg-Haus auf der Werkbund-Ausstellung in Köln, Bremen 1914
Deutscher Werkbund u. a. (Hg.): Die Zwanziger Jahre des Deutschen Werkbundes, Gießen 1982
Dillschneider, K.: Die Ortsplanung, ein Teil der Landesplanung, in: BSW 1940/6
Distel, H.: Rationeller Krankenhausbau, Stuttgart 1932
Doerr, H.: Die „Vierraum-Musterwohnung" in Düsseldorf, in: Der soziale Wohnungsbau in Deutschland 15.10.1941
Doerr, H.: Schönheit des Wohnens – ein politisches Problem, in: Der soziale Wohnungsbau in Deutschland 15.01.1941
Doormann, L.: Eine Stadt will überleben, in: Merian 04.04.1988
Dülffer, J.: Hitlers Städte, Köln 1978
Dülffer, J.: Weimar, Hitler und die Marine, Düsseldorf 1973 (Dissertation)
Dunger, I.: Wilhelmshaven 1870–1914, Wilhelmshaven 1962 (Dissertation)
Duppler, J.: Prinz Adalbert von Preußen, Gründer der deutschen Marine, Herford 1986
Durth, W.: Deutsche Architekten, Braunschweig 1986
Durth, W. u. a.: Träume in Trümmern, Planungen zum Wiederaufbau zerstörter Städte im Westen Deutschlands 1940–1950, Braunschweig 1988
Dustmann, H.: Vom Bauen der Hitlerjugend, in: Baumeister 1940/1
Eckermann, A.: Entstehung der Siedlung Altengroden, in: Festschrift für die Feier des 10jährigen Bestehens der Siedlung Altengroden, Wilhelmshaven 1951
Eckhardt, A. u. a. (Hg.): Geschichte des Landes Oldenburg, Oldenburg 1987
Ehbrecht, W.: Stadtgeschichte als Teil der Kulturarbeit, in: WZ, Lokale Dokumentation 1987
Eigner, G. P.: Mitten entzwei, München 1988
Elsaesser, M. (Hg.): Bauten und Entwürfe aus den Jahren 1924–1932, Berlin 1933
Engels, F.: Die Lage der arbeitenden Klasse in England, in: Marx/Engels Werk, Bd. 2
Ennen, T. u. a.: 50 Jahre Siedlung Voslapp, Wilhelmshaven 1988
Estermann-Juchler, M.: Faschistische Staatsbaukunst, Köln 1982 (Dissertation)
Feder, G.: Die neue Stadt, Berlin 1939
Fest, J. C.: Hitler, Eine Biographie, Frankfurt 1976, 2. Bd.
Feuchter, G. W.: Die Geschichte des Luftkriegs, Bonn 1954
Fischer-Dieskau, J.: Die Förderung des Baues von Volkswohnungen, in: Bauwelt 1935/35
Fischer-Dieskau, J.: Einführung in die Wohnungs- und Siedlungspolitik, Berlin 1938
Fissen, K.: Jeversche Volkskunde, Jever 1936
Frank, H. (Hg.): Faschistische Architekturen, Hamburg 1985
Frede, G.: 100 Jahre Hafenbau in Wilhelmshaven, Wilhelmshaven 1953 (Typoskript)
Friedrich Wilhelm IV., König von Preußen: Kabinettsorder v. 25.06.1856, in: Grundig S. 6
Fritsch, T.: Die Stadt der Zukunft, Leipzig 1896
Fritzsch, R.: Nürnberg unterm Hakenkreuz, Düsseldorf 1983
Gabriels, H.: Siedlung Kniphausersiel Wilhelmshaven, undatiert (Typoskript)
Gamm, H.-J.: Der braune Kult, Hamburg 1962
Gemeinnützige Deutsche Wohnungsbaubgesellschaft mbH: 30 Jahre Gemeinnützige Deutsche Wohnungsbaugesellschaft mbH, Berlin 1953
Gemeinschaft III: 50 Jahre Voslapp, Wilhelmshaven 1988
Gerdes, H.: Die Seeschleuse der III. Einfahrt in Wilhelmshaven und ihre gründliche Instandsetzung in den Jahren 1934–1937, in: Jahrbuch der hafenbautechnischen Gesellschaft 1937
Gesamthochschule Kassel (Hg.): Ausstellung Leberecht Migge 1981, Ausstellungskatalog, Kassel 1981
Gesamthochschule Kassel (Hg.): Leberecht Migge 1881–1935, Worpswede 1981
GEW (Hg.): 50 Jahre Gas- und Elektrizitätswerke Wilhelmshaven, Wilhelmshaven 1970
Gießerei Sande: Festschrift 25 Jahre Gießerei Sande, Sande 1965
Gildemeister, H.: Baumethoden des Nachkriegswohnungsbaus, Ahlhorn 1944 (Typoskript)
Gildemeister, H.: Grundformen der Stadtentwicklung, Bremen 1943 (Typoskript)
Giordano, R.: Wenn Hitler den Kreig gewonnen hätte, Die Pläne der Nazis nach dem Endsieg, Hamburg 1989
Glaser, H.: Das Dritte Reich, Anspruch und Wirklichkeit, Freiburg 1961
Goecke, T.: Der Bebauungsplan für Wilhelshaven, in: Der Städtebau, 1918, 7/8
Gontermann, H.: Land zwischen Weser und Ems, in: Der Türmer, 1938/10
Graubner, G.: Bonatz und seine Schüler, Stuttgart 1930
Graul, J.: Siebethsburg – von der „Colonie" zum Stadtteil, in: Graul, J. u. a.: 600 Jahre Siebetsburg, Wilhelmshaven 1983
Greiner, U.: Söhne und ihre Väter, in: Die Zeit, 20.04.1988
Grundig, E.: Chronik der Stadt Wilhelmshaven, Bd. II, Wilhelmshaven 1953
Grunewald, A. (Hg.): Wilhelmshaven, Tidekurven einer Seestadt, Wilhelmhaven 1969
Günter, S.: Das deutsche Heim, Gießen 1984
Günter, W.: Die Revolution von 1918/89 in Oldenburg, Oldenburg 1979
Gutschow, K. u. a.: Neuzeitlicher Krankenhausbau, in: Baumeister Sept. 1931
Gutschow, N.: Die ersehnte Katastrophe, Wie Europas Stadtplaner ihre große Chance gekommen sahen, in: Die Zeit 1988/2
Gwildies, P.: Das Wilhelmshavener Handwerk in Vergangenheit, Gegenwart und Zukunft, Wilhemshaven, undatiert (Typoskript)
Hackelsberger, C.: Deutsch sein als Auftrag und Sendung, in: Bauwelt 1985/3
Hagen, G.: Beschreibung neuerer Wasserbauwerke in Deutschland, Frankreich, den Niederlanden und Schweiz, Königsberg 1926
Hahn, W.: Aufgaben kommunaler Bauämter im Kriege, in: Technisches Gemeindeblatt 20.12. 1914
Hahn, W.: Die Organisation des Siedelungswesens in den deutschen Städten, Stuttgart 1919 (Dissertation)

Hahn, W.: Vom Aufbau einer neuen Stadt, Hamburg 1920
Hahn, W.: Eine Schrebergartenanlage in Kriegszeit, in: Technisches Gemeindeblatt 20.03.1915
Haker, H.: Die Partei führt den Wohnungsbau, in BSW 1940/23
Hallbauer, W.: Die Stadt von morgen, Bad Nauheim 03.10.1957 (Typoskript)
Hallbauer, W.: Erläuterungsbericht zum Übersichtsplan der neuen Planung für Wilhelmshaven, Wilhelmshaven 06.08.1937 (Typoskript)
Hallbauer, W.: Gedanken und Erinnerungen, Holzminden 10.02. 1969 (Typoskript)
Hallbauer, W.: Grundsätzliche Gedanken zum Raumproblem Lodsch, Lodsch 1940
Hallbauer, W.: Lex Wilhemshaven, Bad Nauheim 02.11.1954 (Typoskript)
Hallbauer, W.: Raumordnung – eine Lebensfrage Wilhelmshavens, in: Institut für Raumforschung Bad Godesberg, Informationen, 1957, Nr. 12
Hallbauer, W.: Wilhelmshaven 1953, Gedanken von Wilhelm Hallbauer, Bad Nauheim 28.02.1953 (Typoskript)
Hallbauer, W.: Wilhelmshaven 1952, Bad Nauheim 1952 (Typoskript)
Hammer-Schenk u. a. (Hg.): Vom Schloß zum Bahnhof, Bauen in Hannover, Ausstellungskatalog, Hannover 1988
Hampe, E.: Der zivile Luftschutz im Zweiten Weltkriege, Frankfurt 1963
Hanffstengel von, H.: Protokoll über die Besprechung des Arbeitsstabes Speer v. 19.–21.08.1944, in: Durth 2 S. 110ff.
Hanke, H.: Luftgefahr und Luftschutz, Berlin 1933
Harlander, T. u. a. (Hg.): Hitlers Sozialer Wohnungsbau 1940–1945, Hamburg 1986
Hartmann, K.: Deutsche Gartenstadtbewegung, München 1976
Has, L. u. a. (Hg.): Wilhelmshaven 1853–1945, Erinnerungen, Wilhelmshaven 1961
Haus der Deutschen Kunst (Hg.): 1. Deutsche Architektur- und Kunsthandwerk-Ausstellung, München 1938, Ausstellungskatalog
Hecker, M.: Die Berliner Mietskaserne, in: Grote L. (Hg.): Die deutsche Stadt im 19. Jahrhundert, München 1974
Hegemann, W.: Das steinerne Berlin, Berlin 1963
Heil, K.: Kommunikation und Entfremdung, Stuttgart 1971
Heinrich-Hampf, V.: Über Gartenidylle und Gartenarchitektur im Dritten Reich, in: Frank S. 271ff.
Heinrichsbauer, A.: Industrielle Siedlung in Vergangenheit, Gegenwart und Zukunft, Essen 1986
Herber, G.: 50 Jahre Kleinsiedlung, Hannover 1982 (Typoskript)
Herde, A.: Der Luftschutzbunker im Wohngebiet, Berlin 1941 (Dissertation)
Herzog, H.-M.: Der Königsplatz in München, in: Bauwelt 1988/28 u. 29
Heuer, F.: Industrielle Massenanfertigung im kommenden Wohnungsbau, Bremen 1944 (Typoskript)
Heuer, F.: Wohnungsbau und Planung von Wohngemeinschaften und Ortschaften vom rassenpolitischen Gesichtspunkt aus, Ahlhorn 1943 (Typoskript)
Heuzeroth, G. (Hg.): Unter der Gewaltherrschaft des Nationalsozialismus 1933–1945, mehrere Bd., Oldenburg ab 1985
Hiller, F.: Deutsche Erziehung im neuen Reich, Berlin 1935
Hinz, B. u. a.: Die Dekoration der Gewalt, Gießen 1979
Hitler, A.: Mein Kampf, 2 Bd., München 1925/27
Höger, F.: Rathaus Wilhelmshaven/Rüstringen, Bekenreihe 1948/49 (Typoskript)
Hofer, W.: Der Nationalsozialismus, Frankfurt 1957
Homann, R.: Die Stadt als Teil des Landschaftsraumes, Bremen 1944 (Typoskript)
Hortleder, G.: Das Gesellschaftsbild des Ingenieurs, Frankfurt 1970
Howard, E.: Gartenstädte in Sicht, Jena 1907
Hümmelchen, G.: Die deutschen Seeflieger 1935–1945, München 1976
Hundt, P.: Die Entstehung des Wilhelmshavener Parks, in: Die Tide 1917/18, 5 u. 7
Jacob, J.: Erprobungstypen, in: Der soziale Wohnungsbau in Deutschland, 01.03.1941
Joedicke, J.: Zum Werk von Paul Schmitthenner, in: Deutsches Architektenblatt 1985/3
Joel, G. u. a. (Hg.): Unser Gauleiter Carl Röver, in: Heimatlese, 1939 Februar
Jürgens, F. W.: Geschichte des Nordseebades Wangerooge 1804–1954, Jever 1954
Kamm, K. U.: Die grüne Stadt am Meer, Wilhelmshaven 1980
Kamphausen, A.: Der Baumeister Fritz Höger, Neumünster 1972
Kayser, G. u. a.: Die wirtschaftliche und technische Umstellung des Reichskriegshafens Wilhelmshaven/Rüstringen, Wilhelmshaven 1921
Kisch, E. E.: Die seltsame Reise der Lanna 8, in: Merian Ostfriesland, April 1988
Kleihues, J. P. (Hg.): 750 Jahre Architektur und Städtebau in Berlin, Stuttgart 1987
Kliche, J.: Vier Monate Revolution in Wilhelmshaven, Rüstringen 1919
Kluke, P.: Hitler und das Volkswagenprojekt, in: Vierteljahreshefte für Zeitgeschichte, 1960/4
Koch, P.: 50 Jahre Wilhelmshaven, Berlin 1919
Kochskämper, M.: Herbergen der neuen Jugend, Berlin 1937
Kogon, E.: Der SS-Staat, Frankfurt 1960
Koller, P.: Die geschichtlichen und rechtlichen Grundlagen der Gründung Wolfsburgs, Wolfsburg 1949
Koop, G. u. a.: Die Marine in Wilhelmshaven, Koblenz 1987
Koop, G. u. a.: Von der Kaiserlichen Werft zum Marinearsenal, München 1982
Krämer, R.: Die Landschafts- und Siedlungsentwicklung im Stadtgebiet von Wilhelmshaven, Wilhelmshaven 1982
Kraft, E.: 80 Jahre Arbeiterbewegung zwischen Meer und Moor, Wilhelmshaven 1952
Kriegsmarine (Hg.): Teilnehmerverzeichnis aller Marine-Fernsprechanlagen in Wilhelmshaven, Wilhelmshaven 1942
Krier, L.: Albert Speer, Architecture 1932–1942, Brüssel 1985
Krohn von, L.: Vierzig Jahre in einem deutschen Kriegshafen, Wilhelmshaven 1905
Krüger, K. H.: Die Entnazifizierung der Steine, in: Der Spiegel 1989/4
Kühnl, R.: Formen bürgerlicher Herrschaft, Liberalismus, Faschismus, Reinbek 1971
Kunstbildbibliothek Berlin (Hg.): Fritz Höger, Baumeister, Zeichnungen, Berlin 1977
Kunsthalle Wilhelmshaven (Hg.): Wilhelmshaven Stadtgesicht, Ausstellungskatalog, Wilhelmshaven 1982
Lampugnani, V. M.: Architektur als Kultur, Köln 1986
Larsson, L. O.: Die Neugestaltung der Reichshauptstadt, Albert Speers Generalbebauungsplan für Berlin, Stuttgart 1978
Linde, G.: Erklärungen zum Raumordnungsplan Wilhelmshaven, Oldenburg 16.12.1937 (Typoskript)
Linde, G.: Neue Wohnflächen für die Durchführung der Wohnungsforderungen der Marine, Oldenburg 27.08.1937 (Typoskript)
Lindner, W. u. a.: Die Stadt, ihre Pflege und Gestaltung, München 1939
Lindner, W.: Heimatschutz – Heimatpflege – Heimatgestaltung, in: BSW 1940/13

Lösche, H.: Münchener Postbauschule damals und heute, in: Bauwelt 14.10.1988
Maas, K.: Bericht über die Wiederherstellungsarbeiten an der durch Feindeinwirkung beschädigten evangelischen Marine, Garnisonskirche, Wilhelmshaven 1943 (Typoskript)
Maas, K.: Die Chronik der Landesbauverwaltung in Schleswig-Holstein, Kiel 1981
Machule, D.: Waldsiedlung Krumme Lanke, in: Bauwelt 1986/27
Marinewerft Wilhelmshaven (Hg.): 75 Jahre Marinewerft Wilhelmshaven, 1956–1931, Oldenburg 1931
Markelíln, A. u. a.: Stadtbaugeschichte Stuttgart, Stuttgart 1985
Marßolek, J.: Bremen im 3. Reich, Bremen 1986
Maser, W.: Adolf Hitler, München 1971
Mattausch, R.: Siedlungsbau und Stadtneugründungen im deutschen Faschismus, Frankfurt 1981 (Dissertation)
Meyer, E.: Menschen zwischen Weser und Ems 1933–1945, Oldenburg 1986
Meyer, E.: Wilhelmshaven, eine nationalsozialistische Aufgabe, in: Die Nordwestmark, Bd. I, Oldenburg 1941
Merker, R.: Die bildenden Künste im Nationalsozialismus, Köln 1983
Middleton, R. u. a.: Klassizismus und Historismus, Weltgeschichte der Architektur, Stuttgart 1981
Migge, L.: Jedermann Selbstversorger, Jena 1918
Migge, L.: Das grüne Maifest 1918, in: Gesamthochschule Kassel 1 S. 59ff.
Migge, L.: Grünaufgaben am Jadebusen – Planung eines Sport- und Lagunenparks, in: Die Tide 1918/12
Miller Lane, B.: Architektur und Politik in Deutschland 1918–1945, Braunschweig 1986
Moderhack, R.: Braunschweigs Stadtgeschichte, 2 Bd., Braunschweig 1985
Möhrle, J.: Das Gesicht der Post, in: Bauwelt 14.10.1988
Mosebach-Tegtmeier, E.: Anfänge und Entwicklung der Sozialdemokratie auf Reichsebene in Rüstringen/Wilhelmshaven (1863/79–1933), in: SPD-Unterbezirk Wilhelmshaven (Hg.): 116 Jahre Sozialdemokratische Politik in Wilhelmshaven, Wilhelmshaven 1988
Mosebach-Tegtmeier, E.: Der andere Weg, Oldenburg 1988
Mosebach-Tegtmeier, E.: Für sich stand die Gestapo, in: Reinhardt, W. u. a. (Hg.): Justiz an der Jade, Wilhelmshaven 1985
Mosebach-Tegtmeier, E.: Von der Arbeiterkolonie Bant zur Stadt Rüstringen 1879–1911, in: Stadt Wilhelmshaven (Hg.): Wilhelmshavener Museumsgespräche, Wilhelmshaven 1988
Müller-Wulckow, W.: Die deutsche Wohnung der Gegenwart, Königstein 1930
Mulitze, E.: Bau der Nordwerft der Kriegsmarine in Wilhelmshaven einst und jetzt, Wilhelmshaven 1973 (Typoskript)
Mulitze, E.: Wilhelmshavens Nordwerft einst und jetzt, in: Nordwestzeitung 02.06.1973
Muntsch, O. u. a.: Die Gefahren der Luft und ihre Bekämpfung im täglichen Leben in der Technik und im Krieg, Berlin 1940
Murken, T.: Haus und Grund in Wilhelmshaven, Wilhelmshaven 1986
Murken, T.: Wilhelmshavens Enttrümmerung, Wilhelmshaven 1979
Muthesius, H.: das englische Haus, Berlin 1904–1905, 3 Bd.
Neidhardt, H. (Hg.): Baudenkmäler im Oldenburger Land, Oldenburg 1980
Neufert, E.: Bau-Entwurfslehre, Berlin 1940
Neupert, K.: Die politischen und organisatorischen Grundlagen der totalen Planung Gestaltung, in: BSW 1940/5
NGBK (Hg.): Inszenierung der Macht, Ästehtische Faszination im Faschismus, Berlin 1987
Nicolaus, F.: Die Aufstellung von einfachen Behelfsheimen in Siedlungsform und ihre gruppenweise Zusammenfassung, in: Der soziale Wohnungsbau in Deutschland, 1944/15 u. 16
Nicolaus, F.: Gutachten über die weitere bauliche Entwicklung von Wilhelmshaven, Berlin 1941 (Typoskript)
Nicolaus, F.: Kolonisatorische Siedlungsplanung, in: Siedlung und Wirtschaft, Berlin 1939/3
Nicolaus, F.: Wehrhafter Städtebau, in: Der soziale Wohnungsbau in Deutschland, 01.03.1941
Niemeyer, R.: Die Entwicklung der Landesplanung in Deutschland, in: Städtebau, Okt. 1936
Niestroj, G. u. a. (Hg.): 50 Jahre Wohnungsbaugesellschaft Jade mbH, 1937–1987, Wilhelmshaven 1987
Notholt, E.: Wilhelmshaven – Stadt und Landschaft am Meer, Wilhelmshaven 1958
Oldewage, H.: Wurten, Deiche und alte Marschenwege im Stadtgebiet Wilhelmshaven, in: Oldenburger Jahrbuch Bd. 68, Oldenburg 1969
Onken, K.: Fedderwardergroden, Planung, Entstehung und Weiterentwicklung eines Stadtteils der Stadt Wilhelmshaven in der NS-Zeit, Göttingen 1985 (Diplomarbeit)
Ottmann, E.: Gotthilf Hagen, Der Altmeister der Wasserbaukunst, Berlin 1934
Otto, K.: Luftkrieg und Städtebau, in: Raumforschung und Raumordnung, 1940/9
Paffrath, F. (Hg.): Wilhelmshaven 1945 bis 1952, Wilhelmshaven 1952
Panerai, P. u. a.: Vom Block zur Zeile, Braunschweig 1985
Pape, W. (Hg.): Joachim Ringelnatz, das Gesamtwerk 7 Bd., Berlin 1982
Pehnt, W.: Die Architektur des Expressionismus, Stuttgart 1981
Peltz-Dreckmann, U.: Nationalsozialistischer Siedlungsbau, München 1978 (Dissertation)
Peters, H.: Von der Revolte zur Restauration, in: Ein Blick zurück, Beiträge zur Geschichte des Jeverlandes, Jever 1986
Petsch, J.: Baukunst und Stadtplanung im Dritten Reich, München 1976
Petsch, J.: Kunst im Dritten Reich, Köln 1983
Pfeiffer, K.: Die Planung von Großsiedlungen, in: Siedlung und Wirtschaft, 1938/2
Poerschke, S.: Heimstättenarbeit in Westfalen, Münster 1952
Posener, J.: Berlin auf dem Wege zu einer neuen Architektur, das Zeitalter Wilhelm II., München 1979
Posener, J.: Ebenezer Howard, Gartenstädte von morgen, Berlin 1968
Prottengeier, H. K.: Zu Hause in Neuengroden, in: WZ 17.07. 1985
Raddatz, F. J.: Die ungeliebte Republik, Serie, in: Die Zeit, 1988 April/Mai
Räcker-Wellnitz, U.: Aufgaben, Funktion und Aufbau einer regionalen Schaltstelle im nationalsozialistischen Herrschaftssystem, dargestellt am Beispiel der Gauleitung Weser-Ems, Münster 1986 (Magisterarbeit)
Raeder, E.: Mein Leben, Bd. 2, Tübingen 1957
Rasch, K.: Aus der städtebaulichen Geschichte Wilhelmshavens, in: Notholt, E.: Wilhelmshaven-Stadt und Landschaft am Meer, Wilhelmshaven 1958, S. 100ff.
Rasch, K.: Das Kleinwohnungsproblem im Stadtgebiet Wilhelmshaven, Borna 1939 (Dissertation)
Rasp, H.-P.: Eine Stadt für tausend Jahre, München 1981
Rave, P. O.: Kunstdiktatur im Dritten Reich, Hamburg 1949
Rehbein, W.: Aufruf an die Katholiken von Wilhelmshaven, Flugblatt, Wilhelmshaven ca. 1952
Reichhardt, H. J. u. a.: Von Berlin nach Germania, Ausstellungskatalog, Berlin 1985
Reichsjugendführung der NSDAP (Hg.): Werkshefte für den Heimbau der Hitlerjugend, Bd. II Leipzig 1938

Reinig, P. u. a.: Architektur in Oldenburg seit der Jahrhundertwende, Oldenburg 1986
Richter, J.: Bauliche Aufgaben des Heimstättenamts, in: BSW 1940/6
Rickels, K.: Die wirtschaftliche und soziale Lage der Wilhelmshavener Werftarbeiter mit besonderer Berücksichtigung der Einwirkungen der Nachkriegszeit, Münster 1921 (Dissertation)
Rittich, W.: Architektur und Bauplastik der Gegenwart, Berlin 1938
Röhl, J. C. G.: Kaiser, Hof und Staat, München 1987
Roh, F.: „Entartete" Kunst, Hannover 1962
Rohlfs, A.: Geschichte des Krankenhauswesens in Wilhelmshaven, Düsseldorf 1978 (Dissertation)
Rosenberg, A.: Der Mythos des 20. Jahrhunderts, München 1933
Rothfels, H. (Hg.): Wilhelm I., Tagebuch meiner Reise nach dem Morgenlande, Frankfurt 1971
Rürup, R. (Hg.): Wissenschaft und Gesellschaft, Berlin 1979
Rüssel, T.: Ebkeriege, eine Standortchronik, Wilhelmshaven 1988 (Typoskript)
Salewski, M: Die Marine und Berlin, in: Marineforum 1987/10
Scarpa, L.: Martin Wagner und Berlin, Braunschweig 1986
Schaap, K.: Die Endphase der Weimarer Republik im Freistaat Oldenburg 1928–1933, Düsseldorf 1978
Schaap, K.: Oldenburgs Weg ins Dritte Reich, Oldenburg 1983
Schachner, B. u. a.: Handbücherei für das gesamte Krankenhauswesen, Berlin 1930
Schäche, W.: Architektur und Städtebau im Nationalsozialismus, in: Bauwelt 1983/4
Schädlich, C.: Die Hochschule für Architektur und Bauwesen Weimar, Weimar 1985
Scharfe, S.: Deutschland über Alles, Ehrenmale des Weltkriegs, Königstein 1940
Schinz, A.: Berlin, Stadtschicksal und Städtebau, Braunschweig 1964
Schmid, P.: Das Niedersächsische Institut für historische Küstenforschung in Wilhelmshaven, in: Mitteilungsblatt der Oldenburgischen Landschaft, Dez. 1988
Schmidt, M.: Albert Speer, Das Ende eines Mythos, Bern 1982
Schmitthenner, P.: Baugestaltung, Das deutsche Wohnhaus, Stuttgart 1940
Schmitthenner, P.: Baukunst im neuen Reich, München 1934
Schneider, C.: Stadtgründungen im Dritten Reich, Wolfsburg und Salzgitter, München 1979 (Dissertation)
Schneider, F. K.: Erläuterung zu der Ideenskizze vom April 1942, Oldenburg 1942 (Typoskript)
Schneider, F. K.: Fragen der städtebaulichen Planung im Lande Oldenburg, in: BSW 1940/6
Schneider, S.: Fliegerhorst Upjever, München 1986
Schönfeldt von, G.: Wilhelm Wortmann – 90 Jahre, in: Bauwelt 1987/14
Scholl, L. U.: Marinemaler in Wilhelmshaven, in: Verein der Kunstfreunde für Wilhelmshaven S. 93ff.
Scholtz: Raumordnung und Städtebau, in: Oldenburgische Staatszeitung, 06.11.1936
Schrape, J.: Städtebauliche Planungen in Oldenburg zwischen 1933 und 1945, in: Oldenburger Jahrbuch, Band 83, 1983
Schulte-Frohlinde, J.: Bauten als Zeugen großer Zeiten, in: Deutscher Beton-Verein: Vorträge 42. Hauptversammlung 1939, Berlin 1940
Schulte-Frohlinde, J.: Über die Aufgabe der Rationalisierung im Wohnungsbau, in: Deutsche Bauzeitung, 1940/16
Schultze-Naumburg, P.: Das Gesicht des deutschen Hauses, München 1929
Schultze-Naumburg, P.: Kunst und Rasse, München 1928
Schulz, W.: Denkschrift über den Bebauungsplan für Wilhelmshaven, Wilhelmshaven 1942 (Typoskript)
Schulz, W.: Erläuterung zu dem Bebauungsplan (Ideenskizze) der Stadt Wilhelmshaven v. Dez. 1941 (Typoskript)
Schulze, F. W. O.: Seehafenbau, Berlin 1911
Schumacher, F.: Das Wesen des neuzeitlichen Backsteinbaus, München 1919
Schuster, P.-K. (Hg.): Die Kunststadt München 1937, München 1987
Seidler, F. W.: Fritz Todt, Baumeister des Dritten Reiches, Frankfurt 1988
Selle, G.: Design-Geschichte in Deutschland, Köln 1987
Senat von Berlin (Hg.): Schinkel in Berlin und Potsdam, Führer zum Schinkel-Jahr 1981, Berlin 1981
Siedlergemeinschaftg Cäciliengroden (Hg.): 50 Jahre Cäciliengroden, Cäciliengroden 1988
Siedlergemeinschaft III (Hg.): 50 Jahre Voslapp, Wilhelmshaven 1988
Siemer, J.: Was wird aus dieser Stadt? Wilhelmshaven, die neue Industriestadt, Bonn 1950
Siemer, J.: Wilhelmshaven, Gestern, Heute, Morgen, Wilhelmshaven 1959
Silbermann, A.: Vom Wohnen der Deutschen, Köln 1963
Sommer, I.: Altengroden, Eine Chronik, Wilhelmshaven 1984
Sommer, I.: Architektur und Stadtgestaltung in Wilhelmshaven, in: Verein der Kunstfreunde für Wilhelmshaven S. 113ff.
Sommer, I.: Auf den Spuren eines hervorragenden Baumeistern, in: WZ, 05.11.1985
Sommer, I.: Badeanlagen in Wilhelmshaven, in: Mitteilungsblatt der Oldenburgischen Landschaft, Sept. 1987
Sommer, I.: Geschichte der Badeanlagen an der Jade, Artikelfolge WZ 1988
Sommer, I.: Martin Wagner, 1911–1914, Stadtplanung und Architektur, Faltblatt zur Ausstellung, Wilhelmshaven 1986
Sommer, I.: Siebethsburger üben gute Nachbarschaft, in: WZ 20.08.1983
Sommer, I.: Vor 100 Jahren geboren: Willy Hahn, in: WZ 31. 10. 1987
Speer, A. (Hg.): Bombensicherer Luftschutz im Wohnungsbau, Berlin 1941
Speer, A.: Der Sklavenstaat, Frankfurt 1984
Speer, A.: Erinnerungen, Frankfurt 1969
Spieker, H.: Totalitäre Architektur, Stuttgart 1981
Staatshochbauamt Wilhelmshaven (Hg.): Chronik über das Staatshochbauamt Wilhelmshaven, Wilhelmshaven 1980 (Typoskript)
Stadt Nürnberg (Hg.): Nürnberg 1933–1945. Nürnberg 1986
Stadt Rüstringen (Hg.): Die Stadt Rüstringen Nordseebad in Oldenburg, Hannover 1927
Stadt Wilhelmshaven (Hg.): Altenheim Lindenhof 1931–1981, Wilhelmshaven 1981
Stadt Wilhelmshaven (Hg.): Die Wilhelmshavener Stadtentwässerung seit 1868, Wilhelmshaven 1979
Stadt Wilhelmshaven (Hg.): Dokumentation, Außenkommando Wilhelmshaven des Konzentrationslagers Neuengamme, Wilhelmshaven 1986
Stadt Wilhelmshaven (Hg.): Flächennutzungsplan 1972, Strukturanalyse – Erläuterungsbericht, Wilhelmshaven 1973
Stadt Wilhelmshaven (Hg.): Krankenhausneubau Wilhelmshaven, Wilhelmshaven 1967
Stadt Wilhelmshaven (Hg.): Stadtteilentwicklungsplan Fedderwardergroden, Kurzfassung, Wilhelmshaven 1982
Stadt Wilhelmshaven (Hg.): Wilhelmshaven, vom preußischen Marinehafen zum deutschen Tiefwasserhafen, Wilhelmshaven 1980
Stadt Wilhelmshaven (Hg.): Wilhelmshaven – wie es wurde, was es ist, Wilhelmshaven 1981
Stadt Wolfsburg (Hg): Wolfsburg 1938–1988, Ausstellungskatalog, Wolfsburg 1988

Stadtschaft für Niedersachsen (Hg.): Wohnen nach dem Kriege, Ein niederdeutscher Architektenwettbewerb, Pinneberg 1941
Staeck, K. (Hg.): Nazi-Kunst ins Museum? Göttingen 1988
Steinhauser, P.: Die Wohnungsfrage, Ein Kernstück unserer Sozialpolitik auch im Kriege, in: BSW 1940/4
Stiftung Niedersachsen (Hg.): Von Laves bis heute, über staatliche Baukultur, Braunschweig 1988
Stommer, R. u. a. (Hg.): Reichsautobahn, Pyramiden des Dritten Reiches, Marburg 1982
Suhren ,W.: Die wirtschaftliche soziale Entwicklung Rüstringens, Rostock 1926 (Dissertation)
Temp, W.: Richtlinien für die weitere Planungsarbeit v. 11.01. 1944, Wilhelmshaven 1944 (Typoskript)
Tessenow, H.: Hausbau und dergleichen, Berlin 1916
Teut, A.: Architektur im Dritten Reich 1933–1945, Frankfurt 1967
Thalenhorst, C. (Hg.): Bremen und seine Bauten 1900–1951, Bremen 1952
Theilen: Die Oldenburgische Verwaltungsreform vom Jahre 1933, Oldenburg 1934
Thiele: Die Entwicklung des deutschen Marinehaushalts 1930–1939, in: Trial fo the Major War Criminals before the International Military tribunal, Bd. XXXV, Nürnberg 1949
Thies, J.: Architekt der Weltherrschaft, Düsseldorf 1980
Toller, E.: Feuer aus den Kesseln, Historisches Schauspiel, in: Spalek, J. M.: Ernst Toller, Gesammelte Werke, Bd. 3, Berlin 1978
Trommer, S.: 50 Jahre Wolfsburg, in: Bauverwaltung 1988/6
Troost, G.: Das Bauen im Neuen Reich, Bd. I, Bayreuth 1938, Bd. II, Bayreuth 1943
Umlauf, J.: Vom Wesen der Stadt und der Stadtplanung, Düsseldorf 1951
Verein der Kunstfreunde für Wilhelmshaven: Kunst an der Jade, Wilhelmshaven 1912–1987, Wilhelmshaven 1987
Verne, P.: Jules Vernes baltische Reise, Husum 1987
Voigt, C.: Das neue Dienstgebäude der obersten Marinebehörden, in: Mitteilungen des Vereins für die Geschichte Berlins, 1914
Voigt, W.: Erinnerungen an einen Unmodernen, in: Die Zeit 04.01. 1985
Wagner, M.: Das sanitäre Grün der Städte, Berlin 1915 (Dissertation)
Wagner, M.: Rede vor dem Deutschen Werkbund am 10.06.1933, in: Bauwelt 1983/4
Wagner, M.: Wohnungswesen und Realkredit nach dem Kriege, in: Technisches Gemeindeblatt, 20.12.1914
Walz, M.: Gegenbilder zur Großstadt, in: Bauwelt 1980/12
Warnke, M.: Politische Architektur in Europa, Köln 1984
Welge, K. (Hg.): Dokument Wilhelmshaven, Wilhelmshaven 1951
Westphal, C. (Hg.): Fritz Höger, Der niederdeutsche Backstein-Baumeister, Wolfshagen 1938
Wickop, W.: Die Baumethoden im Wohnungsbau der Nachkriegszeit, Hannover 1944 (Typoskript)
Wiepking-Jürgensmann, H.: Das Behelfsheim in der Landschaft, in: der soziale Wohnungsbau in Deutschland, 1944/3 u. 4
Wiepking-Jürgensmann, H.: Landschaftsfibel, Berlin 1942
Wiepking-Jürgensmann, H.: Das Grün im Dorf und in der Feldmark, in: BSW 1940/13
Wietelmann, M. u. a.: 100 Jahre Ems-Jade-Kanal, Aurich 1988
Wilhelm-Lehmbruck-Museum (Hg.): Verboten, Verfolgt, Kunstdiktatur im 3. Reich, Ausstellungskatalog, Duisburg 1983
Wilhelmshavener Spar- und Baugesellschaft (Hg.): 75 Jahre Wilhelmshavener Spar- und Baugesellschaft, Wilhelmshaven 1968
Wistrich, R.: Wer war wer im Dritten Reich, München 1983
Wohnungsbaugesellschaft Jade (Hg.): Voslapp, die neue Wohnstadt am Meer, Wilhelmshaven 1938
Wolf, G.: Etwas von der Erziehung zum guten Bauen, in: BSW 1940/3
Wolf, P.: Wohnung und Siedlung, Berlin 1926
Wolters, R.: Albert Speer, Oldenburg 1943
Wortmann, W.: Der Raum Weser-Jade als Beispiel für Landes- und Stadtentwicklung, Oldenburg 1973
Wortmann, W.: Grundlagen für die Gesundung und Neugestaltung des Raumes Bremen, Bremen 1943 (Typoskript)
Wortmann, W.: Planung Wilhemshaven, Oldenburg 06.06.1942 (Typoskript)
Wortmann, W. u. a.: Siedlung und Landschaft in ihren Wechselbeziehungen, Oldenburg, undatiert
Wortmann, W.: Wandel und Kontinuität der Leitvorstellungen in der Stadt- und Regionalplanung, in: Bericht zur Raumforschung und Raumplanung, 1985, Heft 3–4
Wulf, J.: Die bildenden Künste im Dritten Reich, Frankfurt 1983
Wulf, J.: Theater und Film im Dritten Reich, Frankfurt 1983

9.2. Anmerkungen

Wenn sich die Fußnoten auf Schriften beziehen, die in der Bibliographie vorkommen, wird bei Verfassern und Herausgebern, die mit nur einem Werk vertreten sind, lediglich deren Name genannt, bei Verfassern und Herausgebern, die mit zwei oder mehr Werken vertreten oder namensgleich sind, außerdem eine Zahl entsprechend der Reihenfolge dieser Werke in der Bibliographie.

1.

1 Pape Bd. 4 S. 112. Ringelnatz beschrieb seine Abfahrt aus Wilhelmshaven am 29.03.1915.
2 Wortmann 4 S. 15
3 Obwohl sich Hallbauer im Schreiben an Tappe vom 10.02. 1969 nur an 450.000 Einwohner erinnerte, übernimmt der Vf. die Zielzahl von 500.000. s. a.: Grundig S. 358, Niestroj S. 21, Stadt Wilhelmshaven 6 S. 4, Vermerk 13. 11.1941 Temp gegenüber MDF u. LOB STAO 137/VII-182-9
4 Durth 2 S. 1035 ff.
5 s. a. Berndt
6 Sommer 1
7 Durth, Gutschow u. Heike Coldewey, als Mitarbeiterin des Wilhlemshavener Heimatlexikon

2.

1 Verne S. 18. Jules Verne besuchte mit seiner Yacht 1881 Wilhelmshaven.
2 Historische Kommission: Neue deutsche Biographie Bd. 7 S. 472
3 Ottmann S. 5
4 ebd. S. 14
5 Hagen
6 ebd. S. 131
7 Rürup S. 3
8 ebd. S. 8
9 Sommer 2 S. 113 ff.
10 Ottmann S. 20
11 Duppler S. 117
12 Adalbert S. 5
13 Das Handelsministerium war Teil des Ministeriums der öffentlichen Arbeit. Das Gebäude ist zerstört.
14 Ludwig Franzius war Enkel des Auricher Wasserbaudirektors Johann Nikolaus Franzius, des Erbauers der ostfriesischen Kanäle. Er arbeitete bei der Hannoverschen Wasserbauverwaltung und wurde 1866, als Hannover preußisch wurde, von Gotthilf Hagen in das Berliner Handelsministerium und gleichzeitig als Lehrer an die Bauakademie geholt. Ludwig Franzius war ab 1876 Leiter des gesamten bremischen Staatsbauwesens.
15 Börsch-Supan 2 S. 13 ff.
16 Eckhardt S. 327, 344. Am 25.09. 1852 besuchte der König den Großherzog August in Rastede, als die Verhandlungen um das Jadegebiet gerade begonnen hatten. Erst Nikolaus Friedrich Peter (1827–1900), der 1853 an die Regierung kam, schloß den Jadevertrag am 18. bzw. 20.08. 1853 ab.
17 ebd. S. 638
18 Brune Bd. 1 S. 18
19 Ottmann S. 22
20 Friedrich Wilhelm IV. bestimmte mehr als sonst fürstliche Bauherren die Architektur seiner Zeit. Schon als Kronprinz nahm er auf Schinkels Entwürfe Einfluß. Er versuchte, die Staats- und Kirchbauten im Königreich an sich zu ziehen. Er gab schon 1858 wegen Geisteskrankheit die Regierungsgeschäfte an Prinzregent Wilhelm I. ab, blieb aber offiziell preußischer König. s. a.: Börsch-Supan 2 S. 572
21 Middleton Bd. 2 S. 266 ff.
22 Das älteste Fassadenstück Gewerbeschule Weserstr. 78 ist stark an die Fassadenstruktur der Bauakademie Berlin angelehnt.
23 Senat von Berlin S. 16
24 Ottmann S. 22
25 Eckhardt S. 638
26 Marinewerft Wilhelmshaven Anlage 1
27 Hegemann S. 161, 178
28 Schinz S. 92 ff.
29 Börsch-Supan 2 S. 17, 568, 572, 698
30 ebd. S. 21, 547, 697, 821
31 Kleihues S. 117
32 Börsch-Supan 2 S. 582, Posener 1 S. 607
33 Brune Bd. 3 S. 117, Has
34 Krohn
35 Salewski S. 340–348
36 s. a. Röhl
37 Posener 1 S. 11, 106
38 Börsch-Supan 1 S. 346, Baedeker S. 40, 41. Reinhardt u. Süßenguth bauten 1900–1906 den Hauptbahnhof Hamburg.
39 vgl. Entwurf Rathaus Hamburg von 1854 von Stier, Börsch-Supan 2 Abb. 528. Ähnlichkeiten mit der Hafenkaserne Wilhelmshaven.
40 Zinnenkränze und massive Türmchen im Stil mittelalterlicher Burgen wurden zuerst von Persius 1831 beim Schloß Babelsberg verwendet.
41 vgl. Schloß des Baron Oertzen auf Kittendorf, Schloß des Grafen Schwerin zu Göhren. s. a. Börsch-Supan 2 S. 589 Abb. 514, 515
42 Salewski S. 345, 346
43 Grundig S. 357, 358
44 Posener 1 S. 623
45 vgl. Sommer 2 S. 115 ff., Sommer 3, 6, Scarpa, Akademie der Künste
46 Schr. Wagner 13.07.1913 BAW
47 Hahn 3 S. 58 ff., 88 ff.
48 Rosenbaum, E.: Der Vertrag von Versailles, Inhalt und Wirkung gemeinverständlich dargestellt, Leipzig 1920, S. 64–68
49 Wagner 3
50 Beuster
51 Hahn 3 S. 23
52 Goecke S. 71 ff.
53 Miller Lane S. 94 ff.
54 Zuständig war im Magistrat als Dezernent Arthur Kellerhoff.
55 Goecke S. 71 ff.
56 Das Wachstum war ausschließlich durch die Aufrüstung entstanden.
57 Hofer S. 23
58 Kühnl S. 80 ff.
59 Bis 1924
60 WZ 24.05.1932
61 WZ 31. 12.1932
62 Peltz-Dreckmann S. 98
63 Bibliographisches Institut S. 89
64 Grundig S. 526
65 Daniel S. 114 ff. Er vertritt die Auffassung, daß die Flugplatzbauten illegal von der Marine gefördert wurden.
66 Volksblatt 06. 12.1932
67 WZ 30. 12.1932

68 Sämtliche umliegende Gemeinden hatten sich zusammen mit Wilhelmshaven, Rüstringen und Aurich am 13.02.1933 in Friedeburg getroffen.
69 Meyer 1 S. 45
70 WZ 29. 12.1934
71 Bibliographisches Institut S. 539
72 Murken 1 S. 39
73 Rede von Marinestationspfarrer Ronneberger zum 01.05. 1934, Kasernenhof Mühlenweg. WZ 02.05.1934
74 WZ 18.05.1934
75 z. B. Georg Biedermann, Johannes Herrmann
76 Grundig S. 2
77 Stadt Wilhelmshaven 7 S. 4
78 Grundig S. 2
79 Koch S. 13
80 Grundig S. 2
81 Staatshochbauamt Wilhelmshaven S. 2
82 Koch S. 34
83 Adreßbuch 1892 S. 169 STDAW
84 Adreßbuch 1912 S. 528 STDAW
85 Die MI in Kiel und Wilhelmshaven unterstanden den jeweiligen Stationschefs. Sie waren als Mittelinstanz für einen größeren regionalen Raum und für die Bereiche Besoldung, Bekleidung, Wohnungen, Unterkünfte und Verpflegung zuständig.
86 Maas 2
87 Schr. Zenker 15.01.1989, Schr. Tornow 20.03. u. 03.04.1991, Maas 2 S. 38
88 Marinewerft Wilhelmshaven
89 vgl. 3.5.6.
90 Schr. OKM 01. 12.1942 Sammlung Karl Maas
91 Schr. OKM 22.01.1943 Sammlung Karl Maas
92 Viele ihrer Leiter waren Arch. und hatten mit Wilhelmshavens Aufbau zu tun: Franzius, Naud, Maas, Geisenhainer, Lückemeyer, Gerdes, Thees, Zenker.
93 Schr. OKM 24.07.1944 Sammlung Karl Maas
94 Maas 2 S. 45
95 Die Bezeichnungen wechselten häufig. Auch: Luftkreiskommando oder Luftwaffenkommando
96 Petsch 1 Abb. 89, Leiter war Gallwitz.
97 vgl. 4.4.
98 Gespr. Vosshans 06.01.1988
99 Bremer Kinderheim Wangerooge, vgl. Jürgens S. 109, ehem. Internat Langeoog.
100 Stadtschaft für Niedersachsen S. 70ff.
101 Fernsprechverzeichnis Stadt Wilhelmshaven 1940, Gespr. Baur
102 Vermerk 03.01.1939 BAW
103 4 Dipl.-Ing., 8 Hochbautechniker, NOB 08. 11.1937
104 Vermerk 03.01.1939 BAW
105 NOB 17.09.1937
106 s. a. Hinweise auf ungezählten AktenSTAO
107 Vermerk RAM 09.06.1939 BAK R 41/1318
108 WK 01./02.04.1939
109 Schr. 06.04.1939 STAO 137/VII-182-9
110 Gespr. Siedenburg 10. 12.1989
111 Teut S. 251
112 Gespr. Herde 01. 12.1988. Es arbeiteten dort auch Ritter, Michel, Steinbiss, Jessen
113 Gespr. Herde 01. 12.1988
114 Teut S. 309
115 Blöcker
116 WK 31. 10.1936
117 Niederschr. 01.06.1938 STAO 137/VII-182-9
118 Schr.19.01.1939 BAK R 41/93. Umschlagbild oben
119 Schr. LOB an Reichsstelle 06.06.1942 BAK R 41/94
120 ebd.
121 Niederschr. 01.06.1938 STAO 137/VII-182-9
122 Sachbearbeiter waren Thielo und Rudloff.
123 Niederschr. 01.06.1938 STAO 137/VII-182-9
124 Durth 2 S. 112
125 Scholtz
126 Architektenkammer der freien Hansestadt Bremen S. 25
127 Schr. Röver an Reichsstelle 13. 11.1936 BAK R 113/647
128 Niederschr. BAK R 113/648
129 BSW 6/1940
130 Stadtschaft für Niedersachsen
131 Vorträge Wortmann und Heuer 21. 11.1936 Jever
132 Linde 1
133 Umschlagbild oben
134 BSW 5/1940
135 BSW 7/1940 S. 224
136 NOB 05.03.1938
137 Seidler
138 Räcker-Wellnitz
139 vgl. 7.9., 7.10.
140 Vermerk 16. 10.1936 BAW
141 BSW 4/1940 S. 98
142 BSW 23/1940 S. 835
143 NOB 10.06.1941
144 Peltz-Dreckmann S. 125
145 BSW 5/1940 S. 155, Niederschr. über die Tagung der Wohnungsunternehmen 07.02.1940 BAW
146 BSW 6/1940
147 Durth 1 S. 227
148 Maas 2 S. 29
149 Schr. 20.08.1940 STAO 137/VII-182-9
150 NOB 28.07.1941
151 Schr. Kreisleiter 30.04.1941 BAW
152 Schr. RAM 29.03.1939 BAK R 41/1317. Umschlagbild unten
153 ebd.
154 Schädlich S. 39ff.
155 Akademie der bildenden Künste Dresden
156 Kreis war Lehrer an der Kunstgewerbeschule Dresden. Er plante Kommandobauten in Berlin und Ehrenmäler.
157 Durth 1 S. 46

3.

1 Grundig S. 6ff.
2 Engels S. 256ff.
3 Posener 2 S. 17
4 Ottmann S. 7, 22ff.
5 Nach dem Vorbild der 1899 gegründeten engl. Garden-City-Association
6 Kieler Zeitung 14.08.1927
7 Gegründet 1907
8 vgl. Wolf 2 S. 157ff., Posener 1 S. 264ff., 548ff.
9 Migge 1
10 Scarpa S. 27
11 Posener 1 S. 264
12 Kieler Nachrichten 24.03.1961
13 Deutscher Werkbund 2 S. 28ff.
14 Wagner hatte bis 1910 im Büro Muthesius gearbeitet. Dieser hatte 1896–1903 in London Gartenstadt-Ideen kennengelernt.
15 1913 versuchte Wagner auch Peter Behrens und Fritz Höger nach Wilhelmshaven zu ziehen.
16 Mosebach-Tegtmeier 1 S. 37
17 Howard
18 Akademie der Künste S. 25ff.
19 Bauhaus-Archiv
20 Grundig S. 59

21 Vorhanden ist noch das Gebäude Viktoriastraße 26.
22 Muthesius
23 vgl. Häuser Freudenberg (1908) und von Seefeld (1905) von Muthesius.
24 Goecke S. 71, vgl. 3.5.1.
25 Fritsch. Fritsch war heftiger Antisemit.
26 Berndt S. 37
27 Bereits 1902 hatte die Gemeinde Heppens den Delmenhorster Geometer Aschenbeck mit einem Bebauungsplan beauftragt.
28 Skizzen 23. 10.1906 BAW
29 Gesamthochschule Kassel 1 S. 4
30 Beide Kasernen wurden aus angeblichem Platzmangel auf oldenburgischem Gebiet gebaut.
31 Der noch stehende Block I Mühlenweg 48–54
32 Wohnbaracken I u. II. an der Freiligrathstraße Kreuzung Mühlenweg, vgl. Plan vom Juni 1934 ASTHBA, topographische Karte 1919 BAW
33 ebd., Wohnfläche bis zur Arrestanstalt ausgewiesen
34 Posener 1 S. 267, Port Sunlight 1904
35 vgl. Fuggerei Augsburg 1516
36 Cramer 1 S. 15ff.
37 Hecker S. 273ff.
38 Brune Bd. 3 S. 190
39 z. B. Deutsche Nationalbank Bremen, Bauten in der Südstadt 1890
40 Peltz-Dreckmann S. 42
41 1848 Berliner Gemeinnützige Baugesellschaft, 1852 Alexandra Stiftung
42 Bürgerverein Rüstersiel, Jeversches Wochenblatt 25.06.1988, Stadt Rüstringen S. 38
43 Neidhardt S. 240ff.
44 Leiter der Invaliden-Altersversicherungsanstalt Hannover.
45 Wilhelmshavener Spar- und Baugesellschaft
46 Grundig S. 396ff.
47 Gemeinnützige Deutsche Wohnungsbaugesellschaft mbH
48 Republik 02.02.1932
49 Hahn 3 S. 46
50 Bibliographisches Institut S. 1283, 1933 zusammengefaßt in der Nationalsozialistischen Kriegsopferversorgung mit 1,5 Mio. Mitgliedern.
51 Grundig S. 398
52 Heutiges Karl-Hinrichs-Stift, Friedenstraße
53 Reinig S. 155
54 Bibliographisches Institut S. 1283
55 Wilhelmshavener Tageblatt 29.01.1904
56 Wilhelmshavener Tageblatt 23.03.1904, 24.03.1904
57 Niemeyer S. 110
58 Poerschke, S. 206
59 vgl. Verordnung zur Behebung der dringendsten Wohnungsnot vom 19. 12.1919
60 Peltz-Dreckmann S. 63
61 Wilhelmshavener Spar- und Baugesellschaft, Beamten- Spar- und Bauverein, Bauverein Rüstringen
62 u. a. sollte Gelände bei der Kopperhörner Mühle nach Plänen eines Arch. aus Hannover im Landhausstil bebaut werden.
63 Norddeutsches Volksblatt 29.08.1902, 17.03.1903, 04.05.1918
64 Grundig S. 397
65 Republik 31.05.1927
66 Republik 06.11.1926
67 Höchststand der Arbeitslosigkeit 1932: 6,1 Mio., Höchststand in Wilhelmshaven/Rüstringen: 6000.
68 Hahn 4
69 ebd.
70 Norddeutsches Volksblatt 15.10.1911
71 Verträge zu versch. Erwerbshäusern BAW
72 Goecke S. 71, vgl. 3.5.1.
73 Norddeutsches Volksblatt 15. 10.1911, Börsch-Supan 1 S. 373, 377
74 Akazienstraße, Thomas-Mann-Straße usw.
75 Wagner 3
76 Kapitel II, 4. Teil RGBl. I S. 537, 551
77 Deutscher Siedlerbund S. 7
78 Herber
79 Vortrag DSB Vizepräsident Eckart Naumann, Voslapp 20.08.1988
80 Petsch 2 S. 29
81 Wohnungsbaugesellschaft Jade S. 3
82 Deutscher Siedlerbund S. 10
83 Bibliographisches Institut S. 985
84 Leiter war über viele Jahre Johannes Richter
85 Eignungsschein Karl Lüers, Siedlung Sanderbusch, 11.06.1937, Sammlung Almuth Ulbrich
86 Vortrag Joachim Petsch 03.05.1988 Bonn
87 Bundesminister für Raumordnung, Bauwesen und Städtebau S. 13
88 Schr. 13.10.1933 MAF RM 23/1661
89 WZ 30.12.1933. Feder hielt am 18.02.1933 eine Rede im Schützenhof.
90 RGBl. I S. 568
91 Reichssiedlungskommissar Feder wurde wegen seines antikapitalistischen Kurses mit einer Professur an der TH Berlin und der Reichsarbeitsgemeinschaft für Raumforschung „abgefunden".
92 Forderung nach Bodenreform, Abschaffung von Bodenzins und Spekulation.
93 Peltz-Dreckmann S. 102
94 Hahn 3 S. 13
95 ebd. S. 19
96 Cramer 1 S. 15ff. Bremer Häuser wurden zwischen 1859 und 1914 vorwiegend in Bremen gebaut. Sie entstanden auch in Wilhelmshavens Südstadt um 1880, im Villenviertel ab 1909 und in Bant
97 Die Familien hatten 6–14 Kinder. Im Rüstringer Grundstücksamt mußten Familienväter im Losverfahren „ihren" Hausschlüssel ziehen. WZ 01.08.1985
98 Erlasse RAM 12.02.1935 und 19.02.1935
99 Wettbewerb wurde von der Stadt Rüstringen am 18.04.1934 ausgeschrieben, WZ 29.12.1934
100 WZ 29.12.1934
101 WZ 06.05.1935
102 Stadt Rüstringen S. 308. Rüstringen kaufte 1917–1919 3 Ziegeleien.
103 Brune Bd. 3 S. 548
104 Grundig S. 410, 527
105 Nachtrags-Vertrag des Bauarbeiters Johann Schmidt, Himmelreich 103, 24.09.1934, Privatbesitz
106 Schr. Stadtmagistrat Rüstringen 20.09.1934 an Johann Schmidt, Privatbesitz
107 Alting 1
108 vgl. 3.5.2.
109 Gabriels
110 WZ 15.06.1976
111 MAF RM 23/1661
112 Teut S. 378, Durth 2 S. 23ff.
113 Schr. RAM 09.06.1939 BAK R 41/1318
114 BSW 14/1937 S. 368
115 Jeversches Wochenblatt 03.07.1937
116 Goecke S. 72
117 Zu der ganzen Wohnsiedlung Nord-Ost kam es nicht, da die Marine einen Teil des Geländes in den 20er Jahren wieder beanspruchte.
118 Wie Rüstringen (Sport- und Lagunenpark 1918, Projekt Migge) suchte auch Wilhelmshaven zwischen den Kriegen

eine zivile Existenz. Werner Haschenburger plante unter Hermann Zopff umfangreiche Kur- und Badeanlagen im Zusammenhang mit der Wohnsiedlung Nord-Ost.
119 Wilhelmshavener Spar- und Baugesellschaft
120 Nr. 1–13, 2–16
121 Bebauungsplan des Geländes nördlich der Bismarckstraße 13.10.1922 BAW
122 Bauträger vermutlich Gemeinnützige Bau- und Siedlungsgenossenschaft „Leo“
123 Grundig S. 399
124 Kunsthalle Wilhelmshaven S. 46
125 gebaut wurden 1925: Nr. 23–29, 1926–1927: Nr. 15–21, 1927–1930: Nr. 1–13
126 Gebaut 1901 Arch. Ernst Zimmermann, 1938 geschlossen wegen Einsturzgefahr. Im Kriege zerstört.
127 Gebaut 1903 Arch. Alfred Breslauer und Paul Salinger, Berlin
128 Nr. 2–20
129 Grundig S. 51
130 Stadt Wilhelmshaven 1
131 Plan von Rüstringen Wilhelmshaven. Nach dem Stande vom 01.06.1936 zusammengestellt vom Stadtvermessungsamt.
132 BAW
133 vgl. Siedlung Praunheim
134 Scarpa S. 173, 177, Siedlung Lindenhof
135 Scarpa S. 183, Chapman-Projekt
136 Howard
137 1913–1917 Arch. Michel de Klerk
138 Panerai S. 90ff.
139 Scarpa S. 180
140 WK 21.09.1937
141 WZ 29.12.1934
142 WK 01.04.1938
143 Oldewage
144 Hallbauer 2 S. 18
145 Linde 2 S. 3
146 Schr. 11.11.1936 BAW
147 Vermerk 15.03.1937 BAW
148 Vermerk 27.07.1937 BAW
149 Vermerk 30.07.1937 über eine Besprechung mit Staatsminister Pauly
150 WZ 14.07.1937
151 NOB 29.11.1937
152 WK 04.01.1938
153 Gespr. Lübbers 28.12.1983
154 Anfangs Waagestr. danach Hans-Zenker-Str. heute Flutstr. Zeitweise zur Nord-Süd-Achse stilisiert.
155 Oldewage Plananlage
156 Typ G in der heutigen Tiarksstr., Tiefstr., von-Münnich-Str.
157 vgl. 7.7.
158 Ab März 1938 3 Schulen
159 WK 04.01.1938
160 WK 10.02.1938
161 vgl. 7.8.
162 Stellwerke sind noch erhalten. Arch. Hans Klaucke, Brune Bd. 3 S. 404ff.
163 WZ 25./26.06.1938
164 Schr. OKM 22.01.1936 MAF RM 23/1662
165 vgl. 7.4.
166 vgl. 7.3.
167 WK 10.02.1938
168 WK 28.01.1938
169 WK 04.03.1938, Generalunternehmer Büscher, Münster
170 WZ 29.03.1938
171 WZ 25./26.06.1938
172 Gespr. Lübbers 28.12.1983
173 Markelin S. 89
174 WK 01.11.1938
175 Heinrichsbauer S. 157
176 Peltz-Dreckmann S. 139
177 Teut S. 251
178 Fischer-Dieskau 1
179 Wohnungsbaugesellschaft Jade S. 5
180 Plan vom 15.02.1939 BAW
181 Südl. und östl. Marktplatz
182 Schr. Jade 28.07.1939 AJA
183 Schr. Jade 02. 10.1941 AJA
184 Wohnungsbaugesellschaft Jade, Vorwort
185 WK 26.02.1938
186 WK 10.02.1938
187 WK 04.01.1938
188 WK 04.08.1939
189 Nicht identifizierbarer Zeitungsbericht ca. Sept.1939
190 Auch der Flughafen Mariensiel wurde zu der Zeit mit Sand aus dem Westhafen aufgespült.
191 Vertragsmuster AJA
192 Heuer reichte die Pläne am 09.02.1938 bei der Baupolizei des Amtes Friesland ein.
193 Katasterplan 18.05.1989, Plan der KMW 15.05.1941 AJA
194 vgl. 7.3., 7.10.
195 Hallbauer 2 S. 19
196 WK 26.02.1938
197 WZ 10.01.1940
198 ebd.
199 3 Bunker wurden im Krieg gebaut
200 vgl. 7.5.
201 14 Jahre später als versprochen, 30.000 Mark/Siedlerstelle
202 Niederschr. über Bespr. 10. 12.1936 BAW
203 Onken. Onken irrt auf S. 40, wenn er feststellt, daß sich erst 1938 Fedderwardergroden als „künftiger Planungsraum“ herauskristallisierte.
204 Abb. 129
205 Schr. LOB 27.08.1937 S. 3 BAW
206 Hallbauer 2 S. 20
207 ebd. S. 29
208 WK 03. 11.1937
209 WK 04. 01.1938
210 NOB 05.03.1938
211 Niederschr. 01.06.1938 STAO 137/VII-182-9
212 Vom 16.06.1932 bis April 1945 unter Julius Pauly, Dezernent war Fritz Karl Schneider.
213 Schr. MDF an MI 07.03.1938 BAW
214 WK 30.03.1938, 27.08.1938
215 WK 01. 11.1938. Mit Fallersleben ist Wolfsburg gemeint
216 Gespr. Herde 01.12.1988. Nicolaus war im RAM der Manager, Pfeiffer der passionierte Architekt.
217 Anfangs wurde mit 3.000 Arbeitern gerechnet, es wurden dann 4.000, vgl. 7.8.
218 Vermerk 03.01.1939 STAO 137/VII-182-9. Der Plan war nicht auffindbar.
219 Der Europahausplan befindet sich original im STAO. Er wurde allen Beteiligten per Schnellbrief durch den RAM am 23. 12.1938 zugeschickt STAO 137/VII-182-9.
220 Vermerk 03.01.1939 STAO 137/VII-182-9.
221 ebd.
222 Für die Ortsteilverwaltung. Die heutigen Kirchbauten haben die Fläche des Aufmarschplatzes ausgefüllt.
223 Arch. Karl Franzius 1940. Umschlagbild unten
224 Arch. Rudolf Bürgin 1939, NOB 25.06.1942
225 vgl. 7.7.1.
226 Reisebericht STAO 137/VII-182-9
227 Der Wirtschaftsplan vom 03.08.1939 wies 2geschossige Wohnbauten nur innerhalb der Ringstr. aus.

228 z. B. Planung Bürgin August 1939. Planungen RAM 27. 11. 1939 STAO I/6/36 und 11.04.1940 STAO I/7/26 beide unterschrieben von Büge.
229 An der Stelle des heutigen Plauen-Tiefs
230 Der Nordfriedhof sollte nach dem Vorschlag von Wiepking-Jürgensmann mit Sand oder Maadeaushub aufgeschüttet werden.
231 Sommer 1 S. 16
232 Vermerk 19.05.1939 BAK R 41/1317
233 Reichsluftfahrtminister 26.06.1937, Z. L. 5b 9400/37, Hallbauer 2 S. 15
234 Hallbauer 2 S. 15
235 Soldauer Weg, gerade Hausnummern 1940 Typ II
236 Soldauer Weg, ungerade Hausnummern, Typ I
237 Glettkauer Weg
238 Statiker waren Josef Keisel und die belgische Baufirma J. B. van den Broeck.
239 Vermerk RAM 23. 11.1939. Möglicherweise hat Nicolaus Hilgenstock ausgesucht.
240 Gespr. Hilgenstock 10. 11.1988. Hilgenstock war kein Nationalsozialist, er hatte wegen der verweigerten Integration seines Studentenverbandes Konflikte mit dem NS-System.
241 Veeser kam am 16. 12.1938 in die Bauleitung Hilgenstock. Auch er stand der NSDAP ablehnend gegenüber.
242 Vermerk Müller vom 03.01.1939. Müller erhoffte eine Mittelvorlage von 440 Mio. Mark. Umschlagbild unten
243 WZ 10.01.1939
244 Vermerk MI 02.03.1939 BAW
245 Vertrag 29.08.1939, Hilgenstock mußte streng nach VOB vergeben und erhielt 2% der Rechnungsbeträge als Honorar und Auslagenersatz nach Gebührenordnung für Arch.
246 Schr. XIV 1124 vom 02.03.1939 BAW
247 Vermerk RAM IVc, 23.03.1939 BAW
248 Gutachter der Jade
249 Sachverständiger des RAM
250 Das erforderte auch ein Absenken der Maade
251 gem. Vermerk RAM vom 23.03.1939 hatte die Jade 5 weitere Architekten hinzugezogen. Dem Vf. sind nur 3 bekannt geworden.
252 WK 17.06.1939
253 Modernes Trennsystem: Regenwasser in die Gräben zur Maade, Schmutzwaser in Leitungen zur Kläranlage Coldewei und danach Ableitung in die Maade.
254 Stadt Wolfsburg S. 31
255 Kluke S. 352ff.
256 Vermerk OKM 15.03.1940 STAO 137/VII-182-9
257 BAK R 41/1317
258 WZ 26./27. 10.1940
259 Früher Danziger Str., Nord-Süd-Achse
260 In Wolfsburg wurde der Wohnungsbau 1942 stillgelegt.
261 NOB 09.02.1942
262 Grundig S. 401
263 Mattausch S. 146
264 Linde 2 S. 2
265 z. B. Glettkauer Weg oder Typ 136. Typ 150 usw.
266 Die 6 Bauernhöfe wurden erst Baubüros. Danach sollten sie NSV-Heime werden.
267 Mattausch S. 177
268 Gespr. Bork 23. 11.1988
269 AJA
270 Umschlagbild unten
271 WK 01./02.04.1939
272 Sein einziger Baustellenbesuch galt am 01.04.1939 der Nordwerft
273 WK 17.06.1939
274 ebd. An die Trasse erinnern das schnurgerade Fedderwarder Tief und die Grünfläche zwischen Posener- und Steinstr. Der Europahausplan zeigte ein kleines Haltestellengebäude.
275 Vermerk RAM 23.03.1939 STAO 137/VII-182-9
276 Nicolaus 4 S. 153ff.
277 DH 29.06.1940
278 WZ 07. 11.1940
279 DH 11.01.1941
280 NOB 05.05.1941
281 NOB 30.03.1942
282 Sommer 1
283 Aufgegeben zugunsten der Nordwerft, vgl. 4.3
284 Auch Heß und Rosenberg waren im Juni 1937 in Wilhelmshaven.
285 NOB Juni 1937, undatiert
286 Verbindung Fort Schaar mit Fort Rüstersiel
287 Linde 1 S. 4
288 NOB 24.01.1938
289 vgl. 7.3.
290 Schr. RAM IVc 23. 11.1938 STAO 137/VII-182-9
291 NOB 05.03.1938, möglicherweise von Hallbauer
292 Schr. Zenker an Vf. 15.01.1989, 16.09.1989
293 Der Begriff kommt auch bei Voslapp und Fedderwardergroden vor.
294 Schr. Zenker an Vf. 15.01.1989
295 ebd.
296 ebd.
297 vgl. 7.7.5.
298 Niederschr. über Bespr. 03.02.1939 BAW
299 vgl 7.7.2.
300 Tessenow S. 8
301 Zu gleichzeitiger Regenwasser- und Oberflächenentwässerung
302 WK 27.05.1939
303 Bereich der heutigen Werdumer-/Beethovenstr.
304 Schr. Zenker an Vf. 15.01.1989
305 Schr. 25.04.1936 MAF RM 23/1686
306 Schr. MI an OKM 01.04.1936 BAW
307 Der Wunsch des Bauvereins Rüstringen in Altengroden 300 Wohnungen zu bauen, wurde am 23.02.1939 abgelehnt.
308 Gemeinnützige Deutsche Wohnungsbaugesellschaft mbH
309 Lilienburgstr. 31 A–D, 20–26, Arch. Willibald Strempel
310 WZ 05.05.1939
311 Schr. der Wohnungsbaugesellschaft vom 19.06.1939 an die Stadt BAW
312 Schr. Bauverwaltung 06.07.1939 BAW
313 Am Ende der Klinkeburg und der Redlefsstr.
314 Im Verlauf des Lubbeweges
315 Erker und Balkone auf Stützen fanden sich häufig bei Tessenow.
316 Der Wohnungsbau in Deutschland, Juli 1943 Heft 13/14
317 Jacob
318 Harlander S. 38
319 Arch. Tessenow 1929
320 Sie wurden durch MI vergeben.
321 NOB 13.01.1941
322 Vermerk RAM 09.06.1939 BAK R 41/1318
323 Zugeschnittene Sparren. Türen und Fenster kamen aus Bayern.
324 Gespr. Triebel 18. 12.1983
325 NOB 09.03.1942. Die Wohnungen sollten nach dem „gewonnenen" Kriege an die Marine zurückgegeben werden.
326 Prottengeier
327 Stadt Wilhelmshaven 8 S. 58
328 Hahn 3 S. 13, 23
329 u. a. wegen Flugschneisen, Schutzgürtel und Luftschutz

330 Geräumt werden sollten auch Werftkaserne und Altendeichsweg
331 NOB 23. 12.1938
332 Vermerk 27.02.1939 BAW
333 Vermerk 02.01.1939 BAW
334 Im heutigen Grünstreifen Lönsweg sollten u. a. Einfamilienhäuser für städt. „Gefolgschaft" entstehen.
335 Zwischen heutiger Groth- und Raabestr., nach dem Kriege mit Einfamilienhäusern bebaut
336 Am heutigen Neuengrodener Weg
337 Wurde nicht realisiert
338 NOB 30.01.1939
339 Schr. HNBD 23.02.1939 BAW
340 Schr. Stadterweiterungsamt an Möller 25.02.1939 BAW
341 Prottengeier
342 vgl. Katasterplan 02. 11.1940 BAW
343 WZ 07. 11.1940
344 Plan Forster 01.08.1942 BAW
345 Niederschr. über Bespr., 17. 10.1936 BAW
346 Angaben Katasteramt Varel: Salomon Herz. Salomon de Taube, Kurt de Taube, Robert de Taube, Ernst de Taube
347 Linde 2 S. 7
348 Schr. 15.05.1939 STAO 137/VII-182-9
349 Schr. MI 26.05.1939 BAW
350 Schr. 17.06.1939 STAO 137/VII-182-9
351 Möglicherweise die in Gründung befindliche Neue Heimat, Vermerk 09.06.1939 BAK R 41/1318.
352 ebd.
353 DH 20.07.1940
354 Schr. MDF 17.07.1939 STAO 137/4362
355 Es kam zu vielen baupolizeilichen Bemängelungen, Schr. 08.02.1940 STAO 137/4362
356 Schr. 30. 10.1940 STAO 137/4362
357 Brune Bd. 3 S. 27
358 Plan des Präsidenten des Nieders. Verwaltungsbezirks Oldenburg, Abt. I, März 1949 STAO 137/4362
359 Schr. 13. 10.1933 MAF RM 23/1661
360 z. B. Tapken (Peter-/Prinz-Heinrich-Str., 1937–1939, Arch. Peters), Schwarze (Prinz-Heinrich-Str. 22, 1937, Arch. Iwersen), Wilhelmshavener Spar- und Baugesellschaft (Prinz-Heinrich-Str. 61–67, Arch. Peters), Reif, Meyer, Hoeland, Mücke, Mehrtens, Bismarckstr. Baudissinweg, Viktoriastr. (ab 1934, Arch. Höger), Lübbers (Bismarckstr., 1938, Arch. Lübbers)
361 Mit Ausnahmegenehmigung des Schutzbereichsamtes
362 WK 27.04.1938
363 z. B. Günther (Widukindstr. 4, 1935, Arch. Ernst Oppermann), Klein (Widukindstr. 2, 1938, Arch. Karl Richter)
364 vgl. 7.5.
365 Vermerk 14.06.1935 MAF RM 23/1662
366 Niederschr. über Bespr. 10.12.1936 BAW
367 Vermerk Nutzhorn 12.01.1937 BAW
368 vgl. 7.2.
369 Vermerk Nutzhorn 11.02.1937 BAW. Das gesamte Gebiet östlich Gökerstr. fiel 1933 unter die Bausperre.
370 vgl. 7.5.
371 Hallbauer 2 S. 15
372 ebd. S. 17
373 Linde 2 S. 2
374 WK 31. 12.1937
375 Bekanntgeworden durch die Arbeitersiedlung Bremen-Blumenthal, Wilhelmshavener Tageblatt 27.05.1903
376 Norddeutsches Volksblatt 01.09.1903
377 Bauverein Rüstringen 2 S. 20
378 ebd. S. 21
379 Niederschr. über Bespr. 13.05.1936 BAW
380 Graul S. 60ff.
381 Arch. Abbehusen und Blendermann, Bremen/Rüstringen
382 Niederschr. über Bespr. 24.04.1936 BAW
383 Schr. Bauverein Rüstringen 16.07.1938 ABR, Schr. MI 04.01. 1938 ABR
384 vgl. 7.10.
385 Graul S. 85
386 Schr. Bauverein Rüstringen 16.07.1938 ABR
387 Bauverein Rüstringen 4, 1935
388 ebd.
389 Bauverein Rüstringen 4, 1937
390 vgl. 7.5.
391 Bauverein Rüstringen 4, 1937
392 Kunstbibliothek Berlin S. 19
393 ebd. S. 20
394 ebd. S. 42
395 ebd. S. 44
396 ebd. S. 48
397 Bauverein Rüstringen 4, 1937
398 Pehnt S. 127
399 Westphal S. 16ff.
400 Friedrich-Paffrath-Str., Nordabschluß sowie Kreuzung Siebethsburger-/Bismarckstr.
401 Panerei
402 Bauverein Rüstringen 4, 1933
403 Bauverein Rüstringen 4, 1934
404 Cirksenastr. bis Robodesweg
405 z. B. Nordbrock, Apken, Kruse, Dringenberg und Bauhütte des Jadegebietes
406 Bauverein Rüstringen 4, 1937
407 Bauverein Rüstringen 4, 1936
408 WK 12.06.1935
409 WZ Juni 1938, nicht identifizierbar
410 Vermerk RAM 29.06.1939 BAK R 41/1317 und 09.06.1939 BAK R 41/1318
411 Bauverein Rüstringen 4, 1937
412 Bauverein Rüstringen 4, 1938, Aufstellung 11.01.1939 ABR
413 Bauverein Rüstringen 4, 1940 S. 3
414 Pehnt S. 21
415 Seine ehemaligen Mitabeiter Boyken (ab 1930) und Suhr gingen eigene Wege
416 WK 12.06.1937
417 Schr. Bauverein an MI, 12. 12.1939 ABR
418 Bauverein Rüstringen 4, 1939, WK 25.06.1938
419 Er bildete das Fundament für das Nachkriegsgebäude der Arbeiterwohlfahrt.
420 Schr. 28.04.1939 ABR
421 Hahn 3 S. 13
422 ebd. S. 23
423 Heute Paul-Hug-Str.
424 Fluchtlinienplanakte BAW
425 Die Bauarbeiten für Wolfsburg begannen erst 1938
426 Hahn 3 S. 82, runde, ovale, rechteckige und dreieckige Platzformen
427 vgl. 7.7.1.
428 vgl. 7.7.2.
429 Grundig S. 580, seit 1922 nur noch Turnhalle
430 z. B.1939 Schillerstr./Bremer Str.
431 Nur in Fedderwardergroden wurden in geringem Umfang außer Bork noch andere Arch. beauftragt
432 WK 10.07.1937
433 Hallbauer 2 S. 28
434 WK 10.07.1937
435 WK 06.08.1937
436 vgl. 7.7.5.
437 WK 06.08.1937
438 Werftstr./Peterstr. sowie Peterstr. 95/99

439 WK 17.09.1937
440 Typ I G, I F, III, II G usw.
441 Typ I B
442 WK 01./02.04.1939
443 WK 25.02.1938
444 Hahn 3 S. 23.
445 NOB 26.01.1933, Protokoll Baukommission 23.01.1933
446 Schr. MI 01.09.1937 AJA
447 Schr. Hallbauer 21.09.1937 BAW
448 NOB 01. 10.1937
449 Fluchtlinienplanakte BAW
450 vgl. 7.7.1.
451 vgl. 7.7.3.
452 vgl. 7.7.4.
453 Eine Teilbebauung erfolgte durch die Gesellschaft 1965 mit Wohnhauszeilen am Bojenweg. Nördlich des Jadeviertels baute 1954 die Jade die Olympiasiedlung.
454 WK 14.09.1938
455 Frank S. 242
456 vgl. 7.10.
457 WK 16.03.1938
458 Karl Hitzbleck (Duisburg), Fritz Spieker (Oldenburg), F. C. Reincke (Berlin), Deutsche Bau-Aktiengesellschaft (Bremen)
459 WK 23.06.1938
460 WZ 14.07.1939, eingezeichnet im Wirtschaftsplan vom 03.08. 1939.
461 Begutachtung des Oberbürgermeisters von Wilhelmshaven 10.08.1939 BAK R 41/1318
462 Schr. 07.09.1940 BAK R 41/94
463 Linde 2 S. 7
464 Niederschr. über Sitzung 24.04.1936 BAW
465 Linde 2 S. 5
466 Schr. MDF 07.03.1938 STAO 137/VII-182-9, Burg Kniphausen stand seit 1938 unter Landschaftsschutz.
467 Schr. MI an Gemeinde Kniphausen 19.09.1937 BAW
468 NOB 17.09.1937
469 NOB 05.03.1938
470 Schr. 10.03.1938 STAO 137/VII-182-9
471 Brune Bd. 3 S. 52
472 Schr. 25.05.1938 ABR
473 Erläuterungsbericht v. März 1938 STAO 137/VII-182-9
474 Schr. OKM 09. 12.1940 BAK R 41/93
475 Baur 6
476 Vermerk RAM 23. 11.1938 STAO 137/VII-182-9
477 Feder war Wilhelmshaven durch verschiedene Besuche seit 1933 verbunden.
478 Zu Putlitz baute seit 1938 für die Wilhelmshavener-Rüstringer Wohnungsbaugesellschaft. Forster hatte 1937–1938 bei zu Putlitz gearbeitet.
479 Schr. Old. Staatsministerium Nr. II 205 04.01.1939 STAO, vgl. Anm. 480
480 Schr. MI an OKM 09.01.1939 STAO 137/VII-182-9
481 Baur 6, bestätigt durch den früheren Feder-Mitarbeiter Heinz Killus
482 WZ 26./27. 10.1940
483 Vermerk über Bespr. 26. 11.1940 BAW
484 Schr. RAM 04.08.1941 BAW
485 Vermerk MDF 15.03.1941 BAW
486 Erläuterungsbericht Akte Gutschow E 302 02.01.1945 S. 5 BAW
487 Schr. RAM 04.08.1941 BAW
488 Schneider 1 S. 74
489 Feder S. 472
490 Der Reichsstatthalter in Hamburg S. 2
491 Begutachtung des Oberbürgermeisters von Wilhelmshaven v. 10.08.1939 BAK R 41/1318
492 vgl. 3.5.5.
493 Schr. MI an OKM 09.01.1939 STAO 137/VII-182-9
494 Verm. Müller 03.01.1939 STAO 137/VII-182-9
495 vgl. 3.7.3.
496 Bespr. 16. 12.1938, Schr. 09.01.1939 BAW
497 Seit 01.06.1937 Stammkapitalmehrheit bei der Stadt Wilhelmshaven, Aufsichtsratsvorsitzend. seit 06.04.1937 Oberbürgerm. Müller. Umschlagbild unten
498 Schr. MI an OKM 09.01.1939 STAO 137/VII-182-9
499 STAO 137/4363
500 Schr. OKM 13.01.1939 STAO 137/VII-182-9
501 Schr. Staatsministerium 18.01.1939 STAO 137/VII-182-9
502 Baur 4
503 Erläuterungsbericht zum Bebauungsplan-Vorschlag der Stadt Wilhelmshaven für den Ortsteil Breddewarden 28.02.1939 BAW
504 Baur 4
505 Schr. RAM an OKM 14.03.1939 STAO 137/VII-182-9
506 Vermerk RAM 17.05.1939 STAO 137/VII-182-9
507 Vermerk RAM 20.06.1939 STAO 137/VII-182-9
508 Baur 4
509 Der Reichsstatthalter in Hamburg S. 2
510 Vermerk OKM 15.03.1940 BAK R 41/1318
511 Reisebericht zur Dienstreise nach Berlin 26.–29. 11.1941 BAW
512 Schr. Temp an MDF 08.08.1939 STAO 137/VII-182-9
513 Schr. Temp an MDF 04.04.1940 STAO 137/VII-182-9
514 Schr. RAM an MDF 27.07.1940 STAO 137/VII-182-9, vgl. WZ 26./27. 10.1940
515 NOB 10.06.1941
516 NOB 22. 12.1941
517 WZ 26./27. 10.1940
518 NOB 10.06.1941
519 Gespr. Jahn 05.06.1989
520 Schneider 3

4.

1 Kisch S. 63
2 Grundig S. 8 ff.
3 Marinewerft Wilhelmshaven S. 58
4 Grundig S. 8
5 ebd.
6 Brune Bd. 2 S. 640
7 Koch S. 43 ff.
8 Dülffer 2, vgl. 2.2.
9 Ausnahmsweise wurden unentbehrliche Spezialisten beauftragt, zumeist Bauingenieure oder Hafenbauer wie Agatz.
10 vgl. 2.3.
11 Dülffer 2 S. 563
12 Ahner 2 S. 77
13 vgl. 2.1.
14 Koop 1, 2, Teilnehmerverzeichnis der Marine-Fernsprechanlage in Wilhelmshaven, Stand 1942 ARK, Marinestandortbauamt Wilhelmshaven: Standortspläne Wilhelmshaven Band I, Stichtag 01. Juli 1941 ASTHBA
15 Auch als Flak-Zentrale bezeichnet mit Flakleitstelle, II. Marine-Flak-Brigade. Umfangreiche Bunkeranlagen unterstanden der Festungskommandantur. Annähernd 10 Baracken mit Exerzierplatz ergänzten die vom Marinestandortbauamt 1936 erbauten Anlagen, vgl. Blatt 57 Standortspläne Wilhelmshaven Bd. I, Stichtag 01. Juli 1941 ASTHBA a. a. O.
16 vgl. 7.4., bezogen auf die Einrichtung
17 Ab 1910 als UTO-Werft gebaut, 1935–1937 für 20 Mio. Mark umgebaut und ergänzt, vgl. Thiele
18 Z. B. wurden nach Abriß des Nordteiles der Manteuffelstr. an der damaligen Oldenburger Str. ab 1937 weitere Werkstätten und Geschützlagerhäuser gebaut, vgl. Grunewald S. 283

19 Etwa heutige Landesstr. 10, die westlich in die Reichsstr. 69 münden sollte.
20 Gespr. Geisenhainer 01. 12.1988
21 Schmitthenner 1 S. 154
22 ebd.
23 vgl. Montagehalle Heinkel-Werke Oranienburg 1936–1938 (Rimpl), Flugzeughallen Upjever 1936 (Schütte), vgl. 4.2.6., 4.4.
24 Alting 2
25 Paffrath S. 88
26 Bis 1874 im Besitz der Familie von Karl Jaspers.
27 Am 01.04.1948 wurde Sande wieder selbständige Gemeinde.
28 Zwischen Roon- und Kaiserstr., heute Weserstr. 45–47
29 Rohlfs S. 30
30 Stadt Wilhelmshaven 5 S. 14
31 vgl. 7.7.5.
32 Hilfskrankenhäuser waren auch: 9 Holzbaracken in Fedderwardergroden, in der Hafenkaserne, in der Wasserturmschule, im Keller der Schulen Voslapp, Salzastr. und Albrechtstr., im Lager Sudetenstr. und in den Lazarettsbaracken am Banter Weg.
33 Neufert S. 257
34 1967, nach Inbetriebnahme des neuen Stadtkrankenhauses wurden die Gebäude in der Südstadt funktionslos.
35 Mitarbeiter waren Pitz und Maas. Maas schrieb während der Zeit eine Dissertation über das Gut Sanderbusch.
36 Neufert S. 257, Schachner, Distel, Gutschow 1
37 Spieker S. 180
38 Gespr. Geisenhainer 01. 12.1988
39 Wird heute als Altenheim genutzt.
40 Wie Marinegerätelager Roffhausen, vgl. 4.2.1.
41 Das Lazarett wurde nur gering durch Bomben zerstört.
42 vgl. 7.9.
43 Gespr. Gaisenhainer 01. 12.1988
44 Haus der Deutschen Kunst S. 50 Nr. 208
45 Teilnehmerverzeichnis aller Marine-Fernsprechanlagen in Wilhelmshaven, Stand 1942 S. 79ff. ARK
46 Gespr. Thees 06. 12.1988, 06. 1.1989
47 Raeder besuchte häufig die Baustelle.
48 Dülffer 2 S. 213
49 WK 14.05.1937
50 vgl. 7.4.
51 vgl. 2.1.
52 Daniel S. 2
53 Hieß auch Marinekommando West, vgl. Staschen, M.: Die Geschichte der Admiral-Armin-Zimmermann-Kaserne in Wilhelmshaven-Sengwarden, Aurich 1991 (Typoskript)
54 Gespr. Thees 06. 12.1988, 06.01.1989, Gespr. Jahn 05.06. 1989
55 1,6 Mio. Mark Baukosten
56 Rüssel
57 vgl. 3.6.3.
58 Heutiger Bojenweg
59 DH 23. 11.1940
60 DH 18.05.1940
61 Das Maschinenbauressort hatte bis zu 600 Lehrlinge.
62 DH 10.08.1940
63 DH 14. 12.1940
64 DH 05.04.1941
65 DH 30. 11.1940
66 DH 05.04.1941
67 DH 23.08.1941
68 DH 10.08.1940
69 DH 29. 11.1941
70 Gespr. Higgen 21.08.1988
71 DH 23. 11.1940
72 NOB 21.01.1941
73 vgl. 3.6.2.
74 Plan STAO 137/4362
75 Petsch 1 S. 132
76 10 Sportplätze, mehrere Bäder und Turnhallen, vgl. Büsing 1, Sommer 4, 5
77 Feder S. 212
78 vgl. 7.7.1.
79 WZ 29.03.1939
80 WZ 06. 11.1987, DH 01.05.1940
81 Plan 25.01.1937 BAW
82 DH 19. 10.1940
83 Backes S. 135
84 DH 24.08.1940
85 DH 07.09.1940
86 vgl. 3.5.7.
87 Hallbauer 2, S. 7, 8
88 Linde 1 S. 6
89 vgl. 7.8.
90 vgl. 4.2.2.
91 Plan 17.07.1938 AGS
92 Kaysers S. 36
93 Bauunternehmer: Albert Höller, Diepholz, Stahlbau: Norddeutscher Eisenbau, Sande, Verglasung: Eberspächer, Esslingen
94 Gießerei Sande
95 Schr. Oberbürgermeister 03.01.1939 STAO 137/VII-182-9
96 Petsch 1 S. 156
97 Heutiger Anton-Dohrn-Weg
98 Westl. der ehemaligen Berufsschule Weserstr. 78
99 vgl. Hansa-Lloyd-Goliath-Werke Bremen 1935–1936 (Lodders) und Industriebauten Wolfsburg und Salzgitter
100 Frank S. 105
101 Nicht mehr rekonstruierbar, fast alle nach 1945 gesprengt
102 Gespr. Yvonne Klaucke 02.08.1989
103 Bauantrag 28. 10.1935 BAW
104 Koop 1 S. 32ff.
105 Brune Bd. 1 S. 415
106 Brune Bd. 2 S. 641
107 Anfangs offenbarten die Pläne noch Land- und Seeflughafen sowie Luftwerft in den Grodenflächen. Die späteren Luftwaffenplanungen wurden strikt geheimgehalten – auch vor den städtischen Planern.
108 Koop 1 S. 70ff.
109 Sie unterstand als Mittelinstanz direkt der Bauabteilung des OKM.
110 Die Gesamttonnage durfte jetzt 35% der englischen Marine erreichen.
111 Arndt 2
112 Koop 1 S. 98ff.
113 Thalenhorst S. 60, 61
114 Brune Bd. 1. S. 227
115 Goecke, Verfügung BV III 3195 28.05.1917
116 je 350 x 52 m
117 Die Kammern wurden vergrößert auf 390 x 60 m
118 vgl. 3.5.1.
119 vgl. 3.5.6.
120 Alle technischen Angaben Mulitze 1
121 Zeugnis Hellmut Suhr 10.03.1941, Mulitze 1 S. 43a, 42, Westphal S. 31
122 vgl. 7.8.
123 U-Bootbunker wurden auch in Farge bei Bremen von KZ-Häftlingen betoniert.
124 Raeder S. 68
125 WK 01./02.04.1939
126 vgl. 7.10.
127 Marinewerft Wilhelmshaven S. 38, 40

128 Breyer S. 98 ff.
129 vgl. Hümmelchen
130 Daniel S. 115
131 Der Marine blieben nur 30 Bordflugzeuge der großen Einheiten.
132 Bauunternehmer war u. a. Hermann Möller, Wilhelmshaven.
133 Fissen S. 168
134 Ministerpräsident Joel weihte die ersten Kleinsiedlungen 25.08.1935 ein. Im Rathaus Jever warben DAF und Gau-Heimstättenamt mit fertigen Plänen für Kleinsiedlungen. Sammlung Hartmut Peters.
135 Kunstbibliothek Berlin S. 5
136 Jürgens S. 110
137 Nur ein Teil der Maadelinie wurde 1940–144 gebaut, vgl. 7.6.
138 Hallbauer 7 S. 4, Schr. Hallbauer an Tappe 10.02.1969 BAW
139 Der Flughafen Mariensiel wurde ab 1939 für die Lufthansa gesperrt und dem Seefliegerhorst Wilhelmshaven unterstellt.

5.

1 Fest Bd. 2 S. 793, 794
2 Der Plan ist nur als Nachzeichnung des Stadtplanungsamtes „Großraum Wilhelmshaven, Planung 1939" vorhanden.
3 Dülffer 1 S. 13
4 Haus der Deutschen Kunst S. 49 Nr. 202
5 Umschlagbild unten
6 Gespr. Forster 26.01.1988
7 Gespr. Fanzius 10.02.1984, 26. 10.1985
8 Gespr. Lübbers 28. 12.1988
9 Gespr. Lisa Baummann 08. 11.1988. Die Pläne sind nicht mehr vorhanden.
10 WZ 28. 12.1935
11 WZ 03.04.1937
12 Hallbauer 5 S. 3
13 WZ 02.04.1937
14 Hallbauer 5 S. 7
15 Backes S. 120, Giordano S. 88
16 Hallbauer 2 S. 29
17 NOB 29. 11.1937
18 Schr. Oberbürgermeister 20. 12.1937 an Reichsstatthalter STAO 137/VII-182-9
19 Vermerk MDF 24. 12.1937 mit handschriftlicher Notiz Pauly und Sichtvermerk Joel
20 NOB 05.03.1938
21 WZ 31. 12.1938
22 Backes S. 122
23 WK 30.07.1938
24 Durth 2 S. 13 ff.
25 BAK R 43 II 1012a
26 Schrape S. 209 ff.
27 Thies S. 131, BAK R 120/7
28 Bau-Rundschau 02. 10.1913
29 Vermerk 02.05.1935 BAW
30 WK 26.09.1936
31 WZ 29. 10.1987
32 Vermerk Neumann 13.03.1935 BAW
33 Siemer u. Neumann machten am 11.05.1935 Vorschläge BAW
34 NOB 27.05.1935
35 Vermerk Kellerhoff 07.06.1935 BAW
36 NOB 22.07.1935
37 NOB 22. 10.1935
38 WZ 28.09.1936
39 ebd.
40 WK 29.09.1936
41 WK 31. 12.1937
42 WZ 15.04.1937
43 vgl. 7.10.
44 WK 23.04.1938
45 NOB 27.05.1935
46 NOB 17.02.1941
47 WZ 16.08.1989
48 Antrag Stadt Wilhelmshaven 13.02.1939 STAO 137/5929
49 Umschlagbild unten
50 Schr. Oberbürgermeister 20. 12.1937 an Reichsstatthalter STAO 137/VII-182-9
51 WK 11.01.1939
52 WZ 17.03.1939
53 Backes S. 152
54 Hallbauer 8 S. 32
55 ebd. S. 33
56 NOB 05.03.1938
57 ebd.
58 Schr. Oberbürgermeister 20. 12.1937 an Reichsstaathalter STAO 137/VII-182-9
59 Noch am 17.03.1939 wartete die Stadt Wilhelmshaven auf die Entscheidung des Reichsverkehrsministers zur Bahnhofsfestlegung. Besprechungsvermerk Nicolaus 23.03 1939/21.03.1939 STAO 137/VII-182-9
60 Gespr. Lisa Baumann 08. 11.1988
61 Baur 5
62 Die Reichsstraße 69 wurde 1940 gebaut. Am Ortsausgang sollten noch Polizeikasernen und Hauptfeuerwache gebaut werden. Die ab 1939 geplante Ringschnellbahn sollte das gesamte Kulturforum unterirdisch kreuzen und einen Umsteigebahnhof unter dem Hauptbahnhof bekommen.
63 Roth hatte sich mit dem Theater Dessau hervorgetan (1935–1937). Bis 1941 waren 85.000 Mark Planungskosten für von Arent u. Roth angefallen. BAK R 43/II 1025 Bl. 40 ff.
64 Schr. Bartning 12.04.1940 SHB
65 Baur 5
66 von Arent u. Roth lehnten sich formal an das Gautheater Saarbrücken (Paul Baumgarten 1936–1938) und das Projekt der Münchener Oper (Woldemar Klein 1938).
67 Meyer 2 S. 67
68 Vermerk RAM 18. 12.1940 BAW
69 NOB 15. 12.1941
70 vgl. 6.2., 6.3.
71 Wortmann u. Gutschow hatten Ende der 20er Jahre bei Schumacher gearbeitet.
72 Versch. Gespr. Franzius, Thees, Geisenhainer
73 NOB 18. 10.1937
74 NOB 29. 11.1937
75 WK 31. 12.1938
76 WK 23.04.1938
77 WK 31. 12.1938
78 WZ 10.01.1939. Die Diagonalstraße sollte entfallen.
79 Vermerk RAM 23.03.1939 STAO 137/VII-182-9
80 WZ 30.08.1989
81 WK 31. 12.1937
82 vgl. Eingänge Neue Reichskanzlei Berlin (Speer), Wachhäuschen Großer Stern Berlin (Speer), Führerbau München (Troost), Feldherrnhalle München usw.
83 Schr. Oberbürgermeister 20. 12.1937 an Reichsstatthalter STAO 137/VII-182-9
84 Feder S. 179
85 Schr. MDF 11. 11.1940 BAW
86 WK 17.06.1939
87 Backes S. 135
88 Schrape S. 216, 217

89 Mattausch S. 189
90 Schneider 1 S. 2

6.

1 Rothfels S. 62
2 Hampe S. 3
3 Schr. des Oberbefehlshabers der Kriegsmarine 10. 12.1935, 22. 12.1936 MAF 23/1662
4 vgl. Anm. 4.32
5 Speer 3 S. 197
6 Hampe S. 191 ff.
7 ebd. S. 292
8 Brune Bd. 2 S. 192
9 WZ 30. 07.1986, Aufstellung über Schutzbauten im Stadtgebiet 22.05.1981 BAW
10 Moderhack Bd. 1 S. 101
11 Grundig S. 163
12 Speer 1
13 Gemeinschaft III
14 Archivunterlagen des Amtes für Zivilschutz der Stadt Wilhelmshaven
15 Der Reichsstatthalter in Hamburg, Plan: Luftschutz
16 Die Hilfskrankenhaus-Bunker Virchowstr. und Willehad-Krankenhaus fügten sich lückenlos in die Block-Randbebauung der Südstadt.
17 Hampe S. 274
18 Herde
19 Bauwelt 2 S. 425
20 Dem Vf. ist in dieser Region nur der Scheinflughafen Uplengen bekannt geworden.
21 Herde S. 50
22 Mulitze 1 S. 39 ff.
23 U-Bootbunker gab es in fast allen Marinestützpunkten, vor allem an der Altlantikküste, vgl. Breyer S. 80, 138, 141
24 Koop 1 S. 111
25 Farge war wie Wilhelmshaven KZ-Außenstelle von Neuengamme, ARD, 3. Programm 26./27. 12.1988
26 Breyer S. 141
27 Schneider 4 S. 108
28 Hamburger Zeitung 03.08.1943
29 Speer 3 S. 363
30 BSW 6/1940
31 Stadtschaft für Niedersachsen. Aus dem Raum Wilhelmshaven waren unter den 354 Teilnehmern: Walter Sommer (2. Preis); Klaus Rasch (2. Ankauf), Friedrich Heuer (2. Ankauf), Karl Franzius (9. Ankauf).
32 Stadtschaft für Niedersachsen S. VIII ff.
33 STDAW 5610
34 Larsson S. 48 Abb. 34
35 Schulz 2
36 Schulz 1
37 Dort war der Boden tragfähiger als in der für Langewerth vorgesehenen Siedlungsfläche.
38 Akte Gutschow E 302/28 c, E 302/28 d BAW
39 Akte Gutschow E 302/28 d BAW
40 Schr. OKM 11. 8.1942 BAK R 41/94, Schr. LOB 06.06.1942 an Reichsstelle für Raumordnung BAK R 41/94.
41 Akte Gutschow E 302/28 f, g, h, vgl. Schneider 2
42 Schr. LOB 06.06.1942 an Reichsstelle für Raumordnung BAK R 41/94
43 Durth 2 S. 27, 28
44 Schr. 26.06.1943 BAK R 41/1318
45 Abschrift, eingegangen bei der Stadt Wilhelmshaven 12.01. 1944 BAW.
46 Durth 2 S. 55 ff.
47 Temp
48 Schr. Reichsstatthalter an die Stadt Wilhelmshaven 14.03. 1944, BAW, NOB 12.01.1944
49 Am 16.05.1944 teilten Klaje u. Wortmann in Wilhelmshaven Oberbürgermeister Müller mit, daß Speer für die Wilhelmshavener Planungen Gutschow vorgesehen habe.
50 Gespr. Lübbers 12.09.1989
51 Akte Gutschow E 302/4–7 24. 10.1944 BAW
52 Akte Gutschow E 302/23–28 und Erläuterungsbericht 02.03. 1945. Kurz vor Kriegsende unternahm man noch eine Dienstreise nach Holland, um Grachten zu besichtigen.
53 Gespr. Franzius 10.02.1984, 26. 10.1985. Er stellte dem Arbeitsstab Speer noch nach dem Kriege eine Rechnung.
54 Hanffstengel
55 Geführt vom NSDAP-Hauptamt für Technik
56 Schr. 15. 10.1943, Akte Arbeitskreise Stadt- u. Landesplanung Gau Weser-Ems BAW.
57 Verschiedene Typoskripte ebd.
58 Murken 2 S. 49
59 Koop 1, S. 32 ff.
60 Brune Bd. 1 S. 468
61 Murken 2 S. 42, Gespr. Higgen 21.07.1988
62 Murken 2 S. 64, Gespr. Aichinger 16. 12.1988
63 Stadtbaudirektor Temp wurde 1945 entlassen. Er arbeitete danach bei verschiedenen Behörden und ließ sich – obwohl Bauingenieur – als Architekt nieder.
64 Credé S. 30
65 Brüning S. 57
66 Paffrath S. 90 ff.
67 Murken 2 S. 77
68 Credé S. 49
69 vgl. 7.10.
70 Tätigkeitsbericht des Stadtplanungsamtes BAW
71 Credé S. 62, NWZ 15.02.1949
72 Brüning S. 69 ff., Stadt Wilhelmshaven 4 S. 18
73 Brüning S. 69 ff.
74 Brüning S. 72
75 Brune Bd. 1 S. 410
76 Nordwestdeutsche Rundschau 18.07.1948, Niederschr. des Preisgerichts 08./09.07.1948
77 Protokoll, Bespr. beim Nieders. Minister für Arbeit, Aufbau u. Gesundheit 19./20.08.1949 BAW
78 Durth 2 S. 96, 104
79 Zitate im folgenden aus verschiedenen Hausakten BAW
80 Niestroj S. 43
81 Änderungseintragungen Viktoriastr. 20. 12.1949 BAW
82 Beschluß 25.03.1953
83 Rehbein, Gespr. Rehbein 03. 11.1988

7.

1 Bauwelt 3 S. 111, Rede Wagner vor dem Deutschen Werkbund 10.06.1933, mit der er sein Amt als Vorstandsmitglied niederlegte
2 Hallbauer 2 S. 17, 18
3 Vermerk RAM 18.12.1940 BAK R 41/93
4 Nicolaus 2
5 Schr. Wortmann an Reichsstelle für Raumordnung 27.01.1941 BAR R 113/1534
6 Schr. Oberbefehlshaber der Kriegsmarine 10.02.1939 BAW
7 ebd.
8 Schr. LOB 06.06.1942 BAK 41/94
9 Schr. Temp 26.06.1943 BAK R 41/1318
10 Umschlagbild oben
11 Umlauf S. 40
12 Mitteilungen des MDF an die Reichsstelle für Raumordnung v. 21.02.1936, 16.10.1936, 06.09.1938 BAK R 113/854
13 Scholtz
14 Einladungsschr. 11.03.1937 BAW

15 Hallbauer 2. Der Plan liegt nicht vor, er dürfte jedoch dem Plan Wirtschaftsgebiet Wilhelmshaven vom Juni 1937 entsprechen.
16 Nur als Nachzeichnung des Stadtplanungsamtes bekannt, Anm. 5.2
17 Möglicherweise entsprach er dem Wirtschaftsplan Wilhelmshaven u. Umgegend, Sept. 1937.
18 Umschlagbild oben
19 Nicolaus 4 S. 159
20 Schr. RAM 30.10.1936 BAW
21 WK 27.08.1938
22 Tätigkeitsbericht der LOB, abgechlossen April 1938 S. 1 BAW
23 Stommer S. 27
24 Tätigkeitsbericht der LOB, abgeschlossen April 1938 S. 3 BAW
25 Wilhelmshavener Rundschau 30.07.1965, Grundig S. 435, 486, 674, 675
26 Nicolaus 4 S. 153
27 vgl. 2.1., 4.1.
28 Nicolaus 2
29 Wolf 1 S. 447
30 Heuer 2 S. 18
31 BSW 10/1940 S. 353ff.
32 Lindner 1 S. 72
33 Hallbauer 2 S. 12
34 Schneider 1 S. 44, Wortmann 4 S. 15
35 Otto S. 341ff.
36 Schr. Reichsluftfahrtminister 26.06.1937, Hallbauer 2 S. 15
37 Vermerk über Bespr. im RAM 16.08.1938 BAK R 41/95. Ein Schutzbereichsplan war nicht mehr aufzufinden.
38 Schneider 1 S. 59
39 Gespr. Wortmann 28.01.1984, Gespr. Lübbers 28.12.1983. Auch in den verschiedenen Veröffentlichungen Wortmanns fehlen jegliche Hinweise auf Luftschutz.
40 RGBl. I S. 659
41 Bundesgesetzblatt Teil I 12.12.1986. Nach § 1 (5) 9. sind die Belange der Verteidigung und des Zivilschutzes besonders zu berücksichtigen.
42 Peltz-Dreckmann S. 116ff.
43 Otto S. 341
44 ebd. S. 342
45 Der Personenbahnhof Schaar war bis Juni 1937 Planungsziel, danach wurde der Bahnhof Kirchreihe erwogen, ab 1938 wieder der Hauptbahnhof Schaar.
46 Otto S. 342
47 Grundig S. 19ff., Koop 1 S. 98ff.
48 Hallbauer 2 S. 7
49 ebd. S. 8
50 ebd. S. 9
51 ebd. S. 9
52 ebd. S. 9
53 vgl. 6.2., 6.3.
54 Schulz 1 S. 6
55 Schulz 2
56 Zeugnis des Oberbürgermeisters 15.06.1942 für Walther Schulz SHB
57 Akte Gutschow, Schriftsatz zu den zweiten Skizzen 29.12.1944 BAW
58 Otto S. 342
59 vgl. Nieders. Bauordnung §§ 7–13
60 vgl. Bautechnische Grundsätze für Hausschutzräume des Grundschutzes 4.1.2, Bundesminister für Städtebau u. Wohnungswesen 10.02.1972
61 Mattausch S. 3ff.
62 Der Reichsstatthalter in Hamburg S. 1
63 Gespr. Herde 01.12.1988, Herde
64 Hallbauer 2 S. 12
65 ebd. S. 13
66 Linde 1 S. 3
67 Versch. Gespr.
68 Insgesamt wurde Wilhelmshaven im Zweiten Weltkrieg 04.09.1939–30.03.1945 etwa 100 mal bombardiert.
69 vgl. Direktive Churchills 14.02.1942, Marßolek S. 357
70 WILHEMSHAVEN ZONE MAP INFORMATION SHEET, Secret ca. 1942, National Archivs Washington DC, beschafft von Rolf Uphoff.
71 Am 22.01.1947 stellte die Gemeinnützige Gesellschaft mbH zur Schaffung von Wohngelegenheiten für Reichsangehörige in Altengroden nur 5 zerbombte Häuser fest.
72 Block, A.: Der Mann der Wilhelmshaven einnebelte, in: WZ 10.08.1985
73 vgl. Credé S. 18
74 Paffrath S. 5ff.
75 Grundig S. 32ff.
76 Beutz S. 39ff.
77 Baur 1
78 Brüning S. 71, 72
79 Baur 1
80 Brüning S. 71, 72
81 Baur 3 S. 8
82 Rasch S. 100
83 Stadt Wilhelmshaven 4 S. 20
84 Siemer 2 S. 10. Siemer war 1942–1950 Leiter des Arbeitsamtes Wilhelmshaven
85 Wolf 2 S. 150ff.
86 Migge 2 S. 59ff.
87 Schmitthenner 1 S. 175
88 Voigt 2
89 Schultze-Naumburg 1 S. 176
90 ebd. S. 188. Der gepflanzte Baum, möglichst eine Fichte sollte auch den gezeugten Sohn als „Stammhalter" symbolisieren.
91 Heinrich-Hampf S. 271ff.
92 Schr. DSB Gaugrupe Weser-Ems an die Jade 30.04.1942 AJA
93 Vertrag 26.05.1943 AJA
94 WK 10.07.1937
95 WK Sept. 1938, undatiert
96 WK 10.07.1937, 04.08.1939
97 Nicolaus 4 S. 168
98 Lindner 1 S. 144ff., 232ff.
99 Gemeint sind die Volksparkideen von Wagner u. Migge.
100 Lindner 1 S. 233ff.
101 WK 10.02.1938
102 WK 28.04.1938
103 WZ 29.03.1938
104 WK 28.07.1938
105 Die Privatisierung erfolgte erst ab 1956.
106 WK 28.07.1938
107 WK 30.07.1938
108 WK 29.08.1938
109 vgl. Städtebauliche Planung mit Verkehrs- u. Grünplanung für Wilhelmshaven Altstadt u. Nord, abgeschlossen 01.06.1941 (Unterschrift Baur) AHB, Bebauungsplan Wilhelmshaven Dez. 1941 (Unterschriften Temp u. Schulz) BAW
110 WK 10.07.1937. Der Adalbertplatz wiederum leitete sich von „Unter den Linden", Berlin ab.
111 Erläuterungsbericht Däke 25.11.1938 BAW
112 Berndt S. 25, 35
113 Sommer 7
114 Wagner 1
115 Gesamthochschule Kassel 2

116 Hahn 3 S. 58ff.
117 WK 25.06.1938
118 Die zentrierte Anlage des Torbogenplatzes tom-Brok-Str. blieb eine Ausnahme.
119 Westphal S. 99
120 WZ 29.03.1938
121 Stadt Rüstringen S. 44, Sommer 4
122 Entworfen von Leberecht Migge 1915
123 Kamm S. 12ff.
124 Stadt Rüstringen S. 23ff.
125 Hahn 3 S. 66
126 Gesamthochschule Kassel 2 S. 80
127 Wiepking-Jürgensmann 2, vgl. Bericht über Bespr. beim RAM 26./29. 11. 1941 STAO 137/VII-182-9
128 WK 17.06.1939
129 Pfeiffer S. 154ff.
130 WZ 29.12.1939
131 WK 17.09.1938
132 Nicolaus 4 S. 153
133 z. B. um den Krähenbusch, WK 21.01.1938 vgl. Oldewage.
134 Krämer
135 Schmid
136 Vermerk OKM 15.03.1940 über Bespr. bei der MI 07.03.1940 BAK R 41/1318, Schr. Temp 28.06.1940 am MDF STAO 137/VII-182-9
137 Gespr. Fresenius 16.12.1983
138 Heintz war am 17.01.1940 vom RAM empfohlen worden, Schneider 1 S. 81, Stadt Wolfsburg S. 46
139 Durth 2 S. 1041, 1047, 1051, Brüning S. 48ff.
140 Wiepking-Jürgensmann 3 S. 442ff. Gemeint ist Riehl, W. H.: Die Naturgeschichte des Volkes als Grundlage einer deutschen Sozial-Politik, versch. Bd. Stuttgart 1853
141 WK 23.04.1938
142 ebd.
143 Schr. MI 27.01.1938 ABR
144 Schr. RAM 28.09.1939 ABR
145 Schr. MI 24.06.1940 ABR
146 Selle S. 24ff.
147 Troost war 1914 am Bremen-Oldenburg-Haus der Werkbund-Ausstellung Köln beteiligt. Er wurde meistbeschäftigter Architekt für Schiffsausbauten im nordwestdeutschen Raum u. Hitlers Lieblingsarchitekt bis 1934, vgl. Deutscher Werkbund 1, Merker S. 216
148 Asche Bd. I S. 155ff., Bd. II. S. 120ff.
149 Adolf Rauchheld hatte den Bund Oldenburger Werkkünstler gegründet, der 1913 dem Deutschen Werkbund beitrat. Rauchheld war eng mit Wagner verbunden und bekam Aufträge aus Rüstringen.
150 Verein der Kunstfreunde für Wilhelmshaven S. 182ff.
151 Deutscher Werkbund 1 S. 21, Asche Bd. II S. 119. Beteiligt waren auch Abbehusen, Blendermann, Stoffregen, Dieter u. a.
152 Schr. 07.09.1933, 10.01.1934 MAF RM 23/1661
153 Hahn 3 S. 50
154 Günther 1 S. 105
155 Hahn S. 58
156 Schr. Marinewerft 20.01.1934 MAF RM 23/1661
157 Schr. MI 23.10.1933 MAF RM 23/1662, Aufstellung 01.02.1936
158 Schr. Oberbefehlshaber der Kriegsmarine 22.05.1936 MAF RM 23/1662
159 Vermerk Bauverwaltung 19.09.1941 BAW
160 Vermerke 15.–18.07.1940 STDAW
161 Vermerk RAM 04.02.1936 MAF RM 23/1686
162 Selle S. 136
163 Müller-Wulckow
164 WK 12.06.1937
165 DH 18.05.1940
166 Teut S. 280ff.
167 Petsch 2 S. 79, Selle S. 229
168 Doerr 1, 2, Haus der Deutschen Kunst S. 65ff.
169 WZ 25./26.06.1938
170 Hitlers Reden wurden auf Plätze, in Gemeinschaftssäle, HJ-Heime usw. übertragen.
171 WK 10.07.1937. Die Wohnung war in der Uhlandstr. gelegen.
172 WK 24.09.1938
173 WZ 11.07.1939
174 Doerr 2
175 DH 29.06.1940
176 ebd.
177 DH 27.07.1940
178 DH 01.03.1940
179 Thalenhorst S. 97ff. Troost war 1912–1934 Chefeinrichter des Norddeutschen Lloyd.
180 Erläuterungsbericht Däke 17.08.1938 BAW
181 WZ 20.12.1935, s. a. den Urzustand in Stadt Rüstringen: Die Burg am Meer, Rüstringen 1931
182 DH 20.07.1940
183 Die Zeichnungen befinden sich beim Verwaltungsleiter des Nordwest-Krankenhauses Sanderbusch.
184 Ernst Tietjen, Varel, gehörte zu den renommiertesten Innenarch. des nordwestdeutschen Raumes. Er war Mitglied des Deutschen Werkbundes u. am Bremen-Oldenburg-Haus auf der Kölner Werkbund-Ausstellung 1914 beteiligt, vgl. Deutscher Werkbund 1 S. 18.
185 Petsch 2 S. 83
186 Bremer Str. nahe Gökerstr.
187 Bohlen S. 48, 49
188 Grundrißzeichnung August 1938 SHT
189 Überwiegend von den Vereinigten Werkstätten ausgebaut
190 Feder S. 19
191 ebd.
192 ebd. S. 16
193 ebd. S. 52
194 Versch. Gespr., Oberbürgermeister Müller wohnte Birkenweg 18. Umschlagbild unten
195 Neufert S. 137
196 Haus der Deutschen Kunst
197 Gespr. Franzius, Forster, Murken
198 Z. B. auch bis 1935 als Logenmitglieder oder bis 1937 als Rotarier
199 Hallbauer 2 S. 31
200 Linde 2 S. 1ff.
201 Baubeschreibung Rasch 22.07.1940 BAW
202 Der Reichsstatthalter in Hamburg S. 2
203 Rasch 2 S. 35
204 WK 23.06.1938
205 Schr. Marinestandortbauamt 05.09.1938 BAW
206 Das Haus wurde zugunsten des Amtsgerichtsneubaus 1982 abgerissen.
207 Dachneigung 43°, Abwalmung 65°
208 Balkone wurden ohne Rücksicht auf Himmelsrichtungen ausschließlich nach repräsentativen Gesichtspunkten angeordnet.
209 Rasch 2 S. 2
210 ebd. S. 1
211 Meyer war auch Notar für den Bauverein Rüstringen und hatte sich 1934–1935 sein Privathaus von Höger bauen lassen.
212 WK 30.07.1938
213 Bültemann S. 48ff.
214 Forster war ab März 1938 bei der Hochbauabteilung der Stadt Wilhelmshaven

215 Gespr. Kraak 02.10.1988
216 Nr. 58–62
217 WK 31. 12. 1938, SS-Heim Kirchreihe 18 A, 1938 als Sturmbanndienststelle eingeweiht
218 Gespr. Lübbers 28.12.1983, Bauakten z. B. Herbartstr. 88, 1935
219 Hegelstr. 62, Arch. Fimmen
220 WK 03.11.1937
221 Hunzinger ließ sich das Haus von seiner Hochbauabteilung umbauen. Arch. Forster, Gespr. Forster 26.01.1988.
222 Schr. MI 11.11.1941 ABR
223 vgl. 4.2.6.
224 Feder S. 75
225 WZ 05.05.1937, 07.05.1937, 08.05.1937, vgl. 4.3.
226 Agatz, Gespr. Lackner 28.07.1989
227 NOB 23.05.1939
228 WZ 10.07.1971
229 Hallbauer 2 S. 11
230 NOB 23.05.1939
231 Vermerk RAM 12.12.1940 BAW
232 Stadt Wilhelmshaven 2
233 vgl. 3.7.5.
234 Grundig S. 435
235 Nicolaus 4 S. 164
236 GEW
237 Gespr. 22.05.1935, Grundig S. 472
238 Schr. RAM 22.05.1939 an MDF STAO 137/VII-182-9, vgl. Troost Bd. I S. 14
239 vgl. 5.
240 Folgende Hinweise waren zu finden:
03.07.1933 Richtfest SS-Heim
09.07.1933 Einweihung des Heimes SS-Sturm 3/4/24
05.10.1933 Einweihung Feldpolizei-Heim
08.10.1933 Einweihung Heim „Dinklage" des Sturmes 2/5 Genossenschaftsstr.
15.11.1934 Eröffnung Wärmehalle des Vaterländischen Frauenvereins
Mai 1936 Erweiterung Erholungsheim Kriegsmarinewerft Bohlenbergerfeld
02.04.1937 NSDAP-Kreisleitung (Robert-Koch-Haus) eingeweiht, ab 1940 auch Gestapo
Nov. 1937 Baubeginn Spielhalle Erholungsheim Kriegsmarinewerft Bohlenbergerfeld
22.05.1938 SS-Heim Kirchreihe eingeweiht
24.05.1938 Kreisschulungsheim Ruttelerfeld der NSDAP eingeweiht
02.06.1938 Richtfest Strandhalle (KDF-Reisen)
22.10.1939 Richtfest NSV-Heim Siebethsburg, Papingastr.
17.10.1940 Soldatenheim Montsstr. eröffnet
241 Hiller S. 5
242 Benze S. 5
243 Elsaesser S. 162 ff.
244 Erlaß Reichserziehungsminister 27.03.1936
245 Neufert S. 148 ff.
246 Feder S. 181
247 Nicolaus 4 S. 164
248 Schr. RAM 22.10.1938, Archiv Gemeinde Sande
249 Schr. 29.06.1936 BAW
250 Schr. 11.11.1936
251 Vermerk 30.07.1937 über Bespr. bei Staatsminister Pauly 30.07.1937 BAW
252 Runderlaß des Reichs- u. Preuß. Ministers für Ernährung u. Landwirtschaft
253 Vermerk 17.02.1938 BAW. Später wurde auch noch der Standort Hunrichsstr. als Hilfsschule geplant.
254 Schr. MDF 19.03.1938 BAW
255 Baubeschreibung Lübbers 28.11.1938 BAW
256 Schr. RAM 17.09.1938, Archiv Gemeinde Sande
257 Schr. RAM, Unterschrift Büge 18.10.1938, Archiv Gemeinde Sande
258 Schr. Heuer 10.11.1938, Archiv Gemeinde Sande
259 WZ 08.08.1963
260 Schr. Ott 07.11.1938 an MDF, Archiv Gemeinde Sande
261 DH 19.10.1940
262 Genehmigt durch Amtshauptmann des Amtes Friesland 26.06.1938, Archiv Landkreis Friesland
263 vgl. 4.4.
264 vgl. 3.7.5.
265 WK 26.10.1937, heutige Freiherr-vom-Stein-Mittelschule, Paul-Hug-Str.
266 WK 11.01.1939
267 Feder S. 183
268 WK 17.06.1939
269 Schr. MDF 06.04.1939 STAO 137/VII-182-9
270 Max Büge 27.11.1939, 11.04.1940, vgl. Onken S. 60
271 Rasch, der der NSDAP bereits 1928 beigetreten war, hatte zu Fricke (NSDAP-Beitritt 1931) auch nach 1945 beste Beziehungen. Er lud ihn 1957 als Preisrichter zum Wettbewerb Altengroden-Süd ein.
272 WZ 24.06.1958, Barackentyp der Fa. Horgass, Magdeburg
273 WZ 16.09.1959
274 Bültemann S. 20 ff., Petsch 1 S. 125 ff.
275 WK 03.11.1937
276 ebd.
277 Reichsjugendführung der NSDAP
278 Dustmann
279 WK 28.04.1938
280 Erläuterungsbericht Däke 17.08.1938 BAW
281 WK 30.03.1938
282 Petsch 1 Abb. 74
283 Feder S. 194, 197
284 Kochskämper
285 Albrechtstr. neben der Schule A, Neuteicher Str., Kniprodestr. neben der Schule B.
286 Lindner 1 S. 274 ff.
287 Petsch 1 S. 153
288 Nicolaus 4 S. 164
289 Der soziale Wohnungsbau in Deutschland 4/1941 S. 132
290 In seiner Funktion als Leiter des Amtes „Schönheit der Arbeit" der DAF-Abt. KDF
291 Feder S. 110
292 Der Reichsstatthalter in Hamburg S. 1
293 Altengroden, Bebauungsplan Franzius vom 04.10.1940. In den früheren Plänen von Altengroden ist die Kirche noch nicht enthalten. Am 25.08.1939 hatte der ev. Oberkirchenrat in Oldenburg vom RAM einen Bauplatz für Fedderwardengroden u. Voslapp erbeten. Gemeindehaus, Pfarrhaus u. Friedhof sollten entstehen, später auch eine Kirche mit 500 Plätzen. Ein gleicher Wunsch bestand für Langewerth. Der RAM verwies den Oberkirchenrat an die MI. Kirchliches Leben in den Gemeinschaftssiedlungen entwickelte sich in zurückgelassenen Baubaracken u. wechselnden Räumlichkeiten, vgl. NOB 09.04.1942.
294 Gespr. Lübbers 28.12.1983
295 Gespr. beim RAM mit Meyer, Lübbers, Temp, Auf dem Hövel, BAK R 41/1320
296 WZ 07.11.1940
297 vgl. 3.5.4., WZ 26./27.10.1940
298 NOB 25.06.1941
299 Schr. Oberbürgermeister 20.08.1940 an MDF STAO 137/VII-182-9. Die Stadt leistete der NSDAP „Verwaltungshilfe".

300 NOB 10.02.1941. Müller erkannte, daß die NSDAP niemals für alle Stadtteile Gemeinschaftshäuser finanzieren konnte. Er wies das Bauamt an, sich in der Entwurfskritik zurückzuhalten.
301 DH 20.07.1940
302 Akte Gutschow Plan 302/22 16.01.1945
303 Brockhaus S. 274, 273
304 Feder S. 14
305 Feder S. 249
306 Feder S. 250
307 Tessenow
308 Lindner 1 S. 112ff., 192ff.
309 Der Reichsstatthalter in Hamburg S. 4
310 Grundig S. 379
311 Schmitthenner 1 S. 154
312 Erläuterungsbericht 13.05.1938 BAW
313 vgl. 7.10.
314 Schr. Bauverwaltung 02.10.1940 BAW
315 Siedlergemeinschaft III
316 Nur der südliche Trakt mit 5 Läden wurde gebaut.
317 Korrekturhinweise der Baupolizei Wilhelmshaven, März 1941 BAW
318 Geplant mit Schule D, NSDAP-Gemeinschaftshaus, HJ-Heim
319 Schr. MDF 05.12.1940 BAW
320 Der 1. Gewerbetrakt war Mitte 1940 begonnen worden. Unternehmer war Karl Hitzbleck (Duisburg). Der 2. Gewerbetrakt wurde durch Peter Büscher (Münster) am 28.10.1940 begonnen.
321 Vermerk MDF 07.01.1941 BAW
322 Schr. Bauverwaltung 20.12.1940 an MDF BAW
323 Nicolaus 4 S. 164
324 Schr. Linde 29.05.1986 SHB
325 Feder S. 416
326 Stellungnahme 32b 06.12.1940 BAW
327 Friedrich Tamms, Konstanty Gutschow, Rudolf Hillebrecht u. Paul Bonatz waren als Arch. auch für den Autobahnbau eingesetzt, Petsch 1 S. 142, Durth 1 S. 119, Stommer S. 84
328 Möhrle S. 1663ff. Eine Postbauverwaltung wurde bereits Ende des vorigen Jahrhunderts durch Heinrich von Stephan in Berlin gegründet.
329 Lösche S. 1681ff.
330 Schr. Stadt Wilhelmshaven 09.12.1939 an MDF BAW
331 Vermerk über einen Ortstermin 11.09.1940 BAW
332 Baubeschreibung Reichspostdirektion Oldenburg 30.09.1940 BAW
333 Gwildies, Protokolle der Vorstands- und Obermeistersitzungen der Kreishandwerkerschaft Wilhelmshaven. Die Bedeutung des Handwerks in Wilhelmshaven war nicht wie in anderen Städten abhängig von der Wirtschaftskonjunktur im Reich, sondern abhängig von der Marine als einziger Wirtschaftsgrundlage. Nach dem Ersten Weltkrieg mußten viele Handwerksbetriebe um ihre Existenz kämpfen und gaben auf. Ab 1933 gelang es den Nationalsozialisten durch Schuldzuweisungen an die „privatkapitalistisch geleitete Großindustrie", daß Handwerker Hoffnungen in das „III. Reich" setzten. Die Nationalsozialisten organisierten das Handwerk 1934 neu auf der Grundlage von Pflichtinnungen und Führerprinzip. Der „Reichsstand des Deutschen Handwerks" unterstand dem „Reichshandwerksmeister". In der DAF bildete das Handwerk ein eigenes „Fachamt das deutsche Handwerk". Regional wurde am 16.08.1934 der Kreis- und Amtshandwerkerbund Wilhelmshaven-Rüstringen e. V. unter Wilhelm Völker liquidiert, und es entstand die Kreishandwerkerschaft unter Eduard Dobberkau mit 29 Betrieben und 18 Innungen. Handwerksmeister waren über die sich abzeichnende Baukonjunktur erleichtert. Stolz wurden Kreis-Handwerker-Tage abgehalten und erstmals wurden am 02.04.1935 Wilhelmshavener Handwerkslehrlinge nach altem Brauchtum vor der offenen Handwerkslade „freigesprochen". In der Wilhelmshavener Kunsthalle wurde vom 30.07.1939–03.08.1939 die Ausstellung „Deutsche Werkstoffe im Handwerk" gezeigt. Die NSDAP benutzte die Handwerkertradition für ihre Ziele und stellte die Arbeitskraft der „Volksgenossen" in jährl. Arbeitsschlachten heraus.
334 Vermerk 23.07.1938 BAW. Oberbürgermeister Nutzhorn hatte sich ursprünglich für 2½ Geschosse entschieden.
335 NOB 09.09.1935
336 WZ 28.12.1935, Richtfest war am 29.02.1936
337 Schr. 30.10.1935 BAW
338 Schr. Oberbürgermeister 07.12.1938 über MDF an Reichsinnenminister STAO 137/5930.
339 vgl. 4.2.2.
340 Schr. RAM 14.03.1939 STAO 137/5930. Hunzinger war im Febr. 1939 entlassen worden und die Stadt schien als Nachfolger den Bauingenieur Temp zu bevorzugen.
341 Kleihues S. 198, Larsson S. 73, 182
342 Gespr. Pitz 15.02.1989
343 NOB 19.05.1941
344 vgl. 6.1. Bunker sollten städtebaulich integriert und wehrhaft verkleidet werden.
345 Erläuterungsbericht Distel 07.10.1941 STAO 137/5930
346 Leiter des Hochbauwesens war wie schon bei der ersten Beauftragung von Schachner (1941) Klaus Rasch.
347 WK 23.04.1938
348 Theaterkomplex Lehniner Platz, Berlin (1926–1931), Kaufhaus Schocken, Stuttgart und Chemnitz von Mendelsohn.
349 Der Reichsstatthalter in Hamburg S. 2
350 Koop 1 S. 73. Die Beschäftigtenzahlen waren „geheim".
351 STAO 137/4362, 4363. Barackenlager gebaut durch das Marinestandortbauamt: Mozartstr./Baubüros/OT, Lazarettbaracken Marinelazarett Weserstr., Bürobaracke Zedeliusstr., Baracke Marine-Observatorium, Bürobaracke Marineverpflegungsamt, 2 Baracken Funkempfangsstelle Totenweg, Wohn- und Nebenbaracken Grodenbatterie, Büro-Offiziers- und Fluko-Baracken am Fortbunker Schaar, Baracke Fort Rüstersiel, Barackenlager beim Marinegerätelager Roffhausen, 1000-Mann-Lager bei der Marinekraftfahrabteilung Ebkeriege, Marinekartoffellager Kettenstr., Bürobaracken zwischen Kieler-/Mozart-/Viktoriastr., Propagandakompanie-Baracke, Wehrmeldeamt-Baracke und Festungsbaudienststelle Mozart-/Kieler-/Bremer Str. nördlich Marinestationsgebäude, Wasserwerksbaracken Mühlenweg/Holtermannstr., Baracken Luftverteidigungskommando Rosenhügel, Barackenlager Altengroden, Barackenlager Breddewarden, Gemeinschaftslager Sengwarden-Anzetel, Frauenlager Langewerth, Baracken Mennhauser Weg Fedderwarden. Entnommen: Marinestandortbauamt Wilhelmshaven: Standortspläne Wilhelmshaven Bd. I, Stichtag 01.07.1941 ASTHBA.
352 Art. 163 Reichsverfassung
353 Bibliographisches Institut S. 86
354 Brockhaus S. 26
355 Gespr. Magdalene Veeser 25.11.1987, Gespr. Murken 16.03.1984
356 Ministerpräsident Joel 14.11.1939 in Fedderwardergroden
357 Verordnung zur Sicherung des Kräftebedarfs für Arbeiten an bestimmten Orten 22.06.1938
358 Ab 1933 Rüstersiel für Männer und ehemalige Jugendherberge Kanalstr. für Frauen, WZ 30.12.1933, ab 1943 Lager Sengwarden-Anzetel, ab 1942 Lager Sudetenstr.
359 Reichsarbeitsdienstgesetz 26.06.1935
360 WK 21.06.1937

361 NOB 30.07.1937
362 DH 01.06.1940
363 NOB 10.06.1941. Die Einrichtung von Bordellen für ausländische, insbesondere polnische Arbeiter erfolgte reichsweit und gehörte zu den sogenannten „Polen-Erlassen" vom März 1940. Eine gleiche Anordnung der NSDAP-Zentrale vom Dez. 1940 sollte der nur verzögert oder gar nicht erfolgten Einrichtung Druck machen. Im Sept. 1941 verfügte der RAM die Errichtung von Bordellen in Ausländerlagern. Geschlechtsverkehr zwischen Ausländern und deutschen Frauen sollte verhindert werden, vgl. Herbert, U.: Fremdarbeiter, Bonn 1985 S. 127.
364 Wilhelmshaven besaß keinen leistungsfähigen Ortsgüterbahnhof.
365 vgl. 7.1., NOB 10.02.1941, 25.08.1941
366 WK 27.07.1938
367 Gabbey, D.: 440 Jahre Hooksiel, Hooksiel 1986, STAO 137/5967.
368 Vermerk RAM 09.06.1939, 07.11.1939 BAK R 41/1318
369 MAF RM 23/1686
370 Schr. NS HAGO Kreis Friesland an Stadtmagistrat Jever 06.07.1934, Sammlung Hartmut Peters
371 Die Honorierung der Ingenieure war ab 1935 in der Gebührenordnung für Ingenieure geregelt.
372 Gespr. Hilgenstock jun. 10.11.1988
373 Gespr. Lübbers 28.12.1983. Es kam wegen der Arbeitsweise von Lübbers zu heftigen persönlichen Vorwürfen und schließlich am 10.10.1938 zu seiner Entlassung als technischer Geschäftsführer der Jade.
374 WK 18.03.1938
375 WK 08.04.1938
376 1. Vierjahresplan 01.02.1933 (nationaler Wiederaufbau)
2. Vierjahresplan/Parteitag der Ehre 1936 (Rohstoffunabhängigkeit)
3. Vierjahresplan/Verlängerung vom 18.10.1940 (Anpassung an den Krieg)
377 WK 04.03.1938
378 Wulf 1 S. 6
379 WK 07.12.1934
380 Verein der Kunstfreunde für Wilhelmshaven S. 182
381 Schr. Amtshauptmann Ott an Stadt Jever 06.11.1936
382 WK 28.11.1936
383 Meyer 1 S. 15, Aljets S. 13ff.
384 Wulf 2 S. 187. Rosenberg kam mit Darré zur 1. Aufführung der Stedinger am 27.05.1934 nach Berne. Am 19.10.1934 legte er zusammen mit Himmler den Grundstein zur Stedingsehre und weihte die Kultstätte am 13.07.1935 ein.
385 Räcker-Wellnitz S. 15, 73
386 BSW 2/1940 S. 62
387 Teut S. 288
388 WZ 28.11.1934
389 WZ 13.05.1935
390 WZ 20.12.1935
391 DH 14.09.1940, 25.01.1941, 01.02.1941
392 Teut S. 289
393 Petsch 2 S. 34
394 Vermerk Bauverwaltung 19.09.1941 BAW
395 DH 19.10.1940
396 DH 29.06.1940
397 Vermerk Bauverwaltung 19.09.1941 BAW. Scheuernstuhl produzierte die Plastiken am Maschsee Hannover.
398 WZ 07.11.1940
399 Schmitthenner 1 S. 74
400 Scholl S. 93
401 WZ 29.12.1939, Beispiele aus dem Jahre 1939
402 Gamm S. 156
403 WZ 15.09.1933
404 WZ 13.05.1935
405 Schr. Oberbürgermeister 20.12.1937 an Reichsstatthalter STAO 137/VII-182-9
406 Ronneberger (1886–1968) förderte auch den Heldenfriedhof, das Ehrenmal Laboe, das Marine- und Kolonialmuseum (1935) und das Grab des unbekannten Seemannes in der Garnisonkirche (1957).
407 Maas 1
408 Petsch 2 S. 87
409 Speer 3 S. 40
410 Merker S. 224
411 WZ 20.12.1933
412 WK 28.04.1938
413 WZ 02.05.1934
414 Frank S. 107
415 Von Gustav u. Willi Lübbers sind durchaus moderne Bauten bekanntgeworden (AOK, Kuhlmann, Möller), Forster hatte noch 1930 erwogen, zu Ernst May in die UDSSR zu gehen. Der Bauhaus-Schüler Fricke konvertierte erst 1933 zum Heimatschutzstil.
416 Gespr. Aribert Schemm 01.03.1939, AJA
417 Versch. Schr. AJA
418 Schr. DSB 10.10.1938 AJA
419 Schr. Eisenbarth 26.10.1938 AJA
420 Schr. 06.10.1938 AJA
421 Kamphausen S. 14ff.
422 Höger
423 Bauverein Rüstringen 4, 1937
424 Weitere Verbindungen bestanden über die Loge Schlaraffia.
425 Kamphausen S. 38
426 Frank S. 65
427 Petsch 1 S. 72
428 Kamphausen S. 43
429 Vermerk Köhne 16.10.1936 BAW
430 NOB 01.11.1937
431 Kamphausen S. 39
432 Appelius 1 S. 56
433 vgl. 7.5.
434 Kamphausen S. 45
435 Schr. Hochbauamt 06.03.1934 BAW, Gespr. Siedenburg 13.12.1989
436 Gespr. Prof. Dr. Heinfried Tiburtius 13.07.1989, 22.07.1989. Interview NDR (Ingrid Hohenschurz) 11.10.1989 Berlin
437 WZ 05.05.1937
438 Gespr. Tiburtius a.a.O.
439 Auch Beck stand der NSDAP kritisch gegenüber, Gespr. Rosa Manthey, Tochter von Beck, 12.07.1989.
440 Gespr. Ilse Gerhold, Tochter von Frede 12.07.1989
441 Göderitz war 1921–1933 bei der Stadt Magdeburg; ab 1827 dort Stadtbaurat, 1945–1953 Stadtbaurat von Braunschweig.
442 Gespr. Yvonne Klaucke 21.08.1989
443 Karl Tillessen förderte Klaucke trotz KPD-Mitgliedschaft, Gespr. Tillessen 10.07.1989
444 Die Wilhelmshavener Logen wurden 1935 verboten.
445 Alle Angaben Niedersächsisches Staatsarchiv Aurich Rep. 21, 5265, Baur 2.
446 Bartelt war von 1906–1933 Bürgermeister bzw. Oberbürgermeister von Wilhelmshaven. Er stand der DNVP nahe und wurde 1933 auf Betreiben der NSDAP kurzfristig inhaftiert, vgl. Appelius S. 163
447 Herausgegeben von dem radikalen Bauhausgegner Konrad Nonn, vgl. Miller Lane S. 131
448 Gespr. Fricke 01.12.1988, BDA Mitglieder in Wilhelmshaven waren u. a.: Richard Freygang, Ernst Peters, Franz Iwersen.

Die Wilhelmshavener/Rüstringer Mitglieder gehörten zu Oldenburg.

449 RDBK-Mitglieder im Raum Wilhelmshaven waren nach dem Verzeichnis vom 01.08.1935: Gustav Lübbers, Willi Lübbers, Karl Tuchscherer, Ernst Peters, Franz Iwersen, Robert Fuchs, Ludwig Lehmann, Eilert Kleen, Alfred Tönig u. Hans Ahmels (Jever), vgl. Reichskammer der bildenden Künste: Deutschlands Architekten, Berlin 1935.

450 Verordnung über baupolizeiliche Behandlung von öffentlichen Bauten 20.11.1938 RGBl. I S. 1677.

451 Beide Gesetze gültig bis 1973

452 Gespr. Forster 26.01.1988

453 Schr. MDF 13.09.1940 BAW

454 vgl. 7.7.5.

455 Schr. Bauverwaltung 11.04.1941 an die Jade BAW

456 vgl. 3.5.6.

457 Schr. Bauverwaltung 14.07.1939 BAW

458 Schr. MDF 19.07.1939 BAW

459 Schr. Stadt Wilhelmshaven 28.03.1939 BAW

460 Bauantragszeichnungen GEW 28.06.1937 mit Korrekturhinweisen Maas und Christians AGEW

461 Schr. 07.04.1941 BAW

8.

1 Toller S. 143

9.3. Abbildungsnachweis

Schutzumschlag

oben: Entwurf Raumordnungsplan Dezember 1938, BAW
unten: Vor Wilhelmshavens Stadtmodellen 01.04. 1939, v. l. Halbauer, Franzius, Hitler, Müller, Himmler, SKF

1 Vf.
2 Landesbildstelle Berlin II 9456
3 Schaap 2 S. 122
4 Landesbildstelle Berlin 02 Reg. Ministerien 164 162
5 Larsson, Abb. 60 S. 153
6 Marinewerft Wilhelmshaven Anlage 12
7 STDAW
8 Grundig Anlage
9 SIS
10 ASTHBA
11 Nachzeichnung Hochbauamt
12–13 BAW
14 Rieger, K.: „nu mal man schön, mien Engel", Mönchengladbach 1981, S. 22
15 BAW
16 STDAW
17–21 BAW
22 Sammlung Hans Gabriels
23 Ansichtspostkarte ca. 1940
24 STDAW 5040
25 BAW, Ausschnitt „Großraum Wilhelmshaven, Planung 1939" nach dem Plan „Wilhelmshaven 1950" vom 15./16.6.1939. Nachzeichnung Stadtplanungsamt
26 BAW, Bebauungsplan aus einem Bauantrag von 1935
27 Goecke
28 BAW
29 STDAW
30 AJA
31 BAW
32–35 AJA
36 WZ 10.01.1940
37 BAW
38 SIS
39–42 Archiv Bürgerverein Fedderwardergroden
43 BAW
44 Ansichtspostkarte
45 SIS, Foto ca. 1945
46 BAW
47 STDAW
48–50 SIS
51 ADB
52 BAW, Ausschnitt aus dem Plan „Wilhelmshaven, Städtebauliche Neuplanung", 29.03.1939
53 AJA, vom Vf. montiert aus den Bestands- und Typenplänen, ergänzt um die Gemeinschaftsbauten
54 Sammlung Aribert Schemm, Foto ca. 1940
55 Ansichtspostkarte
56 BAW
57 BAW, Kartengrundlage ca. 1940. Bauten des Bauvereins Rüstringen 1935–1945 wurden vom Vf. schwarz markiert unter Verwendung des Lageplans WK 25.06.1938 sowie verschiedener Lagepläne aus den Geschäftsberichten der Jade.
58 Kunstbibliothek Berlin
59–61 ABR
62–63 SRB
64–65 Dia-Sammlung Stadtplanungsamt
66 BAW, Ausschnitt aus Stadtplan Wilhelmshaven vom Juni 1940
67 BSW 31.03.1940
68 Schneider 4 S. 21
69 SIS, Foto 1988
70 Ansichtspostkarte
71–72 ASTHBA
73–74 AOL
75 ASTHBA
76 SIS, Foto 1988
77 Verwaltung Nordwest-Krankenhaus Sanderbusch
78–80 SHT
81 STDAW 5040
82–88 SHT
89 ASTHBA
90 SHT
91 ASTHBA
92 SEM
93–94 AGS
95–97 SYK
98 STAO 137/4362 (zu 5/1), Arbeiterlager westlich Altengroden
99 DH 23.11.1940
100 Has
101 SEM
102 STDAW 8082
103 SEM
104 Schneider 4 S. 19
105 Sammlung Elisabeth Rasch
106 STDAW
107 SHB
108 SKF
109 STDAW
110 ARK
111 Sammlung Gustav Bork
112 Has
113 SIS, Foto 1988
114 SYK
115 SIS, Foto 1988
116 STADAW 5610
117 STAO 137/4424b (H. 23 [zu 19]) VII-182-9
118 Akte Gutschow BAW W 302/5 26.10.1944
119 SKF
120 BAW
121 NWZ 15.02.1949
122 Brüning S. 52
123 BAW, Wettbewerbsbeitrag 240548
124 AJA
125 Sammlung Wilhelm Rehbein
126 Hallbauer 8 Anlage 2, Skizze A
127 Hallbauer 8 Anlage 3, Skizze B
128 STDAW, Original Maßstab 1:25.000
129 STDAW, Original Maßstab 1:25.000
130 Schneider 4 S. 105
131 WZ Bilddienst
132 BAW
133–134 SHT
135 BAW
136 Iwersen + Kucan (Hg.): Bauplanung seit 50 Jahren, Wilhelmshaven 1982
137 AGEW

138 STDAW
139 AJA
140 WK 11.01.1939
141 SRB
142 BAW
143 Archiv Landkreis Friesland
144–145 BAW
146 SRB
147–150 BAW
151 STAO 137/5930 (zu 50), 01.10.1941
152 SHT
153 Marßolek S. 167
154 Sammlung Magdalene Veeser, 09.06.1941
155 Welge S. 60
156 Schneider 4 S. 45
157 ARK
158 SYK
159 Koop 2 S. 39, Künstler unbekannt
160 SIS, Foto 1988
161 Foto Christa Kruse
162 SIS, Foto 1988

9.4. Archivmaterialien

Niedersächsisches Staatsarchiv Oldenburg (STAO)
131/305
136/2801–2802
137/4328, 4361–4370, 4424, 5025, 5044, 5870, 5871, 5890–5892, 5902, 5908, 5914, 5916, 5926, 5930, 5936, 5946, 5951, 5961, 5967, 5972, 5976, 5978, 5983, 6003, 6035, 6201, VII-182-9
262/6–22

Niedersächsisches Staatsarchiv Aurich
Rep. 21, 5265, 5353

Bundesarchiv Koblenz (BAK)
R 3
R 41/89–96, 1316–1331
R 43 I u. II/378 (II/1025), 1158 (I/608), 1429 (I/894), 1737 (I/1952), 1642 (I/2163), (I/955)
R 113/344, 646–648, 854, 1534
R 120

Militärarchiv Freiburg (MAF)
RM 1
RM 3
RM 7
RM 9
RM 20
RM 23
RM 45

Stadtarchiv Wilhelmshaven (STDAW)
Plansammlung, verschiedene Bestände

Fritz-Küster-Archiv der Universität Oldenburg

Bauakten der Stadt Wilhelnmshaven (BAW)
Hochbauamt:
Akten städtischer Gebäude
Stadtplanungsamt:
5 Tätigkeitsberichte bis 1966
17 Arbeitskreis Stadt- u. Landesplanung Gau Weser-Ems
19 Landesplanungsgemeinschaft Oldenburg-Bremen
21 Landesplanungsgemeinschaft Oldenburg-Bremen
23 Schriftwechsel mit Ministerien
34 Akte Gutschow
52 Nachkriegsindustrie
108 Postgebäude Fedderwardergroden
214 Akte Gutschow
269 Bahnhofsplanung
Bauordnungsamt:
Verschiedene Gebäudeakten, nach Straßen u. Nr. geordnet
Sachgebiet Vermessung des Stadtplanungsamtes:
Stadtpläne, Luftbilder, Planunterlagen
Abteilung Stadtentwässerung des Tiefbauamtes:
Akten von Stadtentwässerungsmaßnahmen

Darüberhinaus wurden benutzt: Adreßbücher, Telefonverzeichnisse, Geschäftsberichte der Wohnungsbaugesellschaften, Tätigkeitsberichte der Ämter, Organisationspläne der Behörden, Sitzungsprotokolle der politischen Gremien sowie Archive, Sammlungen, Fotoarchive, Zeitungen, Fachzeitschriften wie unter Abkürzungen aufgeführt.

9.5. Kurzbiographien

Im folgenden sind Kurzbiographien von Personen zusammengestellt, die mit Architektur und Stadtplanung des Raumes Wilhelmshaven im „III. Reich" zu tun hatten. Schwerpunktmäßig geht es um: Architekten, Stadtplaner, Ingenieure, Juristen, Künstler und Politiker. Bei Studium an einer Technischen Hochschule wurde nur der Ort genannt, Fachhochschulen wurden als solche bezeichnet.

AGATZ, ARNOLD
(*23.08.1891 Hannover, †27.03.1980 Bremen)

1911–1918	Bauingenieurstudium Hannover
1919	Promotion
1919–1922	Bauindustrie Berlin, Bremerhaven
1922–1931	Stadt Bremen, zuletzt Hafenbaudirektor, u. a. Nordschleuse Bremerhaven (1928–1933), Ausbau Columbuskaje Bremerhaven
ab 1931	Professor für Grundbau, Wasser- u. Hafenbau, TH Berlin
1935–1945	Ingenieurbüro Agatz und Bock Berlin/Köln, meistbeschäftigtes Ingenieurbüro des OKM u. a. Nordwerft und IV. Einfahrt Wilhelmshaven (ab 1935), Häfen Bangkok, Nürnberg, Salzgitter
1947–1954	Präsident Hafenbauverwaltung Bremen
1954–1964	Berater Bremer Senat für Häfen

AHMELS, HANS

bis 1920	Architekturstudium Karlsruhe, Stuttgart
ab 1920	Freier Architekt Jever
1938–1939	Zusammenarbeit mit Adalbert Jahn, u. a. Privatbauten in Wittmund, Esens, Jever, Schule Cleverns (1938), Offiziershäuser Mooshütter Weg, Hohnholzstr., Anton-Günther-Str., Sophienstr. (1938), Unteroffiziershäuser Schützenhofstr. 58–62 (1937), Zusammenarbeit mit Wohnungsbaugesellsch. Friesland
ab 1940	Konflikte mit NS-System, im Kriege umgekommen (vermutlich Warschau)

AHNER, HERMANN
(*09.07.1895 Wilhelmshaven, †22.01.1982 Hameln)

1919–1931	Journalist Wilhelmshavener Zeitung, NSDAP-Eintritt (1930)
1931–1945	Journalist Wilhelmshavener Kurier, Leiter NSDAP-Kreisamt für Kultur, NSDAP-Kreishauptstellenleiter
ab 1933	Ortsgruppenleiter KDK Wilhelmshaven, DAF-Betriebsobmann
ab 1934	Kreisobmann NS-Kulturgemeinde Wilhelmshaven, Vorsitzender Ring Volkstum und Heimat, SS-Eintritt
1938–1945	Vorstandsmitglied Verein der Kunstfreunde für Wilhelmshaven e. V., KDF-Funktionär
ab 1945	Büchereileiter Olympia AG, Heimatschriftsteller

AICHINGER, WILHELM
(*09.06.1907 Nürnberg, lebt in Bremerhaven)

bis 1931	Bauingenieur- und Städtebaustudium München
1931–1935	Baubeamtenausbildung Nürnberg
1935–1939	Reichsbahndirektion Nürnberg, Dezernent für Brückenbau, Oberste Bauleitung der Reichsautobahn Nürnberg
ab 02.05.1939	Bauverwaltung Stadt Wilhelmshaven, Planungsabteilung Straßenbauamt, danach Leiter Kanalbauamt unter Lehn, u. a. Maadebegradigung (1941), Erschließungsmaßnahmen Wilhelmshavener Gemeinschaftssiedlungen
1943–1945	Kriegsdienst
1945–1960	Leiter Tiefbauamt Stadt Wilhelmshaven, zuständig für Enttrümmerungsarbeiten, stellvertretender Baudezernent, Preisrichter Ideenwettbewerb 1948
1960–1972	Stadtbaurat Bremerhaven

ARENT, BENNO VON
(*19.06.1898 Görlitz)

ab 1916	Kostümzeichner
ab 1923	Bühnenbildner-Volontär Berlin, meistbeschäftigter Bühnenbildner im Reich, erste architektonische Arbeiten
1931	Gründung Bund nationalsozialistischer Bühnen- und Filmkünstler
ab 1933	Staatliche Aufträge an Theatern
ab 1935	Persönliche Aufträge Hitlers für Festaufführungen, Feierlichkeiten, Theaterbauten usw., u. a. Festgestaltung Berlin zum Mussolini-Besuch (1937), Planung Theater im Kulturforum Wilhelmshaven zusammen mit Werry Roth
ab 1942	Reichsbeauftragter für die Mode

BARTNING, HANS
(*17.03.1908 Karlsruhe, †27.08.1943 Orel)

	Neffe von Otto Bartning
bis 1936	Architekturstudium Karlsruhe (Schweizer), zwischendurch Stadtplanungsamt Stockholm, Assistent TH Karlsruhe
1936–1938	Baubeamtenausbildung Karlsruhe
ab 09.11.1938	Leiter Stadterweiterungsamt Wilhelmshaven unter Hallbauer, u. a. Bebauungspläne für Gemeinschaftssiedlungen in Wilhelmshaven, Breddewarden (1939), Nördliche Stadtparkkolonie (1939)
ab Juni 1940	Leiter Stadtplanungsamtes Lódź (Litzmannstadt), Zuletzt Oberbaurat, u. a. Bebauungsplan Stockhof
1942–1943	Kriegsdienst, gefallen

BAUMANN, HEINZ
(*10.08.1909 Brünn, †14.01.1969 Kiel)

1927–1937	Architekturstudium Brünn
1933–1938	Assistent TH Brünn, u. a. Theater Belgrad (1. Preis Wettbewerb)
1938–1956	Stadterweiterungsamt (bis 1939), danach Neubauamt Wilhelmshaven, NSDAP-Eintritt (1940), u. a. Planungen für Kulturforum und Sportforum, Schulen und andere öffentliche Bauten im Rahmen der Stadterweiterung, Bunkerbau, Gewerbestrukturplanungen (1940)
1943–1945	Kriegsdienst
1946	Internierung Luxemburg
1947–1956	Baurat Hochbauamt Stadt Wilhelmshaven, u. a. Schulbauten, Kulturbauten, Rathauserweiterungsplanungen, Universitätsplanung (1947), Preisrichter Ideenwettbewerb 1948
1956–1969	Schleswig-Holsteinisches Sozialministerium Kiel, zuletzt Oberregierungsbaurat

BAUR, HELMUTH
(*26.01.1911, lebt in Friedeburg)

1934–1938	Architekturstudium Karlsruhe (Schweizer)
1938	Büro Looser Freiburg
ab 24.07.1939	Stadterweiterungsamt Wilhelmshaven unter Hallbauer und Bartning
1939–1941	Leiter Stadterweiterungsamt Wilhelmshaven, u. a. Bebauungsentwürfe für Gemeinschaftssiedlungen, Nördliche Stadtparkkolonie (1939), Breddewarden (1940)
1941–1945	Kriegsdienst
1945–1946	Leiter Stadtplanungsamt Wilhelmshaven
1946–1947	Büro Peters Wilhelmshaven
1947–1973	Abteilungsleiter, ab 1959 Amtsleiter Stadtplanungsamt Wilhelmshaven, zuletzt Baudirektor, u. a. Wirtschaftspläne, Flächennutzungspläne, Bebauungspläne, Stadtgestaltung, Verkehrsplanung, Wiederaufbauplanungen, Preisrichter Ideenwettbewerb 1948, Vorprüfer Wettbewerb Altengroden-Süd (1957)
ab 1973	Planungsgeschichtliche sowie bauhistorische Arbeiten

BECK, ERNST
(*07.09.1877 Baunach, †08.08.1947 Oldenburg)

1895–1899	Studium Wasser-, Kultur- und Straßenbau
1899–1903	Baubeamtenausbildung
1903–1906	Straßenbauämter Deggendorf, Würzburg, Ansbach
1906–1910	Marinehafenbaumeister Wilhelmshaven, u. a. III. Einfahrt, Hafenerweiterung
1910–1913	Marinehafenbaumeister Helgoland, Zusammenarbeit mit Linde
1913–1914	Marinehafenbaumeister Cuxhaven
1914–1915	Leiter Marinebauamt Namur
1915–1919	Leiter Marinebauamt Ahlhorn/Oldenburg, u. a. Marineluftschiffhafen Ahlhorn
1919–1920	Leiter Reichsvermögensamt IV Wilhelmshaven
1920–1924	Landesfinanzamt München
1924–1935	Wartestandsbeamter
1936–1939	Marineoberbaurat HNBD Wilhelmshaven unter Tiburtius
1939–1943	Leiter HNBD Wilhelmshaven, Gesamtleitung IV. Einfahrt und Nordwerft, Hafenbaudirektor
1943–1944	Stellvertretender Leiter Marineoberbauamt Wilhelmshaven unter Triebel
1946	Pensionierung

BELLWINKEL, LUDWIG

	Architekturstudium, Lehrer Polierschule Varel
bis 1937	Marinestandortbauamt Bremerhaven, Wohnungen Gelände Tecklenborg-Werft (1934–1935)
1938–1940	Hochbauabteilung OKM unter Kelm, u. a. Planungen für Wilhelmshaven
1941–1943	Leiter Marinestandortbauamt Wilhelmshaven, umfangreiche Bautätigkeit für die Marine
1943–1945	Hochbauabteilung OKM
ab 1945	Kreisbaurat in einer niederrheinischen Stadt
ca. 1960	Verurteilung zu Zuchthaus wegen einer Frauenmordaffäre

BENDA, HANS
(*1877, †1952)

	Jurastudium
1939–1944	Ministerialdirektor, Leiter Marineverwaltungsamt des OKM, ihm unterstand die Unterkunftsabteilung mit deren Leiter Ernst Fitzler und die Hochbauabteilung mit deren Leiter Walter Kelm

BERG, PAUL
(*03.02.1886 Magdeburg)

	Architekturstudium, Baubeamtenausbildung
1935–1948	Leiter Reichsbauamt Wilhelmshaven, NSDAP-Eintritt (1937), u. a. zuständig für Hochbauten von Staatsdienststellen, NSDAP und Finanzverwaltung
1949–1950	Leiter Staatshochbauamt Wilhelmshaven
1952	Nach Koblenz verzogen

BLOCK, HEINZ
(*19.11.1911 Bremerhaven, lebt in Kiel)

1929–1934	Architekturstudium Hannover (Fischer, Kanold)
1935–1937	Baubeamtenausbildung Stade, Bremerhaven, u. a. Technische Marineschule und Schiffsstammabteilung Gelände Tecklenborg-Werft Bremerhaven, Schiffsstammabteilung Brake
1938	Beendigung Baubeamtenausbildung beim OKM Berlin
1938–1945	Referent Hochbauabteilung Marineverwaltungsamt des OKM unter Walter Kelm, u. a. Durchführung des reichsoffenen Architektenwettbewerbes für das neue OKM (Gelände Invalidenfriedhof) Berlin
1945–1956	Freier Architekt Kiel
1956–1966	Landesbauverwaltung Schleswig-Holstein Kiel
1966–1976	Leitender Ministerialrat Landesrechnungshof Kiel

BLUM, OTTO
(*01.09.1876 Neunkirchen, †26.10.1944 Hannover)

	Bauingenieurstudium
1903	Promotion
1904–1906	Assistent TH Berlin
ab 1906	Professor für Verkehrspolitik, Eisenbahnwesen und Städtebau TH Hannover
1924–1928	Verwaltungsrat der Deutschen Reichsbahngesellschaft, u. a. Gutachter für Eisenbahnfragen Wilhelmshaven (1941–1944)

BORK, GUSTAV
(*09.01.1905 Berlin, lebt in Hannover)

bis 1923	Ausbildung bei Bruno Paul Berlin
1923–1926	Büro Straumer Berlin
1926–1930	Büro Breuhaus Düsseldorf, u. a. Innenraumgestaltung Schnelldampfer „Bremen"
1930–1943	Freier Architekt Hannover, seit 1937 Zusammenarbeit mit Fritz Hilgenstock, u. a. Gewerbe- und Privatbauten Hannover, Kino Salzgitter (1939), Marktplatz Voslapp (ab 1939), Wohnungsbau Fedderwardergroden (ab 1939), Bebauungsplan Pillau (1940–1941), Siedlung Neukieritsch/Leipzig (1942), Gewerbeplanung Posener Str., Fedderwardergroden (1940)
1943–1946	Freier Architekt Bad Harzburg, u. a. Barackenbau OT

ab 1946	Weiterführung des Büros Hannover, u. a. Messebauten, Conti-Block, Kröpcke

BOYKEN, ERNST
(*14.03.1900 Nordenham, †05.08. 1984 Oldenburg)

	Architekturstudium Darmstadt, Hannover
1925–1930	Büro Fritz Höger Hamburg, u. a. Mitarbeit Krankenhaus Delmenhorst (1926), Rathaus Wilhelmshaven (1928–1929), Kirche Hohenzollernplatz Berlin (1929–1930)
ab 1930	Freier Architekt Oldenburg, u. a. Wohnhaus Esser Nordenham (1930), Wohnungsbau Hansaviertel Wilhelmshaven (1937), Otto-Meentz-Platz Wilhelmshaven (1937), Planung Volksschule Breddewarden (1941), Projekte für Zivil- und U-Bootbunker Wilhelmshaven, Wohnungsbau Weichselstraße/Fedderwardergroden (1941–1944), Ideenwettbewerb 1948, Gemeindezentrum Schortens (1950–1951), Wettbewerb Sparkasse Wilhelmshaven (1955), Weser-Ems-Halle Oldenburg zusammen mit Meckseper (1950–1954/1974), Park-Hotel Oldenburg (1959), Restaurierungsplanung Kirche Heppens (1964)

BOYSEN, FRITZ
(*29.05.1900 Kiel, †14.02.1956 Wilhelmshaven)

	Architekturstudium
1935–1938	Freier Architekt Bremerhaven
1938–1945	Hafenbauressort KMW Wilhelmshaven, Mitarbeit an Hochbauten und techn. Gebäuden unter Klaucke
1945–1947	Dienststelle Land- und Wasserbau Wilhelmshaven, u. a. Universitätsplanung Wilhelmshaven (1947)
ab 1948	Freier Architekt Wilhelmshaven, u. a. Haus Brook, Wasseraufbereitungsanlage Rodenkirchen, Gemeindehaus Bant, Haus Frier, Wettbewerbe

BRAEUER, WALTER
(*22.01.1880 Nieder-Großmannsdorf)

	Jurastudium
1935–1939	Marinestationsintendant Wilhelmshaven, zuständig u. a. für Wohnungsbau und Siedlungsangelegenheiten der Marine
ab 1939	Berlin, während des Krieges reaktiviert und im besetzten Ausland eingesetzt

BRAUNSCHWEIG, ADOLF

	Architekturstudium
1938–1940	Techn. Geschäftsführer Wohnungsbaugesellsch. Jade, zuständig für Abwicklung Fedderwardergroden
ab 1940	Freier Architekt zusammen mit seiner Frau in Wilhemshaven, Zusammenarbeit mit dem Arbeitsstab Sperr und Konstanty Gutschow (1944–1945)

BRINKMANN, KARL
(*1891 Bielefeld, †1954 Sande)

	Architekturstudium, Baugeschäft Essen
1937–1945	Hafenbauressort KMW Wilhelmshaven, Mitarbeiter von Klaucke, Barackenbau, Werfthochbau
1950–1954	Freier Architekt Sande, u. a. Schule Sande (1950), Schule Bockhorn (1951)

BÜRGIN, RUDOLF
(*1909 Karlsruhe, lebt in Düsseldorf)

	Architekturstudium Wien, Karlsruhe (Schweizer, Billing), Assistent TH Karlsruhe
ab 1937	Freier Architekt Oldenburg, u. a. Volksschule Eversten Oldenburg (1938–1948), Siedlung Burg Kniphausen (1937), Wohnungsbau Hansaviertel Wilhelmshaven (1937), Handelslehranstalten Oldenburg (1938), Siedlung Schilldeich Wilhelmshaven (1938), Planung Zentrum Nord Fedderwardergroden (1939), Planungen „Langer Markt" Fedderwardergroden (1941), Gewerbebau Kniprodestraße (1941), Wohnungsbau Fedderwardergroden
bis 1945	Kriegsdienst
	Fachhochschullehrer Ingenieurschule Oldenburg, u. a. Ideenwettbewerb 1948, Wettbewerb Altengroden-Süd (1957)
ab 1953	Büro Düsseldorf, u. a. Wohnungsbauprogramme in Düsseldorf

CARL, FRIEDRICH

	Präsident Landesfinanzamt Weser-Ems Bremen, Oberfinanzpräsident Bremen
1937–1945	Beirat LOB

DÄKE, HEINZ
(*13. 06. 1909 Wanne-Eickel, †08. 03. 1962 Georgsmarienhütte)

	Architekturstudium
	Freier Architekt Oldenburg, Gebietsführung HJ Gau Weser-Ems unter Hogrefe, u. a. Jugendherbergen Carolinensiel, Leer, Wilhelmshaven (1936), HJ-Heim Krähenbusch (1938)
bis 1945	Kriegsdienst
ab 1946	Freier Architekt Georgsmarienhütte

DIECKELMANN, BRUNO

	Kaufmann
1925	Gründungsmitglied NSDAP Oldenburg
ab 1933	Gaubetriebszellenleiter, Gleichschaltung der Gewerkschaften, Gauobmann DAF, zuständig für Reichsheimstättenamt Gau Weser-Ems, Gauwalter der DAF Weser-Ems
ab 1937	Beirat LOB
ab 1939	Leiter des Wohnungsbauwesens Gau Weser-Ems, Mitglied des Reichstages
1943	Von Gauleiter Wegener entmachtet

DILLSCHNEIDER, KARL
(*08. 12. 1904 Berlin, lebt in Bremen)

1924–1928	Architekturstudium Berlin, München, Zwischenpraktika Stadtbauamt Wilhelmshaven unter Zopff, u. a. Südstrandplanungen
1928	Stadtbauamt Velbert
1938–1944	Sachbearbeiter Städtebau LOB, stellvertr. Landesplaner unter Linde und Wortmann, u. a. Wirtschaftspläne Wilhelmshaven, insbesondere Wirtschaftsplan Wilhelmshaven-Friesland Mai 1940
1944–1945	Kriegsdienst
ab 1945	Denkmalpfleger Stadtplanungsamt Bremen, u. a. Sanierung Schnoor, Ortsstatut Bremen

DISTEL, HERMANN
(*05.9.1875 Weinsberg)

	Freier Architekt Hamburg, u. a. Krankenhausbauten, Universtität Hamburg (1911), Standardwerk „Rationeller Krankenhausbau" (1922), Planung Stadtkrankenhaus Wilhelmshaven (1939–1941), Planung Universitätsklinik Berlin für Albert Speer (1941)
1944–1945	Ständiger Berater Arbeitsstab Speer

DÖNITZ, KARL
(*16. 09. 1891 Berlin, †24. 12. 1980 Aumühle)

1910	Eintritt Kaiserliche Marine
1917–1918	Dienst U-Bootflotte
ab 1918	Mehrere Jahre Admiralstabsoffizier Kommando der Marinestation der Nordsee Wilhelmshaven
ab 1935	Chef U-Boot-Flottille, Bauherr von Kasernenanlagen in Wilhelmshaven
ab 1939	Befehlshaber U-Boote
1943–1945	Oberbefehlshaber Kriegsmarine, Nachfolger von Raeder, nahm verschiedentlich Stellung zu Planungen Wilhelmshaven
Mai 1945	Bildung einer amtierenden Reichsregierung in Flensburg
1946	In Nürnberg zu 10 Jahren Haft verurteilt
02. 06. 1957	Gemeinsam mit Raeder Einweihung Ehrenmal Christuskirche Wilhelmshaven

DÖSCHER, HELMUT

	Bauingenieurstudium Berlin, Gutachter des RAM zu bodenmechanischen Fragen
ab 1938	Mitarbeiter der Prüfstelle des RAM, beteiligt an Planungen Salzgitter und Wilhelmshaven
1939	Bewerber um den Posten des Stadtbaudirektors Wilhelmshaven
1941–1943	Eingeschaltet für Gutachten zu Wiederaufbauplanungen, u. a. Wilhelmshaven, Hamburg, Lübeck

DRENCKHAN, ERICH
(*13. 02. 1901 Lüneburg, †03. 09. 1971 Lüneburg)

1920–1925	Bauingenieurstudium Hannover
1925–1929	Baubeamtenausbildung
1929–1933	Kanalbauamt Magdeburg
1933–1943	Hafenbauressort KMW Wilhelmshaven, zuständig auch für Hochbauten unter Frede
1943–1945	Leiter der Marinebaudirektion Oslo, Marineoberbaudirektor
ab 1945	Freies Ingenieurbüro Lüneburg
1961–1964	Oberbürgermeister Lüneburg

ECKERMANN, ALFRED
(*21. 08. 1891 Equard)

	Bauamtmann Marinestandortbauamt Wilhelmshaven, NSDAP-Eintritt (1937)
1939–1945	Bauleiter Marinesiedlung Altengroden für die Gemeinnützige Gesellschaft zur Schaffung von Wohngelegenheiten für Reichsangehörige Berlin
ab 1945	Leiter der Baugruppe Wilhelmshaven der Gemeinnützigen Deutschen Wohnungsbaugesellschaft
1952	Nach Bückeburg verzogen

ECKHARDT, ALFRED
(*30.05.1872 Kassel, †10.02.1960 Bad Wiessee)

	Bauingenieurstudium, Baubeamtenausbildung
1901–1919	Baurat Hafenbauressort Kaiserliche Werft Wilhelmshaven, u. a. Mitarbeit an III. Einfahrt und Hafenerweiterung, Planung IV. Einfahrt (1917)
1919–1922	Leiter Strombauressort Marinewerft Wilhelmshaven
1922–1933	Direktor Strom- und Hafenbauressort Marinewerft Wilhelmshaven
1933–1943	Leiter Hafenbauabteilung OKM
1943–1944	Chef Amt Bauwesen OKM

EISENBARTH, FRIEDRICH

	Studium Gartenarchitektur, Dipl.-Gartenbauinspektor, Bauverwaltung Bremen, entlassen, Freier Gartenarch. Bremen, u. a. Gartenplanung Voslapp (1937), Gartenplanung Cäciliengroden (1938)
1939	Vertragskündigung durch Jade, Konflikte mit NS-System

FEDER, GOTTFRIED
(*27.01.1883 Würzburg, †24.09.1941 Murnau)

bis 1905	Bauingenieurstudium Berlin, Zürich, München
1908–1917	Selbständiger Bauunternehmer, Bau von Flugzeughallen
1917	Gründung des Kampfbundes zur Brechung der Zinsknechtschaft
ab 1924	Reichstagsmitglied, Kampfschriften für die NSDAP, Gründungsmitglied der NSDAP
ab 1928	Verschiedene Besuche in Wilhelmshaven und Weser-Ems
ab 1932	Vorsitz KDAI
1934	Reichssiedlungskommissar, Staatssekretär im Reichswirtschaftsministerium
1935–1941	Professor TH Berlin, Leiter der Reichsarbeitsgemeinschaft für Raumforschung
1939	Herausgabe des Standardwerkes für NS-Stadtplanung „Die neue Stadt", u. a. Planung Stadtteil Langewerth Wilhelmshaven (1939)

FIEDERLING, OTTO
(*30.11.1891 Wiesbaden)

	Freier Architekt Hannover, Professor TH Hannover, u. a. Planung Gemeinschaftsbauten „Langer Markt" Fedderwardergroden (1941), Ideenwettbewerb 1948

FITZLER, ERNST
(*1893, †1971)

	Jurastudium
	Promotion
ab 1939	Ministerialdirigent, Leiter der Unterkunftsabteilung des Marineverwaltungsamtes des OKM unter Hans Benda, Nachfolger von Friedrich Schaller, zuständig für Hochbau der Marine einschl. Wohnungsbau
ab 1955	Ministerialdirigent im Bundesverteidigungsministerium

FORSTER, EGON
(*13.07.1907 Hamburg, †18.09.1989 Hamburg)

bis 1928	Maurerlehre, Baugewerkschule Hamburg
1928–1930	Büro Vogl München

1930–1932	Väterlicher Baubetrieb Hamburg
1932–1933	Strom- und Hafenbau Hamburg
1933	Büro Hallbauer Hamburg
1933–1936	Studium Hochschule für Baukunst Weimar (Schultze-Naumburg, Norkauer)
1936	Büro Zerbe/Harder Hamburg
1937–1938	Büro zu Putlitz/Schoch Hamburg, u. a. Mitarbeit Akademie für Deutsche Jugendführung Braunschweig
1937	NSDAP-Eintritt
ab 01.03.1938	Hochbauabteilung Stadt Wilhelmshaven, Vertreter des Gaubeauftragten für die Regelung der Bauwirtschaft, u. a. Mitarbeit Strandhalle (1938), Seewasseraquarium (1938), Kulturforum, Verwaltungsforum, Schule am Krähenbusch (1938), Luftschutzbauten, Barackenlager
1943–1945	Kriegsdienst, Kriegsbauvorhaben, Waffen-SS
1946–1948	Baulenkungsamt Wilhelmshaven
1948–1970	Leiter Hochbauamt/Hochbauabteilung Wilhelmshaven, umfangreiche kommunale Bautätigkeit, u. a. Verwaltungsgebäude Wohnungsbaugesellschaft Jade (1953), Ideenwettbewerb 1948, Schulen, Kulturbauten, Sportanlagen

FRANZIUS, KARL
(*25.12.1905, lebt in Bremerhaven)

bis 1930	Architekturstudium Wien, Danzig, Stuttgart (Bonatz), Hannover (Hölscher)
1930–1933	Baubeamtenausbildung Danzig, Hannover
1933–1935	Assistent TH Hannover
1935	Stadterweiterungsamt Hannover
1935–1939	Marinestandortbauamt Wilhelmshaven unter Lükkemeyer, u. a. Marinestationsgebäude (1937), Dekoration Ehrenfriedhof (1937), Wettbewerb OKM, Marineinstandsetzungsamt (1937), Schiffsmaschineninspektion Marineforum (1937), Marine-Observatorium (1937), Wohnhäuser Villenviertel (1938)
1939–1943	Marineintendantur Wilhelmshaven Abteilung XXIV. (Siedlungsabteilung), u. a. Überwachung der Siedlungstätigkeit für die Marinedienststellen, Gemeinschaftsbauten, Mitarbeit an Gewerbeplanung Gemeinschaftssiedlungen, Freizeithalle Fedderwardergroden (1940)
1943–1944	Leiter Marinebauamt Bremerhaven
1944–1945	OT Berlin, Oberitalien, u. a. Wiederaufbauplanung Wilhelmshaven für Arbeitsstab Speer (1945)
1945–1986	Freier Arch. Bremerhaven, u. a. Kirchen, Wohnungsbau, öffentl. Bauten, Ideenwettbewerb 1948

FREDE, GEORG
(*23.02.1982 Braunschweig, †24.04.1970 Wilhelmshaven)

1901–1905	Bauingenieurstudium Braunschweig
1905–1906	Wehrdienst
1906–1910	Baubeamtenausbildung Wilhelmshaven
1911–1918	Hafenbaumeister Kaiserliche Werft Wilhelmshaven, u. a. Hafenerweiterung, Hallen Fliegerdeich, III. Einfahrt
1919–1922	Leiter Reichsbauamt I Wilhelmshaven
1925–1930	Referent für Land- und Wasserbau Marineleitung Berlin
1930–1936	Betriebsdirektor Hafenbauressort Werft Wilhelmshaven
1936–1939	Leiter Strombauressort KMW Wilhelmshaven, u. a. Strombauarbeiten im Jadegebiet und auf den Ostfriesischen Inseln
1939–1943	Leiter Hafenbauressort KMW Wilhelmshaven, umfangreiche Werftbauten für die Kriegsmarine in Wilhelmshaven, zuständig auch für Hochbauten der KMW
1943–1944	Leiter Marineoberbaudirektion Wilhelmshaven, zuständig für Hafen- und Hochbau an der gesamten deutschen Nordseeküste in Holland und Dänemark
1945–1949	Leiter Dienststelle Land- und Wasserbau Wilhelmshaven

FRESENIUS, WOLFGANG
(*20.06.1910 Michelbach, lebt in Wilhelmshaven)

bis 1936	Studium Gartenarchitektur Landwirtschaftl. und Techn. Hochschule Berlin (Wiepking-Jürgensmann)
1936–1939	Freier Gartenarchitekt
ab Febr. 1939	Leiter Garten- und Friedhofsamt Stadt Wilhelmshaven
1939–1945	Kriegsdienst
1945–1952	Freier Gartenarchitekt u. a. Hannover
1952–1974	Leiter Gartenamt Wilhelmshaven, verantwortlich für öffentliches Grün, städtische Gartenanlage, Friedhöfe, Spielplätze
ab 1974	Pensionierung, Arbeit für Landschafts-, Natur- und Vogelschutz

FREYGANG, RICHARD
(*07.07.1882 Halle, †09.12.1973 Wilhelmshaven)

bis 1904	Königlich-preußische Baugewerkschule Magdeburg
1907	Baumeisterprüfung Plauen
1907–1908	TH Darmstadt, ohne Diplom
1908–1912	Büro Schultze-Naumburg
ab 1912	Übernahme des Architekturbüros Beißel und Jung Wilhelmshaven, Freundschaft mit Fritz Höger, u. a. Synagoge (1915), Leiteweg (1926), Projekt Leo (1929), Konsumbauten (1931), Parkstraße 12–22, Bauten GEW, Wohnhaus Oldebrüggestraße (1933), Wohnungsbau Hansaviertel (1937)
bis 1933	DNVP-Mitglied
1939–1945	Amtsbaumeister Tepel
ab 1945	Freier Architekt Wilhelmshaven, Ideenwettbewerb 1948

FRICKE, HANS
(*11.09.1906 Oldenburg, lebt in Oldenburg)

1922–1925	Bauhausstudium Weimar, Dessau, (Gropius, Itten, Klee, Moholy-Nagy, Kandinski, Werkmeister, Breuer)
1925–1927	Studium Ingenieurakademie Oldenburg
1928–1929	Stadtbauamt Bremen
1929–1930	Niedersächsische Heimstätte Bremen
1930–1932	Büro Rudolf Jacobs Bremen
ab 1933	BDA Vorstandstätigkeit
1933–1976	Freier Architekt Oldenburg
1935–1945	Landesleiter Weser-Ems der RDBK, Kreisamtsleiter des Amtes für Technik, Gaustellenleiter Kultur, Bauausschußmitglied Stadt Oldenburg, Vorsitzender des Oldenburger Kunstvereins, Mitglied Museumsvorstand, u. a. Einfamilienhäuser Schleusenstraße Oldenburg, Planung Luftflottenkom-

mando 2 Braunschweig (1936), Entwürfe Gauwaltung der DAF Oldenburg zusammen mit Gildemeister (1938), Wohnungsbau Fedderwardergroden Soldauer Weg (1939), Planung Schule Altengroden (1940), Einfamilienhäuser und öffentliche Bauten

1942–1943 Kriegsdienst
1943–1945 Stabsfrontführer OT
1953–1976 Mit Unterbrechung BDA-Vorsitzender Oldenburg-Ostfriesland, Preisrichter Wettbewerb Altengroden-Süd (1957), u. a. Bürogebäude Ludwig Freytag Oldenburg (1954/1955), Haus Koch Oldenburg (1951), Wiederaufbauten Wilhelmshaven, Schöpfwerke Wesermarsch

FUCHS, ROBERT
(*22.05.1885 Mühlhausen, †25.11.1963 Wilhelmshaven)

Architekturstudium
Freier Architekt Wilhelmshaven/Rüstringen, u. a. Kaufhaus EPEGE Gökerstraße (1930), 10 Wohnhäuser der Jadestädtischen Wohnungsbaugesellschaft Peterstraße „Kaninchenhausen" (1933), Wohnbauten Hansaviertel/Werftstraße (1937)

GATTNAR, D. J.

Architekturstudium, Architekt Oldenburg
ab 1937 Kreisplaner Amt Friesland, u. a. Bebauungspläne Sande und Gödens

GEISENHAINER, KURT
(*10.02.1907 Hameln, lebt in Hannover)

1926–1931 Architekturstudium Stuttgart (Schmitthenner, Wetzel, Bonatz), Hannover (Kanold)
1931–1935 Baubeamtenausbildung
1935–1936 Preußischer Staatsdienst
1936–1939 Marinestandortbauamt Wilhelmshaven, u. a. Mitarbeit Marinestationsgebäude (1937), Marinelazarett Sanderbusch (1937), Marinegerätelager Roffhausen (1937), Offizierswohnungen Villenviertel (1938)
ab 1939 Hochbauabteilung OKM, Marinesportschule Lanke/Bernau, Bauten in Straßburg, Linz, Hannover, Privathaus Raeder Berlin
1943–1945 Leiter Marineoberbauamt Berlin
1945–1956 Volkseigenes Entwurfsbüro Halle, u. a. Volkspolizeikasernen Marienborn
1956 Übersiedlung in die Bundesrepublik, angestellter Architekt, Baurat
1960–1963 Leiter Staatshochbauamt Celle
1963–1972 Zuletzt Oberfinanzdirektion Hannover, Landesbauabteilung, Baudirektor

GERDES, HANS
(*08.03.1898)

Bauingenieurstudium
Promotion
Baubeamtenausbildung
Abteilungsleiter HNBD Wilhelmshaven unter Tiburtius und Beck, u. a. Instandsetzung III. Einfahrt Wilhelmshaven (1934–1937), Familienlager Neuengroden zusammen mit Generalunternehmer Möller (1939)
ab 1943 Leiter Marinebaudirektion Paris

GILDEMEISTER, HERMANN

Freier Architekt Bremen, u. a. Planung Gemeinschaftsbauten „Langer Markt" Fedderwardergroden (1941), Haus des Reiches Bremen (1928–1930)
Mitglied Ahlhorner Architektenkreis, Leiter Fachgruppe Bauwesen des NSBDT Gau Weser-Ems

GILGES, FRANZ
(*1907)

Jurastudium
Intendanturrat, Leiter der Siedlungs- und Wohnungsbauabteilung der Marineintendantur Wilhelmshaven
ab 1949 Bundeswirtschaftsministerium, zuletzt Ministerialdirigent

GRAU, RUDOLF
(*15.10.1882 Gotha, †27.04.1964 Wilhelmshaven)

Kam aus Hamburg
Freier Bildhauer Wilhelmshaven, u. a. „Harmonikaspieler" Tom-Brok-Straße (1935), Plastischer Schmuck und Reliefs Marinelazarett Sanderbusch (1937), Kunst am Bau Bonte-Kasernen (1937), „Mutter und Kind" Paul-Hug-Straße (1938), Kunst am Bau Schule Altengroden (1961)
Modellbauer Stadterweiterungsamt Wilhelmshaven

GRAUBNER, GERHARD
(*29.01.1899 Dorpat, †24.07.1970 Hannover)

bis 1923 Architekturstudium Stuttgart
1923–1932 Assistent TH Stuttgart (Bonatz)
1932–1939 Freier Architekt
1939–1942 Technischer Leiter Stadtplanungs-GmbH Düsseldorf
1940–1967 Professor TH Hannover
1948–1955 Städtebaulicher Berater des Rates der Stadt Wilhelmshaven, u. a. 1. Preis Ideenwettbewerb 1948, Innenstadtkonzepte Wilhelmshaven (1950), Chirurgietrakt Stadtkrankenhaus (1951)

GRIESBACH, HEINRICH VON
(*11.04.1899 Wolfenbüttel, †1939)

Architekturstudium
Luftwaffenbauamt Wilhelmshaven, u. a. Seefliegerhorst Wilhelmshaven (1936–1938)
1939 Nach Berlin verzogen

GUTHER, MAX
(*12.02.1909 Neu-Ulm)

Architekturstudium Stuttgart (Bonatz, Schmitthenner, Wetzel)
1931–1934 Freier Architekt Stuttgart, Colmar
1934–1938 Freier Architekt Hamburg, Zusammenarbeit mit Konstanty Gutschow
1938–1940 Freier Architekt Schwerin
1940–1945 Kriegsdienst
1945–1947 Stadtbaurat Wismar
1947–1954 Stadtbaurat Ulm
1954–1974 Professor für Städtebau und Siedlungswesen TH Darmstadt, u. a. City-Rahmenkonzept Wilhelmshaven zusammen mit Ferdinand Stracke (1970)

GUTSCHOW, KONSTANTY
(* 10.12.1902 Hamburg, †08.06.1978)

1921–1926 Architekturstudium Danzig, Stuttgart (Wetzel, Bonatz, Schmitthenner)
1926–1928 Hochbaudirektion Hamburg unter Fritz Schumacher
1928–1929 Stadtplanungsamt Wandsbek
ab 1930 Freier Architekt Hamburg, u. a. Stadtrandhäuser Hamburg-Horn, Planten und Blomen (1936), Wohnungsbau
1936 Ernennung zum Vertrauensarchitekten des Generalbauinspektors für das deutsche Straßenwesen
ab 01.04.1939 Architekt des Elbufers Hamburg
ab 08.01.1941 Architekt für die Neugestaltung der Hansestadt Hamburg
1943 Beigeordneter Stadt Hamburg
1943 Stellvertretender Leiter des Arbeitsstabs Speer zum Wiederaufbau zerstörter Städte, u. a. Wiederaufbauplanungen für Hamburg, Wilhelmshaven und Kassel (1944–1945)
1943–1945 Ahlhorner Architektenkreis
ab 1945 Freier Architekt Hamburg, umfangreiche Nachkriegsbauten, Krankenhausbau

HAARNAGEL, WERNER
(*07.12.1907 Torgau, †03.08.1984 Wilhelmshaven)

Geographie- und Geologiestudium
1934 Promotion
1938 Gründung Provinzialstelle für Marschen- und Wurtenforschung mit Außenstelle Wilhelmshaven, umfangreiche Bohrungen und Bodengutachten für Stadterweiterung und Neugestaltung Wilhelmshavens
1939–1945 Dienstverpflichtet bei der Kriegsmarine für Militärbauten, gleichzeitig Bodenuntersuchungsstelle der Stadt Wilhelmshaven
1947–1972 Direktor des Nieders. Landesinstituts für Marschen- und Wurtenforschung Wilhelmshaven, Honorarprofessor Universität Göttingen

HACKE, BRUNO
(*30.07.1900 Braunschweig)

Architekturstudium
Baubeamtenausbildung
bis 1938 Hafenbauressort KMW Wilhelmshaven, umfangreiche Hochbautätigkeit, Werftbauten
1938 Nach Bremerhaven verzogen

HAEFNER, CARL

Bauingenieurstudium
ca. 1913 Oberingenieur Stadtbauamt Rüstringen unter Martin Wagner
ab 1914 Leiter Tiefbauamt Rüstringen
1921–1937 Nachfolger von Willy Hahn als Stadtbaurat Rüstringen unter Bürgermeister Arthur Kellerhoff als Dezernent für das Bauwesen

HALLBAUER, WILHELM
(*30.03.1889 Straßburg, †26.10.1969 Holzminden)

Architekturstudium
Marineoffizier Erster Weltkrieg
bis 1936 Freier Architekt Hamburg
Okt. 1936–Dez. 1937 Leiter der Außenstelle Wilhelmshaven LOB bzw. Bezirksstelle LOB
01.04. 1937–Nov. 1939 Leiter der Stadtplanungs-, Garten- und Friedhofsabteilung Stadt Wilhelmshaven unter den Stadtbaudirektoren Walter Hunzinger und Walter Temp, umfangreiche stadtplanerische Tätigkeit, Raumpläne, Wirtschaftspläne, Bebauungspläne, einflußreichster Stadtplaner des „III. Reiches" in Wilhelmshaven
1940–1945 Stadtoberbaudirektor Lódź (Litzmannstadt), Landesplaner, Stadtbaurat und Stadthauptmann Lemberg (Lwow)
1950–1960 Mitglied der Deutschen Akademie für Städtebau und Landesplanung, Mitglied der Akademie für Bauforschung, verschiedene Denkschriften zum Wiederaufbau Wilhelmshaven
bis 1969 Sachverständiger für Standortfragen und Raumordnung, Holzminden

HARMS, GEORG HEINRICH
(*1894 Rüstringen, †1955 Wilhelmshaven)

1911–1915 Kunststudium Weimar, München, Schüler von Fritz Mackensen
ab 1916 Freier Künstler Rüstringen, viele öffentliche Aufträge im „III. Reich", Porträtist von Hitler, Röver, Joel
1945 Nach Rastede verzogen

HASCHENBURGER, WERNER
(*12.07.1899 Wilhelmshaven, †22.01.1989 Gödens)

1920–1925 Architekturstudium Karlsruhe, Dresden
1925–1927 Marinestandortbauamt Wilhelmshaven, u. a. Offizierswohnhaus Parkstr. 21 (1925), Strandhalle zusammen mit Gustav Lübbers (1925), Wohnhäuser Bismarckstraße (1925–1927)
1927–1929 Baubeamtenausbildung Norden, Berlin
1929–1930 Stadtbauamt Wilhelmshaven, Planung Nordstrand
1930–1932 Arbeitslos
1932–1935 NS-Ortsgruppenleiter Wilhelmshaven, Gauamtsleiter
1933–1935 Leiter Tiefbauamt Wilhelmshaven
1935–1939 Leiter Neubauabteilung Wohnungsbauamt Bremen, u. a. Fritz-Reuter-Siedlung, Kolonial-Siedlung, Osterholz-Tenever
1939 Thüringer Zellwollekonzern
1940–1945 Kriegsdienst
1945–1955 Kriegsgefangenschaft UDSSR
1957–1967 Bezirksregierung Stade, u. a. Wettbewerb Altengroden-Süd (1957)

HEILMANN, GEORG

Jurastudium
Promotion
Ministerialrat RAM, zuständig für Wilhelmshavens Stadterweiterung
ab 1940 Generalreferent beim Reichskommissar für den sozialen Wohnungsbau (ab 1942 Reichswohnungskommissar) Robert Ley
ab 1944 Ministerialdirigent und Leiter der Abteilung Verwaltung und Recht beim Reichswohnungskommissar

HERDE, ALEXANDER
(*19.03.1910 Leipzig, lebt in Oldenburg)

1930–1936	Architekturstudium Berlin (Tessenow, Jansen)
1936–1939	Baubeamtenausbildung Berlin, u. a. Bauleitung Preußische Staatsbank Gendarmenmarkt, Studentenheim Steinplatz
1939–1942	Prüfstelle des RAM für Gemeinschaftssiedlungen und städtebauliche Planungen unter Scholtz, Büge, Nicolaus, u. a. Bearbeitung städtebaulicher Maßnahmen in Kiel, Danzig, Pillau, Wilhelmshaven, Wettbewerb „Alarm" des Reichsluftfahrtministeriums (1940)
1941	Dissertation „Der Luftschutzbunker im Wohngebiet"
1942–1945	Kriegsdienst
1945–1950	Bezirksregierung Hildesheim
1950–1955	Staatshochbauamt Hameln
1955–1975	Bezirksregierung Oldenburg, Dezernent für Städtebau, zuständig f. d. Stadtplanung Wilhelmshaven

HERMEKING, LUDWIG
(*10.03.1876 Berlin)

	Architekturstudium Berlin
	Baubeamtenausbildung
ab 1918	Leiter Reichsvermögensamt II Wilhelmshaven
ab 1922	Reichsbauamt II Wilhelmshaven, Initiator der Industrieansiedlung in der Weimarer Republik
1939	Nach Kiel verzogen

HEUER, FRIEDRICH
(*15.12.1897 Drakenburg, †13.02.1960 Bremen)

1914–1918	Kriegsdienst
1918–1923	Architekturstudium Hannover, Stuttgart (Schmitthenner, Wetzel, Bonatz)
1923–1925	Büro Rudolf Jacobs Bremen
ab 1925	Freier Architekt Bremen, u. a. Kath. Kirche Osternburg (1928), Siedlung Bremen-Grolland (1936), Siedlung Cäciliengroden (1937), Bebauungsskizzen für Altengroden (1938), Planung Breddewarden (1939), Nachkriegsplanung Wilhelmshaven zusammen mit Schneider (1942), Altenwohnungen Bremen (1954), Altengroden-Süd Wilhelmshaven zusammen mit Schwerdtfeger (1957–1959), Sternenviertel Bremen (1955), Einfamilienhäuser Kuhlenkampffallee Bremen (1966), Erweiterung Siedlung Bremen-Grolland (1961–1965), Wiederaufbau Stadtwaage Bremen (1958), Hochhaus Doventorsviertel Bremen (1958)
1943–1945	Ahlhorner Architektenkreis, enge Verbindungen zu LOB und Gauleitung, Zusammenarb. mit Gutschow

HIGGEN, KARL
(*26.02.1913, lebt in Wilhelmshaven)

bis 1933	Ingenieurschule Oldenburg
1933–1935	Freiwilliger Arbeitsdienst, Landgewinnungsmaßnahmen Rüstersiel
1935–1936	Luftwaffenbauamt Wilhelmshaven, Mitarbeiter am Fliegerhorst Upjever unter Schütte, Entwurf verschiedener Hochbauten, Pförtnerhäuser usw.
ab 01.03.1936	Baufirma Hermann Möller, verschiedene Bauleitungen, u. a. Lager Sudetenstraße (1938), Lager Norderneystraße (1937–1938), Kasernen Ebkeriege (1939–1940), Bauleitung U-Bootstützpunkt (1937), Bauleitung Kasernen Fliegerdeich (ab 1936), Wohnungsbauten für Wilhelmshaven-Rüstringer Wohnungsbaugesellschaft, Hilfskrankenhäuser Virchowstraße und Willehad-Krankenhaus (1940)
1942–1945	Kriegsdienst
ab 1945	Baufirma Hermann Möller, Leitung der Wilhelmshavener Enttrümmerungsmaßnahmen (1946)

HILGENSTOCK, FRITZ
(*08.09.1898 Wuppertal, †1961 Salzburg)

1916–1918	Kriegsdienst
1918–1924	Bauingenieurstudium Hannover
1924–1928	Elterlicher Baubetrieb Wuppertal
1928–1944	Büro für Bauberatung und -organisation Berlin, Hannover, Wilhelmshaven, Salzgitter, u. a. Bauleitung Reichswerke Hermann Göring Salzgitter zusammen mit Rimpl (1937), Bauleitung Gemeinschaftssiedlung Fedderwardergroden zusammen mit Bork (ab 1938), Wohnhäuser Sande-Neufeld (1939), Gewerbebauten Voslapp zusammen mit Bork (1940), Fabrikanlagen Focke-Wulf Bremen, Schieferölwerke in Estland (1941), Industriebauten in Niederschlesien und Posen, Stollenbau im Harz
1944–1945	Büro von OT verpflichtet
ab 1945	Bautätigkeit für Amerikaner, Siedlungsbau

HILLEBRECHT, RUDOLF
(*26.02.1910 Hannover, lebt in Hannover)

1928–1930	Architekturstudium Hannover, Berlin (Tessenow, Jansen)
1929–1930	Büro Walter Gropius
1934–1935	Reichsverband der Deutschen Luftfahrtindustrie Travemünde
1936–1937	Baubeamtenausbildung Berlin
ab 1937	Büro Konstanty Gutschow Hamburg, Bürochef von Gutschow, Mitarbeit an den Planungen für die Neugestaltung Hamburgs, Wiederaufbauplanungen für Wilhelmshaven (1944–1945)
1946–1948	Sekretär für Bau- und Wohnungswesen Zonenbeirat der britischen Zone Hamburg
1948–1975	Stadtbaurat Hannover
ab 1951	Professor TU Hannover
ab 1975	Freier Architekt Hannover

HITLER, ADOLF
(*20.04.1889 Braunau, †30.04.1945 Berlin)

	Verhinderter Architekt und Kunstmaler
	Vorsitzender und Neugründer NSDAP
ab 30.01.1933	Reichskanzler
1934–1945	Regierungs- und Staatschef im „III. Reich", Oberster Befehlshaber der Wehrmacht, griff verschiedentlich in Wilhelmshavens Stadtplanung ein, besuchte Wilhelmshaven u. a.:

1929	Reichsmarine zusammen mit Heß und Himmler
1931–1936	Hitler viermal in Horumersiel
12.05.1931	Jever
23.05.1932	Massenversammlung Rüstringen
02.06.1932	Besuch des Kreuzers Köln
12.12.1933	Besuch des Kreuzers Köln
Juni 1936	Besichtigung Marinestützpunkt, Entscheidungen zum Militärbau
03.10.1936	Stapellauf Scharnhorst

17.06.1937	Beisetzung der Toten der „Deutschland“ Ehrenfriedhof
01.04.1939	Besuch der gesamten Reichsregierung in Wilhelmshaven, Stapellauf Tirpitz, Entscheidungen zu Monumentalplanungen, Massenversammlung
27. 09.1939	Gast der U-Bootfahrer
09.04.1940	Besuch U-Bootstützpunkt

HÖGER, FRITZ
(*12.06.1877 Bekenreihe, †21.06.1949 Segeberg)

1896–1899	Baugewerkschule Hamburg
1899–1901	Militärdienst, Gasthörer TU Berlin
1901–1905	Büro Lund und Kallmorgen Hamburg
1907–1949	Freier Architekt Hamburg, Mitglied Deutscher Werkbund (1910), Aufforderung zum Wettbewerb Rüstringer Zentralanlagen (1913), Klöpperhaus Hamburg (1913)
1914–1918	Kriegsdienst
	u. a. Hapaghaus Hamburg (1913–1919), Chilehaus Hamburg (1922–1924), Broschekhaus Hamburg (1925), Krankenhaus und Friedhofskapelle Delmenhorst (1926)
1926	Erster Besuch in Wilhelmshaven, Treffen mit Arthur Kellerhoff und Reinhard Nieter, u. a. Hannoverscher Anzeiger (1926), Reemtsmabau Wandsbek (1926–1927), Wasserturm Norderney (1928), Rathaus Rüstringen (1928–1929), Kirche Hohenzollernplatz Berlin (1929–1930), Wassertürme Zwischenahn und Hohenkirchen (1930), Wohnsiedlung Meppen (1935), Konsumverein Leipzig (1930)
1929–1937	Städtebaulicher Berater Stadt Rüstringen, u. a. Häuser Reif, Meyer und Hoeland Wilhelmshaven (1934–1935), Siedlung Neu-Siebethsburg (1934–1941), Siedlung Meppen (1935), Entwürfe „Haus des Handwerks“ Wilhelmshaven (1935), Häuser Mücke und Mehrtens Wilhelmshaven (1936), Fassade Finanzamt Wilhelmshaven (1936–1937), Arbeitsamt Wilhelmshaven (1936–1937), Kreishaus Meppen (1937)
1934–1936	Professor Nordische Kunsthochschule Bremen, Mitglied des Verwaltungsrates der RDBK
1936	Bruch mit dem NS-Regime, u. a. Wohnhäuser Mühlenweg für Friedrich Möller (1935–1936), Entwürfe Tausend-Wohnungsprojekt Schaarreihe (1937–1938), Wohnungsprojekt Arngaststraße (1939–1941), Polizeipräsidium Aussig (1941), NSV-Heim Siebethsburg (1939), Entwürfe Rathauserweiterung Wilhelmshaven (1938), Entwürfe für den Rüstringer Bauverein (1947), Ideenwettbewerb 1948

HÖGG, HANS

	Aufgewachsen Bremen
	Architekturstudium
1935	Promotion
	Bauamt Hildesheim
1939	Bewerber um den Posten des Stadtbaudirektors Wilhelmshaven
bis 1945	Stadtbaurat Stettin
ab 1947	Professor für Wohnungswesen und Städtebau Hannover, u a. Frankfurter Innenstadt, Gutachten Hannover, Kassel (1947)

HÖROLD, LUDWIG
(*19.03.1892 Wiesbaden)

	Architekturstudium
	Freier Architekt Wiesbaden
bis 1938	Freier Architekt Hamburg
1938–1943	Projekte in Wilhelmshaven, u. a. Schule Albrechtstraße (1939), Wohnbebauung Glettkauer Weg (1941)
	Gewerbebauten Albrechtstraße (1941), Bauten für den Bauverein Rüstringen
ab 1943	Freier Architekt Wiesbaden

HOFFMANN, WERNER
(*12.03.1898)

	Architekturstudium
	Baubeamtenausbildung
	Marinebaurat, SA-Eintritt (1934)
bis 1943	HNBD Wilhelmshaven, zuständig für Hochbauten, u. a. Familienlager Neuengroden zusammen mit Baufirma Möller (1939–1940), Hochbauten der Nordwerft und IV. Einfahrt
1943–1944	Leiter Marinebauamt Emden

HOMANN, RICHARD

	Gartenbaudirektor Bremen
ab 1939	Fortsetzung der Gartenplanungen von Friedrich Eisenbarth in Wilhelmshaven, Mitarbeit Wiederaufbauplanungen Wilhelmshaven (bis 1945)
1943–1945	Ahlhorner Architektenkreis, Zusammenarbeit mit Wortmann und Schwarz

HUCKE, ALFRED
(*27.01.1898 Berlin, lebt in Heide)

bis 1922	Architekturstudium Berlin (Jansen)
1922–1925	Baubeamtenausbildung
1925–1935	Bauabteilung Stettiner Vororteverband
ab 1935	Marinestandortbauamt Wilhelmshaven
1943–1944	stellvertretender Leiter Marineoberbauamt Wilhelmshaven unter Triebel
1944–1945	OT-Bauleitung Wilhelmshaven
	Kriegsgefangenschaft
bis 1963	Bauverwaltung Kreis Dithmarschen

HÜBOTTER, WILHELM
(*16.06.1895 Neu-Garge, †28.07.1976 Hannover)

	Gärtnerlehre Hannover
1919–1922	Studium Gartenarchitektur Geisenheim
ab 1922	Freier Gartenarchitekt Prag, Berlin, Hannover, u. a. Gartenplanungen für Gemeinschaftssiedlungen Wilhelmshaven (ab 1938), Wiederaufbauplanungen Hamburg (1943–1945), Berater für den Wiederaufbau Hannover (1944)
ab 1938	Landschaftsanwalt OT
ab 1945	Professor TU Hannover, u. a. Grünplanung Helgoland (ab 1952)

HUNZINGER, WALTER
(*17.07.1889 Köln)

	Kein Architekturstudium
	Freier Architekt Hösel
ab 1935	Bauabteilung der Luftwaffe, u. a. Flugplatz Delmenhorst/Adelheide, Flugplatzplanungen in Dortmund und Rotenburg

ab 01.11.1937	Stadtbaudirektor Wilhelmshaven, u. a. Umbau Strandhalle (1938), Beteiligung an Bebauungsplänen Wilhelmshavener Gemeinschaftssiedlungen, stadtplanerisch wenig hervorgetreten
Febr. 1939	Entlassung, verzogen nach Kleve/Pfalzdorf

ISENBERG, GERHARD

	Architekturstudium
ab 1935	Reichsstelle für Raumordnung, zuständig für den Bereich LOB, beteiligt an Wiederaufbauplanungen Wilhelmshaven des Arbeitsstabs Speer (1944–1945)

IWERSEN, FRANZ
(*20.02.1906 Wilhelmshaven, †20.09.1971 Wilhelmshaven)

1929–1931	Architekturstudium Ingenieurakademie Oldenburg
1932–1971	Freier Architekt Wilhelmshaven
1937–1944	Bürogemeinschaft mit Hanns Iwersen
1945–1967	Bürogemeinschaft mit Carl Peters, u. a. Wohnhäuser Hansaviertel (ab 1933), Wohnhäuser Baltrumstraße (1936), Wohnhäuser Peter-/Werftstraße (1937), Gewerbetrakt Banter Weg (1938), Schule B Fedderwardergroden (1939), Umbau Marinegerätelager Roffhausen/Olympia Werke (ab 1946), Olympiasiedlung Roffhausen (1950), Volksbank Theaterplatz (1950), Bunkerumbauten (1950/52), Senckenberg-Institut (1960), Förderanlagen und Maschinenbau

JAHN, ADALBERT
(*09.07.1909 Schortens, lebt in Jever)

1929–1932	Architekturstudium Ingenieurakademie Oldenburg
1933–1935	Hafenbauressort KMW Wilhelmshaven, u. a. Schnürbodengebäude, Schiffsinnenausbauten (Admiral Scheer, Meteor usw.)
1935–1938	Luftwaffenkommando See, Bauleitung Fliegerhorst Upjever unter Schütt, u. a. Mitarbeit an Wohngebäuden, Offiziersledigenheim, Offizierskasino, Siedlung Addernhausen (1935–1936)
1938–1939	Zusammenarb. m. Hans Ahmels, u. a. Wohnungsbau f. Luftwaffenpersonal Jever, Schule Cleverns (1938)
1939–1941	Marinestandortbauamt Wilhelmshaven, u. a. Mitarbeit an Kasernen Varel, Sengwarden, Funkstelle Dangast, Marinesportanlage
1941–1945	Kriegsdienst
1947–1971	Technischer Leiter Bauamt Stadt Jever, Stadtbauingenieur

JOCHEM, FRIEDRICH
(*20.02.1905 Herlisheim, †03.05.1984)

1925–1930	Architekturstudium Aachen
1930–1934	Baubeamtenausbildung
1934–1936	Bauabteilung Reichsluftfahrtministerium, Entwurf von Fliegerhorstanlagen
1936–1938	Reichspostdirektion Köln
1938–1939	Reichspostdirektion Münster
1939–1941	Reichspostdirektion Oldenburg, zuständig für Postbauten im Raum Wilhelmshaven, u. a. Fernsprechamt Schillerstraße (1939), Entwurf Postgebäude Fedderwardergroden (1940), Notpostämter, Postamt Voslapp
1941–1945	Reichspostdirektion Königsberg
1945–1970	Oberpostdirektion Hamburg, Post- und Fernmeldebauten in Hamburg und Lübeck

JOEL, GEORG
(*08.08.1898 Wilhelmshaven, †10.10.1981 Rastede)

	Oberrealschule Wilhelmshaven
1917–1918	Kriegsdienst
ab 1919	Reichsbahnobersekretär
1925	NSDAP-Eintritt
ab 1931	NSDAP-Mitglied Oldenburgischer Landtag
ab 1932	Landtagspräsident
1933–1945	Ministerpräsident Land Oldenburg, stellvertretender Gauleiter
ab 1936	Reichstagsabgeordneter der NSDAP, umfangreiche Mitwirkung an Landes- und Stadtplanung in Wilhelmshaven, Beirat LOB (1936–1945), 1948 entnazifiziert: $2\frac{1}{2}$ Jahre Gefängnis
1955–1959	DRP-Landtagsabgeordneter Land Niedersachsen

KELLERHOFF, ARTHUR
(*25.01.1886 Oldenburg, †20.12.1950 Wilhelmshaven)

	Jurastudium
	Promotion
ab 1912	Juristischer Hilfsarbeiter Stadt Rüstringen
	Kriegsdienst
1918–1920	Stadtsyndikus Rüstringen
1920–1937	Bürgermeister Rüstringen, im Magistrat für das Bauwesen zuständig, befreundet mit Fritz Höger, machte Höger zum städtebaulichen Berater der Stadt Rüstringen (1929–1937), förderte den Bau des Rüstringer Rathauses
1937–1945	Bürgermeister Wilhelmshaven, NSDAP-Eintritt (1937)
1947–1950	Stadtkämmerer Wilhelmshaven, Veröffentlichungen zur Geschichte der Stadt Rüstringen

KELM, WALTER
(*21.08.1883, †12.10.1954)

	Ältester Sohn des Geh. Baurats der Kaiserlichen Marine Adalbert Kelm (u a. Marineschule Mürwik)
1903–1907	Architekturstudium Darmstadt, Braunschweig
	Baubeamtenausbildung, Kriegsdienst
1919–1934	Marineintendantur Kiel und Marinestandortbauamt Kiel
1934–1945	OKM Berlin, zunächst Ministerialrat und Referent Hochbau in der Unterkunftsabteilung des Marineverwaltungsamtes, ab 1939 Ministerialdirigent, unter Hans Benda, ranghöchster Hochbaubeamter der Marine, Aufsichtsratsmitglied der Gemeinnützigen Gesellschaft zur Schaffung von Wohngelegenheiten für Reichsangehörige Berlin, u. a. Siedlung Kiel Schillstr., Marinesiedlungen, Kasernen Marinelazarette, Offizierswohnhäuser, Leiter Hochbauabteilung OKM unter Schreiber, zuständig f. Hochbauangelegenheiten der Marine in Wilhelmshaven, Ministerialinstanz über Marineintendantur

KLATT, OTTO
(*18.04.1886 Bülz, †15.09.1974 Wilhelmshaven)

	Architekturstudium
	Freier Architekt Wilhelmshaven/Rüstringen, u. a. Brandkassenverwaltung, später Adolf-Hitler-Haus

Oldenburg (1921–1923), Wohnungsbauten Hansaviertel (1937)

KLAUCKE, HANS
(*18.03.1901 Magdeburg, †25.09.1975 Norden)

1920–1925	Architekturstudium, Braunschweig (Mühlenpfordt), Darmstadt (Hofmann, Roth, Hummel)
1925–1938	Hochbauamt Stadt Magdeburg unter Göderitz
1938–1945	Hafenbauressort KMW Wilhelmshaven, Abteilungsleiter Hochbauplanung, u. a. Gießerei Maschinenbauressort KMW Sande (1938–1940), Leuchtturm Mellum-Plate (1939–1942), Leuchtfeuer Minsener Oog (1942–1943), Bürohaus Ressort II der KMW (1940), Werkstatt für Elektrogeräte der KMW (1940) Motorenbauwerkstatt Westwerft (1940), Planung Erweiterung Hafenbauressort Virchowstraße (1943), Bunkerbauten der KMW (1940–1941), Stellwerke und Bahnhöfe der Werftbahn nach Fedderwardergroden und Voslapp (1941), Torpedowerftgebäude Verbindungshafen (1940), Heizhaus Emden (1941), Umspannstationen Cuxhaven
1945	Kriegsdienst
1945–1965	Freier Architekt Wilhelmshaven, u a. Maadesiel (1948–1951), Dangastersiel (1954–1956), Volksbank Bordumplatz (1950), Schweiburger Siel (1957), Schöpfwerk Harlesiel (1953–1957), Mündungsschöpfwerk Jade (1962–1965), Schöpfwerk Petershörn (1959–1962), Anker-Denkmal Heppenser Groden (1953), Wangersiel (1960–1963), Schöpfwerk Wangerland (1971–1974), Stapelmoorer Siel und Schöpfwerk (1958), Schöpfwerk Ditzum, Kläranlage Zetel, Kläranlage Bockhorn (1962–1964)

KÖHLER

Jurastudium
Promotion
Oberregierungsrat OKM, zuständig für Wohnungsbau der Marine, vielfach mit Wilhelmshavens Stadterweiterung verbunden

KÖHNE, ADOLF

	Leiter Reichsheimstättenamt Gau Weser-Ems Oldenburg
ab 1945	Freier Architekt Oldenburg

KRETSCHMER, ERNST
(*30.06.1892 Berlin)

	Architekturstudium Berlin
bis 1939	Marineintendantur Wilhelmshaven, Marineoberbaurat
1943–1944	Stellvertr. Leiter Marineoberbaudirektion Wilhelmshaven unter Frede

KRÜGER, WILHELM
(*15.02.1871 Oldenburg, †29.02.1940 Bad Zwischenahn)

bis 1895	Bauingenieurstudium Hannover, Berlin
1896–1901	Eisenbahndirektion Oldenburg
1901–1933	Strombauressort Werft Wilhelmshaven, u. a. Leuchtturm Voslapp (1906–1907), Leuchtturm Arngast (1909–1910), Wasserbau-Versuchsanstalt Wilhelmshaven (1908), umfangreiche Forschungen zur Küstengeologie, Bau einer zweiten Wasserbauversuchsanstalt (1926), Jade-Korrektion (bis 1931)
1926	Ehrenpromotion
1933–1936	Leiter Hafenbauressort KMW Wilhelmshaven, u. a. Vergrößerung Insel Wangerooge, Schutzbauwerke

KÜHN, ERICH

Architekturstudium Berlin
Promotion
Stadtbaurat a. D.
Freier Architekt Berlin, umfangreiche Bautätigkeit für die Gemeinnützige Gesellschaft zur Schaffung von Wohngelegenheiten für Reichsangehörige Berlin, u. a. Siedlung Berlin-Schlachtensee, Ladentrakt Altengroden-Süd Wilhelmshaven (1940)

KUÖHL, RICHARD EMIL
(*31.05.1880 Meißen)

	Lehre in Modellatelier Meißen
1902–1905	Kunstgewerbeschule
seit 1919	Freier Künstler Hamburg, u. a. Baukeramiken an vielen Hamburger Staatsbauten, Plastiken Chilehaus (1923–1924), Kriegerdenkmal Dammtor-Bahnhof, Relief und Bauplastik Marinestationsgebäude Wilhelmshaven (1937)

KUSSIN
(*ca. 1885, †1945)

	Architekturstudium
	Baubeamtenausbildung
bis 1940	Marineintendantur Kiel
1940–1945	Referent in der Hochbauabteilung des Marineverwaltungsamtes des OKM, Stellvertreter von Abteilungsleiter Walter Kelm zuletzt Ministerialrat

LADWIG, HANNS

1937–1945	Amtsbaumeister Amt Friesland
ab 1940	Ortsgruppenleiter und Propagandaleiter NSDAP Ortsgruppe Jever, zuständig für Hochbauten und Bauplanungen Amt Friesland und Gemeinschaftssiedlungen soweit sie außerhalb der Wilhelmshavener Stadtgrenzen lagen
ab 1945	Staatshochbauamt Wilhelmshaven, u. a. Arbeitsamt Jever (1955)

LANGEWAND, PHILIPP
(*16.07. 1884)

Architekturstudium
Baubeamtenausbildung
Reichsbahnrat, Dezernent Reichsbahndirektion Münster, zuständig für Bahnhofsbauten im Raum Wilhelmshaven, u. a. Hauptbahnhof Oldenburg zusammen mit Friedrich Mettegang (1912–1916), Bahnhof Sande (1940–1941), Planung des neuen Ortsgüterbahnhofs Wilhelmshaven (1940–1941), Bau von Stellwerken und Betriebsgebäude für die Bahnhöfe Wilhelmshaven

LEHN, OTTO
*28.01. 1883 Cuxhaven, †30.01.1969 Wilhelmshaven)

	Bauingenieurstudium, Gasthörer an einer TH ohne Examen
ab 1908	Bauamt Stadt Wilhelmshaven unter Paul Wolf

ab 1913	Leiter des Tiefbauamtes Stadt Wilhelmshaven unter Hermann Zopff, Stellvertr. des Baudezernenten
1933–1935	Von Nationalsozialisten dienstenthoben
1937–1945	Leiter Tiefbauabteilung Stadt Wilhelmshaven, Magistratsbaurat
ab 01.09.1944	Vertreter von Stadtbaurat Temp, nach dessen Einberufung zum Kriegsdienst beteiligt an Wiederaufbauplanungen
1945–1949	Baudezernent, Stadtbaurat von Wilhelmshaven, Preisrichter Ideenwettbewerb 1948

LEY, ROBERT
(*15.02.1890 Gummersbach, †25.10.1945 Nürnberg)

	Chemiestudium
	Promotion
1925	Gauleiter NSDAP Rheinland
ab 1932	Reichsorganisationsleiter der NSDAP, Nachfolger von Gregor Strasser
1934–1945	Gründer und Leiter der DAP, Schöpfer der Ordensburgen und des KDF-Wagens
ab 15.11.1940	Reichskommissar für den sozialen Wohnungsbau
ab 1942	Reichswohnungskommissar, besuchte häufig Wilhelmshaven, hielt Propagandareden vor den Arbeitern der KMW und Bauarbeitern der Gemeinschaftssiedlungen (17.08.1935, 01.04.1939, 24.11.1939, 09.06.1941)

LINDE, GEORG
(*14. 12.1878 Kottbus, †1966 Berlin)

ab 1898	Bauingenieurstudium Berlin
bis 1914	Marinehafenbaumeister Pillau
1914–1918	Kriegshafenbau Frankreich
ab 1918	Marineoberbaurat im Wartestand, private Ingenieurtätigkeit Jever, Holland, Belgien
bis 1929	Stellvertr. Landesvorsitzender der DNVP, NSDAP-Beitritt, Stahlhelm-Führer Jever (1923), verschiedentlich inhaftiert
ab 1932	Wirtschafts-Ortsgruppenleiter für das Amt Friesland, Gauamtsleiter
ab 1933	Leiter der Zentrale des Arbeitsamtes der NSDAP, Leiter des Arbeitsamtes Northeim
ab 1934	Gauwalter NSV, Gauführer Winterhilfswerk
1936–1938	Landesplaner LOB Oldenburg, eng mit Wilhelmshavens Stadterweiterung verbunden
1939–1945	Kriegshafenbau Frankreich
ab 1950	Ruhestand in Berlin

LINDE, HORST
(*06.04.1912 Heidelberg, lebt in Freiburg)

1931–1936	Architekturstudium Karlsruhe (Schweizer)
1936–1939	Baubeamtenausbildung Emmendingen, Baden-Baden, Karlsruhe
ab 1940	Kriegsdienst
1941	Wiederaufbauplanungen Elsaß
ab 1942	Eingezogen zur Marine, Marineintendantur Abteilung XXIV. (Siedlungsabteilung), Mitarbeit an Gewerbeplanung Fedderwardergroden unter Franzius
1942–1945	Rouen, Boulogne, Dünkirchen
1947	Wiederaufbauplanungen Freiburg, Karlsruhe
ab 1957	Staatliche Hochbauverwaltung Baden-Württemberg
1961–1977	Professor TH Stuttgart

LÜBBERS, FRIEDRICH
(*04.06.1911 Hamburg, lebt in Bad Kreuznach)

bis 1936	Architekturstudium Stuttgart (Bonatz, Wetzel)
1936–1939	Büro Kraemer, Braunschweig
1939–1940	Architekturbüro der DAF unter Julius Schulte-Frohlinde
1940–1941	Kriegsdienst
1941–1942	Bauabteilung der DAF Berlin, Wien, u. a. Theaterbauten
1942–1944	Referent für Städtebau bei der Landesregierung Linz, Sonderbeauftragter des Reichsstatthalters für die Neugestaltung von Braunau
1944–1945	Büro Gutschow Hamburg (Empfehlung Bonatz), u. a. Wiederaufbauplanungen Wilhelmshaven für den Arbeitsstab Speer
1946	Promotion
1945–1954	Stadtplanungsamt Braunschweig unter Kraemer und Göderitz
1954–1976	Leiter Stadtplanungs- und Vermessungsamt Bad Kreuznach

LÜBBERS, GUSTAV
(*05.01.1878 Jeringhave, †23.01.1958 Wilhelmshaven)

bis 1900	Baugewerkschule Holzminden
ab 1897	Freier Architekt Wilhelmshaven/Rüstringen
ab 1913	Bürogemeinschaft mit Hermann Dieter, u. a. Bebauung Parkstraße (1910–1911), Kaufhaus Leffers (1912), Wettbewerb Zentralanlagen Rüstringen (1913)
1916–1918	Kriegsdienst
ab 1925	Bürogemeinschaft mit Alwin Zeitler, u. a. Tierseucheninstitut Oldenburg (1925), Wohnungsbauten
1933	NSDAP-Eintritt
1933–1937	Bürogemeinschaft mit Sohn Willi Lübbers, Wohnungsbau Hansaviertel, Mitscherlichstraße/Paul-Hug-Straße/1937), Ideenwettbewerb 1948

LÜBBERS, WILLI
(*06.04.1906 Wilhelmshaven, lebt in Wilhelmshaven)

bis 1933	Architekturstudium Darmstadt, Stuttgart (Schmitthenner, Bonatz)
ab 01.04.1934	Freier Architekt in Wilhelmshaven, Beauftragter der NSDAP Wilhelmshaven f. Bauangelegenheiten
01.02.1937–10.10.1938	Technischer Geschäftsführer Wohnungsbaugesellschaft Jade mit eigenem Architekturbüro
ab 01.04.1937	Nach der Zusammenlegung von Wilhelmshaven und Rüstringen Ratsherr und ehrenamtliches Magistratsmitglied der NSDAP, einflußreicher Baupolitiker in Wilhelmshaven während des „III. Reiches", u. a. Bürogebäude Baufirma Möller (1934), Privathäuser Herbartstraße (1935), Gewerbegebäude Kuhlmann (1936), Wohnungs- und Gemeinschaftsbauten Voslapp (ab 1937), Wohnungsbauten Hansaviertel (1937), Unteroffizierswohnhäuser Inselviertel (1937), Schule Voslapp (1938), Stadttheater im Seemannshaus (1938), NSDAP-Gemeinschaftshaus Voslapp (1938), Hauptumspannstelle Alter Banter Weg (1938)
1939–1940	Kriegsdienst Fliegerhorst Upjever

1940–1945	OT, Marinebaurat, u. a. Bauvorhaben Posen, Brückenbau an der Mosel (1944), OKM-Quartiere Eberswalde (1944)
1945	Leitung Staatshochbauamt Stettin
ab 1945	Freier Architekt Wilhelmshaven, Wiederaufbauten, viele Neubauten, u. a. Sparkasse Theaterplatz (1957), Kreuzkirche Schulstraße, Versicherungsanstalt Paul-Hug-Str., Dresdner Bank, Commerzbank

LÜCKEMEYER, OTTO
(*19.09.1988 Osede, †27.12.1979 Kiel)

1909–1914	Architekturstudium München, Berlin (Caesar, Koch, Hartung)
1914–1916	Kriegsdienst
1916–1920	Marine-Garnison-Bauamt bzw. Reichsvermögensamt bzw. Marinestandortbauamt Wilhelmshaven
1920–1925	Möglicherweise Marinestandortbauamt Cuxhaven
1925–1941	Leiter Marinestandortbauamt Wilhelmshaven, u. a. Offizierswohnhäuser Parkstr. 21 (1925), Wohnbebauung Bismarckstr. 1–29 (1925–1927), Marinesiedlung Athenstraße, Wohnbebauung Lilienburgstr. 31 A–D, 20–26 (1925–1927), umfangreiche Militär-Kasernen Wohnungs- und Stützpunktbauten im „III. Reich"
1943–1944	Leiter Marineoberbaudirektion Kiel
1946–1953	Oberfinanzdirektion Kiel

MAAS, KARL
(*1911 Bremen, lebt in Kiel)

1930–1936	Architekturstudium Stuttgart (Schmitthenner), Braunschweig (Petersen, Flesche, Diekmann)
ab 1936	Marinestandortbauamt Wilhelmshaven, Mitarbeit an u. a. Marinestationsgebäude (1937), Marinelazarett Sanderbusch (1937), Wiederherstellung Christus-Kirche (1942)
1943	Promotion
1944	Bauleiter OT
1945	Kriegsdienst
1947–1954	Freier Architekt an verschiedenen Orten
1954–1976	Landesbauverwaltung Schleswig-Holstein, zuletzt Ltd. Regierungsbaudirektor

MARKWORTH, FRIEDRICH
(*01.08.1887 Straßburg, †1951)

	Bauingenieurstudium
	Baubeamtenausbildung
1939–1943	Leiter Strombauressort KMW Wilhelmshaven, zuständig für Fahrwasser, Leuchttürme, Uferschutz, Buhnen usw.
1943–1944	Stellvertr. Leiter Marineoberbaudirektion Wilhelmshaven unter Frede
1945	Verzogen nach Stettin

MEICHNER, WILHELM
(*31.01.1863 Osterwald, †15.01.1940 Wilhelmshaven)

	Architekturstudium
	Baumeister Wilhelmshaven
ab 1891	Übernahme des Baugeschäfts Groß, beteiligt an vielen öffentlichen Bauten, Funktionen in Bauwirtschaft und Gewerbevereinen, baugewerblich tätiger Architekt, u. a. Haus des Handwerks (1935)

MELLIN, FRIEDRICH

	Architekturstudium
	Baubeamtenausbildung
bis 1945	Techn. Geschäftsführer der Gemeinnützigen Gesellschaft zur Schaffung von Wohngelegenheiten für Reichsangehörige Berlin, u. a. Wohnbebauung Altengroden Wilhelmshaven (ab 1939)

MINKE, ALEXANDER
(*17.05.1894 Wilhelmshaven, †07.07.1956 Wilhelmshaven)

	Freier Architekt in Wilhelmshaven, u. a. Wohnhaus Freiligrathstraße (1932), Reihenhaustypen Kettenstraße (1936), Graf-Galen-Siedlung Wilhelmshaven (1952)

MÖLLER, FRIEDRICH
(13.10.1888 Ihlbruck, †12.06.1951 Wilhelmshaven)

	Bauingenieurstudium
ab 1916	Übernahme der väterlichen Baufirma Hermann Möller, viele Niederlassungen im Reich
1931	Gründung des Norddeutschen Eisenbau Sande
1923–1931	Abgeordneter der DDP Oldenburgischer Landtag, u. a. beteiligt an: Rüstringer Rathaus (1928–1929), Hindenburg-Damm Sylt, Deichanlangen Voslapp (1928), Kasernen Ebkeriege (1939), Lager Sudetenstraße (1936), Marinedienstgebäude (1937), Familienlager Neuengroden (1939), Befehlstelle Marinekommando Nord (1938–1942), Flugzeughallen Upjever (ab 1936), Luftschutzbauten in Wilhelmshaven (ab 1940), U-Bootstützpunkt Wilhelmshaven (1936–1941), Kasernen Fliegerdeich (1936–1938), Olympia-Stadion Berlin (1936), Leuchtturm- und Hafenbauten, Enttrümmerungsarbeiten, umfangreiche Nachkriegsbautätigkeit
1945–1946	FDP-Abgeordneter Oldenburgischer Landtag, Funktionen in Industrie- und Handelskammer
1951	Baufirma an Söhne übergeben, 1976 eingestellt.

MÜLLER, PETER AUGUST

	Architekturstudium
	Büro Fritz Höger, Hamburg, Zweigniederlassung Wilhelmshaven
ab 1936	Luftwaffenbauamt Wilhelmshaven unter Schütt, u. a. Siedlung Addernhausen (1935–1936), Flugplatzbauten Upjever

MÜLLER, WILHELM
(*04.02.1889 Kiel, †30.11.1965 Rastede)

	Studium Deutsch, Geschichte, Philosophie, Religion, Göttingen, Kiel
1917	Promotion über Goethe
	Kriegsdienst
1919–1932	Schuldienst Rüstringen, Brake, Delmenhorst, NSDAP-Eintritt (1930), Ortsgruppenleiter Hude (ab 1930)
1933–1937	Staatskommissar bzw. Oberbürgermeister in Delmenhorst, NSDAP-Kreisleiter Delmenhorst (1933–1934)
März 1937	Überleitungskommissar zur Zusammenlegung von Wilhelmshaven und Rüstringen
ab 01.04. 1937	NSDAP-Oberbürgermeister Groß-Wilhelmshaven, vielfach mit Stadterweiterung und Bauplanungen verbunden, Aufsichtsratsvorsitzender Wohnungsbaugesellschaft Jade, Beirat LOB (1937–

	1945), Abstimmung der Stadterweiterung mit Berliner Dienststellen RAM, OKM, DAF, Initiative zur Neugestaltung Wilhelmshavens nach dem Gesetz zur Neugestaltung deutscher Städte vom 04.10.1937, Motor der Monumental- und Wiederaufbauplanungen
1945	Entnazifiziert

NAUD, WILHELM
(*12.12.1878 St. Wendel, †23.06.1957 Wilhelmshaven)

	Architekturstudium
	Baubeamtenausbildung
	Zeitweise Marinestandortbauamt Wilhelmshaven
ab 1937	Vertreter des Marine-Chefintendanten in Bauangelegenheiten, Beirat für Hochbauwesen der KMW, Mitglied des Bauausschusses beim Kommando der Marinestation der Nordsee
ab 1943	Marineoberbaudirektion Wilhelmshaven, stellvertr. Leiter Marineoberbauamt Hamburg

NEUMANN, MAX
(*18.03.1880 Waldenburg, †05.08.1960 Wilhelmshaven)

ab 1906	Bauführer Gemeinde Heppens
1911–1914	Arbeit unter Martin Wagner, Stadtbauamt Rüstringen
ab 1918	Stadtbaumeister Rüstringen unter Willy Hahn
1921–1937	Leiter Hochbauamt Rüstringen
1937–1939	Leiter Hochbauamt Stadt Wilhelmshaven
1939–1945	Leiter Baupolizeiabteilung Wilhelmshaven, Magistratsbaurat, u. a. Bauleitung Rüstringer Rathaus für Fritz Höger (1928–1929), Erschließung Villenviertel (ab 1909), Bauleitung Wasserturmschule (1911), Paul-Hug-Kinderheim (zus. mit Heinrich Siemer 1928), Bebauungsplan Hansaviertel (1936)

NICOLAUS, FRIEDRICH
(*vor 1900, †1945)

	Architekturstudium
	Baubeamtenausbildung
	Bezirksregierung Leipzig
bis 1942	Oberregierungsrat RAM, zuständig für Stadtneugründungen des „III. Reiches", intensive Mitwirkung an der Wilhelmshavener Stadterweiterung, Theoretiker des „Wehrhaften Städtebaus"
ab 1942	Oberregierungsrat beim Reichswohnungskommissar für städtebauliche und baupolizeiliche Fragen, zuletzt Ministerialrat
Mai 1945	Nach Ermordung seiner Frau Selbstmord Berlin

NOBE, FRITZ
(*1880, †1945)

	Jurastudium
	Werftverwaltungsdirektor
ab 1942	Generalintendant bzw. Marine-Chefintendant, umfangreich mit Wilhelmshavens Marinebauten und Stadterweiterung verflochten
1945	Selbstmord in Norwegen, danach Selbstmord der Familie

NORDECK, WILLY VON
(*26.01.1888 Brandenburg, †12.10.1956 Berlin)

1903	Eintritt Kaiserliche Marine, Artillerie-Offizier
1919–1930	Marineleitung Berlin
1930–1932	Linienschiff-Kommandant
1932–1935	Leiter der Zentralabteilung Marinewerft Wilhelmshaven
1935–1942	Oberwerftdirektor KMW Wilhelmshaven, seit 1940 als Admiral, wichtigste Schlüsselfigur der Marineaufrüstung des „III. Reiches", als Leiter der KMW verantwortlich u. a. für Werftausbau, Werftarbeiterwohnungsbau, Arbeiterlager der Werft, Hafen- und Strombau, Werftbahn, Deichbau, Wohnungsfürsorge, Gemeinschaftsbauten der Werftarbeiter, technische Hochbauten der KMW
1943–1945	Landrat, Rügen
1945	Internierung Dänemark

NOTT, ERICH
(*1883, †1954)

	Jurastudium
1938–1945	Marineintendant, Vertreter der Marine-Chefintendanten Braeuer und Nobe, zuständig für Stadterweiterungsfragen

NUTZHORN, GUSTAV
(*08.08.1886 Oldenburg, †29.06.1981 Bad Zwischenahn)

ab 1920	Studienrat Rüstringen, NSDAP-Eintritt (1930), ab 1931 NSDAP-Ratsherr Rüstringen
ab 1932	NSDAP-Landtagsabgeordneter
1933–1937	Oberbürgermeister Rüstringen, Förderer der Adolf-Hitler-Siedlung
1936–1937	Beirat LOB
1937–1945	Polizeipräsident Aussig (Sudetenland), SS-Führer, 1945 als Mitläufer entnazifiziert

OSTENDORF (I), HANS
(*26.02.1883 Eutin, †19.04.1972 Oldenburg)

	Jurastudium
ab 1909	Landesregierung Oldenburg
1915–1918	Kriegsdienst
1920–1923	Amtshauptmann, Jever
1923–1946	MDF Oldenburg, zuletzt Ministerialrat, zuständig für Grundstücksangelegenheiten, Domänen, vielfach mit Wohnsiedlungstätigkeit im Raum Wilhelmshaven 1933–1945 verbunden

OSTENDORF (II), WILHELM
(*20.05.1885 Ahrensböck, †07.11.1975 Oldenburg)

bis 1909	Bauingenieurstudium München, Hannover
1909–1913	Baubeamtenausbildung Oldenburgische Eisenbahnverwaltung
1913–1914	Bauunternehmen Bremen
ab 1914	Wege- und Wasserbauverwaltung Oldenburg
ab 1918	Bezirksbaumeister Wege- und Wasserbauverwaltung Nordenham
ab ca. 1920	Landesregierung Oldenburg, Leiter der gesamten technischen Abteilungen, zuletzt Ministerialrat beim MDF, vielfach mit Stadterweiterung und Neugestaltung Wilhelmshavens im „III. Reich" verbunden, Beirat LOB (1936–1945)
1946–1951	Regierungsdirektor bei der Bezirksregierung Oldenburg
ab 1951	Freier Gutachter, Berater der Baufirma Möller Wilhelmshaven

OTT, HERMANN
(*1895 Brake, †in den 50er Jahren)

1914–1921	Jurastudium Jena, München, Göttingen, unterbrochen durch Kriegsdienst
1921–1924	Beamtenausbildung
1924–1927	Ministerium des Innern Berlin
1927–1928	Mitglied der Regierung Eutin
1928–1932	Amtshauptmann Westerstede
1932–1933	Amtshauptmann Butjadingen
1933–1945	Amtshauptmann Amt Friesland Jever, zuständig für Kreisplanung und Siedlungsgebiete im Wohnsiedlungsgebiet Wilhelmshaven-Jeverland, umfangreich mit Planung und Wohnungsbau verbunden, u. a. Cäciliengroden, Sande, Schortens, Upjever, Jever, Gödens, Cleverns, Schilldeich, Burg Kniphausen usw.

PAULS, HEINRICH

	Architekturstudium
ab 1938	Luftwaffenkommando See, Bauverwaltungsbezirk Wilhelmshaven
ab 1939	Luftwaffenbauamt Wilhelmshaven, zuständig für Hochbauplanungen im Raum Wilhelmshaven, Bremerhaven und Ostfriesische Inseln unter Baurat Schäfer

PAULY, JULIUS
(*16.02.1901 Birkenfeld)

	Jurastudium
ab 1927	Amtsgerichtsrat
ab 1929	Landgerichtsrat
1932–1945	Ortsgruppenleiter NSDAP Oldenburg, Abgeordneter Oldenburgischer Landtag, Oldenburgischer Staatsminister, als MDF zuständige Aufsichtsbehörde für Planung und Hochbau in Wilhelmshaven

PEIN, RICHARD
(*23.11.1883 Harburg)

	Hafenbaudirektor
1936–1939	Leiter Hafenbauressort KMW Wilhelmshaven
ab 1939	Ministerialrat OKM Abteilung Hafenbau

PETERS, ERNST
(*28.03.1886 Marne, †02.01.1975 Wilhelmshaven)

	Baugewerksschule Buxtehude, TH Darmstadt (ohne Diplom)
ab 1912	Übernahme Büro des Bruders Heinrich Peters, Wilhelmshaven
1914–1918	Kriegsdienst
ab 1918	Freier Architekt, BDA-Mitglied, befreundet mit dem Fotografen August Iwersen und dem Maler Georg Siehl-Feystett, u. a. Wohnbebauung Leiteweg (1924–1927), umfangreiche Wohnungsbautätigkeit im Inselviertel (1924–1937), Konsumgebäude Marktstraße (1931), Bebauung Admiral-Klatt-Str., Wohnhaus Freiligrathstr. 169 (1933), private Wohnbauten, umfangreiche Wiederaufbauten für Wilhelmshavener Spar- und Baugesellschaft, Geschäftshausbauten.

PFEIFFER, KARL
(*11.10.1902 Solnhofen, †30.07.1957 Bayreuth)

1922–1927	Architekturstudium München
1927–1931	Büro Bestelmeyer München
1931–1934	Assistent TH München (Bestelmeyer), u. a. Haus Ridder München (1934), Wettbewerbe Reichstagserweiterung Berlin (1929), Reichsehrenmal Berka (1930), Glaspalast München (1933), Theater Dessau (1934)
1934	Deutsches Siedlungswerk Berlin
1935–1939	RAM, Leitung Referat Städtebau, Theoretiker des Siedlungsbaus des „III. Reiches", u. a. Entwurf Haus Riefenstahl Berlin (1935), Wettbewerbe: Olympiabauten Berlin (1936), Museum Kiel (1936), TH Erweiterung München (1937), Bebauungspläne: Voslapp (1937), Fedderwardergroden (1938), Brandenburg (1939), Salzgitter (1939)
ab 1939	Professor Staatliche Kunstakademie Düsseldorf, u. a. Bebauungspläne: Blumberg (1940), Blexersande (1940), Behrsande (1941), Kiel-Raisdorf (1941), Lünen (1941), Speyer (1942), Reydt (1942), Dormagen (1942), Wiederaufbauplanungen im Elsaß (1941)
1943–1945	Kriegsdienst
1947–1957	Freier Architekt Bayreuth
ab 1949	Mitglied der Akademie für Städtebau und Landesplanung
ab 1952	Mitglied des Bayerischen Landes-Baukunstausschusses, umfangreiche Nachkriegsbautätigkeit

PICKER, DANIEL
(*10.04.1876 Brake, †02.12.1952 Wilhelmshaven)

ab 1900	Großhandelsgeschäft Wilhelmshaven
1914–1918	Kriegsdienst
1929–1943	Senator bzw. Stadtrat (1937–1943) in den Dezernaten für Wirtschaft und Verkehr, verfügte über guten Kontakt zu Adolf Hitler, ab 1929 NSDAP Bürgervorsteher Wilhelmshaven

PICKER, HENRY
(*06.12.1912 Wilhelmshaven, †02.05.1988 Starnberg)

	Jurastudium
	Promotion
	Regierungsrat Landratsämter Brake, Jever, Chef des obersten Gerichts der Reichsjugendführung
ab 1938	Regierungsrat bei der Vertretung Oldenburgs in Berlin, verschiedentlich eingeschaltet wegen Stadterweiterung und Neugestaltung Wilhelmshavens, NSDAP-Eintritt (1939), NSDAP-Gaustellenleiter
1942	Oberregierungsrat im Führerhauptquartier, Landrat in Norden
1943–1945	Kriegsdienst
nach 1945	Industriebeauftragter für Wilhelmshaven in Bonn
seit 1951	Historiker, Herausgeber von Hitlers Tischgesprächen

PITZ, KLEMENS
(*18.06.1907 Kleve, lebt in Ganderkesee)

bis 1934	Architekturstudium München (Göschel), Aachen (Gruber, von Schöver)
1934–1937	Baubeamtenausbildung
1937–1938	Luftfahrtministerium Berlin, Abteilung Lazarettbau
1938–1939	Marinestandortbauamt Bremerhaven

1939–1944	Marinestandortbauamt Wilhelmshaven, u. a. Mitarbeit am Marinelazarett Sanderbusch, Marinegerätelager Roffhausen
1944–1945	Leiter Marinebauamt Emden
1945–1949	Freier Architekt, Wilhelmshaven
1949–1972	Oberpostdirektion Bremen, zuletzt Abteilungspräsident Hochbau

PUTLITZ, ERICH ZU
(*01.02.1892 Brahlstorf, †28.01.1945)

1913–1914	Kunstgewerbeschule Magdeburg Bühnenbildner Düsseldorf
1919–1924	Büro Lasso und Kühne, Dresden
1924–1928	Büro Klophaus und Schoch, Hamburg, u. a. Wohnungen Kaiser-Friedrich-Ufer, Siedlungen am Weißen Berge, Bürohaus Mohlendorf, Siedlungen am Hasenberge, erster Preis Völkerbundpalast Genf
1932–1945	Freier Architekt Hamburg zusammen mit August Schoch, u. a. Wettbewerb Reichsführerschule der NSDAP München-Grünwald (1934), Wettbewerb Schlageterforum Düsseldorf, Siedlung am Rübenkamp Hamburg (1935), Wettbewerb Adolf-Hitler-Forum Dresden (1935), Akademie für deutsche Jugendführung Braunschweig (1936), HJ-Heim Tilsit (1938), Wohnungsbau in Wilhelmshaven (ab 1938), Wettbewerb Langewerth/Wilhelmshaven (1939)

QUEISNER, JOACHIM
(*12.07.1898 Königsberg)

	Architekturstudium
bis 1945	Hafenbauressort KMW, Mitarbeiter von Hans Klaucke, Werfthochbau
1951	verzogen nach Kaiserlautern

RAEDER, ERICH
(*24.04.1876 Wandsbek, †06.11.1960 Kiel)

1894	Eintritt Kaiserliche Marine
1912–1917	Admiralstabsoffizier und Chef des Stabes der Aufklärungsstreitkräfte
1914–1918	Erster Weltkrieg, Teilnahme an mehreren Seeschlachten, Skagerrakschlacht
ab 1925	Vizeadmiral
ab 1928	Admiral, Chef der Marineleitung
ab 1934	Generaladmiral, Aufbau der deutschen Kriegsflotte
01.04.1939	Von Hitler anläßlich des Stapellaufs der Tirpitz zum Großadmiral befördert
07.11.1942	Einweihung der IV. Einfahrt, verantwortlich für den uneingeschränkten U-Bootkrieg, setzte sich für die Gemeinschaftssiedlungen und Militärbauten in Wilhelmshaven ein, besuchte die Großsiedlungen und förderte die Monumentalplanungen
30.01.1943	Nach Unstimmigkeiten mit Hitler durch Dönitz ersetzt
1946	Zu lebenslänglicher Haft verurteilt
1955	Aus Haft entlassen
02.06.1957	Gemeinsam mit Dönitz Einweihung Ehrenmal Christuskirche Wilhelmshaven

RAMBACHER
(*ca. 1885, †1948)

	Architekturstudium
	Baubeamtenausbildung
bis 1940	Leiter Marinestandortbauamt Kiel
1940–1945	Referent in der Hochbauabteilung des Marineverwaltungsamtes des OKM, zuletzt Ministerialrat
1945	Nach Baden-Württemberg gezogen

RASCH, KLAUS
(*18.07.1910 Weimar, †28.04.1971 Wilhelmshaven)

	Sohn des Kunstmalers Prof. Otto Rasch und der Malerin Else Kaiser
1928–1936	Architekturstudium Darmstadt, Hannover, NSDAP-Eintritt (1930)
1936–1939	Baubeamtenausbildung Weimar, Upjever, Wilhelmshaven, Aurich, Hannover u. a. beim Luftwaffenbauamt, Siedlung Fliegerhost Upjever (1936), Luftwaffenhäuser Siebethsburg (1938)
1939	Promotion
ab 16.06.1936	Leiter Hochbauabteilung Stadt Wilhelmshaven, u. a. zuständig für Schulen und öffentliche Bauten, Kulturforum und Gestaltung Wilhemshavens im „III. Reich"
1942–1945	Kriegsdienst
1945–1953	Freier Architekt Wilhelmshaven, BDA-Vorsitzender Ostfriesland (1946–1953), u. a. Schlickbad (1948), Wettbewerbe Emden, Friesoythe, Wilhelmshaven, Preisrichter Ideenwettbewerb 1948, viele Wiederaufbauten, Dennoch-Programm (1949), Stadttheater (1952)
1953–1959	Leiter des städtischen Hochbauamtes Wilhelmshaven, verantwortlich für Hochbauten, Stadtteilerweiterungen, u. a. Altengroden-Süd, Reinhard-Nieter-Krankenhaus, Volkshochschule, Schulbauten, Preisrichter Wettbewerb Altengroden-Süd (1957)
1959–1966	Stadtbaurat Remscheid

RECH, TONI
(*04.03.1900 Bad Neuenahr, †29.08.1988 Wilhelmshaven)

bis 1928	Staatsbauschule Köln
1929	Meisterprüfung Maurerhandwerk, Gründung Bauunternehmen Wilhelmshaven, umfangreiche Bautätigkeit in Wilhelmshaven/Rüstringen
1934–1937	Obermeister der Baugewerkeinnung Wilhelmshaven
1956–1972	CDU-Ratsmitglied
1960–1971	Kreishandwerksmeister

REEKEN, OTTO VAN

	Architekturstudium
	Luftwaffenbauamt Wilhelmshaven, Mitarbeit Flugplatz Upjever unter Schütte (ab 1936)

REICHOW, HANS BERNHARD
(*25.11.1899, †07.05.1974 Bad Mergentheim)

1919–1923	Architekturstudium Danzig, München
1923–1925	Assistent TH Danzig
1926	Promotion
1925–1928	Büro Erich Mendelsohn Berlin
1928–1934	Stadtplaner Dresden unter Paul Wolf
1934–1936	Baurat Braunschweig
1936–1945	Baudirektor Stettin
1939	Bewerber um den Posten des Stadtbaudirektors Wilhelmshaven
ab 1945	Freier Architekt Hamburg, u. a. Sennestadt
ab 1964	Professor

RENKEN, CARL
(*31.03.1893 Sandel, †12.11.1954 Wilhelmshaven)

	Kaufmann, Mitglied des Bürgervorsteherkollegiums Wilhelmshaven
1931	NSDAP-Eintritt
1932	Senator Wilhelmshaven
1933–1937	Oberbürgermeister Wilhelmshaven, förderte den Bau Inselviertel und Marineforum
1937–1945	Oberbürgermeister Emden

RICHARD

Ministerialrat im Reichsverkehrsministerium, Professor, Gutachter für Reichsbahnangelegenheiten in Wilhelmshaven, Mitarb. an Bahnhofsplanung Schaar und Ortsgüterbahnhof Ebkeriege

RICHTER, JOHANNES

Architekturstudium
Leiter Reichsheimstättenamt Gau Weser-Ems Oldenburg, zuständig für Kleinsiedlungen im Land Oldenburg, Gegner von Fritz Höger, u. a. Planungen für Voslapp (1937), Kleinsiedlungen in Schortens (1940)

RODENBERG, WILHELM
(*09.08.1892 Westerstede)

	Jurastudium
1922	Regierungsrat Vechta
1924–1938	Legationsrat Berlin Ministerialdirektor Oldenburger Vertretung beim Reich Berlin, vielfach wegen Wilhelmshavens Stadterweiterung eingeschaltet, NSDAP-Mitglied seit 1932
ab 1938	Regierungspräsidium Osnabrück

RÖVER, CARL
(*12.02.1889 Lemwerder, †19.05.1942 Berlin)

	Kaufmann Oldenburg
1914–1918	Kriegsdienst
ab 1928	NSDAP Landtagsabgeordneter Oldenburg
1928–1942	Gauleiter Weser-Ems der NSDAP
ab 1930	Mitglied des Reichstages
1932–1933	Ministerpräsident Land Oldenburg
05.05.1933–19.05.1942	Reichsstatthalter in Oldenburg und Bremen, Vorsitzender LOB (1936–1942)
ab Sept. 1939	Reichsverteidigungskommissar Gau Weser-Ems
1940–1942	Gauwohnungskommissar, Reichskommissar für die Erledigung von Beihilfeanträgen
ab 1940	Beauftragter für die Neugestaltung im Gau Weser-Ems, vielfach mit Stadterweiterung und Neugestaltung Wilhelmshavens verflochten, Initiator Wettbewerb Wohnen nach dem Kriege (1941)

ROSENBERG, ALFRED
(*21.01.1893 Reval, †16.10.1946 Nürnberg)

bis 1916	Architekturstudium Riga, Moskau
ab 1923	Hauptschriftleiter Völkischer Beobachter, rassistisch-antisemitischer Schriftsteller
1927	Gründung Nationalsozialistische Gesellschaft für deutsche Kunst
1928	Gründung KDK
ab 1930	Reichstagsabgeordneter, Chefideologe der NSDAP, Freundschaft mit Carl Röver
1933–1945	Leiter außenpolitisches Amt der NSDAP, Plünderung von Kunstgalerien
27.05.1934	Besuch der Erstaufführung von „De Stedinger" Berne mit Walter Darré
19.10.1934	Grundsteinlegung Kultstätte Stedingsehre zusammen mit Himmler
13.07.1935	Einweihung Kultstätte Stedingsehre
1935	Beauftragte Höger mit dem Projekt einer Reichsführerschule in Hohen-Lychen
31.05.1937	Zusammen mit Heß in Wilhelmshaven
ab 1941	Minister für die besetzten Ostgebiete
Mai 1942	Totenrede für Röver Berlin
ab 1943	Beziehung zum Ahlhorner Architektenkreis
1946	In Nürnberg hingerichtet

ROTH, WERRY
(*17.05.1885 Berlin, †03.01.1985 Berlin)

Freier Architekt Berlin-Schöneberg, u. a. Theater Breslau, Rostock, Stolp, Dessau (1935–1937), Entwürfe für Theater Wilhelmshaven zusammen mit Benno von Arent (1939), Großsiedlung Berlin-Britz (1953–1959), Bauverwaltung Berlin (1957), Professor

SCHACHNER, BENNO
(*20.01.1902 Freising, †02.08.1987 Aachen)

bis 1925	Architekturstudium München
1925–1928	Baubeamtenausbildung
ab 1928	Zusammenarbeit mit seinem Vater Richard Schachner, Krankenhausbauten
1932	Habilitation TH Aachen
1938–1940	Professor TH Aachen
1940–1948	Professor TH Brünn, u. a. Planung städtisches Krankenhaus Wilhelmshaven zusammen mit Hermann Distel (1941)
ab 1948	Professor TH Aachen, u. a. umfangreiche Krankenhausbauten in der Bundesrepublik, Reinhard-Nieter-Krankenhaus Wilhelmshaven (1967)

SCHALLER, FRIEDRICH
(*1878, †1942)

	Jurastudium
bis 1939	Ministerialdirigent, Leiter der Unterkunftsabteilung des Marineverwaltungsamtes des OKM unter Theodor Schreiber

SCHEMM, WILLY
(*02.06.1897 Wuppertal, †1960 Wilhelmshaven)

	Lehre bei einem Architekten in Wuppertal
1916–1918	Kriegsdienst, Pilot bei der Luftwaffe
1918–1925	Abendkurse zur Vorbereitung auf ein Studium, Architekturstudium Aachen als Externer
1925–1937	Freier Architekt Wuppertal, u. a. Brotfabrik, Aufenthalt in Bulgarien im Auftrag des Reichsluftfahrtministeriums
ab 01.12.1937	Wohnungsbaugesellschaft Jade Wilhelmshaven, Leiter des Entwurfsbüros, u. a., Wohnungsbau Jadeviertel (1938), Volkswohnungen Voslapp (1939–1942), Abwicklung Cäciliengroden, Fedderwardergroden, Voslapp, Hansaviertel, Gewerbebauten, Fedderwardergroden

ab 1946	Techn. Geschäftsführer Wohnungsbaugesellsch. Jade, u. a. Lagerräumprogramme, Wiederaufbauten

SCHIPPER, GEORG
(*12.06.1878 Connhausen, †27.07.1972 Schortens)

ab 1911	Landtagsabgeordneter Oldenburg
1920–1926	Abteilungsleiter Staatliches Siedlungsamt Oldenburg
ab 1926	Oldenburgisches Domänenamt
ab 1932	Leiter Oldenburgisches Domänenamt, Landesökonomierat, für Kleinsiedlungen im Wirtschaftsgebiet Wilhelmshaven-Jeverland zuständig, u. a. Wohnkolonie Kniphausersiel (1929–1937), Siedlung Sanderbusch, Wattkultivierungen, Arbeitsdienstlager Kniphausersiel und Hooksiel, Deichbauten

SCHNEIDER, FRITZ KARL
(*02.03.1901 Worms, †30.05.1983 Oldenburg)

bis 1929	Architekturstudium München
1930–1935	Freier Architekt Nürnberg
1935–1955	Dezernent MDF für Hochbau und Stadtplanung, u. a. Bebauungsentwürfe für fast alle Wilhelmshavener Stadtteile, Planungsaufsicht des Landes Oldenburg über Stadterweiterung und Wiederaufbauplanung, beteiligt an den Bebauungsplänen Voslapp (1937), Fedderwardengroden (1938), Nachkriegsplanungen (1942)
1955–1967	Bezirksregierung Hildesheim, u. a. Preisrichter Ideenwettbewerb 1948, DRK-Kurheim Schillig

SCHOCH, AUGUST
(*12.03.1881 Herisau/Schweiz, †1957)

	Zimmermannslehre
1897–1902	Baugewerbeschule Winterthur, TH Zürich
1902–1920	Büros in Köln und Hamburg
ab 1920	Freier Architekt Hamburg
1932–1945	Freier Architekt zusammen mit Erich zu Putlitz, u. a. Wohnungsbau in Wilhelmshaven (1938), Wettbewerb Langewerth (1939)

SCHOLTZ

	Jurastudium
	Promotion
	Abteilungsdirigent RAM, zuständig für Städtebau und Raumordnung, griff verschiedentlich in Wilhelmshavens Stadterweiterung ein

SCHREIBER, THEODOR
(*1874, †1947)

	Jurastudium
bis 1939	Ministerialdirektor, Chef des Marineverwaltungsamtes des OKM, ihm unterstand die Unterkunftsabteilung mit deren Leiter Friedrich Schaller und die Hochbauabteilung mit deren Leiter Walter Kelm

SCHÜTT, FERDINAND
(*ca. 1890)

	Architekturstudium Berlin
	Promotion
ab 1934	Architekt Luftgaukommando See bzw. Luftwaffenbauamt Wilhelmshaven, Chef der Bauleitung der Luftwaffe Fliegerhorst Upjever (1935–1937)
ab 1937	Bauvorhaben an der Ostsee, KDF-Bäder, Bauvorhaben in Frankfurt für die OT
ab 1945	Freier Architekt Jever

SCHULTE-FROHLINDE, JULIUS
(*ca. 1894, †1968)

	Architekturstudium Stuttgart
1929	Stadtbaurat Nürnberg
1936–1939	Mustersiedlung Braunschweig Mascherode
1936–1940	Leiter Bauabteilung der DAF und Amt „Schönheit der Arbeit" unter Albert Speer
1938	Betrieb mit der Stadt Wilhelmshaven Gaumustersiedlungen im Wohnsiedlungsgebiet Wilhelmshaven-Jeverland
1940–1943	Hauptreferent beim Reichskommissar für den sozialen Wohnungsbau
ab 1943	Professor TH München
1944	Berater Arbeitsstab Speer
ab 1951	Stadtbaudirektor Düsseldorf

SCHULZ, WALTHER
(*17.07.1891 Weinheim, †06.1.1980 Freiburg)

	Architekturstudium
1933–1934	Leiter der Arbeitsbeschaffung Stadt Göttingen
1936–1941	Leiter des Amtes für Stadtgestaltung Göttingen, Stadtbaudirektor, kurzzeitiger Kriegsdienst, u. a. Entwürfe f. Gesamtbebauungsplan Göttingen, Umgestaltung der Innenstadt, Universitätsforum
06.09.1941–14.05.1942	Leiter Stadterweiterungsamt Wilhelmshaven unter Temp, u. a. Gesamtplan und Wiederaufbauplanungen Wilhelmshaven
1942–1946	Bauverwaltung Stadt Göttingen, auf Anordnung der Militärregierung entlassen

SCHULZE, ERNST

1925	Gründungsmitglied NSDAP Oldenburg
ab 1933	Gaugeschäftsführer NSDAP Gau Weser-Ems, Landeskulturwalter, zuständig für Angelegenheit der RDBK und der Nordischen Kunsthochschule Bremen, Gaupropagandaleiter und Landesstellenleiter des Reichsministeriums für Volksaufklärung und Propaganda

SCHUMACHER, FRITZ
(*1869 Bremen, †1947 Hamburg)

bis 1893	Architekturstudium München, Berlin
1893–1899	Büros München, Leipzig
1899–1909	Professor TH Dresden, Mitbegründer Deutscher Werkbund (1907)
1909–1933	Leiter Hochbauwesen Stadt Hamburg, u. a. zahlreiche öffentliche Bauten in Hamburg, Zusammenarbeit mit Gutschow und Wortmann (1926–1928)
1933	Entlassung aus dem öffentlichen Dienst, empfahl 1936 Landesplaner Wortmann Hallbauer als Leiter der Landesplanungsstelle Wilhelmshaven, Gutachter im Wettbewerb Marinestationsgebäude Wil-

helmshaven (1936), Preisrichter Wettbewerb Wohnen nach dem Kriege (1941)

SCHUSTER, RUDOLF

(†1950)

	Maschinenbaustudium
bis 1945	Mitarbeiter LOB unter Wortmann
ab 1945	Bauverwaltung Bremen
1948–1950	Mitarbeit am Gutachten über die Grundlagen für den Neuaufbau der Stadt Wilhelmshaven

SCHWARTING, THEODOR

(*17.10.1902)

	Architekt bei der Baufirma Hermann Möller, Entwurf verschiedener Arbeiterlager
ab 1937	Planungen Neuengroden
1951	nach Bonn verzogen

SCHWARZ, MAX KARL

(*07.11.1895 Mannheim, †04.10. 1963 Worpswede)

1914–1918	Kriegsdienst
1919–1922	Gartenarchitekturstudium Proskau
ab 1922	Gartenarchitekt Worpswede, anfangs bei Leberecht Migge
ab 1933	Landschaftsanwalt für Autobahnbau unter Todt
1941–1945	Berater Arbeitsstab Speer
1944–1945	Berater für den Wiederaufbau Wilhelmshaven, Zusammenarbeit mit Gutschow, Ahlhorner Architektenkreis (1943–1945)
1948–1950	Gutachter über die Grundlagen für den Neuaufbau der Stadt Wilhelmshaven

SEETZEN, FRITZ

(*31.10.1897 Wittmund, †01.01.1957 Wilhelmshaven)

	Elektroingenieurstudium
	Vermutlich Elektroingenieur KMW, NSDAP-Eintritt (1932), Kreisobmann des NSBDT, Leiter des Kreisamtes für Technik der NSDAP (ab 1934)

SIEDENBURG, KURT

(*03.06.1910, lebt in Oldenburg)

bis 1935	Architekturstudium München, Hannover (Hölscher, Fiederling, Wickop)
1935–1938	Baubeamtenausbildung
ab 1938	Oldenburgisches Staatsministerium, Landesbeauftragter für den Neuaufbau Wilhelmshavens
ab 1940	Kriegsdienst
1943	Marineoberbaudirektion Wilhelmshaven unter Naud und Kretschmer
1943–1944	Leiter Marinebauamt Esbjerg
1944–1945	Marinepionierbauwesen
1945–1949	Land bzw. Verwaltungsbezirk Oldenburg
1949–1975	Leiter Staatshochbauamt I bzw. II Oldenburg, nebenamtlich Denkmalpfleger Land Oldenburg

SIEMER, HEINRICH

(*10.02.1891 Bant, †18.01.1960 Wilhelmshaven)

bis 1911	Baugewerkschule Bremen
1911–1937	Hochbauamt Rüstringen unter Martin Wagner, Willy Hahn, Carl Haefner, u. a. Wettbewerbe mit Martin Wagner, Paul-Hug-Kinderheim (1928), Bauleitung Rüstringer Rathaus für Fritz Höger (1928–1929), Adolf-Hitler-Siedlung (ab 1934), Bauleitung Arbeitsamt für Fritz Höger (1937)
1929–1930	Zwischenzeitlicher Gewerbeschuldienst
1937–1954	Bauverwaltung Stadt Wilhelmshaven, Hochbau, Baupflege, Stadtplanung, u. a. Bebauungspläne Hansaviertel und Jadeviertel (1937), umfangreiche Planungs- und Stadtgestaltungsaufgaben

SIEPERMANN, JOHANN

(*14.10.1898 Elberfeld)

	Architekturstudium
ab 1938	Wohnungsbaugesellschaft Jade
1959–1964	Technischer Direktor Wohnungsbaugesellschaft Jade, u. a. Olympia-Siedlung Bismarckstr. (1954), Nachkriegswohnungsbau
1974	Nach Schweinfurt verzogen

SPEER, ALBERT

(*19.03.1905 Mannheim, †01.09.1981 London)

1924–1928	Architekturstudium München, Berlin
1928–1932	Assistent TH Berlin (Tessenow)
ab 1932	Freier Architekt Berlin, Aufträge der NSDAP: Reichskanzlei (1938–1939), Planungen für NS-Massenkundgebungen, Reichsparteitagsgelände Nürnberg (ab 1934)
ab 03.01.1937	Generalbauinspektor für die Reichshauptstadt und andere deutsche Städte (GBI), griff verschiedentlich in Wilhelmshavener Monumentalplanungen ein, Amt „Schönheit der Arbeit"
ab 1938	Professor
ab 1941	Reichstagsabgeordneter
ab 08.02.1942	Rüstungsminister, Generalbevollmächtigter für die Regelung der Bauwirtschaft, Leiter OT (Nachfolger von Fritz Todt)
1943–1945	Arbeitsstab Speer, Planung des Wiederaufbaus bombenzerstörter Städte
1944	Marinebauwesen Speer und der OT unterstellt
1946	Zu 20 Jahren Haft verurteilt
1966	Aus Haft entlassen, schriftstellerische Arbeit

SUHR, HELLMUT

(*08.05.1909 Rüstringen, †21.02.1968 Wilhelmshaven)

	Musikstudium Weimar
1933–1936	Staatsbauschule Oldenburg
1936	Bauleitung der Luftwaffe Nordenham
1937–1940	Büro Fritz Höger, Leiter der Außenstelle Wilhelmshaven, u. a. Mitarbeit Neu-Siebethsburg, Rathauserweiterungsplanung, möglicherweise Hochbauten für die Nordwerft
1940–1945	Kriegsdienst
1945–1950	Gefangenschaft
1950–1955	Bauverein Rüstringen, u. a. Wohnhäuser Störtebeker Platz (1950), Flüchtlingssonderprogramm Siebethsburg (1950–1952), Bebauung Berliner Platz (1954)
1955–1968	Freier Architekt Wilhelmshaven, u. a. Wiederaufbauten, Wohnungsbau

TEMP, WALTER

(*15.06.1902 Flatow, †04.08.1974 Soltau)

bis 1924	Bauingenieurstudium Danzig
1924–1928	Baubeamtenausbildung Berlin
1928–1933	Kulturbauverwaltung Allenstein, NSDAP-Eintritt (1932)
ab 1933	Aufbau des Arbeitsdienstes in Ostpreußen

1933–1936	Leiter der städtischen Betriebswerke Allenstein, verurteilt zu 1 Jahr Gefängnis wegen Landfriedensbruch
1936–1937	Bauverwaltung Danzig, Kreisamtsleiter
1937–1939	HNBD Wilhelmshaven, Mitarbeit am Bau der IV. Einfahrt
ab 04.05. 1939	Stadtbaudirektor Wilhelmshaven
1940–1945	Gebietsbeauftragter für die Regelung der Bauwirtschaft im Lande Oldenburg, verantwortlich für Stadterweiterung und Monumentalplanungen in Wilhelmshaven, umfangreiche Planungen in Absprache mit den Berliner Behörden
01.09.1944	Leiter des Kreisamtes für Technik der NSDAP, Ahlhorner Architektenkreis (1943–1945)
bis 1945	Kriegsdienst
ab 1945	Wasser- und Schiffahrtsamt, Jade-Wasser-Werke
ab ca. 1950	Freier Architekt Wilhelmshaven
1956–1960	DP-Ratsherr Wilhelmshaven

THEES, HEINZ-GÜNTHER
(*24.04.1911 Lehe, lebt in Bremerhaven)

1932–1935	Architekturstudium Danzig (Gruber, Fendrich), Hannover (Hölscher, Fischer, Kanold, Fiederling, Wickop)
1935–1936	Assistent TH Hannover (Hölscher)
1936–1942	Marinestandortbauamt Wilhelmshaven unter Otto Lückemeyer, u. a. Kasernen-U-Bootstützpunkt (1936–1941), Bonte-Kasernen am Großen Hafen (1937–1938), Standortschwimmanstalt (1937–1938), Marinenachrichtenstelle Langenargen/Bodensee (1937–1938), Marinenachrichtenstelle Soest (1937–1938), Umbau Marinekleiderkammer (1937), Innenausbau Marineoffiziersheim (1938–1939), Kasernen Sengwarden/Befehlsstelle Marinekommando-Nord (1938–1942), verschiedene Ausweichvermittlungsstellen (1939–1941), Hauptvermittlungsstelle Langewerth (1939–1941), Kasernen Varel/6. Schiffsstammabteilung (ab 1939), Flak-Schießplätze Dangast (1941), Frauenlager Langwarden (1942)
1943–1945	Leiter Marinebauämter Paris, Nantes, La Rochelle
1945–1949	Gefangenschaft, Entminung des Hafens La Rochelle
1949–1950	Büro Heinrich Jäger Bremerhaven, Gründungsmitglied BDA
ab 1950	Freier Architekt Bremerhaven, zweitweise Bürogemeinschaft mit Franzius, u. a. Museumsbauten, Denkmalpflege, viele Nachkriegsbauten
ab 1953	FDP-Ratsmitglied Bremerhaven

TIBURTIUS, CARL
(*20.04.1874 Berlin, †24.02.1953 Berlin)

1892–1898	Architekturstudium Berlin, Braunschweig, München
1898–1901	Baubeamtenausbildung
bis 1914	Hafenbauressort der Kaiserlichen Werft Wilhelmshaven, u. a. Leiter der Schleusenbauabteilung, III. Einfahrt, Hafen-Süderweiterung
1914–1918	Baudirektor der baltischen Gewässer, u. a. Hafenbauten Riga, Libau
1918–1920	Hafenbaudirektor Kiel
1920–1922	Leiter Reichsbauamt Kiel II
1922–1935	Vorsteher des Hafenbauamtes Marinearsenal Kiel, Referent für Hafenbaufragen beim Kommandierenden Admiral der Marinestation der Ostsee
1935–1939	Hafenbaudirektor, Leiter der HNBD Wilhelmshaven, Bau von Nordwerft und IV. Einfahrt
1941–1942	Hafenbauten Nordafrika, Schwarzes Meer
1942–1953	Ruhestand Berlin

TIETJEN, ERNST
(*03.02.1882 Varel, †29.11.1964 Varel)

	Tischlerfachschule Detmold, Meisterprüfung, Innenarchitekt, Mitglied des Bundes Oldenburger Werkkünstler, Übernahme des väterlichen Tischlerbetriebes Varel
ab 1914	Mitglied des Deutschen Werkbundes
1914–1918	Kriegsdienst, u. a. Schiffsausbauten, Arbeiten für Möbelhäuser, Zusammenarbeit mit den Vereinigten Werkstätten, Innenausbau Bahnhof Zoo Berlin, Innenausbau Reichskanzlei (1938–1939), Innenausbau Spanische Botschaft Berlin, Möbelausstattung Marinelazarett Sanderbusch (1937), Innenausbau Offiziersheime
ab 1945	Möbelproduktion, Baukastenmöbel

TILLESSEN, KARL
(*09.08.1907 Flensburg, lebt in Wilhelmshaven)

bis 1932	Bauingenieurstudium Danzig, Hannover (Franzius)
1932–1936	Baubeamtenausbildung
1936–1944	Abteilungsleiter Strombauressort KMW Wilhelmshaven unter Frede und Markworth, zahlreiche Bauten für Fahrwasser und Uferschutz an der Jade, u. a. Leuchtturm Mellum-Plate (1939–1942), Leuchtfeuer Minsener Oog (1942–1943), Seebuhne Wangerooge
1944–1945	Bauabteilung OKM
ab 1945	Abteilungsleiter Gruppe Strombau Wilhelmshaven
1955–1962	Leiter Wasserwirtschaftsamt Wilhelmshaven, zahlreiche Sielbauwerke zusammen mit Hans Klaucke
1962–1972	Technischer Leiter des Oldenburgisch-Ostfriesischen Wasserverbandes

TODT, FRITZ
(*1891 Pforzheim, †09.02.1942)

1911–1920	Bauingenieurstudium Karlsruhe, München
1917–1918	Kriegsdienst
1921–1933	Bauunternehmung Sager und Woerner
ab 1931	Leiter der Fachgruppe Bauingenieure und Landesleiter KDAI, Brauner Bericht (1932)
1931	Promotion
ab 1933	Generalinspektor für das deutsche Straßenwesen
ab 1934	Leiter NSBDT und Amt für Technik der NSDAP
1938–1942	Generalbevollmächtigter für die Regelung der Bauwirtschaft, Aufbau OT
1938	Generalinspektor für Sonderaufgaben im Vierjahresplan
1940–1942	Reichsminister für Bewaffnung und Munition
ab 1941	Generalinspektor für Wasser und Energie, verschiedentlich wegen Verkehrsbauten und Autobahnanbindungen für Wilhelmshaven eingeschaltet

TORNOW, WALTER
(*1907, lebt in Kiel)

	Jurastudium
	Promotion
1934–1938	Marineintendantur Kiel, zuständig für Marinewohnungsbau im Ostseebereich
1938–1944	Oberintendanturat in der Unterkunftsabteilung des Marineverwaltungsamtes des OKM unter Friedrich Schaller und Ernst Fitzler, Referent für den gesamten Marinewohnungsbau
bis 1945	Aufsichtsratsmitglied der Gemeinnützigen Gesellschaft zur Schaffung von Wohngelegenheiten für Reichsangehörige Berlin
1948–1953	Geschäftsführer des BDA Schleswig-Holstein Kiel
1953–1973	Wirtschaftsministerium Kiel zuletzt Leitender Ministerialrat
1965–1983	Vorsitzender des Eintragungsausschusses der Architektenkammer Schleswig-Holstein

TRIEBEL, WALTER
(*03.10. 1898 Ingolstadt)

	Bauingenierstudium
	Baubeamtenausbildung
1943–1944	Leiter Marineoberbauamt Wilhelmshaven, die Hochbauaufgaben des aufgelösten Marinestandortbauamtes übernahm der Hochbauer Alfred Hukke
1944	Verließ Wilhelmshaven

TRITTLER, GUIDO

	Architekturstudium
	Promotion
	Hafenbauressort KMW Wilhelmshaven, zuständig für Werftbauten
	Nach dem Kriege Ministerialbeamter, zuletzt Ministerialdirigent Stuttgart

TUCHSCHERER, KARL
(*21.01.1903 Heppens, †12.12.1969 Wilhelmshaven)

	Jurastudium
bis 1928	Architekturstudium Braunschweig
1928–1930	Arbeitslos
1930–1935	Baubüro Karstadt-Konzern Hamburg
1935–1937	Bauverwaltung Stadt Rüstringen, NSDAP-Eintritt (1937)
1937–1945	Hafenbauressort KMW Wilhelmshaven, u. a. Bebauungsplan Sande-Neufeld (1939–1940), Arbeiterunterkünfte Ebkeriege (1939), HJ-Wohnheim für Jungarbeiter Schillerstr., viele techn. Marinebauten
1945–1949	Freier Architekt Wilhelmshaven, umfangreiche Nachkriegsbautätigkeit

VEESER, JOSEPH
(*22.01.1906 Ostrach, †29.11. 1981 Wilhelmshaven)

	Praktikum Bauhütte Berlin unter Martin Wagner
bis 1930	Architekturstudium Baugewerkschule Berlin-Neukölln
1930–1933	Planungsabteilung Bauhütte Berlin, u. a. Verwaltungsgebäude des Baugewerksbundes Berlin, Schulungsheim Werlsee, Bauausstellung Berlin (1931), Planungsaufträge Paris für den Verband sozialer Baubetriebe, Schulung von arbeitslosen Bauhandwerkern, Mitglied der Eisernen Front
ab 1933	Nach Verbot der gewerkschaftseigenen Betriebe Arbeit bei der Bau- und Bodenbank und kurzzeitige Tätigkeit bei der Gemeinnützigen Siedlungs- und Wohnungsbaugesellschaft Berlin
ab 16.12.1938	Oberbauleiter Büro für Bauberatung und Organisation Fritz Hilgenstock, Fedderwordergroden
ab 1946	Architekt in Wilhelmshaven, umfangreiche Wiederaufbau- und Nachkriegsbautätigkeit in Wilhelmshaven und im Umland, Ideenwettbewerb 1948, Wettbewerb Altengroden-Süd (1957)

VOSSHANS, JOSEF
(*12.12.1906 Dortmund, lebt in Bremerhaven)

bis 1927	Kunstgewerbeschule Dortmund, Bildhauerei, Architektur, Stukkateur im väterlichen Betrieb
1930–1933	Staatsbauschule Holzminden
ab 1933	Bauverwaltung Stadt Dortmund, Hüro Holzapfel Bochum
1936–1937	Hafenbauressort KMW Wilhelmshaven unter Trittler und Hacke
1937–1941	Luftwaffenbauamt Wilhelmshaven, u. a. Mitarbeit Seefliegerhorst Wilhelmshaven (1937–1938), Luftwaffenhäuser Siebethsburg (1938), Flugplatz Bremerhaven-Weddewarden (1940), Flugplatz Helgoland (1941)
1941–1945	Luftwaffenbauverwaltung Norwegen, OT
1945–1947	Büro Rump Bremerhaven
ab 1947	Freier Architekt Bremerhaven

WEGENER, PAUL
(*01.10.1908 Varel)

	Kolonialwirt
1930	NSDAP-Eintritt, Ortsgruppenleiter Varel
1933	NSDAP Kreisleiter Bremen, Reichstagsmitglied, SS-Gruppenführer Adjudant von Heß und Bormann
1939	Stellvertreter des Reichskommissars für Norwegen
1942–1945	Gauleiter Weser-Ems, Reichsstatthalter, Gauwohnungskommissar, Reichsverteidigungskommissar Gau Weser-Ems, Vorsitzender LOB, Beauftragter für die Neugestaltung im Gau Weser-Ems, Verbindungen zum Ahlhorner Architektenkreis (1943–1945), Förderer der Neugestaltung Wilhelmshavens
Mai 1945	Mitglied der Regierung Dönitz
23.5.1945	Festnahme in Flensburg, 1949 zu 6½ Jahren Gefängnis verurteilt

WIDEKIND, KARL HASSO VON
(*1910, †1982)

	Jurastudium
1938–1943	Abteilungsleiter Siedlungswesen und Wohnungsbau der Marineintendantur Wilhelmshaven, zuständig auch für Landesplanung, Gemeinschaftseinrichtungen und Gewerbebauten
bis 1975	Bundesverkehrsministerium, zuletzt Ministerialrat

WIEPKING-JÜRGENSMANN, HEINRICH
(*1891, †17.06.1973)

	Gärtnerlehre Hannover
	Städtebaustudium London, Paris

	Expeditionen Sibirien, Kaukasus, Süd-Ost-Europa
	Freier Gartenarchitekt Berlin, u. a. Tannenbergdenkmal (1927), Reichssportfeld und Olympiabauten Berlin (1936), Großaufträge der Nationalsozialisten
ab 1934	Professor Berlin, empfahl Fresenius als Gartenamtsleiter nach Wilhelmshaven (1939), von Himmler zum Sonderbeauftragten für Fragen der Landschaftsgestaltung berufen
ab 1938	Grünplanungen für Wilhelmshavener Gemeinschaftssiedlungen Altengroden, Fedderwardergroden, Breddewarden in Zusammenarbeit mit Hübotter, Heintz, Homann
1945	Gründung Höhere Lehranstalt für Gartenbau Osnabrück
1949–1959	Professor für Landschaftspflege und Naturschutz TU Hannover

WOHLSCHLÄGER, JOSEF
(†ca. 1950)

	Architekturstudium München
	Büro Bonatz
1914–1916	Bauleitung Oldenburger Landtag für Bonatz
ab 1916	Hochbaudezernat Land Oldenburg, MDF, u. a. Amtsgebäude Friesland Jever (1937–1938)

WOLF, GUSTAV
(*1887, †1954 Münster)

bis 1913	Architekturstudium München (Fischer), Zusammenarbeit mit Schmitthenner
1922–1927	Westfälische Heimstätte Münster
1928–1935	Leiter Kunstgewerbeschule Breslau, Kontakte zur LOB
1937–1951	Leiter Landesamt für Baupflege Münster, Ahlhorner Architektenkreis (1943–1945), Vors. Preisrichter Ideenwettbewerb 1948, Professor

WORTMANN, WILHELM
(*15.03.1897 Bremen, lebt in Hannover)

1914–1924	Architekturstudium München, Dresden
1916–1919	Kriegsdienst und Kriegsgefangenschaft
1922	Praktikum bei Paul Wolf Dresden
1924–1925	Büro Emil Högg
1926–1927	Hochbaudirektion Hamburg unter Fritz Schumacher, Zusammenarbeit mit Gutschow
1928–1932	Stadtplanungsamt Bremen, Generalsiedlungsplan mit Schumacher und Blum
1932–1934	Freier Architekt Bremen
1934–1945	Baurat im Stadtplanungsamt Bremen, Landesplaner LOB, empfahl Hallbauer nach Wilhelmshaven, starker Einfluß auf Raumordnung, Kommunalpolitik und Stadtplanung im Wohnsiedlungsgebiet Wilhelmshaven-Jeverland, Preisrichter Wettbewerb Wohnen nach dem Kriege (1941), Mitarbeiter Arbeitsstab Speer, Wiederaufbauplanungen Wilhelmshaven zusammen mit Gutschow (1944–1945), Ahlhorner Architektenkreis (1943–1945), Leiter Arbeitskreis Stadt- und Landesplanung NSBDT Gau Weser-Ems
August 1939–Juli 1940	Kriegsdienst
1945–1958	Freier Architekt Bremen, umfangreiche stadtplanerische Arbeiten für viele Städte, Gutachten über die Grundlagen für den Neuaufbau der Stadt Wilhelmshaven (1948–1950), Preisrichter Wettbewerb Altengroden-Süd (1957)
1956–1963	Professor TU Hannover

ZENKER, EWALD
(*28.06.1910 Berlin, lebt bei Augsburg)

	Sohn von Admiral Hans Zenker (1870–1932), aufgewachsen Berlin und Wilhelmshaven (1916–1917, 1919–1922)
1929–1933	Architekturstudium Stuttgart (Schmitthenner), Berlin (Tessenow)
1934–1935	Büro Mebes und Emmerich Berlin
1935–1938	Bauabteilung der Luftwaffe, u. a. Flughafenbauten Burg, Utersen, Pinneberg
1938–1942	Hochbauabteilung OKM unter Kelm, zuständig für Marinehochbauten an Nord- und Ostsee, u. a. Bebauungsplan Altengroden (1938–1939)
1942–1945	Stellv. Leiter Marineoberbauamt Drontheim (Norwegen)
1945–1947	Stadt Hannover, zuletzt Ltd. Baudirektor, Zusammenarbeit mit Otto Meffert, Walter Kleffner, Rudolf Hillebrecht
1947–1975	Geschäftsführer Gemeinnützige Baugesellschaft Hannover

ZOPFF, HERMANN
(*15.11.1871 Plauen, † wahrscheinlich 13.02.1945 Dresden

bis 1899	Architekturstudium Stuttgart, Dresden
1899–1903	Baubeamtenausbildung
1904–1905	Studienaufenthalt Holland, Belgien, England, Frankreich, Spanien, Italien
1905–1910	Baubeamter Sächsischer Staatsdienst
1910–1937	Stadtbaurat Wilhelmshaven, ständiger Mitarbeiter der Zeitschrift „Der Städtebau", u. a. Erweiterung Königin-Luise-Schule (1911), Städtisches Lagerhaus (1912), Gewerbeschule Rheinstraße (1912), Kath. Volksschule Mozartstraße (1913), Herrenbadeanstalt Südstrand (1918), Bebauungsplan Wilhelmshaven, Bahnhofs- und Inselviertel (1918), Ratskellerumbau (1924), Südstrandanlagen (1928), Altenheim Lindenhof (1931), Adolf-Hitler-Tor (1933), Sparkasse Goekerstraße (1933), Kurparkgestaltung (1933), Marine- und Kolonialmuseum (1935), Schlickbad Rheinstraße (1936)
1937–1945	Ruhestand in Dresden, Tod vermutlich durch Bombenangriff

9.6. Abbildungen

1 Heute noch vorhandene Spuren der NS-Architektur in Wilhelmshaven, außer Bunker, Straßen, Deiche, Kartengrundlage um 1940
2 Admiralität Berlin, Leipziger Platz, zerst., 1850 (Hitzig)
3 Oldenburger Ministerien mit Hoheitszeichen, 1914–17 (Bonatz)
4 Reichsmarineamt Berlin, Reichpietsch Ufer, 1914 (Reinhardt/Süssenguth)
5 Projekt OKM Berlin, 1937/43 (Dübbers/Bonatz/Speer)
6 Hagen-Plan mit Befestigung nach Fischer, 25.06.1856
7 Klotz-Plan der Befestigung, vermutl. 1863
8 Stoltz-Plan der Befestigung, vermutl. 1865/68
9 Übersichtsplan der marinefiskalischen Gebäude, Borrmann-Plan 1891
10 Werfttor 1, Gökerstr., 1876 (Hafenbauressort)
11 Marinestationsgebäude, Adalbertplatz, zerst., 1870 (Adler/Hitzig/Klingenberg)
12 Kasernen 2. Seebatallion, Gökerstr., zerst., 1904/07 (Balfanz/Marineintendantur)
13 Kasernen II. Matrosendivision, Mühlenweg, 1912/13 (Strempel/Balfanz/Marineintendantur)
14 Marineintandentur, Peterstr., teilzerst., 1904 (Schubert/Fleinert)
15 Kaufhaus EPEGE, Gökerstr., zerst., 1930 (Fuchs)
16 Werftspeisehaus, Gökerstr., zerst., 1902 (Marineintendantur)
17 Projekt Leo, Gökerstr., 1929 (Freygang)
18 Kaufhaus Karstadt, Bahnhofsvorplatz, 1921/24 (Schäfer)
19 Bürogebäude Konsum, Marktstr., 1931 (Peters)
20 Allgemeine Ortskrankenkasse Wilhelmshaven, Kieler Str., 1929/31 (Lübbers)
21 Wohnhäuser Freiligrathstr., 1932 (Minke)
22 Wohnkolonie Kniphauser Siel, 1936 (Domänenamt)
23 Wohnhäuser Widukindstr., 1938 (Stadterweiterungsamt/verschiedene Architekten)
24 Richtfest Adolf-Hitler-Siedlung, 04.05.1935
25 Siedlung Sanderbusch, 1935 (Domänenamt)
26 Planung Adolf-Hitler-Siedlung, 50 Wohnungen, 1934 (Siemer)
27 Planung Inselviertel und Wohnsiedlung Nordost, 1918 (Zopff)
28 Bestandsplan Inselviertel mit Arbeitslagern und Bunkern, um 1941 (Zopff)
29 Wohnhäuser Wangeroogestr., um 1934 (Peters)
30 Planung Voslapp, 1235 Wohnungen, 1937, Gemeinschaftsskizze (Schneider)
31 Volkswohnungen Voslapp, Geniusbankstr., Typ VIII a, Vorentwurf, 1939 (Schemm)
32 Kleinsiedlungstypen Voslapp, 1938 (Lübbers)
33 Kleinsiedlungstypen Voslapp, 1938 (Lübbers)
34 Planung Cäciliengroden, 360 Wohnungen, 1938 (Heuer)
35 Wohnhäuser Cäciliengroden, 1938 (Heuer)
36 Cäciliengroden, Visite des Großadmirals, 09.01.1940
37 Planung Fedderwardergroden, 4100 Wohnungen, 1938 (Pfeiffer/Schneider/Hallbauer), überarbeitet 1940
38 Modell Fedderwardergroden für die Besprechung mit Hitler, 01.04. 1939
39 Bauarbeiten Fedderwardergroden, 1941
40 Wohnhäuser Fedderwardergroden, 1940 (Bork)
41 Wohnhäuser Fedderwardergroden, Zoppoter Str., 1940 (Bork)
42 Fedderwardergroden, Preußenstr. mit Notpostamt, 1940
43 OKM-Bebauungsplan Altengroden, 14.02.1939 (Zenker)
44 Ladenstr. 1, Altengroden, Goedenser Weg, 1943 (Kühn)
45 Wohnhäuser Altengroden, Ubbostr. Typ H, 1945 (Mellin/Kühn)
46 2. Bebauungsplan Neuengroden, 600 Wohnungen, 03.03.1939 (Schwarting)
47 Neuengrodener Weg, 1939 (Schwarting)
48 Planung Sande-Neufeld, 1939 (Tuchscherer)
49 Wohnblocks Sande-Neufeld, 1939 (Hilgenstock)
50 Gemeinschaftshalle Sande-Neufeld, 1939 (Hilgenstock)
51 Empfangsgebäude Bahnhof Sande, 1941 (Langewand/Reichsbahndirektion Münster)
52 Neugestaltung Wilhelmshaven, 29.03.1939 (Hallbauer/Stadterweiterungsamt)
53 Bebauungsplan Hansa- und Jadeviertel, 1937 (Stadterweiterungsamt) Bauten der Jade u. der Jadestädtischen Wohnungsbaugesellschaft sind schwarz markiert.
54 Wohnhäuser Jadeviertel, zerst., 1938 (Schemm)
55 Hansaviertel, Paul-Hug-Str., 1937 (Lübbers)
56 Wohnhäuser Hansaviertel, Typ Ib; Otto-Meentz-Platz, 1937 (Boyken)
57 Neu-Siebethsburg, 1600 Wohnungen, 1935–38 (Höger)
58 Wohnhäuser Meyer und Hoeland, Baudissinweg, teilzerst. 1934–35 (Höger)
59 Wohnhäuser Werftstr., 1938 (Höger)
60 Wohnhaus tom-Brok-Str. im NS-Schmuck, 1938 (Höger)
61 Tausend-Wohnungsprojekt Schaarreihe, Typ RIII, 1935 (Höger)
62 Projekt Siedlung Burg Kniphausen, 1800 Wohnungen, 1937 (Bürgin)
63 Projekt Siedlung Schilldeich, 1500 Wohnungen, 1938 (Bürgin)
64 Wettbewerbsentwurf Langewerth, 15–19000 Wohnungen, 1939 (Feder)
65 Wettbewerbsentwurf Langewerth, 15–19000 Wohnungen, 1939 (zu Putlitz)
66 Projekt Breddewarden, 1939 (Heuer)
67 Projekt Schortens-Klosterneuland, 1940 (Richter)
68 Werfthalle III, Fliegerhost Upjever, 1937 (Schütt)
69 Marineinstandsetzungsamt Ulmenstr., 1937 (Franzius)
70 Marinestationsgebäude Marineforum, Haupteingang, 1937 (Franzius/Geisenhainer)
71 Marinestationsgebäude, geplante Fassade Mozartstr., 1937 (Franzius/Geisenhainer)
72 Marinegerätelager und Arbeiterlager Roffhausen, Lageplan, 1944 (Geisenhainer)
73 Marinegerätelager Roffhausen, 1937 (Geisenhainer), später AEG/Olympia
74 Marinegerätelager Roffhausen und Reste des Arbeiterlagers, Luftbild, 1950
75 Marinelazarett Sanderbusch, Lageplan, 1937 (Geisenhainer)
76 Marinelazarett Sanderbusch, 1937 (Geisenhainer)
77 Marinelazarett Sanderbusch, 1937 (Geisenhainer)
78 U-Bootstützpunkt, Modell, 1939 (Thees)
79 U-Bootstützpunkt Wilhelmshaven für die U-Flottille „Saltzwedel“, 1936–41(Thees)
80 U-Bootstützpunkt, Terrasse Wirtschaftsgebäude, 1939 (Thees)
81 U-Bootstützpunkt, Hitlerbesuch 09.04.1940
82 Bonte-Kasernen, zerst., 1937–38 (Thees)
83 Eingangsschmuck Bonte-Kasernen, zerst., 1937–38 (Grau)
84 Mannschaftsraum Bonte-Kasernen, zerstr., 1937–38 (Thees)
85 Gedenkstätte Bonte-Kasernen, zerst., 1937–38 (Thees)
86 Marinekommando Nord, 1938–42 (Thees)
87 Marinekommando Nord, 1938–42 (Thees)

88 Kasernen der 6. Schiffsstammabteilung Varel, 1939–42 (Thees)
89 Torpedowerftgebäude, 1940 (Klaucke)
90 Standortschwimmanstalt, 1937–38 (Thees)
91 Navigationsressort der Kriegsmarinewerft, Büro- u. Plangebäude, Südufer Zwischenhafen, 1939–42 (Hafenbauressort der Kriegsmarinewerft)
92 Modell der geplanten Werkshallen der Nordwerft mit Panzerschiff, 1938 (Hafenneubaudirektion)
93 Gießerei der Kriegsmarinewerft, Schiff 2, 1938–40 (Klaucke)
94 Werftbahnhof, Gießerei, Norddeutscher Eisenbau, um 1940
95 Minsener Oog, 1942–43 (Tillessen/Klaucke)
96 Bürohaus Ressort II der Kriegsmarinewerft, zerstr., um 1940 (Klaucke)
97 Entwurf Motorenbauwerkstatt Westwerft zerst. um 1940 (Klaucke)
98 Normentwurf Arbeiterlager für 1000 Mann, 1939 (Marinestandortbauamt)
99 Planung Arbeiterunterkünfte Ebkeriege, 1939 (Tuchscherer)
100 Schleusenkammer IV. Einfahrt, 1936 (Agatz)
101 Planung Nordwerft u. IV. Einfahrt vom 19.07.1938 (Agatz/Hafenneubaudirektion)
102 Geplante Gestaltung IV. Einfahrt, 1940 (Agatz/Hafenneubaudirektion)
103 Projekt Gesenkschmiede der Nordwerft, 1939 (Hafenneubaudirektion)
104 Werfthalle Fliegerhorst Upjever, 1936 (Schütt)
105 Siedlung Fliegerhorst Upjever, 1936 (Rasch)
106 Modell Kulturforum, 3. Fassung, 1940 (Temp/Baumann)
107 Projekt Kulturforum 3. Fassung, 1940 (Temp/Baumann)
108 Modell Marineforum, 1939 (Hallbauer/Franzius)
109 Sport- und Kulturforum, 1. Fassung, 24.01.1939 mit Änder. 29.03.1939 (Hallbauer/Bartning/Baumann)
110 Strandalle, 1938 (Hunzinger)
111 Projekt Langer Markt und Aufmarschplatz Fedderwardergroden, 1940 (Bork)
112 NS-Schmuck Rathausplatz, 01.04.1939
113 Detail, Bunker Metzer Weg, zerst., 1940 (Hochbauamt)
114 Befehlsbunker der Kriegsmarinewerft, zerst., um 1941 (Klaucke)
115 Zu Wohnungen umgebauter Bunker, Hamburger Str. (Iwersen)
116 Nachkriegsplanung Wilhelmshaven, 1942 (Schulz)
117 Nachkriegsplanung Wilhelmshaven, 1942 (Heuer/Schneider)
118 Wiederaufbauplanung des Arbeitsstabes Speer, 1944 (Gutschow)
119 Projekt Bandstadt, 1945 (Franzius)
120 Flüchtlingsstoßprogramm, Virchowstr., 1950 (zu Putlitz/Kraak)
121 Nachkriegskonzept Werft (1949)
122 Bebauungsplan Stadtpark, 1949 (Schwarz)
123 Wettbewerb Stadtzentrum, Rathaus, 1948 (Graubner)
124 Wettbewerb Verwaltungsgebäude Jade, Lessingstr., 1952 (Forster)
125 Graf-Galen-Siedlung, 1952 (Minke)
126 Neuordnungsplan, 20.11.1936 (Hallbauer)
127 Projekt Die Stadt der Dreihunderttausend, 15.03.1937 (Hallbauer)
128 Wirtschaftsgebiet Wilhelmshaven, Juni 1937 (Hallbauer/Landesplanungsgemeinschaft)
129 Wirtschaftsplan Wilhelmshaven-Friesland, Mai 1940 (Dillschneider/Landesplanungsgemeinschaft)
130 Eßsaal Fliegerhorst Upjever, 1939 (Luftgaukommando)
131 Strandhalle Wilhelmshaven, 1938 (Hunzinger)
132 Geplante Siedl. für städtische Angestellte, Ernst-Moritz-Arndt-Str., 1938 (Forster)
133 Gesellschaftsraum Offiziersheim Wilhelmshaven, zerst., 1939 (Thees)
134 Mannschaftskantine U-Bootstützpunkt, 1939 (Thees)
135 Unternehmerhaus Wilhelmshaven, 1935 (Lübbers)
136 Privatwohnhäuser Hansaviertel, 1933 (Iwersen)
137 Hauptumspannstelle, Alter Banter Weg, 1938 (Lübbers)
138 Schule A Fedderwardergroden, 1939 (Hörold)
139 Schule Cäciliengroden, 1941 (Heuer)
140 Schule am Krähenbusch, 1938–39 (Forster/Hochbauamt)
141 Projekt Schule D Fedderwardergroden, 1939 (Bürgin)
142 Schule B Fedderwardergroden, 1939 (Iwersen)
143 Amtsgebäude Friesland, 1936–38 (Wohlschläger)
144 Projekt Freizeithalle Fedderwardergroden, 1940 (Franzius)
145 Projekt Schule Altengroden-Nord, 1940 (Fricke)
146 Projekt HJ-Heim und NSDAP-Haus Fedderwardergroden, 1939 (Bürgin)
147 HJ-Heim, Krähenbusch, 1938 (Däke)
148 Projekt Ladentrakt, Aufmarschplatz Neuengroden, 1939 (Schwarting)
149 Gewerbebau Altengroden, Goedenser Weg, 1940 (Kühn)
150 Projekt Postamt Fedderwardergroden, Vorentwurf, 1940 (Jochem/West)
151 Projekt Städtisches Krankenhaus Fedderwarden, 1941 (Distel)
152 Baustelleneinrichtung der Fa. Hermann Möller, Marinekommando Nord, um 1939
153 NS-Baupropaganda, 1934
154 DAF-Führer Ley in Fedderwardergroden, 09.06.1941
155 KZ-Häftlinge bei Bauarbeiten in Wilhelmshaven
156 Dekoration Flugzeugwerft Upjever, 1936 (Luftwaffenbauamt)
157 NS-Schmuck Rathausplatz, 01.04.1939
158 Relief Kriegsmarinewerft, um 1939 (Grau)
159 Relief Lehrlingshochhaus Kriegsmarinewerft, zerst.
160 „Schmied", tom-Brok-Str., 1939 (Oehler)
161 „Mutter und Kind", Paul-Hug-Str., 1938 (Grau)
162 „Zimmermann", tom-Brok-Str., 1939 (Oehler)

9.6. Bildteil

1 Heute noch vorhandene Spuren der NS-Architektur in Wilhelmshaven, außer Bunker, Straßen, Deiche, Kartengrundlage um 1940

2 Admiralität Berlin, Leipziger Platz, zerst.,1850 (Hitzig)

3 Oldenburger Ministerien mit Hoheitszeichen, 1914–17 (Bonatz)

4 Reichsmarineamt Berlin, Reichpietsch Ufer, 1914 (Reinhardt/Süssenguth)

5 Projekt OKM Berlin, 1937/43 (Dübbers/Bonatz/Speer)

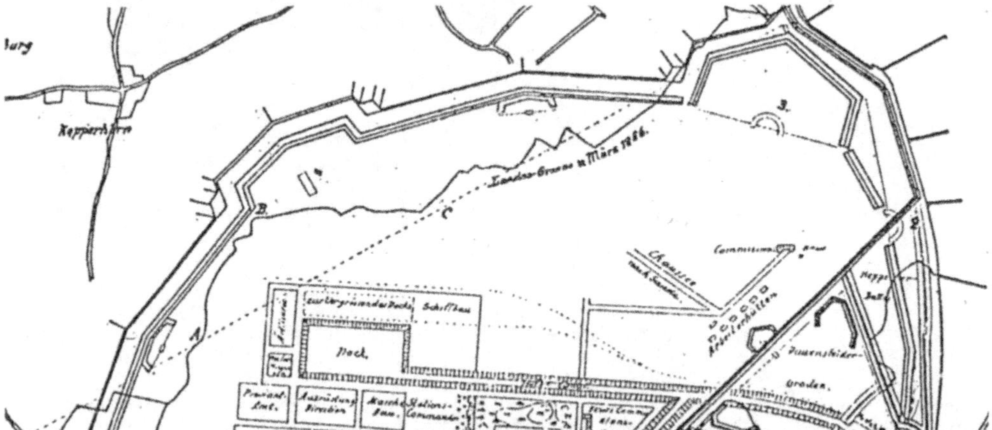

6 Hagen-Plan mit Befestigung nach Fischer, 25. 6. 1856

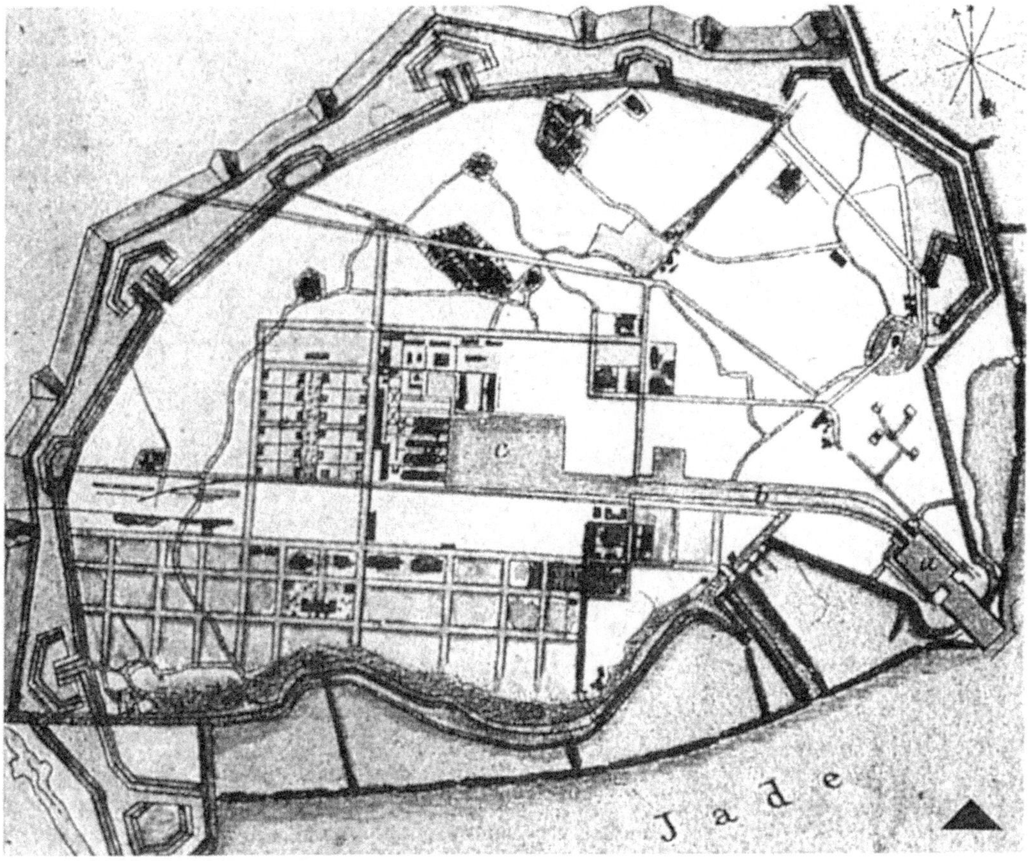

7 Klotz-Plan der Befestigung, vermutl. 1863

8 Stoltz-Plan der Befestigung, vermutl. 1865/68

9 Übersichtsplan der marinefiskalischen Gebäude, Borrmann-Plan 1891

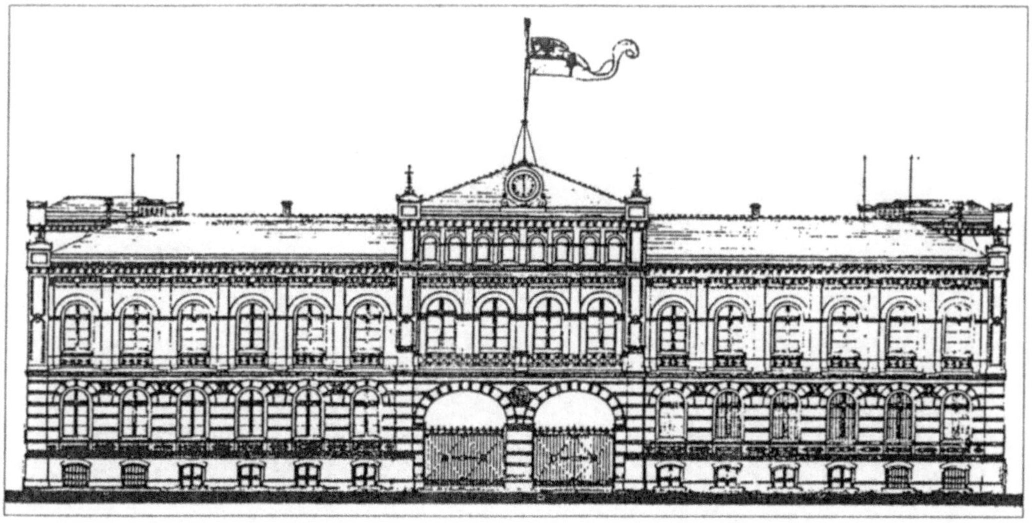

10 Werfttor 1, Gökerstr., 1876 (Hafenbauressort)

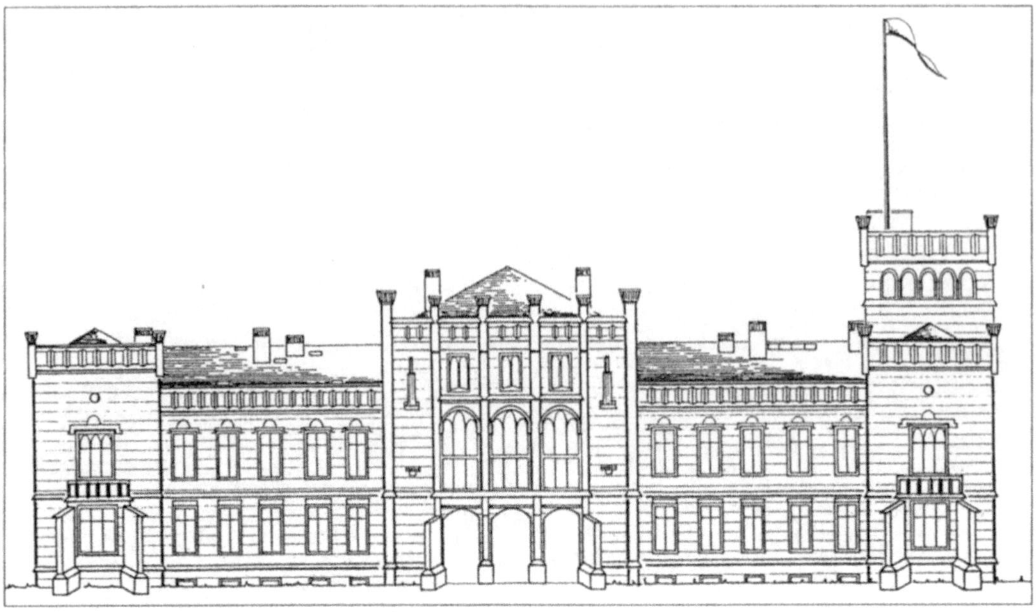

11 Marinestationsgebäude, Adalbertplatz, zerst., 1870 (Adler/Hitzig/Klingenberg)

12 Kasernen 2. Seebataillon, Gökerstr., zerst., 1904/07 (Balfanz/Marineintendantur)

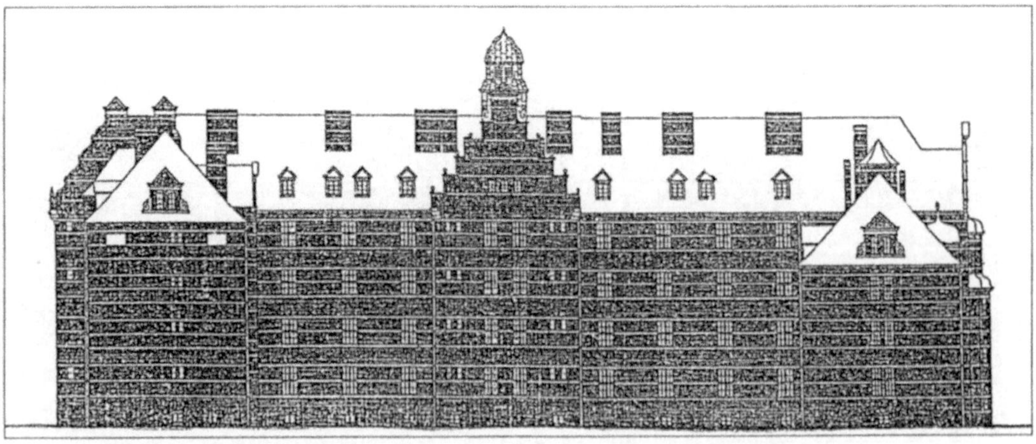

13 Kasernen II. Matrosendivision, Mühlenweg, 1912/13 (Strempel/Balfanz/Marineintendantur)

14 Marineintendantur, Peterstr., teilzerst., 1904 (Schubert/Fleinert)

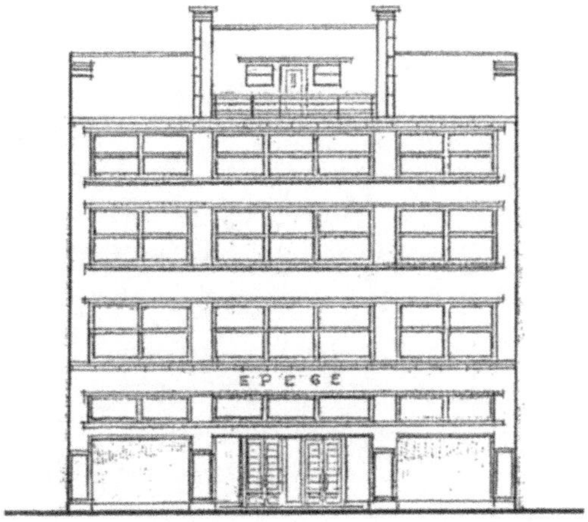

15 Kaufhaus EPEGE, Gökerstr., zerst., 1930 (Fuchs)

16 Werftspeisehaus, Gökerstr., zerst., 1902 (Marineintendantur)

17 Projekt Leo, Gökerstr., 1929 (Freygang)

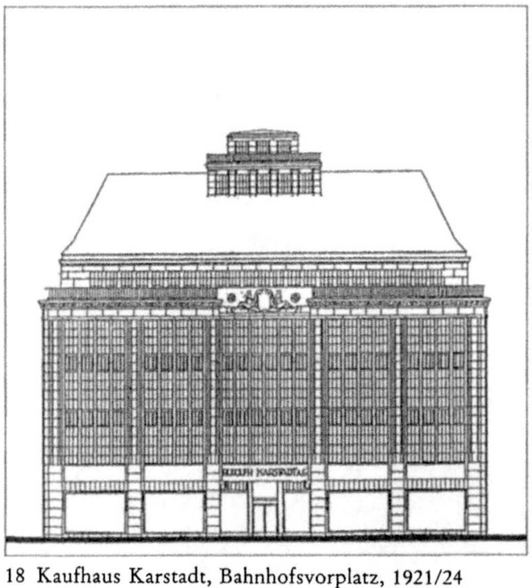

18 Kaufhaus Karstadt, Bahnhofsvorplatz, 1921/24 (Schäfer)

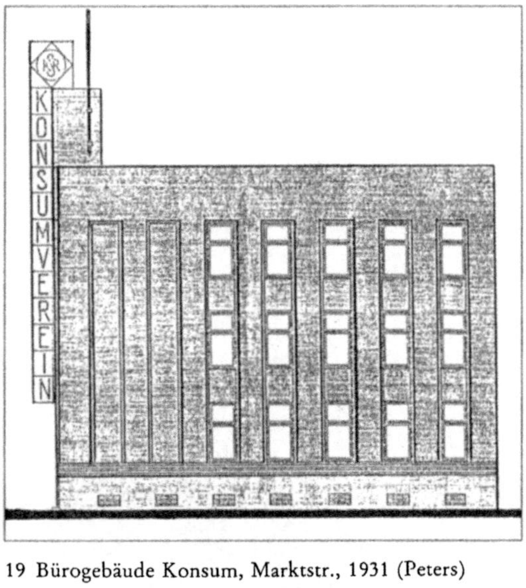

19 Bürogebäude Konsum, Marktstr., 1931 (Peters)

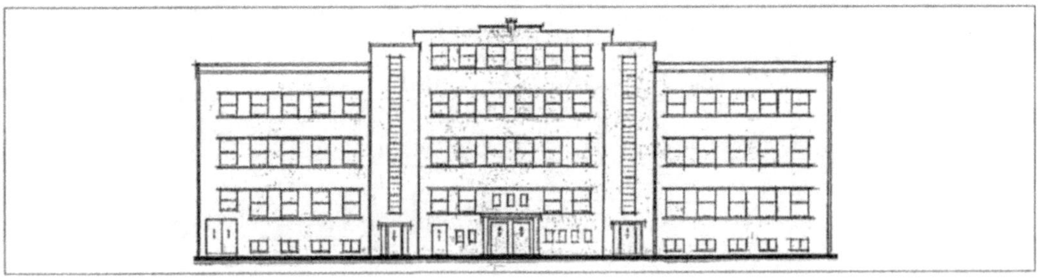

20 Allgemeine Ortskrankenkasse Wilhelmshaven, Kieler Str., 1929/31 (Lübbers)

21 Wohnhäuser Freiligrathstr., 1932 (Minke)

22 Wohnkolonie Kniphausersiel, 1936 (Domänenamt)

23 Wohnhäuser Widukindstr., 1938 (Stadterweiterungsamt/verschiedene Architekten)

24 Richtfest Adolf-Hitler-Siedlung, 4. 5. 1935

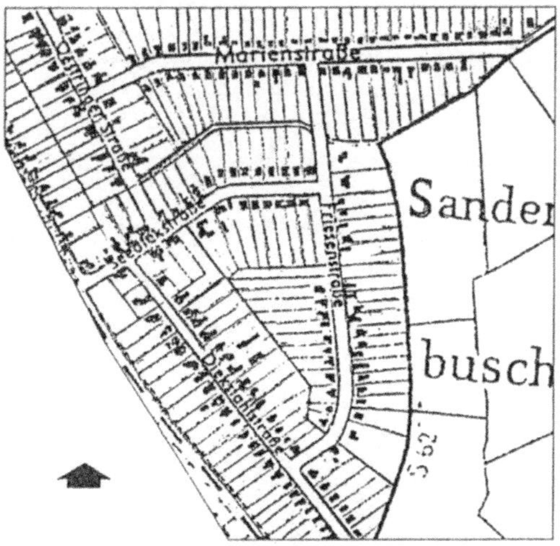

25 Siedlung Sanderbusch, 1935 (Domänenamt)

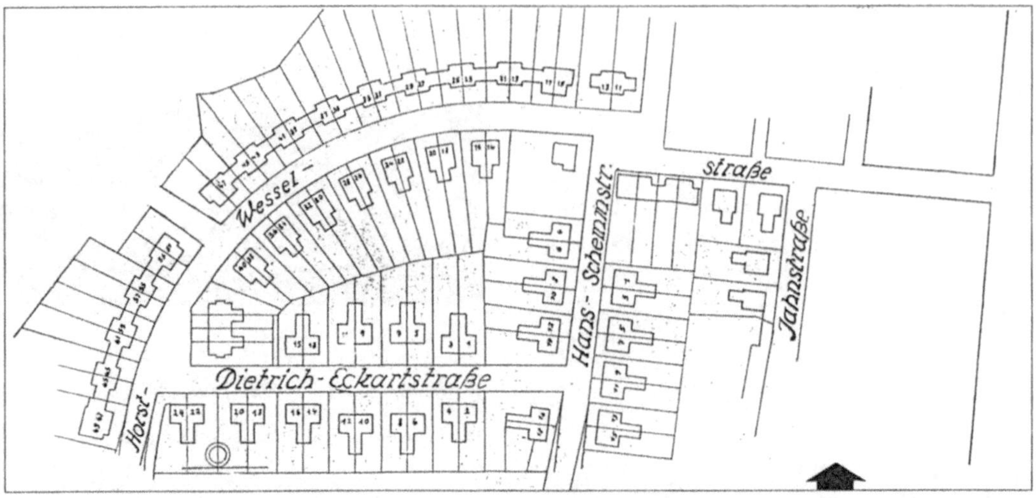

26 Planung Adolf-Hitler-Siedlung, 50 Wohnungen, 1934 (Siemer)

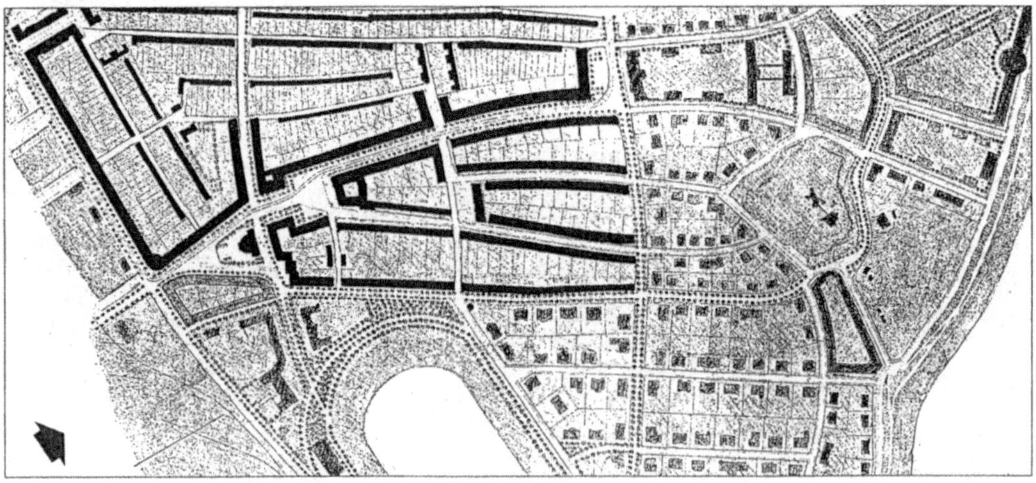

27 Planung Inselviertel und Wohnsiedlung Nordost, 1918 (Zopff)

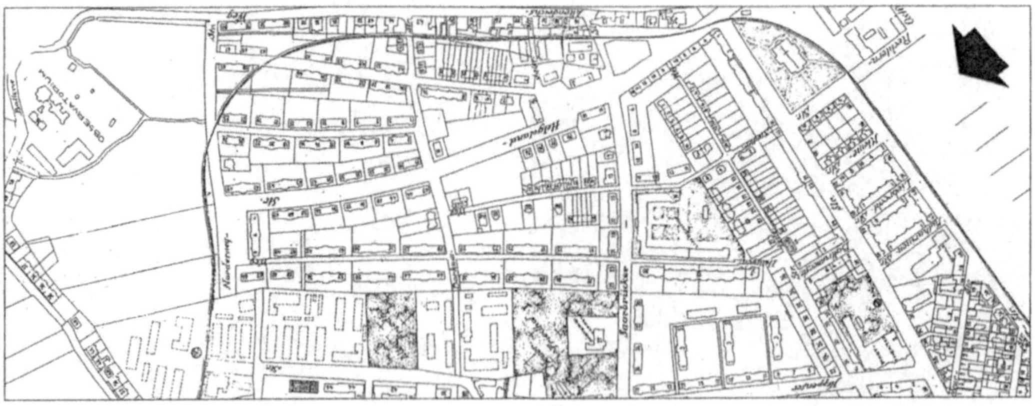

28 Bestandsplan Inselviertel mit Arbeiterlagern und Bunkern, um 1941 (Zopff)

29 Wohnhäuser Wangeroogestr., um 1934 (Peters)

30 Planung Voslapp, 1235 Wohnungen, 1937, Gemeinschaftsskizze (Schneider)

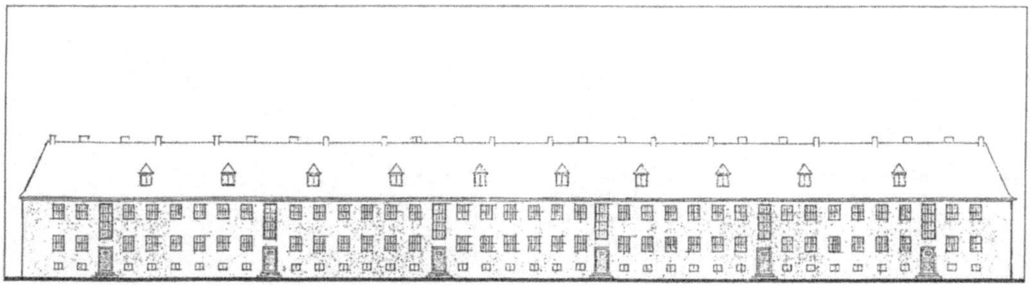

31 Volkswohnungen Voslapp, Geniusbankstr., Typ VIIIa, Vorentwurf, 1939 (Schemm)

32 Kleinsiedlungstypen Voslapp, 1938 (Lübbers)

33 Kleinsiedlungstypen Voslapp, 1938 (Lübbers)

34 Planung Cäciliengroden, 360 Wohnungen, 1938 (Heuer)

35 Wohnhäuser Cäciliengroden, 1938 (Heuer)

36 Cäciliengroden, Visite des Großadmirals, 9.1.1940

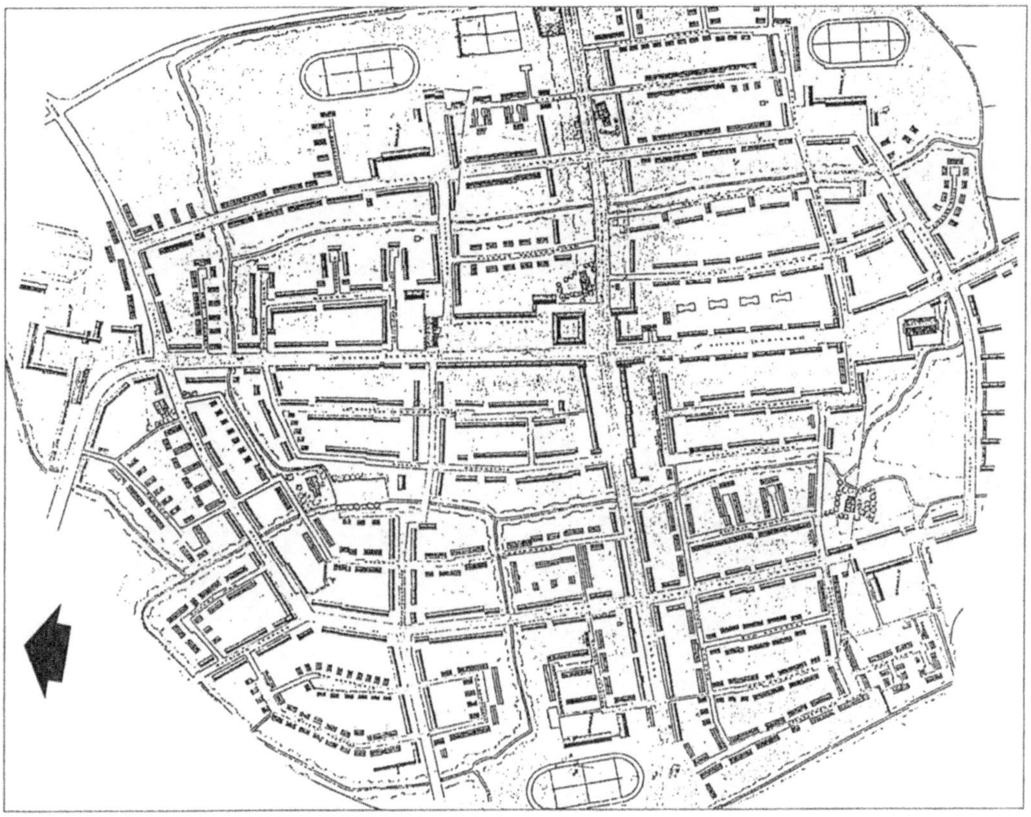

37 Planung Fedderwardergroden, 4100 Wohnungen, 1938 (Pfeiffer/Schneider/Hallbauer), überarbeitet 1940

38 Modell Fedderwardergroden für die Besprechung mit Hitler, 1.4.1939

39 Bauarbeiten Fedderwardergroden, 1941

40 Wohnhäuser Fedderwardergroden, 1940 (Bork)

41 Wohnhäuser Fedderwardergroden, Zoppoter Str., 1940 (Bork)

42 Fedderwardergroden, Preußenstr. mit Notpostamt, 1940

44 Ladenstr. 1, Altengroden, Goedenser Weg, 1943 (Kühn)

45 Wohnhäuser Altengroden, Ubbostr. Typ H, 1945 (Mellin/Kühn)

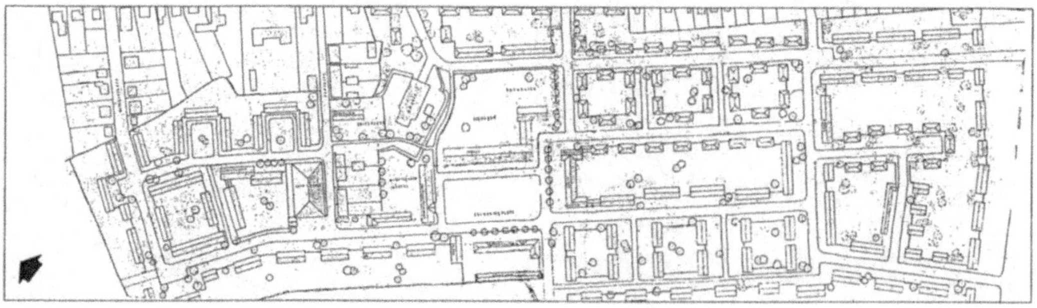

46 2. Bebauungsplan Neuengroden, 600 Wohnungen, 3.3.1939 (Schwarting)

43 OKM-Bebauungsplan Altengroden, 14.2.1939 (Zenker)

48 Planung Sande-Neufeld, 1939 (Tuchscherer)

47 Neuengrodener Weg, 1939 (Schwarting)

49 Wohnblocks Sande-Neufeld, 1939 (Hilgenstock)

50 Gemeinschaftshalle Sande-Neufeld, 1939 (Hilgenstock)

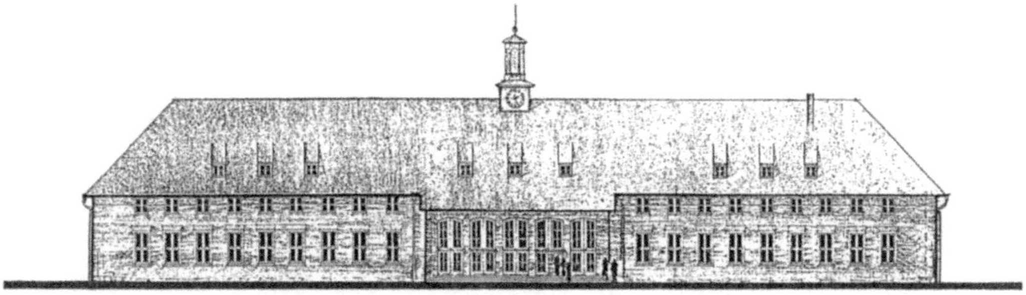

51 Empfangsgebäude Bahnhof Sande, 1941 (Langewand/Reichsbahndirektion Münster)

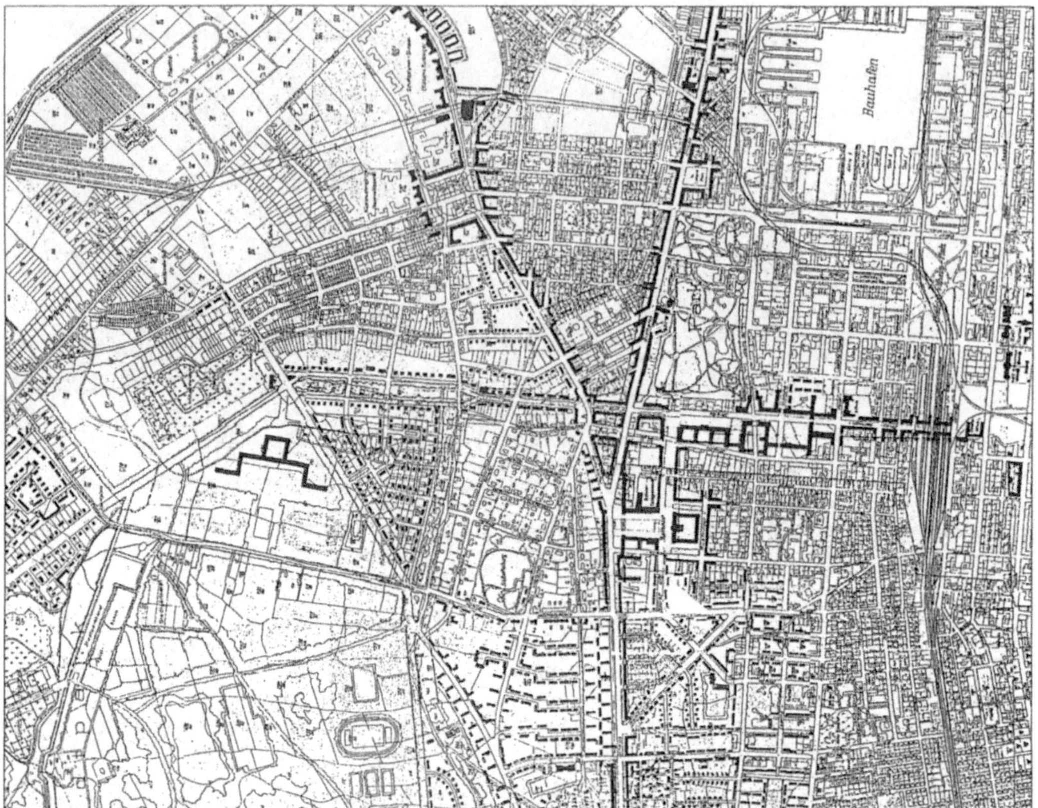

52 Neugestaltung Wilhelmshaven, 29.3.1939 (Hallbauer/Stadterweiterungsamt)

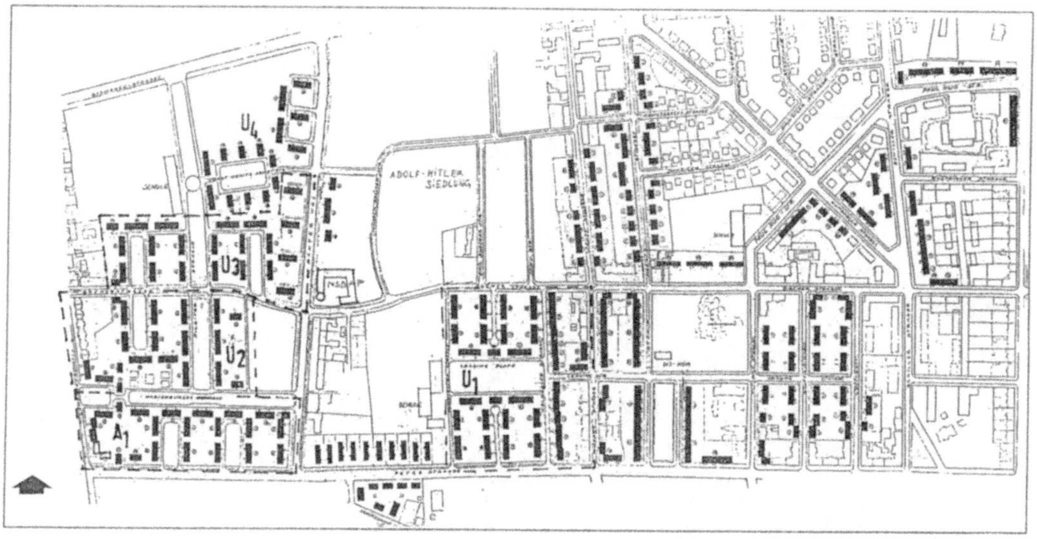

53 Bebauungsplan Hansa- und Jadeviertel, 1937 (Stadterweiterungsamt) Bauten der Jade und der Jadestädtischen Wohnungsbaugesellschaft sind schwarz markiert.

54 Wohnhäuser Jadeviertel, zerst., 1938 (Schemm)

55 Hansaviertel, Paul-Hug-Str., 1937 (Lübbers)

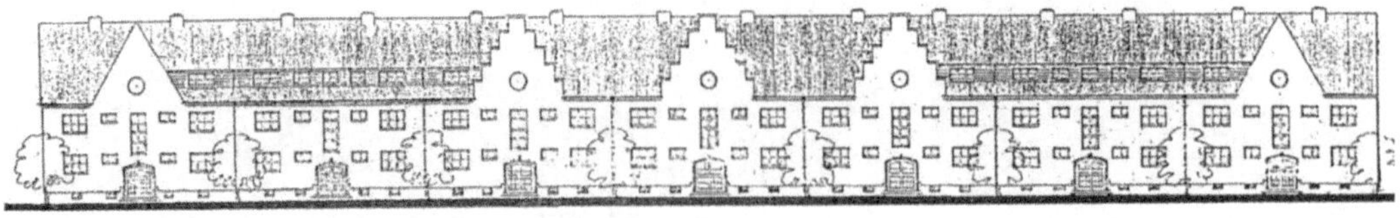

56 Wohnhäuser Hansaviertel, Typ Ib, Otto-Meentz-Platz, 1937 (Boyken)

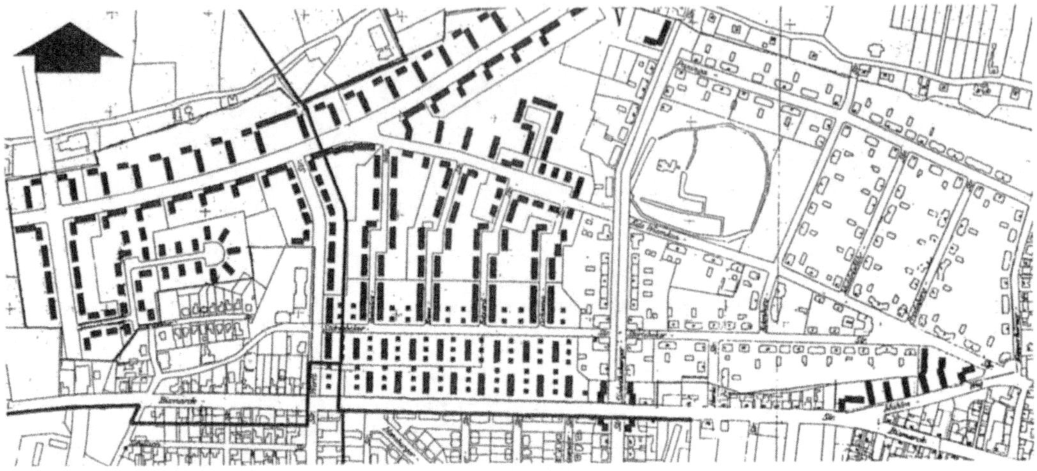

57 Neu-Siebethsburg, 1600 Wohnungen, 1935–38 (Höger)

58 Wohnhäuser Meyer und Hoeland, Baudissinweg, teilzerst., 1934–35 (Höger)

59 Wohnhäuser Werftstr., 1938 (Höger)

60 Wohnhaus tom-Brok-Str. im NS-Schmuck, 1938 (Höger)

61 Tausend-Wohnungsprojekt Schaarreihe, Typ RIII, 1935 (Höger)

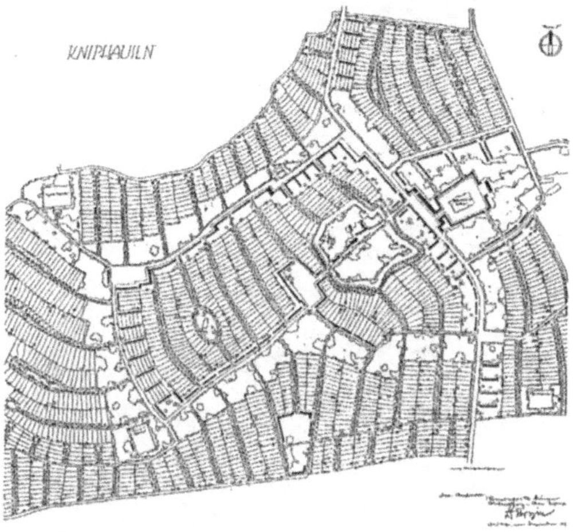

62 Projekt Siedlung Burg Kniphausen, 1800 Wohnungen, 1937 (Bürgin)

63 Projekt Siedlung Schilldeich, 1500 Wohnungen, 1938 (Bürgin)

64 Wettbewerbsentwurf Langewerth, 15–19000 Wohnungen, 1939 (Feder)

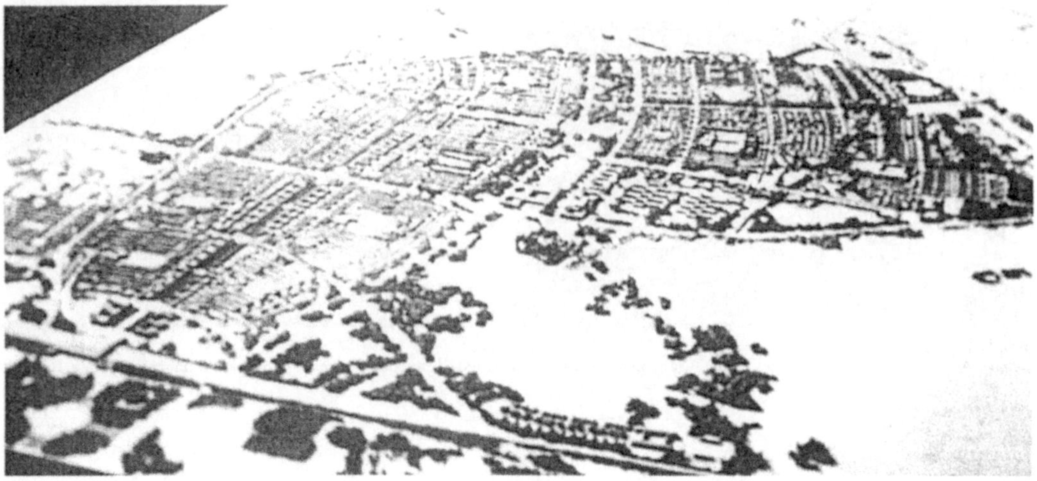

65 Wettbewerbsentwurf Langewerth, 15–19000 Wohnungen, 1939 (zu Putlitz)

66 Projekt Breddewarden, 1939 (Heuer)

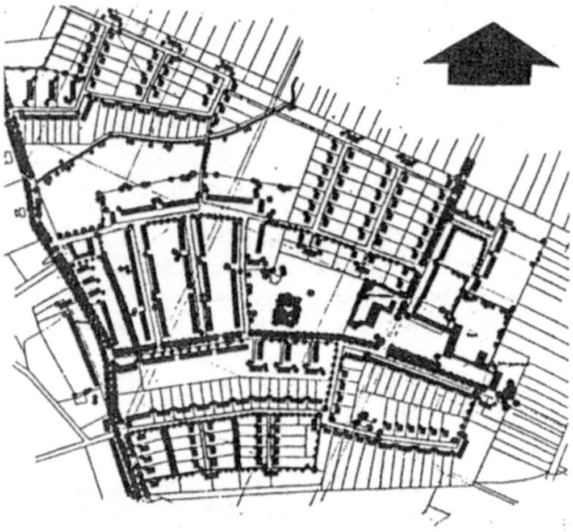

67 Projekt Schortens-Klosterneuland, 1940 (Richter)

68 Werfthalle III, Fliegerhorst Upjever, 1937 (Schütt)

69 Marineinstandsetzungsamt Ulmenstr., 1937 (Franzius)

70 Marinestationsgebäude Marineforum, Haupteingang, 1937 (Franzius/Geisenhainer)

71 Marinestationsgebäude, geplante Fassade Mozartstr., 1937 (Franzius/Geisenhainer)

73 Marinegerätelager Roffhausen, 1937 (Geisenhainer), später AEG/Olympia

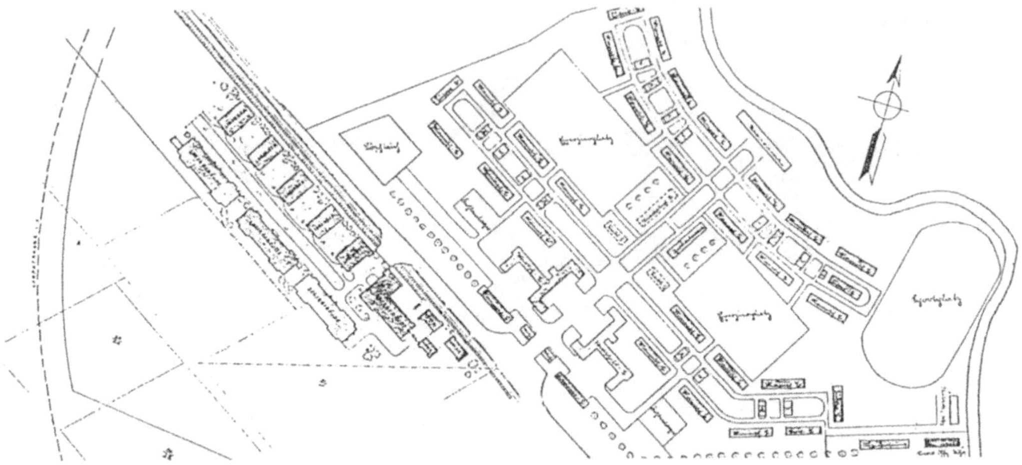

72 Marinegerätelager und Arbeiterlager Roffhausen, Lageplan, 1944 (Geisenhainer)

74 Marinegerätelager Roffhausen und Reste des Arbeiterlagers, Luftbild, 1950

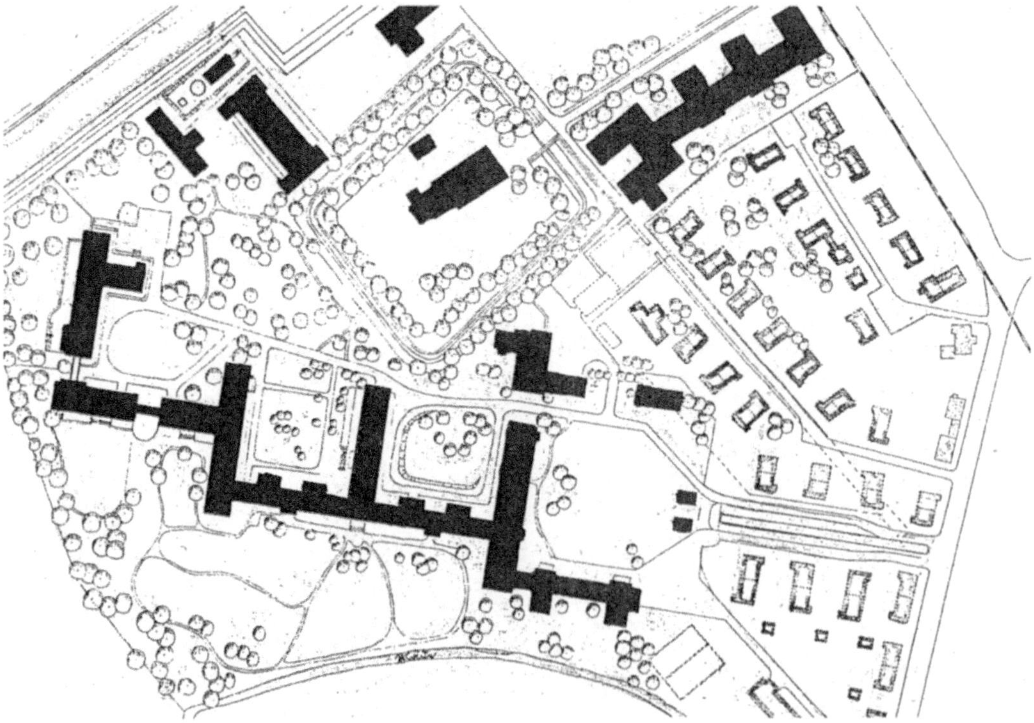

75 Marinelazarett Sanderbusch, Lageplan, 1937 (Geisenhainer)

76 Marinelazarett Sanderbusch, 1937 (Geisenhainer)

77 Marinelazarett Sanderbusch, 1937 (Geisenhainer)

78 U-Bootstützpunkt, Modell, 1939 (Thees)

79 U-Bootstützpunkt Wilhelmshaven für die U-Flottille „Saltzwedel“, 1936–41 (Thees)

80 U-Bootstützpunkt, Terrasse Wirtschaftsgebäude, 1939 (Thees)

81 U-Bootstützpunkt, Hitlerbesuch 9.4.1940

82 Bonte-Kasernen, zerst., 1937–38 (Thees)

83 Eingangsschmuck Bonte-Kasernen, zerst., 1937–38 (Grau)

84 Mannschaftsraum Bonte-Kasernen, zerst., 1937–38 (Thees)

85 Gedenkstätte Bonte-Kasernen, zerst., 1937–38 (Thees)

86 Marinekommando Nord, 1938–42 (Thees)

87 Marinekommando Nord, 1938–42 (Thees)

88 Kasernen der 6. Schiffsstammabteilung Varel, 1939–42 (Thees)

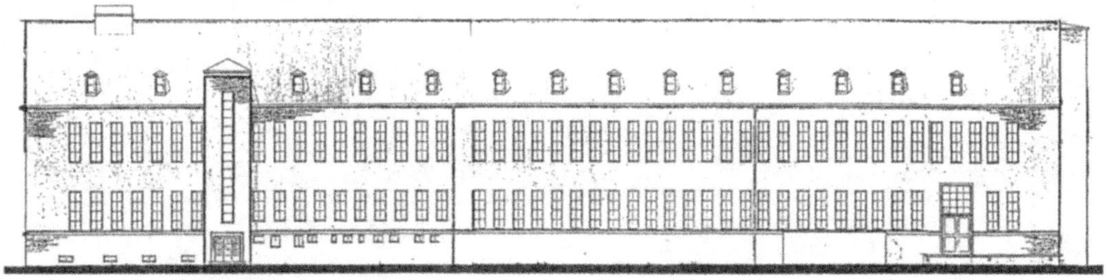

89 Torpedowerftgebäude, 1940 (Klaucke)

90 Standortschwimmanstalt, 1937–38 (Thees)

91 Navigationsressort der Kriegsmarinewerft, Büro- u. Plangebäude, Südufer Zwischenhafen, 1939–42 (Hafenbauressort der Kriegsmarinewerft)

92 Modell der geplanten Werkshallen der Nordwerft mit Panzerschiff, 1938 (Hafenneubaudirektion)

93 Gießerei der Kriegsmarinewerft, Schiff 2, 1938–40 (Klaucke)

94 Werftbahnhof, Gießerei, Norddeutscher Eisenbau, um 1940

95 Minsener Oog, 1942–43
(Tillessen/Klaucke)

96 Bürohaus Ressort II der Kriegsmarinewerft, zerst., um 1940 (Klaucke)

97 Entwurf Motorenbauwerkstatt Westwerft, zerst., um 1940 (Klaucke)

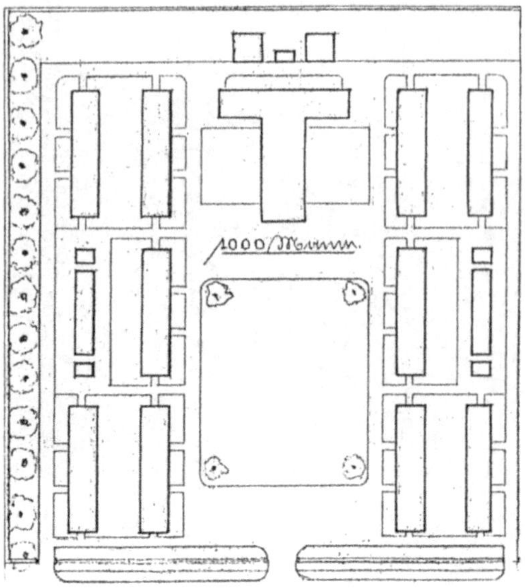

98 Normentwurf Arbeiterlager für 1000 Mann, 1939 (Marinestandortbauamt)

99 Planung Arbeiterunterkünfte Ebkeriege, 1939 (Tuchscherer)

100 Schleusenkammer IV. Einfahrt, 1936 (Agatz)

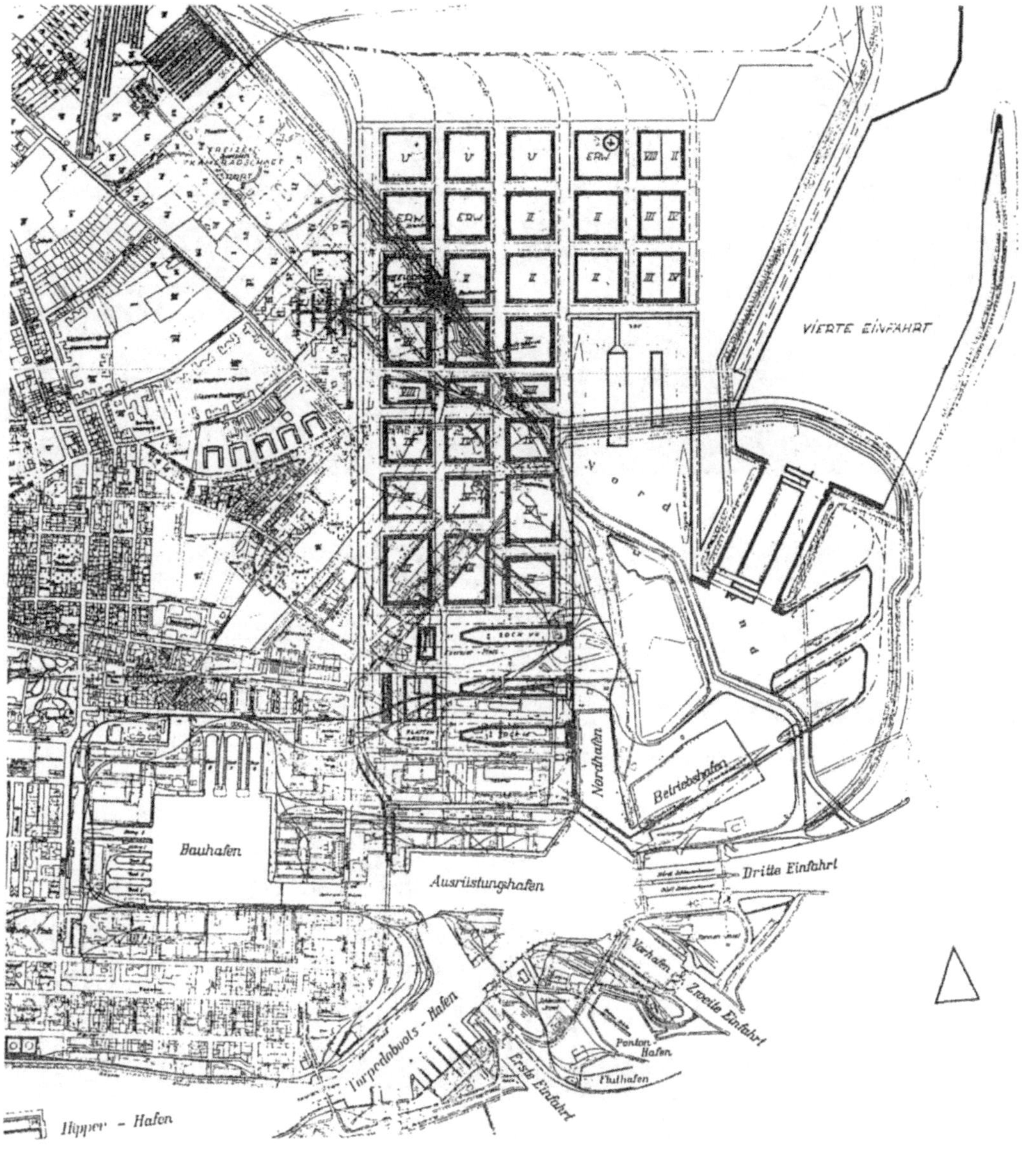

101 Planung Nordwerft u. IV. Einfahrt vom 19.7.1938 (Agatz/Hafenneubaudirektion)

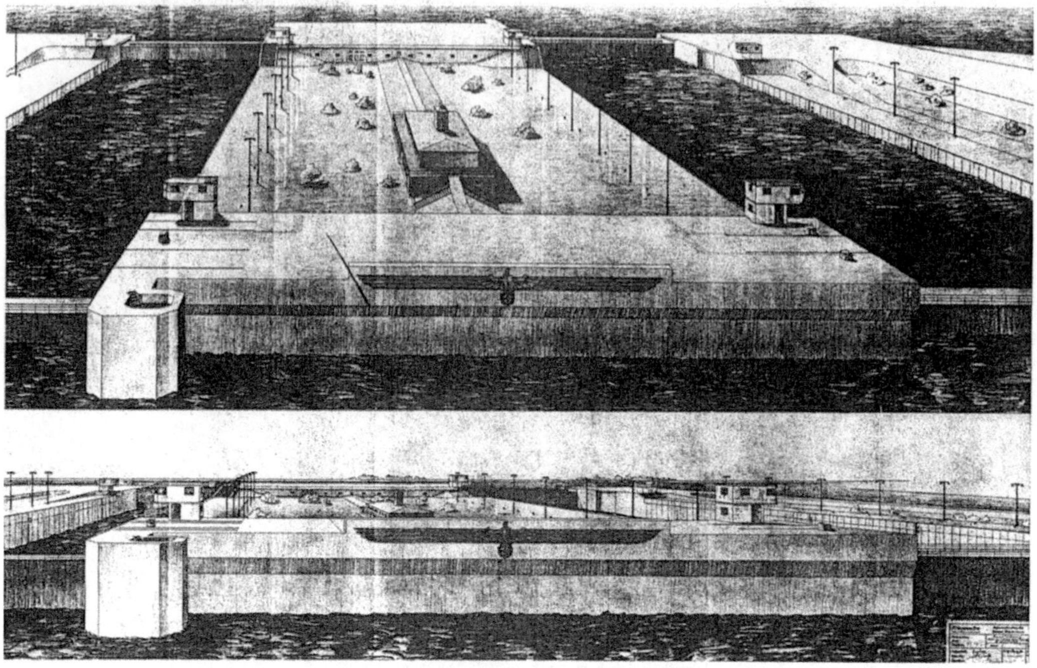

102 Geplante Gestaltung IV. Einfahrt, 1940 (Agatz/Hafenneubaudirektion)

103 Projekt Gesenkschmiede der Nordwerft, 1939 (Hafenneubaudirektion)

104 Werfthalle Fliegerhorst Upjever, 1936 (Schütt)

105 Siedlung Fliegerhorst Upjever, 1936 (Rasch)

106 Modell Kulturforum, 3. Fassung, 1940 (Temp/Baumann)

107 Projekt Kulturforum 3. Fassung, 1940 (Temp/Baumann)

108 Modell Marineforum, 1939 (Hallbauer/Franzius)

109 Sport- und Kulturforum, 1. Fassung, 24.1.1939 mit Änder. 29.3.1939 (Hallbauer/Bartning/Baumann)

110 Strandhalle, 1938 (Hunzinger)

111 Projekt Langer Markt und Aufmarschplatz Fedderwardergroden, 1940 (Bork)

112 NS-Schmuck Rathausplatz, 1.4.1939

113 Detail, Bunker Metzer Weg, zerst., 1940 (Hochbauamt)

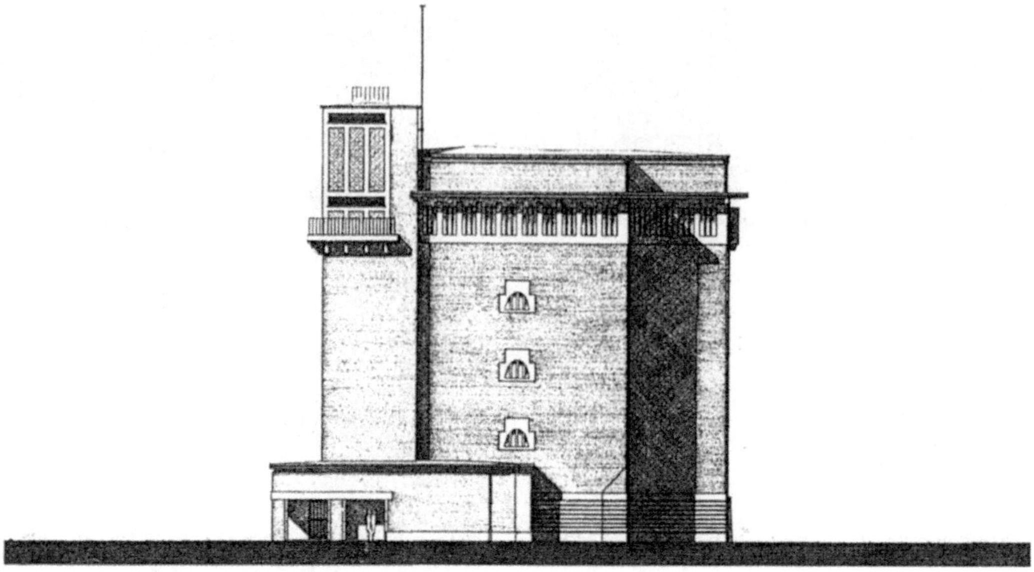

114 Befehlsbunker der Kriegsmarinewerft, zerst., um 1941 (Klaucke)

115 Zu Wohnungen umgebauter Bunker, Hamburger Str. (Iwersen)

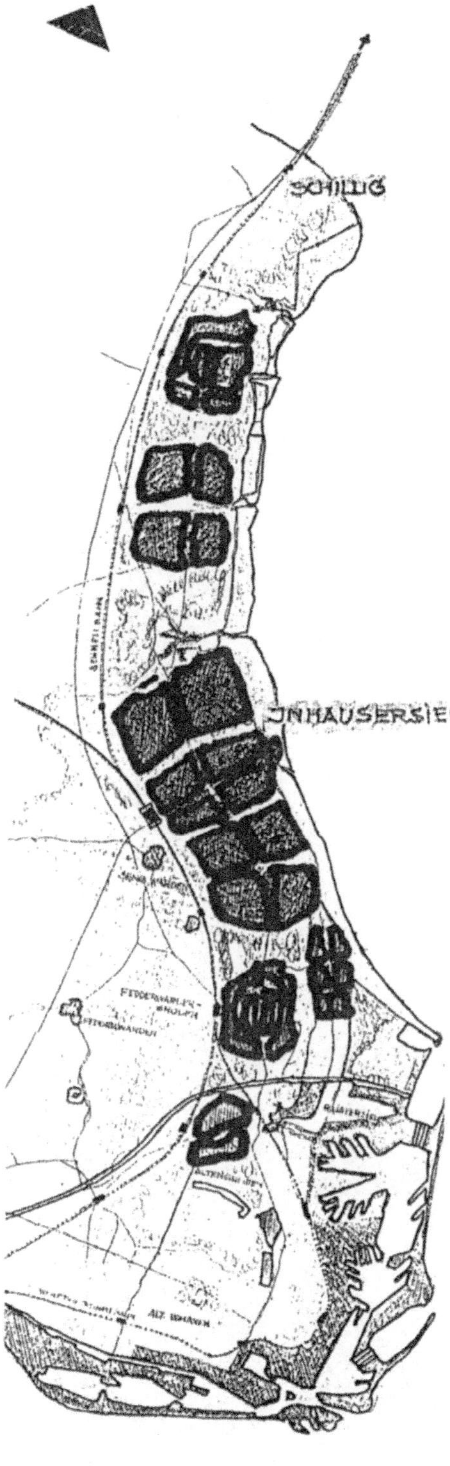

119 Projekt Bandstadt, 1945 (Franzius)

116 Nachkriegsplanung Wilhelmshaven, 1942 (Schulz)

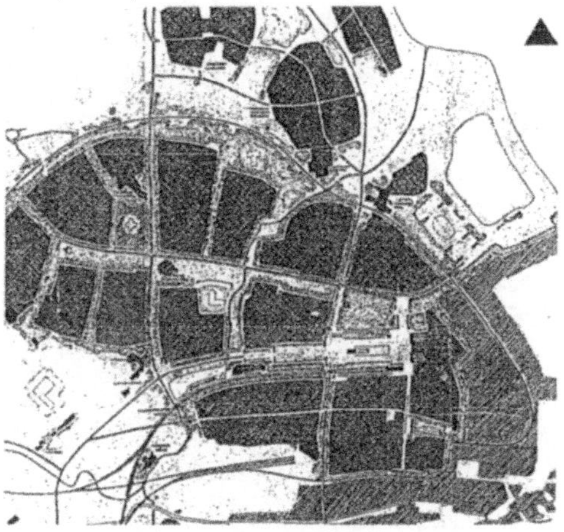

117 Nachkriegsplanung Wilhelmshaven, 1942 (Heuer/Schneider)

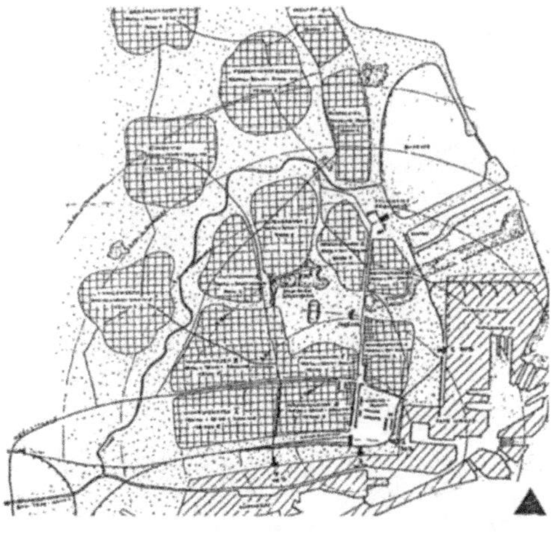

118 Wiederaufbauplanung des Arbeitsstabes Speer, 1944 (Gutschow)

120 Flüchtlingsstoßprogramm, Virchowstr., 1950 (zu Putlitz/Kraak)

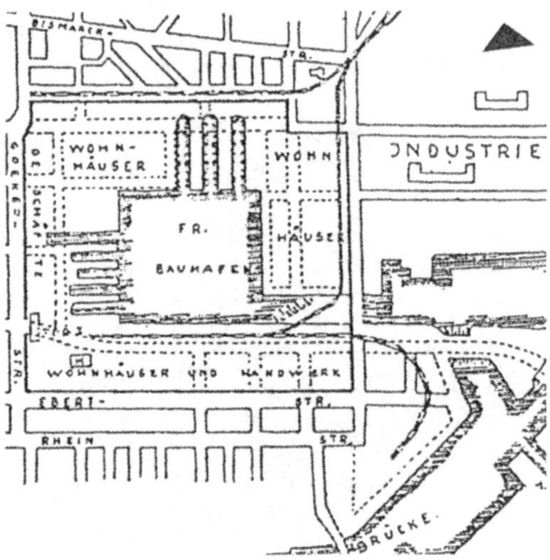

121 Nachkriegskonzept Werft (1949)

122 Bebauungsplan Stadtpark, 1949 (Schwarz)

123 Wettbewerb Stadtzentrum, Rathaus, 1948 (Graubner)

124 Wettbewerb Verwaltungsgebäude Jade, Lessingstr., 1952 (Forster)

125 Graf Galen Siedlung, 1952 (Minke)

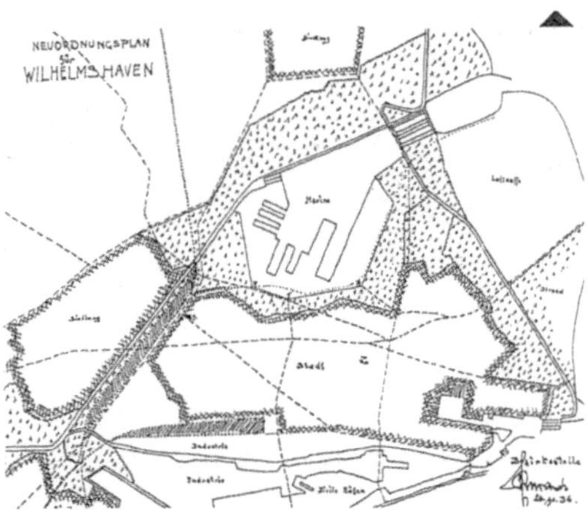

126 Neuordnungsplan, 20.11.1936 (Hallbauer)

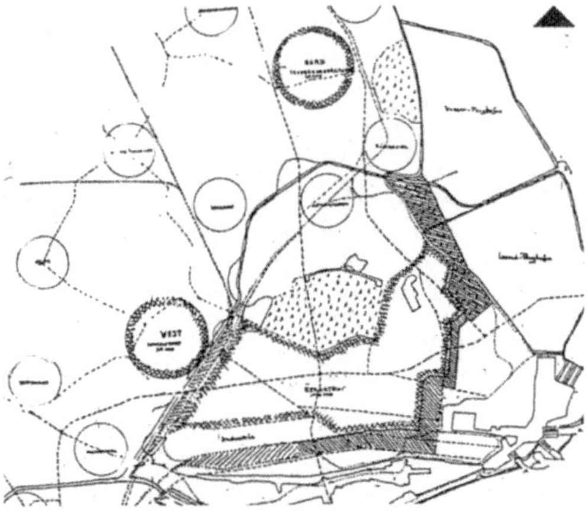

127 Projekt Die Stadt der Dreihunderttausend, 15.3.1937 (Hallbauer)

128 Wirtschaftsgebiet Wilhelmshaven, Juni 1937 (Hallbauer/Landesplanungsgemeinschaft)

129 Wirtschaftsplan Wilhelmshaven-Friesland, Mai 1940 (Dillschneider/Landesplanungsgemeinschaft)

130 Eßsaal Fliegerhorst Upjever, 1939 (Luftgaukommando)

131 Strandhalle Wilhelmshaven, 1938 (Hunzinger)

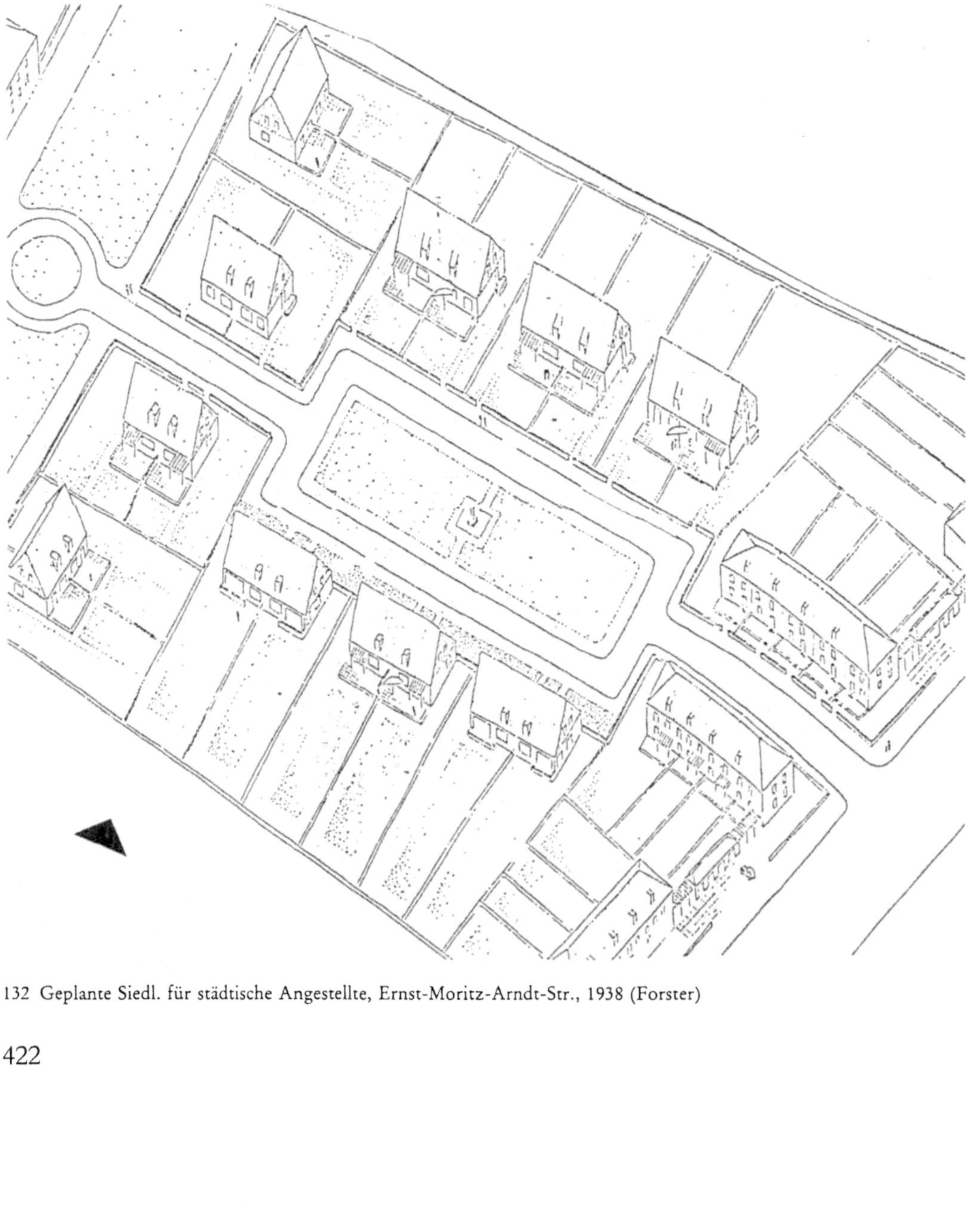

132 Geplante Siedl. für städtische Angestellte, Ernst-Moritz-Arndt-Str., 1938 (Forster)

133 Gesellschaftsraum Offiziersheim Wilhelmshaven, zerst., 1939 (Thees)

134 Mannschaftskantine U-Bootstützpunkt, 1939 (Thees)

135 Unternehmerhaus Wilhelmshaven, 1935 (Lübbers)

136 Privatwohnhäuser Hansaviertel, 1933 (Iwersen)

137 Hauptumspannstelle, Alter Banter Weg, 1938 (Lübbers)

138 Schule A Fedderwardergroden, 1939 (Hörold)

139 Schule Cäciliengroden, 1941 (Heuer)

140 Schule am Krähenbusch, 1938–39 (Forster/Hochbauamt)

141 Projekt Schule D Fedderwardergroden, 1939 (Bürgin)

142 Schule B Fedderwardergroden, 1939 (Iwersen)

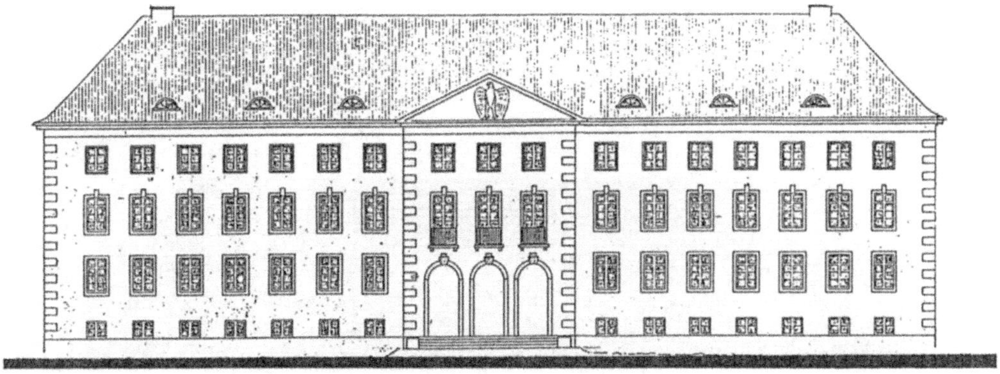

143 Amtsgebäude Friesland, 1936–38 (Wohlschläger)

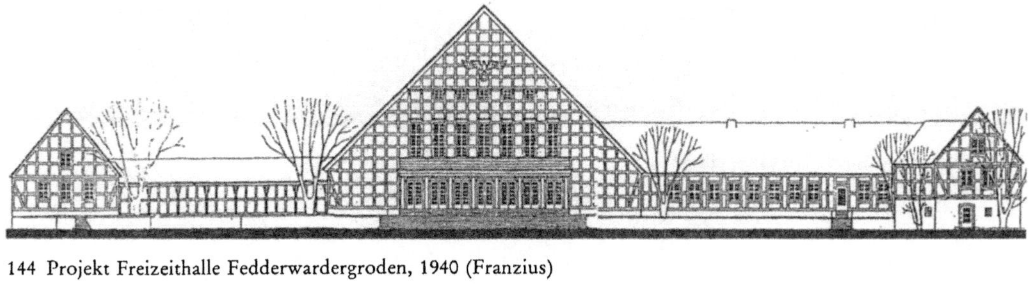

144 Projekt Freizeithalle Fedderwardergroden, 1940 (Franzius)

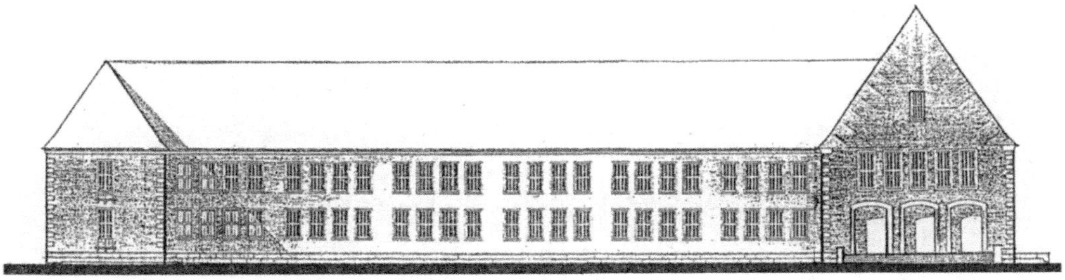

145 Projekt Schule Altengroden-Nord, 1940 (Fricke)

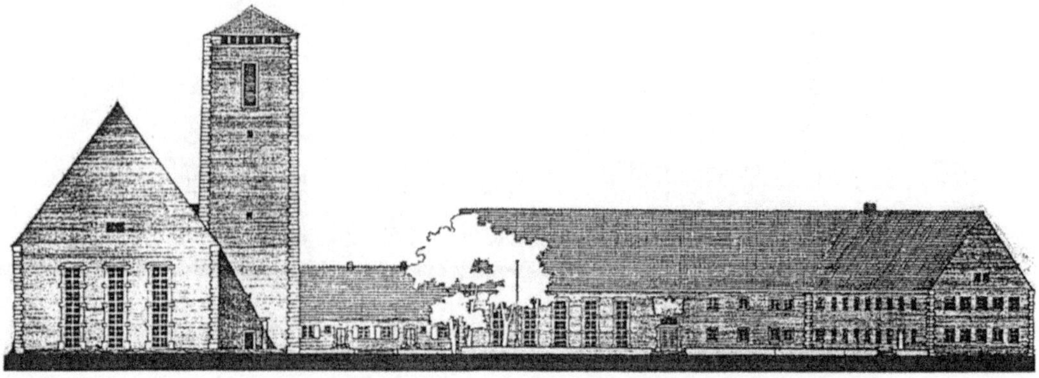

146 Projekt HJ-Heim und NSDAP-Haus Fedderwardergroden, 1939 (Bürgin)

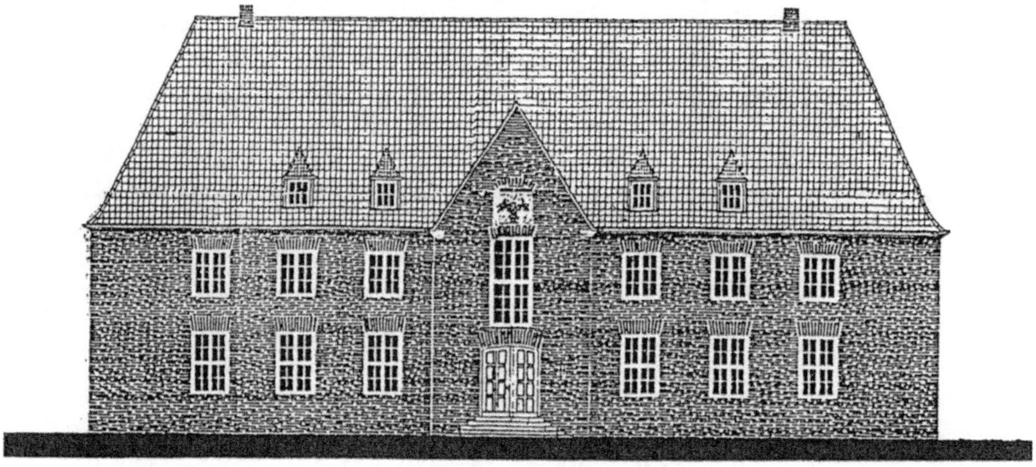

147 HJ-Heim, Krähenbusch, 1938 (Däke)

148 Projekt Ladentrakt, Aufmarschplatz Neuengroden, 1939 (Schwarting)

149 Gewerbebau Altengroden, Goedenser Weg, 1940 (Kühn)

150 Projekt Postamt Fedderwardergroden, Vorentwurf, 1940 (Jochem/West)

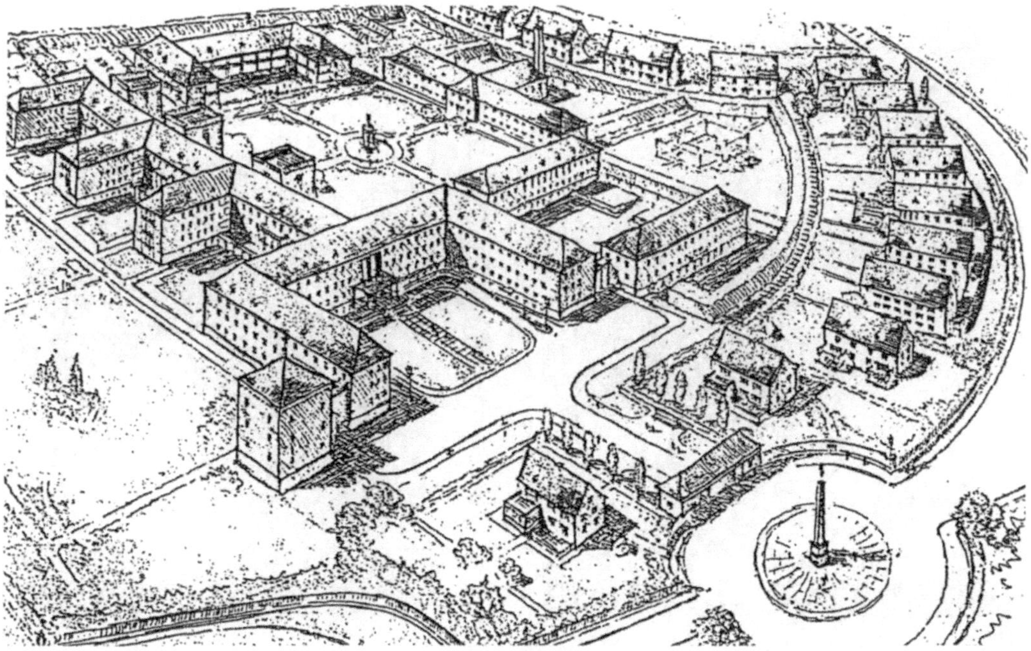

151 Projekt Städtisches Krankenhaus Fedderwarden, 1941 (Distel)

152 Baustelleneinrichtung der Fa. Hermann Möller, Marinekommando Nord, um 1939

153 NS-Baupropaganda, 1934
154 DAF-Führer Ley in Fedderwardergroden, 9.6.1941

155 KZ-Häftlinge bei Bauarbeiten in Wilhelmshaven

156 Dekoration Flugzeugwerft Upjever, 1936
(Luftwaffenbauamt)

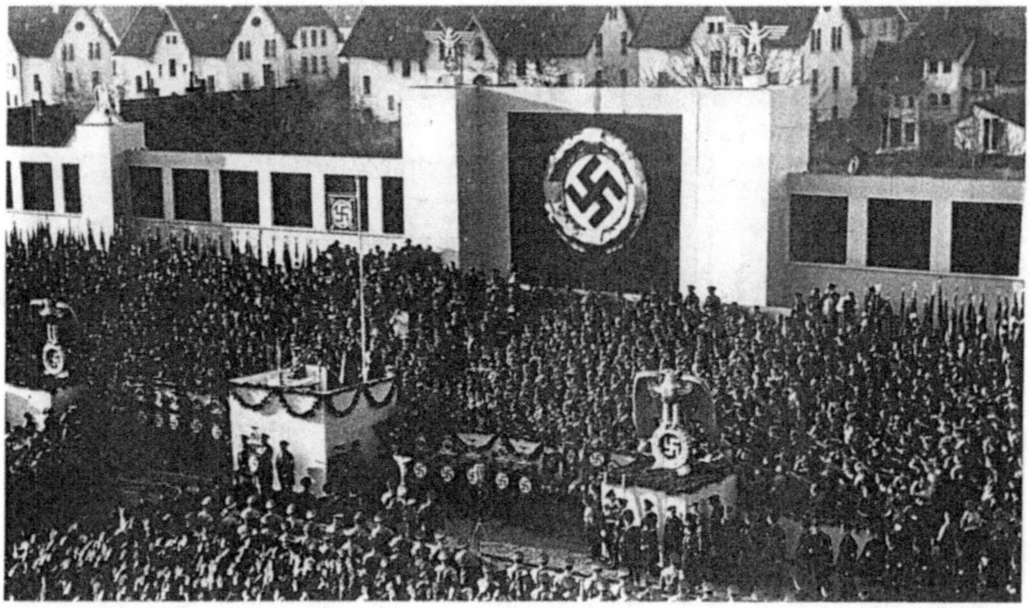

157 NS-Schmuck Rathausplatz, 1.4.1939

158 Relief Kriegsmarinewerft, um 1939 (Grau)

159 Relief Lehrlingshochhaus Kriegsmarinewerft, zerst.

160 „Schmied", tom-Brok-Str., 1939 (Oehler)

161 „Mutter und Kind", Paul-Hug-Str., 1938 (Grau)

162 „Zimmermann", tom-Brok-Str., 1939 (Oehler)

9.7. Tabelle

Wohnungsbauunternehmen in Wilhelmshaven/Rüstringen

	Baugesellschaften	Bemerkungen	Wohnungs-produktion
ca. 1868–1875	Consortium de Cousser	Gegr. von Gutsbesitzer Adolph de Cousser aus Hahn/Ol., 1873 Bau der Arbeitersiedlung Bant, ca. 1875 an Marinefiskus verkauft	452
ca. 1870–	Bremer Baugesellschaft	Baute um 1880 in der Südstadt „Bremer Häuser", Roonstraße, Kronprinzenstraße	200
1893–heute	Wilhelmshavener Spar- und Baugesellschaft	Gefördert durch Landesrat Liebrecht, Genossenschaft mit fast 7.000 Mitgliedern, Mietwohnungen, auch Betreuung und Eigenheime	über 3.000 (Bestand)
1897–1941	Beamten-Spar- und Bauverein	Baute Wohnungen im Inselviertel und in Rüstringen, 1941 mit Wilhelmshavener Spar- und Baugesellschaft zusammengeschlossen, Mietwohnungen	300
1902–1918	Gartenbaukolonie Lebensborn, Bant/Rüstringen	32 Genossen, mit Konsumverein verflochten, wollte Familien auf dem Land ansiedeln	4 (in Schortens)
1903–1907	Verein für ländliche Bauten in Wilhelmshaven	Abspaltung des Beamten-Spar- und Bauvereins, geplant wurden engl. Landhäuser bei der Kopperhörner Mühle	–
1903–heute	Bauverein Rüstringen	Gefördert durch den Staatssekretär des Innern, Graf Posadowsky und Oberwerftdirektor Wodrig, 5500 Mitglieder, Mietwohnungen, auch Betreuung und Eigenheime	3.100 (Bestand)
1904–1941	Bauverein Selbsthilfe	Gefördert von der Gemeinde Bant, baute an der Luisenstraße, mit Wilhelmshavener Spar- und Baugesellschaft zusammengeschlossen	47
1907–1908	Bauverein Siebethsburg	Wollte Wohnungen für Werftfeuerwehrleute schaffen	–
1911–1912	Gemeinnütziger Bauverein	Gefördert durch die Stadt Heppens, wollte Miets- und Erwerbshäuser bauen	90
1913–1943	Baugenossenschaft Daheim	Gegr. von Werftangestellten, wollte am Rüstringer Schützenhof 120 Wohnungen bauen, baute 1923 12 Häuser an der Freiligrathstraße, mit Wilhelmshavener Spar- und Baugesellschaft zusammengeschlossen	26
1916–1933	Bauverein Schaar	Baute Erwerbshäuser beim Schützenhof Schaar	?
1917–	Eigenheimbaugenossenschaft des evangelisch-sozialen Vereins Wilhelmshaven-Rüstringen	Wollte Wohnungen für Minderbemittelte bauen	–
ca. 1917–1933	Bauverein Eigenheim	Baute im Villenviertel für Beamte	?
1917–heute	Gemeinnützige Aktiengesellschaft für Angestelltenheimstätten (GAGFAH), Berlin	Baute 1932 30 Wohnungen in Wilhelmshaven, rege Bautätigkeit in der Nachkriegszeit	?
1918–	Kriegerheimstätten- und Kleinwohnungsbaugesellschaft	Gefördert durch Admiral Scheer, Mietwohnungen und Eigenheime für Kriegsversehrte, Hinterbliebene, Witwen und Kinderreiche, Gesellschaft löste sich auf	–
1920–ca. 1925	Eigenhandbauverein, Reichsbund Wilhelmshaven-Rüstringen	Gegr. von Kriegsbeschädigten unter Einfluß von Stadtbaurat Willy Hahn, große Projekte in Neuengroden und Inselviertel	50
1921–1933	Reichsheimstättenbaugesellschaft Rüstringen	Baute in der nördlichen Herbartstraße, 20 Reihenhäuser	20

	Baugesellschaften	Bemerkungen	Wohnungsproduktion
1923–heute	Gemeinnützige Gesellschaft zur Schaffung von Wohngelegenheiten für Reichsangehörige, Berlin	Gefördert durch Marine und Reich, Mietwohnungen in den Küstenstädten und Berlin, baute ab 1939 Altengroden und auch nach dem Kriege, heute Gemeinnützige Deutsche Wohnungsbaugesellschaft	1.200
1924–1941	Gemeinnützige Bau- und Siedlungsgenossenschaft der Jadestädte	Baute am Alten Deichsweg und in der Adolf-Hitler-Siedlung, 1941 mit der Wilhelmshavener Spar- und Baugesellschaft zusammengeschlossen	250
1926–1933	Gemeinnütziger Bau- und Siedlungsverein Wilhelmshaven		30
1926–1933	Gemeinnütziger Heimstättenverein des Mietervereins der Jadestädte		3
1927–1934	Baugesellschaft Mieterfreunde	34 Genossen, wurde mit der Wilhelmshavener Spar- und Baugesellschaft zusammengeschlossen	16
1927–1937	Gemeinnützige Bau- und Siedlungsgenossenschaft „Leo", Rüstringen	Baute in der Adolf-Hitler-Siedlung, Leiteweg sowie Projekte in der Gökerstraße, ging in Wilhelmshavener Spar- und Baugesellschaft auf	11
1933–1971	Jadestädtische Wohnungsbaugesellschaft GmbH	Gegr. von Wilhelmshavener Unternehmen, baute Wohnungen in der Peterstr. 131–169	100
1934–heute	Wilhelmshavener-Rüstringer Wohnungsbaugesellschaft	Gegr. von Bauunternehmer Friedrich Möller und versch. Handwerkern, baute Mietwohnungen für öffentl. Bedienstete, Gesellschaftsform verschiedentlich geändert	ca. 2.000
1937–heute	Wohnungsbaugesellschaft Jade	Gegr. von den an der Stadterweiterung der 30er Jahre beteiligten Gebietskörperschaften, Mietwohnungen, auch Betreuung und Eigenheime	ca. 10.000
ab 1938 in Wilhelmshaven	Niedersächsische Bauträgergesellschaft, Hannover	Sollte ab 1939 die Großsiedlung Breddewarden bauen	–
1939–1945	Neue Heimat	Gegr. von DAF im Gau Weser-Ems	–
ca. 1949–	Baugenossenschaft der Ostvertriebenen	Plante Wohngebäude in der Parkstraße 1950	?
ca. 1949–1972	Gemeinschaft für Wohnungsbau	Baute 1950–1952 ehem. Luftschutzbunker zu Wohnungen und Läden um	40
1952–1958	Siedlungswerk Kardinal Graf Galen	Baute die Graf-Galen-Siedlung für kath. Kinderreiche und Flüchtlinge	70
1953–heute	GEFA, Gesellschaft für Aufbau	Gegr. von Unternehmen und Handwerkern, Wiederaufbau, Eigenheime, Betreuung	2.200
1954–	Neue Heimat	Verschiedene Regionalniederl.	200
1961–heute	Haertle Wohnungsbaugesellschaft	Verschiedene Gesellschaftsformen, baute ausschl. Eigentumswohnungen als Bauträger	400
60er und 70er Jahre	Gemeinnützige Siedlungsgesellschaft mbH des ev. Siedlungswerks in Deutschland und der Leonberger Bausparkasse (GSG)	Baute im Wiesenhof	?
1967–heute	Wohnungsunternehmen Toni Rech		1.000
70er Jahre	NEBA, Neue Baugesellschaft Wilhelmshaven	Private Wohnungsbautätigkeit in Wilhelmshaven	?

9.8. Personenverzeichnis

Aufgeführt sind nur Personen aus dem Textteil 1.–8.

Abbehusen, August 44, 183
Adalbert, Prinz von Preußen 10, 11, 23, 146
Adena, Hugo 251
Adler, Johann Heinrich Friedrich 12, 14, 240, 311
Agatz, Arnold 72, 170, 171, 260
Ahmels, Hans 176, 270
Ahner, Hermann 38, 126, 306
Aichinger, Wilhelm 29, 214, 261
Arent, Benno von 190, 311
Aschenbeck, Diedrich 16
August Wilhelm, Prinz von Preußen 60
Augustin, Karl 267
Aven, Heinrich 168

Backes, Klaus 2
Balfanz, Friedrich 24
Balfanz, Ulrich 106
Barlage, Bernhard 29
Bartelt, Emil 21, 45, 320
Bartetzko, Dieter 2
Bartning, Hans 29, 40, 65, 105, 106, 140, 136, 139, 140, 141, 189, 190, 220, 270, 279, 329
Bartning, Otto 40, 72
Bauer, Ernst 319
Baumann, Georg-Emil 310
Baumann, Heinz 29, 180, 189, 190, 191, 196, 202, 217, 218, 284
Baur, Helmuth 7, 29, 40, 103, 135, 136, 137, 139, 190, 207, 216, 217, 235, 323
Beck, Ernst 25, 26, 169, 170, 172, 173, 318, 319
Behrendt, Hubert 11, 170
Behrens, Peter 16, 40, 42, 183, 244, 245, 253
Bellwinkel, Ludwig 25, 97, 98, 99, 285
Benda, Hans 25
Berg, Paul 25, 38, 184, 185
Berghaus, Jan 320
Berlage, Hendrik Petrus 72
Bernoulli, Hans 16, 42, 44, 183
Bestelmeyer, German 39, 40, 149, 174
Beuth, Peter Christian Wilhelm 10
Bieske, Georg 11
Billing, Hermann 40, 132
Bischof, Josef 294
Blendermann, Otto 44, 183
Block, Heinz 25, 285
Blomberg, Werner von 22, 60, 61
Blum, Otto 209, 211
Bochert, Kurt 128, 314
Böckler, Erich 94, 321
Böckmann (Regierungsrat) 267
Bodelschwingh, Friedrich von 58
Bohländer, Friedrich 29
Bollermann (Architekt) 269
Bonatz, Paul 26, 39, 40, 94, 149, 321
Bonte, Friedrich 159
Borchers (Architekt) 165
Bork, Gustav 87, 90, 91, 93, 94, 134, 218, 285, 286, 287, 288, 289, 323
Bormann, Martin 307
Boswau (Bauunternehmer) 116
Boyken, Ernst 116, 118, 124, 125, 126, 142, 217, 270, 287, 288
Boysen, Fritz 165, 218, 319, 320
Brahms, Albert 212
Brandes, Paul 46, 69, 115, 116, 118, 240, 316
Bräuning, Fritz 16, 42, 57, 183
Braunschweig, Adolf 90, 128, 211, 314
Breitschwert (Architekt) 27, 28
Breker, Arno 309
Brennecke, Ludwig 11
Brenner, Hermann 174
Breslauer, Alfred 12
Brinkmann, Karl 164, 165
Brix, Joseph 42
Bröcker, Paul 316
Brune, Ernst 18
Brune, Siegmar 46, 47
Brüning, Kurt 216
Büge, Max 31, 39, 85, 182, 266
Bugge, Valentin 24
Bürde, Georg Heinrich 10
Bürgin, Rudolf 37, 40, 90, 93, 124, 125, 126, 131, 132, 133, 134, 188, 217, 272, 273, 274, 280, 287
Buscher, Heinrich 307
Büscher, Peter 158, 304

Carl, Friedrich 183, 184
Christensen (Offizier) 169
Cousser, Adolph de 48
Czermak, Wilhelm 98

Däke, Heinz 38, 124, 249, 276
Damaschke, Adolf 58
Darré, Richard Walter 35, 236
Daun, Redmar 154
Dieckelmann, Bruno 35, 37, 143
Diersen, Wilhelm 161
Dieter, Hermann 44
Dillschneider, Karl 34, 39, 153, 154, 224, 294, 295
Distel, Hermann 153, 154, 294, 295
Dönitz, Karl 156, 211, 297
Döscher, Helmut 31, 39, 91, 209, 260
Drenckhan, Erich 26, 164
Drewitz, Wilhelm 12, 14
Dübbers, Kurt 26
Durth, Werner 2
Dustmann, Hanns 276, 285
Duvendack, Erich 314, 315

Eberstadt, Rudolf 42
Eckart, Dietrich 190, 308
Eckermann, Alfred 101, 102
Eckhardt, Alfred 25, 26, 170, 171, 173, 260, 319
Eiermann, Egon 166
Eilts, Bernhard 29
Eisenbarth, Friedrich 75, 77, 82, 83, 237, 238, 239, 243, 315
Elkart, Karl 207, 213
Elsaesser, Martin 265
Emmerich, Paul 43
Engel, Sigmund 293

Feder, Gottfried 31, 32, 39, 60, 74, 79, 112, 134, 135, 136, 162, 195, 196, 213, 226, 252, 253, 265, 272, 277, 279, 283, 284, 288, 294, 316, 321, 322
Ferse, Walter 165
Fiederling, Otto 217

Fink (Grundbesitzer) 123
Fischer (Generalmajor) 146
Fischer, Theodor 39, 42
Fischer-Dieskau, Joachim 31
Fitzler, Ernst 25
Fleinert, Friedrich 24
Fleischinger, August Ferdinand 12, 14
Folkerts, Heinz 220
Folkerts, Johann 21
Forster, Egon 36, 40, 106, 118, 124, 129, 180, 188, 191, 217, 219, 255, 256, 271, 272, 273, 274, 275, 310, 317, 318, 323, 324
Fourier, Charles 42
Frank, Hans 60
Frank, Hartmut 2
Franzius, Karl 39, 80, 97, 98, 99, 100, 115, 140, 149, 171, 180, 192, 193, 195, 212, 216, 217, 255, 280, 281, 286, 287, 288, 289, 309, 313, 323
Franzius, Ludwig 11
Frede, Georg 26, 111, 165, 170, 282, 318, 319, 320
Fresenius, Wolfgang 29, 243
Freygang, Richard 18, 19, 37, 70, 124, 134, 313, 321
Freytag, Ludwig 215
Frick, Wilhelm 60, 181, 307
Fricke, Hans 36, 90, 93, 103, 273, 274, 275, 307, 315, 316, 322
Friedrich Wilhelm III. 10
Friedrich Wilhelm IV. 10, 11, 23, 49, 146
Fritsch, Theodor 45, 240, 283
Fuchs, Robert 18, 37, 63, 113, 124, 125, 127, 134, 313

Galen, Clemens August Graf von 220
Gattnar, D.J. 110, 165
Geisenhainer, Kurt 25, 39, 97, 151, 154, 155, 192, 193, 250, 255, 309, 310
Genzmer, Felix 42
Gerdes, Hans 106, 108
Gereke 20
Gerking, Friedrich 36
Giesler, Hermann 186, 191
Gildemeister, Hermann 38, 213
Gilly, David 9
Göderitz, Johannes 320
Goebbels, Joseph 60, 306, 307, 308, 309, 322
Goecke, Theodor 16, 39, 42, 46, 62, 68, 104, 115, 122, 123
Göker, Heinrich Wilhelm 11
Göring, Hermann 20, 27, 30, 89, 90, 135, 174, 197, 221, 222, 247, 252, 316
Grau, Rudolf 155, 159, 308, 309
Graubner, Gerhard 215, 217, 219
Griesbach, Heinrich von 27, 28, 177
Groeneveldt, Jaques 35
Gropius, Martin 10
Gropius, Walter 44
Gummels (Architekt) 269
Gurlitt, Cornelius 16, 39
Gutschow, Konstanty 36, 39, 112, 115, 118, 135, 136, 137, 141, 177, 191, 197, 208, 210, 211, 212, 213, 218, 224, 231, 232, 282, 317
Gutschow, Niels 2

Haarnagel, Werner 29, 168
Hacke, Bruno 26
Haefner, Carl 18, 29, 44, 186, 316, 321
Hagen, Gotthilf 9, 10, 11, 23, 41, 48, 145, 146, 169
Hahn, Willy 1, 17, 18, 19, 29, 42, 44, 46, 50, 52, 55, 56, 62, 63, 64, 69, 78, 104, 115, 116, 122, 123, 127, 183, 192, 233, 240, 245, 316, 321
Hakenholz, Paul 46, 69, 115, 116, 118, 240, 316
Hallbauer, Wilhelm 29, 34, 38, 66, 74, 82, 83, 86, 87, 89, 92, 96, 97, 98, 104, 105, 106, 109, 110, 114, 115, 118, 124, 125, 127, 139, 140, 152, 163, 169, 176, 177, 180, 181, 188, 189, 190, 192, 193, 196, 221, 223, 224, 230, 233, 234, 253, 294, 317, 323, 329
Häring, Hugo 40, 98
Harlander, Tilman 2
Harms, Georg Heinrich 96, 242, 308
Harmsen, Heinrich 83, 111, 269, 305
Harris, Arthur 234
Haschenburger, Werner 19, 40, 320
Häuser, Hanns 276
Häußler, Albert 24
Heeren (Architekt) 164, 165
Heiligenthal, Roman 40
Heilmann, Georg 85, 182
Heinrich, Prinz von Preußen 12
Heintz, Wilhelm 243
Held, Richard 186
Hentrich, Helmut 276
Herde, Alexander 31, 39, 203, 204, 232
Herde, Gertraude 31, 204, 232
Herdejürgen, Wilhelm 83, 111, 269, 305
Herenbrock (Architekt) 322
Hermeking, Ludwig 25
Heß, Rudolf 35, 60, 180, 181, 278, 279, 280, 307, 315, 317
Heuer, Friedrich 37, 38, 39, 82, 84, 97, 98, 112, 115, 134, 135, 139, 140, 141, 160, 191, 195, 197, 202, 209, 211, 212, 213, 224, 226, 232, 239, 254, 268, 269, 270, 279, 284, 288, 305
Hierl, Konstantin 299
Higgen, Karl 27, 175, 214
Hilberseimer, Ludwig 253
Hilgenstock, Fritz 87, 89, 90, 91, 92, 95, 102, 111, 112, 285, 287, 304, 324
Hillebrecht, Rudolf 211, 218
Himmler, Heinrich 60
Hindenburg, Paul von 252, 311, 312
Hinrichs, Karl 52
Hinz, Karl-Heinrich 162, 188
Hipper, Franz von 159
Hitler, Adolf 20, 22, 25, 26, 31, 32, 33, 35, 39, 60, 65, 67, 73, 81, 92, 94, 96, 101, 110, 111, 112, 114, 134, 136, 140, 149, 156, 157, 162, 163, 169, 171, 173, 174, 179, 180, 181, 182, 184, 185, 186, 187, 188, 189, 190, 191, 192, 193, 194, 200, 201, 206, 209, 210, 213, 214, 215, 222, 223, 225, 226, 247, 252, 259, 263, 264, 269, 272, 277, 279, 284, 290, 296, 298, 305, 308, 309, 311, 312, 313, 316, 329
Hitzbleck, Fritz 304
Hitzig, Friedrich 10, 12, 14, 240
Hoeland, Franz 316
Hoffmann, Ludwig 16, 183
Hoffmann, Werner 105, 106, 171
Höger, Fritz 18, 37, 44, 113, 115, 116, 117, 118, 119, 120, 121, 122, 123, 125, 126, 127, 134, 171, 175, 176, 183, 184, 185, 186, 187, 188, 192, 212, 213, 217, 219, 240, 249, 256, 258, 292, 293, 308, 309, 313, 315, 316, 317, 318, 321
Högg, Hans 39
Homann, Richard 38, 213, 243, 315
Hopf (Gauamtsleiter) 36, 60
Hoppe, Jeanette 211
Horgass, Paul 273
Hörold, Ludwig 37, 90, 93, 272, 286, 323, 324
Horstmann, Bernhard 38
Howard, Ebenezer 41, 42, 43, 45, 58, 283
Hübbe, Heinrich 145
Huber, Victor Aimé 58
Hübotter, Wilhelm 99, 237, 243
Hucke, Alfred 25

Hug, Paul 44
Humboldt, Alexander von 9, 10
Hunzinger, Walter 27, 29, 38, 74, 85, 86, 88, 89, 98, 105, 123, 127, 180, 182, 186, 249, 258, 295, 296, 303, 317, 320, 323
Hyltén-Cavallius (Kapitän zur See) 145

Ibeling, Albert 70
Imken, Adolf 167
Isenberg, Gerhard 32
Iwersen, Franz 18, 37, 71, 124, 125, 134, 152, 257, 272, 273, 290, 313
Iwersen, Hanns 124, 257, 273, 290

Jacob, Johannes 102
Jaeger (Architekt) 128, 314
Jahn, Adalbert 27, 175, 176, 270
Jahn, Karl 54, 142
Jansen, Hermann 39, 97
Janssen-Jennelt, Diedrich 310
Janßen, Georg 27, 115
Jaunisch (Baurat) 27
Jensen, Herbert 213
Jochem, Friedrich 291
Joel, Georg 34, 38, 95, 114, 308
Johann, Erzherzog von Österreich 10

Kämmerer, Otto 29
Kampen, Peter 305
Kampffmeyer, Bernhard 42
Kampffmeyer, Hans 42
Kellerhoff, Arthur 108, 183, 185, 212, 316, 317, 318
Kelm, Walter 25, 75, 97, 98, 100
Keppler, Wilhelm 262
Kerrl, Hanns 32, 34
Kimpel, Dieter 7
Klaje, Hans-Hermann 210
Klatt, Otto 124
Klaucke, Hans 26, 40, 149, 164, 165, 166, 167, 171, 172, 205, 318, 319, 320
Klett (Baurat) 25
Klette, Otto 164
Kleyhauer, Gerhard Hinrich Diedrich 49
Klotz (Major) 146
Knauer (Bauunternehmer) 116
Kniphausen, Wilhelm Edzard Graf von 130, 131, 132
Koch, Wilhelm 29
Köhler (Oberregierungsrat) 74, 100, 189
Kohlscheen, Julius 220
Köhne, Adolf 37, 317
Köllmann, Johann 36
König (Marinebaurat) 11
Kraak, Karl-Heinz 219, 256
Kreft, Paul 101
Kreis, Wilhelm 39, 174, 310, 317
Krüger, Herbert 305
Krüger, Wilhelm 26, 168, 242
Kühn, Erich 99, 101, 102, 103, 289, 323
Kuöhl,Richard Emil 193, 309
Kussin (Baurat) 25
Küster, Wilhelm 37

Lackner, Erich 260
Lademann (Verkehrsgutachter) 209, 211
Ladwig, Hanns 305
Lammers, Hans Heinrich 60
Lang, Richard 294
Lehmann, Ludwig 266
Lehn, Otto 29, 211, 214, 216, 217, 261, 303, 320, 321
Ley, Robert 31, 60, 79, 95, 163, 207, 209, 278, 298
Lichtwark, Alfred 316
Liebrecht, Wilhelm 50
Linde, Georg 32, 33, 35, 38, 39, 74, 85, 110, 130, 132, 224, 253, 315
Linde, Horst 40, 288
Lindner, Werner 94, 112, 238, 259, 321
Link (Amtsleiter) 29
Loye, Fritz zur 35
Lübbers, Friedrich 36, 39, 211, 212
Lübbers, Gustav 18, 124, 125, 205, 295, 323
Lübbers, Willi 18, 33, 38, 39, 71, 73, 75, 76, 77, 78, 79, 82, 107, 124, 125, 128, 180, 192, 217, 218, 238, 254, 255, 257, 262, 267, 268, 277, 279, 285, 290, 293, 304, 305, 314, 315, 317
Lückemeyer, Otto 18, 25, 26, 39, 70, 101, 103, 149, 151, 154, 156, 158, 192
Luckhardt, Hans 295
Luckhardt, Wassili 295
Ludowici, Johann Wilhelm 35
Lueken, Emil 42, 43, 44, 46, 64, 115, 116
Luttmer (Schlachter) 285
Lutze, Viktor 61

Maas, Karl 29, 39, 192, 311
Mackensen, Fritz 308, 316, 317
Mainzer, Josef 38
Mangel (Gaugruppenleiter) 213
March, Werner 190
Markworth, Friedrich 167
Marquardsen, Harald 119, 120, 316, 318
Martens, Werner 167
Martini (Oberpostbaurat) 257
Mattausch, Roswitha 2
May, Ernst 40, 44, 72, 98
Mebes, Paul 42, 43
Mehrtens 192, 316
Meichner, Wilhelm 292, 293
Mellin, Friedrich 98, 101, 102, 103
Mendelsohn, Erich 40, 118, 295
Merker, Reinhard 2
Metzendorf, Georg 42
Meurer, Franz 165
Meyer, Ernst 38, 39, 81, 190, 277, 319, 325
Meyer, Hellmut 256, 316
Mies van der Rohe, Ludwig 40
Migge, Leberecht 16, 19, 21, 42, 43, 44, 78, 123, 161, 233, 236, 240, 241, 310
Milch, Erhard 28
Minke, Alexander 18, 37, 113, 134, 220, 313
Moeller, Paul 11
Möhring, Bruno 16, 46, 57, 72
Möller, Friedrich 104, 105, 106, 107, 108, 156, 158, 161, 167, 214, 300, 316, 318, 324, 326
Möller, Hermann 18, 51, 106, 271
Moltke, Helmuth von 146
Moraller, Franz Karl 307
Mücke, Willy 192, 316
Müller, Ludwig 60, 311
Müller, Peter August 175, 176
Müller, Richard 24
Müller, Wilhelm 24, 29, 34, 38, 39, 75, 82, 85, 87, 89, 90, 96, 97, 101, 120, 127, 131, 132, 136, 138, 181, 182, 186, 187, 188, 189, 191, 192, 210, 211, 212, 243, 261, 277, 281, 293, 294, 311, 314
Müller-Schiedmayer, Clemens 251
Müller-Wulckow, Walter 246
Mussolini, Benito 312

Muthesius, Hermann 16, 42, 45, 316

Nathenstedt (Grundbesitzer) 123
Naud, Wilhelm 25
Naumann, Friedrich 43, 58
Nehl, Walther 275
Neufert, Ernst 102, 202, 203, 211, 213, 253, 265, 276
Neumann, Max 29, 44, 62, 63, 107, 113, 122, 123, 162, 281, 288, 292, 321
Nevermann, Arthur 27, 175
Nicolaus, Friedrich 31, 86, 91, 92, 95, 112, 134, 140, 141, 221, 225, 232, 237, 242, 266
Niemeyer, Reinhold 208, 213
Nieter, Reinhard 183, 316
Nikolaus Friedrich Peter von Oldenburg 11
Nobe, Fritz 84, 138
Norkauer, Fritz 40
Nutzhorn, Gustav 34, 60, 62, 64, 65, 114, 116, 184, 293, 308, 318

Ober-Bloibaum, Gerhard 29
Oehler, Emmerich 309
Offenberg, Gerd 207
Olbrich, Joseph Maria 18
Oldewage, Heinrich 96, 242
Oppenheimer, Franz 42
Osborn, Frederic 41
Ostendorf, Hans 30
Ostendorf, Wilhelm 30, 34, 105, 108, 140
Ott, Hermann 74, 84, 112, 132, 138, 165, 175, 266, 267, 270
Otto, Karl 227, 228
Owen, Robert 42

Paffrath, Friedrich 21, 65, 127, 214, 235
Paul, Bruno 90
Pauls, Heinrich 27, 258
Paulsmeier (Oberingenieur) 91, 141
Pauly, Julius 30, 31, 47, 132
Pein, Richard 26, 165
Peltz-Dreckmann, Ute 2
Penger, Zoltan 128, 314
Peters, Ernst 18, 37, 63, 70, 71, 73, 113, 128, 134, 219, 220, 313
Petsch, Joachim 2
Pfeffer (Baurat) 11
Pfeiffer, Karl 31, 39, 75, 86, 87, 91, 97, 140, 242, 269
Picker, Daniel 260, 320, 321
Picker, Henry 215
Pinnau, Cäsar 210
Pitz, Klemens 39
Plate, Ludwig 212
Poelzig, Hans 16, 39, 40, 44, 183
Pollet, Julius 53
Posadowsky, Graf von 50, 115
Prinz, Ernst 130, 134, 135, 196
Pront, Abel 29
Putlitz, Erich zu 107, 118, 130, 134, 135, 196, 218, 219, 256, 257, 318

Queisner, Joachim 165, 319

Raeder, Erich 22, 60, 61, 77, 84, 100, 120, 132, 156, 157, 173, 182, 192, 193, 260, 296, 319
Rahn, (Architekt) 101
Rambacher (Ministerialrat) 25
Rasch, Klaus 27, 29, 187, 190, 191, 217, 218, 235, 254, 258, 273, 274, 294, 323, 324
Rastede, Richard 266
Rauch, Christian Daniel 10
Rauchheld, Adolf 44
Rech, Toni 292
Rechenberg, Fritz 135
Rechtern, Balduin Emil 11
Reeken, Otto van 175
Refardt, Heinrich 320, 321
Reh (Bauunternehmer) 304
Rehbein, Wilhelm 220
Reichow, Hans-Bernhard 39, 211
Reif, Wilhelm 316
Reinhard 20
Reinhardt, Heinrich 13
Rendel, Sir John 145, 146
Renken, Carl 64, 73, 262, 320, 321
Renner, Paul 101
Richard (Ministerialrat) 209
Richter, Johannes 37, 38, 75, 142, 143
Riehl, Wilhelm Heinrich 240, 243
Riemerschmid, Richard 42
Rimpl, Herbert 90, 135, 211, 213
Riphahn, Wilhelm 43
Rodenberg, Wilhelm 181
Rohlfs, Werner 128, 314
Rollmann, Julius 11
Rosenberg, Alfred 35, 213, 293, 306, 307, 316, 322
Roth, Werry 190
Röver, Carl 21, 22, 31, 33, 34, 35, 36, 37, 39, 101, 140, 143, 180, 181, 182, 191, 207, 211, 224, 225, 274, 276, 278, 280, 307, 308, 316, 319, 321
Rudolphi, Joachim 100
Ruep, J. 70

Saalwächter, Alfred 34
Sagebiel, Ernst 174
Salinger, Paul 12
Salvisberg, Otto Rudolf 117
Schäche, Wolfgang 2
Schachner, Benno 39, 137, 153, 294, 295
Schäfer (Baurat) 27, 175
Schäfer, Philipp 18
Schaller, Friedrich 25
Scharoun, Hans 117
Scheeb (Architekt) 185
Scheer, Reinhard 50, 52, 156, 243, 308
Schemm, Willy 79, 80, 81, 128, 129, 218, 314, 315
Schikore (Baurat) 184, 185
Schinkel, Karl Friedrich 9, 10, 11, 23, 41
Schipper, Georg 66, 142
Schirach, Baldur von 256, 275, 276, 277
Schleicher, Gustav 211
Schmidt, Friedrich 31
Schmitthenner, Paul 39, 40, 42, 62, 77, 82, 94, 97, 121, 125, 128, 149, 151, 154, 236, 253, 264, 285, 309, 321, 328
Schnedler (Vermessungsrevisor) 16
Schneider, Christian 2
Schneider, Fritz Karl 30, 38, 39, 75, 76, 79, 82, 86, 87, 88, 108, 110, 112, 115, 140, 141, 191, 197, 208, 209, 210, 217, 224, 267, 269, 281
Schoch, August 118, 256
Scholtz (Abteilungsdirigent) 31
Schöning, Klaus-Georg 217
Schreber, Daniel 58
Schreiber, Theodor 25, 74, 132, 189, 267
Schubert, Friedrich 24, 45
Schulte-Frohlinde, Julius 35, 36, 39, 85, 211
Schulte-Ostrup, Egon 156
Schultze, Otto 34, 193

Schultze-Naumburg, Paul 40, 42, 94, 191, 213, 236, 253, 259, 264, 306, 316, 321
Schulz, Walther 32, 115, 136, 177, 191, 197, 207, 208, 209, 224, 230, 231, 260
Schulze, Ernst 36, 307
Schumacher, Fritz 29, 39, 118, 192, 207, 217, 223, 261, 315, 317
Schumacher, Heinz 261
Schürmann, Arthur 32
Schuster, Rudolf 34
Schütt, Ferdinand 27, 39, 175
Schütte, Heinrich 242
Schwarting, Theodor 105, 106, 107, 324
Schwarz, Franz Xaver 35
Schwarz, Max Karl 211, 215, 216, 217, 243
Schweizer, Otto Ernst 40, 132, 136
Schwerin von Krosigk, Lutz Graf 185
Schwichow, Friedemann 215
Scotland, Eduard 38, 213
Seetzen, Fritz 36
Seldte, Franz 31
Semler, Johannes 53
Siedenburg, Kurt 31, 318
Siemer, Heinrich 29, 32, 44, 62, 63, 113, 123, 162, 321
Siepermann, Johann 128, 314
Sievers (Bauunternehmer) 269
Soissons, Louis de 42
Sommer, Walter 27
Speer, Albert 26, 27, 28, 33, 36, 39, 112, 134, 177, 180, 182, 185, 188, 189, 190, 191, 194, 195, 196, 197, 201, 202, 203, 207, 209, 210, 212, 213, 216, 217, 225, 231, 232, 243, 246, 260, 278, 282, 294, 298, 309, 311, 317
Spiegel, Hans 207
Springer, Peter 7
Spröde (Reg. Baumeister a.D.) 140
Stein, Friedrich 29
Steinbrecher (Baurat) 27
Steinhauser, Paul 37
Stoffregen, Heinz 44, 183
Stoltz (Hauptmann) 103, 146
Strack, Johann Heinrich 10
Streicher, Julius 307
Strempel, Willibald 13, 18, 70
Stuhlmann (Architekt) 281
Stüler, Friedrich August 10, 11, 12
Suhr, Dietrich 134
Suhr, Hellmut 118, 219
Süßenguth, Georg 13

Tamms, Friedrich 203, 211, 213, 290
Tappe, Ulrich 260
Taube, Samuel de 132
Taut, Bruno 39, 40, 42, 44, 72, 98, 118
Taut, Max 40
Temp, Walter 29, 36, 38, 39, 80, 92, 95, 101, 134, 136, 139, 141, 182, 190, 191, 197, 209, 210, 211, 216, 222, 272, 280, 285, 294, 297
Tessenow, Heinrich 39, 40, 42, 49, 94, 97, 100, 102, 121, 149, 195, 203, 213, 264, 283, 321
Teut, Anna 2
Thees, Heinz-Günther 149, 156, 157, 158, 159, 162, 192, 203, 251
Thiele, Kurt 35, 36, 303
Thorak, Joseph 309
Tiburtius, Carl 26, 39, 105, 169, 170, 171, 173, 260, 318, 319
Tiburtius, Irina 173, 319
Tietjen, Ernst 250
Tillessen, Karl 167, 215, 320
Tirpitz, Alfred von 12, 13, 15, 53, 308
Todt, Fritz 36, 61, 189, 201, 213, 225, 296, 302, 303
Tornow, Walter 25, 100
Tribius (Obersturmbannführer) 22
Triebel, Walter 26
Triebel, Wolfgang 213
Trittler, Guido 26
Troost, Paul-Ludwig 180, 252, 317
Trotha, Adolf von 16
Tuchscherer, Karl 110, 111, 112, 118, 160, 161, 195, 220, 320

Unwin, Raymond 42

Veeser, Joseph 90, 304
Völker, Wilhelm 292, 293
Vosshans, Josef 27, 28, 177, 258

Wagner, Martin 1, 15, 16, 17, 29, 40, 42, 43, 44, 46, 47, 56, 57, 62, 72, 78, 98, 116, 119, 123, 136, 183, 192, 233, 236, 240, 243, 244, 245, 310, 312, 316
Walkenhorst, Heinrich 35
Wallbaum (Baudirektor) 169
Weddigen, Otto 156, 276
Wedel, Haro Graf von 109
Wegener, Paul 33, 36, 37, 209, 210, 212, 213
Wentzel (Gutachter) 209
West (Architekt) 291
Wetzel, Heinz 38, 39, 82, 149
Weving (Baurat) 27
Wickop, Walter 213
Widekind, Karl Hasso von 25
Wiepking-Jürgensmann, Heinrich 88, 98, 99, 103, 141, 237, 242, 243
Wilhelm I. 11
Wilhelm II. 12, 15, 244
Wohlschläger, Josef 31, 264
Wolf, Gustav 38, 39, 217, 226
Wolf, Paul 15, 29, 42, 43, 45, 236
Wortmann, Wilhelm 32, 33, 34, 38, 39, 82, 110, 118, 141, 165, 192, 207, 208, 210, 211, 213, 216, 217, 222, 223, 224, 235, 317
Wucherpfennig, Wilhelm 211

Zenker, Ewald 25, 39, 97, 98, 99, 100, 102, 136, 195, 203, 274, 289
Zieb, Paul-Willy 216
Zimmermann, Ernst 24
Zinsser, Ernst 217
Zopff, Hermann 15, 17, 18, 19, 29, 39, 45, 56, 60, 68, 69, 71, 72, 73, 113, 123, 180, 186, 295, 312, 320, 321
Zorn, Richard 211

9.9. Abkürzungen

Arch.	Architekt
BDA	Bund deutscher Architekten
BDM	Bund deutscher Mädel
DAF	Deutsche Arbeitsfront
DSB	Deutscher Siedlerbund
FAD	Freiwilliger Arbeitsdienst
GBI	Generalbauinspektor (Albert Speer)
Gestapo	Geheime Staatspolizei
HJ	Hitler-Jugend
HNBD	Hafenneubaudirektion Wilhelmshaven
KDAI	Kampfbund deutscher Architekten und Ingenieure
KDF	NS-Gemeinschaft Kraft durch Freude
KDK	Kampfbund für deutsche Kultur
KMW	Kriegsmarinewerft Wilhelmshaven
KZ	Konzentrationslager
LOB	Landesplanungsgemeinschaft Oldenburg-Bremen
MDF	Minister der Finanzen Oldenburg
MI	Marineintendantur
NS	nationalsozialistisch
NSBDT	Nationalsozialistischer Bund deutscher Techniker
NSDAP	Nationalsozialistische deutsche Arbeiterpartei
NSV	Nationalsozialistische Volkswohlfahrt
OKM	Oberkommando der Kriegsmarine Berlin
OT	Organisation Todt
RAD	Reichsarbeitsdienst
RAM	Reichsarbeitsminister Berlin
RDBK	Reichskammer der bildenden Künste
RUGES	Reichsumsiedlungsgesellschaft
SA	Sturmabteilung der NSDAP
SS	Schutzstaffel der NSDAP
TH	Technische Hochschule

ABR	Archiv Bauverein Rüstringen
ADB	Archiv Bundesbahndirektion Hannover
AGEW	Archiv Gas- und Elektrizitätswerke Wilhelmshaven
AGS	Archiv Gießerei Sande
AJA	Archiv Wohnungsbaugesellschaft Jade
ANE	Archiv Norddeutscher Eisenbau Sande
AOL	Archiv Olympia-Werke Roffhausen
ARK	Archiv Ralf Kampen
ASTHBA	Archiv Staatshochbauamt Wilhelmshaven
BAK	Bundesarchiv Koblenz
BAW	Bauakten der Stadt Wilhelmshaven
BSW	Zeitschrift Bauen, Siedeln, Wohnen
DH	Der Hammer, Werkszeitung der KMW Wilhelmshaven
MAF	Militärarchiv Freiburg
NOB	Niederschriften des Wilhelmshavener Oberbürgermeisters mit seinen Beigeordneten
SEM	Sammlung Erich Mulitze
SHB	Sammlung Helmuth Baur
SHT	Sammlung Heinz-Günther Thees
SIS	Sammlung Ingo Sommer
SKF	Sammlung Karl Franzius
SRB	Sammlung Rudolf Bürgin
STAO	Niedersächsisches Staatsarchiv Oldenburg
STDAW	Stadtarchiv Wilhelmshaven
SYK	Sammlung Yvonne Klaucke
WK	Wilhelmshavener Kurier
WZ	Wilhelmshavener Zeitung

a.a.O.	am angegebenen Ort
Anm.	Anmerkung
Bd.	Band
Bespr.	Besprechung
ebd.	ebenda
ff.	folgende
gem.	gemäß
Gespr.	persönliches Gespräch
Hg.	Herausgeber
Niederschr.	Niederschrift
S.	Seite
s.a.	siehe auch
Schr.	Schreiben
u.a.	und andere/unter anderem
v.	von/vom
Vf.	Verfasser
vgl.	vergleiche
z.B.	zum Beispiel

GPSR Compliance

The European Union's (EU) General Product Safety Regulation (GPSR) is a set of rules that requires consumer products to be safe and our obligations to ensure this.

If you have any concerns about our products, you can contact us on ProductSafety@springernature.com

In case Publisher is established outside the EU, the EU authorized representative is:

Springer Nature Customer Service Center GmbH
Europaplatz 3
69115 Heidelberg, Germany

Batch number: 08567489

Printed by Printforce, the Netherlands